The Algebra Pyramid

Equations and Inequalities
$2 + 5(7) = 37$ or
$x + 2y > 5$

Expressions
$2 + 5(7)$ or $x + 2y$

Constants and Variables
$2, 5, 7, x, y$

The Algebra Pyramid illustrates how variables, constants, expressions, equations, and inequalities relate. At the foundation of algebra, and this pyramid, are constants and variables, which are used to build expressions, which, in turn, are used to build equations and inequalities.

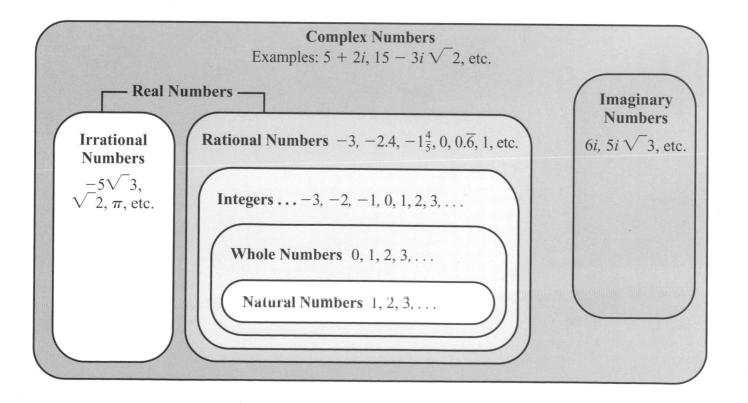

Complex Numbers
Examples: $5 + 2i$, $15 - 3i\sqrt{2}$, etc.

Real Numbers

Irrational Numbers
$-5\sqrt{3}$, $\sqrt{2}$, π, etc.

Rational Numbers $-3, -2.4, -1\frac{4}{5}, 0, 0.\overline{6}, 1$, etc.

Integers ... $-3, -2, -1, 0, 1, 2, 3, \ldots$

Whole Numbers $0, 1, 2, 3, \ldots$

Natural Numbers $1, 2, 3, \ldots$

Imaginary Numbers
$6i$, $5i\sqrt{3}$, etc.

Arithmetic Summary

Each operation has an inverse operation. In the diagram below, the operations build from the top down. Addition leads to multiplication, which leads to exponents. Subtraction leads to division, which leads to roots.

PROPERTIES OF ARITHMETIC

In each of the following, a, b, and c represent real numbers.

Additive Identity
$$a + 0 = a$$

Commutative Property of Addition
$$a + b = b + a$$

Associative Property of Addition
$$(a + b) + c = a + (b + c)$$

Multiplicative Identity
$$a \cdot 1 = a$$

Multiplicative Property of 0
$$a \cdot 0 = 0$$

Commutative Property of Multiplication
$$ab = ba$$

Associative Property of Multiplication
$$(ab)c = a(bc)$$

Distributive Property
$$a(b + c) = ab + ac$$

INVERSE OPERATIONS

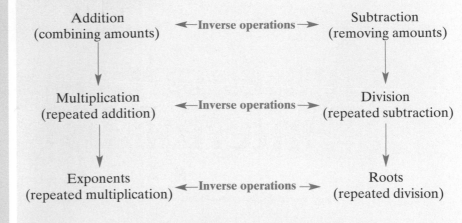

ORDER OF OPERATIONS

1. Grouping symbols
2. Exponents or roots from left to right
3. Multiply or divide from left to right
4. Add or subtract from left to right

SOLVING LINEAR EQUATIONS

To solve linear equations in one variable:

1. Simplify both sides of the equation as needed.
   a. Distribute to clear parentheses.
   b. Clear fractions or decimals by multiplying through by the LCD. In the case of decimals, the LCD is a power of 10 determined by the decimal number with the most places. (Clearing fractions and decimals is optional.)
   c. Combine like terms.
2. Use the addition principle so that all variable terms are on one side of the equation and all constants are on the other side. (Clear the variable term with the lesser coefficient.) Then combine like terms.
3. Use the multiplication principle to clear the remaining coefficient.

USING THE QUADRATIC FORMULA

To solve a quadratic equation in the form $ax^2 + bx + c = 0$, where $a \neq 0$, use the quadratic formula:

$$x = \frac{-b \pm \sqrt{b^2 - 4ac}}{2a}$$

FACTORING A POLYNOMIAL

To factor a polynomial, first factor out any monomial GCF, then consider the number of terms in the polynomial. If the polynomial has:

I. **Four terms,** then try to factor by grouping.

II. **Three terms,** then determine if the trinomial is a perfect square or not.

   A. If the trinomial is a perfect square, then consider its form.
      1. If in the form $a^2 + 2ab + b^2$, then the factored form is $(a + b)^2$.
      2. If in the form $a^2 - 2ab + b^2$, then the factored form is $(a - b)^2$.

   B. If the trinomial is not a perfect square, then consider its form.
      1. If in the form $x^2 + bx + c$, then find two factors of c whose sum is b, and write the factored form as $(x + \text{first number})(x + \text{second number})$.
      2. If in the form $ax^2 + bx + c$, where $a \neq 1$, then use trial and error. Or, find two factors of ac whose sum is b; write these factors as coefficients of two like terms that, when combined, equal bx; and then factor by grouping.

III. **Two terms,** then determine if the binomial is a difference of squares, a sum of cubes, or a difference of cubes.

   A. If given a binomial that is a difference of squares $a^2 - b^2$, then the factors are conjugates and the factored form is $(a + b)(a - b)$. Note that a sum of squares cannot be factored.
   B. If given a binomial that is a sum of cubes, $a^3 + b^3$, then the factored form is $(a + b)(a^2 - ab + b^2)$.
   C. If given a binomial that is a difference of cubes, $a^3 - b^3$, then the factored form is $(a - b)(a^2 + ab + b^2)$.

Note: Always check to see if any of the factors can be factored.

Elementary and Intermediate Algebra

Second Edition

TOM CARSON
Midlands Technical College

ELLYN GILLESPIE
Midlands Technical College

BILL JORDAN
Seminole Community College

PEARSON

Addison
Wesley

Boston   San Francisco   New York
London   Sydney   Tokyo   Singapore   Madrid
Mexico City   Paris   Cape Town   Hong Kong   Montreal

Publisher: Greg Tobin

Editor in Chief: Maureen O'Connor

Executive Editor: Jennifer Crum

Executive Project Manager: Kari Heen

Project Editor: Lauren Morse

Editorial Assistants: Elizabeth Bernardi and Emily Ragsdale

Managing Editor: Ron Hampton

Senior Designer: Dennis Schaefer

Cover Designer: Dennis Schaefer

Photo Researcher: Beth Anderson

Supplements Supervisor: Emily Portwood

Media Producer: Sharon Smith

Software Development: Rebecca Williams, MathXL; Marty Wright, TestGen

Marketing Manager: Jay Jenkins

Marketing Coordinator: Alexandra Waibel

Senior Prepress Supervisor: Caroline Fell

Senior Manufacturing Buyer: Carol Melville

Production Coordination: Pre-Press Company, Inc.

Composition: Pre-Press Company, Inc.

Artwork: Pre-Press Company, Inc.

Cover photo: © Jeremy Woodhouse/Masterfile—The Louvre at sunset; Paris, France

Photo credits can be found on page PC-1 in the back of the book.

Many of the designations used by manufacturers and sellers to distinguish their products are claimed as trademarks. Where those designations appear in this book, and Addison-Wesley was aware of a trademark claim, the designations have been printed in initial caps or all caps.

Library of Congress Cataloging-in-Publication Data

Carson, Tom, 1967–
    Elementary and Intermediate Algebra—2nd ed./Tom Carson, Ellyn Gillespie,
    Bill E. Jordan.
        p. cm.
    Includes index.
    ISBN 0-321-36854-1 (Student's Edition)
    1. Albegra I. Gillespie, Ellyn. II. Jordan, Bill E. III. Title.

QA152.3.C374 2005b
512.9—dc22   2005050923

4 5 6 7 8 9 10—VH—10 09 08

Contents

Preface

Welcome to the second edition of *Elementary and Intermediate Algebra* by Carson, Gillespie, and Jordan! Revising this series has been both exciting and rewarding. It has given us the opportunity to respond to valuable instructor and student feedback and suggestions for improvement. It is with great pride that we share with you both the improvements and additions to this edition as well as the hallmark features and style of the Carson/Gillespie/Jordan series.

Elementary and Intermediate Algebra, Second Edition, is the fourth book in a series that includes *Elementary Algebra*, Second Edition, *Prealgebra*, Second Edition, *Elementary Algebra with Early Systems of Equations*, and *Intermediate Algebra*, Second Edition. This text is designed to be versatile enough for use in a standard lecture format, a self-paced lab, or even in an independent study format. Written in a relaxed, nonthreatening style, *Elementary and Intermediate Algebra* takes great care to ensure that students who have struggled with math in the past will be comfortable with the subject matter. Explanations are carefully developed to provide a sense of why an algebraic process works the way it does, instead of just an explanation of how to follow the process. In addition, problems from science, engineering, accounting, health, the arts, and everyday life link algebra to the real world. A complete study system beginning with a Learning Styles Inventory and supported by frequent Learning Strategy boxes, is also provided to give students extra guidance and to help them be successful. (See page xxiii.)

Changes to the Second Edition

This revision includes refinements to the presentation of the material as well as the addition of many more examples and applications throughout the text. However, the primary focus of this revision is the exercise sets. The section-level exercise sets have been scrutinized and reworked to create a gradation that slowly progresses from easy to more difficult. There is also better pairing between odd and even exercises sets, and many more midlevel problems have been added.

In addition to the exercise sets, the Learning Strategy boxes and Algebra Pyramid references have been enhanced and increased in number to provide students with even more guidance.

Interval notation is introduced in Section 2.6 instead of Section 8.1 and is now used throughout the text.

A review chapter called Chapter R, which reviews basic Elementary Algebra concepts, has been added.

Small versions of the Algebra Pyramid have been added to the Chapter Review Exercises and the Cumulative Review Exercises to help students distinguish groups of expression exercises from groups of equation or inequality exercises.

Finally, the number of exercises included in MyMathLab and MathXL has been increased dramatically for an even stronger correlation between the book and the technology that supports it.

Key Features

Study System   A study system is presented in the *To the Student* section on pages xvii–xxii. This system is then reinforced throughout the text. The system recommends color codes for taking notes. The color codes are consistent in the text itself: red for definitions, blue for procedures and rules, and black for notes and examples. In addition, the study system presents strategies for succeeding in the course. These learning strategies have been expanded and are revisited in the chapter openers and throughout the body of the text.

Learning Styles Inventory   A Learning Styles Inventory is presented on page xxiii to help students assess their particular learning style. Learning Strategy boxes are then presented throughout the book with different learning styles in mind.

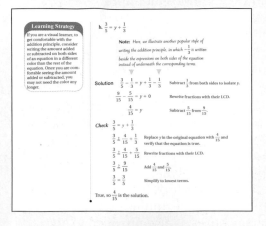

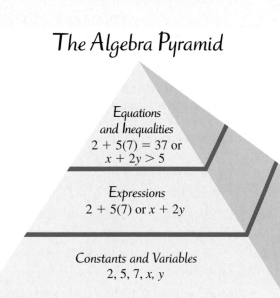

Learning Styles Inventory

What is your personal learning style?

A learning style is the way in which a person processes new information. Knowing your learning style can help you make choices in the way you focus on and study new material. Below are fifteen statements that will help you assess your learning style. After reading each statement, rate your response to the statement using the scale below. There are no right or wrong answers.

3 = Often applies   2 = Sometimes applies   1 = Never or almost never applies

_____ **1.** I remember information better if I write it down or draw a picture of it.

_____ **2.** I remember things better when I hear them instead of just reading or seeing them.

_____ **3.** When I receive something that has to be assembled, I just start doing it. I don't read the directions.

_____ **4.** If I am taking a test, I can "visualize" the page of text or lecture notes where the answer is located.

Learning Strategy Boxes   Learning Strategy boxes appear where appropriate in the text to offer advice on how to effectively use the study system and how to study specific topics based on a student's individual learning style (see pages 3, 121, and 394).

The Algebra Pyramid   An Algebra Pyramid is used throughout the text to help students see how the topic they are learning relates to the big picture of algebra—particularly focusing on the relationship between constants, variables, expressions, and equations (see pages 3, 102, and 387). In Chapter Review Exercises and Cumulative Review Exercises, an Algebra Pyramid icon indicates the level of the pyramid that correlates to a particular group of exercises to help students determine what actions are appropriate with these exercises, for example, whether to "simplify" or "solve" (see pages 265, 364, and 630).

Equations and Inequalities

Exercises 11–16

The Algebra Pyramid

Equations and Inequalities
$2 + 5(7) = 37$ or
$x + 2y > 5$

Expressions
$2 + 5(7)$ or $x + 2y$

Constants and Variables
$2, 5, 7, x, y$

Chapter Openers   Like the Algebra Pyramid, chapter openers are designed to help students see how the topics in the upcoming chapter relate to the big picture of the entire course. The chapter openers give information about the importance of the topics in each chapter and how they fit into the overall structure of the course (see pages 1, 101, and 189).

Connection Boxes Connection boxes bridge concepts and ideas that students have learned elsewhere in the text so they see how the concepts are interrelated and build on each other (see pages 102, 237, and 404).

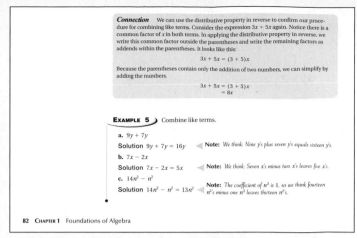

Your Turn Practice Exercises Your Turn practice exercises are found after most examples to give students an opportunity to work problems similar to the examples they have just seen. This practice step makes the text more interactive and provides immediate feedback so students can build confidence in what they are learning (see pages 6, 110, and 331).

Real, Relevant, and Interesting Applications
A large portion of application problems in examples and exercise sets are taken from real situations in science, engineering, health, finance, the arts, or just everyday life. The real-world applications illustrate the everyday use of basic algebraic concepts and encourage students to apply mathematical concepts to solve problems (see pages 110, 236, and 334).

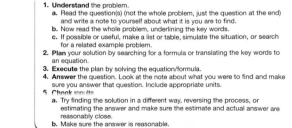

Thorough Explanations Great care is taken to explain not only how to do the math, but also why the math works the way it does, where it comes from, and how it is relevant to students' everyday lives. Knowing all of this gives students a context in which to remember the concept.

Problem-Solving Outline A five-step problem-solving outline is introduced on page 106 of Section 2.1 with the following headings:

1. Understand
2. Plan
3. Execute
4. Answer
5. Check

Application examples throughout the rest of the text follow the steps given in this outline, presenting the headings to model the thinking process clearly (see pages 127, 193, and 212).

PROCEDURE Problem-Solving Outline

1. **Understand** the problem.
   a. Read the question(s) (not the whole problem, just the question at the end) and write a note to yourself about what it is you are to find.
   b. Now read the whole problem, underlining the key words.
   c. If possible or useful, make a list or table, simulate the situation, or search for a related example problem.
2. **Plan** your solution by searching for a formula or translating the key words to an equation.
3. **Execute** the plan by solving the equation/formula.
4. **Answer** the question. Look at the note about what you were to find and make sure you answer that question. Include appropriate units.
5. **Check** results.
   a. Try finding the solution in a different way, reversing the process, or estimating the answer and make sure the estimate and actual answer are reasonably close.
   b. Make sure the answer is reasonable.

Warning Boxes   Warning boxes alert students to common mistakes and false assumptions that students often make and explain *why* these are incorrect (see pages 226, 344, and 393).

The word FOIL is a popular way to remember the process of multiplying two binomials. FOIL stands for **First Outer Inner Last**. We will use Example 2 to demonstrate.

Warning: FOIL only helps keep track of products when multiplying two binomials. When multiplying larger polynomials, just remember to multiply every term in the second polynomial by every term in the first polynomial.

First terms: $x \cdot x = x^2$
$(x + 4)(x + 3)$

Outer terms: $x \cdot 3 = 3x$
$(x + 4)(x + 3)$

Inner terms: $4 \cdot x = 4x$
$(x + 4)(x + 3)$

Last terms: $4 \cdot 3 = 12$
$(x + 4)(x + 3)$

Of Interest

After exiting the plane, a skydiver accelerates from 0 to approximately 110 miles per hour (terminal velocity) in about 10 seconds and falls about 1000 feet. Terminal velocity means that air resistance balances out gravitational acceleration so that the skydiver falls at a constant 110 miles per hour until deploying the parachute. At terminal velocity, the skydiver falls at a rate of about 1000 feet every 6 seconds. Of course, the speed of the fall can be changed by varying body position to increase or decrease air resistance.

Of Interest Boxes   Of Interest boxes are positioned throughout the text to offer a unique perspective on content that some students might otherwise consider to be ho-hum mathematics. Sometimes containing trivia and other times historical notes, Of Interest boxes are designed to enhance the learning process by making concepts fun, interesting, and memorable (see pages 23, 145, and 392).

Puzzle Problems   These mathematical brainteasers, often solved without a formulaic approach, appear at the end of selected exercise sets to encourage critical thinking (see pages 117, 422, and 450).

PUZZLE PROBLEM

Fill in each square with a number 1 through 9 so that the sum of the numbers in each row, each column, and the two diagonals is the same.

Collaborative Exercises   These exercises, which appear once per chapter, encourage students to work in groups to discuss mathematics and use the topics from a particular section or group of sections to solve a problem (see pages 152, 220, and 312).

Collaborative Exercises OCCUPATION GROWTH

Complete the following table by calculating the amount of change and the percent of the increase for each occupation, then answer the questions. Round to the nearest tenth of a percent.

The 10 Fastest-Growing Occupations, 2002–2012

Occupation	Employment		Change	
	2 002	2012	Amount	Percent
Medical assistants	364,600	579,400		
Network systems and data communications analysts	186,000	292,000		
Physician assistants	63,000	93,800		
Social and human service assistants	305,200	453,900		
Home health aides	579,700	858,700		
Medical records and health information technicians	146,900	215,600		
Physical therapist aides	37,000	54,100		
Computer software engineers, applications	394,100	573,400		
Computer software engineers, systems software	281,100	408,900		
Physical therapist assistants	50,200	72,600		

Source: Bureau of Labor Statistics, Office of Occupational Statistics and Employment Projections

1. In 2002, in which occupation were the greatest number of people employed?

2. By 2012, which occupation is projected to have the greatest number of people employed?

3. Explain why the two occupations you listed in Exercises 1 and 2 are not at the top of the list.

4. Which is a better indicator of the demand for people in a particular occupation: the number of people employed in a particular year, the amount of change projected from 2002 to 2012, or the percent of increase in employment from 2002 to 2012?

5. Based on your conclusions in Exercise 4, what occupation will have the greatest demand? What college majors might have the greatest potential for employment in that occupation? Write your conclusions and present them to the class.

Calculator Tips   The relevant functions of calculators (scientific or graphing, depending on the topic) are explained and illustrated throughout the text in the optional Calculator Tips feature. In addition, an occasional calculator icon ▦ in the exercise sets indicates that the problem is designed to be solved using a calculator, though one is not required (see pages 18, 49, and 306).

▦ Calculator
TIPS

To enter scientific notation on a graphing or scientific calculator, we can use the [EE] function. For example, to enter 4.2×10^{-3}, type

[4] [.] [2] [EE] [(−)] [3] (You may have to press [2nd] to get to the [EE] function.)

Notice that the base 10 is understood and is not entered when using the [EE] function. We can also enter scientific notation without using [EE] like this:

[4] [.] [2] [×] [1] [0] [^] [(−)] [3]

After entering scientific notation, if you press [ENTER], the calculator will change the number to standard form. However, if the number has too many digits for the calculator to display, then it will leave it in scientific notation. Similarly, if the answer to a calculation is too large for the calculator to display, then it will show the result in scientific form.

REVIEW EXERCISES

[1.1]   **1.** Write a set containing the last names of the first four presidents of the United States.

[1.1]   **2.** Simplify: $|-7|$

[1.2]   **3.** Find the missing number that makes $\frac{40}{48} = \frac{?}{6}$ true.

[1.2]   **4.** Use $<$, $>$, or $=$ to make a true statement: $-\frac{4}{5}$ ■ $-\frac{8}{20}$

[1.2]   **5.** Find the prime factorization of 100.

[1.2]   **6.** A company report indicates that 78 employees out of 91 have medical flexible spending accounts. What fraction of the employees have medical flexible spending accounts? Express the fraction in lowest terms.

Review Exercises   Since continuous review is important in any mathematics course, this text includes Review Exercises at the end of each exercise set. These exercises review previously learned concepts not only to keep the material fresh for students, but also to serve as a foundational review for the discussion in the upcoming section (see pages 40, 236, and 399).

Chapter Summaries and Review Exercises   An extensive Summary at the end of each chapter provides a list of defined terms referenced by section and page number, a two-column summary of key concepts, and a list of important formulas appearing in that chapter. A set of Review Exercises is also provided with answers to all Review Exercises provided in the back of the book (see pages 177–183, 252–261, and 355–369).

Chapter 5   Summary

Defined Terms

Section 5.1
Scientific notation (p. 378)

Section 5.2
Monomial (p. 387)
Coefficient (p. 389)
Degree of a monomial (p. 389)
Polynomial (p. 389)

Polynomial in one variable (p. 390)
Binomial (p. 390)
Trinomial (p. 390)
Degree of a polynomial (p. 391)

Section 5.5
Conjugates (p. 428)

Procedures, Rules, and Key Examples

Procedures/Rules	Key Examples
Section 5.1 Exponents and Scientific Notation If the base of an exponential form is a negative number and the exponent is even, then the product is positive. If the base is a negative number and the exponent is odd, then the product is negative.	**Example 1:** Evaluate. **a.** $(-2)^4 = (-2)(-2)(-2)(-2) = 16$ **b.** $(-2)^5 = (-2)(-2)(-2)(-2)(-2) = -32$
…numbers, where $b \neq 0$ and n is a natural number, …ber and $a \neq 0$, then $a^0 = 1$. …ber, where $a \neq 0$ and n is a natural number, then …ber, where $a \neq 0$ and n is a natural number, then …numbers, where $a \neq 0$ and $b \neq 0$ and n is a natural $\left(\frac{}{} \right)^{-n} = \left(\frac{b}{a} \right)^n$.	**Example 2:** Simplify. **a.** $\left(\frac{3}{4} \right)^2 = \frac{3^2}{4^2} = \frac{9}{16}$ **b.** $7^0 = 1$ **c.** $2^{-4} = \frac{1}{2^4} = \frac{1}{16}$ **d.** $\frac{1}{3^{-2}} = 3^2 = 9$ **e.** $\left(\frac{2}{3} \right)^{-4} = \left(\frac{3}{2} \right)^4 = \frac{81}{16}$

Chapter 5   Review Exercises

For Exercises 1–6, answer true or false.

[5.1]   **1.** A negative exponent will always make the coefficient negative.

[5.2]   **2.** The degree of a binomial is two.

[5.4]   **3.** To raise an exponential form to a power, multiply the exponents.

[5.5]   **4.** Conjugates occur when both binomials contain subtraction.

[5.5]   **5.** FOIL can be used for all types of polynomial multiplication.

[5.5]   **6.** $(x - 4)^2 = x^2 + 16$

For Exercises 7–10, complete the rule.

[5.3]   **7.** To add polynomials, _____ like terms.

[5.4]   **8.** To multiply monomials:
   a. Multiply the _____.
   b. Add the _____ of the like bases.
   c. Rewrite any unlike bases in the product.

Chapter Practice Tests A Practice Test follows each set of chapter review exercises. The problem types in the practice tests correlate to the short-answer tests in the *Printed Test Bank*. This is especially comforting for students who have math anxiety or who experience test anxiety (see pages 99, 186, and 262).

Cumulative Reviews Cumulative Review Exercises appear after Chapters 3, 6, 9, and 13. These exercises help students stay current with all the material they have learned and help prepare them for midterm and final exams (see pages 269 and 534).

Chapter 5 Practice Test

For Exercises 1 and 2, evaluate the exponential form.

1. 2^{-3}

2. $\left(\dfrac{2}{3}\right)^{-2}$

3. Write 6.201×10^{-3} in standard form.

4. Write 275,000,000 in scientific notation.

For Exercises 5 and 6, identify the degree.

5. $-7x^2y$

6. $4x^2 - 9x^4 + 8x - 7$

7. Evaluate $-6mn - n^3$, where $m = 4$ and $n = -2$.

8. Combine like terms and write the resulting polynomial in descending order of degree.
   $-5x^4 + 7x^2 + 6x^2 - 5x^4 + 12 - 6x^3 + x^2$

For Exercises 9 and 10, add or subtract and write the resulting polynomial in descending order of degree.

9. $(3x^2 + 4x - 2) + (5x^2 - 3x - 2)$

10. $(7x^4 - 3x^2 + 4x + 1) - (2x^4 + 5x - 7)$

For Exercises 11–18, multiply.

11. $(4x^2)(3x^5)$

12. $(2ab^3c^7)(-a^5b)$

13. $(4xy^3)^2$

14. $3x(x^2 - 4x + 5)$

15. $-6t^2u(4t^3 - 8tu^2)$

16. $(n - 1)(n + 4)$

17. $(2x - 3)^2$

18. $(x + 2)(x^2 - 4x + 3)$

19. Write an expression for the area of the shape shown.

   $2n - 5$ □ $3n + 4$

For Exercises 20–25, simplify.

20. $x^9 \div x^4$

21. $\dfrac{(x^3)^{-2}}{x^4 \cdot x^{-5}}$

22. $\dfrac{(3y)^{-2}}{(x^3y^2)^{-3}}$

24. $\dfrac{x^2 - x - 12}{x + 3}$

25. $\dfrac{15x^2 - 22x + 14}{3x - 2}$

Chapters 1–6 Cumulative Review Exercises

For Exercises 1–6, answer true or false.

[5.4] 1. $x^3 \cdot x^4 = x^{12}$

[6.1] 2. The GCF of 12 and 5 is 1.

[1.3] 3. The commutative property can be used for both addition and subtraction.

[2.2] 4. $2x - y^2 = 1$ is a linear equation.

[1.7] 5. The expression $5 + x$ can be simplified to equal $5x$.

[6.1] 6. $4x + 12 = 2(2x + 6)$ is factored completely.

For Exercises 7–10, fill in the blank.

[6.3] 7. To factor a trinomial of the form $ax^2 + bx + c$, where $a \neq 1$, by grouping:

   a. Look for a monomial _____ in all the terms. If there is one, factor it out.

   b. Multiply a and c.

   c. Find two factors of this product whose sum is b.

   d. Write a four-term polynomial in which bx is written as the sum of two like terms whose coefficients are the two numbers you found in step c.

   e. Factor by _____.

[5.2] 8. To write a polynomial in descending order, place the _____ degree term first, then the next highest degree, and so on.

[6.4] 9. To factor a difference of squares, we use the rule $a^2 - b^2 =$ _____.

[6.6] 10. To solve a quadratic equation:

   a. Manipulate the equation as needed so that one side is an expression and the other side is _____.

   b. Write the expression in _____ form.

   c. Use the zero-factor theorem to solve.

[4.3] 11. What are the x- and y-intercepts on a graph?

[5.2] 12. What is a binomial?

Supplements for *Elementary and Intermediate Algebra,* Second Edition

Student Supplements

STUDENT'S SOLUTIONS MANUAL
- By Doreen Kelly, *Mesa Community College.*
- Contains complete solutions to the odd-numbered section exercises and solutions to all of the section-level Review Exercises, Chapter Review Exercises, Practice Tests, and Cumulative Review Exercises.
ISBN: 0-321-37496-7

DIGITAL VIDEO TUTOR
- Complete set of digitized videos on CD-ROM for student use at home or on campus.
- Ideal for distance learning or supplemental instruction.
ISBN: 0-321-42249-X

MATHXL® TUTORIALS ON CD
- Provides algorithmically generated practice exercises that correlate to the exercises in the textbook.
- Every exercise is accompanied by an example and a guided solution, and selected exercises may also include a video clip to help students visualize concepts.
- The software provides helpful feedback for incorrect answers and can generate printed summaries of students' progress.
ISBN: 0-321-37580-7

ADDISON-WESLEY MATH TUTOR CENTER
- Staffed by qualified mathematics instructors.
- Provides tutoring on examples and odd-numbered exercises from the textbook through a registration number with a new textbook or purchased separately.
- Accessible via toll-free telephone, toll-free fax, e-mail, or the Internet.
- White Board technology allows tutors and students to actually see problems worked while they "talk" in real time over the Internet during tutoring sessions.
www.aw-bc.com/tutorcenter

Instructor Supplements

ANNOTATED INSTRUCTOR'S EDITION
- Includes answers to all exercises, including Puzzle Problems and Collaborative Exercises, printed in bright blue near the corresponding problems.
- Useful teaching tips are printed in the margin.
- A ★ icon, found in the AIE only, indicates especially challenging exercises in the exercise sets.
ISBN: 0-321-36855-X

INSTRUCTOR'S SOLUTIONS MANUAL
- By Doreen Kelly, *Mesa Community College.*
- Contains complete solutions to all even-numbered section exercises, Puzzle Problems, and Collaborative Exercises.
ISBN: 0-321-37504-1

INSTRUCTOR AND ADJUNCT SUPPORT MANUAL
- Includes resources designed to help both new and adjunct faculty with course preparation and classroom management.
- Offers helpful teaching tips specific to the sections of the text.
ISBN: 0-321-37498-3

PRINTED TEST BANK
- By Laura Hoye, *Trident Technical College.*
- Contains one diagnostic test per chapter, four free-response tests per chapter, one multiple-choice test per chapter, a mid-chapter check-up for each chapter, one midterm exam, and two final exams.
ISBN: 0-321-37497-5

VIDEOTAPES
- A series of lectures presented by author-team member Ellyn Gillespie, among others, correlated directly to the chapter content of the text. A video symbol at the beginning of each exercise set references the videotape or CD (*see also* Digital Video Tutor).
- Include a pause-the-video feature that encourages students to stop the videotape, work through an example, and resume play to watch the video instructor work through the same example.
ISBN: 0-321-37500-9

Instructor Supplements

TESTGEN®

- Enables instructors to build, edit, print, and administer tests.
- Features a computerized bank of questions developed to cover all text objectives.
- Alogrithmically based, allowing instructors to create multiple, but equivalent, verisions of the same questions or test with the click of a button.
- Instructors can also modify test bank questions or add new questions.
- Tests can be printed or administrated online.
- Available on a dual-platform Windows/Macintosh CD-ROM.
  ISBN: 0-321-37501-7

MATHXL®

MathXL® is a powerful online homework, tutorial, and assessment system that accompanies your Addison-Wesley textbook in mathematics or statistics. With MathXL, instructors can create, edit, and assign online homework and tests using algorithmically generated exercises correlated at the objective level to your textbook. They can also create and assign their own online exercises and import TestGen tests for added flexibility. All student work is tracked in MathXL's online gradebook. Students can take chapter tests in MathXL and receive personalized study plans based on their test results. The study plan diagnoses weaknesses and links students directly to tutorial exercises for the objectives they need to study and retest. Students can also access supplemental video clips directly from selected exercises. MathXL® is available to qualified adopters. For more information visit our Web site at www.mathxl.com, or contact your Addison-Wesley sales representative.

MYMATHLAB

MyMathLab is a series of text-specific, easily customizable online courses for Addison-Wesley textbooks in mathematics and statistics. MyMathLab is powered by CourseCompass™—Pearson Education's online teaching and learning environment—and by MathXL®—our online homework, tutorial, and assessment system. MyMathLab gives instructors the tools they need to deliver all or a portion of their course online, whether students are in a lab setting or working from home. MyMathLab provides a rich and flexible set of course materials, featuring free-response exercises that are algorithmically generated for unlimited practice and mastery. Students can also use online tools, such as video lectures, animations, and a multimedia textbook, to independently improve their understanding and performance. Instructors can use MyMathLab's homework and test managers to select and assign online exercises correlated directly to the textbook, and they can also create and assign their own online exercises and import TestGen tests for added flexibility. MyMathLab's online gradebook—designed specifically for mathematics and statistics—automatically tracks students' homework and test results and gives the instructor control over how to calculate final grades. Instructors can also add offline (paper-and-pencil) grades to the gradebook. MyMathLab is available to qualified adopters. For more information, visit our Web site at www.mymathlab.com or contact your Addison-Wesley sales representative.

Acknowledgments

Many people gave of themselves in so many ways during the development of this text. Mere words cannot contain the fullness of our gratitude. Though the words of thanks that follow may be few, please know that our gratitude is great.

We would like to thank the following people who gave of their time in reviewing the text. Their thoughtful input was vital to the development of the text.

Khadija Ahmed, *Monroe County Community College*
Frank Attanucci, *Scottsdale Community College*
Daniel Bacon, *Massasoit Community College*
Kerry Bailey, *Laramie Community College*
Debra Bryant, *Tennessee Technological University*
Baruch Cahlon, *Oakland University*
Pat Cook, *Weatherford College*
Patrick S. Cross, *University of Oklahoma*
Cheryl B. Davids, *Central Carolina Technical College*
Elias Deeba, *University of Houston–Downtown Campus*
Stephan DeLong, *Tidewater Community College–Virginia Beach Campus*
Laura Ferguson, *Weatherford College*
Margret Hathaway, *Kansas City Kansas Community College*
Allen Miller, *South Plains College*
Carol Murphy, *Miramar College*
Joanne Peeples, *El Paso Community College*
Larry Pontaski, *Pueblo Community College*
Jack Sharp, *Floyd College*
Linda Shoesmith, *Scott Community College*
James Vicich, *Scottsdale Community College*
Linda J. Wagner, *Indiana University Purdue University Fort Wayne*
Walter Wang, *Baruch College*

We would like to extend a heartfelt thank-you to everyone at Addison-Wesley for giving so much to this project. We would like to offer special thanks to Jennifer Crum and Greg Tobin, who believed in us and gave us the opportunity; to Elizabeth Bernardi, Emily Ragsdale, Lauren Morse, and Kari Heen, for keeping us on track; and also to Jay Jenkins, Tracy Rabinowitz, and Alexandra Waibel for the encouragement and working so hard to get us "out there."

A very special thank-you to Dennis Schaefer, who created the beautiful, student-friendly text design and cover; to Ron Hampton, whose keen eyes and editorial sense were invaluable during production; and to Lisa Laing, Gordon Laws, Sam Blake, and all of the fabulous people at Pre-Press Company, Inc. for working so hard to put together the finished pages.

To Sharon Smith, Ruth Berry, Mary Ann Perry, and all the people involved in developing the media supplements package, we are so grateful for all that you do. A special thank-you to Laura Hoye, who created the excellent *Printed Test Bank,* and to Doreen Kelly for her work on the solutions manuals. Thank you to Cheryl Davids, Perian Herring, Elizabeth Morrison, and Vince Koehler for their wonderful job of accuracy checking the manuscript and page proofs. A big thank-you goes to Lisa Sims, Cheryl Cantwell, and Laura Wheel for their help keeping the application problems fresh and up to date.

Finally, we'd like to thank our families for their support and encouragement during the process of developing and revising this text.

Tom Carson

Ellyn Gillespie Stewart

Bill Jordan

To the Student

Why do I have to take this course?

Often this is one of the first questions students ask when they find out they must take an algebra course, especially when they believe that they will never use the math again. You may think that you will not use algebra directly in daily life, and you may assume that you can get by knowing enough arithmetic to balance a checkbook. So, what is the real point of education? Why don't colleges just train students for the jobs they want? The purpose of education is not just job training but also exercise—mental exercise. An analogy that illustrates this quite well is the physical training of athletes.

During the off-season, athletes usually develop an exercise routine that may involve weight lifting, running, swimming, aerobics, or maybe even dance lessons. Athletes often seek out a professional trainer to push them further than they might push themselves. The trainer's job is not to teach an athlete better technique in his or her sport, but to develop the athlete's raw material—to work the body for more strength, stamina, balance, etc. Educators are like physical trainers, and going to college is like going to the gym. An educator's job is to push students mentally and work the "muscle" of the mind. A college program is designed to develop the raw material of the intellect so the student can be competitive in the job market. After the athlete completes the off-season exercise program, he or she returns to the coach and receives specific technique training. Similarly, when students complete their college education and begin a job, they receive specific training to do that job. If the trainer or teacher has done a good job with hardworking clients, the coaching or job training should be absorbed easily.

Taking this analogy a step further, a good physical trainer finds the athlete's weaknesses and designs exercises that the athlete has never performed before, and then pushes him or her accordingly. Teachers do the same thing—their assignments are difficult in order to work the mind effectively. If you feel "brain-strained" as you go through your courses, that's a good sign that you are making progress, and you should keep up the effort.

The following study system is designed to help you in your academic workouts. As teachers, we find that most students who struggle with mathematics have never really *studied* math. A student may think, "Paying attention in class is all I need to do." However, when you watch a teacher do math, keep in mind that you are watching a pro. Going back to the sports analogy, you can't expect to shoot a score of 68 in golf by watching Tiger Woods. You have to practice golf yourself in order to learn and improve. The study system outlined in the following pages will help you get organized and make efficient use of your time so that you can maximize the benefits of your course work.

What do I need to do to succeed?

We believe there are four prerequisites one must have or acquire in order to succeed in college:

1. **Positive Attitude**
2. **Commitment**
3. **Discipline**
4. **Time**

A **Positive Attitude** is most important because commitment and discipline flow naturally from it. Consider Thomas Edison, inventor of the lightbulb. He tried more than 2000 different combinations of materials for the filament before he found the successful combination. When asked by a reporter about all his failed attempts, Edison replied, "I didn't fail once, I invented the lightbulb. It was just a 2000-step process." Recognize that learning can be uncomfortable and difficult, and mistakes are part of the process. So, embrace the learning process with its discomforts and difficulties, and you'll see how easy it is to be committed and disciplined.

Commitment means giving everything you've got with no turning back. Consider Edison again. Imagine the doubts and frustrations he must have felt trying material after material for the filament of his lightbulb without success. Yet he forged ahead. In Edison's own words, "Our greatest weakness lies in giving up. The most certain way to succeed is always to try just one more time."

Discipline means doing things you should be doing even when you don't want to. According to author W. K. Hope, "Self-discipline is when your conscience tells you to do something and you don't talk back." Staying disciplined can be difficult given all the distractions in our society. The best way to develop discipline is to create a schedule and stick to it.

Make sure you have enough **Time** to study properly, and make sure that you manage that time wisely. Too often, students try to fit school into an already full schedule. Take a moment to complete the exercise that follows under "How do I do it all?" to make sure you haven't overcommitted yourself. Once you have a sense of how much time school requires, read on about the study system that will help you maximize the benefits of your study time.

How do I do it all?

Now that we know a little about what it takes to be successful, let's make sure that you have enough time for school. In general, humans have a maximum of 60 hours of productivity per week. Therefore, as a guide, let's set the maximum number of work hours, which means time spent at your job(s) and at school combined, at 60 hours per week. Use the following exercise to determine the time you commit to your job and to school.

Exercise: Calculate the time that you spend at your job and at school.

1. Calculate the total hours you work in one week.
2. Calculate the number of hours you are in class each week.
3. Estimate the number of hours you should expect to spend outside of class studying.
   A general rule is to double the number of hours spent in class.
4. Add your work hours, in-class hours, and estimated out-of-class hours to get your total time commitment.
5. Evaluate the results. *See below.*

Evaluating the Results
   a. If your total is greater than 60 hours, you will probably find yourself feeling overwhelmed. This feeling may not occur at first, but doing that much for an extended period of time will eventually catch up with you, and something may suffer. It is in your best interest to cut back on work or school until you reduce your time commitment to under 60 hours per week.
   b. If your total is under 60 hours, good. Be sure you consider other elements in your life, such as your family's needs, health problems, commuting, or anything that could make demands on your time. Make sure that you have enough time for everything you put in your life. If you do not have enough time for everything, consider what can be cut back. It is important to note that it is far better to pass fewer classes than to fail many.

How do I make the best use of my time? How should I study?

We've seen many students who had been making D's and F's in mathematics transform their grades to A's and B's by using the study system that follows.

The Study System

Your Notebook

1. Get a loose-leaf binder so that you can put papers in and take them out without ripping any pages.
2. Organize the notebook into four parts:
   a. Class notes
   b. Homework
   c. Study sheets (a single piece of paper for each chapter onto which you will transfer procedures from your notes)
   d. Practice tests

In Class

Involve your mind completely.

1. **Take good notes.** Use three different colors. Most students like using red, blue, and black (pencil).

   - Use the red pen to write *definitions*. Also, use this color to mark problems or items that the instructor indicates will be covered on a test.
   - Use the blue pen to write procedures and rules.
   - Use the pencil to write problems and explanations.

When taking notes, don't just write the solutions to the problems that the instructor works out, but write the explanations as well. To the side of the problem, make notes about each step so that you remember the significance of the steps. Pay attention to examples or issues the instructor emphasizes: they will usually appear on a test, so make an effort to include them in your notes. Include common errors that the instructor points out or any words of caution. If you find it is difficult to write and pay attention at the same time, ask your instructor if you can record the lectures with a tape recorder. If your instructor follows the text closely, when he or she points out definitions or procedures in the text, highlight them or write a page reference in your notes. You can then write these referenced items in their proper place in your notes after class.

2. **Answer the instructor's questions.** This does not mean you have to answer every question verbally, but you should think through every question and answer in your mind, write an answer in your notes, or answer out loud.
3. **Ask questions.** You may find it uncomfortable to ask questions in front of other people, but keep in mind that if you have a question, then it is very likely that someone else has the same question. If you still don't feel like asking in class, then be sure to ask as soon as class is over. The main thing is to get that question answered as soon as possible because in mathematics, one misconception can grow and cause confusion in the future.

After Class

Prepare for the next class meeting as if you were going to have a test on everything covered so far. To make the most of your time, set aside a specific time that is reserved for math. Since there are often many distractions at home, study math while on campus in a quiet place such as the library or tutorial lab. Staying on campus also allows you to visit your instructor or tutorial services if you have a question that you cannot resolve. Here is a systematic approach to organizing your math study time outside of class:

1. As soon as possible, go over your notes. Clarify any sentences that weren't quite complete. Fill in any page-referenced material.

2. Read through the relevant section(s) in the text again, and make sure you understand all the examples.

3. Transfer each new procedure or rule to your study sheet for that chapter. You might also write down important terms and their definitions. Make headings for each objective in the section(s) you covered that day. Write the procedures and definitions in your own words.

4. Study the examples worked in class. Transfer each example (without the solution) to the practice test section of your notebook, leaving room to work it out later.

5. Use your study sheet to do the assigned practice problems. As soon as you finish each problem, check your answer in the back of the book or in the *Student's Solutions Manual*. If you did not get it correct, then immediately revisit the problem to determine your error (see the box on troubleshooting). If you are asked to do even-numbered problems, then work odd-numbered problems that mirror the even problems. This way you can check your answers for the odd-numbered problems and then work the even-numbered problems with confidence.

> **Troubleshooting:** For the problems that you do not get correct, first look for simple arithmetic errors. If you find no arithmetic errors, then make sure you followed the procedure or rules correctly. If you followed the or rules correctly, then you have likely interpreted something incorrectly, either with the problem or the rules. Read the instructions again carefully and try to find similar examples in your notes or in the book. If you still can't find the mistake, go on to something else for a while. Often after taking a fresh look you will see the mistake right away. If all these tips fail to resolve the problem, then mark it as a question for the next class meeting.

6. After completing the homework, prepare a quiz for yourself. Select one of each type of homework problem. Don't just pick the easy ones! Set the quiz aside for later.

7. After making the quiz, study your study sheet. To test your understanding, write the rules and procedures in your own words. Do not focus on memorizing the wording in the textbook.

8. Now it is time to begin preparing for the next class meeting. Read the next section(s) to be covered. Don't worry if you do not understand everything. The idea is to get some feeling for the topics to be discussed so that the class discussion will actually be the second time you encounter the material, not the first. While reading, you might mark points that you find difficult so that if the instructor does not clear them up, you can ask about them. Also, attempt to work through the examples. The idea is for you to do as much as possible on your own before class so that the in-class discussion merely ties together loose ends and solidifies the material.

9. After you have finished preparing for the next day, go back and take the quiz that you made. If you get all the answers correct, then you have mastered the material. If you have difficulty, return to your study sheet and repeat the exercise of writing explanations for each objective.

How do I ace the test?

Preparing for a Test

If you have followed all of the preceding suggestions, then preparing for a test should be quite easy.

1. **Read.** In one sitting, read through all of your notes on the material to be tested. In the same sitting, read through the book, observing what the instructor has highlighted in class. To guide your studies, look at any information or documents provided by your instructor that address what will be on the test. The examples given by the instructor will usually reflect the test content.
2. **Study.** Compare your study sheet to the summary in the book at the end of the chapter. Use both to guide you in your preparation, but keep in mind that the sheet you made from your notes reflects what the instructor has emphasized. Make sure you understand everything on your study sheet. Write explanations of the objectives until you eliminate all hesitation about how to approach an objective. The rules and procedures should become second nature.
3. **Practice.** Create a game plan for the test by writing the rule, definition, or procedure that corresponds to each problem on your practice tests. (Remember, one practice test is in your book and the other you made from your notes.) Next, work through the practice tests without referring to your study sheet or game plan.
4. **Evaluate.** Once you have completed the practice tests, check them. The answers to the practice tests in the book are in the answer section in the back of the book. Check the practice test that you made using your notes.
5. **Repeat.** Keep repeating steps 2, 3, and 4 until you get the right answer for every problem on the practice tests.

Taking a Test

1. When the test hits your desk, don't look at it. Instead, do a memory dump. On paper, dump everything you think you might forget. Write out rules, procedures, notes to yourself, things to watch out for, special instructions from the instructor, and so on. This will help you relax while taking the test.
2. If you get to a problem that you cannot figure out, skip it and move on to another problem. First do all the problems you are certain of and then return to the ones that are more difficult.
3. Use all the time given. If you finish early, check to make sure you have answered every problem. Even if you cannot figure out a problem, at least guess. Use any remaining time to check as many problems as possible by doing them over on separate paper.

If You Are Not Getting Good Results

Evaluate the situation. What are you doing or not doing in the course? Are you doing all the homework and taking the time to prepare as suggested? Sometimes people misjudge how well they have prepared. Just like an athlete, to excel, you will need to prepare beyond the minimum requirements.

Here are some suggestions:

1. **Go to your instructor.** Ask your instructor for help to evaluate what is wrong. Make use of your instructor's office hours, because this is your opportunity for individual attention.
2. **Get a tutor.** If your school has a tutorial service, and most do, do your homework there so you can get immediate help when you have a question.

3. **Use Addison-Wesley's support materials.** Use the support materials that are available with your text, which include a *Student's Solutions Manual*, the Addison-Wesley Math Tutor Center, videotapes (available on CD-ROM as well), and tutorial software available on CD and online. Full descriptions of these supplements are provided on page xiii of this book.

4. **Join a study group.** Meet regularly with a few people from class and go over material together. Quiz each other and answer questions. Meet with the group only after you have done your own preparation so you can then compare notes and discuss any issues that came up in your own work. If you have to miss class, ask the study group for the assignments and notes.

We hope you find this study plan helpful. Be sure to take the Learning Styles Inventory that follows to help determine your primary learning style. Good luck!

Learning Styles Inventory

What is your personal learning style?

A learning style is the way in which a person processes new information. Knowing your learning style can help you make choices in the way you study and focus on new material. Below are fifteen statements that will help you assess your learning style. After reading each statement, rate your response to the statement using the scale below. There are no right or wrong answers.

3 = Often applies   **2** = Sometimes applies   **1** = Never or almost never applies

_____ **1.** I remember information better if I write it down or draw a picture of it.

_____ **2.** I remember things better when I hear them instead of just reading or seeing them.

_____ **3.** When I receive something that has to be assembled, I just start doing it. I don't read the directions.

_____ **4.** If I am taking a test, I can "visualize" the page of text or lecture notes where the answer is located.

_____ **5.** I would rather the professor explain a graph, chart, or diagram to me instead of just showing it to me.

_____ **6.** When learning new things, I want to do it rather than hear about it.

_____ **7.** I would rather the instructor write the information on the board or overhead instead of just lecturing.

_____ **8.** I would rather listen to a book on tape than read it.

_____ **9.** I enjoy making things, putting things together, and working with my hands.

_____ **10.** I am able to conceptualize quickly and visualize information.

_____ **11.** I learn best by hearing words.

_____ **12.** I have been called hyperactive by my parents, spouse, partner, or professor.

_____ **13.** I have no trouble reading maps, charts, or diagrams.

_____ **14.** I can usually pick up on small sounds like bells, crickets, frogs, or distant sounds like train whistles.

_____ **15.** I use my hands and gesture a lot when I speak to others.

Write your score for each statement beside the appropriate statement number below.

Then add the scores in each column to get a total score for that column.

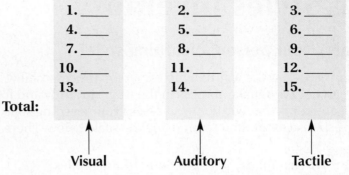

1. ____	2. ____	3. ____
4. ____	5. ____	6. ____
7. ____	8. ____	9. ____
10. ____	11. ____	12. ____
13. ____	14. ____	15. ____

Total:

Visual **Auditory** **Tactile**

The largest total of the three columns indicates your dominant learning style.

Visual learners learn best by seeing. If this is your dominant learning style, you should focus on learning strategies that involve seeing. The color coding in the study system (see page xix) will be especially important. The same color coding is used in the text. Draw lots of diagrams, arrows, and pictures in your notes to help you see what is happening. Reading your notes, study sheets, and text repeatedly will be an important strategy.

Auditory learners learn best by hearing. If this is your dominant learning style, you should use learning strategies that involve hearing. After getting permission from your instructor, bring a tape recorder to class to record the discussion. When you study your notes, play back the tape. Also, when you learn rules, say the rule over and over. As you work problems, say the rule before you do the problem. You may also find the video tapes beneficial in that you can hear explanations of problems taken from the text.

Tactile (also known as kinesthetic) learners learn best by touching or doing. If this is your dominant learning style, you should use learning strategies that involve doing. Doing lots of practice problems will be important. Make use of the Your Turn exercises in the text. These are designed to give you an opportunity to do problems like the examples as soon as the topic is discussed. Writing out your study sheets and doing your practice tests repeatedly will be important strategies for you.

Note that the study system developed in this text is for all learners. Your learning style will help you decide what aspects and strategies in the study system to focus on, but being predominantly an auditory learner does not mean that you shouldn't read the textbook, do lots of practice problems, or use the color-coding system in your notes. Auditory learners can benefit from seeing and doing, and tactile learners can benefit from seeing and hearing. In other words, do not use your dominant learning style as a reason for not doing things that are beneficial to the learning process. Also, remember that the Learning Strategy boxes presented throughout the text provide tips to help you use your personal learning style to your advantage.

This Learning Styles Inventory is adapted from *Cornerstone: Building on Your Best*, Third Edition, by Montgomery/Moody/Sherfield © 2002. Reprinted by permission of Prentice-Hall, Inc., Upper Saddle River, NJ.

CHAPTER

1

Foundations of Algebra

"Three people were at work on a construction site. All were doing the same job, but when each was asked what the job was, the answers varied. 'Breaking rocks,' the first replied. 'Earning my living,' the second said. 'Helping to build a cathedral,' said the third."

—Peter Schultz, German businessman

"Build up your weaknesses until they become your strengths."

—Knute Rockne, Notre Dame football coach 1888–1931

In Peter Schultz's quote, each of the three people answered the question correctly. Notice that the first two people were rather narrowly focused, whereas the third person had a greater vision and appreciation of the finished structure and his or her place in the construction process. Your curriculum and the courses in it are like a building, with the units or chapters like construction materials. Try to keep the big picture in mind as your courses and curriculum build. In the chapter openers of this book, we will discuss how the mathematics in the chapter fits into the bigger picture of the entire course structure to help you see the "cathedral" we are building.

In this chapter, we review the foundation of algebra, which is arithmetic. More specifically, we will review number sets, operations of arithmetic, properties of real numbers, and the evaluation and simplification of expressions. This review is by no means a complete instruction of arithmetic. Since a solid foundation is important to support the rest of our building, if you encounter a topic in this chapter that is a particular weakness for you, then it is important that you consult with your instructor or other more complete sources for extra practice.

1.1 Number Sets and the Structure of Algebra

OBJECTIVES

1. Understand the structure of algebra.
2. Classify number sets.
3. Graph rational numbers on a number line.
4. Determine the absolute value of a number.
5. Compare numbers.

OBJECTIVE 1. Understand the structure of algebra. Learning mathematics is like learning a language. When we learn a language we must learn the alphabet, vocabulary, and sentence structure. Similarly, mathematics has its own alphabet, vocabulary, and sentence structure. In this section we begin the development of the foundation of algebra with an overview of its components and structure. The basic components are **variables** and **constants**.

DEFINITIONS   *Variable:* A symbol that can vary in value.
Constant: A symbol that does not vary in value.

Variables are usually letters of the alphabet, like x or y. Usually constants are symbols for numbers, like $1, 2, \frac{3}{4}, 6.74$, and so on. However, constants can sometimes be symbols like e or the Greek letter π, which each have special numeric values. Variables and constants are used to make **expressions**, **equations**, and **inequalities**.

DEFINITION   *Expression:* A constant, variable, or any combination of constants, variables, and arithmetic operations that describes a calculation.

Examples of expressions:

$$2 + 6 \quad \text{or} \quad 4x - 5 \quad \text{or} \quad \frac{1}{3}\pi r^2 h$$

DEFINITION   *Equation:* A mathematical relationship that contains an equal sign.

Examples of equations:

$$2 + 6 = 8 \quad \text{or} \quad 4x - 5 = 12 \quad \text{or} \quad V = \frac{1}{3}\pi r^2 h$$

> **Connection**   Think of expressions as phrases and equations as complete sentences. The expression $2 + 6$ is read "two plus six," which is not a complete sentence. The equation $2 + 6 = 8$ is read "two plus six is eight." Notice the equal sign translates to the verb *is*, which makes the sentence a complete sentence.

DEFINITION *Inequality:* A mathematical relationship that contains an inequality symbol ($\neq$, $<$, $>$, $\leq$, or $\geq$).

Inequality Symbols and Their Translations

Symbolic form	Translation
$8 \neq 3$	Eight is not equal to three.
$5 < 7$	Five is less than seven.
$7 > 5$	Seven is greater than five.
$x \leq 3$	x is less than or equal to three.
$y \geq 2$	y is greater than or equal to two.

The algebra pyramid shown illustrates how variables, constants, expressions, equations, and inequalities relate. At the foundation of algebra and our pyramid are constants and variables, which are used to build expressions, which in turn are used to build equations and inequalities.

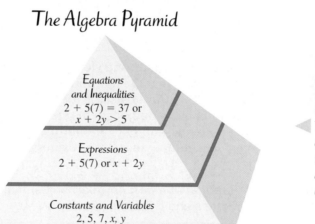

The Algebra Pyramid

Equations and Inequalities
$2 + 5(7) = 37$ or
$x + 2y > 5$

Expressions
$2 + 5(7)$ or $x + 2y$

Constants and Variables
$2, 5, 7, x, y$

Note: *During this course, we will move back and forth between expressions, equations, and inequalities. When we transition from one to the other we will use the algebra pyramid as a quick visual so that it's clear what we are working on.*

OBJECTIVE 2. Classify number sets.   Numbers can be placed into different categories using **sets**.

DEFINITION *Set:* A collection of objects.

Braces are used to indicate a set. For example, the set containing the numbers 1, 2, 3, and 4 would be written $\{1, 2, 3, 4\}$. The numbers 1, 2, 3, and 4 are called the *members* or *elements* of this set.

PROCEDURE Writing Sets

To write a set, write the members or elements of the set separated by commas within braces, $\{\ \}$

EXAMPLE 1 Write the set containing the first five letters of the alphabet.

Answer {A, B, C, D, E}

YOUR TURN Write the set containing the first four months of the year.

Of Interest

There are symbols for the number sets. In particular the set of real numbers is often written as $\mathbb{R}$ or $\mathscr{R}$

Numbers are classified using number sets. The set of *natural numbers* contains the counting numbers 1, 2, 3, 4, ... and is written {1, 2, 3, ...}. The three dots are called *ellipses* and indicate that the numbers continue forever in the same pattern. The set of *whole numbers* contains all of the natural numbers and 0 and is written {0, 1, 2, 3, ...}. The set of *integers* contains all the whole numbers and the opposite (or negative) of every natural number and is written {..., −3, −2, −1, 0, 1, 2, 3, ...}. Notice the integers continue forever in both directions.

A number line is often useful in mathematics. The following number line is marked with the integers.

$$\longleftarrow \overset{}{\underset{-4}{|}} \ \overset{}{\underset{-3}{|}} \ \overset{}{\underset{-2}{|}} \ \overset{}{\underset{-1}{|}} \ \overset{}{\underset{0}{|}} \ \overset{}{\underset{1}{|}} \ \overset{}{\underset{2}{|}} \ \overset{}{\underset{3}{|}} \ \overset{}{\underset{4}{|}} \longrightarrow$$

Though we can mark and view only a portion of the number line, the arrows at the ends indicate that the line and the numbers on it continue on forever in both directions. If we traveled along the number line forever in both directions we would encounter every number in the set of *real numbers*.

Some real numbers are not integers. The set of **rational numbers** contains every real number that can be expressed as a ratio of integers.

DEFINITION *Rational number:* Any real number that can be expressed in the form $\frac{a}{b}$, where *a* and *b* are integers and $b \neq 0$.

Note: *In the definition for a rational number, the notation $b \neq 0$ is important because if the denominator b were to equal 0, then the fraction would be undefined. The reason it is undefined will be explained later.*

For example, $\frac{3}{4}$ is a rational number because 3 and 4 are integers. Note that numbers like 0.75 and 75% are also rational numbers because they can be written as $\frac{3}{4}$.

It is important to realize that the definition of a rational number does not state that the number *must* be expressed in the form $\frac{a}{b}$. Rather, if a number *can* be expressed in the form $\frac{a}{b}$, it is a rational number. For example, the number 5 is a rational number because it can be expressed as $\frac{5}{1}$.

ANSWER

{January, February, March, April}

EXAMPLE 2 Determine whether the given number is a rational number.

a. $\dfrac{2}{3}$

Answer Yes, because 2 and 3 are integers.

b. 0.4

Answer Yes, 0.4 is a rational number because it can be expressed as $\dfrac{4}{10}$, and 4 and 10 are integers.

c. $0.\overline{6}$

Answer A bar written over a decimal digit indicates that the digit repeats without end, so $0.\overline{6} = 0.66666\ldots$, and we say these decimal numbers are nonterminating decimal numbers. Usually, we encounter these numbers as quotients in certain division problems, such as when we write certain fractions as decimals. In this case, $0.\overline{6}$ is the decimal equivalent of $\dfrac{2}{3}$. Since $0.\overline{6}$ can be expressed as the fraction $\dfrac{2}{3}$, it is a rational number.

Note: *All nonterminating decimal numbers with repeating digits are rational numbers because they can be expressed as fractions with integers in both the numerator and denominator.*

YOUR TURN Determine whether the given number is a rational number.

a. 0.56 **b.** -7

Not all numbers can be expressed as a ratio of integers. One such number is π (pronounced "pie"). Because the exact value of π cannot be expressed as a ratio of integers, it is categorized as an **irrational number**.

DEFINITION *Irrational number:* Any real number that is not rational.

Some other irrational numbers are $\sqrt{2}$ or $\sqrt{3}$. (Square roots will be explained in more detail in Section 1.5.) Because an irrational number cannot be written as a ratio of integers, if a calculation involves an irrational number, we must leave it in symbolic form or use a rational number approximation. We can approximate π with rational numbers like 3.14 or $\dfrac{22}{7}$.

The following figure illustrates how the number sets relate in the real number system.

Real Numbers

Irrational Numbers: Any real number that is not rational, such as $-\sqrt{2}$, $-\sqrt{3}$, $\sqrt{0.8}$, and π.

Rational Numbers: Real numbers that can be expressed in the form $\frac{a}{b}$, where a and b are integers and $b \neq 0$, such as $-4\frac{3}{4}$, $-\frac{2}{3}$, 0.018, $0.\overline{3}$, and $\frac{5}{8}$.

Integers: $\ldots, -3, -2, -1, 0, 1, 2, 3, \ldots$

Whole Numbers: $0, 1, 2, 3, \ldots$

Natural Numbers: $1, 2, 3, \ldots$

OBJECTIVE 3. Graph rational numbers on a number line. Number lines can be useful tools when comparing numbers or solving certain arithmetic problems. Let's review how to graph a number on a number line.

EXAMPLE 3 ▶ Graph on a number line.

a. $1\frac{3}{5}$

Solution The number $1\frac{3}{5}$ is located $\frac{3}{5}$ of the way between 1 and 2, so we divide the space between 1 and 2 into 5 equal divisions and place a dot on the 3rd mark.

b. -2.56

Solution Because -2.56 means $-2\frac{56}{100}$, we could divide the space in between -2 and -3 into 100 divisions and count to the 56th mark. Because this is tedious, we will gradually zoom in on smaller and smaller sections of the number line.

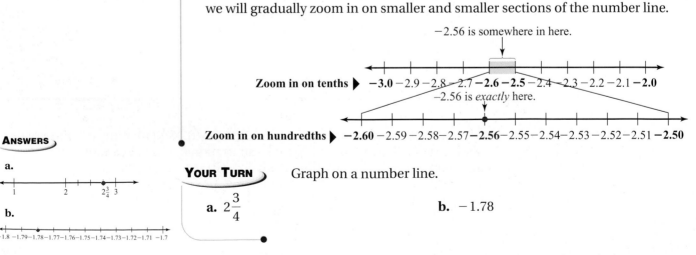

-2.56 is somewhere in here.

Zoom in on tenths ▶ $-3.0 \ -2.9 \ -2.8 \ -2.7 \ -2.6 \ -2.5 \ -2.4 \ -2.3 \ -2.2 \ -2.1 \ -2.0$

-2.56 is *exactly* here.

Zoom in on hundredths ▶ $-2.60 \ -2.59 \ -2.58 \ -2.57 \ -2.56 \ -2.55 \ -2.54 \ -2.53 \ -2.52 \ -2.51 \ -2.50$

YOUR TURN ▶ Graph on a number line.

a. $2\frac{3}{4}$

b. -1.78

ANSWERS

a.

b.

OBJECTIVE 4. Determine the absolute value of a number. The word *value* indicates how much something is worth, or its size. In mathematics, **absolute value** indicates the size of a number.

DEFINITION *Absolute value:* A given number's distance from 0 on a number line.

For example, the absolute value of 5 is 5 because the number 5 is five units from 0 on a number line. Likewise, the absolute value of -5 is 5 because -5 is also five units from 0 on a number line.

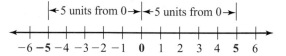

Learning Strategy

In the To the Student section, we suggested that when taking notes, you should use a red pen for definitions and a blue pen for rules or procedures. Notice that we have used those colors in the design of the text to connect with your notes.

The absolute value of a number n is written $|n|$. The examples just mentioned would translate this way:

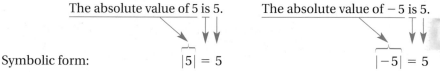

Symbolic form: $|5| = 5$ $|-5| = 5$

What about the absolute value of 0? Since there are no units between 0 and itself, the absolute value of 0 is 0.

RULE Absolute Value

The absolute value of every real number is either positive or 0.

EXAMPLE 4 Simplify.

a. $|-4.5|$

Answer $|-4.5| = 4.5$ because -4.5 is 4.5 units from 0 on a number line.

b. $\left|\dfrac{3}{8}\right|$

Answer $\left|\dfrac{3}{8}\right| = \dfrac{3}{8}$ because $\dfrac{3}{8}$ is $\dfrac{3}{8}$ of a unit from 0 on a number line.

YOUR TURN Simplify.

a. $\left|-2\dfrac{4}{5}\right|$ **b.** $|12.8|$

ANSWERS

a. $2\dfrac{4}{5}$ **b.** 12.8

OBJECTIVE 5. Compare numbers. Number lines can also be used to determine which of two numbers is greater. Because numbers increase from left to right on a number line, the number farthest to the right will be the greater of two numbers.

> **RULE** **Comparing Numbers**
>
> For any two real numbers a and b, a is greater than b if a is to the right of b on a number line. Equivalently, b is less than a if b is to the left of a on a number line.

Consider the following number line where we compare 2 and 8.

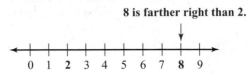

8 is farther right than 2.

Because the number 8 is farther to the right on a number line than the number 2, we say that 8 is greater than 2, or in symbols, $8 > 2$. Or, we could say that since 2 is to the left of 8, $2 < 8$.

EXAMPLE 5 Use $=$, $<$, or $>$ to write a true statement.

a. $4 \ \blacksquare \ -4$

Answer $4 > -4$ because 4 is farther right on a number line than -4.

b. $-2.7 \ \blacksquare \ -2.5$

Answer $-2.7 < -2.5$ because -2.7 is farther left on a number line than -2.5.

c. $\left| 3\frac{5}{6} \right| \ \blacksquare \ 3\frac{5}{6}$

Answer $\left| 3\frac{5}{6} \right| = 3\frac{5}{6}$ because the absolute value of $3\frac{5}{6}$ is equal to $3\frac{5}{6}$.

d. $|-0.7| \ \blacksquare \ -1.5$

Answer $|-0.7| > -1.5$ because the absolute value of -0.7 is equal to 0.7, which is farther to the right on a number line than -1.5.

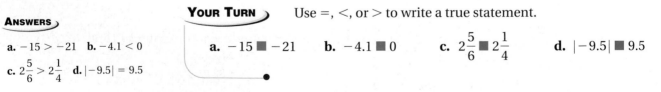

ANSWERS

a. $-15 > -21$ **b.** $-4.1 < 0$
c. $2\frac{5}{6} > 2\frac{1}{4}$ **d.** $|-9.5| = 9.5$

YOUR TURN Use $=$, $<$, or $>$ to write a true statement.

a. $-15 \ \blacksquare \ -21$ **b.** $-4.1 \ \blacksquare \ 0$ **c.** $2\frac{5}{6} \ \blacksquare \ 2\frac{1}{4}$ **d.** $|-9.5| \ \blacksquare \ 9.5$

1.1 Exercises

For Extra Help

MyMathLab
MyMathLab

Videotape/DVT

InterAct Math

Tutor Center
Math Tutor Center

Math XL.com

1. Define a set in your own words.

2. Define rational numbers in your own words.

3. Explain the difference between a rational number and an irrational number.

4. Explain why it is incorrect to say that absolute value is always positive.

For Exercises 5–10, answer true or false.

5. Every rational number is a real number.

6. Every real number is a rational number.

7. Every whole number is an integer.

8. Every real number is a natural number.

9. A number exists that is both rational and irrational.

10. For every real number n, $|n| = n$.

For Exercises 11–20, write a set representing each description.

11. The days of the week

12. The last ten letters of the alphabet

13. The vowels of the English alphabet

14. The states in the United States that do not share a border with any other state

15. The natural numbers that are multiples of 5

16. The even natural numbers

17. The odd natural numbers greater than 7

18. The even integers greater than or equal to 16

19. The integers greater than or equal to -6 but less than -2

20. The integers greater than -2.1 and less than $\frac{3}{4}$

For Exercises 21–30, determine whether each of the following is a rational number or an irrational number.

21. $-\dfrac{4}{5}$

22. $\dfrac{1}{4}$

23. $\sqrt{5}$

24. $-\sqrt{17}$

25. π

26. $\dfrac{\pi}{4}$

27. -0.21

28. -0.8

29. $0.\overline{6}$

30. $0.\overline{13}$

For Exercises 31–38, graph each of the following on a number line.

31. $-\dfrac{5}{6}$ **32.** $5\dfrac{1}{2}$ **33.** $2\dfrac{3}{8}$ **34.** $-\dfrac{2}{5}$

35. -3.5 **36.** 7.4 **37.** 2.45 **38.** -7.62

For Exercises 39–48, simplify.

39. $|23|$ **40.** $|6|$ **41.** $|-2|$ **42.** $|-8|$ **43.** $|-5.7|$

44. $|4.5|$ **45.** $\left|-3\dfrac{1}{8}\right|$ **46.** $\left|2\dfrac{3}{5}\right|$ **47.** $|0|$ **48.** $|-67.8|$

For Exercises 49–76, use =, <, or > to write a true statement.

49. $9 \;\blacksquare\; 3$ **50.** $2 \;\blacksquare\; 7$ **51.** $4 \;\blacksquare\; -3$ **52.** $-6 \;\blacksquare\; 5$

53. $-7 \;\blacksquare\; -8$ **54.** $-19 \;\blacksquare\; -7$ **55.** $-8 \;\blacksquare\; 0$ **56.** $0 \;\blacksquare\; -5$

57. $6.2 \;\blacksquare\; 4.5$ **58.** $2.63 \;\blacksquare\; 3.75$ **59.** $-4.3 \;\blacksquare\; -7.6$ **60.** $-3.5 \;\blacksquare\; -3.1$

61. $2\dfrac{2}{3} \;\blacksquare\; 2\dfrac{3}{4}$ **62.** $3\dfrac{5}{6} \;\blacksquare\; 3\dfrac{1}{4}$ **63.** $5.8 \;\blacksquare\; |-5.8|$ **64.** $|4.1| \;\blacksquare\; 4.1$

65. $7.3 \;\blacksquare\; |-8.7|$ **66.** $|-10.4| \;\blacksquare\; 3.2$ **67.** $|4.31| \;\blacksquare\; |-4.31|$ **68.** $|-0.59| \;\blacksquare\; |0.59|$

69. $\left|-6\dfrac{5}{8}\right| \;\blacksquare\; 5\dfrac{3}{8}$ **70.** $4\dfrac{2}{9} \;\blacksquare\; \left|4\dfrac{5}{9}\right|$ **71.** $|-4| \;\blacksquare\; |-2|$ **72.** $|-10| \;\blacksquare\; |8|$

73. $|29.5| \;\blacksquare\; |-29.7|$ **74.** $|-5.36| \;\blacksquare\; |5.76|$ **75.** $\left|-\dfrac{2}{3}\right| \;\blacksquare\; \left|-\dfrac{4}{3}\right|$ **76.** $\left|-\dfrac{9}{11}\right| \;\blacksquare\; \left|-\dfrac{7}{11}\right|$

For Exercises 77–80, list the given numbers in order from least to greatest.

77. $-2.56, 5.4, |8.3|, \left|-7\dfrac{1}{2}\right|, -4.7$ **78.** $2.9, 1, -12.6, |-1.3|, -9.6, \left|-2\dfrac{3}{4}\right|$

79. $0.4, -0.6, 0, 3\dfrac{1}{4}, |-0.02|, -0.44, \left|1\dfrac{2}{3}\right|$ **80.** $1.02, -0.13, -4\dfrac{1}{8}, -2\dfrac{1}{4}, |-1.06|, -2, |0.1|$

1.2 Fractions

OBJECTIVES

1. Write equivalent fractions.
2. Write equivalent fractions with the LCD.
3. Write the prime factorization of a number.
4. Simplify a fraction to lowest terms.

In this section, we review some basic principles of **fractions**.

DEFINITION **Fraction:** A quotient of two numbers or expressions a and b having the form $\frac{a}{b}$, where $b \neq 0$.

Connection We learned in Section 1.1 that if a number can be expressed in the form $\frac{a}{b}$, where a and b are integers, it is a rational number, so fractions that have integer numerators and denominators are rational numbers.

For example, $\frac{3}{4}$ is a fraction. The top number in a fraction is called the *numerator* and the bottom number is called the *denominator*.

$$\frac{3}{4} \longleftarrow \text{Numerator} \atop \longleftarrow \text{Denominator}$$

We can use fractions to indicate a part of a whole. For example, the following rectangle has been divided into four equal pieces, three of which are shaded, so the shaded region represents $\frac{3}{4}$ of the rectangle.

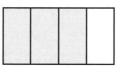

OBJECTIVE 1. Write equivalent fractions. Now, let's review how to write *equivalent fractions*, which are fractions that represent the same amount. For example, $\frac{1}{2}$ and $\frac{2}{4}$ are equivalent. We can see that they represent the same amount on a ruler.

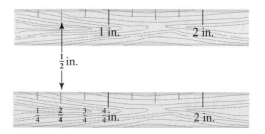

We can rewrite $\frac{1}{2}$ as $\frac{2}{4}$ by multiplying its numerator and denominator by 2.

$$\frac{1}{2} = \frac{1 \cdot 2}{2 \cdot 2} = \frac{2}{4}$$

We could also begin with $\frac{2}{4}$ and get back to $\frac{1}{2}$ by dividing both the numerator and denominator by 2.

$$\frac{2}{4} = \frac{2 \div 2}{4 \div 2} = \frac{1}{2}$$

RULE   Writing Equivalent Fractions

For any fraction, we can write an equivalent fraction by multiplying or dividing both its numerator and denominator by the same nonzero number.

This rule makes use of the fact that multiplying or dividing both the numerator and denominator by the same nonzero number is equivalent to multiplying or dividing by 1, which does not change a number.

EXAMPLE 1   Find the missing number that makes the fractions equivalent.

a. $\dfrac{7}{12} = \dfrac{?}{48}$

Solution $\dfrac{7 \cdot 4}{12 \cdot 4} = \dfrac{28}{48}$ Multiply the numerator and denominator by 4.

Note: *When a fraction is negative, the sign can be placed to the left of the numerator, to the left of the fraction line, or to the left of the denominator.*

b. $-\dfrac{30}{42} = \dfrac{?}{7}$

Solution $-\dfrac{30 \div 6}{42 \div 6} = -\dfrac{5}{7}$ Divide the numerator and denominator by 6. ◀ $-\dfrac{30}{42} = -\dfrac{5}{7} = \dfrac{-5}{7} = \dfrac{5}{-7}$

YOUR TURN   Find the missing number that makes the fractions equivalent.

a. $\dfrac{-9}{16} = \dfrac{?}{32}$

b. $\dfrac{54}{72} = \dfrac{3}{?}$

OBJECTIVE 2. Write equivalent fractions with the LCD.   It is sometimes necessary to write fractions that have different denominators, such as $\frac{1}{2}$ and $\frac{1}{3}$, as equivalent fractions with a common denominator. A common denominator can be any number that is a **multiple** of the denominators.

DEFINITION   *Multiple:* A multiple of a given integer *n* is the product of *n* and an integer.

We can generate multiples of a given number by multiplying the given number by the integers. For example, to generate the positive multiples of 2 and 3, we multiply 2 and 3 by 1, 2, 3, 4, and so on.

Multiples of 2	Multiples of 3
$2 \cdot 1 = 2$	$3 \cdot 1 = 3$
$2 \cdot 2 = 4$	$3 \cdot 2 = 6$
$2 \cdot 3 = 6$	$3 \cdot 3 = 9$
$2 \cdot 4 = 8$	$3 \cdot 4 = 12$
$2 \cdot 5 = 10$	$3 \cdot 5 = 15$
$2 \cdot 6 = 12$	$3 \cdot 6 = 18$

Note: *Every multiple of 2 is evenly divisible by 2. Similarly, every multiple of 3 is evenly divisible by 3. In general, every multiple of a number n is evenly divisible by n.*

Notice some common multiples of 2 and 3 in our lists are 6 and 12. When we rewrite fractions with a common denominator, it is usually preferable to use the **least common multiple** of the denominators. Such a denominator is called the **least common denominator**. From our lists, we see that the least common multiple of 2 and 3 is 6, so 6 is the least common denominator of $\frac{1}{2}$ and $\frac{1}{3}$.

DEFINITIONS   *Least common multiple (LCM):* The smallest number that is a multiple of each number in a given set of numbers.
Least common denominator (LCD): The least common multiple of the denominators of a given set of fractions.

We can now write equivalent fractions for $\frac{1}{2}$ and $\frac{1}{3}$ with their LCD, 6.

$$\frac{1}{2} = \frac{1 \cdot 3}{2 \cdot 3} = \frac{3}{6} \quad \text{and} \quad \frac{1}{3} = \frac{1 \cdot 2}{3 \cdot 2} = \frac{2}{6}$$

EXAMPLE 2   Write $\frac{5}{8}$ and $\frac{1}{6}$ as equivalent fractions with the LCD.

Solution   The LCD of 8 and 6 is 24.

$$\frac{5}{8} = \frac{5 \cdot 3}{8 \cdot 3} = \frac{15}{24} \quad \text{and} \quad \frac{1}{6} = \frac{1 \cdot 4}{6 \cdot 4} = \frac{4}{24}$$

YOUR TURN   Write each pair as equivalent fractions with the LCD.

   a. $\frac{4}{9}$ and $\frac{5}{6}$ **b.** $-\frac{3}{4}$ and $-\frac{5}{8}$

ANSWERS

a. $\frac{8}{18}$ and $\frac{15}{18}$

b. $-\frac{6}{8}$ and $-\frac{5}{8}$

OBJECTIVE 3. Write the prime factorization of a number. We now consider simplifying fractions. When simplifying fractions, we work with **factors**, which are numbers or variables that are multiplied together.

DEFINITION *Factors:* If $a \cdot b = c$, then a and b are factors of c.

In the multiplication statement $20 = 2 \cdot 10$, the 2 and 10 are factors of 20. Notice that a factor of a given number divides the given number evenly. The numbers 1, 2, 4, 5, 10, and 20 are all factors of 20 because they all divide 20 evenly. When a number is expressed as a product of its factors, it is said to be in *factored form*. Factored form is sometimes referred to as a *factorization*. Here are some factorizations for 20:

$$20 = 1 \cdot 20 \quad \text{or} \quad 20 = 2 \cdot 10 \quad \text{or} \quad 20 = 4 \cdot 5$$

Some natural numbers have only two distinct factors, 1 and the number itself. These numbers are called **prime numbers**.

DEFINITION *Prime number:* A natural number that has exactly two different factors, 1 and the number itself.

Note that 0 is not a prime number because it is not a natural number (0 is a whole number). Also, 1 is not a prime number because the only factor for 1 is itself and, by definition, a prime number must have *two different* factors.

The first prime number is 2, because it is the first natural number greater than 1 that has exactly two factors, 1 and 2. The number 4 is not a prime number because it is divisible by a number other than 1 and 4, namely 2. Here is a list of prime numbers:

2, 3, 5, 7, 11, 13, 17, 19, 23, 29, 31, 37, 41, 43, . . .

Note: *A natural number that has factors other than 1 and itself is called a* **composite number***. Some examples are 4, 6, 8, and 9. Notice that 0 and 1 are not composite, so they are neither prime nor composite.*

When simplifying a fraction it is useful to write the **prime factorization** of the numerator and denominator. For example, the prime factorization for 20 is $2 \cdot 2 \cdot 5$.

DEFINITION *Prime factorization:* A factorization that contains only prime factors.

We can find the prime factorization of a number using any factorization of the number as a starting point. Consider 20 again and suppose we factored it as $4 \cdot 5$. Since 4 is not a prime number, we can factor 4 as $2 \cdot 2$. It looks like this:

$$20 = 4 \cdot 5$$
$$= 2 \cdot 2 \cdot 5$$

We will use factor trees to find prime factorizations. The idea is to draw two branches beneath any composite number and write two factors of that number at the end of the branches. Following are two ways we could use a factor tree to find the prime factorization of 20. Notice that when a branch ends in a prime factor, we circle the prime factor. This is optional, but many people find it helpful in making the prime factors stand out.

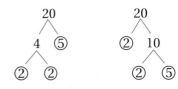

We usually write the prime numbers in a prime factorization in ascending order, so we would write $20 = 2 \cdot 2 \cdot 5$ (instead of $20 = 5 \cdot 2 \cdot 2$ or $20 = 2 \cdot 5 \cdot 2$). Note that no matter how we write the prime factorization, the product of two 2's and one 5 is always 20. No two numbers have the same prime factorization. Consequently, we say that a number's prime factorization is unique.

EXAMPLE 3 Find the prime factorization of 360.

Solution Use a factor tree.

```
            360
           /   \
         10      36
        / \     /  \
       ②  ⑤   4     9
             / \   / \
            ②  ②  ③  ③
```

Factor 360 to 10 and 36. (Any two factors will work.)

Factor 10 to 2 and 5, which are primes. Then factor 36 to 4 and 9.

4 is then factored to 2 and 2, which are primes, and 9 is factored to 3 and 3, which are primes.

Answer $360 = 2 \cdot 2 \cdot 2 \cdot 3 \cdot 3 \cdot 5$

YOUR TURN Find the prime factorization.

a. 84 **b.** 280

OBJECTIVE 4. Simplify a fraction to lowest terms. We can use prime factorizations to simplify fractions to **lowest terms**.

DEFINITION *Lowest terms:* Given a fraction $\frac{a}{b}$ and $b \neq 0$, if the only factor common to both a and b is 1, then the fraction is in lowest terms.

For example, $\dfrac{3}{4}$ is in lowest terms because the only factor common to both 3 and 4 is the number 1, whereas $\dfrac{6}{8}$ is not in lowest terms because 6 and 8 have a common factor of 2.

To develop the method of reducing fractions using primes, we must understand that if the numerator and denominator are identical and not both 0, then the fraction is equivalent to 1.

$$1 = \frac{1}{1} = \frac{2}{2} = \frac{3}{3} = \frac{4}{4} = \frac{5}{5} = \cdots$$

This suggests the following rule.

RULE   **Simplifying a Fraction with the Same Nonzero Numerator and Denominator**

$\dfrac{n}{n} = \dfrac{1}{1} = 1$, when $n \neq 0$.

This rule can be extended to common factors within a fraction. It allows us to eliminate factors that are identical in the numerator and denominator of a fraction.

RULE   **Eliminating a Common Factor in a Fraction**

$\dfrac{an}{bn} = \dfrac{a \cdot 1}{b \cdot 1} = \dfrac{a}{b}$, when $b \neq 0$ and $n \neq 0$.

These rules allow us to write fractions in lowest terms using prime factorization. The idea is to replace the numerator and denominator with their prime factorizations and then eliminate the prime factors that are common to both the numerator and denominator. Consider the fraction $\dfrac{10}{15}$. Replacing the 10 and 15 with their prime factorizations, we have:

$$\frac{10}{15} = \frac{2 \cdot 5}{3 \cdot 5}$$

$$= \frac{2 \cdot 1}{3 \cdot 1}$$

Note: *The common factor of 5 is replaced by 1. In Example 4 we will leave this step out.*

$$= \frac{2}{3}$$

This process of eliminating common factors is called *dividing out* the common factors and is sometimes written this way:

$$\frac{10}{15} = \frac{2 \cdot \cancel{5}}{3 \cdot \cancel{5}} = \frac{2}{3}$$

Though there are different styles of showing which common factors are eliminated, we will simply highlight the common factors that are eliminated.

To simplify a fraction to lowest terms:

1. Replace the numerator and denominator with their prime factorizations.
2. Eliminate (divide out) all prime factors common to the numerator and denominator.
3. Multiply the remaining factors.

Connection   You may remember a method for simplifying fractions in which you divide out the greatest common factor. For example, to simplify $\frac{28}{70}$, we recognize that 14 is the greatest common factor of 28 and 70 and divide both by 14 to get lowest terms: $\frac{28}{70} = \frac{28 \div 14}{70 \div 14} = \frac{2}{5}$. This method is a perfectly good way of simplifying fractions. However, we show the prime method to prepare for simplifying rational expressions in Chapter 7.

EXAMPLE 4   Simplify to lowest terms.

a. $\frac{28}{70}$

Solution $\frac{28}{70} = \frac{2 \cdot 2 \cdot 7}{2 \cdot 5 \cdot 7} = \frac{2}{5}$ Replace the numerator and denominator with their prime factorizations, then eliminate the common prime factors.

b. $\frac{156}{210}$

Solution $\frac{156}{210} = \frac{2 \cdot 2 \cdot 3 \cdot 13}{2 \cdot 3 \cdot 5 \cdot 7} = \frac{26}{35}$

YOUR TURN   Simplify to lowest terms.

a. $\frac{36}{48}$ b. $\frac{126}{315}$ c. $\frac{312}{378}$

EXAMPLE 5   A newspaper reports that over the last year, 160 households per 1000 households were victims of a crime. Write a fraction in simplest form representing the fraction of households that were victims of crime.

Solution If 160 out of 1000 households were victims of crime, then the fraction of households that were victims of crime is $\frac{160}{1000}$. Now we can simplify the fraction.

$$\frac{160}{1000} = \frac{2 \cdot 2 \cdot 2 \cdot 2 \cdot 2 \cdot 5}{2 \cdot 2 \cdot 2 \cdot 5 \cdot 5 \cdot 5} = \frac{4}{25}$$

Answer The fraction in simplest form of households that were victims of crime is $\frac{4}{25}$, which means 4 out of 25 households were victims of crime.

ANSWERS

a. $\frac{3}{4}$ b. $\frac{2}{5}$ c. $\frac{52}{63}$

ANSWER

$\frac{3}{35}$

YOUR TURN   Researchers gave 560 volunteers a new medication and found that 48 of them developed a headache within 30 minutes of taking the medication. Write a fraction in simplest form representing the portion of the volunteers that developed a headache.

1.2 Exercises

For Extra Help   *MyMathLab*   MyMathLab   Videotape/DVT   InterAct Math   Tutor Center Math Tutor Center   Math XL.com

1. What are two ways to generate fractions equivalent to a given fraction?

2. Is 4 a multipleof 12? Explain.

3. Is $2 \cdot 2 \cdot 4 \cdot 7$ a prime factorization? Explain.

4. Is it possible for two different numbers to have the same prime factorization? Explain.

5. Is every whole number a composite number?

6. How do you know if a fraction is in lowest terms?

For Exercises 7–10, answer true or false, then explain.

7. Every prime number is a whole number.

8. Prime numbers are rational numbers.

9. Every composite number has at least one factor that is a prime number.

10. The number 1 is neither prime nor composite.

For Exercises 11–14, identify the fraction represented by the shaded region.

11.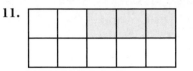

12.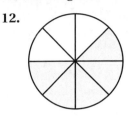

13.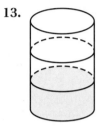

14.

For Exercises 15–18, write the length of each line segment in lowest terms.

15.
1 in.

16.
1 in.

17.
1 in.

18.
1 in.

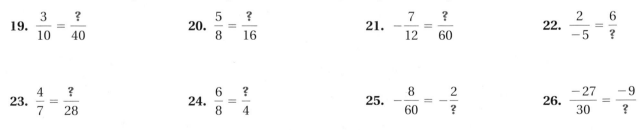

For Exercises 19–26, find the missing number that makes the fractions equivalent.

19. $\dfrac{3}{10} = \dfrac{?}{40}$

20. $\dfrac{5}{8} = \dfrac{?}{16}$

21. $-\dfrac{7}{12} = \dfrac{?}{60}$

22. $\dfrac{2}{-5} = \dfrac{6}{?}$

23. $\dfrac{4}{7} = \dfrac{?}{28}$

24. $\dfrac{6}{8} = \dfrac{?}{4}$

25. $-\dfrac{8}{60} = -\dfrac{2}{?}$

26. $\dfrac{-27}{30} = \dfrac{-9}{?}$

For Exercises 27–34, write the fractions as equivalent fractions with the LCD.

27. $\frac{4}{5}$ and $\frac{2}{3}$

28. $\frac{5}{7}$ and $\frac{3}{11}$

29. $\frac{4}{9}$ and $\frac{7}{6}$

30. $\frac{5}{8}$ and $\frac{7}{12}$

31. $-\frac{11}{18}$ and $-\frac{17}{24}$

32. $-\frac{9}{20}$ and $-\frac{7}{15}$

33. $-\frac{1}{16}$ and $-\frac{15}{36}$

34. $-\frac{13}{21}$ and $-\frac{9}{14}$

For Exercises 35–42, write the prime factorization for each number.

35. 44

36. 33

37. 36

38. 42

39. 64

40. 48

41. 250

42. 810

For Exercises 43–50, simplify to lowest terms.

43. $\frac{72}{90}$

44. $\frac{48}{84}$

45. $\frac{63}{99}$

46. $\frac{42}{91}$

47. $-\frac{78}{104}$

48. $-\frac{30}{54}$

49. $-\frac{48}{90}$

50. $-\frac{24}{162}$

For Exercises 51 and 52, determine whether the simplification is correct. If not, explain why.

51. $\dfrac{\overset{1}{\cancel{2}} + 3}{\underset{1}{\cancel{2}}} = \dfrac{1 + 3}{1} = 4$

52. $\dfrac{84}{240} = \dfrac{\overset{1}{\cancel{2}} \cdot \overset{1}{\cancel{2}} \cdot \overset{1}{\cancel{3}} \cdot 7}{2 \cdot \underset{1}{\cancel{2}} \cdot \underset{1}{\cancel{2}} \cdot \underset{1}{\cancel{3}} \cdot 3 \cdot 5} = \dfrac{7}{30}$

For Exercises 53–56, write each fraction in lowest terms.

53. A student scores 294 points out of a total of 336 points in a college course. What fraction of the total points is the student's score?

54. The nutrition label on a package of frozen fish sticks indicates that a serving of fish sticks has 250 calories with 130 of those calories coming from fat. What fraction of the calories in a serving of fish sticks comes from fat?

55. At a company, 300 of the 575 employees have optional life insurance coverage as part of their benefits package. What fraction of the employees have optional life insurance coverage?

56. Laura uses 120 square feet of a room in her home as an office for her business. The total living space of her home is 1830 square feet. For tax purposes, she needs to compute the fraction of her home that is used as an office. What is that fraction?

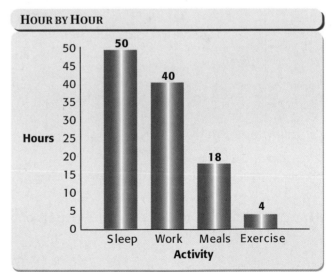

For Exercises 57–60, use the following bar graph, which shows the number of hours that Carla spends each week in each activity. Write all fractions in lowest terms.

57. What fraction of a week does Carla spend working?

58. What fraction of a week does Carla spend sleeping?

59. What fraction of a week does Carla spend in all of the listed activities combined?

60. What fraction of the week does Carla have for free time, which is time away from all the listed activities?

For Exercises 61–68, use the following table that shows the number of households per 1000 that were victims of property crime in each year. Write all fractions in lowest terms.

National Crime Victimization Survey, Property Crime Trends, 1973–2003

Year	Total Property Crime*	Burglary	Thefts	Motor Vehicle Thefts
1973	520	110	391	19
1974	552	112	421	19
1975	554	110	424	20
1976	544	107	421	17
1977	544	106	421	17
1978	533	103	412	18
1979	532	101	413	18
1980	496	101	378	17
1981	497	106	374	17
1982	468	94	358	16
1983	428	84	330	15
1984	399	77	307	15
1985	385	75	296	14
1986	373	74	284	15
1987	380	75	289	16
1988	378	74	287	18
1989	373	68	287	19
1990	349	65	264	21
1991	354	65	267	22
1992	325	59	248	19
1993	319	58	242	19
1994	310	56	235	19
1995	291	49	224	17
1996	266	47	206	14
1997	248	45	190	14
1998	217	39	168	11
1999	198	34	154	10
2000	178	32	138	9
2001	167	29	129	9
2002	159	28	122	9
2003	163	30	124	9

*Victimizations per 1000 households

Source: U.S. Department of Justice, Bureau of Crime Statistics

61. What fraction of households were victims of property crime in 1973?

62. What fraction of households were victims of property crime in 1992?

63. In 1987 what fraction of households were not victims of property crime?

64. In 1994 what fraction of households were not victims of property crime?

65. a. What year had the highest number of property crimes?

    b. What fraction of households were victims in that year?

66. a. What year had the lowest number of burglaries?

    b. What fraction of households were victims of burglary in that year?

67. In 1974, what fraction of the total property crimes were burglaries?

68. In 1985, what fraction of total property crimes were motor vehicle theft?

For Exercises 69–76, write each fraction in lowest terms.

69. What fraction of a year is 30 days? Use 365 days for the number of days in a year.

70. What fraction of an hour is 8 minutes?

71. What fraction of a minute is 40 seconds?

72. What fraction of a foot is 4 inches?

Of Interest

It takes Earth one year to complete one revolution around the Sun, which is actually $365\frac{1}{4}$ days. Since it is impractical to have $\frac{1}{4}$ days on the calendar, every four years, those $\frac{1}{4}$ days add up to a full day, which is why we have a leap year with 366 days.

73. In the first session of 2005, the U.S. House of Representatives had 201 Democrats, 232 Republicans, 1 Independent and 1 vacant seat. What fraction of the House of Representatives was Republican? (*Source:* Official list of United States House of Representatives)

74. In 2005, the U.S. Senate had 44 Democrats, 55 Republicans, and 1 Independent. What fraction of the Senate was not Democrat? (*Source:* Official list of United States Senators)

Of Interest

The chemical formula for a molecule of lactose is $C_{12}H_{22}O_{11}$. Notice the subscripts in a chemical formula indicate the number of atoms of each element that are present in the molecule.

75. One molecule of lactose contains 12 carbon atoms, 22 hydrogen atoms, and 11 oxygen atoms. What fraction of the atoms that make up the molecule are carbon?

76. One molecule of glucose contains 6 carbon atoms, 12 hydrogen atoms, and 6 oxygen atoms. What fraction of the atoms that make up the molecule are other than carbon?

REVIEW EXERCISES

[1.1]   **1.** Is $5x + 2$ an expression or an equation? Explain.

[1.1]   **2.** Write a set containing the names of the four planets closest to the Sun.

[1.1]   **3.** Is 0.8 a rational or irrational number? Explain.

[1.1]   **4.** Graph 3.7 on a number line.

[1.1]   **5.** Simplify: $|27|$

[1.1]   **6.** Use $<$, $>$, or $=$ to make a true statement: $|-6|$ ■ 6

1.3 Adding and Subtracting Real Numbers; Properties of Real Numbers

OBJECTIVES

1. Add integers.
2. Add rational numbers.
3. Find the additive inverse of a number.
4. Subtract rational numbers.

OBJECTIVE 1. Add integers.   We now turn our attention to adding and subtracting numbers. We begin with addition. First, consider the parts of an addition statement. The numbers added are called *addends* and the answer is called a *sum*.

$$2 + 3 = 5$$

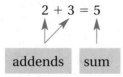

addends   sum

There are three important properties of addition that are true for all real numbers. First, the sum of a number and 0 is that number. Because adding 0 to a number does not change the identity of the number, 0 is called the *additive identity*.

$$2 + 0 = 2$$

Second, the order of addends can be changed without affecting the sum. This is known as the *commutative property of addition*.

$$2 + 3 = 5 \quad \text{or} \quad 3 + 2 = 5$$

Third, when adding three or more numbers, we can group the addends differently without affecting the sum. This is known as the *associative property of addition*.

$$2 + (3 + 4) \quad \text{or} \quad (2 + 3) + 4$$
$$= 2 + 7 \qquad\qquad = 5 + 4$$
$$= 9 \qquad\qquad\quad = 9$$

Following is a summary of these properties.

Properties of Addition	Symbolic Form	Word Form
Additive Identity	$a + 0 = a$	The sum of a number and 0 is that number.
Commutative Property of Addition	$a + b = b + a$	Changing the order of addends does not affect the sum.
Associative Property of Addition	$a + (b + c) = (a + b) + c$	Changing the grouping of three or more addends does not affect the sum.

Adding Numbers with the Same Sign

With these properties in mind, let's consider how to add signed numbers. First we learn how to add numbers that have the same sign. It is helpful to relate adding signed numbers to money. For example, adding two positive numbers is like adding credits, whereas adding two negative numbers is like adding debts.

$$15 + 10 = 25$$

This addition is like having $15 in an account and depositing another $10 into the same account, bringing the total to $25.

$$-20 + (-8) = -28$$

This addition is like having a debt balance of $20 on a credit card and charging another $8 on the same card, bringing the balance to a total debt of $28.

We can also illustrate each of these cases using a number line. The first addend is the starting point and the second addend indicates the direction and distance to travel on the number line so that we finish at the sum.

These two cases are similar and suggest the following procedure.

PROCEDURE Adding Numbers with the Same Sign

To add two numbers that have the same sign, add their absolute values and keep the same sign.

EXAMPLE 1 Add.

a. $15 + 13$

Solution $15 + 13 = 28$

b. $-14 + (-23)$

Solution $-14 + (-23) = -37$

Note: *In terms of money, part a illustrates two credits, whereas part b illustrates two debts.*

YOUR TURN Add.

a. $32 + 19$ **b.** $-62 + (-13)$

Adding Numbers with Different Signs

Now consider adding two numbers with different signs, which is like making a payment toward a debt. Consider the following cases:

$-15 + 3 = -12$ $-9 + 20 = 11$

| This addition is like having a debt of \$15 and a payment of \$3 is made toward that debt. This decreases the debt to \$12. | This addition is like having a debt of \$9 and a payment of \$20 is made toward that debt. Because the payment is more than the debt, we now have a credit of \$11. |

Using number lines:

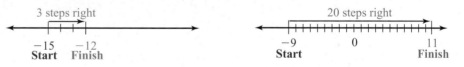

Note that the commutative property of addition indicates that in the examples, the addends can be rearranged without affecting the sum.

$$3 + (-15) = -12 \qquad 20 + (-9) = 11$$

These examples suggest the following procedure.

PROCEDURE Adding Numbers with Different Signs

To add two numbers that have different signs, subtract their absolute values and keep the sign of the number with the greater absolute value.

EXAMPLE 2 Add.

a. $22 + (-14)$

Solution $22 + (-14) = 8$

b. $-13 + 20$

Solution $-13 + 20 = 7$

c. $12 + (-18)$

Solution $12 + (-18) = -6$

d. $-24 + 5$

Solution $-24 + 5 = -19$

Note: *In terms of money, parts a and b illustrate situations in which the credit is greater than the debt. Therefore, the result is a credit, or positive.*

Note: *In terms of money, parts c and d illustrate situations in which a payment toward a debt is not enough to pay off the debt. Therefore, the result is still debt, or negative.*

YOUR TURN Add.

a. $15 + (-6)$ **b.** $27 + (-35)$ **c.** $-13 + 19$ **d.** $-29 + 14$

OBJECTIVE 2. Add rational numbers.

Adding Fractions with the Same Denominator

Recall that the set of rational numbers contains any number that can be expressed as a ratio of integers. This set includes the integers themselves because every integer can be expressed as a ratio with 1 as the denominator. Let's now turn our attention to the rest of the rational number set, namely fractions and decimals. Consider the intuitive fact that adding two quarters equals a half (in money this would be a half-dollar). The following circle shows this sum.

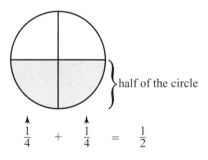

$$\frac{1}{4} \quad + \quad \frac{1}{4} \quad = \quad \frac{1}{2}$$

Notice if the numerators are added and the denominator remains the same, the result simplifies to $\frac{1}{2}$.

$$\frac{1}{4} + \frac{1}{4} = \frac{1+1}{4} = \frac{2}{4} = \frac{1}{2}$$

PROCEDURE   To add fractions with the same denominator, add the numerators and keep the same denominator, then simplify.

EXAMPLE 3 Add.

a. $\dfrac{5}{8} + \dfrac{1}{8}$

Solution $\dfrac{5}{8} + \dfrac{1}{8} = \dfrac{5+1}{8} = \dfrac{6}{8}$

Add the numerators and keep the same denominator. Because $\dfrac{6}{8}$ is not in lowest terms, we must simplify.

$= \dfrac{2 \cdot 3}{2 \cdot 2 \cdot 2} = \dfrac{3}{4}$

Replace 6 and 8 with their prime factorizations, divide out the common factor, 2, and then multiply the remaining factors.

b. $-\dfrac{3}{10} + \left(-\dfrac{1}{10} \right)$

Solution $-\dfrac{3}{10} + \left(-\dfrac{1}{10} \right) = \dfrac{-3 + (-1)}{10} = -\dfrac{4}{10}$

Note: *Because the fractions have the same sign, we add and keep the same sign.*

$= -\dfrac{2 \cdot 2}{2 \cdot 5} = -\dfrac{2}{5}$

Simplify to lowest terms by dividing out the common factor, 2.

c. $\dfrac{7}{9} + \left(-\dfrac{4}{9} \right)$

Solution $\dfrac{7}{9} + \left(-\dfrac{4}{9} \right) = \dfrac{7 + (-4)}{9} = \dfrac{3}{9}$

Note: *Because the two addends have different signs, we subtract the absolute values and keep the sign of the number with the greater absolute value.*

$= \dfrac{3}{3 \cdot 3} = \dfrac{1}{3}$

Simplify to lowest terms by dividing out the common factor, 3.

YOUR TURN Add.

a. $\dfrac{3}{8} + \dfrac{1}{8}$

b. $-\dfrac{5}{9} + \left(-\dfrac{2}{9} \right)$

c. $-\dfrac{7}{12} + \dfrac{5}{12}$

ANSWERS

a. $\dfrac{1}{2}$ **b.** $-\dfrac{7}{9}$ **c.** $-\dfrac{1}{6}$

Adding Fractions with Different Denominators

If the denominators are different, then we must first find a common denominator. Recall that in Section 1.2 we wrote equivalent fractions with the LCD. Consider the addition $\frac{3}{4} + \frac{1}{6}$. The LCD of 4 and 6 is 12. We now write each fraction as an equivalent fraction with the LCD.

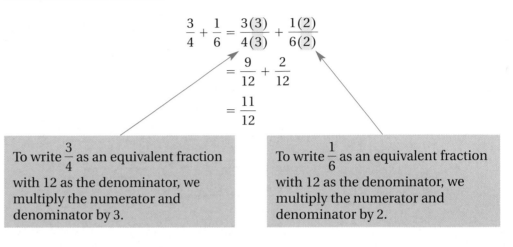

$$\frac{3}{4} + \frac{1}{6} = \frac{3(3)}{4(3)} + \frac{1(2)}{6(2)}$$

$$= \frac{9}{12} + \frac{2}{12}$$

$$= \frac{11}{12}$$

To write $\frac{3}{4}$ as an equivalent fraction with 12 as the denominator, we multiply the numerator and denominator by 3.

To write $\frac{1}{6}$ as an equivalent fraction with 12 as the denominator, we multiply the numerator and denominator by 2.

PROCEDURE Adding Fractions

To add fractions with different denominators:

1. Write each fraction as an equivalent fraction with the LCD.
2. Add the numerators and keep the LCD.
3. Simplify.

EXAMPLE 4 Add.

a. $-\frac{1}{6} + \left(-\frac{5}{9}\right)$

Solution The LCD of 6 and 9 is 18.

$$-\frac{1}{6} + \left(-\frac{5}{9}\right) = -\frac{1(3)}{6(3)} + \left(-\frac{5(2)}{9(2)}\right)$$ Write equivalent fractions with 18 in the denominator.

$$= -\frac{3}{18} + \left(-\frac{10}{18}\right)$$ Add numerators and keep the common denominator.

$$= \frac{-3 + (-10)}{18}$$ Because the addends have the same sign, we add and keep the same sign.

$$= -\frac{13}{18}$$

b. $-\dfrac{4}{5} + \dfrac{2}{3}$

Solution The LCD of 5 and 3 is 15.

$$-\frac{4}{5} + \frac{2}{3} = -\frac{4(3)}{5(3)} + \frac{2(5)}{3(5)}$$ Write equivalent fractions with 15 in the denominator.

$$= -\frac{12}{15} + \frac{10}{15}$$ Add numerators and keep the common denominator.

$$= \frac{-12 + 10}{15}$$ Because the addends have different signs, we subtract and keep the sign of the number with the greater absolute value.

$$= -\frac{2}{15}$$

YOUR TURN Add.

a. $\dfrac{5}{6} + \dfrac{1}{4}$ **b.** $-\dfrac{1}{15} + \left(-\dfrac{5}{12}\right)$ **c.** $\dfrac{7}{8} + \left(-\dfrac{1}{5}\right)$

Adding Decimal Numbers

Let's now review adding decimal numbers, which are also a part of the set of rational numbers. Recall that to add decimal numbers, we add like place values.

EXAMPLE 5 Angela has a balance of $578.26 and incurs the following debts:

$$\$25.67, \$92.45, \$78.50, \$15.00$$

What is Angela's new balance?

Solution First, we calculate the total debts.

$$-25.67 + (-92.45) + (-78.50) + (-15.00) = -211.62$$

Now we reconcile the total debt with the current balance.

$$578.26 + (-211.62) = \$366.64$$

YOUR TURN Daryl has a balance of $-\$452.75$ on a credit card. In one month he charges $25.80, $65.75, and $35.21 using the same credit card. If the finance charge is $7.47 and he makes a payment of $120, what is his new balance?

OBJECTIVE 3. Find the additive inverse of a number. What happens if we add two numbers that have the same absolute value but different signs, such as $5 + (-5)$? In money terms, this is like making a $5 payment toward a debt of $5. Notice the payment pays off the debt so that the balance is 0.

$$5 + (-5) = 0$$

Because their sum is 0, we say 5 and -5 are **additive inverses**, or opposites.

DEFINITION | *Additive inverses:* Two numbers whose sum is 0.

EXAMPLE 6 Find the additive inverse of the given number.

a. 9

Answer -9 because $9 + (-9) = 0$

b. -7

Answer 7 because $-7 + 7 = 0$

c. 0

Answer 0 because $0 + 0 = 0$

Calculator TIPS

To enter a negative number or find an additive inverse on a graphing or scientific calculator, use the $\boxed{(-)}$ *key. Note that this key is different from the subtraction key.*

To find the additive inverse of 9, type: $\boxed{(-)}$ $\boxed{9}$

To find the additive inverse of -7, *type:* $\boxed{(-)}$ $\boxed{(-)}$ $\boxed{7}$ $\boxed{\text{ENTER}}$

On older scientific calculators, the $\boxed{+/-}$ *key is used to find the additive inverse of a number. You enter the value first, then press the* $\boxed{+/-}$ *key. Each time you press the* $\boxed{+/-}$ *key the sign will change.*

To find the additive inverse of 9, type: $\boxed{9}$ $\boxed{+/-}$

To find the additive inverse of -7, *type:* $\boxed{7}$ $\boxed{+/-}$ $\boxed{+/-}$

YOUR TURN Find the additive inverse of the given number.

a. -12 **b.** 0.67   **c.** $-\dfrac{3}{8}$

ANSWERS

a. 12   **b.** -0.67 **c.** $\dfrac{3}{8}$

OBJECTIVE 4. Subtract rational numbers. Now let's turn our attention to subtraction. The parts of subtraction are

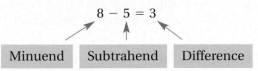

$$8 - 5 = 3$$

| Minuend | Subtrahend | Difference |

Recall that when we add two numbers that have different signs, we actually subtract. This implies that subtraction and addition are related. In fact, every subtraction statement can be written as an equivalent addition statement. Notice that the subtraction statement $8 - 5 = 3$ and the addition statement $8 + (-5) = 3$ are equivalent: They are both ways of finding the balance given an \$8 credit with a \$5 debt. Note that in writing $8 - 5$ as an equivalent addition $8 + (-5)$, we change the operation sign from a minus sign to a plus sign and also change the 5 to its additive inverse, -5.

Any subtraction statement can be written as an equivalent addition statement using this procedure. Consider $8 - (-5)$, which represents an \$8 credit subtracting a debt of \$5. What does it mean to subtract a debt? If a bank records a debt and then realizes it made a mistake, then it must reverse that debt. Subtracting, or reversing, a debt means that the bank must record a credit (or deposit) to correct the mistake. In math terms, this means that $8 - (-5)$ is equivalent to $8 + 5$, which equals 13. Notice this still follows the process just discussed.

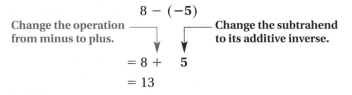

$$8 - (-5)$$

Change the operation from minus to plus.   Change the subtrahend to its additive inverse.

$$= 8 + \ 5$$
$$= 13$$

PROCEDURE   Rewriting Subtraction

To write a subtraction statement as an equivalent addition statement, change the operation symbol from a minus sign to a plus sign, and change the subtrahend to its additive inverse.

EXAMPLE 7   Subtract.

a. $-15 - (-6)$

Solution Write the subtraction as an equivalent addition.

$$-15 - (-6)$$

Change the operation from minus to plus.   Change the subtrahend to its additive inverse.

$$= -15 + \ 6$$
$$= -9$$

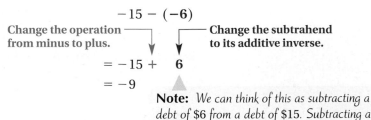

Note: *We can think of this as subtracting a debt of \$6 from a debt of \$15. Subtracting a debt from a debt is equivalent to making a deposit or payment against the existing debt balance, so this decreases the amount of debt.*

b. $-\dfrac{2}{3} - \dfrac{3}{4}$

Solution Write an equivalent addition, then find a common denominator and continue the process of adding fractions.

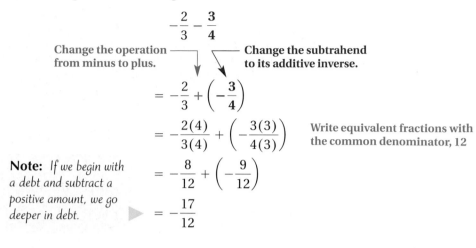

$$-\dfrac{2}{3} - \dfrac{3}{4}$$

Change the operation from minus to plus. ⟶ ⟵ Change the subtrahend to its additive inverse.

$$= -\dfrac{2}{3} + \left(-\dfrac{3}{4}\right)$$

$$= -\dfrac{2(4)}{3(4)} + \left(-\dfrac{3(3)}{4(3)}\right) \quad \text{Write equivalent fractions with the common denominator, 12}$$

Note: *If we begin with a debt and subtract a positive amount, we go deeper in debt.*

$$= -\dfrac{8}{12} + \left(-\dfrac{9}{12}\right)$$

$$= -\dfrac{17}{12}$$

Calculator TIPS

To enter a problem like Example 7 (b) into a scientific calculator, type:

[(−)] [2] [$a^{b/c}$] [3] [−] [3] [$a^{b/c}$] [4] [ENTER]

On a graphing calculator, type:

[(−)] [2] [÷] [3] [−] [3] [÷] [4] [ENTER]

To see the answer in fraction form, press [MATH] *and select the 1:Frac option from the menu, then press* [ENTER].

---

c. $5.04 - 8.01$

Solution Write an equivalent addition statement.

$5.04 - \mathbf{8.01}$
$= 5.04 + (\mathbf{-8.01})$
$= -2.97$

Note: *We begin with a positive balance of $5.04 and subtract $8.01. Because we are subtracting more than we have, the result is negative.*

Note: *To actually calculate the difference, the number with the greater absolute value must be placed on top.*

$$\begin{array}{r} 8.01 \\ -5.04 \\ \hline 2.97 \end{array}$$

We then use the equivalent addition to figure whether 2.97 is a positive or negative difference. Because the number with greater absolute value is negative, this result is negative.

YOUR TURN Subtract.

a. $25 - (-17)$ **b.** $-\dfrac{3}{8} - \left(-\dfrac{1}{3}\right)$ **c.** $0.06 - 4.02$

ANSWERS

a. 42 **b.** $-\dfrac{1}{24}$ **c.** -3.96

EXAMPLE 8 On Mars, the daytime high temperature can reach $-15°C$, while at night the temperature can plummet to $-140°C$. What is the difference between the high and low temperatures?

Solution *Difference* indicates that we subtract the temperatures. Because the wording in the problem was to find the "difference between the high and low temperatures," we arrange the subtraction with the low temperature subtracted from the high temperature.

$$-15 - (-140) = -15 + 140 \quad \text{Write as an equivalent addition.}$$
$$= 125 \quad \text{Add.}$$

Answer The difference between the high and low temperatures is 125°C.

YOUR TURN In an experiment, a mixture begins at a temperature of 5.8°C. The mixture is then cooled to a temperature of $-28.6°C$. Find the difference of the initial and final temperatures.

ANSWER

34.4°C

1.3 Exercises

For Extra Help   *MyMathLab*   MyMathLab   Videotape/DVT   InterAct Math   Math Tutor Center   Math XL.com

1. Explain the difference between the commutative property of addition and the associative property of addition.

2. Explain why 4 and -4 are additive inverses.

3. Why is 0 called the additive identity?

4. In your own words, explain the process of adding or subtracting two fractions with different denominators.

5. In your own words, explain how to add two numbers that have the same sign.

6. In your own words, explain how to add two numbers that have different signs.

7. Explain how to write a subtraction as an equivalent addition.

8. Explain why $6 - 5$ and $5 - 6$ have different answers, whereas $-6 - 5$ and $-5 - 6$ have the same answer. (*Hint:* Think about the properties of addition.)

For Exercises 9–24, indicate whether the given equation illustrates the additive identity, commutative property of addition, associative property of addition, or additive inverse.

9. $5 + 6 = 6 + 5$

10. $-3 + 7 = 7 + (-3)$.

11. $0 + 8 = 8$

12. $-5 + 0 = -5$

13. $0.8 + (-0.8) = 0$

14. $-\dfrac{4}{9} + \dfrac{4}{9} = 0$

15. $(3 + 5) + 2 = 3 + (5 + 2)$

16. $6 + (2 + 3) = (6 + 2) + 3$

17. $-8 + (7 + 3) = (7 + 3) + (-8)$

18. $(5 + 3) - 4 = -4 + (5 + 3)$

19. $\dfrac{5}{11} - \dfrac{8}{9} = -\dfrac{8}{9} + \dfrac{5}{11}$

20. $-\dfrac{2}{7} + \dfrac{3}{4} = \dfrac{3}{4} + \left(-\dfrac{2}{7}\right)$

21. $6.3 + (2.1 - 2.1) = 6.3 + 0$

22. $(-4.6 + 4.6) + 9.5 = 0 + 9.5$

23. $\left(\dfrac{1}{2} + \dfrac{3}{5}\right) + \dfrac{1}{6} = \left(\dfrac{3}{5} + \dfrac{1}{2}\right) + \dfrac{1}{6}$

24. $-12.5 + (-6.8) + (-9) = -12.5 + (-9) + (-6.8)$

For Exercises 25–50, add.

25. $8 + 13$

26. $15 + 7$

27. $6 + (-12)$

28. $-5 + (-7)$

29. $27 + (-13)$

30. $29 + (-7)$

31. $-8 + 4$

32. $-16 + 19$

33. $\dfrac{5}{8} + \dfrac{1}{8}$

34. $\dfrac{9}{16} + \dfrac{5}{16}$

35. $-\dfrac{4}{9} + \left(-\dfrac{2}{9}\right)$

36. $-\dfrac{3}{5} + \left(-\dfrac{1}{5}\right)$

37. $\dfrac{1}{6} + \left(-\dfrac{5}{6}\right)$

38. $-\dfrac{9}{14} + \dfrac{3}{14}$

39. $\dfrac{3}{4} + \dfrac{1}{6}$

40. $\dfrac{1}{4} + \dfrac{7}{8}$

41. $-\dfrac{1}{12} + \left(-\dfrac{2}{3}\right)$ **42.** $-\dfrac{2}{5} + \left(-\dfrac{3}{20}\right)$ **43.** $-\dfrac{5}{6} + \dfrac{4}{21}$ **44.** $-\dfrac{5}{16} + \left(-\dfrac{3}{12}\right)$

45. $0.21 + 0.05$ **46.** $0.06 + 0.17$ **47.** $-0.18 + 6.7$ **48.** $-15.81 + 4.28$

49. $-0.28 + (-4.1)$ **50.** $-7.8 + (-9.16)$

For Exercises 51–64, find the additive inverse.

51. 5 **52.** 7 **53.** -12 **54.** -6

55. 0 **56.** -9 **57.** $\dfrac{5}{6}$ **58.** $-\dfrac{6}{17}$

59. -0.29 **60.** 2.8 **61.** $-x$ **62.** b

63. $\dfrac{m}{n}$ **64.** $-\dfrac{a}{b}$

For Exercises 65–84, subtract.

65. $6 - 15$ **66.** $8 - 20$ **67.** $-4 - 9$ **68.** $-7 - 15$

69. $4 - (-3)$ **70.** $6 - (-7)$ **71.** $-8 - (-2)$ **72.** $-13 - (-6)$

73. $\dfrac{7}{10} - \left(-\dfrac{3}{5}\right)$ **74.** $\dfrac{3}{8} - \left(-\dfrac{5}{6}\right)$ **75.** $-\dfrac{4}{5} - \left(-\dfrac{2}{7}\right)$ **76.** $-\dfrac{1}{2} - \left(-\dfrac{1}{3}\right)$

77. $-\dfrac{1}{5} - \left(-\dfrac{1}{5}\right)$ **78.** $-\dfrac{3}{4} - \left(-\dfrac{3}{4}\right)$ **79.** $4.01 - 3.65$ **80.** $8.1 - 4.76$

81. $0.07 - 5.82$ **82.** $0.107 - 5.802$ **83.** $-6.1 - (-4.5)$ **84.** $-7.1 - (-2.3)$

For Exercises 85–96, add or subtract.

85. $-18 + |-12|$

86. $|-24| + 11$

87. $-42 - |-14|$

88. $31 - |-54|$

89. $|-2.4| - |-0.78|$

90. $-0.6 - |9.1|$

91. $\left|-\dfrac{3}{8}\right| - \left|\dfrac{5}{6}\right|$

92. $\left|-\dfrac{4}{5}\right| - \dfrac{3}{4}$

93. $-|-4| + |6|$

94. $-|-9| - |-12|$

95. $-|8| - (-5)$

96. $-(-14) - |-6|$

For Exercises 97–106, solve.

97. The Disney corporation's fourth-quarter financial report for 2004 contains the following information. Find the net profit or loss.

Fourth-quarter report for the year ending September 30, 2004

Income (in millions)	Expenditures (in millions)
Revenue = $7,160	Costs and expenses = $6,458
Equity in the income of investors = $102	Restructuring and impairment charges = $5
	Interest expense = $149
	Income taxes = $67

98. Following is a balance sheet of income and expenditures for a small business. Find the net profit or loss.

June 2005 balance sheet

Income	Expenditures
Revenue = $24,572.88	Materials = $1545.75
Dividends = $1284.56	Lease = $2700
	Utilities = $865.45
	Employee wages = $21,580.50

99. In engineering, the resultant force on an object is the sum of all the forces acting on the object. The diagram shows the forces acting on a steel beam. Find the resultant force. What does the sign of the resultant force indicate?

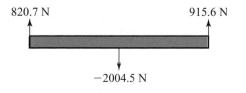

820.7 N   915.6 N

-2004.5 N

100. A family's taxable income is their income less deductions. Following is a table that shows the Brendel family's income and deductions for the 2000 tax year. What is the Brendel family's taxable income?

Income	Deductions
Mr. Brendel = $31,672.88	Mortgage interest = $6545.75
Mrs. Brendel = $32,284.56	Charitable donations = $1200
Dividends = $124.75	Medical expenses = $165.45
Miscellaneous income = $2400	Exemptions = $10,800

101. On January 1, 2004, the Dow Jones Industrial Average closed at 10,409.85. On December 31, 2004, the DJIA closed at 10,783.01. Find the difference in closing value from January 1 to December 31.

102. On January 13, 2005, the closing price of 7-Eleven Corporation's stock was $22.55. On January 12, 2005, the closing price was $22.20. Find the difference in closing price from January 12 to January 13.

103. On January 13, 2005, the closing price of Airtran Corporation's stock was $8.27. If the price had a change of −$0.60 from the previous date's closing price, what was the closing price on January 12, 2005?

104. On January 5, 2005, an investor found that the NASDAQ closed at 2,091.24, which was a change of −16.62 from the previous day's closing value. What was the closing price on January 4, 2005?

105. The temperature of liquid nitrogen is −208°C. An apple is placed in the liquid, and the apple then raises the temperature of the liquid to its boiling point of −196°C. Write an expression that describes the difference between the boiling point and the initial temperature, then calculate the difference.

Of Interest

Absolute zero refers to 0 on the Kelvin scale, which is the scale most scientists use to measure temperatures. Increments on the Kelvin scale are equal to increments on the Celsius scale, so a change of 1 K corresponds to a change of 1°C.

106. Absolute zero is the temperature at which molecular motion is at a minimum. This temperature is −273.15°C. A piece of metal is cooled to −256.5°C. Write an expression that describes the difference between the piece of metal's current temperature and absolute zero, then calculate the difference.

Use the following line graph to answer Exercises 107 and 108. The graph shows the mean composite ACT score of entering college freshmen from 1995 to 2000.

107. a. Write an expression that describes the difference between the mean composite scores in 1989 and 1986.

   b. Calculate the difference.

   c. What does the sign of the difference indicate about how the scores changed?

108. a. Write an expression that describes the difference between the mean composite scores in 1997 and 1986.

b. Calculate the difference.

c. What does the sign of the difference indicate about how the scores changed?

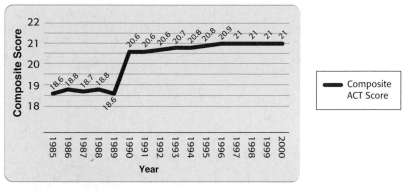

Source: U.S. Department of Education

Use the following bar graph to answer Exercises 109 and 110. The graph shows the median annual income of year-round full-time workers 25 years old and over, by level of education completed and gender, from 1992 to 2001.

109. In what year was the difference between salaries for males and females with a bachelor's degree the greatest? How much was that difference?

110. In what year was the difference in salaries greatest between a person with a high school degree and a person with a bachelor's degree? How much was that difference?

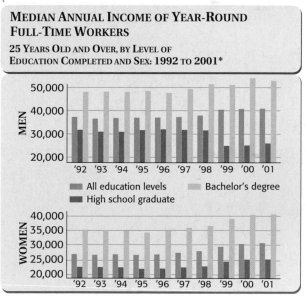

*In constant 2001 dollars

Source: U.S. Department of Education

PUZZLE PROBLEM
Fill in each square with a number 1 through 9 so that the sum of the numbers in each row, each column, and the two diagonals is the same.

Collaborative Exercises DOLLARS AND SENSE

Your group is to prepare a simple monthly household budget. The budget should be for a "typical" student or for a new employee in your field of study. (However, you should not use any of your personal information.) Remember to be realistic about both income and expenditures.

1. Discuss and agree upon an amount of income you think an average working student makes each month (or use a starting salary for a person in your field of study).
2. Estimate average expense amounts for each of the following. If your group feels that an expense does not apply, then explain why.

Taxes	Transportation/Gasoline
Housing	Child care
Electricity	Loan repayments
Water	Hair care
Telephone (home and cellular)	Clothing
Cable TV	Magazine/Newspaper subscriptions
Internet service	Allowance
Food	Movies
Household supplies	Other entertainment (video rentals, concerts, eating out)
Medicine	

3. Using your estimates in Exercises 1 and 2, calculate the monthly net.
4. Discuss and list ways that a person could cut expenses to improve his or her monthly net.

REVIEW EXERCISES

[1.1] **1.** Write a set containing the last names of the first four presidents of the United States.

[1.1] **2.** Simplify: $|-7|$

[1.2] **3.** Find the missing number that makes $\dfrac{40}{48} = \dfrac{?}{6}$ true.

[1.2] **4.** Use $<, >$, or $=$ to make a true statement: $-\dfrac{4}{5} \blacksquare -\dfrac{8}{20}$

[1.2] **5.** Find the prime factorization of 100.

[1.2] **6.** A company report indicates that 78 employees out of 91 have medical flexible spending accounts. What fraction of the employees have medical flexible spending accounts? Express the fraction in lowest terms.

1.4 Multiplying and Dividing Real Numbers; Properties of Real Numbers

OBJECTIVES

1. Multiply integers.
2. Multiply more than two numbers.
3. Multiply rational numbers.
4. Find the multiplicative inverse of a number.
5. Divide rational numbers.

OBJECTIVE 1. Multiply integers. In a multiplication statement, *factors* are multiplied to equal a *product*.

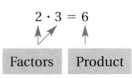

$$2 \cdot 3 = 6$$

Factors Product

In Section 1.3, we discussed properties of addition. Multiplication has similar properties to addition. The following table shows the properties of multiplication.

Properties of Multiplication	Symbolic Form	Word Form
Multiplicative Property of 0	$0 \cdot a = 0$	The product of a number multiplied by 0 is 0.
Multiplicative Identity	$1 \cdot a = a$	The product of a number multiplied by 1 is the number.
Commutative Property of Multiplication	$ab = ba$	Changing the order of factors does not affect the product.
Associative Property of Multiplication	$a(bc) = (ab)c$	Changing the grouping of three or more factors does not affect the product.
Distributive Property of Multiplication over Addition	$a(b + c) = ab + ac$	A sum multiplied by a factor is equal to the sum of that factor multiplied by each addend.

Multiplying Numbers with Different Signs

To determine the rules for multiplying signed numbers, consider the following pattern. Pay attention to the product as we decrease the second factor.

$2 \cdot 4 = 8$

$2 \cdot 3 = 6$ ◄—— As we decrease the second factor by 1, the product decreases by 2, from 8 to 6.

Notice that the same pattern continues with the product decreasing by 2 each time we decrease the factor by 1.

$2 \cdot 2 = 4$

$2 \cdot 1 = 2$

$2 \cdot 0 = 0$

$2 \cdot (-1) = -2$ ◄—— Notice that when we decrease the 0 factor to -1, we continue the pattern and decrease the product by 2 so that it is -2. Decreasing the factor by 1 again decreases the product by 2, so that we get -4.

$2 \cdot (-2) = -4$

This pattern suggests that multiplying a positive number by a negative number equals a negative product. Note that by the commutative property of multiplication we can exchange the factors and conclude that multiplying a negative number times a positive number also equals a negative product.

$$2 \cdot (-2) = (-2) \cdot 2 = -4$$

RULE **Multiplying Two Numbers with Different Signs**

When multiplying two numbers that have different signs, the product is negative.

EXAMPLE 1 Multiply.

a. $9(-8)$

Solution $9(-8) = -72$

b. $(-12)3$

Solution $(-12)3 = -36$

Warning: Make sure that you see the difference between $9(-8)$, which indicates multiplication, and $9 - 8$, which indicates subtraction.

YOUR TURN Multiply.

a. $13 \cdot (-7)$

b. $-5 \cdot 9$

Multiplying Numbers with the Same Sign

Now let's examine multiplying numbers with the same sign. Consider the following pattern to determine the rule.

$(-2) \cdot 4 = -8$ From the rule we already established, a negative number times a positive number is a negative product.

$(-2) \cdot 3 = -6$ ← As we decrease the positive factor by 1, the product *increases* by 2, from −8 to −6.

$(-2) \cdot 2 = -4$ Notice that the same pattern continues with the product increasing by 2 each time we decrease the factor by 1.

$(-2) \cdot 1 = -2$

$(-2) \cdot 0 = 0$

$(-2) \cdot (-1) = 2$ ← To continue the same pattern, when we decrease the 0 factor to −1, we must continue to increase the product by 2. This means the product must become positive!

$(-2) \cdot (-2) = 4$

This pattern indicates that the product of two negative numbers is a positive number. We have already seen that the product of two positive numbers is a positive number, so we can draw the following conclusion.

ANSWERS

a. -91 **b.** -45

RULE Multiplying Two Numbers with the Same Sign

When multiplying two numbers that have the same sign, the product is positive.

EXAMPLE 2 Multiply.

a. $-6(-8)$

Solution $-6(-8) = 48$

b. $(-9)(-7)$

Solution $(-9)(-7) = 63$

YOUR TURN Multiply.

a. $-6(-9)$ **b.** $(-3)(-13)$

OBJECTIVE 2. Multiply more than two numbers. Suppose we have to multiply more than two numbers, such as $(-2)(-3)(5)$. The associative property of multiplication allows us to group the factors any way we wish.

$$(\mathbf{-2})(\mathbf{-3})(5) = \mathbf{6}(5)$$ **Multiply the first two factors.**
$$= 30$$

or

$$(-2)(-3)(5) = (-2)(\mathbf{-15})$$ **Multiply the second and third factors.**
$$= 30$$

Notice that no matter which way we choose to multiply the factors, the result is the same. Also notice that the result is positive because there are two negative factors involved in the multiplication. What if we have three negative factors, as in $(-2)(-3)(-5)$?

$$(\mathbf{-2})(\mathbf{-3})(-5) = \mathbf{6}(-5)$$ **Multiply the first two factors.**
$$= -30$$

or

$$(-2)(\mathbf{-3})(\mathbf{-5}) = (-2)(\mathbf{15})$$ **Multiply the second and third factors.**
$$= -30$$

With three negative factors, the result is negative. This suggests the following rule.

RULE Multiplying with Negative Factors

The product of an even number of negative factors is positive, whereas the product of an odd number of negative factors is negative.

EXAMPLE 3 Multiply.

a. $(-1)(-1)(-2)(6)$

Solution Because there are three negative factors (an odd number of negative factors), the result is negative.

$$(-1)(-1)(-2)(6) = -12$$

b. $(-1)(-3)(-7)(2)(-4)$

Solution Because there are four negative factors (an even number of negative factors), the result is positive.

$$(-1)(-3)(-7)(2)(-4) = 168$$

YOUR TURN Multiply.

a. $(-3)(4)(-5)(2)$ **b.** $(6)(-5)(-1)(2)(-1)$

OBJECTIVE 3. Multiply rational numbers.

Multiplying Fractions

On a ruler, note that half of a fourth is an eighth.

When preceded by a fraction, the word *of* means multiply, so the sentence would translate as follows:

Half of a fourth is an eighth.

$$\frac{1}{2} \cdot \frac{1}{4} = \frac{1}{8}$$

Notice that the numerators multiply to equal the numerator and the denominators multiply to equal the denominator. This suggests the following rule for multiplying fractions.

RULE Multiplying Fractions

$\dfrac{a}{b} \cdot \dfrac{c}{d} = \dfrac{ac}{bd}$, where $b \neq 0$ and $d \neq 0$.

Multiplying and Simplifying

In mathematics, it is expected that we always simplify results to lowest terms. Consider $\dfrac{3}{4} \cdot \dfrac{5}{6}$. We can multiply first, then simplify.

$$\frac{3}{4} \cdot \frac{5}{6} = \frac{15}{24} = \frac{3 \cdot 5}{2 \cdot 2 \cdot 2 \cdot 3} = \frac{5}{8}$$

Note: *After multiplying, we divide out the common factor, 3.*

Or, as an alternative, we can actually divide out the common factors before multiplying the fractions. In our example, we can divide out the common factor of 3 and then multiply.

$$\frac{3}{4} \cdot \frac{5}{6} = \frac{3}{2 \cdot 2} \cdot \frac{5}{2 \cdot 3} = \frac{5}{8}$$

Note: *Some people prefer to divide out the common factors without breaking down to primes, like this:*

$$\frac{\overset{1}{\cancel{3}}}{4} \cdot \frac{5}{\underset{2}{\cancel{6}}} = \frac{5}{8}$$

We can divide out common factors before or after multiplying because every factor in the original numerators becomes a factor in the product's numerator and every factor in the original denominators becomes a factor in the product's denominator. Now let's put this together with the sign rules for multiplication.

EXAMPLE 4 Multiply.

a. $-\frac{2}{3} \cdot \left(\frac{7}{8}\right)$

Solution $-\frac{2}{3} \cdot \left(\frac{7}{8}\right) = -\frac{2}{3} \cdot \left(\frac{7}{2 \cdot 2 \cdot 2}\right)$ Divide out the common factor, 2.

$= -\frac{7}{12}$ Because we are multiplying two numbers that have different signs, the product is negative.

Calculator TIPS

To use a calculator for Example 4 (a), type:

$\boxed{(-)}$ $\boxed{2}$ $\boxed{a^{b/c}}$ $\boxed{3}$ $\boxed{\times}$ $\boxed{7}$ $\boxed{a^{b/c}}$ $\boxed{8}$ $\boxed{=}$

b. $-\frac{4}{15} \cdot \left(-\frac{9}{16}\right)$

Solution $-\frac{4}{15} \cdot \left(-\frac{9}{16}\right) = -\frac{2 \cdot 2}{3 \cdot 5} \cdot \left(-\frac{3 \cdot 3}{2 \cdot 2 \cdot 2 \cdot 2}\right)$ Divide out the common factors.

$= \frac{3}{20}$ Because we are multiplying two numbers that have the same sign, the product is positive.

Connection Using the slash marks style, Example 4(b) looks like this:

$$-\frac{\overset{1}{\cancel{4}}}{\underset{5}{\cancel{15}}} \cdot \left(-\frac{\overset{3}{\cancel{9}}}{\underset{4}{\cancel{16}}}\right) = \frac{3}{20}$$

Notice that we are still dividing out a factor of 4 in the 4 and 16, and a common factor of 3 in the 9 and 15.

YOUR TURN Multiply.

a. $\frac{5}{8} \cdot \left(-\frac{3}{10}\right)$

b. $-\frac{4}{5} \cdot \left(\frac{1}{6}\right) \cdot \left(-\frac{3}{9}\right)$

ANSWERS

a. $-\frac{3}{16}$ **b.** $\frac{2}{45}$

Multiplying Decimal Numbers

Recall that decimal numbers name fractions with denominators that are powers of 10. In the problem $0.3 \cdot 0.7$, we could multiply by replacing the decimal factors with their fraction equivalents.

$$0.3 \cdot 0.7 = \frac{3}{10} \cdot \frac{7}{10} = \frac{21}{100} = 0.21$$

Because we multiply the denominators, if we multiply tenths times tenths, then the product will be hundredths. Notice that if we count the number of decimal places in the factors, this corresponds to the total places in the product.

$$
\begin{array}{l}
0.7 \longrightarrow \quad 1 \text{ place} \\
\underline{\times\ 0.3} \longrightarrow \underline{+\ 1 \text{ place}} \\
0.21 \longrightarrow \quad 2 \text{ places}
\end{array}
$$

Likewise, hundredths (2 places) times tenths (1 place) would equal thousandths (3 places).

$$0.03 \cdot 0.7 = \frac{3}{100} \cdot \frac{7}{10} = \frac{21}{1000} = 0.021 \quad \text{or}$$

$$
\begin{array}{l}
0.7 \longrightarrow \quad 1 \text{ place} \\
\underline{\times\ 0.03} \longrightarrow \underline{+\ 2 \text{ place}} \\
0.021 \longrightarrow \quad 3 \text{ places}
\end{array}
$$

PROCEDURE   Multiplying Decimal Numbers

To multiply decimal numbers:

1. Multiply as if they named whole numbers.
2. Place the decimal in the product so that it has the same number of decimal places as the total number of decimal places in the factors.

EXAMPLE 5   Multiply. $(-5.6)(0.03)$

Solution   First, we calculate the value and disregard the signs for now.

$$
\begin{array}{r}
0.03 \qquad 2 \text{ places} \\
\underline{\times\ \ 5.6} \qquad \underline{+\ 1 \text{ place}} \\
0\,1\,8 \\
\underline{+\ 0\,1\,5\ \ } \\
0.1\,6\,8 \longleftarrow 3 \text{ places}
\end{array}
$$

Answer -0.168 ◀   **Note:** *When we multiply two numbers with different signs, the product is negative.*

YOUR TURN   Multiply.

a. $(-0.07)(-2.65)$ b. $(-1)(-0.9)(-24)(0.2)$

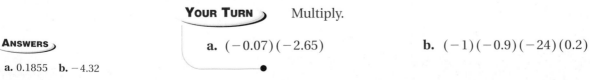

OBJECTIVE 4. Find the multiplicative inverse of a number. Two numbers that multiply to equal 1 are called **multiplicative inverses**.

DEFINITION *Multiplicative inverses:* Two numbers whose product is 1.

For example, $\frac{2}{3}$ and $\frac{3}{2}$ are multiplicative inverses because their product is 1.

$$\frac{2}{3} \cdot \frac{3}{2} = \frac{6}{6} = 1$$

Notice that to write a number's multiplicative inverse, we simply invert the numerator and denominator. Multiplicative inverses are also known as *reciprocals*.

EXAMPLE 6 Find the multiplicative inverse.

a. $\frac{5}{8}$

Answer The multiplicative inverse of $\frac{5}{8}$ is $\frac{8}{5}$ because $\frac{5}{8} \cdot \frac{8}{5} = 1$.

b. $-\frac{1}{4}$

Answer The multiplicative inverse of $-\frac{1}{4}$ is -4 because
$-\frac{1}{4} \cdot (-4) = -\frac{1}{4} \cdot \left(-\frac{4}{1}\right) = 1$.

c. -6

Answer The multiplicative inverse of -6 is $-\frac{1}{6}$ because
$-6 \cdot \left(-\frac{1}{6}\right) = -\frac{6}{1} \cdot \left(-\frac{1}{6}\right) = 1$.

YOUR TURN Find the multiplicative inverse.

a. $\frac{4}{5}$　　　　　b. $\frac{1}{7}$　　　　　c. -9

OBJECTIVE 5. Divide rational numbers. Let's now turn our attention to division. The parts of a division statement are shown next.

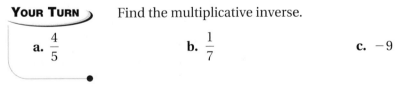

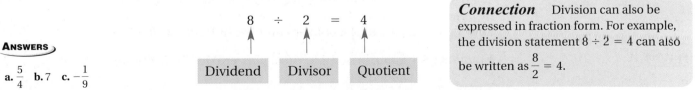

$$8 \div 2 = 4$$

Dividend　　Divisor　　Quotient

> **Connection** Division can also be expressed in fraction form. For example, the division statement $8 \div 2 = 4$ can also be written as $\frac{8}{2} = 4$.

Sign Rules for Division

The sign rules for division are the same as for multiplication. We can see this by looking at the relationship between division and multiplication. We can use one operation to check the other. For example, if $8 \cdot 2 = 16$, then $16 \div 2 = 8$ or $16 \div 8 = 2$. In checking a multiplication, we divide the product by one of the factors to equal the other factor. If we place signs in the problem, we could say that if $8 \cdot (-2) = -16$, then $-16 \div (-2) = 8$ or $-16 \div 8 = -2$.

RULE   Dividing Signed Numbers

When dividing two numbers that have the *same* sign, the quotient is positive.
When dividing two numbers that have *different* signs, the quotient is negative.

EXAMPLE 7 Divide.

a. $48 \div (-6)$

Solution $48 \div (-6) = -8$

b. $-35 \div (-7)$

Solution $-35 \div (-7) = 5$

YOUR TURN   Divide.

a. $(-60) \div 12$ **b.** $-42 \div -6$

Division Involving 0

What if 0 is involved in division? First, consider when the *dividend* is 0 and the *divisor* is not 0, as in $0 \div 9$. The result should be $0 \div 9 = 0$. Remember that we can check division by multiplying the quotient and the divisor to equal the dividend, so we can verify that $0 \div 9 = 0$ by writing the following multiplication:

$$\text{Quotient} \cdot \text{Divisor} = \text{Dividend}$$
$$0 \quad \cdot \quad 9 \quad = \quad 0$$

Conclusion   If the dividend is 0 and the divisor is not zero, then the quotient is 0.

What if the divisor is 0 as in $8 \div 0$? Again, think about the check statement:

$$\text{Quotient} \cdot \text{Divisor} = \text{Dividend}$$
$$? \quad \cdot \quad 0 \quad = \quad 8$$

There is no number that can make $? \cdot 0 = 8$ true because any number multiplied by 0 would equal 0, not 8. So there is also no number that can make $8 \div 0 = ?$ true, and we say that dividing a nonzero number by 0 is *undefined*.

Conclusion   If the divisor is 0 with a nonzero dividend, the statement is undefined.

ANSWERS

a. -5 **b.** 7

What if both the dividend and divisor are 0 as in $0 \div 0$? Again, think about the check statement.

$$\text{Quotient} \cdot \text{Divisor} = \text{Dividend}$$
$$? \quad \cdot \quad 0 \quad = \quad 0$$

Notice any number would work for this quotient because any number times 0 equals 0. Because we cannot determine a unique quotient here, we say that $0 \div 0$ is *indeterminate*.

Conclusion If both the dividend and divisor are 0, then the statement is indeterminate.

Following is a summary of these rules.

RULE   Division Involving 0

> $0 \div n = 0$ when $n \neq 0$.
> $n \div 0$ is undefined when $n \neq 0$.
> $0 \div 0$ is indeterminate.

Calculator TIPS

Try dividing any number by 0 on your calculator. When you press the equal key the screen will display error, which is your calculator's way of indicating undefined or indeterminate.

Dividing Fractions

To determine how to divide when the divisor is a fraction, consider this problem: How many quarters are in \$5? A quarter is $\frac{1}{4}$ of a dollar, so this question translates to the division problem:

$$\text{Number of quarters} = 5 \div \frac{1}{4}$$

Now, since there are 4 quarters in each of those 5 dollars, then there must be 20 quarters in \$5. Notice in analyzing the situation we translated the division problem into a multiplication problem. We can say:

$$\text{Number of quarters} = 5 \div \frac{1}{4} = 5 \cdot 4 = 20$$

Notice that $\frac{1}{4}$ and 4 are multiplicative inverses. We can write a division problem as an equivalent multiplication problem by changing the divisor to its multiplicative inverse.

RULE   Dividing Fractions

> $\frac{a}{b} \div \frac{c}{d} = \frac{a}{b} \cdot \frac{d}{c}$, where $b \neq 0$, $c \neq 0$, and $d \neq 0$.

EXAMPLE 8 Divide. $-\dfrac{7}{8} \div \dfrac{5}{6}$

Solution $-\dfrac{7}{8} \div \dfrac{5}{6} = -\dfrac{7}{8} \cdot \dfrac{6}{5}$ Write an equivalent multiplication.

$\qquad\qquad = -\dfrac{7}{2 \cdot 2 \cdot 2} \cdot \dfrac{2 \cdot 3}{5}$ Divide out the common factor, 2.

$\qquad\qquad = -\dfrac{21}{20}$ Because we are dividing two numbers that have different signs, the result is negative.

> ***Connection***   Using slash marks, Example 8 looks like this:
>
>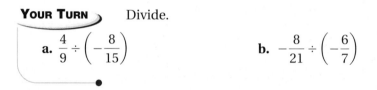

YOUR TURN   Divide.

   a. $\dfrac{4}{9} \div \left(-\dfrac{8}{15}\right)$ **b.** $-\dfrac{8}{21} \div \left(-\dfrac{6}{7}\right)$

Dividing Decimal Numbers

Division with decimal numbers can be broken into two categories: one, if the divisor is an integer, and two, if the divisor contains a decimal point. The following procedure summarizes how to divide decimal numbers.

PROCEDURE   Dividing Decimal Numbers

To divide decimal numbers, set up a long division and consider the divisor.

Case 1: If the divisor is an integer, then divide as if the dividend were a whole number and place the decimal point in the quotient directly above its position in the dividend.

Case 2: If the divisor is a decimal number, then

   1. Move the decimal point in the divisor to the right enough places to make the divisor an integer.
   2. Move the decimal point in the dividend the same number of places.
   3. Divide the divisor into the dividend as if both numbers were whole numbers. Be sure to align the digits in the quotient properly.
   4. Write the decimal point in the quotient directly above its position in the dividend.

In either of the two cases, continue the division process until you get a remainder of 0 or a repeating digit or block of digits in the quotient.

ANSWERS

a. $-\dfrac{5}{6}$ **b.** $\dfrac{4}{9}$

EXAMPLE 9 Divide. $-12.8 \div (-0.09)$

Solution Because the divisor is a decimal number, we move the decimal point enough places to the right to create an integer—in this case, two places. Then move the decimal point two places to the right in the dividend. Because we are dividing two numbers with the same sign, the result is positive.

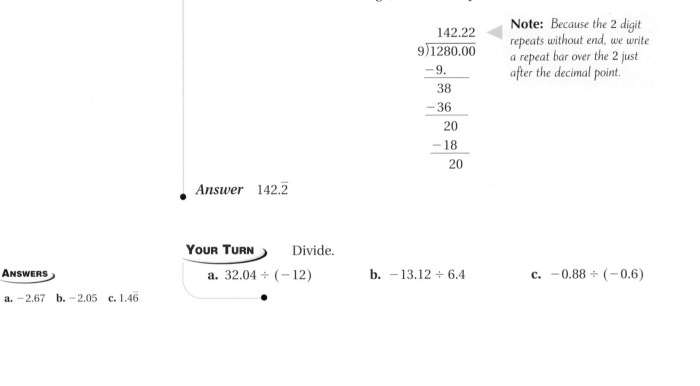

$$
\begin{array}{r}
142.22 \\
9{\overline{\smash{\big)}\,1280.00}} \\
-9. \\
\hline
38 \\
-36 \\
\hline
20 \\
-18 \\
\hline
20
\end{array}
$$

Note: *Because the 2 digit repeats without end, we write a repeat bar over the 2 just after the decimal point.*

Answer $142.\overline{2}$

YOUR TURN Divide.

ANSWERS
a. -2.67 **b.** -2.05 **c.** $1.4\overline{6}$

a. $32.04 \div (-12)$ **b.** $-13.12 \div 6.4$ **c.** $-0.88 \div (-0.6)$

1.4 Exercises

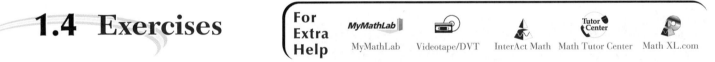

For Extra Help   MyMathLab   MyMathLab   Videotape/DVT   InterAct Math   Math Tutor Center   Math XL.com

For Exercises 1–4, state whether the result is positive or negative.

1. When multiplying or dividing two numbers that have the same sign, the sign of the result is _____.

2. When multiplying or dividing two numbers that have different signs, the sign of the result is _____.

3. When a multiplication problem contains an odd number of negative factors, the sign of the product is _____.

4. When a multiplication problem contains an even number of negative factors, the sign of the product is _____.

5. Why is $5 \div 0$ undefined?

6. Why is $0 \div 0$ indeterminate?

7. Why are 5 and $\frac{1}{5}$ multiplicative inverses?

8. Explain how to divide fractions.

9. Explain why $\dfrac{-a}{b} = \dfrac{a}{-b} = -\dfrac{a}{b}$, where a and b are any real numbers and $b \neq 0$.

10. Explain the difference between $6(-7)$ and $6 - 7$.

For Exercises 11–26, indicate whether the equation illustrates the multiplicative property of 0, the multiplicative identity, the commutative property of multiplication, the associative property of multiplication, or the distributive property.

11. $6(3 + 4) = 6 \cdot 3 + 6 \cdot 4$

12. $7(2 - 5) = 7 \cdot 2 - 7 \cdot 5$

13. $1 \cdot \left(-\dfrac{5}{6}\right) = -\dfrac{5}{6}$

14. $1 \cdot (-7) = -7$

15. $4(3 \cdot 2) = (4 \cdot 3)2$

16. $(3.4)(7.8) = (7.8)(3.4)$

17. $0 \cdot (-6.3) = 0$

18. $\dfrac{2}{3} \cdot 0 = 0$

19. $\dfrac{3}{5} \cdot \left(-\dfrac{1}{4}\right) = -\dfrac{1}{4} \cdot \dfrac{3}{5}$

20. $-\dfrac{2}{3} \cdot \dfrac{4}{5} = \dfrac{4}{5} \cdot \left(-\dfrac{2}{3}\right)$

21. $[4(-2)] \cdot 3 = 4[-2 \cdot 3]$

22. $-6 \cdot (5 \cdot 4) = (-6 \cdot 5) \cdot 4$

23. $(-4.5)(2 + 0.6) = (2 + 0.6)(-4.5)$

24. $-6.2(-4.3 + 7.1) = (-4.3 + 7.1)(-6.2)$

25. $\dfrac{2}{3} \cdot 5 + \dfrac{2}{3} \cdot 4 = \dfrac{2}{3}(5 + 4)$

26. $-(5 + 9) = -5 - 9$

For Exercises 27–48, multiply.

27. $6(-3)$

28. $(-10)(5)$

29. $-4(-5)$

30. $(-4)(-3)$

31. $6(-9)(-1)$

32. $(-5)(-3)(-2)$

33. $(-1)(-6)(-40)(-3)$

34. $12(-6)(-2)(-1)$

35. $(-2)(3)(-4)(-1)(2)$

36. $(-1)(-1)(4)(-5)(-3)$

37. $\dfrac{1}{2} \cdot \left(-\dfrac{1}{2}\right)$

38. $\dfrac{4}{5} \cdot \left(\dfrac{20}{3}\right)$

39. $\left(-\dfrac{2}{3}\right)\left(-\dfrac{3}{4}\right)$

40. $\left(-\dfrac{5}{6}\right)\left(-\dfrac{6}{5}\right)$

41. $\left(-\dfrac{5}{6}\right)\left(\dfrac{8}{15}\right)$

42. $\left(\dfrac{2}{9}\right)\left(-\dfrac{21}{26}\right)$

43. $3.8(-10)$

44. $8(-2.5)$

45. $-1.6(-4.2)$

46. $-7.1(-0.5)$

47. $-4.3(1.52)$

48. $(-8.1)(-2.75)$

For Exercises 49–56, find the multiplicative inverse.

49. $\dfrac{2}{3}$

50. $\dfrac{20}{3}$

51. $-\dfrac{5}{2}$

52. $-\dfrac{6}{7}$

53. -5

54. 17

55. 0

56. -1

For Exercises 57–80, divide.

57. $-6 \div 2$

58. $42 \div (-7)$

59. $-56 \div (-4)$

60. $-12 \div (-4)$

61. $\dfrac{-18}{-9}$

62. $\dfrac{75}{-3}$

63. $\dfrac{-8}{0}$

64. $\dfrac{0}{0}$

65. $0 \div 0$

66. $-21 \div 0$

67. $6 \div \dfrac{2}{3}$

68. $-8 \div \dfrac{3}{4}$

69. $\dfrac{3}{5} \div \left(-\dfrac{9}{10}\right)$

70. $-\dfrac{4}{5} \div \dfrac{4}{5}$

71. $-\dfrac{2}{7} \div \left(-\dfrac{8}{21}\right)$

72. $-\dfrac{1}{3} \div \left(-\dfrac{3}{2}\right)$

73. $-\dfrac{4}{9} \div \dfrac{10}{21}$

74. $\dfrac{7}{15} \div \left(-\dfrac{35}{24}\right)$

75. $9.03 \div 4.3$

76. $8.1 \div 0.6$

77. $-36.72 \div (-0.4)$

78. $-10.65 \div (-7.1)$

79. $-14 \div 0.3$

80. $-19 \div (-0.06)$

For Exercises 81–86, solve.

81. In a poll where respondents can agree, disagree, or have no opinion, $\frac{3}{4}$ of the respondents said they agreed and $\frac{2}{3}$ of those that agreed were women. What fraction of all respondents were women who agreed with the statement in the poll?

82. On a standard-size guitar, the length of the strings between the saddle and nut is $25\frac{1}{2}$ inches. Guitar makers must place the 12th fret at exactly half of the length of the string. How far from the saddle or nut should one measure to place the 12th fret?

Saddle

Nut

12th fret

83. A financial planner estimates that at the rate one of his clients is increasing debt she will have $3\frac{1}{2}$ times her current debt in five years. If her current debt is represented by $-\$2480$, how can her debt in five years be represented?

84. Mario finds that $\frac{2}{3}$ of his credit card debt is from dining out. If his total credit card debt is represented by $-\$858$, how much of the debt is from dining out?

85. In 1995, the Garret family's only debt was credit card balance represented by $-\$258.75$. In 2005, they have a mortgage, two car loans, and three credit cards with a total debt represented by $-\$158,572.85$. How many times greater is their debt in 2005 than it was in 1995?

86. A company's stock loses value by an amount represented by $-\$\frac{3}{8}$ each day for four days. Write a representation of the total loss in value at the close of the fourth day.

Of Interest

Three components affect the pitch of a string on a stringed instrument: diameter, tension, and length. Given two strings of the same diameter and same tension, if one string is half the length of the other, it will have a pitch that is one octave higher than the longer string. On a guitar, placing our finger at the 12th fret cuts the string length in half, thereby creating a tone that is an octave higher than the string's open tone.

For Exercises 87–90, use the fact that an object's weight is a downward force due to gravity and is calculated by multiplying the object's mass by the acceleration due to gravity, which is a constant (force = mass × acceleration). The following table indicates the units.

	Force	=	Mass	×	Acceleration
American measurement	pounds (lb)		slugs (s)		-32.2 ft./sec.^{2}
Metric measurement	newtons (N)		kilograms (kg)		-9.8 m/sec.^{2}

87. Find the force due to gravity on an object with a mass of 12.5 slugs. What does the sign of the weight indicate?

88. Find the force due to gravity on a person with a mass of 70.4 kilograms.

89. The blue whale is the largest animal on Earth. The largest blue whale ever caught weighed 1,658,181.8 newtons, which means the downward force due to gravity was −1,658,181.8 newtons. What was its mass in kilograms? (*Source: The Whale Watcher's Guide: Whale Watching Trips in North America*, Corrigan, 1999)

90. The Liberty Bell weighs 2080 pounds, which means the downward force due to gravity is −2080 pounds. What is its mass in slugs? (*Source: Ring in the Jubilee: The Epic of America's Liberty Bell*, Boland, 1973)

For Exercises 91 and 92, use the fact that in an electrical circuit, voltage (V) is equal to the product of the current, measured in amperes (A), and the resistance of the circuit, measured in ohms (Ω) (voltage = current × resistance).

91. Suppose the current in a circuit is −6.4 amperes and the resistance is 8 ohms. Find the voltage.

92. An electrical technician measures the voltage in a circuit to be −15 volts and the current to be −8 amperes. What is the resistance of the circuit?

For Exercises 93 and 94, use the fact that in an electrical circuit, power, which is measured in watts (W), is the product of the voltage and the current (power = voltage × current).

93. An engineer measures the voltage in a circuit to be −120 volts and the current to be −30 amperes. What is the power?

94. An electrical technician determines that the power in a circuit is 400 watts. If the current is measured to be −6.5 amperes, what is the resistance in the circuit?

REVIEW EXERCISES

[1.1] **1.** Is π a rational or irrational number?

[1.1] **2.** Graph −7.2 on a number line.

[1.1] **3.** Simplify: $|-6.8|$

[1.3] **4.** Add: $-7 + (-15)$

[1.3] **5.** Find the additive inverse of $\frac{2}{3}$.

[1.3] **6.** Subtract: $-\frac{5}{8} - \left(-\frac{1}{6}\right)$

1.5 Exponents, Roots, and Order of Operations

OBJECTIVES

1. Evaluate numbers in exponential form.
2. Evaluate square roots.
3. Use the order-of-operations agreement to simplify numerical expressions.
4. Find the mean of a set of data.

OBJECTIVE 1. Evaluate numbers in exponential form. Sometimes problems involve repeatedly multiplying the same number. In such problems, we can use an **exponent** to indicate that a **base** number is repeatedly multiplied.

DEFINITIONS *Exponent:* A symbol written to the upper right of a base number that indicates how many times to use the base as a factor.
Base: The number that is repeatedly multiplied.

When we write a number with an exponent, we say the expression is in *exponential form*. The expression 2^4 is in exponential form, where the base is 2 and the exponent is 4, it is read "two to the fourth power," or simply "two to the fourth." To *evaluate* 2^4, write 2 as a factor 4 times, then multiply.

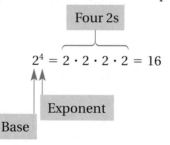

Four 2s

$$2^4 = 2 \cdot 2 \cdot 2 \cdot 2 = 16$$

Exponent

Base

PROCEDURE Evaluating an Exponential Form

To evaluate an exponential form raised to a natural number exponent, write the base as a factor the number of times indicated by the exponent, then multiply.

EXAMPLE 1 Evaluate. $(-7)^2$

Solution The exponent 2 indicates we have two factors of -7. Because we multiply two negative numbers, the result is positive.

$$(-7)^2 = (-7)(-7) = 49$$

◄ **Note:** *If a base is raised to the second power, we can say it is* **squared.** *The expression* $(-7)^2$ *can be read as "negative seven* **squared.**"

EXAMPLE 2 Evaluate. $\left(-\dfrac{2}{3}\right)^3$

Solution The exponent 3 means we must multiply the base by itself three times.

$$\left(-\dfrac{2}{3}\right)^3 = \left(-\dfrac{2}{3}\right)\left(-\dfrac{2}{3}\right)\left(-\dfrac{2}{3}\right)$$

$$= -\dfrac{8}{27}$$

◀ **Note:** *If a base is raised to the third power, we can say it is* cubed. *The expression* $\left(-\dfrac{2}{3}\right)^3$ *can be read as "negative two-thirds* cubed.*"*

Examples 1 and 2 suggest the following sign rules.

RULE **Evaluating Exponential Forms with Negative Bases**

If the base of an exponential form is a negative number and the exponent is even, then the product is positive.
If the base is a negative number and the exponent is odd, then the product is negative.

YOUR TURN Evaluate.

a. $\left(-\dfrac{1}{2}\right)^5$ **b.** $(-0.3)^4$

Calculator TIPS

Use the $\boxed{\wedge}$ *key to indicate an exponent. To evaluate* $\left(-\dfrac{2}{3}\right)^3$, *type:*

$\boxed{(}\ \boxed{(-)}\ \boxed{2}\ \boxed{\div}\ \boxed{3}\ \boxed{)}\ \boxed{\wedge}\ \boxed{3}\ \boxed{\text{ENTER}}$

The answer will be given as a decimal. To change the decimal answer to a fraction, press the $\boxed{\text{MATH}}$ *key, select 1:Frac, then press* $\boxed{\text{ENTER}}$

On some scientific calculators, you use a $\boxed{y^x}$ *key. To evaluate* $\left(-\dfrac{2}{3}\right)^3$ *on one of these, type:*

$\boxed{2}\ \boxed{a^{b}/c}\ \boxed{3}\ \boxed{+/-}\ \boxed{y^x}\ \boxed{3}\ \boxed{=}$

OBJECTIVE 2. Evaluate square roots. Roots are inverses of exponents. More specifically, a square root is the inverse of a square, so a square root of a given number is a number that, when squared, equals the given number. For example, 5 is a square root of 25 because $5^2 = 25$. Notice -5 is also a square root of 25 because $(-5)^2 = 25$.

Conclusion For every real number there are *two* square roots, a positive root and a negative root.

For convenience, if asked to find all square roots of 25, we can write both the positive and negative answers in a compact expression like this: ± 5.

EXAMPLE 3 Find all square roots of the given number.

a. 36

Answer ±6

b. −25

Answer No real-number square roots exist.

Note: *We do not say that this situation is undefined or indeterminate, because we eventually will define the square root of a negative number. Such roots will be defined using imaginary numbers.*

Explanation Because of the sign rules for multiplication, there is no way to square a real number and get anything other than a positive result.

Our work so far suggests the following rules.

RULE Square Roots

Every positive number has two square roots, a positive root and a negative root. Negative numbers have no real-number square roots.

YOUR TURN Find all square roots of the given number.

a. 144

b. −81

The Principal Square Root

A symbol, $\sqrt{}$, called the *radical*, is used to indicate finding only the *positive* (or *principal*) square root of a given number. The given number or expression inside the radical is called the *radicand*.

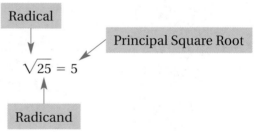

What about the square root of a fraction like $\sqrt{\dfrac{4}{9}}$? Since $\left(\dfrac{2}{3}\right)^2 = \dfrac{2}{3} \cdot \dfrac{2}{3} = \dfrac{4}{9}$, we can say that $\sqrt{\dfrac{4}{9}} = \dfrac{2}{3}$. Notice that we could evaluate $\sqrt{\dfrac{4}{9}}$ by finding the square root of the numerator and denominator separately, or $\sqrt{\dfrac{4}{9}} = \dfrac{\sqrt{4}}{\sqrt{9}} = \dfrac{2}{3}$. We now have the following rules for square roots involving the radical sign.

ANSWERS

a. ±12

b. No real-number roots exist.

Calculator TIPS

To calculate a square root, use the $\sqrt{}$ function, which may require pressing [2nd] *and then the* [x²] *key. To find the square root of 144, type:*

[2nd] [x²] [1] [4] [4] [ENTER]

Some scientific calculators have a [√x] *key. To find the square root of 144 on one of these, type:*

[1] [4] [4] [√x]

EXAMPLE 4 Evaluate the square root.

a. $\sqrt{144}$

Solution $\sqrt{144} = 12$

b. $\sqrt{\dfrac{49}{81}}$

Solution $\sqrt{\dfrac{49}{81}} = \dfrac{\sqrt{49}}{\sqrt{81}} = \dfrac{7}{9}$

c. $\sqrt{0.81}$

Solution $\sqrt{0.81} = 0.9$

d. $\sqrt{-81}$

Solution not a real number

YOUR TURN Evaluate the square root.

a. $\sqrt{121}$

b. $\sqrt{\dfrac{81}{169}}$

c. $\sqrt{-0.09}$

OBJECTIVE 3. **Use the order-of-operations agreement to simplify numerical expressions.** Some expressions are too complicated to evaluate using only the associative and commutative properties. If an expression contains a mixture of operations, then we must be careful about the order in which we perform the operations. Performing the operations in a different order can change the results. Consider the expression $2 + 5 \cdot 3$.

Adding first: $2 + 5 \cdot 3$
$= 7 \cdot 3$
$= 21$

Multiplying first: $2 + 5 \cdot 3$
$= 2 + 15$
$= 17$

Changing the order of the operations changes the outcome. To ensure that everyone arrives at the same answer, mathematicians developed an order of operations that we all agree to follow.

ANSWERS

a. 11 **b.** $\dfrac{9}{13}$

c. not a real number

PROCEDURE   Order-of-Operations Agreement

Perform operations in the following order:

1. Grouping symbols: parentheses ( ), brackets [ ], braces { }, absolute value | |, and radicals $\sqrt{}$.
2. Exponents/Roots from left to right, in order as they occur.
3. Multiplication/Division from left to right, in order as they occur.
4. Addition/Subtraction from left to right, in order as they occur.

If we follow this agreement to simplify $2 + 5 \cdot 3$, the correct order is to multiply first and then add so that the answer is 17.

Embedded Parentheses

When one set of grouping symbols contains another set of grouping symbols, we say they are *nested* or *embedded*. Work from the innermost set of grouping symbols outward.

EXAMPLE 5   Simplify.

a. $-34 + 12 \div (-3)(5)$

Solution $-34 + 12 \div (-3)(5)$

$= -34 + (-4)(5)$ **Divide $12 \div (-3) = -4$.**

$= -34 + (-20)$ **Multiply $-4(5) = -20$.**

$= -54$ **Add $-34 + -20 = -54$.**

Note: *The division and multiplication must be calculated before adding. As we read from left to right, we see the division before the multiplication, so we divide first.*

b. $(-2)^3 + 4|18 - 24|$

Solution $(-2)^3 + 4|18 - 24|$

$= (-2)^3 + 4|-6|$ **Subtract inside the absolute value: $18 - 24 = -6$.**

$= (-2)^3 + 4 \cdot 6$ **Simplify the absolute value $|-6| = 6$.**

$= -8 + 4 \cdot 6$ **Evaluate the exponential form $(-2)^3 = -8$.**

$= -8 + 24$ **Multiply $4 \cdot 6 = 24$.**

$= 16$ **Add $-8 + 24 = 16$.**

Warning: The expressions $(-5)^2$ and -5^2 are different. The expression $(-5)^2$ means the square of the number -5, whereas -5^2 means the additive inverse of the square of 5.

$(-5)^2 = (-5)(-5) = 25$
whereas
$-5^2 = -(5 \cdot 5) = -25$

c. $(-5)^2 - 8[4 - (6 + 3)] - \sqrt{36}$

Solution $(-5)^2 - 8[4 - (6 + 3)] - \sqrt{36}$

$= (-5)^2 - 8[4 - 9] - \sqrt{36}$ **Calculate within the innermost parentheses: $6 + 3 = 9$.**

$= 25 - 8(-5) - 6$ **Evaluate the exponential form, brackets, and square root.**

$= 25 - (-40) - 6$ **Multiply $8(-5) = -40$.**

$= 25 + 40 - 6$ **Write $25 - (-40)$ as an equivalent addition.**

$= 65 - 6$ **Add $25 + 40 = 65$.**

$= 59$ **Subtract $65 - 6 = 59$.**

YOUR TURN   Simplify.

  a. $28 - 36 \div 9(-5)$ **b.** $6|-5 - 4| + 2(-3)^3$

  c. $15.8 - 0.2[7^2 - (12 + 8) \div 4]$

Radical Symbols

Radical symbols can be grouping symbols. Consider $\sqrt{36 \cdot 4}$. Following the order of operations, we multiply first, then find the root of the product. However, if a radical contains multiplication, we can find the roots of the factors and multiply those roots to get the same answer.

Multiplying first:		Multiplying the roots:
$\sqrt{36 \cdot 4} = \sqrt{144}$	or	$\sqrt{36 \cdot 4} = 6 \cdot 2$
$= 12$		$= 12$

Either approach yields the same result. The same holds for division under a radical.

$\sqrt{36 \div 4} = \sqrt{9}$	or	$\sqrt{36 \div 4} = 6 \div 2$
$= 3$		$= 3$

RULE   **Square Root of a Product or Quotient**

> If a square root contains multiplication or division, we can multiply or divide first, then find the square root of the result, or we can find the square roots of the individual numbers, then multiply or divide the square roots.

Note that this rule does *not* hold for addition or subtraction under a radical, such as in $\sqrt{9 + 16}$.

Adding first:		Adding the roots:
$\sqrt{9 + 16} = \sqrt{25}$		$\sqrt{9 + 16} = 3 + 4$
$= 5$		$= 7$

Because the results are different, we do not have a choice of how to approach the situation when the radical contains addition. The correct approach is to treat the radical as a grouping symbol and add first. Then find the root of the sum. Thus, the correct answer for $\sqrt{9 + 16}$ is 5. The same procedure applies when a radical contains subtraction.

RULE   **Square Root of a Sum or Difference**

> When a radical contains addition or subtraction, we must add or subtract first, then find the root of the sum or difference.

ANSWERS

a. 48   **b.** 0   **c.** 7

EXAMPLE 6) Simplify.

a. $12.4 \div 5(-3)^2 + 4\sqrt{169 - 25}$

Solution $12.4 \div 5(-3)^2 + 4\sqrt{169 - 25}$

$= 12.4 \div 5(-3)^2 + 4\sqrt{144}$ Subtract within the radical: $169 - 25 = 144$.

$= 12.4 \div 5(9) + 4(12)$ Evaluate the exponential form and root.

$= 2.48(9) + 4(12)$ Divide $12.4 \div 5 = 2.48$.

$= 22.32 + 48$ Multiply $2.48(9) = 22.32$ and $4(12) = 48$

$= 70.32$ Add $22.32 + 48 = 70.32$.

◀ **Note:** *We divide before multiplying here because the order is to multiply or divide from left to right in the order in which those operations occur.*

b. $\left(-\dfrac{1}{4}\right)^3 - \left(\dfrac{1}{2} + \dfrac{3}{8}\right) \div \sqrt{\dfrac{48}{3}}$

Solution $\left(-\dfrac{1}{4}\right)^3 - \left(\dfrac{1(4)}{2(4)} + \dfrac{3}{8}\right) \div \sqrt{\dfrac{48}{3}}$

$= \left(-\dfrac{1}{4}\right)^3 - \left(\dfrac{4}{8} + \dfrac{3}{8}\right) \div \sqrt{\dfrac{48}{3}}$ Write equivalent fractions with a common denominator in order to add within the parentheses.

$= \left(-\dfrac{1}{4}\right)^3 - \dfrac{7}{8} \div \sqrt{16}$ Add within parentheses and divide within the radical.

$= -\dfrac{1}{64} - \dfrac{7}{8} \div 4$ Evaluate the exponential form and square root.

$= -\dfrac{1}{64} - \dfrac{7}{8} \cdot \dfrac{1}{4}$ Write an equivalent multiplication using the reciprocal of the divisor.

$= -\dfrac{1}{64} - \dfrac{7(2)}{32(2)}$ Multiply $\dfrac{7}{8} \cdot \dfrac{1}{4}$ to get $\dfrac{7}{32}$, then write it as an equivalent fraction with the common denominator 64 in order to subtract.

$= -\dfrac{1}{64} - \dfrac{14}{64}$

$= -\dfrac{15}{64}$ Subtract.

YOUR TURN) Simplify.

a. $5 + (0.2)^3 - 3(8 - 14)$ **b.** $-\dfrac{3}{5} \div \dfrac{1}{10} \cdot 4 + \sqrt{64 + 36}$

ANSWERS)

a. 23.008   **b.** -14

Fraction Lines

Sometimes fraction lines are used as grouping symbols. When they are, we simplify the numerator and denominator separately, then divide the results.

Connection The expression in Example 7 is equivalent to the expression $[(-2)^3 + 5(-6)] \div [3(7) - 2]$.

EXAMPLE 7 Simplify.

a. $\dfrac{(-2)^3 + 5(-6)}{3(7) - 2}$

Solution $\dfrac{(-2)^3 + 5(-6)}{3(7) - 2}$

$= \dfrac{-8 + 5(-6)}{21 - 2}$ Evaluate the exponential form in the numerator and multiply in the denominator.

$= \dfrac{-8 + (-30)}{19}$ Multiply in the numerator and subtract in the denominator.

$= \dfrac{-38}{19}$ Add in the numerator.

$= -2$ Divide.

b. $\dfrac{9(3) + 15}{4^2 + 2(-8)}$

Solution $\dfrac{9(3) + 15}{4^2 + 2(-8)}$

$= \dfrac{27 + 15}{16 + 2(-8)}$ Multiply in the numerator and evaluate the exponential form in the denominator.

$= \dfrac{42}{16 + (-16)}$ Add in the numerator and multiply in the denominator.

$= \dfrac{42}{0}$ Add in the denominator.

Because the denominator or divisor is 0, the answer is undefined.

YOUR TURN Simplify.

a. $\dfrac{7(3 - 7) + 1}{(-2)^3 - 1}$

b. $\dfrac{2[9 - 4(3 + 5)]}{25 - (6 - 1)^2}$

ANSWERS

a. 3   **b.** undefined

OBJECTIVE 4. Find the mean of a set of data. Fraction line notation can be used to indicate the proper sequence of operations in calculating an arithmetic mean, or average, of a set of data.

PROCEDURE Finding the Arithmetic Mean

Note: *The subscripts indicate that each x represents a different given number, so x_1 represents the first given number, x_2 represents the second, and so on until x_n, which represents the last given number.*

To find the arithmetic mean, or average, of n numbers, divide the sum of the numbers by n.

$$\text{Arithmetic mean} = \frac{x_1 + x_2 + \cdots + x_n}{n}$$

EXAMPLE 8 Jacky has the following test scores: 86, 95, 78, 82, 84. Find the average of her test scores.

Solution $\dfrac{86 + 95 + 78 + 82 + 84}{5} = \dfrac{425}{5}$ Divide the sum of the 5 scores by 5.

$$= 85$$

YOUR TURN This table has the daily rainfall accumulations for the month of April in a certain city. Find the average daily accumulation for April. (Remember, April has 30 days.)

Date	Accumulation
April 2	0.5 in.
April 7	1.25 in.
April 9	1.0 in.
April 15	1.25 in.
April 21	0.25 in.
April 26	1.0 in.

ANSWER

0.175 in./day

1.5 Exercises

For Extra Help

MyMathLab   Videotape/DVT   InterAct Math   Math Tutor Center   Math XL.com

1. What is another way to say "two to the third power"?

2. What is another way to say "three to the second power"?

3. Explain the difference between squaring a number and finding its square root.

4. If the base of an exponential form is a negative number and the exponent is even, will the sign of the result be positive or negative?

5. When simplifying a square root of a product of two numbers, what is the proper order of operations?

6. When simplifying a square root of a sum of two numbers, what is the proper order of operations?

For Exercises 7–12, identify the base and the exponent, then translate the expression to words.

7. 7^2

8. 9^4

9. $(-5)^3$

10. $(-8)^2$

11. -2^7

12. -3^8

For Exercises 13–28, evaluate.

13. 3^4

14. 2^5

15. -8^2

16. -5^3

17. $(-2)^3$

18. $(-3)^2$

19. $(-1)^6$

20. $(-1)^5$

21. $\left(-\dfrac{1}{5}\right)^2$

22. $\left(-\dfrac{2}{7}\right)^2$

23. $\left(-\dfrac{3}{4}\right)^3$

24. $\left(-\dfrac{1}{3}\right)^4$

25. $(0.2)^3$

26. $(0.3)^4$

27. $(-4.1)^3$

28. $(-0.2)^4$

For Exercises 29–36, find all square roots of the given number.

29. 121 **30.** 49 **31.** -81 **32.** -36

33. 196 **34.** 169 **35.** 256 **36.** 225

For Exercises 37–48, evaluate the square root.

37. $\sqrt{16}$ **38.** $\sqrt{36}$ **39.** $\sqrt{144}$ **40.** $\sqrt{289}$

41. $\sqrt{0.49}$ **42.** $\sqrt{0.01}$ **43.** $\sqrt{-64}$ **44.** $\sqrt{-25}$

45. $\sqrt{\dfrac{64}{81}}$ **46.** $\sqrt{\dfrac{9}{100}}$ **47.** $\sqrt{\dfrac{50}{2}}$ **48.** $\sqrt{\dfrac{48}{3}}$

For Exercises 49–86, simplify using the order of operations.

49. $5 \cdot 3 + 4$ **50.** $8 \div 2 + 4$ **51.** $4 \cdot 6 - 7 \cdot 5$

52. $-3 \cdot 4 - 2 \cdot 7$ **53.** $8 - 3(-4)^2$ **54.** $16 - 5(-2)^2$

55. $4^2 - 36 \div (12 - 8)$ **56.** $3^2 - 18 \div (6 - 3)$ **57.** $-4 + 3(-1)^4 + 18 \div 3$

58. $12 - 2(-2)^3 - 64 \div 4$ **59.** $(-2)^3 + 8 - 7(4 - 3)$ **60.** $(-3)^3 - 16 - 5(7 - 2)$

61. $24 \div (-9 + 3)(3 + 4)$ **62.** $18 \div (-6 + 3)(4 + 1)$ **63.** $13.02 \div (-3.1) + 6^2 - \sqrt{25}$

64. $-15.54 \div 3.7 + (-2)^4 + \sqrt{49}$ **65.** $18.2 - 3.4[6^2 - (5 + 2) \div 7]$ **66.** $16.3 + 2.8[(8 + 7) \div 5 - 4^2]$

67. $-4^3 - 3^2 + 4|7 - 12|$ **68.** $-2|9 - 15| + 5^2 - 3^2$ **69.** $-\dfrac{3}{4} \div \dfrac{1}{8} + \left(-\dfrac{2}{5}\right)(-3)(-4)$

70. $\dfrac{5}{6} \div \left(-\dfrac{2}{3}\right) + \left(-\dfrac{2}{7}\right)(5)(-14)$ **71.** $-36 \div (-2)(3) + \sqrt{169 - 25} + 12$ **72.** $\sqrt{100 - 64} + 18 \div (-3)(-2)$

73. $\dfrac{9}{8} \cdot \left(-\dfrac{2}{3}\right) + \left(\dfrac{1}{5} - \dfrac{2}{3}\right) \div \sqrt{\dfrac{125}{5}}$

74. $\left(\dfrac{3}{4} - \dfrac{2}{3}\right) \div \sqrt{\dfrac{9}{81}} - \dfrac{16}{27} \div \dfrac{4}{9}$

75. $\dfrac{2}{5} \div \left(-\dfrac{1}{10}\right) \cdot (-3) + \sqrt{64 + 36}$

76. $\dfrac{5}{6}(-18) \div \dfrac{3}{2} - \sqrt{9 + 16}$

77. $-16 \cdot \dfrac{3}{4} \div (-2) + |9 - 3(4 + 1)|$

78. $18 \cdot \left(-\dfrac{5}{6}\right) \div (-3) + 2|4 + 2(7 - 3)|$

79. $\dfrac{7 - |0 - 7|}{6^2 - 3(2 - 14)}$

80. $\dfrac{|6(-3) + 7| - 11}{5^3 - 2(6 - 12)}$

81. $\dfrac{4[5 - 8(2 + 1)]}{3 - 6 - (-4)^2}$

82. $\dfrac{3[24 - 4(6 - 2)]}{-3^3 + 4^2 + 3}$

83. $\dfrac{5^3 - 3(4^3 - 41)}{19 - (3 - 10)^2 + 38}$

84. $\dfrac{6^2 - 3(4 + 2^5)}{5 + 20 - (2 + 3)^2}$

85. $\dfrac{4(5^2 - 10) + 3}{\sqrt{25 - 16} - 3}$

86. $\dfrac{5(4 - 9) + 1}{7 - 2^3 - \sqrt{49}}$

In Exercises 87–90, a property of arithmetic was used as an alternative to the order of operations. Determine what property of arithmetic was applied, and explain how it is different from the order-of-operations agreement.

87. $14 - 2 \cdot 6 \cdot 3 + 8^2$

$= 14 - 2 \cdot 18 + 64$

$= 14 - 36 + 64$

$= 42$

88. $2(3 + 8) - \sqrt{81}$

$= 6 + 16 - 9$

$= 13$

89. $-6[5 + 3^2] - \sqrt{14 + 11}$

$= -6[5 + 9] - \sqrt{25}$

$= -30 + (-54) - 5$

$= -89$

90. $(-4)^3 + 2[-10 + 8 + (-3)]$

$= -64 + 2[-13 + 8]$

$= -64 + 2[-5]$

$= -64 + (-10)$

$= -74$

For Exercises 91–94, explain the mistake. Then simplify correctly.

91. $24 \div 4 \cdot 2 - 11$

$= 24 \div 8 - 11$

$= 3 - 11$

$= -8$

92. $19 - 6(10 - 8)$

$= 19 - 6(2)$

$= 13(2)$

$= 26$

93. $40 \div 2 + \sqrt{25 - 9}$

$= 20 + \sqrt{25 - 9}$

$= 20 + 5 - 3$

$= 22$

94. $-3^4 + 20 \div 5 - (16 - 24)$

$= 81 + 4 - (-8)$

$= 85 + 8$

$= 93$

For Exercises 95–102, solve.

95. Tomeka has the following test scores in a history course: 82, 76, 64, 90, 74. What is the average of her test scores?

96. Will's math instructor gives quizzes worth 10 points each. Will has the following quiz scores: 9, 8, 4, 8, 7, 7, 6, 9, 8. If his instructor drops the lowest quiz score, what is Will's quiz average?

97. Michael can exempt his chemistry exam if he has a test average greater than or equal to 90 on the five tests in the course. His current test scores are 96, 88, 86, 84. Using trial and error, determine the minimum test score on the last test that will give him a test average of 90.

98. To get an A in her psychology course, Lisa must have a test average greater than or equal to 90 on four out of five tests (the lowest test score is dropped). Her test scores on the first four tests are 98, 68, 84, 86. What is the minimum score on the last test that will give her a test average of 90?

99. In November of 2004, Reuters News Service reported the highest gas prices ever for Thanksgiving. Following are some cities from the report and their respective gas prices. What was the average price of gas for the listed cities? (*Source:* Reuters News Service, "Gas Prices Highest Ever for Thanksgiving," November 22, 2004)

San Francisco	$2.32
Houston	$1.81
Miami	$2.05
Seattle	$1.99
New York City	$1.99
Chicago	$1.95
Cleveland	$1.85

Of Interest

Though the prices reported in 2004 were the highest to date, with inflation factored in, they still fell short of the 1981 gas prices. Using the November 2004 value of the dollar, the 1981 price would be equivalent to $3.04.

100. The following graphic shows the number of people 20 years of age or older who were employed in the United States for each month during 2004. Find the average number of people employed during 2004.

ALL IN A MONTH'S WORK		YEAR 2004	
Month	Employees (in thousands)	Month	Employees (in thousands)
January	146,785	July	147,823
February	146,529	August	147,676
March	146,737	September	147,531
April	146,788	October	147,843
May	147,018	November	148,313
June	147,386	December	148,203

Source: Bureau of Labor Statistics, *Labor Force Statistics from Current Population Survey*

101. The following bar graph shows the daily closing price of Wal-Mart stock over a one-week period from January 10 to January 14, 2005. Find the average of the daily prices for that week.

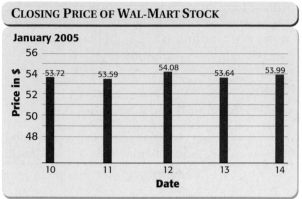

Source: Yahoo Finance

102. The following bar graph shows the Dow Jones Industrial Average at the close of each of the days listed. What is the average of the DJIA for that week?

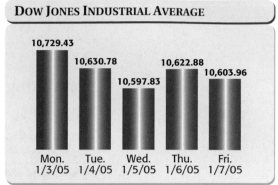

Source: Yahoo Finance

REVIEW EXERCISES

[1.1] **1.** Is $5x - 4y$ an equation or an expression? Explain.

[1.1] **2.** Write a set containing the whole numbers through 10.

[1.1] **3.** Graph $-\dfrac{4}{5}$ on a number line.

[1.4] **4.** Multiply: $-4(3)(-2)$

[1.4] **5.** Divide: $\dfrac{-15}{-5}$

[1.4] **6.** Find the multiplicative inverse of -9.

1.6 Translating Word Phrases to Expressions

OBJECTIVE 1. Translate word phrases to expressions.   An important technique in solving problems is to translate the words in the problem to math symbols. In this section, we focus on how to translate word phrases to expressions containing variables. The first step in translating is to identify the unknown amount. If a variable is not already given, then select a variable to represent that unknown amount.

Translating Basic Phrases

The following table contains some basic phrases and their translations.

Addition	Translation
The sum of x and 3	$x + 3$
h plus k	$h + k$
7 added to t	$t + 7$
3 more than a number	$n + 3$
y increased by 2	$y + 2$

Note: *Since addition is a commutative operation, it does not matter in what order we write the translation.*

For "the sum of x and 3" we can write $x + 3$ or $3 + x$.

Subtraction	Translation
The difference of x and 3	$x - 3$
h minus k	$h - k$
7 subtracted from t	$t - 7$
3 less than a number	$n - 3$
y decreased by 2	$y - 2$

Note: *Subtraction is not a commutative operation; therefore, the way we write the translation matters. We must translate each key phrase exactly as it was presented above. Notice when we translate "less than" or "subtracted from," the translation is in reverse order from what we read.*

Multiplication	Translation
The product of x and 3	$3x$
h times k	hk
Twice a number	$2n$
Triple the number	$3n$
$\frac{2}{3}$ of a number	$\frac{2}{3}n$

Note: *Like addition, multiplication is a commutative operation. This means we can write the translation order any way we wish.*

h times k can be hk or kh.

Division	Translation
The quotient of x and 3	$x \div 3$
h divided by k	$h \div k$
h divided into k	$k \div h$
The ratio of a to b	$a \div b$

Note: *Division is like subtraction in that it is not a commutative operation; therefore, we must translate division phrases exactly as presented above. Notice how "divided into" is translated in reverse order of what we read.*

Exponents	Translation
c squared	c^2
The square of b	b^2
k cubed	k^3
The cube of b	b^3
n to the fourth power	n^4
y raised to the fifth power	y^5

Roots	Translation
The square root of x	$\sqrt{x}$

The key words *sum, difference, product,* and *quotient* indicate the answer for their respective operations. Notice that they all involve the word *and*. In the translation, the word *and* separates the parts and can therefore be translated to the operation symbol indicated by the key word *sum, difference, product,* or *quotient.*

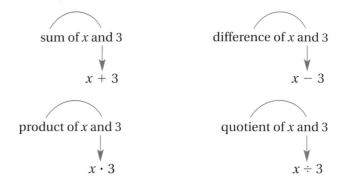

sum of *x* and 3

$x + 3$

difference of *x* and 3

$x - 3$

product of *x* and 3

$x \cdot 3$

quotient of *x* and 3

$x \div 3$

Combinations of Basic Phrases

Now that we have seen the basic phrases, let's translate phrases that involve combinations of the basic phrases.

EXAMPLE 1 Translate to an algebraic expression.

a. four more than three times a number

Note: *The commutative property of addition allows us to write the expression in either order.*

Translation $4 + 3n$ or $3n + 4$

b. six less than the square of a number

Translation $n^2 - 6$

Note: *Because subtraction is not commutative, we must translate the subtraction in a specific order.* Less than *indicates that the* 6 *is the subtrahend and the* square of a number *is the minuend.*

c. the sum of *h* raised to the fifth power and fifteen

Translation $h^5 + 15$

d. the ratio of *m* times *n* to *r* cubed

Note: *When coupled with the word* ratio, *the word* to *translates to the fraction line. The amount to the left of the word* to *goes in the numerator and the amount to the right of the word* to *goes in the denominator.*

Translation $\dfrac{mn}{r^3}$

e. one-half of *v* divided by the square root of *t*

Translation $\dfrac{1}{2}v \div \sqrt{t}$

Note: *When the word* of *is preceded by a fraction, it means multiply.*

YOUR TURN Translate to an algebraic expression.

a. two-thirds subtracted from the product of nine and a number

b. -6.2 increased by 9.8 times a number

ANSWERS

a. $9n - \dfrac{2}{3}$ **b.** $-6.2 + 9.8n$

Translating Phrases Involving Parentheses

Sometimes the word phrases imply an order of operations that would require us to use parentheses in the translation. These situations arise when the phrase indicates that a sum or difference is to be calculated before performing a higher-order operation such as multiplication, division, exponent, or root.

EXAMPLE 2) Translate to an algebraic expression.

a. five times the sum of x and y

Translation $5(x + y)$

Note: *Without the parentheses, the expression is $5x + y$, which indicates that we are to multiply 5 times x first, and then add y to the result. The parentheses indicate that we add x and y first, then multiply the resulting sum by 5.*

b. the square root of the difference of the square of x and the square of y

Translation $\sqrt{x^2 - y^2}$

Note: *The square root of the difference indicates that we are to calculate the difference before calculating the square root. We indicate this symbolically by placing the entire subtraction under the radical.*

c. the product of x and y divided by the sum of x^2 and 5

Translation $xy \div (x^2 + 5)$ or $\dfrac{xy}{x^2 + 5}$

YOUR TURN)   Translate to an algebraic expression.

a. a number minus twice the sum of the number and seven

b. the difference of m and n, all raised to the fourth power

EXAMPLE 3)   Isaac Newton developed an expression that describes the gravitational effect between two objects. The relationship is the product of the masses of the two objects divided by the square of the distance between the objects. Translate the relationship to an expression.

Translation $Mm \div d^2$ or $\dfrac{Mm}{d^2}$

Note: *If the same letter is desired to describe two different quantities, uppercase and lowercase can be used to distinguish the quantities. Subscripts can also be used, as in $m_1 m_2 \div d^2$.*

YOUR TURN)   Two asteroids are on a collision course with each other. We can calculate the time to impact by dividing the distance that separates them by the sum of their velocities. Translate the relationship to an expression.

1.6 Exercises

MyMathLab   Videotape/DVT   InterAct Math   Math Tutor Center   Math XL.com

1. List three key words that indicate addition.

2. List three key words that indicate subtraction.

3. List three key words that indicate multiplication.

4. List three key words that indicate division.

5. The phrase "9 subtracted from n" translates to $n - 9$, which is in reverse order of what we read. What other key words for subtraction translate in reverse order?

6. What key words for division translate in reverse order?

For Exercises 7–40, translate the phrase to an algebraic expression.

7. four times a number

8. the product of a number and four

9. the sum of four times x and sixteen

10. the difference of seven times x and eight

11. five more than y

12. six less than T

13. the quotient of negative four and the cube of y

14. the ratio of seven to the square of m

15. eight times p decreased by four

16. thirteen subtracted from twice a number

17. fourteen divided into m

18. r divided by six

19. x to the fourth power increased by five

20. the sum of b cubed and seven

21. one-fifth subtracted from the product of seven and a number

22. two-thirds added to the product of four and a number

23. the product of negative three and the difference of a number and two

24. a number plus triple the sum of the number and four

25. the difference of two and l, all raised to the third power

26. the sum of four and n, all raised to the fifth power

27. five less than the product of m and n

28. seven added to the quotient of x and y

29. the quotient of four and a number, decreased by two

30. the product of three and a number, increased by five

31. negative twenty-seven decreased by the sum of a and b

32. the difference of m and n subtracted from negative eight

33. six-tenths decreased by the product of four and the difference of y and two

34. eighty-one hundredths increased by the product of eight and the sum of x and three tenths

35. the difference of p and q decreased by the sum of m and n

36. the sum of a and b subtracted from the difference of c and d

37. the product of m and n subtracted from the square root of y

38. the square root of x subtracted from the product of a and b

39. a number minus three times the difference of the number and six

40. the product of five and a number minus the sum of the number and two

For Exercises 41–44, explain the mistake. Then translate the phrase correctly.

41. nine times the sum of x and y
Translation: $9x + y$

42. seventeen less than three times t
Translation: $17 - 3t$

43. four subtracted from the square of m
Translation: $4 - m^2$

44. Nineteen divided into the product of h and k
Translation: $19 \div hk$

For Exercises 45–54, translate the indicated phrase.

45. The length of a rectangle is five more than the width. If the width is represented by w, then write an expression that describes the length.

46. The width of a rectangle is four less than the length. If the length is represented by l, then write an expression that describes the width.

47. The length of a rectangle is three times the width. If the width is represented by the variable w, then write an expression that describes the length.

48. The width of a rectangle is one-fourth of the length. If the length is represented by l, then write an expression that describes the width.

49. The radius of a circle is one-half of the diameter. If d represents the diameter, then write an expression for the radius.

50. The diameter of a circle is twice the radius. If r represents the radius, then write an expression for the diameter.

51. Lindsey has 42 coins in her change purse that are all either dimes or quarters. If n represents the number of quarters she has, then write an expression in terms of n that describes the number of dimes.

52. Sherice owns a total of 60 shares of stock in two companies. If n represents the number of the higher-priced stock, then write an expression in terms of n for the number of shares of the lower-priced stock.

53. Don passes mile marker 51 on the highway. One-fourth of an hour later, a state trooper traveling in the same direction passes the same mile marker. If t represents the amount of time it takes the trooper to catch up to Don, write an expression in terms of t that describes the amount of time Don has traveled since passing marker 51.

54. Berry is jogging along a trail and passes a sign. One-third of an hour later, Dedra, who is traveling in the same direction, passes the same sign. If t represents the amount of time it takes Dedra to catch up to Berry, write an expression in terms of t that describes the amount of time Berry has traveled since passing the sign.

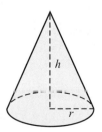

Exercises 55–62 contain word descriptions of expressions from mathematics and physics. Translate the descriptions to symbolic form.

55. The perimeter of a rectangle is twice the width plus twice the length.

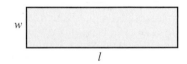

56. The area of a circle can be found by multiplying the square of the radius by π.

57. The volume of a cone is one-third of the product of π, the square of the radius, and the height of the cone.

58. The volume of a sphere is four-thirds of the product of π and the cube of the radius.

59. René Descartes developed an expression for the distance between two points in the coordinate plane. The distance between two points is the square root of the sum of the square of the difference of x_2 and x_1, and the square of the difference of y_2 and y_1.

60. The centripetal acceleration of an object traveling in a circular path is found by an expression that is the ratio of the square of the velocity, v, to the radius, r, of the orbital path.

61. Albert Einstein developed an expression that describes the energy of a particle at rest, which is the product of the mass, m, of the particle and the square of the speed of light, which is represented by c.

62. Albert Einstein's theory of relativity includes a mathematical expression that is the square root of the difference of one and the ratio of the square of the velocity, v, of an object in motion to the square of the speed of light, c.

Of Interest

Though Albert Einstein (1879–1955) is considered one of the greatest intellects of all time, he actually struggled academically. As a teen, he was expelled from school because his apathy was a bad influence on his classmates. He failed his first attempt at the technical college entrance exam. He later passed the exam after taking a year of remedial classes at a secondary school. His marginal academic performance did not deter him from pursuing on his own his love of math and science, and in 1905 while working as a clerk in the Swiss patent office, he submitted three papers to a German science journal that outlined quantum theory and his theory of relativity. The genius of these papers was quickly recognized and he was asked to lecture at several European universities. Ironically, he never left academia thereafter, accepting a position at Princeton University to escape persecution by the Nazis. He remained at Princeton until his death in 1955.

For Exercises 63–70, translate the expression to a word phrase.

63. The area of a triangle: $\dfrac{1}{2}bh$

64. The area of a trapezoid: $\dfrac{1}{2}h(a + b)$

65. The volume of a box: lwh

66. The volume of a cylinder: $\pi r^2 h$

67. The surface area of a box: $2(lw + lh + wh)$

68. The surface area of a cylinder: $2\pi r(r + h)$

69. The slope of a line: $\dfrac{y_2 - y_1}{x_2 - x_1}$

70. A quadratic expression: $ax^2 + bx + c$

PUZZLE PROBLEM

Consecutive integers follow one another in a pattern like 1, 2, 3, . . .
Let n represent the first integer in a set of consecutive integers.

a. Write expressions for the next two consecutive integers.

b. Suppose n is odd. Write expressions for the next two consecutive odd integers.

c. Suppose n is even. Write expressions for the next two consecutive even integers.

REVIEW EXERCISES

[1.1] **1.** What is a rational number?

[1.2] **2.** What property of addition is represented by $6 + 0 = 6$?

[1.4] **3.** What property is represented by $(3 \cdot 5) \cdot 6 = 3 \cdot (5 \cdot 6)$?

[1.5] *For Exercises 4–6, simplify.*

4. $-6^2 \div 3 + 4(5 - 9)$

5. $-4|3 - 8| + 2^5$

6. -3^2

1.7 Evaluating and Rewriting Expressions

OBJECTIVES

1. Evaluate an expression.
2. Determine all values that cause an expression to be undefined.
3. Rewrite an expression using the distributive property.
4. Rewrite an expression by combining like terms.

Because algebraic expressions form equations, we must learn how to manipulate those expressions. There are two actions we can perform with algebraic expressions. We can *evaluate* them or *rewrite* them.

OBJECTIVE 1. Evaluate an expression. First we learn how to evaluate an expression.

PROCEDURE Evaluating an Algebraic Expression

To evaluate an algebraic expression:

1. Replace the variables with their corresponding given numbers.
2. Calculate the numerical expression using the order of operations.

EXAMPLE 1 Evaluate $y^2 - 0.2xy + 5$, when $x = 6$ and $y = -4$.

Solution $y^2 - 0.2xy + 5$

$(-4)^2 - 0.2(6)(-4) + 5$ Replace x with 6 and y with -4.

$= 16 - 0.2(6)(-4) + 5$ Begin calculating by simplifying the exponential form.

$= 16 - (-4.8) + 5$ Multiply.

$= 16 + 4.8 + 5$ Write the subtraction as an equivalent addition.

$= 25.8$ Add from left to right.

YOUR TURN Evaluate $m^3 - 6n^2$ when $m = -2$ and $n = 3$.

EXAMPLE 2 Evaluate $\dfrac{5r^2}{d - 3}$ when $r = 2$ and $d = -5$.

Solution $\dfrac{5r^2}{d - 3}$

$\dfrac{5(2)^2}{-5 - 3}$ Replace r with 2 and d with -5.

$= \dfrac{5(4)}{-8}$ Calculate the top and bottom expressions separately.

$= \dfrac{20}{-8}$

$= -\dfrac{5}{2}$ or -2.5

ANSWER

-62

OBJECTIVE 2. Determine all values that cause an expression to be undefined. In Example 2, what if $d = 3$? Notice that the denominator would become 0: $\dfrac{5r^2}{3-3} = \dfrac{5r^2}{0}$, which is undefined if $r \neq 0$ and indeterminate if $r = 0$. When asked to evaluate a division expression in which the divisor or denominator contains a variable or variables, we must be careful about what values replace the variable(s). Often we'll need to know what values could replace the variable(s) and cause the expression to be undefined or indeterminate.

EXAMPLE 3 Determine all values that cause the expression to be undefined.

a. $\dfrac{5}{x+7}$

Answer If $x = -7$, we have $\dfrac{5}{-7+7} = \dfrac{5}{0}$, which is undefined because the denominator is 0.

b. $\dfrac{-y}{(y+2)(y-6)}$

Solution Note that because $(y+2)(y-6)$ is a product, if either factor is 0, then the entire denominator is 0.

Notice if $y = -2$, we have: $\dfrac{-(-2)}{(-2+2)(-2-6)}$

$= \dfrac{2}{(0)(-8)}$

$= \dfrac{2}{0}$, which is undefined

Also, if $y = 6$, we have: $\dfrac{-6}{(6+2)(6-6)}$

$= \dfrac{-6}{(8)(0)}$

$= \dfrac{-6}{0}$, which is undefined

Connection This principle, that if either factor is 0 then the entire product is 0, is known as the *zero-factor theorem*. We will use this theorem extensively in Chapter 6 when solving equations like $(y+2)(y-6) = 0$.

Answer Both $y = -2$ and $y = 6$ cause the expression to be undefined.

YOUR TURN Determine all values that cause the expression to be undefined.

a. $\dfrac{9}{n+6}$

b. $\dfrac{2m}{(m-3)(m+4)}$

OBJECTIVE 3. Rewrite an expression using the distributive property. Recall from Section 1.4 the distributive property of multiplication over addition.

> **RULE** The Distributive Property of Multiplication over Addition
> $$a(b + c) = ab + ac$$

This property gives us an alternative to the order of operations. Look at the numerical expression $2(5 + 6)$ and compare using the order of operations versus using the distributive property.

Following order of operations:

$$2(\mathbf{5 + 6}) = 2(\mathbf{11})$$
$$= 22$$

or

Using the distributive property:

$$2(5 + 6) = 2 \cdot 5 + 2 \cdot 6$$
$$= 10 + 12$$
$$= 22$$

Note: *Using the order of operations, we add within the parentheses first, then multiply.*

Note: *Using the distributive property, we multiply each addend by the factor outside the parentheses, then add the products.*

The result is the same either way, so the distributive property allows us to say that $2(5 + 6)$ and $2 \cdot 5 + 2 \cdot 6$ are equivalent expressions.

Conclusion We can use the distributive property to rewrite an expression in another form that is equivalent to the original form.

EXAMPLE 4 Use the distributive property to write an equivalent expression.

a. $2(x + 5)$

Solution $2(x + 5) = 2 \cdot x + 2 \cdot 5$
$$= 2x + 10$$

b. $-6(n - 9)$

Solution $-6(n - 9) = -6 \cdot n - (-6) \cdot 9$
$$= -6n - (-54)$$
$$= -6n + 54$$

Connection We can apply the distributive property to $-6(n - 9)$ even though the parentheses contain subtraction instead of addition because subtraction can be expressed as addition like this:

$$-6(n - 9) = -6(n + (-9))$$

In fact, it is helpful to think of the expression in this way because when we multiply the -6 by the -9, we get positive 54.

c. $\dfrac{3}{8}\left(2m + \dfrac{4}{5}\right)$

Solution $\dfrac{3}{8}\left(2m + \dfrac{4}{5}\right) = \dfrac{3}{8} \cdot 2m + \dfrac{3}{8} \cdot \dfrac{4}{5}$ Use the distributive property.

$$= \dfrac{3}{\overset{}{\underset{4}{8}}} \cdot \dfrac{\overset{1}{2}}{1}m + \dfrac{3}{\overset{}{\underset{2}{8}}} \cdot \dfrac{\overset{1}{4}}{5}$$ Divide out common factors.

$$= \dfrac{3}{4}m + \dfrac{3}{10}$$

YOUR TURN   Use the distributive property to write an equivalent expression.

a. $-9(7 - 3x)$

b. $\dfrac{4}{5}\left(\dfrac{1}{2}y - 10\right)$

OBJECTIVE 4. Rewrite an expression by combining like terms. Many expressions are sums of expressions called **terms**. For example, the expression $5x + 3$ is a sum of the terms $5x$ and 3.

DEFINITION   *Terms:* Expressions that are the addends in an expression that is a sum.

If an expression contains subtraction, as in $4x^2 - 6x + 8$, we can identify its terms by writing the subtraction as addition. Since $4x^2 - 6x + 8 = 4x^2 + (-6x) + 8$, its terms are $4x^2$, $-6x$, and 8. Notice that we can see that the second term in $4x^2 - 6x + 8$ is $-6x$ without writing the equivalent addition by recognizing that the sign to the left of the term is its sign.

The numerical factor in a term is called the numerical **coefficient**, or simply the coefficient of the term.

DEFINITION   *Coefficient:* The numerical factor in a term.

The coefficient of $4x^2$ is 4.
The coefficient of $-6x$ is -6.
The coefficient of 8 is 8.

Another way that we can rewrite an expression is to combine **like terms**.

DEFINITION   *Like terms:* Variable terms that have the same variable(s) raised to the same exponents, or constant terms.

Examples of like terms:

$3x$ and $5x$
$4y^2$ and $9y^2$
$7xy$ and $3xy$
6 and 15

Examples of unlike terms:

$2x$ and $8y$ (different variables)
$4t^2$ and $4t^3$ (different exponents)
x^2y and xy^2 (different exponents)
12 and $12x$ (different variables)

ANSWERS

a. $-63 + 27x$ **b.** $\dfrac{2}{5}y - 8$

Combining Like Terms

Now, consider an expression that is a sum of like terms, such as $3x + 5x$. We can rewrite this expression in a more compact form. Multiplication of 3 times x means that x is repeatedly added three times. Likewise 5 times x means that x is repeatedly added five times. We can expand the expression out like this:

$$
\begin{array}{c}
\overbrace{3x}\ \ +\ \ \overbrace{5x} \\
= \underbrace{x + x + x\ \ +\ \ x + x + x + x + x} \\
= \qquad\qquad 8x
\end{array}
$$

Note: *After expanding, we see that there are a total of eight x's repeatedly added. We can write those eight x's as a single term, $8x$.*

Notice that when we combine the like terms, we get an expression that is more compact. When we rewrite an expression in a more compact form, we say that we are *simplifying* the expression. Also notice that when combining like terms, we simply add the coefficients and keep the variable the same.

PROCEDURE   Combining Like Terms

To combine like terms, add or subtract the coefficients and keep the variables and their exponents the same.

> ***Connection***   We can use the distributive property in reverse to confirm our procedure for combining like terms. Consider the expression $3x + 5x$ again. Notice there is a common factor of x in both terms. In applying the distributive property in reverse, we write this common factor outside the parentheses and write the remaining factors as addends within the parentheses. It looks like this:
>
> $$3x + 5x = (3 + 5)x$$
>
> Because the parentheses contain only the addition of two numbers, we can simplify by adding the numbers.
>
> $$
> \begin{aligned}
> 3x + 5x &= (3 + 5)x \\
> &= 8x
> \end{aligned}
> $$

EXAMPLE 5   Combine like terms.

a. $9y + 7y$

Solution $9y + 7y = 16y$ **Note:** *We think: Nine y's plus seven y's equals sixteen y's.*

b. $7x - 2x$

Solution $7x - 2x = 5x$ **Note:** *We think: Seven x's minus two x's leaves five x's.*

c. $14n^2 - n^2$

Solution $14n^2 - n^2 = 13n^2$ **Note:** *The coefficient of n^2 is 1, so we think fourteen n^2's minus one n^2 leaves thirteen n^2's.*

YOUR TURN   Combine like terms.

     a. $6.5x + 2.3x$ **b.** $\dfrac{1}{3}n^3 - \dfrac{3}{4}n^3$

Collecting Like Terms

Sometimes expressions are more complex and contain different sets of like terms. In such cases, we combine the like terms and copy any unlike terms in the final expression. Many people like to use the commutative property of addition and rearrange the expression first so that the like terms are together, then combine them. This type of manipulation is called *collecting* the like terms.

EXAMPLE 6   Combine like terms in $4y^2 + 5 + 3y^2 - 9$.

Solution $4y^2 + 5 + 3y^2 - 9$

         $= 4y^2 + 3y^2 + 5 - 9$ **Collect the like terms.**

         $= 7y^2 - 4$ **Combine like terms.**

Collecting like terms is optional. Alternatively, many people mark through the terms as they combine them as a way of keeping track of what has been combined. Using this technique with the expression from Example 6 looks like this:

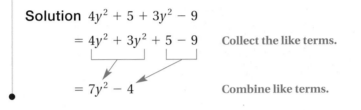

$$= 7y^2 - 4$$

EXAMPLE 7   Combine like terms in $15y + 8x - y - 8x$.

Solution   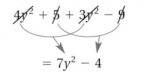   **Note:** *The term $-y$ is equivalent to $-1y$.*

         $= 14y + 0$

         $= 14y$

Sometimes collecting the like terms is beneficial, as in Example 8, where there are like terms with fraction coefficients. Some people find it easier to "see" the steps in writing the equivalent fractions when the fractions are closer together.

ANSWERS

a. $8.8x$ **b.** $-\dfrac{5}{12}n^3$

EXAMPLE 8 Combine like terms in $\frac{1}{3}n - 9m + 2 - m + \frac{1}{4}n$.

Solution $\frac{1}{3}n - 9m + 2 - m + \frac{1}{4}n$

$= \frac{1}{3}n + \frac{1}{4}n - 9m - m + 2$ Collect the like terms.

$= \frac{1(4)}{3(4)}n + \frac{1(3)}{4(3)}n - 9m - m + 2$ Write fraction coefficients as equivalent fractions with their LCD, 12.

$= \frac{4}{12}n + \frac{3}{12}n - 9m - m + 2$

$= \frac{7}{12}n - 10m + 2$ Combine like terms.

YOUR TURN Combine like terms.

a. $\frac{3}{8}y - x + 2 - \frac{3}{4}y + 5x$ **b.** $4.2x^2 - 6x + 9 - 0.7x - 12$

Connection One way to verify that two expressions are equivalent is to evaluate them both using the same value(s) for the variable(s). Consider the expression in Example 6. If we evaluate the original expression $4y^2 + 5 + 3y^2 - 9$ and the simplified expression $7y^2 - 4$ using the same chosen value for y, we should get the same answer. Let's choose the number 2 for y.

Original Expression:

$4y^2 + 5 + 3y^2 - 9$

$4(2)^2 + 5 + 3(2)^2 - 9$

$= 4(4) + 5 + 3(4) - 9$

$= 16 + 5 + 12 - 9$

$= 24$

Simplified expression:

$7y^2 - 4$

$7(2)^2 - 4$

$= 7(4) - 4$

$= 28 - 4$

$= 24$

Warning: Some expressions that are not equivalent may give the same answer using this method, so this "check" does not guarantee that the expressions are equivalent. It is, however, a good sign.

Notice the simplified expression gave the same answer, yet was easier to work with. This is one of the reasons we simplify expressions. If we can simplify an expression, then we have an easier time evaluating it.

ANSWERS

a. $-\frac{3}{8}y + 4x + 2$

b. $4.2x^2 - 6.7x - 3$

1. In your own words, explain how to evaluate an expression.

2. Explain the difference between evaluating and rewriting an expression.

3. What causes an expression to be undefined?

4. What are like terms?

5. What is the coefficient in a term?

6. Explain how to combine like terms.

For Exercises 7–28, evaluate the expression using the given values.

7. $3(a + 5) - 4b$; $a = 2, b = 3$

8. $8n - (m + 1)$; $m = 5, n = 3$

9. $-0.2(x + 3) + 4$; $x = 1$

10. $6 - 0.4(y - 2)$; $y = 5$

11. $2p^2 - 3p - 4$; $p = -3$

12. $n^2 - 8n + 1$; $n = -1$

13. $2y^2 - 4y + 3$; $y = -\dfrac{1}{2}$

14. $3r^2 - 9r + 6$; $r = -\dfrac{1}{3}$

15. $8 - 2(y + 4)$; $y = -2.3$

16. $-6 - 2(l - 5)$; $l = -0.4$

17. $-|3x| + |y^3|$; $x = 5, y = -1$

18. $-|2m^2| - |4n|$; $m = 3, n = -2$

19. $|2r^2 - 3q|$; $r = 3, q = -5$

20. $|m^2 + 2n|$; $m = 4, n = -5$

21. $\sqrt{c} + 4ab^2$; $a = -1, b = -2, c = 16$

22. $-2x^3y + \sqrt{z}$; $x = -2, y = -3, z = 4$

23. $-5\sqrt{a} + 2\sqrt{b}$; $a = 25, b = 4$

24. $\sqrt{h} + \sqrt{k}$; $h = 16, k = 9$

25. $\dfrac{6x^3}{y - 8}$; $x = 2, y = 7$

26. $\dfrac{4m^2}{n + 4}$; $m = 2, n = 6$

27. $\dfrac{3b^3 - 5}{2\sqrt{c} + d}$; $b = 3, c = 25, d = 144$

28. $\dfrac{5 - 2a^2}{3\sqrt{x} + y}$; $a = 1, x = 64, y = 36$

29. The expression $b^2 - 4ac$ is called the *discriminant* and is used to determine the types of solutions for quadratic equations. Find the value of the discriminant given the following values.

  a. $a = 2, b = 1, c = -3$ **b.** $a = 1, b = -2, c = 4$

> **Connection**   You will learn all the details about each expression presented in Exercises 29–32 in future chapters (and future courses). Notice that even without a full understanding of the context in which the expressions are used, you can still evaluate them.

30. The expression $ad - bc$ is used to calculate the determinant of a matrix. Find the determinant given the following values.

  a. $a = 1, b = 0.5, c = -4, d = 6$ **b.** $a = -3, b = \dfrac{4}{5}, c = 2, d = \dfrac{1}{2}$

31. The expression $\dfrac{y_2 - y_1}{x_2 - x_1}$ is used to calculate the slope of a line. Find the slope given each set of the following values.

  a. $x_1 = 2, y_1 = 1, x_2 = 5, y_2 = 7$ **b.** $x_1 = -1, y_1 = 2, x_2 = -7, y_2 = -2$

32. The expression $\sqrt{(x_2 - x_1)^2 + (y_2 - y_1)^2}$ is used to calculate the distance between two points in the coordinate plane. Evaluate the expression using the following values.

  a. $x_1 = 2, y_1 = 1, x_2 = 5, y_2 = 7$ **b.** $x_1 = -1, y_1 = 2, x_2 = -7, y_2 = -2$

For Exercises 33–40, determine all values that cause the expression to be undefined.

33. $\dfrac{-7}{5 + y}$

34. $\dfrac{8}{x + 3}$

35. $\dfrac{6m}{(m + 1)(m - 3)}$

36. $\dfrac{-5a}{(a - 4)(a - 2)}$

37. $\dfrac{6 + x^2}{x}$

38. $\dfrac{7 - y}{y}$

39. $\dfrac{x + 1}{3x - 2}$

40. $\dfrac{3y}{2y + 1}$

For Exercises 41–48, use the distributive property to write an equivalent expression.

41. $6(a + 2)$

42. $4(b - 5)$

43. $-8(4 - 3y)$

44. $-7(3 - 2m)$

45. $\dfrac{7}{8}\left(\dfrac{1}{2}c - 16\right)$

46. $\dfrac{4}{5}\left(-10h + \dfrac{2}{9}\right)$

47. $0.2(3n - 8)$

48. $-1.5(6x + 7)$

For Exercises 49–58, identify the coefficient of each term.

49. $-6x^3$

50. $-1.4y$

51. m^9

52. y^5

53. $-b$

54. $-n$

55. $-\dfrac{2}{3}m$

56. $\dfrac{5}{8}a$

57. $\dfrac{y}{5}$

58. $-\dfrac{u}{3}$

For Exercises 59–78, simplify by combining like terms.

59. $3y + 5y$

60. $6m + 7m$

61. $\dfrac{1}{2}b^2 - \dfrac{5}{6}b^2$

62. $\dfrac{3}{4}z - \dfrac{7}{5}z$

63. $6.3n - 8.2n$

64. $5.1x^4 + 3.4x^4$

65. $-11c - 10c - 5c$

66. $-15w - 6w - 11w$

67. $6x + 7 - x - 8$

68. $5y^2 + 6 + 3y^2 - 8$

69. $-8x + 3y - 7 - 2x + y + 6$

70. $4a + 9b - a + 5 + 7c + 2b - 8 - 7c$

71. $-10m + 7n + 1 + m - 2n - 12 + y$

72. $-3h + 7k - 5 - 8h - 7k + 19$

73. $\frac{1}{6}c + 3d + \frac{2}{3}c + \frac{1}{7} - \frac{1}{2}d$

74. $\frac{5}{8}y + \frac{1}{2} - \frac{3}{4}x + \frac{2}{3} - \frac{1}{4}y$

75. $7a + \frac{4}{5}b^2 - 5 - \frac{3}{4}a - \frac{2}{7}b^2 + 9$

76. $\frac{1}{2}m - 3n + 14 - \frac{3}{8}m - \frac{9}{10}n - 5$

77. $1.5x + y - 2.8x + 0.3 - y - 0.7$

78. $0.4t^2 + t - 2.8 - t^2 + 0.9t - 4$

79. a. Translate to an algebraic expression: fourteen plus the difference of six times a number and eight times the same number.

   b. Simplify the expression.

   c. Evaluate the expression when the number is -3.

80. a. Translate to an algebraic expression: the sum of negative five times a number and eight minus two times the same number.

   b. Simplify the expression.

   c. Evaluate the expression when the number is 0.2.

PUZZLE PROBLEM

Each letter in the following addition problem represents a different whole number, 0–9. What number does each letter represent?

```
  FORTY
    TEN
+   TEN
-------
  SIXTY
```

REVIEW EXERCISES

[1.1]   **1.** Arrange in order from least to greatest.

      $4.2, |-6|, 4\frac{5}{8}, -2.5, -3$

[1.2]   **2.** Simplify: $\dfrac{72}{420}$

For Exercises 3–6, perform the indicated operation(s).

[1.3]   **3.** $-8 + (-14)$

[1.3]   **4.** $1.2 - (-4.9)$

[1.4]   **5.** $\frac{1}{4}(-2)(-3)$

[1.4]   **6.** $0 \div (-9.8)$

Chapter 1   Summary

Defined Terms

Section 1.1
Variable (p. 2)
Constant (p. 2)
Expression (p. 2)
Equation (p. 2)
Inequality (p. 3)
Set (p. 3)
Rational number (p. 4)
Irrational number (p. 5)
Absolute value (p. 7)

Section 1.2
Fraction (p. 11)
Multiple (p. 12)
LCM (p. 13)
LCD (p. 13)
Factors (p. 14)
Prime number (p. 14)
Prime factorization (p. 14)
Lowest terms (p. 15)

Section 1.3
Additive inverses (p. 31)

Section 1.4
Multiplicative inverses (p. 47)

Section 1.5
Exponent (p. 56)
Base (p. 56)

Section 1.7
Terms (p. 81)
Coefficient (p. 81)
Like terms (p. 81)

The Real Number System

Real Numbers

Irrational Numbers: Any real number that is not rational, such as $-\sqrt{2}$, $-\sqrt{3}$, $\sqrt{0.8}$, and π.

Rational Numbers: Real numbers that can be expressed in the form $\frac{a}{b}$, where a and b are integers and $b \neq 0$, such as $-4\frac{3}{4}$, $-\frac{2}{3}$, 0.018, $0.\overline{3}$, and $\frac{5}{8}$.

Integers: $\ldots, -3, -2, -1, 0, 1, 2, 3, \ldots$

Whole Numbers: $0, 1, 2, 3, \ldots$

Natural Numbers: $1, 2, 3, \ldots$

Arithmetic Summary Diagram

Each operation has an inverse operation. In the diagram the operations build from the top down. Addition leads to multiplication, which leads to exponents. Subtraction leads to division, which leads to roots.

Properties of Arithmetic

In each of the following, a, b, and c represent real numbers.

Additive Identity
$$a + 0 = a$$

Commutative Property of Addition
$$a + b = b + a$$

Associative Property of Addition
$$(a + b) + c = a + (b + c)$$

Multiplicative Identity
$$a \cdot 1 = a$$

Multiplicative Property of 0
$$a \cdot 0 = 0$$

Commutative Property of Multiplication
$$ab = ba$$

Associative Property of Multiplication
$$(ab)c = a(bc)$$

Distributive Property
$$a(b + c) = ab + ac$$

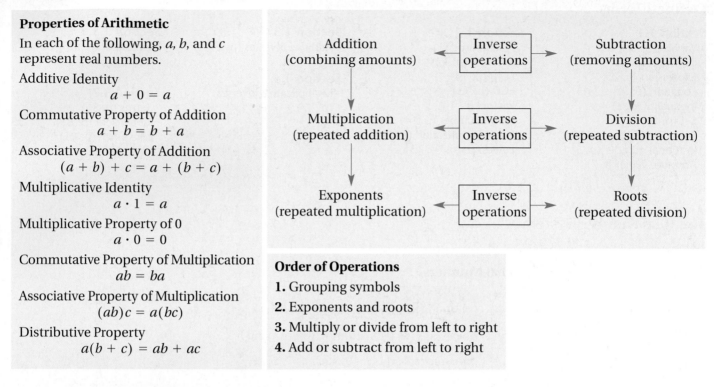

Order of Operations
1. Grouping symbols
2. Exponents and roots
3. Multiply or divide from left to right
4. Add or subtract from left to right

Procedures, Rules, and Key Examples

Procedures/Rules	Key Examples
Section 1.1 Number Sets and the Structure of Algebra To write a set, write the elements or members of the set within braces.	**Example 1:** Write the set containing the natural numbers divisible by 3. Answer: $\{3, 6, 9, 12, \ldots\}$
The absolute value of a number is positive or 0.	**Example 2:** Find the absolute value. **a.** $\lvert 8 \rvert = 8$ **b.** $\lvert -15 \rvert = 15$ **c.** $\lvert 0 \rvert = 0$
For any two real numbers a and b, a is greater than b if a is to the right of b on a number line.	**Example 3:** Use $<$ or $>$ to make a true sentence. **a.** $-8\ \blacksquare\ -10$ Answers: **a.** $-8 > -10$ **b.** $-6\ \blacksquare\ 9$ **b.** $-6 < 9$ *continued*

Procedures/Rules	Key Examples

Section 1.2 Fractions

For any fraction, we can write an equivalent fraction by multiplying or dividing both its numerator and denominator by the same nonzero number.

Example 1: Find the missing number in $\dfrac{5}{6} = \dfrac{?}{24}$ so that the fractions are equivalent.

Answer: $\dfrac{5}{6} = \dfrac{20}{24}$

To simplify a fraction to lowest terms:
1. Replace the numerator and denominator with their prime factorizations.
2. Eliminate (divide out) all prime factors common to the numerator and denominator.
3. Multiply the remaining factors.

Example 2: Reduce $\dfrac{54}{60}$ to lowest terms.

$$\dfrac{54}{60} = \dfrac{2 \cdot 3 \cdot 3 \cdot 3}{2 \cdot 2 \cdot 3 \cdot 5} = \dfrac{9}{10}$$

Section 1.3 Adding and Subtracting Real Numbers; Properties of Real Numbers

When adding two numbers that have the same sign, add their absolute values and keep the same sign.

When adding two numbers that have different signs, subtract their absolute values and keep the sign of the number with the greater absolute value.

To write a subtraction statement as an equivalent addition statement, change the operation symbol from a minus sign to a plus sign and change the subtrahend to its additive inverse.

Example 1: Add.
a. $4 + 9 = 13$
b. $-4 + (-9) = -13$
c. $-4 + 9 = 5$
d. $4 + (-9) = -5$

Example 2: Subtract.
a. $4 - 9 = 4 + (-9) = -5$
b. $-4 - 9 = -4 + (-9) = -13$
c. $4 - (-9) = 4 + 9 = 13$
d. $-4 - (-9) = -4 + 9 = 5$

To add fractions with the same denominator, add the numerators and keep the same denominator, then simplify.

To add fractions with different denominators,
1. Write each fraction as an equivalent fraction with the LCD.
2. Add the numerators and keep the LCD.
3. Simplify.

Example 3: Add.
a. $\dfrac{1}{10} + \dfrac{3}{10} = \dfrac{4}{10} = \dfrac{2}{5}$

b. $-\dfrac{3}{4} + \dfrac{5}{6} = -\dfrac{3(3)}{4(3)} + \dfrac{5(2)}{6(2)}$

$$= -\dfrac{9}{12} + \dfrac{10}{12}$$

$$= \dfrac{1}{12}$$

continued

Procedures/Rules	Key Examples

Section 1.4 Multiplying and Dividing Real Numbers; Properties of Real Numbers

The product of an even number of negative factors is positive, whereas the product of an odd number of negative factors is negative.

Example 1: Multiply.
a. $(5)(7) = 35$
b. $(-5)(-7) = 35$
c. $(-1)(-2)(-5)(-7) = 70$
d. $(-5)(7) = -35$
e. $(5)(-7) = -35$
f. $(-2)(-5)(-7) = -70$

Rule for multiplying fractions: $\dfrac{a}{b} \cdot \dfrac{c}{d} = \dfrac{ac}{bd}$, where $b \neq 0$ and $d \neq 0$.

Example 2: Multiply.
a. $\dfrac{3}{4} \cdot \dfrac{5}{7} = \dfrac{15}{28}$
b. $\dfrac{5}{6} \cdot \dfrac{4}{9} = \dfrac{5}{2 \cdot 3} \cdot \dfrac{2 \cdot 2}{9} = \dfrac{10}{27}$

To multiply decimal numbers:
1. Multiply as if they named whole numbers.
2. Place the decimal in the product so that it has the same number of decimal places as the total number of decimal places in the factors.

Example 3: Multiply $(3.6)(2.4)$.

$$
\begin{array}{r}
2.4 \\
\times\, 3.6 \\
\hline
144 \\
+\,72 \\
\hline
8.64
\end{array}
$$

When dividing two numbers that have the same sign, the quotient is positive.

When dividing two numbers that have different signs, the quotient is negative.

Rules for division involving 0: $0 \div n = 0$, when $n \neq 0$

$n \div 0$ is undefined, when $n \neq 0$

$0 \div 0$ is indeterminate

Example 4: Divide.
a. $24 \div 8 = 3$
b. $-24 \div (-8) = 3$
c. $-24 \div 8 = = -3$
d. $24 \div (-8) = -3$

e. $0 \div 5 = 0$
f. $14 \div 0$ is undefined

Rule for dividing fractions: $\dfrac{a}{b} \div \dfrac{c}{d} = \dfrac{a}{b} \cdot \dfrac{d}{c}$, where $b \neq 0$, $c \neq 0$, and $d \neq 0$.

Example 5: Divide $\dfrac{5}{8} \div \dfrac{3}{4}$.

$$\frac{5}{8} \div \frac{3}{4} = \frac{5}{\underset{2}{\cancel{8}}} \cdot \frac{\overset{1}{\cancel{4}}}{3} = \frac{5}{6}$$

continued

Procedures/Rules	Key Examples

Section 1.4 Multiplying and Dividing Real Numbers; Properties of Real Numbers (continued)

When dividing decimal numbers, set up a long division and consider the divisor.

Case 1: If the divisor is an integer, then divide as if the dividend were a whole number, and place the decimal point in the quotient directly above its position in the dividend.

Case 2: If the divisor is a decimal number, then:

1. Move the decimal point in the divisor to the right enough places to make the divisor an integer.
2. Move the decimal point in the dividend the same number of places.
3. Divide the divisor into the dividend as if both numbers were whole numbers. Be sure to align the digits in the quotient properly.
4. Write the decimal point in the quotient directly above its position in the dividend.

In either of the two cases, continue the division process until you get a remainder of 0 or a repeating digit or block of digits in the quotient.

Example 6: Divide $12.88 \div 0.06$.

$$
\begin{array}{r}
214.66 \\
6\overline{)1288.00} \\
-12 \\
\overline{08} \\
-6 \\
\overline{28} \\
-24 \\
\overline{40} \\
-36 \\
\overline{40}
\end{array}
$$

Answers: $214.\overline{6}$

Section 1.5 Exponents, Roots, and Order of Operations

To evaluate an exponential form raised to a natural number exponent, write the base as a factor the number of times indicated by the exponent, then multiply.

If the base of an exponential form is a negative number and the exponent is even, then the product is positive.

If the base is a negative number and the exponent is odd, then the product is negative.

Every positive number has two square roots, a positive root and a negative root.

Negative numbers have no real-number square roots.

The radical symbol $\sqrt{}$ indicates to find only the positive (principal) square root.

$\sqrt{\dfrac{a}{b}} = \dfrac{\sqrt{a}}{\sqrt{b}}$, where $a \geq 0$ and $b > 0$.

Example 1: Evaluate.

a. $3^4 = 3 \cdot 3 \cdot 3 \cdot 3 = 81$

b. $(-2)^4 = (-2)(-2)(-2)(-2) = 16$

c. $(-2)^5 = (-2)(-2)(-2)(-2)(-2)$
$ = -32$

Example 2: Find all square roots.

a. 49

Answer: ± 7

b. -36

Answer: No real-number square roots exist.

Example 3: Simplify.

a. $\sqrt{81} = 9$

b. $\sqrt{-81}$ is not a real number.

c. $\sqrt{\dfrac{25}{64}} = \dfrac{\sqrt{25}}{\sqrt{64}} = \dfrac{5}{8}$

continued

Procedures/Rules	Key Examples

Section 1.5 Exponents, Roots, and Order of Operations (continued)
Order of Operations

Perform operations in the following order:
1. Grouping symbols.
2. Exponents/Roots from left to right, in order as they occur.
3. Multiplication/Division from left to right, in order as they occur.
4. Addition/Subtraction from left to right, in order as they occur.

Example 4: Simplify.
$$35 - [29 - 2(12 + 4)] + 4^3$$
$$= 35 - [29 - 2(16)] + 4^3$$
$$= 35 - [29 - 32] + 4^3$$
$$= 35 - (-3) + 4^3$$
$$= 35 - (-3) + 64$$
$$= 35 + 3 + 64$$
$$= 102$$

To find the arithmetic mean, or average, of n numbers, divide the sum of the numbers by n.
$$\text{Arithmetic mean} = \frac{x_1 + x_2 + \ldots + x_n}{n}$$

Example 5: Carl has the following test scores: 68, 76, 82, 78. Find the average of his test scores.
$$\frac{68 + 76 + 82 + 78}{4} = \frac{304}{4} = 76$$

Section 1.6 Translating Word Phrases to Expressions
To translate a word phrase to an expression, identify the variables, constants, and key words, then write the corresponding symbolic form.

Example 1: Translate to an algebraic expression:
a. Five more than seven times a number
   Answer: $7n + 5$
b. Six times the difference of a number and nine
   Answer: $6(n - 9)$
c. Four divided by a number cubed
   Answer: $4 \div n^3$

Section 1.7 Evaluating and Rewriting Expressions
To evaluate an algebraic expression:
1. Replace the variables with the corresponding given numbers.
2. Calculate following the order of operations.

Example 1: Evaluate $4x^2 - 5x$ when $x = -3$.
$$4(-3)^2 - 5(-3)$$
$$= 4(9) - 5(-3)$$
$$= 36 - (-15)$$
$$= 36 + 15$$
$$= 51$$

The Distributive Property of Multiplication over Addition
$$a(b + c) = ab + ac$$

Example 2: Use the distributive property to write an equivalent expression.
$$3(x + 7) = 3 \cdot x + 3 \cdot 7$$
$$= 3x + 21$$

To combine like terms, add or subtract the coefficients and keep the variables and their exponents the same.

Example 3: Combine like terms.
$$15x^2 + 8x + 3x^2 + 7 - 9x$$
$$= 15x^2 + 3x^2 + 8x - 9x + 7$$
$$= 18x^2 - x + 7$$

Chapter 1   Review Exercises

For Exercises 1–6, answer true or false.

[1.1]   **1.** π is a rational number.

[1.1]   **2.** $-\sqrt{2}$ is a rational number.

[1.7]   **3.** $-6x$ and $2x^3$ are like terms.

[1.7]   **4.** $8xy$ and yx are like terms.

[1.5]   **5.** $-6^2 = -36$

[1.5]   **6.** $(-6)^2 = 36$

For Exercises 7–10, fill in the blank.

[1.1]   **7.** The absolute value of every number is either _____ or 0.

[1.2]   **8.** A natural number that has exactly two different factors, 1 and the number itself, is a(n) _____ number.

[1.3]   **9.** When adding two addends that have the same sign, add their absolute values and _____ the sign.

[1.4]   **10.** If there are a(n) _____ number of negative factors, then the product is positive.

[1.1] *For Exercises 11–14, write a set representing each description.*

11. The months beginning with the letter *M*

12. The letters in the word *algebra*

13. The even natural numbers

14. The even integers

[1.1] *For Exercises 15–18, graph each on a number line.*

15. $-3\dfrac{2}{5}$

16. $2\dfrac{1}{4}$

17. 8.2

18. -4.6

[1.1] *For Exercises 19–22, simplify.*

19. $|-6.3|$

20. $|8.46|$

21. $\left|2\dfrac{1}{6}\right|$

22. $\left|-4\dfrac{3}{8}\right|$

Expressions

Exercises 19–22

[1.1] *For Exercises 23–26, use =, +, <, or > to write a true sentence.*

23. $|-6.4|$ ▨ 6.4

24. -6.9 ▨ 0

25. -14 ▨ -9

26. $\left|5\dfrac{6}{7}\right|$ ▨ $\left|-5\dfrac{6}{7}\right|$

Equations and Inequalities

Exercises 23–26

For Exercises 27 and 28, name the fraction represented by the shaded region.

27.

28.

[1.2] *For Exercises 29–32, find the missing number that makes the fractions equivalent.*

29. $-\dfrac{6}{7} = -\dfrac{?}{14}$

30. $\dfrac{2}{5} = \dfrac{10}{?}$

31. $\dfrac{15}{30} = \dfrac{?}{10}$

32. $\dfrac{-5}{2} = \dfrac{?}{4}$

[1.2] *For Exercises 33–36, write the prime factorization for each number.*

33. 51

34. 100

35. 108

36. 84

[1.2] *For Exercises 37–40, simplify to lowest terms.*

37. $\dfrac{152}{200}$

38. $\dfrac{61}{122}$

39. $\dfrac{250}{360}$

40. $\dfrac{26}{39}$

[1.3–1.4] *For Exercises 41–46, indicate whether the expression illustrates the commutative property of addition, the associative property of addition, the commutative property of multiplication, the associative property of multiplication, or the distributive property of multiplication over addition.*

41. $4 + (7 + 3) = (4 + 7) + 3$

42. $3(9 + 1) = 27 + 3$

43. $\dfrac{1}{3} \cdot \dfrac{5}{6} = \dfrac{5}{6} \cdot \dfrac{1}{3}$

44. $5.6 + (11.2 + 4.3) = 5.6 + (4.3 + 11.2)$

45. $(n - 8)(-3) = -3(n - 8)$

46. $(2 \cdot 3) \cdot 7 = 2 \cdot (3 \cdot 7)$

[1.3–1.5] *For Exercises 47–68, perform the indicated operation(s).*

47. $-8 - (-4.2)$

48. $6 - (-9.1)$

49. $\dfrac{2}{5} + \dfrac{1}{3}$

50. $-\dfrac{1}{4} + \dfrac{2}{3}$

51. $-7(-13)$

52. $6(-8)$

53. $-25 \div 5$

54. $-30 \div (-3)$

55. -6^2

56. $(-3)^2$

57. $7(6 + 2) - 48 \div 12$

58. $6(9 - 4) \div 3 - 1$

59. $(-2)^3(4)(-5)$

60. $3^5 \div 3^2 \div 3^2 \div 3$

61. $-\dfrac{1}{3} - \dfrac{3}{2} \div \dfrac{1}{6}$

62. $-6.8 + (-4.1 + 2.3)$

63. $4|-2| \div (-8)$

64. $10 \div 5 + (-5)(-5)$

65. $6 + \{3(4 - 5) + 2[6 + (-4)]\}$

66. $-1 - 3[4^2 - 3(-6)]$

67. $\dfrac{-6 + 3^2(8 - 10)}{2^3 - 16}$

68. $\dfrac{3(4 - 5) - 4 \cdot 3 + (11 - 20)}{(3 - 4)^3}$

[1.5] *For Exercises 69 and 70, find all square roots of the given number.*

69. 49

70. -36

[1.6] *For Exercises 71–74, translate to an algebraic expression.*

71. twice a number subtracted from fourteen

72. the ratio of y to seven

73. a number minus twice the sum of the number and four

74. seven times the difference of m and n

[1.7] *For Exercises 75–80, evaluate the expression using the given values.*

75. $b^2 - 4ac$; $a = -3, b = -1, c = 5$

76. $b^2 - 4ac$; $a = 2, b = 6, c = -2$

77. $-|4x| + |y^2|$; $x = -3, y = -2$

78. $\sqrt{m} + \sqrt{n}$; $m = 16, n = 9$

79. $\sqrt{m + n}$; $m = 16, n = 9$

80. $\dfrac{3l^2}{4 - n}$; $l = -4, n = 16$

[1.7] *For Exercises 81 and 82, determine the value that could replace the variable and cause the expression to be undefined.*

81. $\dfrac{n}{n + 6}$

82. $\dfrac{y}{(y - 4)(y + 3)}$

[1.7] *For Exercises 83–86, use the distributive property to write an equivalent expression.*

83. $5(x + 6)$

84. $-3(5n - 8)$

85. $\dfrac{1}{4}(8y + 3)$

86. $0.6(4.5m - 2.1)$

[1.7] *For Exercises 87–92, simplify by combining like terms.*

87. $6x + 3y - 9x - 6y - 15$

88. $5y^2 - 6y + 4y - y^2 + 9y$

89. $-6xy + 9xy - xy$

90. $8x^3 - 4x - 6x^2 - 10x^3 + 4x$

91. $-4m - 4m - 4n + 4n$

92. $14 - 6x + 4y - 8 - 10y - 8x$

[1.3] **93.** The credit card statement to the right shows when each transaction was posted to the account. Find the new balance.

4-01-01	Beginning balance	−$685.92
4-03-01	Dillard's	−$45.80
4-05-01	Payment	$250.00
4-08-01	Applebee's	−$36.45
4-16-01	CVS Pharmacy	−$12.92
4-24-01	Target credit	$32.68
4-30-01	Finance charge	−$5.18

[1.3] **94.** The bar graph to the right shows the closing price for Merck and Company, Inc. stock each day for the week beginning September 27, 2004. This is the week in which it was announced that the pain-relieving drug Vioxx, which was manufactured by Merck, had been linked to an increased risk of heart attack.

CLOSING PRICE OF MERCK STOCK

$44.46 $44.92 $45.07 $33.00 $33.31

9/27/04   9/28/04   9/29/04   9/30/04   10/1/04

    a. Find the difference between the closing price at the beginning of the week and the closing price at the end of the week.

    b. If a person bought shares at the close of 9/28 and sold at the close of 10/1, what would that person's net profit or loss be on each share?

[1.4] **95.** Refer to the graph in Exercise 92.

    a. If Kaye bought 200 shares at the close of 9/28, how much did she spend?

    b. If she sold all 200 shares on 10/1, what was her loss?

[1.4] **96.** The voltage in a circuit can be calculated by multiplying the current and the resistance. Suppose the current in a circuit is −9.5 amperes and the resistance is 16 ohms. Find the voltage.

[1.4] **97.** In 1995, Larry has a loan balance of −$2405.80. If he makes no payment for ten years, his balance will become −$14,896.08. How many times greater is his debt in 2005 than it was in 1995?

[1.5] **98.** A company reports the following revenues for the first six months of 2001. Find the average revenue.

Month	Revenue
January	$45,320
February	$38,250
March	$61,400
April	$42,500
May	$74,680
June	$62,800

Chapter 1 Practice Test

1. Graph $-3\frac{2}{7}$ on a number line.

2. Simplify: $|-3.67|$

3. Write the prime factorization of 100.

4. Simplify: $\dfrac{17}{51}$

5. Find the missing number: $-\dfrac{6}{5} = -\dfrac{?}{10}$

For Exercises 6 and 7, indicate whether the expression illustrates the commutative property of addition, the associative property of addition, the commutative property of multiplication, the associative property of multiplication, or the distributive property of multiplication over addition.

6. $3(2 + 5) = 6 + 15$

7. $4(7 + 1) = 4(1 + 7)$

For Exercises 8–12, calculate.

8. $8 + (-4)$

9. $\dfrac{7}{8} - \left(-\dfrac{5}{6}\right)$

10. $(-1.5)(-0.4)$

11. $-\dfrac{5}{6} \div \dfrac{2}{3}$

12. $(-4)^3$

13. Find all square roots of 81.

For Exercises 14–17, simplify.

14. $-12 \div 3 \cdot 2$

15. $\dfrac{6^2 + 14}{(2 - 7)^2}$

16. $-8 \div |4 - 2| + 3^2$

17. $\sqrt{25 - 16} + [(14 + 2) - 3^2]$

For Exercises 18 and 19, translate the indicated phrase.

18. twice the sum of m and n

19. five less than three times w

20. After making a payment, Karen's balance on her credit card is $-\$854.80$. If her balance prior to making the payment was $-\$1104.80$, how much did she pay toward the balance?

21. On six days over a three-month period, a city receives 6 inches, 10 inches, 4 inches, 3 inches, 2.5 inches, and 8 inches of snow. Calculate the average daily amount of snowfall over those days.

22. Evaluate $-|3x^2 + 2y|$, when $x = -1$ and $y = 4$.

23. Evaluate $\sqrt{a - b}$, when $a = 64$ and $b = -36$.

24. Use the distributive property to write an equivalent expression: $-5(4y + 9)$

25. Simplify: $3.5x - 8 + 2.1x + 9.3$

Solving Linear Equations and Inequalities

"The difficulties you meet will resolve themselves as you advance. Proceed, and light will dawn, and shine with increasing clearness on your path."

—D'Alembert, French mathematician (1717–1783)

In Chapter 1, we reviewed numbers, arithmetic, and expressions, which form a foundation for studying equations. Remember from Chapter 1 that our most basic components are variables and constants, which make up expressions. We saw that we can perform two actions with expressions: (1) evaluate or (2) rewrite. In this chapter, we will build on that foundation by creating equations and inequalities out of those expressions, and then we will *solve* those equations and inequalities. This development will be repeated throughout the course. We will study and classify new types of expressions, evaluate and rewrite those expressions, and then solve equations and inequalities made from the new expressions.

Remember, as we build algebra concepts, they become more complex, just as a building becomes more complex during its construction. Try not to get discouraged if a new concept does not make sense right away. Learning new concepts is like the construction of a building: during the early stages, the building is not as solid as it will be upon completion. This chapter is like the frame of the walls of a house: The frame is not as solid as the walls will be after the drywall and roof are put in place. Similarly, as we revisit equations and inequalities throughout the course, we will reinforce and solidify the concepts you will learn in this chapter.

2.1 Equations, Formulas, and the Problem-Solving Process

OBJECTIVES

1. Verify solutions to equations.
2. Determine whether an equation is an identity.
3. Use formulas to solve problems.

We now move to the top tier of the Algebra Pyramid and focus on equations. Recall from Section 1.1 the definition of an **equation**.

DEFINITION   *Equation:* Two expressions set equal.

For example, $4x + 5 = 9$ is an equation made from the expressions $4x + 5$ and 9.

Connection We can think of an equation as a complete sentence and the equal sign as the verb "is." We can read $4x + 5 = 9$ as "Four x plus five is equal to nine" or "Four x plus five is nine."

$$4x + 5 = 9$$

$$\updownarrow$$

"Four x plus five is 9."

The Algebra Pyramid

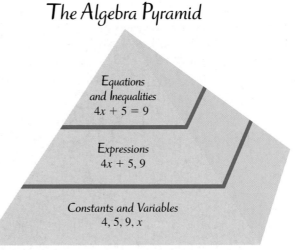

Equations
and Inequalities
$4x + 5 = 9$

Expressions
$4x + 5, 9$

Constants and Variables
$4, 5, 9, x$

OBJECTIVE 1. Verify solutions to equations. An equation can be either true or false. For example, the equation $3 + 1 = 4$ is true, whereas the equation $3 + 1 = 5$ is false. When given an equation that contains a variable for a missing number, such as $x + 2 = 7$, our goal is to solve it, which means to find every number that can replace the variable and make it true. Such a number is called a **solution** to the equation.

DEFINITION   *Solution:* A number that makes an equation true when it replaces the variable in the equation.

For example, the number 5 is a solution to the equation $x + 2 = 7$ because when 5 replaces x, it makes the equation true.

$$x + 2 = 7$$

$$\downarrow$$

$$5 + 2 = 7 \qquad \text{True.}$$

By showing that the equation $x + 2 = 7$ is true statement when x is replaced with 5, we verify or *check* that 5 is the solution to the equation.

PROCEDURE **Checking a Possible Solution**

To determine whether a value is a solution to a given equation, replace the variable(s) in the equation with the value. If the resulting equation is true, then the value is a solution.

EXAMPLE 1 Check to see if -2 is a solution to $4x + 9 = 1$.

Solution $4x + 9 = 1$

$$4(-2) + 9 \stackrel{?}{=} 1 \qquad \text{Replace } x \text{ with } -2 \text{ and see if the equation is true.}$$
$$-8 + 9 \stackrel{?}{=} 1 \qquad \text{Note: } \textit{The symbol}$$
$$1 = 1 \qquad \stackrel{?}{=} \textit{indicates we are}$$

Note: *The symbol $\stackrel{?}{=}$ indicates we are asking, "How does the left side compare with the right side? Are they equal or not equal?"*

Since -2 makes the equation true, it is a solution to $4x + 9 = 1$.

EXAMPLE 2 Check to see if 0.4 is a solution to $n^2 + 0.8 = 3n - 0.1$.

Solution $n^2 + 0.8 = 3n - 0.1$

$$(0.4)^2 + 0.8 \stackrel{?}{=} 3(0.4) - 0.1 \qquad \text{Replace it with 0.4 and see if the}$$
$$0.16 + 0.8 \stackrel{?}{=} 1.2 - 0.1 \qquad \text{equation is true.}$$
$$0.96 \neq 1.1.$$

Since 0.4 does not make the equation true, it is not a solution to $n^2 + 0.8 = 3n - 0.1$.

EXAMPLE 3 Check to see if $5\frac{1}{3}$ is a solution to $\frac{3}{4}y - 1 = \frac{y}{2} + \frac{1}{3}$.

Solution $\frac{3}{4}y - 1 = \frac{1}{2}y + \frac{1}{3}$

> **Note:** *It is convenient to write $\frac{y}{2}$ as $\frac{1}{2}y$, which we can do because $\frac{y}{2} = y \div 2 = y \cdot \frac{1}{2} = \frac{1}{2}y$.*

$\frac{3}{4}\left(5\frac{1}{3}\right) - 1 \stackrel{?}{=} \frac{1}{2}\left(5\frac{1}{3}\right) + \frac{1}{3}$ Replace y with $5\frac{1}{3}$ and see if the equation is true.

$\frac{\cancel{3}^{1}}{\cancel{4}_{1}}\left(\frac{\cancel{16}^{4}}{\cancel{3}_{1}}\right) - 1 \stackrel{?}{=} \frac{1}{\cancel{2}_{1}}\left(\frac{\cancel{16}^{8}}{3}\right) + \frac{1}{3}$ Rewrite $5\frac{1}{3}$ as an improper fraction, divide out common factors and then multiply.

$4 - 1 \stackrel{?}{=} \frac{8}{3} + \frac{1}{3}$

$3 = 3$ Therefore, $5\frac{1}{3}$ is a solution to $\frac{3}{4}y - 1 = \frac{y}{2} + \frac{1}{3}$.

YOUR TURN

  a. Check to see if 6 is a solution to $5n - 8 = 3n + 4$.

  b. Check to see if -1.5 is a solution to $4y - 3.5 = 2y + 3(y + 0.2)$.

  c. Check to see if $\frac{1}{6}$ is a solution to $\frac{2}{3}x + \frac{1}{4} = \frac{1}{5} - x^2$.

OBJECTIVE 2. Determine whether an equation is an identity. Consider $3x + 9 = 3x + 9$. Because the expressions on each side of the equal sign are identical, they will each produce the same result no matter what number replaces the variable x. This means that every real number is a solution. Such an equation is called an **identity**.

DEFINITION *Identity:* An equation that has every real number as a solution (excluding any numbers that cause an expression in the equation to be undefined).

The equation $5y - 4 = 7y - 3$ is not an identity because not every real number is a solution. Consider the number 2, for example. If we replace y with 2 in the equation, we see that it does not check:

$5(2) - 4 \stackrel{?}{=} 7(2) - 3$

$10 - 4 \stackrel{?}{=} 14 - 3$ The real number we picked, 2, is not a solution for

$6 \neq 11$ $5y - 4 = 7y - 3$, so the equation is not an identity.

Sometimes it is not obvious that an equation is an identity. Simplifying the expressions on each side of the equal sign can make the identity apparent.

ANSWERS

a. yes **b.** no **c.** no

To determine whether an equation is an identity, simplify the expressions on each side of the equal sign. If, after simplifying, the expressions are identical, then the equation is an identity.

EXAMPLE 4   Determine whether the equation is an identity.

a. $5x - 8x + 7 = 13 - 3x - 6$

Solution Simplify the expressions on each side of the equal sign. If the simplified expressions are identical, then the equation is an identity.

$$5x - 8x + 7 = 13 - 3x - 6$$
$$-3x + 7 = -3x + 7 \qquad \text{Combine like terms.}$$

The expressions on each side of the equal sign are identical. Therefore, the equation is an identity.

b. $2(3y - 4) - 10y = 15 - 4(y + 2)$

Solution $2(3y - 4) - 10y = 15 - 4(y + 2)$

$$6y - 8 - 10y = 15 - 4y - 8 \qquad \text{Distribute to clear parentheses.}$$
$$-4y - 8 = -4y + 7 \qquad \text{Combine like terms.}$$

The expressions on each side of the equal sign are not identical. Therefore, this equation is not an identity.

YOUR TURN   Determine whether the equation is an identity.

a. $0.5(3n - 8) = n - 4 + 0.5n$ **b.** $10 + \dfrac{1}{4}t^2 - 9 = t^2 - 1 - 0.75t^2$

OBJECTIVE 3. Use formulas to solve problems.   A primary purpose of studying mathematics is to develop and improve problem-solving skills. George Polya proposed the idea that all problem solving follows a four-step outline: 1. Understand the problem, 2. devise a plan for solving the problem, 3. execute the plan, and 4. check the results. Following is an outline for problem solving based on Polya's four stages. We'll see Polya's four stages illustrated throughout the rest of the text in application problems.

ANSWERS

a. yes   **b.** no

1. **Understand** the problem.
   a. Read the question(s) (not the whole problem, just the question at the end) and write a note to yourself about what it is you are to find.
   b. Now read the whole problem, underlining the key words.
   c. If possible or useful, make a list or table, simulate the situation, or search for a related example problem.
2. **Plan** your solution by searching for a formula or translating the key words to an equation.
3. **Execute** the plan by solving the equation/formula.
4. **Answer** the question. Look at the note about what you were to find and make sure you answer that question. Include appropriate units.
5. **Check** results.
   a. Try finding the solution in a different way, reversing the process, or estimating the answer and make sure the estimate and actual answer are reasonably close.
   b. Make sure the answer is reasonable.

As the course develops, we will explore the various strategies listed in the outline. In this section, we focus on using **formulas** to solve problems. Later in this chapter, we will solve problems by translating key words to an equation.

DEFINITION   *Formula:* An equation that describes a mathematical relationship.

First, let's consider using formulas from geometry. The following table lists some common geometric formulas. Before we examine the formulas, let's review a few terms.

DEFINITIONS   *Perimeter:* The distance around a figure.
Area: The total number of square units that fill a figure.
Volume: The total number of cubic units that fill a space.
Circumference: The distance around a circle.
Radius: The distance from the center of a circle to any point on the circle.
Diameter: The distance across a circle through its center.

Geometric Formulas

Plane Figures

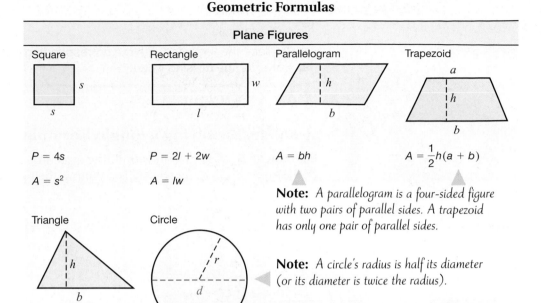

Square

$P = 4s$

$A = s^2$

Rectangle

$P = 2l + 2w$

$A = lw$

Parallelogram

$A = bh$

Trapezoid

$A = \frac{1}{2}h(a + b)$

Note: *A parallelogram is a four-sided figure with two pairs of parallel sides. A trapezoid has only one pair of parallel sides.*

Triangle

$A = \frac{1}{2}bh$

Circle

$C = \pi d$

$A = \pi r^2$

Note: *A circle's radius is half its diameter (or its diameter is twice the radius).*

Note: *Recall that π represents an irrational number. The π key on most calculators approximates the value as* **3.141592654.** *Some people prefer to round the value to* **3.14,** *which is simpler but less accurate.*

Solids

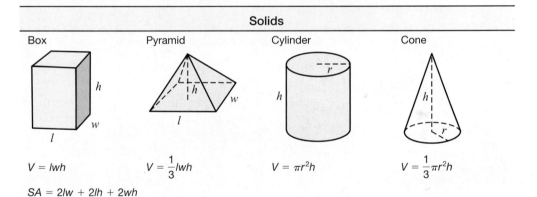

Box

$V = lwh$

$SA = 2lw + 2lh + 2wh$

Pyramid

$V = \frac{1}{3}lwh$

Cylinder

$V = \pi r^2 h$

Cone

$V = \frac{1}{3}\pi r^2 h$

Sphere

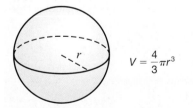

$V = \frac{4}{3}\pi r^3$

Connection Notice that the volume formulas for a box and pyramid are similar. This is because a pyramid has $\frac{1}{3}$ the volume of a box with the same-size base and same height. The same relationship is true for cylinders and cones with the same-size base and same height.

To use a formula:
1. Replace the variables with the corresponding given values.
2. Solve for the missing value.

Problems Involving a Single Formula

EXAMPLE 5) A mason is building a rectangular foundation wall that is to be 75 feet by 40 feet. What is the total distance around the wall?

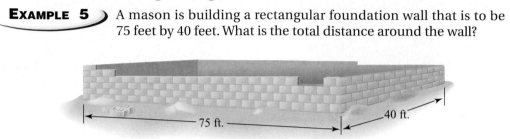

Understand   The "total distance around the wall" is the perimeter.

Plan   Because the shape is a rectangle, we can use the formula $P = 2l + 2w$.

Execute   Replace l with **75** feet and w with **40** feet, then calculate.

$$P = 2l + 2w$$
$$P = 2(\mathbf{75}) + 2(\mathbf{40})$$
$$P = 150 + 80$$
$$P = 230 \text{ ft.}$$

Answer   The mason must have enough supplies to build a wall that is 230 feet long.

Check   In this case, the solution can be verified using an alternative method. One could add all four side lengths.

$$P = 75 + 75 + 40 + 40 = 230 \text{ ft.}$$

Problems Involving Combinations of Formulas

Sometimes a problem requires a combination of formulas. This occurs if we are finding the area of a figure that is composed of two or more simpler figures.

EXAMPLE 6 Following is a drawing of a room that is to be carpeted. Calculate the area of the room.

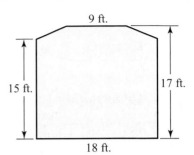

Understand This figure can be viewed as a trapezoid and a rectangle. The height of the trapezoid is found by subtracting 15 feet from 17 feet.

Plan Because the figure consists of a trapezoid and rectangle combined, we can add the areas of these shapes to find the total area of the figure. The area of a rectangle is found using the formula $A = lw$, and the area of a trapezoid is found using the formula $A = \frac{1}{2}h(a + b)$. The total area is

$$A = \text{Area of the rectangle} + \text{area of the trapezoid}$$
$$A = \qquad lw \qquad + \qquad \frac{1}{2}h(a + b)$$

Execute Replace the variables with the corresponding values and calculate.

$$A = lw + \frac{1}{2}h(a + b)$$
$$A = (15)(18) + \frac{1}{2}(2)(18 + 9)$$
$$A = (15)(18) + \frac{1}{2}(2)(27)$$
$$A = 270 + 27$$
$$A = 297 \text{ ft.}^2$$

Answer The total area is 297 square feet.

Check We can verify the reasonableness of the answer using estimation. Suppose the figure had been a rectangle measuring 18 feet by 17 feet. We would expect the area of this rectangle to be slightly greater than the area of the actual figure. The area of an 18-foot by 17-foot rectangle is $A = (18)(17) = 306 \text{ ft.}^2$, which indicates that 297 square feet is reasonable.

Sometimes a figure may be placed within another figure and we must calculate the area of the region between the outside figure and inside figure.

EXAMPLE 7 A portion of the front of a house is to be covered with siding (shaded in the figure). The window is 2.8 feet by 4.8 feet. Calculate the area to be covered.

Understand The shaded area is the entire triangle excluding the rectangular window.

Plan We can exclude the area of the window by calculating the area of the triangle and subtracting the area of the window.

A = Area of the triangle − area of the rectangle

$$A = \frac{1}{2}bh \quad - \quad lw$$

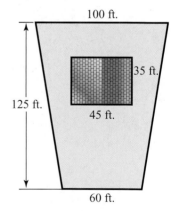

8 ft.

12.5 ft.

Execute $A = \frac{1}{2}(12.5)(8) - (2.8)(4.8)$

$A = 50 - 13.44$

$A = 36.56 \text{ ft.}^2$

Answer The area to be covered is 36.56 square feet.

Check Estimate the area by rounding the decimal numbers. Rounding the measurements of the triangle to the nearest ten, we have a base of 10 feet and a height of 10 feet, which means an area of 50 square feet. Rounding the window dimensions to the nearest whole, we have a 3-foot by 5-foot window, which has an area of 15 square feet. Subtracting these areas, we have a final area of 35 square feet, which indicates our answer of 36.56 square feet is reasonable.

YOUR TURN A house is to be situated on a lot as shown. Once the house is built, the remaining area will be landscaped. Calculate the area to be landscaped.

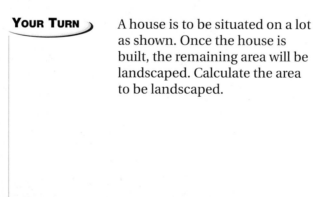

100 ft.

35 ft.

125 ft.

45 ft.

60 ft.

Problems Involving Nongeometric Formulas

Following is a list of other formulas you may encounter.

The distance, d, an object travels given its rate, r, and the time of travel, t: $d = rt$

The average rate of travel, r, given the total distance, d, and total time, t: $r = \dfrac{d}{t}$

The voltage, V, in a circuit with a current, i, in amperes (A), and a resistance, R, in ohms (Ω): $V = iR$

The temperature in degrees Celsius given degrees Fahrenheit:

$$C = \frac{5}{9}(F - 32)$$

The temperature in degrees Fahrenheit given degrees Celsius:

$$F = \frac{9}{5}C + 32$$

EXAMPLE 8 A truck driver begins a delivery at 9 A.M. and travels 150 miles. He then takes a 30-minute break. He then travels another 128 miles, arriving at his destination at 2:00 P.M. What was his average driving rate?

Understand We are given travel distances and times and we are to find the average driving rate.

Plan To find the average rate, we first need the total distance traveled and the total time spent driving. We can then use the formula $r = \dfrac{d}{t}$.

Execute Total distance = 150 + 128 = 278 miles

Total time driving = 5 − 0.5 = 4.5 hours

Note: *From 9 A.M. to 2:00 P.M. is 5 hours, but we must deduct the 0.5-hour (30-minute) break because he was not traveling during that time.*

Now we can calculate the average rate:

$$r = \frac{278}{4.5}$$

$$r = 61.\overline{7} \text{ mph (miles per hour)}$$

Answer His average driving rate was $61.\overline{7}$ miles per hour.

Check We can use $d = rt$ to verify that if he traveled an average rate of $61.\overline{7}$ miles per hour for 4.5 hours, he would travel 278 miles.

$$d \approx 61.8(4.5)$$
$$d \approx 278.1$$

Note: *We rounded $61.\overline{7}$ so the distance is an approximation.*

YOUR TURN A person hikes 10 miles then rests for 20 minutes, after which she hikes another 8 miles. If she began her hike at 10:30 A.M. and stopped at 4:30 P.M., what was her average rate?

ANSWER

$3\dfrac{3}{17}$ or ≈ 3.2 mph

2.1 Exercises

For Extra Help

MyMathLab

MyMathLab   Videotape/DVT   InterAct Math   Math Tutor Center   Math XL.com

1. What symbol is present in an equation, but not in an expression?

2. What is a solution for an equation?

3. Explain how to check a value to see if it is a solution for a given equation.

4. If an equation is also an identity, what does this indicate about its solutions?

For Exercises 5–20, check to see if the given number is a solution for the given equation.

5. $3x - 5 = 41; x = 2$

6. $4a + 7 = 51; a = 11$

7. $7y - 1 = y + 3; y = \dfrac{2}{3}$

8. $-8t - 3 = 2t - 15; t = -\dfrac{6}{5}$

9. $-4(x - 5) + 2 = 5(x - 1) + 3; x = -2$

10. $2(3m + 2) = 5m - 1; m = -3$

11. $-\dfrac{1}{2} + \dfrac{1}{3}y = \dfrac{3}{2}; y = 6$

12. $\dfrac{1}{2}p - 1 = \dfrac{2}{5}p + 3; p = 20$

13. $4.3z - 5.71 = 2.1z + 0.07; z = 2.2$

14. $12.7a + 12.6 = a + 5.4a; a = -2$

15. $b^2 - 4b = b^3 + 4; b = -1$

16. $-x^3 - 9 = 2x^2; x = -3$

17. $|x^2 - 9| = -x + 3; x = 3$

18. $-|2u - 3| = -3u + 8; u = 5$

19. $\dfrac{4x - 3}{x - 4} = \sqrt{x + 8}; x = 17$

20. $\dfrac{-y}{10 + y} = \dfrac{\sqrt{4 - y}}{3}; y = -5$

For Exercises 21–30, determine whether the equation is an identity.

21. $6x + 2 - 4x = 2x + 7 - 5$

22. $-5x + 8x - 1 = 3x - 4 + 2$

23. $3(y + 1) - 2y = y + 3$

24. $2(2 + m) = 6m - 4m + 4$

25. $12 - 2(x + 8) = 3x + 4 - 2x$

26. $5n - 20 - 9n = 8 - 4(n + 5)$

27. $\frac{1}{3}(t + 6) - t = \frac{5}{6}t + 2 - \frac{1}{2}t$

28. $\frac{1}{2}x + 3 - \frac{4}{5}x = -3\left(\frac{1}{10}x + 1\right)$

29. $0.27p + 3p = 7(p + 1) - (3.73 + 1)$

30. $0.4(n^2 - 6) - 3.6 = 1.9n^2 - 1.5(n + 4)$

For Exercises 31–54, solve using the geometric formulas. (Answers to exercises involving π were calculated using the π key on a calculator, which approximates the value as 3.141592654.)

31. Janet wants to put a wallpaper border around her den. The room is 13 feet by 20 feet.

   a. What is the total length of wallpaper border she needs?

   b. If the wallpaper border comes in packages of 12 feet, how many packages must she buy?

   c. If the packages are priced at $8.99 each, what will be the total cost of the wallpaper?

32. Jamal is planning to install crown molding where the walls and ceiling of his living room meet. The room is 16.5 feet by 22 feet.

   a. What is the total length of crown molding he needs?

   b. If the crown molding comes in strips of 8 feet, how many strips must he purchase?

   c. If the strips cost $9.99 each, what will be the cost of the crown molding?

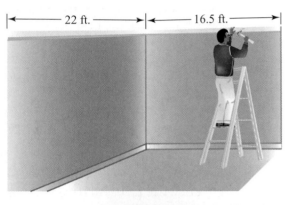

33. Fermi National Accelerator Laboratory is a circular tunnel that is used to accelerate elementary particles. The radius of the tunnel is about 2 kilometers. If a particle travels one complete revolution through the tunnel, what distance does it travel? (*Source:* Cesare Emiliani, *The Scientific Companion*, 2nd ed., Wiley, 1995)

34. The Chicxulub crater, which is buried partly beneath the Yucatan peninsula and partly beneath the Gulf of Mexico, is circular and about 180 kilometers in diameter. What is the circumference of this crater?

35. At one time, the Venetian Hotel and Casino in Las Vegas boasted having the largest hotel rooms in the world. If one of the rectangular rooms is 70 feet by 100 feet, what is the total square footage in the room?

36. The world's largest oil production barge was built by Kvaernel Oil and Gas, Inc. of Norway and Single Bouy Moorings, Inc. of Switzerland. The rectangular deck of the barge is 273 meters long and 50 meters wide. What is the area of the deck?

Of Interest

The Chicxulub crater was created when Earth was hit by an asteroid about 65 million years ago. The devastation of the impact is thought to have caused the extinction of about 50 percent of the species on Earth, including the dinosaurs.

37. A large wall in a home is to be painted with a custom accent color. The rectangular wall measures 32.5 feet by 12 feet. A painter says he charges $2.50 per square foot to paint a wall with a custom color. How much will it cost to have the painter paint the wall?

38. Hardwood floors are to be installed in a bedroom that measures 14 feet by 15 feet. If the contractor charges $34.50 per square foot to install the floors, what will be the cost of this improvement?

39. Tina needs to buy enough pine straw to cover a rectangular flower bed that measures 6 feet by 13 feet.

   a. What is the total area she must cover?

   b. A landscape consultant indicates that she should use 1 bale of pine straw for every 10 square feet to be covered. If she follows this advice, how many bales must Tina purchase?

   c. If Tina's local garden center charges $4.50 per bale, what will be the cost of covering her flower bed?

40. Juan is considering installing carpet in his new office space. The office is rectangular and measures 42 feet by 36 feet. Juan has a budget of $3000.

   a. What is the total area that will be carpeted?

   b. How many square yards is this area? (There are 9 square feet in 1 square yard.)

   c. If the contractor quotes Juan a price of $22.50 per square yard to install the carpet, how much will it cost to carpet the space? Is this feasible with Juan's budget?

41. A city planning committee decides to install 20 historic markers around the city. The signs are triangular, measuring 26 inches wide at the base and 20 inches high.

   a. What is the area of each sign?

   b. If a contractor charges $0.75 per square inch to build and install the signs, how much will it cost to have them installed?

42. A parks department is going to install a flower bed in a triangular space created by three trails, as shown. The flower bed measures 32 feet at the base and 24 feet high.

   a. What is the area of the flower bed?

   b. If a landscaper charges $6.50 per square foot to design and install the bed, what will the total cost be?

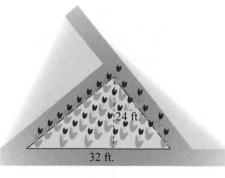

43. A kitchen floor is to be re-covered with granite tiles. There is a fixed island in the center of the floor (see figure). If the contractor charges $28.80 per square foot to install the new floor, what will be the installation cost?

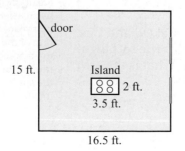

44. Conner and Ellyn purchase a new home. The area around an 8-foot by 9-foot storage building in the rectangular back-yard is to be landscaped (see figure).

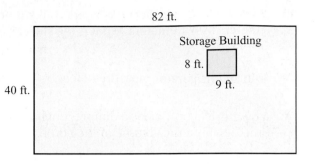

   a. What is the total area that is to be landscaped?

   b. If $\frac{2}{3}$ of this area is to be grass, what amount will be covered with grass?

   c. If each pallet of grass sod covers 504 square feet, how many pallets will be needed?

   d. Each pallet cost $106. What is the total cost of the sod?

45. The wall shown is to be painted. The door, which will not be painted, is 2.5 feet by 7.5 feet. What area will be painted?

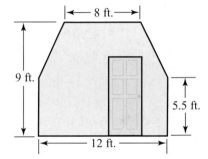

46. The side of a house needs new siding (see figure). The window measures 3 feet by 4.5 feet. Find the area to be covered in siding.

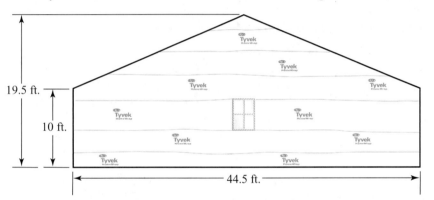

47. A company cuts two circular pieces out of a sheet of metal and the rest of the sheet is recycled. The sheet is 4 feet by 8 feet and the diameter of each circle is 4 feet. Find the area of the sheet that is recycled.

48. A compact disc has a diameter of $5\frac{3}{4}$ inches. On a CD, no information is placed within a circle in the center of the CD with a diameter of $1\frac{3}{4}$ inches. Find the area on a CD that can contain information.

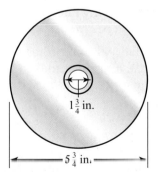

49. A small storage building is shaped like a box that measures 2 meters by 5 meters by 4.75 meters. What is the storage building's volume?

50. A dorm refrigerator measures 2 feet by $1\frac{1}{2}$ feet by 4 feet. What is its volume?

51. A cylindrical bottle has a diameter of 8 centimeters and a height of 20 centimeters. Find the volume of the bottle.

52. Earth is roughly a sphere with a radius of approximately 6370 kilometers. What is the approximate volume of Earth?

53. The base of the Great Pyramid at Giza, Egypt, is 745 feet by 745 feet and its height is 449 feet. Find the volume of the pyramid.

54. If a funnel is approximately a cone, approximate the volume of a funnel with a diameter of $8\frac{1}{2}$ inches and a height of 6 inches.

For Exercises 55–60 use the formulas relating distance, rate, and time.

55. A flight departs Atlanta at 8:30 A.M. EST to arrive in Las Vegas at 10:00 A.M. PST. If the plane flies at an average rate of $388\frac{1}{3}$ mph, what distance does it travel? (*Hint:* There is a 3-hour time difference between EST and PST.)

56. A flight departs at 7:30 A.M. EST from Philadelphia and travels to Dallas–Fort Worth, arriving at 10:10 A.M. CST. If the plane is averaging a speed of 368.2 miles per hour, what distance does it travel? (*Hint:* There is a 1-hour-time difference between EST and CST. Also, express the time using fractions.)

57. A family began a trip of 516 miles at 8 A.M. They arrived at their final destination at 5:30 P.M. If they took two 15-minute breaks and took an hour for lunch, what was their average driving rate?

58. A family began a trip at 7:30 A.M. At the beginning of the trip, the car's odometer read 45,362.6. They arrived at their final destination at 6:00 P.M. and the odometer read 45,785.2. If they took three 15-minute breaks and took an hour for lunch, what was their average driving rate?

59. A truck driver is to make a delivery that requires him to drive 285 miles by 4:00 P.M. If he starts at 10:00 A.M. and must make three stops of 15 minutes each at weigh stations along the route, what must his average driving rate be to make the delivery on time?

60. In 2005, Lance Armstrong won the Tour de France, which covered a distance of 3606 kilometers, with a time of 82 hours, 34 minutes, and 5 seconds. What was his average rate in kilometers per hour for the race? (*Source: The World Almanac and Book of Facts 2005*)

For Exercises 61 and 62, use the formula relating voltage, current, and resistance.

61. A technician measures the current in a circuit to be −6.5 amperes and the resistance is 8 ohms. Find the voltage.

62. The current in a circuit is measured to be 4.2 amperes. If the resistance is 16 ohms, find the voltage.

For Exercises 63–68, use the formulas for converting degrees Fahrenheit to degrees Celsius or degrees Celsius to degrees Fahrenheit.

63. On a drive to work, Tim notices that the temperature shown on a bank's digital thermometer is 34°C. What is this temperature in degrees Fahrenheit?

64. Liquid oxygen boils at a temperature of −183°C. What is this temperature in degrees Fahrenheit?

65. Iron melts at a temperature of 1535°C. What is this temperature in degrees Fahrenheit?

66. The average temperature on Neptune is approximately −360°F. What is this temperature in degrees Celsius?

67. At the South Pole, temperatures can get down to −76°F. What is this temperature in degrees Celsius?

68. The average temperature on Venus is approximately 890°F. What is this temperature in degrees Celsius?

Of Interest

Because of the tilt of Earth the South Pole experiences daylight for six months, then night for six months. The Sun rises in mid-September and sets in mid-March.

PUZZLE PROBLEM

A contest has people guess how many marbles are inside a jar. The jar is a cylinder with a diameter of 9 inches and a height of 12 inches. Suppose the marbles have a diameter of 0.5 inches. Calculate the number of marbles that might fit inside the jar. Discuss the accuracy of your calculation.

REVIEW EXERCISES

[1.5] *For Exercises 1–2, simplify.*

1. $14 + 9(7 − 2)$

2. $6(2 + 3^2) − 4(2 + 3)^2$

[1.7] *For Exercises 3–4, multiply using the distributive property.*

3. $\frac{1}{3}(x + 6)$

4. $−(4w − 6)$

[1.7] *For Exercises 5–6, combine like terms.*

5. $5x − 14 + 3x + 9$

6. $6.2y + 7 − 1.5 − 0.8y$

2.2 The Addition Principle

OBJECTIVES

1. Determine whether a given equation is linear.
2. Solve linear equations in one variable using the addition principle.
3. Solve equations with variables on both sides of the equal sign.
4. Solve application problems.

OBJECTIVE 1. Determine whether a given equation is linear. There are many types of equations. In this section, we introduce and begin solving **linear equations**.

DEFINITION *Linear equation:* An equation in which each variable term contains a single variable raised to an exponent of 1.

Equations that are not linear are *nonlinear equations*.

EXAMPLE 1 Determine whether the equation is linear or nonlinear.

a. $3x + 5 = 12$

Answer This equation is linear because the variable x has an exponent of 1.

b. $7x^3 + x = 1$

Answer This equation is nonlinear because the variable in the term $7x^3$ has an exponent of 3.

Connection Every equation has a corresponding graph. The names *linear* and *nonlinear* refer to the graphs of the equations that they describe. The graph of a linear equation is a line; the graph of a nonlinear equation is not.

YOUR TURN Determine whether the given equation is linear.

a. $3x - 4 = 11$ **b.** $-6x + y = 30$ **c.** $n^2 = 5m + 4$ **d.** $y = 3x^2 + 1$

In this chapter, we will focus on equations with the same variable throughout the equation. If a linear equation has the same variable throughout, then we say it is a **linear equation in one variable**.

DEFINITION *Linear equation in one variable:* An equation that can be written in the form $ax + b = c$, where a, b, and c are real numbers and $a \neq 0$.

$5x + 9 = -6$ is a linear equation in one variable.
$x^2 - 2x + 5 = 8$ is a nonlinear equation in one variable.

In Chapter 4, we will consider linear equations in two variables, such as $x + y = 3$.

ANSWERS

a. yes **b.** yes **c.** no **d.** no

OBJECTIVE 2. Solve linear equations in one variable using the addition principle. In Section 2.1, we discussed solutions to equations and how to check an equation's solution. We now consider a technique for solving an equation (which means to find its solution[s]). This technique is called the *balance technique.* Imagine an equation as scales with the equal sign as the fulcrum.

When weight is added or removed on one side of a balanced set of scales, the scales tip out of balance. In this case, we add 4 to the left side.

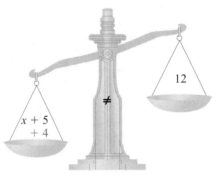

To maintain balance, the same weight must be added or removed on *both* sides of the scale. Adding 4 to the right side balances the scale.

In the language of mathematics, this balance technique is stated in two principles: the *addition principle of equality* and the *multiplication principle of equality.* In this section, we will consider the addition principle of equality. In Section 2.3, we will consider the multiplication principle of equality. The addition principle of equality says that we can add the same amount to (or subtract the same amount from) both sides of an equation without affecting its solution(s).

The Addition Principle of Equality

Note: *We say that* $a = b$ *and* $a + c = b + c$ *are equivalent equations because they have the same solution.*

▶ If $a = b$, then $a + c = b + c$ is true for all real numbers a, b, and c.

In solving an equation, the goal is to write an equivalent equation in a simpler form, $x = c$, where the variable is alone, or isolated, on one side of the equal sign and the solution appears on the other side. For example, to isolate x in $x + 5 = 12$, we need to clear $+5$ from the left-hand side. To clear $+5$, we add -5 (or subtract 5) because $+5$ and -5 are additive inverses, which means their sum is 0.

To isolate x, we clear $+5$ by adding -5 (or subtracting 5) to the left side.	$\begin{aligned} x + 5 &= 12 \\ \underline{-5} \quad &\underline{-5} \\ x + 0 &= 7 \\ x &= 7 \end{aligned}$	To keep the equation balanced, we add -5 (or subtract 5) to the right side as well.

The statement $x = 7$ is the solution statement.

PROCEDURE Using the Addition Principle

To use the addition principle of equality to clear a term in an equation, add the additive inverse of that term to both sides of the equation (that is, add or subtract appropriately so that the term you want to clear becomes 0).

EXAMPLE 2 Solve and check.

a. $x - 13 = -22$

Solution To isolate x, we need to clear -13.

Add 13 to the left-hand side so that $-13 + 13 = 0$.	$\begin{aligned} x - 13 &= -22 \\ \underline{+13} \quad &\underline{+13} \\ x + 0 &= -9 \\ x &= -9 \end{aligned}$	Since we added 13 to the left side, we *must* add 13 to the right side as well.

Check Recall from Section 2.1 that to check, we replace x in the original equation with -9 and verify that the equation is true.

$$x - 13 = -22$$
$$-9 - 13 \stackrel{?}{=} -22 \qquad \text{Replace } x \text{ with } -9.$$
$$-22 = -22$$

True, so -9 is the solution.

b. $\dfrac{3}{5} = y + \dfrac{1}{3}$

Note: *Here, we illustrate another popular style of writing the addition principle, in which $-\dfrac{1}{3}$ is written beside the expressions on both sides of the equation instead of underneath the corresponding terms.*

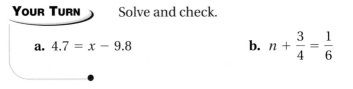

Solution $\quad \dfrac{3}{5} - \dfrac{1}{3} = y + \dfrac{1}{3} - \dfrac{1}{3} \qquad$ Subtract $\dfrac{1}{3}$ from both sides to isolate y.

$\qquad\qquad \dfrac{9}{15} - \dfrac{5}{15} = y + 0 \qquad\qquad$ Rewrite fractions with their LCD.

$\qquad\qquad\qquad\quad \dfrac{4}{15} = y \qquad\qquad\qquad$ Subtract $\dfrac{5}{15}$ from $\dfrac{9}{15}$.

Check $\quad \dfrac{3}{5} = y + \dfrac{1}{3}$

$\qquad \dfrac{3}{5} \overset{?}{=} \dfrac{4}{15} + \dfrac{1}{3} \qquad$ Replace y in the original equation with $\dfrac{4}{15}$ and verify that the equation is true.

$\qquad \dfrac{3}{5} \overset{?}{=} \dfrac{4}{15} + \dfrac{5}{15} \qquad$ Rewrite fractions with their LCD.

$\qquad \dfrac{3}{5} \overset{?}{=} \dfrac{9}{15} \qquad\qquad$ Add $\dfrac{4}{15}$ and $\dfrac{5}{15}$.

$\qquad \dfrac{3}{5} = \dfrac{3}{5} \qquad\qquad$ Simplify to lowest terms.

True, so $\dfrac{4}{15}$ is the solution.

YOUR TURN) $\qquad$ Solve and check.

$\qquad$ **a.** $4.7 = x - 9.8$ $\qquad\qquad\qquad$ **b.** $n + \dfrac{3}{4} = \dfrac{1}{6}$

Simplifying Before Isolating the Variable

Some equations have expressions that can be simplified. If like terms are on the same side of the equation, we will combine the like terms before isolating the variable.

ANSWERS)

a. 14.5 $\quad$ **b.** $-\dfrac{7}{12}$

EXAMPLE 3 Solve and check. $8y - 2.1 - 7y = -3.7 + 9$

Solution Simplify the expressions, then isolate y.

$$8y - 2.1 - 7y = -3.7 + 9$$

Combine $8y$ and $-7y$ to equal y. $y - 2.1 = 5.3$ Combine -3.7 and 9 to equal 5.3.

$$\underline{+2.1 \quad +2.1} \qquad \text{Add 2.1 to both sides to isolate } y.$$
$$y + 0 = 7.4$$
$$y = 7.4$$

Check   Replace y in the original equation with 7.4 and verify that the equation is true. We will leave this to the reader.

Using the Distributive Property

If the equation to be solved contains parentheses, we use the distributive property to clear the parentheses before isolating the variable.

EXAMPLE 4 Solve and check. $6(n - 3) - 5n = -25 + 4$

Solution Use the distributive property, simplify, and then isolate n.

$$6(n - 3) - 5n = -25 + 4$$

Distribute 6 to clear parentheses. $6n - 18 - 5n = -21$ Combine -25 and 4 to equal -21.

Combine $6n$ and $-5n$ to equal n. $n - 18 = -21$

$$\underline{+18 \qquad +18} \qquad \text{Add 18 to both sides}$$
$$n + 0 = \quad -3 \qquad \text{to isolate } n.$$
$$n = -3$$

Check   Replace n with -3 in the original equation and verify that the equation is true. We will leave this to the reader.

YOUR TURN   Solve and check.

    a. $9x + 4 - 8x = 1 - 6$ **b.** $6.2 - 0.4 = 1.5(m - 6) - 0.5m$

OBJECTIVE 3. Solve equations with variables on both sides of the equal sign.   Some equations have variable terms on both sides of the equal sign. In order to isolate the variable, we must first get the variable terms together on the same side of the equal sign. To do this, we can use the addition principle.

ANSWERS

   a. -9 **b.** 14.8

122   **CHAPTER 2**   Solving Linear Equations and Inequalities

EXAMPLE 5 Solve and check. $5x - 8 = 4x - 13$

Solution Use the addition principle to get the variable terms together on the same side of the equal sign. Then isolate the variable.

$$5x - 8 = 4x - 13$$
$$\underline{-4x \qquad\quad -4x} \qquad \text{Subtract } 4x \text{ from both sides.}$$
$$x - 8 = 0 - 13$$
$$x - 8 = -13$$
$$\underline{+8 \qquad\quad +8} \qquad \text{Add 8 to both sides to isolate } x.$$
$$x + 0 = -5$$
$$x = -5$$

Note: *By choosing to subtract $4x$ on both sides, we clear the $4x$ term from the right side and combine it with the $5x$ on the left side. It does not matter which term you move first (see the following explanation).*

Check $5x - 8 = 4x - 13$
$$5(-5) - 8 \overset{?}{=} 4(-5) - 13$$
$$-25 - 8 \overset{?}{=} -20 - 13$$
$$-33 = -33$$

Replace x in the original equation with -5 and verify that the equation is true.

True, so -5 is the solution.

When variable terms appear on both sides of the equal sign, it does not matter which term you choose to clear first, although some choices make the process easier. Suppose in Example 5 we clear the $5x$ term first:

$$5x - 8 = 4x - 13$$
$$\underline{-5x \qquad\quad -5x}$$
$$0 - 8 = -x - 13$$
$$-8 = -x - 13$$
$$\underline{+13 \qquad\qquad +13}$$
$$5 = -x + 0$$
$$5 = -x$$
$$-5 = x$$

Note: *Clearing the $5x$ term first gives $-x$, whereas when we cleared the $4x$ term, we got x. Most people prefer working with positive coefficients.*

Note: *$5 = -x$ means we must find a number whose additive inverse is 5, so x must be equal to -5. In effect, we simply changed the signs of both sides. In Section 2.3, we will see another way to interpret $-x$.*

Notice that the term we chose to clear in the first step did not affect the solution, but it did affect our approach to the rest of the problem.

Conclusion In solving an equation that has variable terms on both sides of the equal sign, when we select a variable term to clear, we can avoid negative coefficients by clearing the term with the lesser coefficient.

YOUR TURN Solve and check.

a. $7y - 3 = 6y + 10$ 　　　　　 b. $2.6 + 3n = 9.5 + 4n$

ANSWERS

a. 13　b. -6.9

EXAMPLE 6 Solve and check. $y - (3y + 11) = 5(y - 2) - 8y$

Solution Simplify both sides of the equation. Then isolate y.

Note: *When we distribute a minus sign in an expression like* $-(3y + 11)$*, we can think of the minus sign as* -1*, so that we have:*

$$-(3y + 11) = -1(3y + 11)$$
$$= -1(3y) + (-1)(11)$$
$$= -3y - 11$$

$$y - (3y + 11) = 5(y - 2) - 8y$$

Distribute the minus sign.	$y - 3y - 11 = 5y - 10 - 8y$	Distribute 5.
	$-2y - 11 = -3y - 10$	Combine like terms.
	$\underline{+3y \qquad\qquad +3y}$	Add $3y$ to both sides.
	$y - 11 = \quad 0 - 10$	
	$y - 11 = -10$	
	$\underline{+11 \quad\quad +11}$	Add 11 to both sides to isolate y.
	$y + 0 = \quad 1$	
	$y = 1$	

Note: *The coefficient of* $-3y$ *is less than the coefficient of* $-2y$*, so we chose to clear* $-3y$ *by adding* $3y$*.*

Check Replace y in the original equation with 1 and verify that the equation is true. We will leave this to the reader.

Following is an outline for solving equations based on everything we've learned so far.

PROCEDURE Solving Linear Equations

To solve linear equations requiring the addition principle only:

1. Simplify both sides of the equation as needed.
  a. Distribute to clear parentheses.
  b. Combine like terms.
2. Use the addition principle so that all variable terms are on one side of the equation and all constants are on the other side. Then combine like terms.

Tip: Clear the variable term that has the lesser coefficient to avoid negative coefficients.

YOUR TURN Solve and check.

a. $7n - (n - 3) = 4(n + 4) + n$ **b.** $11 + 3(5m - 2) = 7(2m - 1) + 9$

ANSWERS

a. 13   **b.** -3

Number of Solutions

In general, a linear equation in one variable has only one real-number solution. However, there are two special cases that we need to consider. First, if an equation is an identity, as in Example 7, every real number is a solution. Second, as we'll see in Example 8, some linear equations have no solution.

Equations with an Infinite Number of Solutions

EXAMPLE 7 Solve and check. $5(2t + 1) - 8 = 4t - 3(1 - 2t)$

Solution Simplify both sides of the equation. Then isolate the variable.

$$5(2t + 1) - 8 = 4t - 3(1 - 2t)$$

Distribute 5.

$$10t + 5 - 8 = 4t - 3 + 6t$$

Distribute −3.

$$10t - 3 = 10t - 3$$

Combine like terms.

Note: *We can stop here because the equation is obviously an identity. However, if we continue and apply the addition principle, the equation is still an identity.*

$$\begin{array}{r} 10t - 3 = 10t - 3 \\ \underline{-10t \qquad -10t} \\ 0 - 3 = \quad 0 - 3 \\ -3 = -3 \end{array}$$

Because the equation is an identity, every real number is a solution.

Check   Every real number is a solution for an identity, so any number we choose should check in the original equation. We will choose to test the number 1.

$$5(2t + 1) - 8 = 4t - 3(1 - 2t)$$
$$5(2(1) + 1) - 8 \stackrel{?}{=} 4(1) - 3(1 - 2(1))$$

Replace *t* with 1.

$$5(3) - 8 \stackrel{?}{=} 4 - 3(-1)$$
$$15 - 8 \stackrel{?}{=} 4 + 3$$
$$7 = 7$$

The equation is true, so 1 is a solution, which supports, but does not prove, our conclusion that the equation is an identity. To be more certain, we could choose another number to check in the equation (remember, every number should work). We will leave this to the reader.

Conclusion   If, after simplifying, the equation is an identity, then every real number is a solution to the equation.

Equations with No Solution

Some equations, called *contradictions*, have no solution. Linear equations of this type can be recognized by the fact that after the expressions on each side of the equation have been simplified, the variable terms will match but the constant terms will not. For example, $2x - 5 = 2x - 3$ is a contradiction. In Example 8, watch what happens when we try to solve the equation.

EXAMPLE 8 Solve and check. $2x - 5 = 2x - 3$

Solution
$$2x - 5 = 2x - 3$$
$$\underline{-2x \qquad -2x} \qquad \text{Subtract } 2x \text{ from both sides.}$$
$$0 - 5 = 0 - 3$$
$$-5 = -3$$

Note: *The variable terms are eliminated and the resulting numeric equation is false, which indicates that the given equation is a contradiction.*

Because the equation is a contradiction, it has no solution.

Check Because the variable terms, $2x$, are identical on both sides of the equal sign, replacing x with any number will yield an identical product. Since we subtract 5 from that product on the left side and we subtract 3 from that product on the right side, the equation can never be true. Therefore, it has no solution.

Conclusion When solving a linear equation in one variable, if applying the addition principle of equality causes the variable terms to be eliminated from the equation and the resulting equation is false, then the equation is a contradiction and has no solution.

OBJECTIVE 4. Solve application problems. Now consider some problems that can be solved using the addition principle.

EXAMPLE 9 Laura wants to buy a car stereo that costs $275. She currently has $142. How much more is needed?

Understand We are given the total required and the amount she currently has, and we must find how much she needs.

Plan Let x represent the amount Laura needs. We will write an equation, then solve.

Execute Current amount + needed amount = 275
$$142 \qquad + \qquad x \qquad = 275$$
$$142 + x = 275$$
$$\underline{-142 \qquad -142} \qquad \text{Subtract 142 from both sides}$$
$$0 + x = 133 \qquad \text{to isolate } x.$$
$$x = 133$$

Answer Laura needs $133 to buy the stereo.

Check Does $142 plus the additional $133 equal $275?

$$142 + 133 \stackrel{?}{=} 275$$
$$275 = 275 \qquad \text{It checks.}$$

YOUR TURN Daryl has a balance of $-\$568$ on a credit card. How much must he pay to bring his balance to $-\$480$?

EXAMPLE 10 The following figure shows the front-view design of a bookcase. Find the missing length.

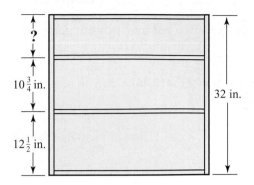

Understand In the drawing, the sum of the dimensions on the left side must be equal to 32 inches.

Plan Let d represent the unknown distance. We will write an equation and solve for d.

Execute

$$12\frac{1}{2} + 10\frac{3}{4} + d = 32$$

$$23\frac{1}{4} + d = 32 \qquad \text{Combine like terms.}$$

$$23\frac{1}{4} - 23\frac{1}{4} + d = 32 - 23\frac{1}{4} \qquad \text{Subtract } 23\frac{1}{4} \text{ from both sides to isolate } d.$$

$$d = 8\frac{3}{4}$$

Answer The unknown length is $8\frac{3}{4}$ inches.

Check Verify that the sum of the three lengths on the left side is equal to 32 inches. We will leave this to the reader.

Note: *As in Example 2(b), since we are subtracting a fraction form on both sides, we are writing the subtraction horizontally instead of vertically. Use whichever approach, horizontal or vertical, you prefer.*

YOUR TURN In researching the grades given in a particular course during one semester, a department chair finds that 89 students received an A, 154 received a B, and 245 received a C. If the department does not give Ds and there were 645 students that initially enrolled in the course, how many students did not receive a passing grade in the course?

ANSWER

157 students

2.2 Exercises

1. In your own words, explain what the addition principle of equality says.

2. In the equation $5x - 9 = 4x + 1$, how do you use the addition principle to clear the $4x$ term?

3. In the equation $5x - 9 = 4x + 1$, how do you use the addition principle to clear the -9 term?

4. If an equation has like terms on the same side of the equation, what should you do first?

5. If an equation has variable terms on both sides of the equation, what should you do first?

6. How do you handle parentheses in an equation?

For Exercises 7–22, determine whether the given equation is linear.

7. $4y + 7 = 2y - 8$

8. $6x - 5 = 3y + 64$

9. $m^2 + 4 = 20$

10. $-1 = w^2 - 5w$

11. $7u^4 + u^2 = 16$

12. $6n^3 - 9n^2 = 4n + 6$

13. $4x - 8 = 2(x + 3)$

14. $5(t - 2) + 7 = 3t - 1$

15. $6x - y = 11$

16. $3x + 2y = 12$

17. $x^2 + y^2 = 1$

18. $3x^2 + 2y^2 = 24$

19. $y = 8$

20. $x = -3$

21. $y = \dfrac{1}{4}x - 3$

22. $y = -0.5x + 2$

For Exercises 23–66, solve and check.

23. $x - 7 = 2$

24. $a + 8 = 30$

25. $-16 = y + 9$

26. $-24 = n - 11$

27. $m + \dfrac{7}{8} = \dfrac{4}{5}$

28. $\dfrac{3}{4} = c - \dfrac{2}{3}$

29. $-\dfrac{5}{8} = y - \dfrac{1}{6}$

30. $k + \dfrac{5}{9} = -\dfrac{1}{3}$

31. $15.8 + y = 7.6$

32. $b + 4.8 - 9.2 = 5.4$

33. $-2.1 = n - 7.5 - 0.8$

34. $x + 0.4 - 1.6 = -12.5$

35. $18 = 7x - 3 - 6x$ **36.** $2z + 6 - z = 5 - 9$ **37.** $5m = 4m + 7$ **38.** $-6y = -7y - 8$

39. $7x - 9 = 6x + 4$ **40.** $12y + 22 = 11y - 3$ **41.** $-2y - 11 = -3y - 5$ **42.** $-4t + 9 = -5t + 1$

43. $7x - 2 + x = 10x - 1 - x$ **44.** $3t + 6 + 4t = 9t - 2 - t$

45. $8n + 7 - 12n = 4n + 5 - 9n$ **46.** $-10x - 9 + 8x = -4x - 5 + 3x$

47. $2.6 + 7a + 5 = 8a - 5.6$ **48.** $9c + 4.8 = 7.5 + 4.8 + 8c$

49. $4 + 3y + y - 12 = 5y + 9 - y - 17$ **50.** $-9 - 4v - 1 + v = -2v + 5 - v - 15$

51. $7.3 - 0.2x + 1.3 - 0.6x = 12 - 0.8x - 3.6$ **52.** $2.5y - 3.4 - 1.2y = 6.7 - 9.1 + 1.3y$

53. $8(h + 6) - 7h = 14 - 8$ **54.** $19 - 3(m + 4) + 4m = 42$

55. $5 - \dfrac{1}{2}(4x + 6) = -3x - 8$ **56.** $\dfrac{2}{3}(6b - 9) + 6 = 3b - 21$

57. $5y - (3y + 7) = y + 9$ **58.** $-15 - 2x = 16 - (3x - 9)$

59. $3(3x - 5) - 2(4x - 3) = 5 - 10$ **60.** $5(5x - 3) - 6(4x - 2) = 12 - 15$

61. $1.5(x + 8) - (6.2 + 0.5x) = -7 + 4$ **62.** $0.5(3.8x - 6.2) - (0.9x - 4) = 2.9 - 4.7$

63. $20z + \dfrac{2}{3}(6z - 48) - 9z = 14z + 0.2(50z - 250) - 9z$

64. $6b - 1.5(8 + 2b) + 4 = 6b - \dfrac{1}{8}(24b + 48)$

65. $8(2m - 4) + 20(m - 5) = 4m + 5(5m - 20) - 32 + 7m$

66. $-3(2x + 5) + 8(x + 2) - 7 = 6(x - 5) - 4(x - 6)$

For Exercises 67–78, translate to an equation, then solve.

67. Latonia is planning to buy a new car. The down payment is $2373. She has $1947 saved. How much more does she need?

68. Kent owes $12,412 on his Visa card. What payment should he make to get the balance to −$10,500?

69. Robert knows the distance from his home to work is 42 miles. Unfortunately, he gets a flat tire 16 miles from work. How far did Robert drive before his flat tire?

70. A patient must receive 350 cc of a medication in three injections. He has received two injections at 110 cc each. How much should the third injection be?

71. Susan is playing Yahtzee. The "chance" score is found by adding the value shown on five dice. If she has a total of 23 on four of the dice, what does the fifth die need to be so that her score will be 28?

72. On May 16, Nikki's balance in her checking account was $1741.62. Afterwards, she writes four checks and doesn't record the amount of the last check. The first is for $16.82, the second is for $150.88, and the third is for $192.71. On May 21, she finds her balance to be $1286.65. If the checks were the only transactions that could have cleared the bank during that time, what was the amount of the fourth check?

73. The perimeter of the trapezoid shown is 67.2 centimeters. Find the length of the missing side.

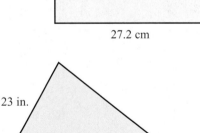

12.4 cm   16.3 cm

27.2 cm

74. The perimeter of the triangle shown is $84\frac{1}{2}$ inches. Find the length of the missing side.

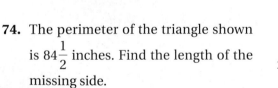

23 in.

$35\frac{1}{4}$ in.

75. In the following blueprint, what is the distance *x*?

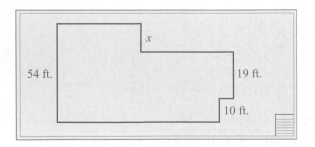

x

54 ft.   19 ft.

10 ft.

76. John sells scanning devices. The company quota is set at $8500 each month. The spreadsheet that follows shows John's sales as of the end of the second week of April. How much more does John need to sell to make the quota? Do you think John will make the quota? Why?

Date	Item no.	Quantity	Price per Unit
4/2	32072	1	$900
4/9	17032	3	$1500
4/13	48013	2	$1245

77. In a survey, respondents were given a statement, and they could agree, disagree, or have no opinion regarding the statement. The results indicate that $\frac{1}{3}$ of the respondents agree with the statement while $\frac{2}{5}$ disagree. What fraction had no opinion?

78. In a *Time*/CNN poll, $\frac{19}{50}$ of the respondents said that they felt that the increase in the number of divorces was due more to changes in women's attitude toward marriage, while $\frac{9}{50}$ said the increase was due more to changes in men's attitude toward marriage. The rest felt the increase was a result of both men and women equally. What fraction of the respondents felt the increase was due to men and women equally?

PUZZLE PROBLEM

In a study of an experimental medication, half of the participants are given the medication and half are given a placebo. The results indicate that $\frac{1}{3}$ of the participants had improvement in their condition, $\frac{1}{8}$ had no improvement in their condition but experienced known side effects of the medication, and the rest showed no effect at all. What fraction of the group that received the actual medication showed no discernible effects from the medication? (Hint: Assume all the participants that saw improvement in their condition or experienced known side effects took the medication.)

REVIEW EXERCISES

For Exercises 1–4, simplify.

[1.4] **1.** $4(-8.3)$

[1.4] **2.** $27 \div (-3)$

[1.5] **3.** $16 - (7 + 4) + 32$

[1.5] **4.** $\dfrac{20 + 3(2 + 4)}{-6(2) + 10}$

[1.7] **5.** Combine like terms:
$3x^2 - 6x + 5x^2 + x - 9$

[1.7] **6.** Use the distributive property to rewrite:
$4(3x - 5)$

2.3 The Multiplication Principle

OBJECTIVES

1. Solve linear equations using the multiplication principle.
2. Solve linear equations using both the addition and the multiplication principles.
3. Use the multiplication principle to clear fractions and decimals from equations.
4. Solve application problems.

OBJECTIVE 1. Solve linear equations using the multiplication principle. In Section 2.2, when we isolated the variable, the coefficient of that variable was always 1. What if we ended up with a coefficient other than 1? Look at the equation $6n - 11 = 4n + 3$. Using the methods of Section 2.2, we have:

$$6n - 11 = 4n + 3$$
$$\underline{-4n \qquad\qquad -4n}$$
$$2n - 11 = 0 + 3$$
$$\underline{+11 \qquad +11}$$

Note: *The coefficient is not 1.* ▶
$$2n + 0 = \qquad 14$$
$$2n = 14$$

How can we solve $2n = 14$? To isolate n we must clear the coefficient 2 so that we have $1n$, or simply n. Because coefficients multiply variables, we must undo that multiplication using its inverse operation, division. If we are going to divide, we must divide on both sides to keep the equation balanced. We write:

$$2n \div 2 = 14 \div 2 \text{ or the more popular form: } \frac{2n}{2} = \frac{14}{2}$$

Notice the coefficient 2 divides out.

> The 2s divide out, leaving $1n$, which can be simplified as n. $\longrightarrow$

$$\frac{\cancel{2}n}{\cancel{2}} = \frac{14}{2}$$
$$1n = 7$$
$$n = 7$$

You might remember in Chapter 1 we discussed that multiplicative inverses are numbers whose product is 1. Multiplying the coefficient by its multiplicative inverse would be another way to clear the coefficient. In the previous case, if we multiply the coefficient 2 by its multiplicative inverse $\frac{1}{2}$, we get the desired $1n$.

$$\frac{1}{2} \cdot 2n = 14 \cdot \frac{1}{2}$$

Note: *Throughout the rest of the book, we will omit this step of writing the resulting 1 coefficient.* ▶

$$\frac{1}{\cancel{2}} \cdot \frac{\overset{1}{\cancel{2}}n}{1} = \frac{\overset{7}{\cancel{14}}}{1} \cdot \frac{1}{\cancel{2}}$$
$$1n = 7$$
$$n = 7$$

> **Connection** Remember that when we divide by a fraction, we write an equivalent multiplication using the multiplicative inverse. Multiplying by the multiplicative inverse is the same as dividing.

If we manipulate an equation using the addition principle and we still have a coefficient other than 1, we can clear that coefficient by dividing (or multiplying by its multiplicative inverse). This is the main purpose of the *multiplication principle*. The multiplication principle says that we can multiply (or divide) both sides of an equation by the same amount without affecting its solution(s).

RULE **The Multiplication Principle of Equality**

If $a = b$, then $ac = bc$ is true for all real numbers a, b, and c, where $c \neq 0$.

PROCEDURE **Using the Multiplication Principle**

To use the multiplication principle of equality to clear a coefficient in an equation, multiply both sides of the equation by the multiplicative inverse of that coefficient, or divide both sides by the coefficient.

EXAMPLE 1 Solve and check. $-\dfrac{4}{5}m = \dfrac{6}{7}$

Solution $-\dfrac{4}{5}m = \dfrac{6}{7}$

$$-\frac{\overset{1}{\cancel{5}}}{\underset{1}{\cancel{4}}} \cdot -\frac{\overset{1}{\cancel{4}}}{\underset{1}{\cancel{5}}}m = \frac{\overset{3}{\cancel{6}}}{7} \cdot -\frac{5}{\underset{2}{\cancel{4}}}$$ Clear the coefficient $-\dfrac{4}{5}$ by multiplying both sides by its multiplicative inverse, $-\dfrac{5}{4}$.

$$m = -\frac{15}{14}$$

Check $-\dfrac{4}{5}m = \dfrac{6}{7}$

$$-\frac{\overset{2}{\cancel{4}}}{\underset{1}{\cancel{5}}}\left(-\frac{\overset{3}{\cancel{15}}}{\underset{7}{\cancel{14}}}\right) \overset{?}{=} \frac{6}{7}$$ Replace m in the original equation with $-\dfrac{15}{14}$ and verify that the equation is true.

$$\frac{6}{7} = \frac{6}{7}$$

True, therefore $-\dfrac{15}{14}$ is correct.

YOUR TURN Solve and check.

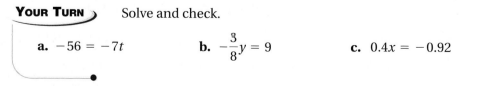

a. $-56 = -7t$

b. $-\dfrac{3}{8}y = 9$

c. $0.4x = -0.92$

OBJECTIVE 2. Solve linear equations using both the addition and multiplication principles. Now let's put the multiplication principle together with the addition principle. We follow the same outline as in Section 2.2, with one new step: using the multiplication principle to clear any remaining coefficient at the end.

PROCEDURE   Solving Linear Equations

To solve linear equations in one variable:

1. Simplify both sides of the equation as needed.
   a. Distribute to clear parentheses.
   b. Combine like terms.
2. Use the addition principle so that all variable terms are on one side of the equation and all constants are on the other side. (Clear the variable term with the lesser coefficient.) Then combine like terms.
3. Use the multiplication principle to clear the remaining coefficient.

EXAMPLE 2   Solve and check. $-3x - 11 = 7$

Solution Use the addition principle to separate the variable term and constant terms, then use the multiplication principle to clear the remaining coefficient.

Note: *Multiplying both sides by* $-\dfrac{1}{3}$

would also work:

$$\left(-\frac{1}{\overset{1}{\cancel{3}}}\right)\left(-\frac{\overset{1}{\cancel{3}}x}{1}\right) = \left(\frac{\overset{6}{\cancel{18}}}{1}\right)\left(-\frac{1}{\overset{1}{\cancel{3}}}\right)$$

$$x = -6$$

$$
\begin{array}{rl}
-3x - 11 = & 7 \\
\underline{+11 \quad +11} & \\
-3x + 0 = & 18 \\
\end{array}
$$ Add 11 to both sides to isolate the $-3x$ term.

$$
\begin{array}{rl}
\dfrac{-3x}{-3} = & \dfrac{18}{-3} \\
x = & -6
\end{array}
$$ Divide both sides by -3 to clear the -3 coefficient.

Check

$$-3x - 11 = 7$$
$$-3(-6) - 11 \overset{?}{=} 7$$ Replace x in the original equation with -6 and
$$18 - 11 \overset{?}{=} 7$$ verify that the equation is true.
$$7 = 7$$

True, therefore -6 is correct.

YOUR TURN   Solve and check.

a. $6y - 19 = -22$ **b.** $-12 = 20 - 8t$

Solving Equations with Variable Terms on Both Sides

Recall that when variable terms appear on both sides of the equal sign, we use the addition principle to get the variable terms on one side of the equal sign and the constant terms on the other side.

a. $-\dfrac{1}{2}$ **b.** 4

EXAMPLE 3) Solve and check. $8y - 5 = 2y - 29$

Solution Use the addition principle to get the variable terms on one side of the equation and the constant terms on the other side, then use the multiplication principle to clear the remaining coefficient.

$$8y - 5 = 2y - 29 \qquad \text{Subtract } 2y \text{ from both sides (} 2y \text{ has the lesser coefficient).}$$
$$\underline{-2y \qquad\;\; -2y}$$
$$6y - 5 = 0 - 29$$
$$6y - 5 = -29$$
$$\underline{+5 \qquad +5} \qquad \text{Add 5 to both sides to isolate the } 6y \text{ term.}$$
$$6y + 0 = -24$$
$$\frac{6y}{6} = \frac{-24}{6} \qquad \text{Divide both sides by 6 to clear the 6 coefficient.}$$
$$y = -4$$

Check $\quad 8y - 5 = 2y - 29$
$$8(-4) - 5 \stackrel{?}{=} 2(-4) - 29 \qquad \text{Replace } y \text{ in the original equation with } -4 \text{ and verify that the equation is true.}$$
$$-32 - 5 \stackrel{?}{=} -8 - 29$$
$$-37 = -37$$

True, therefore -4 is correct.

Simplifying First

Recall that when an equation contains parentheses or like terms that appear on the same side of the equal sign, we first clear the parentheses and combine like terms. Then we use the addition principle to separate the variable terms and constant terms.

EXAMPLE 4) Solve and check. $17 - (4n - 5) = 9n - 3(n + 7)$

Solution $17 - (4n - 5) = 9n - 3(n + 7)$
$$17 - 4n + 5 = 9n - 3n - 21 \qquad \text{Distribute to clear parentheses.}$$
$$22 - 4n = 6n - 21 \qquad \text{Combine like terms.}$$
$$22 - 4n = 6n - 21$$
$$\underline{+4n \quad +4n} \qquad \text{Add } 4n \text{ to both sides (} -4n \text{ has the lesser coefficient).}$$
$$22 + 0 = 10n - 21$$
$$22 = 10n - 21$$
$$\underline{+21 = \qquad +21} \qquad \text{Add 21 to both sides to isolate } 10n.$$
$$43 = 10n + 0$$
$$\frac{43}{10} = \frac{10n}{10} \qquad \text{Divide both sides by 10 to clear the 10 coefficient.}$$
$$4.3 = n$$

Check
$$17 - (4n - 5) = 9n - 3(n + 7)$$
$$17 - (4(4.3) - 5) \overset{?}{=} 9(4.3) - 3(4.3 + 7)$$
$$17 - (17.2 - 5) \overset{?}{=} 38.7 - 3(11.3)$$
$$17 - 12.2 \overset{?}{=} 38.7 - 33.9$$
$$4.8 = 4.8$$

Replace *n* in the original equation with 4.3 and verify that the equation is true.

True, therefore 4.3 is correct.

YOUR TURN Solve and check.

a. $3t - 10 + 9t = 17 + 4t - 3$ **b.** $13 - 6(x + 2) = 3x - (11x + 5)$

OBJECTIVE 3. Use the multiplication principle to clear fractions and decimals from equations. To clear fractions and decimals from an equation, we can use the multiplication principle of equality. Though equations can be solved without clearing the fractions or decimals, most people find equations that contain only integers easier to solve.

Clearing Fractions in an Equation

If the equation contains fractions, we multiply both sides by a number that will clear all the denominators. We could multiply both sides by any multiple of the denominators; however, using the LCD (least common denominator) results in the simplest equations.

EXAMPLE 5 Solve and check. $\dfrac{1}{3}x - \dfrac{3}{4} = \dfrac{5}{6}x + 2$

Solution Clear the fractions by multiplying through by the LCD. The LCD of 3, 4, and 6 is 12.

Note: *We have chosen to clear the fractions in our first step. Remember, you can clear the fractions at any step in solving the equation, or not at all.*

$$12\left(\frac{1}{3}x - \frac{3}{4}\right) = \left(\frac{5}{6}x + 2\right)12$$

Clear the fractions by multiplying both sides by the LCD, 12.

$$\frac{\overset{4}{\cancel{12}}}{1} \cdot \frac{1}{\underset{1}{\cancel{3}}}x - \frac{\overset{3}{\cancel{12}}}{1} \cdot \frac{3}{\underset{1}{\cancel{4}}} = \frac{\overset{2}{\cancel{12}}}{1} \cdot \frac{5}{\underset{1}{\cancel{6}}}x + 12 \cdot 2$$

Distribute 12, then divide out the denominators.

$$4x - 9 = 10x + 24$$
$$4x - 9 = 10x + 24$$
$$\underline{-4x \qquad\qquad -4x}$$

Subtract 4x from both sides.

$$0 - 9 = 6x + 24$$
$$-9 = 6x + 24$$
$$\underline{-24 \qquad\qquad -24}$$

Subtract 24 from both sides.

$$-33 = 6x + 0$$

$$\frac{-33}{6} = \frac{6x}{6}$$

Divide both sides by 6 to clear the coefficient.

$$-\frac{11}{2} = x$$

Simplify.

ANSWERS

a. 3 **b.** −3

136 **CHAPTER 2** Solving Linear Equations and Inequalities

Check

$$\frac{1}{3}x - \frac{3}{4} = \frac{5}{6}x + 2$$

$$\frac{1}{3}\left(-\frac{11}{2}\right) - \frac{3}{4} \overset{?}{=} \frac{5}{6}\left(-\frac{11}{2}\right) + 2$$

Replace x in the original equation with $-\frac{11}{2}$ and verify that the equation is true.

$$-\frac{11}{6} - \frac{3}{4} \overset{?}{=} -\frac{55}{12} + 2$$

$$-\frac{22}{12} - \frac{9}{12} \overset{?}{=} -\frac{55}{12} + \frac{24}{12}$$

Write equivalent fractions with their LCD.

$$-\frac{31}{12} = -\frac{31}{12}$$

True, therefore $-\frac{11}{2}$ is correct.

Clearing Decimals in an Equation

The multiplication principle can also be used to clear decimals in an equation. Because multiplying by a power of 10 will cause the decimal point to move to the right, we can clear decimals by multiplying both sides of the equation by an appropriate power of 10. The power of 10 we use depends on the decimal number with the most decimal places.

EXAMPLE 6 Solve and check. $0.2(y - 6) = 0.48y + 3$

Solution Decimals can be cleared at any step in the process of solving an equation. In this case, we will distribute to clear the parentheses first, then clear the decimals.

$$0.2y - 1.2 = 0.48y + 3$$

Distribute to clear parentheses.

$$100(0.2y - 1.2) = (0.48y + 3)100$$

Note: *The number 0.48 has two decimal places, which is more decimal places than any of the other decimal numbers, so we multiply both sides by 100 to clear the decimals.*

$$20y - 120 = 48y + 300$$

$$20y - 120 = 48y + 300$$

$$\underline{-20y \qquad\quad -20y}$$

$$0 - 120 = 28y + 300$$

$$-120 = 28y + 300$$

$$\underline{-300 \qquad\quad -300}$$

$$-420 = 28y + 0$$

$$\frac{-420}{28} = \frac{28y}{28}$$

$$-15 = y$$

Connection Clearing decimals and fractions uses the same process. Decimal numbers represent fractions with denominators that are powers of 10. For example, $0.48 = \frac{48}{100}$ and $0.2 = \frac{2}{10}$. The LCD for these fractions is 100.

Check $0.2(y - 6) = 0.48y + 3$

$0.2(-15 - 6) \overset{?}{=} 0.48(-15) + 3$ **Replace *y* in the original equation with**
$0.2(-21) \overset{?}{=} -7.2 + 3$ **−15 and verify that the equation is true.**
$-4.2 = -4.2$

True, therefore -15 is correct.

We can amend the outline for solving equations to include clearing fractions and decimals. Remember that this process is optional.

PROCEDURE   **Solving Linear Equations**

To solve linear equations in one variable:

1. Simplify both sides of the equation as needed.
   a. Distribute to clear parentheses.
   b. Clear fractions or decimals by multiplying through by the LCD. In the case of decimals, the LCD is the power of 10 with the same number of zero digits as decimal places in the number with the most decimal places. (Clearing fractions and decimals is optional.)
   c. Combine like terms.
2. Use the addition principle so that all variable terms are on one side of the equation and all constants are on the other side. (Clear the variable term with the lesser coefficient.) Then combine like terms.
3. Use the multiplication principle to clear the remaining coefficient.

YOUR TURN   Solve and check.

a. $\dfrac{1}{3}(y - 2) = \dfrac{3}{5}y + \dfrac{1}{6}$ **b.** $2.5n - 1.04 = 0.15n - 0.1$

OBJECTIVE 4. Solve application problems.   In Section 2.1, we began solving problems using formulas. Recall that to use a formula, we replace the variables with the corresponding given numbers, then solve for the unknown amount.

EXAMPLE 7   The perimeter of the figure shown is 188 feet. Find the width and length.

Understand   The width is represented by w and the length is represented by $w + 7$. We are given the perimeter, so we can use the formula $P = 2l + 2w$.

Plan   In the perimeter formula, replace P with **188**, l with **$w + 7$**, and solve for w.

ANSWERS

a. $-\dfrac{25}{8}$ **b.** 0.4

Execute

$$P = 2l + 2w$$
$$188 = 2(w + 7) + 2w$$
$$188 = 2w + 14 + 2w \qquad \text{Distribute.}$$
$$188 = 4w + 14 \qquad \text{Combine like terms.}$$
$$188 = 4w + 14$$
$$\underline{-14 \qquad\quad -14} \qquad \text{Subtract 14 from both sides.}$$
$$174 = 4w + 0$$
$$\frac{174}{4} = \frac{4w}{4} \qquad \text{Divide both sides by 4.}$$
$$43.5 = w$$

Answer The width is 43.5 feet. To find the length, we evaluate the expression that represents the length, $w + 7$, with $w = 43.5$.

$$\text{Length} = 43.5 + 7 = 50.5 \text{ feet}$$

Check First, verify that the length is 7 more than the width: $50.5 = 7 + 43.5$. Second, verify that a rectangle with a length of 50.5 feet and a width of 43.5 feet has a perimeter of 188 feet.

$$P = 2l + 2w$$
$$P = 2(50.5) + 2(43.5)$$
$$P = 101 + 87$$
$$P = 188$$

It checks.

EXAMPLE 8 The total material allotted for the construction of a metal box is 4754 square inches. The length is to be 30 inches and the width is to be 28 inches. Find the height.

Understand The material will be used to create the outer shell that is the box, which means that 4754 square inches is the surface area of the box. The formula for the surface area of a box is $SA = 2lw + 2lh + 2wh$.

Plan Replace SA with **4754**, l with **30**, w with **28**, and solve for h.

Execute

$$SA = 2lw + 2lh + 2wh$$
$$4754 = 2(\mathbf{30})(\mathbf{28}) + 2(\mathbf{30})h + 2(\mathbf{28})h$$
$$4754 = 1680 + 60h + 56h \qquad \text{Simplify.}$$
$$4754 = 1680 + 116h \qquad \text{Combine like terms.}$$
$$4754 = 1680 + 116h$$
$$\underline{-1680 \quad -1680} \qquad \text{Subtract 1680 from both sides.}$$
$$3074 = \quad 0 + 116h$$
$$\frac{3074}{116} = \frac{116h}{116} \qquad \text{Divide both sides by 116.}$$
$$26.5 = h$$

Answer The height must be 26.5 inches.

Check Does a 30-inch by 28-inch by 26.5-inch box have a surface area of 4754 square inches?

$$4754 \stackrel{?}{=} 2(30)(28) + 2(30)(26.5) + 2(28)(26.5)$$
$$4754 \stackrel{?}{=} \quad 1680 \quad + \quad 1590 \quad + \quad 1484$$
$$4754 = 4754$$

It checks.

Your Turn The desired surface area for a box is 98 square feet. If the length of the box is to be 4 feet and the height is to be 6 feet, what will the width be?

Answer

2.5 ft.

2.3 Exercises

For Extra Help MyMathLab | Videotape/DVT | InterAct Math | Math Tutor Center | Math XL.com

1. In your own words, explain what the multiplication principle of equality says.

2. How do you use the multiplication principle to clear a coefficient that is an integer?

3. How do you use the multiplication principle to clear a coefficient that is a fraction?

4. To clear a negative coefficient, do you multiply/divide by a positive or negative number? Why?

5. If an equation contains fractions, how do you transform the equation so that it contains only integers?

6. If an equation contains decimal numbers, how do you transform the equation so that it contains only integers?

For Exercises 7–18, solve and check.

7. $3x = 12$

8. $8x = -24$

9. $-6x = -18$

10. $-5y = 20$

11. $\dfrac{n}{2} = 8$

12. $\dfrac{t}{3} = -4$

13. $\dfrac{3}{4}y = -15$

14. $-\dfrac{5}{6}x = 20$

15. $\dfrac{4}{7}t = -\dfrac{2}{3}$

16. $\dfrac{3}{8}a = \dfrac{5}{6}$

17. $-\dfrac{2}{5}t = \dfrac{8}{15}$

18. $-\dfrac{7}{9}t = -\dfrac{5}{12}$

For Exercises 19–34, solve and check.

19. $5n - 2n = 21$

20. $7t - 3t = 20$

21. $4x + 1 = 21$

22. $3x - 8 = 10$

23. $9 - 2n = 12$

24. $1 - 7y = -8$

25. $\dfrac{5}{8}x - 2 = 8$

26. $7 = \dfrac{3}{4}x + 13$

27. $2(x - 3) = -6$

28. $4(5x + 7) = 13$

29. $3a - 2a + 7 + 6a = -28$

30. $c + 2c + 3 + 4c = 24$

31. $-12b + 4(2b - 3) = 16$

32. $2x - 6(x + 8) = -12$

33. $2(y - 33) + 3(y - 2) = 8$

34. $4(r - 8) + 2(r + 3) = -8$

For Exercises 35–52, solve and check.

35. $8x + 5 = 2x + 17$

36. $10t + 1 = 6t + 13$

37. $5y - 7 = 2y + 13$

38. $9m + 1 = 3m - 14$

39. $9 - 4k = 15 - k$

40. $6 - 12m = -20m + 22$

41. $9x + 12 - 3x - 2 = x + 8 + 5x$

42. $12 - 6r - 14 = 9 - 4r - 7 - 2r$

43. $14(w - 2) + 13 = 4w + 5$

44. $2x + 2(3x - 4) = -23 + 5x$

45. $2n - (6n + 5) = 9 - (2n - 1)$

46. $2 - (17 - 5m) = 9m - (m + 7)$

47. $4 - (6x + 5) = 7 - 2(3x + 4)$

48. $-6 - (3z - 2) = 5z - 4(2z + 1)$

49. $5(a - 1) - 9(a - 2) = -3(2a + 1) - 2$

50. $4(k + 4) + 13(k - 1) = 3(k - 1) + 20$

51. $3(2x - 5) - 4x = 2(x - 3) - 9$

52. $4(2x - 1) - 3(x + 5) = 5(x - 2) + 7$

For Exercises 53–72, use the multiplication principle of equality to eliminate the fractions or decimals, then solve and check.

53. $\dfrac{2}{3}n - 1 = \dfrac{1}{4}$

54. $3 - \dfrac{5}{8}n = \dfrac{3}{2}$

55. $\dfrac{3}{4}n - 2 = -\dfrac{7}{8}n + 4$

56. $-\dfrac{2}{5}t + 1 = \dfrac{3}{10}t - 3$

57. $\dfrac{3}{4}x + \dfrac{5}{6} = \dfrac{1}{4} + \dfrac{4}{3}x$

58. $\dfrac{7}{9}w + 3w = \dfrac{7}{2}w + \dfrac{5}{9}$

59. $\dfrac{1}{2}(y + 5) = \dfrac{3}{2} - y$

60. $\dfrac{2}{3}(x - 4) = \dfrac{4}{3} + 2x$

61. $\dfrac{1}{6}(m - 4) = \dfrac{2}{3}(m + 1) - \dfrac{1}{3}m$

62. $\dfrac{1}{5}(y - 3) = \dfrac{3}{10}(y + 5) - \dfrac{2}{5}y$

63. $-2.9u + 3.6u = 6.3$

64. $-4.6z + 2.2z = 4.8$

65. $0.2 - 1.3t - 0.7t = 1 - 2.9t + 0.9t$

66. $4.2y - 8.2 + 2.3y = 0.9 + 6.5y - 9.1$

67. $0.3(a - 18) = 0.9a$

68. $0.6(w - 12) = 0.2w$

69. $0.08(15) + 0.4x = 0.05(36 + 7x)$

70. $0.06(25) + 0.27x = 0.3(4 + x)$

71. $9 - 0.2(v - 8) = 0.3(6v - 8)$

72. $0.5 - (t - 6) = 10.4 - 0.4(1 - t)$

For Exercises 73–76, the check indicates a mistake was made. Find and correct the mistake.

73.

$$6x - 11 = 8x - 15$$

$$\underline{-6x \qquad\quad -6x}$$

$$11 = 2x - 15$$

$$\underline{+15 \qquad\quad +15}$$

$$26 = 2x$$

$$\frac{26}{2} = \frac{2x}{2}$$

$$13 = x$$

Check: $6x - 11 = 8x - 15$

$$6(13) - 11 \stackrel{?}{=} 8(13) - 15$$

$$78 - 11 \stackrel{?}{=} 104 - 15$$

$$67 \neq 89$$

74.

$$6n + 1 = -2n + 21$$

$$\underline{\quad -1 \qquad\qquad -1}$$

$$6n + 0 = -2n + 20$$

$$6n = -2n + 20$$

$$\underline{+2n \qquad\; +2n}$$

$$8n = 0 + 20$$

$$\frac{8n}{8} = \frac{20}{8}$$

$$n = \frac{5}{2}$$

Check: $6n + 1 = -2n + 21$

$$\overset{3}{\cancel{6}}\!\left(\frac{5}{\underset{1}{\cancel{2}}}\right) + 1 \stackrel{?}{=} -\overset{1}{\cancel{2}}\!\left(\frac{5}{\underset{1}{\cancel{2}}}\right) + 21$$

$$3 + 1 \stackrel{?}{=} -5 + 21$$

$$4 \neq 16$$

75.

$$4 - (x + 2) = 2x + 3(x - 8)$$

$$4 - x + 2 = 2x + 3x - 24$$

$$6 - x = 5x - 24$$

$$\underline{+x \quad +x}$$

$$6 + 0 = 6x - 24$$

$$6 = 6x - 24$$

$$\underline{+24 \qquad\quad +24}$$

$$30 = 6x + 0$$

$$\frac{30}{6} = \frac{6x}{6}$$

$$5 = x$$

Check: $4 - (x + 2) = 2x + 3(x - 8)$

$$4 - (5 + 2) \stackrel{?}{=} 2(5) + 3(5 - 8)$$

$$4 - 7 \stackrel{?}{=} 10 + 3(-3)$$

$$-3 \stackrel{?}{=} 10 + (-9)$$

$$-3 \neq 1$$

76. $\frac{1}{2}n - 3 = \frac{2}{3}n + \frac{1}{4}$

$\frac{\cancel{12}^{6}}{1} \cdot \frac{1}{\cancel{2}_{1}}n - 3 = \frac{\cancel{12}^{4}}{1} \cdot \frac{2}{\cancel{3}_{1}}n + \frac{\cancel{12}^{3}}{1} \cdot \frac{1}{\cancel{4}_{1}}$

$6n - 3 = 8n + 3$

$\underline{-6n \qquad -6n}$

$0 - 3 = 2n + 3$

$\underline{-3 \qquad -3}$

$-6 = 2n + 0$

$\frac{-6}{2} = \frac{2n}{2}$

$-3 = n$

Check: $\frac{1}{2}n - 3 = \frac{2}{3}n + \frac{1}{4}$

$\frac{1}{2}\left(-\frac{3}{1}\right) - 3 \stackrel{?}{=} \frac{2}{\cancel{3}_{1}}\left(-\frac{\cancel{3}^{1}}{1}\right) + \frac{1}{4}$

$-\frac{3}{2} - 3 \stackrel{?}{=} -2 + \frac{1}{4}$

$-\frac{3}{2} - \frac{6}{2} \stackrel{?}{=} -\frac{8}{4} + \frac{1}{4}$

$-\frac{9}{2} \neq -\frac{7}{4}$

For Exercises 77–90, solve for the unknown amount.

77. The area of a rectangular deck is known to be 140 square feet. Find the length if the width must be 10 feet. (Use $A = lw$)

78. A crate manufacturer receives an order for a crate with a volume of 128 cubic feet. The crate must be 5 feet, 4 inches by 6 feet. What must be the crate's height? (Use $V = lwh$)

79. The area of the metal plate shown in the figure is 45.6 square centimeters. What is the length of the side labeled h? (Use $A = \frac{1}{2}bh$)

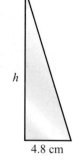
h

4.8 cm

80. The Worthy family is planning a 465-mile trip. If they travel at an average speed of 62 miles per hour, what will be their travel time? (Use $d = rt$)

81. A fence is to be installed around a rectangular field. The field's perimeter is 212 feet. Find the width if the length of the field is 54 feet. (Use $P = 2l + 2w$)

82. The base of a building has a perimeter of 241 feet. The length is 16 feet more than the width, as shown in the figure. Find the dimensions of the building.

w
$w + 16$

83. The area of the trapezoidal plot of land shown in the figure is 7500 square feet. What is the length of the side labeled b? (Use $A = \frac{1}{2}h(a + b)$)

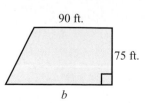
90 ft.
75 ft.
b

84. The area of the L-shaped room shown is 321 square feet. Find all the missing dimensions.

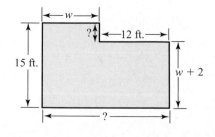
w
?
12 ft.
15 ft.
$w + 2$
?

144 **CHAPTER 2** Solving Linear Equations and Inequalities

85. A small storage box is designed so that its surface area is 106 square inches. If the length is 7.5 inches and the width is 4 inches, find the height. (Use $SA = 2lw + 2lh + 2wh$)

86. Jerry has 1992 square inches of wood to build a chest for storing musical equipment. He wants the length to be 22 inches and the width to be 18 inches. What will be the height? (Use $SA = 2lw + 2lh + 2wh$)

87. The circumference of Earth along the equator is approximately 40,053.84 kilometers. What is the equatorial radius? (Use $C = 2\pi r$)

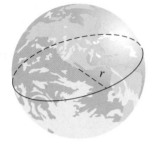

88. A giant sequoia is measured to have a circumference of 26.4 meters. What is the diameter? (Use $C = \pi d$)

> **Of Interest**
> The giant sequoia tree is found along the western slopes of the Sierra Nevada range. These trees reach heights of nearly 100 meters and some are over 4000 years old.

89. An engineer is designing a chemical storage tank in the shape of a cylinder. The volume of the tank is to be 60 cubic meters and the radius is to be 2 meters. Find the height. (Use $V = \pi r^2 h$)

90. The liquid in a standard soda can occupies a volume of 355 cubic centimeters. If the diameter is 6 centimeters, find the height of the liquid inside the can. (Use $V = \pi r^2 h$)

The formula $V = iR$ describes the voltage, V, in a circuit when the current, i, passes through a resistance, R. Voltage is measured in volts (V), current is measured in amps (A), and resistance is measured in ohms (Ω). Use the formula for voltage to solve Exercises 91 and 92.

91. The voltage in a circuit is measured to be -76 V. If the resistance is 8 Ω, find the current.

92. A technician applies 15 V to a circuit. The current is measured to be 2.5 A. What is the resistance?

The formula $F = ma$ describes the force that an object with mass m experiences when accelerated an amount a. If acceleration is measured in meters per second squared ($m/sec.^2$), and the mass is measured in kilograms (kg), then the force is in units of newtons (N). Use the force formula to solve Exercises 93–96.

93. A car with a mass of 1200 kilograms exerts a force of 4800 newtons. Find the acceleration.

> **Of Interest**
> Earth's gravity accelerates all objects at the same rate, -9.8 m/sec.^{2} (or -32.2 ft./sec.^{2}). The accelerations are negative because the direction is downward.

94. An object weighs 44.1 newtons, that is, it exerts a downward force of -44.1 newtons. The object is dropped so that it accelerates at -9.8 meters per second squared. Find the mass of the object.

95. The Liberty Bell weighs 9265.9 newtons, so its downward force is -9265.9 newtons. If the acceleration due to gravity is about -9.8 meters per second squared, what is the mass of the bell?

96. Jacob weighs 160 pounds (a downward force of -160 pounds). If the acceleration due to gravity is about -32.2 feet per second squared, what is his mass? (The mass will be in terms of the unit *slugs*.)

97. The formula $w = \dfrac{s}{f}$ describes the wavelength w of a musical note if f is the frequency and s is the speed of sound. A note is sounded with a wavelength of 0.94 meters/cycle and a frequency of 440 Hertz. Find the speed of sound.

98. The formula $C = 0.3m + 25$ describes the total cost of using a digital phone over 100 minutes, where m represents the number of minutes beyond 100 minutes. Pam's bill shows that she used the phone for 187 minutes. What is the total cost of her use of the phone?

99. A rental car agency rents a van for a weekend using the formula $C = 0.27m + 200$, where m represents the number of miles. Laura rents a van from the company over a weekend that the company is offering 100 free miles. If she returns the van and has driven 421 miles, what is the total cost of the rental?

100. The formula $d = v_i t + \dfrac{1}{2}at^2$ describes the distance, d, an object travels if the object has an initial velocity (speed) v_i, an acceleration of a, for a time t. An object travels 135 feet after being accelerated for 3 seconds with an acceleration of 20 feet per second squared. Find the object's initial velocity.

REVIEW EXERCISES

[1.5] **1.** Simplify: $-6 + 3(5^2 + 2) - (6 - 1)$

[1.7] **2.** Combine like terms:
$$-3.2x^2 + 0.4x + 7x^2 - 8.03x + 1.5$$

[1.7] **3.** Use the distributive property to rewrite:
$$-\frac{3}{4}(6x + 8)$$

For Exercises 4–6, solve and check.

[2.2] **4.** $6 + x = -14 + 2x$

[2.3] **5.** $-\dfrac{2}{5}x = 10$

[2.3] **6.** $2x + 3(x - 3) = 10x - (3x - 11)$

2.4 Applying the Principles to Formulas

OBJECTIVE 1. Isolate a variable in a formula using the addition and multiplication principles. So far when using formulas, we have substituted given numbers for all the variables except one. We then used the addition and multiplication principles of equality to isolate that variable. However, we can actually isolate a particular variable in a formula without being given numbers for the other variables. We follow the same steps as if we had been given numbers.

Consider the area formula, $A = lw$. If we were given values for A and l and asked to isolate w, or solve for w, we would divide both sides by the value of l. We can perform this manipulation symbolically.

$$\frac{A}{l} = \frac{\cancel{l}w}{\cancel{l}} \qquad \text{To isolate } w, \text{ we clear } l \text{ by dividing both sides by } l.$$

$$\frac{A}{l} = w$$

This development suggests the following procedure.

PROCEDURE   Isolating a Variable in a Formula

To isolate a particular variable in a formula:

1. Treat all other variables like constants.
2. Isolate the desired variable using the outline for solving equations.

Isolate a Variable Using the Addition Principle

Recall that the addition principle is used to clear terms that are added or subtracted.

EXAMPLE 1   Isolate R in the formula $P = R - C$.

Solution

$$\begin{array}{rcl} P &=& R - C \\ \underline{+\,C} & & \underline{+\,C} \\ P + C &=& R + 0 \\ P + C &=& R \end{array}$$

To isolate R we must clear C. Because C is subtracted from R, we add C to both sides.

◀ **Note:** *Because P and C are not like terms, they cannot be combined.*

Of Interest
The formula $P = R - C$ is used to calculate profit, given revenue, R, and cost, C.

YOUR TURN   Isolate v_i in the formula that describes the velocity of an object after being accelerated from an initial velocity (v_i), $v = v_i + at$.

ANSWER

$v - at = v_i$

Isolate a Variable Using the Multiplication Principle

Now let's use the multiplication principle to isolate a variable in a formula.

EXAMPLE 2 **a.** Isolate w in the formula for the volume of a box, $V = lwh$.

Solution

$$\frac{V}{lh} = \frac{\cancel{l}w\cancel{h}}{\cancel{l}\cancel{h}}$$

To isolate w, we must clear l and h. Because l and h are multiplying w, we divide both sides by l and h.

$$\frac{V}{lh} = w$$

b. Isolate P in the formula $B = P(1 + rt)$.

Solution

$$\frac{B}{1 + rt} = \frac{P\cancel{(1 + rt)}}{1 \cancel{+ rt}}$$

To isolate P, we must clear $(1 + rt)$. Since $(1 + rt)$ is multiplying P, we divide both sides by $1 + rt$.

$$\frac{B}{1 + rt} = P$$

Of Interest

The formula $B = P(1 + rt)$ is used in finance to calculate the balance for a simple-interest investment or loan. The variable B represents the final balance, P represents the principal, r represents the interest rate, and t represents time.

YOUR TURN

a. Isolate b in the formula for the area of a parallelogram, $A = bh$.

b. Isolate i in the voltage formula for a series circuit with two resistors, $V = i(R + r)$.

Isolate a Variable Using Both Principles

Now consider formulas that require the use of both principles to isolate the desired variable.

EXAMPLE 3 **a.** Isolate l in the perimeter formula for rectangles, $P = 2l + 2w$.

Solution First, use the addition principle to isolate the term that contains l. Then use the multiplication principle to clear the coefficient.

$$P = 2l + 2w$$
$$\underline{-2w \qquad\quad -2w}$$
$$P - 2w = 2l + 0$$

Isolate $2l$ by subtracting $2w$ from both sides.

$$\frac{P - 2w}{2} = \frac{2l}{2}$$

Isolate l by dividing both sides by 2.

$$\frac{P - 2w}{2} = l$$

ANSWERS

a. $\dfrac{A}{h} = b$ **b.** $\dfrac{V}{R + r} = i$

b. Isolate d in the formula from physics $v^2 = v_0^2 + \dfrac{2Fd}{m}$.

Solution First, use the addition principle to isolate the term that contains d. Then use the multiplication principle to clear the coefficient.

$$v^2 = v_0^2 + \frac{2Fd}{m} \qquad \text{Isolate } \frac{2Fd}{m} \text{ by subtracting } v_0^2 \text{ on both sides.}$$

$$\underline{ -v_0^2 \quad -v_0^2}$$

$$v^2 - v_0^2 = 0 + \frac{2Fd}{m}$$

$$v^2 - v_0^2 = \frac{2Fd}{m}$$

$$\frac{m}{2F} \cdot \frac{v^2 - v_0^2}{1} = \frac{\overset{1}{\cancel{m}}}{\underset{1}{\cancel{2F}}} \cdot \frac{\overset{1}{\cancel{2F}}d}{\underset{1}{\cancel{m}}} \qquad \text{Clear the coefficient } \frac{2F}{m} \text{ by multiplying both sides by its reciprocal.}$$

$$\frac{m(v^2 - v_0^2)}{2F} = d$$

> **Of Interest**
> In this example, the 0 subscript is used to indicate that v_0 is an initial velocity. We have also seen v_i used to indicate initial velocity. Both forms are used in physics and engineering.

YOUR TURN

a. Isolate t in the formula that describes the velocity of an object after being accelerated from an initial velocity, $v = v_i + at$.

b. Isolate n in the formula $T = \dfrac{nt}{r} - R$.

ANSWERS

a. $\dfrac{v - v_i}{a} = t$

b. $\dfrac{r}{t}(T + R) = n$ or $\dfrac{r(T + R)}{t} = n$

2.4 Exercises

For Extra Help   MyMathLab   Videotape/DVT   InterAct Math   Math Tutor Center   Math XL.com

1. What would you do to isolate w in the formula $l = w - 2$? Why?

2. What would you do to isolate p in the formula $C = np$? Why?

3. To isolate y in $2x + 3y = 8$, what would you do first? Why?

4. How can you verify that your answer is correct after isolating a variable in a formula?

For Exercises 5–24, solve for the indicated variable.

5. $t - 4u = v; t$

6. $x = a + 3y; a$

7. $-5y = x; y$

8. $2n = a; n$

9. $2x - 3 = b; x$

10. $3m + b = y; m$

11. $y = mx + b; m$

12. $ab + c = d; b$

13. $3x + 4y = 8; y$

14. $19 = 2l + 2w; w$

15. $\dfrac{mn}{4} - Y = f; Y$

16. $q = \dfrac{rs}{2} - p; p$

17. $6(c + 2d) = m - np; c$

18. $5(2n + a) = bn - c; a$

19. $\dfrac{x}{3} + \dfrac{y}{5} = 1; y$

20. $\dfrac{a}{4} + \dfrac{b}{6} = 3; a$

21. $\dfrac{3}{4} + 5a = \dfrac{n}{c}; n$

22. $\dfrac{x}{6} + 5y = \dfrac{m}{n}; m$

23. $t = \dfrac{A - P}{pr}; p$

24. $S = \dfrac{C}{1 - M}; M$

For Exercises 25–40, solve the geometric formula for the indicated variable.

25. $P = a + b + c; a$

26. $a + b + c = 180; b$

27. $A = lw; l$

28. $C = 2\pi r; r$

29. $\dfrac{C}{d} = \pi; d$

30. $\dfrac{A}{b} = h; b$

31. $V = \dfrac{1}{3}\pi r^2 h; r^2$

32. $V = \dfrac{4}{3}\pi r^3; r^3$

33. $A = \dfrac{1}{2}bh; b$

34. $S = \dfrac{1}{2}gt^2; t^2$

35. $P = 2l + 2w; w$

36. $S = 2\pi r^2 - 2\pi rh; h$

37. $S = \dfrac{n}{2}(a + l); l$

38. $A = \dfrac{h}{2}(a + b); b$

39. $V = h(\pi r^2 - lw); w$

40. $V = h\left(\dfrac{1}{3}\pi r^2 + lw\right); l$

For Exercises 41–48, solve the financial formula for the indicated variable.

41. $P = R - C; C$

42. $P = R - C; R$

43. $I = Prt; r$

44. $I = Prt; t$

45. $P = \dfrac{C}{n}; C$

46. $P = \dfrac{C}{n}; n$

47. $A = P + Prt; r$

48. $A = P + Prt; t$

For Exercises 49–60, solve the physics formula for the indicated variable.

49. $d = rt$; t **50.** $F = ma$; m **51.** $W = Fd$; d **52.** $K = \dfrac{1}{2}mv^2$; m

53. $P = \dfrac{W}{t}$; t **54.** $D = \dfrac{M}{V}$; V **55.** $v = -32t + v_0$; t **56.** $x = x_0 + vt$; v

57. $F = \dfrac{9}{5}C + 32$; C **58.** $C = \dfrac{5}{9}(F - 32)$; F **59.** $F = G\dfrac{Mm}{R^2}$; m **60.** $\dfrac{P_1 V_1}{T_1} = \dfrac{P_2 V_2}{T_2}$; V_2

For Exercises 61–64, find and explain the mistake in each solution.

61. $3n + 7t = 54$; isolate t

$$3n + 7t = 54$$
$$\underline{-3n \qquad\quad -3n}$$
$$7t = 54 - 3n$$
$$7t = 54 - 3n$$
$$\underline{-7 \quad\; -7}$$
$$t = 47 - 3n$$

62. $\dfrac{1}{4}kt = r$; isolate t

$$\frac{1}{4}kt = r$$
$$\frac{4}{1} \cdot \frac{1}{4}kt = 4r$$
$$\frac{1}{k} \cdot kt = 4r\frac{k}{1}$$
$$t = 4kr$$

63. $\dfrac{3m - 2}{5} = nk$; isolate m

$$\frac{3m - 2}{5} = nk$$
$$\frac{5}{1} \cdot \frac{3m - 2}{5} = nk \cdot 5$$
$$3m - 10 = 5nk$$
$$3m - 10 = 5nk$$
$$\underline{+10 \;\; +10}$$
$$3m = 5nk + 10$$
$$\frac{3m}{3} = \frac{5nk + 10}{3}$$
$$m = \frac{5nk + 10}{3}$$

64. $7(y - 5) = xv$; isolate y

$$7(y - 5) = xv$$
$$7y - 5 = xv$$
$$7y - 5 = xv$$
$$\underline{+5 \qquad +5}$$
$$\frac{7y}{7} = \frac{xv + 5}{7}$$
$$y = \frac{xv + 5}{7}$$

Collaborative Exercises WHERE DOES SPEEDING GET YOU?

1. The formula that relates distance, rate, and time is $d = rt$. Solve this formula for t.

2. Use the formula you found in Problem 1 to determine how long it would take you to travel 20 miles (an average daily commute) at a rate of 55 miles per hour. Note that your result will be in hours. Convert the time to minutes by multiplying by 60 and round to the nearest tenth of a minute.

3. Now use the formula from Problem **1** to determine how long it would take you to travel 20 miles at 65 miles per hour. Again, convert the time to minutes rounded to the nearest tenth of a minute.

4. How much time does speeding save you? To find this, subtract the time found in Problem 3 from the time found in Problem 2.

5. Discuss in your group whether the additional time is worth the additional risk that comes with increasing the speed by 10 miles per hour.

6. Complete the table, rounding all values to the nearest tenth. (You've already calculated the values for the 20-mile trip.)

Trip Distance	Time at 55 mph	Time at 65 mph	Time Saved
20 miles			
30 miles			
40 miles			

7. Is the time saved on longer trips worth the additional risk of increasing speed by 10 miles per hour? Why or why not? Does the longer distance also increase risk? Why or why not?

REVIEW EXERCISES

[1.1]   1. Use $<$, $>$, or $=$ to make a true sentence:
$-7 \quad -9$

[1.5] *For Exercises 2–4, evaluate.*

2. $(0.6)^2$

3. -4^2

4. $-|-6|$

[1.6] *For Exercises 5 and 6, translate the phrase to an expression.*

5. four more than seven times a number

6. nine less than three times the sum of a number and two

2.5 Translating Word Sentences to Equations

OBJECTIVE 1. Translate sentences to equations using key words, then solve. In previous sections, we have translated phrases to expressions (Section 1.6) and solved problems using formulas (Section 2.1). In this section, we will further develop problem-solving strategies by translating sentences to equations. Since we have not learned how to solve equations containing exponents and roots, we will explore translations that involve only addition, subtraction, multiplication, and division. See the following table, which we first saw in Section 1.6.

Key Words and Their Translations

Addition	Translation		Subtraction	Translation
The sum of x and 3	$x + 3$		The difference of x and 3	$x - 3$
h plus k	$h + k$		h minus k	$h - k$
7 added to t	$t + 7$		7 subtracted from t	$t - 7$
3 more than a number	$n + 3$		3 less than a number	$n - 3$
y increased by 2	$y + 2$		y decreased by 2	$y - 2$

Note: *Since addition is a commutative operation, it does not matter in what order we write the translation.*

For "the sum of x and 3" we can write $x + 3$ or $3 + x$.

Note: *Subtraction is not a commutative operation. Therefore, the way we write the translation matters. We must translate each key phrase exactly as presented. Particularly note how* less than *and* subtracted from *are translated.*

Multiplication	Translation		Division	Translation
The product of x and 3	$3x$		The quotient of x and 3	$x \div 3$
h times k	hk		h divided by k	$h \div k$
Twice a number	$2n$		h divided into k	$k \div h$
Triple the number	$3n$		The ratio of a to b	$a \div b$
$\frac{2}{3}$ of a number	$\frac{2}{3}n$			

Note: *Like addition, multiplication is a commutative operation. This means we can write the translation order any way we wish.*

"h times k" can be hk or kh.

Note: *Division is like subtraction in that it is not a commutative operation. We must translate division phrases exactly as presented. Particularly note how* divided into *is translated.*

In order to translate a sentence to an equation, you must also know the key words that indicate an equal sign.

Key words for an equal sign:	is equal to	is	yields
	is the same as	produces	results in

PROCEDURE   Translating Word Sentences

To translate a word sentence to an equation, identify the variable(s), constants, and key words, then write the corresponding symbolic form.

Problems Involving Addition or Subtraction

EXAMPLE 1) Translate to an equation, and then solve.

a. The sum of twenty-five and a number is equal to fourteen. Translate to an equation, then solve for the number.

Understand The key word *sum* indicates addition, *is equal to* indicates an equal sign, and *a number* indicates a variable.

Plan Translate the key words to an equation, and then solve the equation. We will use *n* as the variable.

Execute Translate: The sum of twenty-five and a number is equal to fourteen.

$$25 + n = 14$$

Solve: $25 + n = 14$

$$\underline{-25 \qquad -25}$$ **Subtract 25 from both sides to isolate *n*.**

$$0 + n = -11$$

Answer $n = -11$

Check In the original sentence, replace the unknown amount with -11 and verify that it makes the sentence true. Verify that the sum of twenty-five and a negative eleven is equal to fourteen.

$$25 + (-11) \overset{?}{=} 14$$
$$14 = 14$$

Yes, the sum of 25 and -11 is equal to 14.

b. Nineteen less than a number is seven. Translate to an equation and then solve.

Understand The key words *less than* indicate subtraction. The key word *is* means an equal sign.

Plan Translate the key words and then solve. We'll use *n* for the variable.

Execute Translate: Nineteen less than a number is seven.

$$n - 19 = 7$$

Note: *The key words* **less than** *require careful translation. In the sentence, the word* **nineteen** *comes before the words* **less than** *and the words* **a number** *come after. In the translation, this order is reversed.*

Solve: $n - 19 = 7$

$$\underline{+19 \quad +19}$$ **Add 19 to both sides.**

$$n + 0 = 26$$

Answer $n = 26$

Check   Verify that 19 less than 26 is 7.

$$26 - 19 \overset{?}{=} 7$$
$$7 = 7$$

Yes, 19 less than 26 is 7.

YOUR TURN   Translate to an equation, and then solve.

    a. Fifteen more than a number is negative seven.

    b. The difference of x and thirty-five is negative nine.

Problems Involving Multiplication

EXAMPLE 2   Three-fourths of a number is negative five-eighths. Translate to an equation and then solve.

Understand   When *of* is preceded by a fraction, it means multiply. The word *is* means an equal sign.

Plan   Translate the key words and then solve.

Execute   Translate: Three-fourths of a number is negative five-eighths.

$$\frac{3}{4} \cdot n = -\frac{5}{8}$$

Solve: $\dfrac{\cancel{4}^{1}}{\cancel{3}^{1}} \cdot \dfrac{\cancel{3}^{1}}{\cancel{4}^{1}} n = -\dfrac{5}{\cancel{8}^{2}} \cdot \dfrac{\cancel{4}^{1}}{3}$ Clear the coefficient $\dfrac{3}{4}$ by multiplying both sides by its reciprocal, $\dfrac{4}{3}$.

Answer $n = -\dfrac{5}{6}$

Check   Verify that $\dfrac{3}{4}$ of $-\dfrac{5}{6}$ is $-\dfrac{5}{8}$.

$$\frac{\cancel{3}^{1}}{4} \cdot -\frac{5}{\cancel{6}^{2}} \overset{?}{=} -\frac{5}{8}$$
$$-\frac{5}{8} = -\frac{5}{8}$$

Yes, $\dfrac{3}{4}$ of $-\dfrac{5}{6}$ is $-\dfrac{5}{8}$, so $-\dfrac{5}{6}$ is correct.

YOUR TURN   Translate to an equation and then solve. The product of 0.6 and a number is -25.08.

Problems Involving More Than One Operation

EXAMPLE 3 Seven less than the product of four and a number is equal to negative five. Translate to an equation, and then solve.

Understand *Less than* indicates subtraction in reverse order, *product* indicates multiplication, and *is equal to* indicates an equal sign.

Plan Translate to an equation using the key words and then solve the equation.

Execute Translate:

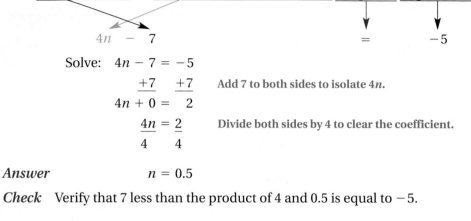

Seven less than the product of four and a number is equal to negative five.

$$4n - 7 \qquad = \qquad -5$$

Solve: $4n - 7 = -5$

$$\underline{+7 \qquad +7} \qquad \text{Add 7 to both sides to isolate } 4n.$$

$$4n + 0 = 2$$

$$\frac{4n}{4} = \frac{2}{4} \qquad \text{Divide both sides by 4 to clear the coefficient.}$$

Answer $n = 0.5$

Check Verify that 7 less than the product of 4 and 0.5 is equal to -5.

$$4(0.5) - 7 \stackrel{?}{=} -5$$
$$2 - 7 \stackrel{?}{=} -5$$
$$-5 = -5$$

Yes, 7 less than the product of 4 and 0.5 is equal to -5.

YOUR TURN Translate to an equation and then solve. Thirteen more than the product of four and a number yields negative seven.

Translations Requiring Parentheses

Sometimes a translation requires parentheses. This occurs when a sum or difference is multiplied or divided. When we multiply or divide a sum or difference, the addition or subtraction is to be calculated first. Because addition and subtraction follow multiplication and division in the order of operations, parentheses must be inserted around the addition or subtraction so that these operations are performed first.

EXAMPLE 4 Two-thirds of the difference of six and a number is the same as the number divided by four. Translate to an equation and then solve.

Understand *Of* means to multiply, and the word *difference* indicates subtraction. Because the difference is being multiplied, we write the difference expression in parentheses. *Divided by* indicates division.

Plan Translate the key words to an equation and then solve.

Execute Translate:

Two-thirds of the difference of six and a number is the same as the number divided by four.

$$\frac{2}{3} \quad \cdot \quad (6 - n) \qquad\qquad = \qquad \frac{n}{4}$$

Solve: $\dfrac{2}{3}(6 - n) = \dfrac{n}{4}$

$\dfrac{2}{\cancel{3}}\cdot\dfrac{\cancel{6}^{2}}{1} - \dfrac{2}{3}n = \dfrac{n}{4}$ **Distribute to clear the parentheses.**

$4 - \dfrac{2}{3}n = \dfrac{n}{4}$ **Simplify.**

$12\left(4 - \dfrac{2}{3}n\right) = \left(\dfrac{n}{4}\right)12$ **Multiply both sides by 12 to clear the fractions.**

$12 \cdot 4 - \dfrac{\cancel{12}^{4}}{1}\cdot\dfrac{2}{\cancel{3}}n = \dfrac{n}{\cancel{4}}\cdot\dfrac{\cancel{12}^{3}}{1}$ **Distribute 12 and simplify.**

$$48 - 8n = 3n$$
$$\underline{+8n \qquad +8n}$$
$$48 + 0 = 11n$$

Add 8n to both sides to separate the constant term and variable terms.

$\dfrac{48}{11} = \dfrac{11n}{11}$ **Clear the coefficient by dividing both sides by 11.**

Answer $\dfrac{48}{11} = n$

Check Verify that $\frac{2}{3}$ of the difference of 6 and $\frac{48}{11}$ is the same as $\frac{48}{11}$ divided by 4.

$$\frac{2}{3}\left(6 - \frac{48}{11}\right) \stackrel{?}{=} \frac{1}{4}\left(\frac{48}{11}\right)$$

Write 6 as a fraction with a denominator of 11.
$$\frac{2}{3}\left(\frac{66}{11} - \frac{48}{11}\right) \stackrel{?}{=} \frac{1}{\overset{}{\underset{1}{\cancel{4}}}}\left(\frac{\overset{12}{\cancel{48}}}{11}\right)$$ Divide out 4.

Subtract.
$$\frac{2}{3}\left(\frac{18}{11}\right) \stackrel{?}{=} \frac{12}{11}$$ Multiply.

Divide out 3.
$$\frac{2}{\overset{}{\underset{1}{\cancel{3}}}}\left(\frac{\overset{6}{\cancel{18}}}{11}\right) \stackrel{?}{=} \frac{12}{11}$$

Multiply.
$$\frac{12}{11} = \frac{12}{11}$$

Yes, $\frac{2}{3}$ of the difference of 6 and $\frac{48}{11}$ is the same as $\frac{48}{11}$ divided by 4.

ANSWERS

a. $3(n - 5) = 7n - 12;$
$$n = -\frac{3}{4}$$

b. $\frac{1}{4}n + 6 = \frac{1}{5}(n + 10);$
$$n = -80$$

YOUR TURN Translate to an equation and then solve.

a. Three times the difference of a number and five is equal to seven times the number, minus twelve.

b. The sum of one-fourth of a number and six is equal to one-fifth of the sum of the number and ten.

2.5 Exercises

For Extra Help MyMathLab Videotape/DVT InterAct Math Math Tutor Center Math XL.com

1. List three key words that indicate addition.

2. List three key words that indicate subtraction.

3. List three key words that indicate multiplication.

4. List three key words that indicate division.

5. Why are "four minus n" and "four less than n" different?

6. Write a sentence that would require parentheses when translated and explain why your sentence requires parentheses.

For Exercises 7–44, translate each sentence to an equation and then solve.

7. Six decreased by the number x is equal to negative three.

8. c divided by four is seven.

9. Eleven times m is negative ninety-nine.

10. Six added to p is negative two

11. Five less than y is negative one.

12. Eighteen subtracted from a number is equal to negative three.

13. Four divided into m is 1.6.

14. The quotient of a number and -6.5 is 4.2.

15. Four-fifths of a number is five-eighths.

16. Three-sevenths of a number is negative nine-eighths.

17. Seven more than three times a number is equal to thirty-four.

18. Nineteen more than triple a number is one hundred.

19. Nine less than five times a number is seventy-six.

20. Twenty-five less than four times a number is eleven.

21. Eight times the sum of eight and a number is equal to one hundred sixty.

22. Four times the sum of a number and three is equal to sixteen.

23. Negative three times the difference of x and two is twelve.

24. Tripling the difference of x and five produces negative fifteen.

25. One-third of the sum of a number and two is one.

26. Half of the difference of a number and two results in four.

27. Eleven less than three times a number is equal to that number added to five.

28. Five more than four times a number is equal to seven subtracted from that number.

29. Ten is the result when one is subtracted from the ratio of a number to four.

30. Two added to the quotient of a number and five is six.

31. The product of three and the sum of a number and four added to twice the number yields negative three.

32. Seven times the sum of a number and six subtracted from four times the number results in negative twenty-one.

33. The difference of a number and twelve is subtracted from the difference of twice the number and eight so that the result is eleven.

34. The sum of a number and six is added to the difference of the number tripled and four so that the result is fourteen.

35. Five times the sum of a number and one-third is equal to three times the number increased by two-thirds.

36. The product of three and the sum of a number and two-fifths is four-fifths less than negative two times the number.

37. Two times a number decreased by four is equal to triple the number added to sixteen.

38. Six less than three times a number is equal to the difference of fourteen and twice the number.

39. Two-fifths of a number is equal to two less than one-half the number.

40. One-third of a number is the same as half of the number increased by one.

41. The quotient of three less than a number and three is the same as the number divided by six.

42. The quotient of one more than a number and two is the same as the quotient of six more than twice the number and eight.

43. The product of negative four and the difference of two and a number is six less than the product of two and the difference of four and three times the number.

44. Negative two times the difference of one and three times a number is equal to four added to five times the difference of one and the number.

For Exercises 45–54, translate the equation to a word sentence.

45. $4x + 3 = 7$

46. $-3y + 8 = 10$

47. $6(y + 4) = -10y$

48. $8(w - 2) = 3w$

49. $\dfrac{1}{2}(x - 3) = \dfrac{2}{3}(x - 8)$

50. $\dfrac{1}{2}(t + 1) = \dfrac{1}{3}(t - 5)$

51. $0.05m + 0.06(m - 11) = 22$

52. $0.05v + 0.03(v - 4.5) = 0.465$

53. $\dfrac{2}{3}x + \dfrac{3}{4}x + \dfrac{1}{2}x = 10$

54. $\dfrac{1}{2}a + \dfrac{1}{3}a + \dfrac{1}{6}a = 5$

For Exercises 55–60, explain the mistake in the translation.

55. Ten less than a number is forty.

　　Translation: $10 - n = 40$

56. Twelve divided into a number is negative eight.

　　Translation: $12 \div n = -8$

57. Five times the difference of a number and six is equal to negative two.

　　Translation: $5x - 6 = -2$

58. Nine times the sum of a number and eight is the same as three subtracted from the number.

　　Translation: $9(y + 8) = 3 - y$

59. Twice a number minus the sum of the number and three is equal to negative six.

　　Translation: $2t - t + 3 = -6$

60. Four times the sum of a number and two is equal to the number minus the difference of the number and seven.

　　Translation: $4(y + 2) = y - (7 - y)$

For Exercises 61–68, translate to a formula, then use the formula to solve the problem.

61. The perimeter of a rectangle is equal to twice the sum of its length and width. Find the perimeter with the following length and width.

Length	Width
a. 24 ft.	18 ft.
b. 12.5 cm	18 cm
c. $8\frac{1}{4}$ in.	$10\frac{3}{8}$ in.

62. The surface area of a box is equal to twice the sum of its length times its width, its length times its height, and its width times its height. Find the surface area of a box with the following length, width, and height.

Length	Width	Height
a. 15 in.	6 in.	4 in.
b. 9.2 cm	12 cm	6.5 cm.
c. 8 ft.	$2\frac{1}{2}$ ft.	3 ft.

63. An isosceles triangle has two sides that are equal in length. The perimeter of an isosceles triangle is the sum of its base and twice the length of one of the other sides. Find the perimeter of each listed isosceles triangle.

Base	Sides
a. 5 in.	11 in.
b. 1.5 m	0.6 m
c. $15\frac{1}{4}$ in.	$9\frac{1}{2}$ in.

64. The perimeter of a semicircle is the sum of the diameter and π times the radius. Find the perimeter of each semicircle listed.

　　a. Diameter = 18 in.

　　b. Radius = 4.5 m

　　c. Diameter = $2\frac{3}{4}$ ft.

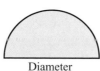
Diameter

65. The simple interest earned after investing an amount of money, called principal, is equal to the product of the principal, the interest rate, and the time in years that the money remains invested. Use the formula to calculate the interest of each of the following investments.

	Principal	Rate	Time
a.	$4000	0.05	2 years
b.	$500	0.03	$\frac{1}{2}$ year
c.	$2000	0.06	$\frac{1}{4}$ year

66. Investors often consider a stock's price-to-earnings ratio to help them decide whether a stock is worth investing in. The price-to-earnings ratio is equal to the stock's price divided by its company's earnings per share over the last 12 months. Use your formula to calculate each company's price-to-earnings ratio. (*Source:* David and Tom Gardner, *The Motley Fool Investment Guide*, Simon & Schuster, 1996)

	Company	Price	Earnings per Share
a.	Microsoft	$47.98	$1.41
b.	Intel	$18.79	$0.29
c.	General Motors	$46.55	$4.01

67. Skydivers often use a formula to estimate free-fall time. The amount of free-fall time in seconds is approximately equal to the difference of the exit altitude and the parachute deployment altitude divided by 153.8. Use this formula to calculate free-fall times if a skydiver deploys at each of the following altitudes after falling from an exit altitude of 12,500 feet.

    a. 5000 feet
    b. 4500 feet
    c. 3000 feet

68. A person's body mass index (BMI) is a measure of the amount of body fat based on height, in inches, and weight, in pounds. A person's BMI is equal to the ratio of the product of his or her weight and 704.5 to his or her height in inches multiplied by itself. Use your formula to find the BMI of each person listed below. (*Source:* National Institutes of Health)

	Height	Weight
a.	5'4"	138 lbs.
b.	5'9"	155 lbs.
c.	6'2"	220 lbs.

Fill in each blank with a whole number so that the resulting rational number has the following properties:

The rational number is between 300 and 500.
The tenths placeholder is less than the tens placeholder.
The sum of the tens place and the tenths place is 8.
The product of all four placeholders is 42.

$$\underline{\quad}\ \underline{\quad}\ \underline{\quad}\ .\ \underline{\quad}$$

REVIEW EXERCISES

[1.1]   **1.** Is $19 \geq 19$ true or false? Explain.

[1.1, 1.5] *For Exercises 2–3, use <, >, or = to make a true sentence.*

2. $-|-6| \quad -2(3)$

3. $5(4) - 23 \quad 2(3 - 8)$

[2.3] *For Exercises 4–5, solve and check.*

4. $-6z = 72$

5. $\dfrac{1}{3}(y - 2) = \dfrac{3}{5}y + 6$

Find the missing number that makes the fractions equivalent.

[1.2]   **6.** $-\dfrac{6}{8} = \dfrac{\quad}{40}$

2.6 Solving Linear Inequalities

OBJECTIVES

1. Represent solutions to inequalities graphically and using set notation.
2. Solve linear inequalities.
3. Solve problems involving linear inequalities.

OBJECTIVE 1. Represent solutions to inequalities graphically and using set notation. Not all problems translate to equations. Sometimes a problem can have a range of values as solutions. In mathematics we can write inequalities to describe situations where a range of solutions is possible. Following are the inequality symbols and their meanings:

$<$ is less than
$>$ is greater than
$\leq$ is less than or equal to
$\geq$ is greater than or equal to

Throughout the chapter, we have explored techniques for solving linear equations. In this section, we will explore how to solve **linear inequalities**.

DEFINITION **_Linear inequality:_** An inequality containing expressions in which each variable term contains a single variable with an exponent of 1.

Following are some examples of linear inequalities:

$$x > 5 \qquad n + 2 < 6 \qquad 2(y - 3) \leq 5y - 9 \qquad 2x + 3y \geq 6$$

A solution for an inequality is any number that can replace the variable(s) in the inequality and make it true. Since inequalities will have a range of solutions, we often write those solutions in set-builder notation, which we introduced in Section 1.1. For example, the solution set for the inequality $x \geq 5$ contains 5 and every real number greater than 5, which we write as $\{x \mid x \geq 5\}$. We read this set-builder notation as shown here.

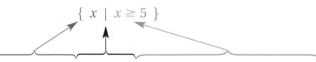

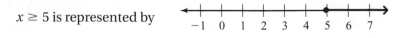

The set of all x such that x is greater than or equal to five.

We can graph solution sets for inequalities on a number line. Since the solution set for $x \geq 5$ contains 5 and every real number to the right of 5, we draw a dot (or solid circle) at 5 and shade to the right of 5.

$x \geq 5$ is represented by

The solution set for $x < 2$ contains every real number to the left of 2, but not 2 itself, so it is written $\{x \,|\, x < 2\}$. To graph this solution set, we draw an open circle at 2, indicating that 2 is not included, and shade to the left of 2.

$x < 2$ is represented by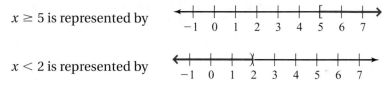

Another popular notation used to indicate ranges of values is *interval notation,* which uses parentheses and brackets to indicate whether end values are included or not. Parentheses are used for end values that are not included in the interval and brackets are used for end values that are included. To write interval notation, we imagine traveling from left to right on the number line.

For example, as we travel from left to right on the graph for $x \geq 5$ we encounter 5 and then every real number to the right of 5, so the interval notation is $[5, \infty)$. The symbol ∞ means infinity and is used to indicate the positive extreme. Since ∞ can never be reached, it will always have a parenthesis indicating that it is not included as an end value. As we travel from left to right on the graph for $x < 2$ we go from the negative extreme up to 2, but not including 2, so the interval notation is $(-\infty, 2)$. Notice $-\infty$ represents the left-most extreme.

The parentheses and brackets from interval notation can be used on the graphs instead of open and solid circles.

$x \geq 5$ is represented by

$x < 2$ is represented by

Your instructor may have a preference of style. Because interval notation is a preferred notation in college algebra courses, we will draw our graphs with the parentheses and brackets.

PROCEDURE   Graphing Inequalities

To graph an inequality on a number line,

1. If the symbol is $\leq$ or $\geq$, draw a bracket (or solid circle) on the number line at the indicated number. If the symbol is $<$ or $>$, draw a parenthesis (or open circle) on the number line at the indicated number.
2. If the variable is greater than the indicated number, shade to the right of the indicated number. If the variable is less than the indicated number, shade to the left of the indicated number.

EXAMPLE 1 Write the solution set in set-builder notation and interval notation, then graph the solution set.

a. $x \le -6$

Solution Set-builder notation: $\{x \mid x \le -6\}$

Interval notation: $(-\infty, -6]$

Note: *When writing interval notation, imagine traveling from left to right so this range of values is from $-\infty$ to -6, including -6.*

Graph:

b. $n > 0$

Solution Set-builder notation: $\{n \mid n > 0\}$

Interval notation: $(0, \infty)$

Connection $\{n \mid n > 0\}$ is a way of expressing the set of all positive real numbers.

Graph:

YOUR TURN Write the solution set in set-builder notation and interval notation, then graph the solution set.

a. $x < 2$

b. $t \ge -1$

Compound Inequalities

Inequalities containing two inequality symbols are called *compound inequalities*. Compound inequalities are useful in writing a range of values between two numbers. For example, $1 < x < 6$ is a compound inequality meaning x can be any number that is greater than 1 and less than 6. Its solution set contains every real number between 1 and 6, but not 1 and 6.

Set-builder notation: $\{x \mid 1 < x < 6\}$

Interval notation: $(1, 6)$

Graph:

Note: *Because the end values 1 and 6 are not included in the solution set, we use parentheses in the interval notation and on the graph.*

ANSWERS

a. Set-builder notation: $\{x \mid x < 2\}$

Interval notation: $(-\infty, 2)$

Graph:

b. Set-builder notation: $\{t \mid t \ge -1\}$

Interval notation: $[-1, \infty)$

Graph:

EXAMPLE 2 Write the solution set for $-5 \leq x < 1$ in set-builder notation and interval notation, then graph the solution set.

Solution Set-builder notation: $\{x \mid -5 \leq x < 1\}$

Interval notation: $[-5, 1)$

Graph:

Note: *Because the end value -5 is included in the solution set, we use a bracket in the interval notation and on the graph. Since 1 is not included, we use a parenthesis.*

YOUR TURN Write the solution set for $-4 < y \leq -1$ in set-builder notation and interval notation, then graph the solution set.

OBJECTIVE 2. Solve linear inequalities. To solve inequalities like $n + 2 < 6$ and $2(x - 3) \leq 5x - 9$, we will follow essentially the same process as for solving equations. The addition and multiplication principles of inequalities are similar to the principles of equality. First, let's examine the addition principle of inequality.

Consider the inequality $3 < 7$. According to the addition principle, if we add or subtract the same amount on both sides, then the inequality should still be true. Let's test this by choosing some numbers to add and subtract on both sides.

Add 2 to both sides:	**Subtract 9 from both sides:**
$3 < 7$	$3 < 7$
$\underline{+2 \quad +2}$	$\underline{-9 \quad -9}$
$5 < 9$ Still true.	$-6 < -2$ Still true.

RULE   **The Addition Principle of Inequality**

If $a < b$, then $a + c < b + c$ is true for all real numbers a, b, and c.

Note: *This principle indicates that we can add (or subtract) the same amount on both sides of an inequality without affecting its solution(s). Although the principle is written in terms of the $<$ symbol, the principle is true for any inequality symbol.*

The multiplication principle, on the other hand, does not work as neatly as it did for equations. As we shall see, sometimes multiplying or dividing both sides of an inequality by the same number can turn a true inequality into a false inequality. Let's multiply both sides of $3 < 7$ by a positive number and then by a negative number.

Multiply both sides by 4:	**Multiply both sides by -2:**
$4(3) < 4(7)$	$-2(3) < -2(7)$
$12 < 28$ Still true	$-6 < -14$ Not true!

Conclusion An inequality remains true when we multiply both sides by a positive number, but not when we multiply both sides by a negative number.

ANSWER

Set-builder notation: $\{y \mid -4 < y \leq -1\}$

Interval notation: $(-4, -1]$

Graph:

RULE **The Multiplication Principle of Inequality**

Note: *We can multiply (or divide) both sides of an inequality by the same positive amount without affecting its solution(s). However, if we multiply (or divide) both sides of an inequality by the same negative amount, we must reverse the direction of the inequality symbol to maintain the truth of the inequality. Although we have written the principle in terms of the < symbol, the principle is true for any inequality symbol.*

▶ If a and b are real numbers, where $a < b$, then $ac < bc$ is true if c is a positive real number.

If a and b are real numbers, where $a < b$, then $ac > bc$ is true if c is a negative real number.

EXAMPLE 3 Solve and write the solution set in set-builder notation and in interval notation. Then graph the solution set.

a. $3x > -7$

Solution

Note: *Dividing both sides by a positive number does not affect the inequality.*

$$\frac{3x}{3} > \frac{-7}{3}$$

$$x > -\frac{7}{3}$$

To isolate x, we clear the 3 coefficient by dividing by 3 on both sides.

Set-builder notation: $\left\{ x \,|\, x > -\frac{7}{3} \right\}$

Interval notation: $\left(-\frac{7}{3}, \infty \right)$

Graph:

b. $-8x \geq 24$

Solution

Note: *Because we divided both sides by a negative number, we reversed the direction of the inequality symbol.*

$$\frac{-8x}{-8} \leq \frac{24}{-8}$$

$$x \leq -3$$

To isolate x, we clear the -8 coefficient by dividing by -8 on both sides.

Set-builder notation: $\{ x \,|\, x \leq -3 \}$

Interval notation: $(-\infty, -3]$

Graph:

c. $\dfrac{5}{6} \le -\dfrac{3}{4}n$

Solution

Note: *You may find it helpful to rewrite the inequality so that the variable is on the left side.*

The inequality $-\dfrac{10}{9} \ge n$ *is the same as* $n \le -\dfrac{10}{9}.$

$$-\dfrac{\cancel{4}}{3}\left(\dfrac{5}{\cancel{6}}\right) \ge -\dfrac{\cancel{4}}{\cancel{3}}\left(-\dfrac{\cancel{3}}{\cancel{4}}n\right)$$

$$-\dfrac{10}{9} \ge n$$

To isolate n**, we clear the** $-\dfrac{3}{4}$ **coefficient by multiplying by its reciprocal** $-\dfrac{4}{3}$ **on both sides. Multiplying by a negative number means we must reverse the direction of the inequality symbol.**

Set-builder notation: $\left\{ n \mid n \le -\dfrac{10}{9} \right\}$

Interval notation: $\left(-\infty, -\dfrac{10}{9} \right]$

Graph:

YOUR TURN Solve $-\dfrac{2}{5}t < \dfrac{9}{10}$ and write the solution set in set-builder notation and in interval notation. Then graph the solution set.

We can solve more complex linear inequalities using an outline similar to that for solving linear equations.

PROCEDURE **Solving Linear Inequalities**

To solve linear inequalities:

1. Simplify both sides of the inequality as needed.
   a. Distribute to clear parentheses.
   b. Clear fractions or decimals by multiplying through by the LCD just as we did for equations. (Clearing fractions and decimals is optional.)
   c. Combine like terms.
2. Use the addition principle so that all variable terms are on one side of the inequality and all constants are on the other side. Then combine like terms.
3. Use the multiplication principle to clear any remaining coefficient. If you multiply (or divide) both sides by a negative number, reverse the direction of the inequality symbol.

ANSWERS

a. $t > -\dfrac{9}{4}$ or $-2\dfrac{1}{4}$

Set-builder notation: $\left\{ t \mid t > -\dfrac{9}{4} \right\}$

Interval notation: $\left(-\dfrac{9}{4}, \infty \right)$

Graph:

Notice that clearing the term with the smaller coefficient results in a positive coefficient, so you won't have to reverse the direction of the inequality.

EXAMPLE 4 Solve $7x + 9 > 4x - 6$ and write the solution set in set-builder notation and interval notation. Then graph the solution set.

Solution Use the addition principle to separate the variable terms and constant terms, then clear the remaining coefficient.

$$7x + 9 > 4x - 6$$

Note: *Clearing the 4x term results in a positive coefficient after combining like terms, so we won't have to reverse the inequality when we clear the 3 coefficient.*

$$\underline{-4x \qquad -4x} \qquad \text{Subtract } 4x \text{ from both sides.}$$
$$3x + 9 > 0 - 6$$
$$3x + 9 > -6$$
$$\underline{-9 \qquad -9} \qquad \text{Subtract 9 from both sides.}$$
$$3x + 0 > -15$$
$$\frac{3x}{3} > \frac{-15}{3} \qquad \text{Divide both sides by 3 to isolate } x.$$
$$x > -5$$

Set-builder notation: $\{x \mid x > -5\}$

Interval notation: $(-5, \infty)$

Graph:

YOUR TURN Solve $7(x - 3) \le 4x + 3$ and write the solution set in set-builder notation and in interval notation. Then graph the solution set.

OBJECTIVE 3. Solve problems involving linear inequalities. Problems requiring inequalities can be translated using key words much like those we used to translate sentences to equations. The following table lists common key words that indicate inequalities.

Less Than:		Greater Than:	
A number is less than 7.	$n < 7$	A number is greater than 2.	$n > 2$
A number must be smaller than 5.	$n < 5$	A number must be greater than 3.	$n > 3$
		A number must be more than -6.	$n > -6$

Less Than or Equal to:		Greater Than or Equal to:	
A number is at most 9.	$n \le 9$	A number is at least 2.	$n \ge 2$
The maximum is 14.	$n \le 14$	The minimum is 18.	$n \ge 18$

ANSWER

$x \le 8$

Set-builder notation: $\{x \mid x \le 8\}$

Interval notation: $(-\infty, 8]$

Graph:

EXAMPLE 5 Three-fourths of a number is at least eighteen. Translate to an inequality, then solve.

Understand Because the word *of* is preceded by a fraction, it means multiplication. The key words *at least* indicate a greater-than or equal-to symbol.

Plan Translate the key words, then solve. We'll use *n* for the variable.

Execute Translate.

Three-fourths of a number is at least eighteen.

$$\frac{3}{4} \cdot n \geq 18$$

Solve: $\frac{3}{4} \cdot n \geq 18$

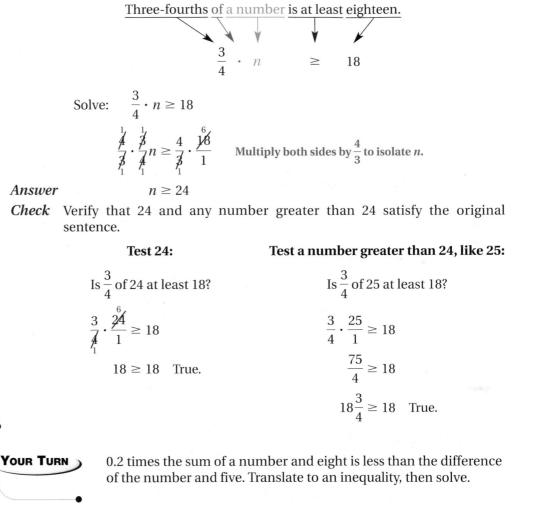

Multiply both sides by $\frac{4}{3}$ to isolate *n*.

Answer $n \geq 24$

Check Verify that 24 and any number greater than 24 satisfy the original sentence.

Test 24:

Is $\frac{3}{4}$ of 24 at least 18?

$$\frac{3}{\overset{1}{\underset{1}{4}}} \cdot \frac{\overset{6}{24}}{1} \geq 18$$

$$18 \geq 18 \quad \text{True.}$$

Test a number greater than 24, like 25:

Is $\frac{3}{4}$ of 25 at least 18?

$$\frac{3}{4} \cdot \frac{25}{1} \geq 18$$

$$\frac{75}{4} \geq 18$$

$$18\frac{3}{4} \geq 18 \quad \text{True.}$$

YOUR TURN 0.2 times the sum of a number and eight is less than the difference of the number and five. Translate to an inequality, then solve.

EXAMPLE 6 Darwin is planning a garden area, which he will enclose with a fence. Cost restricts him to a total of 240 feet of fencing materials. He wants the garden to span the entire width of his lot, which is 80 feet. What is the maximum length of the garden?

Understand The fence surrounds the garden, so that 240 feet is the maximum perimeter for the garden. The width of the garden is to be 80 feet, and we must find the length.

ANSWERS

$0.2(n + 8) < n - 5;$
$n > 8.25$

Plan The formula for the perimeter of a rectangle is $P = 2l + 2w$. Because 240 feet is a *maximum* perimeter, we write the perimeter formula as an inequality so that the expression used to calculate perimeter ($2l + 2w$) is less than or equal to 240. Because the width is to be 80 feet, we replace w in the formula with 80.

Execute Translate:

The calculation for perimeter must be less than or equal to 240.

$$2l + 2(80) \quad \leq \quad 240$$

Solve: $\quad 2l + 160 \leq 240 \qquad$ **Subtract 160 from both sides.**

$$\underline{\quad -160 \quad -160\quad}$$

$$2l + 0 \leq 80$$

$$\frac{2l}{2} \leq \frac{80}{2} \qquad \textbf{Divide both sides by 2 to isolate } l.$$

$$l \leq 40$$

Answer The length must be less than or equal to 40 feet, which means the maximum length is 40 feet.

Check Verify that a garden with a length of 40 feet or less and a width of 80 feet has a perimeter that is 240 feet or less. This will be left to the reader.

YOUR TURN It is desired that the surface area of a box be at least 678 square centimeters. If the length is to be 12 centimeters and the width is to be 9 centimeters, find the range of values for the height that satisfies the minimum surface area of 678 square centimeters.

ANSWER

$h \geq 11$ cm

2.6 Exercises

For Extra Help MyMathLab · Videotape/DVT · InterAct Math · Math Tutor Center · Math XL.com

1. What is a solution for an inequality?

2. Explain the difference between $x < 8$ and $x \leq 8$.

3. In your own words, explain the meaning of $x \geq -5$.

4. What action causes the direction of an inequality symbol to change?

For Exercises 5–16: a. Write the solution set in set-builder notation.
b. Write the solution set in interval notation.
c. Graph the solution set.

5. $x \geq -3$ **6.** $n \leq -5$ **7.** $h < 6$ **8.** $x > 0$

9. $n < -\dfrac{2}{3}$ **10.** $a \geq \dfrac{4}{5}$ **11.** $t \geq 2.4$ **12.** $p \leq -0.6$

13. $-3 < x < 6$ **14.** $4 < c \leq 10$ **15.** $0 \leq n \leq 5$ **16.** $-1 < a < 5$

For Exercises 17–48: a. Solve.
b. Write the solution set in set-builder notation.
c. Write the solution set in interval notation.
d. Graph the solution set.

17. $m - 5 < 20$ **18.** $x + 8 \geq -12$ **19.** $16y \geq 32$ **20.** $4z \leq -20$

21. $-3x \geq -12$ **22.** $-4x < -16$ **23.** $\dfrac{2}{3}x \geq 4$ **24.** $\dfrac{3}{5}a < -6$

25. $-\dfrac{3}{4}m < -6$ **26.** $-\dfrac{5}{3}p > -10$ **27.** $5y + 1 > 16$ **28.** $3y - 2 > 10$

29. $1 - 6x < 25$ **30.** $-2c + 3 \geq 17$ **31.** $\dfrac{a}{2} + 1 < \dfrac{3}{2}$ **32.** $2 > \dfrac{2}{3} - \dfrac{k}{3}$

33. $1 - f \geq 5 + f$

34. $9 + 2n < 3 - 4n$

35. $3 - 6u \geq -5 - 2u$

36. $4 - 9k < -4k + 19$

37. $2(c - 3) < 3(c + 2)$

38. $4(d + 4) \geq 5(d + 2)$

39. $10(9w + 13) - 5(3w - 1) \leq 8(11w + 12)$

40. $6(n + 1) + 2(4n - 2) > 3(5n - 1)$

41. $\frac{1}{2}(x + 1) \leq \frac{1}{3}(x - 5)$

42. $\frac{1}{5}(6m - 7) > \frac{1}{2}(3m - 1)$

43. $\frac{1}{6}(5 + n) - \frac{1}{3}(10 - n) < 1$

44. $-\frac{1}{3}(2y - 5) + \frac{1}{6}(5y - 2) \geq 1$

45. $0.4l - 0.37 < 0.3(l + 2.1)$

46. $0.05 + 0.03(v - 4.5) \geq 0.465$

47. $0.6t + 0.2(t - 8) \leq 0.5t - 0.6$

48. $2.1s + 3(1.4s - 1.5) \geq 0.09s + 20.34$

For Exercises 49–60, translate to an inequality, then solve.

49. Four more than a number is greater than twenty-four.

50. Six less than a number is less than or equal to negative four.

51. Four-ninths of a number is less than or equal to negative eight.

52. Three-fifths of a number is greater than negative twenty-one.

53. Eight times a number less thirty-six is at least sixty.

54. Three subtracted from a number times eleven is at least forty-one.

55. Five added to half of a number is at most two.

56. The sum of two-fifths of a number and two is at most eight.

57. The difference of four times a number and eight is less than two times the number.

58. Triple a number is less than the number increased by eight.

59. Twenty-five is greater than or equal to seven more than six times a number.

60. Negative five is greater than fifteen less than five times a number.

For Exercises 61–72, solve.

61. A school is planning a playground. The area of a rectangular playground may not exceed 170 square feet. If the width is 17 feet, what range of values can the length have?

62. Scotty needs to make a rectangular pool cover that is at least 128 square feet. If the pool length is 16 feet, what range of values must the width be?

63. The design of a storage box calls for a width of 27 inches and a length of 41 inches. If the surface area must be at least 4254 square inches, find the range of values for the height.

64. Andre builds tables. One particular circular table requires him to glue a strip of veneer around the circumference of the top. He is designing a new circular table and has a strip of veneer that is 150 inches long that he wishes to use. What range of values can the radius have?

65. Jon never exceeds 65 miles per hour when driving on the highway. If he is traveling 410 miles, what range of values can his time spent driving have?

66. A truck driver must make a delivery 298 miles away in 5 hours or less. What range of values can his average rate be so that he meets the schedule?

67. The design of a circuit specifies that the voltage cannot exceed 12 volts. If the resistance of the circuit is 8 ohms, find the range of values that the current can have.

68. A label in an elevator indicates that the maximum load is 9000 newtons, that is, the downward force should not exceed -9000 newtons. If the acceleration due to gravity is -9.8 meters per second squared, find the range of values of mass that can be loaded onto the elevator.

69. The fourth-quarter profit goal for a software company is at least $850,000. If it is known that the cost for the quarter will be $625,000, find the revenue that must be generated to achieve the goal.

70. Silver is a solid up to a temperature of 960.8°C. Write an inequality to describe this range of temperatures in degrees Fahrenheit.

71. To get an A in her math course, Tina needs the average of five tests to be at least 90. Her current test scores are 82, 91, 95, and 84. What range of scores on the fifth test would get her an A in the course?

72. In Aaron's English course, the final grade is determined by the average of five papers. The department requires any student whose average falls below 75 to repeat the course. Aaron's scores on the first four papers are 68, 78, 80, and 72. What range of scores on the fifth paper would cause him to have to repeat the course?

PUZZLE PROBLEM

Write the set of all values such that
$|x| = -x.$

REVIEW EXERCISES

[1.1] **1.** Write a set representing the natural numbers up to 10.

[1.3] **2.** What property of addition is represented by $-6.5 + 3.4 = 3.4 + (-6.5)$?

[1.4] **3.** Simplify: $-6 \div 0$

[1.6] **4.** Translate to an expression: six more than twice the difference of a number and five.

[1.7] **5.** Evaluate $b^2 - 4ac$ when $a = 1$, $b = -2$, and $c = -3$.

[1.7] **6.** Evaluate $-|3x| + |y^3|$ when $x = 5$ and $y = -1$.

Chapter 2   Summary

Defined Terms

Section 2.1
Equation (p. 102)
Solution (p. 102)
Identity (p. 104)
Formula (p. 106)
Perimeter (p. 106)

Area (p. 106)
Volume (p. 106)
Circumference (p. 106)
Radius (p. 106)
Diameter (p. 106)

Section 2.2
Linear equation (p. 118)
Linear equation in one
    variable (p. 118)

Section 2.6
Linear inequality (p. 164)

Learning Strategy

The summaries in the textbook are like the study sheet suggested in the To the Student section (see page XIX). Remember that your study sheet is a list of the rules, procedures, and formulas that you need to know. If you are a tactile or visual learner, spend a lot of time reviewing and writing the rules or procedures. Try to get to the point where you can write the essence of each rule and procedure from memory. If you are an audio learner, record yourself saying each rule and procedure, then listen to the recording over and over. Consider developing clever rhymes or songs for each rule or procedure to help you remember them.

PROCEDURE   Problem-Solving Outline

1. **Understand** the problem.
   a. Read the question(s) (not the whole problem, just the question at the end) and write a note to yourself about what it is you are to find.
   b. Now read the whole problem, underlining the key words.
   c. If possible or useful, make a list or table, simulate the situation, or search for a related example problem.
2. **Plan** your solution by searching for a formula or translating the key words to an equation.
3. **Execute** the plan by solving the equation/formula.
4. **Answer** the question. Look at the note about what you were to find and make sure you answer that question. Include appropriate units.
5. **Check** results.
   a. Try finding the solution in a different way, reversing the process, or estimating the answer and make sure the estimate and actual answer are reasonably close.
   b. Make sure the answer is reasonable.

Procedures, Rules, and Key Examples

Procedures/Rules	Key Examples

Section 2.1 Equations, Formulas, and the Problem-Solving Process

To determine whether a value is a solution for a given equation, replace the variable(s) in the equation with the value. If the resulting equation is true, the value is a solution.

Example 1: Check to see if the given number is a solution for the given equation.

a. $7n - 13 = 3n + 11; n = 6$

$$7(6) - 13 \overset{?}{=} 3(6) + 11$$
$$42 - 13 \overset{?}{=} 18 + 11$$
$$29 = 29 \quad \text{Yes, 6 is a solution.}$$

b. $2(x - 3) = 5x + 4; x = 1$

$$2(1 - 3) \overset{?}{=} 5(1) + 4$$
$$2(-2) \overset{?}{=} 5 + 4$$
$$-4 \neq 9 \quad \text{No, 1 is not a solution.}$$

To determine whether an equation is an identity, simplify the expressions on each side of the equal sign. If, after simplifying, the expressions are identical, the equation is an identity.

Example 2: Is $5x - 9 - x = 4(x - 2) - 1$ an identity?

$$4x - 9 = 4x - 8 - 1$$
$$4x - 9 = 4x - 9 \quad \text{Yes, it is an identity.}$$

To use a formula:
1. Replace the variables with the corresponding given values.
2. Solve for the missing value.

Example 3: Find the volume of a box with a length of 4 cm, width of 5 cm, and height of 6 cm.

$$V = lwh$$
$$V = (4)(5)(6)$$
$$V = 120 \text{ cm}^3$$

For composite figures:
1. To calculate the area of a figure composed of two or more figures that are next to each other, add the areas of the individual figures.
2. To calculate the area of a region defined by a smaller figure within a larger figure, subtract the area of the smaller figure from the area of the larger figure.

Example 4: Find the shaded area in the figure shown.

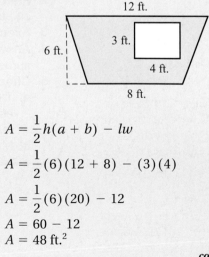

$$A = \frac{1}{2}h(a + b) - lw$$
$$A = \frac{1}{2}(6)(12 + 8) - (3)(4)$$
$$A = \frac{1}{2}(6)(20) - 12$$
$$A = 60 - 12$$
$$A = 48 \text{ ft.}^2$$

continued

Procedures/Rules	Key Examples

Section 2.2 The Addition Principle

If $a = b$, then $a + c = b + c$ is true for all real numbers a, b, and c. To use the addition principle of equality to clear a term in an equation, add the additive inverse of that term to both sides of the equation (that is, add or subtract appropriately so that the term you want to clear becomes 0).

To solve equations requiring the addition principle only:
1. Simplify both sides of the equation as needed.
   a. Distribute to clear parentheses.
   b. Combine like terms.
2. Use the addition principle so that all variable terms are on one side of the equation and all constants are on the other side. Then combine like terms.

Tip: Clear the variable term that has the lesser coefficient to avoid negative coefficients.

If applying the addition principle of equality causes the variable terms to be eliminated from the equation and the resulting equation is false, then the equation is a contradiction and has no solution.

Example 1: Solve.

a.
$$x + 9 = 15$$
$$\underline{-9 \quad -9}$$
$$x + 0 = 6$$
$$x = 6$$

b.
$$y - 5 = 7$$
$$\underline{+5 \quad +5}$$
$$y + 0 = 12$$
$$y = 12$$

c.
$$5(n - 2) + 3n = 7n - 4$$
$$5n - 10 + 3n = 7n - 4$$
$$8n - 10 = 7n - 4$$
$$\underline{-7n \qquad\qquad -7n}$$
$$n - 10 = 0 - 4$$
$$n - 10 = -4$$
$$\underline{+10 \qquad +10}$$
$$n + 0 = \quad 6$$
$$n = 6$$

d.
$$3n + 7 = 3n + 5$$
$$\underline{-3n \qquad\quad -3n}$$
$$0 + 7 = 0 + 5$$
$$7 = 5$$

The equation is a contradiction so it has no solution.

Section 2.3 The Multiplication Principle

If $a = b$, then $ac = bc$ is true for all real numbers a, b, and c, where $c \neq 0$.
To use the multiplication principle of equality to clear a coefficient in an equation, multiply both sides of the equation by the multiplicative inverse of that coefficient, or divide both sides by the coefficient.

To solve linear equations in one variable:
1. Simplify both sides of the equation as needed.
   a. Distribute to clear parentheses.
   b. Clear fractions or decimals by multiplying through by the LCD. In the case of decimals, the LCD is the power of 10 with the same number of zero digits as decimal places in the number with the most decimal places. (Clearing fractions and decimals is optional.)
   c. Combine like terms.
2. Use the addition principle so that all variable terms are on one side of the equation and all constants are on the other side. (Clear the variable term with the lesser coefficient.) Then combine like terms.
3. Use the multiplication principle to clear the remaining coefficient.

Example 1: Solve.

a.
$$-5x = 15$$
$$\frac{-5x}{-5} = \frac{15}{-5}$$
$$x = -3$$

b.
$$\frac{3}{4}n = \frac{5}{6}$$
$$\frac{\cancel{4}}{3} \cdot \frac{\cancel{3}}{\cancel{4}} n = \frac{5}{\cancel{6}} \cdot \frac{\cancel{4}^2}{3}$$
$$n = \frac{10}{9}$$

c.
$$7n - (n - 3) = 2n - 5$$
$$7n - n + 3 = 2n - 5$$
$$6n + 3 = 2n - 5$$
$$\underline{-2n \qquad\quad -2n}$$
$$4n + 3 = 0 - 5$$
$$4n + 3 = -5$$
$$\underline{-3 \qquad -3}$$
$$4n + 0 = -8$$
$$\frac{4n}{4} = \frac{-8}{4}$$
$$n = -2$$

continued

Summary **179**

Procedures/Rules	Key Examples

Section 2.4 Applying the Principles to Formulas

To isolate a particular variable in a formula:
1. Imagine all other variables to be constants.
2. Isolate the desired variable using the outline for solving equations.

Example 1: Isolate w in the formula $P = 2l + 2w$.

$$P = 2l + 2w$$
$$\underline{-2l \quad -2l}$$
$$P - 2l = 0 + 2w$$
$$\frac{P - 2l}{2} = \frac{2w}{2}$$
$$\frac{P - 2l}{2} = w$$

Section 2.5 Translating Word Sentences to Equations

To translate a word sentence to an equation, identify the variable(s), constants, and key words, then write the corresponding symbolic form.

Example 1: Translate to an equation.
a. Five more than seven times a number is twelve.
   Answer: $7n + 5 = 12$
b. Six times the difference of a number and nine is equal to four times the sum of a number and five.
   Answer: $6(n - 9) = 4(n + 5)$

Section 2.6 Solving Linear Inequalities

To graph an inequality:
1. If the symbol is $\leq$ or $\geq$, draw a bracket (or solid circle) on the number line at the indicated number. If the symbol is $<$ or $>$, draw a parenthesis (or open circle) on the number line at the indicated number.
2. If the variable is greater than the indicated number, shade to the right of the indicated number. If the variable is less than the indicated number, shade to the left of the indicated number.

Example 1: Write the solution set in set-builder notation and interval notation, then graph the solution set.
a. $x \leq 6$
   Set-builder notation: $\{x \mid x \leq 6\}$
   Interval notation: $(-\infty, 6]$
   Graph:

b. $x > -2$
   Set-builder notation: $\{x \mid x > -2\}$
   Interval notation: $(-2, \infty)$
   Graph:

c. $-3 \leq x < 1$
   Set-builder notation: $\{x \mid -3 \leq x < 1\}$
   Interval notation: $[-3, 1)$
   Graph:

continued

Procedures/Rules	Key Examples

Section 2.6 Solving Linear Inequalities (Continued)

To solve linear inequalities:

1. Simplify both sides of the inequality as needed.
   a. Distribute to clear parentheses.
   b. Clear fractions or decimals by multiplying through by the LCD just as we did for equations. (Clearing fractions and decimals is optional.)
   c. Combine like terms.
2. Use the addition principle so that all variable terms are on one side of the inequality and all constants are on the other side. Then combine like terms. (Remember, moving the term with the lesser coefficient results in a positive coefficient.)
3. Use the multiplication principle to clear any remaining coefficient. If you multiply (or divide) both sides by a negative number, reverse the direction of the inequality symbol.

Example 2: Solve and write the solution set in set-builder notation and in interval notation. Then graph the solution set.

a. $-6x > 24$

$$\frac{-6x}{-6} < \frac{24}{-6}$$
$$x < -4$$

Set-builder notation: $\{x \mid x < -4\}$
Interval notation: $(-\infty, -4)$

Graph:

$$\overset{\longleftarrow}{\underset{-7\ \ -6\ \ -5\ \ -4\ \ -3\ \ -2\ \ -1\ \ \ \ 0}{\mid\ \ \mid\ \ \mid\ \ )\ \ \mid\ \ \mid\ \ \mid\ \ \mid\ \ \mid}}$$

b. $8y - 2 \geq 5y + 7$

$$8y - 2 \geq 5y + 7$$
$$\underline{-5y \qquad -5y}$$
$$3y - 2 \geq 0 + 7$$
$$3y - 2 \geq 7$$
$$\underline{+2 \quad +2}$$
$$3y + 0 \geq 9$$
$$\frac{3y}{3} \geq \frac{9}{3}$$
$$y \geq 3$$

Set-builder notation: $\{y \mid y \geq 3\}$
Interval notation: $[3, \infty)$

Graph:

$$\overset{\longrightarrow}{\underset{-1\ \ \ 0\ \ \ 1\ \ \ 2\ \ \ 3\ \ \ 4\ \ \ 5\ \ \ 6}{\mid\ \ \mid\ \ \mid\ \ \mid\ \ [\ \ \mid\ \ \mid\ \ \mid}}$$

Formulas

Perimeter of a rectangle: $P = 2l + 2w$

Circumference of a circle: $C = \pi d$ or $C = 2\pi r$

Area of a parallelogram: $A = bh$

Area of a triangle: $A = \dfrac{1}{2}bh$

Area of a trapezoid: $A = \dfrac{1}{2}h(a + b)$

Area of a circle: $A = \pi r^2$

Surface area of a box: $SA = 2lw + 2lh + 2wh$

Volume of a box: $V = lwh$

Volume of a pyramid: $V = \dfrac{1}{3}lwh$

Volume of a cylinder: $V = \pi r^2 h$

Volume of a cone: $V = \dfrac{1}{3}\pi r^2 h$

Volume of a sphere: $V = \dfrac{4}{3}\pi r^3$

Distance, d, an object travels given its rate, r, and the time of travel, t: $d = rt$

The temperature in degrees Celsius given degrees Fahrenheit: $C = \dfrac{5}{9}(F - 32)$

The temperature in degrees Fahrenheit given degrees Celsius: $F = \dfrac{9}{5}C + 32$

The profit, P, after cost, C, is deducted from revenue, R: $P = R - C$

Chapter 2 Review Exercises

For Exercises 1–6, answer true or false.

[2.1]   **1.** $6x + 3y = 7x^2 - 4y$ is an equation.

[2.1]   **2.** $5(x + 2) = 10 - 3x + 8x$ is an identity.

[2.1]   **3.** 1 is a solution for $b^2 - 4b = 4$.

[2.5]   **4.** "Five less than a number is fourteen" translates to $5 - n = 14$.

[2.2]   **5.** $5x - 3y = 2$ is a linear equation.

[2.6]   **6.** $3 \le x < 5$ is a compound inequality.

For Exercises 7–9, fill in the blank.

[2.1]   **7.** If an equation is an _____, then every real number for which its expressions are defined is a solution.

[2.2]   **8.** An equation that is made of expressions in which each variable term contains a single variable raised to an exponent equal to 1 is a _____ equation.

[2.3]   **9.** To solve equations:

    1. _____ both sides of the equation as needed.

      a. _____ to clear parentheses.

      b. Clear fractions or decimals by _____ by their _____.

      c. _____ like terms.

    2. Use the _____ principle so that all variable terms are on one side of the equation and all constants are on the other side.

    3. Use the _____ principle to clear any remaining coefficient.

[2.1]   **10.** In your own words, explain how to check a potential solution.

[2.1] For Exercises 11–16, check to see if the given number is a solution for the given equation.

11. $y + 6 + 3y - 4 = 14; y = 3$

12. $\frac{1}{4}(n - 1) = \frac{1}{2}(n + 1); n = 5$

Equations and Inequalities

13. $x^2 + 6.1 = 3x - 0.1; x = -0.1$

14. $\frac{5}{6}m - 1 = \frac{m}{3} + \frac{1}{2}; m = 3$

Exercises 11–64

15. $\frac{3}{5}(a + 10) = -2; a = -20$

16. $6.1(7.2 + b) = 43.92; b = 0$

[2.2–2.3] *For Exercises 17–30, solve and check.*

17. $-6x = 30$

18. $-\dfrac{2}{3}y = -12$

19. $4n - 9 = 15$

20. $5(2a - 3) - 6(a + 2) = -7$

21. $12t + 9 = 4t - 7$

22. $2(c + 4) - 5c = 17 - 8c$

23. $1 - 2(3c - 5) = 10 - 6c$

24. $5(b - 1) - 3b = 7b - 5(1 + b)$

25. $5(u - 1) - (u - 2) = 10 - (2u + 1)$

26. $\dfrac{3}{4}m - \dfrac{1}{2} = \dfrac{2}{3}m + 1$

27. $\dfrac{1}{12}(4x + 9) = \dfrac{2}{3}x$

28. $\dfrac{5}{3} + 7x - 2x = \dfrac{1}{3}(9x + 2)$

29. $2 - \dfrac{2}{3}m = 6\left(m - \dfrac{1}{3}\right) - 4m + 2$

30. $1.4x - 0.5(9 - 6x) = 6 + 2.4x$

[2.4] *For Exercises 31–40, solve each for the indicated variable.*

31. $x + y = 1$; for x

32. $a^2 + b^2 = c^2$; for a^2

33. $C = \pi d$; for d

34. $E = mc^2$; for m

35. $A = \dfrac{1}{2}bh$; for h

36. $F = \dfrac{mv^2}{r}$; for m

37. $V = \dfrac{1}{3}\pi r^2 h$; for h

38. $y = mx + b$; for m

39. $P = 2l + 2w$; for w

40. $A = \dfrac{1}{2}h(a + b)$; for h

[2.5] *For Exercises 41–44, translate to an equation and solve.*

41. Six times a number is the same as negative eighteen.

42. Three subtracted from half of a number is equal to nine less than one-fourth of the number.

43. One less than two times the sum of a number and four results in one.

44. Twice the difference of a number and one added to triple the number yields twenty less than six times the number.

[2.6] *For Exercises 45–50, a. Solve.*
           b. Write the solution set in set-builder notation.
           c. Write the solution set in interval notation.
           d. Graph the solution set.

45. $-6x > 18$

46. $-3x + 2 \le 5$

47. $1 - 2z + 4 \le -3(z + 1) + 7$

48. $\dfrac{1}{4}v < 3 + v$

49. $\dfrac{1}{2}(m - 1) > 5 - \dfrac{7}{2}$

50. $2c - (5c - 7) + 4c \ge 8 - 10$

For Exercises 51–54, translate to an inequality and solve.

51. Thirteen is greater than three minus ten times a number.

52. Three increased by twice a number is less than three times the difference of a number and five.

53. Negative one-half of a number is greater than or equal to four.

54. Negative two is less than or equal to one minus one-fourth of a number.

For Exercises 55–64, solve.

[2.1] **55.** Best Buy is selling an oak entertainment center. Elijah is concerned it will take up too much room in his den. Using the figure shown, determine the total area.

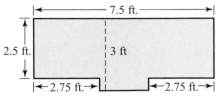

[2.1] **56.** Find the area of the figure shown.

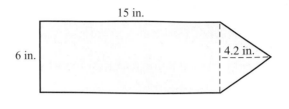

[2.1] **57.** A long-distance company uses the formula $C = 3 + 0.05m$ to describe a calling plan where C is the total cost of a call lasting m minutes. If a customer uses 415 minutes, what is the total bill?

[2.3] **58.** Bonnie is making a shower curtain. The pattern says the curtain will use 5760 square inches. If the length is to be 72 inches, how wide will the curtain be?

[2.3] **59.** The area of the trapezoid shown is 58 square inches. Find the missing length.

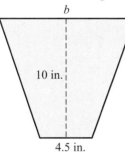

[2.3] **60.** Randy is snowskiing at Lake Tahoe. He knows the trail is 1.8 miles long and he is one-third of the way down. How far does he have left to ski?

[2.3] **61.** The surface area of a storage trunk is 2610 square inches. If the height of the trunk is 15 inches and the width is 36 inches, what is the length?

[2.6] **62.** The cost of border material limits the circumference of a circular flower garden to 10 feet. What is the range of values for the diameter of the garden?

[2.6] **63.** Terrel has to travel at least 350 miles each day to keep on his schedule. If he averages 60 miles per hour, what is the range of values for his travel time?

[2.6] **64.** A company has a first-quarter profit goal of at least $350,000. If the projected cost is $475,000, what range of values can revenue have to meet the profit goal?

Chapter 2 Practice Test

1. Is $5m - 14 = 21$ an expression or an equation? Why?

2. Is $5y + 14 = y^2 - 9$ linear or nonlinear? Why?

3. Check to see if 3 is a solution for $3x^2 + x = 10x$.

4. Check to see if $\dfrac{2}{3}$ is a solution for $4 - 3(x + 2) = 0$.

For Exercises 5–10, solve and check.

5. $6x + 9 = -9$

6. $-5.6 + 8a = 7a + 7.6$

7. $8 + 5(x - 3) = 9x - (6x + 1)$

8. $1 - 3(c + 4) + 8c = 1 - 42 + 5(c + 7)$

9. $\dfrac{1}{2}x + \dfrac{5}{2} = \dfrac{3}{2} - x$

10. $-\dfrac{1}{8}(6m - 32) = \dfrac{3}{4}m - 2$

For Exercises 11–14, solve for the indicated variable.

11. $2x + y = 7;\ x$

12. $I = Prt;\ r$

13. $A = \dfrac{1}{2}bh;\ h$

14. $C = \dfrac{5}{9}(F - 32);\ F$

For Exercises 15 and 16, write the solution set in set-builder notation and interval notation, then graph the solution set.

15. $x \geq 3$

16. $-1 \leq x < 4$

For Exercises 17–20: a. Solve.
b. Write the solution set in set-builder notation.
c. Write the solution set in interval notation.
d. Graph the solution set.

17. $-5m + 3 > -12$

18. $-2(2 - x) \geq -20$

19. $6(p - 2) - 5 > 3 - 4(p + 6)$

20. $\frac{1}{5}(l + 10) \geq \frac{1}{2}(l + 5)$

For Exercises 21-24, translate and then solve.

21. Two-thirds of a number minus one-sixth is the same as twice the number.

22. The product of negative seven and a number added to twelve yields five.

23. Three subtracted from five times the difference of a number and two is equal to ten minus four times the difference of the number and one.

24. One minus a number is greater than twice the number.

For Exercises 25–30, solve.

25. A carpenter needs to determine the amount of wood required to trim a window. If the window is 4 feet by 6 feet, what is the perimeter of the window?

26. Find the area of the figure shown.

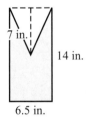

7 in.
14 in.
6.5 in.

27. The perimeter of a trapezoid shown is 63 centimeters. Find the length of the missing side.

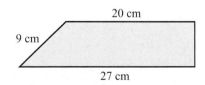

20 cm
9 cm
27 cm

28. A rent-to-own store advertises a VCR for $10 down with $10 payments each week. The formula $c = 10 + 10w$ represents the cost, C, after making payments for w weeks. If the customer signs a contract for one year, what is the total cost of the VCR?

29. The area of a sandbox cannot exceed 9 square feet. If the width of the sandbox is to be 3 feet, what range of values can the length have?

30. In Dan's psychology class, if his average on five tests is at least 92, then he can be exempt from the final exam. His current scores are 90, 84, 89, and 96. What range of scores can he have on the last test that would allow him to be exempt from the final exam?

Problem Solving

"Difficulty, my brethren, is the nurse of greatness—
a harsh nurse, who roughly rocks her foster-children
into strength and athletic proportion."

—William Cullen Bryant (1794–1878), U.S. poet, editor

"It is one of man's curious idiosyncrasies to create
difficulties for the pleasure of resolving them."

—Joseph de Maistre (1753–1821),
French diplomat, philosopher

In Chapter 2, we learned to solve linear equations and developed some techniques for solving problems, namely using formulas and translating key words. In this chapter, we will expand the problem-solving techniques introduced in Chapter 2 and solve problems that are a little more complex. These problems will involve proportions, percents, and two unknown amounts.

Some of the exercises you encounter in this chapter will target a specific strategy such as translating key words or using a table to organize information. In the same way that an athlete uses exercise to target specific muscles, we use rather contrived problems to exercise specific problem-solving strategies and techniques. Most of the exercises an athlete performs in weight training are not sport-specific, meaning they are generic exercises designed to strengthen raw muscle tissue. Similarly, many of the problems in the chapter will not be what you would encounter in your job or everyday life. Rather, they are designed to exercise the strategies and techniques of problem solving.

3.1 Ratios and Proportions

OBJECTIVES
1. Solve problems involving ratios.
2. Solve for a missing number in a proportion.
3. Solve proportion problems.
4. Use proportions to solve for missing lengths in figures that are similar.

OBJECTIVE 1. Solve problems involving ratios. In this section we will solve problems involving **ratios**.

DEFINITION *Ratio:* A comparison of two quantities using a quotient.

Ratios are usually expressed in fraction form. When a ratio is written in English, the word *to* separates the numerator and denominator quantities.

The ratio of **12** to **17** translates to $\dfrac{12}{17}$.

↑ Numerator ↑ Denominator

Note: *The quantity preceding the word* **to** *is written in the numerator and the quantity following* **to** *is written in the denominator.*

EXAMPLE 1 A small college has 2450 students and 140 faculty. Write the ratio of students to faculty in simplest form.

Solution The ratio of students to faculty $= \dfrac{2450}{140} = \dfrac{35}{2}$.

The ratio in Example 1 indicates that there are 35 students for every 2 faculty members. Expressing a ratio as a **unit ratio** often makes the ratio easier to interpret.

DEFINITION *Unit ratio:* A ratio with a denominator of 1.

To express the ratio from Example 1 as a unit ratio, we divide.

$$\frac{2450}{140} = \frac{35}{2} = \frac{17.5}{1}$$

The unit ratio $\dfrac{17.5}{1}$ indicates that there are 17.5 students for every faculty member at the college.

ANSWER

$\dfrac{\$0.228}{1\text{ oz.}}$

YOUR TURN The price of a 12.5-ounce box of cereal is $2.85. Write the unit ratio of price to weight.

OBJECTIVE 2. Solve for a missing number in a proportion. Some problems involve two equal ratios in a **proportion**.

DEFINITION   *Proportion:* An equation in the form $\dfrac{a}{b} = \dfrac{c}{d}$, where $b \neq 0$ and $d \neq 0$.

In Section 1.2 we discussed equivalent fractions and saw that fractions like $\dfrac{3}{8}$ and $\dfrac{6}{16}$ are equivalent because $\dfrac{3}{8}$ can be rewritten as $\dfrac{6}{16}$ and $\dfrac{6}{16}$ can be reduced to $\dfrac{3}{8}$. We also saw that when fractions are equivalent, the cross products are equal.

$$16 \cdot 3 = 48 \qquad\qquad 8 \cdot 6 = 48$$

$$\dfrac{3}{8} \;\text{\Large$\times$}\; \dfrac{6}{16}$$

Let's see why all proportions have equal cross products. We begin with the general form of a proportion from the definition $\dfrac{a}{b} = \dfrac{c}{d}$, where $b \neq 0$ and $d \neq 0$. We can use the multiplication principle of equality to clear the fractions by multiplying both sides by the LCD. In this case, the LCD is bd.

$$\dfrac{\overset{1}{\cancel{b}}d}{1} \cdot \dfrac{a}{\underset{1}{\cancel{b}}} = \dfrac{b\overset{1}{\cancel{d}}}{1} \cdot \dfrac{c}{\underset{1}{\cancel{d}}}$$

$$ad = bc$$

Note that ad and bc are the cross products.

Conclusion   If two ratios are equal, then their cross products are equal.

RULE   **Proportions and Their Cross Products**

If $\dfrac{a}{b} = \dfrac{c}{d}$, where $b \neq 0$ and $d \neq 0$, then $ad = bc$.

We can use cross products to solve for a missing number in a proportion.

PROCEDURE   **Solving a Proportion**

To solve a proportion using cross products:

1. Calculate the cross products.
2. Set the cross products equal to one another.
3. Use the multiplication principle of equality to isolate the variable.

EXAMPLE 2 Solve.

a. $\dfrac{7}{16} = \dfrac{x}{10}$

Solution $10 \cdot 7 = 70$ $16 \cdot x = 16x$ Calculate the cross products.

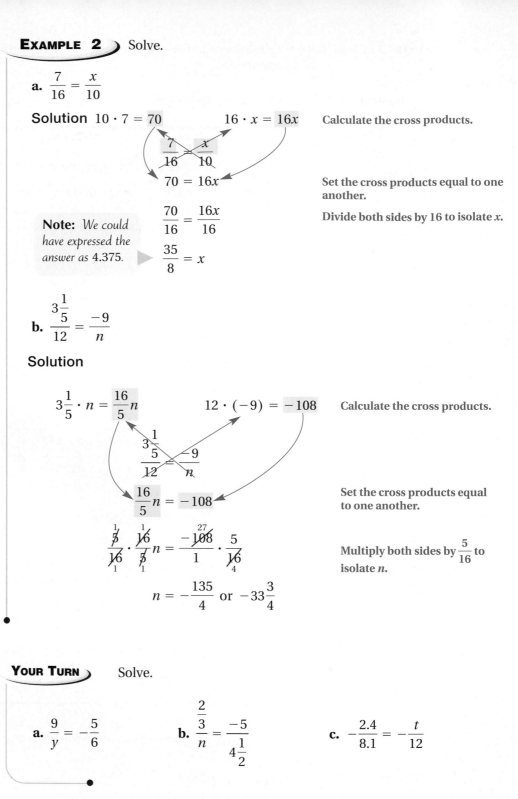

$70 = 16x$ Set the cross products equal to one another.

Note: *We could have expressed the answer as 4.375.*

$\dfrac{70}{16} = \dfrac{16x}{16}$ Divide both sides by 16 to isolate x.

$\dfrac{35}{8} = x$

b. $\dfrac{3\frac{1}{5}}{12} = \dfrac{-9}{n}$

Solution

$3\dfrac{1}{5} \cdot n = \dfrac{16}{5}n$ $12 \cdot (-9) = -108$ Calculate the cross products.

$\dfrac{16}{5}n = -108$ Set the cross products equal to one another.

$\dfrac{\cancel{5}^{1}}{\cancel{16}} \cdot \dfrac{\cancel{16}^{1}}{\cancel{5}}n = \dfrac{-\cancel{108}^{27}}{1} \cdot \dfrac{5}{\cancel{16}_{4}}$ Multiply both sides by $\dfrac{5}{16}$ to isolate n.

$n = -\dfrac{135}{4}$ or $-33\dfrac{3}{4}$

YOUR TURN Solve.

a. $\dfrac{9}{y} = -\dfrac{5}{6}$ b. $\dfrac{\frac{2}{3}}{n} = \dfrac{-5}{4\frac{1}{2}}$ c. $-\dfrac{2.4}{8.1} = -\dfrac{t}{12}$

ANSWERS

a. $-\dfrac{54}{5}$ or -10.8 b. $-\dfrac{3}{5}$ c. $3.\overline{5}$

OBJECTIVE 3. Solve proportion problems. In a typical proportion problem, you are given a ratio and then asked to find an equivalent ratio. The key to translating problems to a proportion is to write the proportion so that the numerators and denominators correspond to each other in a logical way.

EXAMPLE 3 A company determines that it spends $254.68 every 5 business days on equipment failure. At this rate, how much would the company spend over 28 business days?

Understand We are to find the amount spent on equipment failure over 28 days if the company spends $254.68 every 5 days. Translating $254.68 every 5 days to a ratio, we have $\dfrac{\$254.68}{5 \text{ days}}$.

Plan *At this rate* indicates that $\dfrac{\$254.68}{5 \text{ days}}$ stays the same for the 28 days, so we can use a proportion. If we let n represent the unknown amount spent over the 28 days, we have a ratio of $\dfrac{\$n}{28 \text{ days}}$. We create the proportion by setting those ratios equal to each other.

Execute $\dfrac{\$254.68}{5 \text{ days}} = \dfrac{\$n}{28 \text{ days}}$

Note: *The units in the numerators and denominators correspond.*
Tip: *One approach to setting up a proportion is to write the ratios so that the numerator units match and the denominator units match.*

$28 \cdot 254.68 = 7131.04 \qquad 5 \cdot n = 5n$

$$\frac{254.68}{5} \diagdown \diagup \frac{n}{28}$$

$7131.04 = 5n$ — **Set the cross products equal.**

$\dfrac{7131.04}{5} = \dfrac{5n}{5}$ — **Divide both sides by 5 to isolate n.**

$1426.208 = n$

Answer Because the answer is in dollars, we round to the nearest hundredth. The company spends $1426.21 over a 28-day cycle.

Check Do a quick estimate to verify the reasonableness of the answer. The company spends $254.68 every 5 business days. Since 28 days is a little less than 6 times the number of days, the company must spend a little less than 6 times the $254.68 spent in the 28 business days. To estimate the calculation, we round $254.68 to $250. We then have $6 \cdot \$250 = \$1500$, which is reasonably close to $1426.21.

Checking Proportion Problems

One of the nice things about proportion problems is that if you set the proportion up incorrectly, you will usually get an unreasonable answer. A quick estimate check can be a fast way to recognize when you have set up a proportion incorrectly. For example, suppose we had incorrectly set up Example 4 this way:

$$\frac{\$254.68}{5 \text{ days}} = \frac{28 \text{ days}}{\$n}$$

Warning: This setup is incorrect because the numerators and denominators do not correspond.

$254.68 \cdot n = 254.68n \qquad\qquad 5 \cdot 28 = 140$ **Cross multiply.**

$$\frac{254.68}{5} \times \frac{28}{n}$$

$254.68n = 140$ **Set the cross products equal.**

$$\frac{254.68n}{254.68} = \frac{140}{254.68}$$

Divide both sides by 254.68 to isolate n.

$n \approx 0.55$ **Warning:** This answer indicates that the company spends only $0.55 over 28 business days based on spending $254.68 every 5 business days, which is clearly too low.

There are many correct ways to set up a proportion for a particular problem. For example, following are some other correct ways we could have set up the proportion for Example 4. Note that the proportions involve rates that correspond with each other.

$$\frac{5 \text{ days}}{\$254.68} = \frac{28 \text{ days}}{\$n} \qquad \frac{5 \text{ days}}{28 \text{ days}} = \frac{\$254.68}{\$n} \qquad \frac{28 \text{ days}}{5 \text{ days}} = \frac{\$n}{\$254.68}$$

Think:

| In 5 days, $254.68 is spent. | In 28 days, $n is spent. |

Think of this version as an analogy:

If 5 days means $254.68, then 28 days means $n.

This version is a rearrangement of the previous version.

PROCEDURE   Solving Proportion Application Problems

To solve proportion problems:

1. Set up the given ratio any way you wish.
2. Set the given ratio equal to the other ratio with the unknown so that the numerators and denominators correspond logically.
3. Solve using cross products.

YOUR TURN   Pam drove 351.6 miles using 16.4 gallons of gasoline. At this rate, how much gasoline would it take for her to drive 1000 miles?

ANSWER

≈ 46.6 gal.

OBJECTIVE 4. Use proportions to solve for missing lengths in figures that are similar. Consider triangles *ABC* and *DEF*.

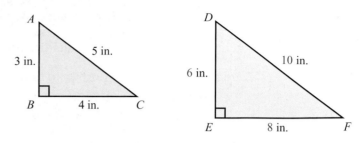

Though the triangles are different in size, their corresponding angle measurements are **congruent**.

DEFINITION   *Congruent angles:* Angles that have the same measure. The symbol for congruent is ≅.

In symbols, we write: $\angle B \cong \angle E$, $\angle A \cong \angle D$, and $\angle C \cong \angle F$.

    Besides having all angles congruent, notice that the sides of triangle *DEF* are twice as long as the corresponding sides of triangle *ABC*. Or, we could say that the sides of triangle *ABC* are all half as long as the corresponding sides of triangle *DEF*. In other words, the corresponding side lengths are proportional.

Ratio of Triangle *DEF*
to
Triangle *ABC* → $\dfrac{6}{3} = \dfrac{8}{4} = \dfrac{10}{5} = \dfrac{2}{1}$

or

Ratio of Triangle *ABC*
to
Triangle *DEF* → $\dfrac{3}{6} = \dfrac{4}{8} = \dfrac{5}{10} = \dfrac{1}{2}$

    Because the corresponding angles in the two triangles are congruent and the side lengths are proportional, we say these are **similar figures**.

DEFINITION   *Similar figures:* Figures with congruent angles and proportional side lengths.

EXAMPLE 4 The two trapezoids shown are similar. Find the missing lengths.

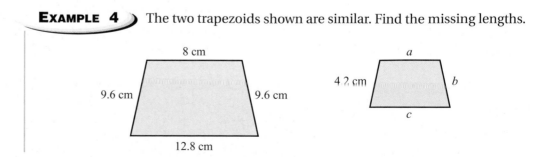

Understand   Since the figures are similar, the lengths of the corresponding sides are proportional. The corresponding sides can be shown as follows:

Ratio of larger trapezoid →
to
Smaller trapezoid → $\dfrac{9.6}{4.2} = \dfrac{8}{a} = \dfrac{9.6}{b} = \dfrac{12.8}{c}$

Plan   Write a proportion to find each missing side length.

Execute   It is not necessary to write a proportion to find *b*. Since two sides in the larger trapezoid measure 9.6 cm, the corresponding sides in the smaller trapezoid must be equal in length, so we can conclude that $b = 4.2$ cm.

To find *a*:

$$\dfrac{9.6}{4.2} = \dfrac{8}{a}$$

$$a \cdot 9.6 = 9.6a \qquad 4.2 \cdot 8 = 33.6$$

$$\dfrac{9.6}{4.2} \diagup\!\!\!\!\diagdown \dfrac{8}{a}$$

$$9.6a = 33.6$$

$$\dfrac{9.6a}{9.6} = \dfrac{33.6}{9.6}$$

$$a = 3.5$$

To find *c*:

$$\dfrac{9.6}{4.2} = \dfrac{12.8}{c}$$

$$c \cdot 9.6 = 9.6c \qquad 4.2 \cdot 12.8 = 53.76$$

$$\dfrac{9.6}{4.2} \diagup\!\!\!\!\diagdown \dfrac{12.8}{c}$$

$$9.6c = 53.76$$

$$\dfrac{9.6c}{9.6} = \dfrac{53.76}{9.6}$$

$$c = 5.6$$

Answer   The missing lengths are: $a = 3.5$ cm, $b = 4.2$ cm, and $c = 5.6$ cm.

Check   We can verify the reasonableness of the answers by doing quick estimates of the missing side lengths. Comparing the smaller trapezoid to the larger, we see that the 4.2-cm side is a little less than half of the corresponding 9.6-cm side in the larger trapezoid. Each of the corresponding sides should follow this same rule. The 8-cm side of the larger trapezoid should correspond to a length of a little less than 4 for side *a* of the smaller trapezoid, and the 12.8-cm side should correspond to a length of a little less than 6.4 for side *c* on the smaller trapezoid. These do check out.

YOUR TURN   Find the missing lengths in the similar shapes.

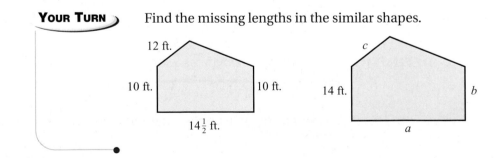

3.1 Exercises

1. In the phrase "the ratio of the longest side to shortest side," how do you determine the numerator and denominator of the ratio?

2. In your own words, define a proportion.

3. If $\dfrac{a}{b} = \dfrac{c}{d}$, where $b \neq 0$ and $d \neq 0$, is true, then what can we say about the cross products ad and bc?

4. Explain how to solve for a missing number in a proportion.

5. How do you set up and solve a word problem that translates to a proportion?

6. Why are $\dfrac{5}{8} \cdot \dfrac{x}{10}$ and $\dfrac{5}{8} = \dfrac{x}{10}$ different?

For Exercises 7–16, write the ratio in simplest form.

7. One molecule of lactose contains 12 carbon atoms, 22 hydrogen atoms, and 11 oxygen atoms.

   a. Write the ratio of carbon atoms to total atoms in the molecule.

   b. Write the ratio of carbon atoms to hydrogen atoms.

   c. Write the ratio of hydrogen atoms to oxygen atoms.

Of Interest

Lactose is a molecule found in milk. Its chemical formula is $C_{12}H_{22}O_{11}$. Notice that the subscripts indicate the number of atoms of each element in the molecule.

8. One molecule of hematcin contains 16 carbon atoms, 12 hydrogen atoms, and 6 oxygen atoms.

    a. What is the ratio of hydrogen atoms to the total number of atoms in the molecule?

    b. What is the ratio of carbon atoms to hydrogen atoms?

    c. What is the ratio of carbon atoms to oxygen atoms?

9. The roof of a house rises 0.75 feet for every 1.5 feet of horizontal length. Write the ratio of rise to horizontal length as a fraction in simplest form.

10. A steep mountain road has a grade of 8%, which means that the road rises 8 feet vertically for every 100 feet it travels horizontally. Write the grade as a ratio in simplest form.

11. In a certain gear on a bicycle, the back wheel rotates $1\frac{1}{3}$ times for every 2 rotations of the pedals. Write the ratio of rotations of the back wheel to pedal rotations as a fraction in simplest form.

12. A recipe calls for $2\frac{1}{2}$ cups of flour and $1\frac{1}{3}$ cups of milk. Write the ratio of milk to flour as a fraction in simplest form.

13. In 2004, La-Z-Boy's stock earned $0.92 per share. The current price of La-Z-Boy stock is $15.15. Write a unit ratio that expresses the price to earnings. What does this ratio indicate?

14. The price of a 10.5-ounce can of soup is $1.68. Write the unit ratio that expresses the price to weight. What does this ratio indicate?

15. An 8-ounce container of sour cream costs $1.42. A 24-ounce container of the same brand of sour cream costs $3.59. Which is the better buy? Why?

16. The price of a 12-ounce box of corn flakes is $3.59, and an 18-ounce box of the same cereal is $4.59. Which is the better buy? Why?

For Exercises 17 and 18, use the following graph.

17. **a.** Write the unit ratio of convictions to total cases in 1994.

    b. Write the unit ratio of convictions to total cases in 2002.

    c. Compare the ratios for 1994 to 2002. What do these ratios indicate?

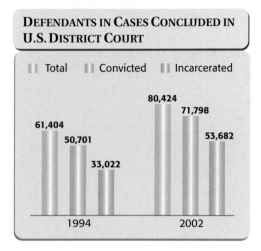

Source: Bureau of Federal Justice Statistics

18. **a.** Write the unit ratio of incarcerated to convictions in 1994.

    b. Write the unit ratio of incarcerated to convictions in 2002.

    c. Compare the ratios for 1994 to 2002. What do these ratios indicate?

For Exercises 19 and 20, use the following table.

Median Annual Income of Year-Round Full-Time Workers 25 Years Old and Over, by Level of Education Completed and Sex: 1990 to 2001 (constant 2001 dollars)

	Men				Women		
Year	All Education Levels	High School Graduate	Bachelor's Degree	Year	All Education Levels	High School Graduate	Bachelor's Degree
1990	41,644	36,115	53,168	1990	28,959	24,822	37,963
1992	40,465	34,435	52,202	1992	29,208	24,523	38,280
1994	39,961	33,504	52,178	1994	29,157	24,346	37,931
1996	40,208	34,663	51,748	1996	29,131	23,901	37,841
1998	41,195	34,200	55,852	1998	30,374	24,751	39,721
2000	42,227	35,279	57,937	2000	31,190	25,681	41,565
2001	41,617	34,723	55,929	2001	31,356	25,303	40,994

Source: U.S. Department of Education, National Center for Education Statistics. *Digest of Education Statistics*, 2003

19. a. In 1990, what was the unit ratio of women's median income to men's median income for all educational levels? What does this ratio indicate?

b. Calculate the same unit ratio for the year 2001.

c. Compare the ratio you found in 2001 to the ratio you found in 1990. What do these ratios indicate?

20. a. In 1992, what was the unit ratio of women's median income to men's median income for high school graduates? What does this ratio indicate?

b. Calculate the same unit ratio for the year 1996.

c. Compare the ratio you found in 1992 to the ratio you found in 1996. What do these ratios indicate?

For Exercises 21–30, determine whether the ratios are equal.

21. $\dfrac{15}{25} \stackrel{?}{=} \dfrac{3}{5}$

22. $\dfrac{6}{8} \stackrel{?}{=} \dfrac{15}{20}$

23. $\dfrac{20}{25} \stackrel{?}{=} \dfrac{25}{30}$

24. $\dfrac{12}{20} \stackrel{?}{=} \dfrac{18}{30}$

25. $\dfrac{7}{4} \stackrel{?}{=} \dfrac{15}{6.2}$

26. $\dfrac{15}{21} \stackrel{?}{=} \dfrac{130.2}{93}$

27. $\dfrac{10}{14} \stackrel{?}{=} \dfrac{2.5}{3.5}$

28. $\dfrac{6.5}{14} \stackrel{?}{=} \dfrac{10.5}{42}$

29. $\dfrac{\frac{2}{5}}{\frac{3}{10}} \stackrel{?}{=} \dfrac{15}{10\frac{1}{2}}$

30. $\dfrac{3\frac{1}{4}}{5\frac{1}{2}} \stackrel{?}{=} \dfrac{4\frac{2}{3}}{6\frac{1}{2}}$

For Exercises 31–46, solve for the missing number.

31. $\dfrac{2}{3} = \dfrac{7}{x}$

32. $\dfrac{7}{8} = \dfrac{y}{40}$

33. $\dfrac{-9}{2} = \dfrac{63}{n}$

34. $\dfrac{11}{m} = \dfrac{132}{-24}$

35. $\dfrac{21}{5} = \dfrac{h}{2.5}$

36. $\dfrac{29}{7} = \dfrac{k}{1.75}$

37. $\dfrac{-17}{b} = \dfrac{8.5}{4}$

38. $\dfrac{1.6}{4.96} = \dfrac{-4.8}{c}$

39. $\dfrac{-1.5}{60} = \dfrac{-2\frac{7}{8}}{t}$

40. $\dfrac{u}{-4\frac{7}{16}} = \dfrac{-60}{1.5}$

41. $\dfrac{2\frac{1}{4}}{6\frac{2}{3}} = \dfrac{d}{10\frac{2}{3}}$

42. $\dfrac{\frac{2}{5}}{4\frac{1}{2}} = \dfrac{7\frac{2}{3}}{j}$

43. $\dfrac{2}{3} = \dfrac{8}{x+8}$

44. $\dfrac{x+5}{6} = \dfrac{7}{3}$

45. $\dfrac{4x-4}{8} = \dfrac{2x+1}{5}$

46. $\dfrac{3}{4} = \dfrac{x+1}{x+3}$

For Exercises 47–70, solve.

47. LaTonia drives 110 miles in 3 hours. At this rate, how far will she go in 5 hours?

48. If 90 feet of wire weighs 18 pounds, what will 110 feet of the same type of wire weigh?

49. Jodi pays taxes of $1040 on her house, which is valued at $154,000. She is planning to buy a house valued at $200,000. At the same rate, how much will her taxes be on the new house?

50. Gary notices that his water bill was $24.80 for 600 cubic feet of water. At that rate, what would the charges be for 940 cubic feet of water?

51. A company determines that it spends $1575 every 20 business days (one month) for equipment maintenance. How much does the company spend in 120 business days (six months)?

52. A quality-control manager samples 1200 units of her company's product and finds seven defects. If the company produces about 50,000 units each year, how many of those units would she expect to have a defect?

53. Carson is planning a trip to London. He knows that it takes 1.8946 U.S. dollars to equal a British pound. If he has $800 to take with him, how much will he have in British pounds?

54. Sydney is planning a trip to Mexico. It takes 11.083 Mexican pesos to equal 1 U.S. dollar. If she has $400 to take with her, how many pesos will she have?

55. Fletcher is building a model of an F-16 jet. If the scale is $1\frac{1}{2}$ inches to 1 foot, what length on the model represents 3.25 feet?

56. The legend on a map indicates that 1 inch = 500 feet. If a road measures $3\frac{3}{4}$ inches on the map, how long is the road in feet?

57. A recipe for pancakes calls for $1\frac{1}{2}$ teaspoons of sugar to make 12 pancakes. How many teaspoons of sugar should be used to make 20 pancakes?

58. Mary Kennon learns in science class that baking powder can be made by mixing 21 parts of sodium bicarbonate with 47 parts of cream of tartar. She has already used 5 ounces of cream of tartar. How much sodium bicarbonate does she need to mix?

59. Ford estimates that its 2005 Expedition will travel 532 highway miles on one tank of gas. If the gas tank of the Expedition holds 28 gallons, how far can a driver expect to travel on 20 gallons?

60. Chevrolet estimates that its 2005 Tahoe will travel 520 highway miles on one tank of gas. If the gas tank of the Tahoe holds 26 gallons, how far can a driver expect to travel on 20 gallons?

61. Gregory Arakelian of the United States holds the world record for fastest typing. In 1991, he typed 158 words per minute in the Key Tronic World Invitational Type-Off. If he could maintain that rate, how long would it take him to type a term paper with 2500 words? (*Source: Guinness Book of World Records*, 2002)

62. Sean Shannon of Canada holds the world record for the fastest recital of Hamlet's soliloquy "To be or not to be." On August 30, 1995, he recited the 260-word soliloquy in 23.8 seconds. If he could maintain that rate, how long would it take him to recite Shakespeare's entire script of Hamlet, which contains 32,241 words. (*Sources: Guinness Book of World Records*, 2002, and William-Shakespeare.info)

63. A consultant recommends that about 9 pounds of rye grass seed be used for every 1000 square feet. If the client's lawn is about 25,000 square feet, how many pounds of seed should he buy?

64. Mark used 3 gallons of paint to cover 1200 square feet of wall space in his home. Mark's neighbor wants to paint 2000 square feet. Based on Mark's rate, how many gallons should the neighbor buy?

65. A carpet-cleaning company charges $49.95 to clean a rectangular living room that is 20 feet by 15 feet. The client asks the company to also clean the dining room, which measures 18 feet by 12 feet. If the company charges the same rate, how much should it cost to clean both the living room and dining room?

66. A painter charged $320 to paint two walls that measured 12 feet by 9 feet and two walls that measured 10 feet by 9 feet. The client asks him to return to paint two walls that measure 15 feet by 12 feet and two walls that measure 18.5 feet by 12 feet. If the painter charges the same rate, what should the price be?

67. A state wildlife department wants to estimate the number of deer in a state forest. In one month, it captures 25 deer, tags them, and releases them. Later, it captures 45 deer and finds that 17 are tagged. Assuming that the ratio of tagged deer to total deer in the forest remains constant, estimate the number of deer in the preserve.

68. A marine biologist is studying the population of trout in a mountain lake. During one month she catches 225 trout, tags them, and releases them. Later, she captures 300 trout and finds that 48 are tagged. Assuming that the ratio of tagged trout to total trout in the lake remains constant, estimate the population of trout in the lake.

69. The Hoover Dam has 17 generators and can supply all the electricity needed by a city of 750,000 people. If only 14 generators are running, how many people will be supplied with electricity?

70. Until the early 1950s the erosion rate for Niagara Falls was an average of 3 feet per year. The suggested rate of erosion may now be as low as 1 foot per 10 years for Niagara Falls. Given this rate, how much erosion may be expected over the next 50 years?

For Exercises 71–74, find the missing lengths in the similar figures.

71.

72.

73.

74.

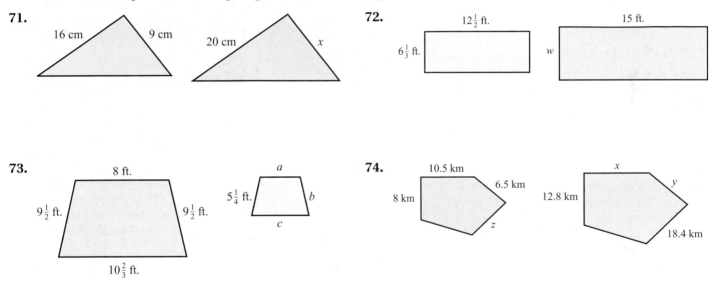

75. To estimate the height of the Eiffel Tower, a tourist uses the concept of similar triangles. The Eiffel Tower is found to have a shadow measuring 200 meters in length. The tourist has his own shadow measured at the same time. His shadow measures 1.2 meters. If he is 1.8 meters tall, how tall is the Eiffel Tower?

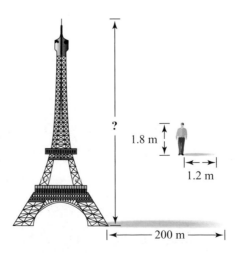

76. To estimate the height of the Great Pyramid in Egypt, the Greek mathematician Thales performed a procedure similar to that in Exercise 75, except that he supposedly used a staff instead of his body. Suppose the shadow from the staff was measured to be 6 feet and at

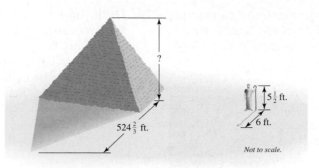

the same time the shadow of the pyramid was measured to be $524\frac{2}{3}$ feet. If the

staff was $5\frac{1}{2}$ feet tall, then what was the height of the pyramid?

77. To estimate the distance across a river, an engineer creates similar triangles. The idea is to create two similar right triangles as shown. Calculate the width of the river.

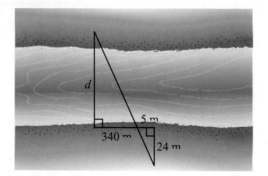

78. An engineer wants to estimate the distance across the Amazon River at his current location. The engineer creates two similar right triangles as shown. Calculate the distance across the river.

REVIEW EXERCISES

[1.1]   **1.** Is 0.58 a rational number? Why?

For Exercises 2–4, simplify.

[1.2]   **2.** $\dfrac{42}{350}$ [1.4]   **3.** $(0.15)(20.4)$ [1.4]   **4.** $45.2 \div 100$

For Exercises 5 and 6, solve.

[2.3]   **5.** $0.75n = 240$ [2.3]   **6.** $60p = 24$

3.2 Percents

OBJECTIVES 1. Write a percent as a fraction or decimal number.

2. Write a fraction or decimal number as a percent.

3. Translate and solve percent sentences.

4. Solve problems involving percents.

OBJECTIVE 1. Write a percent as a fraction or decimal number. The word *percent* is a compound word made from the prefix *per* and the suffix *cent*. The prefix *per* means "for each." The suffix *cent* comes from the Latin word *centum*, which means "100." Therefore, **percent** means "for each 100." In other words, percent is a ratio with the denominator 100.

DEFINITION *Percent:* A ratio representing some part out of 100.

The symbol for percent is %. For example, 20 percent, which means 20 out of 100, is written 20%. Note that the definition can be used to write percents as fractions or decimal numbers.

$$20\% = 20 \text{ out of } 100$$

$$= \frac{20}{100}$$

$$= \frac{1}{5} \text{ or } 0.2$$

PROCEDURE Rewriting a Percent

To write a percent as a fraction or decimal:

1. Write the percent as a ratio with 100 in the denominator.
2. Simplify to the desired form.

Note: When simplifying, remember that dividing a decimal number by 100 moves the decimal point two places to the left.

EXAMPLE 1) Write each percent as a decimal number and a fraction.

a. 20.5%

Solution Write 20.5 over 100, then simplify. In this case it might be helpful to write the decimal number first, then write the equivalent fraction from that decimal number.

$$20.5\% = \frac{20.5}{100} \qquad \text{Write as a ratio with 100 in the denominator.}$$

$$= 0.205 \qquad \text{Write the decimal form.}$$

$$= \frac{205}{1000} \qquad \text{Write the fraction form.}$$

$$= \frac{41}{200} \qquad \text{Simplify to lowest terms.}$$

b. $30\frac{1}{4}\%$

Solution Write $30\frac{1}{4}$ over 100, then simplify. In contrast to part a, you could find the fraction form first, and then find the decimal form.

$$30\frac{1}{4}\% = \frac{30\frac{1}{4}}{100} \qquad \text{Write as a ratio with 100 in the denominator.}$$

$$= 30\frac{1}{4} \div 100 \qquad \text{Rewrite the division.}$$

$$= \frac{121}{4} \cdot \frac{1}{100} \qquad \text{Write an equivalent multiplication.}$$

$$= \frac{121}{400} \text{ or } 0.3025 \qquad \text{Multiply.}$$

YOUR TURN) Write each percent as a fraction in simplest form and as a decimal.

a. 26%   **b.** 9.25%   **c.** $30\frac{2}{3}\%$

OBJECTIVE 2. Write a fraction or decimal number as a percent. We have seen that $20\% = \frac{20}{100} = \frac{1}{5}$ or

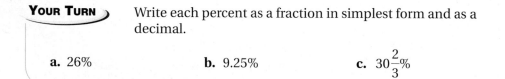

0.2. To write a percent as a fraction or a decimal, we replace the % sign with division by 100. To write a fraction or decimal as a percent, we should multiply by 100% and simplify.

ANSWERS

a. $\frac{13}{50}$ or 0.26

b. $\frac{37}{400}$ or 0.0925

c. $\frac{23}{75}$ or $0.30\overline{6}$

PROCEDURE Writing a Fraction or Decimal as a Percent

To write a fraction or decimal number as a percent:
1. Multiply by 100%.
2. Simplify.

EXAMPLE 2 Write as a percent.

a. $\dfrac{5}{8}$

Solution $\dfrac{5}{8} = \dfrac{5}{\overset{}{\underset{2}{8}}} \cdot \dfrac{\overset{25}{\cancel{100}}}{1}\%$ **Multiply by 100%.**

$= \dfrac{125}{2}\%$ **Simplify.**

$= 62\dfrac{1}{2}\%$ or 62.5%

b. 0.267

Solution $0.267 = 0.267 \cdot 100\%$ **Multiply by 100%.**

$= 26.7\%$

> **Connection** Multiplying an amount by 100% is a way of rewriting the amount because multiplying by 100% is equivalent to multiplying by 1. The percent symbol means $\dfrac{1}{100}$, so that 100% literally means:
>
> $$100 \cdot \% = 100 \cdot \dfrac{1}{100} = 1$$

YOUR TURN Write as a percent.

a. $\dfrac{5}{9}$

b. 2.4

OBJECTIVE 3. Translate and solve percent sentences. To solve problems involving percents, it is often helpful to reduce the problem to the following simple percent sentence:

A **percent** of a **whole** is a **part** of the whole.

Note that there are three pieces in the simple sentence: the *percent*, the *whole*, and the *part*. Any one of those three pieces could be unknown in a problem.

	A **percent** of	a **whole**	is a **part**.
Unknown part:	42% of	68	is what amount?
Unknown whole:	76% of	what number	is 63.84?
Unknown percent:	What percent of	72	is 63?

To solve for an unknown piece of a simple percent sentence, we can translate the sentence to an equation. We will look at two methods of translation: 1. translating word for word, and 2. translating to a proportion.

Method 1. Translate the sentence word for word:

1. Select a variable for the unknown.
2. Translate the word *is* to an equal sign.
3. If *of* is preceded by the percent, then it is translated to multiplication.
If *of* is preceded by a whole number, translate it to division.

Method 2. Translate to a proportion by writing the following form:

$$\text{Percent} = \frac{\text{Part}}{\text{Whole}}$$

where the percent is expressed as a fraction with a denominator of 100.

First, let's consider the case where the part is unknown.

Simple Percent Sentence with an Unknown Part

EXAMPLE 3   42% of 68 is what number?

Note: *This question can also be worded this way: "What is 42% of 68?"*

Solution   Notice the part is the unknown.

42%	of	68	is	what number?
↑		↑		↑
Percent	of	the whole	is	the part

Method 1.   Word for word translation: Since *of* is preceded by the percent, it means multiply, and *is* means equals.

Note: *To calculate, we write the percent as a decimal (or fraction).*

42% of 68 is what number?

$$0.42 \cdot 68 = n$$
$$28.56 = n$$

Method 2.   Proportion: The percent, which is a ratio, is equivalent to the ratio of the part to the whole, so we can write a proportion in the following form:

$$\text{Percent} = \frac{\text{Part}}{\text{Whole}}$$

Note: *This is 42% expressed as a ratio.*

$$\frac{42}{100} = \frac{n}{68} \begin{matrix} \leftarrow \text{Part} \\ \leftarrow \text{Whole} \end{matrix}$$

Note: *The whole amount will always follow* of.

$$2856 = 100n \qquad \text{Equate the cross products.}$$

$$\frac{2856}{100} = \frac{100n}{100} \qquad \text{Divide both sides by 100 to isolate } n.$$

$$28.56 = n$$

ANSWER

5.4

YOUR TURN   What number is 15% of 36?

Simple Percent Sentence with an Unknown Whole

Now consider the situation in which the whole amount is unknown.

EXAMPLE 4 76% of what number is 63.84?

Note: *This question can be worded: "63.84 is 76% of what number?"*

Solution Note the three pieces:

76%	of	what number	is	63.84?
↑		↑		↑
Percent	of	the whole	is	the part

Method 1. Word-for-word translation:

76% of what number is 63.84?

$$0.76 \quad \cdot \quad n \quad = 63.84$$

$$0.76n = 63.84$$

$$\frac{0.76n}{0.76} = \frac{63.84}{0.76}$$

$$n = 84$$

Method 2. Proportion:

Note: *This is 76% expressed as a ratio.*

$$\text{Percent} = \frac{\text{part}}{\text{whole}}$$

$$\frac{76}{100} = \frac{63.84}{n} \quad \begin{array}{l} \longleftarrow \boxed{\text{Part}} \\ \longleftarrow \boxed{\text{Whole}} \end{array}$$

$$76n = 6384 \qquad \text{Equate the cross products.}$$

$$\frac{76n}{76} = \frac{6384}{76} \qquad \text{Divide both sides by 76 to isolate } n.$$

$$n = 84$$

YOUR TURN 48 is 120% of what number?

ANSWER

40

Simple Percent Sentence with an Unknown Percent

Finally, consider the case in which the percent is the unknown.

EXAMPLE 5 What percent of 72 is 63?

Note: *This question can be worded:* "63 is what percent of 72?"
or
"What percent is 63 out of 72?"
or
"What percent is 63 of 72?"
Note how 72, the whole amount, always follows of.

Solution Note the pieces.

What percent   of   72   is   63?

Percent   of   the whole   is   the part

Method 1. Word-for-word translation:

What percent of 72 is 63?

$$p \cdot 72 = 63$$
$$72p = 63$$
$$\frac{72p}{72} = \frac{63}{72}$$
$$p = 0.875$$

Note: *The division yields a decimal number that must be written as a percent.*

To write 0.875 as a percent, we multiply by 100%.

Answer $0.875 \cdot 100\% = 87.5\%$

Method 2. Proportion:

Note: *This is the unknown percent written as a ratio. We use an upper case P here because the 100 in the proportion causes the solution to be slightly different from the solution to the equation in the word-for-word translation.*

$$\frac{P}{100} = \frac{63}{72}$$ ← Part
← Whole

$$72P = 6300$$ **Equate the cross products.**

$$\frac{72P}{72} = \frac{6300}{72}$$ **Divide both sides by 72 to isolate** *p.*

$$P = 87.5$$

Note: *In calculating the cross products, we multiplied by 100, so the solution is the percent value and we only need to add the percent sign.*

Answer   87.5%

As Example 5 shows, when the percent is unknown, the proportion method gives the answer as a percent, so you may want to consider using proportions in this case.

YOUR TURN   21.6 is what percent of 90?

ANSWER

24%

OBJECTIVE 4. Solve problems involving percents. If a problem involves a percent, it can be helpful to try to write the information in the form of a simple percent sentence.

PROCEDURE Solving Percent Application Problems

To solve problems involving percent:

1. Determine whether the percent, the whole, or the part is unknown.
2. Write the problem as a simple percent sentence (if needed).
3. Translate to an equation (word for word or proportion).
4. Solve for the unknown.

EXAMPLE 6 Twelve of the 28 students in an English class received an A. What percent of the class received an A?

Understand The unknown is the percent of the class that received an A. The number of students in the class is 28, which is the whole. The number of students that received an A, which is the part, is 12.

Plan Write a simple percent sentence, then solve. We will let P represent the unknown percent and we will translate to a proportion because the proportion's solution will be the percent value.

Execute What percent of 28 is 12?

$$\frac{P}{100} = \frac{12}{28} \xleftarrow{} \boxed{\text{Part}}$$
$$\xleftarrow{} \boxed{\text{Whole}}$$

$28P = 1200$ **Equate the cross products.**

$\dfrac{28P}{28} = \dfrac{1200}{28}$ **Divide both sides by 28 to isolate P.**

$P \approx 42.9$ ◁ **Note:** *We rounded to the nearest tenth.*

Answer About 42.9% of the students in the class received an A on their first paper.

Check Verify that 42.9% of the 28 students is about 12 students.

$$0.429 \cdot 28 = 12.012$$

Since 12.012 is reasonably close to 12, we can conclude that 42.9% is correct.

YOUR TURN On Monday, January 17, 2005, the *Morning Sentinel* reported it planned to cut 100 full-time jobs in a bid to return to profitability. This is 6% of the newspaper's workforce. How many employees does the *Sentinel* have?

ANSWER

1667

Percent Problems Involving Increase or Decrease

Sometimes a percent problem involves an increase or decrease of some initial amount. Sales tax, interest, and discount are situations in which an initial amount of money is increased or decreased by an amount of money that is a percent of the initial amount.

EXAMPLE 7 If the sales tax rate in a certain city is 6%, what is the total cost of:

Raisin Bran	$3.79
Milk	$1.29
Orange Juice	$1.49
Yogurt	$2.59

Understand The total cost is the purchase price plus the sales tax, which is a percent of the purchase price.

Plan We must first calculate the purchase price. We can then write a simple percent sentence and solve for the sales tax. Finally, we add the tax to the purchase price.

Execute Purchase price = 3.79 + 1.29 + 1.49 + 2.59 = $9.16.

Simple percent sentence: The sales tax is 6% of $9.16. We will translate word for word, letting x represent the unknown sales tax.

$$x = 0.06 \cdot 9.16$$
$$x = 0.5496$$

Because the tax is a dollar amount, we round to the nearest hundredth. The sales tax is $0.55.

Answer The total cost is $9.16 + $0.55 = $9.71.

Check Determine whether the total cost of $9.71 represents a 6% increase in the total sale of $9.16. To determine the percent of increase, we would need to calculate the amount of the increase by subtracting 9.16 from 9.71.

$$9.71 - 9.16 = 0.55 \qquad \text{This verifies that the final addition was correct.}$$

Now, to determine the percent of the increase we can set up a proportion.

$$\frac{P}{100} = \frac{0.55}{9.16} \longleftarrow \text{Part}$$
$$\longleftarrow \text{Whole}$$

$$9.16P = 55 \qquad \textbf{Equate the cross products.}$$

$$\frac{9.16P}{9.16} = \frac{55}{9.16} \qquad \textbf{Divide both sides by 9.16 to isolate } P.$$

$$P \approx 6.004 \qquad \blacktriangleleft \textbf{Note:} \textit{ This does not match 6\% exactly because we used the rounded sales tax amount, 0.55, instead of the exact value, 0.5496.}$$

Connection The steps in this check will help us understand how to approach a type of problem in which the percent of increase or decrease is unknown. We will see this type of problem in Example 8.

YOUR TURN An advertisement indicates that a coat is on sale at 30% off the marked price. If the marked price is $79.95, what will be the price after the discount?

ANSWER

$55.97

> **Connection** A clever shortcut for Example 7 is to recognize that we pay 100% of the initial price plus 6% in sales tax, so the total is 106% of the initial price.
>
> Total cost $= 1.06 \cdot 9.16 = 9.7096$, which rounds to $9.71
>
> A similar shortcut exists for the Your Turn problem involving discount. If we take 30% off the marked price, then we pay 70% of the marked price.
>
> Price after discount $= 0.7 \cdot 79.95 = 55.965$, which rounds to $55.97

The check in Example 7 is actually a common type of percent problem in which the percent of increase or decrease is unknown.

EXAMPLE 8 The Air Transport Association reports that airline fuel prices jumped from 96.2 cents in January of 2004 to 135 cents in January of 2005. What is the percent of the increase?

Understand We are to determine the percent of the increase given the initial price and the price after the increase.

Plan To determine the percent of the increase, we need the amount of the increase, which is found by subtracting the initial amount from the final amount.

Execute Amount of increase $= 135 - 96.2 = 38.8$.

Simple percent sentence: What percent of 96.2 is 38.8?

Since the percent is unknown, we will translate to a proportion letting P represent the unknown percent.

$$\frac{P}{100} = \frac{38.8}{96.2} \quad \overset{\longleftarrow \;\; \boxed{\text{Part}}}{\underset{\longleftarrow \;\; \boxed{\text{Whole}}}{}}$$

Note: *The part is the amount of the increase and the whole is the initial amount.*

$$96.2P = 3880 \quad \text{Equate the cross products.}$$

$$\frac{96.2P}{96.2} = \frac{3880}{96.2} \quad \text{Divide both sides by 72.1 to isolate } P.$$

$$P \approx 40.3$$

Answer The percent of the increase in fuel price is about 40.3%.

Check Determine whether increasing the initial price of 96.2 cents by 40.3% results in a price of 135 cents.

$$0.403(96.2) = n$$
$$38.8 \approx n$$

Note: *When rounded to the nearest tenth, this amount of the increase corresponds to our earlier calculation.*

Now add the amount of the increase: $96.2 + 38.8 = 135$.

In general, when solving for a percent of increase or decrease, the *part* will be the amount of increase or decrease and the *whole* will be the initial amount.

$$\frac{P}{100} = \frac{\text{Amount of increase/decrease}}{\text{Initial amount}}$$

Your Turn

On February 8, 2005, Wal-Mart's stock closed at a price of $53.20. On June 8, 2005, the stock closed at $47.57. Find the percent of the decrease in the price of Wal-Mart's stock. (*Source:* Motleyfool.com)

Answer

≈ 10.6%

3.2 Exercises

For Extra Help — MyMathLab — MyMathLab — Videotape/DVT — InterAct Math — Math Tutor Center — Math XL.com

1. Explain how to write a percent as a decimal or fraction.

2. Explain how to write a decimal or fraction as a percent.

3. When preceded by a percent, the word *of* indicates what operation?

4. Explain how to translate a simple percent sentence to a proportion.

5. When using word-for-word translation to find a percent, the calculation will not be a percent, whereas if you use the proportion method, the result will be a percent. Why?

6. When given an initial amount and a final amount after an increase, how do you find the percent of the increase?

For Exercises 7–18, write each percent as a decimal and as a fraction in simplest form.

7. 20%

8. 30%

9. 15%

10. 85%

11. 14.8%

12. 18.6%

13. 3.75%

14. 6.25%

15. $45\frac{1}{2}\%$

16. $65\frac{1}{4}\%$

17. $33\frac{1}{3}\%$

18. $18\frac{1}{6}\%$

For Exercises 19–38, write as a percent rounded to the nearest tenth.

19. $\dfrac{3}{5}$ **20.** $\dfrac{1}{5}$ **21.** $\dfrac{3}{8}$ **22.** $\dfrac{5}{8}$

23. $\dfrac{5}{6}$ **24.** $\dfrac{4}{9}$ **25.** $\dfrac{2}{3}$ **26.** $\dfrac{5}{11}$

27. 0.96   **28.** 0.42   **29.** 0.8   **30.** 0.7

31. 0.09   **32.** 0.01   **33.** 1.2   **34.** 3.58

35. 0.028   **36.** 0.065   **37.** 4.051   **38.** 0.007

For Exercises 39–58, translate word for word or to a proportion, then solve.

39. 80% of 35 is what number?

40. What number is 40% of 90?

41. 2.5% of 124 is what number?

42. 13.5% of 940 is what number?

43. What number is $9\dfrac{1}{4}$% of 64?

44. $3\dfrac{3}{4}$% of 24 is what number?

45. 120% of 86 is what number?

46. What number is 250% of 62.8?

47. 30% of what number is 15?

48. $45\dfrac{3}{5}$ is 5% of what number?

49. 16.4 is 20.5% of what number?

50. 30.2% of what number is 18.12?

51. 7.8 is $12\dfrac{1}{2}$% of what number?

52. $5\dfrac{1}{4}$% of what number is 1.26?

53. What percent of 45 is 9?

54. 15 is what percent of 40?

55. What percent of 38 is 39.9?

56. 73.44 is what percent of 68?

57. What percent is 18 out of 27?

58. 50 out of 60 is what percent?

For Exercises 59–98, solve.

59. Angela answers 86% of the questions on a psychology multiple choice test correctly. If there were 50 questions on the test, how many did she answer correctly?

60. Terra is a server at a restaurant. She serves a table of seven people, for which the restaurant automatically adds a service charge of 15% of the cost of the meal. If the meal costs $127.40, how much is the service charge?

61. A county's annual property taxes are 4% of the market value of the property. How much is the tax on a property valued at $120,000? If the tax is paid in monthly installments, what amount must be paid each month?

62. How much is the tax on a house valued at $230,000 in the same county as that in Exercise 61? If the tax is paid in monthly installments, what amount must be paid each month?

63. In general, lenders do not want home buyers to spend more than 28% of their gross monthly income on a house payment. If a buyer has a monthly gross income of $3210, what is the most a lender will allow for a mortgage payment?

64. In general, lenders do not want home buyers to spend more than 36% of gross income a month on debts. Using the buyer in Exercise 63, what is the most a lender will accept in total debt payments per month? If she makes the highest mortgage payment allowed each month, what does she have left to spend on other debt and still be within the lender's guidelines?

> *Of Interest*
> In finance, the ratio of the house payment to the gross monthly income is known as the *front-end ratio*. The ratio of the monthly debts to the gross monthly income is known as the *back-end ratio*. The front-end and back-end ratios, along with a person's credit history, are primary factors involved in determining loan qualification.

For Exercises 65 and 66, use the following circle graph.

65. How many teragrams of carbon dioxide were emitted by the United States in 2002?

66. How many teragrams of methane were emitted by the United States in 2002?

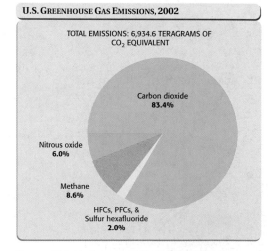

Source: U.S. Environmental Protection Agency

67. Terra, the server from Exercise 60, receives a tip of $8.50. If this was 20% of the cost of the meal, what was the cost of the meal?

68. When an item is sold on ebay, the seller must pay ebay 2% of the selling price. If ebay charges $22.50 for the sale of a guitar amplifier, what was the selling price of the amplifier?

69. A theater charges 2.5% of the ticket price as a handling fee to hold tickets at the "Will Call" booth. If the handling fee for four tickets is $7.50, what is the total cost of the tickets?

70. Carol's checking account earns 0.025% of her average daily balance as a dividend every month. If she receives a dividend of $0.15 one month, what was her average daily balance?

71. In the 2003–2004 season, Detroit Pistons player Richard Hamilton made 530 field goals out of 1170 attempts. What percent of his field goal attempts did he score?

72. As of June 1, 2004, Ichiro Suzuki of the Seattle Mariners had the number 1 batting average in the American League with 262 hits out of 704 at-bats. What percent of his at-bats are hits?

73. According to the 2000 census, the total population of the U.S. was 281,421,906. If there were 196,899,193 people 21 years of age or older, what percent of the total population was 21 years of age or older? *Source:* Bureau of the Census, Department of Commerce, 2000 Census)

74. In 2002, there were 56,747,000 households in the United States headed by married couples out of 109,297,000 households. What percent of U.S. households was headed by married couples in 2002? (*Source:* Bureau of the Census, Department of Commerce)

Of Interest
Batting averages are expressed as decimal numbers rounded to the nearest thousandth. For example, a player's batting average might be 0.285, which means that he has had a hit 28.5% of the number of times he's been at bat.

For Exercises 75–78, use the following table, which shows revenue (in thousands) for public elementary and secondary schools in the 2003–2004 school year for several states in New England. (Source: National Education Association)

	Total	Federal	State	Local
Massachusetts	13,692,829	709,402	3,864,414	9,119,013
New Hampshire	2,065,830	103,789	1,545,110	416,931
Vermont	1,182,553	75,966	843,280	263,307

75. What percent of New Hampshire's revenue was funded by state government?

76. What percent of Massachusetts' total revenue was funded by state government?

77. For each state, find the percent of the total funding that came from local sources. Which state had the greatest percentage?

78. For each state, find the percent of the total funding that came from the federal government. Which state had the greatest percentage?

79. A small company has the following costs in one quarter:

Category	Cost (in thousands)
Plant operations	124.2
Employee wages	248.6
Research and development	115.7

What percent of the company's total cost is spent on research and development?

80. A pharmaceutical company conducts a study on the side effects of a new drug. Following is a report of the data from the study.

	Number of Subjects
No side effects	1462
Headache	35
Rash	24
Fever	65

What percent of the subjects in the study experienced fever?

81. Margaret mixes 50 milliliters of a solution that is 10% HCl with a 100 milliliters of a solution that is 25% HCl. What percent of the resulting solution is HCl?

82. Leon invests $3400 in a stock that returns 12% in the first year and $2200 in stock that returns 15% during the same year. What is his total return as a percent of his total investment?

83. Carl purchases $124.45 in groceries. If the sales tax rate is 5%, what is the amount of the tax and the cost after tax is added?

84. Juanita purchases several books over the Internet that have a total price of $85.79. The online store charges a handling fee of 2.5% of the total cost. What is the handling fee and the price after the fee is added?

85. An appliance store has all refrigerators discounted 10% off the regular price. If the regular price of a refrigerator is $949, what is the amount of the discount and the price after the discount?

86. A furniture store has all its sofas discounted 40% off. If the original price of a particular sofa is $785.99, what is the amount of the discount and the price after the discount?

87. Video game sales in a department store increased from $2600 in November to $4550 in December. What was the amount of increase? What was the percent of increase?

88. Fifteen years ago, James inherited a marble-top table. It was appraised at $420. The insurance company has asked that he have it reappraised. It is now worth $1200. What was the amount of the increase? What is the percent of increase?

89. In 1989–90, about 580,000 bachelor's degrees were awarded to women in the United States. In 2003–04, about 780,000 bachelor's degrees were awarded to women. What was the percent of increase in degrees awarded to women? (*Source:* National Center of Education Statistics, U.S. Department of Education)

90. At 9:30 A.M. on February 18, 2005, the Dow Jones Industrial Average (DJIA) was 10,755. At 4:30 P.M., the DJIA was at 10,785. What was the percent of the increase? (*Source: USA Today,* February 18, 2005)

91. The price of a TI-83 calculator dropped from $120 to $79. What is the percent of decrease?

92. After the Vallina family purchased a smaller car, they realized their monthly gasoline bill dropped from $225 per month to $170 per month. What is the percent of decrease in the monthly gasoline bill?

93. In 1993, Los Angeles had 134 days in which the area failed to meet acceptable air quality standards. In 2002, the area had 80 days (an improvement) that did not meet acceptable air-quality standards. What was the percent of the decrease in the number of days that Los Angeles did not meet acceptable air-quality standards? (*Source:* U.S. Environmental Protection Agency, Office of Air Quality Planning and Standards)

94. In 1990, the population of Decatur, Illinois was 83,900. In 2003, the population was 79,285. What was the percent of decrease in population? (*Source:* U.S. Bureau of the Census, Department of Commerce)

95. In 1938, the federal hourly minimum wage was $0.25. In 2005, the hourly minimum wage was $5.15. What is the percent of increase in the minimum wage from 1938 to 2005? (*Source:* U.S. Department of Labor)

96. A small business owner has an audit performed and finds that the net worth of the company is $58,000. After two years of poor business, an audit shows the net worth is −$20,000. What is the percent of the decrease in the net worth of the business?

97. A report indicates that on Monday November 15, 2004, the price of Microsoft's stock closed at $27.39, which was a change of −8.6% from the previous day's closing price. What was the price of Microsoft's stock at closing on Friday November 12?

98. The U.S. Bureau of the Census reported a 61.8% increase in the number of single fathers from 1990 to 2000. If the number of single fathers in 2000 was 2,190,989, how many single fathers were there in 1990? (*Source: USA Today,* May 18, 2001)

Collaborative Exercises (Occupation Growth)

Complete the following table by calculating the amount of change and the percent of the increase for each occupation, then answer the questions. Round to the nearest tenth of a percent.

The 10 Fastest-Growing Occupations, 2002–2012

Occupation	Employment		Change	
	2002	2012	Amount	Percent
Medical assistants	364,600	579,400		
Network systems and data communications analysts	186,000	292,000		
Physician assistants	63,000	93,800		
Social and human service assistants	305,200	453,900		
Home health aides	579,700	858,700		
Medical records and health information technicians	146,900	215,600		
Physical therapist aides	37,000	54,100		
Computer software engineers, applications	394,100	573,400		
Computer software engineers, systems software	281,100	408,900		
Physical therapist assistants	50,200	72,600		

Source: Bureau of Labor Statistics, Office of Occupational Statistics and Employment Projections

1. In 2002, in which occupation were the greatest number of people employed?

2. By 2012, which occupation is projected to have the greatest number of people employed?

3. Explain why the two occupations you listed in Exercises 1 and 2 are not at the top of the list.

4. Which is a better indicator of the demand for people in a particular occupation: the number of people employed in a particular year, the amount of change projected from 2002 to 2012, or the percent of increase in employment from 2002 to 2012?

5. Based on your conclusions in Exercise 4, what occupation will have the greatest demand? What college majors might have the greatest potential for employment in that occupation? Write your conclusions and present them to the class.

REVIEW EXERCISES

[1.5]   **1.** Simplify: -3^0

[1.7]   **2.** Multiply: $-3(x + 7)$

[1.7]   **3.** Combine like terms: $7xy + 4x - 8xy + 3y$

For Exercise 4, translate to an equation and then solve.

[2.5]   **4.** Three times the sum of a number and nine is equal to eleven less than the number.

[3.1]   **5.** Solve: $\dfrac{1}{5} = \dfrac{n}{10.5}$

[3.1]   **6.** The figures shown are similar. Find the missing side length.

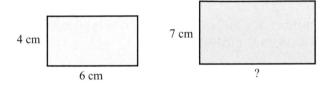

4 cm   6 cm   7 cm   ?

3.3   Problems with Two or More Unknowns

OBJECTIVES

1. Solve problems involving two unknowns.
2. Use a table in solving problems with two unknowns.

OBJECTIVE 1. Solve problems involving two unknowns. Sometimes problems have two or more unknown amounts. In general, if there are two unknowns, then there will be two relationships. Our approach to solving these problems will be to use the relationships to write an equation that we can solve. The following outline gives a more specific approach.

PROCEDURE   Solving Problems with Two or More Unknowns

To solve problems with two or more unknowns:
1. Determine which unknown will be represented by a variable.
   Tip: Let the unknown that is acted on be represented by the variable.
2. Use one of the relationships to describe the other unknown(s) in terms of the variable.
3. Use the other relationship to write an equation.
4. Solve the equation.

First, let's look at problems in which we simply translate key words.

EXAMPLE 1 Lindsey eats a snack consisting of 1 ounce of almonds and one oatmeal fruit bar. She is surprised to see that the ounce of almonds has 40 more calories than the fruit bar. If her snack has a total of 300 calories, how many calories are in the almonds and how many in the fruit bar?

Understand There are two unknowns in the problem: We must find the number of calories in 1 ounce of almonds and the number of calories in an oatmeal fruit bar. We are given two relationships:

Relationship 1: The number of calories in the almonds is 40 more than the number of calories in the fruit bar

Relationship 2: The total calories in the snack are 300.

Plan Translate each relationship to an equation, and then solve.

Execute Use the first relationship to determine which unknown will be represented by a variable.

Relationship 1: The number of calories in the almonds is 40 more than the number of calories in the fruit bar

$$\text{Calories in almonds} = 40 \text{ calories} + \text{Calories in the fruit bar}$$
$$\text{Calories in almonds} = \quad 40 \quad + \quad n$$

Note: *In general, we choose the unknown amount that is acted on to be the variable. Because 40 is added to the calories in the fruit bar, we will let n represent the calories in a fruit bar.*

Now use the second relationship to write an equation that we can solve for n.

Relationship 2: The total calories are 300.

$$\text{Calories in the fruit bar} + \text{Calories in the almonds} = 300$$
$$n \quad + \quad 40 + n \quad = 300$$

$$2n + 40 = 300 \qquad \text{Combine like terms.}$$
$$2n + 40 = 300$$
$$\underline{ -40 \quad -40} \qquad \text{Subtract 40 from both}$$
$$2n + 0 = 260 \qquad \text{sides.}$$
$$\frac{2n}{2} = \frac{260}{2} \qquad \text{Divide both sides by 2.}$$
$$n = 130$$

Answer Since n represents the number of calories in the fruit bar, 1 fruit bar has 130 calories. The almonds contain 40 more calories than the fruit bar, so the almonds contain $40 + 130 = 170$ calories.

Check Verify that the total calories is 300, which it is because $170 + 130 = 300$.

YOUR TURN One positive number is one-third of another positive number. The larger number minus the smaller number is equal to 15.

ANSWER

22.5 and 7.5

Geometry Problems

Sometimes problems do not give all the needed relationships in an obvious manner. This often occurs in geometry problems in which the definition of a geometry term provides the needed relationship. For example, we must know that *perimeter* means the sum of the lengths of the sides in order to solve a problem that involves perimeter.

EXAMPLE 2 A rectangular frame is to be built so that the length is 4 inches more than the width. The perimeter of the frame is to be 72 inches. Find the dimensions of the frame.

Understand Draw a picture and list the relationships.

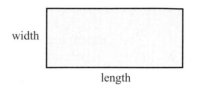

width

length

Relationship 1: The length is 4 inches more than the width.
Relationship 2: The perimeter must be 72 inches.

Plan Translate to an equation using the key words, and then solve the equation.

Execute Relationship 1: The length is 4 inches more than the width.

$$\text{Length} = 4 + \text{Width}$$
$$\text{Length} = 4 + w$$

Note: *Since 4 is added to the width and the length is isolated, we let w represent width.*

Now use the other given relationship to write an equation to solve.

Relationship 2: The perimeter is 72 inches. Recall that the perimeter is the sum of the lengths of all the sides.

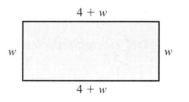

$4 + w$

w w

$4 + w$

$$\text{Perimeter} = 72$$

$$\text{Length} + \text{Width} + \text{Length} + \text{Width} = 72$$
$$4 + w + \quad w \quad + 4 + w + \quad w \quad = 72$$

$$4w + 8 = 72 \qquad \text{Combine like terms.}$$

$$\underline{\quad -8 \qquad -8}$$

$$4w + 0 = 64 \qquad \text{Subtract 8 from both sides to isolate the variable term.}$$

$$\frac{4w}{4} = \frac{64}{4} \qquad \text{Divide both sides by 4 to isolate } w.$$

$$w = 16$$

Answer The width, *w*, is 16 inches. To determine the length, use relationship 1. If the length = 4 + *w*, then the length = 4 + 16 = 20 inches.

Check Verify that both conditions in the problem are satisfied. First, the length must be 4 inches more than the width, and 20 inches is 4 inches more than 16 inches. Second, the perimeter must be 72 inches, which is true because 20 + 16 + 20 + 16 = 72 inches.

YOUR TURN Karen is constructing a rectangular flower garden with a wooden border. She wants the length to be twice the width and has 30 feet of border material. Find the dimensions of the garden.

In Example 2, the word *perimeter* was a key word that helped us write the equation that we solved. There are other terms in geometry that, by their definitions, give information to help us translate a problem into an equation. For example, a problem might involve angles that are **complementary** or **supplementary**.

DEFINITION *Complementary angles:* Two angles are complementary if the sum of their measures is 90°.

In the following figure, ∠*ABD* and ∠*DBC* are complementary because 32° + 58° = 90°.

Note: *The small square indicates a 90° angle.* ▶

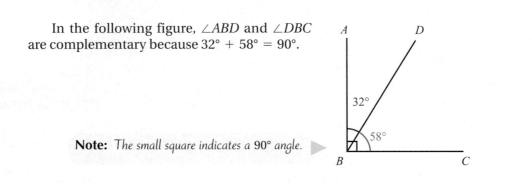

DEFINITION *Supplementary angles:* Two angles are supplementary if the sum of their measures is 180°.

A straight line forms an angle that measures 180°. Any line that intersects a straight line will divide the 180° angle into two angles that are supplementary. In the following figure, line segment *DB* joins line segment *AC*, forming two angles, ∠*ABD* and ∠*DBC*. These two angles are supplementary angles because 20° + 160° = 180°.

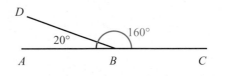

Now look at a problem containing one of these terms.

EXAMPLE 3 A steel beam is to be welded to a cross beam, creating two angles, an inner angle and an outer angle. If the outer angle is to be 15° less than twice the inner angle, what are the angle measurements?

Understand We must find the inner and outer angle measurements. The sketch for this situation is shown (notice it is like the preceding figure).

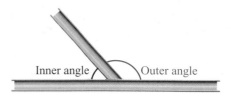

Next, list the relationships:

Relationship 1: The outer angle is 15° less than twice the inner angle.

Relationship 2: inner angle + outer angle = 180.

Note: *The second relationship comes from the sketch. It shows that the inner and outer angles are supplementary, which means the sum of their measures is 180°.*

Plan Translate each relationship to an equation, and then solve.

Execute Relationship 1: The outer angle is 15° less than twice the inner angle.

$$\text{Outer angle} = 2 \cdot \text{Inner angle} - 15$$
$$\text{Outer angle} = 2a - 15$$

Note: *Since the measure of the inner angle is multiplied by 2, we should let the variable represent the inner angle. Let's use the letter a.*

Now use relationship 2 to write an equation that we can solve.

$$\text{Inner angle} + \text{Outer angle} = 180$$
$$a \quad + \quad 2a - 15 \quad = 180$$

$3a - 15 = 180$ **Combine like terms.**

$\underline{+15 \quad +15}$ **Add 15 to both sides.**

$3a + 0 = 195$

$\dfrac{3a}{3} = \dfrac{195}{3}$ **Divide both sides by 3 to isolate a.**

$a = 65$

Answer The inner angle, represented by a, is 65°. According to relationship 1, the outer angle $= 2a - 15$, so the outer angle $= 2(65) - 15 = 115°$.

Check Verify that all the conditions in the problem are satisfied. The outer angle measuring 115° is 15° less than twice the inner angle measuring 65°. The sum of the inner and outer angles is 180°, which verifies that they are supplementary.

YOUR TURN
The wooden joists supporting a roof are connected as shown. The measure of $\angle DBC$ is 6° less than three times the measure of $\angle ABD$. Find the measure of $\angle DBC$ and $\angle ABD$.

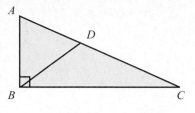

Problems Involving Consecutive Integers

Another group of problems in which the relationships are not obvious involve *consecutive integers*. The word *consecutive* means that the numbers are in sequence. For example, beginning with -3, consecutive integers are $-3, -2, -1, 0, 1, 2 \ldots$.

What relationship exists in the pattern of consecutive integers? Notice to get from one integer in the sequence to the next integer, we add 1. For the sake of finding the general relationship, suppose the first integer in a sequence of consecutive integers is 1. By noting how we get from the number 1 to its consecutive integer, we can find a general pattern when the first integer is an unknown number, n, as the following table illustrates:

	With 1 as the First Integer	With n as the First Integer
First integer:	1	n
Next integer:	$1 + 1 = 2$	$n + 1$
Next integer:	$1 + 2 = 3$	$n + 2$

Now let's determine a general relationship for a sequence of consecutive odd integers, such as $1, 3, 5, 7, \ldots$ As we see in the following table, adding 2 to an odd integer gives the next consecutive odd integer.

	With 1 as the First Odd Integer	With n as the First Odd Integer
First odd integer:	1	n
Next odd integer:	$1 + 2 = 3$	$n + 2$
Next odd integer:	$1 + 4 = 5$	$n + 4$

Warning: The pattern $n, n + 1, n + 3, \ldots$ is not correct for consecutive odd integers.

As we see in the following table, the pattern for consecutive even integers is the same as for consecutive odd integers. Adding 2 to an even integer gives the next consecutive even integer.

	With 2 as the First Even Integer	With n as the First Even integer
First even integer:	2	n
Next even integer:	$2 + 2 = 4$	$n + 2$
Next even integer:	$2 + 4 = 6$	$n + 4$

ANSWERS

$\angle DBC = 66°$

$\angle ABD = 24°$

EXAMPLE 4) The sum of three consecutive even integers is 78. What are the integers?

Understand The key word *sum* means to "add." If we let *n* represent the smallest of the even integers, then the pattern for three unknown consecutive even integers is

Smallest unknown even integer: n

Next even integer: $n + 2$

Third even integer: $n + 4$

Plan Translate to an equation, then solve.

Execute Translate: The sum of three consecutive even integers is 78.

Smallest even integer	+	Next even integer	+	Third even integer	= 78
n	+	$n + 2$	+	$n + 4$	= 78

$$3n + 6 = 78 \quad \text{Combine like terms.}$$
$$\underline{-6 \quad -6} \quad \text{Subtract 6 from both sides.}$$
$$3n + 0 = 72$$
$$3n = 72$$
$$\frac{3n}{3} = \frac{72}{3} \quad \text{Divide both sides by 3 to isolate } n.$$
$$n = 24$$

Answer The smallest of the three unknown integers, represented by n, is 24.

Next even integer: $n + 2 = 24 + 2 = 26$

Third even integer: $n + 4 = 24 + 4 = 28$

Check The numbers 24, 26, and 28 are three consecutive even integers and $24 + 26 + 28 = 78$.

YOUR TURN) The sum of three consecutive integers is 96. What are the integers?

OBJECTIVE 2. Use a table in solving problems with two unknowns. We have seen how using key words, drawing pictures, making lists, and looking for patterns have helped us understand problems. Now we consider a group of problems in which a table is helpful in organizing information.

ANSWER)

31, 32, 33

EXAMPLE 5 A home improvement store sells two sizes of cans of wood stain. A small can sells for $8.95 and a large can sells for $15.95. One day the store sold twice as many of the large cans as the small cans. If the total revenue that day for these cans of wood stain was $694.45, how many cans of each size were sold?

Understand We know that the total revenue is $694.45. From this we can say:

$$\boxed{\begin{array}{c}\text{Revenue from} \\ \text{the small cans}\end{array}} + \boxed{\begin{array}{c}\text{Revenue from} \\ \text{the large cans}\end{array}} = 694.45$$

We can also describe the revenue from the sale of each size can. We know the price of each size. If we knew the number of cans sold, we could multiply that number by the price per can to calculate the revenue from that particular size.

$$\boxed{\begin{array}{c}\text{Price} \\ \text{per can}\end{array}} \cdot \boxed{\begin{array}{c}\text{Number of} \\ \text{cans sold}\end{array}} = \boxed{\begin{array}{c}\text{Revenue for} \\ \text{that size can}\end{array}}$$

With so much information, it is helpful to use a table to list it all. We use a four-column table with the labels categories, price, number of cans, and revenue. The categories are the two sizes of wood stain.

Categories	Price per Can	Number of Cans	Revenue
Small	$ 8.95	n	$8.95n$
Large	$15.95	$2n$	$15.95(2n)$

We were given these.

We selected n to represent the number of small cans, then translated "twice as many of the large cans as the small cans."

We multiplied straight across because

$$\boxed{\begin{array}{c}\text{Price} \\ \text{Per can}\end{array}} \cdot \boxed{\begin{array}{c}\text{Number} \\ \text{of cans}\end{array}} = \boxed{\begin{array}{c}\text{Revenue from} \\ \text{each size}\end{array}}$$

The expressions in the last column in the table give the revenue from the sale of each size can.

Plan Translate the information to an equation, and then solve.

Execute Now we can use our initial relationship:

$$\boxed{\begin{array}{c}\text{Revenue from} \\ \text{the small cans}\end{array}} + \boxed{\begin{array}{c}\text{Revenue from} \\ \text{the large cans}\end{array}} = 694.45$$

$$8.95n \quad + \quad 15.95(2n) \quad = 694.45$$

$8.95n + 31.9n = 694.45$ **Multiply 2n by 15.95.**

$40.85n = 694.45$ **Combine like terms.**

$\dfrac{40.85n}{40.85} = \dfrac{694.45}{40.85}$ **Divide both sides by 40.85 to isolate n.**

$n = 17$

Answer The number of small cans sold, *n*, is 17. The number of large cans sold, 2*n*, is 34 (found by multiplying $2 \cdot 7$).

Check First, verify that twice the number of small cans sold, 17, is 34, which is true. Second, verify that the total revenue from the sale of wood stain is $694.45.

$$8.95(17) + 15.95(34) = 152.15 + 542.30 = 694.45 \qquad \text{It checks.}$$

The table used in Example 5 is a common type of table. In general, when a problem involves two or more categories, we can construct a table with a column labeled *Categories* and a row for each category in the problem. We then create a column for each parameter described in the problem. For example, parameters in a problem might be the *value* of each item, the *number* of items, and the total *amount* of money in each item, and the table would be

Categories	Value	Number	Amount

Since *value · number = amount*, the column labeled *Amount*, contains the products of the expressions in the Value and Number columns.

YOUR TURN Complete a table, write an equation, and then solve.

A computer store sells two different ink cartridges for a particular printer. The black ink cartridge sells for $12.75 and the color cartridge sells for $24.95. A company orders five more black ink cartridges than color cartridges. If the total cost is $629.25, how many of each cartridge were purchased?

Now consider a problem in which a total number of items is given and you must determine how they are split up.

EXAMPLE 6 A company sells two versions of the same software product. The regular version sells for $29.95. The deluxe version sells for $39.95. Due to an inventory error, the number of each version sold was lost. However, the company knows that there were 458 units of sales of the two versions combined and the total revenue was $14,667.10. How many of each version was sold?

Understand We are given the combined income, $14,667.10, which indicates the following relationship:

$$\boxed{\begin{array}{c}\text{Revenue from the}\\\text{regular version}\end{array}} + \boxed{\begin{array}{c}\text{Revenue from the}\\\text{deluxe version}\end{array}} = 14{,}667.10$$

ANSWER

20 black ink and 15 color

Because the problem involves two different items, their value, and the number of each item, we can use a four-column table to organize the information. We begin filling in the table with the categories and the given value of each version.

Categories	Selling Price (Value)	Number	Revenue (Amount)
Regular	29.95		
Deluxe	39.95		

Filling in the number column is tricky. We are told that a total of 458 units were sold, so we can say:

$$\boxed{\text{Number of the regular version}} + \boxed{\text{Number of the deluxe version}} = 458$$

Notice we can isolate one of the two unknowns using related subtraction statements. We will isolate the number of the regular version and let n represent the number of the deluxe version so that we have

$$\boxed{\text{Number of the regular version}} = 458 - \boxed{\text{Number of the deluxe version}}$$

$$= 458 - n$$

◁ **Note:** *By letting the variable represent the number of the larger-valued item, which is the deluxe version, we will avoid negative coefficients in solving the equation.*

Now we can complete the table.

Categories	Selling price (Value)	Number	Revenue (Amount)
Regular	29.95	$458 - n$	$29.95(458 - n)$
Deluxe	39.95	n	$39.95n$

In general, if given a total number of items, select a variable to represent the number of one of the items. The number of the other item will be

Total number − Variable

Since selling price · number = revenue, we multiply straight across the columns to get the expressions in the revenue column.

Plan Translate the information to an equation, and then solve.

Execute Use the initial relationship:

Revenue from the regular version	$+$	Revenue from the deluxe version	$= 14{,}667.10$

$$29.95(458 - n) \quad + \quad 39.95n \quad = 14{,}667.10$$

$$13{,}717.10 - 29.95n + 39.95n = 14{,}667.10 \qquad \textbf{Distribute.}$$

$$13{,}717.10 + \quad 10n = 14{,}667.10 \qquad \textbf{Combine like terms.}$$

$$\underline{-13{,}717.10 \quad -13{,}717.10} \qquad \textbf{Subtract 13,717.10}$$
$$0 + 10n = \quad 950.00 \qquad \textbf{from both sides.}$$

$$\frac{10n}{10} = \frac{950}{10} \qquad \begin{array}{l}\textbf{Divide both sides}\\ \textbf{by 10 to isolate } n.\end{array}$$

$$n = 95$$

Answer The number of the deluxe versions, n, is 95. The number of the regular versions, $458 - n$, is 363 (found by subtracting $458 - 95$).

Check First, verify that the sum of 95 deluxe versions and 363 regular versions gives 458 copies of the software sold, which is true. Second, check the total revenue.

$$29.95(363) + 39.95(95) = 10{,}871.85 + 3795.25 = 14{,}667.10$$

It checks.

YOUR TURN Complete a table, write an equation, and then solve.

A company offers two dental plans for its employees: individual coverage at $22.75 each month, or family coverage at $87.50 each month. The company has 26 employees with dental coverage paying a total of $1757 each month in premiums. How many employees have individual coverage and how many have family coverage?

TIPS FOR TABLE PROBLEMS

You could be given a relationship about the number of items in each category or a total number of items to split up.

- *If you are given a relationship, then use its key words to translate to an expression.*

- *If you are given a total number of items to split up, then choose a variable to represent the number of items for one of the categories. It is better to choose the category with the larger-valued item. The number of items in the other category will be*

Total number − Variable

ANSWER

8 have individual coverage,
18 have family coverage

3.3 Exercises

1. Suppose a problem gives the following information: "one number is five times a second number." Which is easier: letting the *one number* be represented by n or letting the *second number* be represented by n? What expression would describe the other unknown?

2. Suppose you are given that a rectangle has a width that is 5 centimeters less than the length. Which is easier: representing the length or the width with a variable? What expression would describe the other unknown?

3. In describing consecutive even integers or consecutive odd integers, we use the same pattern:

   Consecutive Odd   Consecutive Even

   first odd integer $= n$ first even integer $= n$
   second odd integer $= n + 2$ second even integer $= n + 2$
   third odd integer $= n + 4$ third even integer $= n + 4$

   Explain why the same pattern can be used to describe both odd and even consecutive integers.

4. Consider the following problem: Grant has a jar containing only quarters and half-dollars. If there are a total of 42 coins and their total value is $13.75, how many of each type of coin are there?

   a. Though we could let either unknown be represented by n, what is advantageous about letting n represent the number of half-dollars, which is the greater-valued coin?

   b. If n represents the number of half-dollars, what expression describes the number of quarters?

For Exercises 5–42, translate to an equation, then solve.

5. The larger of two numbers is 5 less than twice the smaller number. The sum of the two numbers is 28. What are the numbers?

6. The second of two numbers is three times the first. Twice the sum of 15 and the first number is equal to 15 plus the second. What are the numbers?

7. The greater of two numbers is 2 more than the lesser. The sum of 25 times the greater number and 50 times the lesser number is 500. What are the numbers?

8. The difference between two numbers is 6. If three times the larger is subtracted from 52, the answer is the same as when 56 is subtracted from seven times the smaller number. Find the numbers.

9. One number is three times another. The sum of the numbers is 52. What are the numbers?

10. A positive number is one-fifth of another positive number. The larger number less the smaller number is 8. What are the numbers?

11. Marie is four times as old as her youngest sister Susan. In eight years, Marie will be only twice as old as Susan. What are their ages now?

12. Jason is three times as old as his brother Michael. In five years, Jason will be twice as old as Michael. What are their ages now?

13. Arianna is 25 years older than Emma. In 15 years, Arianna will be twice as old as Emma. What will their ages be in 15 years?

14. Bob is now twice as old as his son. Sixteen years ago he was four times as old. How old is each now?

15. One cup of canned unsweetened grapefruit juice has 5 fewer grams of carbohydrates than 1 cup of frozen diluted orange juice. If a person drinking a mixture of 1 cup of each consumes 49 total grams of carbohydrates, how many carbohydrates do 1 cup of orange juice and 1 cup of grapefruit juice contain?

16. One slice of white bread has 9 fewer milligrams of sodium than one slice of wheat bread. If a person ate one slice of each, he would consume a total of 267 milligrams of sodium. How many milligrams of sodium are in a slice of each type of bread?

17. A McDonald's Big Mac contains 60 more calories than a large fry. If a person eats both, she consumes a total of 1100 calories. Find the number of calories in a Big Mac and the number of calories in a large fry. (*Source:* McDonald's USA)

18. In one workout, Jennifer walked on a treadmill for 30 minutes, then spent 30 minutes on a stationary bicycle. She burned 75 more calories on the stationary bicycle. If she burned a total of 527 calories during that hour of exercise, how many calories did she burn on each piece of equipment?

Of Interest

Adding a 32-oz. chocolate triple-thick shake to the meal discussed in Exercise 17 brings the total calories for the meal to 2260 calories. A general recommendation for calorie intake is around 1950 for an adult female and 2500 for an adult male per *day*.

19. The brightness of light is measured in a unit called lumens. A 60-watt bulb produces 20 more than four times the number of lumens produced by a 25-watt bulb. Combined they produce 1095 lumens. Find the number of lumens produced by each.

20. The top two states for tourism in the United States are California and Florida. In 2001, the combined domestic traveler spending for those two states was $99.55 billion. If the domestic traveler spending in California was $20.72 billion less than twice the amount spent in Florida, find the amount spent in each state. (*Source:* Travel Industry Association of America)

21. The length of a rectangular garden is 20 feet more than its width and the perimeter is 240 feet. What are the length and width?

22. The length of a rectangular hallway is three times its width. The perimeter is 32 feet. Find the length and width.

23. The perimeter of a rectangular frame is 36 centimeters. If the length is 3 less than twice the width, find the length and width.

24. The width of a rectangular field is 12 meters less than its length. Find the length and width if the perimeter is 400 meters.

25. The width of a rectangular bathroom is $\frac{2}{3}$ of its length. What are its length and width if the perimeter is 25 feet?

26. An artist creates a rectangular sculpture with a width that is $\frac{3}{5}$ of the length. If the perimeter of the sculpture is 32 feet, find the length and width.

27. The base of an isosceles triangle is 9 meters shorter than the other two sides. If the perimeter is 69 meters, find the length of each side.

28. Two sides of an isosceles triangle are each 3 inches less than the base. If the perimeter is 30 inches, find the length of each side.

29. Two angles are supplementary. One of the angles is 20° more than twice the other angle. What are the measures of the two angles?

30. Two angles are supplementary. One of the angles is 30° less than the other angle. What are the measures of the two angles?

31. Two angles are complementary. One angle is 15° less than two times the other angle. What are the measures of the two angles?

32. Two angles are complementary. The smaller angle is $\frac{1}{4}$ of the measure of the larger angle. What are the measures of the two angles?

33. The sum of two consecutive integers is 147. What are the integers?

34. The sum of three consecutive integers is 279. What are the integers?

35. The sum of three consecutive even integers is 234. What are the integers?

36. The sum of three consecutive even integers is 42. What are the integers?

37. The sum of four consecutive odd integers is 48. What are the integers?

38. The sum of three consecutive odd integers is 339. What are the integers?

39. There are three consecutive integers such that the sum of the first and second integers decreased by the third integer will result in 68. What are the integers?

40. Twice the lesser of two consecutive odd integers decreased by the greater integer is 53. What are the integers?

41. Find three consecutive odd integers such that twice the smallest plus three times the largest will result in the middle integer being increased by 70. What are the integers?

42. Find three consecutive odd integers whose sum is 5 more than twice the next odd integer.

For Exercises 43–50, complete a table, write an equation, then solve.

43. The local high school is performing *Oklahoma* as its spring play. The tickets cost $5.00 for students and $8.50 for the general public. Although the organizers lost track of the ticket count, they know that they sold 310 more general public tickets than student tickets. The total sales for the play were $5605. How many of each type of ticket were sold?

44. A vending machine has 12-ounce and 16-ounce drinks. The 12-ounce drinks are $0.50 and the 16-ounce drinks are $1.00. If 3600 total drinks were sold in one month and the total sales were $2225, how many of each size drink were sold?

45. Darryl has some $5 bills and $10 bills in his wallet. If he has a total of 16 bills worth a total of $110, then how many of each bill is in the wallet?

46. Tax preparation software is sold in the standard version for $19.95 and the deluxe version for $10 more. If a store sold 174 copies of the software for total sales of $4011.30, how many of each type were sold?

47. Ember is playing Monopoly, and has two different types of money. The more valuable bill is worth ten times the less valuable bill. She has 12 of the more valuable bills and 23 of the other bills for a total of $715. What is the value of each kind of bill?

48. Kelly is purchasing plants from a garden center where the prices are based on the size of the pot. She is purchasing 13 plants in 1-gallon pots and 7 plants in half-gallon pots. A single 1-gallon and single half-gallon pot have a combined price of $14.50. The total for all the plants she is buying is $155.50. What is the individual price of each size pot?

49. Rachel has two different stocks. She notices in the newspaper that one of her stocks is $5 more valuable than the other. She has 36 shares of the more valuable stock and 21 shares of the other stock. Her total assets in the stocks are $351. How much is each stock worth?

50. Aimee sells textbooks and instructor's materials. The combined price of a tutorial CD-ROM and a study guide is $37. One day she sold 6 CD-ROM and 13 study guides for a total of $372.50. Find the individual prices of the CD-ROM and study guide.

(**PUZZLE PROBLEM**

The product of three integers is 84. Two of the integers are prime numbers and the third is not. If the sum of the composite number and one of the two prime numbers is 17, then what are the three numbers?

REVIEW EXERCISES

[2.1]   **1.** Check to see if $x = -2$ is a solution for $5x + 7 = x - 1$.

[2.2]   **2.** Is $3x - 8 = 12$ a linear equation?

[2.3] *For Exercises 3–5, solve and check.*

3. $-\dfrac{3}{4}x = 12$ **4.** $4x + 5 = -27$ **5.** $9x - 7(x + 2) = -3x + 1$

[2.1]   **6.** Carla knows from running on a treadmill that her average running speed is about 4.5 miles per hour. If she maintains this rate running on roads in her neighborhood for 1.5 hours, what distance will she travel? (Use $d = rt$)

3.4   Rates

OBJECTIVES

1. Solve problems involving two objects traveling in opposite directions.

2. Solve problems involving two objects traveling in the same direction.

In Section 3.3, we developed the use of tables to organize information. In this section, we will use a similar table to organize information in problems that involve distance, rate, and time. Recall that the general formula relating distance, rate, and time is $d = rt$.

OBJECTIVE 1. Solve problems involving two objects traveling in opposite directions. First, consider the situation in which two people or objects are traveling in opposite directions. Whether the two people or objects are moving in opposite directions toward or away from each other does not matter because the math involved will be the same.

EXAMPLE 1 In practicing maneuvers, two fighter jets fly toward each other. One flies east at 582 miles per hour and the other flies west at 625 miles per hour. If the two are 22 miles apart, how much time will it take for them to meet?

Understand   Draw a picture of the situation:

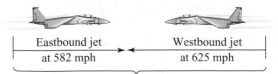

| Eastbound jet at 582 mph | Westbound jet at 625 mph |

22 miles

Note: *The westbound jet will travel the greater distance because it is going faster.*

From the diagram we see that the sum of the individual distances traveled will be the total distance separating the two jets, which is 22 miles.

$$\boxed{\text{Distance traveled by the eastbound jet}} + \boxed{\text{Distance traveled by the westbound jet}} = 22$$

Use a table to find expressions for the individual distances.

Categories	Rate	Time	Distance
Eastbound	582 mph	t	$582t$
Westbound	625 mph	t	$625t$

Both jets start at the same time and meet at the same moment in time, so they traveled the same amount of time, t.

Multiplying the rate value by the time value gives the distance value.

Plan   Use the information to write an equation, then solve.

Execute

$$\boxed{\text{Distance traveled by the eastbound jet}} + \boxed{\text{Distance traveled by the westbound jet}} = 22$$

$$582t \quad + \quad 625t \quad = 22$$
$$1207t = 22$$
$$\frac{1207t}{1207} = \frac{22}{1207}$$
$$t \approx 0.018$$

Connection   When we combined the coefficients in the like terms, we added the individual speeds of the jets, yielding the relative closing speed of the jets.

$$582 \text{ mph} + 625 \text{ mph} = 1207 \text{ mph}$$

The closing speed for these jets is 1207 miles per hour, which is the same as if one of them remained still and the other traveled 1207 miles per hour for the entire 22 miles.

The same reasoning applies if the two jets were going away from each other. They would separate at a rate of 1207 miles per hour.

Note: *If solving for time, the time units will match the unit of time in the rate. In this problem, the rate is in miles per hour, so the time unit is hours.*

Answer   The jets will meet in 0.018 hours, which is approximately 1.09 minutes, or about 1 minute and 6 seconds.

Check   Verify that in 0.018 hours, the jets will travel a combined distance of 22 miles.

Eastbound jet: $d = 582(0.018) = 10.476$ miles
Westbound jet: $d = 625(0.018) = 11.25$ miles

Note: *The westbound jet goes farther because it is traveling faster.*

The combined distance, in miles, is $10.476 + 11.25 = 21.726 \approx 22$. Because we rounded the time, 21.726 is reasonable.

Learning Strategy

If you are a tactile learner, you may find it helpful to simulate the situations in rate problems with toy cars or pencils.

3.4   Rates   **237**

We can summarize how to solve problems involving two objects traveling in opposite directions as follows:

PROCEDURE   Two Objects Traveling in Opposite Directions

To solve for time when two objects are moving in opposite directions:
1. Use a table with columns for categories, rate, time, and distance. Use the fact that rate · time = distance.
2. Write an equation that is the sum of the individual distances equal to the total distance of separation.
3. Solve the equation.

YOUR TURN   Kari is running west at 5 miles per hour along a trail. Vernon is walking east on the same trail at $3\frac{1}{4}$ miles per hour. If they are $\frac{1}{2}$ mile apart, how long will it be until they meet?

OBJECTIVE 2. Solve problems involving two objects traveling in the same direction. If two objects are traveling in the same direction, we can use the same type of four-column table to organize the information. However, we will see that the equation is different.

EXAMPLE 2   Carol and Richard are traveling north in separate cars on the same highway. Carol is traveling at 65 miles per hour and Richard at 70 miles per hour. Carol passes Exit 102 at 1:30 P.M. Richard passes the same exit at 1:45 P.M. At what time will Richard catch up to Carol?

Understand   To determine the time at which Richard catches up to Carol, we must calculate the amount of time it will take him to catch up to her. We can then add that amount of time to 1:45 P.M.

Richard passed the same exit 15 minutes after Carol did. In order to use $d = rt$, the time units must match the time units in the rate. Since 65 and 70 were in miles per hour (mph), the time units must be hours. What portion of an hour is 15 minutes? Since there are 60 minutes in an hour, we can write

$$\frac{15}{60} = \frac{1}{4} = 0.25 \text{ hr.}$$

Now we can begin completing the table. We let t represent the amount of time it takes Richard to catch up. Carol's travel time is a bit tricky. Suppose Richard catches up at 2:30 P.M. Notice this is 1 hour after Carol passed Exit 102 and 45 minutes after Richard passed Exit 102. We see that at the time Richard catches up, Carol's travel time from Exit 102 is 15 minutes (0.25 hours) more than Richard's travel time from the same exit. This will be the case no matter what time he catches up, so Carol's travel time is $t + 0.25$.

Categories	Rate	Time	Distance
Carol	65	$t + 0.25$	$65(t + 0.25)$
Richard	70	t	$70t$

Since $d = rt$, we multiply the rate and time values to equal the distance values.

What can we conclude about the distances they each will have traveled when Richard catches up? Notice Exit 102 is the common reference point from which we describe the time for Richard to catch up. When Richard catches up, he and Carol will be the same distance from that exit, so their distances are equal.

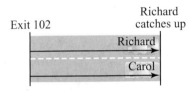

Plan   Set the expressions of their individual distances equal, and solve for t.

Execute   Richard's distance = Carol's distance (at the time he catches up)

Connection   The coefficients of the like terms are the individual rates, so $70 - 65 = 5$ means that Richard's rate relative to Carol's is 5 miles per hour. This means he must make up the 16.25 miles that separates them at a rate of 5 miles per hour.

$$70t = 65(t + 0.25)$$
$$70t = 65t + 16.25$$
$$\underline{-65t \quad -65t}$$
$$5t = 0 + 16.25$$
$$\frac{5t}{5} = \frac{16.25}{5}$$
$$t = 3.25$$

Connection   Since 65 is the rate and 0.25 is the additional amount of time, their product, 16.25 miles, is the distance that Carol is ahead of Richard. Richard must make up this distance to catch her.

Connection   Dividing 16.25 miles by 5 miles per hour calculates the time it takes Richard to make up the distance between Carol and him.

Answer   It will take Richard 3.25 hours, which is 3 hours and 15 minutes, after passing Exit 102 to catch up to Carol. Since he passed Exit 102 at 1:45 P.M., he catches up to Carol at 5:00 P.M.

Check   Verify that Richard and Carol are equal distances from Exit 102 after traveling for their respective times. If Richard travels for 3.25 hours after Exit 102, then Carol travels for $3.25 + 0.25 = 3.5$ hours after Exit 102. Using $d = rt$:

Richard: $d = (70)(3.25)$ Carol: $d = (65)(3.5)$
$d = 227.5$ miles $d = 227.5$ miles.

Following is a summary of the process for problems involving two objects traveling in the same direction in which we are to calculate the time for one of the objects to catch up to the other.

PROCEDURE   Two Objects Traveling in the Same Direction

To solve problems involving two objects traveling in the same direction in which the objective is to determine the time for one object to catch up to the other:

1. Use a table to organize the rates and times. Let t represent the time for the object to catch up, and add the time difference to t to represent the other object's time.
2. Set the expressions for the individual distances equal.
3. Solve the equation.

YOUR TURN   Juan and Angela are bicycling along the same trail. Juan passes a marker at 9:00 A.M. and Angela passes the same marker at 9:05 A.M. Juan is traveling at 8 miles per hour while Angela is traveling at 10 miles per hour. What time will Angela catch up to Juan?

ANSWER

9:25 A.M.

3.4 Exercises

For Extra Help   MyMathLab   Videotape/DVT   InterAct Math   Math Tutor Center   Math XL.com

1. Suppose two objects begin moving toward each other on a collision course. What can we conclude about the amount of time it takes them to meet?

2. Suppose two objects start "back to back" and travel in opposite directions. Object 1 travels $24t$ miles while object 2 travels $45t$ miles. In the following figure, which arrow represents which object? Explain.

?   start   ?

3. Suppose the two objects described in Exercise 2 are separated by 10 miles after traveling t hours. How do you write the equation?

4. A car passes under a bridge on a highway. Forty-five minutes later, a second car passes under the same bridge traveling in the same direction at a speed greater than the first car. Suppose the second car catches up to the first car in t hours. Write an expression for the first car's travel time (in hours) since passing under the bridge.

5. Draw a picture of the situation described in Exercise 4.

6. What can you say about the individual distances of the cars described in Exercise 4?

Exercises 7–18 involve people or objects moving in opposite directions. Complete a four-column table, write an equation, and then solve.

7. An Expedition and a Camaro leave a gas station and travel in opposite directions. The Camaro travels at 50 miles per hour and the Expedition travels at 45 miles per hour. In how many hours will the two cars be 380 miles apart?

8. Two planes leave Denver at the same time, one flying north and the other flying south. If the plane flying south is traveling at 653 miles per hour and the plane flying north is traveling 560 miles per hour, after how many hours will the planes be 2426 miles apart?

9. If a plane flying from Cincinnati travels west at 900 kilometers per hour and another leaving at the same time travels east at 1050 kilometers per hour, in how many hours will the planes be 6825 kilometers apart?

10. Monica and Chandler pass each other along a straight road traveling in opposite directions. Monica is driving at 40 miles per hour and Chandler at 50 miles per hour. In how many hours will the cars be 225 miles apart?

11. Two cars leave a restaurant at the same time with one heading north and the other heading south. If the first car travels at 55 miles per hour and the second travels at 60, how long will it take them to be 230 miles apart?

12. If two planes leave the Dallas airport at the same time with one flying west at 530 miles per hour and the other flying east at 620 miles per hour, how long will it take them to be 1725 miles apart?

13. Two trains started at the same time from Grand Central Station and traveled in opposite directions. One traveled at a rate of 45 miles per hour and the other at the rate of 60 miles per hour. In how many hours were they 210 miles apart?

14. Two trains start at the same time from Victoria Station and travel in opposite directions. One travels at a rate of 40 miles per hour and the other at the rate of 55 miles per hour. After how many hours will they be 190 miles apart?

15. Kim and Karmen begin walking toward each other from opposite directions at 8 A.M. If Kim is walking at 3.5 miles per hour and Karmen is walking at 4 miles per hour, what time will they meet if they are 12 miles apart?

16. Two trains traveling on parallel tracks are going toward each other from a distance of 378 miles. If the freight train is moving at 22 miles per hour and the passenger train is moving at 50 miles per hour, how long will it take for them to pass each other?

17. Two cars pass each other traveling in opposite directions on a highway. One car is going $1\frac{1}{2}$ times as fast as the other. At the end of $3\frac{1}{2}$ hours, they are $192\frac{1}{2}$ miles apart. Find each car's rate.

18. Dustin drives 5 miles per hour faster than Kelly. If he leaves town 1 hour after Kelly and they travel in opposite directions, they will have driven equal distances after Dustin has traveled 7 hours. How far apart are they?

Exercises 19–30 involve people or objects moving in the same direction. Complete a table, write an equation, and then solve.

19. At 9 A.M., a freight train leaves Seattle traveling at 45 miles per hour. At 10 A.M., a passenger train leaves the same station traveling in the same direction at 55 miles per hour. How long will it take the passenger train to overtake the freight train? How far will they be from the station at this time?

20. At 6 A.M., a freight train leaves Washington, D.C., traveling at 50 miles per hour. At 9 A.M., a passenger train leaves the same station traveling in the same direction at 75 miles per hour. How long will it take the passenger train to overtake the freight train? How far will they be from the station at this time?

21. Dan and Kathy are both traveling south for spring break. Dan is taking a bus traveling 60 miles per hour and Kathy is driving her own car at 70 miles per hour. If Dan's bus passes Jacksonville at 2:00 P.M. and Kathy passes through Jacksonville at 2:30 P.M., what time will Kathy catch up to Dan?

22. Two truckers are delivering materials from the same site. One leaves the site at 8 A.M. and the other leaves at 8:15 A.M. If the first trucker travels at 60 miles per hour and the second at 65 miles per hour, at what time will the second catch up to the first?

23. Daphne and Niles are both driving west on Route 66. At 4:00 P.M., Daphne is 5 miles west of Niles. A little later, Niles passes Daphne. If Niles is driving at 55 miles per hour and Daphne at 50 miles per hour, at what time does Niles pass Daphne?

24. Janet and Paul were both traveling west from Charleston, West Virginia to the Rupp Arena in Lexington, Kentucky for a concert. At 6:00 P.M., Janet, traveling at 45 miles per hour, was 10 miles west of Paul. A little later, Paul, traveling at 50 miles per hour, passed Janet. What time did Paul pass Janet?

25. Jane leaves Oklahoma City at 9 A.M. heading for Kansas City. Harrison leaves at 10 A.M. on the same highway, heading toward Kansas City. By driving 25 miles per hour faster, Harrison overtakes Jane at 1 P.M.

   a. How fast is Jane driving?

   b. How far have they traveled when Harrison catches up with Jane?

26. Morrison leaves Cincinnati at 4:00 P.M., heading for New York City. Barbara leaves at 4:30 P.M. on the same highway, heading toward New York City. By driving 10 miles per hour faster, Barbara overtakes Morrison at 7:30 P.M.

   a. How fast is Barbara driving?

   b. How far have they traveled when Barbara catches up with Morrison?

27. If two cars start driving in the same direction from the same place, while one is traveling 45 miles per hour and the other 50 miles per hour, how long will it take them to be 30 miles apart?

28. A car traveling 70 miles per hour passes a truck traveling 62 miles per hour in the same direction on the highway. If they maintain their speeds, how long will it take them to be 1 mile apart?

29. If two joggers leave the same point and run in the same direction, while one is jogging at 1.5 miles per hour and the other jogs at 2.5 miles per hour, how long will it take them to be 4 miles apart?

30. Two brothers start a bicycling race at the same time. If one brother rides at a rate of 15 miles per hour and the other brother rides at a rate of 12 miles per hour, how long will it take them to be 0.75 miles apart?

For Exercises 31–34, use a table and the techniques discussed in this section to solve.

31. On the way to a ski weekend in Vail, Carlita drove 2 hours in a snowstorm. When it finally quit snowing, she was able to increase her speed by 35 miles per hour and drove another 3 hours. If the entire trip was 325 miles, how fast did she drive when it finally quit snowing?

32. While driving from Los Angeles, Tori had a visibility problem due to the smog. After $1\frac{1}{2}$ hours, she hit the open highway outside of the city and was able to increase her speed by 40 miles per hour driving another $5\frac{1}{2}$ hours. If the entire trip was 514 miles, how fast was she traveling in the smog?

33. A trucker traveling along a highway encounters road construction so that traffic is down to a single lane. After driving 30 minutes, the road construction ends and the trucker is able to increase his speed by 64 miles per hour for 4.5 hours. If his entire trip is 303 miles, what was his speed in the road construction? What was his speed after the road construction?

34. An athlete is training for a triathlon. She is working on the running and cycling legs of the event. She runs for 45 minutes then bikes for 1.5 hours. If her combined distance running and cycling is 31.5 miles and her cycling speed is three times her speed running, what is her running speed?

REVIEW EXERCISES

[1.4] **1.** What property of arithmetic is illustrated by $3 \cdot (5 \cdot 7) = 3 \cdot (7 \cdot 5)$?

[1.5] **2.** Simplify: $|14 - 9 \cdot 3| + (4 - 9)^2$

[1.7] **3.** Evaluate $-3x^2 - 4y^3$, where $x = 5$ and $y = -2$.

[1.7] **4.** Simplify: $9x - 5y - 7 + 12y + 13 - x$

[2.3] **5.** Solve: $0.15n + 0.4n = 0.25(200)$

[3.2] **6.** 12% of what number is 84?

3.5 Investment and Mixture

OBJECTIVES
1. Use a table to solve problems involving two investments.
2. Use a table to solve problems involving mixtures.

In Sections 3.3 and 3.4, we began using tables to organize information in problems with two unknowns. In this section, we continue the use of tables in problems involving investments and mixtures.

OBJECTIVE 1. Use a table to solve problems involving two Investments.   When money is invested in an account that earns a percentage of the money invested, the amount earned is called *interest* and is represented by *I*. The amount of money invested is called the *principal* and is represented by *P*. When dealing with interest, the *percentage rate, r*, is usually an annual percentage rate (APR), which is used to calculate the interest if the principal is invested for one year. The relationship for the interest after one year is $I = Pr$. For example, if a person invests $500 at 8% for one year, the interest would be calculated as follows:

$$I = 0.08(500)$$
$$I = \$40$$

> **Connection**   The formula $I = Pr$ comes from the basic percent sentence:
>
> Percent   of a   whole   is a   part of the whole.
>
> Percent   of the   principal   is the   interest.
> r · P = I

Let's explore problems in which an investor invests money in two different accounts.

EXAMPLE 1   Han invests a total of $8000 in two different accounts. The first account earns 5% while the second account earns 8%. If the total interest earned after one year is $565, what principal was invested in each account?

Understand   Since Han invests a total of $8000, we can say

$$\boxed{\text{Principal invested in first account}} + \boxed{\text{Principal invested in second account}} = 8000$$

Note that we can isolate one of the unknown amounts by writing a related subtraction statement. We will isolate the principal invested in the first account and let *P* represent the principal invested in the second account.

$$\boxed{\text{Principal invested in first account}} = 8000 - \boxed{\text{Principal invested in second account}}$$
$$= 8000 - P$$

> **Connection**   This is the same idea that we used in Example 6 of Section 3.3. By letting the variable represent the principal in the account with the larger APR, we will avoid negative coefficients in solving the equation.

We can use a table similar to those used in prior sections. Because $I = Pr$, we need columns for rate, principal, and interest, plus a column for the account names.

Accounts	Rate	Principal	Interest
First account	0.05	$8000 - P$	$0.05(8000 - P)$
Second account	0.08	P	$0.08P$

These rates were given. Don't forget to write the percents as decimals.

From our analysis, if P represents the principal invested in the second account, then the principal invested in the first account is $8000 - P$.

Multiply each principal by the corresponding rate to get the expressions for interest.

Rate · Principal = Interest

Plan Write an equation. Then solve for P.

Execute Since the total interest after one year is $565, we can say

Interest from first account	+	Interest from second account	= Total interest

$$0.05(8000 - P) \quad + \quad 0.08P \quad = 565$$

$$400 - 0.05P + 0.08P = 565 \qquad \text{Distribute.}$$

$$400 + 0.03P = 565 \qquad \text{Combine like terms.}$$

$$\underline{-400 \qquad\qquad -400} \qquad \text{Subtract 400 from both sides.}$$

$$0 + 0.03P = 165$$

$$\frac{0.03P}{0.03} = \frac{165}{0.03} \qquad \text{Divide both sides by 0.03.}$$

$$P = 5500$$

Answer The principal Han invested in the second account, P, is $5500. In the first account, he invested $8000 - P$, or $2500 (found by subtracting $8000 - 5500$).

Check Verify that investing $5500 at 8% and $2500 at 5% results in a total of $565 in interest.

$$5500(0.08) + 2500(0.05) = 440 + 125 = 565 \qquad \text{It checks.}$$

The following procedure can be used as a general guide for solving the type of problems illustrated in Example 1.

PROCEDURE Solving Problems Involving Investment

To solve problems involving two interest rates, in which the total interest is given:

1. Use a table to organize the interest rates and principals. Multiply the individual rates and principals to get expressions for the individual interests.
2. Write an equation that is the sum of the expressions for interest set equal to the given total interest.
3. Solve the equation.

Marvin invests $12,000 in two plans. Plan 1 is at an APR of 6% and plan 2 is at an APR of 9%. If the total interest after one year is $828, what principal was invested in each plan?

OBJECTIVE 2. Use a table to solve problems involving mixtures. Chemicals are often mixed to achieve a solution that has a particular concentration. *Concentration* refers to the portion of a solution that is pure. For example, an 80-milliliter solution may have a 5% concentration of hydrochloric acid (HCl), which means that 5% of that 80 milliliters is pure HCl and the rest is water. To calculate the actual number of milliliters that is pure HCl, we use the simple percent sentence:

Percent of a whole is a part of the whole.
$$0.05 \cdot 80 = 4$$

This means that 4 milliliters out of the 80 milliliters is pure HCl. The general relationship is

Concentration · Whole solution volume = Volume of the particular chemical

Let's consider a problem in which two solutions are mixed to obtain a solution with a particular concentration.

EXAMPLE 2 Margaret has a bottle containing 50 milliliters of 10% HCl solution and a bottle of 25% HCl solution. She wants a 20% HCl solution. How much of the 25% solution must be added to the 10% solution so that a 20% concentration is created?

Understand Because more than one solution is involved, a table is helpful in organizing the information. There are three solutions in this problem: the 10% solution, the 25% solution, and the new 20% solution that is created by mixing the 10% and 25% solutions. The table therefore has a row for each solution. We need columns for the concentration of HCl, the solution volume, and the volume of HCl in each solution. Because we are to find the volume of the 25% solution, we choose a variable, n, to represent this volume.

Solutions	Concentration of HCl	Volume of Solution	Volume of HCl
10% solution	0.10	50	0.10(50)
25% solution	0.25	n	$0.25n$
20% solution	0.20	$50 + n$	$0.20(50 + n)$

These are the concentration percents written in decimal form.

Because the 10% solution and 25% solution are combined to form the 20% solution, we add their volumes to get the volume of the 20% solution.

Multiply straight across the columns to generate the expressions for the volume of HCl in each solution.

$$\text{Concentration} \cdot \begin{array}{c}\text{Volume}\\\text{of solution}\end{array} = \begin{array}{c}\text{Volume}\\\text{of HCl}\end{array}$$

ANSWER

$3600 at 9% and $8400 at 6%

Plan   Write an equation that describes the mixture. Then solve for n.

Execute   The 10% solution and 25% solution combine to equal the 20% solution. We can say

Volume of HCl in the 10% solution	+	Volume of HCl in the 25% solution	=	Volume of HCl in the 20% solution
$0.10(50)$	+	$0.25n$	=	$0.20(50 + n)$

$$5 + 0.25n = 10 + 0.20n \qquad \text{Multiply } 0.10(50) \text{ and distribute.}$$

$$\begin{aligned}5 + 0.25n &= 10 + 0.20n \\ -0.20n & \quad\quad\;\; -0.20n \\ \hline 5 + 0.05n &= 10 \;+\; 0\end{aligned} \qquad \text{Subtract } 0.20n \text{ from both sides.}$$

$$\begin{aligned}-5 & \quad\quad\; -5 \\ \hline 0 + 0.05n &= \;\; 5\end{aligned} \qquad \text{Subtract 5 from both sides.}$$

$$\frac{0.05n}{0.05} = \frac{5}{0.05} \qquad \text{Divide both sides by 0.05.}$$

$$n = 100$$

Answer   100 milliliters of the 25% solution must be added to 50 milliliters of 10% solution to create a solution that is 20% HCl.

Check   Verify that the volume of HCl in the two original solutions combined is equal to the volume of HCl in the combined solution.

$$0.10(50) + 0.25(100) \stackrel{?}{=} 0.20(150)$$

$$5 + 25 \stackrel{?}{=} 30$$

$$30 = 30 \qquad \text{True.}$$

Following is a procedure that can be used as a general guide for solving problems of the type illustrated in Example 2.

PROCEDURE   **Solving Mixture Problems**

To solve problems involving mixing solutions:

1. Use a table to organize the concentrations and volumes. Multiply the individual concentrations and solution volumes to get expressions for the volume of the particular chemical.
2. Write an equation that the sum of the volumes of the chemical in each solution set is equal to the volume in the combined solution.
3. Solve the equation.

YOUR TURN   Andre has 75 milliliters of a 5% sulfuric acid solution. How much of a 20% sulfuric acid solution must be added to create a solution that is 15% sulfuric acid?

ANSWERS

150 ml

3.5 Exercises

For Extra Help

MyMathLab
MyMathLab

Videotape/DVT

InterAct Math

Tutor Center
Math Tutor Center

Math XL.com

1. Suppose an investor has $10,000 to invest in two accounts. The following table lists some ways to split the principal that he is considering.

   a. Complete the table.

   b. How did you determine the unknown principals in the table?

Account 1	Account 2
$9000	$1000
$8500	?
?	$6400
$3200	?

   c. In general, if he invests P in one of the accounts, how do you describe the amount invested in the other account?

2. An investor always invests in two accounts according to the pattern shown in the following table.

   a. Complete the table.

Account 1	Account 2
$500	$1000
$700	$1400
$900	?
?	$2400

   b. In general, if he invests P in account 1, then how would you describe the amount invested in account 2?

   c. If he invests P in account 2, then how would you describe the amount invested in account 1?

3. Given the total interest from an investment in two plans, once you have the expressions for the two interest amounts, how do you write the equation?

4. When completing a table for mixture problems, what are the three categories?

5. The following table shows some possible combinations of two solutions.

   a. Look for a pattern, then complete the table.

Solution 1	Solution 2	Combined Solution
25 ml	50 ml	75 ml
40 ml	80 ml	?
100 ml	?	?

   b. How do you determine the amount of solution 2?

   c. If the amount of solution 1 is represented by n, how would you describe the amount of solution 2?

   d. If solution 1 is represented by n, how would you describe the combined solution?

6. a. If the volume of solution 1 from Exercise 5 is represented by n and the solution is 20% HCl, write an expression for the volume of HCl in solution 1.

   b. Refer to your answer to part c of Exercise 5. If 5% of solution 2 is HCl, write an expression for the volume of HCl in solution 2.

   c. Refer to your answer to part d of Exercise 5. Suppose the combined solution is to be a 10% HCl solution. Write an expression for the volume of HCl in the combined solution.

Exercises 7–22 involve investment. Complete a table, write an equation, then solve.

7. Janice invests in a plan that has an APR of 2%. She invests three times as much in a plan that has an APR of 4%. If the total interest after one year from the investments is $644, how much was invested in each plan?

8. Gayle invests in a plan that has an APR of 3%. She invests twice as much in a plan that has an APR of 5%. If the total interest after one year from the investments is $637, how much was invested in each plan?

9. Tory invests money in two plans. She invests two-fifths of the money at a return rate of 2% in a money market account. The remainder of the money she invests in a short-term mutual fund account with a return rate of 3%. If the total interest from the investments is $104, how much was invested in each plan?

10. Zelda invests money into two plans. She invests two-fifths of the money at an APR of 8% and the rest at an APR of 7.6%. If the total interest after one year from the investments is $620.80, how much was invested in each plan?

11. Boyd invests in a plan that has an APR of 8%. He invests in a 12% APR account $650 more than what he invested in the 8% account. If the total interest after one year from the investments is $328, how much was invested in each plan?

12. Deon invests in an account that has an APR of 2.5%. He invests in a 3.5% APR account $400 more than what he invests in the 2.5% account. If the total interest from the investments is $242, how much was invested in each account?

13. Agnes invests $4500 in two plans. Plan 1 is at an APR of 4%, and plan 2 is at an APR of 6%. If the total interest after one year is $234, what principal was invested in each plan?

14. Juan invests $12,800 in two plans. Plan 1 is at an APR of 7%, and plan 2 is at an APR of 5%. If the total interest after one year is $706, what principal was invested in each plan?

15. Kennon invested $1600. She was able to put some of this into a CD at 5%, and the rest into a CD at only 4%. If she achieved an annual income of $73.40 on these investments, how much was invested in each plan?

16. Shanequa has $4000. She put some of the money into savings that pays 6% and the rest in an account that pays 7%. If her total interest for the year is $264, how much did she invest at each rate?

17. Annette invested $5000. She put some into an investment earning 5.5% and the rest earning 7%. If her income during this year was $335, how much did she invest at each rate?

18. Buddy inherited $12,000. He invested part of the money at 4% and the rest at $5\frac{1}{2}$%. If his annual income from these investments is $532.50, how much did he invest at each rate?

19. Deloris invests in a plan that has an APR of 6%. She invests twice as much in a plan that ends up as a loss of 2% (that is, a −2% APR). If the net gain from the investments is $84, how much was invested in each plan?

20. Dominique invests a total of $10,000 in two plans. Plan 1 is at an APR of 5% while plan 2 ends up returning a loss of 9% (that is, a −9% APR). If the net return on the investments after one year is a net loss of $312, then what principal was invested in each plan?

21. Cody has two investments totaling $10,000. Plan A is at an APR of 4% and plan B is at an APR of 8%. After one year, the interest earned from plan B is $320 more than the interest from plan A. How much did he invest in each plan?

22. Curtis has two investments totaling $12,000. Plan A is at an APR of 5% and plan B is at an APR of 7%. After one year, the interest earned from plan B is $360 more than the interest from plan A. How much did he invest in each plan?

Exercises 23–40 involve mixtures. Complete a table, write an equation, then solve.

23. A cough medicine contains 25% alcohol. How much liquid should a pharmacist add to 120 milliliters of the cough medicine so that it contains only 20% alcohol?

24. How many liters of a 40% solution of HCl are added to 2000 liters of a 20% solution to obtain a 35% solution?

25. A solution of gasoline and oil used for a gas-powered weed trimmer is 8% oil. How much gasoline must be added to 3 gallons of the solution to obtain a new solution that is 5% oil?

26. A 30-milliliter solution of alcohol and water is 10% alcohol. How much water must be added to yield a 5% solution?

27. How many quarts of pure antifreeze must be added to 6 quarts of a 40% antifreeze solution to obtain a 50% antifreeze solution?

28. A pharmacist has a 45% acid solution and a 35% acid solution. How many liters of each must be mixed to form 80 liters of a 40% acid solution?

29. The dairy is making a 30% buttermilk cream. If it mixes a 26% buttermilk cream with a 35% buttermilk cream, how much of each does it need to use to produce 300 pounds of the 30% buttermilk cream?

30. Sharon has a bottle containing 45 milliliters of 15% HCl solution and a bottle of 35% HCl solution. She wants a 25% solution. How much of the 35% solution must be added to the 15% solution so that a 25% concentration is created?

31. Wilson has a bottle containing 35 milliliters of 15% saline solution and a bottle of 40% saline solution. She wants a 30% solution. How much of the 40% solution must be added to the 15% solution so that a 30% concentration is created?

32. Charles has 80 milliliters of a 15% acid solution. How much of a 20% acid solution must be added to create a solution that is 18% acid?

33. To form a 10% copper alloy weighing 75 grams, a 6% copper alloy is combined with an 18% copper alloy. How much of each type should be used?

34. A jeweler is mixing two silver alloys. How many ounces of an alloy containing 25% silver must be mixed with an alloy containing 20% silver to obtain 50 ounces of a 22% silver alloy?

35. A candy store is selling two types of candy, one at $1.20 per pound and the other at 90 cents per pound. To sell a mixture at $1.11 a pound, how many pounds of each must the store mix to make 80 pounds?

36. Pam has a mixture of peanuts and cashews that weighs 12 ounces. Of that, 25% is cashews. If she eats 2 ounces of peanuts, what percent of the remaining mixture is cashews?

37. The Market Street Sweets sells two types of pralines: a caramel version for $1.65 a pound and a chocolate version for $1.25 a pound. A mixture of these pralines can be sold for $1.49 a pound. How many pounds of each kind should be used to make 5 pounds of the $1.49 mixture?

38. Daly's Farm Feed sells two kinds of horse feed. One is 20% oats and the other is 28% oats. How many pounds of each should be mixed to form 50 pounds of a feed that is 25% oats?

39. Cromer's sells salted peanuts for $2.50 per pound and cashews for $6.75 per pound. How many pounds of each are needed to obtain a 10-pound mixture costing $4.20 per pound?

40. The Candy Shoppe wants to mix 115 pounds of candy to sell for $0.80 per pound. How many pounds of $0.60 candy must be mixed with a candy costing $1.20 per pound to make the desired mix?

PUZZLE PROBLEM

Herschal has $18,000 to invest. He invests $\frac{2}{3}$ in one account and the rest in a second account. The first account returned 2% more than the second account. The total return on his investment was $1050. What were the percentage rates of the two accounts?

REVIEW EXERCISES

[1.1]   **1.** Use $<, >$, or $=$ to make a true statement: $-|-14|$ ■ 14

[1.5] *For Exercises 2–3, simplify.*

2. $\dfrac{3}{4} - \dfrac{5}{6} \cdot \left(\dfrac{2}{3}\right)^2$

3. $(-4)^3$

$\begin{bmatrix}\textbf{1.6}\\\textbf{1.7}\end{bmatrix}$ **4.** Write an expression in simplest form that describes the perimeter of the rectangle shown.

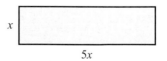

x

$5x$

[2.3] *For Exercises 5–6, solve.*

5. $4(y - 3) + y = 7y - 8$

6. $\dfrac{3}{4}x - 5 = \dfrac{2}{3}x + 1$

Chapter 3   Summary

Defined Terms

Section 3.1
Ratio (p. 190)
Unit ratio (p. 190)
Proportion (p. 191)
Congruent angles
(p. 195)
Similar figures (p. 195)

Section 3.2
Percent (p. 205)

Section 3.3
Complementary angles
(p. 224)
Supplementary angles
(p. 224)

Procedures, Rules, and Key Examples

Procedures/Rules	Key Examples

Section 3.1 Ratios and Proportions

If $\dfrac{a}{b} = \dfrac{c}{d}$, where $b \neq 0$ and $d \neq 0$, then $ad = bc$.

Example 1: Determine whether the ratios are proportional.

$$\frac{5}{6} \, ? \, \frac{15}{18}$$

$18 \cdot 5 = 90 \qquad\qquad 6 \cdot 15 = 90$

$$\frac{5}{6} \diagdown\!\!\!\!\diagup \frac{15}{18}$$

Since the cross products are equal, the ratios are proportional.

To solve a proportion using cross products:
1. Calculate the cross products.
2. Set the cross products equal to one another.
3. Use the multiplication principle of equality to isolate the variable.

Example 2: Solve for the missing number in the proportion.

$$\frac{5}{12} = \frac{x}{18}$$
$$90 = 12x$$
$$\frac{90}{12} = \frac{12x}{12}$$
$$7.5 = x$$

To solve proportion problems:
1. Set up the given ratio any way you wish.
2. Set the given ratio equal to the other ratio with the unknown so that the numerators and denominators correspond logically.
3. Solve using cross products.

Example 3: A car can travel about 350 miles on 16 gallons of gasoline. How many gallons would be needed to travel 600 miles?

$$\frac{350}{16} = \frac{600}{x}$$
$$350x = 9600$$
$$\frac{350x}{350} = \frac{9600}{350}$$
$$x \approx 27.4 \text{ gal.}$$

continued

Procedures/Rules	Key Examples

Section 3.2 Percents

To write a percent as a fraction or decimal:
1. Write the percent as a ratio with 100 in the denominator.
2. Simplify to the desired form.

Note: When simplifying, remember that dividing a decimal number by 100 moves the decimal point two places to the left.

Example 1: Write each percent as a fraction in lowest terms and as a decimal.

a. $42\% = \dfrac{42}{100} = \dfrac{21}{50} = 0.42$

b. $8.5\% = \dfrac{8.5}{100} = \dfrac{85}{1000} = \dfrac{17}{200} = 0.085$

c. $20\frac{1}{2}\% = \dfrac{20\frac{1}{2}}{100} = \dfrac{41}{200} = 0.205$

To write a fraction or decimal number as a percent:
1. Multiply by 100%.
2. Simplify.

Example 2: Write as a percent.

a. $0.453 = 0.453 \cdot 100\% = 45.3\%$

b. $\dfrac{5}{8} = \dfrac{5}{\overset{}{\underset{2}{8}}} \cdot \dfrac{\overset{25}{\cancel{100}}}{1}\% = \dfrac{125}{2}\% = 62\frac{1}{2}\%$

Method 1. Translate the sentence word-for-word:
1. Select a variable for the unknown.
2. Translate the word *is* to an equal sign.
3. If *of* is preceded by the percent, then it is translated to multiplication. If *of* is preceded by a whole number, translate it to division.

Method 2. Translate to a proportion by writing the following form:

$$\text{Percent} = \dfrac{\text{Part}}{\text{Whole}}$$

where the percent is expressed as a fraction with a denominator of 100.

Example 3: 15% of what number is 33?

Method 1: Direct Translation.
$$0.15n = 33$$
$$\dfrac{0.15n}{0.15} = \dfrac{33}{0.15}$$
$$n = 220$$

Method 2: Proportion.
$$\dfrac{15}{100} = \dfrac{33}{n}$$
$$15n = 3300$$
$$\dfrac{15n}{15} = \dfrac{3300}{15}$$
$$n = 220$$

To solve problems involving percent, try to write a basic sentence:
1. Determine whether the percent, whole, or part is unknown.
2. Write the problem as a simple percent sentence (if needed).
3. Translate to an equation (word for word or proportion).
4. Solve for the unknown.

Example 4: A student answered 24 questions out of 30 correctly. What percent of the questions did the student answer correctly?

Simple percent sentence: What percent of 30 is 24?

Direct Translation:
$$p \cdot 30 = 24$$
$$\dfrac{30p}{30} = \dfrac{24}{30}$$
$$p = 0.8$$

Answer:
$0.8 \cdot 100\% = 80\%$

Proportion:
$$\dfrac{P}{100} = \dfrac{24}{30}$$
$$30P = 2400$$
$$\dfrac{30P}{30} = \dfrac{2400}{30}$$
$$P = 80$$

Answer: 80%

continued

Procedures/Rules	Key Examples

with Two or More Unknowns

To solve problems with two or more unknowns:
1. Determine which unknown will be represented by a variable.
   Tip: Let the unknown that is acted on be represented by the variable.
2. Use one of the relationships to describe the other unknown(s) in terms of the variable.
3. Use the other relationship to write an equation.
4. Solve the equation.

Example 1: One number is 15 more than twice another. The sum of the numbers is 39. Find the numbers.

Relationship 1: "One number is fifteen more than twice another."

Translation: $2x + 15$ where x represents the smaller number.

Relationship 2: "The sum of the numbers is 39."

Translation:
$$x + 2x + 15 = 39$$
$$3x + 15 = 39$$
$$\underline{-15 \quad -15}$$
$$3x + 0 = 24$$
$$\frac{3x}{3} = \frac{24}{3}$$
$$x = 8$$

The smaller number is 8 and the other number is $2(8) + 15 = 31$.

Section 3.4 Rates

To solve for time when two objects are moving in opposite directions:
1. Use a table with columns for categories, rate, time, and distance. Use the fact that rate · time = distance. The travel times are equal.
2. Write an equation that is the sum of the individual distances equal to the total distance of separation.
3. Solve the equation.

Example 1: Mark and Latonia are 17 miles apart traveling toward each other. Mark is traveling at a rate of 45 miles per hour and Latonia is traveling at a rate of 40 miles per hour. How long will it be until they meet?

Categories	Rate	Time	Distance
Mark	45	t	$45t$
Latonia	40	t	$40t$

Equation:
$$45t + 40t = 17$$
$$85t = 17$$
$$\frac{85t}{85} = \frac{17}{85}$$
$$t = 0.2 \text{ hr.}$$

continued

Procedures/Rules	Key Examples

Section 3.4 Rates (continued)

To solve problems involving two objects traveling in the same direction in which the objective is to determine the time for one object to catch up to the other:

1. Use a table to organize the rates and times. Let t represent the time for the object to catch up and add the time difference to t to represent the other object's time.
2. Set the expressions for the individual distances equal.
3. Solve the equation.

Example 2: Frank is traveling at a rate of 64 miles per hour and passes Exit 43 at 4:35 P.M. Juanita is traveling at a rate of 72 miles per hour in the same direction and passes the same exit at 4:50 P.M. At what time does she catch up with Frank?

Categories	Rate	Time	Distance
Frank	64	$t + 0.25$	$64(t + 0.25)$
Juanita	72	t	$72t$

Equation:
$$64(t + 0.25) = 72t$$
$$64t + 16 = 72t$$
$$\frac{-64t \qquad -64t}{0 + 16 = 8t}$$
$$\frac{16}{8} = \frac{8t}{8}$$
$$2 = t$$

Answer: Juanita will catch up at 6:50 P.M.

Section 3.5 Investment and Mixture

To solve problems involving two interest rates in which the total interest is given:

1. Use a table to organize the interest rates and principals. Multiply the individual rates and principals to get expressions for the individual interests.
2. Write an equation that is the sum of the expressions for interest set equal to the given total interest.
3. Solve the equation.

Example 1: Dan invests a total of $4000 in two different plans. The first plan returns 9% while the second plan returns 6%. If the total interest earned after one year is $285, what principal was invested in each plan?

Categories	Rate	Principal	Interest
Plan 1	0.09	p	$0.09p$
Plan 2	0.06	$4000 - p$	$0.06(4000 - p)$

Equation:
$$0.09p + 0.06(4000 - p) = 285$$
$$0.09p + 240 - 0.06p = 285$$
$$0.03p + 240 = 285$$
$$\frac{-240 \qquad -240}{0.03p + 0 = \quad 45}$$
$$\frac{0.03p}{0.03} = \frac{45}{0.03}$$
$$p = 1500$$

Answer: $1500 in plan 1 and
$4000 - $1500 = $2500 in
plan 2

continued

Procedures/Rules	Key Examples

Section 3.5 Investment and Mixture (continued)

To solve problems involving mixing solutions:

1. Use a table to organize the concentrations and volumes. Multiply the individual concentrations and solution volumes to get expressions for the volume of the particular chemical.
2. Write an equation that is the sum of the volumes of the chemical in each solution set equal to the volume in the combined solution.
3. Solve the equation.

Example 2: 200 milliliter of a 15% alcohol solution is mixed with a 30% alcohol solution to make a solution that is 20% alcohol. How much of the 30% solution is needed?

Categories	Concentration	Volume of Solution	Volume of Alcohol
15% solution	0.15	200	0.15(200)
30% solution	0.30	n	0.30n
20% solution	0.20	200 + n	0.20(200 + n)

Equation:

$$0.15(200) + 0.30n = 0.20(200 + n)$$
$$30 + 0.30n = 40 + 0.20n$$
$$\underline{-0.20n \qquad -0.20n}$$
$$30 + 0.10n = 40 + 0$$
$$30 + 0.10n = 40$$
$$\underline{-30 \qquad\qquad -30}$$
$$0 + 0.10n = 10$$
$$0.10n = 10$$
$$\frac{0.10n}{0.10} = \frac{10}{0.10}$$
$$n = 100$$

Answer: 100 ml of the 30% solution is needed.

Formulas

Simple percent sentence: A percent of a whole is a part of the whole.

Distance, rate, and time: $d = rt$

Simple interest: $I = Pr$

Chapter 3 Review Exercises

For Exercises 1–6, answer true or false.

[3.1] **1.** The cross products of a proportion are equal.

[3.2] **2.** A percent is a ratio out of 100.

[3.2] **3.** $7.8\% = 0.78$

[3.3] **4.** Consecutive odd integers can be represented as $x, x + 1, x + 3, \ldots$.

[3.3] **5.** Complementary angles have a sum of 180°.

[3.3] **6.** "One less than a number" can be represented as $1 - x$.

For Exercises 7–10, fill in the blank.

[3.2] **7.** To write a percent as a fraction or decimal:
  1. Write the numeral over _____.
  2. Simplify to the desired form.

[3.3] **8.** Angles that have the same measurement are _____.

[3.3] **9.** Angles whose sum is 90° are _____.

[3.1] **10.** An equation in the form $\dfrac{a}{b} = \dfrac{c}{d}$, where $b \neq 0$ and $d \neq 0$, is a _____.

[3.1] **11.** A recipe calls for $\dfrac{3}{4}$ of a cup of sugar and 2 cups of flour. Write the ratio of sugar to flour as a fraction in simplest form.

[3.1] **12.** The price of a 10.5-ounce can of vegetables is $0.89. Write the unit ratio of price to weight rounded to the nearest thousandth.

[3.1] **For Exercises 13–16, determine whether the ratios are equal.**

Equations and
Inequalities

13. $\dfrac{2}{5} \overset{?}{=} \dfrac{10}{20}$

14. $\dfrac{1}{3} \overset{?}{=} \dfrac{13}{39}$

15. $-\dfrac{2}{8} \overset{?}{=} -\dfrac{1}{4}$

16. $\dfrac{15}{21} \overset{?}{=} \dfrac{130.2}{93}$

Exercises 13–32

[3.1] **For Exercises 17–22, solve for the missing number.**

17. $\dfrac{3}{5} = \dfrac{15}{x}$

18. $\dfrac{-11}{12} = \dfrac{n}{-24}$

19. $\dfrac{\frac{2}{5}}{4\frac{1}{2}} = \dfrac{7}{p}$

20. $\dfrac{k}{15.3} = \dfrac{5}{8}$

21. $\dfrac{-60}{w} = \dfrac{20}{40}$

22. $\dfrac{c}{12.7} = \dfrac{5}{3}$

23. Jason has driven 225 miles in 3 hours. If he averages the same speed after lunch, how far can he expect to travel in 4 hours?

24. Gutzon Borglum used a simple ratio of 1:12 (1 inch to 1 foot) when he transferred the plaster models to the granite for Mt. Rushmore. Each of the sculptures is 60 feet tall. How tall was a plaster model?

25. It costs $5.85 to mail a 4.5-pound package. How much will it cost to mail a 7-pound package?

26. Tamika pays property taxes of $1848 on her house, which is assessed at $125,000. She is planning to buy a house assessed at $175,000. At the same rate, how much will her taxes be on the new home?

27. Ferrari estimates that the 2004 612 Scaglietti two-door coupe will travel 241 city miles on one tank of gas. If the gas tank of the automobile holds 28.5 gallons, how far can a driver expect to travel on 25 gallons?

28. Dan notes on his natural gas bill that he paid $195.16 for 159 Therms in November. If he knows he usually uses around 220 Therms in January, how much can he expect his bill to be in January?

> *Of Interest*
>
> Sculptor Gutzon Borglum began work on the series of 60-foot sculptures collectively called Mount Rushmore in 1927. The Washington head was formally dedicated in 1930, followed by Jefferson in 1936, Lincoln in 1937, and Teddy Roosevelt in 1939.

[3.1] *For Exercises 29–32, find the missing lengths in the similar shapes.*

29.

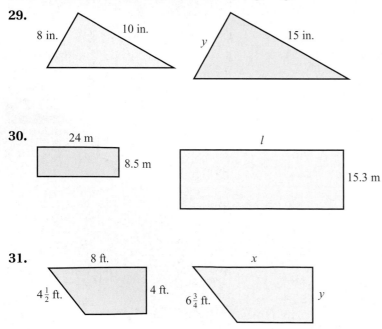

8 in. 10 in. y 15 in.

30. 24 m l 8.5 m 15.3 m

31. 8 ft. x $4\frac{1}{2}$ ft. 4 ft. $6\frac{3}{4}$ ft. y

32.

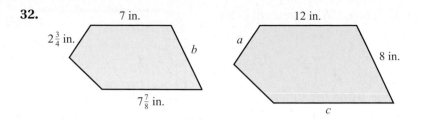

[3.2] *For Exercises 33–38, write each percent as a decimal and as a fraction in simplest form.*

33. 15%

34. 82.5%

35. $12\frac{1}{2}\%$

36. 2.45%

37. 10%

38. $33\frac{1}{3}\%$

[3.2] *For Exercises 39–44, write as a percent.*

39. $\frac{2}{5}$

40. $\frac{1}{3}$

41. $\frac{4}{11}$

42. 0.35

43. 1.2

44. 2.016

[3.2] *For Exercises 45–48, translate the percent sentence to an equation, then solve.*

45. 16% of 91 is what number?

46. 14.5% of what number is 42?

47. 6.5 is what percent of 20?

48. What percent is 18 out of 32?

For Exercises 49–68, solve.

[3.2] **49.** In a recent Gallup poll, 47% of those polled knew someone who lost a job during the past year. If 1016 Americans were polled, how many knew someone who lost a job? (*Source:* Gallup News Service 2/15/01)

[3.2] **50.** Karen buys a new kitchen table for $759.99. If the tax rate in her area is 6%, what is the cost after tax is added?

Expressions

Exercises 33–44

Equations and Inequalities

Exercises 45–68

[3.2] **51.** In 2004, the Boston Red Sox won 98 out of 162 games. What percent of its games did the Red Sox win?

[3.2] **52.** In 1995, cotton farmers received an average price for upland cotton of 75.4 cents per pound. In 2001, the average price of upland cotton was 29.8. What was the percent of the decrease in average price? (*Source:* National Agricultural Statistics Services, U.S. Department of Agriculture)

[3.3] **53.** The larger of two numbers is five more than three times the smaller number. The sum of the two numbers is 15. What are the numbers?

[3.3] **54.** Find three consecutive integers such that twice the sum of the first two is six less than triple the third number.

[3.3] **55.** The length of a rectangle is 15 feet more than twice its width, and the perimeter is 120 feet. What are the length and width?

[3.3] **56.** Two sides of an isosceles triangle are 2 centimeters more than the other side. If the perimeter is 55 centimeters, find the length of each side.

[3.3] **57.** Two angles are supplementary. One of the angles is 32° less than the measure of the other angle. What are the measures of the two angles?

[3.3] **58.** Two angles are complementary. One of the angles is three less than twice the other. What are the measures of the two angles?

[3.3] **59.** While buying pantyhose, Sophie notices that the regular fit costs $6.50 and control top costs $9.50. If she buys five pairs of hose and spends $38.50, how many of each type did she buy?

[3.3] **60.** Candice owns stock in Coca-Cola and Pepsi. She notices that the Coca-Cola stock is worth about $12 more per share. If she has $3646 invested and owns 20 shares of the Pepsi and 47 shares of the Coca-Cola, how much is one share of each stock worth?

> **Of Interest**
>
> Although the New York Yankees won the most games (101) in the American League during the 2004 season, Boston beat the Yankees in the best of seven American League championship series, winning the last four games after being down three games to 0, to advance to the World Series. Boston then won the first four games of the World Series against St. Louis to win the World Series.

[3.4] **61.** In Cincinnati, a Land Rover and an Escort enter I-71 at Exit 2 at the same time and travel in opposite directions. The Land Rover traveling south is caught in traffic and travels at 45 miles per hour. The Escort traveling north encounters light traffic and is able to travel at 65 miles per hour. In how many hours will the cars be 385 miles apart?

[3.4] **62.** If a plane flying from O'Hare travels west at 550 miles per hour and another leaving at the same time travels east at 450 miles per hour, how many hours until the planes are 2600 miles apart?

[3.4] **63.** Wayne and Libby are both driving south. At 3:00 P.M., Wayne was 5 miles south of Libby. A little later, Libby passed Wayne. If Libby is driving at 55 miles per hour and Wayne at 50 miles per hour, what time did Libby pass Wayne?

[3.4] **64.** Two trains started from Victoria Station and traveled in the same direction on parallel tracks. The freight train leaves at 10 A.M. The passenger train leaves at 11 A.M. but, traveling at 25 miles per hour faster, overtakes the freight train at 2 P.M. How fast is the freight train traveling?

[3.5] **65.** Terrell invests in a retirement plan that has an APR of 8%. He invests twice as much in a second plan that has an APR of 10%. If the total interest from the investments is $5600, how much was invested in each plan?

[3.5] **66.** Brian invests in a retirement plan that has an APR of 7%. He invests three times as much in a second plan that ended up as a loss of 2% (that is, at −2% APR). If the net gain from the investments is $102, how much was invested in each plan?

[3.5] **67.** A solution that is 90% alcohol is to be mixed with one that is 40% alcohol. The chemist is preparing 100 liters of a 60% alcohol solution. How many liters of each must he use?

[3.5] **68.** How many ounces of a 20% saline solution must be added to 40 ounces of a 10% solution to produce a 16% solution?

Chapter 3 Practice Test

1. The roof of a building rises 8 inches for every 20 inches of horizontal length. Write the ratio of rise to horizontal length as a fraction in simplest form.

2. A 14.5-ounce box of cereal costs $3.58. Write the unit ratio of price to weight.

For Exercises 3 and 4, solve for the missing number.

3. $\dfrac{m}{8} = \dfrac{5}{12}$

4. $\dfrac{-32}{-9} = \dfrac{8}{b}$

5. On a trip, Michael drives 857.5 miles using 35 gallons of fuel. If he has another trip planned that will cover about 1200 miles, how many gallons of fuel should he expect to use?

6. The following figures are similar. Find the missing side length.

For Exercises 7 and 8, write each percent as a decimal and as a fraction in simplest form.

7. 22%

8. $3\dfrac{1}{3}\%$

For Exercises 9 and 10, write as a percent.

9. 3.2

10. $\dfrac{2}{5}$

For Exercises 11–20, solve.

11. 12 is what percent of 60?

12. 20.5% of 70 is what number?

13. 70 is 40% of what number?

14. According to a survey conducted by the National Center for Health, 37.1% of Americans say they are in excellent health. Based on these findings, in a group of 500 Americans, how many would say they are in excellent health? (*Source:* National Center for Health Statistics)

15. In 2004, Los Angeles was the top seed of the the Western Conference in the WNBA with 25 wins out of 34 games. What percent of its games did Los Angeles win? (*Source: The World Almanac and Book of Facts*, 2005)

16. In March, Andrea notes her electric bill is $148.40. In April, her bill is $166.95. What is the percent of the increase in her bill?

17. If five times a number is reduced by 14, the result is 26. Find the number.

18. The sum of two consecutive even integers is 110. Find the integers.

19. The length of a rectangle is one more than double the width. If the perimeter of the rectangle is 32 kilometers, what are the length and width?

20. Two angles are supplementary. If the larger angle is 24° more than the smaller angle, find the measure of the two angles.

21. Christe owns stock in both Wal-Mart and Target. On May 26, 2005, she notices that the Target stock is worth about $6 more per share. If she has 22 shares of Target stock and 15 shares of Wal-Mart stock for a total investment of $1871, what was the value of each stock?

22. Two runners begin running toward each other along the same path. One person runs at a speed of 4 miles per hour and the other person runs at 4.5 miles per hour. If they begin $4\frac{1}{4}$ miles apart, how long will it be until they meet?

23. An F-18 jet fighter passes a landmark at a speed of 450 miles per hour. A second F-18 traveling in the same direction at 570 miles per hour passes the same landmark 10 minutes later. How long will it take the second jet to catch up with the first jet?

24. Kari invests a total of $4500 in two plans. After one year, she found that plan A earned 8% interest and plan B earned 6% interest. If she earned a total of $330, how much did she invest in each plan?

25. How many ounces of a 15% saline solution must be added to 30 ounces of a 25% solution to produce a 20% solution?

Chapters 1–3 Cumulative Review Exercises

For Exercises 1–6, answer true or false.

[1.1] **1.** A constant is a symbol that can vary in value.

[1.1] **2.** $\sqrt{2}$ is an irrational number.

[1.5] **3.** When evaluating an expression using the order of operations, one must always do multiplication before division.

[1.6] **4.** "Four less than a number" can be translated to $4 - n$.

[2.1] **5.** $\frac{1}{2}x = 10$ is a linear equation.

[2.1] **6.** The equation $2(x - 1) - 3x = 5 - x - 7$ is an identity.

For Exercises 7–10, fill in the blank.

[1.2] **7.** A prime number is a natural number that has exactly two different factors, 1 and _____.

[2.3] **8.** To clear fractions from an equation, we can use the multiplication principle of equality and multiply both sides by the _____.

[3.2] **9.** A percent is a ratio representing some part out of _____.

[3.1] **10.** The _____ products of equivalent fractions are equal.

[1.1] **11.** Write a set representing the natural numbers.

[1.1] **12.** Graph $-3\frac{1}{4}$ on a number line.

[1.2] **13.** Find the prime factorization of 120.

[1.7] **14.** Determine the values that make $\dfrac{4x}{x - 3}$ undefined.

[1.7] **15.** Which property of arithmetic is illustrated by $4(x + 3) = 4x + 12$?

[3.1] **16.** Explain in your own words how to solve a proportion.

[1.5] *For Exercises 17–25, simplify.*

17. $\left(-\dfrac{2}{3}\right)^2$

18. $\sqrt{\dfrac{4}{81}}$

19. $\sqrt{-36}$

Expressions

Exercises 17–27

20. $(-1)^2 - 8.5 + 1$

21. $\sqrt{16} + \sqrt{9}$

22. $\dfrac{5 - 2(-1)^2}{3\sqrt{64} + 36}$

23. $-|3 \cdot (-5)| + |(-1)^3|$

24. $2\left(-\dfrac{1}{2}\right)^2 - 4\left(\dfrac{3}{5}\right) + 3$

25. $-6 - 2(3 + 0.2)^2$

[1.7] **26.** Evaluate $\dfrac{2x^3}{y - 3}$, where $x = -1$ and $y = 3$.

[1.7] **27.** Use the distributive property to simplify $-3\left(\dfrac{1}{6}x - 9\right)$.

[1.3] *For Exercises 28–31, indicate whether the given equation illustrates the additive identity, commutative property of addition, associative property of addition, or additive inverse.*

Equations and
Inequalities

Exercises 28–50

28. $-5 + (4 - 4) = -5 + 0$

29. $6 + (-3) = -3 + 6$

30. $4 + (5 + 6) = (4 + 5) + 6$

31. $-4 + 4 = 0$

[2.2–2.3] *For Exercises 32–37, solve and check.*

32. $4(x + 1) = 2(x - 3)$

33. $3.23c - 8.75 = 1.41c + 7.63$

34. $2m + \dfrac{1}{2} = 4m - 3\dfrac{1}{2} + 8m$

35. $\dfrac{1}{8}w = \dfrac{4}{3} - \dfrac{2}{3}w$

36. $6x - 2 = 3\left(2x - \dfrac{2}{3}\right)$

37. $-7 = -8(p + 2) + 5(2p - 3)$

[2.4] *For Exercises 38–41, solve for the indicated variable.*

38. $A = \frac{1}{2}bh$ for b

39. $A = P + Prt$ for r

40. $P = a + b + c$ for c

41. $d = rt$ for t

For Exercises 42–50, solve.

[2.5] **42.** Twice a number is twelve less than three times the number. Find the number.

[3.2] **43.** What is 15% of 42.3?

[3.3] **44.** The sum of three consecutive odd integers is -3. What are the integers?

[2.6] **45.** Jon made an 82 on his first test and a 70 on his second test. What range of scores does he need to make on his third test to have an average of at least 80?

[3.5] **46.** How many liters of a 60% solution of boric acid should be added to 10 liters of a 30% solution to obtain a 50% solution?

[3.3] **47.** Kathie has $6.50 in dimes and quarters. If there are 35 coins in all, how many of each coin does she have?

[3.4] **48.** Archie and Veronica are both driving east on Riverdale Drive. At 3:00, Archie, driving at 50 miles per hour, was 5 miles east of Veronica. A little later, Veronica, driving at 55 miles per hour, passed Archie. At what time did Veronica pass Archie?

[3.5] **49.** Cromers sells cashews for $7.00 per pound and peanuts for $3.00 per pound. If a 10-pound bag is mixed selling at $4.50 per pound, how many pounds of each type nut should be used?

[3.1] **50.** If a car travels 120 miles on 4 gallons of gasoline, how far can it travel on 10 gallons?

CHAPTER 4

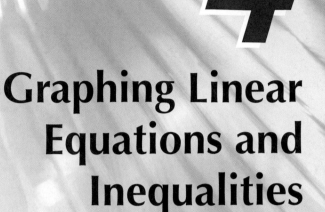

Graphing Linear Equations and Inequalities

There's a compelling reason to master information and news. Clearly there will be better job and financial opportunities. Other high stakes will be missed by people if they don't master and connect information.

—Everette Dennis, Media foundation executive and professor of media and entertainment

Where is the wisdom we have lost in knowledge?
Where is the knowledge we have lost in information?

—T. S. Eliot, American-born British critic and writer (1888–1965)

All equations and inequalities have corresponding pictures, or graphs. The graph of an equation or inequality offers information about the solutions to the equation or inequality. Graphs can be helpful in solving problems. For example, we may be given an equation or inequality that describes a particular problem. The graph of that equation may offer additional insight about possible solutions to the problem. Or, a problem may have some data that we can graph. We can often use such a graph to help us write an equation or inequality that describes the situation. Both approaches can be quite useful in solving problems.

4.1 The Rectangular Coordinate System

1. Determine the coordinates of a given point.

2. Plot points in the coordinate plane.

3. Determine the quadrant for a given coordinate.

4. Determine whether the graph of a set of data points is linear.

OBJECTIVE 1. Determine the coordinates of a given point. In 1619 René Descartes, the French philosopher and mathematician, recognized that positions of points in a plane could be described using two number lines that intersect at a right angle. Each number line is called an **axis**.

DEFINITION   *Axis:* A number line used to locate a point in a plane.

Two perpendicular axes form the *rectangular*, or Cartesian, *coordinate system*, named in honor of René Descartes. Usually, we call the horizontal axis the *x-axis* and the vertical axis the *y-axis*.

Note: *When letters are used to label the axes, they are usually placed so that the horizontal axis is the first letter in alphabetical order. For example, if the letters t and u are used, then the horizontal axis will be the t-axis and the vertical axis will be the u-axis.*

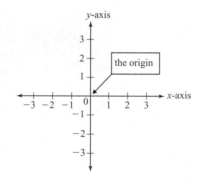

The point where the axes intersect is 0 for both the *x*-axis and the *y*-axis. This position is called the *origin*. The positive numbers are to the right and up from the origin, whereas negative numbers are to the left and down from the origin.

Any point in the plane can be described using two numbers, one number from each axis. To avoid confusion, these two numbers are written in a specific order. The number representing a point's *horizontal distance* from the origin is given *first* and the number representing a point's *vertical distance* from the origin is given *second*. Because the order in which we say or write these two numbers matters, we say that they form an *ordered pair*. Each number in an ordered pair is called a *coordinate* of the ordered pair. The notation for writing ordered pairs is: (horizontal coordinate, vertical coordinate).

Consider the point labeled *A* in the coordinate plane shown. The point is drawn at the intersection of the 3rd line to the right of the origin and the 4th line up from the origin. The ordered pair that describes point *A* is (3, 4).

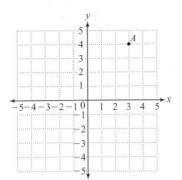

Connection Think of the intersecting lines in the grid as avenues and streets in a city such as New York, where avenues run north-south and streets run east-west. To describe an intersection, a person would say the avenue first, then the street. Point *A* is at the intersection of 3rd avenue and 4th street, or, as a New Yorker would say, "3rd and 4th."

PROCEDURE **Identifying the Coordinates of a Point**

To determine the coordinates of a given point in the rectangular system:

1. Follow a vertical line from the point to the *x*-axis (horizontal axis). The number at this position on the *x*-axis is the first coordinate.
2. Follow a horizontal line from the point to the *y*-axis (vertical axis). The number at this position on the *y*-axis is the second coordinate.

EXAMPLE 1 Write the coordinates for each point shown.

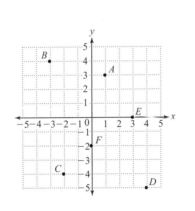

Answers

A: (1, 3)   We can get to point *A* from the origin by moving to the right 1 and then up 3.

B: (−3, 4)   From the origin, point *B* is left 3, then up 4.

C: (−2, −4) Left 2, then down 4.

D: (4, −5)   Right 4, then down 5.

E: (3, 0)   Right 3. Since the point is on the *x*-axis, the *y*-coordinate is 0.

F: (0, −2)   Down 2. Since the point is on the *y*-axis, the *x*-coordinate is 0.

YOUR TURN   Write the coordinates for each point shown.

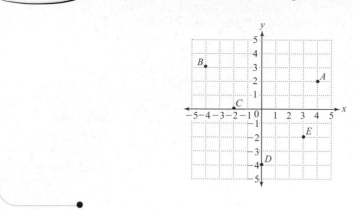

OBJECTIVE 2. Plot points in the coordinate plane.   Any point can be plotted in the coordinate plane using its coordinates. Remember that the first coordinate in an ordered pair indicates the distance to move right or left, and the second coordinate indicates the distance to move up or down. To plot the ordered pair (3, 4), we begin at the origin, move to the right 3, then up 4, and draw a dot to indicate the point.

PROCEDURE   **Plotting a Point**

To graph or plot a point given its coordinates:

1. Beginning at the origin, (0, 0) move to the right or left along the *x*-axis the amount indicated by the first coordinate.
2. From that position on the *x*-axis, move up or down the amount indicated by the second coordinate.
3. Draw a dot to represent the point described by the coordinates.

Learning Strategy

If you are a tactile learner, when plotting a point, move your pencil along the *x*-axis first, then move up or down to the point location, just as we've done with the arrows.

ANSWERS

A: (4, 2)   D: (0, −4)
B: (−4, 3)   E: (3, −2)
C: (−2, 0)

ANSWER

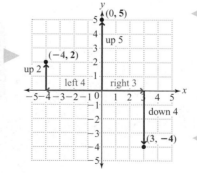

EXAMPLE 2   Plot the point described by the coordinates.

a. $(3, -4)$ **b.** $(-4, 2)$ **c.** $(0, 5)$

Solution

Note: For $(-4, 2)$, we begin at the origin and move to the left 4, then up 2.

Note: For $(0, 5)$, because the first coordinate is 0, we do not move right or left. Since the second coordinate is 5, we move straight up 5, drawing the point on the *y* axis.

Note: For $(3, -4)$, we begin at the origin and move to the right 3, then down 4.

YOUR TURN   Plot the point described by the coordinates.

    A: $(-3, 2)$ B: $(1, 4)$ C: $(-2, -4)$ D: $(3, 0)$

OBJECTIVE 3. Determine the quadrant for a given coordinate. The two perpendicular axes divide the coordinate plane into four regions called *quadrants.*

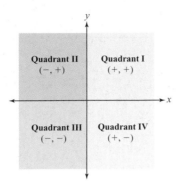

The quadrants are numbered using Roman numerals, as shown to the left. Note that the signs of the coordinates determine the quadrant in which a point lies. For points in quadrant I, both coordinates are positive. Every point in quadrant II has a negative first coordinate (horizontal) and a positive second coordinate (vertical). Points in quadrant III have both coordinates negative. Every point in quadrant IV has a positive first coordinate and a negative second coordinate. Points on the axes are not in any quadrant.

PROCEDURE Identifying Quadrants

To determine the quadrant for a given ordered pair, consider the signs of the coordinates.

$(+, +)$ means the point is in quadrant I
$(-, +)$ means the point is in quadrant II
$(-, -)$ means the point is in quadrant III
$(+, -)$ means the point is in quadrant IV

EXAMPLE 3 State the quadrant in which each point is located.

a. $(-61, 23)$

Answer Quadrant II (upper left), because the first coordinate is negative and the second coordinate is positive.

b. $\left(14, -37\dfrac{2}{3}\right)$

Answer Quadrant IV (lower right), because the first coordinate is positive and the second coordinate is negative.

c. $(0, -12)$

Answer Since the x-coordinate is 0, this point is on the y-axis and is not in a quadrant.

YOUR TURN State the quadrant in which each point is located.

a. $(-42, -109)$ **b.** $(37, -15.9)$ **c.** $(4.7, 0)$ **d.** $\left(-16\dfrac{3}{4}, 124\right)$

ANSWERS

a. III **b.** IV **c.** on the x-axis
d. II

OBJECTIVE 4. Determine whether the graph of a set of data points is linear. In many problems data are listed as ordered pairs. For example, in a report on stock prices, we might have the following listing of the closing price of a particular stock for each day over four days.

Day	Closing Price
1	$24\frac{1}{2}$
2	$25\frac{3}{4}$
3	$27
4	$28\frac{1}{4}$

Note: *Because the vertical axis has dollar values that involve fractions, there are two marks for every whole dollar. This allows halves to be plotted accurately and fourths to be estimated well.*

If we plot each pair of data as an ordered pair in the form (day, closing price), we see that the points lie on a straight line.

Because the points can be connected to form a straight line they are said to be *linear*. Points that do not form a straight line are *nonlinear*.

Note: *The broken line in the axis indicates that values between 0 and 24 were skipped to conserve space.*

EXAMPLE 4 The following data points track the trajectory, or path, of an object over a period of time. Plot the points with the time along the horizontal axis and the height along the vertical axis. Then state whether the trajectory is linear or nonlinear.

Time (in seconds)	Height (in feet)
0	6.00
0.5	7.50
1.0	8.25
1.5	7.50
2.0	6.00
2.5	2.00

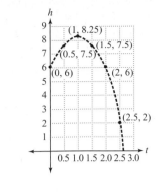

Solution Because the data points do not form a straight line when connected, the trajectory is nonlinear.

◀ **Note:** *It seems here that our axis labels violate the alphabetical convention. However, the problem states to plot the time along the horizontal axis. Also, in general, when a graph involves time, the time values are plotted along the horizontal axis.*

YOUR TURN The following data points show the velocity of an object as time passes. Plot the points with the time along the horizontal axis and the velocity along the vertical axis. Then state whether the points are linear or nonlinear.

ANSWER

The points are linear.

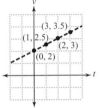

Time (in seconds)	Height (in feet)
0	2.0
1	2.5
2	3.0
3	3.5

4.1 Exercises

1. When writing an ordered pair, which is written first, the horizontal-axis coordinate or the vertical-axis coordinate?

2. Describe how to use the coordinates of an ordered pair to plot a point in the rectangular coordinate system.

3. Draw the axes in the rectangular coordinate system, and then label the quadrants. Include the signs for coordinates in each quadrant.

4. Describe how to determine whether a set of data points is linear.

For Exercises 5–8, write the coordinates for each point.

5.

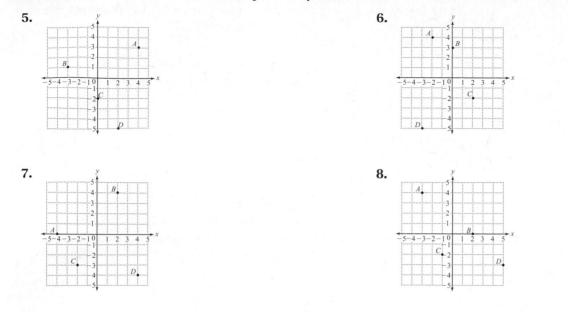

6.

7.

8.

For Exercises 9–12, on a piece of graph paper, draw and label x- and y-axes. Then plot and label the points indicated by the coordinate pairs.

9. $(5, 4), (2, -3), (-1, -3), (2, 0)$

10. $(-3, 0), (0, 1), (-2, -2), (4, -1)$

11. $(-2, 0), (3, -3), (-1, -5), (2, 2)$

12. $(-1, -4), (0, -2), (4, -2), (-4, 2)$

For Exercises 13–24, state the quadrant in which or axis on which the point is located.

13. $(-6.5, 1050)$

14. $\left(-42\frac{1}{3}, -500\right)$

15. $(620, 50)$

16. $(-42, 50)$

17. $(57, -82.71)$

18. $(16, 27)$

19. $\left(-41\frac{1}{2}, -82\right)$

20. $\left(47, -51\frac{5}{8}\right)$

21. $(0, -9)$

22. $(0, 5.3)$

23. $(-0.6, 0)$

24. $(8.2, 0)$

For Exercises 25–30, determine whether the set of points is linear or nonlinear.

25. $(-2, -9), (0, -5), (2, -1), (4, 3), (5, 5)$

26. $(-5, 7), (-1, 3), (2, 0), (4, -2), (6, -4)$

27. $(-4, 8), (-2, 2), (0, 0), (1, 0.5), (3, 4.5)$

28. $(-1, -3), (0, 0), (2, 0), (3, -3)$

29. (pounds of chicken, cost): $(1, 2.5), (2, 5), (3, 7.5), (4, 10)$

30. (time, distance): $(0, 2), (1, 5), (2, 8), (3, 11), (4, 14)$

31. (time, height): $(0, 8.0), (0.5, 9.5), (1.0, 11.0), (1.5, 9.5), (2.0, 8.0), (2.5, 6.5)$

32. (days, stock closing price): $(1, 23), (2, 21), (3, 19), (4, 17)$

33. (copy costs, number of copies): $(15, 300), (18, 400), (21, 500), (24, 600)$

34. (number of hours, plumbing cost): $(1, 120), (2, 160), (3, 200), (4, 240)$

35. List the coordinates of three points on the line shown.

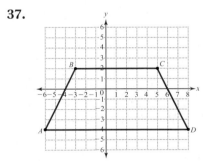

36. List the coordinates of three points on the line shown.

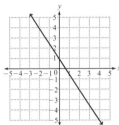

A vertex on a figure is a point where two line segments form a corner or joint on the figure. For Exercises 37 and 38, find the coordinates of each labeled vertex.

37.

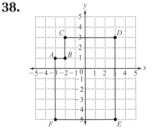

38.

39. A parallelogram has been moved from its original position, shown with the dashed lines, to a new position, shown with solid lines.

   a. List the coordinates of each vertex of the parallelogram in its original position.

   b. List the coordinates of each vertex of the parallelogram in its new position.

   c. Write a rule that describes in mathematical terms how to move each point on the parallelogram in its original position to each point on the parallelogram in its new position.

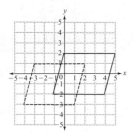

40. A trapezoid has been moved from its original position, shown with the dashed lines, to a new position, shown with solid lines.

   a. List the coordinates of each vertex of the trapezoid in its original position.

   b. List the coordinates of each vertex of the trapezoid in its new position.

   c. Write a rule that describes in mathematical terms how to move each point on the trapezoid in its original position to each point on the trapezoid in its new position.

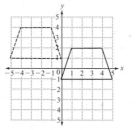

41. The ordered pairs $(-3, 1)$, $(-1, 1)$, $(-1, 3)$, $(2, 3)$, $(2, -2)$, and $(-3, -2)$ form the vertices of a figure.

   a. Plot the points and connect them to form the figure.
   b. Find the perimeter of the figure.

   c. Find the area of the figure.

42. The ordered pairs $(-4, 2)$, $(-2, 2)$, $(-2, 4)$, $(1, 4)$, $(1, 3)$, $(3, 3)$, $(3, -4)$, and $(-4, -4)$ form the vertices of a figure.

   a. Plot the points and connect them to form the figure.
   b. Find the perimeter of the figure.

   c. Find the area of the figure.

REVIEW EXERCISES

[1.1]   **1.** Graph -3 on a number line.

[1.7]   **2.** Evaluate $2x - 3y$, when $x = 4$ and $y = 2$.   [1.7]   **3.** Evaluate $-4x - 5y$, when $x = -3$ and $y = 0$.

[2.3] *For Exercises 4–6, solve.*

  4. $2x - 5 = 10$ **5.** $-\dfrac{5}{6}x = \dfrac{3}{4}$ **6.** $7n - (2n - 9) = 3(4n + 1) - 8$

4.2 Graphing Linear Equations

OBJECTIVES

1. Determine whether a given pair of coordinates is a solution to a given equation with two unknowns.

2. Find solutions for an equation with two unknowns.

3. Graph linear equations.

OBJECTIVE 1. Determine whether a given pair of coordinates is a solution to a given equation with two unknowns. In Chapter 2 we considered linear equations with one variable. We now consider linear equations that have two variables, such as $x + y = 4$ and $y = 2x - 3$.

A solution for an equation with two variables is a pair of numbers, one number for each variable, that can replace the corresponding variables and make the equation true. Since these solutions are ordered pairs, we can write them using coordinates. The ordered pair $(1, 3)$ is a solution for $x + y = 4$ because replacing x with 1 and y with 3 makes the equation true.

Note: *Recall from Section 4.1 that we assume coordinates are written in alphabetical order, so $(1, 3)$ means that $x = 1$ and $y = 3$.*

$$x + y = 4$$
$$\downarrow \quad \downarrow$$
$$1 + 3 = 4$$

The equation is true, so the ordered pair is a solution.

PROCEDURE Checking a Potential Solution for an Equation with Two Variables

To determine whether a given ordered pair is a solution for an equation with two variables:

1. Replace the variables in the equation with the corresponding coordinates.
2. Verify that the equation is true.

EXAMPLE 1 Determine whether the ordered pair is a solution for the equation.

a. $(-4, -10); y = 2x - 3$

Solution $-10 \overset{?}{=} 2(-4) - 3$ **Replace x with -4 and y with -10 and see if the equation is true.**

$-10 \overset{?}{=} -8 - 3$

$-10 \neq -11$ **Because the equation is not true, $(-4, -10)$ is not a solution for $y = 2x - 3$.**

b. $(4, -1); 2m - n = 9$

Solution $2(4) - (-1) \overset{?}{=} 9$ **Unless otherwise specified, ordered pairs are expressed in alphabetical order; therefore, replace m with 4 and n with -1 and see if the equation is true.**

$8 + 1 \overset{?}{=} 9$

$9 = 9$ **The equation is true, so $(4, -1)$ is a solution for $2m - n = 9$.**

c. $\left(-2, -\dfrac{1}{3}\right); y = \dfrac{2}{3}x + 1$

Solution $-\dfrac{1}{3} \overset{?}{=} \dfrac{2}{3}(-2) + 1$ Replace x with -2 and y with $-\dfrac{1}{3}$ and see if the equation is true.

$-\dfrac{1}{3} \overset{?}{=} -\dfrac{4}{3} + \dfrac{3}{3}$

$-\dfrac{1}{3} = -\dfrac{1}{3}$ The equation is true, so $\left(-2, -\dfrac{1}{3}\right)$ is a solution

for $y = \dfrac{2}{3}x + 1$.

YOUR TURN   Determine whether the ordered pair is a solution for the equation.

a. $(-8, -2); x - 3y = -2$ **b.** $(1, -6); q = -5p + 1$ **c.** $\left(\dfrac{3}{4}, \dfrac{9}{4}\right); y = \dfrac{x}{3} + 2$

OBJECTIVE 2. Find solutions for an equation with two unknowns.   Consider the equation $x + y = 4$. We have already seen that $(1, 3)$ is a solution. But there are other solutions. We list some of those solutions in the following table:

x	y	Ordered Pair
0	4	(0, 4)
1	3	(1, 3)
2	2	(2, 2)
3	1	(3, 1)
4	0	(4, 0)
5	−1	(5, −1)

Note: *Every solution for $x + y = 4$ is an ordered pair of numbers whose sum is 4.*

In fact, $x + y = 4$ has an infinite number of solutions. For every x-value there is a corresponding y-value that will add to the x-value to equal 4 and vice versa, which gives a clue about how to find solutions. We can simply choose a value for either x or y and solve for the corresponding value of the other variable. (In this case, our equation was easy enough that we could solve for y mentally.)

PROCEDURE   Finding Solutions to Equations with Two Variables

To find a solution to an equation in two variables:

1. Choose a value for one of the variables (any value).
2. Replace the corresponding variable with your chosen value.
3. Solve the equation for the value of the other variable.

ANSWERS

a. yes   **b.** no   **c.** yes

EXAMPLE 2 Find three solutions for the equation.

a. $3x + y = 2$

Solution To find a solution we replace one of the variables with a chosen value and then solve for the value of the other variable.

For the first solution, we will choose x to be 0.	For the second solution, we will choose x to be 1.	For the third solution, we will choose x to be 2.

$$3x + y = 2$$
$$3(0) + y = 2$$
$$y = 2$$

Solution $(0, 2)$

Note: *Choosing x (or y) to equal 0 usually makes the equation very easy to solve.*

$$3x + y = 2$$
$$3(1) + y = 2$$
$$3 + y = 2$$
$$y = -1$$

Subtract 3 from both sides.

Solution $(1, -1)$

$$3x + y = 2$$
$$3(2) + y = 2$$
$$6 + y = 2$$
$$y = -4$$

Subtract 6 from both sides.

Solution $(2, -4)$

We can summarize the solutions in a table:

If we choose x to be 0, then y is equal to 2.

If we choose x to be 1, then y is equal to -1.

If we choose x to be 2, then y is equal to -4.

x	y	Ordered Pair
0	2	$(0, 2)$
1	-1	$(1, -1)$
2	-4	$(2, -4)$

Keep in mind that there are an infinite number of correct solutions for a given equation in two variables, so your solutions may be different from the solutions someone else finds.

b. $y = -\dfrac{1}{3}x + 2$

Solution Notice this equation has y isolated. If we select values for x, we will not have to isolate y as we did in Example 2. We will simply calculate the y value. Also notice that the coefficient for x is a fraction. Since we can choose any value for x, let's choose values like 3 and 6 that will divide out nicely with the denominator of 3.

For the first solution we will choose x to be 0.	For the second solution, we will choose x to be 3.	For the third solution, we will choose x to be 6.

$$y = -\frac{1}{3}x + 2$$
$$y = -\frac{1}{3}(0) + 2$$
$$y = 2$$

$$y = -\frac{1}{3}x + 2$$
$$y = -\frac{1}{3}(3) + 2$$
$$y = -\frac{1}{\cancel{3}}\left(\frac{\cancel{3}^{1}}{1}\right) + 2$$

$$y = -\frac{1}{3}x + 2$$
$$y = -\frac{1}{3}(6) + 2$$
$$y = -\frac{1}{\cancel{3}}\left(\frac{\cancel{6}^{2}}{1}\right) + 2$$

Solution $(0, 2)$

$y = -1 + 2$
$y = 1$

$y = -2 + 2$
$y = 0$

Solution $(3, 1)$

Solution $(6, 0)$

Solution In summary,

If we choose x to be 0, then y is equal to 2.

If we choose x to be 3, then y is equal to 1.

If we choose x to be 6, then y is equal to 0.

x	y	Ordered Pair
0	2	(0, 2)
3	1	(3, 1)
6	0	(6, 0)

YOUR TURN Find three solutions for each equation. (Answers may vary.)

a. $x + 4y = 8$

b. $3x - 2y = 6$

c. $y = \frac{1}{2}x + 5$

OBJECTIVE 3. Graph linear equations. We have learned that equations in two variables have an infinite number of solutions. Because of this fact, there is no way that all solutions to an equation can be found, much less listed. However, all the solutions can be represented using a graph. Consider some of the solutions for $x + y = 4$, which we listed in the table on page 278: $(0, 4)$ $(1, 3)$ $(2, 2)$ $(3, 1)$ $(4, 0)$ $(5, -1)$.

Notice that when we plot our ordered pair solutions for $x + y = 4$, the corresponding points lie in a straight line. In fact, all solutions for $x + y = 4$ are on that same line, which is why we say it is a *linear* equation. By connecting the points, we are graphically representing every possible solution for $x + y = 4$.

The graph of the solutions of every linear equation will be a straight line. Since two points determine a line in a plane, we need a minimum of two ordered pairs. This means we must find at least two solutions in order to graph the line. However, it is wise to find three solutions, using the third solution as a check. If we plot all three points and they cannot be connected with a straight line, then we know something is wrong.

PROCEDURE **Graphing Linear Equations**

To graph a linear equation:

1. Find at least two solutions to the equation.

2. Plot the solutions as points in the rectangular coordinate system.

3. Connect the points to form a straight line.

ANSWERS

Remember, your solutions may be different.

a. $(0, 2)$ $(4, 1)$ $(8, 0)$

b. $(0, -3)$ $(2, 0)$ $\left(1, -\frac{3}{2}\right)$

c. $(0, 5)$ $(2, 6)$ $(4, 7)$

EXAMPLE 3 Graph.

a. $3x + y = 2$

Solution We found three solutions to this equation in Example 2. Those solutions are listed in the following table. We plot each solution as a point in the rectangular coordinate system, and then connect the points to form a straight line.

x	y	Ordered Pair
0	2	(0, 2)
1	−1	(1, −1)
2	−4	(2, −4)

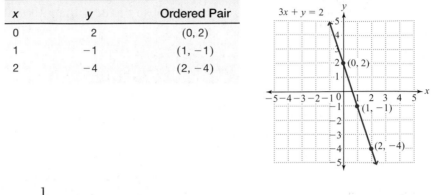

b. $y = -\frac{1}{3}x + 2$

Solution We found three solutions to this equation in Example 3. Those solutions are listed in the following table. We plot each solution as a point in the rectangular coordinate system, and then connect the points to form a straight line.

x	y	Ordered Pair
0	2	(0, 2)
3	1	(3, 1)
6	0	(6, 0)

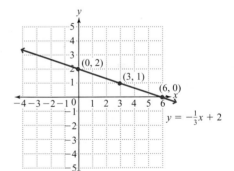

ANSWERS

a.

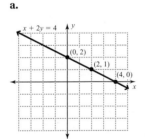

b.

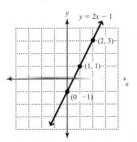

YOUR TURN Graph each equation.

a. $x + 2y = 4$ **b.** $y = 2x - 1$

We have graphed linear equations in two variables. Now let's consider the graphs of equations in one variable, such as $y = 3$ or $x = 4$, in which the variable is equal to a constant.

EXAMPLE 4 ▶ Graph: $y = 3$

Solution The equation $y = 3$ indicates that y is equal to a constant, 3. To establish ordered pairs, we can rewrite the equation $y = 3$ as $0x + y = 3$. Because the coefficient of x is 0, y is always 3 no matter what we choose for x. We could complete a table of solutions like this:

If we choose x to be 0, then y equals 3.

If we choose x to be 2, then y is 3.

If we choose x to be 4, then y is still 3.

x	y	Ordered Pair
0	3	(0, 3)
2	3	(2, 3)
4	3	(4, 3)

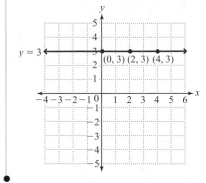

Note: *The graph of* $y = 3$ *is a horizontal line parallel to the x-axis that passes through the y-axis at a point with coordinates* (0, 3).

RULE **Horizontal Lines**

The graph of $y = c$, where c is a real-number constant, is a horizontal line parallel to the x-axis that passes through the y-axis at a point with coordinates (0, c).

EXAMPLE 5 ▶ Graph: $x = 4$

Solution The equation $x = 4$ indicates that x is equal to a constant, 4. If the y-variable is missing from an equation, it means that its coefficient is 0, so we can write $x = 4$ as $x + 0y = 4$. Notice that x is always 4 no matter what we choose for y. We could complete a table of solutions like this:

If we choose y to be 0, then x equals 4.

If we choose y to be 1, then x is 4.

If we choose y to be 2, then x is still 4.

x	y	Ordered Pair
4	0	(4, 0)
4	1	(4, 1)
4	2	(4, 2)

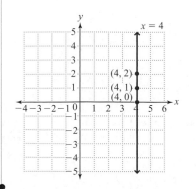

Note: *The graph of* $x = 4$ *is a vertical line parallel to the y-axis that passes through the x-axis at a point with coordinates* (4, 0).

RULE **Vertical Lines**

The graph of $x = c$, where c is a real number constant, is a vertical line parallel to the y-axis that passes through the x-axis at a point with coordinates $(c, 0)$.

a.

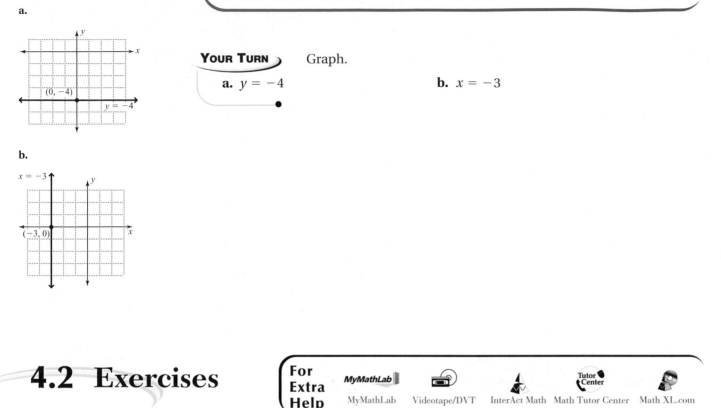

YOUR TURN   Graph.

     a. $y = -4$ **b.** $x = -3$

b.

4.2 Exercises

For Extra Help

MyMathLab   Videotape/DVT   InterAct Math   Math Tutor Center   Math XL.com

1. How do you determine if an ordered pair is a solution for a given equation?

2. If a given ordered pair is not a solution for a linear equation with two variables, does that mean the equation has no solutions? Explain.

3. How do you find a solution for a given equation with two unknowns?

4. What does the graph of an equation represent?

5. What is the minimum number of points needed to draw a straight line? Explain.

6. In your own words, explain the process of graphing a linear equation.

For Exercises 7–22, determine whether the given ordered pair is a solution for the given equation.

7. $(2, 3)$; $x + 2y = 8$

8. $(3, 1)$; $2x - y = 5$

9. $(4, 9)$; $3x - 2y = -6$

10. $(-5, 10)$; $5x + 3y = 5$

11. $(-5, 2)$; $y - 4x = 3$

12. $(4, -6)$; $y - 3x = -1$

13. $(9, 0)$; $y = -2x + 18$

14. $(0, -1)$; $y = 2x + 1$

15. $(6, -2)$; $y = -\dfrac{2}{3}x$

16. $(-10, -4)$; $y = \dfrac{2}{5}x$

17. $(-8, -8)$; $y = \dfrac{3}{4}x - 2$

18. $(-5, 0)$; $y = -\dfrac{3}{5}x + 3$

19. $\left(-1\dfrac{2}{5}, 0\right)$; $y - 3x = 5$

20. $\left(0, -2\dfrac{1}{3}\right)$; $4x - 3y = 7$

21. $(2.2, -11.2)$; $y + 6x = -2$

22. $(-1.5, -1.3)$; $y = 0.2x - 1$

For Exercises 23–58, find three solutions for the given equation. Then graph. (Answers may vary for the three solutions.)

23. $x - y = 8$

24. $x + y = -5$

25. $2x + y = 6$

26. $3x - y = 9$

27. $2x + 3y = 12$

28. $3x + 5y = 15$

29. $4x - 3y = -12$

30. $3x - 2y = -6$

31. $y = x$

32. $y = -x$

33. $y = 2x$

34. $y = 3x$

35. $y = -5x$

36. $y = -2x$

37. $y = x - 3$

38. $y = x + 4$

39. $y = -x - 2$

40. $y = -x + 5$

41. $y = 2x - 5$

42. $y = 3x + 2$

43. $y = -2x + 4$

44. $y = -5x - 1$

45. $y = \dfrac{1}{2}x$

46. $y = \dfrac{1}{3}x$

47. $y = -\dfrac{2}{3}x$

48. $y = -\dfrac{3}{4}x$

49. $y = -\dfrac{2}{3}x + 4$

50. $y = \dfrac{4}{5}x - 1$

51. $x - \dfrac{1}{4}y = 2$

52. $\dfrac{1}{3}x + y = -1$

53. $y = -5$

54. $y = 4$

55. $x = 7$

56. $x = -6$

57. $y = 0.4x - 2.5$

58. $1.2x + 0.5y = 6$

59. Compare the graphs of $y = x$, $y = 2x$, and $y = 3x$ from Exercises 31, 33, and 34. For an equation in the form $y = ax$ where $a > 0$, what effect does increasing a seem to have on the graph?

60. Compare the graphs of $y = x$ and $y = -x$ from Exercises 31 and 32, then compare the graphs of $y = 2x$ and $y = -2x$ from Exercises 33 and 36. For an equation in the form $y = ax$, what can you conclude about the graph when a is positive versus when a is negative?

61. Compare the graphs of $y = 3x$ and $y = 3x + 2$ from Exercises 34 and 42, then compare the graphs of $y = -2x$ and $y = -2x + 4$ from Exercises 36 and 43. For an equation in the form $y = ax + b$ where $b > 0$, what effect does adding b to $y = ax$ seem to have on the graph?

62. Compare the graphs of $y = x$ and $y = x - 3$ from Exercises 31 and 37, then compare the graphs of $y = -5x$ and $y = -5x - 1$ from Exercises 35 and 44. For an equation in the form $y = ax - b$ where $b > 0$, what effect does subtracting b from $y = ax$ seem to have on the graph?

For Exercises 63–70, solve.

63. A plumber charges $80 plus $40 per hour of labor. The equation $c = 40n + 80$ describes the total that she would charge for a service visit, where n represents the number of hours of labor and c is the total cost.

  a. Find the total cost if labor is 2 hours.

  b. If a client's total charges are $240, for how many hours of labor was the client charged?

  c. Graph the equation with n along the horizontal axis and c along the vertical axis. (*Hint:* Let each grid mark on the y-axis represent $20.)

  d. What does the c-intercept (y-intercept) represent?

64. An academic tutor charges $20 for supplies and then $25 per hour of tutoring. The equation $c = 25n + 20$ describes the total that he would charge for tutoring, where n represents the number of hours of tutoring and c is the total cost.

  a. Find the total cost if the tutor works 3 hours.

  b. If a client's total charges are $145, for how many hours of labor was the client charged?

  c. Graph the equation with n along the horizontal axis and c along the vertical axis.

  d. What does the c-intercept (y-intercept) represent?

65. A copy center charges $15 for the first 300 copies plus $0.03 per copy after that. The equation $c = 0.03n + 15$ describes the total that would be charged for a copy service where n represents the number of copies above 300 and c is the total cost.

  a. Find the total cost for 400 copies.

  b. If a client's total charges are $21, for how many copies was the client charged?

  c. Graph the equation with n along the horizontal axis and c along the vertical axis. (*Hint:* Let each grid mark on the x-axis represent 50 copies and each grid mark on the y-axis represent $3.)

66. A wireless company charges $20 per month for 300 anytime minutes plus $0.10 per minute above 300. The equation $c = 0.10n + 20$ describes the total that the company would charge for a monthly bill where n represents the number of minutes used above 300.

  a. Find the total cost if a customer uses 400 anytime minutes.

  b. If a client's total charges are $22, for how many additional minutes was the client charged?

  c. Graph the equation with n along the horizontal axis and c along the vertical axis.

67. It is recommended that a hot tub be emptied and cleaned regularly. The hot tub Benjamin owns holds approximately 500 gallons of water. He can drain out 3.5 gallons of water per minute using a garden hose. The equation $g = -3.5m + 500$ describes the amount of water in the hot tub in gallons, where g represents the number of gallons remaining after pumping water out for m minutes

  a. After 25 minutes, what is the amount of water in the hot tub?

  b. How long does it take to pump half of the water out of the hot tub?

  c. How long does it take to pump all of the water out of the hot tub?

d. Graph the equation.

e. What does the g-intercept (y-intercept) represent?

f. What does the m-intercept (x-intercept) represent?

68. The battery in Jenny's laptop is fully charged (100%). While using the laptop continuously on a flight from Kansas to California, she depletes the battery at a rate of 0.75% per minute. The equation $P = -0.75m + 100$ describes the battery's power, where P represents the percent of the battery's power that remains after using the laptop for m minutes.

a. After 1 hour, what percent of the battery's power remains?

b. How long does it take for the battery to be at one-fourth of its full power?

c. How long does it take for the battery to reach 0%?

d. Graph the equation.

e. What does the P-intercept (y-intercept) represent?

f. What does the m-intercept (x-intercept) represent?

69. A small business owner buys a new computer for $2400. Each year the computer is in use, she can deduct its depreciated value when calculating her taxable income. The equation $c = -300n + 2400$ describes the value of the computer, where c is the value after n years of use.

a. Find the value of the computer after 4 years.

b. In how many years will the computer be worth half of its initial value?

c. After how many years will the computer have a value of $0?

d. Graph the equation with n along the horizontal axis and c along the vertical axis.

70. As of 2005, it is estimated that Earth is being deforested at a rate of about 40,000 square miles per year. The equation $A = -40,000n + 2,000,000$ describes the area of forest remaining, where A represents the area remaining n years after 2005. (*Source:* www.wikipedia.org)

a. Find the area of forest remaining in 2015.

b. If the rate does not change, in what year will Earth have half of the area of the forest that was present in 2005?

c. If the rate does not change, in what year will forest be gone from Earth?

d. Graph the equation with n along the horizontal axis and c along the vertical axis.

Of Interest

Estimates for the rate of deforestation vary. As of 2005, the rates range from 20,000 square miles per year to 50,000 square miles per year.

PUZZLE PROBLEM

Without lifting your pencil from the paper, connect the dots using only four straight lines:

• • •

• • •

• • •

REVIEW EXERCISES

[1.7] **1.** Evaluate $-32t + 70$ when $t = 4.2$.

[2.1] **2.** Calculate the perimeter of a rectangle with a length of 8.5 inches and a width of 7.25 inches.

[2.1] **3.** Calculate the area of a triangle with a base of 4.2 inches and a height of 2.7 inches.

[2.3] **4.** Solve: $2x + 3 = 9$

[2.4] **5.** Isolate x in the formula $v = mx + k$.

4.3 Graphing Using Intercepts

OBJECTIVES

1. Given an equation, find the coordinates of the x- and y-intercepts.
2. Graph linear equations using intercepts.

OBJECTIVE 1. Given an equation, find the coordinates of the x- and y-intercepts. Look at the following graph.

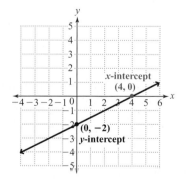

Notice that the line intersects the x-axis at $(4, 0)$ and the y-axis at $(0, -2)$. These points are called *intercepts*. The point where a graph intersects the x-axis is called the **x-intercept**. The point where a graph intersects the y-axis is called the **y-intercept**.

DEFINITIONS   **x-intercept:** A point where a graph intersects the *x*-axis.
y-intercept: A point where a graph intersects the *y*-axis.

Note that the *y*-coordinate of any *x*-intercept will always be 0. Similarly, the *x*-coordinate of any *y*-intercept will always be 0. These facts about intercepts suggest the following procedures.

PROCEDURE   **Finding the *x*- and *y*-intercepts**

To find an *x*-intercept:

1. Replace *y* with 0 in the given equation.
2. Solve for *x*.

To find a *y*-intercept:

1. Replace *x* with 0 in the given equation.
2. Solve for *y*.

EXAMPLE 1   For the equation $5x - 2y = 10$, find the *x*- and *y*-intercepts.

Solution

For the *x*-intercept, replace *y* with 0 and solve for *x*.

$$5x - 2(0) = 10$$
$$5x = 10$$
$$x = 2 \quad \textbf{Divide both sides by 5.}$$

x-intercept: $(2, 0)$

For the *y*-intercept, replace *x* with 0 and solve for *y*.

$$5(0) - 2y = 10$$
$$-2y = 10$$
$$y = -5 \quad \textbf{Divide both sides by −2.}$$

y-intercept: $(0, -5)$

Note: *We will no longer show the steps containing the addition or multiplication principles as we did in Chapters 2 and 3.*

YOUR TURN   For the equation $3x - 4y = 8$, find the *x*- and *y*-intercepts.

The equation in Example 1 is in the form $Ax + By = C$, where *A, B,* and *C* are real-number constants with *A* and *B* not both equal to 0. A linear equation written in this form is said to be in *standard form.* One of the advantages of having a linear equation in standard form is that it is easy to find the intercepts. Throughout this section we introduce other forms for linear equations and explore some of the information that can be discerned from equations written in various forms.

EXAMPLE 2   For the equation $y = \dfrac{2}{3}x$, find the *x*- and *y*-intercepts.

Solution   Replace *x* with 0 and solve for *y*. Then replace *y* with 0 and solve for *x*.

y-intercept: $y = \dfrac{2}{3}(0)$

$$y = 0$$

x-intercept: $0 = \dfrac{2}{3}x$

$$0 = x \quad \textbf{Multiply both sides by } \dfrac{3}{2}.$$

x- and *y*-intercept: $(0, 0)$

ANSWERS

x-intercept: $\left(\dfrac{8}{3}, 0\right)$

y-intercept: $(0, -2)$

The equation in Example 2 is a linear equation in the form $y = mx$, where m is any real number. The graph of any equation in the form $y = mx$ is always a line that passes through the origin, which means that the x- and y-intercepts are both at $(0, 0)$.

RULE Intercepts for $y = mx$

If an equation can be written in the form $y = mx$, where m is a real number other than 0, then the x- and y-intercepts are at the origin, $(0, 0)$.

EXAMPLE 3 For the equation $y = 2x + 1$, find the coordinates of the x- and y-intercepts.

Solution Replace x with 0 and solve for y. Then replace y with 0 and solve for x.

x-intercept: $0 = 2x + 1$

$-1 = 2x$ **Subtract 1 from both sides.**

$-\dfrac{1}{2} = x$ **Divide both sides by 2.**

x-intercept: $\left(-\dfrac{1}{2}, 0\right)$

y-intercept: $y = 2(0) + 1$

$y = 1$

y-intercept: $(0, 1)$

Note: *When we replace x with 0, we are left with the constant 1 for the y-intercept.*

The equation in Example 3 is in the form $y = mx + b$, where m and b can be any real number. In this form, when we replace x with 0 to find the y-intercept, we are left with the constant b.

RULE The y-intercept for $y = mx + b$

If an equation is in the form $y = mx + b$, where m and b are real numbers, then the y-intercept will be $(0, b)$.

YOUR TURN For the equation $y = 3x - 5$, find the x- and y-intercepts.

EXAMPLE 4 For the equation $y = 3$, find the x- and y-intercepts.

Solution Remember from Section 4.2, Example 6, that the graph of $y = 3$ is a horizontal line parallel to the x-axis that passes through the y-axis at the point $(0, 3)$. Notice that this point is the y-intercept. Since the line is parallel to the x-axis, it will never intersect the x-axis. Therefore, there is no x-intercept.

Example 4 suggests the following rule.

RULE Intercepts for $y = c$

The graph of an equation in the form $y = c$, where c is a real-number constant, has no x-intercept and the y-intercept is $(0, c)$.

We can draw a similar conclusion about the graph of an equation in the form $x = c$.

> **RULE** Intercepts for $x = c$
>
> The graph of an equation in the form $x = c$, where c is a real-number constant, has no y-intercept and the x-intercept is $(c, 0)$.

YOUR TURN Find the coordinates of the x- and y-intercepts.

   a. $y = -5$ **b.** $x = -2$

OBJECTIVE 2. Graph linear equations using intercepts. Because two points determine a line in a plane, we can use the x- and y-intercepts to draw the graph of a linear equation.

EXAMPLE 5 Graph using the x- and y-intercepts: $5x - 2y = 10$

Solution In Example 1, the intercepts were found to be $(2, 0)$ and $(0, -5)$. We plot these intercepts and connect them to graph the line. It is helpful to find a third solution and verify that all three points can be connected to form a straight line. For this third point, we will choose $x = 3$ and solve for y.

Third solution: (check) If $x = 3$, then

$$5(3) - 2y = 10$$
$$15 - 2y = 10$$
$$-2y = -5 \qquad \text{Subtract 15 from both sides.}$$
$$y = \frac{5}{2} \qquad \text{Divide both sides by } -2. \qquad \textbf{Solution}\ \left(3, \frac{5}{2}\right)$$

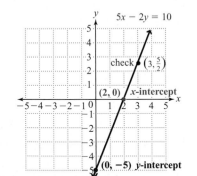

Table of Solutions

	x	y
x-intercept:	2	0
y-intercept:	0	-5
Third solution: (check)	3	$\dfrac{5}{2}$

a. no x-intercept
   y-intercept: $(0, -5)$

b. x-intercept: $(-2, 0)$
   no y-intercept

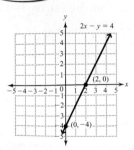

YOUR TURN Graph $2x - y = 4$ using the x- and y-intercepts.

If the x- and y-intercepts are both at the origin, then we must find an additional point in order to graph the linear equation.

EXAMPLE 6 Graph using the x- and y-intercepts.

a. $y = \dfrac{2}{3}x$

Solution In Example 2 we found $(0, 0)$ to be both the x- and the y-intercepts. Because this single point is not enough to determine the line in the plane, we must find at least one more solution. We will also find a third solution as a check.

Second solution:
Choose x to be 3.

$$y = \frac{2}{3}(3)$$

$$y = \frac{2}{\cancel{3}}\left(\frac{\cancel{3}^{\,1}}{1}\right)$$

$$y = 2$$

Second solution $(3, 2)$

Third solution (check):
Choose x to be -3.

$$y = \frac{2}{3}(-3)$$

$$y = \frac{2}{\cancel{3}}\left(-\frac{\cancel{3}^{\,1}}{1}\right)$$

$$y = -2$$

Third solution $(-3, -2)$

Now plot the solutions and connect the points to form the line.

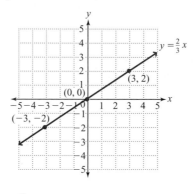

Table of Solutions

	x	y
x- and y-intercept:	0	0
Second solution:	3	2
Third solution: (check)	-3	-2

b. $y = \dfrac{2}{3}x + 2$

Solution Notice that this equation is in the form $y = mx + b$. Because $b = 2$ in this case, we can conclude that the y-intercept is $(0, 2)$. For the x-intercept, let $y = 0$ and solve for x. Also, as a check, we will choose $x = 3$ to generate a third point.

x-intercept: $0 = \dfrac{2}{3}x + 2$

$$-2 = \frac{2}{3}x \qquad \text{Subtract 2 from both sides.}$$

$$\frac{3}{\cancel{2}}\cdot\frac{-\cancel{2}^{\,1}}{1} = \frac{\cancel{3}^{\,1}}{\cancel{2}}\cdot\frac{\cancel{2}^{\,1}}{\cancel{3}^{\,1}}x \qquad \text{Multiply both sides by } \frac{3}{2}.$$

$$-3 = x$$

x-intercept $(-3, 0)$

Third solution: (check) $y = \dfrac{2}{3}(3) + 2$

$$y = \frac{2}{\cancel{3}}\cdot\frac{\cancel{3}^{\,1}}{1} + 2$$

$$y = 2 + 2$$

$$y = 4$$

Third solution $(3, 4)$

Now plot the solutions and connect the points to form the line.

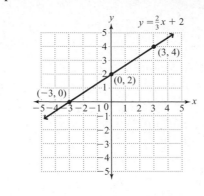

Table of Solutions

	x	y
x-intercept:	-3	0
y-intercept:	0	2
Third solution: (check)	3	4

Compare the graphs in Examples 6a and 6b. If we place both lines on the same grid (shown here), we see that they are parallel, which means that they do not intersect at any point. Further, each point on the graph of $y = \frac{2}{3}x + 2$ is 2 units higher than each point with the same x-coordinate on the graph of $y = \frac{2}{3}x$.

Learning Strategy

If you are a visual learner, imagine the line $y = \frac{2}{3}x$ sliding up the y-axis 2 units to become $y = \frac{2}{3}x + 2$.

Table of Solutions

x	$y = \frac{2}{3}x$	$y = \frac{2}{3}x + 2$
-3	-2	0
0	0	2
3	2	4

From the table we see that the difference in an equation in the form $y = mx$ and an equation in the form $y = mx + b$ is the value of the y-intercept. In other words, the line $y = mx$ shifts upwards or downwards by b units. The graph of $y = \frac{2}{3}x - 2$ is a line parallel to $y = \frac{2}{3}x$, only shifted down 2 units so that its y-intercept is $(0, -2)$. We will explore this further in the next section.

YOUR TURN Graph $y = \frac{2}{3}x - 2$ using the x- and y-intercepts.

ANSWER

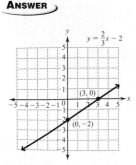

4.3 Exercises

For Extra Help

MyMathLab   Videotape/DVT   InterAct Math   Math Tutor Center   Math XL.com

1. Explain what an x-intercept is.

2. Explain what a y-intercept is.

3. How do you find an x-intercept?

4. How do you find a y-intercept?

5. If an equation is in the form $y = mx$, where m is a real-number constant, what are the coordinates of the x- and y-intercepts? Explain.

6. Compare the graphs of $y = 2x$ and $y = 2x + 5$.

For Exercises 7–22, find the x- and y-intercepts.

7. $x - y = 4$

8. $x + y = 7$

9. $2x + 3y = 6$

10. $2x + 3y = 12$

11. $3x - 4y = -10$

12. $3x - 4y = 6$

13. $y = 3x - 6$

14. $y = 2x + 8$

15. $y = 2x + 5$

16. $y = 3x - 1$

17. $y = 2x$

18. $y = \dfrac{2}{5}x$

19. $2x + 3y = 0$

20. $x + 3y = 0$

21. $x - 5 = 0$

22. $y = 7$

For Exercises 23–50, graph using the x- and y-intercepts.

23. $x + y = 7$

24. $x + y = 10$

25. $3x - y = 6$

26. $x - 5y = 15$

27. $2x + 3y = 6$

28. $3x + 4y = 12$

29. $5x + 2y = 10$

30. $3x - 2y = 18$

31. $y = 2x$

32. $y = -3x$

33. $y = \dfrac{2}{5}x$

34. $-\dfrac{1}{4}x = y$

35. $y = x - 4$

36. $y = x + 2$

37. $y = 2x - 6$

38. $y = 3x + 9$

39. $6x + 3y = 9$

40. $2x - y = -1$

41. $4x + 2y = 5$

42. $3x + 2y = 7$

43. $x = -4$

44. $x = 5$
45. $y = -2$
46. $y = 5$

47. $y - 2 = 0$
48. $y + 4 = 0$
49. $x - 2 = 0$

50. $x + 3 = 0$

51. Which of the following could be the graph of $1.4x - 0.9y = 2.7$? Explain.

  a.   **b.**   **c.**   **d.**

52. Which of the following could be the graph of $0.65x - 2.9y = -3.5$? Explain.

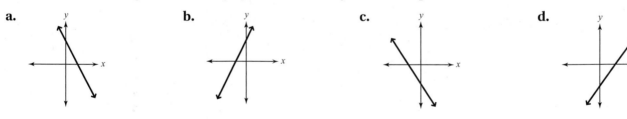

  a.   **b.**   **c.**   **d.**

53. If a circle with a radius of 2.7 units were to be drawn in the coordinate plane with its center at the origin, what would be the coordinates of its x- and y-intercepts?

54. If a circle with a diameter of 6.8 units were to be drawn in the coordinate plane with its center at the origin, what would be the coordinates of its x- and y-intercepts?

55. A salesperson sells two different sizes of facial cleansing lotion. She sells the small size for $15 and the large size for $20.

    a. If x represents the number of units sold of the small size and y represents the number of units of the large size sold, then write an expression that describes her total sales.

    b. If her supervisor sets a goal of selling $2000 worth of the cleansing lotion, use the expression from part a to write an equation describing the sales goal.

    c. Find the x- and y-intercepts.

    d. What do the x- and y-intercepts mean in terms of sales?

    e. Find a third combination of sales numbers that would equal the sales goal.

    f. Graph the equation.

56. A comic book collector will pay $50 for any Superman comic published prior to 1970 if it is in mint condition and $30 if it is not in mint condition.

    a. If x represents the number of comic books in mint condition and y represents the number of comic books not in mint condition, then write an expression that describes the amount he would pay for a combination of comic books in mint and nonmint condition.

    b. Suppose the collector visits a convention for collectors and has a budget of $1800 to spend. Use the expression from part a to write an equation describing his expenditure on a combination of the comics if he spends all of his budget.

    c. Find the x- and y-intercepts.

    d. What do the x- and y-intercepts mean in terms of purchases?

    e. Find a third combination of comic books he could purchase to equal his budget amount.

    f. Graph the equation.

REVIEW EXERCISES

[2.4]   **1.** Solve for w in the equation $P = 2l + 2w$.

[2.5]   **2.** Translate to an equation, then solve. Four more than three times a number is equal to five times the sum of the number and six.

[2.5]   **3.** Translate to an equation, then solve. The sum of three consecutive odd integers is 141. Find the integers.

[3.3] 4. Two steel rods are welded to form a 90° angle. A third rod is welded to the other two as shown. If one of the angles is 15° more than the other, what are the two angles?

[4.1] 5. In what quadrant is (150, 200) located?

[4.2] 6. Determine whether $(-8, -2)$ is a solution for $x + 3y = -5$.

4.4 Slope-Intercept Form

OBJECTIVES

1. Compare lines with different slopes.
2. Graph equations in slope-intercept form.
3. Find the slope of a line given two points on the line.

In Section 4.3, we discussed various forms of linear equations, one of which was $y = mx + b$, where m and b are real-number constants. Further, we discovered that the graph of an equation in this form is a line with its y-intercept at $(0, b)$. In this section, we explore how the coefficient m affects the graph.

OBJECTIVE 1. Compare lines with different slopes. First, we consider graphs of equations in which $b = 0$, so that the equation is $y = mx$ and the graphs all have the origin $(0, 0)$ as the x- and y-intercepts.

EXAMPLE 1 Graph each of the following on the same grid.

$$y = x \qquad y = 2x \qquad y = 3x$$

Solution In the following table, the same choice for x has been substituted into each equation and the corresponding y-coordinate has been found. We plot the ordered pairs and connect the points to graph the lines.

If x is	and y = x, then y is	and y = 2x, then y is	and y = 3x, then y is
0	0	0	0
1	1	2	3
2	2	4	6

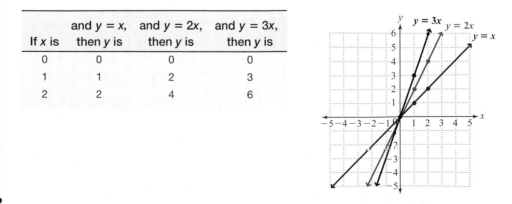

From Example 1 we see that for equations of the form $y = mx$, as the coefficient m increases, the graphs get steeper. Because the coefficient m affects how steep a line is, m is called the *slope* of the line.

For $y = x$, the slope is 1 ($m = 1$).

For $y = 2x$, the slope is 2 ($m = 2$).

For $y = 3x$, the slope is 3 ($m = 3$).

We have seen that if the slope is greater than 1, then the corresponding line has a steeper incline than the graph of $y = x$. What values for m would cause a line to be less inclined than $y = x$? The pattern in the graphs suggests we should explore values less than 1. First consider values less than 1 but greater than 0, such as $\frac{1}{2}$ and $\frac{1}{3}$.

EXAMPLE 2 Graph each of the following on the same grid.

$$y = x \qquad y = \frac{1}{2}x \qquad y = \frac{1}{3}x$$

Solution For consistency, we use the same three choices for x that we used in Example 1. After finding the corresponding y-coordinates for each equation, we plot the ordered pairs and connect the points to graph the lines.

If x is	and $y = x$, then y is	and $y = \frac{1}{2}x$, then y is	and $y = \frac{1}{3}x$, then y is
0	0	0	0
1	1	$\frac{1}{2}$	$\frac{1}{3}$
2	2	1	$\frac{2}{3}$

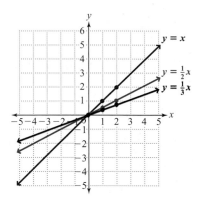

Notice that if the slope is a fraction between 0 and 1, then the smaller the fraction is, the less inclined or flatter the lines get. So far, we have considered only positive slope values. If we "travel" from left to right along lines with positive slopes, we travel uphill. What would a line with a negative slope look like?

EXAMPLE 3 Graph each of the following on the same grid.

$$y = -x \qquad y = -2x \qquad y = -\frac{1}{2}x$$

Solution Again, we make a table of ordered pairs, then plot those ordered pairs and connect the points to graph the lines.

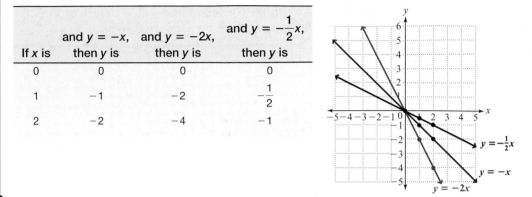

If x is	and $y = -x$, then y is	and $y = -2x$, then y is	and $y = -\frac{1}{2}x$, then y is
0	0	0	0
1	−1	−2	$-\frac{1}{2}$
2	−2	−4	−1

Note: *When we speak of traveling along a graph, the convention is to travel from left to right, just as we do when we read English.*

Notice that if we "travel" from left to right along lines with negative slopes, we travel downhill. As the slope becomes more negative, the downhill incline becomes steeper.

RULE **Graphs of $y = mx$**

Given an equation of the form $y = mx$, the graph of the equation is a line passing through the origin and having the following characteristics:

If $m > 0$, then the graph is a line that slants uphill from left to right.

If $m < 0$, then the graph is a line that slants downhill from left to right.

The greater the absolute value of m, the steeper the line.

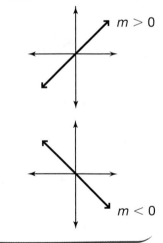

YOUR TURN

a. Is $y = 5x$ steeper or less inclined than $y = x$?

b. Is the graph of $y = -\frac{2}{3}x$ uphill or downhill from left to right?

ANSWERS

a. $y = 5x$ is steeper than $y = x$.

b. downhill

OBJECTIVE 2. Graph equations in slope-intercept form. We have seen that slope determines the incline of a line. What more can we discover about slope? Notice that if we isolate m in $y = mx$, we get $m = \dfrac{y}{x}$, which suggests that **slope** is a ratio of the amount of vertical change to the amount of horizontal change.

Note: *A precise formula for the slope of a line will be given in Objective 3.*

DEFINITION **Slope:** The ratio of the vertical change between any two points on a line to the horizontal change between these points.

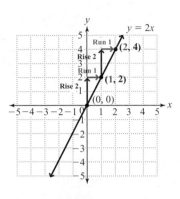

For example, the slope of $y = 2x$ is 2, which can be written as $\dfrac{2}{1}$. This slope value means that rising vertically 2 units and then running horizontally 1 unit from any point on the line locates a second point on the line. For example, if we rise 2 units and run 1 unit from (0, 0), we end up at (1, 2). Similarly, rising 2 units and running 1 unit from (1, 2) puts us at (2, 4).

We have learned that in an equation of the form $y = mx + b$, the slope is m and the y-intercept is (0, b). For this reason, we say that equations of the form $y = mx + b$ are in *slope-intercept* form. We can graph equations in slope-intercept form using the y-intercept as a starting point, and then using the slope to locate other points on the line.

PROCEDURE Graphing Equations in Slope-Intercept Form

To graph an equation in slope-intercept form, $y = mx + b$,

1. Plot the y-intercept, (0, b).
2. Plot a second point by rising the number of units indicated by the numerator of the slope, m, then running the number of units indicated by the denominator of the slope, m.
3. Draw a straight line through the two points.

Note: You can check by locating additional points using the slope. Every point you locate using the slope should be on the line.

EXAMPLE 4 For the equation $y = -\dfrac{1}{3}x + 2$, determine the slope and the y-intercept. Then graph the equation.

Solution Because the equation is in the form $y = mx + b$, where $m = -\dfrac{1}{3}$ and $b = 2$, the slope of the line is $-\dfrac{1}{3}$ and the y-intercept is (0, 2). To graph the line, we can plot the y-intercept and then use the slope to find other points. To interpret $-\dfrac{1}{3}$

in terms of "rise" and "run," we place the negative sign in either the numerator or denominator. We can "rise" -1 and "run" 3, which gives the point $(3, 1)$. Or we can "rise" 1 and "run" -3, which gives the point $(-3, 3)$. Either interpretation gives another point on the same line.

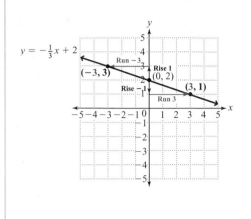

The points we found using the slope can be verified in the equation. We can check the ordered pair $(-3, 3)$ in the equation:

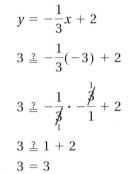

$$y = -\frac{1}{3}x + 2$$

$$3 \stackrel{?}{=} -\frac{1}{3}(-3) + 2$$

$$3 \stackrel{?}{=} -\frac{1}{\cancel{3}} \cdot -\frac{\cancel{3}^{1}}{1} + 2$$

$$3 \stackrel{?}{=} 1 + 2$$

$$3 = 3 \qquad \text{This checks.}$$

We will leave the check of $(3, 1)$ to the reader.

YOUR TURN For the equation $y = \dfrac{2}{5}x - 2$, determine the slope and the y-intercept. Then graph the equation.

If an equation is not in slope-intercept form, as in $3x + 4y = 8$, and we need to determine the slope and y-intercept, then we can write the equation in slope-intercept form by isolating y.

EXAMPLE 5 For the equation $3x + 4y = 8$, determine the slope and the y-intercept. Then graph the equation.

Solution Write the equation in slope-intercept form by isolating y.

$$3x + 4y = 8$$

$$4y = -3x + 8 \qquad \textbf{Subtract } 3x \textbf{ from both sides to isolate } 4y.$$

$$y = -\frac{3}{4}x + \frac{8}{4} \qquad \textbf{Divide both sides throughout by 4 to isolate } y.$$

$$y = -\frac{3}{4}x + 2 \qquad \textbf{Simplify the fraction.}$$

ANSWER

$m = \dfrac{2}{5}$, y-intercept: $(0, -2)$

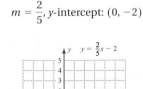

The slope is $-\dfrac{3}{4}$ and the y-intercept is $(0, 2)$. To graph the line we begin at $(0, 2)$ and then rise -3 and run 4, which gives the point $(4, -1)$. Or we could begin at $(0, 2)$ and then rise 3 and run -4, which gives the point $(-4, 5)$.

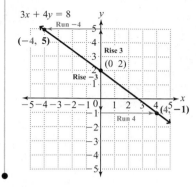

We can check the ordered pairs in the equation:

For $(-4, 5)$:

$$3x + 4y = 8$$
$$3(-4) + 4(5) \stackrel{?}{=} 8$$
$$-12 + 20 \stackrel{?}{=} 8$$
$$8 = 8$$

For $(4, -1)$:

$$3x + 4y = 8$$
$$3(4) + 4(-1) \stackrel{?}{=} 8$$
$$12 + (-4) \stackrel{?}{=} 8$$
$$8 = 8$$

Both ordered pairs check.

YOUR TURN For the equation $3x - 2y = 6$, determine the slope and the y-intercept. Then graph the equation.

OBJECTIVE 3. Find the slope of a line given two points on the line. Given two points on a line, we can determine the slope of the line. Consider the points $(-3, 0)$ and $(3, 4)$ on the graph of $y = \dfrac{2}{3}x + 2$, which we graphed in Example 7 of Section 4.3. In the equation, we see that the slope is $\dfrac{2}{3}$. Recall that slope is the ratio of the vertical change between any two points on a line to the horizontal change between those points. Therefore, the ratio of the vertical distance between $(-3, 0)$ and $(3, 4)$ to the horizontal distance between those points should be $\dfrac{2}{3}$. Let's verify.

ANSWER

$m = \dfrac{3}{2}$, y-intercept: $(0, -3)$

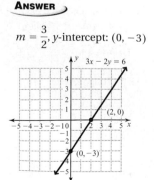

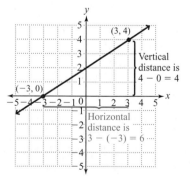

To find the vertical distance between the two points, we can subtract their y-coordinates: $4 - 0 = 4$. Similarly, the horizontal distance between the points is the difference of the x-coordinates: $3 - (-3) = 6$. Now we can write the ratio.

$$\frac{\text{Vertical distance}}{\text{Horizontal distance}} = \frac{4 - 0}{3 - (-3)} = \frac{4}{6} = \frac{2}{3}$$

Our example suggests the following formula.

PROCEDURE   The Slope Formula

Given two points (x_1, y_1) and (x_2, y_2), where $x_2 \neq x_1$, the slope of the line connecting the two points is given by the formula $m = \dfrac{y_2 - y_1}{x_2 - x_1}$.

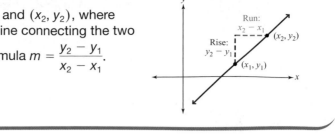

EXAMPLE 6   Find the slope of the line connecting the given points.

a. $(3, 5)$ and $(-1, 7)$

Solution Using $m = \dfrac{y_2 - y_1}{x_2 - x_1}$, replace the variables with their corresponding values, and then simplify. We will let $(3, 5)$ be (x_1, y_1) and $(-1, 7)$ be (x_2, y_2).

$$m = \frac{7 - 5}{-1 - 3} = \frac{2}{-4} = -\frac{1}{2}$$

It does not matter which ordered pair is (x_1, y_1) and which is (x_2, y_2). If we let $(-1, 7)$ be (x_1, y_1) and $(3, 5)$ be (x_2, y_2), we get the same slope.

$$m = \frac{5 - 7}{3 - (-1)} = \frac{-2}{4} = -\frac{1}{2}$$

b. $(-3, -4)$ and $(1, -4)$

Solution We will let $(-3, -4)$ be (x_1, y_1) and $(1, -4)$ be (x_2, y_2).

$$m = \frac{-4 - (-4)}{1 - (-3)} = \frac{0}{4} = 0$$

In Example 6b, we found the slope of the line connecting $(-3, -4)$ and $(1, -4)$ to be 0. What does a line with a slope of 0 look like? Let's plot the points $(-3, -4)$ and $(1, -4)$ and see.

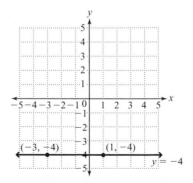

Because the two points have the same y-coordinate, they form a horizontal line parallel to the x-axis. The equation of this line is $y = -4$.

RULE   Zero Slope

Two points with different x-coordinates and the same y-coordinates, (x_1, c) and (x_2, c), will form the line with a slope 0 (a horizontal line) and equation $y = c$.

What is the slope of a vertical line? Two points on a vertical line would have the same x-coordinates but different y-coordinates, as in $(2, 4)$ and $(2, -3)$.

EXAMPLE 7 Find the slope of the line connecting the points $(2, 4)$ and $(2, -3)$.

Solution Note that $(2, 4)$ and $(2, -3)$ connect to form a vertical line whose equation is $x = 2$. The slope of this line is found using the slope formula:

$$m = \frac{-3 - 4}{2 - 2}$$ **Replace the variables in** $m = \dfrac{y_2 - y_1}{x_2 - x_1}$ **with their corresponding values.**

$$m = \frac{-7}{0}$$ **Simplify.**

Because we never divide by 0, the slope is *undefined*.

RULE Undefined Slope

Two points with the same x-coordinates and different y-coordinates, (c, y_1) and (c, y_2) will form a line with a slope that is undefined (a vertical line) and equation $x = c$.

YOUR TURN Find the slope of the line connecting the points with the given coordinates.

a. $(2, 5)$ and $(6, 1)$ **b.** $(-5, -2)$ and $(7, -5)$ **c.** $(4, -1)$ and $(-3, -1)$

▦ Calculator
TIPS

To graph an equation on a graphing calculator, you first enter the equation using the [Y=] *key. When you press the* [Y=] *key, you will see a list of Y's: Y1, Y2, Y3, etc. Each of the Y's can be used to enter an equation, which means you can graph multiple equations on the same graph if desired.*

To enter the equation $y = 3x + 2$*, press* [Y=] *and at Y1, type:* [3] [X, T, θ, n] [+] [2]

Notice that the [X, T, θ, n] *key is used to type the variable x. To have the calculator draw the graph, press* [GRAPH] *.*

Sometimes you may need to adjust or change your viewing window. To change the range of marks on your axes. Press [WINDOW] *. For example, if you want the x-axis to range from* -10 *to 10, then press* [WINDOW] *and at Xmin, enter* [−] [1] [0] *and at Xmax, enter* [1] [0] *. Xscl sets the frequency of the marks in the range. For example, setting the Xscl to 1 places a mark at each integer in the range, so if the range is* -10 *to 10, a mark will appear at* $-10, -9, -8, \ldots, 9, 10$*. If the Xscl is 2 and the range is* -10 *to 10, a mark will appear at* $-10, -8, -6, \ldots, 8, 10$*. Similarly, Ymax and Ymin change the range on the y-axis, and Yscl affects how many marks are shown in that range.*

ANSWERS

a. -1 **b.** $-\dfrac{1}{4}$ **c.** 0

1. In your own words, what is the slope of a line?

2. What is the formula for finding the slope of a line connecting two points (x_1, y_1) and (x_2, y_2)?

3. Does the graph of $y = -2x + 1$ go uphill from left to right or downhill? Why?

4. Does the graph of $3x - 5y = 10$ go uphill from left to right or downhill? Why?

5. Describe a line with a slope of 0. Explain why a slope of 0 produces this type of line.

6. Describe a line with a slope that is undefined. Explain why an undefined slope produces this type of line.

For Exercises 7–14, graph each set of equations on the same grid. For each set of equations, compare the slopes, y-intercepts, and their effects on the graphs.

7. $y = \dfrac{1}{2}x$

   $y = x$

   $y = 2x$

8. $y = \dfrac{1}{3}x$

   $y = x$

   $y = 3x$

9. $y = x$

   $y = \dfrac{1}{2}x$

   $y = \dfrac{1}{5}x$

10. $y = x$

    $y = \dfrac{1}{3}x$

    $y = \dfrac{1}{6}x$

11. $y = -x$

    $y = -2x$

    $y = -4x$

12. $y = -x$

    $y = -\dfrac{3}{2}x$

    $y = -3x$

13. $y = x$

    $y = x + 2$

    $y = x - 2$

14. $y = 2x$

    $y = 2x - 1$

    $y = 2x - 3$

For Exercises 15–38, determine the slope and the y-intercept. Then graph the equation.

15. $y = 2x + 3$

16. $y = 4x - 3$

17. $y = -2x - 1$

18. $y = -3x + 5$

19. $y = \dfrac{1}{3}x + 2$

20. $y = \dfrac{1}{2}x - 4$

21. $y = \dfrac{3}{4}x - 2$

22. $y = \dfrac{2}{5}x - 5$

23. $y = -\dfrac{2}{3}x + 8$

24. $y = -\dfrac{4}{3}x + 7$

25. $2x + y = 4$

26. $3x + y = 6$

27. $3x - y = -1$

28. $2x - y = 5$

29. $2x + 3y = 6$

30. $4x - 3y = 12$

31. $2x + 3y = 7$

32. $x + 2y = -5$

33. $2x - 3y + 8 = 0$

34. $3x - 2y + 7 = 0$

35. $2y = -3x + 5$

36. $3y = 2x - 4$

37. $0.6x - 0.2y = 1$

38. $0.3x + 1.5y = -6$

For Exercises 39–54, find the slope of the line through the given points.

39. $(4, 1), (8, 11)$

40. $(1, 5), (3, 8)$

41. $(2, 4), (5, 2)$

42. $(6, 2), (3, 7)$

43. $(-3, 5), (4, 7)$

44. $(6, -2), (5, 5)$

45. $(2, -5), (-6, 3)$

46. $(-4, 6), (6, -2)$

47. $(-3, 1), (-2, -2)$

48. $(-4, -6), (2, -3)$

49. $(0, 5), (4, 0)$

50. $(-3, 0), (0, 6)$

51. $(-3, 5), (4, 5)$

52. $(-2, -8), (4, -8)$

53. $(6, 1), (6, -8)$

54. $(-3, 2), (-3, 5)$

For Exercises 55–60, match the equation with the appropriate graph.

55. $y = 2x + 4$

56. $3x + 2y = 6$

57. $x = y$

58. $y = 3$

59. $x - 2 = 0$

60. $x - 2y = 8$

a.

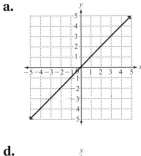

b.

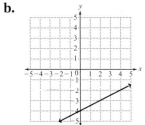

c.

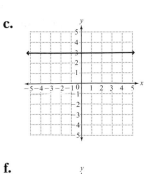

d.

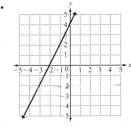

e.

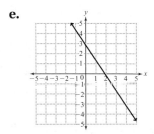

f.

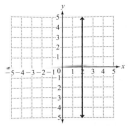

For Exercises 61–64, find the slope of the line, the y-intercept, and write the equation of the line in slope-intercept form.

61.

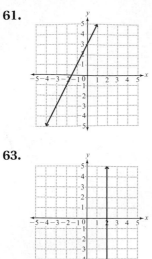

62.

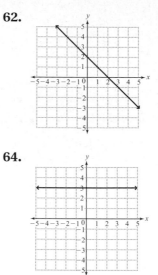

63.

64.

65. What is the equation of a horizontal line through $(0, -2)$?

66. What is the equation of a vertical line through $(-3.5, 0)$?

67. What equation describes the *x*-axis?

68. What equation describes the *y*-axis?

69. A parallelogram has vertices at $(1, -2)$, $(3, 4)$, $(-3, -2)$, and $(-1, 4)$.
   a. Plot the vertices in a coordinate plane, then connect them to form the parallelogram.
   b. Find the slope of each side of the parallelogram.

   c. What do you notice about the slopes of the parallel sides?

70. A right triangle has vertices at $(-2, 3)$, $(2, 1)$, and $(3, 3)$.
   a. Plot the vertices in a coordinate plane, then connect them to form the triangle.
   b. Find the slope of each side of the triangle.

   c. What do you notice about the slopes of the perpendicular sides?

71. In architecture, the slope of a roof is called its *pitch*. The roof of a house rises 3.5 feet for every 6 feet of horizontal distance. Find the pitch of the roof. Write the pitch as a decimal number and a fraction in simplest form.

72. The slope of a hill is referred to as the *grade* of the hill. Heartbreak Hill, which is part of the Boston Marathon, rises from an elevation of 150 feet to an elevation of 225 feet over a horizontal distance of about 5000 feet. Find the grade of Heartbreak Hill. (*Source:* Boston Athletic Association)

73. The Great Pyramid at Giza rises 29.3 meters for every 23 meters of horizontal distance. Find the slope of the pyramid.

74. The steps of the Pyramid of the Sun at Teotihuacan, Mexico, rise 65 meters over a horizontal distance of 112.5 meters. Find the slope of the steps of the Pyramid of the Sun.

75. The graph to the right shows the percentage rate for Treasury bonds from 9 A.M. until 2:00 P.M. on July 17, 2001.

   a. Draw a straight line from the point on the graph at 9 A.M. to the point on the graph at 2 P.M.

   b. Find the slope of this line.

   c. If the same trend continued, predict the T-bond percentage at 4 P.M.

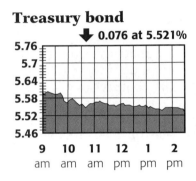

76. The graph shown is a comparison of an average diver's consumption of gases using a standard open-circuit tank versus a closed-circuit breathing system.

   a. What is the slope of the line for the open-circuit tank?

   b. What is the slope of the line for the closed-circuit breathing system?

   c. What are some conclusions you could draw from the graphs?

Of Interest

Closed-circuit breathing systems operate by recycling the exhaled breath of the diver. A special filter absorbs carbon dioxide from the exhaled breath, then sensors detect how much oxygen remains, and a computer regulates the amount of new oxygen that is mixed in from a tank for the diver to inhale.

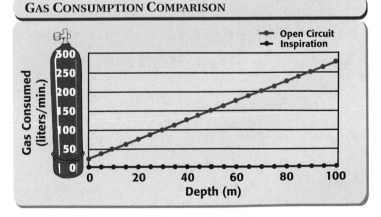

Source: http://ambientpressurediving.com/Graphs.html

Collaborative Exercises POPULATION GROWTH

The following data show the total U.S. population by decade.

Decade	Population (in millions)
1900	76.2
1910	92.2
1920	106.0
1930	123.2
1940	132.2
1950	151.3
1960	179.3
1970	203.3
1980	226.5
1990	248.7
2000	281.4

1. Plot each ordered pair as a point on a graph with the decades along the horizontal axis and the population along the vertical axis.

2. Do the points have a linear relationship?

3. Draw a straight line that connects the point at 1900 to the point at 1950. Find the slope of this line.

4. Draw a straight line that connects the point at 1950 to the point at 2000. Find the slope of this line.

5. Compare the slopes of the two lines you have drawn. Which has the greater slope? What does this indicate about population growth in the United States during the 20th century?

6. Use the line you drew from 1950 to 2000 to predict the population at 2110.

REVIEW EXERCISES

[1.7]   1. Simplify: $12x - 9y + 7 - x + 9y$

[1.7]   2. Multiply: $-2(x + 7)$

[2.3]   3. Solve and check: $(7x - 3) - (2x + 4) - (3x - 7) = 21$

[2.4]   4. Isolate x in the equation $Ax + By = C$.

[4.2]   5. Graph: $2x + 3y = 8$

[4.3]   6. Find the x- and y-intercepts for $3x - 5y = 7$.

4.5 Point-Slope Form

OBJECTIVES

1. Use slope-intercept form to write the equation of a line.
2. Use point-slope form to write the equation of a line.
3. Write the equation of a line parallel to a given line.
4. Write the equation of a line perpendicular to a given line.

In this section, we will explore how to write an equation of a line given information about the line, such as its slope and the coordinates of a point on the line.

OBJECTIVE 1. Use slope-intercept form to write the equation of a line. If we are given the y-intercept $(0, b)$ of a line, to write the equation of the line, we will need either the slope or another point so that we can calculate the slope. First, consider a situation in which we are given the y-intercept and the slope of a line.

EXAMPLE 1 A line with a slope of 5 crosses the y-axis at the point $(0, 3)$. Write the equation of the line.

Solution We use $y = mx + b$, the slope-intercept form of the equation, replacing m with the slope, 5, and b with the y-coordinate of the y-intercept, 3.

$$y = 5x + 3$$

YOUR TURN A line has a slope of -3. If the y-intercept is $(0, -4)$, write an equation of the line in slope-intercept form.

Now suppose we are given the y-intercept and a second point on the line. In order to write the equation in slope-intercept form, we need the slope. So our first step is to calculate the slope using the formula from Section 4.4, $m = \dfrac{y_2 - y_1}{x_2 - x_1}$, where the two points, (x_1, y_1) and (x_2, y_2), are the y-intercept and the other given point.

ANSWER

$y = -3x - 4$

EXAMPLE 2 The following graph shows the relationship between Celsius and Fahrenheit temperatures. The table lists two ordered pairs of equivalent temperatures on the two scales.

a. Write the equation for converting degrees Celsius to degrees Fahrenheit.

b. Use the equation to convert 60°C to degrees Fahrenheit.

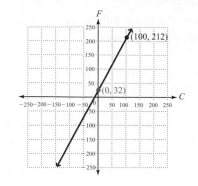

	Celsius (C)	Fahrenheit (F)
Freezing point of water:	0	32
Boiling point of water:	100	212

Solution

a. We are given (0, 32), which is like the *y*-intercept, and a second point, (100, 212). To use the slope-intercept equation, we must first find the slope using the slope formula:

Note: *In this case, the y-coordinates are degrees Fahrenheit and the x-coordinates are degrees Celsius.*

$$m = \frac{212 - 32}{100 - 0} \qquad \text{Use } m = \frac{y_2 - y_1}{x_2 - x_1}.$$

$$m = \frac{180}{100}$$

$$m = \frac{9}{5}$$

Connection A slope of $\frac{9}{5}$ means that for every 9 degrees of change on the Fahrenheit scale, there is a 5-degree change on the Celsius scale.

Since the *F*-intercept is (0, 32) and the slope is $\frac{9}{5}$, we can write an equation in slope-intercept form. We let *C* (degrees Celsius) take the place of the *x*-variable, and *F* (degrees Fahrenheit) take the place of the *y*-variable.

$$F = \frac{9}{5}C + 32$$

b. Now we can use this equation to convert Celsius temperatures to degrees Fahrenheit. We are to convert 60°C to degrees Fahrenheit.

$$F = \frac{9}{5}(60) + 32 \qquad \text{Replace } C \text{ with 60.}$$

$$F = 108 + 32$$

$$F = 140°F$$

Examples 1 and 2 suggest the following procedure.

To write the equation of a line given its *y*-intercept, $(0, b)$, and its slope, *m*, use the slope-intercept form of the equation, $y = mx + b$. If given a second point and not the slope, we must first calculate the slope using $m = \dfrac{y_2 - y_1}{x_2 - x_1}$, then use $y = mx + b$.

YOUR TURN   Write the equation of the line connecting the given points in slope-intercept form.

$$(0, -5), (2, 4)$$

OBJECTIVE 2. Use point-slope form to write the equation of a line.   In our first objective we used the slope-intercept form of the equation of a line because we were given the *y*-intercept. However, we need a more general approach so that we can write the equation of a line given *any* two points on the line. For this more general approach, we can use the slope formula to derive a new form of the equation of a line called the *point-slope* form. Recall the slope formula:

Connection   We can also view the formula for slope as a proportion and cross multiply to get the point-slope form:

$$\frac{m}{1} \diagdown \frac{y_2 - y_1}{x_2 - x_1}$$

$$(x_2 - x_1)m = y_2 - y_1$$

$$m = \frac{y_2 - y_1}{x_2 - x_1}$$

$$(x_2 - x_1) \cdot m = \left(\frac{y_2 - y_1}{x_2 - x_1}\right) \cdot \left(\frac{x_2 - x_1}{1}\right) \qquad \text{Multiply both sides by } x_2 - x_1 \text{ to isolate the } y\text{'s.}$$

$$(x_2 - x_1)m = y_2 - y_1 \qquad \text{Simplify.}$$

$$y_2 - y_1 = m(x_2 - x_1) \qquad \text{Rewrite with } y\text{'s on the left side to resemble slope-intercept form.}$$

To write the equation of a line with this formula, we replace *m* with the slope and substitute one of the given ordered pairs for x_1 and y_1 leaving x_2 and y_2 as variables. To indicate that x_2 and y_2 remain variables, we remove their subscripts so that we have $y - y_1 = m(x - x_1)$, which is called the point-slope form of the equation of a line.

PROCEDURE   Using the Point-Slope Form of the Equation of Line

To write the equation of a line given its slope and any point, (x_1, y_1), on the line, use the point-slope form of the equation of a line, $y - y_1 = m(x - x_1)$. If given a second point, (x_2, y_2), and not the slope, we first calculate the slope using $m = \dfrac{y_2 - y_1}{x_2 - x_1}$, then use $y - y_1 = m(x - x_1)$.

ANSWER

$$y = \frac{9}{2}x - 5$$

EXAMPLE 3 Write the equation of a line with a slope of 2 and passing through the point (3, 5). Write the equation in slope-intercept form.

Solution Because we are given the coordinates of a point and the slope of a line passing through the point, we begin with the point-slope formula. After replacing m, x_1, and y_1 with their corresponding values, we isolate y to get slope-intercept form.

$$y - y_1 = m(x - x_1)$$
$$y - 5 = 2(x - 3) \qquad \text{Replace } m \text{ with 2, } x_1 \text{ with 3, and } y_1 \text{ with 5.}$$
$$y - 5 = 2x - 6 \qquad \text{Simplify.}$$
$$y = 2x - 1 \qquad \text{Add 5 to both sides to isolate } y. \\ \text{This is now slope-intercept form.}$$

YOUR TURN Write the equation of the line in slope-intercept form with the given slope passing through the given point.

$$m = 0.6; (-4, 2)$$

EXAMPLE 4 Write the equation of a line passing through the points (3, 2) and (−3, 6). Write the equation in slope-intercept form.

Solution Since we do not have the slope, we first calculate it.

$$m = \frac{6 - 2}{-3 - 3} = \frac{4}{-6} = -\frac{2}{3} \qquad \text{Use } m = \frac{y_2 - y_1}{x_2 - x_1}.$$

Now we can use the point-slope form, then isolate y to write the slope-intercept form. Since we were given two points, we can use either point for (x_1, y_1) in the point-slope equation. We will select (3, 2) to be (x_1, y_1).

$$y - y_1 = m(x - x_1)$$
$$y - 2 = -\frac{2}{3}(x - 3) \qquad \text{Replace } m \text{ with } -\frac{2}{3}, x_1 \text{ with 3, and } y_1 \text{ with 2.}$$
$$y - 2 = -\frac{2}{3}x + 2 \qquad \text{Simplify.}$$
$$y = -\frac{2}{3}x + 4 \qquad \text{Add 2 to both sides to isolate } y.$$

YOUR TURN Write the equation of the line connecting the given points in slope-intercept form.

$$(-2, -4), (4, -1)$$

ANSWER

$y = 0.6x - 4.4$

ANSWER

$y = \frac{1}{2}x - 3$

Writing Linear Equations in Standard Form

Equations can also be written in standard form. Recall from Section 4.3 that standard form is $Ax + By = C$, where A, B, and C are real numbers. We will manipulate the point-slope form of the equation so that the x and y terms are on the same side of the equation and a constant appears on the other side of the equation.

EXAMPLE 5 The following data points relate the velocity of an object as time passes (You may recall this data from the last Your Turn problem in Section 4.1). The graph shows that the points are in a line. Write the equation of the line in standard form.

Time (in seconds)	Velocity (in ft./sec.)
0	2.0
1	2.5
2	3.0
3	3.5

Solution First, we find the slope of the line using any two ordered pairs in the slope formula. We will use $(0, 2)$ and $(1, 2.5)$.

$$m = \frac{2.5 - 2}{1 - 0} = 0.5 \text{ or } \frac{1}{2} \qquad \text{Use } m = \frac{y_2 - y_1}{x_2 - x_1}.$$

Because $(0, 2)$ is the y-intercept, we can write the equation in slope-intercept form, then manipulate the equation so that x and y appear on the same side of the equal sign to get standard form.

$$y = \frac{1}{2}x + 2 \qquad \text{Slope-intercept form.}$$

$$y - \frac{1}{2}x = \frac{1}{2}x - \frac{1}{2}x + 2 \qquad \text{Subtract } \frac{1}{2}x \text{ from both sides to get } x \text{ and } y \text{ together.}$$

$$y - \frac{1}{2}x = 2$$

When we write an equation in standard form ($Ax + By = C$), it is customary to write the equation so that the x term is first with a positive coefficient and, if possible, so that A, B, and C are all integers. Let's manipulate $y - \frac{1}{2}x = 2$ to this more polished form.

$$-\frac{1}{2}x + y = 2 \qquad \text{Use the commutative property of addition to write the } x \text{ term first.}$$

$$(-2)\left(-\frac{1}{2}x + y\right) = (-2)(2) \qquad \text{Multiply both sides by } -2 \text{ so that the coefficient of } x \text{ is a positive integer.}$$

$$x - 2y = -4 \qquad \text{Standard form with } A, B, \text{ and } C \text{ integers and } A > 0.$$

EXAMPLE 6 A line connects the points $(1, 5)$ and $(-3, 2)$. Write the equation of the line in the form $Ax + By = C$, where A, B, and C are integers and $A > 0$.

Solution Find the slope, then write the equation of the line using point-slope form. To finish, rewrite the equation in standard form.

$$\text{Find the slope:} \quad m = \frac{2 - 5}{-3 - 1} = \frac{-3}{-4} = \frac{3}{4} \qquad \text{Use } m = \frac{y_2 - y_1}{x_2 - x_1}.$$

Because we were not given the y-intercept, we use the point-slope form of the linear equation:

$$y - 5 = \frac{3}{4}(x - 1) \qquad \text{Use } y - y_1 = m(x - x_1).$$

Now manipulate the equation to get the form $Ax + By = C$, where A, B, and C are integers and $A > 0$.

$$y - 5 = \frac{3}{4}x - \frac{3}{4} \qquad \text{Distribute } \frac{3}{4} \text{ to clear parentheses.}$$

$$4(y - 5) = 4\left(\frac{3}{4}x - \frac{3}{4}\right) \qquad \text{Multiply both sides by the LCD, 4.}$$

$$4y - 20 = 3x - 3$$

$$4y - 3x - 20 = -3 \qquad \text{Subtract } 3x \text{ from both sides to get } x \text{ and } y \text{ together.}$$

$$4y - 3x = 17 \qquad \text{Add 20 to both sides to get the constant terms on the right-hand side of the equation.}$$

$$-3x + 4y = 17 \qquad \text{Use the commutative property of addition to write the } x \text{ term first.}$$

$$(-1)(-3x + 4y) = (-1)(17) \qquad \text{Multiply both sides by } -1 \text{ so that the coefficient of } x \text{ is positive.}$$

$$3x - 4y = -17 \qquad \text{Standard form with } A, B, \text{ and } C \text{ integers and } A > 0.$$

YOUR TURN Write the equation of the line through the given points in the form $Ax + By = C$, where A, B, and C are integers and $A > 0$.

$$(-3, 6), (1, -4)$$

ANSWER

$5x + 2y = -3$

OBJECTIVE 3. Write the equation of a line parallel to a given line. Consider the graphs of $y = 2x - 3$ and $y = 2x + 1$.

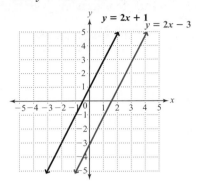

These two lines are parallel, which means that they will never intersect. Notice in the equations that their slopes are equal, which is why they are parallel.

RULE   Parallel Lines

The slopes of parallel lines are equal.

EXAMPLE 7 Write the equation of a line that passes through $(1, -4)$ and parallel to $y = -3x + 5$. Write the equation in slope-intercept form.

Solution In the given equation, we see that the slope of the line is -3, so the slope of the parallel line will also be -3. We can now use the point-slope form to write the equation of the parallel line.

$$y - y_1 = m(x - x_1)$$
$$y - (-4) = -3(x - 1) \qquad \text{Replace } m \text{ with } -3, x_1 \text{ with 1, and } y_1 \text{ with } -4.$$
$$y + 4 = -3x + 3 \qquad \text{Simplify.}$$
$$y = -3x - 1 \qquad \text{Subtract 4 from both sides to isolate } y.$$

YOUR TURN Write the equation of the line in slope-intercept form that passes through the given point and is parallel to the given line.

$$(-4, -2); y = \frac{3}{4}x - 7$$

OBJECTIVE 4. Write the equation of a line perpendicular to a given line. Consider the graphs of $y = \frac{2}{3}x - 4$ and $y = -\frac{3}{2}x + 1$.

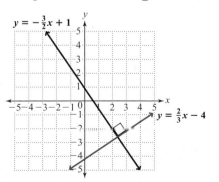

These two lines are perpendicular, which means that they intersect at a 90° angle. Notice in the equations that their slopes are negative reciprocals. In other words $-\frac{3}{2}$ is the negative reciprocal of $\frac{2}{3}$.

ANSWER

$$y = \frac{3}{4}x + 1$$

RULE   Perpendicular Lines

> The slope of a line perpendicular to a line with a slope of $\frac{a}{b}$ will be $-\frac{b}{a}$.
>
> Horizontal and vertical lines are perpendicular.

EXAMPLE 8   Write the equation of a line that passes through $(3, -1)$ and is perpendicular to $3x + 4y = 8$. Write the equation in slope-intercept form.

Solution To write the equation of the perpendicular line, we need its slope. To determine its slope, we need the slope of the given line. Since the equation of the given line is in standard form, to determine its slope, we rearrange it in slope-intercept form.

$$3x + 4y = 8$$

$$4y = -3x + 8 \qquad \text{Subtract } 3x \text{ from both sides.}$$

$$y = -\frac{3}{4}x + 2 \qquad \text{Divide both sides by 4.}$$

We now see that the slope of the given line is $-\frac{3}{4}$, so the slope of the perpendicular line will be $\frac{4}{3}$. We can now use point-slope form to write the equation of the perpendicular line.

$$y - y_1 = m(x - x_1)$$

$$y - (-1) = \frac{4}{3}(x - 3) \qquad \text{Replace } m \text{ with } \frac{4}{3}, x_1 \text{ with 3, and } y_1 \text{ with } -1.$$

$$y + 1 = \frac{4}{3}x - 4 \qquad \text{Simplify.}$$

$$y = \frac{4}{3}x - 5 \qquad \text{Subtract 1 from both sides to isolate } y.$$

ANSWER

$y = -\frac{2}{5}x - 1$

YOUR TURN   Write the equation of a line that passes through $(-10, 3)$ and is perpendicular to $5x - 2y = 6$. Write the equation in slope-intercept form.

4.5 Exercises

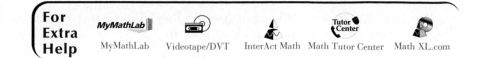

For Extra Help

MyMathLab   Videotape/DVT   InterAct Math   Math Tutor Center   Math XL.com

1. If given the slope and the *y*-intercept and asked to write the equation of the line, which form of the equation would be easiest to use? Why?

2. If given two ordered pairs, neither of which is the *y*-intercept, and asked to write the equation of the line, which form of the equation would be easiest to use? Why?

3. What is your first step when you are given two ordered pairs and asked to write the equation of the line connecting the two points?

4. When given two ordered pairs and asked to write an equation of a line, does it matter which ordered pair is used in the point-slope equation? Explain.

5. How do you rewrite an equation that is in standard form in slope-intercept form?

6. How do you rewrite an equation that is in point-slope form in standard form?

For Exercises 7–18, write the equation of the line in slope-intercept form given the slope and the coordinate of the y-intercept.

7. $m = 2$; $(0, 3)$

8. $m = 3$; $(0, -2)$

9. $m = -3$; $(0, -9)$

10. $m = -4$; $(0, -5)$

11. $m = \dfrac{3}{4}$; $(0, 5)$

12. $m = \dfrac{3}{7}$; $(0, -1)$

13. $m = -\dfrac{2}{5}$; $\left(0, \dfrac{7}{8}\right)$

14. $m = \dfrac{5}{2}$; $\left(0, \dfrac{3}{8}\right)$

15. $m = 0.8$; $(0, -5.1)$

16. $m = -0.75$; $(0, 2.5)$

17. $m = -3$; $(0, 0)$

18. $m = 1$; $(0, 0)$

For Exercises 19–22, write the equation of the line in slope-intercept form given the y-intercept and one other point.

19. $(3, 4)$, $(0, -2)$

20. $(0, 2)$, $(4, 10)$

21. $(0, 3)$, $(5, -3)$

22. $(-3, -1)$, $(0, -3)$

For Exercises 23–34, write the equation of a line in slope-intercept form with the given slope passing through the given point.

23. $m = 3$; $(5, 2)$

24. $m = 2$; $(3, -5)$

25. $m = -1$; $(-1, -1)$

26. $m = -5$; $(2, 0)$

27. $m = \dfrac{2}{3}$; $(6, 3)$

28. $m = \dfrac{3}{2}$; $(-2, 3)$

29. $m = -\dfrac{3}{4}$; $(-1, -5)$

30. $m = -\dfrac{2}{5}$; $(-2, -1)$

31. $m = -\dfrac{4}{3}$; $(-2, 4)$

32. $m = \dfrac{2}{9}$; $(-1, -7)$

33. $m = 2$; $(0, 0)$

34. $m = -3$; $(0, 0)$

For Exercises 35–46, write the equation of a line connecting the given points in slope-intercept form.

35. $(4, -3), (-1, 7)$

36. $(-1, -5), (-4, 1)$

37. $(3, 1), (5, 6)$

38. $(-5, -2), (7, 5)$

39. $(0, 0), (-1, -7)$

40. $(0, 0), (-5, 1)$

41. $(8, -1), (8, -10)$

42. $(-3, 2), (-3, 5)$

43. $(-9, -1), (2, -1)$

44. $(-2, 5), (6, 5)$

45. $(1.4, 5), (3, 9.8)$

46. $(3.9, -2.1), (4.1, -1)$

For Exercises 47–58, write the equation of a line through the given points in the form $Ax + By = C$, where A, B, and C are integers and $A > 0$.

47. $(2, 3), (4, 8)$

48. $(4, 1), (8, 2)$

49. $(2, 0), (0, -3)$

50. $(-6, 0), (0, 6)$

51. $(-4, -1), (7, -5)$

52. $(5, -2), (-3, -3)$

53. $(7, -2), (-1, -5)$

54. $(4, 8), (-4, -8)$

55. $(-1, 0), (-1, 8)$

56. $(-2, 5), (-2, -6)$

57. $(3, -1), (4, -1)$

58. $(2, -4), (3, -4)$

For Exercises 59–68, determine whether the given lines are parallel, perpendicular, or neither.

59. $y = 4x - 7$
$y = 4x + 2$

60. $y = 3x + 2$
$y = 3x - 2$

61. $y = \dfrac{1}{2}x$
$y = -2x + 4$

62. $y = -x + 3$
$y = x - 2$

63. $4x + 6y = 5$
$8x + 12y = 9$

64. $3x + 6y = 10$
$2x + 4y = 9$

65. $5x + 10y = 12$
$4x - 2y = 5$

66. $5x - 3y = 11$
$3x + 5y = 8$

67. $y = 1$
$x = -3$

68. $x = 4$
$x = -2$

For Exercises 69–78, write the equation of a line that passes through the given point and is parallel to the given line.
a. Write the equation in slope-intercept form.
b. Write the equation in the form $Ax + By = C$, where A, B, and C are integers and $A > 0$.

69. $(4, 2); y = 4x + 1$

70. $(-2, 1); y = -3x + 1$

71. $(0, 3); y = -2x + 4$

72. $(0, -1); y = 3x + 2$

73. $(5, 2); y = \dfrac{1}{3}x + 4$

74. $(-1, -3); y = -\dfrac{3}{4}x + 1$

75. $(2, -3); 2x + y = 5$

76. $(5, 1); 6x + 3y = 8$

77. $(-2, 4); 3x - 4y = 6$

78. $(-1, -1); 2x + 3y = 6$

For Exercises 79–88, write the equation of a line that passes through the given point and perpendicular to the given line.
a. Write the equation in slope-intercept form.
b. Write the equation in the form $Ax + By = C$, where A, B, and C are integers and $A > 0$.

79. $(-4, -1); y = 2x + 3$

80. $(3, -3); y = 3x - 1$

81. $(3, -4); y = \dfrac{1}{4}x + 2$

82. $(2, 3); y = \dfrac{1}{5}x - 4$

83. $(2, -4); y = -2x + 7$

84. $(-1, -1); y = -3x + 4$

85. $(3, -4); y = -\dfrac{2}{3}x + 4$

86. $(4, 2); y = \dfrac{3}{4}x - 11$

87. $(-2, 3); 3x - 2y = 5$

88. $(3, 7); 5x + 2y = 3$

89. From Logan Airport in Boston, a taxi charges a $7.75 initial fee plus $0.30 for each one-eighth of a mile.

   a. Let n represent the number of eighths of a mile and c represent the total cost of the taxi. Write an equation in slope-intercept form that describes how much it costs to hire a taxi from Logan Airport.

   b. What will be the total cost of traveling 5 miles from the airport?

   c. Graph the equation.

90. A company manager is examining profits. He sees that the profit in May was $52,000. He then discovers that his company's profits have been declining $1000 each month thereafter.

   a. Let n represent the number of months and p represent the profit. Write an equation in slope-intercept form that describes the profit n months after May.

   b. If the decline continues at the same rate, what will be the profit in nine months?

   c. Graph the equation.

91. In the 1990–91 school year, about 1.89 million females were participating in high school athletics. Over the next ten years, participation increased in a linear pattern so that in 2000–2001, 2.78 million females were participating in high school athletics. (*Source:* 2001 High School Participation Survey)

   a. Consider 1990–91 to be year 0 so that 2000–2001 is year 10. Plot the two ordered pairs with the number of years along the horizontal axis and the number of females participating in athletics along the vertical axis. Then, draw a line through the two points.

   b. Find the slope of the line.

   c. Let n represent the number of years after the 1990–91 school year and p represent the number of female participants in high school athletics. Write an equation of the line in slope-intercept form.

   d. Using your equation, how many females participated in high school athletics in the 1997–98 school year?

   e. If the linear trend continues, how many females will participate in high school athletics in the 2012–13 school year?

92. A report about national health expenditure amounts shows the costs of dental services for senior citizens (age 65 and older) was $53.2 billion in 1998. The costs continued to rise in a nearly linear pattern over the next six years so that in 2004 that cost was $76.0 billion. (*Source:* Centers for Medicare and Medicaid Services, Office of the Actuary)

    a. Consider 1998 to be year 0 so that 2004 is year 6. Plot the two ordered pairs with the year along the horizontal axis and cost along the vertical axis. Then, draw a line through the two points.

    b. Find the slope of the line.

    c. Let n represent the number of years after 1998 and c represent the cost in billions of dollars. Write an equation of the line in slope-intercept form.

    d. Using your equation, what was the cost of dental services for senior citizens in 2000?

    e. If the linear trend continues, what will be the cost of dental care for senior citizens in 2010?

93. A stock begins a decline in price following a linear pattern of depreciation. The stock's initial price before the decline began was $36. On the 5th day, the closing price was $28.50.

    a. Graph the line with the number of days after the decline begins plotted along the horizontal axis and the price along the vertical axis.

    b. Find the slope of the line.

    c. Write an equation of the line in slope-intercept form with n representing the number of days of decline in price and p representing the price.

    d. Find the p-intercept. Explain what the p-intercept represents in the problem.

    e. Find the n-intercept. Explain what the n-intercept represents in the problem.

    f. If the decline in price were to continue at the same rate, what would be the price of the stock after eight days of decline?

94. Keith deducts the value of his computer on his taxes. The amount of the deduction is based on the value of the computer, which depreciates each year. The initial value of the computer was $2100. After two years in service the computer's value is $1500.

a. Graph the line with the number of years plotted along the horizontal axis and the value along the vertical axis.

b. Find the slope of the line.

c. Write an equation of the line in slope-intercept form with n representing the number of years the computer is in service and v representing the value.

d. Find the v-intercept. Explain what the v-intercept represents in the problem.

e. Find the n-intercept. Explain what the n-intercept represents in the problem.

f. If the decline in value were to continue at the same rate, what would be the computer's value after five years in service?

REVIEW EXERCISES

[1.1] **1.** Use $<$, $>$, or $=$ to write a true statement:

$$0.6 \ ? \ \frac{3}{4}$$

[2.3] **2.** Solve: $\frac{3}{4}x - \frac{4}{5} = \frac{1}{2}x + 1$

[2.6] *For Exercises 3 and 4, solve and graph the solution set.*

3. Solve $4x - 9 \geq 6x + 5$ and graph the solution set on a number line.

4. Solve $2(x + 1) - 5x \geq 3x - 6$ and graph the solution set on a number line.

[3.2] **5.** What is 12% of 86?

[4.3] **6.** Find the x- and y-intercepts for $2x - 7y = 14$.

4.6 Graphing Linear Inequalities

OBJECTIVES

1. Determine whether an ordered pair is a solution for a linear inequality with two variables.

2. Graph linear inequalities.

OBJECTIVE 1. Determine whether an ordered pair is a solution for a linear inequality with two variables. Now that we have seen how to graph linear equations, let's turn our attention to linear inequalities. Linear inequalities have the same form as linear equations, except that they contain an inequality symbol instead of an equal sign. Examples of linear inequalities are

$$2x + 3y > 6$$
$$y \le x - 3$$

Remember that a solution for a linear equation in two variables is an ordered pair that makes the equation true. Similarly, a solution to a linear inequality is an ordered pair that makes the inequality true. For example, the ordered pair $(2, 1)$ is a solution for $2x + 3y > 6$:

$$2x + 3y > 6$$
$$2(2) + 3(1) \overset{?}{>} 6 \qquad \text{Replace } x \text{ with 2 and } y \text{ with 1.}$$
$$4 + 3 \overset{?}{>} 6$$
$$7 > 6 \qquad \text{This is true, so } (2, 1) \text{ is a solution.}$$

PROCEDURE Checking an Ordered Pair

To determine whether an ordered pair is a solution for an inequality, replace the variables with the corresponding coordinates and see if the resulting inequality is true. If so, the ordered pair is a solution.

EXAMPLE 1 Determine whether $(-4, 7)$ is a solution for $y \ge -2x + 3$.

Solution $y \ge -2x + 3$

$$7 \overset{?}{\ge} -2(-4) + 3 \qquad \text{Replace } x \text{ with } -4 \text{ and } y \text{ with 7, and then determine whether the inequality is true.}$$
$$7 \overset{?}{\ge} 8 + 3$$
$$7 \ge 11$$

This statement is false, so $(-4, 7)$ is not a solution.

YOUR TURN Determine whether the given ordered pair is a solution for the given inequality.

a. $3x - 5y < 6$; $(-1, -2)$ **b.** $y \le -4x + 1$; $(2, -7)$

OBJECTIVE 2. Graph linear inequalities. Graphing linear inequalities is very much like graphing linear equations. Consider $2x + 3y \geq 6$. The greater than or equal to sign indicates that solutions to the inequality $2x + 3y \geq 6$ will be ordered pairs that satisfy $2x + 3y > 6$ and also $2x + 3y = 6$. The graph of $2x + 3y = 6$ is a line. Let's graph this line using intercepts.

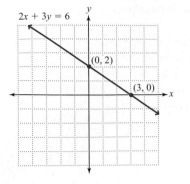

Find the x- and y-intercepts.

x-intercept:	y-intercept:
$2x + 3y = 6$	$2x + 3y = 6$
$2x + 3(0) = 6$	$2(0) + 3y = 6$
$2x = 6$	$3y = 6$
$x = 3$	$y = 2$

x-intercept coordinates: $(3, 0)$
y-intercept coordinates: $(0, 2)$

The "greater than" portion of the inequality consists of ordered pairs that satisfy $2x + 3y > 6$. To find one of these pairs, we select a point on one side of the line and see if its coordinates satisfy $2x + 3y > 6$. Let's select the point with coordinates $(1, 4)$.

$$2x + 3y > 6$$
$$2(1) + 3(4) \overset{?}{>} 6 \qquad \text{Replace } x \text{ with 1 and } y \text{ with 4.}$$
$$2 + 12 \overset{?}{>} 6$$
$$14 > 6 \qquad \text{This statement is true, therefore } (1, 4) \text{ is a solution.}$$

It can be shown that every ordered pair in the region containing the point $(1, 4)$, which is above the line, will satisfy $2x + 3y > 6$. It can also be shown that no ordered pair on the other side of the line will satisfy $2x + 3y > 6$. For example, consider the origin $(0, 0)$.

$$2x + 3y > 6 \qquad \text{Replace } x \text{ with 0 and } y \text{ with 0.}$$
$$2(0) + 3(0) \overset{?}{>} 6$$
$$0 > 6 \qquad \text{This statement is false, so } (0, 0) \text{ is not a solution.}$$

The line that is the graph of $2x + 3y = 6$ is called a *boundary*. The region on one side of this boundary line contains all the ordered pair solutions satisfying $2x + 3y > 6$. The region on the other side contains all ordered pairs that satisfy $2x + 3y < 6$. To indicate that every ordered pair above the boundary line is a solution to the inequality, we shade the region above the line.

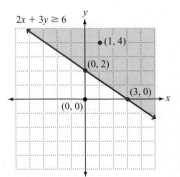

If the inequality had been $2x + 3y > 6$, the ordered pairs that satisfy $2x + 3y = 6$ would not be solutions. Because these ordered pairs form the boundary line, we draw a dashed line instead of a solid line to indicate that those ordered pairs are not solutions. The graph of $2x + 3y > 6$ looks like this:

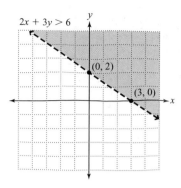

For the inequalities $2x + 3y \leq 6$ and $2x + 3y < 6$, we shade the region on the other side of the boundary line.

The graph of $2x + 3y \leq 6$:   The graph of $2x + 3y < 6$:

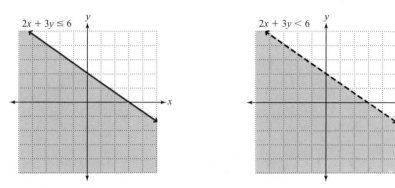

Connection   When graphing inequalities with the symbols $\leq$ or $\geq$ on a number line, we used a bracket (or closed circle) to indicate that equality is included. When the inequalities have $<$ or $>$, we used a parenthesis (or open circle) to indicate that equality is excluded. When graphing linear inequalities in the rectangular coordinate system, a solid line is like a bracket and a dashed line is like a parenthesis.

PROCEDURE   Graphing Linear Inequalities

To graph a linear inequality:

1. Graph the related equation. The related equation has an equal sign in place of the inequality symbol. If the inequality symbol is $\leq$ or $\geq$, then draw a solid line. If the inequality symbol is $<$ or $>$, then draw a dashed line.
2. Choose an ordered pair on one side of the boundary line and test this ordered pair in the inequality. If the ordered pair satisfies the inequality, then shade the region that contains it. If the ordered pair does not satisfy the inequality, then shade the region on the other side of the boundary line.

EXAMPLE 2 Graph: $y > 3x + 1$

Solution First, graph the related equation $y = 3x + 1$. Two ordered pairs that satisfy $y = 3x + 1$ are $(0, 1)$ and $(1, 4)$.

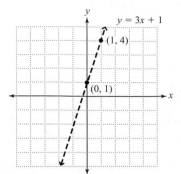

Because the inequality is strictly greater than, we draw a dashed line to indicate that ordered pairs along this boundary line are not solutions.

Now we choose an ordered pair on one side of the line and test this ordered pair in the inequality. We will choose $(3, 0)$.

$$y > 3x + 1$$

$$0 \stackrel{?}{>} 3(3) + 1 \qquad \text{Replace } x \text{ with 3 and } y \text{ with 0.}$$

$$0 > 10 \qquad \text{This statement is false, so } (3, 0) \text{ is not a solution for the inequality.}$$

Since $(3, 0)$ did not satisfy the inequality, we shade the region on the other side of the boundary line.

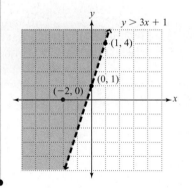

We can confirm that our shading is correct by choosing an ordered pair in this region. We will choose the ordered pair $(-2, 0)$:

$$y > 3x + 1$$

$$0 \stackrel{?}{>} 3(-2) + 1 \qquad \text{Replace } x \text{ with } -2 \text{ and } y \text{ with 0.}$$

$$0 > -5 \qquad \text{This statement is true, which confirms that we have shaded the correct side of the boundary line.}$$

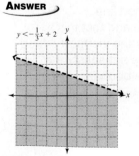

YOUR TURN Graph the linear inequality.

$$y < -\frac{1}{3}x + 2$$

EXAMPLE 3 ▸ Graph: $x - 2y \le 4$

Solution First, graph the related equation $x - $?
isfy $x - 2y = 4$ are $(0, -2)$ and $(4, 0)$.

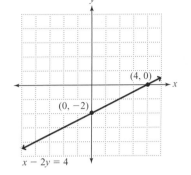

Because the i.
less than or equal tc
line to indicate that ord
the boundary line are solu.

Now we choose an ordered pair on one side of the line and test this ordered pa
the inequality. We will choose the origin $(0, 0)$.

$$x - 2y \le 4$$
$$0 - 2(0) \overset{?}{\le} 4 \qquad \text{Replace } x \text{ with 0 and } y \text{ with 0.}$$
$$0 \le 4 \qquad \text{This statement is true, therefore, } (0, 0)$$
$$\text{is a solution for the inequality.}$$

Since $(0, 0)$ satisfies the inequality, we shade the region that contains it.

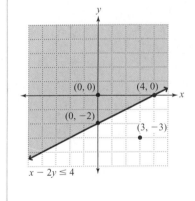

We can confirm that the region on the other
side of the line should not be shaded by choosing
a point in that region, such as $(3, -3)$:

$$x - 2y \le 4$$
$$3 - 2(-3) \overset{?}{\le} 4 \qquad \text{Replace } x \text{ with 3 and } y \text{ with } -3.$$
$$3 + 6 \overset{?}{\le} 4$$
$$9 \le 4$$

This statement is false, which confirms that the region containing the point with coordinates $(3, -3)$
should not be shaded.

YOUR TURN ▸ Graph the linear inequality.

$$2x + 3y \ge 6$$

1. How d

2. Ho

3.

Calculator TIPS

Shading can be added to a graph on a graphing calcula-tor. Using the $\boxed{Y=}$ key, enter the related equation for the given inequality. On the same line where you have entered the equation, use the arrow keys to move the cursor to the left of the variable y. Nor-mally, in this position, you see a slash, which indicates that the graph will be a solid line. Pressing enter will change this slash to various other options for the graph. There are two options for shading:

◥ *means the graph will have the region above the line shaded.*

◣ *means the graph will have the region below the line shaded.*

Notice that it is up to you to determine which region should be shaded.

ANSWER

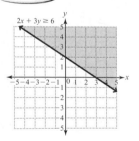

you determine if an ordered pair is a solution to an inequality?

w do you determine the related equation for a given inequality?

In graphing a linear inequality, when do you draw a dashed line and when do you draw a solid line?

4. When graphing a linear inequality, how do you determine which side of the boundary line to shade?

For Exercises 5–12, determine whether the ordered pair is a solution for the linear inequality.

5. $(5, 5); y > -x + 2$

6. $(0, 1); y > -x + 2$

7. $(2, 0); y \geq \dfrac{1}{2}x + 1$

8. $(0, 0); 2x + 3y < 7$

9. $(-1, 2); 3x - y \leq -8$

10. $(3, -1); y > x + 2$

11. $(-1, -2); 2y - x < 7$

12. $(0, -2); x + 4y \geq 8$

For Exercises 13–40, graph the linear inequality.

13. $y \leq -x + 1$

14. $y \leq x - 8$

15. $y > -2x + 5$

16. $y < -3x + 2$

17. $y > 2x$

18. $y < x$

19. $y > \dfrac{1}{3}x$

20. $y < \dfrac{2}{5}x$

21. $y \geq \dfrac{3}{4}x + 2$

22. $y \geq -\dfrac{2}{3}x - 3$

23. $3x + y \leq 9$

24. $2x + y > 6$

25. $5x - 2y < 10$

26. $3x - 2y \geq 6$

27. $2x + 3y < 9$

28. $5x + y \geq -10$

29. $4x + 2y \leq 3$

30. $3x + y \geq 8$

31. $3x + y \geq 0$

32. $-3x - 2y < 0$

33. $x > -4$

34. $x \leq 6$

35. $y \leq 2$

36. $y > -1$

37. $x + 2 \geq 0$

38. $x - 4 < 0$

39. $y - 3 < 0$

40. $y + 2 \geq 0$

41. A company sells two sizes of bottles of hand lotion. The revenue for each unit of the large bottle is $2.00. The revenue for each unit of the small bottle is $1.50. For the company to break even during the first quarter, the company must generate $280,000 in revenue. The inequality $2x + 1.5y \geq 280,000$ describes the amount of revenue that must be generated from the sale of both size bottles in order to break even or turn a profit.

   a. What do x and y represent?

   b. Graph the inequality.

   c. What does the boundary line represent? What does the shaded region represent?

   d. List three combinations of sales numbers of each size that allow the company to break even.

   e. List three combinations of sales numbers of each size that allow the company to turn a profit.

   f. In reality, is every combination of units sold represented by the line and shaded region a possibility? Explain.

42. A company produces two versions of tax-preparation software. The regular version costs $12.50 per unit to produce, and the deluxe version costs $15.00 per unit to produce. Management plans for the cost of production to be a maximum of $300,000 for the first quarter. The inequality $12.50x + 15y \leq 300,000$ describes the total cost as prescribed by management.

   a. What do x and y represent?

   b. Graph the inequality.

   c. What does the line represent?

   d. What does the shaded region represent?

   e. List three combinations of numbers of units produced of each version of software that yield a total cost of $300,000.

   f. List three combinations of numbers of units produced of each version of software that yield a total cost less than $300,000.

   g. In reality, is every combination of units produced represented by the line and shaded region a possibility? Explain.

43. Andrea visits a garden center to purchase some new plants for her garden. She has a gift certificate in the amount of $100. The plants that she is considering come in two sizes and are priced according to size. Plants in 1-gallon containers are priced at $5.50 each, and plants in 5-gallon containers are priced at $12.50 each.

  a. Select a variable for the number of each plant size, and write an inequality that has Andrea's total cost within the amount of the gift certificate.

  b. Graph the inequality.
  c. What does the line represent?

  d. What does the shaded region represent?

  e. What are some possible combinations of containers she could purchase?

  f. In reality, are there any combinations that she could purchase that would yield a total in the exact amount of the gift certificate? Explain.

44. Michelle is coordinating a fund-raiser that will sell two different cookbooks. One cookbook sells for $10 and the other for $12. The goal is to raise at least $30,000.

  a. Select variables for each type of cookbook, and write an inequality in which the total sales is at least $30,000.

  b. Graph the inequality.
  c. What does the line represent?

  d. What does the shaded region represent?

  e. Find two combinations of book sales that yield exactly $30,000 in total sales.

  f. Find two combinations of units sold that yield more than $30,000.

45. A building with a rectangular base is to be designed so that the maximum perimeter is 180 feet.

  a. Select variables for the length and width, and write an inequality in which the maximum perimeter is 180 feet.

  b. Graph the inequality.
  c. What does the line represent?

  d. What does the shaded region represent?

  e. Find three combinations of length and width that yield a perimeter of exactly 180 feet.

  f. Find three combinations of length and width that yield a perimeter of less than 180 feet.

46. Steel beams are to be welded together to form a frame that is an isosceles triangle. The perimeter of the frame cannot exceed 60 feet.

    a. Select a variable for the length of the base and a second variable for the sides of equal length. Then write an inequality in which the maximum perimeter is 60 feet.

    b. Graph the inequality.

    c. What does the line represent?

    d. What does the shaded region represent?

    e. Find sets of dimensions that yield a perimeter of exactly 60 feet.

    f. Find sets of dimensions that yield a perimeter of less than 60 feet.

PUZZLE PROBLEM

The sum of two different whole numbers is equal to 8, while the difference of the two numbers is greater than or equal to 1. Find all possible ordered pairs that satisfy both constraints.

REVIEW EXERCISES

[1.1]   **1.** Write a set containing all vowels in the alphabet.

[1.5]   **2.** Evaluate: 6^0

[2.3]   **3.** Solve and check: $3x - 4(x + 7) = -15$

[3.3]   **4.** The length of a rectangle is four more than the width. If the perimeter is 88 feet, find the dimensions.

[4.3]   **5.** Find the x- and y-intercepts for $3x - 2y = 8$.   [4.3]   **6.** Graph: $-x + y = 7$

4.7 Introduction to Functions and Function Notation

1. Identify the domain and range of a relation.
2. Identify functions and their domains and ranges.
3. Find the value of a function.
4. Graph linear functions.

In this section, we study relations and functions, which develop naturally from our work with graphing because they involve ordered pairs.

OBJECTIVE 1. Identify the domain and range of a relation. In order to formally define a function, we will first discuss **relations**.

DEFINITION *Relation:* A set of ordered pairs.

A linear equation like $y = 2x$ is a relation because it pairs a given x-value with a corresponding y-value. You might recall that with an equation in this form, we thought of the x-values as input values and the y-values as output values. In a relation, the set of all the input values (x-values) is called the **domain** and the set of all the output values (y-values) is called the **range**.

DEFINITIONS *Domain:* The set of all input values for a relation.
Range: The set of all output values for a relation.

The following table lists some values in the domain and range for $y = 2x$:

Domain (x-values)	Range (y-values)
0	0
1	2
2	4
3	6

EXAMPLE 1 Determine the domain and range of the relation $\{(2, -1), (3, 0), (4, 5), (5, 8)\}$.

Solution The domain is the set containing all the x-values $\{2, 3, 4, 5\}$, and the range is the set containing all the y-values $\{-1, 0, 5, 8\}$.

Domain and Range with a Graph

We can determine the domain and range of a relation from its graph. Look at the graph of $y = 2x$.

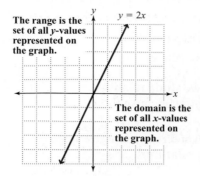

The range is the set of all y-values represented on the graph.

The domain is the set of all x-values represented on the graph.

It is convenient to imagine traveling along a graph from left to right. As we travel along $y = 2x$ from left to right, every x-value along the x-axis has a corresponding y-value. Thus the domain is a set containing all real numbers. Similarly, every y-value is paired with an x-value, so the range also contains all real numbers.

PROCEDURE   Determining Domain and Range

To determine the domain of a relation given its graph, answer the question: What are all the x-values that have a corresponding y-value? To determine the range, answer the question: What are all the y-values that have a corresponding x-value?

EXAMPLE 2 Determine the domain and range of the relation.

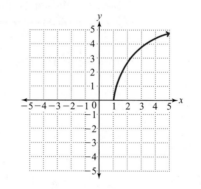

Solution From left to right, this graph "begins" at the point $(1, 0)$. Values along the x-axis begin at 1 and continue infinitely, so the domain is $\{x \mid x \geq 1\}$. For the range, the y-values begin at 0 and continue infinitely, so the range is $\{y \mid y \geq 0\}$.

YOUR TURN   Determine the domain and range of the relation.

a. $\{(-2, -4), (0, -1), (3, 6), (7, 5)\}$ **b.**

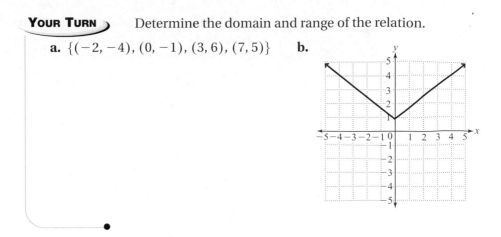

OBJECTIVE 2. Identify functions and their domains and ranges.   Now we turn to a special type of relation called a *function*. Let's reconsider the trajectory data from Example 4 of Section 4.1 (pages 272–273), which we have listed here as a relation in the following table:

Time in Seconds (domain)	Height in Feet (range)
0	6.00
0.5	7.50
1.0	8.25
1.5	7.50
2.0	6.00
2.5	2.00

Note that each value for time is paired with one and only one height value. This is the key feature that makes a relation a **function**.

DEFINITION   *Function:* A relation in which every value in the domain is paired with exactly one value in the range.

From the data in the preceeding table, we can see that a function can have different values in the domain assigned to the same value in the range. For example, domain values 0.5 seconds and 1.5 seconds each correspond to the same range value 7.50 feet. Also, the domain values 0 seconds and 2.0 seconds each correspond to 6.00 feet in the range. Using arrows to match the values in the domain with their corresponding values in the range can be a helpful technique in determining whether a relation is a function. Such a map might look like this:

Domain: {0,   0.5,   1.0,   1.5,   2.0,   2.5}

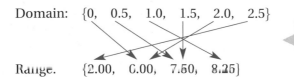

Range.   {2.00,   6.00,   7.50,   8.25}

Note: *Each element in the domain has a single arrow pointing to an element in the range.*

ANSWERS

a. Domain: $\{-2, 0, 3, 7\}$
  Range: $\{-4, -1, 6, 5\}$

b. Domain: All real numbers
  Range: $\{y \mid y \geq 1\}$

4.7   Introduction to Functions and Function Notation   **339**

Note that every function is a relation, but not every relation is a function. For example, consider the following relation that assigns baseball outfield positions (the domain) to the 2002 New York Yankees outfielders (the range).

Domain: {left field, center field, right field}

Range: {Spencer, White, Williams, Mondesi, Rivera, Thames, Vander Wal}

Notice that left field and right field are assigned to more than one player. Since an element in the domain (left or right field) is assigned to more than one element in the range, this relation is not a function.

Conclusion   If any value in the domain is assigned to more than one value in the range, then the relation is not a function.

EXAMPLE 3 ❯ Determine whether the relation is a function.

  a. The following table indicates the percent of people in the United States under 65 years of age with private health care coverage:

Year (domain)	Percent (range)
1984	76.8
1989	75.9
1993	71.5
1994	70.3
1995	71.6
1996	71.4
1997	70.7

Solution This relation is a function because each year value in the domain has only one percent value in the range assigned to it.

  b. The following relation assigns a birthdate to the corresponding person.

Domain: {May 1, June 4, July 8, August 2}

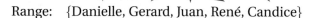

Range: {Danielle, Gerard, Juan, René, Candice}

Solution This relation is not a function because an element in the domain, July 8, is assigned to two people in the range.

YOUR TURN Determine whether the relation is a function.

a.

Index Topic (domain)	Page Number(s) (range)
Celestial sphere	56
Celsius	36
Centigrade	36
Centimeter	6
Centripetal force	239, 305

b. The table shows the top five finishers in the 2005 Daytona 500 race.

Final Position (domain)	Car Number (range)
1st	24 (Jeff Gordon)
2nd	97 (Kurt Busch)
3rd	8 (Dale Earnhardt Jr.)
4th	10 (Scott Riggs)
5th	48 (Jimmie Johnson)

We can determine whether a relation is a function by its graph. Remember that in Example 4 of Section 4.1 (p. 272) we created a graph of the trajectory data (that data also appears in this section on p. 339) by plotting the ordered pairs as points in the coordinate plane.

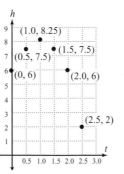

Notice the domain (time in this case) is plotted along the horizontal axis and the range (height) is plotted along the vertical axis.

To determine whether a relation is a function from its graph, we can perform a test called the *vertical line test*. The nature of the test is to draw or imagine a vertical line through every point in the domain.

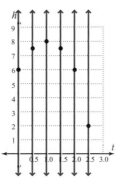

Here, we've drawn lines in blue through each value in the domain. Note that each vertical line intersects the graph at one and only one point, which means that each value in the domain corresponds to exactly one value in the range. This relation is a function.

ANSWERS

a. not a function
b. function

PROCEDURE The Vertical Line Test

To determine whether a relation is a function from its graph, perform a vertical line test:

1. Draw or imagine vertical lines through each point in the domain.
2. If each vertical line intersects the graph at only one point, then the graph is the graph of a function.
3. If any vertical line intersects the graph at two different points, then the graph is not the graph of a function.

EXAMPLE 4 For each graph, determine the domain and range. Then state whether the relation is a function.

a.

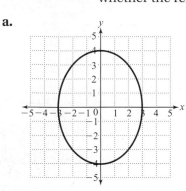

Solution Domain: $\{x \mid -3 \le x \le 3\}$

Range: $\{y \mid -4 \le y \le 4\}$

This relation is not a function because a vertical line can be drawn that intersects the graph at two different points.

For example, the vertical line $x = 2$ (shown in blue) passes through two different points on the graph, $(2, 3)$ and $(2, -3)$.

b.

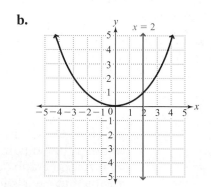

Solution Domain: all real numbers

Range: $\{y \mid y \ge 0\}$

This relation is a function. A vertical line through any value along the horizontal axis will intersect the graph at only one point. We have shown $x = 2$ to illustrate one such line.

c.

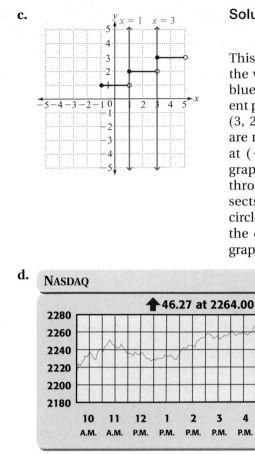

Solution Domain: $\{x \mid -1 \le x < 5\}$

Range: $\{y \mid y = 1, 2, 3\}$

This relation is a function. It may seem that the vertical lines $x = 1$ and $x = 3$ (shown in blue) each intersect the graph at two different points. However, the open circles at $(1, 1)$, $(3, 2)$, and $(5, 3)$ indicate that those points are not part of the graph. The closed circles at $(-1, 1)$, $(1, 2)$, and $(3, 3)$ are part of the graph. The vertical line $x = 1$ "passes through" the open circle at $(1, 1)$ and intersects the graph at only one point, the closed circle at $(1, 2)$. Similarly, $x = 3$ passes through the open circle at $(3, 2)$ and intersects the graph at only one point, $(3, 3)$.

d.

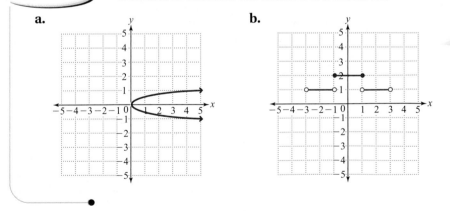

Source: USA Today, June 6, 2001

Solution

Domain:

$\{t \mid 9\!:\!30 \text{ A.M.} \le t \le 4\!:\!30 \text{ P.M.}\}$

Range:

$\{N \mid 2220 \le N \le 2264\}$

This relation is a function. A vertical line through any value along the horizontal axis will intersect the graph at only one point. In the terminology from the graph, at each time during this business day, the Nasdaq had only one value.

YOUR TURN Determine whether the relation is a function.

a.

b.

ANSWERS

a. not a function

b. function

Function Notation

When written as an equation, the notation for a function is a modification of an equation in two variables. Instead of $y = 2x + 5$, in function notation, we would write $f(x) = 2x + 5$ to mean the same equation. Notice that the notation $f(x)$, which is read "a function in terms of x," or "f of x," replaces the variable y. The following table shows how equations in two variables can be written using function notation:

Equation in two variables: Function notation:

$$y = 3x \qquad\qquad\qquad f(x) = 3x$$
$$y = -4x + 1 \qquad\qquad f(x) = -4x + 1$$
$$y = \frac{2}{3}x - 5 \qquad\qquad f(x) = \frac{2}{3}x - 5$$

Warning: The notation $f(x)$ does not mean multiply f and x.

Finding the Value of a Function

Remember that when we found solutions to equations with y isolated, we often spoke of the x-value as an input value and the corresponding y-value as the output value. Finding the value of a function is the same as finding a solution to an equation in two variables: We input a value for x, and calculate the output value, $f(x)$. The following tables list the same ordered pairs. The table on the left is for $y = 2x + 5$, and the table on the right is for $f(x) = 2x + 5$. Notice that the only difference is in labeling the output values.

(input)	(output)	(input)	(output)
x	y	x	$f(x)$
0	$2(0) + 5 = 5$	0	$2(0) + 5 = 5$
1	$2(1) + 5 = 7$	1	$2(1) + 5 = 7$
2	$2(2) + 5 = 9$	2	$2(2) + 5 = 9$

Function notation offers a clever way to indicate that a specific x-value is to be used. For example, given a function $f(x) = 3x$, the notation $f(2)$ means to find the value of the function where $x = 2$, so that $f(2) = 3(2) = 6$.

PROCEDURE   Finding the Value of a Function

Given a function $f(x)$, to find $f(a)$, where a is a real number in the domain of f, replace x in the function with a and calculate the value.

EXAMPLE 5 For the function $f(x) = -4x + 1$, find the following.

a. $f(2)$

Solution $f(2) = -4(2) + 1$ Replace x in $-4x + 1$ with 2, then calculate.

$\quad\quad\quad\quad = -8 + 1$

$\quad\quad\quad\quad = -7$

b. $f(-3)$

Solution $f(-3) = -4(-3) + 1$ Replace x in $-4x + 1$ with -3, then calculate.

$\quad\quad\quad\quad\quad = 12 + 1$

$\quad\quad\quad\quad\quad = 13$

c. $f(a)$

Solution $f(a) = -4a + 1$ Replace x in $-4x + 1$ with a, then simplify as needed.

> **Connection** We use similar procedures to find the value of a function and to evaluate an expression.
>
> **Function language:**
> For $f(x) = -4x + 1$, find $f(2)$.
>
> **Expression language:**
> Evaluate the expression $-4x + 1$ where $x = 2$.

YOUR TURN For the function $f(x) = \dfrac{2}{3}x - 5$, find the following.

   a. $f(-3)$ **b.** $f(6)$ **c.** $f(a)$

Finding the Value of a Function Given Its Graph

We can also find the value of a function given its graph. For a given value in the domain (x-value), we look on the graph for the corresponding value in the range (y-value).

EXAMPLE 6 Using the graph, find the value of the function.

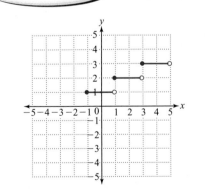

a. $f(0)$

Solution The notation $f(0)$ means to find the value of the function (y-value), where $x = 0$. On the graph we see that when $x = 0$, the corresponding y-value is 1, so we say $f(0) = 1$.

b. $f(1)$

Solution When $x = 1$, we see that $y = 2$, so $f(1) = 2$.

c. $f(-4)$

Solution Remember from Example 4, part c, that the domain is $\{x|-1 \le x < 5\}$. The value -4 is not in the domain, so we say that $f(-4)$ is undefined.

ANSWERS

a. -7 **b.** -1 **c.** $\dfrac{2}{3}a - 5$

Use the graph to find the value of the function.

a. $f(-1)$

b. $f(0)$

c. $f(5)$

d. $f(-3)$

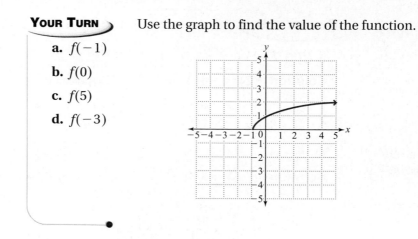

OBJECTIVE 4. Graph linear functions. We create the graph of a function the same way that we create the graph of an equation in two variables. In this chapter, we will graph only linear functions. Linear functions have a form similar to linear equations in slope-intercept form.

Slope-intercept form: $y = mx + b$

Linear function: $f(x) = mx + b$

EXAMPLE 7 Graph: $f(x) = -2x + 3$

Solution Think of the function as the equation $y = -2x + 3$. We could make a table of ordered pairs or use the fact that the slope is -2 and the y-intercept is 3.

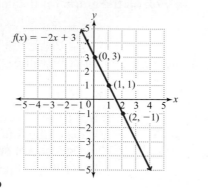

Table of ordered pairs

x	$f(x)$
0	3
1	1
2	-1

ANSWERS

a. 0 **b.** 1 **c.** 2 **d.** undefined

ANSWER

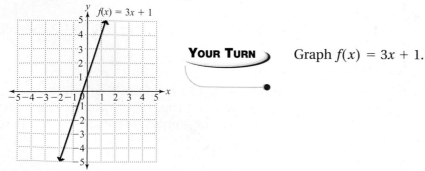

YOUR TURN Graph $f(x) = 3x + 1$.

1. Explain domain of a relation in your own words.

2. Explain range of a relation in your own words.

3. Given a graph of a relation, how do you determine the domain?

4. Given a graph of a relation, how do you determine the range?

5. When using the vertical line test on a graph, how do you determine if the graph is the graph of a function?

6. True or false: All relations are also functions. Explain.

For Exercises 7–12, determine the domain and range of the relation.

7. $\{(2, 1), (3, -2), (1, 4), (-1, -1)\}$

8. $\{(-2, 1), (4, 3), (2, 5), (-3, -2)\}$

9. $\{(4, 1), (2, 1), (3, -5), (-4, 0)\}$

10. $\{(-2, 1), (8, 11), (-3, 5), (4, -4)\}$

11. $\{(0, 0), (-3, -3), (2, 9), (8, -16)\}$

12. $\{(5, 4), (-3, 8), (10, -2), (0, -5)\}$

For Exercises 13–24, determine if the relation is a function and explain your answer.

13. $\{(2, 7), (3, 4), (-1, 4), (0, 5)\}$

14. $\{(-1, 5), (2, 6), (2, 9), (-3, -2)\}$

15. $\{(8, 2), (4, -5), (3, 0), (8, -1)\}$

16. $\{(1, 2), (4, -7), (9, 5), (6, 14)\}$

17. The grades on a recent calculus test:

Student	Score
Jonathon	54
Nathan	79
Rebecca	84
Terri	97
Allen	97

18. The sources of distraction among inattentive drivers:

Distraction	Percentage of Drivers
Outside person, object, or event	29.4
Adjusting radio/cassette/CD	11.4
Other occupant	10.9
Using/dialing cell phone	1.5
Smoking-related	0.9

Source: AAA Foundation for Traffic Safety/UNC Safety Research Center

19. Number of bridge-construction applications for South Carolina coastal counties:

Applications	Year Applied
8	1990
	1998
7	1993
6	1991
5	1994
	1996
	2000
	2001
1	1999

Source: South Carolina Department of Health and Environmental Control, Coastal Division

20. Babysitter prices per hour around the country:

Rate	Location
$10.00	Chappaqua, New York
$ 7.00	Yellowstone National Park
	Menlo Park, California
$ 6.00	Coral Springs, Florida
	St. Louis, Missouri
	Hawi, Hawaii
$ 5.00	North Pole, Alaska
$ 2.00	Las Vegas, Nevada

Source: Good Housekeeping, October 2002

21.

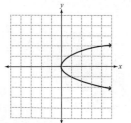

22.

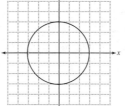

23.

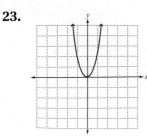

24.

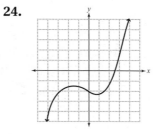

For Exercises 25–32, determine if the relation is a function. In each table, the domain is in the left column and the range is in the right column.

25. Average college tuition in the United States for 2002–2003:

Sector	Costs
Two-year public	$ 1,479
Two-year private	$10,755
Four-year public	$ 4,059
Four-year private	$16,948

Source: National Center for Education Statistics, U.S. Department of Education

26. The top-five money earners on the concert circuit during 2003:

Performer	Revenue in Millions of Dollars
Bruce Springsteen	$115.9
Celine Dion	$ 80.5
The Eagles	$ 69.3
Fleetwood Mac	$ 69.0
Cher	$ 68.2

Source: Pollstar

27. Results of the 2004 Masters golf tournament:

Position	Name
1st	Phil Mickelson
2nd	Ernie Els
3rd	K. J. Choi
4th	Sergio Garcia, Bernhard Langer

Source: www.masters.org

28. Top-five cable TV networks of 2004:

Rank	Channel
1	Discovery Channel
2	C-SPAN and USA Network
3	ESPN
4	CNN, TBS, and TNT
5	A&E

Source: The World Almanac and Book of Facts, 2005

29. Highest average fourth-quarter home prices by metropolitan area in 2004:

Metropolitan Area	Average Price
San Jose–San Francisco–Oakland	$568,900
San Diego–Carlsbad–San Marcos	$532,500
Los Angeles–Long Beach–Riverside	$451,900
Sacramento–Arden-Arcade–Truckee	$414,700
Washington–Baltimore–Northern Virginia	$399,000
New York–Newark–Bridgeport	$394,700
Boston–Worcester–Manchester	$381,400
Las Vegas–Paradise–Pahrump	$320,200
Seattle–Tacoma–Olympia	$309,600
Denver–Aurora–Boulder	$304,800

Source: Federal Housing Finance Board

30. Top brands of ice cream by total sales:

Brand	Sales
Private Label	$837,943,872
Breyers	$525,857,856
Dreyer's Edys Grand	$396,139,488
Blue Bell	$240,826,448
Haagen-Dazs	$195,961,056

Source: Information Resources Inc.

31. Average high temperature for Orlando, Florida:

Average High	Month
92	July, August
91	June
89	September
88	May
84	October
83	April
77	March, November
73	February, December
72	January

32. Average rainfall for Seattle:

Avg. Rain (inches)	Month
6	January, December
4	February, March
3	October
2	April, May, September
1	June, July, August

For Exercises 33–42, determine the domain and range. Then state whether the relation is a function.

33. This graph shows the number of Catholics in the United States from 1900 to 2001.

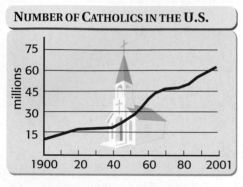

Source: Newsweek, March 4, 2002

34. The line graph represents postage costs from 1991 to 2004.

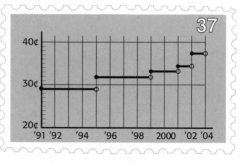

Source: Postal Rate Commission

35. This graph shows the percentage of the population that was foreign-born at each census.

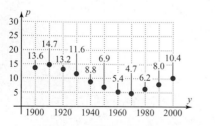

Source: Bureau of the Census, U.S. Department of Commerce

36. This graph shows the total number of people enrolled in a program each month after the program began.

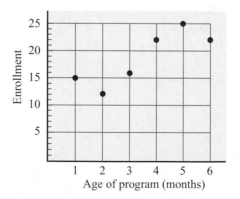

37.

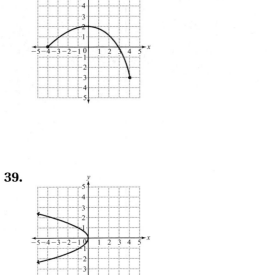

38.

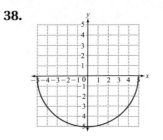

39.

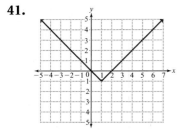

40.

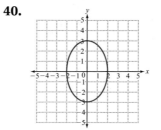

41.

42.

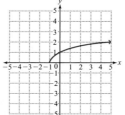

For Exercises 43–62, find the value of the function.

43. $f(x) = 2x - 5$
   a. $f(0)$ **b.** $f(1)$

   c. $f(-2)$ **d.** $f\left(\dfrac{1}{2}\right)$

44. $f(x) = 5x + 3$
   a. $f(1)$ **b.** $f(0)$

   c. $f(-1)$ **d.** $f(t)$

45. $f(x) = x^2 - 3x + 7$
   a. $f(0)$ **b.** $f(1)$

   c. $f(-1)$ **d.** $f(-2)$

46. $f(x) = 3x^2 + 2x - 1$
   a. $f(-2)$ **b.** $f(0.2)$

   c. $f(a)$ **d.** $f(3)$

47. $f(x) = \sqrt{x + 3}$
   a. $f(0)$ **b.** $f(-4)$

   c. $f(-1)$ **d.** $f(a)$

48. $f(x) = \sqrt{x - 5}$
   a. $f(-2)$ **b.** $f(6)$

   c. $f(a)$ **d.** $f(5)$

49. $f(x) = \sqrt{x^2 - 4}$

   a. $f(2)$ **b.** $f(1)$

   c. $f(-2)$ **d.** $f(3)$

50. $f(x) = \sqrt{x^2 + 2x}$

   a. $f(-2)$ **b.** $f(-1)$

   c. $f(0)$ **d.** $f(1)$

51. $f(x) = \dfrac{1}{x - 1}$

   a. $f(0)$ **b.** $f(1)$

   c. $f(-1)$ **d.** $f(-2)$

52. $f(x) = \dfrac{2}{x + 2}$

   a. $f(-2)$ **b.** $f(-1)$

   c. $f(a)$ **d.** $f(3)$

53. $f(x) = x^2 - 2x - 3$

   a. $f(0)$ **b.** $f(0.1)$

   c. $f(a)$ **d.** $f(a - 3)$

54. $f(x) = 2x^2 + x - 3$

   a. $f\left(\dfrac{3}{4}\right)$ **b.** $f(-1)$

   c. $f(a)$ **d.** $f(x + 2)$

55. $f(x) = \dfrac{1}{3}x - 2$

   a. $f(0)$ **b.** $f(1)$

   c. $f(-1)$ **d.** $f(-3)$

56. $f(x) = \dfrac{2}{3}x + 1$

   a. $f(-3)$ **b.** $f(-1)$

   c. $f(t)$ **d.** $f(3)$

57. $f(x) = \dfrac{x + 3}{x - 4}$

   a. $f(0)$ **b.** $f(1)$

   c. $f(-2)$ **d.** $f(3)$

58. $f(x) = \dfrac{9 - x}{x + 2}$

   a. $f(0)$ **b.** $f(-2)$

   c. $f(3)$ **d.** $f(-1)$

59. $f(x) = \dfrac{1}{\sqrt{x + 2}}$

   a. $f(-2)$ **b.** $f(1)$

   c. $f(-3)$ **d.** $f(a)$

60. $f(x) = \dfrac{2}{\sqrt{2 - x}}$

   a. $f(-2)$ **b.** $f(-1)$

   c. $f(a)$ **d.** $f(3)$

61. $f(x) = |4x - 2|$

   a. $f(0)$ **b.** $f(1.5)$

   c. $f(-1)$ **d.** $f(-2)$

62. $f(x) = |3x - 5|$

   a. $f(-2)$ **b.** $f(-1)$

   c. $f(2)$ **d.** $f(3)$

For Exercises 63–66, use the given graph to determine the value of the function.

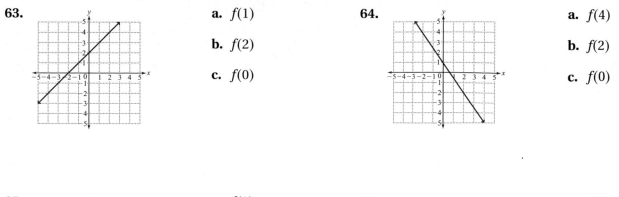

63.

a. $f(1)$

b. $f(2)$

c. $f(0)$

64.

a. $f(4)$

b. $f(2)$

c. $f(0)$

65.

a. $f(0)$

b. $f(2)$

c. $f(3)$

66.

a. $f(0)$

b. $f(2)$

c. $f(-2)$

For Exercises 67–76, graph.

67. $f(x) = 2x + 1$

68. $f(x) = 3x - 5$

69. $f(x) = \dfrac{1}{3}x + 3$

70. $f(x) = \dfrac{1}{2}x - 3$

71. $f(x) = -4x + 1$

72. $f(x) = -x - 1$

73. $f(x) = -\dfrac{2}{3}x$

74. $f(x) = \dfrac{2}{5}x$

75. $f(x) = -\dfrac{1}{4}x - 2$

76. $f(x) = -\dfrac{1}{3}x + 4$

77. The cost, $C(x)$, as a function of the number of items produced, x, is given by $C(x) = 20x + 150$.

   a. Find the cost of producing 10 items.

   b. Find the cost of producing 15 items.

   c. What does $C(30) = 750$ mean?

78. For a rental car, the cost, $C(x)$, as a function of the number of miles driven, x, is given by $C(x) = 0.12x + 22$.

   a. Find the cost of driving 100 miles.

   b. Find the cost of driving 350 miles.

   c. What does $C(250) = 52$ mean?

79. The value, $V(x)$, of a car as a function of the number of years after it was purchased, x, is given by $V(x) = -1500x + 18{,}000$.

   a. Find the value of the car after three years.

   b. Find the value of the car after six years.

   c. What does $V(10) = 3000$ mean?

80. The monthly salary, $S(x)$, of a salesperson as a function of her sales, x, is given by $S(x) = 0.10x + 175$.

   a. Find her salary for sales of $15,000.

   b. Find her salary for sales of $10,000.

   c. What does $S(20{,}000) = 2175$ mean?

REVIEW EXERCISES

[1.3]   **1.** Which property is illustrated in the equation $4(x + 9) = 4(9 + x)$?

For Exercises 2 and 3, simplify.

[1.5]   **2.** $\dfrac{2}{5} - \dfrac{5}{8} \div \dfrac{3}{4}$

[1.7]   **3.** $x + 3y - 5x + 7 - 3y - 9$

[2.3] *For Exercises 4 and 5, solve.*

   4. $3(x + 5) - x = 5x - (x + 1)$

   5. $\dfrac{1}{2}x - \dfrac{4}{5} = \dfrac{1}{4}x + 1$

[3.2]   **6.** Angie graduated in the top 5% of her class. If there were 700 people in her class, how many people were in the top 5%?

Chapter 4　Summary

Defined Terms

Section 4.1
Axis (p. 268)

Section 4.3
x-intercept (p. 290)
y-intercept (p. 290)

Section 4.4
Slope (p. 302)

Section 4.7
Relation (p. 337)
Domain (p. 337)
Range (p. 337)
Function (p. 339)

Procedures, Rules, and Key Examples

Procedures/Rules	Key Examples

Section 4.1 The Rectangular Coordinate System

To determine the coordinates of a given point in the rectangular system:

1. Follow a vertical line from the point to the x-axis (horizontal axis). The number at this position on the x-axis is the first coordinate.
2. Follow a horizontal line from the point to the y-axis (vertical axis). The number at this position on the y-axis is the second coordinate.

To graph or plot a point given its coordinates:

1. Beginning at the origin, move to the right or left along the x-axis the amount indicated by the first coordinate.
2. From that position on the x-axis, move up or down the amount indicated by the second coordinate.
3. Draw a dot to represent the point described by the coordinates.

To determine the quadrant for a given ordered pair, consider the signs of the coordinates.

$(+, +)$ means the point is in quadrant I
$(-, +)$ means the point is in quadrant II
$(-, -)$ means the point is in quadrant III
$(+, -)$ means the point is in quadrant IV

Note: If either coordinate is 0, then the point is on an axis and not in a quadrant.

Example 1: Plot each of the following:
$(-2, 4), (5, 0), (0, -2), (-4, -3),$
$(3, -4), (2, 3)$

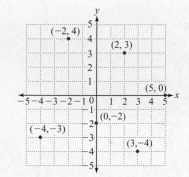

Example 2: State the quadrant in which each point is located.

$(19, 78)$:	quadrant I
$(-67, 45)$:	quadrant II
$(-107, -36)$:	quadrant III
$(58, -92)$:	quadrant IV
$(0, -16)$:	Not in a quadrant. This point is on the y-axis.

continued

Procedures/Rules	Key Examples

Section 4.2 Graphing Linear Equations

To determine whether a given ordered pair is a solution for an equation with two variables:
1. Replace the variables in the equation with the corresponding coordinates.
2. Verify that the equation is true.

Example 1: Determine whether $(3, 5)$ is a solution for $4x - y = 7$.

Solution: $4(3) - 5 \stackrel{?}{=} 7$

$\qquad 12 - 5 \stackrel{?}{=} 7$

$\qquad\qquad 7 = 7$

$\qquad\qquad\qquad (3, 5)$ is a solution.

Example 2: Determine whether $(4, -9)$ is a solution for $4x - y = 7$.

Solution: $4(4) - (-9) \stackrel{?}{=} 7$

$\qquad\qquad 16 + 9 \stackrel{?}{=} 7$

$\qquad\qquad\quad 25 \neq 7$

$\qquad\qquad (4, -9)$ is not a solution.

To find a solution to an equation in two variables:
1. Choose a value for one of the variables (any value).
2. Replace the corresponding variable with your chosen value.
3. Solve the equation for the value of the other variable.

Example 3: Find two solutions for the equation $y = 2x - 3$.

First solution:　　　Second solution:

Let $x = 0$　　　　　Let $x = 1$

$y = 2(0) - 3$　　　　$y = 2(1) - 3$

$y = -3$　　　　　　$y = 2 - 3$

$\qquad\qquad\qquad\qquad y = -1$

Solution: $(0, -3)$　　Solution: $(1, -1)$

To graph a linear equation:
1. Find at least two solutions to the equation.
2. Plot the solutions as points in the rectangular coordinate system.
3. Connect the points to form a straight line.

Example 4: Graph $y = 2x - 3$.

We found two solutions above: $(0, -3)$ and $(1, -1)$

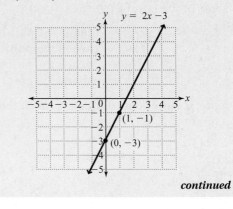

continued

Procedures/Rules	Key Examples

Section 4.2 Graphing Linear Equations (continued)

The graph of $y = c$, where c is a real number constant, is a horizontal line parallel to the x-axis that passes through the y-axis at a point with coordinates $(0, c)$.

Example 5: Graph $y = 4$.

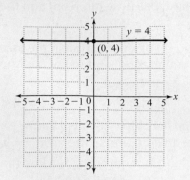

The graph of $x = c$, where c is a real number constant, is a vertical line parallel to the y-axis that passes through the x-axis at a point with coordinates $(c, 0)$.

Example 6: Graph $x = -3$.

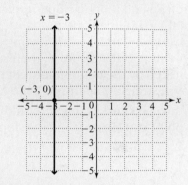

Section 4.3 Graphing Using Intercepts

To find an x-intercept:
1. Replace y with 0 in the given equation.
2. Solve for x.

To find a y-intercept:
1. Replace x with 0 in the given equation.
2. Solve for y.

Example 1: Find the x- and y-intercepts for $2x - 3y = 18$.

Solution:

x-intercept: y-intercept:

$$2x - 3(0) = 18 \qquad 2(0) - 3y = 18$$
$$2x = 18 \qquad -3y = 18$$
$$x = 9 \qquad y = -6$$

x-intercept: $(9, 0)$ y-intercept: $(0, -6)$

If an equation can be written in the form $y = mx$, where m is a real-number constant other than 0, then the x- and y-intercepts are at the origin, $(0, 0)$.

Example 2: The graph of the equation $y = 3x$ has $(0, 0)$ as both x- and y-intercepts.

If an equation is in the form $y = mx + b$, where m and b are nonzero real-number constants, then the y-intercept will be $(0, b)$.

Example 3: The y-intercept of the graph of $y = 3x - 2$ is $(0, -2)$.

continued

Procedures/Rules	Key Examples

Section 4.3 Graphing Using Intercepts (continued)

The graph of an equation in the form $y = c$, where c is a real-number constant, has no x-intercept and the y-intercept is at $(0, c)$.

The graph of an equation in the form $x = c$, where c is a real-number constant, has no y-intercept and the x-intercept is at $(c, 0)$.

Example 4: The graph of the equation $y = 4$ has no x-intercept and its y-intercept is at $(0, 4)$.

Example 5: The graph of the equation $x = -3$ has no y-intercept and its x-intercept is at $(-3, 0)$.

Section 4.4 Slope-Intercept Form

Given an equation of the form $y = mx + b$:

If $m > 0$, then the graph is a line that slants uphill from left to right.

If $m < 0$, then the graph is a line that slants downhill from left to right.

To graph an equation in slope-intercept form, $y = mx + b$,
1. Plot the y-intercept, $(0, b)$.
2. Plot a second point by rising the number of units indicated by the numerator of the slope, m, then running the number of units indicated by the denominator of the slope, m.
3. Draw a straight line through the two points.

Note: You can check by locating additional points using the slope. Every point you locate using the slope should be on the line.

Example 1: Graph.

a. $y = 2x + 1$ **b.** $y = -\dfrac{3}{4}x - 2$

Solution: For $y = 2x + 1$, the slope is 2 and the y-intercept is $(0, 1)$. The positive slope indicates an uphill line from left to right. To graph, we can rise 2, then run 1 from the y-intercept, $(0, 1)$, to get a second point $(1, 3)$.

For $y = -\dfrac{3}{4}x - 2$, the slope is $-\dfrac{3}{4}$ and the y-intercept is $(0, -2)$. The negative slope indicates a downhill line from left to right. To graph, we rise -3, then run 4 from the y-intercept, $(0, -2)$, to get a second point $(4, -5)$.

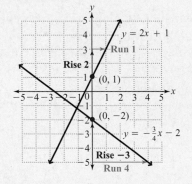

Given two points (x_1, y_1) and (x_2, y_2), the slope of the line connecting the two points is given by the formula

$$m = \frac{y_2 - y_1}{x_2 - x_1}$$

Example 2: Find the slope of a line passing through the points $(2, 7)$ and $(-8, 3)$.

Solution: $m = \dfrac{3 - 7}{-8 - 2} = \dfrac{-4}{-10} = \dfrac{2}{5}$

Two points with different x-coordinates and the same y-coordinates, (x_1, c) and (x_2, c), will form the line with a slope 0 (a horizontal line) and equation $y = c$.

Example 3: The slope of a line connecting the points $(2, 4)$ and $(1, 4)$ is 0 and the equation of the line is $y = 4$.

continued

Procedures/Rules	Key Examples

Procedures/Rules

Section 4.4 Slope-Intercept Form (continued)

Two points with the same x-coordinates and different y-coordinates, (c, y_1) and (c, y_2), will form a line with a slope that is undefined (a vertical line) and equation $x = c$.

Section 4.5 Point-Slope Form

To write the equation of a line given its y-intercept, $(0, b)$, and its slope, m, use the slope-intercept form of the equation, $y = mx + b$. If given a second point and not the slope, we must first calculate the slope using $m = \dfrac{y_2 - y_1}{x_2 - x_1}$, then use $y = mx + b$.

To write the equation of a line given its slope and any point, (x_1, y_1), on the line, use the point-slope form of the equation of a line, $y - y_1 = m(x - x_1)$. If given a second point, (x_2, y_2), and not the slope, we first calculate the slope using $m = \dfrac{y_2 - y_1}{x_2 - x_1}$, then use $y - y_1 = m(x - x_1)$.

Key Examples

Example 4: The slope of a line connecting the points $(-3, 7)$ and $(-3, -2)$ is undefined and the equation of the line is $x = -3$.

Example 1: Write slope-intercept form of the equation for a line with slope $\dfrac{3}{4}$ and y-intercept $(0, -1)$.

Solution: $y = \dfrac{3}{4}x - 1$

Example 2: Write slope-intercept form of the equation for a line with slope -3 and passing through $(2, 5)$.

Solution: $y - y_1 = m(x - x_1)$

$$y - 5 = -3(x - 2)$$
$$y - 5 = -3x + 6$$
$$y = -3x + 11$$

Example 3: Write an equation of a line connecting $(2, 3)$ and $(-3, -1)$ in the form $Ax + By = C$, where A, B, and C are integers and $A > 0$.

Solution: Find the slope:

$$m = \frac{-1 - 3}{-3 - 2} = \frac{-4}{-5} = \frac{4}{5}$$

Write the equation: $y - 3 = \dfrac{4}{5}(x - 2)$

$5(y-3) = 5 \cdot \dfrac{4}{5}(x-2)$ — Multiply by 5 to clear the fraction.

$5y - 15 = 4(x - 2)$

$5y - 15 = 4x - 8$ — Distribute 4.

$5y - 4x - 15 = -8$ — Subtract $4x$ from both sides.

$5y - 4x = 7$ — Add 15 to both sides.

$-4x + 5y = 7$ — Rearrange so that the $-4x$ term is first.

$-1(-4x + 5y) = -1 \cdot 7$ — Multiply by -1 so that the x term is positive.

$4x - 5y = -7$ — Simplify.

continued

Procedures/Rules	Key Examples

Section 4.5 Point-Slope Form (continued)

The slopes of parallel lines are equal.

The slope of a line perpendicular to a line with a slope of $\frac{a}{b}$ will be $-\frac{b}{a}$.

Horizontal and vertical lines are perpendicular.

Example 3: Find the slope of a line parallel and perpendicular to $3x + 2y = 4$.

Solution: Write $3x + 2y = 4$ in slope-intercept form.

$$2y = -3x + 4 \qquad \text{Subtract } 3x \text{ from both sides.}$$

$$y = -\frac{3}{2}x + 2 \qquad \text{Divide both sides by 2.}$$

The slope of $3x + 2y = 4$ is $-\frac{3}{2}$.

The slope of a line parallel to $3x + 2y = 4$ is $-\frac{3}{2}$.

The slope of line perpendicular to $3x + 2y = 4$ is $\frac{2}{3}$.

Section 4.6 Graphing Linear Inequalities

To determine whether an ordered pair is a solution for an inequality, replace the variables with the corresponding coordinates and see if the resulting inequality is true. If so, the ordered pair is a solution.

Example 1: Determine whether $(4, 3)$ is a solution for $2x - 4y \leq 8$.

$$2x - 4y \leq 8$$
$$2(4) - 4(3) \overset{?}{\leq} 8$$
$$8 - 12 \overset{?}{\leq} 8$$
$$-4 \leq 8 \qquad \text{This is true, so } (4, 3) \text{ is a solution.}$$

Example 2: Determine whether $(3, -1)$ is a solution for $2x - 4y \leq 8$.

$$2x - 4y \leq 8$$
$$2(3) - 4(-1) \overset{?}{\leq} 8$$
$$6 + 4 \overset{?}{\leq} 8$$
$$10 \leq 8 \qquad \text{This is false, so } (3, -1) \text{ is not a solution.}$$

continued

Procedures/Rules	Key Examples

Section 4.6 Graphing Linear Inequalities (continued)

To graph a linear inequality:
1. Graph the related equation. The related equation has an equal sign in place of the inequality symbol. If the inequality symbol is $\leq$ or $\geq$, then draw a solid line. If the inequality symbol is $<$ or $>$, then draw a dashed line.
2. Choose an ordered pair on one side of the boundary line and test this ordered pair in the inequality. If the ordered pair satisfies the inequality, then shade the region that contains the chosen ordered pair. If the ordered pair does not satisfy the inequality, then shade the region on the other side of the boundary line.

Example 3: Graph $2x - 4y \leq 8$.

Solution: We found $(4, 3)$ to be a solution, therefore shade the region containing $(4, 3)$.

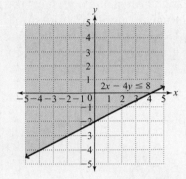

Section 4.7 Introduction to Functions and Function Notation

To determine the domain of a graph, answer the question: What are all the x-values that have a corresponding y-value?

To determine the range, answer the question: What are all the y-values that have a corresponding x-value?

Example 1: Give the domain and range of the function shown.

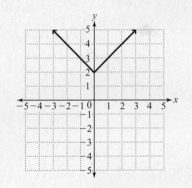

Domain: all real numbers

Range: $\{y \,|\, y \geq 2\}$

continued

Procedures/Rules	Key Examples

Section 4.7 Introduction to Functions and Function Notation (continued)

To determine whether a graph is a graph of a function, perform a vertical line test:

1. Draw or imagine vertical lines through each point in the domain.
2. If each vertical line intersects the graph at only one point, then the graph is the graph of a function.
3. If any vertical line intersects the graph at two different points, then the graph is not the graph of a function.

Example 2: Determine whether the graph is the graph of a function.

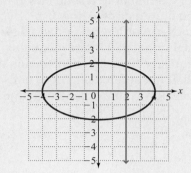

Solution: The relation in the graph is not a function because a vertical line can be drawn that intersects the graph at two points. This means that there is a domain value that corresponds to two values in the range.

Given a function $f(x)$, to find $f(a)$, where a is a real number in the domain of f, replace x in the function with a and calculate the value.

Example 3: For the function $f(x) = x^2 - 3x + 1$, find $f(-2)$.

$$f(-2) = (-2)^2 - 3(-2) + 1$$
$$= 4 + 6 + 1$$
$$= 11$$

Forms of a Linear Equation

Slope-intercept form: $y = mx + b$

Point-slope form: $y - y_1 = m(x - x_1)$

Standard form: $Ax + By = C$

Chapter 4 Review Exercises

For Exercises 1–6, answer true or false.

[4.1] **1.** When writing coordinates, the horizontal coordinate is written first.

[4.3] **2.** An x-intercept is a point where a graph intersects the horizontal axis.

[4.4] **3.** The slope formula is $m = \dfrac{x_2 - x_1}{y_2 - y_1}$.

[4.6] **4.** A linear inequality has just one solution.

[4.5] **5.** Given any two points on a line, the equation of the line may be found.

[4.5] **6.** Given the slope and any point on a line, the equation of the line may be found.

[4.3] *For Exercises 7–10, complete the rule.*

    7. To find an x-intercept:
        1. Replace _____ with 0 in the given equation.
        2. Solve for _____.

[4.3] **8.** To find a y-intercept:
        1. Replace _____ with 0 in the given equation.
        2. Solve for _____.

[4.4] **9.** If an equation is in the form $y = mx + b$, where m and b are nonzero real-numbers, then the _____ will be $(0, b)$.

[4.4] **10.** Given an equation of the form $y = mx + b$,
    if $m > 0$, then the graph will be a line that slants _____ from left to right.
    if $m < 0$, then the graph will be a line that slants _____ from left to right.

[4.1] *For Exercises 11 and 12, write the coordinates for each point.*

11.

12.

[4.1] *For Exercises 13 and 14, plot and label the points indicated by the coordinate pairs.*

13. $(3, 4), (-2, 0), (0, 1), (-1, -5)$

14. $(-2, 2), (0, -3), (4, 1), (5, 0)$

[4.1] *For Exercises 15–18, state the quadrant in which the point is located.*

15. $(-2.8, 1203)$

16. $\left(-\dfrac{1}{8}, -16\right)$

17. $(421, 5300)$

18. $\left(52\dfrac{2}{3}, -0.36\right)$

Equations and Inequalities

[4.2] *For Exercises 19–24, determine whether the given pair of coordinates is a solution for the given equation.*

19. $(-1, 3); x + 3y = 8$

20. $(2, 1); 2x - y = 3$

21. $(0, 0); 6x = 2y + 1$

Exercises 19–85

22. $(-2, -2); y = \dfrac{2}{3}x$

23. $\left(\dfrac{3}{2}, -\dfrac{1}{2}\right); x + y = 2$

24. $(2.1, 3.2); y - 3x = 5$

[4.2] *For Exercises 25–30, find three solutions for the given equation. Then graph.*

25. $y = -2x$

26. $x = y$

27. $y = -x + 7$

28. $x + 4y = 5$

29. $2x + 3y = 6$

30. $y + \dfrac{1}{3}x = 3$

31. $x + 2y = 12$

32. $y = -3x$

33. $2x - 3.1y = 6.2$

34. $y = \dfrac{2}{3}x + 5$

35. $x = -2$

36. $y = 7$

[4.3] *For Exercises 37–42, graph using the x- and y-intercepts.*

37. $3x + 2y = 6$

38. $y = -2x + 1$

39. $y = -\dfrac{1}{5}x + 3$

40. $2x = 5y + 10$

41. $7 = y$

42. $x = -1$

[4.4] *For Exercises 43 and 44, graph each of the following on the same grid.*

43. $y = x$
  $y = 2x$
  $y = 4x$

44. $y = -x$
  $y = -2x$
  $y = -3x$

[4.4] *For Exercises 45–50, determine the slope and the coordinates of the y-intercept, then graph.*

45. $y = -2x + 3$

46. $y = -\dfrac{1}{4}x + 2$

47. $y = 3x$

48. $x + y = 5$

49. $x - 3y = 7$

50. $2x - 3y = 8$

[4.4] *For Exercises 51–56, find the slope of the line through the given points.*

51. $(2, 7), (-1, -2)$

52. $(0, -1), (3, 2)$

53. $(-3, 1), (3, -1)$

54. $(-6, 2), (-1, -1)$

55. $(2, 8), (-1, 8)$

56. $(-7, 3), (-7, -5)$

[4.4] *For Exercises 57–60, write the equation, in slope-intercept form, of the line given the slope and the coordinate of the y-intercept.*

57. $m = -1; (0, 7)$

58. $m = -\frac{1}{5}; (0, -8)$

59. $m = 0.2; (0, 6)$

60. $m = 1; (0, 0)$

[4.5] *For Exercises 61–66, write the equation, in slope-intercept form, of a line with the given slope and passing through the given point.*

61. $m = -2; (1, 7)$

62. $m = 1; (3, -6)$

63. $m = \frac{1}{3}; (-5, 0)$

64. $m = -\frac{2}{5}; (3, 3)$

65. $m = 6.2; (-2, -3)$

66. $m = -0.4; (-1, 2)$

[4.5] *For Exercises 67–70, write the equation of a line connecting the given points in slope-intercept form and in standard form.*

67. $(7, -3), (2, 2)$

68. $(5, -3), (9, 2)$

69. $(3, 2), (5, 1)$

70. $(-3, -2), (-8, -1)$

[4.5] *For Exercises 71–74, write the equations in slope-intercept form.*

71. Find the equation of a line passing through $(0, -2)$ and parallel to the line $y = 2x + 7$.

72. Find the equation of a line passing through $(1, 5)$ and parallel to the line $2x + 3y = 6$.

73. Find the equation of a line passing through $(-1, -2)$ and perpendicular to the line $y = \frac{3}{5}x - 1$.

74. Find the equation of a line passing through $(-2, 4)$ and perpendicular to the line $y = -\frac{2}{5}x - 6$.

[4.2] **75.** A salesperson receives $1000 per month, plus a commission of 5% of his total sales. The equation $p = 0.05s + 1000$ describes the salesperson's gross pay each month, where p represents the gross pay and s represents the total sales in dollars.

   a. Find the gross pay if the person sells $24,000 worth of merchandise.

   b. Find the p-intercept. Explain what it indicates.

   c. Graph the equation.

[4.5] **76.** Andrew purchased a photocopier for his business in the year 2000 for $18,000. In 2004, he considers selling the copier on ebay and notes that others like it are selling for $12,000.

   a. Assuming that the depreciation is linear, plot the two given data points with 2000 being year 0 so that 2004 is year 4. Draw a line connecting the two points.

   b. What is the slope of the line?

   c. Let n represent the number of years the copier is in service and v represent the value of the copier. Write the equation of the line in slope-intercept form.

   d. If the depreciation continues at the same rate, in what year will the copier be worth half of its original value?

   e. In what year will the copier be worth $0?

[4.6] *For Exercises 77–80, determine whether the ordered pair is a solution for the linear inequality.*

77. $(3, 2); x + 2y > 5$ **78.** $(-1, 0); y < x + 2$ **79.** $(-5, 8); 2x - 6y \le 17$ **80.** $(0, 0); x + y \ge 5$

[4.6] *For Exercises 81–85, graph the linear inequality.*

81. $y < 3x - 5$ **82.** $y \ge -\dfrac{2}{3}x$ **83.** $x - y \ge 3$

84. $-3x - 5y < -15$ **85.** $x \ge -1$

[4.6] **86.** A company produces two versions of its product. The lower-priced package costs $6 to make, whereas the higher-priced package costs $8 to make.

    a. Let a represent the number of the lower-priced packages produced and b represent the number of the higher-priced packages produced. Write an inequality in which the total cost is at most $12,000.

    b. Graph the inequality.

    c. Give a combination of the number of each package that the company could produce that has a cost equal to $12,000.

    d. Give a combination of the number of each package that the company could produce that has a cost less than $12,000.

[4.7] *For Exercises 87 and 88, find the domain and range of the relation.*

87. $\{(2, 3), (-1, 5), (3, 6), (-3, -4)\}$

88. Cars holding their value for resale:

Automobile	Percentage of Original Value
Volkswagen	52.2
Honda	49.7
Toyota	49.0
Subaru	47.8
Nissan	45.8

Source: Automotive Lease Guide

[4.7] *For Exercises 89–90, determine whether the relation is a function.*

89. The following relation shows the courses taught by each instructor during one semester:

Domain (instructor name)	Range (courses taught)
Hames	Math 100
Carson	Math 100, Math 102
Pritchard	Math 035
Webb	Math 100, Math 110

90. The following relation shows the price of a particular brand of dog food based on the size of the bag:

Domain (size of bag)	Range (price)
5 lb.	$3.95
10 lb.	$7.90
20 lb.	$15.00
40 lb.	$29.95

[4.7] *For Exercises 91–93, give the domain and range. Then determine whether the graph is the graph of a function.*

91.

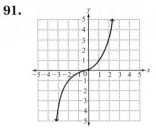

92.

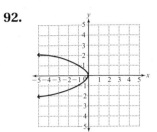

93.

[4.7] **94.** Use the graph in Exercise 93 to find the value of the function at the indicated values.

    a. $f(-3)$ **b.** $f(-1)$

    c. $f(0)$ **d.** $f(3)$

[4.7] *For Exercises 95 and 96, find the value of the function.*

95. $f(x) = x^3 - 5x + 2$

    a. $f(2)$ **b.** $f(0)$ **c.** $f(-3)$

96. $f(x) = \dfrac{x}{x - 3}$

    a. $f(6)$ **b.** $f(3)$ **c.** $f(-5)$

Chapter 4 Practice Test

1. Determine the coordinates for each point.

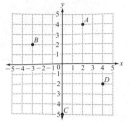

2. Plot and label the points indicated by the coordinate pairs.

$(0, -3), (4, -1), (-3, 2), (2, 0), (-1, -5)$

3. State the quadrant in which $\left(4\frac{2}{3}, 5115 \right)$ is located.

4. Determine whether $(1, 7)$ is a solution for $y = -\frac{2}{5}x - 8$.

For Exercises 5–8, find the coordinates for the x- and y-intercepts, then graph.

5. $x - 2y = 4$

6. $2x + 5y = 10$

7. $y = -2x$

8. $y = \frac{3}{4}x - 1$

For Exercises 9 and 10, determine the slope and the coordinates of the y-intercept.

9. $y = \frac{3}{4}x + 11$

10. $5x - 3y = 8$

For Exercises 11 and 12, determine the slope of the line through the given points.

11. $(-5, 6), (-2, -4)$

12. $(6, 2), (3, 2)$

For Exercises 13–18, write the equation of the line in slope-intercept form.

13. $m = \dfrac{3}{5}$; y-intercept $(0, 4)$

14. Passing through the points $(8, -1)$, $(-7, 5)$

15. Write the equation of a line in the form $Ax + By = C$ through the points $(2, 1)$ and $(5, 3)$.

16. Write the equation of a line in the form $Ax + By = C$ through the point $(4, -2)$ and perpendicular to the line $y = -3x + 4$.

17. From 1999 to 2003, the number of people traveling more than 50 miles for leisure increased at a linear rate. In 1999, the number of people traveling 50 miles or more for leisure was 848.6 million. In 2003, that number was 929.5 million people. (*Source:* Travel Industry Association of America, TravelScope)

  a. Let 1999 be year 0 so that 2003 is year 4. Let x represent the number of years after 1999 and y represent the number of people traveling 50 miles or more. Plot the two data points in the coordinate plane, then draw a line connecting them.

  b. Find the slope of the line.

  c. Write the equation of the line in slope-intercept form.

  d. If the trend continues, predict the number of people traveling more than 50 miles for leisure in 2010. Round to the nearest tenth.

18. Determine whether $(2, 7)$ is a solution for the linear inequality $y \geq 2x - 9$.

For Exercises 19 and 20, graph the linear inequality.

19. $y \geq \dfrac{2}{5}x - 1$

20. $x - 3y < 8$

21. Alex works two part-time jobs. He receives $10 per hour when working as a cook in a restaurant and $8 per hour when working at a music store. To pay all his monthly expenses, he needs to make at least $360 per week.

  a. Let x represent the number of hours that he works at the restaurant and y represent the number of hours that he works at the music store. Write an inequality in which his total income is at least $360.

  b. Graph the inequality. Since he can only work a positive number of hours, restrict the graph to the first quadrant.

  c. Give a combination of hours that provides him an income of exactly $360.

  d. Give a combination of hours that provides him an income of more than $360.

22. The following relation shows U.S. per capita income for each year. Is the relation a function?

Year	Income (in 2001 dollars)
1995	19,871
1996	20,372
1997	21,162
1998	21,821
1999	22,499
2000	22,970
2001	22,851

23. Determine whether the graph is the graph of a function.

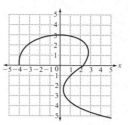

24. Find the indicated value the function $f(x) = \dfrac{x^2}{x - 4}$.

a. $f(2)$ **b.** $f(4)$ **c.** $f(-3)$

25. a. Give the domain and range of the function graphed below.

b. Find $f(1)$.

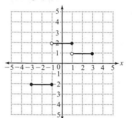

CHAPTER 5

Polynomials

"The ability to simplify means to eliminate the unnecessary so that the necessary may speak."

—Hans Hofmann, early 20th-century teacher and painter

"Everything should be made as simple as possible, but not simpler."

—Albert Einstein, early 20th-century physicist

In Chapter 1, we reviewed real numbers and introduced algebraic expressions. We learned there are two actions we can perform on an expression: (1) We can evaluate an expression, or (2) we can rewrite an expression. In this chapter, we will focus on polynomial expressions, exploring how to evaluate and rewrite these expressions. We will also see that we can add, subtract, multiply, and divide polynomial expressions in much the same way that we add, subtract, multiply, and divide numbers.

5.1 Exponents and Scientific Notation

OBJECTIVES
1. Evaluate exponential forms with integer exponents.
2. Write scientific notation in standard form.
3. Write standard form numbers in scientific notation.

OBJECTIVE 1. Evaluate exponential forms with integer exponents.

Positive Integer Exponents

We learned in Section 1.5 that exponents indicate repeated multiplication of a base number.

$$2^4 = \underbrace{2 \cdot 2 \cdot 2 \cdot 2}_{\text{four factors of two}} = 16 \qquad (-5)^3 = \underbrace{(-5)(-5)(-5)}_{\text{three factors of negative five}} = -125$$

We also discovered some rules for evaluating exponential forms with negative bases.

RULE   **Evaluating Exponential Forms with Negative Bases**

If the base of an exponential form is a negative number and the exponent is even, then the product is positive.

If the base is a negative number and the exponent is odd, then the product is negative.

What if the base is a fraction, as in $\left(\dfrac{2}{3}\right)^4$?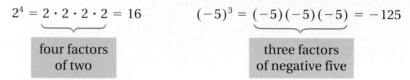

$$\left(\frac{2}{3}\right)^4 = \frac{2}{3} \cdot \frac{2}{3} \cdot \frac{2}{3} \cdot \frac{2}{3} = \frac{2^4}{3^4} = \frac{16}{81}$$

This example suggests the following conclusion and rule:

Conclusion   If a fraction is raised to a power, we can write its numerator and denominator to that power.

RULE   **Raising a Quotient to a Power**

If a and b are real numbers, where $b \neq 0$ and n is a natural number, then
$$\left(\frac{a}{b}\right)^n = \frac{a^n}{b^n}.$$

5.2 Exercises

For Extra Help — MyMathLab · Videotape/DVT · InterAct Math · Math Tutor Center · Math XL.com

1. How do we determine the degree of a monomial?

2. Explain the differences between a monomial, a binomial, and a trinomial.

3. How do we determine the degree of a polynomial with two or more terms?

4. Given a polynomial in one variable, what does *descending order of degree* mean?

5. Explain how to combine like terms.

6. Why do we say that combining the like terms in a polynomial simplifies it?

For Exercises 7–18, determine whether the expression is a monomial.

7. $-3x^2$

8. $5y^3$

9. $\dfrac{1}{5}$

10. $\dfrac{4}{7}$

11. $\dfrac{6m}{4y^3}$

12. $\dfrac{-8t}{5x}$

13. $4x^2 - 3x + 7$

14. $p^2 - q^2$

15. y

16. 0.76

17. $3m^4n^2$

18. $6u^3v^4w$

For Exercises 19–30, identify the coefficient and degree of each monomial.

19. $-5m^3$

20. $8p^2$

21. $-xy^4$

22. $-m$

23. -9

24. 18

25. $4.2n^3p$

26. $-6.7uv^7$

27. $16abc$

28. $-8.1lkm$

29. y

30. w

For Exercises 31–42, determine whether the expression is a monomial, binomial, trinomial, or none of these.

31. $6x + 8y + 3z$

32. $17.3x^2 - 3x + 2.1$

33. $-7m^2n$

34. $18xy^3$

35. $5.2x^3 - 3x^2 + 4.1x - 11$

36. $x^3 - 4x^2 + 4x + 5$

37. $5u^3 - 16u$

38. $7k + 5k^3$

39. -21

40. 36

41. $\dfrac{25}{x} - x^2$

42. $y^2 - \dfrac{16}{y}$

For Exercises 43–48, identify the degree of each polynomial.

43. $19 - 7y^4 + 3y - 2y^3 - 7$

44. $6a^5 - 19a^3 + a^9 + 5a - 14$

45. $16z^4 + 7z^2 - z^5 - 4z^3 + z$

46. $11 + 4t^5 - 5t + 7t^3 - 18t^2 + t^8$

47. $2u^2 + 5u^3 - u^7 + 13u^4 - 3u$

48. $22j^3 + 5j^2 - 16j^4 + 21 - 14j$

For Exercises 49–54, evaluate the polynomial using the given values.

49. $-3xy^2;\, x = -5,\, y = 2$

50. $-2x^2y;\, x = -1,\, y = 4$

51. $x^2 - 6x + 1;\, x = 3$

52. $n^2 - 8n - 3;\, n = -4$

53. $a^3 + 0.5ab + 2.4b;\, a = -1,\, b = -2$

54. $m^3 - 0.2n^2;\, m = -3,\, n = 4$

55. If we neglect air resistance, the polynomial $-16t^2 + h_0$ describes the height of a falling object after falling from an initial height h_0 for t seconds. A marble is dropped from a tower at a height of 50 feet.

  a. What is its height after 0.5 seconds?

  b. What is its height after 1.2 seconds?

56. The polynomial $2lw + 2lh + 2wh$ describes the surface area of a box.

  a. Find the surface area of a box with a length of 8 inches, width of 6.5 inches, and height of 4 inches.

  b. Find the surface area of a box with a length of 4 inches, width of 3 inches, and height of 2.5 inches.

57. The polynomial $5p^2 - 6p + 3$ describes the number of units sold for every 1000 people in a particular region based on the price of the product, which is represented by p.

  a. Find the number of units sold for every 1000 people if the price of each unit is $3.00.

  b. Find the number of units sold for every 1000 people if the price of each unit is $3.50.

58. The polynomial $7r^2 - 2r + 6$ describes the voltage in a circuit, where r represents the resistance in the circuit.

  a. Find the voltage if the resistance is 6 ohms.

  b. Find the voltage if the resistance is 8 ohms.

59. An engineer is designing a chemical-storage tank that is capsule shaped. The polynomial $\frac{4}{3}\pi r^3 + \pi r^2 h$ describes the volume of the tank.

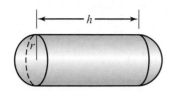

a. Find the volume of the tank if the radius is 4 feet and the height is 20 feet. Round the result to the nearest tenth.

b. Find the volume of the tank if the radius is 3 feet and the height is 15 feet. Round the result to the nearest tenth.

60. The polynomial $lwh - \pi r^2 h$ describes the volume of metal remaining in a block after a cylinder has been bored into the block of metal.

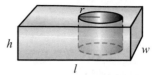

a. Find the volume of metal remaining if the length is 15 inches, the width is 7 inches, the height is 4 inches, and the radius of the cylinder is 3 inches. Round to the nearest tenth.

b. Find the volume of metal remaining if the length is 15 inches, the width is 7 inches, the height is 4 inches, and the radius of the cylinder is 2 inches. Round to the nearest tenth.

61. The polynomial $-0.2x^2 + 4.25x + 26$ describes the number of international visitors (in millions) to the United States each year from 1986 on, where x represents the number of years after 1986 ($x = 0$ means 1986, $x = 1$ means 1987, and so on). (*Source:* Tourism Industries, International Trade Administration, Department of Commerce)

a. How many international visitors came to the United States in 1986?

b. How many international visitors came to the United States in 1993?

c. How many international visitors came to the United States in 2003?

d. If the model holds, predict the number of international visitors in 2010. Do you think the prediction is reasonable? Explain.

62. The polynomial $28x^3 + 25x^2 + 100x + 340$ describes the number of cell phone subscribers in the United States (in thousands of subscribers) each year since 1985, where x represents the number of years since 1985 ($x = 0$ means 1985, $x = 1$ means 1986, and so on). (*Source:* The CTIA Semiannual Wireless Industry Survey)

a. How many subscribers were there in the United States in 1985?

b. How many subscribers were there in the United States in 1995?

c. How many subscribers were there in the United States in 2001?

d. If the model holds, predict the number of subscribers in 2010. Do you think the prediction is reasonable? Explain.

For Exercises 63–68, write each polynomial in descending order.

63. $7x^4 + 5x^7 - 8x + 14 - 3x^5$

64. $-4y^3 + 7y - 8y^5 - 6y^2 + 9 + 2y^4$

65. $4r - 3r^2 + 18r^5 + 7r^3 - 8r^6$

66. $u^4 + 7u^2 - u^9 + 15u + 27 - 3u^5 - 8u^3$

67. $20 - w^3 + 5w + 11w^2 - 12w^4$

68. $a - 6 + 7a^5 + 3a^2 - 4a^3$

For Exercises 69–84, combine like terms and write the resulting polynomial in descending order of degree.

69. $3x + 4 + 2x - 7$

70. $2y - 9 + 5y + 3$

71. $7 - 6a + 2a - 2$

72. $9 + 4b - 8b - 3$

73. $7x^2 + 2x + 4x - 9x^2$

74. $4a^3 + 3a^2 - 2a^2 - 7a^3$

75. $2x^2 + 5x - 7x^2 + 8 - 5x + 6$

76. $3m^5 + 7m^2 - 8m + 9m^5 - 7m - m^2$

77. $15 - 4y + 8y^2 - 4y - 3y^2 + 7 - y$

78. $6l - 5l^3 - 2l^4 + 7l^4 + 5l - l^3 + 4l^3$

79. $9k - 5k^2 + 6k^3 + 2k^2 + k^4 - 3k - 3k^4$

80. $11p^5 - p^2 + 12p^3 + 5p^2 - 7p^3 + 20 - 4p$

81. $7a^2 - 5a^4 + 6a^3 - 5a^4 + 12 - 6a^3 + a + 4$

82. $-6c^9 + 7c^3 - 4c^5 + 8c - 14c^2 + 13c^5 - 7 - c^9$

83. $12v^3 + 19v^2 - 20 - v^5 - 5v^3 + 16v^2 - 20v^5 + v^3$

84. $b^3 - 13b^2 + b^4 + 6 - 2b^3 - 15b^4 - 17b^3 + 8$

For Exercises 85–94, combine like terms.

85. $6a - 3b - 2a + b$

86. $5x - y - 3x - 2y$

87. $3y^2 + 5y - 2y - 7y^2$

88. $2z^3 - 4z - 6z - 10z^3$

89. $2x^2 + 3y - 4x^2 + 6y - 5x^2 - 2x^2$

90. $3a^2 - 5b - 4a^2 + 7b - 5b + 6b$

91. $y^6 + 2yz^4 - 10yz + 3yz^4 - 3z^5 - 7z^3 + 4yz - 10$

92. $x^4 + xy^3 - 6xy + xy^3 + 3y^4 - 4y^2 + 3xy - 8$

93. $-3w^2z - 9 + 6wz^2 + 3w^2 - w^2z + 8 + 9w^2 - 7wz^2$

94. $-m^3n - 8mn + 4 - 5m^2n^2 + 3m^3n - 7 + mn - 16m^2n^2$

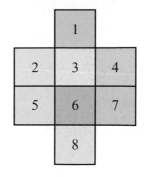
REVIEW EXERCISES

[1.5] **1.** Simplify: $(-2)^3 - 9|12 - 14 \div (-2)|$

[2.3] **2.** Solve and check: $\dfrac{1}{4}x = 3$

[3.2] **3.** Write 14.2% as a fraction in simplest form.

[3.5] **4.** How many liters of a 40% solution must be added to 200 liters of a 20% solution to obtain a 35% solution?

[5.1] **5.** Evaluate: -7^0

[5.1] **6.** The Andromeda galaxy (the closest one to our Milky Way galaxy) contains at least 200,000,000,000 stars. Write this in scientific notation.

5.3 Adding and Subtracting Polynomials

OBJECTIVES
1. Add polynomials.
2. Subtract polynomials.

OBJECTIVE 1. Add polynomials.

Understanding Polynomial Addition

We can add and subtract polynomials in the same way that we add and subtract numbers. In fact, polynomials are like whole numbers that are in an expanded form. Consider the polynomial $4x^2 + 3x + 6$. If we replace the x's with the number 10, then we have the expanded form for the number 436.

$$4x^2 \quad + \quad 3x \quad + 6$$

$$4 \cdot 10^2 \quad + \quad 3 \cdot 10 \quad + 6 \qquad \text{Replacing } x \text{ with 10.}$$
$$= 400 \quad + \quad 30 \quad + 6$$
$$= 436$$

In our base-ten number system each place value is a power of 10. We can think of polynomials as a variable-base number system. In other words, the place values are variables, where x^2 is like the hundreds place (10^2), and x is like the tens place (10^1). To add whole numbers, we add the digits in like place values. Polynomials are added in a similar way. However, instead of adding digits in like place values, we add like terms. Consider the following comparison:

Numeric addition:
$436 + 251 = 687$

$$\begin{array}{r} 436 \\ + 251 \\ \hline 687 \end{array}$$

In numeric addition, like place values are added.

Polynomial addition:
$(4x^2 + 3x + 6) + (2x^2 + 5x + 1) = 6x^2 + 8x + 7$

$$\begin{array}{r} 4x^2 + 3x + 6 \\ + 2x^2 + 5x + 1 \\ \hline 6x^2 + 8x + 7 \end{array}$$

In polynomial addition, like terms are added.

From this simple case, we see that polynomials are added by combining the like terms.

PROCEDURE Adding Polynomials

To add polynomials, combine like terms.

Adding Polynomials

Although we stacked the preceding polynomials, we do not actually have to stack the polynomials in order to add them.

EXAMPLE 1 Add. $(2x^3 + 5x^2 + 4x + 1) + (6x^3 + 3x + 8)$

Solution Combine like terms. Notice that combining in order of degree places the resulting polynomial in descending order of degree.

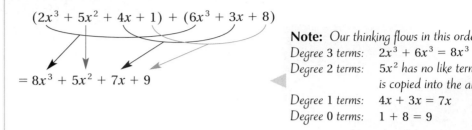

$= 8x^3 + 5x^2 + 7x + 9$

Note: *Our thinking flows in this order:*
Degree 3 terms: $2x^3 + 6x^3 = 8x^3$
Degree 2 terms: $5x^2$ *has no like term, so it is copied into the answer.*
Degree 1 terms: $4x + 3x = 7x$
Degree 0 terms: $1 + 8 = 9$

If we had stacked the polynomials in Example 1, we would have had to leave a blank space under the $5x^2$ because there was no x^2 term in the second polynomial. It is the same as having a 0 in that place.

$$
\begin{array}{r}
2x^3 + 5x^2 + 4x + 1 \\
+\,6x^3 + 0 + 3x + 8 \\
\hline
8x^3 + 5x^2 + 7x + 9
\end{array}
$$

As the polynomials get more complex, these blanks become more prevalent. This is why the stacking method is not a preferred method for adding polynomials.

EXAMPLE 2 Add. $(y^5 + 9y^3 - 3y - 7) + (2y^5 - 4y^3 + 3y + 5)$

Solution $(y^5 + 9y^3 - 3y - 7) + (2y^5 - 4y^3 + 3y + 5)$ **Combine like terms.**

$= 3y^5 + 5y^3 \qquad + 0 \qquad - 2$
$= 3y^5 + 5y^3 - 2$

YOUR TURN Add.

a. $(x^4 + 3x^3 + 4x + 2) + (5x^3 + 2x + 7)$

b. $(14n^5 - 6n^3 - 12n + 15) + (2n^5 - 4n^3 + 8n - 5)$

Sometimes the polynomials may contain terms that have several variables or coefficients that are fractions or decimals.

ANSWERS

a. $x^4 + 8x^3 + 6x + 9$
b. $16n^5 - 10n^3 - 4n + 10$

EXAMPLE 3 Add. $\left(7t^4 + 2.9tu^2 - \dfrac{3}{4}tu + 11\right) + \left(-8t^4 - 7tu^2 + \dfrac{1}{6}tu - 5\right)$

Solution $\left(7t^4 + 2.9tu^2 - \dfrac{3}{4}tu + 11\right) + \left(-8t^4 - 7tu^2 + \dfrac{1}{6}tu - 5\right)$ **Combine like terms.**

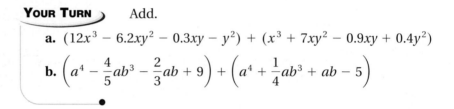

$$= -t^4 - 4.1tu^2 - \dfrac{7}{12}tu + 6$$

Note: *Our thinking flows like this:*
The t^4 terms: $7t^4 + (-8t^4) = -1t^4 = -t^4$
The tu^2 terms: $2.9tu^2 + (-7tu^2) = -4.1tu^2$
The tu terms: $-\dfrac{3}{4}tu + \dfrac{1}{6}tu = -\dfrac{9}{12}tu + \dfrac{2}{12}tu = -\dfrac{7}{12}tu$
The constant terms: $11 + (-5) = 6$

YOUR TURN Add.

a. $(12x^3 - 6.2xy^2 - 0.3xy - y^2) + (x^3 + 7xy^2 - 0.9xy + 0.4y^2)$

b. $\left(a^4 - \dfrac{4}{5}ab^3 - \dfrac{2}{3}ab + 9\right) + \left(a^4 + \dfrac{1}{4}ab^3 + ab - 5\right)$

EXAMPLE 4 Write an expression in simplest form for the perimeter of the rectangle shown.

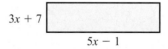

$3x + 7$

$5x - 1$

Understand *Perimeter* means the total distance around the shape. Therefore, we need to add the lengths of all the sides of the shape.

Plan In this case, the lengths of the sides are represented by polynomials. Therefore, we add the polynomials to represent the perimeter.

Execute Perimeter = Length + Width + Length + Width
$= (5x - 1) + (3x + 7) + (5x - 1) + (3x + 7)$
$= 5x + 3x + 5x + 3x - 1 + 7 - 1 + 7$
$= 16x + 12$

Answer The expression for the perimeter is $16x + 12$.

ANSWERS

a. $13x^3 + 0.8xy^2 - 1.2xy - 0.6y^2$

b. $2a^4 - \dfrac{11}{20}ab^3 + \dfrac{1}{3}ab + 4$

Check To check, we (1) choose a value for x and evaluate the original expressions for length and width, (2) determine the corresponding numeric perimeter, and (3) evaluate the perimeter expression using the same value for x and verify that we get the same numeric perimeter. Let's choose $x = 2$.

Length: $5x - 1$ Width: $3x + 7$ Perimeter:

$5(2) - 1$ $3(2) + 7$ The perimeter of the rectangle with a
$= 10 - 1$ $= 6 + 7$ length of 9 and width of 13 is
$= 9$ $= 13$ $9 + 13 + 9 + 13 = 44$.

Now evaluate the perimeter expression where $x = 2$ and we should find that the result is 44.

$$\text{Perimeter expression:} \quad 16x + 12$$
$$16(2) + 12$$
$$= 32 + 12$$
$$= 44 \quad \text{This agrees with our calculation above.}$$

YOUR TURN Write an expression in simplest form for the perimeter of the following shape.

OBJECTIVE 2. Subtract polynomials.

Understanding Polynomial Subtraction

Subtracting polynomials is similar to subtracting signed numbers. When we subtract signed numbers it is often simpler to write the subtraction statement as an equivalent addition statement. Consider the following comparison of numeric subtraction and polynomial subtraction.

Numeric subtraction: **Polynomial subtraction:**

$975 - 621$ $(9x^2 + 7x + 5) - (6x^2 + 2x + 1)$

We can stack by place: We can stack like terms:

$$\begin{array}{r} 975 \\ -621 \\ \hline 354 \end{array}$$ $$\begin{array}{r} 9x^2 + 7x + 5 \\ -(6x^2 + 2x + 1) \\ \hline 3x^2 + 5x + 4 \end{array}$$

In numeric subtraction, we subtract digits in the same place value, whereas in polynomial subtraction, we subtract like terms. Note that subtracting 1 from 5 is equivalent to combining 5 and -1, subtracting $2x$ from $7x$ is equivalent to combining $7x$ and $-2x$, and subtracting $6x^2$ from $9x^2$ is equivalent to combining $9x^2$ and $-6x^2$. This suggests that we can write polynomial subtraction as equivalent

ANSWER

$6x + 3$

polynomial addition by changing the sign of each term in the second polynomial (the subtrahend) like this:

$$(9x^2 + 7x + 5) - (6x^2 + 2x + 1)$$
$$= (9x^2 + 7x + 5) + (-6x^2 - 2x - 1)$$
$$= 3x^2 + 5x + 4$$

Note: *We are allowed to change the signs because of the distributive property. If we disregard the initial polynomial, we have*

$$-(6x^2 + 2x + 1) = -1(6x^2 + 2x + 1)$$

The distributive property tells us we can distribute the -1 *(or minus sign) to each term inside the parentheses.*

$$= -1 \cdot 6x^2 - 1 \cdot 2x - 1 \cdot 1$$
$$= -6x^2 - 2x - 1$$

Subtracting Polynomials

Our exploration suggests the following procedure for subtracting polynomials.

PROCEDURE   Subtracting Polynomials

To subtract polynomials:
1. Write the subtraction statement as an equivalent addition statement.
   a. Change the operation symbol from a minus sign to a plus sign.
   b. Change the subtrahend (second polynomial) to its additive inverse. To get the additive inverse, we change the sign of each term in the polynomial.
2. Combine like terms.

Connection   The procedure for subtracting polynomials is the same as the procedure for subtracting signed numbers (Section 1.3). In both cases, we write subtraction as equivalent addition by changing the subtrahend to its additive inverse.

EXAMPLE 5   Subtract. $(7x^3 + 8x^2 + 6x + 9) - (3x^3 + 6x^2 + 5x + 1)$

Solution   Write an equivalent addition statement, then combine like terms.

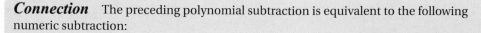

$(7x^3 + 8x^2 + 6x + 9) - (\quad 3x^3 + 6x^2 + 5x + 1)$

Change the minus sign to a plus sign.   **Change all signs in the subtrahend.**

$$= (7x^3 + 8x^2 + 6x + 9) + (-3x^3 - 6x^2 - 5x - 1)$$
$$= 4x^3 + 2x^2 + x + 8 \qquad \text{Combine like terms.}$$

Connection   The preceding polynomial subtraction is equivalent to the following numeric subtraction:

$$\begin{array}{r} 7869 \\ -3651 \\ \hline 4218 \end{array}$$

Notice the numeric result 4218 corresponds to the polynomial result $4x^3 + 2x^2 + x + 8$. We could use the stacking method for subtracting polynomials; however, as we saw with adding polynomials, it is not the best method for all cases because we often have missing terms or terms that are not like terms.

You may have noticed that the polynomials in Example 5 contained only plus signs. It is important to recognize that the polynomials may contain a mixture of signs and multiple variables. This is shown in Examples 6 and 7.

EXAMPLE 6 Subtract. $(15y^3 + 3y^2 + y - 2) - (7y^3 - 6y^2 + 5y - 8)$

Solution Write an equivalent addition statement, then combine like terms.

$$(15y^3 + 3y^2 + y - 2) - (\quad 7y^3 - 6y^2 + 5y - 8)$$

Change the minus sign to a plus sign.   Change all signs in the subtrahend.

$$= (15y^3 + 3y^2 + y - 2) + (-7y^3 + 6y^2 - 5y + 8)$$

$$= 8y^3 + 9y^2 - 4y + 6 \qquad \text{Combine like terms.}$$

Learning Strategy

If you are a tactile learner, collecting the like terms first will probably be more comfortable for you.

If you are a visual learner, striking through the terms as you combine them will probably be more comfortable.

EXAMPLE 7 Subtract. $(9.7x^5 - 2x^2y + xy^2 - 14.6xy - 7y^2) - (x^5 - 5.8xy^2 + 10.3xy - 15y^2)$

Solution We write an equivalent addition statement, then combine like terms.

$$(9.7x^5 - 2x^2y + xy^2 - 14.6xy - 7y^2) - (\quad x^5 - 5.8xy^2 + 10.3xy - 15y^2)$$

Change the minus sign to a plus sign.   Change all signs in the subtrahend.

$$= (9.7x^5 - 2x^2y + xy^2 - 14.6xy - 7y^2) + (-x^5 + 5.8xy^2 - 10.3xy + 15y^2)$$

$$= 8.7x^5 - 2x^2y + 6.8xy^2 - 24.9xy + 8y^2$$

Note: *The* $-2x^2y$ *term had no like term, so we rewrote it in the final expression.*

YOUR TURN Subtract.

a. $(8t^4 + 5t^2 + 9t + 7) - (2t^4 + 5t^2 + t + 4)$

b. $(12.5x^4 - 15x^3 + 2x^2 - 9) - (13.1x^4 + 9x^3 - 6x^2 - 14)$

c. $(7a^5 - a^3b^2 + 9a^2b^2 - 2ab^2 + b - 18) - (2a^5 + a^3b^2 - ab^2 + 6b - 15)$

ANSWERS

a. $6t^4 + 8t + 3$

b. $-0.6x^4 - 24x^3 + 8x^2 + 5$

c. $5a^5 - 2a^3b^2 + 9a^2b^2 - ab^2 - 5b - 3$

5.3 Exercises

For
Extra
Help

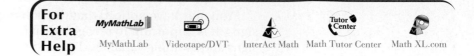

MyMathLab   Videotape/DVT   InterAct Math   Math Tutor Center   Math XL.com

MyMathLab
Tutor Center

1. How do we add two polynomials?

2. What is the function of parentheses in the addition problem $(3x + 9) + (4x - 2)$?

3. How do we find the additive inverse of a polynomial?

4. How do we subtract two polynomials?

For Exercises 5–24, add and write the resulting polynomial in descending order.

5. $(3x + 2) + (5x - 1)$

6. $(5y + 4) + (3y + 1)$

7. $(8y + 7) + (2y + 5)$

8. $(3m - 4) + (5m + 1)$

9. $(2x + 3y) + (5x - 3y)$

10. $(4a - 3b) + (6a - 2b)$

11. $(2x + 5) + (3x^2 - 4x + 7)$

12. $(5p^2 + 3p - 1) + (4p + 8)$

13. $(z^2 - 3z + 7) + (5z^2 - 8z - 9)$

14. $(2w^2 - 5w - 1) + (4w^2 - 5w + 1)$

15. $(3r^2 - 2r + 10) + (2r^2 - 5r - 11)$

16. $(7m^2 + 8m - 1) + (-4m^2 - 3m - 2)$

17. $(4y^2 - 8y + 1) + (5y^2 + 8y + 2)$

18. $(5k^2 - k - 1) + (4k^2 + k + 7)$

19. $(9a^3 + 5a^2 - 3a - 1) + (-4a^3 - 2a^2 + 6a - 2)$

20. $(4u^3 - 6u^2 + u + 11) + (-5u^3 - 3u^2 + u - 5)$

21. $(7p^3 - 9p^2 + 5p - 1) + (-4p^3 + 8p^2 + 2p + 10)$

22. $(12r^4 - 5r^2 + 8r - 15) + (-7r^4 + 3r^3 + 2r - 9)$

23. $(-5w^4 - 3w^3 - 8w^2 + w - 14) + (-3w^4 + 6w^3 + w^2 + 12w + 5)$

24. $(-r^4 + 3r^2 - 12r - 14) + (5r^4 - 2r^3 - 3r^2 + 10r - 5)$

For Exercises 25–30, add.

25. $(a^3b^2 - 6ab^2 - 6a^2b + 5ab + 5b^2 - 6) + (-4a^3b^2 + 4ab^2 - 2a^2b - ab - 2b^2 - 2)$

26. $(x^2y^2 + 8xy^2 - 12x^2y - 4xy + 7y^2 - 9) + (-3x^2y^2 + 2xy^2 + 4x^2y - xy + 6y^2 + 4)$

27. $(-2u^4 + 6uv^3 - 8u^2v^3 + 7u^3v - u + v^2 + 9) + (-5u^4 - 6uv^3 + 9u^2v^3 + 8u - 14v^2 - 12)$

28. $(-13a^6 + a^3b^2 + 3ab^3 - 8b^3) + (11a^6 - 3a^2b^3 - 4ab^3 + 10ab^2 + 4b^3)$

29. $(-8mnp - 6m^2n^2 - 13mn^2p + 14m^2n - 12n + 6) + (-4nmp - 10m^2n^2 + mn^2p - 19n + 3)$

30. $(14abc - 2a^2b^2 - ab^2c + 4b - 9c - 3) + (-15abc + 5a^2b^2 - 13ab^2c + 7a - 12b + 6)$

For Exercises 31–34, write an expression for the perimeter in simplest form.

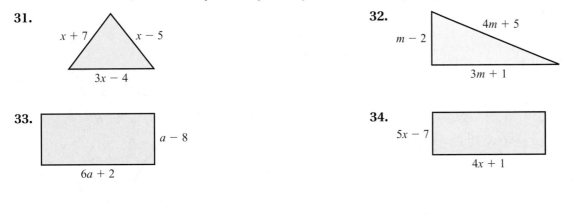

31.

$x + 7$ $x - 5$

$3x - 4$

32.

$4m + 5$

$m - 2$

$3m + 1$

33.

$a - 8$

$6a + 2$

34.

$5x - 7$

$4x + 1$

For Exercises 35–52, subtract and write the resulting polynomial in descending order.

35. $(6x + 3) - (2x + 1)$

36. $(5m + 8) - (2m + 3)$

37. $(18a^2 + 3) - (5a^2 + 4)$

38. $(12b^2 + 4b) - (5b^2 + b)$

39. $(2x^2 - 3x + 4) - (6x + 2)$

40. $(7y^2 - 3y + 2) - (8y + 5)$

41. $(6z^2 - 3z + 1) - (4z^2 + 4z - 8)$

42. $(5y^2 - 3y - 10) - (5y^2 + 2y - 10)$

43. $(8a^2 + 11a - 12) - (-4a^2 + a + 3)$

44. $(y^2 + 3y + 6) - (-5y^2 + 3y - 8)$

45. $(6u^3 + 3u^2 - 8u + 5) - (7u^3 + 5u^2 - 2u - 3)$

46. $(-8t^3 + 3t^2 - 7t - 9) - (-10t^3 - 5t^2 + 3t - 11)$

47. $(-w^5 + 3w^4 + 8w^2 - w - 1) - (10w^5 + 3w^2 - w - 3)$

48. $(-s^4 + 7s^3 + 6s^2 - 13s - 11) - (3s^4 - 2s^2 - s + 5)$

49. $(-5p^4 + 8p^3 - 9p^2 + 3p + 1) - (6p^4 + 5p^3 + 4p^2 - 2p + 7)$

50. $(-6r^5 - 9r^4 + 5r^3 - 2r^2 + 8) - (5r^5 + 2r^4 + 7r^3 - 8r^2 - 9)$

51. $(16v^4 + 21v^2 + 4v - 8) - (5v^4 - 3v^3 - 2v^2 + 9v + 1)$

52. $(12q^4 + 5q^3 - 2q + 15) - (8q^4 - 14q^3 - 3q^2 + 15q + 8)$

For Exercises 53–58, subtract.

53. $(x^2 + xy + y^2) - (x^2 - xy + y^2)$

54. $(a^2 - 4ab + b^2) - (a^2 + 4ab + b^2)$

55. $(4xy + 5xz - 6yz) - (10xy - xz - 8yz)$

56. $(4ab - 5bc + 6ac) - (10ab - bc - ac)$

57. $(18p^3q^2 - p^3q^3 + 6pq^2 - 4pq + 16q^2 - 4) - (-p^3q^2 - p^3q^3 + 4pq^2 - 3pq - 6)$

58. $(15a^3y^4 + a^2y^2 + 5ay^3 - 7ay + 12y^2 - 11) - (-a^3y^4 + a^2y^2 + 2ay^3 - ay + 9)$

For Exercises 59–62, solve. (Hint: Profit = Revenue − Cost)

59. The polynomial $54.95x + 92.20y + 25.35z$ describes the revenue a company generates from the sale of three different portable CD players. The expression $22.85x + 56.75y + 19.34z$ describes the cost of producing each of the CD players. Write a polynomial in simplest form that describes the company's net profit.

60. The polynomial $27.50x + 14.70y + 42.38z$ describes the revenue a company generates from the sale of three printer cartridges. The expression $12.75x + 5.25y + 27.42z$ describes the cost of producing each of the cartridges. Write a polynomial in simplest form that describes the company's net profit.

61. The polynomial $6.50a + 3.38b + 25.00c$ describes the revenue a department store generates from the sale of three different candles. The expression $2.28a + 1.75b + 12.87c$ describes the cost the store pays to sell each of the candles. Write a polynomial in simplest form that describes the store's net profit.

62. The polynomial $10.55m + 14.75n + 27.50p$ describes the revenue a pet store generates from the sale of three different litter boxes. The expression $5.73m + 8.26n + 15.22p$ describes the cost the store pays to sell each of the products. Write a polynomial in simplest form that describes the store's net profit.

For Exercises 63–64, find and explain the mistake, then work the problem correctly.

63. $(3x^2 - 5x + 10) - (7x^2 - 3x + 5) = (3x^2 - 5x + 10) + (-7x^2 - 3x + 5)$
$$= 3x^2 - 5x + 10 - 7x^2 - 3x + 5$$
$$= -4x^2 - 8x + 15$$

64. $(7y^2 - 3y + 1) - (y^2 + 4y - 1) = (-7y^2 + 3y - 1) + (-y^2 - 4y + 1)$
$$= -7y^2 + 3y - 1 - y^2 - 4y + 1$$
$$= -8y^2 - y$$

Collaborative Exercises ⎛BUILDING FURNITURE AND PROFITS⎞

Suppose your group operates a furniture manufacturing company that produces two different sizes of coffee table. The larger coffee table sells for $350 each and the smaller one for $280 each.

1. Write a monomial expression for the revenue if x number of large tables are sold per month.

2. Write a monomial expression for the revenue if y number of small tables are sold per month.

3. Write a polynomial expression that describes the total revenue generated from the sale of the two tables per month.

Suppose that it costs $225 to produce each large table and $135 to produce each small table.

4. Write a monomial expression for the cost of producing x number of large tables per month.

5. Write a monomial expression for the cost of producing y number of small tables per month.

6. Your company also spends $18,500 per month to pay for the lease, utilities, and salaries. Write a polynomial expression that describes the total cost involved in production.

7. Using your polynomials for revenue and cost, write a polynomial in simplest form for the profit if x number of large tables are produced and sold and y number of small tables are produced and sold.

8. What does each coefficient in the polynomial for profit indicate?

9. Suppose in one month, 200 large tables are produced and sold and 400 small tables are produced and sold. Find the revenue, cost, and profit for the month.

(continued)

10. Suppose in one month, 56 large tables are produced and sold and 65 small tables are produced and sold. Find the revenue, cost, and profit for the month. Explain the meaning of your answer.

11. Suppose in one month, only large tables are produced and sold. How many would have to be produced and sold to break even (profit is 0)?

12. Suppose only small tables are produced and sold in one month. How many would have to be produced and sold to break even?

REVIEW EXERCISES

For Exercises 1–4, evaluate.

[1.4] **1.** $2(-341)$ $\begin{bmatrix} \mathbf{1.5} \\ \mathbf{5.1} \end{bmatrix}$ **2.** $(-2)^5$ [5.1] **3.** 10^{-4} [1.5] **4.** $(2^3)(2^4)$

[5.1] **5.** Write 5.89×10^7 in standard form.

[2.1] **6.** The surface of the water in a rectangular swimming pool measures 20 meters by 15 meters. What is the surface area of the water?

5.4 Exponent Rules and Multiplying Monomials

OBJECTIVES
1. Multiply monomials.
2. Multiply numbers in scientific notation.
3. Simplify a monomial raised to a power.

In this section, we will learn how to multiply monomials, multiply numbers in scientific notation, and simplify a monomial raised to a power. These simplifications require some new rules for exponents.

OBJECTIVE 1. Multiply monomials. Because monomials contain variables in exponential form, in order to multiply monomials, we must develop a rule for multiplying exponential forms.

Multiplying Exponential Forms

Consider $2^3 \cdot 2^4$, which is a product of exponential forms. To simplify $2^3 \cdot 2^4$, we could follow the order of operations and evaluate the exponential forms first, then multiply.

$$2^3 \cdot 2^4 = 8 \cdot 16 = 128$$

However, there is an alternative. We can write the result in exponential form by first writing 2^3 and 2^4 in their factored forms.

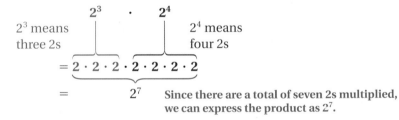

Notice that $2^7 = 128$, which agrees with our earlier calculation. Also notice that in the alternative method the resulting exponent is the sum of the original exponents.

$$2^3 \cdot 2^4 = 2^{3+4} = 2^7$$

Conclusion To multiply exponential forms that have the same base, we can add the exponents and keep the same base.

This suggests the following rule.

RULE Product Rule for Exponents

If a is a real number and m and n are integers, then $a^m \cdot a^n = a^{m+n}$.

We apply this rule when multiplying monomials.

EXAMPLE 1 Multiply. $(x^5)(x^3)$

Solution Because the bases are the same, we can add the exponents and keep the same base.

$$(x^5)(x^3) = x^{5+3} = x^8$$

YOUR TURN Multiply.

a. $y^2 \cdot y^5$ **b.** $(c^7)(c^3)$ **c.** $t^5(t^3)$

Multiplying Monomials

Now let's consider multiplying monomials that have coefficients like $(4x^3)(2x^6)$. Because monomials are products by definition, we can use the commutative property of multiplication and write

$$(4x^3)(2x^6) = 4 \cdot x^3 \cdot 2 \cdot x^6$$
$$= 4 \cdot 2 \cdot x^3 \cdot x^6$$
$$= 8 \cdot x^{3+6}$$
$$= 8x^9$$

We multiply the coefficients and use the product rule for exponents to simplify $x^3 \cdot x^6$.

What if the monomial factors have some variables that are different, as in $(5x^2)(7xy)$? If we expand out the monomials in factored form and use the commutative property of multiplication, we have

$$(5x^2)(7xy) = 5 \cdot x^2 \cdot 7 \cdot x \cdot y$$
$$= 5 \cdot 7 \cdot x^2 \cdot x \cdot y$$
$$= 35 \cdot x^{2+1} \cdot y$$
$$= 35x^3y$$

Notice that the unlike variable base, y, is simply written unchanged in the result. Our examples suggest the following procedure.

PROCEDURE Multiplying Monomials

To multiply monomials:

1. Multiply coefficients.
2. Add the exponents of the like bases.
3. Write any unlike variable bases unchanged in the product.

EXAMPLE 2 Multiply.

a. $(9y^5)(3y^8)$

Solution $(9y^5)(3y^8) = 9 \cdot 3y^{5+8}$ Multiply the coefficients and add the exponents of the like bases.

$\qquad\qquad\qquad\qquad = 27y^{13}$ Simplify the exponents.

b. $\left(-\dfrac{5}{6}a^2bc^3\right)\left(\dfrac{2}{3}a^5b^4\right)$

Solution $\left(-\dfrac{5}{6}a^2bc^3\right)\left(\dfrac{2}{3}a^5b^4\right) = -\dfrac{5}{\overset{}{\underset{3}{6}}} \cdot \dfrac{\overset{1}{2}}{3}a^{2+5}b^{1+4}c^3$ Multiply the coefficients, add the exponents of the like bases, and write the unlike variable base c unchanged in the product.

Note: *Remember that a variable or number with no apparent exponent has an understood exponent of 1.*

$\qquad\qquad\qquad\qquad = -\dfrac{5}{9}a^7b^5c^3$ Simplify.

$-\dfrac{5}{6}a^2bc^3 = -\dfrac{5}{6}a^2b^1c^3$

YOUR TURN Multiply.

a. $(9n^4)(5n^8)$

b. $-2.5x^2y^3(-6xy^2z^3)$

Connection Multiplying distance measurements in area and volume is like multiplying monomials.

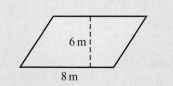

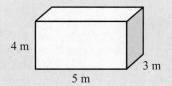

The area of the parallelogram is

$A = bh$
$A = (8\,\text{m})(6\,\text{m})$
$A = 48\,\text{m}^2$

The volume of the box is

$V = lwh$
$V = (5\,\text{m})(3\,\text{m})(4\,\text{m})$
$V = 60\,\text{m}^3$

The area calculation is similar to multiplying two monomials with m as the variable.

Multiply: $(8m)(6m) = 48m^2$

The volume calculation is similar to multiplying three monomials with m as the variable.

Multiply: $(5m)(3m)(4m) = 60m^3$

ANSWERS

a. $45n^{12}$ **b.** $15x^3y^5z^3$

EXAMPLE 3 Write an expression in simplest form for the volume of the box shown.

Understand We are given a box with side lengths that are monomial expressions.

Plan The volume of a box is found by multiplying the length, width, and height.

Execute
$$V = lwh$$
$$V = (7x)(x)(2x)$$
$$V = 14x^3$$

Answer The expression for volume is $14x^3$.

Check Since $(7x)(x)(2x) = 14x^3$ is an identity, we can evaluate $(7x)$, (x), $(2x)$, and $14x^3$ using any chosen value for x and get the same result. Let's choose a value for x such as 3 and try it.

$$(7x)(x)(2x) = 14x^3$$
$$[7(3)](3)[2(3)] \stackrel{?}{=} 14(3)^3$$
$$(21)(3)(6) \stackrel{?}{=} 14(27)$$
$$378 = 378 \qquad \text{This is true.}$$

Note: *Choosing only one value does not guarantee that our simplification is correct, but it is a good indicator. Testing several values would give more conclusive assurance.*

OBJECTIVE 2. Multiply numbers in scientific notation. We multiply numbers in scientific notation using the same procedure we used to multiply monomials. If we replace a with 10 in $(4a^3)(2a^6)$, we have a product of two numbers in scientific notation $(4 \times 10^3)(2 \times 10^6)$. Following the same procedure for multiplying monomials, we multiply the $4 \cdot 2$ to get 8 and then add the exponents of the like bases.

Monomials:
$$(4a^3)(2a^6) = 4 \cdot 2a^{3+6}$$
$$= 8a^9$$

Scientific notation:
$$(4 \times 10^3)(2 \times 10^6) = 4 \times 2 \times 10^{3+6}$$
$$= 8 \times 10^9$$

EXAMPLE 4 Multiply $(2.3 \times 10^5)(7.5 \times 10^6)$. Write the answer in scientific notation.

Solution Multiply 2.3 and 7.5, then add the exponents for base 10s.

$$(2.3 \times 10^5)(7.5 \times 10^6) = 2.3 \times 7.5 \times 10^{5+6}$$
$$= 17.25 \times 10^{11}$$
$$= 1.725 \times 10^{12}$$

Note: *The product 17.25×10^{11} is not in scientific notation because the absolute value of 17.25 is not less than 10. To fix this problem, we must move the decimal point one place to the left and account for this move by increasing the exponent by 1. If you are unconvinced that this is correct, consider the fact that 17.25×10^{11} is equal to 1,725,000,000,000. If we now write that standard form number in scientific notation, the proper position for the decimal point is between the 1 and the 7 digits, which means we must account for 12 places.*

OBJECTIVE 3. Simplify a monomial raised to a power. We now discuss how to simplify an expression like $(3x^2)^4$, which is a monomial raised to a power. In order to simplify such an expression, we need to discuss two new rules of exponents: (1) raising a power to a power and (2) raising a product to a power.

Raising a Power to a Power

Consider a power raised to a power, like $(2^3)^2$. To calculate $(2^3)^2$, we could follow the order of operations and evaluate the exponential form 2^3 within the parentheses first, then square the result.

$$(2^3)^2 = (2 \cdot 2 \cdot 2)^2 = 8^2 = 64$$

However, there is an alternative. The outside exponent, 2, indicates to multiply the inside exponential form 2^3 by itself.

$$(2^3)^2 = 2^3 \cdot 2^3$$
$$= 2^{3+3}$$
$$= 2^6$$

Since this is a multiplication of exponential forms that have the same base, we can add the exponents.

Notice that $2^6 = 64$. Also, in comparing $(2^3)^2$ with 2^6, notice that the exponent in 2^6 is the product of the original exponents.

$$(2^3)^2 = 2^{3 \cdot 2} = 2^6$$

Conclusion If an exponential form is raised to a power, we can multiply the exponents and keep the same base.

RULE **A Power Raised to a Power**

If a is a real number and m and n are integers, then $(a^m)^n = a^{mn}$.

Now let's consider a product raised to a power, like $(3 \cdot 4)^2$. We could write the factored form twice and then use the commutative property to rearrange the like bases.

$$(3 \cdot 4)^2 = (3 \cdot 4)(3 \cdot 4)$$
$$= 3 \cdot 3 \cdot 4 \cdot 4$$
$$= 3^2 \cdot 4^2$$

Since there are two factors of 3 and two factors of 4, we can write each of these in exponential form.

Conclusion If a product is raised to a power, then we can evaluate the factors raised to that power.

ANSWER

1.176×10^{14}

RULE **Raising a Product to a Power**

If a and b are real numbers and n is an integer, then $(ab)^n = a^n b^n$.

Raising a Monomial to a Power

We apply these exponent rules to simplify an expression like $(3x^2)^4$, which is a monomial raised to a power. Because the monomial $3x^2$ expresses a product, we can use the rule for raising a product to a power and raise the factors 3 and x^2 to the 4th power.

$$(3x^2)^4 = (3)^4(x^2)^4 \qquad \text{To simplify } (x^2)^4, \text{ we use the rule for}$$
$$= 81x^{2 \cdot 4} \qquad \text{raising a power to a power.}$$
$$= 81x^8$$

Notice that the outside exponent is distributed to the coefficient and the variable(s) within the monomial. This suggests the following procedure.

PROCEDURE **Simplifying a Monomial Raised to a Power**

To simplify a monomial raised to a power:

1. Evaluate the coefficient raised to that power.
2. Multiply each variable's exponent by the power.

EXAMPLE 5 Simplify.

a. $(5a^6)^3$

Solution $(5a^6)^3 = (5)^3 a^{6 \cdot 3}$ Write the coefficient, 5, raised to the 3rd power, and multiply the variable's exponent by 3.

$$= 125a^{18} \qquad \text{Simplify.}$$

b. $\left(-\dfrac{2}{3}xy^3z^5\right)^4$

Note: *Recall that $x = x^1$.*

Solution $\left(-\dfrac{2}{3}xy^3z^5\right)^4 = \left(-\dfrac{2}{3}\right)^4 x^{1 \cdot 4}y^{3 \cdot 4}z^{5 \cdot 4}$ Write the coefficient, $-\dfrac{2}{3}$, raised to the 4th power, and multiply each variable's exponent by 4.

$$= \frac{16}{81}x^4y^{12}z^{20} \qquad \text{Simplify.}$$

c. $(-0.4mn^6)^3$

Solution $(-0.4mn^6)^3 = (-0.4)^3 m^{1 \cdot 3}n^{6 \cdot 3}$

$$= -0.064m^3n^{18}$$

YOUR TURN Simplify.

a. $(2y^4)^5$ **b.** $(-5t^2u^6)^3$ **c.** $(-0.2a^4b^5c)^6$

ANSWERS

a. $32y^{20}$ **b.** $-125t^6u^{18}$
c. $0.000064a^{24}b^{30}c^6$

Now let's consider expressions that require us to use all of these exponent rules.

EXAMPLE 6 Simplify.

a. $(3x^2)(4x^5)^3$

Solution Since the order of operations is to simplify exponents before multiplying, we will simplify $(4x^5)^3$ first, then multiply the result by $3x^2$.

$$
\begin{aligned}
(3x^2)(4x^5)^3 &= (3x^2)(4^3x^{5\cdot3}) \\
&= (3x^2)(64x^{15}) \quad \textbf{Simplify.} \\
&= 3 \cdot 64x^{2+15} \quad \textbf{Multiply coefficients and add} \\
&= 192x^{17} \quad\quad\quad \textbf{exponents of like variables.}
\end{aligned}
$$

b. $(-0.2a)^3(4ab)(1.5a^3c)^2$

Solution We follow the order of operations and simplify the monomials raised to a power first, then multiply the monomials.

$$
\begin{aligned}
(-0.2a)^3(4ab)(1.5a^3c)^2 &= ((-0.2)^3a^{1\cdot3})(4ab)((1.5)^2a^{3\cdot2}c^{1\cdot2}) \\
&= (-0.008a^3)(4ab)(2.25a^6c^2) \quad \textbf{Simplify.} \\
&= -0.008 \cdot 4 \cdot 2.25a^{3+1+6}bc^2 \quad \textbf{Multiply coefficients} \\
&= -0.072a^{10}bc^2 \quad\quad\quad\quad\quad\quad \textbf{and add exponents of} \\
&\quad\quad\quad\quad\quad\quad\quad\quad\quad\quad\quad\quad\quad\quad\quad \textbf{like variables.}
\end{aligned}
$$

YOUR TURN Simplify.

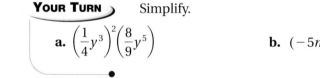

a. $\left(\dfrac{1}{4}y^3\right)^2\left(\dfrac{8}{9}y^5\right)$

b. $(-5mn^4)(-2mp^2)^3(1.5m^2n)$

ANSWERS

a. $\dfrac{1}{18}y^{11}$ **b.** $60m^6n^5p^6$

5.4 Exercises

For Extra Help

MyMathLab MyMathLab   Videotape/DVT   InterAct Math   Math Tutor Center   Math XL.com

1. What rule can be applied when multiplying two exponential forms that have the same base?

2. Explain why $2^3 \cdot 5^2 \neq 10^5$.

3. Explain how to multiply monomials.

4. In multiplying the powers of 10 in $(8 \times 10^4)(6 \times 10^5)$, we add the exponents to get 10^9. However, the correct answer in scientific notation is 4.8×10^{10}. Why is the exponent 10 instead of 9?

5. How do we simplify $(x^3)^8$?

6. Explain how to simplify $(3x^4)^2$.

For Exercises 7–32, multiply.

7. $a^3 \cdot a^4$

8. $b^2 \cdot b^3$

9. $2^3 \cdot 2^{10}$

10. $5^4 \cdot 5^3$

11. $a^3 \cdot a^2 b$

12. $y^2 \cdot y^3 z$

13. $2x \cdot 3x^2$

14. $5b \cdot 2b^3$

15. $(-2mn)(4m^2 n)$

16. $(-3uv^2)(2u^3 v)$

17. $\left(\dfrac{5}{8}st\right)\left(-\dfrac{2}{7}s^2 t^7\right)$

18. $\left(\dfrac{3}{8}x^2 y\right)\left(\dfrac{4}{9}xy^2\right)$

19. $(2.3ab^2 c)(1.2a^2 b^3 c^5)$

20. $(-3.1mn)(2.4m^2 n)$

21. $(r^2 st^2)(2r^2 s^3 t)(-3rs^6 t^4)$

22. $(-2jkl)(-5j^3 kl^4)(7jk^4 l^2)$

23. $3xz^2(-4x^3 tz^2)(tz^5)$

24. $-5mn^4(-m^3 n^9 p)(4m^3 np^5)$

25. $(5xyz)(9x^2 y^4)(2x^3 z^2)$

26. $(6hj^3 k)(7h^5 k)(hkj)$

27. $-5a^2\left(-\dfrac{1}{15}abc\right)\left(-\dfrac{3}{4}a^3 b^2 c\right)$

28. $\left(-\dfrac{1}{4}l^2 m\right)\left(\dfrac{5}{6}lm^3 n\right)\left(-\dfrac{2}{3}l^5\right)$

29. $(0.4wxy)(w^2 y^2)(2.5x^2 y^2)$

30. $(0.4u^2 v)(-3.2u^2 v^3 w^3)(0.2vw^9)$

31. $(5q^2 r^4 s^9)(-q^3 rs^2)(-r^5 s^{10})$

32. $(a^3 b^3 c^3)(-6a^3 b^3 c^3)(-a^4 c^2)$

For Exercises 33–36, find and explain the mistake(s), then work the problem correctly.

33. $7x^3 \cdot 5x^4y = 35x^{12}y$

34. $-3m^2n \cdot 10m^5n = -30m^7n$

35. $(9x^3y^2z)(6xy^4) = 54x^3y^6z$

36. $(a^3b^2c)(ab^5) = a^4b^7$

For Exercises 37 and 38, write an expression in simplest form for the area of the shape.

37.

w
$5w$

38.
h
$7h$

For Exercises 39 and 40, write an expression in simplest form for volume.

39.

$0.5w$
$9w$
w

40.
h
$0.1h$
$0.2h$

41. The height of a trapezoid is half the base. The top side of the trapezoid is one-fourth the base.

    a. If b represents the length of the base, write expressions for the height and the length of the top side of the trapezoid.

    b. Using your expressions from part a, write an expression in simplest form for the area of the trapezoid.

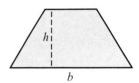
h
b

42. The height of a box is three times the width. The length is five times the width.

    a. If w represents the width, write expressions for the height and the length of the box.

    b. Using your expressions from part a, write an expression in simplest form for the volume of the box.

    c. Using your expressions from part a, write an expression in simplest form for the surface area of the box.

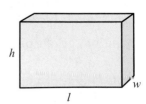
h
l
w

For Exercises 43–48, multiply and write your answer in scientific notation.

43. $(2 \times 10^5)(6 \times 10^3)$

44. $(9 \times 10^6)(5 \times 10^5)$

45. $(3.2 \times 10^5)(7.5 \times 10^8)$

46. $(4.7 \times 10^{10})(8.4 \times 10^7)$

47. $(9.5 \times 10^7)(2.63 \times 10^{11})$

48. $(8.52 \times 10^{12})(3.4 \times 10^9)$

49. Tennessee is shaped like a parallelogram with a base of approximately 350 miles and a height of approximately 120 miles. Calculate the area of Tennessee and write the answer in scientific notation. (Use $A = bh$.)

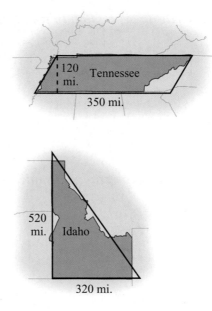

120 mi.   Tennessee

350 mi.

50. Idaho is shaped like a triangle with a base of approximately 320 miles and a height of approximately 520 miles. Calculate the area of Idaho and write the answer in scientific notation.

$$\left(\text{Use } A = \frac{1}{2}bh. \right)$$

520 mi.   Idaho

320 mi.

For Exercises 51 and 52, use the formula for the circumference of a circle: $C = 2\pi r$. (Use 3.14 to approximate π.)

51. The path of Earth's orbit around the Sun is approximately a circle with a radius of about 9.3×10^7 miles. What is the circumference of Earth's orbit?

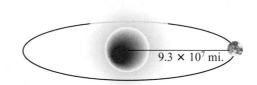

9.3×10^7 mi.

Not to scale.

Of Interest

Earth travels the distance that you calculated in Exercise 51 in one year, which translates to about 18.5 miles every second.

52. The radius of Earth is approximately 6.37×10^6 meters. Calculate the circumference of the equator.

For Exercises 53 and 54, use the following information. The energy, in joules, of a single photon of light can be determined by $E = hv$, where h is the constant 6.626×10^{-34} joule-seconds and v is the frequency of the light in hertz.

53. Find the energy of a photon of red light with a frequency of 4.5×10^{14} Hz.

54. Find the energy of a photon of violet light with a frequency of 8.5×10^{14} Hz.

For Exercises 55–86, simplify.

55. $(x^2)^3$

56. $(y^3)^5$

57. $(3^2)^4$

58. $(2^3)^2$

59. $(-h^2)^4$

60. $(-j^3)^2$

61. $(-y^2)^3$

62. $(-x^4)^5$

63. $(xy)^3$

64. $(ab)^3$

65. $(3x^2)^3$

66. $(4x^4)^2$

67. $(-2x^3y^2)^3$

68. $(-5a^4b^2)^2$

69. $\left(\frac{1}{4}m^2n\right)^3$

70. $\left(\frac{5}{6}u^3v\right)^2$

71. $(-0.2x^2y^5z)^3$

72. $(-0.3r^4st^3)^4$

73. $(-2p^4q^4r^2)^6$

74. $(-6u^2v^4w^2)^3$

75. $(2x^3)(3x^2)^2$

76. $(-3y^2)(2y^2)^3$

77. $(rs^2)(r^2s)^2$

78. $(m^2n)^3(mn)$

79. $(2ab)^2(5a^2b)^2$

80. $(4xy^3)^2(2x^2y)^2$

81. $\left(\frac{1}{4}a^2b^3c\right)^2\left(\frac{1}{4}a^2b^3c\right)^3$

82. $\left(-\frac{1}{2}s^2tu^4\right)^2\left(-\frac{1}{2}s^2tu^4\right)^5$

83. $(0.6u)^2(u^2v)(-3uv)^2$

84. $(-0.2w)^4(-w^2z^3)^3(-4w^4z)^2$

85. $(6u)^2(u^2v)(-3uv)^2$

86. $(3x)^2(xy^3)(2x^3yz)^3$

For Exercises 87–90, find and explain the mistake(s), then work the problem correctly.

87. $(5x^3)^2 = 25x^5$

88. $(-4mn^5)^3 = -12m^3n^{15}$

89. $(2x^4y)^4 = 8x^8y$

90. $(-2ab^5)^3 = -8ab^{15}$

For Exercises 91 and 92, use the formula for the volume of a sphere: $V = \frac{4}{3}\pi r^3$.
(Use 3.14 to approximate π.)

91. The radius of the Sun is approximately 6.96×10^8 meters. Calculate the volume of the Sun, and write the answer using scientific notation.

92. The radius of Earth is approximately 6.37×10^6 meters. Calculate the volume of Earth, and write the answer using scientific notation.

Earth   Sun

For Exercises 93 and 94, use the following information. Albert Einstein discovered that if a mass of m kilograms is converted to pure energy (as in a nuclear reaction), the amount of energy, in joules, that is released is described by $E = mc^2$, where c represents a constant speed of light, which is approximately 3×10^8 meters per second.

93. Suppose 2.4×10^{-10} kilograms of uranium is converted to energy in a nuclear reaction. How much energy is released?

94. Suppose 4.5×10^{-8} kilograms of plutonium is converted to energy in a nuclear reaction. How much energy is released?

PUZZLE PROBLEM

a. *If there are about 10^{11} stars in a galaxy and about 10^{11} galaxies in the visible universe, then about how many stars are in the visible universe?*

b. *Suppose one star out of every billion stars has a solar system with one planet that could support life. How many stars in the visible universe would have a solar system with one planet that could support life? (Hint: One out of every billion is one billionth, or 10^{-9}.)*

REVIEW EXERCISES

[5.3]   1. Add: $(3x + 4y - z) + (-7x - 4y - 3z)$

$\begin{bmatrix} 1.5 \\ 5.1 \end{bmatrix}$ 2. Evaluate: -2^4

[1.3]   3. Add: $(-4) + (-57)$

[3.3]   4. Find three consecutive integers whose sum is 15.

[5.2]   5. What is the degree of the monomial $-2r^5s$?

[3.2]   6. A business report indicates that revenues are $541,200 for the quarter, which is 12% less than the previous quarter. What was the revenue in the previous quarter?

5.5 Multiplying Polynomials; Special Products

OBJECTIVES
1. Multiply a polynomial by a monomial.
2. Multiply binomials.
3. Multiply polynomials.
4. Determine the product when given special polynomial factors.

In this section, we will use the rules for multiplying monomials to help us multiply multiple-term polynomials.

OBJECTIVE 1. Multiply a polynomial by a monomial. To multiply polynomials, we use the distributive property, which we introduced in Section 1.7 and used in solving linear equations and inequalities in Chapter 2. For example, we have applied the distributive property in situations like $2(3x + 4)$.

$$2(3x + 4) = 2 \cdot 3x + 2 \cdot 4$$
$$= 6x + 8$$

Notice that $2(3x + 4)$ is a product of a monomial 2 and a binomial $3x + 4$. This suggests the following procedure:

PROCEDURE Multiplying a Polynomial by a Monomial

To multiply a polynomial by a monomial, use the distributive property to multiply each term in the polynomial by the monomial.

Connection Multiplying a polynomial by a monomial is like multiplying a multidigit number by a single-digit number. Consider $2 \cdot 34$:

$$
\begin{array}{r}
34 \\
\times\ 2 \\
\hline
68
\end{array}
$$

Notice that in the numerical multiplication, we distribute the 2 to each digit in the 34. In fact, if we expand 34 by expressing it as $30 + 4$, the problem takes the same form as $2(3x + 4)$, where $x = 10$:

$$2(30 + 4) = 2 \cdot 30 + 2 \cdot 4 \qquad\qquad 2(3x + 4) = 2 \cdot 3x + 2 \cdot 4$$
$$= 60 + 8 \qquad\qquad\qquad\qquad\quad = 6x + 8$$
$$= 68$$

Now let's multiply a polynomial by a monomial that contains variables.

EXAMPLE 1 Multiply.

a. $2y(3y^2 + 4y + 1)$

Solution $2y \quad (3y^2 + 4y + 1)$ **Multiply each term in the polynomial by 2y.**

$2y \cdot 3y^2$

$2y \cdot 4y$

$2y \cdot 1$

$= 2y \cdot 3y^2 + \mathbf{2y} \cdot \mathbf{4y} + 2y \cdot 1$

$= 6y^3 + 8y^2 + 2y$

Note: *When multiplying the individual terms by the monomial, we multiply the coefficients and add the exponents of the like bases.*

b. $-7x^2(5x^2 + 3x - 4)$

Solution $-7x^2 \quad (5x^2 + 3x - 4)$ **Multiply each term in the polynomial by $-7x^2$.**

$-7x^2 \cdot 5x^2$

$-7x^2 \cdot 3x$

$-7x^2 \cdot (-4)$

$= -7x^2 \cdot 5x^2 - 7x^2 \cdot 3x - 7x^2 \cdot (-4)$

$= -35x^4 - 21x^3 + 28x^2$

Note: *When we multiply a poly-nomial by a negative monomial, the signs of the resulting polynomial will be the opposite of the signs in the original polynomial.*

$$-7x^2(\ 5x^2 \ + \ 3x \ - \ 4)$$

$$= -35x^4 - 21x^3 + 28x^2$$

c. $-0.3t^2u(6t^3u + tu^2 - 3t^3 + 0.2uv)$

Solution $-0.3t^2u \quad (\ 6t^3u \ + \ tu^2 \ - 3t^3 \ + \ 0.2uv)$ **Multiply each term in the polynomial by $-0.3t^2u$.**

$-0.3t^2u \cdot 6t^3u$

$-0.3t^2u \cdot tu^2$

$-0.3t^2u \cdot (-3t^3)$

$-0.3t^2u \cdot 0.2uv$

$= -0.3t^2u \cdot 6t^3u - 0.3t^2u \cdot tu^2 - 0.3t^2u \cdot (-3t^3) - 0.3t^2u \cdot 0.2uv$

$= \quad -1.8t^5u^2 \quad - \quad 0.3t^3u^3 \quad + \quad 0.9t^5u \quad - \quad 0.06t^2u^2v$

Note: *When multiplying multivariable terms, it is helpful to multiply the coefficients first, then the variables in alphabetical order.*

For $-0.3t^2u \cdot 6t^3u$, think: $-0.3 \cdot 6 = -1.8$; $t^2 \cdot t^3 = t^5$; $u \cdot u = u^2$

Result: $-1.8t^5u^2$

ANSWERS

a. $36x^5 - 28x^4 + 12x^3$

b. $-15x^5yz^2 + 3x^3y^4z^3 - 24x^4yz^2$

YOUR TURN Multiply.

a. $4x^3(9x^2 - 7x + 3)$ **b.** $-3x^3yz(5x^2z - y^3z^2 + 8xz)$

Understanding Binomial Multiplication

Now that we have seen how to multiply a polynomial by a monomial, we are ready to multiply two binomials, which is like multiplying a pair of two-digit numbers. For example, consider $(12)(13)$ and compare this with $(x + 2)(x + 3)$.

$$
\begin{array}{r}
12 \\
\times\ 13 \\
\hline
36 \\
+\ 12 \\
\hline
156
\end{array}
$$

Note: *We think to ourselves "3 times 2 is 6, then 3 times 1 is 3," which creates the 36. We then move to the 1 in the tens place of the 13 and do the same thing. Since this 1 digit is in the tens place, it really means 10, so when we multiply this 10 times 12 it makes 120. We usually omit writing the 0 in the ones place and write 12 in the next two places.*

Again, the distributive property is the governing principle. We multiply each digit in one number by each digit in the other number and shift underneath as we move to each new place. The same process applies to the binomials. We can stack them as we just did. Notice, however, that we will only stack this once in order to make the connection to numeric multiplication. As the polynomials get more complex, the stacking method becomes too tedious.

$$
\begin{array}{r}
x + 2 \\
x + 3 \\
\hline
3x + 6 \\
x^2 + 2x \\
\hline
x^2 + 5x + 6
\end{array}
$$

Note: *We think "3 times 2 is 6, then 3 times x is 3x." Now move to the x and think "x times 2 is 2x, then x times x is x^2." Notice how we shifted so that the 2x and 3x line up. This is because they are like terms. It is the same as lining up the tens column when multiplying numbers. Note that the numeric result, 156, is basically the same as the algebraic result, $x^2 + 5x + 6$.*

Notice that each term in $x + 2$ is multiplied by each term in $x + 3$. In general, this is how we multiply two polynomials.

PROCEDURE   Multiplying Polynomials

To multiply two polynomials:

1. Multiply every term in the second polynomial by every term in the first polynomial.
2. Combine like terms.

Multiply Binomials

EXAMPLE 2   Multiply. $(x + 4)(x + 3)$

Solution   Multiply each term in $x + 3$ by each term in $x + 4$.

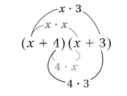

$$(x + 4)(x + 3)$$

Connection   If every term in both binomials is positive, the like terms are added and both signs in the resulting trinomial are plus signs. Noting these sign patterns will help us when factoring in Chapter 6.

$$= x \cdot x + x \cdot \mathbf{3} + 4 \cdot x + 4 \cdot 3$$
$$= x^2 + \underbrace{3x + 4x}_{} + 12$$
$$= x^2 + \quad 7x \quad + 12 \qquad \text{Combine like terms: } 3x + 4x = 7x.$$
$$= x^2 + 7x + 12$$

The word FOIL is a popular way to remember the process of multiplying two binomials. FOIL stands for **First Outer Inner Last**. We will use Example 2 to demonstrate.

First terms: $x \cdot x = x^2$
$$(x + 4)(x + 3)$$

Outer terms: $x \cdot 3 = 3x$
$$(x + 4)(x + 3)$$

$$(x + 4)(x + 3)$$
Inner terms: $4 \cdot x = 4x$

$$(x + 4)(x + 3)$$
Last terms: $4 \cdot 3 = 12$

EXAMPLE 3 Multiply. $(3x + 5)(x - 4)$

Solution Multiply each term in $x - 4$ by each term in $3x + 5$ (think FOIL).

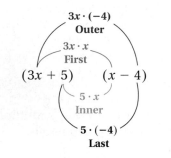

Connection Notice that $3x + 5$ has a plus sign whereas $x - 4$ has a minus sign. When multiplying binomials like these, the last term of the resulting trinomial will always be negative. The middle term could be positive or negative, depending on the coefficients of the like terms.

$$\qquad \text{First} \quad\;\; \textbf{Outer} \quad\;\; \text{Inner} \quad\;\; \text{Last}$$
$$= 3x \cdot x + 3x \cdot (-4) + 5 \cdot x + 5 \cdot (-4)$$
$$= 3x^2 \qquad\;\; - 12x \quad\;\; + 5x \qquad - 20$$
$$= 3x^2 \qquad\qquad\quad - 7x \qquad\qquad - 20 \qquad \text{Combine like terms: } -12x + 5x = -7x.$$
$$= 3x^2 - 7x - 20$$

In Example 4, we will switch the signs in the binomials from Example 3. Look at how this changes the sign of the middle term in the resulting trinomial.

EXAMPLE 4 Multiply. $(3x - 5)(x + 4)$

Solution Multiply each term in $x + 4$ by each term in $3x - 5$.

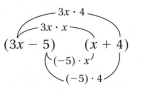

	First	Outer	Inner	Last

$= 3x \cdot x + 3x \cdot 4 + (-5) \cdot x + (-5) \cdot 4$

$= 3x^2 \qquad + 12x \qquad - 5x \qquad - 20$

$= 3x^2 \qquad\qquad + 7x \qquad - 20$ **Combine like terms:** $12x - 5x = 7x$.

$= 3x^2 + 7x - 20$

Now consider a case when the signs between terms in the binomials are both negative.

EXAMPLE 5) Multiply. $(2x - 3)(4x - 5)$

Solution Multiply each term in $4x - 5$ by each term in $2x - 3$ (think FOIL).

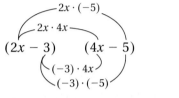

Connection When multiplying binomials like $2x - 3$ and $4x - 5$, notice the last term in the resulting trinomial is positive and the middle term is negative.

	First	Outer	Inner	Last

$= 2x \cdot 4x + 2x \cdot (-5) + (-3) \cdot 4x + (-3) \cdot (-5)$

$= 8x^2 \qquad - 10x \qquad - 12x \qquad + 15$

$= 8x^2 \qquad\qquad - 22x \qquad\qquad + 15$ **Combine like terms:** $-10x - 12x = -22x$.

$= 8x^2 - 22x + 15$

Keep in mind that Examples 2–5 do not illustrate all possible sign combinations that you could be given. The point of our discussion of signs is to note some patterns with the signs in the given binomial factors and see how they affect the resulting polynomial. Noticing these sign patterns now will help us when factoring in Chapter 6.

YOUR TURN) Multiply.

a. $(n + 6)(n - 2)$ **b.** $(2y - 5)(y - 3)$

Connection The product of two binomials can be shown in terms of geometry. Consider the rectangle shown with a length of $x + 7$ and a width of $x + 5$. We can describe its area in two ways: (1) the product of the length and width, $(x + 7)(x + 5)$; or (2) the sum of the areas of the four internal rectangles, $x^2 + 5x + 7x + 35$. Since these two approaches describe the same area, the two expressions must be equal.

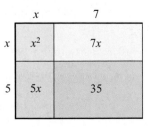

Length · width = Sum of the areas of the four internal rectangles

$(x + 7)(x + 5) = x^2 + 5x + 7x + 35$

$= x^2 + 12x + 35$ **Combine like terms**

OBJECTIVE 3. Multiply polynomials. Now consider an example of multiplication involving a larger polynomial, such as a trinomial. Remember that no matter how many terms are in the polynomials, the process is to multiply every term in the second polynomial by every term in the first polynomial.

> **EXAMPLE 6** Multiply. $(x + 2)(3x^2 + 5x - 4)$

Solution Multiply each term in $3x^2 + 5x - 4$ by each term in $x + 2$.

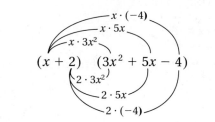

Warning: FOIL does not make sense here because there are too many terms. FOIL handles only the four terms from two binomials.

$$= x \cdot 3x^2 + x \cdot 5x + x \cdot (-4) + 2 \cdot 3x^2 + 2 \cdot 5x + 2 \cdot (-4)$$
$$= 3x^3 \quad\;\; + 5x^2 \quad - 4x \quad\;\;\;\; + 6x^2 \quad\; + 10x \quad - 8$$
$$= 3x^3 + 11x^2 + 6x - 8 \quad \text{Combine like terms.}$$

> **YOUR TURN** Multiply $(x - 3)(x^2 - 6x + 2)$.

OBJECTIVE 4. Determine the product when given special polynomial factors.

Multiplying Conjugates

In Objective 2, we discussed how the signs in the binomial factors affect the signs in the product in order to prepare for factoring in Chapter 6. To further prepare for factoring, it is helpful to note some patterns that occur when multiplying certain special polynomial factors. First, we will multiply **conjugates**.

DEFINITION *Conjugates:* Binomials that differ only in the sign separating the terms.

The following binomial pairs are conjugates:

$$x + 9 \text{ and } x - 9 \qquad 2x + 3 \text{ and } 2x - 3 \qquad -6x + 5 \text{ and } -6x - 5$$

Let's multiply the conjugates $2x + 3$ and $2x - 3$ to see what pattern emerges.

$$\begin{array}{c} \text{First} \quad \text{Outer} \quad \text{Inner} \quad \text{Last} \\ (2x + 3)(2x - 3) = 2x \cdot 2x + 2x \cdot (-3) + 3 \cdot 2x + 3 \cdot (-3) \\ = 4x^2 \qquad - 6x \quad + \quad 6x \qquad - 9 \\ = 4x^2 + 0 - 9 \quad \text{Combine like terms: } -6x + 6x = 0. \\ = 4x^2 - 9 \end{array}$$

◄ **Note:** *The sum of the like terms is 0, so the resulting binomial is the square of the first term minus the square of the last term.*

ANSWER

$x^3 - 9x^2 + 20x - 6$

When multiplying conjugates, the like terms will always be additive inverses, so their sum will always be 0. This suggests the following rule.

RULE **Multiplying Conjugates**

If a and b are real numbers, variables, or expressions, then
$(a + b)(a - b) = a^2 - b^2$.

Using this rule allows us to quickly multiply conjugates.

EXAMPLE 7 Multiply.

a. $(x + 3)(x - 3)$

Solution $(x + 3)(x - 3) = x^2 - 3^2$ Use $(a + b)(a - b) = a^2 - b^2$.

$\qquad\qquad\qquad\qquad\;\; = x^2 - 9$ Simplify.

b. $(3t - 8)(3t + 8)$

Solution $(3t - 8)(3t + 8) = (3t)^2 - 8^2$ Use $(a + b)(a - b) = a^2 - b^2$.

$\qquad\qquad\qquad\qquad\qquad\; = 9t^2 - 64$ Simplify.

YOUR TURN Multiply.

a. $(m + 7)(m - 7)$ **b.** $(5y - 3)(5y + 3)$

Squaring a Binomial

Now let's look for a pattern when we square a binomial, as in $(3x + 4)^2$. Recall that squaring a number or expression means to multiply that number or expression by itself, so $(3x + 4)^2 = (3x + 4)(3x + 4)$.

$$(3x + 4)^2 = (3x + 4)(3x + 4) = \overset{\text{First}}{3x \cdot 3x} + \overset{\text{Outer}}{3x \cdot 4} + \overset{\text{Inner}}{4 \cdot 3x} + \overset{\text{Last}}{4 \cdot 4}$$

$$= 9x^2 + 12x + 12x + 16$$

$$= 9x^2 + 24x + 16 \qquad \text{Combine like terms.}$$
$$12x + 12x = 24x$$

Warning: Notice that $(3x + 4)^2$ does *not* equal $9x^2 + 16$.

Notice the pattern:

$$(3x + 4)^2 = 9x^2 + 24x + 16$$

This first term is the square of the first term in the binomial: $(3x)^2 = 9x^2$.

This last term is the square of the second term in the binomial: $(4)^2 = 16$.

This middle term is twice the product of the two terms in the binomial: $2(3x)(4) = 24x$.

ANSWERS

a. $m^2 - 49$ **b.** $25y^2 - 9$

Squaring a binomial with a minus sign, as in $(3x - 4)^2$, yields a similar pattern.

$$
\begin{array}{ll}
 & \text{First} \qquad \text{Outer} \qquad \text{Inner} \qquad \text{Last} \\
(3x - 4)^2 = (3x - 4)(3x - 4) = 3x \cdot 3x + 3x \cdot (-4) + (-4) \cdot 3x + (-4) \cdot (-4) \\
\qquad\qquad\qquad\qquad\qquad = 9x^2 - 12x - 12x + 16 \\
\qquad\qquad\qquad\qquad\qquad = 9x^2 - 24x + 16 \qquad \text{Combine like terms:} \\
\qquad\qquad\qquad\qquad\qquad\qquad\qquad\qquad\qquad -12x - 12x = -24x.
\end{array}
$$

Warning: Notice that $(3x - 4)^2$ does *not* equal $9x^2 - 16$.

Notice the pattern:

$$(3x - 4)^2 = 9x^2 - 24x + 16$$

This first term is the square of the first term in the binomial: $(3x)^2 = 9x^2$.

This last term is the square of the second term in the binomial: $(-4)^2 = 16$.

This middle term is twice the product of the two terms in the binomial: $2(3x)(-4) = -24x$.

The patterns from our two examples suggest the following rules.

RULES   **Squaring a Binomial**

$$(a + b)^2 = a^2 + 2ab + b^2$$
$$(a - b)^2 = a^2 - 2ab + b^2$$

EXAMPLE 8   Multiply.

a. $(5x + 3)^2$

Solution $(5x + 3)^2 = (5x)^2 + 2(5x)(3) + (3)^2$ Use $(a + b)^2 = a^2 + 2ab + b^2$.

$\qquad\qquad\qquad = 25x^2 + 30x + 9$ Simplify.

b. $(6y - 7)^2$

Solution $(6y - 7)^2 = (6y)^2 - 2(6y)(7) + (7)^2$ Use $(a - b)^2 = a^2 - 2ab + b^2$.

$\qquad\qquad\qquad = 36y^2 - 84y + 49$ Simplify.

YOUR TURN   Multiply.

a. $(3m + 7)^2$ **b.** $(4t - 5)^2$

ANSWERS

a. $9m^2 + 42m + 49$
b. $16t^2 - 40t + 25$

5.5 Exercises

For Extra Help

MyMathLab   Videotape/DVT   InterAct Math   Math Tutor Center   Math XL.com

1. What is the mathematical property applied in the first step when multiplying polynomials?

2. Explain, in general, how to multiply two polynomials.

3. Explain how to recognize that two binomials are conjugates.

4. What happens when we multiply conjugates?

5. Explain why $(a + b)^2 \neq a^2 + b^2$.

6. Explain why $(a - b)^2 \neq a^2 - b^2$.

For Exercises 7–34, multiply the polynomial by the monomial.

7. $5(x + 3)$

8. $2(y + 7)$

9. $-6(2x - 4)$

10. $-4(3y - 5)$

11. $4n(7n - 2)$

12. $6n(2n - 5)$

13. $9a(a - b)$

14. $3y(y - z)$

15. $\dfrac{1}{8}m(2m - 5n)$

16. $\dfrac{1}{4}k(2k - 3l)$

17. $-0.3p^2(p^3 - 2q^2)$

18. $-0.4b^2(b^2 - 3c^2)$

19. $5x(4x^2 + 2x - 5)$

20. $5a(2a^2 - a + 6)$

21. $-3x^2(7x^3 - 5x + 1)$

22. $-4w^3(3w^3 + 7w^2 - 2)$

23. $-r^2s(r^2s^3 + 3rs^2 - s)$

24. $-x^2y^3(-2xy^2 + 3x^2y^5 - y)$

25. $-2a^2b^2(2a^3 - 6b + 3ab - a^2b^2)$

26. $-5hk^3(3k^3 - 2h^2k - 3hk^4 + 4h)$

27. $3abc(4a^2b^2c^2 - 2abc + 5)$

28. $5xy^2z(2xy^3 - 4x^2z^2 + 3)$

29. $4b\left(b^5 - \dfrac{1}{4}b^3 + \dfrac{1}{20}b^2 - \dfrac{1}{8}b + 4\right)$

30. $5r^2\left(r^4 - \dfrac{1}{20}r^3 - \dfrac{1}{5}r^2 + \dfrac{1}{10}r - 1\right)$

31. $-0.4rt^2(2.5r^3t - 8r^2t^2 + 4.1rt^3 - 2)$

32. $-0.25suv^3(10s^2u - 2u^2v - 18sv^4 + 4)$

33. $\dfrac{5}{6}x^2(30x^4y - 18x^3y^2 + 60xy^9)$

34. $\dfrac{2}{3}b^1(-30u^7 + 10ab^3 - 12a^2h^2)$

For Exercises 35 and 36, a larger rectangle is formed out of smaller rectangles:

   a. Write an expression in simplest form for the length (along the top).
   b. Write an expression in simplest form for the width (along the side).
   c. Write an expression that is the product of the length and width that you found in parts a and b.
   d. Write an expression in simplest form that is the sum of the areas of each of the smaller rectangles.
   e. Explain why the expressions in parts c and d are equivalent.

35.
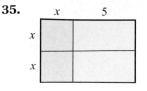

36.
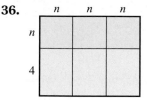

For Exercises 37–54, multiply the binomials (use FOIL).

37. $(x + 3)(x + 4)$ **38.** $(a + 4)(a + 9)$ **39.** $(x - 7)(x + 2)$ **40.** $(n + 7)(n - 5)$

41. $(y - 3)(y - 6)$ **42.** $(r - 4)(r - 5)$ **43.** $(2y + 5)(3y + 2)$ **44.** $(2x + 4)(x + 3)$

45. $(5m - 3)(3m + 4)$ **46.** $(2x + 3)(3x - 2)$ **47.** $(3t - 5)(4t - 2)$ **48.** $(2t - 7)(3t - 1)$

49. $(5q - 3t)(3q + 4t)$ **50.** $(7k - 2j)(3k + 4j)$ **51.** $(7y + 3x)(2y + 4x)$ **52.** $(2a - 5b)(3a - 4b)$

53. $(a^2 + b)(a^2 - 2b)$ **54.** $(x^2 + 4y)(x^2 - 5y)$

For Exercises 55 and 56, a larger rectangle is formed out of smaller rectangles.

   a. Write an expression in simplest form for the length (along the top).
   b. Write an expression in simplest form for the width (along the side).
   c. Write an expression that is the product of the length and width that you found in parts a and b.
   d. Write an expression in simplest form that is the sum of the areas of each of the smaller rectangles.
   e. Explain why the expressions in parts c and d are equivalent.

55.
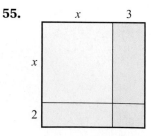

56.
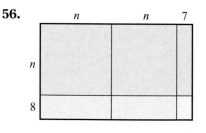

For Exercises 57–70, multiply the polynomials.

57. $(x + 2)(x^2 - 2x + 1)$ **58.** $(x + 3)(x^2 - 3x + 2)$ **59.** $(a - 1)(3a^2 - a - 2)$

60. $(x - 1)(3x^2 - 2x + 1)$ **61.** $(3c + 2)(2c^2 - c - 3)$ **62.** $(2x + 3)(4x^2 + 3x - 5)$

63. $(5p + 2q)(3p^2 + 4pq + 10q^2)$ **64.** $(2f + 3g)(2f^2 - 3fg - 9g^2)$ **65.** $(6y - 3z)(2y^2 - 3yz + 4z^2)$

66. $(2m - 3n)(4m^2 - 6mn + 8)$ **67.** $(x^2 + xy + y^2)(x^2 + 2xy - y^2)$ **68.** $(u^2 - u - 1)(u^2 + 2u + 1)$

69. $(x^2 + 10x + 25)(x^2 + 4x + 4)$ **70.** $(a^2 + 6a + 9)(a^2 - 8a + 16)$

For Exercises 71–78, state the conjugate of the given binomial.

71. $x - 3$ **72.** $y + 8$ **73.** $4x - 2y$ **74.** $2g + 3$

75. $4d - 3c$ **76.** $5m - 2n$ **77.** $-3j - k$ **78.** $-3x - 5y$

For Exercises 79–100, multiply using the rules for special products.

79. $(x - 5)(x + 5)$ **80.** $(y + 4)(y - 4)$ **81.** $(2m - 5)(2m + 5)$ **82.** $(3p + 2)(3p - 2)$

83. $(x + y)(x - y)$ **84.** $(a + b)(a - b)$ **85.** $(8r - 10s)(8r + 10s)$ **86.** $(4b - 5c)(4b + 5c)$

87. $(-2x - 3)(-2x + 3)$ **88.** $(-h + 2k)(-h - 2k)$ **89.** $(x + 3)^2$ **90.** $(y + 5)^2$

91. $(4t - 1)^2$ **92.** $(9k - 2)^2$ **93.** $(m + n)^2$ **94.** $(a - b)^2$

95. $(2u + 3v)^2$ **96.** $(3r + 7s)^2$ **97.** $(9w - 4z)^2$ **98.** $(2q + 11c)^2$

99. $(9 - 5y)^2$ **100.** $(6 - 7t)^2$

For Exercises 101–104, write an expression in simplest form for the area.

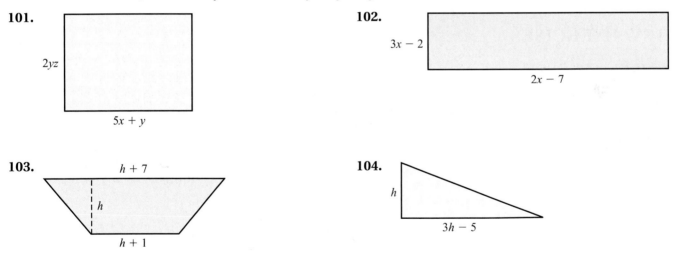

101.

2yz

5x + y

102.

3x − 2

2x − 7

103.

h + 7

h

h + 1

104.

h

3h − 5

105. A rectangular room has a length that is 3 feet more than the width. Using w to represent the width, write an expression in simplest form for the area of the room.

106. A circular metal plate for a machine has radius r. A smaller plate with radius $r - 2$ is to be used in another part of the machine. Write an expression in simplest form for the sum of the areas of the two circles.

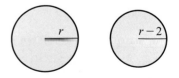

r

r − 2

For Exercises 107 and 108, write an expression in simplest form for the volume.

107.

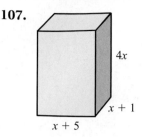

4x

x + 1

x + 5

108.

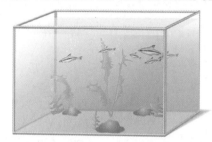

a − 3

4a + 1

a

109. A crate is designed so that the length is twice the width, and the height is 3 feet more than the length. Let w represent the width. Write an expression for the volume in terms of w.

110. A fish tank's length is 8 inches more than its width. The height of the tank is 5 inches greater than its width. Let w represent the width. Write an expression for the volume of the tank in terms of w.

111. A cylinder has a height that is 2 inches more than the radius. Write an expression for the volume of the cylinder in terms of the radius.

112. A cone has a radius that is 5 centimeters less than the height. Write an expression for the volume of the cone in terms of the height.

REVIEW EXERCISES

For Exercises 1 and 2, simplify.

[1.4] **1.** $-42 \div 8$

$\begin{bmatrix} \mathbf{1.5} \\ \mathbf{5.4} \end{bmatrix}$ **2.** -2^0

[2.4] **3.** Isolate r in the formula $\dfrac{5 - d}{r} = m$.

[3.3] **4.** The greater of two positive numbers is four times the smaller number. The difference between the numbers is 27. What are the numbers?

[5.1] **5.** Write 6.304×10^6 in standard form.

[5.4] **6.** Multiply: $x^2 \cdot x^5$

5.6 Exponent Rules and Dividing Polynomials

OBJECTIVES

1. Divide exponential forms with the same base.
2. Divide numbers in scientific notation.
3. Divide monomials.
4. Divide a polynomial by a monomial.
5. Use long division to divide polynomials.
6. Simplify expressions using rules of exponents.

In Section 5.4, we began by learning the product rule of exponents, which we used in multiplying polynomials. In this section, we follow a similar development and begin by learning the quotient rule of exponents so that we can divide polynomials.

OBJECTIVE 1. Divide exponential forms with the same base. In Section 5.4, we found that when we multiply exponential forms that have the same base, we can add the exponents and keep the same base. For example,

$$2^3 \cdot 2^4 = 2^{3+4} = 2^7$$

Now consider how to reverse this process and divide exponential forms with the same base. Remember that multiplication and division are inverse operations, which means they "undo" each other. A related division statement for the multiplication just shown is $2^7 \div 2^4 = 2^3$, which implies that, in the division problem, the exponents can be subtracted. To verify this result, it is helpful to write the division in fraction form and use the definition of exponents.

$$\frac{2^7}{2^4} = \frac{2 \cdot 2 \cdot 2 \cdot 2 \cdot 2 \cdot 2 \cdot 2}{2 \cdot 2 \cdot 2 \cdot 2}$$
$$= \frac{2 \cdot 2 \cdot 2}{1}$$
$$= 2^3$$

◄ **Note:** *We can divide out four of the common factors. There are three 2s left in the top, which can be expressed in exponential form.*

This result supports our thinking that we can subtract the exponents.

$$2^7 \div 2^4 = 2^{7-4} = 2^3$$

Notice that the divisor's exponent is subtracted from the dividend's exponent. This order is important because subtraction is not a commutative operation.

Conclusion   To divide exponential forms that have the same base, we can subtract the divisor's exponent from the dividend's exponent and keep the same base.

Quotient Rule for Exponents

If m and n are integers and a is a real number, where $a \neq 0$, then $a^m \div a^n = a^{m-n}$.

Note: We cannot let a equal 0 because if a were replaced with 0, we would have $0^m \div 0^n$ and in Section 1.4 we concluded that $0 \div 0$ is indeterminate.

Note: *Throughout the remainder of this section, assume that variables will not be replaced by values that cause denominators to be 0.*

EXAMPLE 1 Divide. $x^6 \div x^2$

Solution Because the exponential forms have the same base, we can subtract the exponents and keep the same base.

$$x^6 \div x^2 = x^{6-2} = x^4$$

YOUR TURN Divide.

a. $5^9 \div 5^2$

b. $t^8 \div t^2$

c. $\dfrac{n^5}{n}$

We can apply the quotient rule for exponential forms when the exponents are negative. After simplifying, if the exponent in the result is negative, then we will rewrite the exponential form so that the exponent is positive. Recall the rule from Section 5.1: If n is any real number not equal to 0 and a is a natural number, then

$$a^{-n} = \frac{1}{a^n}.$$

EXAMPLE 2 Divide and write the result with a positive exponent.

a. $x^6 \div x^{-2}$

Solution $x^6 \div x^{-2} = x^{6-(-2)}$ Subtract exponents and keep the same base.

$\phantom{x^6 \div x^{-2}} = x^{6+2}$ **Note:** *To evaluate the subtraction, we write an equivalent addition.*

$\phantom{x^6 \div x^{-2}} = x^8$

b. $n^{-5} \div n^{-3}$

Solution $n^{-5} \div n^{-3} = n^{-5-(-3)}$ Subtract the exponents and keep the same base.

$\phantom{n^{-5} \div n^{-3}} = n^{-5+3}$ Rewrite the subtraction as addition.

$\phantom{n^{-5} \div n^{-3}} = n^{-2}$ Simplify.

$\phantom{n^{-5} \div n^{-3}} = \dfrac{1}{n^2}$ Write with a positive exponent.

ANSWERS

a. 5^7 **b.** t^6 **c.** n^4

ANSWERS

a. $\dfrac{1}{y^4}$ **b.** m^4 **c.** $\dfrac{1}{t^9}$

YOUR TURN Divide and write the result with a positive exponent.

a. $y^3 \div y^7$

b. $m^{-2} \div m^{-6}$

c. $\dfrac{t^{-4}}{t^5}$

OBJECTIVE 2. Divide numbers in scientific notation. We can use the quotient rule of exponents to divide numbers in scientific notation.

EXAMPLE 3 Divide and write the result in scientific notation. $\dfrac{1.08 \times 10^9}{2.4 \times 10^4}$

Solution The decimal factors and powers of 10 can be separated into a product of two fractions. This allows us to calculate the decimal division and divide the powers of 10 separately.

$$\frac{1.08 \times 10^9}{2.4 \times 10^4} = \frac{1.08}{2.4} \times \frac{10^9}{10^4} \qquad \text{Separate the decimal numbers and powers of 10.}$$

$$= 0.45 \times 10^{9-4} \qquad \text{Divide decimal numbers and subtract exponents.}$$

$$= 0.45 \times 10^5$$

$$= 4.5 \times 10^4$$

◀ **Note:** *This is not scientific notation because 0.45 is not a decimal number greater than 1. We must move the decimal point one place to the right. This means we have one less factor of 10 to account for with the exponent, so we subtract 1 from the exponent.*

YOUR TURN Divide $\dfrac{1.82 \times 10^{-9}}{6.5 \times 10^4}$ and write the result in scientific notation.

OBJECTIVE 3. Divide monomials. Dividing monomials is much like dividing numbers in scientific notation. Let's consider $\dfrac{24x^5}{6x^2}$. We can separate the coefficients and variables in the same way that we separated the decimal factors and powers of 10 when dividing numbers in scientific notation:

$$\frac{24x^5}{6x^2} = \frac{24}{6} \cdot \frac{x^5}{x^2}$$

We can now divide the coefficients and subtract the exponents of the like bases.

$$\frac{24x^5}{6x^2} = \frac{24}{6} \cdot \frac{x^5}{x^2} = 4x^{5-2} = 4x^3$$

This suggests the following procedure:

PROCEDURE Dividing Monomials

To divide monomials:

1. Divide the coefficients.
2. Use the quotient rule for the exponents with like bases.
3. Unlike bases are written unchanged in the quotient.
4. Write the final expression so that all exponents are positive.

ANSWER

2.8×10^{-14}

EXAMPLE 4 Divide. $\dfrac{16a^4b^7c}{20a^5b^2}$

Solution $\dfrac{16a^4b^7c}{20a^5b^2} = \dfrac{16}{20} \cdot \dfrac{a^4}{a^5} \cdot \dfrac{b^7}{b^2} \cdot \dfrac{c}{1}$

$= \dfrac{16}{20} \cdot a^{4-5} \cdot b^{7-2} \cdot c$ Divide coefficients and subtract exponents of the like bases. The unlike base, c, will remain unchanged.

$= \dfrac{4}{5}a^{-1}b^5c$

$= \dfrac{4}{5} \cdot \dfrac{1}{a} \cdot \dfrac{b^5}{1} \cdot \dfrac{c}{1}$

$= \dfrac{4b^5c}{5a}$

◀ **Note:** *Since* $a^{-1} = \dfrac{1}{a}$, *the variable* a *is written in the denominator as a factor with the 5. Because the* b^5 *and* c *have positive exponents, they are expressed in the numerator as factors with the 4.*

YOUR TURN Divide.

a. $\dfrac{-32m^2n}{8mn}$

b. $\dfrac{56t^2u^6v}{16t^2u^4}$

OBJECTIVE 4. Divide a polynomial by a monomial. Let's now consider how to divide a polynomial by a monomial, as in $\dfrac{24x^5 + 18x^3}{6x^2}$. Recall that when fractions with a common denominator are added (or subtracted), the numerators are added and the denominator stays the same.

$$\frac{1}{7} + \frac{2}{7} = \frac{1+2}{7}$$

This process can be reversed so that a sum in the numerator of a fraction can be broken into fractions, with each addend over the same denominator.

$$\frac{1+2}{7} = \frac{1}{7} + \frac{2}{7}$$

RULE If a, b, and c are real numbers, variables, or expressions with $c \neq 0$, then

$$\frac{a+b}{c} = \frac{a}{c} + \frac{b}{c}.$$

ANSWERS

a. $-4m$ b. $\dfrac{7}{2}u^2v$

We can apply this rule to the division problem $\dfrac{24x^5 + 18x^3}{6x^2}$.

$$\frac{24x^5 + 18x^3}{6x^2} = \frac{24x^5}{6x^2} + \frac{18x^3}{6x^2}$$

We now have a sum of monomial divisions, which we can simplify separately.

$$= 4x^{5-2} + 3x^{3-2}$$
$$= 4x^3 + 3x$$

Our illustration suggests the following procedure.

PROCEDURE   **Dividing a Polynomial by a Monomial**

To divide a polynomial by a monomial, divide each term in the polynomial by the monomial.

EXAMPLE 5   Divide.

a. $\dfrac{32y^6 + 24y^4 - 48y^2}{8y^2}$

Solution $\dfrac{32y^6 + 24y^4 - 48y^2}{8y^2} = \dfrac{32y^6}{8y^2} + \dfrac{24y^4}{8y^2} - \dfrac{48y^2}{8y^2}$ Divide each term in the polynomial by the monomial.

$$= 4y^4 + 3y^2 - 6y^0$$
$$= 4y^4 + 3y^2 - 6$$

b. $\dfrac{42t^7u^2 - 18t^4u - 10t}{2t^3u}$

Solution $\dfrac{42t^7u^2 - 18t^4u - 10t}{2t^3u} = \dfrac{42t^7u^2}{2t^3u} - \dfrac{18t^4u}{2t^3u} - \dfrac{10t}{2t^3u}$

Note: *When dividing t by t^3, because the denominator's exponent is greater than the numerator's exponent, the result is a negative exponent. Because u has no like base in the third term, it is written unchanged.*

$$= 21t^4u - 9t^1u^0 - \dfrac{5t^{-2}}{u}$$

$$= 21t^4u - 9t - \dfrac{5}{ut^2}$$ Rewrite with positive exponents.

YOUR TURN   Divide.

a. $\dfrac{45x^6 - 27x^5 + 9x^3}{9x^3}$

b. $\dfrac{28m^2n^6 + 36m^4n - 40m^2}{4m^2n}$

ANSWERS

a. $5x^3 - 3x^2 + 1$

b. $7n^5 + 9m^2 - \dfrac{10}{n}$

OBJECTIVE 5. Use long division to divide polynomials. To divide a polynomial by a polynomial, we can use long division. Consider the following numeric long division and think about the process.

Divide: $\dfrac{157}{12}$

Quotient

$$
\begin{array}{r}
13 \\
12\overline{)157} \\
-12 \\
\hline
37 \\
-36 \\
\hline
1
\end{array}
$$

Divisor ⟶ 12)157 ⟵ Dividend

Remainder ⟶ 1

Because the answer has a remainder, we write the result as a mixed number, $13\dfrac{1}{12}$. Notice we could check the answer by multiplying the divisor by the quotient then adding the remainder. The result should be the dividend.

$$\text{Quotient} \cdot \text{Divisor} + \text{Remainder} = \text{Dividend}$$
$$13 \quad\cdot\quad 12 \quad+\quad 1 \quad = 157$$

Now let's consider polynomial division, which follows the same long division process.

EXAMPLE 6 Divide. $\dfrac{x^2 + 5x + 7}{x + 2}$

Solution Use long division. First, we determine what term will multiply by the first term in the divisor to equal the first term in the dividend. A clever way to determine this first term in the quotient is to divide the first term in the dividend by the first term in the divisor.

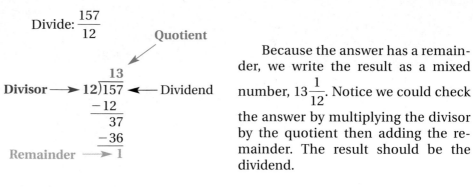

Divide these first terms to determine the first term in the quotient: $x^2 \div x = x$.

Next, we multiply the divisor $x + 2$ by the x in the quotient.

multiply

$$
\begin{array}{r}
x \\
x + 2\overline{)x^2 + 5x + 7} \\
x^2 + 2x
\end{array}
$$

Next we subtract. Note that to subtract the binomial, we change the signs of the terms and then combine like terms. After combining terms, we bring down the next term in the dividend, which is 7.

$$
\begin{array}{r}
x \\
x + 2\overline{)x^2 + 5x + 7} \\
-(x^2 + 2x)
\end{array}
$$

Change signs. ⟶

$$
\begin{array}{r}
x \\
x + 2\overline{)x^2 + 5x + 7} \\
-x^2 - 2x \\
\hline
3x + 7
\end{array}
$$

Combine like terms and bring down the next term.

Now we repeat the process with $3x + 7$ as the dividend. The next term in the quotient is found by dividing the first term of $3x + 7$ by the first term in the divisor, $x + 2$. This gives $3x \div x = 3$. We then multiply the divisor by this 3.

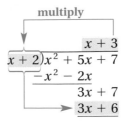

Then we subtract. As before, we change the signs of the binomial to be subtracted.

$$
\begin{array}{r}
x + 3 \\
x + 2 \overline{)\, x^2 + 5x + 7} \\
-x^2 - 2x \\
\hline
3x + 7 \\
-(3x + 6) \\
\hline
\end{array}
\qquad \xrightarrow{\text{Change signs.}} \qquad
\begin{array}{r}
x + 3 \\
x + 2 \overline{)\, x^2 + 5x + 7} \\
-x^2 - 2x \\
\hline
3x + 7 \\
-3x - 6 \\
\hline
1 \qquad \text{Combine like terms.}
\end{array}
$$

Notice that we have a remainder, 1. Recall that in the numeric version, we wrote the answer as a mixed number, $13\dfrac{1}{12}$, which means $13 + \dfrac{1}{12}$ and is in the form:

$\text{quotient} + \dfrac{\text{remainder}}{\text{divisor}}$. With polynomials we write a similar expression:

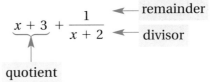

To check, we can multiply the quotient and the divisor, then add the remainder. The result should be the dividend.

$$
\begin{aligned}
(x + 3)(x + 2) + 1 &= x^2 + 2x + 3x + 6 + 1 \\
&= x^2 + 5x + 7 \qquad \text{It checks.}
\end{aligned}
$$

PROCEDURE   Dividing a Polynomial by a Polynomial

To divide a polynomial by a polynomial, use long division. If there is a remainder, write the result in the following form:

$$\text{quotient} + \dfrac{\text{remainder}}{\text{divisor}}$$

In Example 6, the coefficients of the first term in the divisor and dividend were both 1. What if those coefficients are other than 1?

EXAMPLE 7 Divide. $\dfrac{15x^2 - 22x + 14}{3x - 2}$

Solution Begin by dividing the first term in the dividend by the first term in the divisor: $15x^2 \div 3x = 5x$.

$$
\begin{array}{r}
5x \\
3x - 2\overline{)15x^2 - 22x + 14} \\
-(15x^2 - 10x)
\end{array}
$$

Change signs. $\longrightarrow$

$$
\begin{array}{r}
5x \\
3x - 2\overline{)15x^2 - 22x + 14} \\
-15x^2 + 10x \\
\hline
-12x + 14
\end{array}
$$

Combine like terms, then bring down the next term.

Determine the next part of the quotient by dividing $-12x$ by $3x$, which is -4, and repeat the multiplication and subtraction steps.

$$
\begin{array}{r}
5x - 4 \\
3x - 2\overline{)15x^2 - 22x + 14} \\
-15x^2 + 10x \\
\hline
-12x + 14 \\
-(-12x + 8)
\end{array}
$$

Change signs. $\longrightarrow$

$$
\begin{array}{r}
5x - 4 \\
3x - 2\overline{)15x^2 - 22x + 14} \\
-15x^2 + 10x \\
\hline
-12x + 14 \\
+12x - 8 \\
\hline
6
\end{array}
$$

Combine like terms.

Answer $\quad 5x - 4 + \dfrac{6}{3x - 2}$

YOUR TURN Divide.

a. $\dfrac{x^2 + 6x + 8}{x + 2}$

b. $\dfrac{24x^2 + 14x + 4}{6x - 1}$

Using a Place Marker in Long Division

In polynomial division, it is important that we write the terms in descending order of degree. If there is a missing term, then we write that term with a 0 coefficient as a place marker.

EXAMPLE 8 Divide. $\dfrac{-14x^2 - 9 + 8x^4}{2x - 1}$

Solution First, write the dividend in descending order of degree. Also, because degree 3 and degree 1 terms are missing in the dividend, we write those terms with 0 coefficients as place holders.

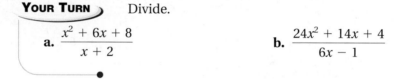

Combine like terms, then bring down the next term.

Determine the next part of the quotient by dividing $4x^3$ by $2x$, and repeat the multiplication and subtraction steps.

$$
\begin{array}{r}
4x^3 + 2x^2 \\
2x - 1\overline{)8x^4 + 0x^3 - 14x^2 + 0x - 9} \\
\underline{-8x^4 + 4x^3} \\
4x^3 - 14x^2 \\
\underline{-(4x^3 - 2x^2)}
\end{array}
$$

$\xrightarrow{\text{Change signs.}}$

$$
\begin{array}{r}
4x^3 + 2x^2 \\
2x - 1\overline{)8x^4 + 0x^3 - 14x^2 + 0x - 9} \\
\underline{-8x^4 + 4x^3} \\
4x^3 - 14x^2 \\
\underline{-4x^3 + 2x^2} \\
-12x^2 + 0x
\end{array}
$$

Combine like terms, then bring down the next term.

Determine the next part of the quotient by dividing $-12x^2$ by $2x$, and repeat the multiplication and subtraction steps.

$$
\begin{array}{r}
4x^3 + 2x^2 - 6x \\
2x - 1\overline{)8x^4 + 0x^3 - 14x^2 + 0x - 9} \\
\underline{-8x^4 + 4x^3} \\
4x^3 - 14x^2 \\
\underline{-4x^3 + 2x^2} \\
-12x^2 + 0x \\
\underline{-(-12x^2 + 6x)}
\end{array}
$$

$\xrightarrow{\text{Change signs.}}$

$$
\begin{array}{r}
4x^3 + 2x^2 - 6x \\
2x - 1\overline{)8x^4 + 0x^3 - 14x^2 + 0x - 9} \\
\underline{-8x^4 + 4x^3} \\
4x^3 - 14x^2 \\
\underline{-4x^3 + 2x^2} \\
-12x^2 + 0x \\
\underline{+12x^2 - 6x} \\
-6x - 9
\end{array}
$$

Combine like terms, then bring down the next term.

Determine the final term by dividing $-6x$ by $2x$, which is -3.

$$
\begin{array}{r}
4x^3 + 2x^2 - 6x - 3 \\
2x - 1\overline{)8x^4 + 0x^3 - 14x^2 + 0x - 9} \\
\underline{-8x^4 + 4x^3} \\
4x^3 - 14x^2 \\
\underline{-4x^3 + 2x^2} \\
-12x^2 + 0x \\
\underline{12x^2 - 6x} \\
-6x - 9 \\
\underline{-(-6x + 3)}
\end{array}
$$

$\xrightarrow{\text{Change signs.}}$

$$
\begin{array}{r}
4x^3 + 2x^2 - 6x - 3 \\
2x - 1\overline{)8x^4 + 0x^3 - 14x^2 + 0x - 9} \\
\underline{-8x^4 + 4x^3} \\
4x^3 - 14x^2 \\
\underline{-4x^3 + 2x^2} \\
-12x^2 + 0x \\
\underline{12x^2 - 6x} \\
-6x - 9 \\
\underline{+6x - 3} \\
-12
\end{array}
$$

Combine like terms.

Answer $4x^3 + 2x^2 - 6x - 3 + \dfrac{-12}{2x - 1}$, or $4x^3 + 2x^2 - 6x - 3 - \dfrac{12}{2x - 1}$

YOUR TURN   Divide. $\dfrac{7 + 8x^3 - 18x}{4x - 2}$

OBJECTIVE 6. Simplify expressions using rules of exponents.   The quotient rule for exponents completes our rules of exponents in this chapter. So far, the problems involving exponent rules have only required that we use one or two rules. Let's now consider problems that require us to use several of the exponent rules. Following is a summary of all of the rules of exponents.

ANSWER

$2x^2 + x - 4 + \dfrac{-1}{4x - 2}$

Exponents Summary

Assume that no denominators are 0, that *a* and *b* are real numbers, and that *m* and *n* are integers.

Zero as an exponent: $a^0 = 1$, where $a \neq 0$

0^0 is indeterminate.

Negative exponents: $a^{-n} = \dfrac{1}{a^n}$

$\dfrac{1}{a^{-n}} = a^n$

$\left(\dfrac{a}{b}\right)^{-n} = \left(\dfrac{b}{a}\right)^n$

Product rule for exponents: $a^m \cdot a^n = a^{m+n}$

Quotient rule for exponents: $a^m \div a^n = a^{m-n}$

Raising a power to a power: $(a^m)^n = a^{mn}$

Raising a product to a power: $(ab)^n = a^n b^n$

Raising a quotient to a power: $\left(\dfrac{a}{b}\right)^n = \dfrac{a^n}{b^n}$

EXAMPLE 9 Simplify. Write all answers with positive exponents.

a. $\left(\dfrac{n^{-5}}{n^2}\right)^3$

Solution Following the order of operations, we simplify within the parentheses first, then consider the exponent outside the parentheses.

$$\left(\dfrac{n^{-5}}{n^2}\right)^3 = (n^{-5-2})^3 \qquad \text{Use the quotient rule for exponents.}$$

$$= (n^{-7})^3$$

$$= n^{-7 \cdot 3} \qquad \text{Use the rule for raising a power to a power.}$$

$$= n^{-21}$$

$$= \dfrac{1}{n^{21}} \qquad \text{Write with a positive exponent.}$$

b. $\dfrac{(m^2)^{-3}}{m^4 \cdot m^{-5}}$

Solution $\dfrac{(m^2)^{-3}}{m^4 \cdot m^{-5}} = \dfrac{m^{2 \cdot (-3)}}{m^{4 + (-5)}}$ In the numerator, use the rule for raising a power to a power. In the denominator, use the product rule for exponents.

$$= \dfrac{m^{-6}}{m^{-1}}$$

$$= m^{-6-(-1)} \qquad \text{Use the quotient rule for exponents.}$$

$$= m^{-6+1} \qquad \text{Write the subtraction as an equivalent addition.}$$

$$= m^{-5}$$

$$= \dfrac{1}{m^5} \qquad \text{Write with a positive exponent.}$$

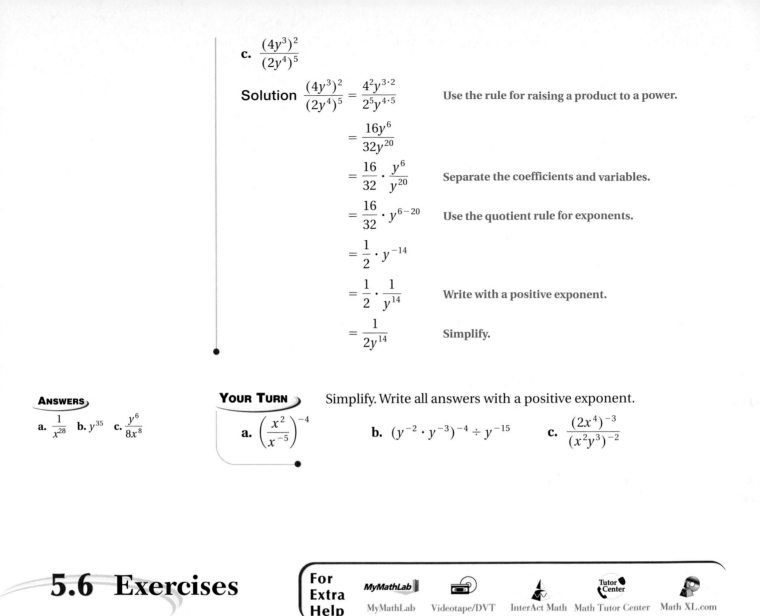

c. $\dfrac{(4y^3)^2}{(2y^4)^5}$

Solution $\dfrac{(4y^3)^2}{(2y^4)^5} = \dfrac{4^2 y^{3 \cdot 2}}{2^5 y^{4 \cdot 5}}$ Use the rule for raising a product to a power.

$= \dfrac{16 y^6}{32 y^{20}}$

$= \dfrac{16}{32} \cdot \dfrac{y^6}{y^{20}}$ Separate the coefficients and variables.

$= \dfrac{16}{32} \cdot y^{6-20}$ Use the quotient rule for exponents.

$= \dfrac{1}{2} \cdot y^{-14}$

$= \dfrac{1}{2} \cdot \dfrac{1}{y^{14}}$ Write with a positive exponent.

$= \dfrac{1}{2 y^{14}}$ Simplify.

ANSWERS

a. $\dfrac{1}{x^{28}}$ **b.** y^{35} **c.** $\dfrac{y^6}{8x^8}$

YOUR TURN   Simplify. Write all answers with a positive exponent.

a. $\left(\dfrac{x^2}{x^{-5}}\right)^{-4}$ **b.** $(y^{-2} \cdot y^{-3})^{-4} \div y^{-15}$ **c.** $\dfrac{(2x^4)^{-3}}{(x^2 y^3)^{-2}}$

5.6 Exercises

For Extra Help   MyMathLab   Videotape/DVT   InterAct Math   Math Tutor Center   Math XL.com

1. Explain how to divide exponential forms that have the same base.

2. What is the rule for rewriting a^{-n} where a is a real number, $a \neq 0$, and n is a natural number so that it has a positive exponent?

3. How do we divide two numbers in scientific notation?

4. How do we divide two monomials?

5. Explain how to divide a polynomial by a monomial.

6. When do we use a 0 place holder in the process of dividing two polynomials by long division?

For Exercises 7–32, simplify. Write all answers with positive exponents.

7. a^{-3}

8. n^{-7}

9. 2^{-5}

10. 3^{-4}

11. $\dfrac{3^4}{3^3}$

12. $\dfrac{6^{11}}{6^5}$

13. $\dfrac{4^3}{4^9}$

14. $\dfrac{5}{5^9}$

15. $\dfrac{y^8}{y^3}$

16. $\dfrac{n^4}{n}$

17. $\dfrac{x^3}{x^5}$

18. $\dfrac{m^4}{m^{10}}$

19. $a^8 \div a^6$

20. $u^5 \div u^3$

21. $w^7 \div w^{10}$

22. $j^6 \div j^{14}$

23. $\dfrac{a^{-2}}{a^5}$

24. $\dfrac{m^{-6}}{m^3}$

25. $\dfrac{r^6}{r^{-4}}$

26. $\dfrac{u^5}{u^{-12}}$

27. $\dfrac{y^{-7}}{y^{-15}}$

28. $\dfrac{x^{-3}}{x^{-8}}$

29. $\dfrac{p^{-5}}{p^{-2}}$

30. $\dfrac{n^{-11}}{n^{-5}}$

31. $\dfrac{t^{-4}}{t^{-4}}$

32. $\dfrac{x^{-6}}{x^{-6}}$

For Exercises 33–40, divide and write your answers in scientific notation.

33. $\dfrac{1.2 \times 10^5}{3.2 \times 10^6}$

34. $\dfrac{3.6 \times 10^6}{2.4 \times 10^9}$

35. $\dfrac{1.4 \times 10^{-4}}{2.1 \times 10^3}$

36. $\dfrac{1.32 \times 10^{-5}}{6.6 \times 10^{15}}$

37. $\dfrac{9.088 \times 10^1}{1.28 \times 10^{-9}}$

38. $\dfrac{4.25 \times 10^4}{1.88 \times 10^{-2}}$

39. $\dfrac{8.75 \times 10^{-3}}{3.2 \times 10^{-7}}$

40. $\dfrac{7.9 \times 10^{-3}}{2.2 \times 10^{-7}}$

For Exercises 41–46, solve.

41. If light travels at 3×10^8 meters per second, how long does it take light to travel the 1.5×10^{11} meters from the Sun to Earth?

42. In 2000, the population of the United States was approximately 2.8×10^8. Calculate the approximate number of people per square mile in the United States in 2000 if the estimated amount of land in the United States is 3.5×10^6 square miles. (*Source:* U.S. Bureau of the Census)

43. As of March 11, 2005, the national debt was about $\$7.754 \times 10^{12}$. If each of the 2.8×10^8 people in the United States contributed an equal share towards the debt, then what would each person have to contribute to pay off the debt? (*Source:* U.S. National Debt Clock)

44. There are about 2.3×10^6 stone blocks in the Great Pyramid. If the total weight of the pyramid is about 1.3×10^{10} pounds and if we assume each block is of equal size, then how much does each block weigh? (*Source:* Nova Online)

For Exercises 45–46, use Einstein's formula $E = mc^2$, which describes the rest energy contained within a mass m, where c represents the speed of light, which is a constant with a value of 3×10^8 meters per second. The units for energy are joules, which are equivalent to $\dfrac{kg \cdot m^2}{s^2}$.

45. The first atom bomb released an energy equivalent of about 8.4×10^{13} joules. What amount of mass was converted to energy?

46. The largest atom bomb detonated was a hydrogen bomb tested by the Soviet Union in 1961. The bomb released 2.4×10^{17} joules of energy. What amount of mass was converted to energy in that explosion? (*Source:* Cesare Emiliani, *The Scientific Companion*; Wiley Popular Science, 1995)

For Exercises 47–58, divide the monomials.

47. $\dfrac{15x^5}{3x^2}$

48. $\dfrac{14a^6}{2a^4}$

49. $\dfrac{24x^2}{-6x^5}$

50. $\dfrac{-18a^6}{9a^{12}}$

51. $\dfrac{9m^3n^5}{-3mn}$

52. $\dfrac{-42x^5y^3}{6xy^2}$

53. $\dfrac{-24p^3q^5}{15p^3q^2}$

54. $\dfrac{48t^4u^7}{-18t^4u^2}$

55. $\dfrac{56x^2y^6}{42x^7y^2}$

56. $\dfrac{60a^8b^4}{66a^2b^9}$

57. $\dfrac{12a^4bc^3}{9a^6bc^2}$

58. $\dfrac{28p^5q^4r}{42p^2q^7}$

For Exercises 59–76, divide the polynomial by the monomial.

59. $\dfrac{7a + 14b}{7}$

60. $\dfrac{4m + 16y}{4}$

61. $\dfrac{12x^3 - 6x^2}{3x}$

62. $\dfrac{18m^4 - 27m^2}{9m}$

63. $\dfrac{12x^3 + 8x}{4x^2}$

64. $\dfrac{24y^3 + 16y}{8y^2}$

65. $\dfrac{5x^2y^2 - 15xy^3}{5xy}$

66. $\dfrac{12k^4l^2 + 15k^2l^3}{3k^2l}$

67. $\dfrac{6abc^2 - 24a^2b^2c}{-3abc}$

68. $\dfrac{30x^3yz^5 - 15xyz^2}{-5xyz^2}$

69. $\dfrac{24x^3 + 16x^2 - 8x}{8x}$

70. $\dfrac{3x^3 + 6x^2 - 3x}{3x}$

71. $\dfrac{6x^3 - 12x^2 + 9x}{3x^2}$

72. $\dfrac{12y^4 - 16y^3 + 8y}{4y^2}$

73. $\dfrac{36u^3v^4 + 12uv^5 - 15u^2v^2}{3u^2v}$

74. $\dfrac{16hk^4 - 28hk - 4h^2k^2}{4hk^2}$

75. $\dfrac{y^5 + y^7 - 3y^8 + y}{y^3}$

76. $\dfrac{x^6 + x^8 - 4x^{10} + x^2}{x^4}$

77. The area of the parallelogram shown is described by the monomial $35mn^2$. Find the height.

78. The area of the triangle shown is $18x^2y$. Find the base.

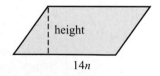

height

14n

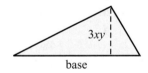

3xy

base

79. An engineer is designing a steel cover plate in the shape of a rectangle. The area of the plate is described by the polynomial $9x^2 - 15x + 12$ and the length must be $3x$.

a. Find an expression for the width.

b. Find the area, length, and width if $x = 2$.

80. The voltage in a circuit is described by the polynomial $36s^3 - 24s^2 + 42s$. The resistance is described by $6s$.

a. Find an expression for the current. (*Hint:* Voltage = Current × Resistance)

b. Find the voltage, current, and resistance if $s = 3$.

For Exercises 81–104, use long division to divide the polynomials.

81. $\dfrac{x^2 + 7x + 12}{x + 3}$

82. $\dfrac{y^2 + 4y + 4}{y + 2}$

83. $\dfrac{m^2 + 8m + 15}{m + 3}$

84. $\dfrac{u^2 - 6u + 9}{u - 3}$

85. $\dfrac{x^2 - 17x + 64}{x - 5}$

86. $\dfrac{x^2 + 6x + 10}{x + 3}$

87. $\dfrac{3x^3 - 73x - 10}{x - 5}$

88. $\dfrac{2x^3 - 35x - 12}{x + 4}$

89. $\dfrac{2x^3 - x^2 + 5}{x - 2}$

90. $\dfrac{2x^3 - 2x^2 + 5}{x + 3}$

91. $\dfrac{p^3 + 125}{p + 5}$

92. $\dfrac{x^3 + 8}{x + 2}$

93. $\dfrac{12x^2 - 7x - 10}{4x - 5}$

94. $\dfrac{6a^2 + 2a - 28}{3a + 7}$

95. $\dfrac{14y^2 - 8y + 17}{7y + 3}$

96. $\dfrac{6x^2 + 7x + 5}{3x - 1}$

97. $\dfrac{4k^3 + 8k^2 - k}{k + 2}$

98. $\dfrac{y^3 - y^2 + 3 - 11y}{y + 2}$

99. $\dfrac{21u^3 - 19u^2 + 14u - 6}{3u + 2}$

100. $\dfrac{2v^3 + 5v^2 - 10v + 2}{2v - 3}$

101. $\dfrac{14b + b^3 - 6b^2 - 12}{b - 3}$

102. $\dfrac{-14x + x^2 + x^3 - 5}{x + 4}$

103. $\dfrac{y^3 - y + 6}{y + 2}$

104. $\dfrac{2b^4 + 3b^3 - 4b - 6}{2b + 3}$

105. The volume of a restaurant's walk-in refrigerator is described by $(6x^2 + 42x + 72)$ cubic feet. The width has to be 6 feet and the length is described by $(x + 3)$ feet.

   a. Find an expression for the height of the refrigerator.

   b. Find the length, height, and volume of the refrigerator if $x = 4$.

106. In an architectural project, the specifications for a rectangular room call for the area to be $(10x^2 + 3x - 18)$ square feet. It is decided that the length should be described by the binomial $(2x + 3)$ feet.

   a. Find an expression for the width.

   b. Find the length, width, and area of the room if $x = 5$.

For Exercises 107–130, simplify. Write all answers with positive exponents.

107. $(6m^2n^{-2})^2$

108. $(-4a^3b^{-2}c)^2$

109. $(4x^{-3}y^2)^{-3}$

110. $(3m^4n^{-3})^{-4}$

111. $(4rs^3)^3(2r^3s^{-1})^2$

112. $(-3abc^{-2})^2(2a^2b^{-3})^3$

113. $\left(\dfrac{m^6}{m^2}\right)^4$

114. $\left(\dfrac{p^7}{p^4}\right)^5$

115. $\left(\dfrac{x^{-3}}{x^4}\right)^2$

116. $\left(\dfrac{r^{-4}}{r^2}\right)^5$

117. $\left(\dfrac{x^2y^7}{x^5y^{-2}}\right)^3$

118. $\left(\dfrac{m^6n^5}{m^{-3}n^8}\right)^2$

119. $\left(\dfrac{x^{-3}}{x^4}\right)^{-2}$

120. $\left(\dfrac{p^{-5}}{p^3}\right)^{-3}$

121. $\left(\dfrac{3x^{-1}y^2}{z^2}\right)^3$

122. $\left(\dfrac{2x^{-2}y}{z^{-1}}\right)^4$

123. $\dfrac{(x^4)^{-3}}{x^{-5}\cdot x^3}$

124. $\dfrac{(y^3)^{-5}}{y^6\cdot y^{-8}}$

125. $\dfrac{4xy^{-2}z^2}{x^{-3}y^3z^{-1}}$

126. $\dfrac{3m^6n^{-1}p^{-2}}{m^{-1}n^{-3}p}$

127. $\dfrac{(2ab)^3}{12a^4b^5}$

128. $\dfrac{(2xy^2)^2}{16x^2y^3}$

129. $\dfrac{(8x^3)^3}{(4x^4)^5}$

130. $\dfrac{(3y^4)^2}{(6y^7)^3}$

PUZZLE PROBLEM

Fill the following blanks with the numbers 1, 2, 3, 4, 5, 6, 7, 8, and 9, using each number once so that the equation is true.

$$\boxed{}\boxed{} \div \boxed{} = \boxed{}\boxed{} \div \boxed{} = \boxed{}\boxed{} \div \boxed{}$$

REVIEW EXERCISES

[1.6] **1.** Translate to an algebraic expression: one number plus half of a second number.

[2.1] **2.** Find the area of a playground that is 17 feet by 20 feet.

[2.3] **3.** Solve and check: $3 - y = 27$

[4.7] **4.** Does the graph to the right represent a function? Explain.

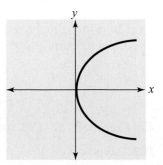

[5.1] **5.** The mass of a proton is 0.00000000000000000000000167 grams. Write this in scientific notation.

[5.1] **6.** The speed of light is approximately 2.998×10^{10} centimeters per second. Write this in standard form.

Chapter 5 Summary

Defined Terms

Section 5.1
Scientific notation
(p. 378)

Section 5.2
Monomial (p. 387)
Coefficient (p. 389)
Degree of a monomial
(p. 389)
Polynomial (p. 389)

Polynomial in one variable
(p. 390)
Binomial (p. 390)
Trinomial (p. 390)
Degree of a polynomial
(p. 391)

Section 5.5
Conjugates (p. 428)

Procedures, Rules, and Key Examples

Procedures/Rules	Key Examples

Section 5.1 Exponents and Scientific Notation

If the base of an exponential form is a negative number and the exponent is even, then the product is positive.

If the base is a negative number and the exponent is odd, then the product is negative.

If a and b are real numbers, where $b \neq 0$ and n is a natural number, then $\left(\dfrac{a}{b}\right)^n = \dfrac{a^n}{b^n}$.

If a is a real number and $a \neq 0$, then $a^0 = 1$.

If a is a real number, where $a \neq 0$ and n is a natural number, then $a^{-n} = \dfrac{1}{a^n}$.

If a is a real number, where $a \neq 0$ and n is a natural number, then $\dfrac{1}{a^{-n}} = a^n$.

If a and b are real numbers, where $a \neq 0$ and $b \neq 0$ and n is a natural number, then $\left(\dfrac{a}{b}\right)^{-n} = \left(\dfrac{b}{a}\right)^n$.

To change from scientific notation with a positive integer exponent to standard notation, move the decimal point to the right the number of places indicated by the exponent.

To write a number expressed in scientific notation with a negative exponent in standard form, move the decimal point to the left the same number of places as the absolute value of the exponent.

To write a number greater than 1 in scientific notation:
1. Move the decimal point so that the number is greater than or equal to 1, but less than 10. (Place the decimal point to the right of the first nonzero digit.)
2. Write the decimal number multiplied by 10^n, where n is the number of places between the new decimal position and the original decimal position.
3. Delete zeros to the right of the last nonzero digit.

Example 1: Evaluate.
a. $(-2)^4 = (-2)(-2)(-2)(-2) = 16$
b. $(-2)^5 = (-2)(-2)(-2)(-2)(-2) = -32$

Example 2: Simplify.
a. $\left(\dfrac{3}{4}\right)^2 = \dfrac{3^2}{4^2} = \dfrac{9}{16}$
b. $7^0 = 1$
c. $2^{-4} = \dfrac{1}{2^4} = \dfrac{1}{16}$
d. $\dfrac{1}{3^{-2}} = 3^2 = 9$
e. $\left(\dfrac{2}{3}\right)^{-4} = \left(\dfrac{3}{2}\right)^4 = \dfrac{81}{16}$

Example 3: Write in standard form.
a. $3.4 \times 10^5 = 340{,}000$
b. $4.2 \times 10^{-5} = 0.000042$

Example 4: Write in scientific notation.
$$7{,}230{,}000 = 7.23 \times 10^6$$

continued

Procedures/Rules	Key Examples

Section 5.1 Exponents and Scientific Notation (continued)

To write a positive decimal number that is less than 1 in scientific notation:

1. Move the decimal point so that the number is greater than or equal to 1, but less than 10. (*Tip:* Place the decimal point to the right of the first nonzero digit.)
2. Write the decimal number multiplied by 10^n, where n is a negative integer whose absolute value is the number of places between the new decimal position and the original decimal position.

Example 5: Write in scientific notation.
$$0.00000056 = 5.6 \times 10^{-7}$$

Section 5.2 Introduction to Polynomials

To write a polynomial in descending order, place the highest degree term first, then the next highest degree, and so on.

Example 1: Write in descending order.
$$2x^2 + 4x^5 + 9 - 6x^3 = 4x^5 - 6x^3 + 2x^2 + 9$$

Section 5.3 Adding and Subtracting Polynomials

To add polynomials, combine like terms.

Example 1: Add.
$$
\begin{aligned}
&(5x^3 + 12x^2 - 9x + 1) + (7x^2 - x - 13) \\
&= 5x^3 + 12x^2 + 7x^2 - 9x - x + 1 - 13 \\
&= 5x^3 + 19x^2 - 10x - 12
\end{aligned}
$$

To subtract polynomials:

1. Write the subtraction statement as an equivalent addition statement.
   a. Change the operation symbol from a minus sign to a plus sign.
   b. Change the subtrahend (second polynomial) to its additive inverse. To get the additive inverse, we change the sign of each term in the polynomial.
2. Combine like terms.

Example 2: Subtract.
$$
\begin{aligned}
&(6y^3 - 5y + 18) - (y^3 - 5y + 7) \\
&= (6y^3 - 5y + 18) + (-y^3 + 5y - 7) \\
&= 6y^3 - y^3 - 5y + 5y + 18 - 7 \\
&= 5y^3 + 11
\end{aligned}
$$

Section 5.4 Exponent Rules and Multiplying Monomials

If a is a real number and m and n are integers, then $a^m \cdot a^n = a^{m+n}$. (Product Rule for Exponents)

To multiply monomials:

1. Multiply coefficients.
2. Add the exponents of the like bases.
3. Write any unlike variable bases unchanged in the product.

If a is a real number and m and n are integers, then $(a^m)^n = a^{mn}$.

If a and b are real numbers and n is an integer, then $(ab)^n = a^n b^n$.

To simplify a monomial raised to a power:

1. Evaluate the coefficient raised to that power.
2. Multiply each variable's exponent by the power.

Example 1: Multiply.
a. $x^3 \cdot x^4 = x^{3+4} = x^7$
b. $-6x^4 yz^2 \cdot 5x^5 y^3 = -30x^{4+5} y^{1+3} z^2$
$$= -30x^9 y^4 z^2$$

Example 2: Simplify.
a. $(x^3)^4 = x^{3 \cdot 4} = x^{12}$
b. $(m^2 n^3)^5 = m^{2 \cdot 5} n^{3 \cdot 5} = m^{10} n^{15}$
c. $(3x^2 y^5)^4 = 3^4 x^{2 \cdot 4} y^{5 \cdot 4} = 81x^8 y^{20}$

continued

Procedures/Rules	Key Examples

Section 5.5 Multiplying Polynomials; Special Products

To multiply a polynomial by a monomial, use the distributive property to multiply each term in the polynomial by the monomial.

To multiply two polynomials:
1. Multiply every term in the second polynomial by every term in the first polynomial.
2. Combine like terms.

Special Products:

Conjugates: $(a + b)(a - b) = a^2 - b^2$

Squaring a sum: $(a + b)^2 = a^2 + 2ab + b^2$

Squaring a difference: $(a - b)^2 = a^2 - 2ab + b^2$

Example 1: Multiply.

a. $3x(x^4 - 7x^2y + 9y^2 - 2)$
$= 3x \cdot x^4 - 3x \cdot 7x^2y + 3x \cdot 9y^2 - 3x \cdot 2$
$= 3x^5 - 21x^3y + 27xy^2 - 6x$

b. $(3x + 5)(2x - 1)$
$= 3x \cdot 2x + 3x \cdot (-1) + 5 \cdot 2x + 5(-1)$
$= 6x^2 - 3x + 10x - 5$
$= 6x^2 + 7x - 5$

c. $(4x + 3)(4x - 3) = 16x^2 - 9$

d. $(3y + 2)^2 = 9y^2 + 12y + 4$

e. $(3y - 2)^2 = 9y^2 - 12y + 4$

Section 5.6 Exponent Rules and Dividing Polynomials

Quotient rule for exponents: If m and n are integers and a is a real number, where $a \neq 0$, then $a^m \div a^n = a^{m-n}$.

To divide monomials:
1. Divide the coefficients.
2. Use the quotient rule for the exponents with like bases.
3. Unlike bases are written unchanged in the quotient.
4. Write the final expression so that all exponents are positive.

If a, b, and c are real numbers, variables, or expressions with $c \neq 0$, then

$$\frac{a + b}{c} = \frac{a}{c} + \frac{b}{c}$$

To divide a polynomial by a monomial, divide each term in the polynomial by the monomial.

To divide a polynomial by a polynomial, use long division. If there is a remainder, write the result in the following form:

$$\text{quotient} + \frac{\text{remainder}}{\text{divisor}}$$

Example 1: Divide.

a. $x^7 \div x^2 = x^{7-2} = x^5$

b. $-39a^6b^5c^2 \div 13a^4b^5c$
$= -3a^{6-4}b^{5-5}c^{2-1}$
$= -3a^2b^0c^1$
$= -3a^2(1)c$
$= -3a^2c$

c. $\dfrac{30y^6 + 45y^3}{5y^2} = \dfrac{30y^6}{5y^2} + \dfrac{45y^3}{5y^2}$
$= 6y^{6-2} + 9y^{3-2}$
$= 6y^4 + 9y$

d. $\dfrac{6x^2 + 7x - 23}{3x - 4}$

$$\begin{array}{r} 2x + 5 \\ 3x - 4 \overline{) 6x^2 + 7x - 23} \\ \underline{-6x^2 + 8x} \\ 15x - 23 \\ \underline{-15x + 20} \\ -3 \end{array}$$

Answer: $2x + 5 + \dfrac{-3}{3x - 4}$, or

$2x + 5 - \dfrac{3}{3x - 4}$

Chapter 5   Review Exercises

For Exercises 1–6, answer true or false.

[5.1]   1. A negative exponent will always make the coefficient negative.

[5.2]   2. The degree of a binomial is two.

[5.4]   3. To raise an exponential form to a power, multiply the exponents.

[5.5]   4. Conjugates occur when both binomials contain subtraction.

[5.5]   5. FOIL can be used for all types of polynomial multiplication.

[5.5]   6. $(x - 4)^2 = x^2 + 16$

For Exercises 7–10, complete the rule.

[5.3]   7. To add polynomials, _____ like terms.

[5.4]   8. To multiply monomials:
  a. Multiply the _____.
  b. Add the _____ of the like bases.
  c. Rewrite any unlike bases in the product.

[5.4]   9. $(a^m)^n = a^?$, where a is a real number and m and n are integers.

[5.5]   10. $(a + b)^2 = a^2 +$ _____ $+ b^2$

Expressions

Exercises 11–112

[5.1] **For Exercises 11–14, evaluate the exponential form.**

11. $\left(\dfrac{2}{5}\right)^3$

12. -4^2

13. 5^{-2}

14. 13^{-1}

[5.1] **For Exercises 15–18, write the number in standard form.**

15. The radioactive half-life of uranium 238 is 4.5×10^9 years.

16. The temperature at the Sun's core is 1.38×10^7 degrees Celsius.

17. The mass of a hydrogen atom is 1.663×10^{-24} grams.

18. The mass of an electron is 2.006×10^{-30} pounds.

[5.1] **For Exercises 19–22, write the number in scientific notation.**

19. Because atoms are so small, their weight is measured in atomic mass units (AMU). 1 atomic mass unit is 0.0000000000000000000001661 grams.

20. One molecule of NaCl, sodium chloride (table salt) is 0.000000000000000000000963 grams.

21. The speed of light is about 300,000,000 meters per second.

22. The approximate radius of Earth is 6,370,000 meters.

[5.2] *For Exercises 23–26, determine whether the expression is a monomial.*

23. $3xy^5$

24. $-\dfrac{1}{2}x$

25. $\dfrac{4}{ab^3}$

26. $2x^2 - 9$

[5.2] *For Exercises 27–30, identify the coefficient and degree of each monomial.*

27. $6x^4$

28. 27

29. $-2.6xy^3$

30. $-m$

[5.2] *For Exercises 31–34, determine whether the expression is a monomial, binomial, trinomial, or none of these.*

31. $4x^2 - 25$

32. $-st^5$

33. $2x^2 - 5x + 7$

34. $5x^3 - 6x^2 - 3x + 11$

[5.2] *For Exercises 35–38, identify the degree of each polynomial and write each polynomial in descending order.*

35. $15 - 2x^9 - 3x + 21x^5 - 19x^3$

36. $22j^2 - 19 - j^4 + 5j$

37. $4u^5 - 18u^3 + 13u^4 - u - 21$

38. $16v^3 + 21v^4 - 2v + 6v^8 - 19 - v^8$

[5.2] *For Exercises 39–49, combine like terms. If possible, write the resulting polynomial in descending order.*

39. $8y^5 - 3y^4 + 2y - 4y^5 - 2y + 8 + 7y^4$

40. $8l + 5l^4 - 2l^3 - l + 18l^5 + 6l^2 + 20$

41. $3m - 4 - m^2 - 2m^3 + 7 - 2m + 8m^2$

42. $-9 + 7y - 2y - 6 + 2y - 3y^2$

43. $a^2bc - 3ab^2c + 2ab^2c - 6$

44. $5xyz^2 - 4x^2yz - 7xyz^2 - 5x^2yz - 2x^2zy$

45. $6cd - 4cd^2 + 3cd^2 + 8c^3 + 5cd - 12c^3 + 2d + 8$

46. $4a^2bc + 5abc^3 - 8abc - 9a^5 - 2abc^3 - 4a^2bc + 8 + a^5$

47. $18jk - 2j^4 + 12jk^3 - j^4 - 8jk - 9jk^3 + 12 + k^2$

48. $4m^2 - 3mn + 2mn^2 - 8n^2 - 6mn - 4m^2 + 7 + mn^2$

49. $2a + 3b - 4ab - c - 18c - 20a + b - 4b + 8$

[5.3] *For Exercises 50–59, add or subtract and write the resulting polynomial in descending order.*

50. $(5y - 2) + (6y - 8)$

51. $(8x^2 - 3) + (2x^2 + 3x - 1)$

52. $(5m - 1) - (2m + 7)$

53. $(5n + 8) - (2n^3 - 3n^2 + n + 5)$

54. $(2y - 4) + (4y + 8) - (6y - 5)$

55. $(3x^3 - 2x + 8) + (4x^2 - 3x - 1) - (4x^3 - 5x - 10)$

56. $(4p^3 + 2p^2 - 8p - 1) + (-5p^3 + 8p^2 - 2p + 10)$

57. $(2x^3 - 3x^2 - 2x - 1) - (18x^3 - 3x^2 + 2x + 1)$

58. $(5x^2 - 2xy + y^2) + (3x^2 + xy + 5y^2)$

59. $(m^2 + 5mn + 6) - (4m^2 - 3mn + 8)$

[5.4, 5.6] *For Exercises 60–74, simplify. Write all answers with positive exponents.*

60. $m \cdot m^4$

61. $2a \cdot 5a^8$

62. $(-3x^2 y)(2x^4 y^5)$

63. $(5x^2 y)^2$

64. $(-6u^3)^2(2u^4)$

65. $(2x^2)(2x^2)^4$

66. $\dfrac{x^3}{x^{-5}}$

67. $\dfrac{u^{-1}}{u^{-8}}$

68. $\dfrac{s^{-5}}{s}$

69. $\left(\dfrac{1}{x^3}\right)^{-2}$

70. $(2a^3 b^{-2})^{-3}$

71. $\dfrac{4hj^4 k^{-2}}{(3hj^{-2})^2}$

72. $-5x^{-2}$

73. $8x^0 - (2x)^0$

74. $(-18x^4 y^{-5})^0$

[5.5] *For Exercises 75–92, multiply.*

75. $2(x - 21)$

76. $-8(2x - 1)$

77. $4a(a - 3)$

78. $-6b(2b^2 - 4b - 1)$

79. $4m^3(-2m^4 - 3m^2 + m + 5)$

80. $-4abc^2(3a - 4ab^3 + 2abc^2 + 8)$

81. $(x + 5)(x - 1)$

82. $(y - 3)(y - 8)$

83. $(2m + 5)(6m - 1)$

84. $(5a + 2b)(3a - 2b)$

85. $(2x + 1)(2x - 1)$

86. $(x - 3)(x^2 + 2x + 5)$

87. $(3y - 1)(y^2 - 2y + 4)$

88. $(a^2 + ab + b^2)(a^2 - 2ab - b^2)$

89. $(4 - x)(4 + x)$

90. $(x - 6)^2$

91. $(3r - 5)^2$

92. $(6s + 2r)(6s - 2r)$

[5.5] 93. a. Write an expression in simplest form for the area of the figure shown.

b. Calculate the area if $x = 3$ feet.

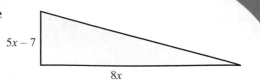

$5x - 7$

$8x$

[5.5] 94. A park ranger is studying the number of trees per square meter in a national forest. The area in the study is rectangular and measures 9×10^3 meters by 2.5×10^4 meters. Find the area of this region. Write your answer in scientific notation.

[5.5] 95. In a circuit, the current is measured to be 3.5×10^{-3} amperes. If the circuit has a resistance of 8.4×10^{-2} ohms, find the voltage. (*Hint:* Voltage = Current × Resistance)

[5.6] *For Exercises 96–101, divide.*

96. $\dfrac{4x - 20}{4}$

97. $\dfrac{5y^2 - 10y + 15}{-5}$

98. $\dfrac{2st + 20s^2t^4 - 4st^5}{2st}$

99. $\dfrac{a^3bc^3 - abc^2 + 2a^2bc^4}{abc^3}$

100. $\dfrac{6a^2b^2 + 3a^2b^3}{3a^2b^2}$

101. $\dfrac{2xyz^2 - 5x^2y^2z^2 + 10xy^2z}{5xy^3z}$

[5.6] 102. a. The area of the parallelogram shown is described by the polynomial $54x^2 - 60x$. Find the height.

b. Find the base, height, and area of this parallelogram if $x = 4$ inches.

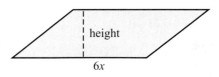

height

$6x$

[5.6] 103. Proxima Centauri is approximately 2.47×10^{13} miles from the Sun. If light travels 5.881×10^{12} miles per year, how long does the light from Proxima Centauri take to reach Earth? (*Hint:* Distance = Rate × Time)

[5.6] *For Exercises 104–111, use long division to divide the polynomials.*

104. $\dfrac{z^2 + 8z + 15}{z + 3}$

105. $\dfrac{3m^2 - 17m + 11}{m - 5}$

106. $\dfrac{2y^2 - 7y - 8}{2y + 1}$

107. $\dfrac{6m^2 - 10m + 5}{3m - 2}$

108. $\dfrac{x^3 - 1}{x + 1}$

109. $\dfrac{8x^3 - 27}{2x - 3}$

110. $\dfrac{2s^3 - 2s - 3}{s - 1}$

111. $\dfrac{10y^4 - 5y^3 - 6y^2 + 7y - 2}{2y - 1}$

[5.6] 112. A storage room is to have an area described by $4x^2 + 23x - 35$. If the length is described by $x + 7$, then find an expression for the width.

Chapter 5 Practice Test

For Exercises 1 and 2, evaluate the exponential form.

1. 2^{-3}

2. $\left(\dfrac{2}{3}\right)^{-2}$

3. Write 6.201×10^{-3} in standard form.

4. Write 275,000,000 in scientific notation.

For Exercises 5 and 6, identify the degree.

5. $-7x^2y$

6. $4x^2 - 9x^4 + 8x - 7$

7. Evaluate $-6mn - n^3$, where $m = 4$ and $n = -2$.

8. Combine like terms and write the resulting polynomial in descending order of degree.
   $-5x^4 + 7x^2 + 6x^2 - 5x^4 + 12 - 6x^3 + x^2$

For Exercises 9 and 10, add or subtract and write the resulting polynomial in descending order of degree.

9. $(3x^2 + 4x - 2) + (5x^2 - 3x - 2)$

10. $(7x^4 - 3x^2 + 4x + 1) - (2x^4 + 5x - 7)$

For Exercises 11–18, multiply.

11. $(4x^2)(3x^5)$

12. $(2ab^3c^7)(-a^5b)$

13. $(4xy^3)^2$

14. $3x(x^2 - 4x + 5)$

15. $-6t^2u(4t^3 - 8tu^2)$

16. $(n - 1)(n + 4)$

17. $(2x - 3)^2$

18. $(x + 2)(x^2 - 4x + 3)$

19. Write an expression in simplest form for the area of the shape shown.

$2n - 5$ [rectangle] $3n + 4$

For Exercises 20–25, simplify.

20. $x^9 \div x^4$

21. $\dfrac{(x^3)^{-2}}{x^4 \cdot x^{-5}}$

22. $\dfrac{(3y)^{-2}}{(x^3y^2)^{-3}}$

23. $\dfrac{24x^5 + 18x^3}{6x^2}$

24. $\dfrac{x^2 - x - 12}{x + 3}$

25. $\dfrac{15x^2 - 22x + 14}{3x - 2}$

6

Factoring

"Discipline is the keynote to learning. Discipline has been the great factor in my life."

—Rose Hoffman, Elementary schoolteacher

"The higher processes are all simplification."

—Willa Cather, U.S. novelist (1873–1947)

In Chapter 5, we learned about a specific class of expressions called *polynomials*. We learned that polynomials can be added, subtracted, multiplied, and divided in much the same way as numbers. In Chapter 6, we will explore polynomials further and in Sections 6.1 through 6.5, we will learn how to factor polynomial expressions. Factoring is another way to rewrite an expression and is one of the most important skills you will acquire in algebra. It is used extensively in simplifying the more complex expressions that we will study in future chapters. In Section 6.6, we will use factoring to solve certain equations that contain polynomial expressions. Finally, In Section 6.7, we will graph equations and functions containing polynomial expressions.

Greatest Common Factor and Factoring by Grouping

OBJECTIVES

1. List all possible factors for a given number.
2. Find the greatest common factor of a set of numbers or monomials.
3. Write a polynomial as a product of a monomial GCF and a polynomial.
4. Factor by grouping.

Often in mathematics we need to consider the factors of a number or expression. A number or expression written as a product of factors is said to be in **factored form**.

DEFINITION *Factored form:* A number or expression written as a product of factors.

Following are some examples of factored form:

An integer written in factored form with integer factors: $28 = 2 \cdot 14$
A monomial written in factored form with monomial factors: $8x^5 = 4x^2 \cdot 2x^3$
A polynomial written in factored form with a monomial factor and a polynomial factor: $2x + 8 = 2(x + 4)$
A polynomial written in factored form with two polynomial factors: $x^2 + 5x + 6 = (x + 2)(x + 3)$

Notice we can check the factored form by multiplying the factors to equal the product. The process of writing an expression in factored form is called *factoring*.

OBJECTIVE 1. List all possible factors for a given number. We begin our exploration of factoring by listing natural number factors when given an integer.

EXAMPLE 1 List all natural number factors of 24.

Solution To list all the natural number factors, we can divide 24 by 1, 2, 3, and so on, writing each divisor and quotient pair as a product until we have all possible combinations.

Note: *The number 5 does not divide 24 evenly and 6 is already in the 4 · 6 pair. When you reach a natural number that you already have listed, you can stop.*

$1 \cdot 24$
$2 \cdot 12$
$3 \cdot 8$
$4 \cdot 6$

The natural number factors of 24 are 1, 2, 3, 4, 6, 8, 12, 24.

ANSWERS

a. 1, 2, 3, 5, 6, 10, 15, 30
b. 1, 2, 3, 6, 7, 14, 21, 42

YOUR TURN List all natural number factors of each given number.

a. 30   **b.** 42

OBJECTIVE 2. Find the greatest common factor of a set of numbers or monomials. When factoring polynomials, the first step is to determine if there is a monomial factor that is common to all the terms in the polynomial. Further, we will want that monomial factor to be the **greatest common factor** of the terms. Let's first consider how to find the greatest common factor of numbers.

DEFINITION *Greatest common factor (GCF):* The largest natural number that divides all given numbers with no remainder.

For example, the greatest common factor of 12 and 18 is 6 because 6 is the largest number that divides into both 12 and 18 evenly.

Using Listing to Find the GCF of Numbers

One way to find the GCF of a given set of numbers is by listing factors.

PROCEDURE Listing Method for Finding GCF

To find the GCF of a set of numbers by listing:

1. List all possible factors for each given number.
2. Search the lists for the largest factor common to all lists.

EXAMPLE 2 Find the GCF of 24 and 60.

Solution Factors of 24: 1, 2, 3, 4, 6, 8, 12, 24
Factors of 60: 1, 2, 3, 4, 5, 6, 10, 12, 15, 20, 30, 60
The GCF of 24 and 60 is 12.

Using Prime Factorization to Find the GCF

The listing method is a good method for smaller numbers, but for larger numbers, it is not the most efficient method to use. It turns out we can use prime factorization, which we explained in Section 1.2, to find the GCF of a given set of numbers. Consider the prime factorizations for the numbers from Example 2.

$$24 = 2 \cdot 2 \cdot 2 \cdot 3 = 2^3 \cdot 3$$
$$60 = 2 \cdot 2 \cdot 3 \cdot 5 = 2^2 \cdot 3 \cdot 5$$

In Example 2, we found the GCF to be 12. Let's look at the prime factorization of 12 to see if we can discover a rule for finding the GCF.

$$12 = 2 \cdot 2 \cdot 3 = 2^2 \cdot 3$$

Notice that the 12 contains two factors of 2 and one factor of 3, which are the factors that are common to 24's and 60's prime factorizations. This suggests that we use only primes that are common to all factorizations involved. The factor 5 is not common to both 24's and 60's prime factorizations, so it is not included in the GCF.

$$24 = 2 \cdot 2 \cdot 2 \cdot 3 = 2^3 \cdot 3$$
$$60 = 2 \cdot 2 \cdot 3 \cdot 5 = 2^2 \cdot 3 \cdot 5$$

$$GCF = 2 \cdot 2 \cdot 3 = 2^2 \cdot 3 = 12$$

Learning Strategy

If you are a visual learner, you may prefer to continue circling the common prime factors.

Notice we could also compare exponents for a given common prime factor to determine how many of that factor should be included in the GCF. For example, 2^2 is included instead of 2^3 because 2^2 has the smaller exponent. This suggests the following procedure.

PROCEDURE   Prime Factorization Method for Finding GCF

To find the greatest common factor of a given set of numbers:

1. Write the prime factorization of each given number in exponential form.
2. Create a factorization for the GCF that includes only those prime factors common to all the factorizations, each raised to its smallest exponent in the factorization.
3. Multiply the factors in the factorization created in Step 2.

Note: If there are no common prime factors, then the GCF is 1.

Learning Strategy

If you are an auditory learner, you might find it easier to remember the procedure by thinking about the words in the procedure: To find the **greatest common** factor, you use the **common** primes raised to their **smallest** exponent. Notice that the words *greatest* and *smallest* are opposites.

EXAMPLE 3   Find the GCF of 3024 and 2520.

Solution Write the prime factorization of 3024 and 2520 in factored form.

$$3024 = 2^4 \cdot 3^3 \cdot 7$$
$$2520 = 2^3 \cdot 3^2 \cdot 5 \cdot 7$$

The common prime factors are 2, 3, and 7. Now we compare the exponents of each of these common factors. The smallest exponent of the factor of 2 is 3. The smallest exponent of 3 is 2. The smallest exponent of 7 is 1. The GCF will be the product of 2^3, 3^2, and 7.

$$GCF = 2^3 \cdot 3^2 \cdot 7 = 8 \cdot 9 \cdot 7 = 504$$

GCF of Monomials

We use a similar approach to find the GCF of a set of monomials that have variables. The variables in the monomials are treated like prime factors.

EXAMPLE 4 Find the GCF of $24x^2y$ and $60x^3$.

Solution Write the prime factorization of each monomial, treating the variables like prime factors.

$$24x^2y = 2^3 \cdot 3 \cdot x^2 \cdot y$$
$$60x^3 = 2^2 \cdot 3 \cdot 5 \cdot x^3$$

The common prime factors are 2, 3, and x. The smallest exponent for the factor of 2 is 2. The smallest exponent of 3 is 1. The smallest exponent of the x is 2. The GCF will be the product of 2^2, 3, and x^2.

$$GCF = 2^2 \cdot 3 \cdot x^2 = 12x^2$$

YOUR TURN Find the GCF.

a. 63, 84, and 105   b. $32a^2b$ and $40abc$ c. $35x^2$ and $18y$

OBJECTIVE 3. Write a polynomial as a product of a monomial GCF and a polynomial. Recall that when we multiply a monomial by a polynomial, we apply the distributive property.

$$2(x + 4) = 2 \cdot x + 2 \cdot 4$$
$$= 2x + 8$$

Now let's reverse this process. Suppose we are given the product, $2x + 8$, and we want to write the factored form, $2(x + 4)$. Notice that terms in $2x + 8$ have 2 as a common factor. When writing factored form, we first determine the GCF of the terms. We then create a missing factor statement like this:

$$2x + 8 = 2(?)$$

Notice that the missing factor must be $x + 4$. To determine this missing factor, we divide the product $2x + 8$ by the known factor, 2.

$$2x + 8 = 2 \, (?)$$
$$\frac{2x + 8}{2} = (?)$$
$$\frac{2x}{2} + \frac{8}{2} = (?)$$
$$x + 4 = (?)$$

Connection In Section 5.6, we learned that we divide a polynomial by a monomial by dividing each term in the polynomial by the monomial.

The factored form of $2x + 8$ is $2(x + 4)$, which suggests the following procedure.

PROCEDURE Factoring a Monomial GCF Out of a Polynomial

To factor a monomial GCF out of a given polynomial:

1. Find the GCF of the terms that make up the polynomial.
2. Rewrite the given polynomial as a product of the GCF and parentheses that contain the result of dividing the given polynomial by the GCF.

$$\text{Given polynomial} = \text{GCF}\left(\frac{\text{Given polynomial}}{\text{GCF}}\right)$$

EXAMPLE 5 Factor. $12x^2 + 18x$

Solution

1. Find the GCF of $12x^2$ and $18x$.

   The GCF of $12x^2$ and $18x$ is $6x$.

   Note: 6 *is the largest number that divides both* 12 *and* 18 *evenly, and* x *has the smaller exponent of the* x^2 *and* x.

2. Write the given polynomial as the product of the GCF and parentheses containing the quotient of the given polynomial and the GCF.

$$12x^2 + 18x = 6x\left(\frac{12x^2 + 18x}{6x}\right)$$

$$= 6x\left(\frac{12x^2}{6x} + \frac{18x}{6x}\right) \quad \text{Separate the terms.}$$

$$= 6x(2x + 3) \quad \text{Divide the terms by the GCF.}$$

Check We can check by multiplying the factored form using the distributive property.

$$6x(2x + 3) = 6x \cdot 2x + 6x \cdot 3 \quad \text{Distribute } 6x.$$

$$= 12x^2 + 18x \quad \begin{array}{l}\text{The product is the}\\\text{original polynomial.}\end{array}$$

Connection Keep in mind that when we factor, we are simply writing the original expression in a different form called *factored form*. When written equal to each other, the factored form and product make an identity, which means that every real number is a solution to the equation. Consider $12x^2 + 18x = 6x(2x + 3)$. Let's choose a value such as $x = 2$ and verify that it satisfies the equation. (Remember, we can select any real number.)

$$12x^2 + 18x = 6x(2x + 3)$$

$$12(2)^2 + 18(2) \stackrel{?}{=} 6(2)(2(2) + 3)$$

$$12(4) + 36 \stackrel{?}{=} 12(4 + 3)$$

$$48 + 36 \stackrel{?}{=} 12(7)$$

$$84 = 84 \quad \text{It checks.}$$

EXAMPLE 6 Factor. $24x^2y - 60x^3$

Solution

1. Find the GCF of $24x^2y$ and $60x^3$.

   We found this GCF in Example 4 to be $12x^2$.

2. Write the given polynomial as the product of the GCF and the parentheses containing the quotient of the given polynomial and the GCF.

$$24x^2y - 60x^3 = 12x^2\left(\frac{24x^2y - 60x^3}{12x^2}\right)$$

$$= 12x^2\left(\frac{24x^2y}{12x^2} - \frac{60x^3}{12x^2}\right)$$

$$= 12x^2(2y - 5x)$$

Connection Remember that when we divide exponential forms that have the same base, we subtract exponents.

Check Multiply the factored form using the distributive property.

$$12x^2(2y - 5x) = 12x^2 \cdot 2y - 12x^2 \cdot 5x$$

$$= 24x^2y - 60x^3$$

Factoring When the First Term Is Negative

Consider factoring the expression $-6x + 10y$. Because the first term of the polynomial is negative, when we factor out the GCF, 2, the first term inside the parentheses is also negative, so that we have $2(-3x + 5y)$. However, it is considered undesirable to have the first term inside parentheses negative. We can avoid a negative first term in the parentheses by factoring the negative of the GCF, so in our example we would factor out -2. This will change the sign of each term inside the parentheses so that we have $-2(3x - 5y)$.

EXAMPLE 7 Factor. $-18x^4y^3 + 9x^2y^2z - 12x^3y$

Solution

1. Find the GCF of $-18x^4y^3$, $9x^2y^2z$, and $-12x^3y$.

   Because the first term in the polynomial is negative, we will factor out the negative of the GCF to avoid a negative first term inside the parentheses. We will factor out $-3x^2y$.

2. Write the given polynomial as the product of the GCF and the parentheses containing the quotient of the given polynomial and the GCF.

$$-18x^4y^3 + 9x^2y^2z - 12x^3y = -3x^2y\left(\frac{-18x^4y^3 + 9x^2y^2z - 12x^3y}{-3x^2y}\right)$$

$$= -3x^2y\left(-\frac{18x^4y^3}{-3x^2y} + \frac{9x^2y^2z}{-3x^2y} - \frac{12x^3y}{-3x^2y}\right)$$

$$= -3x^2y(6x^2y^2 - 3yz + 4x)$$

Check Multiply the factored form using the distributive property.

$$-3x^2y(6x^2y^2 - 3yz + 4x) = -3x^2y \cdot 6x^2y^2 - 3x^2y \cdot (-3yz) - 3x^2y \cdot 4x$$
$$= -18x^4y^3 + 9x^2y^2z - 12x^3y$$

YOUR TURN Factor.

a. $30xy - 45x$ **b.** $-48a^4b^5 - 24a^3b^4 + 16ab^2c$

Before we discuss additional techniques of factoring, it is important to state that no matter what type of polynomial we are asked to factor, we will always first consider whether a monomial GCF (other than 1) can be factored out of the polynomial.

Factoring When the GCF Is a Polynomial

Sometimes, when factoring, the GCF is a polynomial.

EXAMPLE 8 Factor. $y(x + 2) + 7(x + 2)$

Solution Notice that this expression is a sum of two products, y and $(x + 2)$, and 7 and $(x + 2)$. Further, note that $(x + 2)$ is the GCF of the two products.

$$y(x + 2) + 7(x + 2) = (x + 2)\left(\frac{y(x + 2) + 7(x + 2)}{x + 2}\right)$$

$$= (x + 2)\left(\frac{y(x + 2)}{x + 2} + \frac{7(x + 2)}{x + 2}\right)$$

$$= (x + 2)(y + 7)$$

Note: *The parentheses are filled in the same way, by dividing the original expression by the GCF.*

YOUR TURN Factor. $4a(a - 3) - b(a - 3)$

OBJECTIVE 4. Factor by grouping. The process of factoring out a polynomial GCF, as we did in Example 8, is an intermediate step in a process called *factoring by grouping*, which is a technique that we try when factoring a four-term polynomial such as $xy + 2y + 7x + 14$. The method is called *grouping* because we group pairs of terms together and look for a common factor within each group. We begin by pairing the first two terms together as one group and the last two terms as a second group.

$$xy + 2y + 7x + 14 = (xy + 2y) + (7x + 14)$$

Note that the first two terms have a common factor of y and the last two terms have 7 as a common factor. If we factor the y out of the first two terms and the 7 out of the last two terms, we have the same expression that we factored in Example 8:

$$xy + 2y + 7x + 14 = (xy + 2y) + (7x + 14)$$
$$= y(x + 2) + 7(x + 2)$$
$$= (x + 2)(y + 7) \qquad \text{Factor out } (x + 2).$$

ANSWERS

a. $15x(2y - 3)$
b. $-8ab^2(6a^3b^3 + 3a^2b^2 - 2c)$ or $8ab^2(2c - 6a^3b^3 - 3a^2b^2)$

ANSWER

$(a - 3)(4a - b)$

To factor a four-term polynomial by grouping:

1. Factor out any monomial GCF (other than 1) that is common to all four terms.
2. Group together pairs of terms and factor the GCF out of each pair.
3. If there is a common binomial factor, then factor it out.
4. If there is no common binomial factor, then interchange the middle two terms and repeat the process. If there is still no common binomial factor, then the polynomial cannot be factored by grouping.

EXAMPLE 9   Factor.

a. $6x^3 - 8x^2 + 3xy - 4y$

Solution   First, we look for a monomial GCF (other than 1). This polynomial does not have one. Because the polynomial has four terms, we now try to factor by grouping.

$$6x^3 - 8x^2 + 3xy - 4y = (6x^3 - 8x^2) + (3xy - 4y)$$

Factor $2x^2$ out of $6x^3$ and $8x^2$, then factor y out of $3xy$ and $4y$.

$$= 2x^2(3x - 4) + y(3x - 4)$$

$$= (3x - 4)(2x^2 + y)$$

Factor out $3x - 4$.

b. $12mn^2 - 20mn - 24n^2 + 40n$

Solution   Note that in this case there is a monomial GCF, $4n$, so we first factor this GCF out of all four terms.

$$12mn^2 - 20mn - 24n^2 + 40n = 4n(3mn - 5m - 6n + 10)$$

Because the polynomial in the parentheses has four terms, we try to factor by grouping.

$$= 4n[(3mn - 5m) + (-6n + 10)]$$

Factor m out of $3mn$ and $5m$, then factor -2 out of $-6n$ and 10.

$$= 4n[m(3n - 5) - 2(3n - 5)]$$

$$= 4n(3n - 5)(m - 2)$$

Factor out $3n - 5$.

Note: *Now that the expression is in factored form, we no longer need the brackets.*

YOUR TURN   Factor.

a. $6x^2 + 15xz + 2xy + 5yz$

b. $42a^2b - 56a^2 - 12ab + 16a$

ANSWERS

a. $(2x + 5z)(3x + y)$
b. $2a(3b - 4)(7a - 2)$

1. Explain in your own words how to list all possible factors of a number.

2. Define the GCF of a given set of numbers in your own words.

3. Given a set of terms, after finding the prime factorization of each term, how do you use those prime factors to create the GCF's factorization?

4. When factoring a monomial GCF out of the terms of a polynomial, after finding the GCF, how do you rewrite the polynomial?

5. What kinds of polynomials are factored by grouping?

6. When factoring by grouping, after factoring out a monomial GCF, what is the next step?

For Exercises 7–22, list all possible factors of the given number.

7. 9	**8.** 49	**9.** 33	**10.** 21
11. 18	**12.** 12	**13.** 16	**14.** 81
15. 44	**16.** 45	**17.** 60	**18.** 36
19. 56	**20.** 64	**21.** 90	**22.** 84

For Exercises 23–34, find the GCF.

23. 21, 30	**24.** 34, 51	**25.** 72, 80	**26.** 35, 60
27. 12, 42, 60	**28.** 10, 18, 36	**29.** $4xy, 6xy$	**30.** $8x, 2xy$
31. $25h^2, 60h^4$	**32.** $25x^4, 10x^3$	**33.** $6a^5b, 15a^4b^2$	**34.** $7m^6n, 21m^5n^4$

For Exercises 35–66, factor by factoring out the GCF.

35. $5c - 20$	**36.** $8y - 24$	**37.** $8x - 12$	**38.** $15x - 10$

39. $x^2 - x$

40. $y^2 + y$

41. $18z^6 - 12z^4$

42. $15k^3 - 24k^2$

43. $18p^3 - 15p^5$

44. $18r^4 - 27r^6$

45. $6a^2b - 3ab^2$

46. $9m^2n - 3mn^2$

47. $14uv^2 - 7uv$

48. $12x^2y^3 - 6xy^2$

49. $25xy - 50xz + 100x$

50. $8x + 8y - 2xy$

51. $x^2y + xy^2 + x^3y^3$

52. $w^3v^2 + w^2v + wv^2$

53. $28ab^3c - 36a^2b^2c$

54. $30x^3yz^2 - 24x^2y^3z$

55. $-20p^2q - 24pq + 16pq^2$

56. $-6a^2c - 18abc + 12ac^2$

57. $3mn^5p^2 + 18mn^3p - 6mnp$

58. $4axy^2 - 6axyz + 2ayz$

59. $105a^3b^2 - 63a^2b^3 + 84a^6b^4$

60. $21t^4u^3 - 105t^5u^3v + 63t^6u^2v^4$

61. $18x^4 - 9x^3 + 30x^2 - 12x$

62. $10g + 40g^3 - 100g^4 + 120g^5$

63. $2(a - 3) + y(a - 3)$

64. $a(x + y) + b(x + y)$

65. $ay(4x - 3) - 2(4x - 3)$

66. $4m(2x - 3y) - 5(2x - 3y)$

For Exercises 67–90, factor by grouping.

67. $bx + 2b + cx + 2c$

68. $cx + cy + bx + by$

69. $am - an - bm + bn$

70. $xy - xw - yz + wz$

71. $x^3 + 2x^2 - 3x - 6$

72. $y^4 + 4y^3 - by - 4b$

73. $1 - m + m^2 - m^3$

74. $x^2y + y + x^2 + 1$

75. $3xy + 5y + 6x + 10$

76. $2ab + 8a + 3b + 12$

77. $4b^2 - b + 4b - 1$

78. $x^3 - 4x^2 + x - 4$

79. $6 + 2b - 3a - ab$

80. $u^2 - 6u - 4u + 24$

81. $3ax + 6ay + 8by + 4bx$

82. $ac + 2ad + 2bc + 4bd$

83. $x^2y - x^2s - ry + rs$

84. $3x + 3y - x^2 - xy$

85. $w^2 + 3wz + 5w + 15z$

86. $a^2 + 4ab + 3a + 12b$

87. $3st + 3ty - 2s - 2y$

88. $my + 3y - 5m - 15$

89. $ax^2 - 5y^2 + ay^2 - 5x^2$

90. $m^2p - 3n^2 + n^2p - 3m^2$

For Exercises 91–98, factor completely.

91. $2a^2 + 2ab + 6a + 6b$

92. $12pq + 8q + 30p + 20$

93. $3m^2n - 12m^2 - 12m^3 + 3mn$

94. $4a^2b - 4a^2 - 4ab + 4a$

95. $3a^2y - 12a^2 + 9ay - 36a$

96. $10a^2b^2 - 10b^3 + 15a^2b - 15b^2$

97. $2x^3 + 6x^2y + 10x^2 + 30xy$

98. $10uv^2 - 5v^2 + 30uv - 15v$

For Exercises 99–100, write an expression for the area of the shaded region, then factor completely.

99.

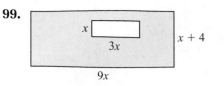

100.

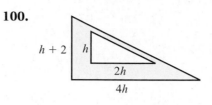

101. The diagram shows the floor plan of a room with the hearth of a fireplace. Write an expression in factored form of the area of the room excluding the hearth, which is 4 feet by 1 foot.

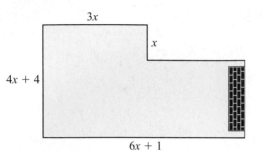

102. Write an expression in factored form for the area of the side of the house shown excluding the window.

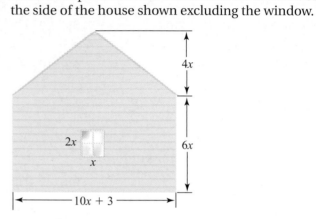

For Exercises 103–104, write an expression for the volume of the object shown, then factor completely.

103.

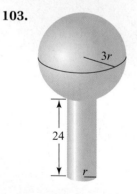

104.

REVIEW EXERCISES

[5.1] **1.** Write 3.74×10^9 in standard form.

[5.1] **2.** Write 45,600,000 in scientific notation.

[5.5] *For Exercises 3–6, multiply.*

3. $(x + 2)(x + 5)$

4. $(x - 4)(x + 3)$

5. $(x - 5)(x - 3)$

6. $x(x - 7)$

6.2 Factoring Trinomials of the Form $x^2 + bx + c$

OBJECTIVES

1. Factor trinomials of the form $x^2 + bx + c$.

2. Factor out a monomial GCF, then factor the trinomial of the form $x^2 + bx + c$.

OBJECTIVE 1. Factor trinomials of the form $x^2 + bx + c$. In this section, we consider trinomials of the form $x^2 + bx + c$. Notice that the coefficient of the squared term is 1. In Section 6.3, we will consider trinomials of this form in which the squared term has a coefficient other than 1. Following are some examples of trinomials of the form $x^2 + bx + c$.

$$x^2 + 5x + 6 \quad \text{or} \quad x^2 - 7x + 12 \quad \text{or} \quad x^2 + 2x - 15 \quad \text{or} \quad x^2 - 5x - 24$$

Trinomials of the form $x^2 + bx + c$ are the result of the product of two binomials. Consider $(x + 2)(x + 3)$. Recall from Section 5.5 that we use FOIL to multiply two binomials. In our use of FOIL to multiply $(x + 2)(x + 3)$, pay attention to how the last term and the middle term of the resulting trinomial are generated.

$$\begin{array}{cccc} \text{F} & \text{O} & \text{I} & \text{L} \\ \end{array}$$
$$(x + 2)(x + 3) = x^2 + 3x + 2x + 6$$
$$= x^2 \quad + 5x \quad + 6$$

Note: *These two numbers multiply to equal the last term, 6, and add to equal the coefficient of the middle term, 5.*

sum of 2 and 3   product of 2 and 3

When we factor a trinomial of the form $x^2 + bx + c$, we reverse the FOIL process, using the fact that b is the sum of the last terms in the binomials and c is the product of the last terms in the binomials. For example, to factor $x^2 + 5x + 6$, we must find two binomials with last terms whose product is 6 and whose sum is 5.

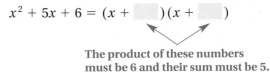

$$x^2 + 5x + 6 = (x + \boxed{})(x + \boxed{})$$

The product of these numbers must be 6 and their sum must be 5.

If unsure of what those two numbers are, we could list all factor pairs of 6, and then find the pair whose sum is 5.

Factors Pairs of 6	Corresponding Sum	
$1 \cdot 6$	$1 + 6 = 7$	
$2 \cdot 3$	$2 + 3 = 5$	← This is the correct pair.

Notice the 2 and 3 become the last terms in the binomials so that the factored form for $x^2 + 5x + 6$ looks like this:

$$x^2 + 5x + 6 = (x + 2)(x + 3)$$

This suggests the following procedure.

PROCEDURE   Factoring $x^2 + bx + c$

To factor a trinomial of the form $x^2 + bx + c$:

1. Find two numbers with a product equal to c and a sum equal to b.
2. The factored trinomial will have the form:
   $(x + \text{first number})(x + \text{second number})$.

Note: The signs in the binomial factors can be minus signs, depending on the signs of b and c.

In the following examples, we explore cases where minus signs occur. Pay attention to how we determine the signs in the binomial factors.

EXAMPLE 1   Factor.

a. $x^2 - 7x + 12$

Solution We must find a pair of numbers whose product is 12 and whose sum is -7. If two numbers have a positive product and negative sum, they must both be negative. Following is a table listing the products and sums:

Product	Sum
$(-1)(-12) = 12$	$-1 + (-12) = -13$
$(-2)(-6) = 12$	$-2 + (-6) = -8$
$(-3)(-4) = 12$	$-3 + (-4) = -7$

This is the correct combination, so -3 and -4 are the second terms in each binomial of the factored form.

Answer $x^2 - 7x + 12 = (x - 3)(x - 4)$

Note: *We could have first written $(x + (-3))(x + (-4))$ to follow the procedure and then simplified.*

Check   We can check by multiplying the binomial factors to see if their product is the original polynomial.

$(x - 3)(x - 4) = x^2 - 4x - 3x + 12$ Multiply the factors using FOIL.
$\qquad\qquad\qquad = x^2 - 7x + 12$ The product is the original polynomial.

b. $x^2 + 2x - 15$

Solution We must find a pair of numbers whose product is -15 and whose sum is 2. Since the product is negative, the two numbers must have different signs. Since the sum is positive, the number with the greater absolute value will be positive.

Product	Sum
$(-1)(15) = -15$	$-1 + 15 = 14$
$(-3)(5) = -15$	$-3 + 5 = 2$

This is the correct combination, so -3 and 5 are the second terms in each binomial of the factored form.

Answer $x^2 + 2x - 15 = (x - 3)(x + 5)$
Check $(x - 3)(x + 5) = x^2 + 5x - 3x - 15$ Multiply the factors using FOIL.
$\qquad\qquad\qquad = x^2 + 2x - 15$ The product is the original polynomial.

c. $x^2 - 5x - 24$

Solution We must find a pair of numbers whose product is -24 and whose sum is -5. Since the product is negative, the two numbers must have different signs. Since the sum is also negative, the number with the greater absolute value will be negative.

Product	Sum	
$(1)(-24) = -24$	$1 + (-24) = -23$	This is the correct combination, so 3 and -8 are the second terms in each binomial of the factored form.
$(2)(-12) = -24$	$2 + (-12) = -10$	
$(3)(-8) = -24$	$3 + (-8) = -5$ ⟵	

Answer $x^2 - 5x - 24 = (x + 3)(x - 8)$

Note: *In Examples 1c–3, we leave the checks to the reader.*

YOUR TURN Factor.

a. $y^2 - 11y + 18$ **b.** $t^2 + 3t - 28$ **c.** $n^2 - n - 30$

Now let's see how to factor trinomials containing two variables.

EXAMPLE 2 Factor.

a. $x^2 - 10xy + 25y^2$

Solution Since y is in the last term, it is helpful to rearrange the order of the variables in the middle term, writing the polynomial as $x^2 - 10yx + 25y^2$, so that we view the "coefficient" of $-10yx$ as $-10y$. We must find a pair of terms whose product is $25y^2$ and whose sum is $-10y$. These terms would have to be $-5y$ and $-5y$.

Answer $x^2 - 10xy + 25y^2 = (x - 5y)(x - 5y)$

b. $a^2 + 2ab - 15b^2$

Solution As in Example 4, we can view the coefficient of a in the middle term as $2b$. We must find a pair of terms whose product is $-15b^2$ and whose sum is $2b$. These terms would have to be $-3b$ and $5b$.

Answer $a^2 + 2ab - 15b^2 = (a - 3b)(a + 5b)$

YOUR TURN Factor.

a. $m^2 + 7mn - 18n^2$ **b.** $x^2 - xy - 12y^2$

OBJECTIVE 2. Factor out a monomial GCF, then factor the trinomial of the form $x^2 + bx + c$. Sometimes there is a monomial GCF (other than 1) in the three terms of a trinomial. Consider $2x^3 + 10x^2 + 12x$. Notice the monomial $2x$ is the GCF of the terms. Factoring out this monomial, we have

$$2x^3 + 10x^2 + 12x = 2x(x^2 + 5x + 6)$$

Now we try to factor the trinomial within the parentheses. We look for two numbers whose product is 6 and whose sum is 5. Note that 2 and 3 work, so we can write:

$$2x^3 + 10x^2 + 12x = 2x(x^2 + 5x + 6)$$
$$= 2x(x + 2)(x + 3) \longleftarrow \boxed{\text{Factored form}}$$

Whenever factoring polynomials, always look for a monomial GCF among the terms.

EXAMPLE 3) Factor.

a. $3mn^3 + 12mn^2 - 96mn$

Solution First, we look for a monomial GCF (other than 1). Notice that the GCF of the terms is $3mn$. Factoring out this monomial, we have

$$3mn^3 + 12mn^2 - 96mn = 3mn(n^2 + 4n - 32)$$

Now try to factor the trinomial to two binomials. We must find a pair of numbers whose product is -32 and whose sum is 4.

Product	Sum	
$(-1)(32) = -32$	$-1 + 32 = 31$	**This is the correct combination, so**
$(-2)(16) = -32$	$-2 + 16 = 14$	-4 **and 8 are the second terms in**
$(-4)(8) = -32$	$-4 + 8 = 4$ $\longleftarrow$	**each binomial.**

Answer $3mn^3 + 12mn^2 - 96mn = 3mn(n - 4)(n + 8)$

b. $x^4 + 6x^3 + 14x^2$

Solution First, we factor out the monomial GCF, x^2.

$$x^4 + 6x^3 + 14x^2 = x^2(x^2 + 6x + 14)$$

Now we try to factor the trinomial to two binomials. We must find a pair of numbers whose product is 14 and whose sum is 6.

Product	Sum
$(1)(14) = 14$	$1 + 14 = 15$
$(2)(7) = 14$	$2 + 7 = 9$
$(-1)(-14) = 14$	$-1 + (-14) = -15$
$(-2)(-7) = 14$	$-2 + (-7) = -9$

Note: *We listed all factor pairs of 14 and found no combination whose sum is 6. This means that $x^2 + 6x + 14$ has no binomial factors with integer terms, so $x^2(x^2 + 6x + 14)$ is the final factored form.*

A polynomial like $x^2 + 6x + 14$ that cannot be factored is like a prime number in that its factors are only 1 and the polynomial itself. We say such a polynomial is *prime*.

ANSWERS

a. $y^2(y - 9)(y - 1)$
b. $4ab(a - 7)(a + 2)$
c. $4x(y^2 - 2y + 3)$

YOUR TURN) Factor.

a. $y^4 - 10y^3 + 9y^2$ **b.** $4a^3b - 20a^2b - 56ab$ **c.** $4xy^2 - 8xy + 12x$

1. If given a trinomial in the form $x^2 + bx + c$, where $c > 0$ and $b > 0$, what can you conclude about the pair of numbers in the binomial factors?

2. If given a trinomial in the form $x^2 - bx + c$, where $c > 0$ and $b > 0$, what can you conclude about the pair of numbers in the binomial factors?

3. If given a trinomial in the form $x^2 + bx - c$, where $c > 0$ and $b > 0$, what can you conclude about the pair of numbers in the binomial factors?

4. If given a trinomial in the form $x^2 - bx - c$, where $c > 0$ and $b > 0$, what can you conclude about the pair of numbers in the binomial factors?

For Exercises 5–12, fill in the missing values in the factors.

5. $x^2 + 7x + 10 = (x + 2)(x +)$

6. $y^2 + 8y + 15 = (y + 5)(y +)$

7. $n^2 - 12n + 20 = (n - 10)(n -)$

8. $t^2 - 9t + 18 = (t -)(t - 3)$

9. $m^2 - 2m - 24 = (m -)(m + 4)$

10. $a^2 + a - 12 = (a + 4)(a -)$

11. $u^2 + 13u - 30 = (u -)(u + 15)$

12. $n^2 - 9n - 36 = (n -)(n + 3)$

For Exercises 13–44, factor.

13. $r^2 + 4r + 3$

14. $t^2 + 8t + 7$

15. $x^2 - 8x + 7$

16. $x^2 - 4x + 3$

17. $z^2 - 2z - 3$

18. $y^2 - 4y - 5$

19. $y^2 + 5y + 6$

20. $n^2 + 5n + 4$

21. $u^2 - 6u + 8$

22. $b^2 - 10b + 21$

23. $u^2 + u - 6$

24. $k^2 - 3k - 10$

25. $a^2 + 7a + 12$

26. $x^2 + 14x + 45$

27. $y^2 - 12y + 36$

28. $x^2 - 10x + 25$

29. $w^2 - w - 12$

30. $b^2 - 3b - 40$

31. $x^2 - x - 30$

32. $b^2 - 6b - 16$

33. $n^2 - 11n + 30$

34. $y^2 - 13y + 12$

35. $r^2 - 9r + 18$

36. $x^2 - 10x + 24$

37. $x^2 - 5x - 24$ **38.** $a^2 - 10a - 24$ **39.** $p^2 - 5p - 36$ **40.** $u^2 - 8u - 20$

41. $m^2 - 4m + 6$ **42.** $z^2 + 7z + 8$ **43.** $x^2 - 6x - 8$ **44.** $a^2 + 3a - 20$

For Exercises 45–52, factor the trinomials containing two variables.

45. $p^2 + 6pq + 9q^2$ **46.** $x^2 - 9xy + 14y^2$ **47.** $a^2 - 6ab - 27b^2$ **48.** $m^2 + mn - 16n^2$

49. $x^2 - 14xy + 24y^2$ **50.** $t^2 - 11tu + 24u^2$ **51.** $r^2 - rs - 30s^2$ **52.** $h^2 - 8hk - 20k^2$

For Exercises 53–70, factor completely.

53. $4x^2 - 40x + 84$ **54.** $3m^2 - 33m + 54$ **55.** $2m^3 - 14m^2 + 12m$

56. $2k^2y - 18ky + 28y$ **57.** $3a^2b - 15ab - 72b$ **58.** $3a^2y + 6ay - 72y$

59. $4x^2 - 24x + 36$ **60.** $6a^2r + 12ar + 6r$ **61.** $n^4 + 5n^3 + 6n^2$

62. $r^4 + 6r^3 + 8r^2$ **63.** $7u^4 + 42u^3 + 35u^2$ **64.** $5x^5 + 20x^4 + 15x^3$

65. $6a^2b^2c - 36ab^2c + 48b^2c$ **66.** $3a^2b^2c - 24ab^2c + 45b^2c$ **67.** $3x^2 - 21xy + 30y^2$

68. $5m^2 - 30mn + 40n^2$ **69.** $2a^2b - 6ab^2 - 36b^3$ **70.** $7h^2k + 14hk^2 - 168k^3$

For Exercises 71–74, find and correct the mistake.

71. $x^2 - 5x - 6 = (x - 3)(x - 2)$ **72.** $x^2 - 5x + 6 = (x - 6)(x + 1)$

73. $x^2 - 3x - 4 = (x - 1)(x + 4)$ **74.** $x^2 + x + 2 = (x + 2)(x + 1)$

For Exercises 75–78, find all natural number values of b that make the trinomial factorable.

75. $x^2 + bx - 21$ **76.** $x^2 + bx - 10$ **77.** $x^2 + bx + 12$ **78.** $x^2 + bx + 18$

For Exercises 79–82, find all natural numbers c that make the trinomial factorable.

79. $x^2 - 7x + c$ **80.** $x^2 - 5x + c$ **81.** $x^2 + 10x + c$ **82.** $x^2 + 11x + c$

83. The expression $h^2 + 6h + 8$ describes the area of the top (or bottom) of the crate shown, where h represents its height. The unknown expression for the length is the sum of h and an integer and the expression for the width is the sum of h and a different integer. Find expressions for the length and width.

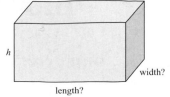
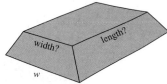

84. The expression $w^2 + w - 2$ describes the area of the top of the bar of gold shown, where w represents the width of the base of the bar. The unknown expression for the length of the top of the bar is the sum of w and an integer. The expression for the width of the top of the bar is the difference of w and an integer. Find the expressions for the length and width of the top of the bar.

85. The expression $w^2 - 16w + 60$ describes the viewing area in the picture frame shown, where w represents the width of the frame. The unknown expression for the length of the viewing area is the difference of w and an integer. The expression for the width of the viewing area is the difference of w and a different integer. Find expressions for the length and width of the viewing area in the frame.

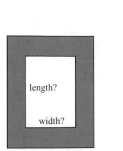

86. The expression $\pi r^2 - 2\pi r + \pi$ describes the area occupied by the circular base of the papasan chair shown, where r represents the radius of the circle on which the chair rests. The expression that describes the radius of the circle that touches the floor is the difference of r and an integer. Find the expression that describes the radius of the circle that touches the floor.

REVIEW EXERCISES

[1.4]   **1.** What property of arithmetic is illustrated by $(6x + 1)(5x - 2) = (5x - 2)(6x + 1)$?

[5.5] *For Exercises 2–4, multiply.*

  2. $(2x + 3)(5x + 1)$ **3.** $(3y - 4)(5y + 6)$ **4.** $4n(6n - 1)(3n - 2)$

[6.1]   **5.** List all factors of 48.   **[6.1]**   **6.** What is the GCF of $6x$ and $35y$?

6.3 Factoring Trinomials of the Form $ax^2 + bx + c$ where $a \neq 1$

OBJECTIVES

1. Factor trinomials of the form $ax^2 + bx + c$, where $a \neq 1$, by trial.
2. Factor trinomials of the form $ax^2 + bx + c$, where $a \neq 1$, by grouping.

OBJECTIVE 1. Factor trinomials of the form $ax^2 + bx + c$, where $a \neq 1$, by trial. In Section 6.2, we factored trinomials in which the coefficient of the squared term was 1. Now we focus on factoring trinomials in which the coefficient of the squared term is other than 1, such as the following:

$$3x^2 + 17x + 10 \qquad 8x^2 + 29x - 12$$

In general, like trinomials of the form $x^2 + bx + c$, trinomials of the form $ax^2 + bx + c$, where $a \neq 1$, also have two binomial factors. When a is not 1, we must consider the factors of the ax^2 term and the factors of the c term. First we will develop a trial-and-error method for finding the factored form. Consider $3x^2 + 17x + 10$. Since all the terms are positive, we know that the terms in the binomial factors will all be positive.

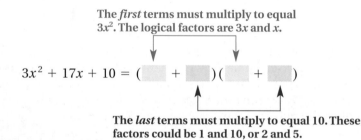

The *first* terms must multiply to equal $3x^2$. The logical factors are $3x$ and x.

$$3x^2 + 17x + 10 = (+)(+)$$

The *last* terms must multiply to equal 10. These factors could be 1 and 10, or 2 and 5.

We now try various combinations of these *first* and *last* terms in binomial factors and consider the resulting trinomial products. If the sum of the inner and outer terms matches the original trinomial, then we know we have the correct combination in the binomial factors.

$$(3x + 10)(x + 1) = 3x^2 + 3x + 10x + 10 = 3x^2 + 13x + 10 \left.\right\}$$
$$(3x + 1)(x + 10) = 3x^2 + 30x + x + 10 = 3x^2 + 31x + 10 \quad \text{Incorrect combinations.}$$
$$(3x + 5)(x + 2) = 3x^2 + 6x + 5x + 10 = 3x^2 + 11x + 10 \left.\right\}$$
$$(3x + 2)(x + 5) = 3x^2 + 15x + 2x + 10 = 3x^2 + 17x + 10 \quad \text{Correct combination.}$$

You may find that drawing lines to connect each product is helpful.

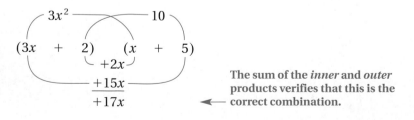

The sum of the *inner* and *outer* products verifies that this is the ← correct combination.

Our example suggests the following procedure.

PROCEDURE **Factoring by Trial and Error**

To factor a trinomial of the form $ax^2 + bx + c$, where $a \neq 1$, by trial and error:

1. Look for a monomial GCF in all the terms. If there is one, factor it out.
2. Write a pair of *first* terms whose product is ax^2.

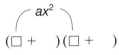

3. Write a pair of *last* terms whose product is c.

$$(\square + \square)(\square + \square)$$

4. Verify that the sum of the *inner* and *outer* products is bx (the middle term of the trinomial).

If the sum of the inner and outer products is not bx, then try the following:
a. Exchange the first terms of the binomials from step 3, then repeat step 4.
b. Exchange the last terms of the binomials from step 3, then repeat step 4.
c. Repeat steps 2–4 with a different combination of first and last terms.

EXAMPLE 1 ⟩ Factor. $8x^2 + 29x - 12$

Solution

The *first* terms must multiply to equal $8x^2$.
These could be x and $8x$, or $2x$ and $4x$.

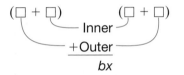

The *last* terms must multiply to equal -12. Because -12 is negative, the last terms in the binomials must have different signs. We have already written the appropriate signs, so these factor pairs could be 1 and 12, 2 and 6, or 3 and 4.

Now we multiply binomials with various combinations of these first and last terms until we find a combination whose inner and outer products combine to equal $29x$.

$$(8x + 1)(x - 12) = 8x^2 - 96x + x - 12 = 8x^2 - 95x - 12$$
$$(8x + 12)(x - 1) = 8x^2 - 8x + 12x - 12 = 8x^2 + 4x - 12$$
$$(8x + 2)(x - 6) = 8x^2 - 48x + 2x - 12 = 8x^2 - 46x - 12$$
$$(8x + 6)(x - 2) = 8x^2 - 16x + 6x - 12 = 8x^2 - 10x - 12$$
$$(8x + 3)(x - 4) = 8x^2 - 32x + 3x - 12 = 8x^2 - 29x - 12$$

Incorrect combinations.

Note: *This middle term has the correct absolute value, but the incorrect sign (we are looking for +29x). Switching the + and − signs in the binomials will give us +29x.*

$$(8x - 3)(x + 4) = 8x^2 + 32x - 3x - 12 = 8x^2 + 29x - 12 \qquad \text{Correct combination.}$$

Answer $8x^2 + 29x - 12 = (8x - 3)(x + 4)$

YOUR TURN Factor. $6y^2 + 13y - 5$

Factoring Out a Monomial GCF First

Remember that we should always look to see if a monomial GCF can be factored out of the terms in the trinomial.

EXAMPLE 2 Factor.

a. $40x^3 - 68x^2 + 12x$

Solution First, we factor out the monomial GCF, $4x$.

$$40x^3 - 68x^2 + 12x = 4x(10x^2 - 17x + 3)$$

Now we factor the trinomial within the parentheses. The fact that 3 is positive and $-17x$ is negative indicates that the signs of the last terms in the binomial factors will both be minus signs.

The *first* terms must multiply to equal $10x^2$.
These could be x and $10x$, or $2x$ and $5x$.

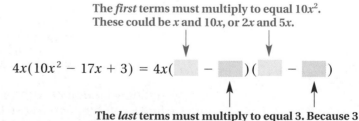

$$4x(10x^2 - 17x + 3) = 4x(\boxed{} - \boxed{})(\boxed{} - \boxed{})$$

The *last* terms must multiply to equal 3. Because 3 is a prime number, its factors are 1 and 3.

ANSWER

$(2y + 5)(3y - 1)$

Now multiply binomials with various combinations of these first and last terms until we find a combination whose inner and outer products combine to equal $-17x$.

$$4x(x - 1)(10x - 3) = 4x(10x^2 - 3x - 10x + 3) = 4x(10x^2 - 13x + 3)$$
$$4x(10x - 1)(x - 3) = 4x(10x^2 - 30x - x + 3) = 4x(10x^2 - 31x + 3)$$
$$4x(2x - 1)(5x - 3) = 4x(10x^2 - 6x - 5x + 3) = 4x(10x^2 - 11x + 3)$$

Incorrect combinations.

$$4x(5x - 1)(2x - 3) = 4x(10x^2 - 15x - 2x + 3) = 4x(10x^2 - 17x + 3)$$

Correct combination.

Answer $\quad 40x^3 - 68x^2 + 12x = 4x(5x - 1)(2x - 3)$

b. $8x^3 - 10x^2y - 18xy^2$

Solution First, factor out the monomial GCF, $2x$.

$$8x^3 - 10x^2y - 18xy^2 = 2x(4x^2 - 5xy - 9y^2)$$

Now we factor the trinomial within the parentheses. The fact that the $-9y^2$ is negative and $-5xy$ is negative indicates that the signs of the last terms in the binomial factors will be different.

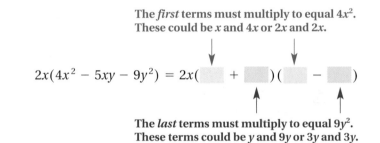

The *first* terms must multiply to equal $4x^2$.
These could be x and $4x$ or $2x$ and $2x$.

$$2x(4x^2 - 5xy - 9y^2) = 2x(\quad + \quad)(\quad - \quad)$$

The *last* terms must multiply to equal $9y^2$.
These terms could be y and $9y$ or $3y$ and $3y$.

Now we multiply binomials with various combinations of these first and last terms until we find a combination whose inner and outer products combine to equal $-5xy$.

$$2x(4x + y)(x - 9y) = 2x(4x^2 - 36xy + xy - 9y^2)$$
$$= 2x(4x^2 - 35xy - 9y^2)$$
$$2x(4x + 9y)(x - y) = 2x(4x^2 - 4xy + 9xy - 9y^2)$$
$$= 2x(4x^2 + 5xy - 9y^2)$$

Incorrect combinations.

Note that the middle term has the *correct absolute value* but the *wrong sign*. As we saw in Example 1, we can change the signs of the second terms in the binomial factors to get the correct middle term.

$$2x(4x - 9y)(x + y) = 2x(4x^2 + 4xy - 9xy - 9y^2) \quad \text{Correct combination.}$$
$$= 2x(4x^2 - 5xy - 9y^2)$$

Answer $\quad 8x^3 - 10x^2y - 18xy^2 = 2x(4x - 9y)(x + y)$

YOUR TURN Factor.

a. $16y^3 - 102y^2 + 36y$ **b.** $36m^2n - 120mn^2 - 21n^3$

OBJECTIVE 2. Factor trinomials of the form $ax^2 + bx + c$, where $a \neq 1$, by grouping. The major drawback about using trial and error to factor is that it can be tedious to find the correct combination. An alternative method is to factor by grouping, which we introduced in Section 6.1. Recall that in the grouping method, we group pairs of terms in a four-term polynomial, then factor out the GCF from each pair of terms. Since a trinomial of the form $ax^2 + bx + c$ has only three terms, we will split the bx term into two like terms to create a four-term polynomial that we can factor by grouping. To determine how to split up the bx term, we use the fact that if $ax^2 + bx + c$ is factorable, then b will always equal the sum of a pair of factors of the product of a and c.

Consider again the trinomial $3x^2 + 17x + 10$, which we factored by trial and error at the beginning of this section. Notice that $a = 3$, $b = 17$, and $c = 10$, so the product of a and c is $(3)(10) = 30$. To split up the bx term, which is $17x$ in this case, we look for a pair of factors of 30 whose sum is 17. It is helpful to list the factor pairs and their corresponding sums in a table:

Factors of ac	Sum of Factors of ac
$(1)(30) = 30$	$1 + 30 = 31$
$(2)(15) = 30$	$2 + 15 = 17$
$(3)(10) = 30$	$3 + 10 = 13$
$(5)(6) = 30$	$5 + 6 = 11$

Notice that 2 and 15 is the only factor pair of 30 whose sum is 17.

Now we can write $17x$ as $15x + 2x$, then factor by grouping.

$$3x^2 + 17x + 10 = 3x^2 + 15x + 2x + 10$$
$$= 3x(x + 5) + 2(x + 5)$$
$$= (x + 5)(3x + 2)$$

Note: *We could have also written $17x$ as $2x + 15x$ to get the same result.*

Every trinomial factorable by grouping has only one factor pair of ac whose sum is b, which suggests the following procedure.

PROCEDURE Factoring $ax^2 + bx + c$, where $a \neq 1$, by Grouping

To factor a trinomial of the form $ax^2 + bx + c$, where $a \neq 1$, by grouping:

1. Look for a monomial GCF in all the terms. If there is one, factor it out.
2. Multiply a and c.
3. Find two factors of this product whose sum is b.
4. Write a four-term polynomial in which bx is written as the sum of two like terms whose coefficients are the two factors you found in step 3.
5. Factor by grouping.

EXAMPLE 3 Factor by grouping.

a. $8x^2 - 22x + 5$

Solution Notice that for this trinomial, $a = 8$, $b = -22$, and $c = 5$. We begin by multiplying a and c: $(8)(5) = 40$. Now we find two factors of 40 whose sum is -22. Notice that these two factors must both be negative. It is helpful to list the combinations in a table.

Factors of ac	Sum of Factors of ac	
$(-1)(-40) = 40$	$-1 + (-40) = -41$	
$(-2)(-20) = 40$	$-2 + (-20) = -22$	◄── Correct
$(-4)(-10) = 40$	$-4 + (-10) = -14$	
$(-5)(-8) = 40$	$-5 + (-8) = -13$	

You do not need to list all possible combinations as we did here. We listed them to illustrate that only one combination is correct. We now write $-22x$ as $-2x - 20x$ and then factor by grouping.

$$8x^2 - 22x + 5 = 8x^2 - 2x - 20x + 5 \qquad \text{Write } -22x \text{ as } -2x - 20x.$$

$$= 2x(4x - 1) - 5(4x - 1) \qquad \begin{array}{l}\text{Factor } 2x \text{ out of } 8x^2 - 2x; \\ \text{factor } -5 \text{ out of } -20x + 5.\end{array}$$

$$= (4x - 1)(2x - 5) \qquad \text{Factor out } (4x - 1).$$

b. $18y^4 + 15y^3 - 12y^2$

Solution Notice that there is a monomial GCF, $3y^2$, that we can factor out.

$$18y^4 + 15y^3 - 12y^2 = 3y^2(6y^2 + 5y - 4)$$

Now we factor the trinomial within the parentheses. For this trinomial, $a = 6$, $b = 5$, and $c = -4$. The product of ac is $(6)(-4) = -24$. Now find two factors of -24 whose sum is 5. Since the product is negative, the two factors will have different signs. Since the sum is positive, the factor with the greater absolute value must be positive.

Factors of ac	Sum of the Factors of ac	
$(-1)(24) = -24$	$-1 + 24 = 23$	
$(-2)(12) = -24$	$-2 + 12 = 10$	
$(-3)(8) = -24$	$-3 + 8 = 5$	◄── Correct

Now write $5y$ as $-3y + 8y$ and then factor by grouping.

$$3y^2(6y^2 + 5y - 4) = 3y^2(6y^2 - 3y + 8y - 4) \qquad \text{Write } 5y \text{ as } -3y + 8y.$$

$$= 3y^2[3y(2y - 1) + 4(2y - 1)] \qquad \begin{array}{l}\text{Factor } 3y \text{ out of } 6y^2 - 3y; \\ \text{factor } 4 \text{ out of } 8y - 4.\end{array}$$

$$= 3y^2(2y - 1)(3y + 4) \qquad \text{Factor out } (2y - 1).$$

YOUR TURN ⟩ Factor by grouping.

a. $6x^2 - 19x + 10$ **b.** $8x^2y - 20xy - 72y$

ANSWERS

a. $(3x - 2)(2x - 5)$
b. $4y(2x - 9)(x + 2)$

1. In factoring a polynomial, your first step should be to look for a
_____.

2. Given a trinomial of the form $ax^2 + bx + c$, where $a \neq 1$, how do you determine the first terms in its binomial factors?

3. Given a trinomial of the form $ax^2 + bx + c$, where $a \neq 1$, how do you determine the last terms in its binomial factors?

4. When factoring a trinomial of the form $ax^2 + bx + c$, assuming there is no monomial GCF, what is the first step?

For Exercises 5–10, fill in the missing values in the factors.

5. $3y^2 + 14y + 8 = (3y + \boxed{})(y + \boxed{})$

6. $6x^2 + 11x + 4 = (2x + \boxed{})(3x + \boxed{})$

7. $4t^2 + 19t - 30 = (\boxed{} + 6)(\boxed{} - 5)$

8. $8m^2 - 2m - 15 = (\boxed{} - 3)(\boxed{} + 5)$

9. $6x^2 - 25x + 14 = (2x - \boxed{})(\boxed{} - 2)$

10. $9n^2 - 56n + 12 = (9n - \boxed{})(\boxed{} - 6)$

For Exercises 11–42, factor completely. If prime, so indicate.

11. $2j^2 + 5j + 2$

12. $2x^2 + 7x + 3$

13. $2y^2 - 3y - 5$

14. $3m^2 - 10m + 3$

15. $3m^2 - 10m + 8$

16. $3y^2 - 10y + 7$

17. $6a^2 + 13a + 7$

18. $4y^2 + 8y + 3$

19. $6p^2 + 2p + 1$

20. $6u^2 + 3u - 7$

21. $4a^2 - 19a + 12$

22. $4u^2 - 17u - 15$

23. $6x^2 + 19x + 15$

24. $8x^2 + 18x + 9$

25. $16d^2 - 14d - 15$

26. $18a^2 + 9a - 35$

27. $3p^2 - 13pq + 4q^2$

28. $2u^2 + 5uv + 3v^2$

29. $12a^2 - 40ab + 25b^2$

30. $4w^2 - 7wv + 3v^2$

31. $5k^2 - 7kh - 12h^2$

32. $3z^2 - 13z - 30$

33. $8m^2 - 27mn + 9n^2$

34. $9x^2 + 9xy - 10y^2$

35. $8x^2y - 4xy - y$

36. $6a^2b + 7ab + 5b$

37. $12y^2 + 24y + 9$

38. $6a^2 - 20a + 16$

39. $6ab^2 - 20ab + 14a$

40. $20xy^2 - 50xy + 30x$

41. $6w^2v^2 + 10wv^2 + 4v^2$

42. $6x^3 + 21x^2y - 90xy^2$

For Exercises 43–72, factor by grouping.

43. $3a^2 + 4a + 1$

44. $5m^2 + 11m + 2$

45. $2t^2 - 3t + 1$

46. $3h^2 - 5h + 2$

47. $3x^2 - 4x - 7$

48. $3l^2 + 2l - 5$

49. $2y^2 - y - 6$

50. $3a^2 + 2a - 8$

51. $10a^2 - 19a + 7$

52. $8m^2 - 10m + 3$

53. $8r^2 - 6r - 9$

54. $6x^2 + 5x - 6$

55. $4k^2 - 10k + 6$

56. $12a^2 + 34a + 20$

57. $20x^2 - 23x + 6$

58. $12c^2 - 8c - 15$

59. $6k^2 + 7jk + 2j^2$

60. $3x^2 + 5xy + 2y^2$

61. $5x^2 - 26xy + 5y^2$

62. $2t^2 - 3tu + u^2$

63. $10s^2 + st - 2t^2$

64. $10x^2 - 33xy - 7y^2$

65. $6u^2 + 5uv - 6v^2$

66. $14u^2 - 6uv - 8v^2$

67. $12m^3 + 10m^2 - 12m$

68. $24x^2 - 64xy - 24y^2$

69. $36x^2y - 78xy + 36y$

70. $28m^2n + 2mn - 8n$

71. $24x^3 - 20x^2y - 24xy^2$

72. $36x^3 - 93x^2y + 60xy^2$

For Exercises 73–76, explain the mistake, then factor the trinomial correctly.

73. $6x^2 + 13x + 5 = (6x + 1)(x + 5)$

74. $2x^2 + x - 3 = (2x - 3)(x + 1)$

75. $4n^2 + 12n + 8 = (4n + 8)(n + 1)$

76. $4h^2 - 3hk - 10k^2 = (h - 2)(4h + 5)$

For Exercises 77–78, given the area of the figure, factor to find possible expressions for the length and width.

77.

Area $= 15x^2 - 11x + 2$

78.

Area $= 2x^2 + 13x + 20$

79. Two adjacent plots of land are for sale. The area of the larger plot is described by $6w^2 - w - 2$ square feet, where w is the width of the smaller plot. Factor to find possible expressions for the dimensions of the larger plot.

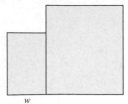

80. A cookie sheet has an area of $5h^2 + 8h - 85$ square inches, where h represents the height of the lip of the sheet. Factor to find possible expressions for the width and length of the cookie sheet.

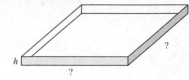

For Exercises 81–86, find all natural numbers that can replace b and make the expression factorable.

81. $2x^2 + bx + 6$

82. $3x^2 + bx + 10$

83. $5x^2 + bx - 9$

84. $2x^2 + bx - 15$

85. $6x^2 + bx - 7$

86. $4x^2 + bx - 5$

REVIEW EXERCISES

[5.1] *For Exercises 1 and 2, simplify.*

1. $(4x)^2$

2. $(2y)^3$

[5.5] *For Exercises 3–6, multiply.*

3. $(3x + 5)(3x - 5)$

4. $(x - 2)(x^2 + x + 4)$

5. $(2x - 5)^2$

6. $(6y + 1)^2$

6.4 Factoring Special Products

OBJECTIVES

1. Factor perfect square trinomials.
2. Factor a difference of squares.
3. Factor a difference of cubes.
4. Factor a sum of cubes.

In Section 5.5, we explored some special products found by multiplying conjugates and squaring binomials. In this section, we see how to factor those special products.

OBJECTIVE 1. Factor perfect square trinomials. First, we consider the product that is a result of squaring a binomial. The trinomial product is a perfect square. Recall the rules for squaring a binomial that we developed in Section 5.5:

$$(a + b)^2 = a^2 + 2ab + b^2$$
$$(a - b)^2 = a^2 - 2ab + b^2$$

Consider $(2x + 3)^2$ and $(2x - 3)^2$. Using the rules, we have

$$(2x + 3)^2 = (2x)^2 + 2(2x)(3) + (3)^2 \qquad (2x - 3)^2 = (2x)^2 - 2(2x)(3) + (3)^2$$
$$= 4x^2 + 12x + 9 \qquad\qquad\qquad = 4x^2 - 12x + 9$$

To use these rules when factoring, we look at the terms in the given trinomial to determine whether the trinomial fits the form of a perfect square. If we were asked to factor $4x^2 + 12x + 9$, we would note that the first and last terms are perfect squares and that twice the product of their square roots equals the middle term.

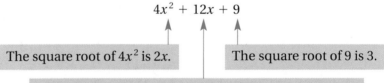

$$4x^2 + 12x + 9$$

The square root of $4x^2$ is $2x$. | The square root of 9 is 3.

Twice the product of $2x$ and 3 is the middle term, $12x$.

Because $4x^2 + 12x + 9$ is a perfect square trinomial fitting the form $a^2 + 2ab + b^2$, where $a = 2x$ and $b = 3$, we can write the factored form as $(a + b)^2$, which is $(2x + 3)^2$.

RULE Factoring Perfect Square Trinomials

$$a^2 + 2ab + b^2 = (a + b)^2$$
$$a^2 - 2ab + b^2 = (a - b)^2$$

EXAMPLE 1 ⟩ Factor.

a. $9y^2 + 30y + 25$

Solution This trinomial is a perfect square because the first and last terms are perfect squares and twice the product of their roots is the middle term.

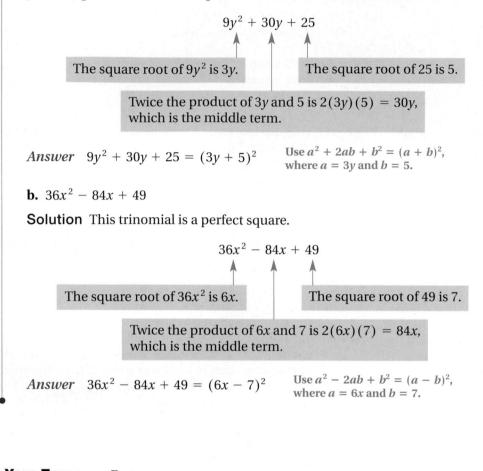

$$9y^2 + 30y + 25$$

The square root of $9y^2$ is $3y$.

The square root of 25 is 5.

Twice the product of $3y$ and 5 is $2(3y)(5) = 30y$, which is the middle term.

Answer $9y^2 + 30y + 25 = (3y + 5)^2$ Use $a^2 + 2ab + b^2 = (a + b)^2$, where $a = 3y$ and $b = 5$.

b. $36x^2 - 84x + 49$

Solution This trinomial is a perfect square.

$$36x^2 - 84x + 49$$

The square root of $36x^2$ is $6x$.

The square root of 49 is 7.

Twice the product of $6x$ and 7 is $2(6x)(7) = 84x$, which is the middle term.

Answer $36x^2 - 84x + 49 = (6x - 7)^2$ Use $a^2 - 2ab + b^2 = (a - b)^2$, where $a = 6x$ and $b = 7$.

YOUR TURN ⟩ Factor.

    a. $25n^2 + 20n + 4$ **b.** $49x^2 - 14x + 1$

          Note that in the perfect square trinomial forms, both a and b can have variables. Also, don't forget that you should always look for a monomial GCF in the terms as a first step.

EXAMPLE 2 ⟩ Factor.

    a. $16x^2 - 8xy + y^2$

    Solution $16x^2 - 8xy + y^2 = (4x - y)^2$ Use $a^2 - 2ab + b^2 = (a - b)^2$, where $a = 4x$ and $b = y$.

b. $4x^3y - 36x^2y + 81xy$

Solution

$$4x^3y - 36x^2y + 81xy = xy(4x^2 - 36x + 81)$$ Factor out the monomial GCF, xy.

$$= xy(2x - 9)^2$$ Factor $4x^2 - 36x + 81$ using $a^2 - 2ab + b^2 = (a - b)^2$, where $a = 2x$ and $b = 9$.

YOUR TURN   Factor.

    a. $9m^2 + 48mn + 64n^2$ **b.** $2xy^4 + 24xy^3 + 72xy^2$

OBJECTIVE 2. Factor a difference of squares. Another special product that we considered in Section 5.5 was found by multiplying conjugates. Recall that conjugates are binomials that differ only in the sign separating the terms. For example, $3x + 2$ and $3x - 2$ are conjugates. Note that multiplying conjugates produces a product that is a difference of squares.

$$(3x + 2)(3x - 2) = 9x^2 - 4$$

| This term is the square of $3x$. | This term is the square of 2. |

The rule for multiplying conjugates that we developed in Section 5.5 is:

$$(a + b)(a - b) = a^2 - b^2$$

To turn this rule into a rule for factoring a difference of squares, we reverse it.

RULE   **Factoring a Difference of Squares**

$$a^2 - b^2 = (a + b)(a - b)$$ **Warning:** A *sum* of squares $a^2 + b^2$ is prime and cannot be factored.

EXAMPLE 3   Factor.

    a. $25x^2 - 36y^2$

Solution This binomial is a difference of squares because $25x^2 - 36y^2 = (5x)^2 - (6y)^2$. To factor it, we use the rule $a^2 - b^2 = (a + b)(a - b)$.

$$a^2 - b^2 = (a + b)(a - b)$$

$$25x^2 - 36y^2 = (5x)^2 - (6y)^2 = (5x + 6y)(5x - 6y)$$ Replace a with $5x$ and b with $6y$.

    b. $9x^4 - 16$

Solution This binomial is a difference of squares because $9x^4 - 16 = (3x^2)^2 - (4)^2$.

$$9x^4 - 16 = (3x^2 + 4)(3x^2 - 4)$$ Use $a^2 - b^2 = (a + b)(a - b)$ with $a = 3x^2$ and $b = 4$.

c. $25y^7 - 64y^9$

Solution The terms in this binomial have a monomial GCF, y^7.

$$25y^7 - 64y^9 = y^7(25 - 64y^2) \qquad \text{Factor out the monomial GCF, } y^7.$$
$$= y^7(5 + 8y)(5 - 8y) \qquad \text{Factor } 25 - 64y^2 \text{ using } a^2 - b^2 = (a + b)(a - b) \text{ with } a = 5 \text{ and } b = 8y.$$

YOUR TURN Factor.

a. $n^2 - 49$ **b.** $9t^6 - 49u^4$ **c.** $50x^3 - 18x$

Sometimes the factors themselves can be factored.

EXAMPLE 4 Factor. $x^4 - 81$

Solution This binomial is a difference of squares, where $a = x^2$ and $b = 9$.

$$x^4 - 81 = (x^2 + 9)(x^2 - 9) \qquad \text{Use } a^2 - b^2 = (a + b)(a - b).$$
$$= (x^2 + 9)(x + 3)(x - 3) \qquad \text{Factor } x^2 - 9, \text{ using } a^2 - b^2 = (a + b)(a - b) \text{ with } a = x \text{ and } b = 3.$$

Note: *The factor $x^2 + 9$ is a sum of squares, which is prime.*

YOUR TURN Factor. $y^4 - 16$

OBJECTIVE 3. Factor a difference of cubes. Another common binomial form that we can factor is a difference of cubes. A difference of cubes is a result of a multiplication like the following:

$$(x - 2)(x^2 + 2x + 4) = x^3 + 2x^2 + 4x - 2x^2 - 4x - 8$$
$$= x^3 - 8 \longleftarrow \boxed{\text{Difference of cubes}}$$

Writing factored form reverses this process so that we have

$$x^3 - 8 = (x - 2)(x^2 + 2x + 4)$$

This suggests the following rule for factoring a difference of cubes.

RULE Factoring a Difference of Cubes
$$a^3 - b^3 = (a - b)(a^2 + ab + b^2)$$

EXAMPLE 5 Factor.

a. $64y^3 - 27$

Solution This binomial is a difference of cubes because $64y^3 - 27 = (4y)^3 - (3)^3$. To factor, we use the rule $a^3 - b^3 = (a - b)(a^2 + ab + b^2)$ with $a = 4y$ and $b = 3$.

$$a^3 - b^3 = (a - b)(a^2 + ab + b^2)$$
$$64y^3 - 27 = (4y)^3 - (3)^3 = (4y - 3)((4y)^2 + (4y)(3) + (3)^2) \quad \text{Replace } a \text{ with } 4y \text{ and } b \text{ with } 3.$$
$$= (4y - 3)(16y^2 + 12y + 9) \quad \text{Simplify.}$$

Note: *This trinomial may seem like a perfect square. However, to be a perfect square, the middle term should be 2ab. In this trinomial, we only have ab, so it cannot be factored.*

b. $27m^4 - 8mn^3$

Solution The terms in this binomial have a monomial GCF, m.

$$27m^4 - 8mn^3 = m(27m^3 - 8n^3) \quad \text{Factor out the monomial GCF, } m.$$
$$= m(3m - 2n)((3m)^2 + (3m)(2n) + (2n)^2) \quad \text{Factor } 27m^3 - 8n^3, \text{ which is a difference of cubes with } a = 3m \text{ and } b = 2n$$
$$= m(3m - 2n)(9m^2 + 6mn + 4n^2) \quad \text{Simplify.}$$

ANSWERS

a. $(3 - u)(9 + 3u + u^2)$
b. $x(y - 4)(y^2 + 4y + 16)$

YOUR TURN Factor.

a. $27 - u^3$

b. $xy^3 - 64x$

OBJECTIVE 4. Factor a sum of cubes. A sum of cubes can be factored using a pattern similar to the difference of cubes. A sum of cubes is a result of a multiplication like the following:

$$(x + 2)(x^2 - 2x + 4) = x^3 - 2x^2 + 4x + 2x^2 - 4x + 8$$
$$= x^3 + 8$$

Writing factored form reverses this process so that we have

$$x^3 + 8 = (x + 2)(x^2 - 2x + 4)$$

In general, we can write the following rule for factoring a sum of cubes.

RULE Factoring a Sum of Cubes

$$a^3 + b^3 = (a + b)(a^2 - ab + b^2)$$

EXAMPLE 6 Factor.

a. $27x^3 + 8$

Solution This binomial is a sum of cubes because $27x^3 + 8 = (3x)^3 + (2)^3$. To factor, we use the rule $a^3 + b^3 = (a + b)(a^2 - ab + b^2)$ with $a = 3x$ and $b = 2$.

$$a^3 + b^3 = (a + b)(a^2 - a\,b + b^2)$$
$$27x^3 + 8 = (3x)^3 + (2)^3 = (3x + 2)((3x)^2 - (3x)(2) + (2)^2) \quad \text{Replace } a \text{ with } 3x \text{ and } b \text{ with 2.}$$
$$= (3x + 2)(9x^2 - 6x + 4) \quad \text{Simplify.}$$

Note: *This trinomial cannot be factored.*

b. $5h + 40hk^3$

Solution The terms in this binomial have a monomial GCF, $5h$.

$$5h + 40hk^3 = 5h(1 + 8k^3) \quad \text{Factor out the monomial GCF, } 5h.$$
$$= 5h(1 + 2k)((1)^2 - (1)(2k) + (2k)^2) \quad \text{Factor } 1 + 8k^3, \text{ which is a sum of cubes with } a = 1$$
$$= 5h(1 + 2k)(1 - 2k + 4k^2) \quad \text{and } b = 2k.$$

ANSWERS

a. $(4 + m)(16 - 4m + m^2)$
b. $x^2(5x - 3)(25x^2 + 15x + 9)$

 **YOUR TURN** Factor.

a. $64 + m^3$

b. $125x^5 - 27x^2$

6.4 Exercises

For Extra Help: MyMathLab, Videotape/DVT, InterAct Math, Math Tutor Center, Math XL.com

1. Explain in your own words how to recognize a perfect square trinomial.

2. How do you determine the sign in the factored form of a perfect square trinomial?

3. The factors of a difference of squares are binomials known as _____.

4. There is only one minus sign in the factors of a difference of cubes. Where is that minus sign placed in the factors?

5. There is only one minus sign in the factors of a sum of cubes. Where is that minus sign placed in the factors?

6. How do you determine the a and b terms in a sum or difference of cubes?

For Exercises 7–24, factor the trinomials that are perfect squares. If the trinomial is not a perfect square, write "not a perfect square."

7. $x^2 + 14x + 49$

8. $y^2 + 6y + 9$

9. $b^2 - 8b + 16$

10. $m^2 - 6m + 9$

11. $n^2 + 12n + 144$

12. $y^2 + 6y + 36$

13. $25u^2 - 30u + 9$

14. $9m^2 - 24m + 16$

15. $100w^2 + 20w + 1$

16. $25a^2 + 10a + 1$

17. $y^2 + 2yz + z^2$

18. $w^2 + 2wv + v^2$

19. $4p^2 - 28pq + 49q^2$

20. $9r^2 - 12rs + 4s^2$

21. $16g^2 + 24gh + 9h^2$

22. $4y^2 - 12by + 9b^2$

23. $16t^2 + 80t + 100$

24. $54q^2 - 72q + 24$

For Exercises 25–44, factor binomials that are the difference of squares. If prime, so indicate.

25. $x^2 - 4$

26. $a^2 - 121$

27. $16 - y^2$

28. $49 - a^2$

29. $p^2 - q^2$

30. $m^2 - n^2$

31. $25u^2 - 16$

32. $16x^2 - 49y^2$

33. $9x^2 - b^2$

34. $16x^2 - y^2$

35. $64m^2 - 25n^2$

36. $121p^2 - 9q^2$

37. $50x^2 - 32y^2$

38. $27a^2 - 48b^2$

39. $4x^2 - 100y^2$

40. $16x^2 - 36y^2$

41. $x^4 - y^4$

42. $a^4 - b^4$

43. $x^4 - 16$

44. $b^4 - 1$

For Exercises 45–54, factor the difference of cubes.

45. $n^3 - 27$

46. $y^3 - 64$

47. $x^3 - 1$

48. $r^3 - 1$

49. $27k^3 - 8$

50. $125a^3 - 64$

51. $c^3 - 64d^3$

52. $a^3 - 8b^3$

53. $27x^3 - 64y^3$

54. $1000a^3 - 27b^3$

For Exercises 55–64, factor the sum of cubes.

55. $x^3 + 27$

56. $y^3 + 8$

57. $m^3 + n^3$

58. $p^3 + q^3$

59. $27k^3 + 8$

60. $125b^3 + 27$

61. $125x^3 + 64y^3$ **62.** $27a^3 + 1000b^3$ **63.** $8p^3 + q^3z^3$

64. $x^3y^3 + 27y^3$

For Exercises 65–80, factor.

65. $2x^2 - 50$ **66.** $5y^2 - 20$ **67.** $16x^2 - \dfrac{25}{49}$ **68.** $25y^2 - \dfrac{1}{36}$

69. $2u^3 - 2u$ **70.** $4 - 16x^2$ **71.** $y^5 - 16y^3b^2$ **72.** $ab^3 - a^3b$

73. $50x^3 - 2x$ **74.** $27m^3 - 363mn^2$ **75.** $3y^3 - 24z^3$ **76.** $2x^3 - 54$

77. $c^3 - \dfrac{8}{27}$ **78.** $8y^3 - \dfrac{1}{125}$ **79.** $16c^4 + 2cd^3$ **80.** $8d^4 - 64c^3d$

For Exercises 81–88, use the rules for a difference of squares, sum of cubes, or difference of cubes to factor completely.

81. $(2a - b)^2 - c^2$ **82.** $(b - 3)^2 - 1$

83. $1 - 9(x - y)^2$ **84.** $25a^2 - 16(b + c)^2$

85. $x^3 + 27(y + z)^3$ **86.** $m^3 - 8(y + b)^3$

87. $64d^3 - (x + y)^3$ **88.** $(t + u)^3 + 64$

For Exercises 89–92, find a natural number b that makes the expression a perfect square trinomial.

89. $16x^2 + bx + 9$ **90.** $25x^2 + bx + 4$ **91.** $4x^2 - bx + 81$ **92.** $36x^2 - bx + 49$

For Exercises 93–96, find the natural number c that completes the perfect square trinomial.

93. $x^2 + 10x + c$ **94.** $x^2 + 8x + c$ **95.** $9x^2 - 24x + c$ **96.** $4x^2 - 28x + c$

For Exercises 97–98, write a polynomial for the area of the shaded region, then factor completely.

97.

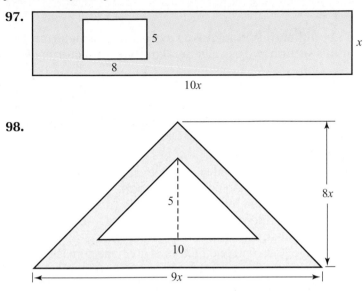

98.

For Exercises 99–100, write a polynomial for the volume of the shaded region, then factor completely.

99.

100.

REVIEW EXERCISES

[6.1]   **1.** List all the factors of 36.

[1.2]   **2.** Find the prime factorization of 100.

[5.3]   **3.** Add: $(4x^3 - 5x^2 + 9x - 11) + (7x^2 - 12x - 15)$

[5.3]   **4.** Subtract: $(10n^4 + n^3 - 15n - 21) - (19n^3 - 12n^2 + n - 15)$

[5.5]   **5.** Multiply: $(4x - 5)^2$

[5.6]   **6.** Divide: $\dfrac{5x^3 + 4x^2 - 5x + 17}{x + 2}$

6.5 Strategies for Factoring

OBJECTIVE 1. Factor polynomials. One of the difficulties in factoring expressions is determining which technique to use. In this section, we use a general outline for factoring. Following is the outline.

PROCEDURE Factoring a Polynomial

To factor a polynomial, first factor out any monomial GCF, then consider the number of terms in the polynomial. If the polynomial has:

I. Four terms, then try to factor by grouping.

II. Three terms, then determine if the trinomial is a perfect square or not.
   A. If the trinomial is a perfect square, then consider its form.
      1. If in the form $a^2 + 2ab + b^2$, then the factored form is $(a + b)^2$.
      2. If in the form $a^2 - 2ab + b^2$, then the factored form is $(a - b)^2$.
   B. If the trinomial is not a perfect square, then consider its form.
      1. If in the form $x^2 + bx + c$, then find two factors of c whose sum is b, and write the factored form as $(x + \text{first number})(x + \text{second number})$.
      2. If in the form $ax^2 + bx + c$, where $a \neq 1$, then use trial and error. Or, find two factors of ac whose sum is b; write these factors as coefficients of two like terms that, when combined, equal bx; and then factor by grouping.

III. Two terms, then determine if the binomial is a difference of squares, sum of cubes, or difference of cubes.
   A. If given a binomial that is a difference of squares, $a^2 - b^2$, then the factors are conjugates and the factored form is $(a + b)(a - b)$. Note that a sum of squares cannot be factored.
   B. If given a binomial that is a sum of cubes, $a^3 + b^3$, then the factored form is $(a + b)(a^2 - ab + b^2)$.
   C. If given a binomial that is a difference of cubes, $a^3 - b^3$, then the factored form is $(a - b)(a^2 + ab + b^2)$.

Note: Always look to see if any of the factors can be factored.

EXAMPLE 1   Factor. $12x^2 + 28x - 5$

Solution There is no monomial GCF. Since this expression is a trinomial, we look to see if it is a perfect square. It is not a perfect square because the first and last terms are not perfect squares. It is a trinomial of the form $ax^2 + bx + c$, where $a \neq 1$. Therefore, we use trial and error or the alternative method that involves factoring by grouping. We will use trial and error here.

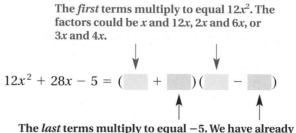

The *first* terms multiply to equal $12x^2$. The factors could be x and $12x$, $2x$ and $6x$, or $3x$ and $4x$.

$$12x^2 + 28x - 5 = (\quad + \quad)(\quad - \quad)$$

The *last* terms multiply to equal -5. We have already written in the appropriate signs, so these factors could be 1 and 5.

Now list possible combinations of these first and last factor pairs and check by multiplying to see if the sum of the products of the inner and outer terms equals $28x$.

$$(x + 5)(12x - 1) = 12x^2 - x + 60x - 5 = 12x^2 + 59x - 5$$
$$(12x + 5)(x - 1) = 12x^2 - 12x + 5x - 5 = 12x^2 - 7x - 5$$ Incorrect combinations.

$$(2x + 5)(6x - 1) = 12x^2 - 2x + 30x - 5 = 12x^2 + 28x - 5$$ Correct combination.

Answer $12x^2 + 28x - 5 = (2x + 5)(6x - 1)$

YOUR TURN Factor. $24y^2 - 53y - 7$

EXAMPLE 2 Factor. $7x^3 + 14x^2 - 105x$

Solution

$$7x^3 + 14x^2 - 105x = 7x(x^2 + 2x - 15)$$ Factor out the monomial GCF, $7x$.

The trinomial in the parentheses is not a perfect square and is in the form $x^2 + bx + c$, so we look for two numbers whose product is -15 and whose sum is 2. Since the product is negative, the two factors must have different signs. Since the sum is positive, the factor with the greater absolute value will be the positive factor.

Product	Sum
$(-1)(15) = -15$	$-1 + 15 = 14$
$(-3)(5) = -15$	$-3 + 5 = 2$

Correct combination.

Answer $7x^3 + 14x^2 - 105x = 7x(x - 3)(x + 5)$

YOUR TURN Factor. $2t^3 - 18t^2 + 36t$

EXAMPLE 3 Factor. $12m^4 - 48n^2$

Solution

$$12m^4 - 48n^2 = 12(m^4 - 4n^2)$$ Factor out the monomial GCF, 12.

The binomial in the parentheses is a difference of squares in the form $a^2 - b^2$, where $a = m^2$ and $b = 2n$. Using $a^2 - b^2 = (a + b)(a - b)$ gives us

$$= 12(m^2 + 2n)(m^2 - 2n)$$ ◀ **Note:** *Neither of these binomial factors can be factored further.*

ANSWER

$(3y - 7)(8y + 1)$

ANSWER

$2t(t - 6)(t - 3)$

EXAMPLE 4 Factor. $10y^5 + 40y^3$

Solution

$$10y^5 + 40y^3 = 10y^3(y^2 + 4) \qquad \text{Factor out the monomial GCF, } 10y^3.$$

The binomial in the parentheses is a sum of squares, which cannot be factored.

YOUR TURN Factor. $3m^2n^3 + 15m^2n^2 + 6m^2n$

EXAMPLE 5 Factor. $45x^3y^3 - 120x^2y^4 + 80xy^5$

Solution

$$45x^3y^3 - 120x^2y^4 + 80xy^5 = 5xy^3(9x^2 - 24xy + 16y^2) \qquad \begin{array}{l}\text{Factor out the monomial}\\ \text{GCF, } 5xy^3.\end{array}$$

The trinomial in the parentheses is a perfect square in the form $a^2 - 2ab + b^2$, where $a = 3x$ and $b = 4y$. Using $a^2 - 2ab + b^2 = (a - b)^2$ gives us

$$= 5xy^3(3x - 4y)^2$$

EXAMPLE 6 Factor. $x^5 - x^3 - 8x^2 + 8$

Solution There is no monomial GCF common to all terms. Since this polynomial has four terms, we try to factor by grouping.

$$\begin{aligned} x^5 - x^3 - 8x^2 + 8 &= x^3(x^2 - 1) - 8(x^2 - 1)\\ &= (x^2 - 1)(x^3 - 8)\end{aligned}$$

The two binomial factors can be factored further. The first binomial, $x^2 - 1$, is a difference of squares with $a = x$ and $b = 1$. The second binomial, $x^3 - 8$, is a difference of cubes with $a = x$ and $b = 2$.

$$= (x + 1)(x - 1)(x - 2)(x^2 + 2x + 4)$$

YOUR TURN Factor. $12x^3 - 8x^2 - 27x + 18$

ANSWER

$3m^2n(n^2 + 5n + 2)$

ANSWER

$(2x + 3)(2x - 3)(3x - 2)$

1. What should you look for first when factoring?

2. What are three different types of factorable binomials?

3. Describe one type of binomial that cannot be factored.

4. What are the two forms a perfect square trinomial can have?

5. When do you factor by grouping?

6. How might you check a factored form to see if it is correct?

For Exercises 7–82, factor completely. If prime, indicate as such.

7. $3xy^2 + 6x^2y$

8. $2a + ab$

9. $7a(x + y) - (x + y)$

10. $(u + v)r + (u + v)s$

11. $2x^2 - 32$

12. $2x^2 - 2y^2$

13. $ax + ay + bx + by$

14. $ax - xy - ay + y^2$

15. $12a^3b^2c + 3a^2b^2c^2 + 5abc^3$

16. $15u^2 - 2u^2v^2 + 2uv^3$

17. $x^2 + 8x + 15$

18. $6b^2 + b - 2$

19. $x^4 - 16$

20. $t^4 - 81$

21. $x^2 + 25$

22. $9a^2 + 49$

23. $15x^2 + 7x - 2$

24. $2a^2 - 5a - 12$

25. $ax^2 + 4ax + 4a$

26. $9a^2x + 12ax + 4x$

27. $6ab - 36ab^2$

28. $5ax - 5a^2x$

29. $x^2 + x + 2$

30. $x^2 + 3x + 4$

31. $x^2 - 49$

32. $1 - 25x^2$

33. $p^2 + p - 30$

34. $y^2 - 6y - 40$

35. $u^3 - u$

36. $5m^2 - 45$

37. $2b^2 + 14b + 24$

38. $xy^2 - xy - 20x$

39. $6r^2 - 15r^3$

40. $y^4 - 6y^3$

41. $7u^2 - 14u - 105$

42. $6r^2 - 26r - 20$

43. $h^4 + h^3 + h^2$

44. $3m^2 + 9m + 27$

45. $5p^2 - 80$

46. $3r^2 - 108$

47. $q^4 - 6q^2 - 16$

48. $r^4 + 4r^2 - 21$

49. $3w^2 - w - 2$

50. $3m^2 - 8m + 5$

51. $12v^2 + 23v + 10$

52. $6x^2 - 5x - 6$

53. $2 - 50x^2$

54. $x^3 - 4xy^2$

55. $80k^2 - 20k^2l^2$

56. $3x^3 - 12x$

57. $2j^2 + j - 3$

58. $3k^2 - 2k - 1$

59. $50 - 20t + 2t^2$

60. $243 - 54x + 3x^2$

61. $3ax - 6ay - 8by + 4bx$

62. $2ax + 6bx - ay - 3by$

63. $x^3 - x^2 - x + 1$

64. $x^3 - 2x^2 - x + 2$

65. $4x^2 - 28x + 49$

66. $4m^2 + 20m + 25$

67. $2x^4 - 162$

68. $a - 81ax^4$

69. $a^2 + 2ab + ab + 2b^2$

70. $x^2 - 3xy + xy - 3y^2$

71. $9 - 4m^2$

72. $64 - 9r^2$

73. $x^5 - 4xy^2$

74. $m^5 - 25mn^2$

75. $20x^2 + 3xy + 2y^2$

76. $20a^2 - 9a + 20$

77. $b^3 + 125$

78. $x^3 + a^3$

79. $3y^3 - 24$

80. $4x^4 - 4x$

81. $54x - 2xy^3$

82. $16m^3 + 250b^3$

For Exercises 83–84, given an expression for the area of each rectangle, find the length and width.

83.

Area = $6x^2 - 11x - 7$

84.

Area = $25n^2 - 49$

For Exercises 85–86, given an expression for the volume of each box, find the length, width, and height.

85.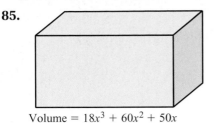

Volume $= 18x^3 + 60x^2 + 50x$

86.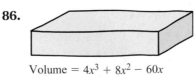

Volume $= 4x^3 + 8x^2 - 60x$

87. A swimmer's goggles fall off of a diving board that is 100 meters high. The height of the goggles after t seconds is given by $100 - 16t^2$. Factor $100 - 16t^2$.

88. An object is dropped from a cliff that is 400 feet high. The expression $400 - 16t^2$ gives the height of the falling object after t seconds. Factor $400 - 16t^2$.

89. The voltage in a circuit is the product of two factors, the resistance in the circuit and the current. If the voltage in a circuit is described by the expression $6ir + 15i + 8r + 20$, find the expressions for the current and resistance. (The expression for current will contain i and the expression for resistance will contain r.)

90. If the voltage in the circuit described in Exercise 89 was changed so that it is now described by the expression $35ir + 15i - 7r - 3$, what would be the expressions for the current and resistance?

REVIEW EXERCISES

[2.1]   **1.** Find the area of a rectangular garden that is 14 feet by 16 feet.

[2.5]   **2.** Translate to an equation, then solve. The product of four and the sum of a number and three is equal to twenty. Find the unknown number.

[3.3]   **3.** Find three consecutive integers whose sum is 54.

[5.6]   **4.** Divide: $\dfrac{2x^2 - 7x - 8}{2x + 1}$

[6.1]   **5.** Find the GCF of $12x^2$ and $-6x^3y^2$.

6.6 Solving Quadratic Equations by Factoring

OBJECTIVES

1. Use the zero-factor theorem to solve equations containing expressions in factored form.
2. Solve quadratic equations by factoring.
3. Solve problems involving quadratic equations.
4. Use the Pythagorean theorem to solve problems.

OBJECTIVE 1. Use the zero-factor theorem to solve equations containing expressions in factored form. So far in Chapters 5 and 6, we have rewritten polynomial expressions. We are now ready to work with polynomial equations, which we will solve using factoring. Notice we are moving to the equation level of our Algebra Pyramid, which is built upon the foundation of expressions, variables, and constants.

Note: *Inequalities are also in the top level of the Algebra Pyramid, but we will not be exploring polynomial inequalities at this point.*

The Algebra Pyramid

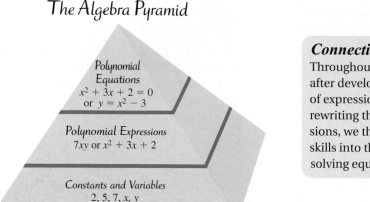

Polynomial Equations
$x^2 + 3x + 2 = 0$
or $y = x^2 - 3$

Polynomial Expressions
$7xy$ or $x^2 + 3x + 2$

Constants and Variables
$2, 5, 7, x, y$

Connection
Throughout this book, after developing new types of expressions and ways of rewriting those expressions, we then put those skills into the context of solving equations.

When solving polynomial equations by factoring, we make use of the **zero-factor theorem**, which states that if the product of two factors is 0, then one or the other of the two factors, or both factors, are equal to 0.

RULE Zero-Factor Theorem

If a and b are real numbers and $ab = 0$, then $a = 0$ or $b = 0$.

EXAMPLE 1 Solve. $(x + 2)(x - 3) = 0$

Solution In this equation, the product of two factors, $x + 2$ and $x - 3$, is equal to 0. According to the zero-factor theorem, one of the two factors, or both factors, must equal 0. In solving, we determine what values for x cause each individual factor to equal 0.

$$x + 2 = 0 \quad \text{or} \quad x - 3 = 0$$

Now solve each of the linear equations to get two solutions.

$$x = -2 \quad x = 3$$

Check Verify that -2 and 3 satisfy the original equation, $(x + 2)(x - 3) = 0$.

For $x = -2$: $(-2 + 2)(-2 - 3) \overset{?}{=} 0$ For $x = 3$: $(3 + 2)(3 - 3) \overset{?}{=} 0$

$(0)(-5) = 0$ $(5)(0) = 0$

Both -2 and 3 check, therefore, they are both solutions.

YOUR TURN   Solve. $(n - 6)(n + 1) = 0$

Example 1 suggests the following procedure.

PROCEDURE   **Solving Equations with Two or More Factors Equal to 0**

To solve an equation in which two or more factors are equal to 0, use the zero-factor theorem:

1. Set each factor equal to zero.
2. Solve each of those equations.

EXAMPLE 2   Solve.

a. $y(3y + 4) = 0$

Solution In this equation, the product of two factors, y and $3y + 4$, is equal to 0. We set each factor equal to 0, and then solve each of those equations.

$$y = 0 \quad \text{or} \quad 3y + 4 = 0$$

Note: *This equation is already solved.*

$3y = -4$ **Subtract 4 from both sides to isolate $3y$.**

$y = -\dfrac{4}{3}$ **Divide both sides by 3 to isolate y.**

Check Verify that 0 and $-\dfrac{4}{3}$ satisfy the original equation, $y(3y + 4) = 0$.

For $y = 0$: $(0)(3(0) + 4) \overset{?}{=} 0$ For $y = -\dfrac{4}{3}$: $\left(-\dfrac{4}{3}\right)\left[\dfrac{\overset{1}{\cancel{3}}}{1}\left(-\dfrac{4}{\cancel{3}_1}\right) + 4\right] \overset{?}{=} 0$

$(0)(4) = 0$ $\left(-\dfrac{4}{3}\right)[-4 + 4] \overset{?}{=} 0$

$\left(-\dfrac{4}{3}\right)(0) = 0$

Both 0 and $-\dfrac{4}{3}$ check, therefore, they are both solutions.

ANSWER

$n = 6 \text{ or } -1$

b. $(x - 4)^2 = 0$

Solution Note that $(x - 4)^2 = 0$ can be written as $(x - 4)(x - 4) = 0$. Because the two factors are identical, there is no need to write two separate equations. We set $x - 4$ equal to 0, then solve.

$$x - 4 = 0$$
$$x = 4 \qquad \text{Add 4 to both sides to isolate } x.$$

Check Verify that 4 satisfies the original equation, $(x - 4)^2 = 0$.

$$(4 - 4)^2 \overset{?}{=} 0$$
$$(0)^2 = 0 \qquad \text{4 checks, therefore it is a solution.}$$

c. $n(n + 6)(5n - 3) = 0$

Solution In this equation, the product of three factors, n, $n + 6$, and $5n - 3$, is 0. We set each factor equal to 0, and then solve each of those equations.

$$n = 0 \quad \text{or} \quad n + 6 = 0 \quad \text{or} \quad 5n - 3 = 0$$
$$n = -6 \qquad \qquad 5n = 3$$
$$n = \frac{3}{5}$$

Check To check, we verify that 0, -6, and $\frac{3}{5}$ satisfy the original equation. We leave this check to the reader.

YOUR TURN Solve.

a. $5x(3x - 2) = 0$ **b.** $(y + 3)^2 = 0$ **c.** $t(4t - 3)(t + 1) = 0$

OBJECTIVE 2. Solve quadratic equations by factoring. Now that we have seen how to use the zero-factor theorem to solve equations containing expressions that are in factored form, we are ready to consider equations that have an expression that is not in factored form. More specifically, we will use factoring to solve **quadratic equations in one variable**.

DEFINITION   *Quadratic equation in one variable:* An equation that can be written in the form $ax^2 + bx + c = 0$, where a, b, and c are all real numbers and $a \neq 0$.

The fact that a cannot equal zero means that the ax^2 term must be present for the equation to be a quadratic equation. A quadratic equation written in the form $ax^2 + bx + c = 0$ is said to be in *standard form*. Following are some examples of quadratic equations.

$$x^2 + 5x + 6 = 0 \qquad\qquad 3x^2 - 48 = 0 \qquad\qquad x^2 = x + 6$$

Note: *This quadratic equation is in standard form.*

Note: *This quadratic equation is in standard form. It has no bx term because $b = 0$.*

Note: *Though not in standard form, this equation is still quadratic. Written in standard form, the equation would be $x^2 - x - 6 = 0$.*

In Objective 1, we learned that to use the zero-factor theorem, we need an expression in factored form set equal to 0, so to solve a quadratic equation using factoring, it needs to be in standard form. If given an equation that is not in standard form, such as $x^2 = x + 6$, we first write it in standard form.

$$x^2 - x - 6 = 0 \qquad \text{Subtract } x \text{ and 6 from both sides to get standard form.}$$

After factoring $x^2 - x - 6$, we have $(x + 2)(x - 3) = 0$, which is the same equation that we solved in Example 1. This suggests the following procedure.

PROCEDURE   Solving Quadratic Equations Using Factoring

To solve a quadratic equation:

1. Write the equation in standard form ($ax^2 + bx + c = 0$).
2. Write the variable expression in factored form.
3. Use the zero-factor theorem to solve.

EXAMPLE 3   Solve. $2x^2 + 9x - 5 = 0$

Solution This equation is in standard form, so we can simply factor the variable expression.

$$(2x - 1)(x + 5) = 0 \qquad \text{Factor } 2x^2 + 9x - 5.$$

Now use the zero-factor theorem to solve.

$$2x - 1 = 0 \quad \text{or} \quad x + 5 = 0$$
$$2x = 1 \qquad\qquad\quad x = -5$$
$$x = \frac{1}{2}$$

Check   To check, we verify that $\frac{1}{2}$ and -5 satisfy the original equation. We leave this check to the reader.

ANSWER

$y = \frac{2}{3} \text{ or } -\frac{1}{4}$

YOUR TURN   Solve. $12y^2 + 11y + 2 = 0$

Sometimes, we may have to factor out a monomial GCF.

EXAMPLE 4 Solve. $x^3 - 3x^2 - 10x = 0$

Solution

$$x(x^2 - 3x - 10) = 0 \qquad \text{Factor out the monomial GCF, } x.$$
$$x(x - 5)(x + 2) = 0 \qquad \text{Factor } x^2 - 3x - 10.$$

Now use the zero-factor theorem to solve.

$$x = 0 \quad \text{or} \quad x - 5 = 0 \quad \text{or} \quad x + 2 = 0$$
$$x = 5 \qquad\qquad x = -2$$

Check Verify that 0, 5, and -2 each satisfy the original equation. We leave the check to the reader.

Note: *Though $x^3 - 3x^2 - 10x = 0$ is not a quadratic equation, after we factor out the x, the parentheses contain a quadratic form:*
$x^2 - 3x - 10.$

YOUR TURN Solve. $12x^3 + 20x^2 - 8x = 0$

Rewrite Quadratic Equations in Standard Form

If a quadratic equation is not in standard form ($ax^2 + bx + c = 0$), we first put it in standard form, then factor and use the zero-factor theorem.

EXAMPLE 5 Solve.

a. $6y^2 - 5y = 4 - 28y$

Solution We first need to manipulate the equation so that it is in standard form.

$$6y^2 + 23y = 4 \qquad\qquad \text{Add } 28y \text{ to both sides.}$$
$$6y^2 + 23y - 4 = 0 \qquad\qquad \text{Subtract 4 from both sides to get standard form.}$$
$$(6y - 1)(y + 4) = 0 \qquad\qquad \text{Factor using trial and error.}$$
$$6y - 1 = 0 \quad \text{or} \quad y + 4 = 0 \qquad \text{Use the zero-factor theorem to solve.}$$
$$6y = 1 \qquad\qquad y = -4$$
$$y = \frac{1}{6}$$

ANSWER

$x = 0, \dfrac{1}{3}, \text{ or } -2$

b. $x(x - 5) = -6$

Solution

Note: *We could have multiplied $x(x - 5)$ first and then moved the 6.*

$$x(x - 5) + 6 = 0$$

Add 6 to both sides so that the right-hand side is 0.

$$x^2 - 5x + 6 = 0$$

Multiply $x(x - 5)$ to get standard form.

$$(x - 2)(x - 3) = 0$$

Factor by looking for two numbers whose product is 6 and sum is -5.

$$x - 2 = 0 \quad \text{or} \quad x - 3 = 0$$

Use the zero-factor theorem to solve.

$$x = 2 \qquad\qquad x = 3$$

YOUR TURN Solve. $15n^2 + 20n = -8 - 6n$

OBJECTIVE 3. Solve problems involving quadratic equations. Many problems and applications are solved using quadratic equations.

EXAMPLE 6 The product of two consecutive odd natural numbers is 195. Find the numbers.

Understand Odd natural numbers are 1, 3, 5, ... Note that adding 2 to a given odd natural number gives a consecutive odd natural number, which suggests the pattern:

First odd natural number: x

Consecutive odd natural number: $x + 2$

The word *product* means that two numbers are multiplied to equal 195.

Connection In Chapter 3, we studied problems that dealt with the sum or difference of consecutive integers. The pattern for consecutive odd natural numbers is the same as for consecutive odd integers because every natural number is an integer.

Plan Translate to an equation, then solve.

Execute $x(x + 2) = 195$

The equation has the product of x and $x + 2$ set equal to 195.

$$x(x + 2) - 195 = 0$$

Subtract 195 from both sides.

$$x^2 + 2x - 195 = 0$$

Multiply $x(x + 2)$ to get the form $ax^2 + bx + c = 0$.

$$(x - 13)(x + 15) = 0$$

Factor by finding a pair of numbers whose product is -195 and sum is 2.

$$x - 13 = 0 \quad \text{or} \quad x + 15 = 0$$

Use the zero-factor theorem to solve.

$$x = 13 \qquad\qquad x = -15$$

Answer Because -15 is not a natural number and 13 is a natural number, the first number is 13. This means that the consecutive odd natural number is 15.

Check 13 and 15 are consecutive odd natural numbers and their product is 195, so the answer is correct.

ANSWER

$n = -\dfrac{2}{5} \text{ or } -\dfrac{4}{3}$

ANSWER

21 and 22

YOUR TURN 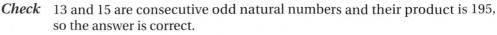 The product of two consecutive natural numbers is 462. Find the numbers.

EXAMPLE 7 ⟩ An architect is designing an addition to a house. The addition will be in the shape of a rectangle and have an area of 270 square feet. The width of the room is to be 3 feet less than the length. Find the dimensions of the room.

Understand The area of a rectangle is found by multiplying the length and width. We are given a relationship about the width and length, which we can translate:

"the width is to be 3 feet less than the length"

Width = $l - 3$, where l represents length.

Plan Translate to an equation, then solve.

Execute $l(l - 3) = 270$ The equation has the product of the length, l, and width, $l - 3$, set equal to 270.

$l(l - 3) - 270 = 0$ Subtract 270 from both sides.

$l^2 - 3l - 270 = 0$ Multiply $l(l - 3)$ to get standard form.

$(l - 18)(l + 15) = 0$ Factor.

$l - 18 = 0$ or $l + 15 = 0$ Use the zero-factor theorem to solve.

$l = 18$ $l = -15$

Answer Because the problem involves room dimensions, only the positive number makes sense as an answer. If the length is 18 feet, and the width is 3 feet less than that, then the width is 15 feet.

Check The width, 15 feet, is 3 feet less than the length, 18 feet. Also, the area of a 15-foot by 18-foot rectangle is 270 square feet.

YOUR TURN ⟩ Two discs are designed so that the smaller of the two has an area of 113.04 square centimeters and a radius that is 8 centimeters less than the larger disc's radius. Find the radius of the larger disc. (Use 3.14 to approximate π.)

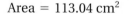

Area = 113.04 cm^2

Note: *This radius is 8 less than the other disc's radius.*

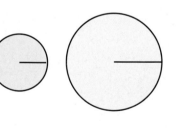

EXAMPLE 8 ⟩ The equation $h = -16t^2 + h_0$ describes the height, h, in feet of an object after falling from an initial height h_0 for a time, t seconds. An object is dropped from a height of 600 feet. When the object reaches 344 feet, how much time has passed?

Understand We are given a formula and we are to find the time it takes an object to drop from an initial height of 600 feet to a height of 344 feet.

ANSWER ⟩

14 cm

Plan Replace the variables in the formula with the given values to get an equation in terms of t, then solve for t.

Execute $344 = -16t^2 + 600$ Replace *h* with 344 and h_0 with 600.

$$-256 = -16t^2 + 0$$ Subtract 600 from both sides.

$$16t^2 - 256 = 0$$ Add $16t^2$ to both sides.

$$16(t^2 - 16) = 0$$ Factor out the monomial GCF, 16.

$$t^2 - 16 = 0$$ Divide out the constant factor, 16.

$$(t - 4)(t + 4) = 0$$ Factor.

$$t - 4 = 0 \quad \text{or} \quad t + 4 = 0$$ Use the zero-factor theorem to solve.

$$t = 4 \qquad\qquad t = -4$$

Answer Because the problem involves the amount of time elapsed after an object was dropped, only the positive value, 4, makes sense. This means it took the object 4 seconds to fall from 600 feet to 344 feet.

Check Verify that after 4 seconds, an object dropped from an initial height of 600 feet will descend to a height of 344 feet.

$$h = -16(4)^2 + 600$$
$$h = -256 + 600$$
$$h = 344$$

YOUR TURN An object is dropped from a height of 225 feet above the ground. When the object is 81 feet from the ground, how much time has passed?

ANSWER

3 sec.

OBJECTIVE 4. Use the Pythagorean theorem to solve problems. One of the most popular theorems in mathematics is the Pythagorean theorem, named after the Greek mathematician Pythagoras. The theorem relates the side lengths of all right triangles. Recall that in a right triangle, the two sides that form the 90° angle are *legs* and the side directly across from the 90° angle is the *hypotenuse.*

RULE **The Pythagorean Theorem**

Given a right triangle, where *a* and *b* represent the lengths of the legs and *c* represents the length of the hypotenuse, then $a^2 + b^2 = c^2$.

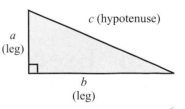

Of Interest

Pythagoras was a Greek mathematician who is believed to have been born around 569 B.C. and died in 500 B.C. The Pythagorean theorem is named after Pythagoras not because he discovered the relationship, as commonly thought, but because he and his followers are the first to have proved that the relationship is true for all right triangles. The relationship was known and used by other cultures prior to the Greeks; however, Pythagoras and others in the Greek culture believed it important to prove mathematical relationships.

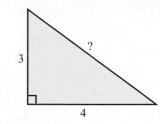

Connection In Chapter 9, we will revisit the Pythagorean theorem. There, we will use another technique involving square roots to solve equations like $c^2 = 25$.

We can use the Pythagorean theorem to find a missing length in a right triangle if we know the other two lengths.

EXAMPLE 9 Find the length of the missing side.

Solution Use the Pythagorean theorem, $a^2 + b^2 = c^2$. The hypotenuse, c, is the missing side length.

$$3^2 + 4^2 = c^2$$ Substitute $a = 3$ and $b = 4$.

$$9 + 16 = c^2$$ Simplify exponential forms.

$$25 = c^2$$ Add.

$$0 = c^2 - 25$$ Subtract 25 from both sides to get standard form.

$$0 = (c - 5)(c + 5)$$ Factor.

$$c - 5 = 0 \quad \text{or} \quad c + 5 = 0$$ Use the zero-factor theorem.

$$c = 5 \qquad\qquad c = -5$$

Because we are dealing with lengths, only the positive solution is sensible, so the missing length is 5.

YOUR TURN In constructing a roof, three beams are used to form a right triangle frame. A 12-foot beam and a 5-foot beam are brought together to form a 90° angle. How long must the third beam be?

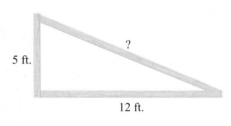

ANSWER

13 ft.

6.6 Exercises

For Extra Help
MyMathLab   Videotape/DVT   InterAct Math   Math Tutor Center   Math XL.com

1. Explain the zero-factor theorem in your own words.

2. How do you use the zero-factor theorem to solve an equation that has an expression in factored form set equal to 0?

3. What is the first step in solving a quadratic equation?

4. Explain why we cannot write $(x + 2)(x - 2) = 32$ as $x + 2 = 32$ and $x - 2 = 32$.

For Exercises 5–16, solve using zero-factor theorem.

5. $(x + 5)(x + 2) = 0$

6. $(x - 7)(x + 3) = 0$

7. $(a + 3)(a - 4) = 0$

8. $(b - 7)(b + 2) = 0$

9. $(3x - 4)(2x + 3) = 0$

10. $(3a + 2)(a - 5) = 0$

11. $x(x - 7) = 0$

12. $y(y + 3) = 0$

13. $m(m - 1)(m + 2) = 0$

14. $r(r - 2)(r + 3) = 0$

15. $b(b - 2)^2 = 0$

16. $m(m - 4)^2 = 0$

For 17–48, solve the quadratic equations.

17. $y^2 - 4y = 0$

18. $p^2 + 3p = 0$

19. $6r^2 + 10r = 0$

20. $12c^2 - 9c = 0$

21. $x^2 - 4 = 0$

22. $y^2 - 25 = 0$

23. $x^2 + x - 6 = 0$

24. $b^2 - 5b + 6 = 0$

25. $n^2 + 6n - 27 = 0$

26. $p^2 - 16p + 64 = 0$

27. $x^2 - 4x = 21$

28. $u^2 + 6u = 27$

29. $p^2 = 3p - 2$

30. $x^2 = 6x - 8$

31. $v^2 = 9$

32. $m^2 = 16$

33. $2a^2 + 18a + 28 = 0$

34. $3x^2 - 6x - 45 = 0$

35. $x^3 - x^2 - 12x = 0$

36. $x^3 + 7x^2 + 10x = 0$

37. $12r^3 + 22r^2 - 20r = 0$

38. $16c^3 - 12c^2 - 18c = 0$

39. $3y^2 + 8 = 14y$

40. $6k^2 + 1 = 7k$

41. $4x^2 = 60 - x$

42. $4n^2 = 5 - 19n$

43. $7t = 12 + t^2$

44. $-9x = -20 - x^2$

45. $b(b - 5) = 14$

46. $m(m + 6) = -9$

47. $4x(x + 7) = -49$

48. $c(2c - 11) = -5$

For Exercises 49–62, translate to an equation, then solve.

49. Find every number such that the square of the number added to 55 is the same as sixteen times the number.

50. Find every number such that triple the square of the number is equal to four times that number.

51. One natural number is four times another. The product of the two numbers is 676. Find both numbers.

52. The product of two consecutive natural numbers is 306. Find the numbers.

53. The sum of the squares of two consecutive natural numbers is 365. Find the integers.

54. The difference of the squares of two consecutive even natural numbers is 60.

55. The length of a rectangular garden is 9 meters more than the width. If the area is 252 square meters, find the dimensions of the garden.

56. The central chamber of the Lincoln Memorial has an area of 4440 square feet. If the length of the chamber is 14 feet more than the width, find the dimensions of the chamber. (*Source:* National Parks Service)

57. The largest billboard in Times Square's history has a length that is 10 feet more than three times its width. The area of the billboard is 9288 square feet. Find the dimensions of the billboard. (*Source:* SOHH.com)

58. The length of a football field is 40 yards less than three times the width. The total football field area is 6400 square yards. Find the dimensions of the football field.

59. The design of a small building calls for a rectangular shape with dimensions of 22 feet by 28 feet. The architect decides to change the shape of the building to a circle, but wants it to have the same area. If we use $\frac{22}{7}$ to approximate π, what would be the radius of the circular building?

60. A design on the front of a marketing brochure calls for a triangle with a base that is 6 centimeters less than the height. If the area of the triangle is to be 216 square centimeters, what are the lengths of the base and height?

61. A steel plate is in the shape of a trapezoid and has an area of 85.5 square inches. The dimensions are shown. Note that the length of the base is equal to the height. Calculate the height.

62. The front elevation of one wing of a house is shown. Because of budget constraints, the total area of the front of this wing must be 352 square feet. The height of the triangular portion is 14 feet less than the base. Find the base length.

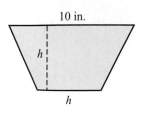

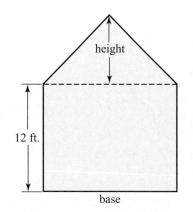

For Exercises 63 and 64, use the formula $h = -16t^2 + v_0t + h_0$, where h is the final height in feet, t is the time of travel in seconds, v_0 is the initial velocity in feet per second, and h_0 is the initial height in feet of an object traveling upwards.

63. A ball is thrown upward at 4 feet per second from a building 29 feet high. When will the ball be 9 feet above the ground?

64. A toy rocket is fired from the ground with an upward velocity of 200 feet per second. How many seconds will it take to return to the ground?

For Exercises 65 and 66, use the formula $B = P(1 + \frac{r}{n})^{nt}$, which is used to calculate the final balance of an investment or a loan after being compounded. Following is a list of what each variable represents:

    B represents the final balance

    P represents the principal, which is the amount invested

    r represents the annual interest rate as a decimal

    t represents the number of years that the principal is compounded

    n represents the number of times per year the principal is compounded

65. Carlita invests $4000 in an account that is compounded annually. If, after two years, her balance is $4840, what was the interest rate of the account?

66. Donovan invests $1000 in an account that is compounded semiannually (every six months). If, after one year, his balance is $1210, what is the interest rate of the account?

For Exercises 67–70, find length of the hypotenuse.

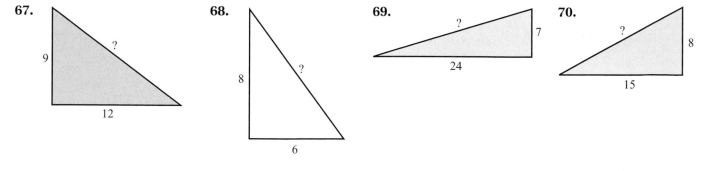

67.

9

?

12

68.

8

?

6

69.

?

7

24

70.

?

8

15

For Exercises 71–74, solve.

71. An artist welds a metal rod measuring 12 centimeters to a second rod measuring 35 centimeters to form a 90° angle. She wants a third rod to be welded to the ends of the first two to form a right triangle. What length must the third rod be?

72. A rectangular screen measures 27 inches by 36 inches. Find the diagonal distance between two opposing corners.

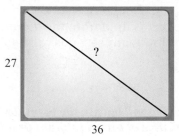

27

?

36

73. A person drives south 18 blocks then turns east and drives 24 blocks. Assuming all the city blocks are equal size and form a grid of streets that intersect at 90°, find the shortest distance in blocks that the person is from her original location.

74. To support a power pole, a wire is to be attached 24 feet from the ground then staked 10 feet from the base of the pole. How long must the wire be?

24 ft.

|←—10 ft.—→|

REVIEW EXERCISES

[4.2] **1.** Is the equation $y = x^2 - 1$ a linear equation? Explain.

[4.2] **2.** Is $(1, -4)$ a solution for $4x + 3y = -8$?

[4.3] **3.** Find the x- and y-intercepts for $2x - 5y = -20$.

[4.5] **4.** Write the equation of the line passing through $(-1, 4)$ and $(-3, -2)$ in the form $Ax + By = C$, where A, B, and C are integers and $A > 0$.

[4.7] **5.** For the function $f(x) = x^2 + 2$, find each of the following.
   a. $f(0)$

   b. $f(-2)$

[4.7] **6.** Graph: $f(x) = 2x - 3$

6.7 Graphs of Quadratic Equations and Functions

OBJECTIVES

1. Graph quadratic equations in the form $y = ax^2 + bx + c$.

2. Graph quadratic functions.

In Chapter 4, we graphed linear equations in two variables. Now we will graph **quadratic equations in two variables**.

DEFINITION *Quadratic equation in two variables:* An equation that can be written in the form $y = ax^2 + bx + c$, where a, b, and c are real numbers and $a \neq 0$.

Following are some examples of quadratic equations in two variables.

$$y = x^2 \qquad y = 3x^2 + 2 \qquad y = x^2 - 4x + 1$$

We learned in Section 2.2 that every term in a linear equation has a degree of 1 or 0. Since quadratic equations always have a term with a degree of 2, they are nonlinear, which means their graphs will not be straight lines. What might their graphs look like?

OBJECTIVE 1. **Graph quadratic equations in the form $y = ax^2 + bx + c$.** Recall that one way to graph an equation is to plot solutions to the equation in the coordinate plane. In Section 4.1, we said that a solution to an equation in two variables is an ordered pair (x, y) that satisfies the equation. For example, $(3, 9)$ is a solution to $y = x^2$ because $9 = (3)^2$ is true. Let's make a table of solutions for $y = x^2$, plot those solutions, and then connect the points to see what the graph looks like.

Note: *Since we know quadratic equations are nonlinear, we find many solutions to get a good sense of the shape of the graph.*

x	y
−3	9
−2	4
−1	1
0	0
1	1
2	4
3	9

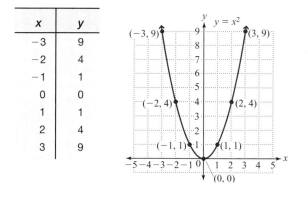

The graph of $y = x^2$ is a *parabola*. In fact, the graph of every quadratic equation in two variables is a parabola.

In the table of solutions for $y = x^2$, notice that when $x = 1$ or -1, $y = 1$; when $x = 2$ or -2, $y = 4$; and when $x = 3$ or -3, $y = 9$. In general, each x-coordinate and its additive inverse have the same y-coordinate. Looking at the graph, we see that the left side of the y-axis is the mirror image of the right side of the y-axis. Consequently, we say that this graph is symmetrical about the line $x = 0$ (the y-axis). Every parabola will have symmetry about a line called its **axis of symmetry**. Further, a parabola's axis of symmetry always passes through a point called the **vertex**.

DEFINITIONS *Axis of symmetry:* A line that divides a graph into two symmetrical halves.
Vertex: The lowest point on a parabola that opens up or highest point on a parabola that opens down.

For $y = x^2$, the axis of symmetry is the line $x = 0$ (y-axis), and the vertex is $(0, 0)$. In Section 10.5, we will learn more about the vertex and axis of symmetry. For now, we will graph by finding enough solutions to see where the parabola turns.

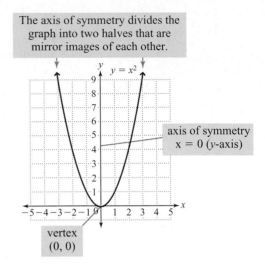

The axis of symmetry divides the graph into two halves that are mirror images of each other.

axis of symmetry
$x = 0$ (y-axis)

vertex
$(0, 0)$

PROCEDURE Graphing Quadratic Equations

To graph a quadratic equation:

1. Find ordered pair solutions and plot them in the coordinate plane. Continue finding and plotting solutions until the shape of the parabola can be clearly seen.
2. Connect the points to form a parabola.

EXAMPLE 1 Graph.

a. $y = 3x^2 + 2$

Solution We complete a table of solutions, plot the solutions in the coordinate plane, and then connect the points to form the graph.

Connection In a quadratic equation in the form $y = ax^2 + bx + c$, the constant c indicates the y-intercept just as the constant b does in a linear equation in the form $y = mx + b$.

Note: *The constant 2 in* $y = 3x^2 + 2$ *indicates the* y-intercept *because replacing* x *with 0 leaves us with* $y = 3(0)^2 + 2 = 2.$

x	y
−2	14
−1	5
0	2
1	5
2	14

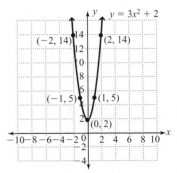

b. $y = x^2 + 5x - 3$

Solution

Note: *Since -3 and -2 have the same y-values, the x-coordinate of the vertex will be halfway between -3 and -2 at -2.5.*

x	y
-5	-3
-4	-7
-3	-9
-2	-9
-1	-7
0	-3

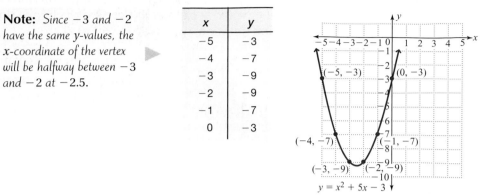

Notice that in each equation that we have graphed so far ($y = x^2$, $y = 3x^2 + 2$, and $y = x^2 + 5x - 3$), the coefficient of x^2 is a positive number. Also notice that the graph of each of those equations was a parabola that opens upwards. Now let's consider quadratic equations in which the coefficient of x^2 is a negative number, as in $y = -2x^2 + 1$. We will see that the graphs of these equations open downwards.

a. $y = x^2 - 5$

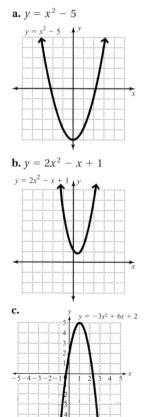

b. $y = 2x^2 - x + 1$

c.

EXAMPLE 2 Graph. $y = -2x^2 + 1$

Solution Complete a table of solutions, plot the solutions, then connect the points to form a parabola.

x	y
-2	-7
-1	-1
0	1
1	-1
2	-7

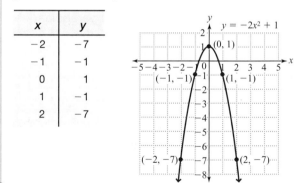

Note: *The coefficient of x^2 is a negative number, -2, and the graph opens downwards. In an equation of the form $y = ax^2 + bx + c$, if the coefficient of x^2 is negative, the parabola opens downwards.*

RULE **Opening of a Parabola**

Given an equation in the form $y = ax^2 + bx + c$, if $a > 0$, then the parabola opens upwards; if $a < 0$, then the parabola opens downwards.

YOUR TURN Graph.

a. $y = x^2 - 5$ **b.** $y = 2x^2 - x + 1$ **c.** $y = -3x^2 + 6x + 2$

OBJECTIVE 2. Graph quadratic functions. In Section 4.7, we learned how to graph linear functions. Now let's graph quadratic functions. A quadratic *equation* in two variables, $y = ax^2 + bx + c$, is a quadratic *function* which can be written in the form $f(x) = ax^2 + bx + c$. Notice that to change the equation in two variables to function notation, we replace y with $f(x)$.

Quadratic equation in two variables: $\quad y = -2x^2 + 1$

Using function notation: $\quad f(x) = -2x^2 + 1$

The graph of $f(x) = -2x^2 + 1$ is the same as $y = -2x^2 + 1$, which we graphed in Example 2. Recall from Section 4.7 that with function notation, we find ordered pairs by evaluating the function using various values of x. For example, the notation $f(1)$ means to find the value of the function when $x = 1$.

> **Connection** Finding $f(1)$ for $f(x) = -2x^2 + 1$ is the same as finding the y-value for $y = -2x^2 + 1$ when $x = 1$.

$$f(1) = -2(1)^2 + 1$$
$$= -2 + 1$$
$$= -1 \qquad \text{The ordered pair is } (1, -1).$$

We would continue finding ordered pairs in this manner to produce the same table and graph that we produced in Example 2, which suggests the following procedure.

PROCEDURE **Graphing Quadratic Functions**

To graph a quadratic function:

1. Find enough ordered pairs by evaluating the function for various values of x so that when those ordered pairs are plotted, the shape of the parabola can be clearly seen.
2. Connect the points to form the parabola.

EXAMPLE 3 Graph.

a. $f(x) = 3x^2 - 5$

Solution Find enough ordered pairs to clearly see the graph, then connect the points to form the parabola.

x	y
-2	7
-1	-2
0	-5
1	-2
2	7

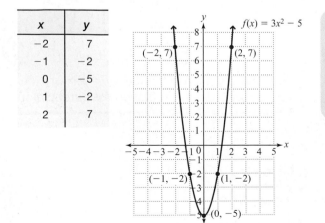

> **Connection** This parabola opens upwards. Given a quadratic function in the form $f(x) = ax^2 + bx + c$, if $a > 0$, then the parabola opens upwards.

b. $f(x) = -x^2 + 2x + 4$

Solution

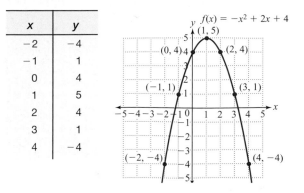

x	y
−2	−4
−1	1
0	4
1	5
2	4
3	1
4	−4

Connection This parabola opens downwards. Given a quadratic function in the form $f(x) = ax^2 + bx + c$, if $a < 0$, then the parabola opens downwards.

Connection In Section 4.7, we learned that a graph represents a function if it passes the vertical line test (a vertical line through any point in the domain touches the graph at only one point). Since graphs of quadratic functions are parabolas that open up or down, they pass the vertical line test. In future courses, you will see parabolas that open right or left, which do not pass the vertical line test and therefore are not functions.

a.

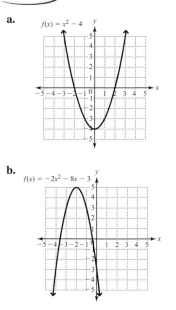

b.

YOUR TURN Graph.

a. $f(x) = x^2 - 4$

b. $f(x) = -2x^2 - 8x - 3$

6.7 Exercises

MyMathLab   Videotape/DVT   InterAct Math   Math Tutor Center   Math XL.com

1. In your own words, describe what is meant by *symmetry* in a parabola.

2. Given an equation in the form $y = ax^2 + bx + c$, what indicates whether the graph opens up or down?

3. Given an equation in the form $y = ax^2 + bx + c$, what indicates the y-coordinate of the y-intercept?

4. Given that the graph of $y = ax^2 + bx + c$, where $a \neq 0$, is a parabola that opens upwards or downwards, explain why $f(x) = ax^2 + bx + c$ is a function. (*Hint:* Refer to the definition of a function and the vertical line test in Section 4.7.)

6.7 Graphs of Quadratic Equations and Functions   **519**

For Exercises 5–10, use the graph of the parabola to determine the coordinates of the vertex.

5.

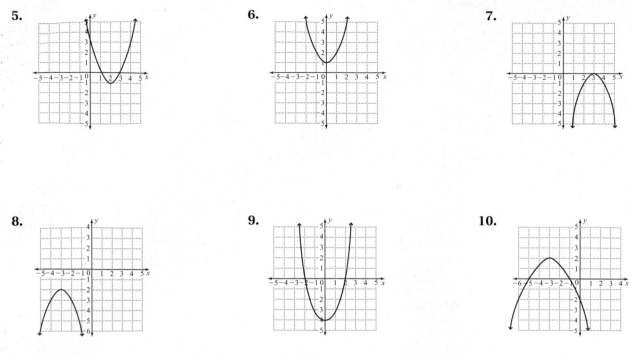

6.

7.

8.

9.

10.

For Exercises 11–16, complete the table of solutions, then graph.

11. $y = 2x^2$

x	y
-2	
-1	
0	
1	
2	

12. $y = -x^2$

x	y
-2	
-1	
0	
1	
2	

13. $y = -x^2 + 2$

x	y
-2	
-1	

14. $y = x^2 - 3$

x	y
-2	
-1	
0	
1	
2	

15. $y = 2x^2 - 4x + 1$

x	y
-1	
0	
1	
2	
3	

16. $y = -3x^2 + 6x + 4$

x	y
-1	
0	
1	
2	
3	

For Exercises 17–32, graph.

17. $y = x^2 + 1$

18. $y = x^2 - 4$

19. $y = -x^2 + 3$

20. $y = -x^2 + 2$

21. $y = 2x^2 - 5$

22. $y = 3x^2 - 4$

23. $y = -3x^2 + 2$

24. $y = -4x^2 + 5$

25. $y = x^2 + 4x - 1$

26. $y = x^2 - 6x + 3$

27. $y = -x^2 + 6x - 5$

28. $y = -x^2 + 4x - 5$

29. $y = 2x^2 - 4x - 3$

30. $y = 3x^2 + 6x - 1$

31. $y = -2x^2 + 8x - 5$

32. $y = -4x^2 - 12x - 3$

For Exercises 33–44, graph the function.

33. $f(x) = -3x^2$

34. $f(x) = 2x^2$

35. $f(x) = -x^2 + 1$

36. $f(x) = x^2 - 1$

37. $f(x) = x^2 + 4x - 1$ **38.** $f(x) = x^2 - 6x + 5$ **39.** $f(x) = -x^2 - 4x - 3$ **40.** $f(x) = -x^2 + 6x - 2$

41. $f(x) = 3x^2 - 12x + 5$ **42.** $f(x) = 2x^2 + 6x - 1$ **43.** $f(x) = -2x^2 - 4x - 1$ **44.** $f(x) = -4x^2 + 8x - 3$

For Exercises 45–48, state whether the graph is the graph of a function. (Hint: Use the vertical line test.)

45.   **46.**   **47.**   **48.**

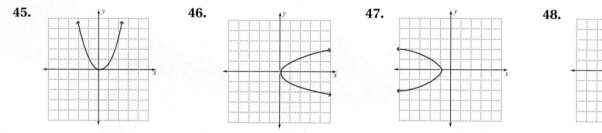

For Exercises 49–52, use a graphing calculator.

49. a. Enter the equation $y = x^2$, then graph. What are the coordinates of the y-intercept?

b. Enter a second equation, $y = x^2 + 2$, then graph. What are the coordinates of the y-intercept?

c. On the same grid, graph $y = x^2 + 4$ and compare all three graphs. What do your explorations suggest about how c affects the graph of an equation in the form $y = x^2 + c$?

d. On the same grid, graph $y = x^2 - 3$. Comparing this graph to the first three graphs, how did changing c to a negative value affect the graph?

50. a. Enter the equation $y = x^2$, then graph. What are the coordinates of the vertex?

b. Enter a second equation, $y = 2x^2$. Compare the graph of $y = 2x^2$ with the graph of $y = x^2$, then discuss their similarities and differences.

c. On the same grid, graph $y = 3x^2$ and compare all three graphs. What does this exploration suggest about how the size of a in $y = ax^2$ affects the graph?

d. In $y = ax^2$, if $a = \dfrac{1}{2}$, how do you think the graph will be different from the graphs in parts a–c? Graph $y = \dfrac{1}{2}x^2$.

e. What do these explorations suggest about how the size of a affects the graph of an equation in the form $y = ax^2$?

51. a. Graph $y = x^2$ and $y = -x^2$ on the same grid. What is different about the two graphs? What does this suggest about the sign of a in an equation of the form $y = ax^2$?

b. Clearing the equations from part a, graph $y = 2x^2$ and $y = -2x^2$ together. Do these graphs confirm your conclusion from part a?

c. On a new grid, graph $y = \dfrac{1}{2}x^2$ and $y = -\dfrac{1}{2}x^2$. What is different about these graphs compared to those in parts a and b?

d. What do these explorations suggest about how the absolute value of a affects the graph of an equation in the form $y = ax^2$?

52. a. Enter the equation $y = x^2 + x + 1$, then graph. What are the coordinates of the y-intercept?

b. On the same grid, graph $y = x^2 + 2x + 1$. Compare the graph of $y = x^2 + x + 1$ with the graph of $y = x^2 + 2x + 1$, then discuss their similarities and differences.

c. Using the same grid, graph $y = x^2 + 3x + 1$. Compare this graph with the graphs of $y = x^2 + x + 1$ and $y = x^2 + 2x + 1$, then discuss similarities and differences.

d. On the same grid, graph $y = x^2 - x + 1$, $y = x^2 - 2x + 1$, and $y = x^2 - 3x + 1$. Compare these graphs with your graphs from parts a–c. What do these explorations suggest about how b affects the graph of an equation in the form $y = ax^2 + bx + c$?

Collaborative Exercises ⟨WHAT GOES UP, MUST COME DOWN⟩

There are many sets of variables that a have a relationship that is roughly modeled by a parabolic shape. The following data give the high temperature and peak power load (in megawatts) for 31 days throughout the calendar year.

1. On a sheet of graph paper, draw and label axes with the temperature along the horizontal axis and the peak load on the vertical axis. Then plot the points.

2. Draw a smooth parabola through the points that might best fit the data.

3. At what temperature does the minimum peak load seem to occur?

4. In your own words, describe what happens to the peak load as the temperature increases and decreases and explain why.

5. Use the model that you drew in part 2 to estimate the peak load on a day when the maximum temperature is 85.

6. Give another example of a situation that might be best modeled by a parabola.

Temp	Peak Load	Temp	Peak Load
69	94	95	141
71	95	46	142
59	100	43	147
65	100	98	150
67	102	96	151
60	104	97	155
74	105	100	158
79	105	43	165
56	112	38	173
84	113	102	176
52	114	36	193
88	115	103	199
50	120	105	204
90	123	106	226
92	132	34	228
92	132		

REVIEW EXERCISES

For Exercises 1 and 2, multiply.

[5.4]   **1.** $3mn(m^2 - 4mn + 5)$ [5.5]   **2.** $(x - 3)(2x + 1)$

For Exercises 3 and 4, factor.

[6.3]   **3.** $3n^2 - 17n + 20$ [6.4]   **4.** $x^2 - 25$

[6.6] *For Exercises 5 and 6, solve.*

  5. $h(h - 3) = 0$ **6.** $y^2 + 3y - 18 = 0$

Chapter 6 Summary

Defined Terms

Section 6.1
Factored form (p. 460)
Greatest common factor
(GCF) (p. 461)

Section 6.6
Quadratic equation in one
variable (p. 504)

Section 6.7
Quadratic equation in two
variables (p. 515)

Axis of symmetry (p. 516)
Vertex (p. 516)

Procedures, Rules, and Key Examples

Procedures/Rules	Key Examples

Section 6.1 Greatest Common Factor and Factoring by Grouping

To find the GCF by listing:
1. List all possible factors for each given number.
2. Search the lists for the largest factor common to all lists.

Example 1: Find the GCF of 54 and 180 by listing.

Factors of 54: 1, 2, 3, 6, 9, 18, 27, 54
Factors of 180: 1, 2, 3, 4, 5, 6, 9, 10, 12, 15, 18, 20, 30, 36, 45, 60, 90, 180
$$GCF = 18$$

To find the greatest common factor of a given set of numbers:
1. Write the prime factorization of each given number in exponential form.
2. Create a factorization for the GCF that includes only those primes common to all factorizations, each raised to its smallest exponent in the factorizations.
3. Multiply the factors in the factorization created in step 2.
Note: If there are no common prime factors, then the GCF is 1.

Example 2: Find the GCF of 54 and 180 using prime factorization.

$$54 = 2 \cdot 3^3$$
$$180 = 2^2 \cdot 3^2 \cdot 5$$
$$GCF = 2 \cdot 3^2 = 18$$

To factor a monomial GCF out of a given polynomial:
1. Find the GCF of the terms that make up the polynomial.
2. Rewrite the given polynomial as a product of the GCF and parentheses that contain the result of dividing the given polynomial by the GCF.

$$\text{Given polynomial} = GCF\left(\frac{\text{Given polynomial}}{GCF}\right)$$

Example 3: Factor. $54x^3y - 180x^2yz$
$54x^3y - 180x^2yz$
$$= 18x^2y\left(\frac{54x^3y - 180x^2yz}{18x^2y}\right)$$
$$= 18x^2y\left(\frac{54x^3y}{18x^2y} - \frac{180x^2yz}{18x^2y}\right)$$
$$= 18x^2y(3x - 10z)$$

To factor a four-term polynomial by grouping:
1. Factor out any monomial GCF (other than 1) that is common to all four terms.
2. Group together pairs of terms and factor the GCF out of each pair.
3. If there is a common binomial factor, then factor it out.
4. If there is no common binomial factor, then interchange the middle two terms and repeat the process. If there is still no common binomial factor, then the polynomial cannot be factored by grouping.

Example 4: Factor by grouping.
$$8x^2 - 20x - 6xy + 15y$$

$8x^2 - 20x - 6xy + 15y$
$$= (8x^2 - 20x) - (6xy - 15y)$$
$$= 4x(2x - 5) - 3y(2x - 5)$$
$$= (2x - 5)(4x - 3y)$$

continued

Procedures/Rules	Key Examples

Section 6.2 Factoring Trinomials of the form $x^2 + bx + c$

To factor a trinomial of the form $x^2 + bx + c$:

1. Find two numbers with a product equal to c and a sum equal to b.
2. The factored trinomial will have the form:
   $(x + \text{first number})(x + \text{second number})$.

Note: The signs in the binomial factors can be minus signs, depending on the signs of b and c.

Example 1: Factor. $x^2 + 9x + 20$
Solution: Find two numbers whose product is 20 and whose sum is 9.

Product:	Sum:
$(1)(20) = 20$	$1 + 20 = 21$
$(2)(10) = 20$	$2 + 10 = 12$
$(4)(5) = 20$	$4 + 5 = 9$ Correct combination.

$$x^2 + 9x + 20 = (x + 4)(x + 5)$$

Example 2: Factor. $x^2 + 2x - 15$

Product:	Sum:
$(-1)(15) = -15$	$-1 + 15 = 14$
$(-3)(5) = -15$	$-3 + 5 = 2$
	Correct combination.

$$x^2 + 2x - 15 = (x - 3)(x + 5)$$

Section 6.3 Factoring Trinomials of the form $ax^2 + bx + c$, where $a \neq 1$

To factor a trinomial of the form $ax^2 + bx + c$, where $a \neq 1$, by trial and error:

1. Look for a monomial GCF in all the terms. If there is one, factor it out.
2. Write a pair of *first* terms whose product is ax^2.
3. Write a pair of *last* terms whose product is c.
4. Verify that the sum of the *inner* and *outer* products is bx (the middle term of the trinomial). If the sum of the inner and outer products is not bx, then try the following:
   a. Exchange the first terms of the binomials from step 3, then repeat step 4.
   b. Exchange the last terms of the binomials from step 3, then repeat step 4.
   c. Repeat steps 2–4 with a different combination of first and last terms.

To factor a trinomial of the form $ax^2 + bx + c$, where $a \neq 1$, by grouping:

1. Look for a monomial GCF in all the terms. If there is one, factor it out.
2. Multiply a and c.
3. Find two factors of this product whose sum is b.
4. Write a four-term polynomial in which bx is written as the sum of two like terms whose coefficients are the two numbers you found in step 3.
5. Factor by grouping.

Example 1: Factor. $4x^2 - 9x + 5$
Factors of 4 are 4 and 1, or 2 and 2.
Factors of 5 are 1 and 5.
Because the middle term $-9x$ is negative and the last term 5 is positive, we know that the second terms in each binomial factor will be negative.

$$(x - 1)(4x - 5) = 4x^2 - 9x + 5$$

Example 2: Factor. $18y^3 + 12y^2 - 48y$
Factor out the monomial GCF, $6y$.

$$18y^3 + 12y^2 - 48y = 6y(3y^2 + 2y - 8)$$

To factor the trinomial, multiply $3(-8) = -24$, then find two factors of -24 whose sum is 2. Note that -4 and 6 work. Write the middle term, $2y$, as $-4y + 6y$, then factor by grouping.

$$= 6y(3y^2 - 4y + 6y - 8)$$
$$= 6y[y(3y - 4) + 2(3y - 4)]$$
$$= 6y(3y - 4)(y + 2)$$

continued

Procedures/Rules	Key Examples

Section 6.4 Factoring Special Products

Rules for factoring special products.

Perfect square trinomials: $\quad a^2 + 2ab + b^2 = (a + b)^2$
$$a^2 - 2ab + b^2 = (a - b)^2$$

Difference of squares: $\quad a^2 - b^2 = (a + b)(a - b)$

Note: A sum of squares cannot be factored.

Difference of cubes: $\quad a^3 - b^3 = (a - b)(a^2 + ab + b^2)$

Sum of cubes: $\quad a^3 + b^3 = (a + b)(a^2 - ab + b^2)$

Example 1: Factor.

a. $9x^2 + 30x + 25 = (3x + 5)^2$
b. $36m^2 - 60mn + 25n^2 = (6m - 5n)^2$

c. $4y^2 - 81 = (2y + 9)(2y - 9)$

d. $8x^3 - 27 = (2x - 3)(4x^2 + 6x + 9)$

e. $n^3 + 64 = (n + 4)(n^2 - 4n + 16)$

Section 6.5 Strategies for Factoring

To factor a polynomial, first factor out any monomial GCF, then consider the number of terms in the polynomial. If the polynomial has:

  I. Four terms, then try to factor by grouping.

 II. Three terms, then determine if the trinomial is a perfect square or not.

    A. If the trinomial is a perfect square, then consider its form.

       1. If in the form $a^2 + 2ab + b^2$, then the factored form is $(a + b)^2$.

       2. If in the form $a^2 - 2ab + b^2$, then the factored form is $(a - b)^2$.

    B. If the trinomial is not a perfect square, then consider its form.

       1. If in the form $x^2 + bx + c$, then find two factors of c whose sum is b and write the factored form as $(x + \text{first number})(x + \text{second number})$.

       2. If in the form $ax^2 + bx + c$, where $a \neq 1$, then use trial and error. Or, find two factors of ac whose sum is b; write these factors as coefficients of two like terms that, when combined, equal bx; and then factor by grouping.

III. Two terms, then determine if the binomial is a difference of squares, sum of cubes, or difference of cubes.

    A. If given a binomial that is a difference of squares $a^2 - b^2$, then the factors are conjugates and the factored form is $(a + b)(a - b)$. Note that a sum of squares cannot be factored.

    B. If given a binomial that is a sum of cubes, $a^3 + b^3$, then the factored form is $(a + b)(a^2 - ab + b^2)$.

    C. If given a binomial that is a difference of cubes, $a^3 - b^3$, then the factored form is $(a - b)(a^2 + ab + b^2)$.

Note: Always look to see if any of the factors can be factored.

The examples shown do not represent every possible type of problem. Each example will contain a monomial GCF other than 1:

Example 1: Factoring by grouping:
$8x^3 + 24x^2 - 2x^2y - 6xy$
$$= 2x(4x^2 + 12x - xy - 3y)$$
$$= 2x[4x(x + 3) - y(x + 3)]$$
$$= 2x(x + 3)(4x - y)$$

Example 2: Factoring a perfect square trinomial:
$75x^4 + 60x^3y + 12x^2y^2$
$$= 3x^2(25x^2 + 20xy + 4y^2)$$
$$= 3x^2(5x + 2y)^2$$

Example 3: Factoring a difference of squares:
$81y^5 - 16y = y(81y^4 - 16)$
$$= y(9y^2 + 4)(9y^2 - 4)$$
$$= y(9y^2 + 4)(3y + 2)(3y - 2)$$

Example 4: Factoring a sum of cubes:
$54n^4 + 16nm^3 = 2n(27n^3 + 8m^3)$
$$= 2n(3n + 2m)(9n^2 - 6mn + 4m^2)$$

continued

Procedures/Rules	Key Examples

Procedures/Rules

Section 6.6 Solving Quadratic Equations by Factoring

Zero-Factor Theorem:

If a and b are real numbers and $ab = 0$, then $a = 0$, or $b = 0$.

To solve an equation in which two or more factors are equal to 0, use the zero-factor theorem:
1. Set each factor equal to zero.
2. Solve each of those equations.

To solve a quadratic equation:
1. Write the equation in standard form ($ax^2 + bx + c = 0$).
2. Write the variable expression in factored form.
3. Use the zero-factor theorem to solve.

Key Examples

Example 1: Solve $(2x - 5)(x + 4) = 0$.
$$2x - 5 = 0 \quad \text{or} \quad x + 4 = 0$$
$$2x = 5 \qquad\qquad x = -4$$
$$x = \frac{5}{2}$$

Example 2: Solve $2x^2 = 3 - x$.

$2x^2 = 3 - x$	
$2x^2 + x = 3$	Add x to both sides.
$2x^2 + x - 3 = 0$	Subtract 3 from both sides.
$(2x + 3)(x - 1) = 0$	Factor.
$2x + 3 = 0 \quad \text{or} \quad x - 1 = 0$	Use the zero-factor theorem.
$2x = -3 \qquad\qquad x = 1$	
$x = -\dfrac{3}{2}$	

Example 3: The length of a small rectangular building is to be 15 feet more than the width. The area is to be 700 square feet. What must the dimensions of the building be?

Let w represent width. The length will be $w + 15$. Writing an equation for area, we have:

$w(w + 15) = 700$	
$w^2 + 15w = 700$	Distribute w.
$w^2 + 15w - 700 = 0$	Subtract 700 from both sides.
$(w - 20)(w + 35) = 0$	Factor.
$w - 20 = 0 \quad \text{or} \quad w + 35 = 0$	Use the zero-factor theorem.
$w = 20 \qquad\qquad w = -35$	

Answer: The width is 20 ft. and the length is 35 ft.

The Pythagorean Theorem:

Given a right triangle where a and b represent the lengths of the legs and c represents the length of the hypotenuse, then $a^2 + b^2 = c^2$.

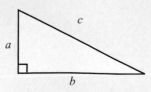

Example 4: Find the unknown length in the following right triangle.

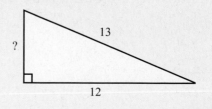

continued

Procedures/Rules	Key Examples

Section 6.6 Solving Quadratic Equations by Factoring (continued)

Solution: Use the Pythagorean theorem, $a^2 + b^2 = c^2$.

$$a^2 + 12^2 = 13^2$$
$$a^2 + 144 = 169$$
$$a^2 - 25 = 0 \quad \text{Subtract 169 from both sides.}$$

$$(a - 5)(a + 5) = 0 \quad \text{Factor.}$$
$$a - 5 = 0 \quad \text{or} \quad a + 5 = 0 \quad \text{Use the zero-factor theorem.}$$

$$a = 5 \qquad a = -5$$

Answer: The unknown length is 5.

Section 6.7 Graphs of Quadratic Equations and Functions

To graph a quadratic equation:
1. Find ordered pair solutions and plot them in the coordinate plane. Continue finding and plotting solutions until the shape of the parabola can be clearly seen.
2. Connect the points to form a parabola.

Given an equation in the form $y = ax^2 + bx + c$, if $a > 0$, then the parabola opens upwards; if $a < 0$, then the parabola opens downwards.

Example 1: Graph $y = x^2 - 3$.

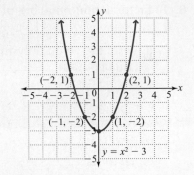

To graph a quadratic function:
1. Find enough ordered pairs by evaluating the function for various values of x so that when those ordered pairs are plotted, the shape of the parabola can be clearly seen.
2. Connect the points to form the parabola.

Example 2: Graph
$f(x) = -x^2 + 4x - 1$.

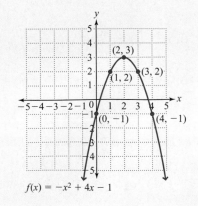

Chapter 6 Review Exercises

For Exercises 1–5, answer true or false.

[6.1] **1.** The GCF of $24x^5$ and $36x^3$ is $6x^3$.

[6.1] **2.** To factor a binomial, we use grouping.

[6.4] **3.** $x^2 + 16 = (x + 4)(x + 4)$

[6.4] **4.** $x^2 + 8x + 16 = (x + 4)^2$

[6.6] **5.** If a and b are real numbers and $ab = 0$, then $a = 0$ or $b = 0$.

[6.1] **For Exercises 6–10, complete the rule.**

    6. To factor a monomial out of a given polynomial:
        1. Find the _____ of the terms that make up the polynomial.
        2. Rewrite the given polynomial as a product of the GCF and parentheses that contain _____.

[6.2] **7.** Procedure: To factor a trinomial of the form $x^2 + bx + c$:
        1. Find two numbers with a product equal to _____ and a sum equal to _____.
        2. The factored trinomial will have the form:
          (x + first number)(x + second number)

[6.4] **8.** $a^2 - b^2 = (\quad)(\quad)$

[6.4] **9.** $a^3 - b^3 = (\quad)(\quad)$

[6.4] **10.** $a^2 - 2ab + b^2 = (\quad)^2$

Constants and
Variables

Exercises 11–18

[6.1] **For Exercises 11–18, list all of the factors.**

11. 60   **12.** 9   **13.** 81   **14.** 44

15. 45   **16.** 49   **17.** 30   **18.** 33

[6.1] *For Exercises 19–26, find the GCF.*

19. $21, 30$

20. $34, 51$

21. $12, 42, 60$

22. $10, 15, 45$

23. $10y^4, 25y^2$

24. $4x, 20xy$

25. $15m^2n, 3mn^5$

26. $8k, 22k^9$

[6.1] *For Exercises 27–34, factor out the GCF.*

27. $4x - 2$

28. $5m - 35m^3$

29. $y^3 - y$

30. $abc + a^2b^2c^2 - a^3b^3c^3$

31. $x^2y + xy^2 + x^3y^3$

32. $105a^3b^2 - 63a^2b^3 + 84a^6b^4$

33. $100k^4 + 120k^5 - 10k + 40k^3$

34. $18ab^3c - 36a^2b^2c$

[6.1] *For Exercises 35–42, factor by grouping.*

35. $ax + ay + bx + by$

36. $ax + 2a + bx + 2b$

37. $y^3 + 2y^2 + 3y + 6$

38. $y^4 + 4y^3 - by - 4b$

39. $xy + y + x + 1$

40. $ax^2 - 5y^2 + ay^2 - 5x^2$

41. $2b^3 - 2b^2 + b - 1$

42. $u^2 - 3u + 4uv - 12v$

[6.2] *For Exercises 43–50, factor.*

43. $x^2 - x - 12$

44. $x^2 + 14x + 45$

45. $n^2 - 6n + 8$

46. $a^2 + 3a - 20$

47. $h^2 + 51h + 144$

48. $y^2 - 10y - 24$

49. $4x^2 - 24x + 36$

50. $3m^2 - 33m + 54$

[6.3] *For Exercises 51–58, factor completely.*

51. $6x^2 + 3x - 7$

52. $2u^2 + 5u + 2$

53. $3m^2 - 10m + 3$

54. $5k^2 - 7kh - 12h^2$

55. $6a^2 - 20a + 16$

56. $8x^2y - 4xy - y$

57. $3p^2 - 13pq + 4q^2$

58. $24x^2 - 64xy - 24y^2$

[6.4] *For Exercises 59–66, factor.*

59. $v^2 - 8v + 16$

60. $u^2 + 6u + 9$

61. $4x^2 + 20x + 25$

62. $9y^2 - 12y + 4$

63. $x^2 - 4$

64. $25 - y^2$

65. $x^3 - 1$

66. $x^3 + 27$

[6.5] *For Exercises 67–76, factor completely.*

67. $6b^2 + b - 2$

68. $4ab - 24ab^2$

69. $y^2 + 25$

70. $3x^2 - 3y^2$

71. $x^4 - 81$ **72.** $7u^2 - 14u - 105$ **73.** $8x^3 - 27y^3$

74. $2 - 50y^2$ **75.** $3am - 6an - 8bn + 4bm$ **76.** $3m^2 + 9m + 27$

Equations and Inequalities

Exercises 77–94

[6.6] *For Exercises 77–84, solve.*

77. $(m + 3)(m - 4) = 0$ **78.** $y^2 - 4 = 0$ **79.** $x^2 - 5x + 6 = 0$ **80.** $x^2 - 4x = 21$

81. $m^2 = 9$ **82.** $x^2 + 3x = 18$ **83.** $y(y + 6) = -9$ **84.** $3n^2 - 11n = 4$

[6.6] *For Exercises 85–90, translate to an equation, then solve.*

85. Find every number such that five times the square of the number is equal to twice the number.

86. The sum of the squares of two consecutive natural numbers is 61. Find the numbers.

87. The product of a number and four times that same number is 100. Find all numbers.

88. Find the dimensions of a rectangle whose length is 6 more than its width and whose area is 91 square inches.

89. The product of two consecutive natural numbers is 110. Find the numbers.

90. Find the length of the missing side of the right triangle shown.

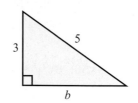

[6.7] *For Exercises 91–94, graph.*

91. $y = -5x^2$ **92.** $y = x^2 - 3$ **93.** $f(x) = x^2 + 2$ **94.** $f(x) = -3x^2 + 6x + 2$

Chapter 6 Practice Test

1. List all possible factors of 80.

2. Find the GCF of $25m^2$ and $40m$.

Factor completely.

3. $5y - 30$

4. $6x^2y - 2y^2$

5. $ax - ay - bx + by$

6. $m^2 + 7m + 12$

7. $r^2 - 8r + 16$

8. $4y^2 + 20y + 25$

9. $3q^2 - 10q + 8$

10. $6x^2 - 23x + 20$

11. $ax^2 - 5ax - 24a$

12. $10n^3 + 38n^2 - 8n$

13. $c^2 - 25$

14. $x^2 + 4$

15. $2 - 50u^2$

16. $-4x^2 + 16$

17. $m^3 + 125$

18. $x^3 - 8$

Solve.

19. $a(a + 3) = 0$

20. $x^2 - 4x - 12 = 0$

21. $2n^2 + 7n = 15$

22. The product of two consecutive natural numbers is 72. Find the numbers.

23. The width of a rectangle is 5 feet less than the length. If the area is 36 square feet, find the length and width.

Graph.

24. $y = -2x^2 + 4$

25. $f(x) = x^2 - 6x + 5$

Chapters 1–6 Cumulative Review Exercises

For Exercises 1–6, answer true or false.

[5.4] **1.** $x^3 \cdot x^4 = x^{12}$

[6.1] **2.** The GCF of 12 and 5 is 1.

[1.3] **3.** The commutative property can be used for both addition and subtraction.

[2.2] **4.** $2x - y^2 = 1$ is a linear equation.

[1.7] **5.** The expression $5 + x$ can be simplified to equal $5x$.

[6.1] **6.** $4x + 12 = 2(2x + 6)$ is factored completely.

For Exercises 7–10, fill in the blank.

[6.3] **7.** To factor a trinomial of the form $ax^2 + bx + c$, where $a \neq 1$, by grouping:

     a. Look for a monomial _____ in all the terms. If there is one, factor it out.

     b. Multiply a and c.

     c. Find two factors of this product whose sum is b.

     d. Write a four-term polynomial in which bx is written as the sum of two like terms whose coefficients are the two numbers you found in step c.

     e. Factor by _____.

[5.2] **8.** To write a polynomial in descending order, place the _____ degree term first, then the next highest degree, and so on.

[6.4] **9.** To factor a difference of squares, we use the rule $a^2 - b^2 =$ _____.

[6.6] **10.** To solve a quadratic equation:

     a. Manipulate the equation as needed so that one side is an expression and the other side is _____.

     b. Write the expression in _____ form.

     c. Use the zero-factor theorem to solve.

[4.3] **11.** What are the x- and y-intercepts on a graph?

[5.2] **12.** What is a binomial?

For Exercises 13–24, simplify.

Expressions

Exercises 13–32

[1.5] **13.** $14 - 2 \cdot 6 \cdot 3 + 8^2$

[1.5] **14.** $-6|5 + 3^2| - \sqrt{14 + 11}$

[5.3] **15.** $(7m^2 - 8) + (5m^2 - 6m + 1)$

[5.3] **16.** $(5x^2 + 6x - 1) - (2x - 3)$

[5.4] **17.** $(6x^2)(3x^2 y)$

[5.4] **18.** $(3x)(4x^3)^2$

[5.5] **19.** $(y - 8)(2y + 1)$

[5.5] **20.** $(2x + 5)^2$

[5.6] **21.** $u^3 \div u^2$

[5.6] **22.** $\dfrac{16h^3 + 4h^2 - 8h}{4h^2}$

[5.6] **23.** $(x^2 + 7x + 12) \div (x + 3)$

[5.6] **24.** $(x^3 - 48) \div (x + 2)$

[6.1–6.5] **For Exercises 25–32, factor completely.**

25. $x^2 - 121$

26. $10x^2 + 40$

27. $x^2 - 8x + 16$

28. $2x^2 + 5x - 12$

29. $x^3 - 5x^2 + 5x - 25$

30. $7m(n - 2) + 6(n - 2)$

31. $8x^3 - 27$

32. $3x^2 + 8x + 4$

Equations and
Inequalities

Exercises 33–50

For Exercises 33–38, solve.

[2.3] **33.** $2x - 9 = 11$

[2.3] **34.** $\dfrac{3}{4} = x - \dfrac{2}{3}$

[2.3] **35.** $7x - 9 = 6x + 4$

[2.3] **36.** $2.6 + 7a + 5 = 8a - 5.6$

[6.6] **37.** $x^2 - 100 = 0$

[6.6] **38.** $x^2 - 3x = 28$

[2.6] **For Exercises 39 and 40, solve.** **a. Write the solution set using set-builder notation.**
b. Write the solution set using interval notation.
c. Graph the solution set.

39. $7x - 5 \leq 3x + 11$

40. $-5x + 6 < 16$

For Exercises 41–44, solve.

[2.4] **41.** Solve for x in the equation $2x - 3y = 7$.

[4.2] **42.** Determine whether $(-1, 3)$ is a solution for $2x + y = 9$.

[4.3] **43.** Find the x- and y-intercepts for $y = 2x - 6$.

[4.4] **44.** Find the slope between the points $(6, 2)$ and $(5, 2)$.

[4.2] *For Exercises 45 and 46, graph.*

45. $y = 3x + 2$ **46.** $2x - y = 6$

For Exercises 47–50, solve.

[3.2] **47.** Using the following poll results, if 1204 husbands were polled, how many said that they often or always told their wives that they were beautiful?

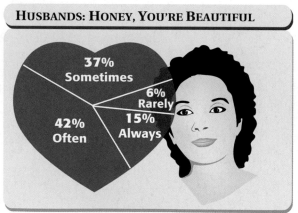

HUSBANDS: HONEY, YOU'RE BEAUTIFUL

37% Sometimes
6% Rarely
42% Often
15% Always

Source: Harris Poll for Clairol

[3.3] **48.** Find three consecutive even integers whose sum is 24.

[3.5] **49.** Janice invests a total of $8000 in two different accounts. The first account returns 5% while the second account returns 8%. If the total interest earned after one year is $565, what principal was invested in each account?

[3.5] **50.** How much of a 35% HCl solution must be added to 50 milliliters of a 10% HCl solution to obtain a 20% solution?

7

Rational Expressions and Equations

"What is the most rigorous law of our being? Growth. No smallest atom of our moral, mental, or physical structure can stand still a year. It grows, it must grow . . . nothing can prevent it."

—Mark Twain, American author and humorist (1835–1910)

Throughout this book, topics are developed in a specific pattern. We first define a new type of expression, then learn how to perform arithmetic operations on those expressions, and finally solve equations that contain those expressions. For example, in Chapter 5 we defined polynomials, then learned how to perform operations of arithmetic with the polynomials. In Chapter 6, we solved equations containing the polynomial expressions. We will now follow a similar pattern of development in Chapter 7. We first define rational expressions, then learn to perform arithmetic operations on these rational expressions, and, finally, solve equations that contain rational expressions.

7.1 Simplifying Rational Expressions

OBJECTIVES
1. Evaluate rational expressions.
2. Find numbers that cause a rational expression to be undefined.
3. Simplify rational expressions containing only monomials.
4. Simplify rational expressions containing multiterm polynomials.

OBJECTIVE 1. Evaluate rational expressions. Now that we have explored polynomials, we are ready to develop a new class of expressions called **rational expressions**.

DEFINITION *Rational expression:* An expression that can be written in the form $\dfrac{P}{Q}$, where P and Q are polynomials and $Q \neq 0$.

Some rational expressions are

$$\frac{3x^5}{18x^2} \qquad \frac{7x}{x^2 - 9} \qquad \frac{x^2 + 2x - 15}{4x + 20}$$

> **Connection** Notice how the definition of a rational expression is like the definition of a rational number. Recall from Chapter 1 that a rational number is a number that can be expressed in the form $\dfrac{a}{b}$, where a and b are integers and $b \neq 0$.

As with any expression, we can evaluate or rewrite rational expressions. First, let's look at evaluating rational expressions. Remember that to evaluate an algebraic expression, we replace the variables with given values and then simplify the resulting numerical expression.

EXAMPLE 1 Evaluate the expression $\dfrac{3x - 5}{x + 1}$, when

a. $x = 2$

Solution $\dfrac{3x - 5}{x + 1} = \dfrac{3(2) - 5}{2 + 1}$ Replace x with 2.

$$= \frac{6 - 5}{3}$$

$$= \frac{1}{3}$$

b. $x = -0.4$

Solution $\dfrac{3x-5}{x+1} = \dfrac{3(-0.4)-5}{-0.4+1}$ **Replace x with -0.4.**

$$= \dfrac{-1.2-5}{0.6}$$

$$= \dfrac{-6.2}{0.6}$$

$$= -10.\overline{3} \quad \text{or} \quad -10\dfrac{1}{3}$$

c. $x = -1$

Solution $\dfrac{3x-5}{x+1} = \dfrac{3(-1)-5}{-1+1}$ **Replace x with -1.**

$$= \dfrac{-3-5}{0}$$

$$= \dfrac{-8}{0}, \quad \text{which is undefined.}$$

YOUR TURN   Evaluate the expression $\dfrac{x^2-9}{x-2}$, when

    a. $x = 4$ **b.** $x = -3$ **c.** $x = 2$

OBJECTIVE 2. Find numbers that cause a rational expression to be undefined.   Note that in Example 1c, the expression $\dfrac{3x-5}{x+1}$ is undefined when $x = -1$, because replacing x with -1 causes the denominator to be 0. We need to avoid values that cause a rational expression to be undefined, which means we need to identify and avoid values that make their denominators 0. This suggests the following procedure.

PROCEDURE   Finding Values That Make a Rational Expression Undefined

To determine the value(s) that make a rational expression undefined:

1. Write an equation that has the denominator set equal to zero.
2. Solve the equation.

EXAMPLE 2   Find every value for the variable that makes the expression undefined.

    a. $\dfrac{5y}{y-2}$

Solution

Note: *We do not consider the numerator because the rational expression is undefined only if the denominator is 0.*

    $y - 2 = 0$ **Set the denominator, $y - 2$, equal to 0.**

       $y = 2$ **Add 2 to both sides.**

The expression $\dfrac{5y}{y-2}$ is undefined if y is replaced with 2 because the denominator would equal 0.

b. $\dfrac{7}{(x+4)(x-3)}$

Solution $(x+4)(x-3) = 0$ Set the denominator equal to 0.

$\qquad\qquad x+4 = 0 \quad \text{or} \quad x-3 = 0$ Use the zero-factor theorem to solve.

$\qquad\qquad\qquad x = -4 \qquad\qquad x = 3$

The original expression is undefined if x is replaced by -4 or 3.

c. $\dfrac{n-6}{3n^3 + n^2 - 4n}$

Solution $3n^3 + n^2 - 4n = 0$ Set the denominator equal to 0.

$\qquad\quad n(3n^2 + n - 4) = 0$ Factor out the monomial GCF, n.

$\qquad\quad n(3n+4)(n-1) = 0$ Factor $3n^2 + n - 4$ by trial.

$\quad n = 0 \quad \text{or} \quad 3n+4 = 0 \quad \text{or} \quad n-1 = 0$ Use the zero-factor theorem.

$\qquad\qquad\qquad\quad 3n = -4 \qquad\qquad n = 1$

$\qquad\qquad\qquad\quad\; n = -\dfrac{4}{3}$

The original expression is undefined if n is replaced by 0, $-\dfrac{4}{3}$, or 1.

YOUR TURN   Find every value that can replace the variable in the expression and cause the expression to be undefined.

a. $\dfrac{y+4}{3y-1}$ **b.** $\dfrac{2x}{x^2+5x}$ **c.** $\dfrac{m}{5m^2+28m-12}$

OBJECTIVE 3. Simplify rational expressions containing only monomials.   Now that we have defined rational expressions, we are ready to begin simplifying them. Recall from Section 1.2 that a fraction is in lowest terms when the only common factor of its numerator and denominator is 1. Also in Section 1.2, we used the rule $\dfrac{an}{bn} = \dfrac{a \cdot 1}{b \cdot 1} = \dfrac{a}{b}$, where $b \neq 0$ and $n \neq 0$, to reduce fractions. We can rewrite this rule so that it applies to rational expressions.

RULE $\dfrac{PR}{QR} = \dfrac{P \cdot 1}{Q \cdot 1} = \dfrac{P}{Q}$, where P, Q, and R are polynomials and Q and R are not 0.

ANSWERS

a. $\dfrac{1}{3}$ **b.** 0 or -5 **c.** $\dfrac{2}{5}$ or -6

The rule indicates that a factor common to both the numerator and denominator can be divided out of a rational expression. Consider the following comparison between simplifying a fraction and simplifying a similar rational expression:

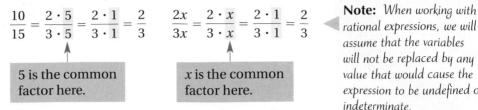

$$\frac{10}{15} = \frac{2 \cdot 5}{3 \cdot 5} = \frac{2 \cdot 1}{3 \cdot 1} = \frac{2}{3} \qquad \frac{2x}{3x} = \frac{2 \cdot x}{3 \cdot x} = \frac{2 \cdot 1}{3 \cdot 1} = \frac{2}{3}$$

5 is the common factor here.

x is the common factor here.

Note: *When working with rational expressions, we will assume that the variables will not be replaced by any value that would cause the expression to be undefined or indeterminate.*

This suggests the following procedure for simplifying rational expressions:

PROCEDURE   **Simplifying Rational Expressions**

To simplify a rational expression to lowest terms:
1. Write the numerator and denominator in factored form.
2. Divide out all common factors in the numerator and denominator.
3. Multiply the remaining factors in the numerator and the remaining factors in the denominator.

We will first simplify rational expressions that are a monomial over a monomial.

EXAMPLE 3) Simplify. $\dfrac{3x^5}{15x^2}$

Solution Write the numerator and denominator in factored form, then divide out all common factors.

Note: *From here on, we will leave out this step with the 1s and simply highlight the common factors that are eliminated.*

$$\frac{3x^5}{15x^2} = \frac{3 \cdot x \cdot x \cdot x \cdot x \cdot x}{3 \cdot 5 \cdot x \cdot x}$$

$$= \frac{1 \cdot x \cdot x \cdot x \cdot 1}{1 \cdot 5 \cdot 1}$$

$$= \frac{x^3}{5}$$

Note: *The common factors are a single 3 and two x's. These form the GCF that we divide out, which is $3x^2$.*

There are different styles for showing the process of dividing out common factors. For example, we could have used cancellation marks to show that the two x's and one 3 divide out in Example 3.

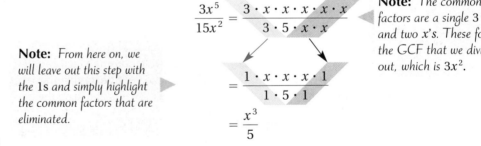

$$\frac{3x^5}{15x^2} = \frac{\cancel{3} \cdot \cancel{x} \cdot \cancel{x} \cdot x \cdot x \cdot x}{\cancel{3} \cdot 5 \cdot \cancel{x} \cdot \cancel{x}} = \frac{x^3}{5}$$

Develop a style that works best for you, keeping in mind that no matter the style, the bottom line is that you will be dividing out the GCF of the numerator and denominator to get a rational expression in lowest terms.

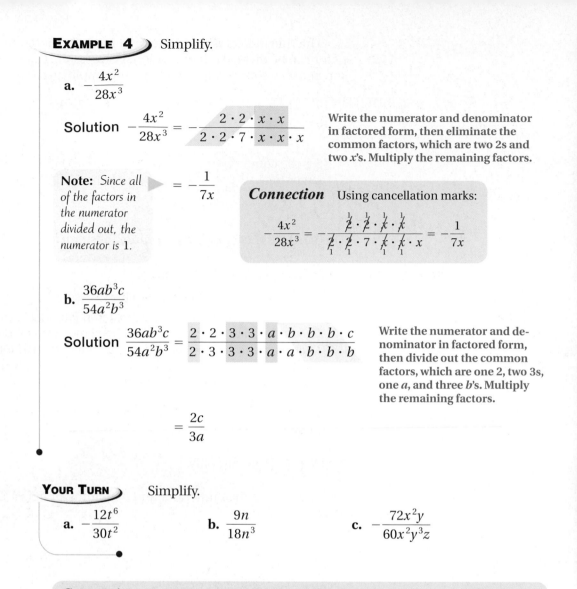

EXAMPLE 4) Simplify.

a. $-\dfrac{4x^2}{28x^3}$

Solution $-\dfrac{4x^2}{28x^3} = -\dfrac{2 \cdot 2 \cdot x \cdot x}{2 \cdot 2 \cdot 7 \cdot x \cdot x \cdot x}$ Write the numerator and denominator in factored form, then eliminate the common factors, which are two 2s and two x's. Multiply the remaining factors.

Note: *Since all of the factors in the numerator divided out, the numerator is 1.* ▶ $= -\dfrac{1}{7x}$

Connection Using cancellation marks:

$$-\dfrac{4x^2}{28x^3} = -\dfrac{\cancel{2} \cdot \cancel{2} \cdot \cancel{x} \cdot \cancel{x}}{\cancel{2} \cdot \cancel{2} \cdot 7 \cdot \cancel{x} \cdot \cancel{x} \cdot x} = -\dfrac{1}{7x}$$

b. $\dfrac{36ab^3c}{54a^2b^3}$

Solution $\dfrac{36ab^3c}{54a^2b^3} = \dfrac{2 \cdot 2 \cdot 3 \cdot 3 \cdot a \cdot b \cdot b \cdot b \cdot c}{2 \cdot 3 \cdot 3 \cdot 3 \cdot a \cdot a \cdot b \cdot b \cdot b}$ Write the numerator and denominator in factored form, then divide out the common factors, which are one 2, two 3s, one a, and three b's. Multiply the remaining factors.

$$= \dfrac{2c}{3a}$$

YOUR TURN) Simplify.

a. $-\dfrac{12t^6}{30t^2}$ **b.** $\dfrac{9n}{18n^3}$ **c.** $-\dfrac{72x^2y}{60x^2y^3z}$

Connection Dividing out common factors when simplifying rational expressions can also be viewed according to the rule for dividing exponential forms with the same base, which is to subtract the exponents and keep the base.

Dividing out common factors:

$$\dfrac{x^5}{x^2} = \dfrac{x \cdot x \cdot x \cdot x \cdot x}{x \cdot x} = \dfrac{x \cdot x \cdot x}{1} = x^3$$

Using rules of exponents:

$$\dfrac{x^5}{x^2} = x^{5-2} = x^3$$

Because we are dividing out two of the five x's in the numerator, we are left with three x's in the numerator. Subtracting the exponents accounts for eliminating the two common factors of x. This connection also applies when the greater exponent is in the denominator.

Dividing out common factors:

$$\dfrac{y^3}{y^7} = \dfrac{y \cdot y \cdot y}{y \cdot y \cdot y \cdot y \cdot y \cdot y \cdot y} = \dfrac{1}{y \cdot y \cdot y \cdot y} = \dfrac{1}{y^4}$$

Using rules of exponents:

$$\dfrac{y^3}{y^7} = y^{3-7} = y^{-4} = \dfrac{1}{y^4}$$

Notice that when dividing out like bases, the position of the base with the greater exponent determines whether we place the result in the numerator or in the denominator.

ANSWERS

a. $-\dfrac{2t^4}{5}$ **b.** $\dfrac{1}{2n^2}$ **c.** $-\dfrac{6}{5y^2z}$

OBJECTIVE 4. Simplify rational expressions containing multiterm polynomials. The rational expressions we have considered so far have had monomials in the both the numerator and denominator. We now consider rational expressions that contain multiterm polynomials. Remember that we can only divide out common factors, so we must first write the polynomials in factored form.

EXAMPLE 5 Simplify. $\dfrac{9ab}{3a + 6}$

Solution $\dfrac{9ab}{3a + 6} = \dfrac{3 \cdot 3 \cdot a \cdot b}{3 \cdot (a + 2)}$

Write the numerator and denominator in factored form, then divide out the common factor, which is 3. Multiply the remaining factors.

$$= \dfrac{3ab}{a + 2}$$

Warning: It may be tempting to try to divide out the a; however, $\dfrac{3ab}{a + 2} \neq \dfrac{3b}{2}$. Though a is a factor in the numerator, it is not a factor in the denominator, so it cannot be divided out.

YOUR TURN Simplify.

a. $\dfrac{6xy}{3x + 12}$

b. $\dfrac{2a + 8}{10a^2}$

We now consider rational expressions with multiterm polynomials in both the numerator and denominator. Again, since we can only divide out common factors, we must first factor the polynomials.

EXAMPLE 6 Simplify.

a. $\dfrac{2x^2 + 8x}{3x^2 + 12x}$

Solution $\dfrac{2x^2 + 8x}{3x^2 + 12x} = \dfrac{2 \cdot x \cdot (x + 4)}{3 \cdot x \cdot (x + 4)}$

Factor the numerator and denominator, then divide out the common factors, x and $x + 4$.

$$= \dfrac{2}{3}$$

b. $\dfrac{x^2 + 2x - 15}{x^2 + x - 12}$

Solution $\dfrac{x^2 + 2x - 15}{x^2 + x - 12} = \dfrac{(x - 3)(x + 5)}{(x - 3)(x + 4)}$

Factor the numerator and denominator, then divide out the common factor, $x - 3$.

$$= \dfrac{x + 5}{x + 4}$$

ANSWERS

a. $\dfrac{2xy}{x + 4}$ b. $\dfrac{a + 4}{5a^2}$

YOUR TURN Simplify.

a. $\dfrac{5x + 10}{7x + 14}$

b. $\dfrac{3n^2 - 23n - 8}{3n^2 + 13n + 4}$

ANSWERS

a. $\dfrac{5}{7}$ b. $\dfrac{n - 8}{n + 4}$

7.1 Simplifying Rational Expressions   **543**

Recall from Chapter 6 that to factor multiterm polynomials, we must sometimes factor the factors. In Example 7, the factors themselves can be factored.

EXAMPLE 7 Simplify.

a. $\dfrac{6y^4 + 2y^3 - 4y^2}{36y^3 - 42y^2 + 12y}$

Solution $\dfrac{6y^4 + 2y^3 - 4y^2}{36y^3 - 42y^2 + 12y} = \dfrac{2y^2(3y^2 + y - 2)}{6y(6y^2 - 7y + 2)}$ Factor out the monomial GCFs.

$= \dfrac{2 \cdot y \cdot y \cdot (y + 1) \cdot (3y - 2)}{2 \cdot 3 \cdot y \cdot (2y - 1) \cdot (3y - 2)}$ Factor the polynomial factors.

$= \dfrac{y(y + 1)}{3(2y - 1)}$ or $\dfrac{y^2 + y}{6y - 3}$ Divide out the common factors 2, y, and $3y - 2$.

b. $\dfrac{3x^2 + 6x - 24}{x^4 - 16}$

Solution Write the numerator and denominator in factored form, then divide out all common factors.

$\dfrac{3x^2 + 6x - 24}{x^4 - 16} = \dfrac{3(x^2 + 2x - 8)}{(x^2 + 4)(x^2 - 4)}$ Factor 3 out of $3x^2 + 6x - 24$ and factor $x^4 - 16$, which is a difference of squares.

$= \dfrac{3 \cdot (x + 4) \cdot (x - 2)}{(x^2 + 4) \cdot (x + 2) \cdot (x - 2)}$ Factor the polynomial factors. Notice that $x^2 - 4$ is a difference of squares.

$= \dfrac{3(x + 4)}{(x^2 + 4)(x + 2)}$ or $\dfrac{3x + 12}{x^3 + 2x^2 + 4x + 8}$ Divide out the common factor, $x - 2$.

YOUR TURN Simplify. $\dfrac{14x^3 + 70x^2 + 84x}{7x^4 - 14x^3 - 105x^2}$

Binomial factors that are additive inverses can be tricky:

EXAMPLE 8 Simplify. $\dfrac{3x^2 - 7x - 20}{8 - 2x}$

Solution Write the numerator and denominator in factored form, then divide out all common factors.

$$\dfrac{3x^2 - 7x - 20}{8 - 2x} = \dfrac{(3x + 5)(x - 4)}{2(4 - x)}$$

ANSWER

$\dfrac{2(x + 2)}{x(x - 5)}$ or $\dfrac{2x + 4}{x^2 - 5x}$

It appears that there are no common factors. However, $x - 4$ and $4 - x$ are additive inverses. We can get matching binomial factors by factoring -1 out of either $x - 4$ or $4 - x$.

$$x - 4 = -1(4 - x) \quad \text{or} \quad 4 - x = -1(x - 4)$$

We can use either form. We will use $x - 4 = -1(4 - x)$.

Note: *We could have factored the -1 out of the denominator:*

$$\frac{(3x + 5)(x - 4)}{2(-1)(x - 4)} = \frac{3x + 5}{-2}$$

$$= \frac{(3x + 5)(-1)(4 - x)}{2(4 - x)}$$

$$= \frac{-3x - 5}{2} \quad \text{or} \quad -\frac{3x + 5}{2}$$

Divide out the common factor, $4 - x$, then multiply the remaining factors.

Example 8 also illustrates the rule of sign placement in a fraction or rational expression. Remember that with a negative fraction (or rational expression), the minus sign can be placed in the numerator, denominator, or aligned with the fraction line. Note, however, that it is generally considered unsightly to write the minus sign in the denominator.

RULE $-\dfrac{P}{Q} = \dfrac{-P}{Q} = \dfrac{P}{-Q}$, where P and Q are polynomials and $Q \neq 0$.

ANSWER

$-\dfrac{x}{x + 6}$

YOUR TURN

Simplify. $\dfrac{2x - x^2}{x^2 + 4x - 12}$

7.1 Exercises

For Extra Help

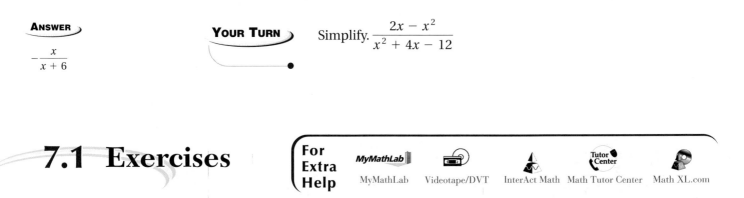

MyMathLab · MyMathLab · Videotape/DVT · InterAct Math · Math Tutor Center · Math XL.com

1. Explain how to evaluate an expression.

2. In general, what causes a rational expression to be undefined?

3. How do you determine what values cause a rational expression to be undefined?

4. When simplifying a rational expression, we divide out all common _____ in the numerator and denominator.

5. Explain why you cannot divide out the y's in $\dfrac{x + y}{2y}$.

6. What is the first step when simplifying rational expressions?

For Exercises 7–14, evaluate the rational expression.

7. $\dfrac{4x^2}{9y}$ **a.** when $x = -2, y = 5$ **b.** when $x = 3, y = -4$ **c.** when $x = 0, y = 7$

8. $\dfrac{3x}{7y^2}$ **a.** when $x = 1, y = 2$ **b.** when $x = -2, y = 3$ **c.** when $x = -4, y = -2$

9. $\dfrac{2x}{x + 4}$ **a.** when $x = 4$ **b.** when $x = -4$ **c.** when $x = -2.4$

10. $\dfrac{4x}{x - 3}$ **a.** when $x = 3$ **b.** when $x = -3$ **c.** when $x = 2.5$

11. $\dfrac{2x + 5}{3x - 1}$ **a.** when $x = 3$ **b.** when $x = -2$ **c.** when $x = -1.3$

12. $\dfrac{3x + 2}{4x - 5}$ **a.** when $x = 2$ **b.** when $x = -1$ **c.** when $x = 1.6$

13. $\dfrac{x^2 - 4}{x + 3}$ **a.** when $x = 0$ **b.** when $x = -1$ **c.** when $x = -4.2$

14. $\dfrac{x^2 + 3}{2x + 1}$ **a.** when $x = 1$ **b.** when $x = -2$ **c.** when $x = -3.5$

For Exercises 15–24, find every value for the variable that makes the expression undefined.

15. $\dfrac{2x}{x - 3}$ 16. $\dfrac{3y}{y + 5}$ 17. $\dfrac{3}{x^2 - 9}$ 18. $\dfrac{4r}{r^2 - 4}$

19. $\dfrac{6}{x^2 - 14x + 45}$ 20. $\dfrac{2x + 3}{x^2 + 5x + 6}$ 21. $\dfrac{2m}{m^2 - 3m}$ 22. $\dfrac{2p + 3}{p^2 + 6p}$

23. $\dfrac{4a}{2a^2 - 3a - 5}$ 24. $\dfrac{t}{2t^2 + 7t + 3}$

For Exercises 25–72, simplify.

25. $\dfrac{9x^2}{12xy}$

26. $\dfrac{14h^3k}{21h}$

27. $-\dfrac{4m^3n}{mn^2}$

28. $-\dfrac{5a^3b}{a^2b^4}$

29. $\dfrac{54x^4y}{36x^6yz^2}$

30. $-\dfrac{48t^5uv}{32tv^4}$

31. $-\dfrac{15m^2n}{5m+10n}$

32. $-\dfrac{7r^3t}{7r+14t}$

33. $\dfrac{7(x+5)}{14}$

34. $\dfrac{9}{18(a-1)}$

35. $\dfrac{3(m-2)}{5(m-2)}$

36. $\dfrac{7(k+2)}{15(k+2)}$

37. $\dfrac{10x-20}{15x-30}$

38. $\dfrac{12y+2}{18y+3}$

39. $\dfrac{30a+15b}{18a+9b}$

40. $\dfrac{8m-24n}{12m-36n}$

41. $\dfrac{x^2+5x}{4x+20}$

42. $\dfrac{ab-b^2}{2a-2b}$

43. $\dfrac{x^2+5x}{x+5}$

44. $\dfrac{m^2-6m}{m-6}$

45. $\dfrac{5y^2-10y-15}{xy^2-2xy-3x}$

46. $\dfrac{3m^3-2m^2+m}{6m^2-4m+2}$

47. $\dfrac{x^2-y^2}{x^2+2xy+y^2}$

48. $\dfrac{a^2-b^2}{a^2+2ab+b^2}$

49. $\dfrac{x^2-5x}{x^2-7x+10}$

50. $\dfrac{x^2-2x}{x^2-6x+8}$

51. $\dfrac{t^2-9}{t^2+5t+6}$

52. $\dfrac{k^2-4}{k^2+k-2}$

53. $\dfrac{4a^2-4a-3}{6a^2-a-2}$

54. $\dfrac{3x^2+16x-35}{5x^2+33x-14}$

55. $\dfrac{x^3-16x}{x^3+6x^2+8x}$

56. $\dfrac{m^3-4m}{m^3-4m^2+4m}$

57. $\dfrac{px+py+qx+qy}{mx-nx+my-ny}$

58. $\dfrac{by-ay+bx-ax}{x^2+ax+xy+ay}$

59. $\dfrac{6u^3-4u^2-3u+2}{3u^2+u-2}$

60. $\dfrac{8n^3+20n^2-2n-5}{4n^2+12n+5}$

61. $\dfrac{4x-4}{1-x}$

62. $\dfrac{12-4m}{m-3}$

63. $\dfrac{y^2-4}{2-y}$

64. $\dfrac{a^2-b^2}{b-a}$

65. $\dfrac{3-w}{w^2-2w-3}$

66. $\dfrac{1-u}{u^2-3u+2}$

67. $\dfrac{3x+6}{x^3+8}$

68. $\dfrac{r^2-s^2}{r^3+s^3}$

69. $\dfrac{x^3 + y^3}{x^3 - x^2y + xy^2}$ 　　　　 **70.** $\dfrac{x^3 - 8}{x^2 + 2x + 4}$ 　　　　 **71.** $\dfrac{3 - m}{m^3 - 27}$ 　　　　 **72.** $\dfrac{x^3 - 64}{4 - x}$

For Exercises 73–78, find and explain the mistake, then correct the mistake.

73. $-\dfrac{5x^2y}{10x^3y} = -\dfrac{x}{2}$ 　　　　　　　　　　　　 **74.** $\dfrac{8x^2y}{4x^3} = 2xy$

75. $\dfrac{x - \overset{3}{\cancel{6}}}{\underset{1}{\cancel{2}y}} = \dfrac{x - 3}{y}$ 　　　　　　　　　　　 **76.** $\dfrac{y - \overset{3}{\cancel{12}}}{\underset{1}{\cancel{4}y} - 3} = \dfrac{y - 3}{y - 3} = 1$

77. $\dfrac{yx - x^2}{x - y} = \dfrac{x(y - x)}{x - y} = x$ 　　　　 **78.** $\dfrac{x^2 - x - 6}{x^3 + 2x^2 - 9x - 18} = \dfrac{(x + 2)(x - 3)}{(x + 2)(x^2 - 9)} = \dfrac{x - 3}{x^2 - 9}$

79. A person's body mass index (BMI) is a measure of the amount of body fat based on height, h, in inches, and weight, w, in pounds. The formula for BMI is

$$\text{BMI} = \frac{704.5w}{h^2}$$

a. Use the formula to find the BMI of each person listed here.

Height	Weight	BMI
5′ 4″	138	
5′ 9″	155	
6′ 2″	220	

b. Use the following table from the National Institutes of Health to classify each person in part (a) as underweight, normal weight, overweight, or obese.

Underweight:	BMI $\leq$ 18.5
Normal Weight:	BMI = 18.5 − 24.9
Overweight:	BMI = 25 − 29.9
Obesity:	BMI $\geq$ 30

Connection　In Section 2.5, Exercise 68, you translated the BMI formula from words to symbols.

80. Recall the formula for slope from Chapter 4:

$$m = \frac{y_2 - y_1}{x_2 - x_1}$$

> **Connection**   Though we hadn't formally defined rational expressions in Chapter 4, we were evaluating the rational expression $\dfrac{y_2 - y_1}{x_2 - x_1}$ to find slope.

a. Find the slope of a line passing through the points $(3, 5)$ and $(-1, -4)$.

b. Find the slope of a line passing through the points $(-4, 7)$ and $(-2, -1)$.

81. Given the diameter in inches, d, and the number of revolutions per minute, N, of a revolving tool such as a circular saw or drill, the cutting speed, C, in feet per minute, of the tool can be found by the formula

$$C = \frac{\pi d N}{12}$$

(*Source:* Robert D. Smith, *Mathematics for Machine Technology*, 4th ed., Delmar Publishers, 1999)

Use the formula to complete the following table for a table saw. Leave answers in terms of π.

Blade Diameter	rpm	Cutting Speed
8 inches	800 rpm	
8 inches	1000 rpm	
10 inches	800 rpm	
10 inches	1000 rpm	

82. The cutting time, T, in minutes, that it takes a drill to cut through something depends on its speed of revolution, N, in rpm, the thickness of the material, L, in inches, and the tool feed, F, in inches per revolution. The formula for calculating cutting time is

$$T = \frac{L}{FN}$$

(*Source:* Robert D. Smith, *Mathematics for Machine Technology*, 4th ed., Delmar Publishers, 1999)

Use the formula to complete the following table for a drill operating at a speed of 480 rpm and a feed of 0.04 inches per revolution.

Material Thickness (inches)	Cut Time (minutes)
0.25	
0.5	
0.75	
1	

83. The formula for the volume of a cone is $V = \dfrac{1}{3}\pi r^2 h$.

   a. Solve this formula for h, so that we have a formula for finding the height of a cone given its volume and radius. (The formula for height of a cone will contain a rational expression.)

   b. Suppose a cone has a volume of approximately 150.8 cubic centimeters with a radius of 4 centimeters. Find the height of the cone.

84. The formula for the volume of a pyramid is $V = \dfrac{1}{3}lwh$.

   a. Solve this formula for h, so that we have a formula for finding the height of a pyramid given its volume, length, and width.

   b. Suppose a decorative onyx pyramid has a volume of 126 cubic inches, a width of 3.5 inches, and a length of 4.5 inches. Find the height of the pyramid.

85. In analyzing circuits, electrical engineers often have to simplify rational expressions. Suppose the following rational expression describes an electrical circuit:

$$\frac{s^2 - 9}{s^2 + 7s + 12}$$

Simplify the expression to lowest terms.

86. An engineer derives the following rational expression, which describes a certain circuit:

$$\frac{4s^2 - 6s}{2s^2 + 7s - 15}$$

Simplify the expression to lowest terms.

Collaborative Exercises Graphs of Rational Expressions

Use a graphing utility.

1. Graph $f(x) = \dfrac{x}{x - 3}$. The vertical line you see is called an *asymptote*.

   a. What is the equation of the asymptote?

   b. What do you think the asymptote indicates?

2. Graph $f(x) = \dfrac{5}{(x + 2)(x - 4)}$.

   a. Why are there two asymptotes?

   b. What are the equations of the two asymptotes?

Note: *The suggested window settings in Exercise 3 are for a TI-83. If a different graphing utility is used, the settings may need to be changed in order to see the hole.*

3. Graph $f(x) = \dfrac{x^2 - x - 6}{x^2 + 3x + 2}$ with these window settings: $x_{min} = -4.7$, $x_{max} = 4.7$, $y_{min} = -10$, and $y_{max} = 10$.

   a. Are there any asymptotes? If so, write the equation for each.

   b. Rewrite $f(x) = \dfrac{x^2 - x - 6}{x^2 + 3x + 2}$ so that the numerator and denominator are in factored form.

   c. Find $f(-2)$.

   d. Look at the graph of $f(x) = \dfrac{x^2 - x - 6}{x^2 + 3x + 2}$ in the second quadrant. The blank space you see in the graph where $x = -2$ is a *removable discontinuity* and is often called a *hole*. How is a hole different from an asymptote?

4. Writing $\dfrac{x^2 - x - 6}{x^2 + 3x - 2}$ in simplest form we have $\dfrac{x - 3}{x + 1}$. Graph $f(x) = \dfrac{x - 3}{x + 1}$.

   a. Compare the graph of $f(x) = \dfrac{x - 3}{x + 1}$ with the graph of $f(x) = \dfrac{x^2 - x - 6}{x^2 + 3x + 2}$. What do you notice about the graphs?

5. Given a rational function, how can you identify its asymptotes and holes?

REVIEW EXERCISES

[1.4] *For Exercises 1 and 2, multiply.*

1. $\dfrac{2}{3} \cdot \dfrac{6}{4}$

2. $-\dfrac{15}{16} \cdot \dfrac{4}{5}$

[1.4] *For Exercises 3 and 4, divide.*

3. $\dfrac{3}{8} \div \dfrac{5}{6}$

4. $-\dfrac{7}{9} \div \left(-\dfrac{5}{12}\right)$

[6.1] *For Exercises 5 and 6, factor completely.*

5. $ax + bx - ay - by$

[6.3]   6. $3x^2 + 13x - 10$

7.2   Multiplying and Dividing Rational Expressions

OBJECTIVES

1. Multiply rational expressions.
2. Divide rational expressions.
3. Convert units of measurement using dimensional analysis.

OBJECTIVE 1. Multiply rational expressions.   In Section 1.4, we used the rule $\dfrac{a}{b} \cdot \dfrac{c}{d} = \dfrac{ac}{bd}$, where $b \neq 0$ and $d \neq 0$, to multiply fractions. We can rewrite this rule so that it applies to multiplying rational expressions.

RULE $\dfrac{P}{Q} \cdot \dfrac{R}{S} = \dfrac{PR}{QS}$, where P, Q, R, and S are polynomials and $Q \neq 0$ and $S \neq 0$.

Also, remember that when multiplying fractions, we can simplify after multiplying, or we can simplify before multiplying by dividing out factors common to both the numerator and denominator. For example,

Multiply, then eliminate common factors:
$$\frac{8}{15} \cdot \frac{3}{4} = \frac{24}{60} = \frac{2 \cdot 2 \cdot 2 \cdot 3}{2 \cdot 2 \cdot 3 \cdot 5} = \frac{2}{5}$$

or

Eliminate common factors, then multiply.
$$\frac{8}{15} \cdot \frac{3}{4} = \frac{2 \cdot 2 \cdot 2}{3 \cdot 5} \cdot \frac{3}{2 \cdot 2} = \frac{2}{5}$$

Multiplying rational expressions works the same way.

Multiplying Rational Expressions

To multiply rational expressions:
1. Write each numerator and denominator in factored form.
2. Divide out any numerator factor with any matching denominator factor.
3. Multiply numerator by numerator and denominator by denominator.
4. Simplify as needed.

EXAMPLE 1 Multiply. $\dfrac{9x}{5y} \cdot \dfrac{15xy^2}{12}$

Solution $\dfrac{9x}{5y} \cdot \dfrac{15xy^2}{12} = \dfrac{3 \cdot 3 \cdot x}{5 \cdot y} \cdot \dfrac{3 \cdot 5 \cdot x \cdot y \cdot y}{2 \cdot 2 \cdot 3}$ Write the numerators and denominators in factored form.

$= \dfrac{3 \cdot 3 \cdot x}{1} \cdot \dfrac{x \cdot y}{2 \cdot 2}$ Divide out the common factors, which are 3, 5, and y.

$= \dfrac{9x^2y}{4}$ Multiply the remaining numerator factors and denominator factors.

YOUR TURN Multiply.

a. $-\dfrac{36m}{8n^4} \cdot \dfrac{14mn^2}{28n}$

b. $-\dfrac{9}{4xy^2} \cdot -\dfrac{20xz}{15y}$

Now consider rational expressions that contain multiterm polynomials.

EXAMPLE 2 Multiply.

a. $-\dfrac{2x}{3x - 9} \cdot \dfrac{5x - 15}{20x^3}$

Solution $-\dfrac{2x}{3x - 9} \cdot \dfrac{5x - 15}{20x^3}$

$= -\dfrac{2 \cdot x}{3 \cdot (x - 3)} \cdot \dfrac{5 \cdot (x - 3)}{2 \cdot 2 \cdot 5 \cdot x \cdot x}$ Write the numerators and denominators in factored form.

$= -\dfrac{1}{3} \cdot \dfrac{1}{2 \cdot x \cdot x}$ Divide out the common factors, which are 2, 5, x, and $(x - 3)$.

$= -\dfrac{1}{6x^2}$ Multiply the remaining numerator factors and denominator factors.

ANSWERS

a. $-\dfrac{9m^2}{4n^0}$ b. $\dfrac{3z}{y^3}$

b. $\dfrac{12 - 4x}{15x} \cdot \dfrac{5x^2}{2x - 6}$

Solution $\dfrac{12 - 4x}{15x} \cdot \dfrac{5x^2}{2x - 6} = \dfrac{4 \cdot (3 - x)}{3 \cdot 5 \cdot x} \cdot \dfrac{5 \cdot x \cdot x}{2 \cdot (x - 3)}$

Note: *We can factor 4 further. Also, since $3 - x$ and $x - 3$ are additive inverses, we can factor -1 out of one of them.*

$= \dfrac{2 \cdot 2 \cdot (-1)(x - 3)}{3 \cdot 5 \cdot x} \cdot \dfrac{5 \cdot x \cdot x}{2 \cdot (x - 3)}$

$= \dfrac{2 \cdot (-1)}{3} \cdot \dfrac{x}{1}$ Divide out the common factors, which are $2, 5, x,$ and $(x - 3)$.

$= \dfrac{-2x}{3}$ Multiply the remaining numerator factors and denominator factors.

c. $\dfrac{x^2 + x - 6}{9x^3} \cdot \dfrac{12x^2 - 6x}{2x^2 - 5x + 2}$

Solution $\dfrac{x^2 + x - 6}{9x^3} \cdot \dfrac{12x^2 - 6x}{2x^2 - 5x + 2}$

$= \dfrac{(x - 2)(x + 3)}{3 \cdot 3 \cdot x \cdot x \cdot x} \cdot \dfrac{2 \cdot 3 \cdot x \cdot (2x - 1)}{(x - 2)(2x - 1)}$ Write the numerators and denominators in factored form.

$= \dfrac{x + 3}{3 \cdot x \cdot x} \cdot \dfrac{2}{1}$ Divide out the common factors, which are $(x - 2), 3, x,$ and $(2x - 1)$.

$= \dfrac{2(x + 3)}{3x^2}$ or $\dfrac{2x + 6}{3x^2}$ Multiply the remaining numerator factors and denominator factors.

YOUR TURN   Multiply.

a. $\dfrac{-8h^2}{2h + 10} \cdot -\dfrac{6h + 30}{18h^4}$ **b.** $-\dfrac{9y - 3}{y^2 + 3y} \cdot \dfrac{2y + 6}{2 - 6y}$ **c.** $\dfrac{m^2 - 25}{20m} \cdot \dfrac{30m + 10}{3m^2 + 16m + 5}$

OBJECTIVE 2. Divide rational expressions. In Section 1.4, we saw how to divide fractions using the rule $\dfrac{a}{b} \div \dfrac{c}{d} = \dfrac{a}{b} \cdot \dfrac{d}{c}$, where $b, c,$ and d are not 0, which indicates to multiply by the reciprocal of the divisor. We can rewrite this rule so that it applies to dividing rational expressions.

RULE $\dfrac{P}{Q} \div \dfrac{R}{S} = \dfrac{P}{Q} \cdot \dfrac{S}{R}$, where $P, Q, R,$ and S are polynomials and $Q \neq 0, R \neq 0,$ and $S \neq 0$.

Following is a procedure for dividing rational expressions.

PROCEDURE Dividing Rational Expressions

To divide rational expressions:
1. Write an equivalent multiplication statement with the reciprocal of the divisor.
2. Write each numerator and denominator in factored form. (Steps 1 and 2 are interchangeable.)
3. Divide out any numerator factor with any matching denominator factor.
4. Multiply numerator by numerator and denominator by denominator.
5. Simplify as needed.

EXAMPLE 3 Divide. $-\dfrac{9x^2y}{28z^3} \div \dfrac{12y}{7z}$

Solution

$$-\frac{9x^2y}{28z^3} \div \frac{12y}{7z} = -\frac{9x^2y}{28z^3} \cdot \frac{7z}{12y}$$

Write an equivalent multiplication statement by changing the division sign to multiplication and changing the divisor to its reciprocal.

$$= -\frac{3 \cdot 3 \cdot x \cdot x \cdot y}{2 \cdot 2 \cdot 7 \cdot z \cdot z \cdot z} \cdot \frac{7 \cdot z}{2 \cdot 2 \cdot 3 \cdot y}$$

Factor the numerators and denominators.

$$= -\frac{3 \cdot x \cdot x}{2 \cdot 2 \cdot z \cdot z} \cdot \frac{1}{2 \cdot 2}$$

Divide out the common factors, which are 3, 7, y, and z.

$$= -\frac{3x^2}{16z^2}$$

Multiply the remaining factors.

YOUR TURN Divide.

a. $\dfrac{15a^3}{28b^2} \div \dfrac{10ab}{7}$

b. $-\dfrac{2t}{33u^3v} \div -\dfrac{8t^3v}{21u^4}$

Now consider dividing rational expressions that contain multiterm polynomials.

EXAMPLE 4 Divide.

a. $\dfrac{2x - 10}{15x^3} \div \dfrac{x^2 - 5x}{12x}$

Solution

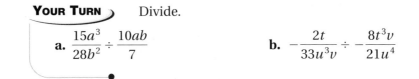

$$\frac{2x - 10}{15x^3} \div \frac{x^2 - 5x}{12x} = \frac{2x - 10}{15x^3} \cdot \frac{12x}{x^2 - 5x}$$

Write an equivalent multiplication statement by changing the division sign to multiplication and changing the divisor to its reciprocal.

$$= \frac{2 \cdot (x - 5)}{3 \cdot 5 \cdot x \cdot x \cdot x} \cdot \frac{2 \cdot 2 \cdot 3 \cdot x}{x \cdot (x - 5)}$$

Factor the numerators and denominators.

$$= \frac{2}{5 \cdot x \cdot x} \cdot \frac{2 \cdot 2}{x}$$

Divide out the common factors, which are 3, x, and $(x - 5)$.

$$-\frac{8}{5x^3}$$

Multiply the remaining factors.

b. $\dfrac{4x^2 - 7x - 2}{2x^2 + 3x - 9} \div \dfrac{4x^2 + x}{6x^3 + 18x^2}$

Solution

$$\dfrac{4x^2 - 7x - 2}{2x^2 + 3x - 9} \div \dfrac{4x^2 + x}{6x^3 + 18x^2}$$

Write an equivalent multiplication statement by changing the division sign to multiplication and changing the divisor to its reciprocal.

$$= \dfrac{4x^2 - 7x - 2}{2x^2 + 3x - 9} \cdot \dfrac{6x^3 + 18x^2}{4x^2 + x}$$

Factor the numerators and denominators.

$$= \dfrac{(4x + 1)(x - 2)}{(2x - 3)(x + 3)} \cdot \dfrac{6x^2(x + 3)}{x(4x + 1)}$$

$$= \dfrac{(4x + 1) \cdot (x - 2)}{(2x - 3) \cdot (x + 3)} \cdot \dfrac{2 \cdot 3 \cdot x \cdot x \cdot (x + 3)}{x \cdot (4x + 1)}$$

Divide out the common factors, which are x, $(4x + 1)$, and $(x + 3)$.

$$= \dfrac{x - 2}{2x - 3} \cdot \dfrac{2 \cdot 3 \cdot x}{1}$$

$$= \dfrac{6x(x - 2)}{2x - 3} \quad \text{or} \quad \dfrac{6x^2 - 12x}{2x - 3}$$

Multiply the remaining factors.

c. $\dfrac{8x^2 - 6x^3}{6xy} \div \dfrac{9x^2 - 16}{5y^2 - 10y}$

Solution

$$\dfrac{8x^2 - 6x^3}{6xy} \div \dfrac{9x^2 - 16}{5y^2 - 10y} = \dfrac{8x^2 - 6x^3}{6xy} \cdot \dfrac{5y^2 - 10y}{9x^2 - 16}$$

Write an equivalent multiplication statement by changing the division sign to multiplication and changing the divisor to its reciprocal.

$$= \dfrac{2x^2(4 - 3x)}{6xy} \cdot \dfrac{5y(y - 2)}{(3x + 4)(3x - 4)}$$

Factor the numerators and denominators.

$$= \dfrac{2 \cdot x \cdot x \cdot (-1) \cdot (3x - 4)}{2 \cdot 3 \cdot x \cdot y} \cdot \dfrac{5 \cdot y \cdot (y - 2)}{(3x + 4) \cdot (3x - 4)}$$

$$= \dfrac{x \cdot (-1)}{3} \cdot \dfrac{5 \cdot (y - 2)}{3x + 4}$$

Divide out the common factors, which are $2, x, y,$ and $(3x - 4)$.

$$= \dfrac{-5x(y - 2)}{3(3x + 4)} \quad \text{or} \quad \dfrac{-5xy + 10x}{9x + 12} \quad \text{or} \quad \dfrac{10x - 5xy}{9x + 12}$$

Multiply the remaining factors.

Note: $4 - 3x$ and $3x - 4$ are additive inverses, so we factored -1 out of $4 - 3x$.

YOUR TURN   Divide.

a. $\dfrac{6yz^3}{2y - y^2} \div \dfrac{8yz}{6 - 3y}$

b. $\dfrac{mn - 3m}{14m^4} \div \dfrac{n}{7n + 7}$

c. $\dfrac{4y^2 - 25}{9y^2 + 18y} \div \dfrac{2y^2 + 3y - 5}{3y^4 - 3y^3}$

d. $\dfrac{3x^2 + 10x + 8}{18 - 6x} \div \dfrac{10 + 5x}{4x^3 - 12x^2}$

ANSWERS

a. $\dfrac{9z^2}{4y}$ **b.** $\dfrac{(n - 3)(n + 1)}{2m^3n}$

c. $\dfrac{y^2(2y - 5)}{3(y + 2)}$ or $\dfrac{2y^3 - 5y^2}{3y + 6}$

d. $-\dfrac{2x^2(3x + 4)}{15}$ or $-\dfrac{6x^3 + 8x^2}{15}$

OBJECTIVE 3. Convert units of measurement using dimensional analysis. We can convert from one unit of a measurement to another using a method called *dimensional analysis*, which will involve writing measurement facts as conversion factors. A conversion factor is a ratio of equivalent measures.

Measurement fact:	**Conversion factor:**	
1 foot = 12 inches	$\dfrac{1 \text{ foot}}{12 \text{ inches}}$ or $\dfrac{12 \text{ inches}}{1 \text{ foot}}$	**Note:** *Because a conversion factor is a ratio of equivalent measures, its ratio will always equal 1.*
1 pound = 16 ounces	$\dfrac{1 \text{ pound}}{16 \text{ ounces}}$ or $\dfrac{16 \text{ ounces}}{1 \text{ pound}}$	

To use dimensional analysis, we multiply a measurement by appropriate conversion factors so that the undesired units divide out, leaving only the desired units of measurement. For example, if we wish to convert 6 feet to inches, we multiply by the conversion factor $\dfrac{12 \text{ inches}}{1 \text{ foot}}$ so that the units of feet divide out, leaving inches.

$$6 \text{ ft.} = \frac{6 \text{ ft.}}{1} \cdot \frac{12 \text{ in.}}{1 \text{ ft.}} = 72 \text{ in.}$$

Connection The ratio of the conversion factor $\dfrac{12 \text{ in.}}{1 \text{ ft.}}$ is 1, so multiplying 6 ft. by this ratio means the result, 72 in., is equal to 6 ft.

PROCEDURE Using Dimensional Analysis to Convert Between Units of Measurement

To convert units using dimensional analysis, multiply the given measurement by conversion factors so that the undesired units divide out, leaving the desired units.

EXAMPLE 5 Convert.

a. 500 ounces to pounds

Solution There are 16 ounces to 1 pound. We write this fact as a unit ratio and multiply so that ounces divides out, leaving pounds.

$$500 \text{ oz.} = \frac{500 \text{ oz.}}{1} \cdot \frac{1 \text{ lb.}}{16 \text{ oz.}} = \frac{500}{16} \text{ lb.} = 31.25 \text{ lb.}$$

Connection Dividing out units in dimensional analysis is like dividing out common factors when multiplying rational expressions.

b. 6 miles to yards

Solution There are 5280 feet to 1 mile and 3 feet to 1 yard. So, we write these facts as unit fractions so that miles and feet divide out, leaving yards.

$$6 \text{ mi.} = \frac{6 \text{ mi.}}{1} \cdot \frac{5280 \text{ ft.}}{1 \text{ mi.}} \cdot \frac{1 \text{ yd.}}{3 \text{ ft.}} = \frac{31{,}680}{3} \text{ yd.} = 10{,}560 \text{ yd.}$$

c. 40 miles per hour to feet per second

Solution First, 40 miles per hour means $\dfrac{40 \text{ miles}}{1 \text{ hour}}$. Notice that we must convert miles to feet and hours to seconds. For the distance conversion, we use the fact that there are 5280 feet to 1 mile. For the time conversion, there are 60 minutes in 1 hour and 60 seconds in 1 minute.

$$\frac{40 \text{ miles}}{1 \text{ hour}} = \frac{40 \text{ mi.}}{1 \text{ hr.}} \cdot \frac{5280 \text{ ft.}}{1 \text{ mi.}} \cdot \frac{1 \text{ hr.}}{60 \text{ min.}} \cdot \frac{1 \text{ min.}}{60 \text{ sec.}} = \frac{211{,}200 \text{ ft.}}{3600 \text{ sec.}} = 58.\overline{6} \text{ ft./sec.}$$

7.2 Exercises

1. When multiplying rational expressions, why is it important to write the numerators and denominators in factored form?

2. If, after factoring, you determine that there are two binomial factors that are additive inverses, how can you get them to be common binomial factors?

3. Why are $\dfrac{x}{y}$ and $\dfrac{y}{x}$ multiplicative inverses (or reciprocals)?

4. To divide rational expressions, we write an equivalent multiplication with the _____ of the divisor.

5. What is a unit fraction? Give two examples.

6. Explain how to convert one unit of measurement to another using dimensional analysis.

For Exercises 7–42, multiply.

7. $\dfrac{m}{n} \cdot \dfrac{2m}{5n}$

8. $\dfrac{x}{y} \cdot \dfrac{3x}{2y}$

9. $\dfrac{r^3}{s^3} \cdot \dfrac{s^5}{r^6}$

10. $\dfrac{a^5}{b^3} \cdot \dfrac{b^2}{a^3}$

11. $\dfrac{3}{y} \cdot \dfrac{y^3}{3}$

12. $\dfrac{y^4}{7} \cdot \dfrac{7}{y^2}$

13. $-\dfrac{n}{rs^2} \cdot \dfrac{rs}{mn^3}$

14. $\dfrac{ab}{xy} \cdot -\dfrac{x}{a^3b^2}$

15. $-\dfrac{6r^2s}{5r^2s^2} \cdot -\dfrac{15r^4s^2}{2r^3s^2}$

16. $-\dfrac{7mn^2}{8m^2n} \cdot -\dfrac{16mz^2}{49n^2z}$

17. $\dfrac{5n^2}{15m} \cdot \dfrac{n}{2} \cdot \dfrac{3m^2n}{n^2}$

18. $\dfrac{a}{b^2} \cdot \dfrac{3b^2}{4a} \cdot \dfrac{ab}{9}$

19. $\dfrac{3b - 12}{14} \cdot \dfrac{21}{5b - 20}$

20. $\dfrac{9}{2a + 4} \cdot \dfrac{3a + 6}{15}$

21. $\dfrac{2x - 6}{3y + 12} \cdot \dfrac{2y + 8}{5x - 15}$

22. $\dfrac{3a + 12}{5b - 30} \cdot \dfrac{3b - 18}{4a + 16}$

23. $\dfrac{4x^2 - 25}{3x - 4} \cdot \dfrac{9x^2 - 16}{2x + 5}$

24. $\dfrac{16x^2 - 9}{4x + 3} \cdot \dfrac{25x^2 - 1}{5x + 1}$

25. $\dfrac{r^2 + 3r}{r^2 - 36} \cdot \dfrac{r^2 - 6r}{r^2 - 9}$

26. $\dfrac{m^2 + 5m}{m^2 - 16} \cdot \dfrac{m^2 - 4m}{m^2 - 25}$

27. $\dfrac{12q}{q^2 - 2q + 1} \cdot \dfrac{q^2 - 1}{24q^2}$

28. $\dfrac{36x^3}{x^2 - 25} \cdot \dfrac{x^2 - 10x + 25}{9x}$

29. $\dfrac{w^2 - 6w + 5}{w^2 - 1} \cdot \dfrac{w - 1}{w^2 - 10w + 25}$

30. $\dfrac{y^2 - 2y - 24}{y^2 - 16} \cdot \dfrac{y^2 + 10y + 25}{y^2 - y - 30}$

31. $\dfrac{n^2 - 6n + 9}{n^2 - 9} \cdot \dfrac{n^2 + 4n + 3}{n^2 - 3n}$

32. $\dfrac{4u^2 + 4u + 1}{2u^2 + u} \cdot \dfrac{u^2 + 3u - 4}{2u^2 - u - 1}$

33. $\dfrac{6x^2 + x - 12}{2x^2 - 5x - 12} \cdot \dfrac{3x^2 - 14x + 8}{9x^2 - 18x + 8}$

34. $\dfrac{2x^2 + 3x - 20}{4x^2 + 11x - 3} \cdot \dfrac{4x^2 - 9x + 2}{2x^2 - 9x + 10}$

35. $\dfrac{ac + 3a + 2c + 6}{ad + a + 2d + 2} \cdot \dfrac{ad - 5a + 2d - 10}{bc + 3b - 4c - 12}$

36. $\dfrac{mn + 2m + 4n + 8}{np - 3n + 2p - 6} \cdot \dfrac{pq + 5p - 3q - 15}{mq + 4m + 4q + 16}$

37. $\dfrac{x^2 + 7xy + 10y^2}{x^2 + 6xy + 5y^2} \cdot \dfrac{x + 2y}{y} \cdot \dfrac{x + y}{x^2 + 4xy + 4y^2}$

38. $\dfrac{2m - 3n}{m^2 + 3mn + 2n^2} \cdot \dfrac{m + 2n}{m - n} \cdot \dfrac{4m^2 - 4n^2}{4m^2 - 9n^2}$

39. $\dfrac{4x^2y^3}{2x - 6} \cdot \dfrac{12 - 4x}{6x^3y}$

40. $\dfrac{6b^3c^3}{70 - 10b} \cdot \dfrac{5b - 35}{8b^2c^5}$

41. $\dfrac{2x^2 - 11x + 12}{2x^2 + 11x - 21} \cdot \dfrac{3x^2 + 20x - 7}{4 - x}$

42. $\dfrac{3x^2 - 17x + 10}{3x^2 + 7x - 6} \cdot \dfrac{2x^2 + x - 15}{5 - x}$

For Exercises 43–70, divide.

43. $\dfrac{x}{y^4} \div \dfrac{x^3}{y^2}$

44. $\dfrac{a^3}{b^3} \div \dfrac{a^5}{b^2}$

45. $\dfrac{2m}{3n^2} \div \dfrac{8m^3}{15n}$

46. $\dfrac{21c^2}{7d^3} \div \dfrac{8c^4}{12d^4}$

47. $-\dfrac{12w^2y}{5z^2} \div -\dfrac{12w^2}{y}$

48. $-\dfrac{7a^2b}{2c^2} \div -\dfrac{7a^2}{b}$

49. $\dfrac{8x^7g^3}{15x^2g^4} \div -\dfrac{3xg^2}{4x^2g^3}$

50. $-\dfrac{10c^5g^5}{12c^2g^4} \div \dfrac{8cg^2}{4c^2g^3}$

51. $\dfrac{4a-8}{15a} \div \dfrac{3a-6}{5a^2}$

52. $\dfrac{6b+24}{7b^2} \div \dfrac{7b+28}{14b^2}$

53. $\dfrac{4p+12q}{3p-6q} \div \dfrac{5p+15q}{6p-12q}$

54. $\dfrac{5r+10s}{3r-12s} \div \dfrac{6r+12s}{4r-16s}$

55. $\dfrac{a^2-b^2}{x^2-y^2} \div \dfrac{a+b}{x-y}$

56. $\dfrac{x+4}{y-3} \div \dfrac{x^2-16}{y^2-9}$

57. $\dfrac{x^2+16x+64}{x^2-16x+64} \div \dfrac{x+8}{x-8}$

58. $\dfrac{w+3}{w-3} \div \dfrac{w^2+6w+9}{w^2-6w+9}$

59. $\dfrac{x^2+7x+10}{x^2+2x} \div \dfrac{x+5}{x^2-4x}$

60. $\dfrac{t^2-2t}{t+1} \div \dfrac{t^2+3t}{t^2+4t+3}$

61. $\dfrac{a^2-b^2}{a^2+2ab+b^2} \div \dfrac{a^2-3ab+2b^2}{a^2+3ab+2b^2}$

62. $\dfrac{3w^2-7w-6}{w^2-9} \div \dfrac{9w^2-4}{3w^2+7w-6}$

63. $\dfrac{3k+2}{5k^2-k} \div \dfrac{6k^2+k-2}{10k^2+3k-1}$

64. $\dfrac{5y-3}{4y^2-y-3} \div \dfrac{5y^3-3y^2}{4y^2+3y}$

65. $\dfrac{x^2+4x+4}{x+1} \div (3x^2+5x-2)$

66. $\dfrac{u^2-2u-8}{u^2+3u+2} \div (u^2-3u-4)$

67. $\dfrac{ab+3a+2b+6}{bc+4b+3c+12} \div \dfrac{ac-3a+2c-6}{bc+4b-4c-16}$

68. $\dfrac{xy-3x+4y-12}{xy+6x-3y-18} \div \dfrac{xy+5x+4y+20}{xy+5x-3y-15}$

69. $\dfrac{b^2-16}{b^2+b-12} \div \dfrac{4-b}{b^2+2b-15}$

70. $\dfrac{y^2-y-20}{y^2-2y-24} \div \dfrac{5-y}{y+3}$

For Exercises 71–76, perform the indicated operations. (Remember that the order of operations is to multiply or divide from left to right.)

71. $\dfrac{y^2 + y - 6}{2y^2 - 3y - 2} \cdot \dfrac{2y^2 + 9y + 4}{y^2 + 7y + 12} \div \dfrac{4y + y^2}{7 + y}$

72. $\dfrac{2m^2 - 7m - 15}{4m^2 - 9} \cdot \dfrac{2m^2 - 5m - 3}{2m^2 - 7m + 3} \div \dfrac{m^2 - 5m}{2m - 1}$

73. $\dfrac{6x^2 - 5x - 6}{x^4 - 81} \div \dfrac{3x - 2x^2}{x^3 + 3x^2 + 9x + 27} \div \dfrac{3x + 2}{x^4}$

74. $\dfrac{12h^2 + 11h - 5}{h^4 - 16} \div \dfrac{h - 3h^2}{h^3 + 4h - 2h^2 - 8} \div \dfrac{4h + 5}{h^3}$

75. $\dfrac{6x^2}{x^2 - 9} \cdot \dfrac{3x^2 + x - 24}{20y - 6xy} \cdot \dfrac{x - 3}{3x - 8} \div \dfrac{3x^2 + 9x}{3x^2 - x - 30}$

76. $\dfrac{8y^3}{3y^2 - 4y - 4} \div \dfrac{8xy - 20x}{2y^2 - 9y + 10} \cdot \dfrac{y + 5}{7 - 4y} \cdot \dfrac{12y^2 - 13y - 14}{2y^2 + 10y}$

For Exercises 77–80, explain the mistake, then correct the mistake.

77. $\dfrac{5y}{9} \div \dfrac{3y}{10} = \dfrac{y}{3} \cdot \dfrac{y}{2} = \dfrac{y^2}{6}$

78. $\dfrac{12x^3y}{5} \cdot \dfrac{3y}{6x^2} = \dfrac{2xy}{5} \cdot \dfrac{3y}{1} = \dfrac{6y^2}{5}$

79.
$$\dfrac{x^2 - 9}{15y} \cdot \dfrac{y - 2}{x + 3} = \dfrac{(x - 3)(x + 3)}{15y} \cdot \dfrac{y - 2}{x + 3}$$
$$= \dfrac{x - 3}{15} \cdot \dfrac{-2}{x + 3}$$
$$= \dfrac{-2x + 6}{15x + 45}$$

80.
$$\dfrac{x^2 + 3x + 2}{x^2 - 1} \cdot \dfrac{x - 1}{x + 1} = \dfrac{(x + 1)(x + 2)}{1} \cdot \dfrac{1}{x + 1}$$
$$= x + 2$$

For Exercises 81–86, use dimensional analysis and the following facts to convert units of length.

$$1 \text{ foot} = 12 \text{ inches}$$
$$1 \text{ yard} = 3 \text{ feet}$$
$$1 \text{ mile} = 5280 \text{ feet}$$

81. 18.5 feet to inches

82. 24.2 miles to feet

83. 6 miles to yards

84. 2400 yards to miles

85. 8 yards to inches

86. 216 inches to yards

For Exercises 87–92, use dimensional analysis and the following values to convert units of weight.

$$1 \text{ pound} = 16 \text{ ounces}$$
$$1 \text{ ton} = 2000 \text{ pounds}$$

87. 5.5 pounds to ounces

88. 36 ounces to pounds

89. 25,400 pounds to tons

90. 2.5 tons to pounds

91. 3 tons to ounces

92. 48,000 ounces to tons

For Exercises 93–100, use dimensional analysis.

93. The fastest winning speed at the Daytona 500 was 177.602 miles per hour by Buddy Baker in 1980. How many feet per second is this? (*Source:* DaytonaInternationalSpeedway.com)

94. The speed of sound is 741.8 miles per hour at 32°F at sea level. How many feet per second is this?

95. A sprinter's best time in the 40-yard dash is 4.3 seconds. What is that speed in miles per hour?

96. A runner knows that she can run 400 yards in 87.2 seconds. If she can maintain that pace, how many minutes will it take her to run a mile?

97. The mean orbital velocity of Earth is 29.78 kilometers per second. How many kilometers does Earth orbit in one year? (Hint: One year is $365\frac{1}{4}$ days.) (*Source:* Windows to the Universe)

98. The mean orbital velocity of Pluto is 4.74 kilometers per second. How many kilometers does Pluto orbit in one year? (*Source:* Windows to the Universe)

99. One light-year is the distance that light travels in a year. If light travels at a speed of approximately 186,000 miles per second, how many miles does light travel in one year?

100. The light from the Sun takes about 8 minutes to reach Earth. Using the fact that light travels at a speed of about 186,000 miles per second, determine Earth's distance from the Sun in miles.

> **Of Interest**
> Pluto is usually the farthest planet from the Sun. However, for 20 years out of its 249-year orbit, its orbital path takes it closer to the Sun than Neptune.

> **Of Interest**
> The nearest star to the Sun is Proxima Centauri, which is about 4.2 light-years away. This means that when we look at Proxima Centauri, we are seeing an image of that star that is 4.2 years old. Some objects we can see in the night sky are much farther. For example, the Andromeda galaxy is about 2,140,000 light-years from Earth, which means that when we look at it, we are seeing it as it was 2,140,000 years ago.

For Exercises 101–106, use dimensional analysis and the following table of exchange rates to convert.

	USD($)	GBP(£)	CAN($)	EUR(€)
USD($)	1	1.9264	0.830702	1.34229
GBP(£)	0.519103	1	0.43122	0.696791
CAN($)	1.2038	2.319	1	1.61586
EUR(€)	0.74499	1.43515	0.618865	1

Source: Federal Reserve Bank of New York

Note: *Each number is the ratio of the row's unit to 1 of the column's unit. For example, 1.61586 means there are 1.61586 Can$ to 1€.*

101. Convert \$500 into £ (Great Britain pound).

102. Convert \$250 into Can\$ (Canadian dollars).

103. Convert 450 € (euros) into U.S. dollars.

104. Convert 700 £ into U.S. dollars.

105. Convert 650 £ to €.

106. Convert 400 Can\$ to €.

PUZZLE PROBLEM

If $\dfrac{1}{1000}$ *of a light year is about* 9.46×10^9 *kilometers, find the speed of light in meters per second.*

REVIEW EXERCISES

For Exercises 1–6, simplify.

[1.3] **1.** $\dfrac{1}{3} + \dfrac{2}{3}$

[1.3] **2.** $\dfrac{1}{4} - \dfrac{3}{4}$

[1.7] **3.** $\dfrac{1}{2}x + \dfrac{3}{2}x$

[1.7] **4.** $\dfrac{3}{8}y - 5 - \dfrac{1}{8}y + 1$

[5.3] **5.** $\left(6x^2 + \dfrac{1}{9}x - 1\right) + \left(2x^2 - \dfrac{7}{9}x - 3\right)$

[5.3] **6.** $\left(\dfrac{3}{5}n^2 - 5n + 7\right) - \left(\dfrac{4}{5}n^2 - 5n + 3\right)$

7.3 Adding and Subtracting Rational Expressions with the Same Denominator

OBJECTIVE 1. Add or subtract rational expressions with the same denominator. Recall that when adding or subtracting fractions with the same denominator, we add or subtract the numerators and keep the same denominator.

$$\frac{3}{7} + \frac{1}{7} = \frac{3+1}{7} = \frac{4}{7} \quad \text{or} \quad \frac{7}{9} - \frac{2}{9} = \frac{7-2}{9} = \frac{5}{9}$$

We add and subtract rational expressions the same way.

$$\frac{3x}{7} + \frac{x}{7} = \frac{3x+x}{7} = \frac{4x}{7} \quad \text{or} \quad \frac{7x}{9} - \frac{2x}{9} = \frac{7x-2x}{9} = \frac{5x}{9}$$

PROCEDURE Adding or Subtracting Rational Expressions (Same Denominator)

To add or subtract rational expressions that have the same denominator:

1. Add or subtract the numerators and keep the same denominator.
2. Simplify to lowest terms (remember to write the numerators and denominators in factored form in order to simplify).

EXAMPLE 1 Add. $\dfrac{5x}{14} + \dfrac{x}{14}$

Solution $\dfrac{5x}{14} + \dfrac{x}{14} = \dfrac{5x + x}{14}$ Since the rational expressions have the same denominator, we add numerators and keep the same denominator.

$\qquad\qquad\quad = \dfrac{6x}{14}$ Combine like terms.

$\qquad\qquad\quad = \dfrac{2 \cdot 3 \cdot x}{2 \cdot 7}$ Factor

$\qquad\qquad\quad = \dfrac{3x}{7}$ Divide out the common factor, 2.

YOUR TURN Subtract. $\dfrac{2a^2}{15} - \dfrac{7a^2}{15}$

Numerators with No Like Terms

In Example 1, the numerators were like terms. If the numerators have no like terms, we write a multiterm polynomial expression in the numerator of the sum or difference.

EXAMPLE 2 Subtract.

a. $\dfrac{y}{y + 5} - \dfrac{3}{y + 5}$

Solution $\dfrac{y}{y + 5} - \dfrac{3}{y + 5} = \dfrac{y - 3}{y + 5}$ Subtract numerators and keep the same denominator. Since y and 3 are not like terms, we express their difference as a polynomial.

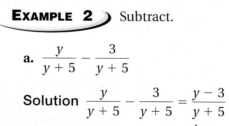

Warning: Though it may be tempting, we cannot divide out the y's because they are terms, not factors.

ANSWER

$-\dfrac{a^2}{3}$

b. $\dfrac{n^2}{n+3} - \dfrac{9}{n+3}$

Solution $\dfrac{n^2}{n+3} - \dfrac{9}{n+3} = \dfrac{n^2-9}{n+3}$

Note: *The numerator can be factored, so we may be able to simplify.*

$= \dfrac{(n-3)(n+3)}{n+3}$ **Factor the numerator.**

$= n - 3$ **Divide out the common factor, $n+3$.**

c. $\dfrac{4m}{15} - \left(-\dfrac{8}{15}\right)$

Solution $\dfrac{4m}{15} - \left(-\dfrac{8}{15}\right) = \dfrac{4m-(-8)}{15}$

$= \dfrac{4m+8}{15}$

Note: $4m + 8$ *can be factored.*

$4m + 8 = 4(m+2)$

However, since there are no common factors in the denominator, we cannot simplify further.

YOUR TURN   Add or subtract.

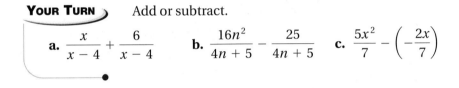

a. $\dfrac{x}{x-4} + \dfrac{6}{x-4}$ **b.** $\dfrac{16n^2}{4n+5} - \dfrac{25}{4n+5}$ **c.** $\dfrac{5x^2}{7} - \left(-\dfrac{2x}{7}\right)$

Numerators That Are Multiterm Polynomials

So far, the numerators in the given rational expressions were monomials. Now look at some cases where the numerators in the given rational expressions are multiterm polynomials. Remember from Chapter 5 that to add polynomials, we combine like terms.

EXAMPLE 3   Add.

a. $\dfrac{7y-13}{5} + \dfrac{y+8}{5}$

Solution $\dfrac{7y-13}{5} + \dfrac{y+8}{5} = \dfrac{(7y-13)+(y+8)}{5}$

Combine $7y$ with y to get $8y$ and combine -13 with 8 to get -5.

$= \dfrac{8y-5}{5}$

Warning: It may be tempting to divide out the 5, but the 5 in $8y - 5$ is not a factor. Because $8y - 5$ cannot be factored, we cannot simplify this rational expression.

b. $\dfrac{x^2 - 5x - 13}{x - 6} + \dfrac{3x - 11}{x - 6}$

Solution $\dfrac{x^2 - 5x - 13}{x - 6} + \dfrac{3x - 11}{x - 6} = \dfrac{(x^2 - 5x - 13) + (3x - 11)}{x - 6}$

Note: *Since the numerator can be factored, we may be able to simplify.* ▶ $= \dfrac{x^2 - 2x - 24}{x - 6}$ **Combine like terms in the numerator.**

$= \dfrac{(x - 6)(x + 4)}{x - 6}$ **Factor the numerator.**

$= x + 4$ **Divide out the common factor, $x - 6$.**

YOUR TURN   Add.

a. $\dfrac{n + 7}{11n} + \dfrac{2 - 5n}{11n}$ **b.** $\dfrac{3t^2 - 5t}{t - 3} + \dfrac{t^2 - 8t + 3}{t - 3}$

Simplifying When the Denominator Is Factorable

Sometimes you may also be able to factor the denominator. If the denominator can be factored, you will need to write it in factored form in order to simplify.

EXAMPLE 4   Add. $\dfrac{4x^2 + 9x + 1}{2x^3 - 18x} + \dfrac{6x^2 + 23x + 5}{2x^3 - 18x}$

Solution $\dfrac{4x^2 + 9x + 1}{2x^3 - 18x} + \dfrac{6x^2 + 23x + 5}{2x^3 - 18x} = \dfrac{(4x^2 + 9x + 1) + (6x^2 + 23x + 5)}{2x^3 - 18x}$

$= \dfrac{10x^2 + 32x + 6}{2x^3 - 18x}$ **Combine like terms in the numerator.**

$= \dfrac{2(5x^2 + 16x + 3)}{2x(x^2 - 9)}$ **Factor 2 out of the numerator and $2x$ out of the denominator.**

$= \dfrac{2(5x + 1)(x + 3)}{2x(x - 3)(x + 3)}$ **Factor the trinomial in the numerator and the difference of squares in the denominator.**

$= \dfrac{5x + 1}{x(x - 3)}$ or $\dfrac{5x + 1}{x^2 - 3x}$ **Divide out the common factors, 2 and $x + 3$.**

YOUR TURN   Add. $\dfrac{x^3 + 3x^2 - 7x}{3x^4 - 12x^2} + \dfrac{5x - 2x^2}{3x^4 - 12x^2}$

Subtracting with Numerators That Are Multiterm Polynomials

Now let's look at subtraction of rational expressions that have the same denominator and multiterm polynomials in the numerators. Recall from Chapter 5 that when subtracting polynomials, we write an equivalent addition with the additive inverse of the subtrahend.

EXAMPLE 5 Subtract.

a. $\dfrac{7t^2 - t + 5}{t + 2} - \dfrac{3t^2 - 6t - 2}{t + 2}$

Solution $\dfrac{7t^2 - t + 5}{t + 2} - \dfrac{3t^2 - 6t - 2}{t + 2} = \dfrac{(7t^2 - t + 5) - (3t^2 - 6t - 2)}{t + 2}$

Note: *To write an equivalent addition, change the operation symbol from a minus sign to a plus sign and change all the signs in the subtrahend (second) polynomial.*

$$= \dfrac{(7t^2 - t + 5) + (-3t^2 + 6t + 2)}{t + 2}$$

$$= \dfrac{4t^2 + 5t + 7}{t + 2} \qquad \text{Combine like terms.}$$

b. $\dfrac{x^3 + 4x - 3}{5x^2 - 10x} - \dfrac{4x + 5}{5x^2 - 10x}$

Solution $\dfrac{x^3 + 4x - 3}{5x^2 - 10x} - \dfrac{4x + 5}{5x^2 - 10x} = \dfrac{(x^3 + 4x - 3) - (4x + 5)}{5x^2 - 10x}$

$$= \dfrac{(x^3 + 4x - 3) + (-4x - 5)}{5x^2 - 10x} \qquad \text{Write the equivalent addition.}$$

Note: *The numerator is a difference of cubes.*

$$= \dfrac{x^3 - 8}{5x^2 - 10x} \qquad \text{Combine like terms.}$$

$$= \dfrac{(x - 2)(x^2 + 2x + 4)}{5x(x - 2)} \qquad \begin{array}{l}\text{Factor the numerator and de-}\\ \text{nominator. Divide out the}\\ \text{common factor, } x - 2.\end{array}$$

$$= \dfrac{x^2 + 2x + 4}{5x}$$

ANSWER

$\dfrac{y^2 + 3y + 9}{3y + 2}$

YOUR TURN Subtract. $\dfrac{y^3 + y - 20}{3y^2 - 7y - 6} - \dfrac{y + 7}{3y^2 - 7y - 6}$

7.3 Exercises

1. Two rational expressions with monomial numerators are added to form a rational expression whose numerator is a multiterm polynomial. Why would this happen?

2. After adding or subtracting two rational expressions, what is the first step in simplifying the resulting rational expression?

3. Two rational expressions that have the same denominator and numerators that are multiterm polynomials are to be subtracted. Explain the process.

4. After adding or subtracting two rational expressions, the resulting rational expression contains a multiterm polynomial in the numerator that can be factored. Does this mean that the rational expression can be simplified? Explain.

For Exercises 5–48, add or subtract. Simplify your answers to lowest terms.

5. $\dfrac{3x}{10} + \dfrac{2x}{10}$

6. $\dfrac{2x}{9} + \dfrac{x}{9}$

7. $\dfrac{4a^2}{9} + \left(-\dfrac{a^2}{9}\right)$

8. $\dfrac{6x^3}{11} + \left(-\dfrac{x^3}{11}\right)$

9. $\dfrac{6}{a} - \left(-\dfrac{3}{a}\right)$

10. $\dfrac{3}{w} - \left(-\dfrac{2}{w}\right)$

11. $\dfrac{6x}{7y^2} - \dfrac{2x}{7y^2} - \dfrac{5x}{7y^2}$

12. $\dfrac{9a}{11x^2} - \dfrac{8a}{11x^2} - \dfrac{3a}{11x^2}$

13. $\dfrac{6}{n+5} + \dfrac{3}{n+5}$

14. $\dfrac{4}{n+2} + \dfrac{7}{n+2}$

15. $\dfrac{2r}{r+2} + \dfrac{r}{r+2}$

16. $\dfrac{5z}{b-6} + \dfrac{7z}{b-6}$

17. $\dfrac{1-2q}{3q} + \dfrac{5q+2}{3q}$

18. $\dfrac{6x-3}{4x} + \dfrac{2x+5}{4x}$

19. $\dfrac{5t+5}{5u} - \dfrac{4t-6}{5u}$

20. $\dfrac{8y+5}{3y} - \dfrac{2y+5}{3y}$

21. $\dfrac{2-2a}{2a-3} + \dfrac{8a-11}{2a-3}$

22. $\dfrac{16x+y}{x-y} + \dfrac{10x-15y}{x-y}$

23. $\dfrac{w^2+5}{w+1} + \dfrac{6-w^2}{w+1}$

24. $\dfrac{x^2+2}{x+1} + \dfrac{6-x^2}{x+1}$

25. $\dfrac{4m+3}{m-1} - \dfrac{m+4}{m-1}$

26. $\dfrac{2x+5}{x+1} - \dfrac{x+4}{x+1}$

27. $\dfrac{1-3t}{2t-3} - \dfrac{8t-2}{2t-3}$

28. $\dfrac{2b}{b+c} - \dfrac{b-c}{b+c}$

29. $\dfrac{2w+3}{w+5} + \dfrac{5w-2}{w+5} - \dfrac{w+6}{w+5}$

30. $\dfrac{6q+21}{q+1} - \dfrac{28-14q}{q+1} + \dfrac{6q-5}{q+1}$

31. $\dfrac{g}{g^2-9} - \dfrac{3}{g^2-9}$

32. $\dfrac{x}{x^2-49} + \dfrac{7}{x^2-49}$

33. $\dfrac{z^2+2z}{z-7} + \dfrac{21-12z}{z-7}$

34. $\dfrac{u^2+5u}{u+1} + \dfrac{2-2u}{u+1}$

35. $\dfrac{2s}{s^2+4s+4} - \dfrac{s-2}{s^2+4s+4}$

36. $\dfrac{v-8}{v^2-16} + \dfrac{4}{v^2-16}$

37. $\dfrac{y^2}{y^2+3y} - \dfrac{9}{y^2+3y}$

38. $\dfrac{x^2}{x^2-5x} - \dfrac{25}{x^2-5x}$

39. $\dfrac{x^2+3x}{x^2-x-12} + \dfrac{2x+6}{x^2-x-12}$

40. $\dfrac{y^2+2y}{y^2+5y+4} + \dfrac{5y+12}{y^2+5y+4}$

41. $\dfrac{t^2 - 4t}{t^2 + 10t + 21} - \dfrac{-6t + 3}{t^2 + 10t + 21}$

42. $\dfrac{v^2 + 6}{v^2 - 3v - 10} - \dfrac{2v + 14}{v^2 - 3v - 10}$

43. $\dfrac{x - 5}{x^2 - 25} + \dfrac{2x + 10}{x^2 - 25} + \dfrac{4}{x^2 - 25}$

44. $\dfrac{3x - 3}{x^2 - 1} + \dfrac{1}{x^2 - 1} + \dfrac{x}{x^2 - 1}$

45. $\dfrac{3x^2 + 2x}{x^2 - 2x - 15} + \dfrac{3x - 8}{x^2 - 2x - 15} + \dfrac{2x + 2}{x^2 - 2x - 15}$

46. $\dfrac{2m^2 - 6m}{m^2 - 5m + 6} + \dfrac{2m - 4}{m^2 - 5m + 6} + \dfrac{4}{m^2 - 5m + 6}$

47. $\dfrac{2x^2 + x - 3}{x^2 + 6x + 5} + \dfrac{x^2 - 2x + 4}{x^2 + 6x + 5} - \dfrac{2x^2 + x + 4}{x^2 + 6x + 5}$

48. $\dfrac{3x^2 - 2x + 3}{x^2 + x - 12} - \dfrac{x^2 + x - 2}{x^2 + x - 12} - \dfrac{x^2 + 4x - 7}{x^2 + x - 12}$

For Exercises 49–52, explain the mistake, then find the correct sum or difference.

49. $\dfrac{4y}{5} + \dfrac{1}{5} = \dfrac{5y}{5} = y$

50. $\dfrac{7}{m} + \dfrac{n}{m} = \dfrac{7 + n}{2m}$

51. $\dfrac{x + 1}{3} - \dfrac{x - 5}{3} = \dfrac{(x + 1) - (x - 5)}{3}$

$\qquad\qquad\qquad = \dfrac{(x + 1) + (-x - 5)}{3}$

$\qquad\qquad\qquad = -\dfrac{4}{3}$

52. $\dfrac{x}{x - 2} + \dfrac{x - 4}{x - 2} = \dfrac{x^2 - 4}{x - 2}$

$\qquad\qquad\qquad = \dfrac{(x + 2)(x - 2)}{x - 2}$

$\qquad\qquad\qquad = x + 2$

For Exercises 53–56, find the perimeter of the figure shown.

53.

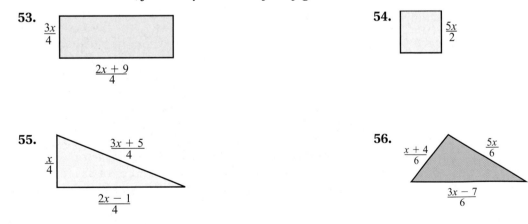

$\dfrac{3x}{4}$

$\dfrac{2x + 9}{4}$

54.

$\dfrac{5x}{2}$

55.

$\dfrac{x}{4}$ $\dfrac{3x + 5}{4}$

$\dfrac{2x - 1}{4}$

56.

$\dfrac{x + 4}{6}$ $\dfrac{5x}{6}$

$\dfrac{3x - 7}{6}$

REVIEW EXERCISES

For Exercises 1–6, simplify.

[1.3]   **1.** $\dfrac{1}{2} + \dfrac{3}{5}$ [1.3]   **2.** $\dfrac{5}{6} - \dfrac{3}{8}$ [1.3]   **3.** $\dfrac{4}{7} - \left(-\dfrac{3}{5}\right)$ [1.7]   **4.** $\dfrac{1}{3}x - \dfrac{1}{6} + \dfrac{2}{5}x + \dfrac{3}{4}$

[5.3]   **5.** $\left(x^2 - 9x - \dfrac{5}{8}\right) + \left(x^2 - 3x + \dfrac{1}{6}\right)$ [5.3]   **6.** $\left(3y^2 + \dfrac{3}{4}y - 8\right) - \left(y^2 + \dfrac{4}{5}y + 2\right)$

7.4   Adding and Subtracting Rational Expressions with Different Denominators

OBJECTIVES

1. Find the LCD of two or more rational expressions.

2. Given two rational expressions, write equivalent rational expressions with their LCD.

3. Add or subtract rational expressions with different denominators.

Adding or subtracting rational expressions with different denominators follows three main stages: (1) finding the LCD, (2) rewriting the expressions with the LCD, and (3) adding or subtracting the numerators. We will explore the first two stages in objectives 1 and 2, then put them back into their proper context in objective 3.

OBJECTIVE 1. Find the LCD of two or more rational expressions.   Remember that when adding or subtracting fractions with different denominators, we must first find a common denominator. It is helpful to use the least common denominator (LCD), which is the smallest number that is evenly divisible by all the denominators. For example, given the fractions $\dfrac{1}{8}$ and $\dfrac{7}{12}$, the LCD is 24.

   To help us find the LCD when denominators are more complex, as in rational expressions, let's consider a method involving prime factorizations. Consider the prime factorizations of the denominators, 8 and 12, compared to the prime factorization of the LCD, 24, in order to determine a procedure.

$$8 = 2^3 \qquad 12 = 2^2 \cdot 3$$

$$\text{LCD} = 2^3 \cdot 3 = 24$$

Both 8 and 12 have factors of 2, but 8 has the greatest number of 2s, which is why 2^3 is included in the factorization for the LCD instead of 2^2. Similarly, 12 has the greatest number of factors of 3 (a single factor, whereas 8 has no factors of 3), so a single factor of 3 is included in the factorization of the LCD. This suggests the following procedure:

To find the LCD of two or more rational expressions:
1. Factor each denominator.
2. For each unique factor, compare the number of times it appears in each factorization. Write a product that includes each unique factor the greatest number of times it appears in the denominator factorizations.

First, let's consider cases where the denominators are monomials.

EXAMPLE 1 Find the LCD. $\dfrac{3}{4x^2}$ and $\dfrac{5}{6x^3}$

Solution We first factor the denominators $4x^2$ and $6x^3$ by writing their prime factorizations.

$$4x^2 = 2^2 \cdot x^2$$
$$6x^3 = 2 \cdot 3 \cdot x^3$$

Note: *We can compare exponents in the prime factorizations to create the LCD. If two factorizations have the same prime factor, we write that prime factor in the LCD with the greater of the two exponents.*

The unique factors are 2, 3, and x. To generate the LCD, include 2, 3, and x the greatest number of times each appears in any of the factorizations.

The greatest number of times that 2 appears is twice (in $2^2 \cdot x^2$).
The greatest number of times that 3 appears is once (in $2 \cdot 3 \cdot x^3$).
The greatest number of times that x appears is three times (in $2 \cdot 3 \cdot x^3$).

$$\text{LCD} = 2^2 \cdot 3 \cdot x^3 = 12x^3$$

YOUR TURN   Find the LCD.

a. $\dfrac{y}{12x^5}$ and $\dfrac{5}{9x^2}$

b. $\dfrac{p}{10mn}$ and $\dfrac{7}{15n^2}$

Now consider denominators that are multiterm polynomials.

EXAMPLE 2 Find the LCD.

a. $\dfrac{7}{5x - 15}$ and $\dfrac{5}{2x - 6}$

Solution Factor the denominators $5x - 15$ and $2x - 6$.

$$5x - 15 = 5(x - 3)$$
$$2x - 6 = 2(x - 3)$$

The unique factors are 2, 5, and $x - 3$.
The greatest number of times that $x - 3$ appears in any of the factorizations is once.
The greatest number of times that 2 appears is once (in $2(x - 3)$).
The greatest number of times that 5 appears is once (in $5(x - 3)$).

$$\text{LCD} = 2 \cdot 5 \cdot (x - 3) = 10(x - 3) \text{ or } 10x - 30$$

ANSWERS

a. $36x^5$ **b.** $30mn^2$

b. $\dfrac{7}{x^2 + 10x + 25}$ and $\dfrac{17}{6x^4 + 30x^3}$

Solution Factor $x^2 + 10x + 25$ and $6x^4 + 30x^3$.

$$x^2 + 10x + 25 = (x + 5)^2$$
$$6x^4 + 30x^3 = 6x^3(x + 5) = 2 \cdot 3 \cdot x^3(x + 5)$$

The factors are 2, 3, x, and $x + 5$.
The greatest number of times that 2 appears is once (in $2 \cdot 3 \cdot x^3(x + 5)$).
The greatest number of times that 3 appears is once (in $2 \cdot 3 \cdot x^3(x + 5)$).
The greatest number of times that $x + 5$ appears is twice (in $(x + 5)^2$).
The greatest number of times that x appears is three times (in $2 \cdot 3 \cdot x^3(x + 5)$).

$$\text{LCD} = 2 \cdot 3 \cdot x^3 \cdot (x + 5)^2 = 6x^3(x + 5)^2$$

YOUR TURN   Find the LCD.

a. $\dfrac{y}{3y + 6}$ and $\dfrac{5}{4y + 8}$

b. $\dfrac{9}{x^2 - 2x + 1}$ and $\dfrac{17}{2x^3 - 2x}$

OBJECTIVE 2. Given two rational expressions, write equivalent rational expressions with their LCD. Writing equivalent rational expressions is much like writing equivalent fractions, which we reviewed in Section 1.2. Remember that to write an equivalent fraction we multiply the numerator and denominator by the same nonzero number.

EXAMPLE 3   Write $\dfrac{3}{4x^2}$ and $\dfrac{5}{6x^3}$ as equivalent rational expressions with their LCD.

Solution In Example 1, we found the LCD to be $12x^3$. For each rational expression, we multiply both the numerator and denominator by an appropriate factor so that the denominator becomes $12x^3$.

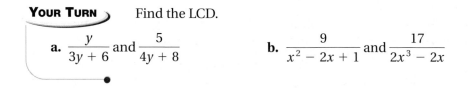

$$\dfrac{3}{4x^2} = \dfrac{3 \cdot 3x}{4x^2 \cdot 3x} = \dfrac{9x}{12x^3}$$

Note: *We can determine the appropriate factor by dividing the LCD by the original denominator. In this case, $12x^3 \div 4x^2 = 3x$.*

$$\dfrac{5}{6x^3} = \dfrac{5 \cdot 2}{6x^3 \cdot 2} = \dfrac{10}{12x^3}$$

Now consider rational expressions with denominators that are multiterm polynomials. Rewriting these rational expressions with their LCD is usually easier if we leave their denominators in factored form.

EXAMPLE 4 Write $\dfrac{7}{x^2 + 10x + 25}$ and $\dfrac{17}{6x^4 + 30x^3}$ as equivalent rational expressions with their LCD.

Solution In Example 2, we found the LCD to be $6x^3(x + 5)^2$.

$$\frac{7}{x^2 + 10x + 25} = \frac{7}{(x + 5)^2}$$

Write the denominator in factored form.

$$= \frac{7 \cdot 6x^3}{(x + 5)^2 \cdot 6x^3}$$

Multiply the numerator and denominator by the same factor, $6x^3$, to get the LCD, $6x^3(x + 5)^2$.

$$= \frac{42x^3}{6x^3(x + 5)^2}$$

Note: *Another way to determine the appropriate factor is to think of it as the factor in the LCD that is missing from the original denominator. In this case, $6x^3$ is in the LCD, but not in the original denominator.*

◀ **Note:** *Though we could multiply out $6x^3(x + 5)^2$, it is simpler to leave it in this form.*

$$\frac{17}{6x^4 + 30x^3} = \frac{17}{6x^3(x + 5)}$$

Write the denominator in factored form.

$$= \frac{17 \cdot (x + 5)}{6x^3(x + 5) \cdot (x + 5)}$$

Multiply the numerator and denominator by the same factor, $x + 5$, to get the LCD, $6x^3(x + 5)^2$.

$$= \frac{17(x + 5)}{6x^3(x + 5)^2}$$

Note: *There are two factors of $x + 5$ in the LCD, whereas only one factor of $x + 5$ appears in the original denominator.*

YOUR TURN Write equivalent rational expressions with their LCD.

a. $\dfrac{p}{10mn}$ and $\dfrac{7}{15n^2}$

b. $\dfrac{9}{x^2 - 2x + 1}$ and $\dfrac{17}{2x^3 - 2x}$

OBJECTIVE 3. Add or subtract rational expressions with different denominators. To add or subtract rational expressions with different denominators, we use the same process that we use to add or subtract fractions.

PROCEDURE Adding or Subtracting Rational Expressions with Different Denominators

To add or subtract rational expressions with different denominators:
1. Find the LCD.
2. Write each rational expression as an equivalent expression with the LCD.
3. Add or subtract the numerators and keep the LCD.
4. Simplify.

ANSWERS

a. $\dfrac{3np}{30mn^2}$ and $\dfrac{14m}{30mn^2}$

b. $\dfrac{18x(x + 1)}{2x(x - 1)^2(x + 1)}$ and

$\dfrac{17(x - 1)}{2x(x - 1)^2(x + 1)}$

EXAMPLE 5) Add or subtract as indicated.

a. $\dfrac{3}{4x^2} + \dfrac{5}{6x^3}$

Solution We found the LCD for these expressions in Example 1. We then wrote equivalent expressions in Example 3. Now we add the numerators and keep the LCD.

$$\dfrac{3}{4x^2} + \dfrac{5}{6x^3} = \dfrac{9x}{12x^3} + \dfrac{10}{12x^3} \qquad \text{Rewrite with the LCD.}$$

$$= \dfrac{9x + 10}{12x^3} \qquad \text{Add numerators.}$$

Note: *Because $9x$ and 10 are not like terms, we must indicate the addition as a polynomial expression. Also, since $9x + 10$ cannot be factored, we cannot simplify this rational expression.*

b. $\dfrac{7}{x^2 + 10x + 25} - \dfrac{17}{6x^4 + 30x^3}$

Solution We found the LCD for these expressions in Example 2 and wrote equivalent expressions in Example 4. Now we combine the numerators and keep the LCD.

$$\dfrac{7}{x^2 + 10x + 25} - \dfrac{17}{6x^4 + 30x^3} = \dfrac{42x^3}{6x^3(x + 5)^2} - \dfrac{17(x + 5)}{6x^3(x + 5)^2} \qquad \begin{array}{l}\text{Rewrite with the}\\ \text{LCD.}\end{array}$$

$$= \dfrac{42x^3 - 17(x + 5)}{6x^3(x + 5)^2} \qquad \begin{array}{l}\text{Combine}\\ \text{numerators.}\end{array}$$

$$= \dfrac{42x^3 - 17x - 85}{6x^3(x + 5)^2} \qquad \begin{array}{l}\text{Simplify in the}\\ \text{numerator.}\end{array}$$

Note: *Leaving the LCD in factored form helps us determine whether the final expression can be reduced. Since $42x^3 - 17x - 85$ cannot be factored, we cannot reduce this rational expression.*

Now that we have seen all the pieces of the process spread out over several examples, let's put them all together using new rational expressions.

EXAMPLE 6) Add or subtract as indicated.

a. $\dfrac{2x}{9y^3} + \dfrac{3}{4xy}$

Solution First, find the LCD.

$$4xy = 2^2 \cdot x \cdot y \qquad 9y^3 = 3^2 \cdot y^3 \qquad \text{LCD} = 2^2 \cdot 3^2 \cdot x \cdot y^3 = 36xy^3$$

$$\dfrac{2x}{9y^3} + \dfrac{3}{4xy} = \dfrac{2x(4x)}{9y^3(4x)} + \dfrac{3(9y^2)}{4xy(9y^2)} \qquad \begin{array}{l}\text{Write equivalent rational expressions with the}\\ \text{LCD, } 36xy^3.\end{array}$$

$$= \dfrac{8x^2}{36xy^3} + \dfrac{27y^2}{36xy^3}$$

$$= \dfrac{8x^2 + 27y^2}{36xy^3} \qquad \text{Add numerators.}$$

Note: *The numerator cannot be factored, so we cannot reduce.*

b. $\dfrac{x+5}{2x} + \dfrac{4x-1}{3x^2}$

Solution The LCD is $6x^2$.

$$\dfrac{x+5}{2x} + \dfrac{4x-1}{3x^2} = \dfrac{(x+5)(3x)}{(2x)(3x)} + \dfrac{(4x-1)(2)}{(3x^2)(2)}$$

Write equivalent rational expressions with the LCD, $6x^2$.

$$= \dfrac{3x^2+15x}{6x^2} + \dfrac{8x-2}{6x^2}$$

$$= \dfrac{3x^2+15x+8x-2}{6x^2}$$

Add numerators.

$$= \dfrac{3x^2+23x-2}{6x^2}$$

◀ **Note:** *Remember that to add polynomials, we combine like terms.*

c. $\dfrac{x}{x+2} - \dfrac{10}{x^2-x-6}$

Solution First, find the LCD by factoring $x+2$ and x^2-x-6.

Factored forms: $x+2$ cannot be factored; $x^2-x-6=(x+2)(x-3)$
LCD $=(x+2)(x-3)$

$$\dfrac{x}{x+2} - \dfrac{10}{x^2-x-6} = \dfrac{x}{x+2} - \dfrac{10}{(x+2)(x-3)}$$

Factor the denominators.

$$= \dfrac{(x)(x-3)}{(x+2)(x-3)} - \dfrac{10}{(x+2)(x-3)}$$

Write equivalent rational expressions with the LCD, $(x+2)(x-3)$.

$$= \dfrac{x^2-3x}{(x+2)(x-3)} - \dfrac{10}{(x+2)(x-3)}$$

Distribute

$$= \dfrac{x^2-3x-10}{(x+2)(x-3)}$$

Combine numerators.

$$= \dfrac{(x+2)(x-5)}{(x+2)(x-3)}$$

Factor the numerator.

$$= \dfrac{x-5}{x-3}$$

Divide out the common factor, $x+2$.

d. $\dfrac{x-1}{x^2+4x} - \dfrac{3x+1}{x^2-16}$

Solution To determine the LCD, we write the factored forms of x^2+4x and x^2-16.

Factored forms: $x^2+4x=x(x+4)$; $x^2-16=(x+4)(x-4)$
LCD $=x(x+4)(x-4)$

$$\frac{x-1}{x^2+4x}-\frac{3x+1}{x^2-16}=\frac{x-1}{x(x+4)}-\frac{3x+1}{(x+4)(x-4)}$$

Factor the denominators.

$$=\frac{(x-1)(x-4)}{x(x+4)(x-4)}-\frac{(3x+1)(x)}{(x+4)(x-4)(x)}$$

Write equivalent rational expressions with the LCD, $x(x+4)(x-4)$.

$$=\frac{x^2-5x+4}{x(x+4)(x-4)}-\frac{3x^2+x}{x(x+4)(x-4)}$$

Multiply in the numerator.

$$=\frac{(x^2-5x+4)-(3x^2+x)}{x(x+4)(x-4)}$$

Combine numerators.

Note: *Remember that to subtract polynomials, we write an equivalent addition.*

$$=\frac{(x^2-5x+4)+(-3x^2-x)}{x(x+4)(x-4)}$$

$$=\frac{-2x^2-6x+4}{x(x+4)(x-4)}$$

Note: *Though the numerator can be factored to $-2(x^2+3x-2)$, none of those factors matches any factor in the denominator, so the result is in lowest terms.*

YOUR TURN Add or subtract as indicated.

a. $\dfrac{5}{6x^2y}+\dfrac{3}{8xy^3}$

b. $\dfrac{2}{x^2+x}-\dfrac{x-4}{x^2-1}$

If denominators are additive inverses, we can multiply the numerator and denominator of either rational expression by -1 to obtain the LCD.

EXAMPLE 7 Add. $\dfrac{5x}{x-3}+\dfrac{2}{3-x}$

Solution $\dfrac{5x}{x-3}+\dfrac{2}{3-x}=\dfrac{5x}{x-3}+\dfrac{2(-1)}{(3-x)(-1)}$

Since $x-3$ and $3-x$ are additive inverses, we obtain the LCD by multiplying the numerator and denominator of one of the rational expressions by -1. We chose the second rational expression.

$$=\dfrac{5x}{x-3}+\dfrac{-2}{x-3}$$

$$=\dfrac{5x-2}{x-3}$$

YOUR TURN Add or subtract as indicated.

a. $\dfrac{3x}{x-2}+\dfrac{6}{2-x}$

b. $\dfrac{7y}{2y-3}-\dfrac{4}{3-2y}$

7.4 Exercises

1. Explain how to find the LCD of two rational expressions.

2. Suppose the LCD of two fractions is $48x^3yz$. If one of the fraction's denominators is $8x^2y$, what factor do you multiply by when writing it as an equivalent fraction with the LCD? Explain how you determine this factor.

3. In general, if you are given rational expressions that contain multiterm polynomials in the denominator(s), what is your first step toward finding the LCD?

4. Instead of using the LCM as a common denominator, a student rewrites $\dfrac{3}{2x-y} + \dfrac{5}{y-2x}$ as $\dfrac{-3}{y-2x} + \dfrac{5}{y-2x}$. Explain what the student did.

5. Suppose you have found the LCD of two rational expressions and have written the equivalent rational expressions. If the problem is now at the stage $\dfrac{(7x+21)-(2x-10)}{(x-5)(x+3)}$, how do you proceed?

6. Suppose you have found the LCD of two rational expressions and have written the equivalent rational expressions. If the problem is now at the stage $\dfrac{(7x+21)-(2x-10)}{(x-5)(x+3)}$, what might the original problem have been? Explain.

For Exercises 7–26, find the least common denominator for the rational expressions and write equivalent rational expressions with the LCD.

7. $\dfrac{2}{3a}, \dfrac{3}{5a^2}$

8. $\dfrac{5}{4b^2}, \dfrac{3}{5b}$

9. $\dfrac{8}{15m^3n^3}, \dfrac{7}{18m^2n}$

10. $\dfrac{5}{10x^2y^3}, \dfrac{3}{14x^2y^2}$

11. $\dfrac{3}{y+2}, \dfrac{4y}{y-2}$

12. $\dfrac{3x}{x-5}, \dfrac{2}{x+1}$

13. $\dfrac{3y}{4y - 20}, \dfrac{7y}{6y - 30}$

14. $\dfrac{2x}{6x - 12}, \dfrac{3x}{8x - 16}$

15. $\dfrac{5}{6x + 18}, \dfrac{2}{x^2 + 3x}$

16. $\dfrac{3}{4w - 8}, \dfrac{5}{w^2 - 2w}$

17. $\dfrac{2}{p^2 - 4}, \dfrac{3 + p}{p + 2}$

18. $\dfrac{5}{y^2 - 36}, \dfrac{y - 2}{y - 6}$

19. $\dfrac{4u}{(u + 2)^2}, \dfrac{2}{3(u + 2)}$

20. $\dfrac{7m}{(x - 3)^2}, \dfrac{4}{7(x - 3)}$

21. $\dfrac{8t^2 - 1}{t^2 - 36}, \dfrac{t}{t^2 - 7t + 6}$

22. $\dfrac{4a + 2}{a^2 - 9}, \dfrac{2a - 1}{a^2 - a - 6}$

23. $\dfrac{x + 2}{x^2 + 3x - 4}, \dfrac{x - 3}{x^2 + 6x + 8}$

24. $\dfrac{b - 4}{b^2 - 2b - 35}, \dfrac{b + 3}{b^2 + 7b + 10}$

25. $\dfrac{2}{x - 1}, \dfrac{3}{x^2 - 3x + 2}, \dfrac{5x}{x - 2}$

26. $\dfrac{3}{y - 3}, \dfrac{2}{y^2 + 2y - 15}, \dfrac{4y}{y + 5}$

For Exercises 27–64, add or subtract as indicated.

27. $\dfrac{3a - 5}{9} - \dfrac{2a + 7}{6}$

28. $\dfrac{3x - y}{6} - \dfrac{3x - 2y}{4}$

29. $\dfrac{5}{2c} + \dfrac{1}{6c}$

30. $\dfrac{5}{8r} + \dfrac{3}{4r}$

31. $\dfrac{5}{x} - \dfrac{2}{x^3}$

32. $\dfrac{5a}{b^2} - \dfrac{3a}{b}$

33. $\dfrac{3}{y + 2} + \dfrac{5}{y - 4}$

34. $\dfrac{2}{c + 4} + \dfrac{3}{c + 3}$

35. $\dfrac{3}{x - 2} - \dfrac{2}{x + 2}$

36. $\dfrac{4}{x + 3} - \dfrac{3}{x - 3}$

37. $\dfrac{4}{x^2 - 25} + \dfrac{2}{x - 5}$

38. $\dfrac{5}{r^2 - 9} + \dfrac{4}{r + 3}$

39. $\dfrac{2p}{p^2 - 2p + 1} + \dfrac{8}{p - 1}$

40. $\dfrac{u}{u - 1} + \dfrac{2u}{u^2 - 2u + 1}$

41. $\dfrac{z + 5}{z^2 + z - 12} - \dfrac{z + 2}{z - 3}$

42. $\dfrac{a + 6}{a^2 + 8a + 15} - \dfrac{a - 3}{a + 3}$

43. $\dfrac{n + 2}{n - 4} + \dfrac{n - 3}{n + 3}$

44. $\dfrac{y - 8}{y + 5} + \dfrac{y - 9}{y - 6}$

45. $\dfrac{t + 2}{t + 4} - \dfrac{t - 1}{t + 6}$

46. $\dfrac{3s + 1}{s - 2} - \dfrac{4s + 1}{s - 3}$

47. $\dfrac{3}{2c + 10} + \dfrac{15}{c^2 - 25}$

48. $\dfrac{k + 3}{k^2 + 6k + 9} - \dfrac{7}{2k + 6}$

49. $\dfrac{5y + 6}{y^2 + 3y} + \dfrac{y}{y + 3}$

50. $\dfrac{2b - 8}{b^2 + 4b} + \dfrac{b}{b + 4}$

51. $\dfrac{3r}{r^2 - 1} - \dfrac{2r + 1}{r^2 - 2r + 1}$

52. $\dfrac{x + 1}{x^2 - 4x + 4} + \dfrac{4}{x^2 + 3x - 10}$

53. $\dfrac{q + 5}{q^2 - 9} - \dfrac{1}{q^2 + 3q}$

54. $\dfrac{x - 2}{x^2 - 16} - \dfrac{1}{x^2 - 4x}$

55. $\dfrac{w - 7}{w^2 + 4w - 5} - \dfrac{w - 9}{w^2 + 3w - 10}$

56. $\dfrac{y - 1}{y^2 - 3y + 2} - \dfrac{y + 2}{y^2 + y - 2}$

57. $\dfrac{2v + 5}{v^2 - 16} - \dfrac{v - 9}{v^2 - v - 12}$

58. $\dfrac{4t}{t^2 - 9} - \dfrac{3t - 2}{t^2 - 8t + 15}$

59. $\dfrac{3}{t - 3} + \dfrac{6}{3 - t}$

60. $\dfrac{10}{2r - 1} - \dfrac{6}{1 - 2r}$

61. $\dfrac{4}{2x - 1} + \dfrac{x}{1 - 2x}$

62. $\dfrac{4}{t - 4} + \dfrac{t}{4 - t}$

63. $\dfrac{y}{3y - 6} - \dfrac{2}{2 - y}$

64. $\dfrac{x}{6x - 2} - \dfrac{3x}{1 - 3x}$

For Exercises 65–70, explain the mistake, then find the correct sum or difference.

65. $\dfrac{3}{x} - \dfrac{x}{2} = \dfrac{3 - x}{x - 2}$

66. $\dfrac{\cancel{5}}{2x} + \dfrac{3}{\cancel{10}x} = \dfrac{1}{2x} + \dfrac{3}{2x}$

$= \dfrac{4}{2x}$

$= \dfrac{2}{x}$

67. $\dfrac{3x}{x + 2} - \dfrac{4x + 7}{x + 2} = \dfrac{3x - (4x + 7)}{x + 2}$

$= \dfrac{3x + (-4x + 7)}{x + 2}$

$= \dfrac{-x + 7}{x + 2}$

68. $\dfrac{3}{x} + \dfrac{2}{x + y} = \dfrac{3(+y)}{x(+y)} + \dfrac{2}{x + y}$

$= \dfrac{3 + y + 2}{x + y}$

$= \dfrac{5 + y}{x + y}$

69. $\dfrac{2w}{x} + \dfrac{5w}{x} = \dfrac{7w}{2x}$

70. $\dfrac{2}{x} + \dfrac{3}{y} = \dfrac{6}{xy}$

For Exercises 71–76, solve.

71. Adam and Candace build and install cabinets in newly constructed homes. When working together, the expression $\dfrac{t}{2}$ represents the portion of the job completed by Adam, and $\dfrac{t}{3}$ represents the portion completed by Candace. Write a rational expression in simplest form that describes their combined work.

72. Two construction teams are combined in an effort to speed construction on a building. The expression $\dfrac{t}{5}$ describes the portion of the job that team 1 can complete and $\dfrac{t}{6}$ describes the portion of the job that team 2 can complete working independently. Write a rational expression in simplest form that describes their combined work.

73. An engineer uses the expression $\dfrac{1}{x + 3}$ to describes the width of a rectangular steel plate she is designing for a machine. The expression $\dfrac{1}{x}$ describes the length. Write a rational expression in simplest form for the perimeter of the steel plate.

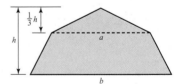

74. Write a rational expression in simplest form for the area of the figure shown.

75. Write a rational expression in simplest form for the area of the shaded region in the figure shown.

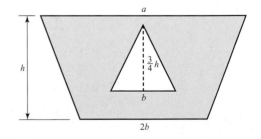

76. An electrical engineer is trying to determine the effects of using a thinner layer of a chemical on a circuit board. Currently, the thickness of the chemical layer is described by $\dfrac{1}{n - 4}$. If the thinner layer is described by $\dfrac{1}{n - 2}$, write a rational expression that describes the difference in the coating thickness?

For Exercises 77–86, use the order-of-operations agreement to simplify.

77. $\dfrac{x}{4x^2 - 1} + \dfrac{2}{4x - 2} - \dfrac{3x + 1}{1 - 2x}$

78. $\dfrac{c}{(c - d)^2} - \dfrac{2}{c + d} - \dfrac{c - 3}{c^2 - d^2}$

79. $\left(\dfrac{a + b}{a - b} + \dfrac{a - b}{a + b}\right) \div \dfrac{a + b}{a - b}$

80. $\left(\dfrac{x}{x - 4} - \dfrac{3}{x + 3}\right) \div \left(\dfrac{1}{x} + \dfrac{2}{x - 4}\right)$

81. $(x + y) \div \left(\dfrac{1}{x} + \dfrac{1}{y}\right)$

82. $\left(\dfrac{25 - 9j^2}{2j + 3}\right) \div (5j + 3j^2)$

83. $\dfrac{1}{m}\left(m + \dfrac{1}{m}\right) \div m^2$

84. $\dfrac{3}{x} + \dfrac{1}{x^2 + 3x} \cdot \dfrac{2x}{3}$

85. $\left(\dfrac{1}{x + 2} - \dfrac{3}{x^2 - 4}\right) \div \dfrac{3}{x - 2}$

86. $\dfrac{6}{x + 4} \div \left(\dfrac{1}{x - 3} - \dfrac{x + 1}{x^2 + x - 12}\right)$

REVIEW EXERCISES

For Exercises 1 and 2, divide.

[1.4]   **1.** $\dfrac{1}{4} \div \dfrac{2}{3}$

[7.2]   **2.** $\dfrac{m}{6} \div \dfrac{18}{m}$

For Exercises 3 and 4, factor completely.

[6.4]   **3.** $4x^2 - 16$

[6.3]   **4.** $6y^3 - 27y^2 - 15y$

[7.1] *For Exercises 5 and 6, simplify.*

5. $\dfrac{x^2 - 7x + 6}{x^2 - 4x - 12}$

6. $\dfrac{x + 3}{x^3 + 27}$

7.5 Complex Rational Expressions

OBJECTIVE 1. Simplify complex rational expressions. Sometimes a rational expression may contain rational expressions in its numerator or denominator. Such an expression is called a **complex rational expression** or a **complex fraction**.

DEFINITION *Complex rational expression:* A rational expression that contains rational expressions in the numerator or denominator.

Here are some complex rational expressions:

$$\dfrac{\dfrac{3}{4}}{\dfrac{5}{6}} \qquad \dfrac{x + \dfrac{x}{3}}{\dfrac{y}{2}} \qquad \dfrac{7}{\dfrac{t+2}{t}} \qquad \dfrac{\dfrac{2}{n} + \dfrac{1}{4}}{\dfrac{1}{n} - \dfrac{n}{3}}$$

There are two common methods for simplifying complex rational expressions. One method is to rewrite the complex rational expression as a horizontal division problem. For example, the complex fraction $\dfrac{\dfrac{3}{4}}{\dfrac{5}{6}}$ can be expressed as the division problem $\dfrac{3}{4} \div \dfrac{5}{6}$, then simplified.

$$\dfrac{\dfrac{3}{4}}{\dfrac{5}{6}} = \dfrac{3}{4} \div \dfrac{5}{6} = \dfrac{3}{\overset{}{\underset{2}{4}}} \cdot \dfrac{\overset{3}{\cancel{6}}}{5} = \dfrac{9}{10}$$

Another method is to multiply the numerator and denominator of the complex fraction by the LCD of the numerator and denominator fractions. In this case, the denominators are 4 and 6, so the LCD is 12.

Connection Notice that the essence of this second method is writing an equivalent fraction by multiplying both the numerator and denominator by the same factor.

$$\dfrac{\dfrac{3}{4} \cdot 12}{\dfrac{5}{6} \cdot 12} = \dfrac{\dfrac{3}{\cancel{4}} \cdot \dfrac{\overset{3}{\cancel{12}}}{1}}{\dfrac{5}{\cancel{6}} \cdot \dfrac{\overset{2}{\cancel{12}}}{1}} = \dfrac{\dfrac{9}{1}}{\dfrac{10}{1}} = \dfrac{9}{10}$$

PROCEDURE Simplifying Complex Rational Expressions

To simplify a complex rational expression, use one of the following methods:

Method 1
1. Simplify the numerator and denominator if needed.
2. Rewrite as a horizontal division problem.

Method 2
1. Multiply the numerator and denominator of the complex rational expression by their LCD.
2. Simplify.

EXAMPLE 1 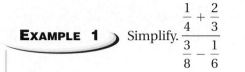 Simplify. $\dfrac{\dfrac{1}{4} + \dfrac{2}{3}}{\dfrac{3}{8} - \dfrac{1}{6}}$

Method 1 Simplify the numerator and denominator, then divide.

$$\dfrac{\dfrac{1}{4} + \dfrac{2}{3}}{\dfrac{3}{8} - \dfrac{1}{6}} = \dfrac{\dfrac{1(3)}{4(3)} + \dfrac{2(4)}{3(4)}}{\dfrac{3(3)}{8(3)} - \dfrac{1(4)}{6(4)}}$$

Write the numerator fractions as equivalent fractions with their LCD, 12, and write the denominator fractions with their LCD, 24.

$$= \dfrac{\dfrac{3}{12} + \dfrac{8}{12}}{\dfrac{9}{24} - \dfrac{4}{24}}$$

$$= \dfrac{\dfrac{11}{12}}{\dfrac{5}{24}}$$

Add in the numerator and subtract in the denominator.

$$= \dfrac{11}{12} \div \dfrac{5}{24}$$

Write the complex fraction as a division problem.

$$= \dfrac{11}{\cancel{12}_{1}} \cdot \dfrac{\cancel{24}^{2}}{5}$$

Write an equivalent multiplication problem, divide out the common factors, and then multiply.

$$= \dfrac{22}{5}$$

Method 2 Multiply the numerator and denominator of the rational expression by the LCD of all the rational expressions involved. In this case, that LCD is 24.

$$\dfrac{\dfrac{1}{4} + \dfrac{2}{3}}{\dfrac{3}{8} - \dfrac{1}{6}} = \dfrac{\left(\dfrac{1}{4} + \dfrac{2}{3}\right) \cdot 24}{\left(\dfrac{3}{8} - \dfrac{1}{6}\right) \cdot 24}$$

Multiply the numerator and denominator by 24.

$$= \dfrac{\dfrac{1}{\cancel{4}_{1}} \cdot \dfrac{\cancel{24}^{6}}{1} + \dfrac{2}{\cancel{3}_{1}} \cdot \dfrac{\cancel{24}^{8}}{1}}{\dfrac{3}{\cancel{8}_{1}} \cdot \dfrac{\cancel{24}^{3}}{1} - \dfrac{1}{\cancel{6}_{1}} \cdot \dfrac{\cancel{24}^{4}}{1}}$$

Distribute the 24 to each fraction, and divide out the common factors.

$$= \dfrac{6 + 16}{9 - 4}$$

Simplify the numerator and denominator.

$$= \dfrac{22}{5}$$

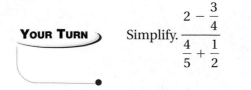

YOUR TURN Simplify. $\dfrac{2 - \dfrac{3}{4}}{\dfrac{4}{5} + \dfrac{1}{2}}$

EXAMPLE 2 Simplify. $\dfrac{5 + \dfrac{x}{3}}{\dfrac{4}{x} - \dfrac{1}{2x}}$

Method 1 Simplify the numerator and denominator, then divide.

$$\dfrac{5 + \dfrac{x}{3}}{\dfrac{4}{x} - \dfrac{1}{2x}} = \dfrac{\dfrac{5(3)}{1(3)} + \dfrac{x(1)}{3(1)}}{\dfrac{4(2)}{x(2)} - \dfrac{1(1)}{2x(1)}}$$

Write the numerator fractions as equivalent fractions with their LCD, 3, and write the denominator fractions with their LCD, $2x$.

$$= \dfrac{\dfrac{15}{3} + \dfrac{x}{3}}{\dfrac{8}{2x} - \dfrac{1}{2x}}$$

$$= \dfrac{\dfrac{15 + x}{3}}{\dfrac{7}{2x}}$$

Add in the numerator, and subtract in the denominator.

$$= \dfrac{15 + x}{3} \div \dfrac{7}{2x}$$

Write the complex fraction as a division problem.

$$= \dfrac{15 + x}{3} \cdot \dfrac{2x}{7}$$

Write an equivalent multiplication problem and then multiply.

$$= \dfrac{30x + 2x^2}{21}$$

Method 2 Multiply the numerator and denominator of the rational expression by the LCD of all the rational expressions involved. In this case, that LCD is $6x$.

$$\dfrac{5 + \dfrac{x}{3}}{\dfrac{4}{x} - \dfrac{1}{2x}} = \dfrac{\left(5 + \dfrac{x}{3}\right) \cdot 6x}{\left(\dfrac{4}{x} - \dfrac{1}{2x}\right) \cdot 6x}$$

Multiply the numerator and denominator by $6x$.

ANSWER

$\dfrac{25}{26}$

$$= \frac{\dfrac{5}{1} \cdot \dfrac{6x}{1} + \dfrac{x}{\cancel{3}_{1}} \cdot \dfrac{\overset{2}{\cancel{6}x}}{1}}{\dfrac{4}{\cancel{7}_{1}} \cdot \dfrac{\overset{1}{6\cancel{x}}}{1} - \dfrac{1}{\cancel{2}x_{1}} \cdot \dfrac{\overset{3}{\cancel{6}x}}{1}}$$

Distribute the $6x$ to each fraction, and divide out the common factors.

$$= \frac{30x + 2x^2}{24 - 3}$$

Simplify the numerator and denominator.

$$= \frac{30x + 2x^2}{21}$$

YOUR TURN   Simplify. $\dfrac{2 - \dfrac{1}{y}}{y - \dfrac{1}{3}}$

EXAMPLE 3   Simplify. $\dfrac{\dfrac{y}{4} - \dfrac{1}{y}}{\dfrac{1}{2} - \dfrac{1}{y}}$

Method 1 Simplify the numerator and denominator, then divide.

$$\frac{\dfrac{y}{4} - \dfrac{1}{y}}{\dfrac{1}{2} - \dfrac{1}{y}} = \frac{\dfrac{y(y)}{4(y)} - \dfrac{1(4)}{y(4)}}{\dfrac{1(y)}{2(y)} - \dfrac{1(2)}{y(2)}}$$

Write the numerator fractions as equivalent fractions with their LCD, $4y$, and write the denominator fractions with their LCD, $2y$.

$$= \frac{\dfrac{y^2}{4y} - \dfrac{4}{4y}}{\dfrac{y}{2y} - \dfrac{2}{2y}}$$

$$= \frac{\dfrac{y^2 - 4}{4y}}{\dfrac{y - 2}{2y}}$$

Subtract in the numerator and in the denominator.

$$= \frac{y^2 - 4}{4y} \div \frac{y - 2}{2y}$$

Write the complex fraction as a division problem.

$$= \frac{y^2 - 4}{4y} \cdot \frac{2y}{y - 2}$$

Write an equivalent multiplication problem.

$$= \frac{(y + 2)(\cancel{y - 2}^{1})}{\cancel{4}\cancel{y}_{2}} \cdot \frac{\overset{1}{\cancel{2}\cancel{y}}}{\cancel{y - 2}_{1}}$$

Factor, divide out the common factors, and then multiply the remaining factors.

$$= \frac{y + 2}{2}$$

ANSWER

$\dfrac{6y - 3}{3y^2 - y}$

Method 2 Multiply the numerator and denominator of the rational expression by the LCD of all the rational expressions involved. In this case, that LCD is $4y$.

$$\frac{\dfrac{y}{4} - \dfrac{1}{y}}{\dfrac{1}{2} - \dfrac{1}{y}} = \frac{\left(\dfrac{y}{4} - \dfrac{1}{y}\right) \cdot 4y}{\left(\dfrac{1}{2} - \dfrac{1}{y}\right) \cdot 4y}$$

Multiply the numerator and denominator by $4y$.

$$= \frac{\dfrac{y}{\cancel{4}} \cdot \dfrac{\cancel{4}y}{1} - \dfrac{1}{\cancel{y}} \cdot \dfrac{4\cancel{y}}{1}}{\dfrac{1}{\cancel{2}} \cdot \dfrac{\cancel{4}y}{1} - \dfrac{1}{\cancel{y}} \cdot \dfrac{4\cancel{y}}{1}}$$

Distribute the $4y$ to each fraction and then divide out the common factors.

$$= \frac{y^2 - 4}{2y - 4}$$

Simplify the numerator and denominator.

$$= \frac{(y + 2)(y - 2)}{2(y - 2)}$$

Factor and then divide out the common factors.

$$= \frac{y + 2}{2}$$

YOUR TURN   Simplify. $\dfrac{\dfrac{3}{x} - \dfrac{1}{x^2}}{\dfrac{1}{4} + \dfrac{1}{x}}$

7.5 Exercises

For Extra Help   MyMathLab   Videotape/DVT   InterAct Math   Math Tutor Center   Math XL.com

1. Identify the numerator and denominator of the complex rational expression
$\dfrac{\dfrac{4}{x + 3}}{\dfrac{3}{x}}.$

2. Given the complex rational expression $\dfrac{2x}{\dfrac{x - 3}{4}}$, is the numerator $2x$ or $\dfrac{2x}{x - 3}$?

   How could the expression be written so that the numerator and denominator are easily identified?

3. Write $\dfrac{2x}{x - 3} \div \dfrac{4 - x}{2x - 6}$ as a complex rational expression.

4. Write $\dfrac{5}{x-1} \div 7x$ as a complex rational expression.

5. Which of the two methods discussed in this section would you use to simplify
$$\dfrac{\dfrac{x}{6} - \dfrac{3}{4}}{\dfrac{2}{3} + x}? \text{ Why?}$$

6. Which of the two methods discussed in this section would you use to simplify
$$\dfrac{\dfrac{x}{7}}{\dfrac{x}{x+1}}? \text{ Why?}$$

For Exercises 7–48, simplify.

7. $\dfrac{\dfrac{3}{4}}{\dfrac{3}{2}}$

8. $\dfrac{\dfrac{2}{3}}{\dfrac{3}{2}}$

9. $\dfrac{\dfrac{m}{n}}{\dfrac{q}{p}}$

10. $\dfrac{\dfrac{a}{b}}{\dfrac{c}{d}}$

11. $\dfrac{\dfrac{1}{2} + \dfrac{1}{3}}{\dfrac{1}{2} - \dfrac{1}{3}}$

12. $\dfrac{\dfrac{2}{3} + \dfrac{1}{4}}{1 + \dfrac{1}{2}}$

13. $\dfrac{5 + \dfrac{2}{a}}{3 + \dfrac{4}{a}}$

14. $\dfrac{4 + \dfrac{1}{x}}{3 + \dfrac{5}{x}}$

15. $\dfrac{\dfrac{1}{y} - 1}{\dfrac{1}{y} + 1}$

16. $\dfrac{x - \dfrac{1}{x}}{1 + \dfrac{1}{x}}$

17. $\dfrac{r - \dfrac{3}{4}}{\dfrac{1}{2} - r}$

18. $\dfrac{\dfrac{5}{6} - x}{x - \dfrac{2}{3}}$

19. $\dfrac{\dfrac{3}{x} - 1}{\dfrac{9}{x^2} - 1}$

20. $\dfrac{1 - \dfrac{2}{p}}{1 - \dfrac{4}{p^2}}$

21. $\dfrac{x + y}{\dfrac{1}{x} + \dfrac{1}{y}}$

22. $\dfrac{\dfrac{2}{a} + \dfrac{3}{b}}{a + b}$

23. $\dfrac{\dfrac{x+2}{x^2-4}}{x}$

24. $\dfrac{\dfrac{a}{b}-1}{a^2-b^2}$

25. $\dfrac{\dfrac{6}{3x-2}}{\dfrac{x}{2x+1}}$

26. $\dfrac{\dfrac{5}{2x-1}}{\dfrac{x}{x+1}}$

27. $\dfrac{\dfrac{a}{a^2-b^2}}{\dfrac{b}{a+b}}$

28. $\dfrac{\dfrac{x}{4x^2-1}}{\dfrac{5}{2x-1}}$

29. $\dfrac{\dfrac{y^2-25}{5y}}{\dfrac{y^2+5y}{2y^3}}$

30. $\dfrac{\dfrac{x^2+4x}{y^2}}{\dfrac{x^2-16}{3y}}$

31. $\dfrac{2+\dfrac{2}{u-1}}{\dfrac{2}{u-1}}$

32. $\dfrac{\dfrac{1}{w+1}-1}{\dfrac{1}{w+1}}$

33. $\dfrac{\dfrac{a}{a-1}-1}{1+\dfrac{a}{a-1}}$

34. $\dfrac{1-\dfrac{x}{x+2}}{\dfrac{x}{x+2}-1}$

35. $\dfrac{\dfrac{x-3}{x}-1}{\dfrac{x^2-9}{x}-x}$

36. $\dfrac{q-\dfrac{q^2-1}{q}}{1-\dfrac{q-1}{q}}$

37. $\dfrac{\dfrac{1}{x-1}-\dfrac{1}{x+1}}{\dfrac{1}{x^2-1}}$

38. $\dfrac{\dfrac{16}{x^2-16}}{\dfrac{2}{x+4}-\dfrac{2}{x-4}}$

39. $\dfrac{1+\dfrac{1}{y}-\dfrac{2}{y^2}}{\dfrac{1}{y}+\dfrac{2}{y^2}}$

40. $\dfrac{1-\dfrac{3}{x}}{1-\dfrac{2}{x}-\dfrac{3}{x^2}}$

41. $\dfrac{1-\dfrac{1}{n}}{n-2+\dfrac{1}{n}}$

42. $\dfrac{y-1-\dfrac{6}{y}}{1+\dfrac{2}{y}}$

43. $\dfrac{1-\dfrac{1}{2y+1}}{\dfrac{1}{2y^2-5y-3}-\dfrac{1}{y-3}}$

44. $\dfrac{\dfrac{1}{f+2}-\dfrac{1}{f-3}}{1+\dfrac{1}{f^2-f-6}}$

45. $\dfrac{\dfrac{x^2-2x-8}{x^2-16}}{\dfrac{x^2+2x}{x^2+3x-4}}$

46. $\dfrac{\dfrac{v^2+v-2}{v^2+4v}}{\dfrac{v^2-4}{v^2+2v-8}}$

47. $\dfrac{\dfrac{2}{x+h}-\dfrac{2}{x}}{h}$

48. $\dfrac{h}{\dfrac{3}{y+h}-\dfrac{3}{y}}$

For Exercises 49–56, explain the mistake, then find the correct answer.

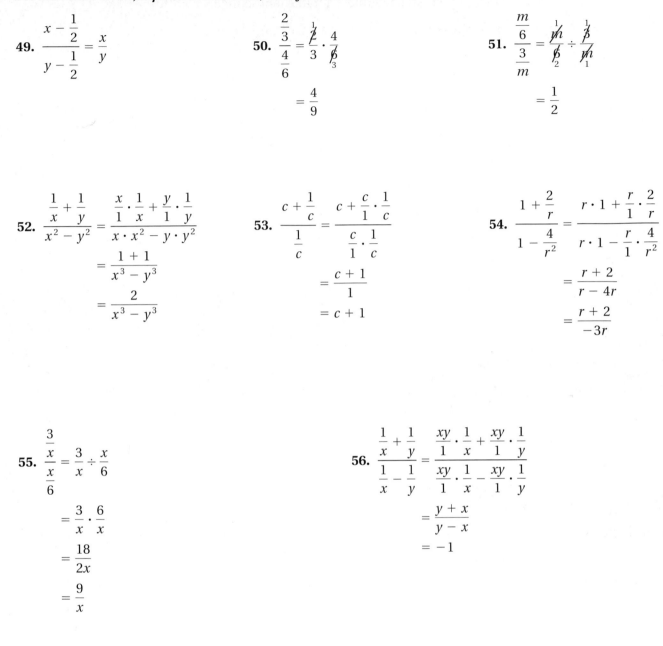

49. $\dfrac{x - \dfrac{1}{2}}{y - \dfrac{1}{2}} = \dfrac{x}{y}$

50. $\dfrac{\dfrac{2}{3}}{\dfrac{4}{6}} = \dfrac{\cancel{2}}{3} \cdot \dfrac{4}{\cancel{6}_{3}}$

$\qquad = \dfrac{4}{9}$

51. $\dfrac{\dfrac{m}{6}}{\dfrac{3}{m}} = \dfrac{\cancel{m}}{\cancel{6}_{2}} \div \dfrac{\cancel{3}}{\cancel{m}_{1}}$

$\qquad = \dfrac{1}{2}$

52. $\dfrac{\dfrac{1}{x} + \dfrac{1}{y}}{x^2 - y^2} = \dfrac{\dfrac{x}{1} \cdot \dfrac{1}{x} + \dfrac{y}{1} \cdot \dfrac{1}{y}}{x \cdot x^2 - y \cdot y^2}$

$\qquad = \dfrac{1 + 1}{x^3 - y^3}$

$\qquad = \dfrac{2}{x^3 - y^3}$

53. $\dfrac{c + \dfrac{1}{c}}{\dfrac{1}{c}} = \dfrac{c + \dfrac{c}{1} \cdot \dfrac{1}{c}}{\dfrac{c}{1} \cdot \dfrac{1}{c}}$

$\qquad = \dfrac{c + 1}{1}$

$\qquad = c + 1$

54. $\dfrac{1 + \dfrac{2}{r}}{1 - \dfrac{4}{r^2}} = \dfrac{r \cdot 1 + \dfrac{r}{1} \cdot \dfrac{2}{r}}{r \cdot 1 - \dfrac{r}{1} \cdot \dfrac{4}{r^2}}$

$\qquad = \dfrac{r + 2}{r - 4r}$

$\qquad = \dfrac{r + 2}{-3r}$

55. $\dfrac{\dfrac{3}{x}}{\dfrac{x}{6}} = \dfrac{3}{x} \div \dfrac{x}{6}$

$\qquad = \dfrac{3}{x} \cdot \dfrac{6}{x}$

$\qquad = \dfrac{18}{2x}$

$\qquad = \dfrac{9}{x}$

56. $\dfrac{\dfrac{1}{x} + \dfrac{1}{y}}{\dfrac{1}{x} - \dfrac{1}{y}} = \dfrac{\dfrac{xy}{1} \cdot \dfrac{1}{x} + \dfrac{xy}{1} \cdot \dfrac{1}{y}}{\dfrac{xy}{1} \cdot \dfrac{1}{x} - \dfrac{xy}{1} \cdot \dfrac{1}{y}}$

$\qquad = \dfrac{y + x}{y - x}$

$\qquad = -1$

For Exercises 57 and 58, average rate can be found using the following formula:

$$\text{Average rate} = \dfrac{\text{Total distance}}{\text{Total time}}$$

57. Suppose a person travels 10 miles in $\dfrac{1}{3}$ of an hour, then returns by traveling that same 10 miles in $\dfrac{1}{4}$ of an hour. What is that person's average rate?

58. Juan travels 50 miles in $\frac{3}{4}$ of an hour and then encounters a traffic jam so that the next 15 miles take $\frac{1}{3}$ of an hour. The last 40 miles of his trip take him $\frac{2}{3}$ of an hour. What was his average rate for the trip?

59. Given the area and length of a rectangle, the width can be found using the formula $w = \frac{A}{l}$. Suppose the area of a rectangle is described by $\frac{5n - 2}{8}$ square feet and the length is described by $\frac{n - 1}{6}$ feet. Find an expression for the width.

60. Given the area and base of a triangle, the height can be found using the formula $h = \frac{2A}{b}$. Suppose the area of a triangle is described by $\frac{n + 1}{6}$ and its base is described by $\frac{5n - 2}{9}$. Find an expression for the height.

61. In electrical circuits, if two resistors R_1 and R_2 are wired in parallel, the resistance of the circuit is found using the complex rational expression $\dfrac{1}{\dfrac{1}{R_1} + \dfrac{1}{R_2}}$.

   a. Simplify this complex rational expression.

   b. If a 100-ohm and 10-ohm resistor are wired in parallel, find the resistance of the circuit.

   c. Two 16-ohm speakers are wired in parallel; find the resistance of the circuit.

62. Suppose three resistors R_1, R_2, and R_3 are wired in parallel. The resistance of the circuit is found using the complex rational expression $\dfrac{1}{\dfrac{1}{R_1} + \dfrac{1}{R_2} + \dfrac{1}{R_3}}$.

   a. Simplify this complex rational expression.

   b. A 20-ohm, a 10-ohm, and a 40-ohm resistor are wired in parallel. Find the resistance of the circuit.

   c. A 300-ohm, a 500-ohm, and a 200-ohm resistor are wired in parallel. Find the resistance of the circuit.

REVIEW EXERCISES

[2.3] *For Exercises 1–4, solve.*

1. $-6x = 42$

2. $9t - 8 = 19$

3. $\dfrac{x}{4} = \dfrac{5}{8}$

4. $\dfrac{3}{4}y - \dfrac{1}{3} = \dfrac{1}{6}y + 2$

[2.1] **5.** A runner travels 1.2 miles in 15 minutes. Find the runner's average rate in miles per hour.

[3.4] **6.** An eastbound car and westbound car pass each other on a highway. The eastbound car is traveling 40 miles per hour and the westbound car is traveling 60 miles per hour. How much time elapses from the time they pass each other until they are 9 miles apart?

7.6 Solving Equations Containing Rational Expressions

OBJECTIVE 1. Solve equations containing rational expressions. In Section 2.3 we learned that we can eliminate fractions from an equation by multiplying both sides of the equation by their LCD. For example, if given the equation $\dfrac{1}{2}x - \dfrac{3}{4} = \dfrac{2}{3}$, we can eliminate the fractions by multiplying both sides by 12, which is the LCD of $\dfrac{1}{2}, \dfrac{3}{4}$, and $\dfrac{2}{3}$:

$$\overset{6}{\cancel{12}} \cdot \dfrac{1}{\underset{1}{\cancel{2}}}x - \overset{3}{\cancel{12}} \cdot \dfrac{3}{\underset{1}{\cancel{4}}} = \overset{4}{\cancel{12}} \cdot \dfrac{2}{\underset{1}{\cancel{3}}}$$

$$6x - 9 = 8$$

Since the rewritten equation contains only integers, it is easier to solve. Similarly, we can use the LCD to eliminate rational expressions in an equation.

EXAMPLE 1 Solve. $\dfrac{x}{5} = \dfrac{x}{2} - \dfrac{1}{4}$

Solution Eliminate the rational expressions by multiplying both sides by the LCD, 20.

$$20\left(\dfrac{x}{5}\right) = 20\left(\dfrac{x}{2} - \dfrac{1}{4}\right) \qquad \text{Multiply both sides by 20.}$$

$$\overset{4}{\cancel{20}} \cdot \dfrac{x}{\underset{1}{\cancel{5}}} = \overset{10}{\cancel{20}} \cdot \dfrac{x}{\underset{1}{\cancel{2}}} - \overset{5}{\cancel{20}} \cdot \dfrac{1}{\underset{1}{\cancel{4}}} \qquad \begin{array}{l}\text{Use the distributive property to multiply every term}\\\text{by 20 and then divide out the common factors.}\end{array}$$

$$4x = 10x - 5$$

$$-6x = -5 \qquad \text{Subtract } 10x \text{ from both sides.}$$

$$\dfrac{-6x}{-6} = \dfrac{-5}{-6} \qquad \text{Divide both sides by } -6.$$

$$x = \dfrac{5}{6} \qquad \text{Simply both sides.}$$

Check   Using the original equation, we replace x with $\dfrac{5}{6}$ and see if the equation is true. It is helpful to write $\dfrac{x}{5}$ as $\dfrac{1}{5}x$ and $\dfrac{x}{2}$ as $\dfrac{1}{2}x$.

$$\frac{1}{5}x = \frac{1}{2}x - \frac{1}{4}$$

$$\frac{1}{\overset{1}{\cancel{5}}}\left(\frac{\overset{1}{\cancel{5}}}{6}\right) \overset{?}{=} \frac{1}{2}\left(\frac{5}{6}\right) - \frac{1}{4} \qquad \text{Replace } x \text{ with } \frac{5}{6} \text{ and then simplify.}$$

$$\frac{1}{6} \overset{?}{=} \frac{5}{12} - \frac{3}{12}$$

$$\frac{1}{6} = \frac{2}{12} \qquad \text{True, therefore } \frac{5}{6} \text{ is a solution.}$$

Now consider equations that contain rational expressions with variables in the denominator.

EXAMPLE 2   Solve. $\dfrac{1}{2} = \dfrac{3}{x} - \dfrac{1}{4}$

Solution Eliminate the rational expressions by multiplying both sides by the LCD, $4x$.

$$4x\left(\frac{1}{2}\right) = 4x\left(\frac{3}{x} - \frac{1}{4}\right) \qquad \text{Multiply both sides by } 4x.$$

$$\overset{2}{\cancel{4}}x \cdot \frac{1}{\underset{1}{\cancel{2}}} = 4\overset{1}{\cancel{x}} \cdot \frac{3}{\underset{1}{\cancel{x}}} - \overset{1}{\cancel{4}}x \cdot \frac{1}{\underset{1}{\cancel{4}}} \qquad \begin{array}{l}\text{Use the distributive property to multiply every term by } 4x \\ \text{and then divide out the common factors.}\end{array}$$

$$2x = 12 - x$$

$$3x = 12 \qquad \text{Add } x \text{ to both sides.}$$

$$x = 4 \qquad \text{Divide both sides by 3.}$$

Check   We will leave the check to the reader.

YOUR TURN   Solve.

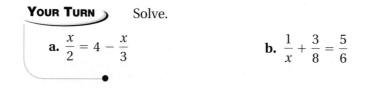

a. $\dfrac{x}{2} = 4 - \dfrac{x}{3}$

b. $\dfrac{1}{x} + \dfrac{3}{8} = \dfrac{5}{6}$

Extraneous Solutions

If an equation contains rational expressions with variables in the denominators, we must make sure that no solution causes any of the expressions in the equation to be undefined. In Example 2, if the solution had turned out to be 0, then the expression $\dfrac{3}{x}$ would be undefined. By inspecting the denominators of each rational expression, you can determine the values that would cause the expressions to be undefined before solving the equation. If by solving an equation, you obtain a number that causes an expression in the equation to be undefined, then we say that number is an ***extraneous solution*** and we discard it.

DEFINITION   ***Extraneous Solution:*** An apparent solution that does not solve its equation.

The following procedure summarizes the process of solving

PROCEDURE   **Solving Equations Containing Rational Expressions**

To solve an equation that contains rational expressions:
1. Eliminate the rational expressions by multiplying both sides of the equation by their LCD.
2. Solve the equation using the methods we learned in Chapters 2 (linear equations) and 6 (quadratic equations).
3. Check your solution(s) in the original equation. Discard any extraneous solutions.

Connection   The equation in Example 3 is a proportion. We could have solved the equation by cross multiplying, which leads to the same equation as in the second step of the solution in Example 3.

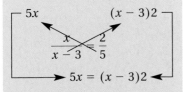

Warning: You can use this method only if the equation is a proportion.

EXAMPLE 3   Solve. $\dfrac{x}{x-3} = \dfrac{2}{5}$

Solution   Notice that if $x = 3$, then $\dfrac{x}{x-3}$ is undefined, so the solution cannot be 3. To solve, we first multiply both sides by the LCD, $5(x-3)$.

$$5(\overset{1}{\cancel{x-3}}) \cdot \dfrac{x}{\underset{1}{\cancel{x-3}}} = \overset{1}{\cancel{5}}(x-3) \cdot \dfrac{2}{\underset{1}{\cancel{5}}} \qquad \text{Divide out the common factors.}$$

$$5x = (x-3)2$$

$$5x = 2x - 6 \qquad \text{Distribute 2 to clear the parentheses.}$$

$$3x = -6 \qquad \text{Subtract } 2x \text{ on both sides.}$$

$$x = -2 \qquad \text{Divide both sides by 3.}$$

Check   We will leave the check to the reader.

YOUR TURN   Solve. $\dfrac{2}{x+5} = \dfrac{4}{x-1}$

In Examples 1 through 3, after eliminating the rational expressions, we were left with linear equations. Now let's look at some equations that transform to quadratic equations after we use the LCD to clear the rational expressions. Remember that when solving quadratic equations, we manipulate the equation so that one side is 0, then factor and use the zero-factor theorem.

ANSWER

-11

EXAMPLE 4 Solve. $\dfrac{3}{x} - 5 = x - 3$

Solution Notice that because of the x in the denominator, the solution cannot be 0.

$$x\left(\dfrac{3}{x} - 5\right) = x(x - 3)$$ Multiply both sides by the LCD, x.

$$\cancel{x}^{1} \cdot \dfrac{3}{\cancel{x}_{1}} - x \cdot 5 = x \cdot x - x \cdot 3$$ Distribute x, then divide out common factors.

$$3 - 5x = x^2 - 3x$$ Since this equation is quadratic, we will set it equal to 0 by subtracting 3 from and adding $5x$ to both sides.

$$0 = x^2 - 3x + 5x - 3$$

$$0 = x^2 + 2x - 3$$ Combine like terms.

$$0 = (x + 3)(x - 1)$$ Factor.

$$x + 3 = 0 \quad \text{or} \quad x - 1 = 0$$ Use the zero-factor theorem.

$$x = -3 \qquad\qquad x = 1$$

Check Verify that both -3 and 1 make the original equation true.

$$\dfrac{3}{-3} - 5 \overset{?}{=} -3 - 3 \qquad\qquad \dfrac{3}{1} - 5 \overset{?}{=} 1 - 3$$

$$-1 - 5 \overset{?}{=} -6 \qquad\qquad\qquad 3 - 5 \overset{?}{=} -2$$

$$-6 = -6 \qquad \text{True} \qquad\qquad -2 = -2 \qquad \text{True}$$

Both -3 and 1 are solutions.

YOUR TURN Solve. $\dfrac{1}{3} - \dfrac{2}{x} = x + 4$

EXAMPLE 5 Solve. $\dfrac{x^2}{x + 2} - 3 = \dfrac{x + 6}{x + 2} - 4$

Solution By inspecting the denominators, notice that the solution cannot be -2.

$$(x + 2)\left(\dfrac{x^2}{x + 2} - 3\right) = (x + 2)\left(\dfrac{x + 6}{x + 2} - 4\right)$$ Multiply both sides by the LCD, $x + 2$.

$$\cancel{(x + 2)}^{1} \cdot \dfrac{x^2}{\cancel{x + 2}_{1}} - (x + 2) \cdot 3 = \cancel{(x + 2)}^{1} \cdot \dfrac{x + 6}{\cancel{x + 2}_{1}} - (x + 2) \cdot 4$$ Distribute $x + 2$ and then divide out the common factors.

$$x^2 - 3x - 6 = x + 6 - 4x - 8$$ Combine like terms.

$$x^2 - 3x - 6 = -3x - 2$$

$$x^2 - 4 = 0$$ Since the equation is quadratic, we set it equal to 0 by adding $3x$ and 2 to both sides.

$$(x + 2)(x - 2) = 0$$ Factor.

$$x + 2 = 0 \quad \text{or} \quad x - 2 = 0$$ Use the zero-factor theorem.

$$x = -2 \qquad\qquad x = 2$$

Check We have already noted that -2 cannot be a solution because it causes an expression in the equation to be undefined, so -2 is extraneous. Therefore, we only need to check 2.

$$\frac{2^2}{2+2} - 3 \overset{?}{=} \frac{2+6}{2+2} - 4 \qquad \text{In the original equation, replace } x \text{ with 2}$$

and then simplify.

$$\frac{4}{4} - 3 \overset{?}{=} \frac{8}{4} - 4$$

$$1 - 3 \overset{?}{=} 2 - 4$$

$$-2 = -2 \qquad \text{True, so 2 is a solution and the only solution.}$$

YOUR TURN Solve. $\dfrac{3x+4}{x-4} = \dfrac{x^2}{x-4} + 3$

EXAMPLE 6 Solve. $\dfrac{5}{x-3} = \dfrac{x}{x-2} + \dfrac{x}{x^2-5x+6}$

Solution To determine the LCD of all the denominators, we first factor quadratic form $x^2 - 5x + 6$.

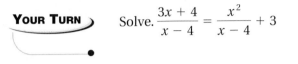

$$\frac{5}{x-3} = \frac{x}{x-2} + \frac{x}{(x-2)(x-3)} \qquad \begin{array}{l}\textbf{Factor the denominator}\\ x^2 - 5x + 6.\end{array}$$

$$(x-2)(x-3)\left(\frac{5}{x-3}\right) = (x-2)(x-3)\left(\frac{x}{x-2} + \frac{x}{(x-2)(x-3)}\right) \qquad \begin{array}{l}\textbf{Multiply both}\\ \textbf{sides by the LCD,}\\ (x-2)(x-3).\end{array}$$

Note: *Inspecting the denominators after factoring, we see that neither 2 nor 3 can be a solution.*

$$(x-2)\overset{1}{(\cancel{x-3})} \cdot \frac{5}{\underset{1}{(\cancel{x-3})}} = \overset{1}{(\cancel{x-2})}(x-3) \cdot \frac{x}{\underset{1}{(\cancel{x-2})}}$$

$$+ \overset{1}{(\cancel{x-2})}\overset{1}{(\cancel{x-3})} \cdot \frac{x}{\underset{1}{(\cancel{x-2})}\,\underset{1}{(\cancel{x-3})}} \qquad \begin{array}{l}\textbf{Distribute } (x-2)(x-3), \textbf{then divide out}\\ \textbf{the common factors.}\end{array}$$

$$(x-2)5 = (x-3)x + x$$

$$5x - 10 = x^2 - 3x + x \qquad \textbf{Distribute to clear the parentheses.}$$

$$5x - 10 = x^2 - 2x \qquad \textbf{Combine like terms.}$$

$$0 = x^2 - 7x + 10 \qquad \begin{array}{l}\textbf{Since the equation is quadratic, we set it}\\ \textbf{equal to 0 by subtracting } 5x \textbf{ and adding 10}\\ \textbf{on both sides.}\end{array}$$

$$0 = (x-2)(x-5) \qquad \textbf{Factor.}$$

$$x - 2 = 0 \quad \text{or} \quad x - 5 = 0 \qquad \textbf{Use the zero-factor theorem.}$$

$$x = 2 \qquad\qquad x = 5$$

Check In the original equation, note that $\dfrac{x}{x-2}$ is undefined when $x = 2$, so 2 is an extraneous solution. Consequently, 5 is the only solution. We will leave the check for 5 to the reader.

ANSWER

-4, (4 is extraneous)

YOUR TURN

Solve. $\dfrac{2}{x + 5} - \dfrac{6}{x^3 + 5x^2} = \dfrac{1}{x}$

Warning: Be sure you understand the difference between performing operations with rational expressions and solving equations containing rational expressions. For example, $\dfrac{5}{x} + \dfrac{1}{3}$ is an *expression* that means two rational expressions are to be added.

Remember, we cannot solve expressions, we can only evaluate or rewrite them. To rewrite this expression, we begin by writing each rational expression with the LCD, $3x$.

$$\frac{5}{x} + \frac{1}{3} = \frac{5(3)}{x(3)} + \frac{1(x)}{3(x)}$$

$$= \frac{15}{3x} + \frac{x}{3x}$$

$$= \frac{15 + x}{3x}$$

To solve an *equation* like $\dfrac{3}{x} + \dfrac{1}{4} = \dfrac{1}{2}$, we first eliminate the rational expressions by multiplying both sides of the equation by the LCD, $4x$.

$$4x\left(\frac{3}{x} + \frac{1}{4}\right) = 4x\left(\frac{1}{2}\right).$$

$$12 + x = 2x$$

$$12 = x \qquad \text{Subtract } x \text{ from both sides.}$$

ANSWER

6 and -1

7.6 Exercises

For Extra Help

MyMathLab   Videotape/DVT   InterAct Math   Math Tutor Center   Math XL.com

1. How are rational expressions eliminated from an equation?

2. What is the LCD for the expressions in $\dfrac{x}{x^2 + 4x - 5} = \dfrac{5}{x} - 2x$?

3. What is an extraneous solution for an equation that contains rational expressions?

4. Why might an equation with rational expressions lead to one or more extraneous solutions?

5. What values might be extraneous solutions in the equation $\dfrac{3}{x-4} = \dfrac{5x}{x+2}$? Why?

6. What values might be extraneous solutions in the equation $\dfrac{x}{x^2+4x-5} = \dfrac{5}{x} - 2x$? Why?

For Exercises 7–12, check the given values to see if they are solutions to the equation.

7. $\dfrac{3}{x} + \dfrac{2}{3} = \dfrac{5}{6}; x = 9$

8. $\dfrac{3}{4} - \dfrac{1}{x} = \dfrac{2}{3}; x = 12$

9. $\dfrac{4}{m} - \dfrac{2}{5} = \dfrac{3}{4m}; m = \dfrac{65}{8}$

10. $\dfrac{5}{6n} + \dfrac{1}{4} = \dfrac{2}{3n}; n = -\dfrac{2}{3}$

11. $\dfrac{1}{t+2} + \dfrac{1}{4} = \dfrac{t}{16}; t = -4$

12. $\dfrac{1}{y-3} - \dfrac{5}{6} = -\dfrac{y}{12}; y = -7$

For Exercises 13–62, solve and check.

13. $\dfrac{2n}{3} + \dfrac{3n}{2} = \dfrac{13}{3}$

14. $\dfrac{3x}{5} - \dfrac{x}{2} = \dfrac{19}{5}$

15. $3y - \dfrac{4y}{5} = 22$

16. $\dfrac{3t}{4} - 2 = \dfrac{t}{4}$

17. $\dfrac{r-7}{2} - 1 = \dfrac{r+9}{9} + \dfrac{1}{3}$

18. $\dfrac{d+5}{3} - 4 = \dfrac{d-8}{4} - \dfrac{1}{2}$

19. $\dfrac{4}{x} + \dfrac{3}{2x} = \dfrac{11}{6}$

20. $\dfrac{3}{2u} - \dfrac{1}{3} = \dfrac{5}{6u}$

21. $\dfrac{4-6t}{2t} + \dfrac{3}{5} = -\dfrac{2}{5t}$

22. $\dfrac{2x+1}{2x} = \dfrac{7}{12} + \dfrac{3x+2}{3x}$

23. $\dfrac{5}{20} = \dfrac{x-7}{x+2}$

24. $\dfrac{y-2}{y+3} = \dfrac{3}{8}$

25. $\dfrac{6}{y-2} = \dfrac{5}{y-3}$

26. $\dfrac{4}{2d+9} = \dfrac{3}{4d-8}$

27. $3 + \dfrac{9-2x}{8x} = \dfrac{5}{2x}$

28. $\dfrac{m+5}{m-3} - 5 = \dfrac{4}{m-3}$

29. $\dfrac{6f-5}{6} = \dfrac{2f-1}{2} - \dfrac{f+2}{2f+5}$

30. $\dfrac{2x^2}{3x-4} + \dfrac{x}{3} = \dfrac{9x-2}{9}$

31. $\dfrac{2}{x^2+x-2} - \dfrac{1}{x+2} = \dfrac{3}{x-1}$

32. $\dfrac{5}{t^2-9} = \dfrac{3}{t+3} - \dfrac{2}{t-3}$

33. $\dfrac{1}{u-4} + \dfrac{2}{u^2-16} = \dfrac{3}{u+4}$

34. $\dfrac{5}{h+5} - \dfrac{2}{h^2+2h-15} = \dfrac{2}{h-3}$

35. $\dfrac{x}{2-3x} = \dfrac{1}{3x+2}$

36. $\dfrac{5p}{p+4} = \dfrac{p}{p-1}$

37. $\dfrac{5}{2x} + \dfrac{15}{2x+16} = \dfrac{6}{x+8}$

38. $\dfrac{4}{3x} - \dfrac{2}{x+4} = \dfrac{2}{3x+12}$

39. $\dfrac{2}{x^2-1} = \dfrac{-3}{7x+7}$

40. $\dfrac{3}{y^2-4} = \dfrac{-2}{5y+10}$

41. $\dfrac{3}{x+5} - \dfrac{x-1}{2x} = \dfrac{-3}{2x^2+10x}$

42. $\dfrac{-1}{3x^2+9x} = \dfrac{1}{x+3} - \dfrac{x-3}{3x}$

43. $\dfrac{3m+1}{m^2-9} - \dfrac{m+3}{m-3} = \dfrac{1-5m}{m+3}$

44. $\dfrac{2n+10}{n^2-1} - \dfrac{n-1}{n+1} + \dfrac{2n+1}{n-1} = 0$

45. $\dfrac{x+3}{x-1} + \dfrac{2}{x+3} = \dfrac{5}{3}$

46. $\dfrac{m+2}{m-2} - \dfrac{2}{m+2} = -\dfrac{7}{3}$

47. $1 - \dfrac{3}{x-2} = -\dfrac{12}{x^2-4}$

48. $\dfrac{1}{x-3} + \dfrac{1}{3} = \dfrac{6}{x^2-9}$

49. $\dfrac{2r}{r-2} - \dfrac{4r}{r-3} = -\dfrac{7r}{r^2-5r+6}$

50. $\dfrac{3a}{a^2-2a-15} - \dfrac{a}{a+3} = \dfrac{2a}{a-5}$

51. $\dfrac{4u^2+3u+4}{u^2+u-2} - \dfrac{3u}{u+2} = \dfrac{-2u-1}{u-1}$

52. $\dfrac{2z}{2z-3} - \dfrac{3z}{2z+3} = \dfrac{15-32z^2}{4z^2-9}$

53. $\dfrac{2k}{k+7} - 1 = \dfrac{1}{k^2+10k+21} + \dfrac{k}{k+3}$

54. $\dfrac{p-5}{p+5} - 2 + \dfrac{p+15}{p-5} = \dfrac{-25}{p^2-25}$

55. $\dfrac{6}{t^2+t-12} = \dfrac{4}{t^2-t-6} + \dfrac{1}{t^2+6t+8}$

56. $\dfrac{7}{v^2-6v+5} - \dfrac{2}{v^2-4v-5} = \dfrac{3}{v^2-1}$

57. $\dfrac{x + 1}{x^2 - 2x - 15} - \dfrac{6}{x^2 - 3x - 10} = \dfrac{2}{x^2 + 5x + 6}$

58. $\dfrac{x + 1}{x^2 + 2x - 24} - \dfrac{3}{x^2 - x - 12} = \dfrac{3}{x^2 + 10x + 24}$

59. $\dfrac{1}{d^2 - 5d + 6} + \dfrac{d - 2}{d^2 - d - 6} - \dfrac{d}{d^2 - 4} = 0$

60. $\dfrac{1}{y^2 - 5y + 6} + \dfrac{y}{y^2 - 2y - 3} = \dfrac{y + 2}{y^2 - y - 2}$

61. $\dfrac{1}{a - 3} - \dfrac{6}{a^2 - 1} = \dfrac{12}{a^3 - 3a^2 - a + 3}$

62. $\dfrac{1}{b + 2} - \dfrac{2}{b^2 - 1} = \dfrac{-2}{b^3 + 2b^2 - b - 2}$

For Exercises 63 and 64, explain the mistake, then find the correct solution(s).

63.
$$\frac{y}{3} + \frac{y}{y - 3} = \frac{3}{y - 3}$$
$$3(y - 3) \cdot \frac{y}{3} + 3(y - 3) \cdot \frac{y}{y - 3} = 3(y - 3) \cdot \frac{3}{y - 3}$$
$$y^2 - 3y + 3y = 9$$
$$y^2 - 9 = 0$$
$$(y + 3)(y - 3) = 0$$
$$y = -3, 3$$

64.
$$\frac{x}{x + 3} + \frac{3}{x + 3} = 0$$
$$(x + 3) \cdot \frac{x}{x + 3} + (x + 3) \cdot \frac{3}{x + 3} = (x + 3) \cdot 0$$
$$x + 3 = 0$$
$$x = -3$$

65. Explain how the LCD is used differently in $\dfrac{3x}{4} - \dfrac{5}{6}$ and in $\dfrac{3x}{4} - \dfrac{5}{6} = \dfrac{x}{2}$.

66. Explain how the LCD is used differently in $\dfrac{5x}{8} + \dfrac{1}{3}$ and in $\dfrac{5x}{8} + \dfrac{1}{3} = \dfrac{x}{6}$.

For Exercises 67–70, use the formula for the total resistance R in an electrical circuit with resistors R_1, R_2, R_3, ..., R_n that are wired in parallel.

$$\frac{1}{R} = \frac{1}{R_1} + \frac{1}{R_2} + \frac{1}{R_3} + \cdots + \frac{1}{R_n}$$

67. Two resistors are wired in parallel, one of which is 100 ohms. If the total resistance is to be 80 ohms, what must be the value of the other resistor?

68. Three resistors are wired in parallel, two of which are equal in value. If the third resistor is 200 ohms and the total resistance is to be 20 ohms, what are the values of the other two resistors?

69. Two resistors are wired in parallel. One resistor is to have a value that is 10 ohms more than the other resistor and the total resistance is to be 12 ohms. Find the value of both resistors.

70. Two resistors are wired in parallel. One resistor is to have a value that is 20 ohms less than the other resistor and the total resistance is to be 24 ohms. Find the value of both resistors.

In optics, a lens can be used to bend light from an object through a focal point to reproduce an inverted image that is larger (or smaller) than the original object. The following formula describes how the image of an object is affected by a lens, where o represents the object's distance from the lens, i represents the image's distance from the lens, and f represents the focal length of the lens.

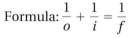

$$\text{Formula: } \frac{1}{o} + \frac{1}{i} = \frac{1}{f}$$

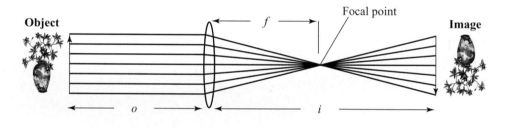

71. An object is 40 feet from a lens with a focal length of 2 feet. What will be the image's distance from the lens?

72. An object is placed 50 centimeters from your eye. Your lens focuses the image of the object onto your retina, which is about 2.5 centimeters from your lens. What is the focal length of your lens?

73. A lens is being designed so that the focal length will be 5 millimeters less than the the image length. Suppose an object is placed 60 millimeters from the lens. What are the focal length and image length?

74. A lens is being designed so that the image length is 9 millimeters more than the the focal length. Suppose an object is placed 40 millimeters from the lens. What are the focal length and image length?

REVIEW EXERCISES

[2.1]   **1.** Andrew notes that he traveled a distance of 120 miles in $1\frac{3}{4}$ hours. What was his average rate in miles per hour?

[3.1]   **2.** A grocery store has a 10.5-ounce can of soup on sale for $0.88. Write a unit ratio in simplest form of the price to capacity.

[3.1]   **3.** Daniel knows that he can paint 800 square feet of wall space in 4 hours. If he were able to maintain that rate, how long would it take him to paint 1200 square feet?

[3.3]   **4.** The sum of two positive consecutive even integers is 146. What are the numbers?

[3.3]   **5.** A rectangular eraser has a perimeter of 36 millimeters. If the length is three less than twice the width, find the length and width.

[3.4]   **6.** Two joggers pass each other going in opposite directions on a path. If one jogger is traveling at a rate of 6 miles per hour and the other is traveling at a rate of 4 miles per hour, how long will it take them to be 1.5 miles apart?

7.7   Applications with Rational Expressions

OBJECTIVES

1. Use tables to solve problems with two unknowns involving rational expressions.

2. Solve problems involving direct variation.

3. Solve problems involving inverse variation.

OBJECTIVE 1. Use tables to solve problems with two unknowns involving rational expressions.   In Chapter 3, we used tables to organize information involving two unknown amounts. Let's now look at some similar problems that lead to equations containing rational expressions.

Problems Involving Work

Tables are helpful in problems involving two or more people (or machines) working together to complete a task. In these problems, we would be given each person's rate of work and asked to find the time for them to complete the task if they work together. For each person involved, the rate of work, time at work, and amount of the task completed are related as follows:

$$\boxed{\text{Person's rate of work}} \cdot \boxed{\text{Time at work}} = \boxed{\text{Amount of the task completed by that person}}$$

Since the people are working together, the sum of their individual amounts of the task completed equals the whole task.

$$\boxed{\text{Amount completed by one person}} \cdot \boxed{\text{Amount completed by the other person}} = \boxed{\text{Whole task}}$$

EXAMPLE 1 Karen and Jeff own a cleaning business. Karen can clean an average-size house in about 2 hours. Jeff can clean a similar house in 3 hours. How long would it take them to clean an average-size house working together?

Understand Karen cleans at a rate of 1 house in 2 hours, or $\frac{1}{2}$ of a house per hour.

Jeff cleans at a rate of 1 house in 3 hours, or $\frac{1}{3}$ of a house per hour.

Category	Rate of Work	Time at Work	Amount of Task Completed
Karen	$\frac{1}{2}$	t	$\frac{1}{2}t$ or $\frac{t}{2}$
Jeff	$\frac{1}{3}$	t	$\frac{1}{3}t$ or $\frac{t}{3}$

Because they are working together on the same job for the same amount of time, we let t represent that amount of time.

Multiplying the rate of work and the time at work gives an expression of the amount of work completed. For example, if Karen works for 4 hours at a rate of $\frac{1}{2}$ of a house every hour, she can clean 2 houses.

The total job in this case is 1 house, so we can write an equation that combines their individual expressions for the task completed and set this sum equal to 1 house.

Plan and Execute   Karen's amount completed + Jeff's amount completed = 1 house

$$\frac{t}{2} + \frac{t}{3} = 1$$

$$6\left(\frac{t}{2} + \frac{t}{3}\right) = 6(1) \qquad \text{Multiply both sides by the LCD, 6.}$$

$$\overset{3}{\cancel{6}} \cdot \frac{t}{\underset{1}{\cancel{2}}} + \overset{2}{\cancel{6}} \cdot \frac{t}{\underset{1}{\cancel{3}}} = 6 \qquad \text{Distribute and then divide out common factors.}$$

$$3t + 2t = 6$$

$$5t = 6 \qquad \text{Combine like terms.}$$

$$t = \frac{6}{5} \qquad \text{Divide both sides by 5.}$$

Answer   Working together, it takes Karen and Jeff $\frac{6}{5}$ or $1\frac{1}{5}$ hours to clean an average-size house.

Check   Karen cleans $\frac{1}{2}$ of a house per hour, so if she works alone $1\frac{1}{5}$ hours, she cleans $\frac{1}{2} \cdot \frac{6}{5} = \frac{3}{5}$ of a house. Jeff cleans $\frac{1}{3}$ of a house per hour, so in $1\frac{1}{5}$ hours, he cleans $\frac{1}{3} \cdot \frac{6}{5} = \frac{2}{5}$ of a house. Combining their individual amounts, we see that in $1\frac{1}{5}$ hours, they clean $\frac{3}{5} + \frac{2}{5} = \frac{5}{5} = 1$ house.

YOUR TURN   Terrell owns a landscaping company. Every Monday he services the same group of clients, and it takes him 5 hours to complete the service. On one occasion he hired a student, Jarod, to service those same clients, and it took him 6 hours. How much time would it take them to service those clients working together?

Motion Problems

Recall that the formula for calculating distance, given the rate of travel and time of travel, is $d = rt$. If we isolate r, we have $r = \frac{d}{t}$. If we isolate t, we have $t = \frac{d}{r}$. These equations suggest that we will use rational expressions when describing rate or time.

ANSWER

$2\frac{8}{11}$ hr.

EXAMPLE 2 Warren runs 3 miles away from the gym, turns around, and runs an average of 2 miles per hour faster on the return trip. If the total time of his run is $1\frac{1}{4}$ hours, what is his speed in the outbound leg and the inbound leg of his run?

Understand We are to find the rates for each leg of Warren's run. Because this situation involves two different rates, we will use a table to organize the distance, rate, and time of each leg of the run.

Category	Distance	Rate	Time
Outbound	3 miles	r	$\dfrac{3}{r}$
Inbound	3 miles	$r + 2$	$\dfrac{3}{r+2}$

Warren runs 2 mph faster during the inbound leg of his run, so we let r represent the outbound rate and add 2 mph to r for the inbound rate.

Since $d = rt$, to describe time, we divide the distance by the rate: $t = \dfrac{d}{r}$.

Because the total time of the trip is $1\frac{1}{4}$ $\left(\text{or } \frac{5}{4}\right)$ hours, we can write an equation that is the sum of the outbound and inbound times.

Plan and Execute Outbound time + Inbound time $= \dfrac{5}{4}$ hours

$$\frac{3}{r} + \frac{3}{r+2} = \frac{5}{4}$$

$$4r(r+2)\left(\frac{3}{r} + \frac{3}{r+2}\right) = 4r(r+2)\left(\frac{5}{4}\right)$$
Multiply both sides by the LCD, $4r(r+2)$.

$$4r(r+2) \cdot \frac{3}{r} + 4r(r+2) \cdot \frac{3}{r+2} = 4r(r+2) \cdot \frac{5}{4}$$
Distribute, then divide out common factors.

$$12(r+2) + 12r = 5r(r+2)$$

$$12r + 24 + 12r = 5r^2 + 10r$$ **Distribute.**

$$24r + 24 = 5r^2 + 10r$$ **Combine like terms.**

$$0 = 5r^2 - 14r - 24$$ **Subtract 24r and 24 from both sides.**

$$0 = (5r + 6)(r - 4)$$ **Factor.**

$$5r + 6 = 0 \quad \text{or} \quad r - 4 = 0$$ **Use the zero-factor theorem.**

$$5r = -6 \qquad\qquad r = 4$$

$$r = -\frac{6}{5}$$

Answer   Though $-\dfrac{6}{5}$ is a solution to the equation, a negative rate does not make sense in this situation, so we do not consider it as an answer. Therefore, the rate of the outbound leg is 4 miles per hour. Since the rate of the inbound leg is 2 miles per hour faster, then it must be 6 miles per hour.

Check   Verify that traveling 3 miles at 4 miles per hour, then 3 miles at 6 miles per hour takes a total of $1\dfrac{1}{4}$ hours.

$$\text{Outbound time} = \frac{3}{4} \text{ hour} \qquad \text{Inbound time} = \frac{3}{6} = \frac{1}{2} \text{ hour}$$

$$\text{Total time} = \frac{3}{4} + \frac{1}{2} = \frac{3}{4} + \frac{2}{4} = \frac{5}{4} = 1\frac{1}{4} \text{ hour}$$

YOUR TURN   A plane travels 1200 miles against the jet stream, causing its airspeed to be decreased by 20 miles per hour. On the return flight, the plane travels with the jet stream so that its airspeed is increased by 20 miles per hour. If the total flight time of the round trip is $6\dfrac{1}{3}$ hours, what would be the plane's rate in still air?

OBJECTIVE 2. Solve problems involving direct variation.   Suppose a vehicle travels at a constant rate of 30 miles per hour. Using the formula $d = rt$, with r replaced by 30, we have $d = 30t$. In the table below, we use $d = 30t$ to determine distances for various values of time.

Time t	Distance Traveled $d = 30t$
1 hour	30 miles
2 hours	60 miles
3 hours	90 miles

From the table we see that as the time of travel increases, so does the distance traveled. Or, more formally, as values of t increase, so do values of d. In $d = 30t$, the two variables, d and t, are said to be in **direct variation** or are *directly proportional*.

DEFINITION   *Direct variation:* Two variables y and x are in direct variation if $y = kx$, where k is a constant.

ANSWER

380 mph

In words, direct variation is written as "y varies directly as x" or "y is directly proportional to x" and these phrases translate to $y = kx$. Often, the constant, k, is not given in problems involving variation, so the first objective is to find its value.

EXAMPLE 3 Suppose y varies directly as x. When $y = 9$, $x = 4$. Find y when $x = 7$.

Solution Translating "y varies directly as x," we have $y = kx$. We need to find the value of the constant k. We replace y with 9 and x with 4 in $y = kx$, then solve for k.

$$9 = k \cdot 4 \qquad \text{Replace } y \text{ with 9 and } x \text{ with 4.}$$
$$2.25 = k \qquad \text{Divide both sides by 4.}$$

Now, we can replace k with 2.25 in $y = kx$ so that we have $y = 2.25x$. We can now use this equation to find y when $x = 7$.

$$y = 2.25(7) = 15.75$$

EXAMPLE 4 The distance a vehicle can travel varies directly with the amount of fuel it carries. A family traveled 207 miles in their van using 9 gallons of fuel. How many gallons are required to travel 368 miles?

Understand Translating "the distance a vehicle travels varies directly with the amount of fuel," we write $d = kf$, where d represents distance and f represents the amount of fuel.

Plan Use $d = kf$, replacing d with 207 miles and f with 9 gallons, in order to solve for the value of k. Then use that value in $d = kf$ to solve for the number of gallons required to travel 368 miles.

Execute $207 = k \cdot 9$

$ 23 = k \qquad$ **Divide both sides by 9.**

> **Connection** Consider the units of measurement. Because miles $= k \cdot$ gallons, when we isolate k, we have $k = \dfrac{\text{miles}}{\text{gallon}}$. The constant k represents the miles per gallon of this van, which is 23 miles per gallon. The number of miles a vehicle can travel per gallon of fuel is also called its *fuel economy*.

Replacing k with 23 in $d = kf$, we have $d = 23f$. We use $d = 23f$ to solve for f when d is 368 miles.

$$368 = 23f$$
$$16 = f \qquad \text{Divide both sides by 23.}$$

Answer To travel 368 miles, the van needs 16 gallons of fuel.

Check If the van's fuel economy is 23 miles per gallon, then with 16 gallons, the van would travel $23(16) = 368$ miles.

At a meat market, the price per pound, or unit price, of ham-burger is constant. The amount you pay is directly proportional to the quantity purchased. Juan notes that a 2.5-pound package is priced at $3.58. If he wishes to buy 8 pounds of hamburger, how much will it cost?

OBJECTIVE 3. Solve problems involving inverse variation. Suppose we know that the distance to a particular place is 60 miles. Given a rate, we can calculate the travel time to that place using the formula $t = \dfrac{d}{r}$. For example, if we travel at a rate of 10 miles per hour, it will take $\dfrac{60 \text{ mi}}{10 \text{ mph}} = 6$ hours to drive the 60 miles. In the following table, we use $t = \dfrac{60}{r}$ to see the relationship between rate and travel time for a fixed distance of 60 miles.

Rate r	Travel time $t = \dfrac{60}{r}$
10 mph	6 hours
20 mph	3 hours
30 mph	2 hours

From the table above, we see that as we increase speed, the travel time decreases. More formally, as values of r increase, values of t decrease. In $t = \dfrac{60}{r}$, the two variables, t and r, are said to be in **inverse variation** or are *inversely proportional.*

DEFINITION *Inverse variation:* Two variables, y and x, are in inverse variation if $y = \dfrac{k}{x}$, where k is a constant.

In words, inverse variation is written as "y varies inversely as x" or "y is inversely proportional to x," and these phrases translate to $y = \dfrac{k}{x}$.

EXAMPLE 5 A technician applies a constant voltage to a circuit that has a variable resistor. Varying the resistance causes the current to vary inversely. At a resistance of 4 ohms, the current is measured to be 8 amperes. Find the current when the resistance is 12 ohms.

Understand Because the current and resistance vary inversely, we can write $I = \dfrac{k}{R}$, where I represents current, R represents the resistance, and k represents the constant voltage.

ANSWER

$11.46

Plan Use the fact that when the resistance is 4 ohms, the current is 8 amperes to determine the value of the constant k. Then use the constant to find the current when the resistance is 12 ohms.

Execute $8 = \dfrac{k}{4}$

$4 \cdot 8 = \overset{1}{\cancel{4}} \cdot \dfrac{k}{\underset{1}{\cancel{4}}}$ **Multiply both sides by 4 to eliminate the 4 in the denominator.**

$32 = k$ ◀ **Note:** *This means that the voltage in the circuit is* **32** *volts.*

Replacing k with 32 in $I = \dfrac{k}{R}$, we have $I = \dfrac{32}{R}$. We use $I = \dfrac{32}{R}$ to solve for I when R is 12 ohms.

$$I = \dfrac{32}{12}$$

$$I = 2\dfrac{2}{3} \text{ or } 2.\overline{6}$$

Answer With a resistance of 12 Ω, the current is $2\dfrac{2}{3}$ or $2.\overline{6}$ A.

Check Recall from the exercises in Section 2.3 (page 145) that the formula for voltage is $V = IR$. Using this formula, we can verify that a current of $2\dfrac{2}{3}$ A with a resistance of 12 Ω yields a voltage of 32 V.

$$V = \left(2\dfrac{2}{3} \right)12 = 32$$

YOUR TURN ⟩ If the wavelength of a wave remains constant, then the velocity, v of a wave is inversely proportional to its period, T. In an experiment, waves are created in a fluid so that the period is 5 seconds and the velocity is 6 centimeters per second. If the period is increased to 8 seconds, what is the velocity?

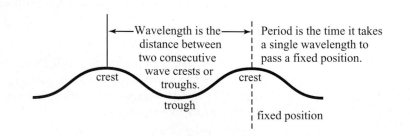

←——Wavelength is the——→ Period is the time it takes
distance between   a single wavelength to
two consecutive   pass a fixed position.
wave crests or
crest   troughs.   crest

trough

fixed position

ANSWER ⟩

3.75 cm/sec.

7.7 Exercises

For Extra Help

MyMathLab   Videotape/DVT   InterAct Math   Math Tutor Center   Math XL.com

1. If a person can complete a task in x hours, write an expression that describes the portion of the task she completes in 1 hour.

2. If x represents the amount of time one person takes to complete a task, and y represents the amount of time for a second person to complete the same task, write an expression that describes the portion of the task completed while working together for one unit of time.

3. If a vehicle travels a distance of 100 miles at a particular rate, r, write an expression that describes the amount of time it takes the vehicle to travel that 100 miles.

4. A cyclist travels 10 miles at an average rate, r, then travels 5 miles per hour faster during the return trip. Write an expression that describes the total time of the trip.

5. If two quantities vary directly, then as one of the quantities increases, what happens to the other?

6. If two quantities are inversely proportional, then as one quantity increases, what happens to the other?

For Exercises 7–22, use a table to organize the information, then solve.

7. Juan can paint a 3600-square-foot house in 12 days, and Carissa can do the same job in only 8 days. How long will it take them to paint a 3600-square-foot house if they work together?

8. After a large snowfall, Franklin can shovel the snow from the sidewalk in 6 minutes and Shawnna can do it 8 minutes. How long will it take them to clear the sidewalk if they work together?

9. The Smith children are responsible for mowing the lawn. If it takes Anna 20 minutes and Christina 30 minutes, assuming they have two lawn mowers, how long will it take them to mow the lawn working together?

10. A large field is being cleared to begin construction on a new library. To remove the dirt, two bulldozers are being used. The larger one could do the job in 6 days, while the smaller one would take 9 days. Working together, how long should the clearing take?

11. When filling a new swimming pool, the builder knows that it will take 6 hours using the smaller pipe. The larger pipe can do the job in only 4 hours. How long will it take the builder to fill the swimming pool if both pipes are used?

12. Karmin is planting vincas for her summer garden. Last summer, it took her 4 hours to complete the job. Her sister Kim planted the fall garden in only 3 hours. If the sisters work together, how long will the planting take?

13. To earn money for college, Peter and Susan address wedding invitations using calligraphy. Peter can complete a job of 150 invitations in 6 hours and Susan can complete the same number in 5 hours. Working together, how long will it take them to address 150 invitations?

14. Normally, it takes Professor Gunther 3 hours to grade a class of final exam essays. If his graduate assistant grades the essays, it takes her 5 hours. How long will it take them to grade the essays together?

15. During an afternoon practice, two track team members took the same amount of time to run 18 miles and 12 miles, respectively. If one of the track team members ran 2 miles per hour faster than the other, find the rate of each team member.

16. If an airplane travels 1050 miles in the same amount of time that an automobile travels 150 miles, and the speed of the airplane is 50 mph more than six times the speed of the car, how fast is each moving?

17. A jet flies 90 miles with a 20 mph tailwind in the same amount of time as it takes to fly against the wind 80 miles. What would the jet's speed be in still air?

18. The river current is 8 miles per hour. A jet ski can travel 34 miles downstream in the same time that it takes to go 18 miles upstream. What is the speed of the jet ski with no current?

19. A car can travel 75 miles in the same amount of time that it takes a moped to travel 25 miles. If the moped is traveling at 40 mph less than the car, what is the rate of each?

20. The Quad Cities offer tours on the Mississippi River. If the *Quad City Queen* riverboat travels upstream 120 miles and then returns, and the river current is measured at 4 miles per hour, what is the speed of the *Quad City Queen* if the total travel time is 22.5 hours?

21. After training for a year, a cyclist discovers that she has increased her average rate by 4 miles per hour and can travel 15 miles in the same time that it used to take her to travel 12 miles. What was her old average rate? What is her new average rate?

22. During target practice, it is determined that a heavily armed bomber flying from base to a target 600 miles away required the same flight time as it did to fly 720 miles at a rate 30 miles per hour faster. What was the rate for each trip?

23. If y varies directly as x and $y = 45$ when $x = 9$, what is the value of y when $x = 6$?

24. If a varies directly as b and $a = 15$ when $b = 27$, what is a when $b = 54$?

25. If x varies directly as y and $x = 4$ when $y = 1$, what is the value of y when $x = 18$?

26. If m varies directly as n and $m = 18$ when $n = 4$, what is the value of n when $m = 20$?

27. At the Piggly Wiggly, the cost of center-cut chuck roast is constant. The purchase price increases with the quantity purchased. Darlene notes a 1.5-pound package is priced at $6.44. If she wished to buy 6 pounds of center-cut chuck roast, how much will it cost?

28. At Kroger, the cost of corn is constant. The purchase price increases with the quantity purchased. Uviv notes that 5 ears of corn cost $0.99. If she wished to buy 8 ears of corn, how much will it cost?

29. The distance a car can travel varies directly with the amount of gas it carries. On the highway, a 2001 Jeep Wrangler travels 112 miles using 7 gallons of fuel. How many gallons are required to travel 240 miles?

30. The distance a car can travel varies directly with the amount of gas it carries. On the highway, a 2005 BMW M3 convertible travels 92 miles using 4 gallons of fuel. How many gallons are required to travel 207 miles?

31. The weight of an object is directly proportional to its mass. A person who weighs 161 pounds has a mass of 5 slugs. Find the weight of a person with a mass of 4.5 slugs.

32. The weight of an object is directly proportional to its mass. An object with a mass of 55 kilograms has a weight of 539 newtons. Find the weight of a person with a mass of 60 kilograms.

33. On a stringed instrument, such as a guitar, the frequency of the vibrating string is inversely proportional to the string's length. Suppose the length of a string on a guitar is 25.5 inches with a frequency of 440 vibrations per second (vps). Now, if we shorten the string to a length of 17 inches by placing a finger on the fret board, what will be the frequency of the string?

Saddle

17"

Nut

34. The frequency of the vibrating string on a stringed instrument is directly proportional to the tension of the string. Suppose the tension of a string is 300 pounds with a frequency of 440 vibrations per second (vps). Now, if we increase the string tension to 340 pounds by turning one of the tuning keys, what will be the frequency of the string?

35. The volume of a gas is directly proportional to temperature. Suppose a balloon containing air has a volume of 525 cubic inches at a temperature of 75°F. If the temperature drops to 60°F, what will be the volume of the balloon?

36. In physics, the energy, E, of a photon of light is directly proportional to the frequency of the light, represented by v (pronounced "nu"). A "hard" X-ray with a frequency of 2×10^{19} hertz has energy of 1.325×10^{-14} joules. What energy would a gamma ray with a frequency of 4×10^{22} hertz have?

37. If y varies inversely as x and $y = 4$ when $x = 10$, what is y when x is 8?

38. If m varies inversely as n and m is 6 when n is 1.5, what is m when n is 9?

39. If a varies inversely as b and b is 4 when a is 7, what is a when b is 2?

40. If x varies inversely as y and $y = 5$ when $x = 2$, what is y when x is 5?

41. The pressure that a gas exerts against the walls of a container is inversely proportional to the volume of the container. A gas is inside a cylinder with a piston at one end that can vary the volume of the cylinder. When the volume of the cylinder is 60 cubic inches, the pressure inside is 20 pounds per square inch (psi). Find the pressure of the gas if the piston compresses the gas to a volume of 40 cubic inches.

20 psi at   Compressed to
60 cubic inches   40 cubic inches

42. Find the pressure in the cylinder from Exercise 41 if the piston compresses the gas to a volume of 20 cubic inches.

43. A constant voltage is applied to a circuit that has a variable resistor. Varying the resistance causes the current to vary inversely. At a resistance of 6 ohms, the current is measured to be 10 amperes. Find the current when the resistance is 15 ohms.

44. Find the current of the circuit in Exercise 43 when the resistance is 20 ohms.

45. The weight of an object is inversely proportional to the square of the distance of the object from the center of Earth. At the surface of Earth, the space shuttle weighs 4.5 million pounds. It is then 4000 miles from the center of Earth. How much will the space shuttle weigh when it is 50 miles from the surface of Earth?

46. The intensity of light is inversely proportional to the square of the distance from the source. A light meter 6 feet from a lightbulb measures the intensity to be 16 foot-candles. What is the intensity at 20 feet from the bulb?

47. If the wavelength of a wave remains constant, then the velocity, v, of a wave is inversely proportional to its period, T. In an experiment, waves are created in a fluid so that the period is 7 seconds and the velocity is 4 centimeters per second. If the period is increased to 12 seconds, what is the velocity?

48. In quantum physics, the de Broglie wavelength of an object, represented by λ (pronounced "lambda"), is inversely proportional to the product of mass and velocity (momentum) of the object. A particle with a mass of 10^{-7} grams is moving at a velocity of 10^6 centimeters per second and has a de Broglie wavelength of 6.6×10^{-26} centimeters. Find the de Broglie wavelength of a particle with a mass of 2.5×10^{-6} grams moving at a velocity of 5×10^7 centimeters per second.

49. The following table lists data from a business of the number of units it supplied to the market and the corresponding price per unit:

Supply (in thousands)	Price per Unit
150	$1.25
175	$1.50
200	$2.00
250	$2.75

a. Plot the ordered pairs of data as points in the coordinate plane. Connect the points to form a graph.

b. Are supply and price directly proportional or inversely proportional? Explain.

c. Do the data represent a function? Explain.

50. The following table lists data for a business of the demand for its product and the price per unit:

Demand (in thousands)	Price per Unit
150	$2.25
175	$1.50
200	$1.00
250	$0.75

a. Plot the ordered pairs of data as points in the coordinate plane. Connect the points to form a graph.

b. Are demand and price directly proportional or inversely proportional? Explain.

c. Do the data represent a function? Explain.

REVIEW EXERCISES

[3.3]   **1.** Find two consecutive integers so that twice the first plus four times the second is equal to 34.

[4.2]   **2.** Determine if $(3, -1)$ is a solution to the equation $-2x + 3y = -1$.

[4.2]   **3.** Graph: $y = -3x - 5$

[4.3]   **4.** Find the x- and y-intercepts of $y - 4x = 6$.

[4.7]   *For Exercises 5 and 6, given $f(x) = -3x + 2$, find each of the following.*

5. $f\left(-\dfrac{3}{4}\right)$ **6.** $f(2 - a)$

Chapter 7 Summary

Defined Terms

Procedures, Rules, and Key Examples

Procedures/Rules	Key Example(s)

Section 7.1 Simplifying Rational Expressions
To determine the value(s) that make a rational expression undefined:
1. Write an equation that has the denominator set equal to zero.
2. Solve the equation.

Example 1: Find every value for the variable that makes the expression undefined.

a. $\dfrac{5x}{x-6}$

Solution: $x - 6 = 0$

$$x = 6$$

b. $\dfrac{n}{n^2 - 4n - 5}$

Solution: $n^2 - 4n - 5 = 0$

$$(n - 5)(n + 1) = 0$$

$$n - 5 = 0 \quad \text{or} \quad n + 1 = 0$$

$$n = 5 \qquad\qquad n = -1$$

Example 2: Simplify.

To simplify a rational expression to lowest terms:
1. Write the numerator and denominator in factored form.
2. Divide out all common factors in the numerator and denominator.
3. Multiply the remaining factors in the numerator and the remaining factors in the denominator.

a. $\dfrac{18x^3y}{24x^2yz} = \dfrac{2 \cdot 3 \cdot 3 \cdot x \cdot x \cdot x \cdot y}{2 \cdot 2 \cdot 2 \cdot 3 \cdot x \cdot x \cdot y \cdot z}$

$$= \dfrac{3 \cdot x}{2 \cdot 2 \cdot z}$$

$$= \dfrac{3x}{4z}$$

b. $\dfrac{5m^2 - 15m}{m^2 - 9} = \dfrac{5 \cdot m \cdot (m - 3)}{(m + 3)(m - 3)}$

$$= \dfrac{5m}{m + 3}$$

Procedures/Rules	Key Example(s)

Section 7.2 Multiplying and Dividing Rational Expressions

To multiply rational expressions:
1. Write each numerator and denominator in factored form.
2. Divide out any numerator factor with any matching denominator factor.
3. Multiply numerator by numerator and denominator by denominator.
4. Simplify as needed.

Example 1: Multiply.

a. $-\dfrac{8u^3}{3t} \cdot \dfrac{9t^2}{10u^3} =$

$$-\frac{2 \cdot 2 \cdot 2 \cdot u \cdot u \cdot u}{3 \cdot t} \cdot \frac{3 \cdot 3 \cdot t \cdot t}{2 \cdot 5 \cdot u \cdot u \cdot u}$$

$$= -\frac{2 \cdot 2}{1} \cdot \frac{3 \cdot t}{5}$$

$$= -\frac{12t}{5}$$

b. $\dfrac{x^2 + 4x}{x^2 - 3x - 10} \cdot \dfrac{3x + 6}{12x}$

$$= \frac{x \cdot (x + 4)}{(x + 2) \cdot (x - 5)} \cdot \frac{3 \cdot (x + 2)}{2 \cdot 2 \cdot 3 \cdot x}$$

$$= \frac{(x + 4)}{(x - 5)} \cdot \frac{1}{2 \cdot 2}$$

$$= \frac{x + 4}{4(x - 5)} \text{ or } \frac{x + 4}{4x - 20}$$

To divide rational expressions:
1. Write an equivalent multiplication statement with the reciprocal of the divisor.
2. Write each numerator and denominator in factored form. (Steps 1 and 2 are interchangeable.)
3. Divide out any numerator factor with any matching denominator factor.
4. Multiply numerator by numerator and denominator by denominator.
5. Simplify as needed.

Example 2: Divide.

a. $\dfrac{3n^3}{14m} \div \dfrac{12mn^2}{7m}$

$$= \frac{3n^3}{14m} \cdot \frac{7m}{12mn^2}$$

$$= \frac{3 \cdot n \cdot n \cdot n}{2 \cdot 7 \cdot m} \cdot \frac{7 \cdot m}{2 \cdot 2 \cdot 3 \cdot m \cdot n \cdot n}$$

$$= \frac{n}{2} \cdot \frac{1}{2 \cdot 2 \cdot m}$$

$$= \frac{n}{8m}$$

b. $\dfrac{r - 4}{3r^2 - 17r + 10} \div \dfrac{r^2 - 3r - 4}{3r^2 + r - 2}$

$$= \frac{r - 4}{3r^2 - 17r + 10} \cdot \frac{3r^2 + r - 2}{r^2 - 3r - 4}$$

$$= \frac{r - 4}{(3r - 2) \cdot (r - 5)} \cdot \frac{(3r - 2) \cdot (r + 1)}{(r - 4) \cdot (r + 1)}$$

$$= \frac{1}{r - 5}$$

continued

Procedures/Rules	Key Example(s)

Section 7.2 Multiplying and Dividing Rational Expressions (Continued)

To convert units using dimensional analysis, multiply the given measurement by conversion factors so that the undesired units divide out, leaving the desired units.

Example 3: Convert.
a. 20 feet to inches

$$20 \text{ ft.} = \frac{20 \text{ ft.}}{1} \cdot \frac{12 \text{ in.}}{1 \text{ ft.}} = 240 \text{ in.}$$

b. 6 miles to yards

$$6 \text{ mi.} = \frac{6 \text{ mi.}}{1} \cdot \frac{5280 \text{ ft.}}{1 \text{ mi.}} \cdot \frac{1 \text{ yd.}}{3 \text{ ft.}} = 10{,}560 \text{ yd.}$$

Section 7.3 Adding and Subtracting Rational Expressions with the Same Denominator

To add or subtract rational expressions that have the same denominator:
1. Add or subtract the numerators and keep the same denominator.
2. Simplify to lowest terms (remember to write the numerators and denominators in factored form in order to simplify).

Example 1: Add or subtract.

a.
$$\frac{7y}{12} + \frac{y}{12} = \frac{7y + y}{12}$$
$$= \frac{8y}{12}$$
$$= \frac{2y}{3}$$

b.
$$\frac{u}{u + 2} - \frac{3}{u + 2} = \frac{u - 3}{u + 2}$$

c.
$$\frac{7x^2 + 5x}{x + 1} - \frac{x + 3}{x + 1}$$
$$= \frac{(7x^2 + 5x) - (x + 3)}{x + 1}$$
$$= \frac{(7x^2 + 5x) + (-x - 3)}{x + 1}$$
$$= \frac{7x^2 + 4x - 3}{x + 1}$$
$$= \frac{(7x - 3)(x + 1)}{x + 1}$$
$$= 7x - 3$$

Section 7.4 Adding and Subtracting Rational Expressions with Different Denominators

To find the LCD of two or more rational expressions:
1. Factor each denominator.
2. For each unique factor, compare the number of times it appears in each factorization. Write a product that includes each unique factor the greatest number of times it appears in the denominator factorizations.

Example 1: Find the LCD.

a. $\frac{5}{6t}$ and $\frac{1}{8t^2}$
$$6t = 2 \cdot 3 \cdot t$$
$$8t^2 = 2^3 \cdot t^2$$
$$\text{LCD} = 24t^2$$

b. $\frac{2}{5t - 15}$ and $\frac{t}{t^2 - 9}$
$$5t - 15 = 5(t - 3)$$
$$t^2 - 9 = (t - 3)(t + 3)$$
$$\text{LCD} = 5(t - 3)(t + 3)$$

continued

Procedures/Rules	Key Example(s)

Section 7.4 Adding and Subtracting Rational Expressions with Different Denominators (Continued)

To add or subtract rational expressions with different denominators:
1. Find the LCD.
2. Write each rational expression as an equivalent expression with the LCD.
3. Add or subtract the numerators and keep the LCD.
4. Simplify.

Example 2: Add or subtract.

a. $\dfrac{5}{6t} + \dfrac{1}{8t^2} = \dfrac{5(4t)}{6t(4t)} + \dfrac{1(3)}{8t^2(3)}$

$\qquad = \dfrac{20t}{24t^2} + \dfrac{3}{24t^2}$

$\qquad = \dfrac{20t + 3}{24t^2}$

b. $\dfrac{2}{5t - 15} - \dfrac{t}{t^2 - 9}$

$= \dfrac{2}{5(t - 3)} - \dfrac{t}{(t - 3)(t + 3)}$

$= \dfrac{2(t + 3)}{5(t - 3)(t + 3)} - \dfrac{t(5)}{(t - 3)(t + 3)(5)}$

$= \dfrac{2t + 6}{5(t - 3)(t + 3)} - \dfrac{5t}{5(t - 3)(t + 3)}$

$= \dfrac{2t + 6 - 5t}{5(t - 3)(t + 3)}$

$= \dfrac{-3t + 6}{5(t - 3)(t + 3)}$

Section 7.5 Complex Rational Expressions

To simplify a complex rational expression, use one of the following methods:

Method 1
1. Simplify the numerator and denominator if needed.
2. Rewrite as a horizontal division problem.

Example 1: Simplify $\dfrac{\dfrac{5t}{t + 1}}{\dfrac{10}{t^2 - 1}}$.

$\dfrac{\dfrac{5t}{t + 1}}{\dfrac{10}{t^2 - 1}} = \dfrac{5t}{t + 1} \div \dfrac{10}{t^2 - 1}$

$\qquad = \dfrac{5t}{t + 1} \cdot \dfrac{t^2 - 1}{10}$

$\qquad = \dfrac{5 \cdot t}{t + 1} \cdot \dfrac{(t + 1)(t - 1)}{2 \cdot 5}$

$\qquad = \dfrac{t}{1} \cdot \dfrac{t - 1}{2}$

$\qquad = \dfrac{t^2 - t}{2}$

continued

Procedures/Rules	Key Example(s)

Section 7.5 Complex Rational Expressions (Continued)

Method 2

1. Multiply the numerator and denominator of the complex rational expression by their LCD.
2. Simplify.

Example 2: Simplify $\dfrac{\dfrac{5}{6} - \dfrac{t}{2}}{\dfrac{t}{4} + \dfrac{1}{3}}$.

$$\dfrac{\dfrac{5}{6} - \dfrac{t}{2}}{\dfrac{t}{4} + \dfrac{1}{3}} = \dfrac{12\left(\dfrac{5}{6} - \dfrac{t}{2}\right)}{12\left(\dfrac{t}{4} + \dfrac{1}{3}\right)}$$

$$= \dfrac{\dfrac{\overset{2}{\cancel{12}}}{1} \cdot \dfrac{5}{\cancel{6}} - \dfrac{\overset{6}{\cancel{12}}}{1} \cdot \dfrac{t}{\cancel{2}}}{\dfrac{\overset{3}{\cancel{12}}}{1} \cdot \dfrac{t}{\cancel{4}} + \dfrac{\overset{4}{\cancel{12}}}{1} \cdot \dfrac{1}{\cancel{3}}}$$

$$= \dfrac{10 - 6t}{3t + 4}$$

Section 7.6 Solving Equations Containing Rational Expressions

To solve an equation that contains rational expressions:

1. Eliminate the rational expressions by multiplying both sides of the equation by their LCD.
2. Solve the equation using the methods we learned in Chapters 2 (linear equations) and 6 (quadratic equations).
3. Check your solution(s) in the original equation. Discard any extraneous solutions.

Example 1: Solve:

$$\dfrac{x^2 - 6}{x - 3} + 2 = \dfrac{x^2 - 2x}{x - 3} - x.$$

Note that x cannot be 3 because it would cause the denominators to be 0, making those expressions undefined.

$$(x - 3)\left(\dfrac{x^2 - 6}{x - 3} + 2\right)$$

$$= (x - 3)\left(\dfrac{x^2 - 2x}{x - 3} - x\right)$$

$$\cancel{(x - 3)} \cdot \dfrac{x^2 - 6}{\cancel{x - 3}} + (x - 3) \cdot 2$$

$$= \cancel{(x - 3)} \cdot \dfrac{x^2 - 2x}{\cancel{x - 3}} - (x - 3) \cdot x$$

$$x^2 - 6 + 2x - 6 = x^2 - 2x - x^2 + 3x$$

$$x^2 + 2x - 12 = x$$

$$x^2 + x - 12 = 0$$

$$(x - 3)(x + 4) = 0$$

$$x - 3 = 0 \quad \text{or} \quad x + 4 = 0$$

$$x = 3 \qquad\qquad x = -4$$

continued

Procedures/Rules	Key Example(s)

Section 7.6 Solving Equations Containing Rational Expressions (Continued)

3 is extraneous. Check -4 by substituting into the original equation:

$$\frac{(-4)^2 - 6}{(-4) - 3} + 2 \stackrel{?}{=} \frac{(-4)^2 - 2(-4)}{(-4) - 3} - (-4)$$

$$\frac{16 - 6}{-7} + 2 \stackrel{?}{=} \frac{16 + 8}{-7} + 4$$

$$-\frac{10}{7} + 2 \stackrel{?}{=} -\frac{24}{7} + 4$$

$$-\frac{10}{7} + \frac{14}{7} \stackrel{?}{=} -\frac{24}{7} + \frac{28}{7}$$

$$\frac{4}{7} \stackrel{?}{=} \frac{4}{7}$$

True, therefore -4 is the solution.

Section 7.7 Applications with Rational Expressions

If y varies directly as x, then we translate to an equation $y = kx$, where k is a constant.

Example 1: The price of photocopies is directly proportional to the number of copies produced. If 20 copies cost $1.40, then how much will 150 copies cost?

Direct variation means $y = kx$. In this case, the price of photocopies is y and the number of copies is x. We can use the fact that 20 copies cost $1.40 in $y = kx$ to find k.

$$1.40 = k(20)$$

$$\frac{1.40}{20} = \frac{20k}{20}$$

$$0.07 = k$$

Now use $y = kx$ again with $k = 0.07$ to solve for y when $x = 150$ copies.

$$y = (0.07)(150)$$

$$y = \$10.50$$

continued

Procedures/Rules	Key Example(s)
Section 7.7 Applications with Rational Expressions (Continued) If y varies inversely as x, then we translate to an equation $y = \dfrac{k}{x}$, where k is a constant.	**Example 2:** Given that $y = 20$ when $x = 8$, if y varies inversely as x, then find x when $y = 9$. Inverse variation means $y = \dfrac{k}{x}$. We can use the fact that $y = 20$ and $x = 8$ to find k. $$20 = \frac{k}{8}$$ $$8(20) = 8\left(\frac{k}{8}\right)$$ $$160 = k$$ Now use $y = \dfrac{k}{x}$ again with $k = 160$ to solve for x when $y = 9$. $$9 = \frac{160}{x}$$ $$x(9) = x\left(\frac{160}{x}\right)$$ $$\frac{9x}{9} = \frac{160}{9}$$ $$x = 17\frac{7}{9}$$

Chapter 7 Review Exercises

For Exercises 1–5, answer true or false.

[7.1] **1.** The rational expression $\dfrac{2x-1}{x}$ is undefined when $x = \dfrac{1}{2}$.

[7.1] **2.** A fraction is in lowest terms when the GCF of the numerator and denominator is 1.

[7.1] **3.** The expression $\dfrac{2-x}{x-2}$ simplifies to -1.

[7.1] **4.** The expression $-x + 5$ can be rewritten as $-1(x-5)$.

[7.3] **5.** $\dfrac{2x}{3} + \dfrac{x}{3} = \dfrac{3x}{3} = x$

For Exercises 6–10, complete the rule.

[7.2] **6.** To divide rational expressions:
1. Write an equivalent _____ statement with the reciprocal of the divisor.
2. Write each numerator and denominator in factored form. (Steps 1 and 2 are interchangeable.)
3. _____ out any numerator factor with any matching denominator factor.
4. Multiply numerator by numerator and denominator by denominator.
5. Simplify as needed.

[7.3] **7.** To add or subtract rational expressions that have the same denominator:
1. Add or subtract the _____ and keep the same _____.
2. Simplify to lowest terms (remember to write the numerators and denominators in factored form in order to reduce).

[7.5] **8.** A _____ rational expression is a rational expression that contains rational expressions in the numerator or denominator.

[7.5] **9.** One method of simplifying a complex rational expression is the following:
1. Simplify the _____ and _____ of the complex rational expression.
2. Divide.

[7.6] **10.** To solve an equation that contains rational expressions:
1. Eliminate the rational expressions by multiplying through by their _____.
2. Solve the equation using the methods we learned in Chapters 2 (linear equations) and 6 (quadratic equations).
3. Check your solution(s) in the original equation.

[7.1] For Exercises 11–12, evaluate the rational expression.

11. $\dfrac{4x + 5}{2x}$ when   **a.** $x = 3$ **b.** $x = -1.3$ **c.** $x = -2$ **Expressions**

12. $\dfrac{5x}{x + 2}$ when   **a.** $x = 3$ **b.** $x = -3$ **c.** $x = 1.2$ Exercises 11–70

[7.1] For Exercises 13–16, find every value that can replace the variable in the expression and cause the expression to be undefined.

13. $\dfrac{9y}{5 - y}$ 14. $\dfrac{2x}{x^2 - 4}$ 15. $\dfrac{6}{x^2 + 5x - 6}$ 16. $\dfrac{2x + 3}{x^2 + 6x}$

[7.1] For Exercises 17–24, simplify to lowest terms.

17. $\dfrac{3(x + 5)}{9}$ 18. $\dfrac{2y + 12}{3y + 18}$ 19. $\dfrac{x^2 + 6x}{4x + 24}$ 20. $\dfrac{xy - y^2}{3x - 3y}$

21. $\dfrac{a^2 - b^2}{a^2 - 2ab + b^2}$ 22. $\dfrac{3x + 9}{x^4 - 81}$ 23. $\dfrac{x^2 - y^2}{y - x}$ 24. $\dfrac{1 - w}{w^2 + 2w - 3}$

[7.2] For Exercises 25–32, multiply.

25. $\dfrac{x}{y} \cdot \dfrac{7x}{4y}$ 26. $-\dfrac{2n^2}{8m^2 n} \cdot \dfrac{24mp}{10n^2 p^5}$ 27. $\dfrac{6}{2m + 4} \cdot \dfrac{3m + 6}{15}$ 28. $\dfrac{x^2 + 3x}{x^2 - 16} \cdot \dfrac{x^2 - 4x}{x^2 - 9}$

29. $\dfrac{n^2 - 6n + 9}{n^2 - 9} \cdot \dfrac{n^2 + 4n + 3}{n^2 - 3n}$ 30. $\dfrac{4r^2 + 4r + 1}{r + 2r^2} \cdot \dfrac{2r}{2r^2 - r - 1}$

31. $\dfrac{4m}{m^2 - 2m + 1} \cdot \dfrac{m^2 - 1}{16m^2}$ 32. $\dfrac{w^2 - 2w - 24}{w^2 - 16} \cdot \dfrac{(w + 5)(w + 4)}{w^2 - w - 30}$

[7.2] For Exercises 33–40, divide.

33. $\dfrac{y^2}{ab} \div \dfrac{x}{y}$ 34. $-\dfrac{7y^2}{10b^2} \div -\dfrac{21y^2}{25b^2}$ 35. $\dfrac{v + 1}{3} \div \dfrac{3v + 3}{18}$

36. $\dfrac{2a + 4}{5} \div \dfrac{4a + 8}{25a}$ 37. $\dfrac{x^2 - y^2}{c^2 - d^2} \div \dfrac{x - y}{c + d}$ 38. $\dfrac{b^3 - 6b^2 + 8b}{6b} \div \dfrac{2b - 4}{10b + 40}$

39. $\dfrac{3j + 2}{5j^2 - j} \div \dfrac{6j^2 + j - 2}{10j^2 + 3j - 1}$ 40. $\dfrac{u^2 - 2u - 8}{u^2 + 3u + 2} \div (u^2 - 3u - 4)$

41. $\dfrac{5x}{18} + \dfrac{x}{18}$

42. $\dfrac{2r}{r+3} + \dfrac{r}{r+3}$

43. $\dfrac{8m+5}{6m^2} - \dfrac{10m+5}{6m^2}$

44. $\dfrac{4x+5}{x+1} - \dfrac{x+4}{x+1}$

45. $\dfrac{2x}{x+y} - \dfrac{x-y}{x+y}$

46. $\dfrac{2r-5s}{r^2-s^2} - \dfrac{2r-6s}{r^2-s^2}$

47. $\dfrac{9x^2-24x+16}{2x+1} + \dfrac{2x^2+3x-9}{2x+1}$

48. $\dfrac{q^2+2q}{q-7} - \dfrac{12q-21}{q-7}$

49. $\dfrac{2}{4c}, \dfrac{3}{3c^3}$

50. $\dfrac{5}{xy^3}, \dfrac{3}{x^2y^2}$

51. $\dfrac{4}{m-1}, \dfrac{4y}{m+1}$

52. $\dfrac{3}{4h-8}, \dfrac{5}{h^2-2h}$

53. $\dfrac{x}{x^2-1}, \dfrac{2+x}{x+1}$

54. $\dfrac{2}{p^2-4}, \dfrac{p}{p^2+4p+4}$

55. $\dfrac{3a-4}{18} - \dfrac{2a+5}{12}$

56. $\dfrac{4}{x+3} - \dfrac{5}{x-3}$

57. $\dfrac{3}{y-2} + \dfrac{5}{y-4}$

58. $\dfrac{5}{x^2-4} + \dfrac{4}{x+2}$

59. $\dfrac{6t}{(t-3)^2} - \dfrac{3t}{2t-6}$

60. $\dfrac{a+6}{a^2+7a+12} - \dfrac{a-3}{a+3}$

61. $\dfrac{10}{2y-1} - \dfrac{5}{1-2y}$

62. $\dfrac{3}{3x-12} + \dfrac{15}{x^2-16}$

63. $\dfrac{\frac{a}{b}}{\frac{x}{y}}$

64. $\dfrac{\frac{1}{3} - \frac{1}{2}}{\frac{1}{2} - \frac{1}{3}}$

65. $\dfrac{\frac{5}{3x+5}}{\frac{x}{x-2}}$

66. $\dfrac{\frac{x+3}{x^2-9}}{x}$

67. $\dfrac{\dfrac{x}{4x^2-1}}{\dfrac{5}{2x+1}}$

68. $\dfrac{r-\dfrac{r^2-1}{r}}{1-\dfrac{r-1}{r}}$

69. $\dfrac{\dfrac{1}{x^2}+\dfrac{1}{y^2}}{\dfrac{7}{xy}}$

70. $\dfrac{\dfrac{1}{h+1}-1}{\dfrac{1}{h+1}}$

Equations and Inequalities

Exercises 71–84

[7.6] *For Exercises 71–78, solve and check.*

71. $\dfrac{3x}{2}-\dfrac{x}{5}=\dfrac{19}{5}$

72. $\dfrac{3t}{6}-2=\dfrac{t}{6}$

73. $\dfrac{2x+1}{4x}=\dfrac{5}{12}+\dfrac{3x-2}{3x}$

74. $\dfrac{y+5}{y-3}-5=\dfrac{4}{y-3}$

75. $\dfrac{x}{x-2}=\dfrac{6}{x-1}$

76. $\dfrac{6g-5}{6}=\dfrac{2g-1}{2}-\dfrac{g+2}{2g+5}$

77. $\dfrac{1}{p-4}+\dfrac{1}{4}=\dfrac{8}{p^2-16}$

78. $\dfrac{3}{r-2}-\dfrac{4}{r+3}=-\dfrac{6}{r^2+r-6}$

[7.7] *For Exercises 79–84, solve.*

79. After giving a math test, Yolonda can grade 20 tests in 60 minutes and Shawn can do it 68 minutes. How long will it take them to grade the tests if they work together?

80. During an afternoon practice, two football team members took the same amount of time to run sprints of 40 yards and 50 yards, respectively. If one of the football team members ran 2 yards per second faster than the other, find the rate of each team member.

81. If a varies directly as b and $a = 15$ when $b = 27$, what is a when $b = 54$?

82. If x varies inversely as y and $y = 5$ when $x = 2$, what is y when x is 5?

83. The distance a car can travel varies directly with the amount of gas it carries. On the highway, a car travels 80 miles using 4 gallons of fuel. How many gallons are required to travel 200 miles?

84. A constant voltage is applied to a circuit that has a variable resistor. Varying the resistance causes the current to vary inversely. At a resistance of 8 ohms, the current is measured to be 12 amperes. Find the current when the resistance is 20 ohms.

Chapter 7 Practice Test

For Exercises 1–2, find the value(s) that can replace the variable in the expression and cause the expression to be undefined.

1. $\dfrac{2x}{x - 7}$

2. $\dfrac{8 - m}{m^2 - 16}$

For Exercises 3–4, simplify to lowest terms.

3. $\dfrac{12 - 3x}{x^2 - 8x + 16}$

4. $\dfrac{x - y}{x^2 - y^2}$

For Exercises 5–6, write equivalent rational expressions with their LCD.

5. $\dfrac{6}{fg^3}, \dfrac{2}{f^2}$

6. $\dfrac{5x}{x^2 - 9}, \dfrac{2}{x^2 + 6x + 9}$

For Exercises 7–18, perform the indicated operation.

7. $\dfrac{4a^2b^3}{15x^3y} \cdot \dfrac{25x^5y}{16ab}$

8. $-\dfrac{9x^3y^4}{16ab^2} \div \dfrac{45x^5y^2}{14a^7b^9}$

9. $\dfrac{4ab - 8b}{x^2} \div \dfrac{2b - ab}{x^3}$

10. $\dfrac{12x^2 - 6x}{x^2 + 6x + 5} \cdot \dfrac{2x^2 + 10x}{4x^2 - 1}$

11. $\dfrac{2x}{2x + 3} + \dfrac{5x}{2x + 3}$

12. $\dfrac{x}{x^2 - 9} - \dfrac{3}{x^2 - 9}$

13. $\dfrac{6}{x^2 - 4} + \dfrac{x - 3}{x^2 - 2x}$

14. $\dfrac{4}{x - 1} + \dfrac{5}{x + 2}$

15. $\dfrac{5}{x - 4} - \dfrac{2}{x + 1}$

16. $\dfrac{x}{2x + 4} - \dfrac{2}{x^2 + 2x}$

17. $\dfrac{2 + \dfrac{1}{y}}{3 - \dfrac{1}{y}}$

18. $\dfrac{\dfrac{x - y}{2}}{\dfrac{x^2 - y^2}{4}}$

For Exercises 19–22, solve the equation.

19. $\dfrac{3x - 1}{4} - \dfrac{7}{6} = \dfrac{2}{3}$

20. $\dfrac{4}{5m - 1} = \dfrac{2}{2m - 1}$

21. $\dfrac{y}{y + 2} - \dfrac{y + 6}{y^2 - 4} + \dfrac{2}{y - 2} = 0$

22. $\dfrac{g}{g - 1} = \dfrac{8}{g + 2}$

For Exercises 23–25, solve.

23. If Erin can paint a room in 3 hours and her husband can paint the same room in 2, how long will it take them to paint the room together?

24. Jake runs 3 miles away from the gym, turns around, and runs an average of 2 miles per hour slower on the return trip. If the total time of his run is $1\dfrac{1}{4}$ hours, what was his speed in the outbound leg and the inbound leg of his run?

25. The weight of an object is directly proportional to its mass. Suppose an object with a mass of 40 kilograms weighs 392 newtons. How much would an object with a mass of 54 kilograms weigh?

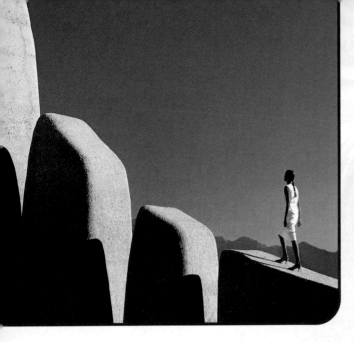

More on Inequalities, Absolute Value, and Functions

"Inequality is a fact. Equality is a value."

—Mason Cooley, U.S. aphorist

"Thinking is the function. Living is the functionary."

—Ralph Waldo Emerson,
U.S. essayist, poet, and philosopher (1803–1882)

In this chapter, we will build upon what we have learned about inequalities, absolute value, and functions. We have solved linear inequalities in which only a single inequality symbol is involved. Now, we will learn to solve compound inequalities, which involve two inequality symbols. We will then learn to solve absolute value equations and inequalities. We will see that compound inequalities are used in solving absolute value inequalities. Finally, after a quick review of functions and graphing, we will learn to perform function operations.

8.1 Compound Inequalities

OBJECTIVES 1. Solve compound inequalities involving "and."

2. Solve compound inequalities involving "or."

In Section 2.6, we learned to solve linear inequalities and represent their solution sets with set notation, interval notation, and graphically on number lines. For example, the solution set for $x > 3$ is represented in those three ways here.

Set-builder notation: $\{x \mid x > 3\}$

Interval notation: $(3, \infty)$

Graph:

$$\begin{array}{ccccccccccc} \leftarrow\!\!\!\!\!+ & + & + & + & (& + & + & + & + & + & \!\!\!\!\!\rightarrow \\ 0 & 1 & 2 & 3 & 4 & 5 & 6 & 7 & 8 & 9 \end{array}$$

Note: *Remember that the parenthesis indicates that 3 is not included in the solution set.*

In this section, we will build on this foundation and explore **compound inequalities.**

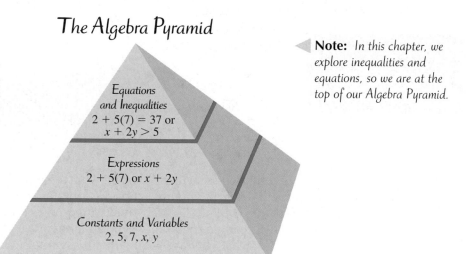

The Algebra Pyramid

Equations
and Inequalities
$2 + 5(7) = 37$ or
$x + 2y > 5$

Expressions
$2 + 5(7)$ or $x + 2y$

Constants and Variables
$2, 5, 7, x, y$

Note: *In this chapter, we explore inequalities and equations, so we are at the top of our Algebra Pyramid.*

DEFINITION   ***Compound inequality:*** *Two inequalities joined by either "and" or "or."*

Some examples of compound inequalities are

$$x > 3 \ \text{and} \ x \leq 8 \qquad\qquad -2 \geq x \ \text{or} \ x > 4$$

Interpreting Compound Inequalities with "and"

First, let's consider inequalities involving "and," such as $x > 3$ *and* $x \leq 8$. The word *and* indicates that the solution set contains only values that satisfy *both* inequalities. Therefore, the solution set is the **intersection**, or overlap, of the two inequalities' solution sets.

DEFINITION *Intersection:* For two sets A and B, the intersection of A and B, symbolized by $A \cap B$, is a set containing only elements that are in both A and B.

For example, if $A = \{1, 2, 3, 4, 5\}$ and $B = \{3, 4, 5, 6, 7\}$, then $A \cap B = \{3, 4, 5\}$ because 3, 4, and 5 are the only elements in both A and B. Let's look at our compound inequality $x > 3$ and $x \leq 8$. First we will graph the two inequalities separately, then consider their intersection.

Note: *We use the blue field to indicate the intersection and the dashed lines to indicate the intersection boundaries.*

$x > 3$

$x \leq 8$

Intersection of $x > 3$ and $x \leq 8$:

Note: *The intersection of the two inequalities contains elements that are only in **both** of their solution sets. Notice that 3 is excluded in the intersection because it was excluded in one of the individual graphs.*

So $x > 3$ and $x \leq 8$ means that x is any number greater than 3 *and* less than or equal to 8. Since this inequality indicates x is between two values, we can write it without the word *and*: $3 < x \leq 8$. In set-builder notation, we write $\{x \mid 3 < x \leq 8\}$. Using interval notation, we write $(3, 8]$.

EXAMPLE 1 For the compound inequality $x > -2$ and $x < 3$, graph the solution set and write the compound inequality without "and," if possible. Then write in set-builder notation and in interval notation.

Solution The solution set is the region of intersection.

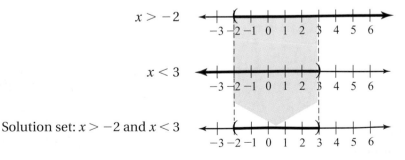

$x > -2$

$x < 3$

Solution set: $x > -2$ and $x < 3$

Without "and": $-2 < x < 3$

Set-builder notation: $\{x \mid -2 < x < 3\}$

Interval notation: $(-2, 3)$

Warning: Be careful not to confuse the interval notation $(-2, 3)$ with an ordered pair.

YOUR TURN

For each compound inequality, graph the solution set and write the compound inequality without "and," if possible. Then write in set-builder notation and in interval notation.

a. $x > 4$ and $x < 7$ **b.** $x \geq -8$ and $x \leq -2$

Solving Compound Inequalities with "and"

Now let's consider solution sets for more complex inequalities involving "and."

EXAMPLE 2 For the inequality $3x - 1 > 2$ and $2x + 4 \leq 14$, graph the solution set. Then write the solution set in set-builder notation and in interval notation.

Solution First, we solve each inequality in the compound inequality.

$3x - 1 > 2$	and	$2x + 4 \leq 14$	
$3x > 3$	and	$2x \leq 10$	Use the addition principle to isolate the x term.
$x > 1$	and	$x \leq 5$	Divide out each coefficient.

The solution set is the intersection of the two individual solution sets.

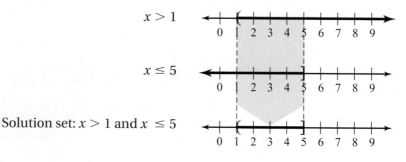

$x > 1$

$x \leq 5$

Solution set: $x > 1$ and $x \leq 5$

Set-builder notation: $\{x \mid 1 < x \leq 5\}$

Interval notation: $(1, 5]$

Example 2 suggests the following procedure.

PROCEDURE Solving Compound Inequalities Involving "and"

To solve a compound inequality involving "and,"

1. Solve each inequality in the compound inequality.
2. The solution set will be the intersection of the individual solution sets.

Finally, we consider some special situations with compound inequalities involving "and."

ANSWERS

a.
0 1 2 3 4 5 6 7 8 9

Without "and": $4 < x < 7$
Set-builder: $\{x \mid 4 < x < 7\}$
Interval: $(4, 7)$

b.
−9 −8 −7 −6 −5 −4 −3 −2 −1 0

Without "and": $-8 \leq x \leq -2$
Set-builder: $\{x \mid -8 \leq x \leq -2\}$
Interval: $[-8, -2]$

EXAMPLE 3 For the compound inequality, graph the solution set. Then write the solution set in set-builder notation and in interval notation.

a. $-2 \le -3x + 7 \le 1$

Solution Note that $-2 \le -3x + 7 \le 1$ means $-3x + 7 \ge -2$ and $-3x + 7 \le 1$. Solving the two inequalities gives us

$$-3x + 7 \ge -2 \quad \text{and} \quad -3x + 7 \le 1$$
$$-3x \ge -9 \quad \text{and} \quad -3x \le -6 \qquad \text{Subtract 7 from both sides of each inequality.}$$
$$x \le 3 \quad \text{and} \quad x \ge 2 \qquad \text{Divide both sides of each inequality by } -3.$$

Remember, we can write $x \le 3$ and $x \ge 2$ as $2 \le x \le 3$. Graphing this solution set we have

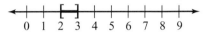

We could have also solved the compound inequality in its original form.

$$-2 \le -3x + 7 \le 1$$
$$-9 \le -3x \le -6 \qquad \text{Subtract 7 from all three parts of the compound inequality.}$$
$$3 \ge x \ge 2 \qquad \text{Divide all three parts of the compound inequality by } -3.$$

Note that $3 \ge x \ge 2$ is the same as $2 \le x \le 3$.

Set-builder notation: $\{x \mid 2 \le x \le 3\}$

Interval notation: $[2, 3]$

b. $2x + 11 \ge 5$ and $2x + 11 > 3$

Solution To graph the solution set, we first solve each inequality.

$$2x + 11 \ge 5 \quad \text{and} \quad 2x + 11 > 3$$
$$2x \ge -6 \quad \text{and} \quad 2x > -8 \qquad \text{Subtract 11 from both sides of each inequality.}$$
$$x \ge -3 \quad \text{and} \quad x > -4 \qquad \text{Divide both sides of each inequality by 2.}$$

$x \ge -3$

$x > -4$

Solution set:

Set-builder notation: $\{x \mid x \ge -3\}$

Interval notation: $[-3, \infty)$

c. $-4x - 3 > 1$ and $-4x - 3 < -11$

Solution First, we solve each inequality.

$$-4x - 3 > 1 \quad \text{and} \quad -4x - 3 < -11$$
$$-4x > 4 \quad \text{and} \quad -4x < -8 \qquad \text{Add 3 to both sides of each inequality.}$$
$$x < -1 \quad \text{and} \quad x > 2 \qquad \text{Divide both sides of each inequality by } -4.$$

$x < -1$

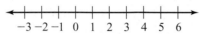

$x > 2$

Note: *There is no region of intersection for these graphs, so the solution set contains no values and is said to be empty.*

To graph an empty set, we draw a number line with no shading.

Set-builder notation: { } or $\varnothing$

Interval notation: We do not write interval notation because there are no values in the solution set.

YOUR TURN For each compound inequality, graph the solution set and write the compound inequality without "and," if possible. Then write in set-builder notation and in interval notation.

a. $-9 \le -5x + 11 < 21$ **b.** $-4x - 3 > 1$ and $-4x - 3 < -5$

OBJECTIVE 2. Solve compound inequalities involving "or."

Interpreting Compound Inequalities with "or"

In a compound inequality such as $x > 2$ or $x \le -1$, the word *or* indicates that the solution set contains values that satisfy *either* inequality. Therefore, the solution set is the **union**, or joining, of the two inequalities' solution sets.

DEFINITION   *Union:* For two sets A and B, the union of A and B, symbolized by $A \cup B$, is a set containing every element in A or in B.

For example, if $A = \{1, 2, 3, 4, 5\}$ and $B = \{3, 4, 5, 6, 7\}$, then $A \cup B = \{1, 2, 3, 4, 5, 6, 7\}$. Consider our compound inequality $x > 2$ or $x \leq -1$. Let's graph the two inequalities separately, then consider their union.

$x > 2$

$x \leq -1$

Note: *This graph is the union of the two individual solution sets because it contains every element in **either** of the individual solution sets.*

Solution set:
$x > 2$ or $x \leq -1$

Using set-builder notation, we write the solution set as $\{x \mid x > 2 \text{ or } x \leq -1\}$. Interval notation is a little more challenging for compound inequalities like $x > 2$ or $x \leq -1$ because the solution set is a combination of two intervals. As we scan from left to right, the first interval is $(-\infty, -1]$. The second interval is $(2, \infty)$. To indicate the union of these two intervals, we write $(-\infty, -1] \cup (2, \infty)$.

Solving Compound Inequalities with "or"

Now let's solve more complex inequalities involving "or." The process is similar to the process we used for solving compound inequalities involving "and."

PROCEDURE   Solving Compound Inequalities Involving "or"

To solve a compound inequality involving "or,"

1. Solve each inequality in the compound inequality.
2. The solution set will be the union of the individual solution sets.

EXAMPLE 4   For each compound inequality, graph the solution set. Then write the solution set in set-builder notation and in interval notation.

a. $\dfrac{2}{3}x - 5 \leq -7$ or $\dfrac{2}{3}x - 5 \geq -1$

Solution First, we solve each inequality in the compound inequality.

$$\frac{2}{3}x - 5 \leq -7 \quad \text{or} \quad \frac{2}{3}x - 5 \geq -1$$

$$\frac{2}{3}x \leq -2 \quad \text{or} \quad \frac{2}{3}x \geq 4 \qquad \text{Add 5 to both sides of each inequality.}$$

$$x \leq -3 \quad \text{or} \quad x \geq 6 \qquad \text{Multiply both sides by } \frac{3}{2} \text{ in each inequality.}$$

The solution set is the union of the two individual solution sets.

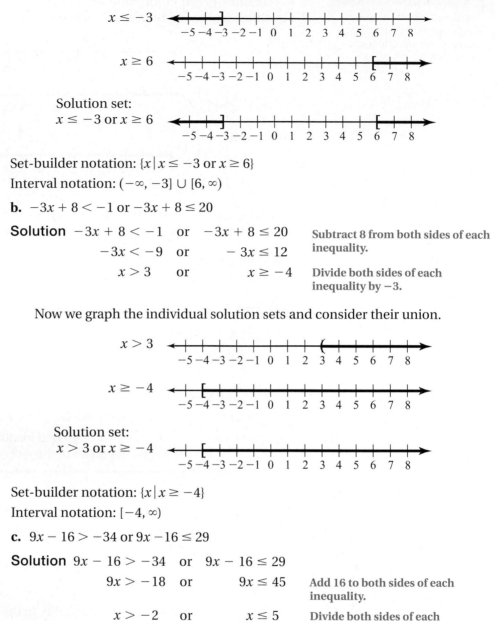

$x \le -3$

$x \ge 6$

Solution set:
$x \le -3$ or $x \ge 6$

Set-builder notation: $\{x \,|\, x \le -3 \text{ or } x \ge 6\}$

Interval notation: $(-\infty, -3] \cup [6, \infty)$

b. $-3x + 8 < -1$ or $-3x + 8 \le 20$

Solution

$-3x + 8 < -1$	or	$-3x + 8 \le 20$	Subtract 8 from both sides of each inequality.
$-3x < -9$	or	$-3x \le 12$	
$x > 3$	or	$x \ge -4$	Divide both sides of each inequality by -3.

Now we graph the individual solution sets and consider their union.

$x > 3$

$x \ge -4$

Solution set:
$x > 3$ or $x \ge -4$

Note: *When we join the two graphs, the graph of $x \ge -4$ covers all of the graph of $x > 3$, so the union of the two graphs is actually $x \ge -4$.*

Set-builder notation: $\{x \,|\, x \ge -4\}$

Interval notation: $[-4, \infty)$

c. $9x - 16 > -34$ or $9x - 16 \le 29$

Solution

$9x - 16 > -34$	or	$9x - 16 \le 29$	Add 16 to both sides of each inequality.
$9x > -18$	or	$9x \le 45$	
$x > -2$	or	$x \le 5$	Divide both sides of each inequality by 9.

Now we graph the individual solution sets and consider their union.

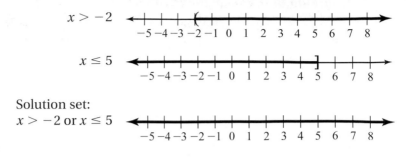

$x > -2$

$x \le 5$

Solution set:
$x > -2$ or $x \le 5$

Note: *When we join the two graphs, the entire number line is covered, so the union of the two graphs is the set of real numbers.*

Set-builder notation: $\{x \mid x \text{ is a real number}\}$, or $\mathbb{R}$.

Interval notation: $(-\infty, \infty)$

ANSWERS

a.

Set-builder: $\{x \mid x \le -4 \text{ or } x \ge -1\}$

Interval: $(-\infty, -4) \cup [-1, \infty)$

b.

Set-builder: $\{x \mid x < 4\}$

Interval: $(-\infty, 4)$

c.

Set-builder: $\{x \mid x \text{ is a real number}\}$, or $\mathbb{R}$

Interval: $(-\infty, \infty)$

YOUR TURN For each compound inequality, graph the solution set. Then write the solution set in set-builder notation and in interval notation.

a. $-5x - 8 > 12$ or $-5x - 8 \le -3$ **b.** $\dfrac{3}{4}x + 5 < -1$ or $\dfrac{3}{4}x + 5 < 8$

c. $5 - 3x \ge 11$ or $5 - 3x < 17$

8.1 Exercises

For Extra Help MyMathLab Videotape/DVT InterAct Math Math Tutor Center Math XL.com

1. What is a compound inequality?

2. Compound inequalities involve what two key words?

3. Describe the intersection of two sets.

4. Describe the union of two sets.

5. Describe how to graph a compound inequality involving "and."

6. Describe how to graph a compound inequality involving "or."

For Exercises 7–14: **a. Find the intersection of the given sets.**
b. Find the union of the given sets.

7. $A = \{1, 3, 5\}$
$B = \{1, 3, 5, 7, 9\}$

8. $A = \{2, 4, 6\}$
$B = \{2, 4, 6, 8, 10\}$

9. $A = \{5, 6, 7\}$
$B = \{7, 8, 9\}$

10. $A = \{8, 10, 12, 14\}$
$B = \{14, 16, 18\}$

11. $A = \{c, a, t\}$
$B = \{d, o, g\}$

12. $A = \{l, o, v, e\}$
$B = \{m, a, t, h\}$

13. $A = \{x, y, z\}$
$B = \{w, x, y, z\}$

14. $A = \{m, n, o, p, q\}$
$B = \{a, b, c, \ldots x, y, z\}$

For Exercises 15–22, write each inequality without "and."

15. $x > -4$ and $x < 5$

16. $n \geq 3$ and $n < 7$

17. $y > -2$ and $y \leq 0$

18. $m \geq 0$ and $m < 3$

19. $w > -7$ and $w < 3$

20. $r > -1$ and $r \leq 1$

21. $u \geq 0$ and $u \leq 2$

22. $t > 7$ and $t < 15$

For Exercises 23–30, graph the compound inequality.

23. $x > 2$ and $x < 7$

24. $x > 3$ and $x < 9$

25. $x > -1$ and $x \leq 5$

26. $x \geq -4$ and $x < 0$

27. $1 \leq x \leq 10$

28. $-2 < x < -1$

29. $-3 \leq x < 4$

30. $0 < x \leq 8$

For Exercises 31–50: **a. Graph the solution set.**
b. Write the solution set in set-builder notation.
c. Write the solution set in interval notation.

31. $x > -3$ and $x < -1$

32. $x > 4$ and $x < 7$

33. $x + 2 > 5$ and $x - 4 \leq 2$

34. $x - 3 \geq 1$ and $x + 2 < 10$

35. $-x > 1$ and $-2x \leq -10$

36. $-4x \geq 8$ and $3x \leq 15$

37. $2x + 6 \geq -4$ and $3x - 1 < 8$

38. $3x + 5 \geq -1$ and $5x - 2 < 13$

39. $-3x - 8 > 1$ and $-4x + 5 \leq -3$

40. $-6x + 4 < -14$ and $-x - 3 \geq -2$

41. $-3 < x + 4 < 1$

42. $-2 \leq x - 2 \leq 2$

43. $-2 \leq 5x + 3 \leq 13$

44. $7 < 2x - 1 < 11$

45. $0 \leq 2 + 3x < 8$

46. $0 \leq -2 + 5x \leq 13$

47. $-6 < -3x + 3 \leq 3$

48. $-3 < 5 - 2x < 1$

49. $3 \leq 6 - x \leq 6$

50. $-4 \leq -4 - x \leq 2$

For Exercises 51–58, graph each inequality.

51. $x < -2$ or $x > 6$

52. $n \leq 3$ or $n > 4$

53. $y < -3$ or $y \geq 0$

54. $m \leq 2$ or $m > 8$

55. $w > -3$ or $w > 2$

56. $r > -3$ or $r \geq -1$

57. $u \geq 0$ or $u \leq 2$

58. $t > 6$ or $t < 13$

For Exercises 59–76: *a. Graph the solution set.*
b. Write the solution set in set-builder notation.
c. Write the solution set in interval notation.

59. $y + 2 < -7$ or $y + 2 > 7$

60. $a - 4 \leq -2$ or $a - 4 \geq 2$

61. $3r + 2 < -4$ or $3r - 3 > 0$

62. $4t + 5 \leq -7$ or $4t + 5 \geq -3$

63. $-w + 2 \leq -5$ or $-w + 2 \geq 3$

64. $-2x - 3 \leq -1$ or $-2x - 3 \geq 5$

65. $7 - 3k \leq -2$ or $7 - 3k \geq 1$

66. $8 - 2q \leq 2$ or $8 - 2q \geq 4$

67. $5x + 2 \leq -1$ or $5x + 2 \geq 3$

68. $2m - 1 < 0$ or $2m - 1 > 5$

69. $-2c + 2 < -4$ or $-2c + 4 < -6$

70. $5d + 3 < 8$ or $5d + 3 < 18$

71. $6 - 3m \leq -2$ or $6 - 3m \geq 3$

72. $3 - 4n > 8$ or $3 - 4n < -3$

73. $2x + 9 \leq 1$ or $2x + 9 \leq -3$

74. $5 + y \geq 4$ or $5 + y \geq 6$

75. $-3x + 2 \le -1$ or $-3x + 2 \ge -2$

76. $4k - 7 \le 1$ or $4k - 7 \ge -5$

For Exercises 77–88: a. Graph the solution set.
b. Write the solution set in set-builder notation.
c. Write the solution set in interval notation.

77. $-3 < x + 4$ and $x + 4 < 7$

78. $0 < x - 1$ and $x - 1 \le 3$

79. $2x + 3 < -1$ or $2x + 3 > 7$

80. $3x - 2 < -5$ or $3x - 2 > 7$

81. $4 < -2x < 6$

82. $9 \le -3x \le 15$

83. $1 \le 2x - 5 \le 7$

84. $-7 < 3x - 4 < 2$

85. $-5 < 1 - 2x < -1$

86. $4 \le -3x + 1 < 7$

87. $x - 3 < 2x + 1 < 3x$

88. $2x - 2 \le x + 1 \le 2x + 5$

For Exercises 89–98, solve. Then: a. Graph the solution set.
b. Write the solution set in set-builder notation.
c. Write the solution set in interval notation.

89. If Andrea's current long-distance bill is between $30 and $45 in a month, then she can save money by switching to another company. Her current rate is $0.09 per minute. What is the range of minutes that Andrea must use long distance to justify switching to the other company?

90. A mail-order music club offers one bonus CD if the total of an order is from $75 to $100. Dayle decides to buy the lowest-priced CDs, which are $12.50 each. In what range would the number of CDs he orders have to be in order to receive a bonus CD?

91. Juan has taken four of the five tests in his history course. To get a B in the course, his average needs to be at least 80 and less than 90. If his scores on the first four tests are 95, 80, 82, 88, in what range of values can his score on the fifth test be so that he has a B average?

92. Students in a chemistry course receive a C if the average of four tests is at least 70 and less than 80. Suppose a student has the following scores: 60, 72, and 70. What range of scores on the fourth test would cause the student to receive a C?

93. To conserve energy, it is recommended that a home's thermostat be set at 5° above 73°F during summer or 5° below 73° during winter. If the heat pump does not run when the temperature is at or between those values, in what range of temperatures is the heat pump off?

94. The house thermostat is set so that the heat pump/air conditioner comes on if the temperature is 2° or more above or below 70°. In what range of temperatures is the heat pump off?

95. To maintain a saltwater aquarium, the temperature should be within 4° of 76°. What range of temperatures would be acceptable for saltwater fish? (*Source: The Conscientious Marine Aquarist*, Robert M. Fenner © 2001.)

96. When driving in the state of Mississippi, motorists risk a ticket if they drive more than 15 miles per hour over or under the 55 miles per hour speed limit on a state highway. In what range of speeds would a motorist not risk receiving a ticket? (*Source:* Mississippi State Code 63-3-509.)

97. A building is to be designed in the shape of a trapezoid as shown. Find the range of values for the length of the back side of the building so that the square footage is from 6000 to 8000 square feet.

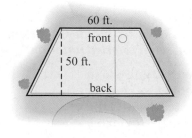

(Use $A = \dfrac{1}{2} h (a + b)$.)

98. At 32°F and below, water is a solid (ice). At 212°F and above, water is a gas (steam). Use a compound inequality to describe the range of values in degrees Celsius for water in its liquid state. (Use $C - \dfrac{5}{9} (F - 32)$.)

For Exercises 99–104, use the following graph and write each solution set using set-builder notation and interval notation.

99. If x represents time, write a solution set that describes the range of years in which there were at least 20,000 drug defendants and 15,000 or more property defendants.

100. If x represents time, write a solution set that describes the range of years in which there were at least 5,000 and at most 10,000 drug defendants.

101. If x represents time, write a solution set that describes the range of years in which there were fewer drug defendants than property and public order defendants.

102. If x represents time, write a solution set that describes the range of years in which there were fewer drug defendants than property defendants or fewer drug defendants than public order defendants.

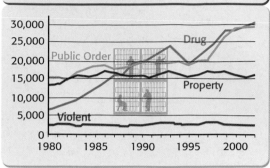

Source: Compendium of Federal Justice Statistics.

103. If x represents time, write a solution set that describes the range of years in which there were fewer public order defendants than drug defendants but more public order defendants than property defendants.

104. If x represents time, write a solution set that describes the range of years in which there were at most 10,000 drug defendants or fewer property defendants than drug defendants.

PUZZLE PROBLEM

A cyclist is involved in a multiple-day race. She feels she needs to complete today's 40 kilometers part in somewhere between 1 hour and 45 minutes and 2 hours. Find the range of values her rate can be to complete this leg of the race in her desired time frame.

REVIEW EXERCISES

[1.1]   1. Is the absolute value of a number always positive? Explain.

[1.5] *For Exercises 2 and 3, simplify.*

2. $-|2 - 3^2| - |-4|$ 3. $-|-|-16||$

[2.3] *For Exercises 4 and 5, solve.*

4. $3x - 14x = 17 - 12x - 11$ 5. $\frac{3}{5}(25 - 5x) = 15 - \frac{3}{5}$

[2.5]   6. One-half the sum of a number and 2 is zero. Find the number.

8.2 Equations Involving Absolute Value

OBJECTIVE 1. Solve equations involving absolute value. We learned in Section 1.1 that the absolute value of a number is its distance from zero on a number line. For example, $|4| = 4$ and $|-4| = 4$ because both 4 and -4 are 4 units from 0 on a number line.

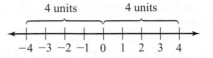

Now we will solve equations in which a variable appears within the absolute value symbols, as in $|x| = 4$. Notice the solutions are 4 and -4, which suggests the following rule.

RULE   **Absolute Value Property**

If $|x| = a$, where x is a variable or an expression and $a \geq 0$, then $x = a$ or $x = -a$.

EXAMPLE 1   Solve.

a. $|2x - 5| = 9$

Solution Using the absolute value property, we separate the equation into a positive case and a negative case, then solve the two cases separately.

$$2x - 5 = 9 \quad \text{or} \quad 2x - 5 = -9$$
$$2x = 14 \quad \text{or} \quad 2x = -4$$
$$x = 7 \quad \text{or} \quad x = -2$$

The solutions are 7 and -2.

b. $|4x + 1| = -11$

Solution This equation has the absolute value equal to a negative number, -11. Because the absolute value of every real number is a positive number or zero, this equation has no solution.

From Example 1b we can say that an absolute value equation in the form $|x| = a$, where $a < 0$, has no solution.

YOUR TURN   Solve.

a. $|3x + 4| = 8$ **b.** $|2x - 1| = 0$ **c.** $|5x - 2| = -3$

ANSWERS

a. $\dfrac{4}{3}$ and -4 **b.** $\dfrac{1}{2}$

c. no solution

To use the absolute value property, we need the absolute value isolated.

EXAMPLE 2) Solve. $|4x - 3| + 9 = 16$

Solution $|4x - 3| = 7$ Subtract 9 from both sides to isolate the absolute value.

$$4x - 3 = 7 \quad \text{or} \quad 4x - 3 = -7 \quad \text{Use the absolute value property.}$$
$$4x = 10 \quad \text{or} \quad 4x = -4$$
$$x = \frac{5}{2} \quad \text{or} \quad x = -1$$

The solutions are $\frac{5}{2}$ and -1.

Examples 1 and 2 suggest the following procedure.

PROCEDURE   **Solving Equations Containing a Single Absolute Value**

To solve an equation containing a single absolute value,

1. Isolate the absolute value so that the equation is in the form $|ax + b| = c$. If $c > 0$, proceed to steps 2 and 3. If $c < 0$, the equation has no solution.
2. Separate the absolute value into two equations, $ax + b = c$ and $ax + b = -c$.
3. Solve both equations.

YOUR TURN)   Solve.

    a. $|6x - 3| - 5 = 4$ **b.** $|2x + 1| + 6 = 4$

More Than One Absolute Value

Some equations contain more than one absolute value expression, such as $|x + 3| = |2x - 8|$. If two absolute values are equal, they must contain expressions that are *equal* or *opposites*. Some simple examples follow:

Equal   **Opposites**

$|7| = |7|$ or $|7| = |-7|$

$|-5| = |-5|$ or $|-5| = |5|$

For the equation $|x + 3| = |2x - 8|$, therefore, the expressions $x + 3$ and $2x - 8$ must be equal or opposites.

Equal   **Opposites**

$x + 3 = 2x - 8$ or $x + 3 = -(2x - 8)$

This suggests the following procedure.

ANSWERS)

a. 2 and -1 **b.** no solution

PROCEDURE Solving Equations in the Form $|ax + b| = |cx + d|$

To solve an equation in the form $|ax + b| = |cx + d|$,

1. Separate the absolute value equation into two equations: $ax + b = cx + d$ and $ax + b = -(cx + d)$.
2. Solve both equations.

EXAMPLE 3 Solve. $|3x + 5| = |x - 9|$

Solution Separate the absolute value equation into two equations.

Equal		**Opposites**
$3x + 5 = x - 9$	or	$3x + 5 = -(x - 9)$
$2x + 5 = -9$	or	$3x + 5 = -x + 9$
$2x = -14$	or	$4x + 5 = 9$
$x = -7$	or	$4x = 4$
		$x = 1$

◄ **Note:** *Think of* $-(x - 9)$ *as* $-1(x - 9)$.

The solutions are -7 and 1.

Sometimes when we separate an equation containing two absolute values into its two equivalent equations, we find that only one of them yields a solution.

EXAMPLE 4 Solve. $|3x - 7| = |5 - 3x|$

Solution

Equal		**Opposites**
$3x - 7 = 5 - 3x$	or	$3x - 7 = -(5 - 3x)$
$6x - 7 = 5$	or	$3x - 7 = -5 + 3x$
$6x = 12$	or	$-7 = -5$
$x = 2$		

Separate into two equations.

◄ **Note:** *Subtracting* $3x$ *from both sides gives us a false equation, so the "opposites" equation has no solution.*

This absolute value equation has only one solution, 2.

YOUR TURN Solve.

a. $|9 + 4x| = |1 - 4x|$ b. $|5 - 2x| = -|x + 3|$

8.2 Exercises

For
Extra
Help   *MyMathLab*   MyMathLab   Videotape/DVT   InterAct Math   Tutor Center   Math Tutor Center   Math XL.com

1. What does absolute value mean in terms of a number line?

2. How do you interpret $|x| = 5$?

3. State the absolute value property in your own words.

4. If given an equation in the form $|ax + b| = c$ and $c < 0$, what can you conclude about its solution(s)?

5. Explain the first step in solving an equation containing two absolute values, such as $|2x + 1| = |3x - 5|$.

6. How can you tell that an absolute value equation containing two absolute values has only one solution?

For Exercises 7–22, solve using the absolute value property.

7. $|x| = 2$

8. $|y| = 5$

9. $|a| = -4$

10. $|r| = -1$

11. $|x + 3| = 8$

12. $|w - 1| = 4$

13. $|2m - 5| = 1$

14. $|3s + 7| = 10$

15. $|6 - 5x| = 1$

16. $|3x - 2| = 5$

17. $|4 - 3w| = 6$

18. $|2x - 1| = 2$

19. $|4m - 2| = -5$

20. $|6p + 5| = -1$

21. $|4w - 3| = 0$

22. $|6 - 3m| = 0$

For Exercises 23–38, isolate the absolute value, then use the absolute value property.

23. $|2y| - 3 = 5$

24. $|3r| + 7 = 10$

25. $|y - 1| + 2 = 4$

26. $|x + 3| - 1 = 5$

27. $|b + 4| - 6 = 2$

28. $|v - 2| + 2 = 4$

29. $3 + |5x - 1| = 7$

30. $2 + |3t - 2| = 5$

31. $1 - |2k + 3| = -4$

32. $-3 = 1 - |4u - 2|$

33. $4 - 3|z - 2| = -8$

34. $3 - 2|x - 5| = -7$

35. $6 - 2|3 - 2w| = -18$

36. $15 - 2|5 - 2x| = -5$

37. $|3x - 2(x + 5)| = 10$

38. $|4y + 2(7 - y)| = 5$

For Exercises 39–48, solve by separating the two absolute values into equal and opposite cases.

39. $|2x + 1| = |x + 5|$

40. $|2p + 5| = |3p + 10|$

41. $|x + 3| = |2x - 4|$

42. $|p - 1| = |2p + 8|$

43. $|3v + 4| = |1 - 2v|$

44. $|3 - 6c| = |2c - 3|$

45. $|n - 3| = |3 - n|$

46. $|2r - 1| = |1 - 2r|$

47. $|2k + 1| = |2k - 5|$

48. $|4 + 2q| = |2q + 8|$

For Exercises 49–56, solve.

49. $|10 - (5 + h)| = 8$

50. $|7 - (2 - n)| = 17$

51. $\left|\dfrac{b}{2} - 1\right| = 4$

52. $\left|\dfrac{u}{3} + 2\right| = 3$

53. $\left|\dfrac{4 - 3x}{2}\right| = \dfrac{3}{4}$

54. $\left|\dfrac{6 - 5w}{6}\right| = \dfrac{2}{3}$

55. $\left|2y + \dfrac{3}{2}\right| - \dfrac{1}{2} = 5$

56. $\left|p + \dfrac{2}{3}\right| - 1 = 7$

REVIEW EXERCISES

[2.6] *For Exercises 1 and 2, solve.*

1. $6 - 4x > -5x - 1$

2. $-4x \le 12$

[2.6] **3.** Translate to a linear inequality and solve: Two less than a number is less than five.

[8.1] **4.** Use interval notation to represent the solution set shown on the graph to the right.

[8.1] **5.** Graph the compound inequality: $-1 < x \le 0$

[8.1] **6.** Solve the compound inequality: $-1 < \dfrac{3x + 2}{4} < 4$

8.3 Inequalities Involving Absolute Value

OBJECTIVES
1. Solve absolute value inequalities involving less than.
2. Solve absolute value inequalities involving greater than.

OBJECTIVE 1. Solve absolute value inequalities involving less than. Now let's solve inequalities that contain absolute value. Think about $|x| \leq 3$. A solution for this inequality is any number whose absolute value is less than or equal to 3. This means the solutions are a distance of 3 units or less from 0 on a number line.

Note: *Solutions for the equal to part of $|x| \leq 3$ are 3 and -3. The **less than** part of $|x| \leq 3$ means all numbers in between 3 and -3 because their absolute values are less than 3.*

Notice the solution region corresponds to the compound inequality $x \geq -3$ and $x \leq 3$, which we can write as $-3 \leq x \leq 3$. In set-builder notation, the solution is $\{x \mid -3 \leq x \leq 3\}$ and in interval notation, we write $[-3, 3]$.

Our examples suggest the following procedure.

PROCEDURE   **Solving Inequalities in the Form $|x| < a$, where $a > 0$**

To solve an inequality in the form $|x| < a$, where $a > 0$,

1. Rewrite as a compound inequality involving "and": $x > -a$ and $x < a$.
   (We can also use $-a < x < a$.)
2. Solve the compound inequality.

Similarly, to solve $|x| \leq a$, we would write $x \geq -a$ and $x \leq a$ (or $-a \leq x \leq a$).

EXAMPLE 1   For each inequality, solve, graph the solution set, and write the solution set in both set-builder and interval notation.

a. $|x - 4| \leq 3$

Solution $x - 4 \geq -3$ and $x - 4 \leq 3$ Rewrite as a compound inequality.
              $x \geq 1$ and $x \leq 7$ Add 4 to both sides of each inequality.

Recall that $x \geq 1$ and $x \leq 7$ means $1 \leq x \leq 7$, so our graph is as follows:

Note: *The solution set for $|x - 4| \leq 3$ contains every number whose distance from 4 is 3 units or less.*

Set-builder notation: $\{x \mid 1 \leq x \leq 7\}$
Interval notation: $[1, 7]$

b. $|3x + 1| < 5$

Solution   Instead of $3x + 1 > -5$ and $3x + 1 < 5$, we will use the more compact form.

$$-5 < 3x + 1 < 5$$ Rewrite as a compound inequality.
$$-6 < 3x < 4$$ Subtract 1 from all parts of the inequality.
$$-2 < x < \frac{4}{3}$$ Divide all parts of the inequality by 3 to isolate x.

The image at the top shows a number line from -3 to 2 with an open interval.

Set-builder notation: $\left\{ x \mid -2 < x < \dfrac{4}{3} \right\}$

Interval notation: $\left(-2, \dfrac{4}{3} \right)$

c. $|-0.5x + 1| - 2 < 1$

Solution Notice that the equation is not in the form $|x| < a$, so our first step is to isolate the absolute value.

$|-0.5x + 1| < 3$	Add 2 to both sides to isolate the absolute value.
$-3 < -0.5x + 1 < 3$	Rewrite as a compound inequality.
$-4 < -0.5x < 2$	Subtract 1 from all parts of the inequality.
$8 > x > -4$	Divide all parts of the inequality by -0.5 to isolate x.

Note: *Because we divided by a negative, we changed the direction of the inequalities.*

A number line from -6 to 10 shows an open interval between -4 and 8.

Set-builder notation: $\{x \mid -4 < x < 8\}$

Interval notation: $(-4, 8)$

d. $\left| \dfrac{1}{3}x - 4 \right| + 6 < -1$

Solution $\left| \dfrac{1}{3}x - 4 \right| < -7$ Subtract 6 from both sides to isolate the absolute value.

Because absolute values cannot be negative, this inequality has no solution, so the solution set is empty.

A number line from -5 to 5 with no solution marked.

Set-builder notation: { } or $\varnothing$

Interval notation: We do not write interval notation because there are no values in the solution set.

ANSWERS

a.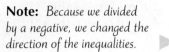

A number line from -8 to 2 showing open interval between -7 and -3.

Set-builder: $\{x \mid -7 < x < -3\}$
Interval: $(-7, -3)$

b. A number line from -4 to 6 showing closed interval between -2 and 5.

Set-builder: $\{x \mid -2 \le x \le 5\}$
Interval: $[-2, 5]$

c. A number line from -8 to 14 showing closed interval between -6 and 12.

Set-builder: $\{x \mid -6 \le x \le 12\}$
Interval: $[-6, 12]$

d. A number line from -5 to 5 with no solution.

Set-builder: { } or $\varnothing$
no interval notation

YOUR TURN For each inequality, solve, graph the solution set, and write the solution set in both set-builder notation and interval notation.

a. $|x + 5| < 2$

b. $|2x - 3| \le 7$

c. $\left| -\dfrac{1}{3}x + 1 \right| - 2 \le 1$

d. $\left| \dfrac{1}{2}x - 1 \right| + 3 < -2$

OBJECTIVE 2. Solve absolute value inequalities involving greater than. Now we consider inequalities with greater than, such as $|x| \geq 5$. A solution for $|x| \geq 5$ is any number whose absolute value is greater than or equal to 5. As the graph shows, the solution set contains all values that are a distance of 5 units or more from 0.

Note: *Solutions for the equal to part of* $|x| \geq 5$ *are 5 and* -5. *The* **greater than** *part means all values that are farther from 0 than 5 and* -5 *because their absolute values are greater than 5.*

Notice that the solutions are equivalent to $x \leq -5$ or $x \geq 5$.

An inequality such as $|x| > 2$ uses parentheses:

The solutions are equivalent to $x < -2$ or $x > 2$. Our examples suggest that we can split these inequalities into a compound inequality involving "or."

PROCEDURE **Solving Inequalities in the Form** $|x| > a$, **where** $a > 0$

To solve an inequality in the form $|x| > a$, where $a > 0$,

1. Rewrite as a compound inequality involving "or": $x < -a$ or $x > a$.
2. Solve the compound inequality.

Similarly, to solve $|x| \geq a$, we would write $x \leq -a$ or $x \geq a$.

EXAMPLE 2 For each inequality, solve, graph the solution set, and write the solution set in both set-builder notation and interval notation.

a. $|x - 2| \geq 5$

Solution $\quad x - 2 \leq -5 \quad$ or $\quad x - 2 \geq 5 \qquad$ Rewrite as a compound inequality.

$\qquad\qquad\quad x \leq -3 \quad$ or $\qquad x \geq 7 \qquad$ Add 2 to both sides of each inequality.

Set-builder notation: $\{x \mid x \leq -3 \text{ or } x \geq 7\}$
Interval notation: $(-\infty, -3] \cup [7, \infty)$

b. $|-3x - 4| > 5$

Solution $\quad -3x - 4 < -5 \quad$ or $\quad -3x - 4 > 5 \qquad$ Rewrite as a compound inequality.

$\qquad\qquad -3x < -1 \quad$ or $\qquad -3x > 9 \qquad$ Add 4 to both sides of each inequality.

$\qquad\qquad\quad x > \dfrac{1}{3} \quad$ or $\qquad\quad x < -3 \qquad$ Divide both sides of each inequality by -3.

Set-builder notation: $\left\{ x \mid x < -3 \text{ or } x > \dfrac{1}{3} \right\}$

Interval notation: $(-\infty, -3) \cup \left(\dfrac{1}{3}, \infty \right)$

c. $\left| \dfrac{3}{4}x - 2 \right| + 1 \geq 6$

Solution $\left| \dfrac{3}{4}x - 2 \right| \geq 5$ Subtract 1 from both sides to isolate the absolute value.

$\dfrac{3}{4}x - 2 \leq -5$ or $\dfrac{3}{4}x - 2 \geq 5$ Rewrite as a compound inequality.

$\dfrac{3}{4}x \leq -3$ or $\dfrac{3}{4}x \geq 7$ Add 2 to both sides of each inequality.

$x \leq -4$ or $x \geq \dfrac{28}{3}$ Multiply both sides of each inequality by $\dfrac{4}{3}$.

Set-builder notation: $\left\{ x \,|\, x \leq -4 \text{ or } x \geq \dfrac{28}{3} \right\}$

Interval notation: $(-\infty, -4] \cup \left[\dfrac{28}{3}, \infty \right)$

d. $|0.2x + 1| - 3 > -7$

Solution $|0.2x + 1| > -4$ Add 3 to both sides to isolate the absolute value.

This inequality indicates that the absolute value is greater than a negative number. Since the absolute value of every real number is either positive or 0, the solution set is $\mathbb{R}$. The graph is the entire number line.

Set-builder notation: $\{ x \,|\, x \text{ is a real number} \}$

Interval notation: $(-\infty, \infty)$

YOUR TURN   For each inequality, solve, graph the solution set, and write the solution set in both set-builder notation and interval notation.

a. $|x - 3| > 2$

b. $|4x + 1| \geq 9$

c. $|-0.4x + 1| + 2 \geq 3$

d. $\left| \dfrac{1}{3}x - 4 \right| - 2 > -9$

*For Exercises 1–4, we assume **a** is a positive number.*

1. What compound inequality is related to $|x| \leq a$, where $a > 0$?

2. What compound inequality is related to $|x| \geq a$, where $a > 0$?

3. How would you characterize the graph of $|x| < a$, where $a > 0$?

4. How would you characterize the graph of $|x| > a$, where $a > 0$?

5. Under what conditions does $|x| < a$ have no solution?

6. Under what conditions does the solution set for $|x| > a$ contain all real numbers?

For Exercises 7–20, solve the inequality. Then: a. Graph the solution set.
b. Write the solution set in set-builder notation.
c. Write the solution set in interval notation.

7. $|x| < 5$

8. $|y| \leq 3$

9. $|x + 3| \leq 7$

10. $|m - 5| < 2$

11. $|s + 3| + 2 < 5$

12. $|p - 3| + 4 \leq 8$

13. $|2m - 5| + 1 < 10$

14. $|4x + 7| + 2 < 5$

15. $|-3k + 5| + 7 \leq 8$

16. $|-5h - 1| - 6 < 8$

17. $2|x| - 7 \leq 3$

18. $3|u| + 2 < 8$

19. $2|w - 3| + 4 < 10$

20. $4|n + 2| - 3 \leq 9$

For Exercises 21–34, solve the inequality. Then: a. Graph the solution set.
            b. Write the solution set in set-builder notation.
            c. Write the solution set in interval notation.

21. $|c| > 12$

22. $|h| \geq 6$

23. $|y + 2| \geq 7$

24. $|a - 4| > 3$

25. $|p - 6| + 2 > 10$

26. $|x + 5| - 3 \geq 6$

27. $|3x + 6| - 3 \geq 9$

28. $|2x - 5| - 1 \geq 10$

29. $|-4n - 5| + 3 > 8$

30. $|-6h + 1| - 7 > 4$

31. $4|v| - 3 \geq 1$

32. $5|m| + 2 > 12$

33. $4|y + 2| - 1 > 3$

34. $3|x - 2| + 5 \geq 11$

For Exercises 35–54, solve the inequality. Then: *a. Graph the solution set.*
b. Write the solution set in set-builder notation.
c. Write the solution set in interval notation.

35. $|4m + 8| - 2 > 10$

36. $|2x + 4| - 2 \geq 10$

37. $|-3x + 6| < 6$

38. $|-2y + 3| \leq 3$

39. $|2r - 3| > -3$

40. $|3b + 7| > -2$

41. $2 - |x + 3| > 1$

42. $5 - 2|u + 4| \geq 1$

43. $6|2x - 1| + 3 < 9$

44. $4|3p + 6| - 2 < 22$

45. $5 - |w + 4| > 10$

46. $4 - |5 + k| > 7$

47. $\left|2 - \dfrac{3}{2}k\right| \leq 5$

48. $\left|3 - \dfrac{1}{2}x\right| \geq 7$

49. $|0.25x - 3| + 2 > 4$

50. $|0.5y - 3| + 4 < 5$

51. $\left|2.4 - \dfrac{3}{4}y\right| \leq 7.2$

52. $\left|5.3 - \dfrac{2}{3}w\right| \geq 5.3$

53. $|2b - 8| + 5 < 1$

54. $|8p + 7| + 4 > 3$

For Exercises 55–62, write an inequality involving absolute value that describes the graph shown.

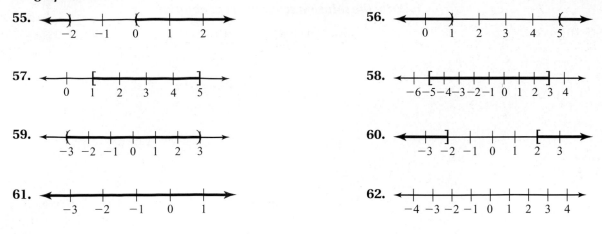

55.

56.

57.

58.

59.

60.

61.

62.

REVIEW EXERCISES

[1.1] *For Exercises 1 and 2, indicate whether the statement is true or false.*

1. $-\dfrac{2}{3} \in \{x \mid x \text{ is a rational number}\}$

2. $\{1, 3, 5, 7\} \subseteq N$

[2.3] *For Exercises 3 and 4, solve.*

3. $23 - 5(2 - 3x) = 12x - (4x + 1)$

4. $\dfrac{5}{6}x - \dfrac{3}{4} = \dfrac{2}{3}x + \dfrac{1}{2}$

[2.4] **5.** Solve $Ax + By = C$ for y.

[3.3] **6.** The sum of three consecutive integers is 141. Find the three integers.

8.4 Functions and Graphing

OBJECTIVES

1. Identify the domain and range of a relation and determine if the relation is a function.
2. Find the value of a function.
3. Graph functions.

In this section, we review and expand on what we have learned about functions and graphing.

OBJECTIVE 1. Identify the domain and range of a relation and determine if the relation is a function. Let's review **relations**, **domain**, **range**, and **functions**, which we introduced in Section 4.7.

DEFINITIONS

Relation: A set of ordered pairs.
Domain: The set of all input values for a relation.
Range: The set of all output values for a relation.
Function: A relation in which every value in the domain is paired with exactly one value in the range.

For example, the linear equation $y = 2x$ is a relation because it pairs a given x-value with a corresponding y-value. We list here some values in the domain and range for $y = 2x$ and draw its graph.

Domain (x-values)	Range (y-values)
-1	-2
0	0
1	2
2	4

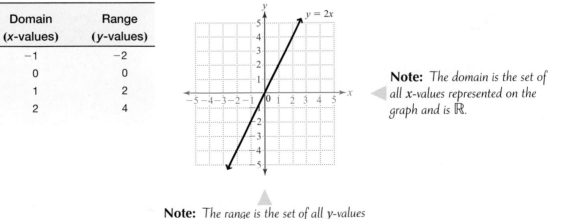

Note: *The domain is the set of all x-values represented on the graph and is* $\mathbb{R}$.

Note: *The range is the set of all y-values represented on the graph and is* $\mathbb{R}$.

We can also conclude that $y = 2x$ is a function because every value in its domain is paired with only one value in its range. In Section 4.7, we developed the vertical line test to determine whether a graph is a graph of a function. With a function, a vertical line drawn through any point of its graph intersects the graph only once.

EXAMPLE 1 Identify the domain and range of each relation and determine whether it is a function.

a. The following relation pairs players on the Philadelphia Eagles football team with their position(s).

Player	Position
David Akers	Kicker
Dirk Johnson	Punter
Brian Westbrook	Running Back
Mike Bartrum	Long Snapper
Thomas Tapen	Full Back and Running Back

Solution Domain: {David Akers, Dirk Johnson, Brian Westbrook, Mike Bartrum, Thomas Tapen}

Range: {Kicker, Punter, Running Back, Long Snapper, Full Back}

It is not a function because an element in the domain, Thomas Tapen, is paired with two elements in the range.

b. $\{(-2, -4), (-1, -1), (0, 0), (1, -1), (2, -4)\}$

Solution Domain: $\{-2, -1, 0, 1, 2\}$
Range: $\{-4, -1, 0, -1, -4\}$

It is a function because every value in the domain is paired with only one value in the range.

c.

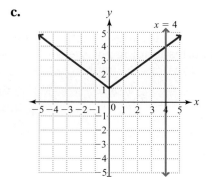

Solution Domain: all real numbers
Range: $\{y \mid y \geq 1\}$

It is a function because it passes the vertical line test: A vertical line drawn through any point of the graph intersects it only once. We have drawn one such line, $x = 4$, in blue.

d.

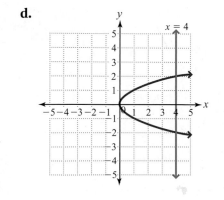

Solution Domain: $\{x \mid x \geq 0\}$
Range: all real numbers

It is a not a function because there are values in the domain paired with more than one value in the range, as the vertical line test confirms. For example, the line $x = 4$ (in blue) intersects the graph twice so the value 4 in the domain is paired with two values in the range, 2 and -2.

a. The following relation shows the forecasted high temperatures for Yellowstone National Park from April 24 to April 28, 2005.

Day	High Temperature
Sunday	51°
Monday	44°
Tuesday	45°
Wednesday	31°
Thursday	30°

b. $\{ (0, 2), (0, -2), (3, 0), (-3, 0) \}$

c.

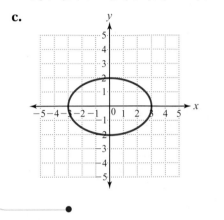

d.

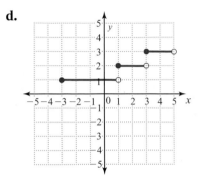

OBJECTIVE 2. Find the value of a function. In Section 4.7, we introduced the function notation $f(x)$, which is read "a function in terms of x" or "f of x." Using function notation, we can write the equation $y = 3x$ as $f(x) = 3x$. Recall that we evaluate functions by replacing x with the given value. For example, the notation $f(2)$ means the value of f when $x = 2$, so if $f(x) = 3x$, then $f(2) = 3(2) = 6$.

EXAMPLE 2 For the function $f(x) = -5x + 3$, find

a. $f(4)$

Solution $f(4) = -5(4) + 3$ Replace x with 4, then calculate.
$$= -20 + 3$$
$$= -17$$

b. $f(-2)$

Solution $f(-2) = -5(-2) + 3$ Replace x with −2, then calculate.
$$= 10 + 3$$
$$= 13$$

Connection $f(4) = -17$ is the same as the ordered pair $(4, -17)$.

a. $f(-6)$ **b.** $f(1)$

OBJECTIVE 3. Graph functions. Now let's review how to graph functions. In general, we can graph a function by finding enough ordered pairs to determine its shape.

Linear Functions

Given a linear function in the form $f(x) = mx + b$, where m is the slope and $(0, b)$ is the y-intercept, we need to find only two ordered pairs. Note that $(0, b)$ can be one of those two ordered pairs.

EXAMPLE 3 Graph. $f(x) = -\dfrac{3}{4}x + 2$

Solution We can use $(0, 2)$ as one ordered pair, then use the slope, $-\dfrac{3}{4}$, to find a second ordered pair. Recall that slope indicates the "rise" and "run" from any point on the line to another point on the line. So, $-\dfrac{3}{4}$ indicates we rise -3 and run 4 from the point $(0, 2)$ to reach a second point, $(4, -1)$.

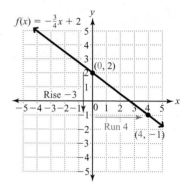

YOUR TURN Graph. $f(x) = 2x - 3$

Quadratic Functions

In Section 6.7, we introduced quadratic functions, which have the form $f(x) = ax^2 + bx + c$, and we saw their graphs are parabolas. To graph a parabola, we have to find many ordered pairs.

EXAMPLE 4 Graph. $f(x) = -2x^2 + 3$

Solution We create a table of ordered pairs, plot the points, and connect them with a smooth curve.

x	f(x)
-2	-5
-1	1
0	3
1	1
2	-5

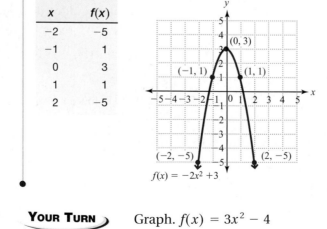

Connection Recall from Section 6.7 that the sign of a in $f(x) = ax^2 + bx + c$ indicates whether the parabola opens up or down. If a is *positive,* the parabola opens *up,* and if a is *negative,* the parabola opens *down.*

YOUR TURN Graph. $f(x) = 3x^2 - 4$

ANSWER

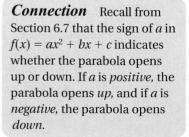

Absolute Value Functions

Now let's consider how to graph functions containing absolute value. In general, absolute value functions have a V shape.

EXAMPLE 5) Graph each function.

a. $f(x) = |x|$

Solution We find ordered pair solutions, plot those ordered pairs, and connect the points.

x	f(x)
−4	4
−2	2
0	0
2	2
4	4

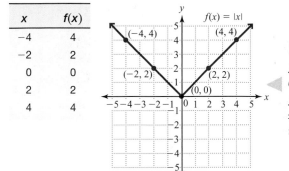

Note: *$f(x) = |x|$ is equivalent to $f(x) = -x$ when $x < 0$ and $f(x) = x$ when $x \geq 0$. Notice $f(x) = -x$ and $f(x) = x$ are linear functions with opposite slopes, which explains the V shape of the graph.*

b. $f(x) = |x| + 1$

Solution

x	f(x)
−4	5
−2	3
0	1
2	3
4	5

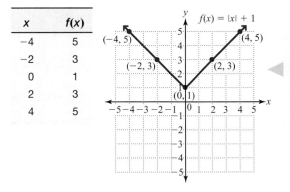

Note: *Compare the graphs in Examples 5a and 5b. Notice that every point on $f(x) = |x| + 1$ is one higher than every point on $f(x) = |x|$. Adding 1 to $|x|$ shifts the graph of $f(x) = |x|$ up 1 unit.*

c. $f(x) = |x - 2| + 1$

Solution

x	f(x)
−4	7
−2	5
0	3
2	1
4	3

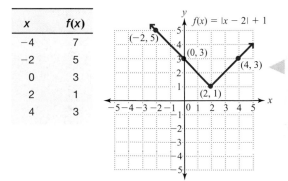

Note: *As we saw in Example 5b, changing $f(x) = |x|$ to $f(x) = |x| + 1$ shifted the graph of $f(x) = |x|$ up 1 unit. Now we see that replacing x in $f(x) = |x| + 1$ with $x - 2$ shifts the graph of $f(x) = |x| + 1$ right 2 units.*

d. $f(x) = -|x - 2| + 1$

Solution

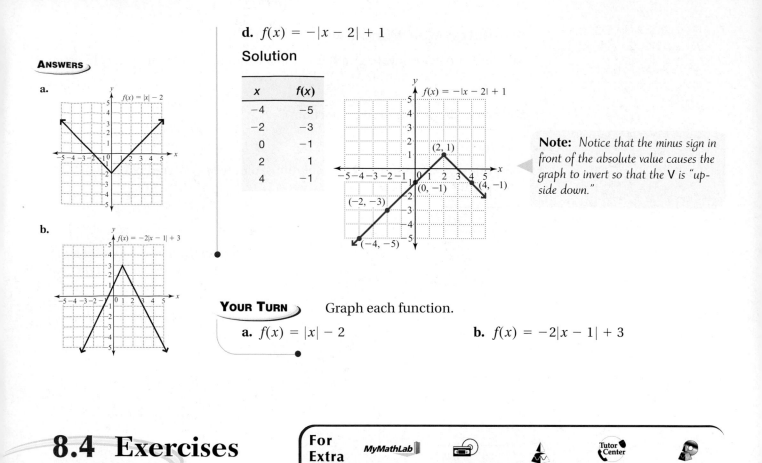

x	f(x)
−4	−5
−2	−3
0	−1
2	1
4	−1

Note: *Notice that the minus sign in front of the absolute value causes the graph to invert so that the V is "upside down."*

a. $f(x) = |x| - 2$

b. $f(x) = -2|x - 1| + 3$

YOUR TURN   Graph each function.

a. $f(x) = |x| - 2$ **b.** $f(x) = -2|x - 1| + 3$

8.4 Exercises

For Extra Help   MyMathLab   Videotape/DVT   InterAct Math   Math Tutor Center   Math XL.com

1. Explain what makes a relation a function.

2. How do you determine if a graph represents a function?

3. In your own words, define *domain* and *range*.

4. What is the shape of the graph of every linear function? How do you graph a linear function?

5. What is the shape of the graph of every quadratic function? How do you graph a quadratic function?

6. What is the shape of the graph of every absolute value function? How do you graph an absolute value function?

For Exercises 7–14, identify the domain and range of each relation and determine whether it is a function.

7. Favorite vacation countries.

Rank	State
1	Australia
2	Italy
3	U.K.
4	France
5	Ireland

Source: Harris Poll, 2003.

8. Best-mannered cities in the United States for 2005 according to etiquette expert Marjabelle Young Stewart.

Rank	City
1	Charleston, SC
2	San Diego, CA
3	Seattle, WA
4	Peoria, IL
5	Omaha, NE

9. Percent of the workforce leaving for work commute.

Percent (%)	Time of departure
2	Work at home
3	Midnight to 4:59 A.M. *and* 10 A.M. to 11:59 A.M.
5	9 A.M. to 9:59 A.M.
14	Noon to 11:59 P.M.
15	8 A.M. to 8:59 A.M.
26	5 A.M. to 6:59 A.M.
33	7 A.M. to 7:59 A.M.

Source: U.S. Census 2000.

10. The percent of resale value to job cost for home repair/remodeling projects according to *Remodeling* magazine's 2001 Cost vs. Value Report.

Cost recouped (%)	Project
128	Bathroom remodel
87	Exterior paint
82	Attic bedroom
69	Minor kitchen remodel, Reroofing
59	Deck addition
58	Major kitchen remodel
37	Home office

11. $\{(2, 1), (-3, 2), (0, 5), (1, 8)\}$

12. $\{(1, 1), (2, 2), (-3, -3), (4, 4)\}$

13. $\{(-3, 2), (4, 1), (-3, 5), (2, 8)\}$

14. $\{(-1, 0), (-3, 0), (4, 5), (2, 8)\}$

For Exercises 15–24, identify the domain and range of each relation and determine whether it is a function.

15.

WOMEN ARE WAITING

The average age of American mothers giving birth to their first child:

21.4 21.8 22.7 23.7 24.2 24.5 24.9

'70 '75 '80 '85 '90 '95 '00

Source: Centers for Disease Control and Prevention.

16.

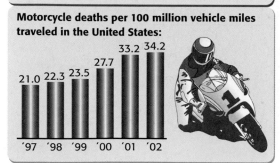

DEATH ON TWO WHEELS

Motorcycle deaths per 100 million vehicle miles traveled in the United States:

33.2 34.2
27.7
21.0 22.3 23.5

'97 '98 '99 '00 '01 '02

17.

$x + 2y = 4$

18.

$y = 4x - 2$

19.

$y = x^2 - 4$

20.

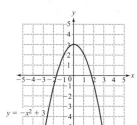

$y = -x^2 + 3$

21.

22.

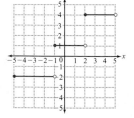

23.

$x = y^2 - 4$

24.

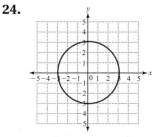

For Exercises 25–28, given $f(x) = 2x^2 - 3x + 1$, *find each of the following.*

25. $f(-2)$ **26.** $f(0)$ **27.** $f\left(\dfrac{1}{3}\right)$ **28.** $f(1.5)$

For Exercises 29–32, given $f(x) = |x - 4| + 2$, *find each of the following.*

29. $f(1)$ **30.** $f(-3)$ **31.** $f\left(\dfrac{2}{5}\right)$ **32.** $f(0.5)$

For Exercises 33–36, given $f(x) = x^3 - 2x^2 + 3$, *find each of the following.*

33. $f(0)$ **34.** $f(-2)$ **35.** $f(0.2)$ **36.** $f\left(\dfrac{1}{4}\right)$

For Exercises 37–40, given $f(x) = \dfrac{2x}{x - 1}$, *find each of the following.*

37. $f(1)$ **38.** $f(-2)$ **39.** $f(1.5)$ **40.** $f\left(\dfrac{2}{3}\right)$

For Exercises 41–46, use the graph to determine the value of the function.

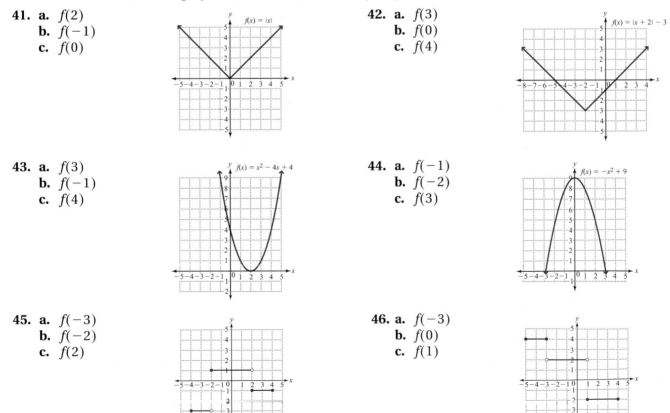

41. **a.** $f(2)$
     b. $f(-1)$
     c. $f(0)$

42. **a.** $f(3)$
     b. $f(0)$
     c. $f(4)$

43. **a.** $f(3)$
     b. $f(-1)$
     c. $f(4)$

44. **a.** $f(-1)$
     b. $f(-2)$
     c. $f(3)$

45. **a.** $f(-3)$
     b. $f(-2)$
     c. $f(2)$

46. **a.** $f(-3)$
     b. $f(0)$
     c. $f(1)$

For Exercises 47–72, graph.

47. $f(x) = 2x + 3$

48. $f(x) = 3x - 1$

49. $f(x) = -x + 4$

50. $f(x) = -2x + 5$

51. $f(x) = \dfrac{1}{4}x - 2$

52. $f(x) = -\dfrac{2}{5}x + 1$

53. $f(x) = x^2 + 4$

54. $f(x) = x^2 - 3$

55. $f(x) = -x^2 + 1$

56. $f(x) = -x^2 + 9$

57. $f(x) = \dfrac{1}{4}x^2 - 2$

58. $f(x) = -\dfrac{1}{3}x^2 + 3$

59. $f(x) = x^2 - 12x + 20$

60. $f(x) = x^2 + 5x + 6$

61. $f(x) = |x| - 4$

62. $f(x) = |x| + 1$

63. $f(x) = |x + 3| - 2$

64. $f(x) = |x - 1| + 3$

65. $f(x) = 2|x - 1| + 3$

66. $f(x) = 3|x + 1| - 2$

67. $f(x) = -|x + 4|$ **68.** $f(x) = -|x - 3|$ **69.** $f(x) = -3|x + 2| - 1$ **70.** $f(x) = -2|x - 1| + 4$

71. $f(x) = -\dfrac{1}{2}|x + 1| + 1$ **72.** $f(x) = -\dfrac{1}{4}|x - 2| + 2$

Collaborative Exercises

In the following exercises, you will use a graphing utility to explore and discuss how numbers can **shift,** **reflect,** *or* **stretch** *a basic function. We will use the graph of* $f(x) = |x|$ *as our basic function.*

1. Graph $f(x) = |x + 2|$ and compare it with the graph of $f(x) = |x|$. What effect does adding 2 to x within the absolute value have on the graph?

2. Graph $f(x) = |x - 3|$ and compare it with the graph of $f(x) = |x|$. What effect does subtracting 3 from x within the absolute value have on the graph?

3. Graph $f(x) = |x| + 4$ and compare it with the graph of $f(x) = |x|$. What effect does adding 4 to $|x|$ have on the graph?

4. Graph $f(x) = |x| - 1$ and compare it with the graph of $f(x) = |x|$. What effect does subtracting 1 from $|x|$ have on the graph?

5. Notice that adding or subtracting in a function *shifts* the position of the basic function. Without graphing, how would you expect the graph of $f(x) = |x + 1| - 3$ to be shifted as compared with the basic function $f(x) = |x|$?

6. Graph $f(x) = -|x|$. How is the graph of $f(x) = -|x|$ different from that of $f(x) = |x|$?

7. Recall that $f(x) = -|x|$ is the same as $f(x) = -1 \cdot |x|$. Multiplying the basic function by a negative number causes a *reflection* of the function about the x-axis. Without graphing, how would you describe the differences between the graph of $f(x) = -|x - 2| + 1$ and the basic function $f(x) = |x|$?

(continued)

8. Graph $f(x) = 2|x|$. How is the graph of $f(x) = 2|x|$ different from that of $f(x) = |x|$?

9. Graph $f(x) = \frac{1}{2}|x|$. How is the graph of $f(x) = \frac{1}{2}|x|$ different from that of $f(x) = |x|$?

10. Notice that multiplying the basic function by a number other than 1 causes the graph to be *stretched*, so that it is wider or narrower than the basic function's graph. Without graphing, describe the differences in each of the following compared to the graph of $f(x) = |x|$.
   a. $f(x) = 3|x| + 1$

   b. $f(x) = -2|x - 4|$

   c. $f(x) = \frac{1}{3}|x + 4| - 2$

For further investigation, change the basic function to $f(x) = x^2$ and work through Exercises 1–10 again. For example, Exercise 1 would read: Graph $f(x) = (x + 2)^2$ and compare it with the graph of $f(x) = x^2$. What effect does adding 2 to x have on the graph? Exercise 3 would read: Graph $f(x) = (x^2 + 4)$ and compare it with the graph of $f(x) = x^2$. What effect does adding 4 to x^2 have on the graph?

REVIEW EXERCISES

For Exercises 1–6, perform the indicated operation.

[5.3] **1.** $(2x^2 + 3x - 1) + (3x^2 - 3x + 7)$

[5.3] **2.** $(5x^2 - 3x + 1) - (-2x^2 + 8)$

[5.5] **3.** $(3x - 4)(2x + 1)$

[5.5] **4.** $(2x - 4)^2$

[5.6] **5.** $(12x^4 + 18x^3 - 30x^2) \div 6x^2$

[5.6] **6.** $(x^4 - 81) \div (x - 3)$

8.5 Function Operations

1. Add or subtract functions.
2. Multiply functions.
3. Divide functions.

Note: *Letters other than f and g can be used to name functions, such as h, p, or q.*

So far, we have seen how to evaluate and graph functions. Now we will consider how to add, subtract, multiply, and divide functions. Because we will be considering two different functions in the same problem, we will write one function as f and the second as g.

OBJECTIVE 1. Add or subtract functions. Given two functions f and g, their sum or difference is found using the following rules.

RULE Adding or Subtracting Functions

> The sum of two functions, $f + g$, is found by $(f + g)(x) = f(x) + g(x)$.
> The difference of two functions, $f - g$, is found by $(f - g)(x) = f(x) - g(x)$.

EXAMPLE 1 Given $f(x) = 2x + 1$ and $g(x) = 7x + 3$, find

a. $f + g$

b. $f - g$

c. $(f - g)(-4)$

Solution

a. $(f + g)(x) = f(x) + g(x)$

$\qquad = (2x + 1) + (7x + 3)$ Write a sum of the two functions.

$\qquad = 9x + 4$ Combine like terms.

b. $(f - g)(x) = f(x) - g(x)$

$\qquad = (2x + 1) - (7x + 3)$ Write a difference of the two functions.

$\qquad = (2x + 1) + (-7x - 3)$ Write as an equivalent addition.

$\qquad = -5x - 2$ Combine like terms.

c. In part b, we found $(f - g)(x) = -5x - 2$, so to find $(f - g)(-4)$, we replace x with -4.

$$(f - g)(-4) = -5(-4) - 2 \quad \text{Replace } x \text{ with } -4.$$
$$= 20 - 2$$
$$= 18$$

ANSWERS

a. $2x^2 - 5x - 4$

b. $-x + 14$

c. 14

YOUR TURN Given $h(x) = x^2 - 3x + 5$ and $k(x) = x^2 - 2x - 9$, find

a. $h + k$ **b.** $h - k$ **c.** $(h + k)(-2)$

OBJECTIVE 2. Multiply functions. Given two functions f and g, their product is found using the following rule.

> **RULE**   Multiplying Functions
>
> The product of two functions, $\boldsymbol{f \cdot g}$, is found by $(f \cdot g)(x) = f(x)\, g(x)$.

EXAMPLE 2 Given $f(x) = 3x - 4$ and $g(x) = x + 5$, find $f \cdot g$.

Solution $(f \cdot g)(x) = \boldsymbol{f(x)\, g(x)}$

$\qquad\qquad = (3x - 4)(x + 5)$ **Write a product of the functions.**

$\qquad\qquad = 3x^2 + 15x - 4x - 20$ **Use FOIL to multiply the binomials.**

$\qquad\qquad = 3x^2 + 11x - 20$ **Combine like terms.**

YOUR TURN Given $p(x) = 7x - 2$ and $q(x) = 5x - 1$, find

a. $p \cdot q$ **b.** $(p \cdot q)(2)$

OBJECTIVE 3. Divide functions. Given two functions f and g, their quotient is found using the following rule.

> **RULE**   Dividing Functions
>
> The quotient of two functions, $\boldsymbol{f \,/\, g}$, is found by $(f/g)(x) = \dfrac{f(x)}{g(x)}$, where $g(x) \neq 0$.

a. $35x^2 - 17x + 2$

b. 108

EXAMPLE 3 Find f/g.

a. $f(x) = 9x^3 - 12x^2 + 6x$, $g(x) = 3x$

Note: *Because we divided by $3x$, we are assuming that $x \neq 0$ in $(f/g)(x)$.*

Solution $(f/g)(x) = \dfrac{f(x)}{g(x)} = \dfrac{9x^3 - 12x^2 + 6x}{3x}$ **Write a quotient of the functions.**

$\qquad\qquad = \dfrac{9x^3}{3x} - \dfrac{12x^2}{3x} + \dfrac{6x}{3x}$ **Since the divisor is a monomial, we separate the terms.**

$\qquad\qquad = 3x^2 - 4x + 2$ **Divide.**

b. $f(x) = 3x^2 - 13x - 6$, $g(x) = x - 5$

Solution $(f/g)(x) = \dfrac{f(x)}{g(x)} = \dfrac{3x^2 - 13x - 6}{x - 5}$ **Write a quotient of the functions.**

To find the quotient, we use long division.

$$\begin{array}{r} 3x \\ x - 5 \overline{)\, 3x^2 - 13x - 6} \\ -(3x^2 - 15x) \end{array}$$

$\xrightarrow{\text{Change signs.}}$

$$\begin{array}{r} 3x \\ x - 5 \overline{)\, 3x^2 - 13x - 6} \\ -3x^2 + 15x \\ \hline 2x - 6 \end{array}$$

Combine like terms and bring down the next term.

$$\begin{array}{r} 3x + 2 \\ x - 5\overline{)3x^2 - 13x - 6} \\ \underline{-3x^2 + 15x} \\ 2x - 6 \\ \underline{-(2x - 10)} \end{array}$$

$$\begin{array}{r} 3x + 2 \\ x - 5\overline{)3x^2 - 13x - 6} \\ \underline{-3x^2 - 15x} \\ 2x - 6 \\ \underline{-2x + 10} \\ 4 \end{array}$$

Change signs. → **Combine like terms.**

Note: *The domain of the quotient function contains all real numbers except 5, so we would write the domain as $\{x \mid x \neq 5\}$ or, using interval notation, $(-\infty, 5) \cup (5, \infty)$.*

▶ **Answer** $3x + 2 + \dfrac{4}{x - 5}$

ANSWERS

a. $f/g = 4x^2 + 12x - 7$;
$(f/g)(-2) = -15$

b. $m/n = 6x - 5 + \dfrac{-1}{x + 4}$

or $6x - 5 - \dfrac{1}{x + 4}$;

$(m/n)(1) = \dfrac{4}{5}$

YOUR TURN

a. Given $f(x) = 12x^4 + 36x^3 - 21x^2$ and $g(x) = 3x^2$, find f/g and $(f/g)(-2)$.

b. Given $m(x) = 6x^2 + 19x - 21$ and $n(x) = x + 4$, find m/n and $(m/n)(1)$.

8.5 Exercises

For Extra Help

MyMathLab   Videotape/DVT   InterAct Math   Math Tutor Center   Math XL.com

1. If $p(x)$ and $q(x)$ are both polynomial functions, how do you find $p + q$?

2. If $p(x)$ and $q(x)$ are both polynomial functions, how do you find $p - q$?

3. If $p(x)$ and $q(x)$ are both binomial functions, how do you find $p \cdot q$?

4. If $p(x)$ is a trinomial function and $q(x)$ is a monomial function, how do you find p/q?

For Exercises 5–14, find $f + g$ and $f - g$.

5. $f(x) = 2x, g(x) = 3x - 1$

6. $f(x) = 3x, g(x) = 4x + 2$

7. $f(x) = x - 5, g(x) = 2x - 3$

8. $f(x) = 5x + 1, g(x) = x - 8$

9. $f(x) = x^2 - 4x + 7, g(x) = x + 1$

10. $f(x) = x^2 - 3x + 1, g(x) = -x - 2$

11. $f(x) = 2x^2 - 5x - 3, g(x) = x^2 + 5$

12. $f(x) = -x^2 - 8x + 2, g(x) = x^2 + 2$

13. $f(x) = -5x^2 + 4x + 8, g(x) = 3x^2 - 2x - 1$

14. $f(x) = -2x^2 + 5x + 3, g(x) = 4x^2 + 2x - 9$

For Exercises 15–24, find $f \cdot g$.

15. $f(x) = x, g(x) = x - 5$

16. $f(x) = -2x, g(x) = x + 4$

17. $f(x) = x + 1, g(x) = x - 2$

18. $f(x) = x - 5, g(x) = x + 4$

19. $f(x) = 3x + 2, g(x) = x + 2$

20. $f(x) = 2x - 5, g(x) = 3x - 7$

21. $f(x) = x^2 - 3x + 2, g(x) = x^2 + 2x - 7$

22. $f(x) = x^2 - 2x - 1, g(x) = x^2 + 3x + 4$

23. $f(x) = 2x^2 - 3x + 1, g(x) = 3x^2 - x - 1$

24. $f(x) = 3x^2 - 2x - 1, g(x) = 2x^2 - 2x - 3$

For Exercises 25–34, find f/g.

25. $f(x) = 2x^2 - 6x, g(x) = 2x$

26. $f(x) = 3x^2 - 9x, g(x) = 3x$

27. $f(x) = 6x^2 - 3x + 6, g(x) = 3x$

28. $f(x) = -4x^2 + 4x - 8, g(x) = 2x$

29. $f(x) = x^2 - 14x + 45, g(x) = x - 9$

30. $f(x) = x^2 - x - 20, g(x) = x + 4$

31. $f(x) = 2x^2 - 3x - 5, g(x) = 2x - 5$

32. $f(x) = 3x^2 - 4x - 7, g(x) = 3x - 7$

33. $f(x) = 2x^2 - 5x - 7, g(x) = x + 5$

34. $f(x) = 6x^2 - 5x + 4, g(x) = 2x - 1$

For Exercises 35–38, find: a. $p + q$, b. $p - q$, c. $p \cdot q$, d. p/q

35. $p(x) = 5x - 1, q(x) = 2x + 3$

36. $p(x) = 3x - 4, q(x) = 7x + 2$

37. $p(x) = 2x^2 - 3x + 5, q(x) = x - 1$

38. $p(x) = 4x^2 - 4x - 1, q(x) = 2x + 3$

39. Using the functions in Exercise 35, find

    a. $(p + q)(-2)$ **b.** $(p - q)(5)$ **c.** $(p \cdot q)(-1)$ **d.** $(p/q)(7)$

40. Using the functions in Exercise 36, find

    a. $(p + q)(-1)$ **b.** $(p - q)(3)$ **c.** $(p \cdot q)(-2)$ **d.** $(p/q)(2)$

41. Using the functions in Exercise 37, find

    a. $(p + q)(3)$ **b.** $(p - q)(-4)$ **c.** $(p \cdot q)(2)$ **d.** $(p/q)(-3)$

42. Using the functions in Exercise 38, find

    a. $(p + q)(-2)$ **b.** $(p - q)(3)$ **c.** $(p \cdot q)(-2)$ **d.** $(p/q)(4)$

For Exercises 43–44: a. Find f + g.
                    b. Complete the table shown.
                    c. Graph f, g, and f + g on the same graph.
                    d. What do the table and graphs suggest about the sum of
                        two functions?

43. $f(x) = 3x + 1, g(x) = x + 5$

x	f(x)	g(x)	(f + g)(x)
−4			
−2			
0			
2			
4			

44. $f(x) = -x + 3, g(x) = 4x + 3$

x	f(x)	g(x)	(f + g)(x)
−4			
−2			
0			
2			
4			

For Exercises 45–46: **a.** *Find h − k.*

                 b. *Complete the table shown.*

                 c. *Graph h, k, and h − k on the same graph.*

                 d. *What do the table and graphs suggest about the difference of two functions?*

45. $h(x) = 2x - 9, k(x) = 3x + 5$
                                            46. $h(x) = 4x + 1, k(x) = 2x - 7$

x	h(x)	k(x)	(h − k)(x)
−4			
−2			
0			
2			
4			

x	h(x)	k(x)	(h − k)(x)
−4			
−2			
0			
2			
4			

For Exercises 47 and 48, the function w(x) describes a company's monthly costs in wages and p(x) describes the cost of production, where x represents the number of units produced.

47. $w(x) = 3x + 2, p(x) = x + 2$

    a. Find the function, $t(x)$, that describes the total cost.

    b. Calculate the total cost if $x = 200$.

    c. Graph all three functions.

48. $w(x) = 2x + 1, p(x) = x - 1$

    a. Find the function, $t(x)$, that describes the total cost.

    b. Calculate the total cost if $x = 175$.

    c. Graph all three functions.

For Exercises 49 and 50, the function r(x) describes a company's monthly revenue and c(x) describes the monthly costs where x represents the number of units produced and then sold. Use the formula Net Profit = Revenue − Cost.

49. $r(x) = x^2 - 3x - 4, c(x) = x + 8$

    a. Find the function, $p(x)$, that describes the net profit or loss.

    b. Calculate the net profit or loss if $x = 100$. Is this value a profit or a loss?

    c. Graph all three functions.

    d. How many units must be sold for the company to break even (which means the net = 0)?

50. $r(x) = x^2 - 4x - 5, c(x) = x + 1$

    a. Find the function, $p(x)$, that describes the net profit or loss.

    b. Calculate the net profit or loss if $x = 300$. Is this value a profit or a loss?

    c. Graph all three functions.

    d. How many units must be sold for the company to break even (which means the net = 0)?

For Exercises 51 and 52, the function *l*(*x*) describes the length of a box and *h*(*x*) describes the height, where *x* represents the width.

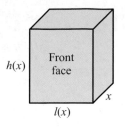

51. $l(x) = 3x + 2, h(x) = 2x$

    a. Find the function that describes the area of the front face of the box.

    b. Calculate the area of the front face if $x = 3$ feet.

52. $l(x) = 5x - 1, h(x) = 3x$

    a. Find the function that describes the area of the front face of the box.

    b. Calculate the area of the front face if $x = 2$ meters.

For Exercises 53 and 54, the function *V*(*x*) describes the volume of a box, *l*(*x*) describes its length, and *w*(*x*) describes its width.

53. $V(x) = 4x^3 - 7x^2 - 14x - 3, l(x) = x - 3, w(x) = x + 1$

    a. Find the function, $h(x)$, that describes the height of the box.

    b. Calculate the height of the box if $x = 4$ centimeters.

54. $V(x) = 6x^3 + 11x^2 - 47x + 20, l(x) = x + 4, w(x) = 2x - 1$

    a. Find the function, $h(x)$, that describes the height of the box.

    b. Calculate the height of the box if $x = 5$ inches.

REVIEW EXERCISES

For Exercises 1 and 2, factor completely.

[6.1]   **1.** $3x - 3y - xy + y^2$ [6.4]   **2.** $-3x^2 + 27$

For Exercises 3 and 4, solve.

[6.6]   **3.** $x^2 - 16 = 0$ [8.2]   **4.** $|x + 5| = -8$

For Exercises 5 and 6, graph.

[4.2]   **5.** $y = 2x - 3$ [6.7]   **6.** $y = x^2 - 9$

Chapter 8 Summary

Defined Terms

Section 8.1
Compound inequality (p. 630)
Intersection (p. 631)
Union (p. 634)

Section 8.4
Relation (p. 657)
Domain (p. 657)
Range (p. 657)
Function (p. 657)

Procedures, Rules, and Key Examples

Procedures/Rules	Key Examples

Section 8.1 Compound Inequalities

To solve a compound inequality involving "and,"
1. Solve each inequality in the compound inequality.
2. The solution set will be the intersection of the individual solution sets.

For each example, solve the compound inequality, then
a. Graph the solution set.
b. Write the solution set in set-builder notation.
c. Write the solution set in interval notation.

Example 1: $-3 < -2x - 5 \leq 1$
Solution:
a. $-2x - 5 > -3$ and $-2x - 5 \leq 1$
$-2x > 2$ and $-2x \leq 6$
$x < -1$ and $x \geq -3$

Graph:

b. Set-builder notation: $\{x \mid -3 \leq x < -1\}$
c. Interval notation: $[-3, -1)$

To solve a compound inequality involving "or,"
1. Solve each inequality in the compound inequality.
2. The solution set will be the union of the individual solution sets.

Example 2: $5x - 1 \leq -16$ or $5x - 1 > 9$
Solution:
a. $5x - 1 \leq -16$ or $5x - 1 > 9$
$5x \leq -15$ or $5x > 10$
$x \leq -3$ or $x > 2$

Graph:

b. Set-builder notation:
$\{x \mid x \leq -3 \text{ or } x > 2\}$
c. Interval notation: $(-\infty, -3] \cup (2, \infty)$

continued

Procedures/Rules	Key Examples

Section 8.2 Equations Involving Absolute Value

If $|x| = a$, where x is a variable or an expression and $a \geq 0$, then $x = a$ or $x = -a$.

To solve an equation containing a single absolute value,
1. Isolate the absolute value so that the equation is in the form $|ax + b| = c$. If $c > 0$, proceed to steps 2 and 3. If $c < 0$, the equation has no solution.
2. Separate the absolute value into two equations, $ax + b = c$ and $ax + b = -c$.
3. Solve both equations.

To solve an equation in the form $|ax + b| = |cx + d|$,
1. Separate the absolute value equation into two equations: $ax + b = cx + d$ and $ax + b = -(cx + d)$.
2. Solve both equations.

Example 1: Solve $|4x + 1| - 3 = 2$.
Solution: $|4x + 1| = 5$ Isolate the absolute value.

$$
\begin{array}{llll}
4x + 1 = 5 & \text{or} & 4x + 1 = -5 & \text{Use the} \\
4x = 4 & \text{or} & 4x = -6 & \text{absolute} \\
 & & & \text{value} \\
x = 1 & \text{or} & x = -\dfrac{3}{2} & \text{property.}
\end{array}
$$

Example 2: Solve $|5x - 7| = |3x + 1|$.
Solution: Separate into two equations.

Equal		Opposites
$5x - 7 = 3x + 1$	or	$5x - 7 = -(3x + 1)$
$2x - 7 = 1$	or	$5x - 7 = -3x - 1$
$2x = 8$	or	$8x = 6$
$x = 4$	or	$x = \dfrac{3}{4}$

Section 8.3 Inequalities Involving Absolute Value

For each example, solve the inequality, then
a. Graph the solution set.
b. Write the solution set in set-builder notation.
c. Write the solution set in interval notation.

To solve an inequality in the form $|x| < a$ where $a > 0$,
1. Rewrite as a compound inequality involving "and": $x > -a$ and $x < a$.
2. Solve the compound inequality.

Similarly, to solve $|x| \leq a$, we would write $x \geq -a$ and $x \leq a$.

Example 1: $|2x - 3| - 1 < 4$
Solution:
a. $|2x - 3| < 5$ Isolate the absolute value.

$$
\begin{array}{lll}
2x - 3 > -5 & \text{and} & 2x - 3 < 5 \quad \text{Separate} \\
2x > -2 & \text{and} & 2x < 8 \quad\;\; \text{using} \\
x > -1 & \text{and} & x < 4 \quad\;\;\; \text{"and."}
\end{array}
$$

b. Set-builder notation: $\{x \mid -1 < x < 4\}$
c. Interval notation: $(-1, 4)$

continued

Procedures/Rules	Key Examples

Section 8.3 Inequalities Involving Absolute Value (continued)

To solve an inequality in the form $|x| > a$,

1. Rewrite as a compound inequality involving "or": $x < -a$ or $x > a$.
2. Solve the compound inequality.

Similarly, to solve $|x| \geq a$, we would write $x \leq -a$ or $x \geq a$.

Example 2: $|3x - 5| - 7 \geq 4$
Solution:
a. $|3x - 5| \geq 11$ Isolate the absolute value.

$$3x - 5 \leq -11 \quad \text{or} \quad 3x - 5 \geq 11 \qquad \text{Separate}$$
$$3x \leq -6 \quad\quad \text{or} \quad\quad\quad 3x \geq 16 \qquad \text{using}$$
$$x \leq -2 \quad\quad \text{or} \quad\quad\quad x \geq \frac{16}{3} \qquad \text{"or."}$$

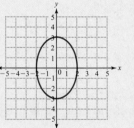

b. Set-builder notation:
$$\left\{ x \,|\, x \leq -2 \text{ or } x \geq \frac{16}{3} \right\}$$

c. Interval notation: $(-\infty, -2] \cup \left[\frac{16}{3}, \infty \right)$

Section 8.4 Functions and Graphing

Relation: A set of ordered pairs.
Domain: The set of all input values for a relation.
Range: The set of all output values for a relation.
Function: A relation in which every value in the domain is paired with exactly one value in the range.

Example 1: For the relation shown, determine the domain, range, and if it is a function.

Domain:
$\{x \,|\, -2 \leq x \leq 2\}$
Range:
$\{y \,|\, -3 \leq y \leq 3\}$
It is not a function because it does not pass the vertical line test.

To find the value of a function replace x with the given value.

Example 2: Given $f(x) = 3x^2 - 4$, find $f(2)$.
$$f(2) = 3(2)^2 - 4$$
$$= 12 - 4$$
$$= 8$$

The graph of a linear function is a line.

A linear function has the form $f(x) = mx + b$, where m is the slope and $(0, b)$ is the y-intercept. To graph, we need to find only two ordered pairs.

Example 3: Graph.
a. $f(x) = -2x + 3$

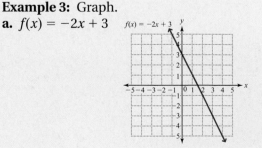

continued

Procedures/Rules	Key Examples

Section 8.4 Functions and Graphing (continued)

The graph of a quadratic function is a parabola.

Quadratic functions have the form $f(x) = ax^2 + bx + c$, and we saw their graphs are parabolas. To graph a parabola we have to find many ordered pairs.

b. $f(x) = x^2 - 2$

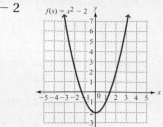

The graph of an absolute value function is a V shape.

To graph, find enough ordered pairs to generate the V shape.

c. $f(x) = |x + 2|$

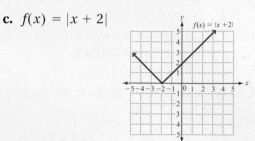

Section 8.5 Function Operations

The sum of two functions, $\boldsymbol{f + g}$, is found by $(f + g)(x) = f(x) + g(x)$.

The difference of two functions, $\boldsymbol{f - g}$, is found by $(f - g)(x) = f(x) - g(x)$.

The product of two functions, $\boldsymbol{f \cdot g}$, is found by $(f \cdot g)(x) = f(x)\, g(x)$.

The quotient of two functions, $\boldsymbol{f / g}$, is found by $(f/g)(x) = \dfrac{f(x)}{g(x)}$, where $g(x) \neq 0$.

Example 1: Given $f(x) = 2x - 5$ and $g(x) = 3x + 4$, find
a. $f + g$ **b.** $f - g$ **c.** $f \cdot g$

a. $(f + g)(x) = (2x - 5) + (3x + 4)$
$= 5x - 1$

b. $(f - g)(x) = (2x - 5) - (3x + 4)$
$= (2x - 5) + (-3x - 4)$
$= -x - 9$

c. $(f \cdot g)(x) = (2x - 5)(3x + 4)$
$= 6x^2 + 8x - 15x - 20$
$= 6x^2 - 7x - 20$

Example 2: Given $f(x) = 12x^3 - 20x^2 + 8x$ and $g(x) = 4x$, find f/g.

Solution:

$$(f/g)(x) = \frac{f(x)}{g(x)} = \frac{12x^3 - 20x^2 + 8x}{4x}$$
$$= \frac{12x^3}{4x} - \frac{20x^2}{4x} + \frac{8x}{4x}$$
$$= 3x^2 - 5x + 2, \, x \neq 0$$

Chapter 8   Review Exercises

For Exercises 1–6, answer true or false.

[8.1]   **1.** The notation $(2, 5)$ can represent an ordered pair or interval notation.

[8.2]   **2.** To solve absolute value equations, the absolute value must be isolated.

[8.4]   **3.** The notation $f(x)$ means $f \cdot x$.

[8.4]   **4.** All functions are also relations.

[8.4]   **5.** The domain of a function $f(x)$ is a set containing all values of x for which the function is defined.

[8.4]   **6.** Functions always pass the vertical line test.

For Exercises 7–10, fill in the blank.

[8.1]   **7.** For two sets A and B, the intersection of A and B, symbolized by _____, is a set containing only those elements that are in both A and B.

[8.2]   **8.** To solve an equation containing a single absolute value,
  1. Isolate the absolute value so that the equation is in the form $|ax + b| = c$.
  2. Separate the absolute value into two equations, _____ and _____.
  3. Solve both equations.

[8.4]   **9.** The set of all output values for a relation is _____.

[8.5]   **10.** The sum of two functions $f + g$ is shown by $(f + g)(x) =$ _____.

[8.1] For Exercises 11–12, find the intersection and union of the given sets.

11. $A = \{1, 5, 9\}$
$B = \{1, 5, 7, 9\}$

12. $A = \{1, 2, 3\}$
$B = \{4, 5, 6, 7\}$

Equations and
Inequalities

Exercises 11-44

[8.1] For Exercises 13–24, solve the compound inequality. Then: a. Graph the solution set.
b. Write the solution set in set-builder notation.
c. Write the solution set in interval notation.

13. $4 < -2x < 6$

14. $9 \le -3x \le 15$

15. $-3 < x + 4 < 7$

16. $0 < x - 1 \le 3$

17. $2x + 3 \ge -1$ and $2x + 3 < 3$

18. $3x - 2 > 1$ and $3x - 2 < -8$

19. $w + 4 \le -2$ or $w + 4 \ge 2$

20. $4w - 3 < 1$ or $4w - 3 > 0$

21. $2m - 5 < 0$ or $2m - 5 > 5$

22. $3x + 2 \le -2$ or $3x + 2 \ge 8$

23. $-x - 6 \le -2$ or $-x - 6 \ge 3$

24. $-4w + 1 \le -3$ or $-4w + 1 \ge 5$

[8.2] *For Exercises 25–26, solve. Then: a. Graph the solution set.*
b. Write the solution set using set-builder notation.
c. Write the solution set using interval notation.

25. Students in a biology course receive a B if the average of four tests is 80 or higher and less than 90. Suppose a student has the following scores: 80, 89, and 83. What range of scores on the fourth test would cause the student to receive a B?

26. The width of a rectangular building is to be 80 feet. What range of values can the length have so that the area is between 12,000 and 16,000 square feet?

[8.2] *For Exercises 27–34, solve.*

27. $|x| = 4$

28. $|x - 4| = 7$

29. $|2w - 1| = 3$

30. $|5r + 8| = -3$

31. $|q - 4| - 3 = 8$

32. $|5w| - 2 = 13$

33. $2|3x - 4| = 8$

34. $4 - 2|r - 5| = -8$

[8.3] *For Exercises 35–44, solve the inequality. Then: a. Graph the solution set.*

                                                     b. Write the solution set in set-builder notation.

                                                     c. Write the solution set in interval notation.

35. $|x| < 5$

36. $|p| \geq 4$

37. $|x - 3| > 7$

38. $|2m + 6| < 4$

39. $|3s - 1| \leq -2$

40. $5|b| - 2 > 3$

41. $7|m + 3| \leq 21$

42. $-2|t - 5| < -10$

43. $5 - 2|2k - 3| \leq -15$

44. $3 - 7|2p + 4| \leq 24$

[8.4] *For Exercises 45–50, identify the domain and range of the relation, then determine whether it is a function.*

45. Top-Selling Comic Books

Rank	Title
1	Superman #204
2	New Avengers #1
3	Superman Batman #8
4	Identity Crisis #1
5	Astonishing X-Men #1
6	Superman #205

Source: Diamond Comics Distributors. Rankings are based on orders placed by retailers and reflect total comics ordered as of 4/8/04.

46. **Top Business Schools for Graduate Studies as Ranked by** *U.S. News and World Report* **2004**

Rank	Business School
1	Harvard University (MA)
2	Stanford University (CA), University of Pennsylvania (Wharton)
3	Massachusetts Institute of Technology (Sloan), Northwestern University (Kellogg) (IL)
4	Columbia University (NY)

Source: U.S. News and World Report

47. $\{(2, 1), (4, -3), (2, 4), (3, -3)\}$

48. $\{(0, 3), (-2, 3), (1, 4)\}$

49.

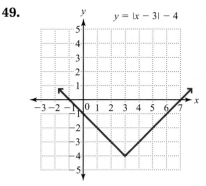

$y = |x - 3| - 4$

50.

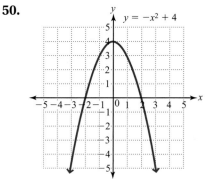

$y = -x^2 + 4$

Equations and Inequalities

Exercises 49-64

[8.4] *For Exercises 51 and 52, find the specified value of the function.*

51. Find $f(3)$ given $f(x) = 3x^2 - 4x + 2$.

52. Find $f(-2)$ given $f(x) = -|2x - 5|$.

[8.4] *For Exercises 53–56, graph.*

53. $f(x) = -3x + 2$

54. $f(x) = 2x^2 - 3$

55. $f(x) = |x - 2|$

56. $f(x) = -2|x + 1| + 3$

$$\text{b. } f - g$$
$$\text{c. } f \cdot g$$

57. $f(x) = -3x + 9, g(x) = 4x + 1$

58. $f(x) = -x - 2, g(x) = 8x + 2$

59. $f(x) = 3x^2 - x + 5, g(x) = x + 2$

60. $f(x) = x^2 - 2x - 1, g(x) = 3x + 1$

For Exercises, 61–62, find f/g.

61. $f(x) = 6x^4 - 15x^3 + 12x^2, g(x) = 3x^2$

62. $f(x) = 2x^3 - 17x^2 + 36x - 17, g(x) = 2x - 5$

[8.5] *For Exercises 63–64, solve.*

63. The function $w(x) = 3x + 5$ describes a company's monthly expenditure on wages. The function $p(x) = x - 1$ describes the total cost of production. Note that x represents the number of units produced.

   a. Find the function, $c(x)$, that describes the total cost.

   b. If $x = 225$, then calculate the total cost.

   c. Graph all three functions.

64. The function $V(x) = 6x^3 + 41x^2 + 26x - 24$ describes the volume of a box, $l(x) = 3x + 4$ describes its length, and $w(x) = x + 6$ describes its width.

   a. Find the function, $h(x)$, that describes the height of the box.

   b. Calculate the height of the box if $x = 9$ inches.

Chapter 8 Practice Test

1. Find the intersection and union of the given sets. $A = \{h, o, m, e\}$
$$B = \{h, o, u, s, e\}$$

For 2–5, solve.

2. $|x + 3| = 5$

3. $3 - |2x - 3| = -6$

4. $|2x + 3| = |x - 5|$

5. $|5x - 4| = -3$

For 6–15, solve. Then: a. Graph the solution set.
  b. Write the solution set in set-builder notation.
  c. Write the solution set in interval notation.

6. $-3 < x + 4 \leq 7$

7. $4 < -2x \leq 6$

8. $5x + 2 \leq -3$ or $5x + 2 \geq 12$

9. $6 - 2n < 2$ or $6 - 2n < 4$

10. $|x + 4| < 9$

11. $2|x - 1| > 4$

12. $3 - 2|x + 4| > -3$

13. $|3y - 2| < -2$

14. $|8t + 4| \geq -12$

15. $2|3x - 4| \leq 10$

For Exercise 16, solve. Then: a. Graph the solution set.
  b. Write the solutions set using set-builder notation.
  c. Write the solution set using interval notation.

16. A mail-order book club offers one bonus book if the total of an order is from $40 to $50. If a person were to order only books that cost $5, what range would the number of books ordered have to be in order to receive a bonus book?

17. Identify the domain and range of the relation shown in the graph to the right, then determine whether it is a function.

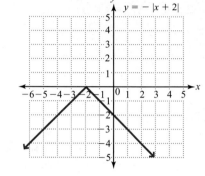

18. Given $f(x) = x^2 - 5x + 4$, find $f(-2)$.

For Exercises 19–21, graph.

19. $f(x) = -\dfrac{3}{4}x + 1$

20. $f(x) = x^2 - 3$

21. $f(x) = |x + 1| - 2$

For Exercises 22 and 23, find a. f + g
  b. f − g
  c. f · g

22. $f(x) = 6x - 9, g(x) = x + 2$

23. $f(x) = x^2, g(x) = 3x^2 - 2$

24. Given $f(x) = 20x^4 + 36x^3 - 28x^2$ and $g(x) = 4x^2$, find f/g.

25. The function $A(x) = 36x^2 - x - 2$ describes the area of the front face of a box and $l(x) = 4x - 1$ describes its length, where x represents its width.

   a. Find the function, $h(x)$, that describes the height of the box.
   b. Calculate the height of the box if $x = 15$ cm.

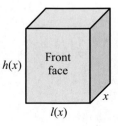

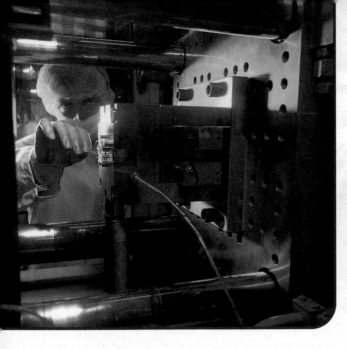

Systems of Equations

"Everybody's a mad scientist, and life is their lab. We're all trying to experiment to find a way to live, to solve problems, to fend off madness and chaos."

—David Cronenberg

"The problem is that we attempt to solve the simplest questions cleverly, thereby rendering them unusually complex. One should seek the simple solution."

—Anton Chekhov

In Chapter 3, we encountered problems involving two unknown amounts, such as finding two consecutive integers whose sum is 63. At that time, we only knew how to solve equations with a *single* variable, so we had to let one of the unknown amounts be represented by a variable, then describe the other unknown amount in terms of that variable. For example, in the consecutive integer problem just mentioned, we would let x represent the first integer and then describe the consecutive integer as $x + 1$.

In this chapter, we will again solve problems involving two or more unknowns; however, we will now assign a different variable to each unknown and translate the problem to several equations, called a *system of equations*. As we will see, it is easier to translate a problem involving two or more unknowns to a system of equations than trying to translate the problem to a single equation in terms of a single variable. The challenge with systems of equations is in solving them. We will learn several techniques for solving a system of equations: graphing (Section 9.1), substitution (Section 9.2), elimination (Section 9.3), matrices (Section 9.5), and Cramer's Rule (Section 9.6). Finally, in Section 9.7, we will solve systems of linear inequalities.

9.1 Solving Systems of Linear Equations Graphically

OBJECTIVES
1. Determine if an ordered pair is a solution for a system of equations.
2. Solve a system of linear equations graphically.
3. Classify systems of linear equations in two unknowns.

OBJECTIVE 1. Determine if an ordered pair is a solution for a system of equations. Problems involving multiple unknowns can be represented by a group of equations called a **system of equations**.

> **DEFINITION** **System of equations:** A group of two or more equations.

For example, look at the following problem: The sum of two numbers is 3. Twice the first number plus three times the second number is 8. What are the two numbers? If x represents the first number and y represents the second number, then we can translate each sentence to an equation.

Sentence in the problem:	Translation:	
The sum of two numbers is 3.	$x + y = 3$	(Equation 1)
Twice the first number plus three times the second number is 8.	$2x + 3y = 8$	(Equation 2)

The two equations together form a system of equations that describes the problem.

$$\text{System of equations} \quad \begin{cases} x + y = 3 & \text{(Equation 1)} \\ 2x + 3y = 8 & \text{(Equation 2)} \end{cases}$$

A **solution for a system of equations** is an ordered set of numbers that satisfies *all* equations in the system.

> **DEFINITION** **Solution for a system of equations:** An ordered set of numbers that makes all equations in the system true.

For example, in the preceding system of equations, $x = 1$ and $y = 2$ make both equations true, so they form a solution to the system of equations. We can check by substituting the values in place of the corresponding variables.

Equation 1:

$x + y = 3$

$1 + 2 \stackrel{?}{=} 3$

$3 = 3$ True

Equation 2:

$2x + 3y = 8$

$2(1) + 3(2) \stackrel{?}{=} 8$

$2 + 6 = 8$ True

We can write a solution to a system of equations in two variables as an ordered pair, in this case: (1, 2). This suggests the following procedure for checking solutions.

PROCEDURE Checking a Solution to a System of Equations

To verify or check a solution to a system of equations:

1. Replace each variable in each equation with its corresponding value.
2. Verify that each equation is true.

EXAMPLE 1 Determine whether each ordered pair is a solution to the system of equations.

$$\begin{cases} x + y = 5 & \text{(Equation 1)} \\ y = 2x - 4 & \text{(Equation 2)} \end{cases}$$

a. $(-1, 6)$

Solution

$x + y = 5$ (Equation 1) $y = 2x - 4$ (Equation 2)

$-1 + 6 \stackrel{?}{=} 5$

$5 = 5$ True.

$6 \stackrel{?}{=} 2(-1) - 4$ **In both equations, replace x with -1 and y with 6.**

$6 \stackrel{?}{=} -2 - 4$

$6 = -6$ False.

Because $(-1, 6)$ does not satisfy both equations, it is not a solution for the system.

b. $(3, 2)$

Solution

$x + y = 5$ (Equation 1) $y = 2x - 4$ (Equation 2)

$3 + 2 \stackrel{?}{=} 5$

$5 = 5$ True.

$2 \stackrel{?}{=} 2(3) - 4$ **In both equations, replace x with 3 and y with 2.**

$2 \stackrel{?}{=} 6 - 4$

$2 = 2$ True.

Because $(3, 2)$ satisfies both equations, it is a solution for the system.

YOUR TURN Determine whether each ordered pair is a solution to the system of equations.

$$\begin{cases} 2x + y = 5 & \text{(Equation 1)} \\ y = 1 - x & \text{(Equation 2)} \end{cases}$$

a. $(3, -1)$ **b.** $(4, -3)$

OBJECTIVE 2. Solve a system of linear equations graphically. A system of two linear equations in two variables can have one solution, no solution, or an infinite number of solutions. To see why, let's look at the graphs of the equations in three different systems. First, let's graph the equations in the system from Example 1.

$$\begin{cases} x + y = 5 \\ y = 2x - 4 \end{cases}$$

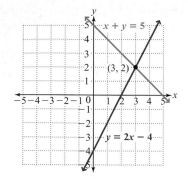

Notice that the graphs of $x + y = 5$ and $y = 2x - 4$ intersect at the point (3, 2), which is the solution for the system. If two linear graphs intersect at a single point, then the system has a *single solution* at the point of intersection.

Note: *These lines also have different slopes, which is always the case when a system of two linear equations has a single solution.*

Now let's see a system with no solution. Look at the graphs of the equations in the system $\begin{cases} y = 3x + 1 \\ y = 3x - 2 \end{cases}$.

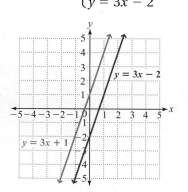

Notice the equations $y = 3x + 1$ and $y = 3x - 2$ have the same slope, 3, so their graphs are parallel lines. Since the graphs of the equations in this system have no point of intersection, the system has no solution.

Last, let's examine a system with an infinite number of solutions, such as

$$\begin{cases} x + 2y = 4 & \text{(Equation 1)} \\ 2x + 4y = 8 & \text{(Equation 2)} \end{cases}$$

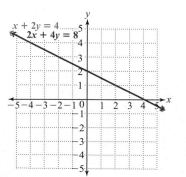

Notice that the graphs of the two equations are identical. Consequently, an infinite number of solutions to this system of equations lie on that line.

We can see without graphing that the two equations describe the same line by writing them in slope-intercept form.

$$x + 2y = 4 \quad \text{(Equation 1)} \qquad 2x + 4y = 8 \quad \text{(Equation 2)}$$

Subtract x from both sides.

$$2y = -x + 4 \qquad\qquad 4y = -2x + 8$$

Subtract $2x$ from both sides.

Divide both sides by 2 to isolate y.

$$y = -\frac{1}{2}x + \frac{4}{2} \qquad\qquad y = -\frac{2}{4}x + \frac{8}{4}$$

Divide both sides by 4 to isolate y.

Simplify the fraction.

$$y = -\frac{1}{2}x + 2 \qquad\qquad y = -\frac{1}{2}x + 2$$

Simplify the fractions.

Our work suggests the following graphical method for solving systems of equations.

PROCEDURE **Solving Systems of Equations Graphically**

To solve a system of linear equations graphically:

1. Graph each equation.
   a. If the lines intersect at a single point, then the coordinates of that point form the solution.
   b. If the lines are parallel, then there is no solution.
   c. If the lines are identical, then there are an infinite number of solutions, which are the coordinates of all the points on that line.
2. Check the solution in the original equations.

EXAMPLE 2 Solve the system of equations graphically.

a. $\begin{cases} y = -3x + 1 & \text{(Equation 1)} \\ 2x - y = 4 & \text{(Equation 2)} \end{cases}$

Solution Graph each equation.

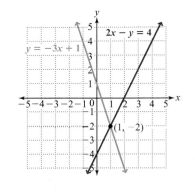

The lines intersect at a single point, which appears to be $(1, -2)$. We can verify that it is the solution by substituting into the equations.

$$y = -3x + 1 \qquad\qquad 2x - y = 4$$
$$-2 \overset{?}{=} -3(1) + 1 \qquad 2(1) - (-2) \overset{?}{=} 4$$
$$-2 \overset{?}{=} -3 + 1 \qquad\qquad 2 + 2 \overset{?}{=} 4$$
$$-2 = -2 \quad \text{True} \qquad\qquad 4 - 4 \quad \text{True}$$

Warning: Graphing by hand can be imprecise and not all solutions have integer coordinates, so you should always check your solutions by substituting them into the original equations.

Answer Because $(1, -2)$ makes both equations true, it is the solution.

b. $\begin{cases} 2x - 3y = 6 & \text{(Equation 1)} \\ y = \dfrac{2}{3}x + 1 & \text{(Equation 2)} \end{cases}$

Solution Graph each equation.

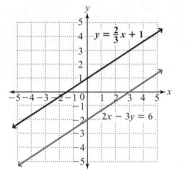

The lines appear to be parallel, which we can verify by comparing the slopes. The slope of equation 2 is $\dfrac{2}{3}$. To determine the slope of Equation 1, we rewrite it in slope-intercept form.

$$2x - 3y = 6$$
$$-3y = -2x + 6 \qquad \text{Subtract } 2x \text{ from both sides.}$$
$$y = \dfrac{2}{3}x - 2 \qquad \begin{array}{l}\text{Divide both sides by } -3 \text{ to} \\ \text{isolate } y.\end{array}$$

The slope of equation 1 is also $\dfrac{2}{3}$. Since the slopes are the same, the lines are indeed parallel.

Answer The system has no solution.

c. $\begin{cases} 6x - 2y = 2 & \text{(Equation 1)} \\ y = 3x - 1 & \text{(Equation 2)} \end{cases}$

Solution Graph each equation.

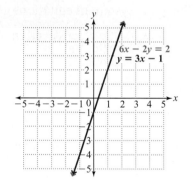

The lines appear to be identical, which we can verify by rewriting $6x - 2y = 2$ in slope-intercept form.

$$6x - 2y = 2$$
$$-2y = -6x + 2 \qquad \text{Subtract } 6x \text{ from both sides.}$$
$$y = 3x - 1 \qquad \begin{array}{l}\text{Divide both sides by } -2 \text{ to} \\ \text{isolate } y.\end{array}$$

The equations are identical.

Answer All ordered pairs along the line $6x - 2y = 2$ (or $y = 3x - 1$).

YOUR TURN Solve the system of equations graphically.

a. $\begin{cases} y = \dfrac{3}{4}x + 1 \\ 2x - y = -6 \end{cases}$

b. $\begin{cases} x + y = -3 \\ y = -x + 2 \end{cases}$

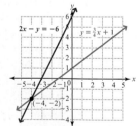
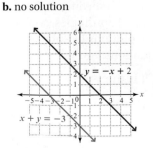

Calculator TIPS

To solve a system of equations using a graphing calculator, begin by graphing each equation. Remember that you must write the equations in slope-intercept form in order to enter them in the calculator. Using the $\boxed{\text{Y=}}$ *key, enter one equation in Y1 and the second equation in Y2, then press* $\boxed{\text{GRAPH}}$ *to see both graphs.*

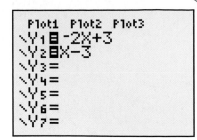

You can find the coordinates of the point of intersection by using the CALC feature found by pressing $\boxed{\text{2nd}}$ $\boxed{\text{TRACE}}$. *Select INTERSECT from the menu. The calculator will then prompt you to verify which lines you want it to find the intersection of, so press* $\boxed{\text{ENTER}}$ *for each line. It will then prompt you to indicate which point of intersection you want. Because these straight lines will have only one point of intersection, you can simply press* $\boxed{\text{ENTER}}$. *The coordinates of the point of intersection will appear at the bottom of the screen.*

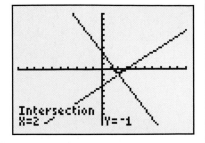

OBJECTIVE 3. Classify systems of linear equations in two unknowns. Systems can be classified as **consistent** or **inconsistent**.

DEFINITIONS **Consistent system of equations:** A system of equations that has at least one solution.
Inconsistent system of equations: A system of equations that has no solution.

We can classify the equations in a system as **dependent** or **independent**. Complete definitions of dependent and independent equations are beyond the scope of this course. For our purposes, we will supply the following definitions that apply for two linear equations in two unknowns.

DEFINITIONS **Dependent linear equations in two unknowns:** Equations with identical graphs.
Independent linear equations in two unknowns: Equations that have different graphs.

We summarize our discussion of classifying systems and equations in those systems with the following procedure.

To classify a system of two linear equations in two unknowns, write the equations in slope-intercept form and compare the slopes and y-intercepts.

Consistent system: The system has a single solution at the point of intersection.

Independent equations: The graphs are different and intersect at one point. They have different slopes.

Consistent system: The system has an infinite number of solutions.

Dependent equations: The graphs are identical. They have the same slope and same y-intercept.

Inconsistent system: The system has no solution.

Independent equations: The graphs are different and are parallel. Though they have the same slopes, they have different y-intercepts.

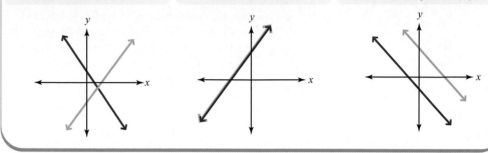

EXAMPLE 3 For each of the systems of equations in Example 2, is the system consistent or inconsistent, are the equations dependent or independent, and how many solutions does the system have?

Solution When we solved the system $\begin{cases} y = -3x + 1 \\ 2x - y = 4 \end{cases}$ in Example 2(a), we found the graphs intersected at a single point. Therefore, the system is consistent (it has a solution), the equations are independent (different graphs), and the system has one solution: $(1, -2)$.

When we solved the system $\begin{cases} 2x - 3y = 6 \\ y = \dfrac{2}{3}x + 1 \end{cases}$ in Example 2(b), we found the graphs to be parallel lines. Therefore, the system is inconsistent (no solution), the equations are independent (different graphs), and the system has no solution.

When we solved the system $\begin{cases} 6x - 2y = 2 \\ y = 3x - 1 \end{cases}$ in Example 2(c), we found the graphs to coincide. Therefore, the system is consistent (it has a solution), the equations are dependent (same graph), and the system has an infinite number of solutions.

YOUR TURN **a.** Determine whether the following system is consistent with independent equations, consistent with dependent equations, or inconsistent with independent equations.

b. How many solutions does the system have?

ANSWERS

a. inconsistent with independent equations

b. no solution

$$\begin{cases} 5x + y = 1 \\ y = -5x - 2 \end{cases}$$

9.1 Exercises

For Extra Help

MyMathLab

MyMathLab

Videotape/DVT

InterAct Math

Tutor Center

Math Tutor Center

Math XL.com

1. How do you check a solution for a system of equations?

2. If a system of linear equations has at least one solution (consistent), what does this indicate about the graphs of the equations?

3. If a system of linear equations has no solution (inconsistent), what does this indicate about the graphs of the equations?

4. If two equations are dependent, what does this mean about their graphs?

5. Can a system of linear equations have more than one solution? Explain.

6. What are some weaknesses in using graphing to find a solution to a system of equations?

For Exercises 7–16, determine whether the given ordered pair is a solution to the given system of equations.

7. $(1, 1);$ $\begin{cases} x = y \\ 9y - 13 = -4x \end{cases}$

8. $(3, -2);$ $\begin{cases} 2x = -3y \\ x - 2y = 7 \end{cases}$

9. $(2, -3);$ $\begin{cases} 4x + 3y = -1 \\ 2x - 5y = -11 \end{cases}$

10. $(4, 3);$ $\begin{cases} 3x + 4y = 24 \\ 3x + 2y = 18 \end{cases}$

11. $\left(\dfrac{2}{3}, \dfrac{4}{3}\right);$ $\begin{cases} x + y = 2 \\ 2x - y = 0 \end{cases}$

12. $\left(\dfrac{3}{2}, -\dfrac{3}{2}\right);$ $\begin{cases} x - y = 3 \\ x + y = 0 \end{cases}$

13. $(2, -4);$ $\begin{cases} 0.5x + 1.25y = 4 \\ x + y = -2 \end{cases}$

14. $(1, 5);$ $\begin{cases} 0.25x + 0.75y = 4 \\ x + y = -7 \end{cases}$

15. $(5, -2);$ $\begin{cases} x + 0.5y = 4 \\ \dfrac{1}{5}x - \dfrac{1}{2}y = 2 \end{cases}$

16. $(-1, -2);$ $\begin{cases} \dfrac{2}{3}x - y = 7 \\ 0.25x + y = -1 \end{cases}$

For Exercises 17–38, solve the system graphically.

17. $\begin{cases} x + y = 5 \\ 2x - y = 7 \end{cases}$

18. $\begin{cases} x - y = 4 \\ x + y = 8 \end{cases}$

19. $\begin{cases} y = x - 1 \\ 3x + y = 11 \end{cases}$

20. $\begin{cases} y = x - 5 \\ 2x - y = 8 \end{cases}$

21. $\begin{cases} x + y = 1 \\ y = 7 + x \end{cases}$

22. $\begin{cases} x + y = 4 \\ y - x = 6 \end{cases}$

23. $\begin{cases} x = y - 4 \\ x + y = 2 \end{cases}$

24. $\begin{cases} x = y + 4 \\ x + 4y = -1 \end{cases}$

25. $\begin{cases} 4x + y = 8 \\ 5x + 3y = 3 \end{cases}$ **26.** $\begin{cases} 2x + 3y = 6 \\ x + 4 = 2y \end{cases}$ **27.** $\begin{cases} 2x + y = 5 \\ x + 2y = -2 \end{cases}$ **28.** $\begin{cases} 2x + y = 1 \\ 2x + 3y = 3 \end{cases}$

29. $\begin{cases} 2x + 2y = -2 \\ 3x - 2y = 12 \end{cases}$ **30.** $\begin{cases} 2x - 4y = 2 \\ 3x - 2y = -1 \end{cases}$ **31.** $\begin{cases} y = 3 \\ x = -2 \end{cases}$ **32.** $\begin{cases} x = 1 \\ y = -4 \end{cases}$

33. $\begin{cases} y = -\dfrac{2}{5}x \\ x - y = 7 \end{cases}$ **34.** $\begin{cases} y = \dfrac{1}{4}x \\ 3x + 2y = 14 \end{cases}$ **35.** $\begin{cases} 3y = -2x + 6 \\ 4x + 6y = 18 \end{cases}$ **36.** $\begin{cases} 2x + y = 0 \\ 2y = 3 - 4x \end{cases}$

37. $\begin{cases} 2x + 3y = 2 \\ 6x + 9y = 6 \end{cases}$ **38.** $\begin{cases} x + y = 4 \\ -2x - 2y = -8 \end{cases}$

For Exercises 39–44, a. determine whether the graph shows a consistent system with independent equations, a consistent system with dependent equations, or an inconsistent system with independent equations. b. How many solutions does the system have?

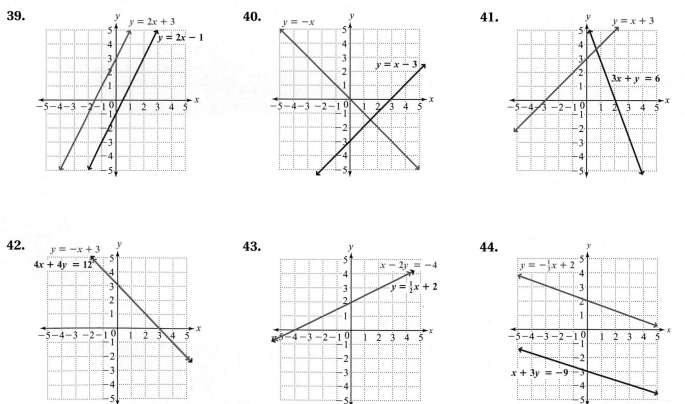

39. $y = 2x + 3$, $y = 2x - 1$

40. $y = -x$, $y = x - 3$

41. $y = x + 3$, $3x + y = 6$

42. $y = -x + 3$, $4x + 4y = 12$

43. $x - 2y = -4$, $y = \frac{1}{2}x + 2$

44. $y = -\frac{1}{3}x + 2$, $x + 3y = -9$

For Exercises 45–52, **a.** *determine whether the system of equations is consistent with independent equations, consistent with dependent equations, or inconsistent with independent equations.* **b.** *How many solutions does the system have?*

45. $\begin{cases} x - 4y = 9 \\ 2x - 3y = 8 \end{cases}$

46. $\begin{cases} -4y = 2x - 12 \\ x - 2y = -4 \end{cases}$

47. $\begin{cases} 3x + 2y = 5 \\ -6x - 4y = 1 \end{cases}$

48. $\begin{cases} 3x + 4y = 15 \\ 9x + 12y = 8 \end{cases}$

49. $\begin{cases} x - y = 1 \\ 2x - 2y = 2 \end{cases}$

50. $\begin{cases} 2x - y = 1 \\ -4x + 2y = -2 \end{cases}$

51. $\begin{cases} 2x - y = 1 \\ x + y = 4 \end{cases}$

52. $\begin{cases} x - y = 5 \\ x + 5y = 11 \end{cases}$

53. A business breaks even when its costs and revenue are equal. To the right is a graph of the cost of a product based on the number of units produced and the revenue based on the number of units sold.

   a. How many units must be produced and sold in order to break even?

   b. What amount of revenue is needed for the business to break even?

   c. Write an inequality that describes the number of units that must be sold for the business to make a profit.

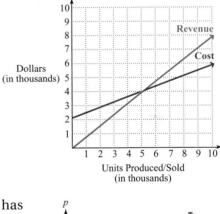

54. John's job requires that he drive between two cities using a toll road. He has two choices for toll plans. The first plan is to stop and pay cash at each booth, which amounts to $15 per trip. The second plan is to purchase a toll pass for $12 per trip, which allows him to drive through the toll booths without stopping. However, to use the pass he must also purchase a transponder for a one-time fee of $24. To the right, the cost of using each plan has been graphed, where x represents the number of trips taken.

   a. Which plan costs less if John makes 6 trips?

   b. How many trips must he make for the costs to be equal?

   c. What other factors should he consider in choosing a plan?

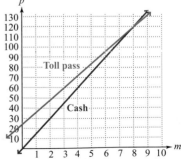

55. The graph shows two long-distance plans. The Unlimited plan allows you to call anywhere in the United States at any time for $30 per month. The One-Rate plan costs $3.75 per month plus 7¢ per minute for each long-distance call.

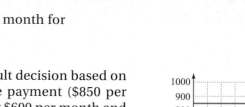

a. Which plan costs less if a person never used more than 350 minutes of long-distance service each month?

b. How many minutes would a person have to talk in a month for the plans to cost the same amount?

c. How many minutes would a person need to talk per month for the Unlimited plan to cost less?

56. Deciding whether to purchase or rent a home is a difficult decision based on many factors. The graph shows the monthly mortgage payment ($850 per month) compared to the monthly rent payment (initially $600 per month and raised $50 every six months) for a 1200-square-foot house.

a. If a customer plans to stay only two years (ignoring any tax benefits), which costs less per month, renting or buying?

b. After how many months will the monthly rent be the same as the monthly mortgage payment?

c. In the fourth year, which costs less per month?

d. What other factors should a person consider when deciding whether to purchase or rent a home?

REVIEW EXERCISES

[1.7] **1.** Simplify: $2y - 5(y + 7)$ [2.4] **2.** Solve for y: $x + 3y = 10$

[2.4] **3.** Solve for x: $\frac{1}{3}x - 2y = 10$

[2.1] **4.** A cone made of marble is used as a decorative accent piece. It is 10 inches tall and the radius of its base is 2 inches. Find its volume.

(Use $V = \frac{1}{3}\pi r^2 h$.)

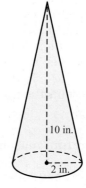

[3.3] **5.** One number exceeds another number by 62. If five times the smaller is subtracted from two times the larger, the difference is 91. What are the numbers?

[4.7] **6.** Given $f(x) = 2x + 9$, find $f(x + 4)$.

9.2 Solving Systems of Linear Equations by Substitution

OBJECTIVES

1. Solve systems of linear equations using substitution.

2. Solve applications involving two unknowns using a system of equations.

In Section 9.1, we solved systems of equations by graphing. However, if the solution to a system of equations contains fractions or decimal numbers, it could be difficult to determine those values using the graphing method. For example, look at the following system of equations:

$$\begin{cases} x + 3y = 10 & \text{(Equation 1)} \\ y = x + 4 & \text{(Equation 2)} \end{cases}$$

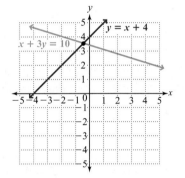

In the graph, notice that the point of intersection is between grid lines, which means the solution contains fractions. We could make a guess at the solution, then check our guess in the equations, but it would be better to use a method that does not require guessing. One such method is the *substitution* method.

OBJECTIVE 1. Solve systems of linear equations using substitution. Remember that in a system's solution, the same x and y values satisfy both equations. In the preceding system, notice Equation 2 indicates that y is equal to $x + 4$, so y must be equal to $x + 4$ in Equation 1 as well. Therefore, we can substitute $x + 4$ for y in equation 1.

$$x + 3y = 10$$
$$x + 3\overbrace{(x + 4)} = 10$$

Now we have an equation in terms of a single variable, x, which allows us to solve for x.

$$x + 3(x + 4) = 10$$
$$x + 3x + 12 - 10 \qquad \text{Distribute 3.}$$
$$4x + 12 = 10 \qquad \text{Combine like terms.}$$
$$4x = -2 \qquad \text{Subtract 12 from both sides.}$$
$$x = -\frac{1}{2} \qquad \text{Divide both sides by 4 and simplify.}$$

We can now find the y-value by substituting $-\frac{1}{2}$ for x in either of the original equations. Equation 2 is easier because y is isolated.

$$y = x + 4$$

$$y = -\frac{1}{2} + 4 \qquad \text{Substitute } -\frac{1}{2} \text{ for } x.$$

$$y = 3\frac{1}{2}$$

The solution to the system of equations is $\left(-\frac{1}{2}, 3\frac{1}{2}\right)$. Notice in our graph the lines do appear to intersect at $\left(-\frac{1}{2}, 3\frac{1}{2}\right)$, which supports our solution graphically.

PROCEDURE   Solving Systems of Two Equations Using Substitution

To find the solution of a system of two linear equations using the substitution method:

1. Isolate one of the variables in one of the equations.
2. In the other equation, substitute the expression you found in step 1 for that variable.
3. Solve this new equation. (It will now have only one variable.)
4. Using one of the equations containing both variables, substitute the value you found in step 3 for that variable and solve for the value of the other variable.
5. Check the solution in the original equations.

EXAMPLE 1   Solve the system of equations using substitution.

$$\begin{cases} x = 3 - y \\ 5x + 3y = 5 \end{cases}$$

Solution

Step 1: We isolate a variable in one of the equations. Since x is already isolated in the first equation, we proceed to step 2

Step 2: Substitute the expression $3 - y$ for x in the second equation.

$$5x + 3y = 5$$

$$5(3 - y) + 3y = 5 \qquad \text{Substitute } 3 - y \text{ for } x.$$

Step 3: We now have a linear equation in terms of a single variable, y, so we solve for y.

$$15 - 5y + 3y = 5 \qquad \text{Distribute 5.}$$

$$15 - 2y = 5 \qquad \text{Combine like terms.}$$

$$-2y = -10 \qquad \text{Subtract 15 from both sides.}$$

$$y = 5 \qquad \text{Divide both sides by } -2 \text{ to isolate } y.$$

Step 4: Now we solve for x by substituting 5 for y in one of the equations containing both variables. We will use $x = 3 - y$.

$$x = 3 - y$$
$$x = 3 - 5 \qquad \text{Substitute 5 for } y.$$
$$x = -2$$

Note: *We chose* $x = 3 - y$ *because x is isolated.*

The solution is $(-2, 5)$. We will let the reader check that $(-2, 5)$ makes both equations true.

YOUR TURN Solve the system of equations using substitution.

$$\begin{cases} x - 3y = 11 \\ y = 2x + 3 \end{cases}$$

In step 1, if no variable is isolated in any of the given equations, then we select an equation and isolate one of its variables. It is easiest to isolate a variable that has a coefficient of 1, as this will avoid fractions.

EXAMPLE 2 Solve the system of equations using substitution.

$$\begin{cases} x + y = 6 \\ 5x + 2y = 8 \end{cases}$$

Solution

Step 1: Isolate a variable in one of the equations. Since both x and y have a coefficient of 1 in $x + y = 6$, isolating either variable is easy. We will isolate y.

$$x + y = 6$$
$$y = 6 - x \qquad \text{Subtract } x \text{ from both sides to isolate } y.$$

Step 2: Substitute $6 - x$ for y in the second equation.

$$5x + 2y = 8$$
$$5x + 2(6 - x) = 8 \qquad \text{Substitute } 6 - x \text{ for } y.$$

Step 3: Solve for x using the equation we found in step 2.

$$5x + 12 - 2x = 8 \qquad \text{Distribute 2.}$$
$$3x + 12 = 8 \qquad \text{Combine like terms.}$$
$$3x = -4 \qquad \text{Subtract 12 from both sides.}$$
$$x = -\frac{4}{3} \qquad \text{Divide both sides by 3 to isolate } x.$$

ANSWER

$(-4, -5)$

Step 4: Find the value of y by substituting $-\dfrac{4}{3}$ for x in one of the equations containing both variables. Since we isolated y in $y = 6 - x$, we will use this equation.

$$y = 6 - x$$

$$y = 6 - \left(-\frac{4}{3}\right) \qquad \text{Substitute } -\frac{4}{3} \text{ for } x.$$

$$y = \frac{18}{3} + \frac{4}{3} \qquad \text{Rewrite as addition of equivalent fractions with a common denominator.}$$

$$y = \frac{22}{3}$$

The solution is $\left(-\dfrac{4}{3}, \dfrac{22}{3}\right)$.

Step 5: We will leave the check to the reader.

YOUR TURN Solve the system of equations by substitution.

$$\begin{cases} x - y = 1 \\ 3x + y = 11 \end{cases}$$

If none of the variables in the system of equations has a coefficient of 1, then for step 1, select the equation that seems easiest to work with and the variable in that equation that seems easiest to isolate. When selecting that variable, recognize that you'll eventually divide out its coefficient, so choose the variable whose coefficient will divide evenly into most, if not all of the other numbers in its equation.

EXAMPLE 3 Solve the system of equations using substitution.

$$\begin{cases} 3x + 4y = 6 \\ 2x + 6y = -1 \end{cases}$$

Note: *We chose to isolate x because its coefficient, 3, divides evenly into one of the other numbers in the equation, 6, whereas y's coefficient, 4, does not divide evenly into any of the other numbers.*

Solution

Step 1: We will isolate x in $3x + 4y = 6$.

$$3x + 4y = 6$$

$$3x = 6 - 4y \qquad \text{Subtract } 4y \text{ from both sides.}$$

$$x = 2 - \frac{4}{3}y \qquad \text{Divide both sides by 3 to isolate } x.$$

Step 2: Substitute $2 - \dfrac{4}{3}y$ for x in $2x + 6y = -1$.

$$2x + 6y = -1$$

$$2\left(2 - \frac{4}{3}y\right) + 6y = -1 \qquad \text{Substitute } 2 - \frac{4}{3}y \text{ for } x.$$

ANSWER

$(3, 2)$

Step 3: Solve for y using the equation we found in step 2.

$$4 - \frac{8}{3}y + 6y = -1 \qquad \text{Distribute to clear the parentheses.}$$

$$3 \cdot 4 - 3 \cdot \frac{8}{3}y + 3 \cdot 6y = 3 \cdot (-1) \qquad \text{Multiply both sides by 3 to clear the fraction.}$$

$$12 - 8y + 18y = -3$$

$$12 + 10y = -3 \qquad \text{Combine like terms.}$$

$$10y = -15 \qquad \text{Subtract 12 from both sides.}$$

$$y = -\frac{15}{10} \qquad \text{Divide both sides by 10 to isolate } y.$$

$$y = -\frac{3}{2} \qquad \text{Simplify.}$$

Step 4: Find the value of x by substituting $-\frac{3}{2}$ for y in one of the equations containing both variables. Since x is isolated in $x = 2 - \frac{4}{3}y$, we will use that equation.

$$x = 2 - \frac{4}{3}y$$

$$x = 2 - \frac{\overset{2}{\cancel{4}}}{\cancel{3}} \cdot \left(-\frac{\overset{1}{\cancel{3}}}{\cancel{2}} \right) \qquad \text{Substitute } -\frac{3}{2} \text{ for } y \text{ and simplify.}$$

$$x = 2 + 2$$

$$x = 4$$

The solution is $\left(4, -\frac{3}{2} \right)$.

Step 5: We will leave the check to the reader.

YOUR TURN Solve the system of equations using substitution.

$$\begin{cases} 5x - 2y = 10 \\ 3x - 6y = 2 \end{cases}$$

Inconsistent Systems of Equations

In Section 9.1, we learned that inconsistent systems of linear equations have no solution because the graphs of the equations are parallel lines. Let's see what happens when we solve an inconsistent system of equations using the substitution method.

ANSWER

$\left(\frac{7}{3}, \frac{5}{6} \right)$

EXAMPLE 4 Solve the system of equations using substitution.

$$\begin{cases} x + 2y = 4 \\ y = -\dfrac{1}{2}x + 1 \end{cases}$$

Solution Since y is isolated in the second equation, we substitute $-\dfrac{1}{2}x + 1$ in place of y in the first equation.

$$x + 2y = 4$$

$$x + 2\overbrace{\left(-\dfrac{1}{2}x + 1\right)} = 4 \qquad \text{Substitute } -\dfrac{1}{2}x + 1 \text{ for } y.$$

Now, solve for x.

$$x - x + 2 = 4 \qquad \text{Distribute to clear the parentheses.}$$
$$2 = 4 \qquad \text{Combine like terms.}$$

Notice that $2 = 4$ no longer has a variable and is false. This false equation with no variable indicates that the system is inconsistent and has no solution.

> **Connection** If we were to solve the system in Example 4 using graphing, we would see that the lines are parallel.

Consistent Systems with Dependent Equations

Recall from Section 9.1 that if the equations in a system of equations are dependent, then there are an infinite number of solutions. Let's see what happens when we use the substitution method to solve a system of dependent equations.

EXAMPLE 5 Solve the system of equations using substitution.

$$\begin{cases} x = 3y + 4 \\ 2x - 6y = 8 \end{cases}$$

Solution Since x is isolated in the first equation, we substitute $3y + 4$ in place of x in the second equation.

$$2x - 6y = 8$$

$$2\overbrace{(3y + 4)} - 6y = 8 \qquad \text{Substitute } 3y + 4 \text{ for } x.$$

Now, solve for x.

$$6y + 8 - 6y = 8 \qquad \text{Distribute to clear the parentheses.}$$
$$8 = 8 \qquad \text{Combine like terms.}$$

> **Connection** If we were to solve the system in Example 5 using graphing, we would see that the lines are identical.

Notice that $8 = 8$ no longer has a variable and is true. This true equation with no variable indicates that the equations in the system are dependent, so there are an infinite number of solutions which are all the ordered pairs along $x = 3y + 4$ (or $2x - 6y = 8$).

OBJECTIVE 2. Solve applications involving two unknowns using a system of equations. In order to solve a problem involving two unknowns, we must be given two relationships about those unknowns. In Chapter 3, we learned to solve these problems by selecting a variable for one of the unknowns and using one relationship to write the other unknown in terms of that variable. We then translated the second relationship to a single equation in terms of the variable.

We will now solve the same types of problems using systems of equations. Using a system of equations is easier because we select two variables to represent the two unknowns, then translate each of the two given relationships to an equation in terms of those variables. We can then solve the system of equations by substitution (or graphing).

PROCEDURE   **Solving Applications Using a System of Equations**

To solve a problem with two unknowns using a system of equations:

1. Select a variable for each of the unknowns.
2. Translate each relationship to an equation.
3. Solve the system.

EXAMPLE 6 Aaron is designing a frame so that the length is three times the width. The perimeter is to be 96 inches. Find the length and width of the frame.

Understand   We are given two relationships about this frame, and we are to find the length and width.

Plan   Select a variable for the length and another variable for the width, translate the relationships to a system of equations, and then solve the system.

Execute   Let l represent the length and w represent the width.

Relationship 1:   The length is three times the width.

Translation: $l = 3w$

Relationship 2:   The perimeter is 96 inches.

Translation: $2l + 2w = 96$

Our system: $\begin{cases} l = 3w \\ 2l + 2w = 96 \end{cases}$

Since the first equation has l isolated, we will substitute $3w$ in place of l in the second equation.

$$2l + 2w = 96$$
$$2(3w) + 2w = 96 \quad \text{Substitute } 3w \text{ for } l.$$
$$6w + 2w = 96$$
$$8w = 96 \quad \text{Combine like terms.}$$
$$w = 12 \quad \text{Divide both sides by 8 to isolate } w.$$

Now we can find the value of l using $l = 3w$.

$$l = 3(12)$$
$$l = 36 \quad \text{Substitute 12 for } w.$$

Answer The length should be 36 inches and the width 12 inches.

Check Verify that the length is three times the width: $3(12) = 36$. Also verify that the perimeter is 96: $2(36) + 2(12) = 72 + 24 = 96$.

> ***Connection*** Prior to learning about systems of equations, when we solved problems involving two unknowns we actually used the concepts of substitution. For example, in Section 3.3, we would have set up Example 6 as follows. We would have selected a variable, w, for the width and translated the first relationship like this:
>
> $$\text{The length} = 3w$$
>
> Then we would have translated the perimeter relationship like this:
>
> $$2 \cdot \text{Length} + 2 \cdot \text{Width} = 96$$
> $$2(3w) + 2w = 96$$
>
> Notice that we substituted $3w$ in place of the word *length* in the perimeter equation.

YOUR TURN One number is four times a second number. The larger number minus the smaller number equals 45. Find the two numbers.

ANSWER

15, 60

9.2 Exercises

For Extra Help   MyMathLab   MyMathLab   Videotape/DVT   InterAct Math   Math Tutor Center   Math XL.com

1. What advantages does the method of substitution have over graphing for solving systems of equations?

2. Suppose you are given the system of equations $\begin{cases} y = x - 6. \\ 4x + 3y = 1 \end{cases}$

   a. Which variable would you replace in the substitution process?

   b. What expression would replace that variable?

3. Suppose you are given the system of equations $\begin{cases} 5x + y = 4. \\ 3x - 2y = 9 \end{cases}$

   Which variable in which equation would you isolate first? Why?

4. Suppose you are given the system of equations $\begin{cases} 3x + 4y = 1. \\ x - 5y = 12 \end{cases}$

   Which variable would you isolate first? Why?

5. How do you know a system has no solution (inconsistent system) when using substitution?

6. How do you know a system has an infinite number of solutions (dependent equations) when using substitution?

For Exercises 7–34, solve the system of equations using substitution. Note that some systems may be inconsistent or consistent with dependent equations.

7. $\begin{cases} x + y = 6 \\ x = 2y \end{cases}$

8. $\begin{cases} x - y = 8 \\ y = 2x \end{cases}$

9. $\begin{cases} x = -3y \\ 2x - 5y = 44 \end{cases}$

10. $\begin{cases} x + 2y = 9 \\ y = -2x \end{cases}$

11. $\begin{cases} 2x - y = 4 \\ y = 3x + 6 \end{cases}$

12. $\begin{cases} x = 3y - 2 \\ 2x - 3y = 2 \end{cases}$

13. $\begin{cases} 2x + y = -4 \\ 3x + 2y = -5 \end{cases}$

14. $\begin{cases} 5x + y = -4 \\ 3x + 2y = -1 \end{cases}$

15. $\begin{cases} 5x - y = 1 \\ 3x + 2y = 24 \end{cases}$

16. $\begin{cases} x + 3y = 5 \\ 2y - x = 10 \end{cases}$

17. $\begin{cases} 5x - 3y = 4 \\ y - x = -2 \end{cases}$

18. $\begin{cases} 2x - y = 39 \\ 14 + 3x = 74y \end{cases}$

19. $\begin{cases} 2x = 10 - y \\ 3x - y = 5 \end{cases}$

20. $\begin{cases} 3y = 7 - x \\ 2x - y = 0 \end{cases}$

21. $\begin{cases} -x + 2y = 1 \\ 3x - 2y = 1 \end{cases}$

22. $\begin{cases} 2x + 3y = -1 \\ 5x - y = 6 \end{cases}$

23. $\begin{cases} 3x + 2y = 8 \\ x + 4y = 1 \end{cases}$

24. $\begin{cases} 4x + y = -2 \\ 8x - y = 11 \end{cases}$

25. $\begin{cases} 4x + 3y = -2 \\ 8x - 2y = 12 \end{cases}$

26. $\begin{cases} 2x + 3y = -1 \\ 8x - 3y = -19 \end{cases}$

27. $\begin{cases} 5x + 4y = 12 \\ 7x - 6y = 40 \end{cases}$

28. $\begin{cases} 4x - 3y = 2 \\ 3x + 5y = 16 \end{cases}$

29. $\begin{cases} 5x + 6y = 2 \\ 10x + 3y = -2 \end{cases}$

30. $\begin{cases} 7x - 6y = 1 \\ 4x - 6y = 3 \end{cases}$

31. $\begin{cases} y = -2x \\ 4x + 2y = 0 \end{cases}$

32. $\begin{cases} x = 2y - 6 \\ -2x + 4y = 12 \end{cases}$

33. $\begin{cases} x - 3y = -4 \\ -5x + 15y = 6 \end{cases}$

34. $\begin{cases} y = 2x + 1 \\ 6x - 3y = 3 \end{cases}$

For Exercises 35–38, solve using substitution, then verify the solution by graphing the equations on a graphing utility.

35. $\begin{cases} y = 2x + 1 \\ 3x + y = 11 \end{cases}$

36. $\begin{cases} y = -4x + 3 \\ y - 2x = -3 \end{cases}$

37. $\begin{cases} 3x + 4y = 11 \\ x + 2y = 5 \end{cases}$

38. $\begin{cases} 4x - y = 12 \\ x - 3y = 14 \end{cases}$

For Exercises 39 and 40, find the mistake.

39. $\begin{cases} 3x + 2y = 5 \\ x - 3y = 6 \end{cases}$

Rewrite: $x = 6 + 3y$

Substitute: $3(6 + 3y) + 2y = 5$

$$18 + 3y + 2y = 5$$
$$18 + 5y = 5$$
$$5y = -13$$
$$y = -\frac{13}{5}$$

$x = 6 + 3y$

$x = 6 + 3\left(-\dfrac{13}{5}\right)$

$x = 6 - \dfrac{39}{5}$

$x = -\dfrac{9}{5}$

Solution: $\left(-\dfrac{9}{5}, -\dfrac{13}{5}\right)$

40. $\begin{cases} 2x + 3y = 8 \\ x + 2y = 6 \end{cases}$

Rewrite: $x = 6 - 2y$

Substitute: $x + 2y = 6$

$$6 - 2y + 2y = 6$$
$$6 - 6 - 2y + 2y = 6 - 6$$
$$0 = 0$$

Solution: Infinite solutions along $x + 2y = 6$

For Exercises 41–56, translate the problem to a system of equations, then solve.

41. Find two numbers whose sum is 40 and whose difference is 6.

42. Find two numbers whose sum is 18 and whose difference is 4.

43. Three times a number added to twice a smaller number is 4. Twice the smaller number less than twice the larger number is 6. Find the numbers.

44. Find two numbers whose sum is 14 if one number is three times as large as the other number.

45. The greater of two integers is 10 more than twice the smaller. The sum of the two integers is -8. Find the integers.

46. One integer is six less than three times another. The sum of the integers is -42. Find the integers.

47. Jon is 13 years older than Tony. The sum of their ages is 27. How old are they?

48. Patrick is twice as old as his sister, Sherry. The sum of their ages is 12. How old are they?

49. At Ann Arundel Community college in Maryland, a difference between a professor's salary and an associate professor's salary if $14,800. The sum of the salaries is $132,000. Find the salary of an associate professor at AACC. (*Source: Chronicle of Higher Education*)

50. The sum of the semester hour fees for both in-state and out-of-state students at Austin Community College is $252. The difference between the two tuitions is $90. What is the semester hour fee for an out-of-state student at ACC? (*Source:* Austin Community College)

51. The length of a rectangle is 1 meter more than twice the width. What are the dimensions of the rectangle if the perimeter is 32 meters?

52. The length of a rectangle is 1 meter more than triple the width. The length is also five less than five times the width. Find the length and width of the rectangle.

53. The width of a rectangular garden plot is 4 feet less than the length. The perimeter of the plot is 48 feet. Find the length and width.

54. A rectangular garden is 6 feet longer than it is wide. If the perimeter is 132 feet, what are the dimensions of the garden?

55. The length of a volleyball court is twice the width. If the perimeter of the court along the doubles lines is 54 meters, what are the dimensions? (*Source: USA Volleyball Rule Book*)

56. The length of the Reflecting Pool at the National Mall in Washington D.C. is 800 feet more than ten times the width. What are the dimensions if the perimeter is 4900 feet? (*Source:* answers.com)

REVIEW EXERCISES

[1.7] *For Exercises 1 and 2, multiply.*

1. $\frac{1}{4}(20x - 16y)$

2. $-3(x - 7y)$

[5.3] *For Exercises 3 and 4, add or subtract.*

3. $(6x + 2y) + (3x - 2y)$

4. $(5x + 3y) - (5x - 10y)$

[2.3] *For Exercises 5 and 6, solve.*

5. $\frac{1}{4}x = 20$

6. $-8y = 12$

9.3 Solving Systems of Linear Equations by Elimination

OBJECTIVES

1. Solve systems of linear equations using elimination.
2. Solve applications using elimination.

We have seen two methods for solving a system of equations: graphing and substitution. Each method has advantages and disadvantages. The substitution method is advantageous over the graphing method when solutions involve fractions. Also, substitution is easy when a variable's coefficient is 1. However, substitution can be tedious when no coefficients are 1. For this reason, we turn to yet a third method, the *elimination* method.

OBJECTIVE 1. Solve systems of linear equations using elimination. The elimination method uses the addition principle of equality to add equations so that a new equation emerges with one of the variables eliminated. We will work through some examples to get a sense of the method before stating it formally.

EXAMPLE 1 Solve the system of equations.

$$\begin{cases} x + y = 5 & \text{(Equation 1)} \\ 2x - y = 7 & \text{(Equation 2)} \end{cases}$$

Solution In Chapter 2, we learned that the addition principle of equality says that adding the same amount to both sides of an equation will not affect its solution(s). Since Equation 1 indicates that $x + y$ and 5 are the same amount, if we add $x + y$ to the left side of Equation 2 and 5 to the right side of Equation 2, then we are applying the addition principle of equality. Most people prefer to stack the equations and combine like terms vertically like this:

$$\begin{array}{ll} x + y = 5 & \text{(Equation 1)} \\ \underline{2x - y = 7} & \text{(Equation 2)} \\ 3x + 0 = 12 & \text{Add Equation 1 to Equation 2.} \end{array}$$

> **Connection** The terms y and $-y$ are additive inverses, or opposites, which means their sum is 0.

Notice that y is eliminated in this new equation, so we can easily solve for the value of x.

$$\begin{array}{ll} 3x = 12 & \\ x = 4 & \text{Divide both sides by 3 to isolate } x. \end{array}$$

Now that we have the value of x, we can find y by substituting 4 for x in one of the original equations. We will use $x + y = 5$:

$$\begin{array}{ll} x + y = 5 & \\ 4 + y = 5 & \text{Substitute 4 for } y. \\ y = 1 & \text{Subtract 4 from both sides to isolate } y. \end{array}$$

The solution is (4, 1). We can check by verifying that (4, 1) makes both of the original equations true. We will leave this to the reader.

Solve the system of equations using elimination.

$$\begin{cases} 4x + 3y = 8 \\ x - 3y = 7 \end{cases}$$

Multiplying One Equation by a Number to Create Additive Inverses

You may have noted that the expressions $x + y$ and $2x - y$ conveniently contained the additive inverses y and $-y$, which eliminated the y's when the expressions were combined. If no such pairs of additive inverses appear in a system of equations, we will use the multiplication principle of equality to multiply both sides of one equation by a number in order to create additive inverse pairs.

> **Connection**   In Chapter 2, we learned that the multiplication principle of equality says that we can multiply both sides of an equation by the same amount without affecting its solution(s).

EXAMPLE 2   Solve the system of equations.

$$\begin{cases} x + y = 6 & \text{(Equation 1)} \\ 2x - 5y = -16 & \text{(Equation 2)} \end{cases}$$

Solution   Because no variables are eliminated when $x + y = 6$ and $2x - 5y = -16$ are added, we will rewrite one of the equations so that it has a term that is the additive inverse of one of the terms in the other equation. We will multiply both sides of Equation 1 by 5 so that its y term becomes $5y$, which is the opposite of the $-5y$ term in Equation 2.

$$x + y = 6$$
$$5 \cdot x + 5 \cdot y = 5 \cdot 6 \qquad \text{Multiply both sides by 5.}$$
$$5x + 5y = 30 \qquad \text{(Equation 1 rewritten)}$$

Note: *We could have multiplied both sides of equation 1 by -2. We would get a $-2x$ term in Equation 1 that is the opposite of the $2x$ term in Equation 2.*

Now we add the rewritten Equation 1 to Equation 2 to eliminate a variable, just as we did in Example 1.

Note: *Multiplying Equation 1 by 5 made the elimination of the y terms possible.*

$$\begin{array}{ll} 5x + 5y = 30 & \text{(Equation 1 rewritten)} \\ 2x - 5y = -16 & \text{(Equation 2)} \\ \hline 7x + 0 \;\; = 14 & \end{array}$$

Add rewritten Equation 1 to Equation 2 to eliminate y.

We can now solve for x.

$$7x = 14$$
$$x = 2 \qquad \text{Divide both sides by 7 to isolate } x.$$

$$\left(3, -\frac{4}{3}\right)$$

To finish, we substitute 2 for x in one of the equations containing both variables. We will use $x + y = 6$.

$$x + y = 6$$
$$2 + y = 6 \qquad \text{Substitute 2 for } x.$$
$$y = 4 \qquad \text{Subtract 2 from both sides to isolate } y.$$

The solution is $(2, 4)$. We will leave the check to the reader.

YOUR TURN Solve the system of equations using elimination.

$$\begin{cases} -3x - 5y = 6 \\ x + 2y = -1 \end{cases}$$

Multiplying Each Equation by a Number to Create Additive Inverses

If every coefficient in a system of equations is other than 1, then we may have to multiply each equation by a number to generate a pair of additive inverses.

EXAMPLE 3 Solve the system of equations.

$$\begin{cases} 4x - 3y = -2 & \text{(Equation 1)} \\ 6x - 7y = 7 & \text{(Equation 2)} \end{cases}$$

Solution We will choose to eliminate x, so we must multiply both equations by numbers that make the x terms additive inverses. We will multiply Equation 1 by 3 and Equation 2 by -2.

$$4x - 3y = -2 \xrightarrow{\text{Multiply by 3.}} 12x - 9y = -6$$
$$6x - 7y = 7 \xrightarrow{\text{Multiply by } -2.} -12x + 14y = -14$$

Note: *The x terms are now additive inverses, $12x$ and $-12x$.*

Now we can add the rewritten equations to eliminate the x term.

$$\begin{array}{r} 12x - 9y = -6 \\ -12x + 14y = -14 \\ \hline 0 + 5y = -20 \qquad \text{Add the rewritten equations to eliminate } x. \\ y = -4 \qquad \text{Divide both sides by 5 to isolate } y. \end{array}$$

To finish, we substitute -4 for y in one of the equations and solve for x. We will use $4x - 3y = -2$.

$$4x - 3y = -2$$
$$4x - 3(-4) = -2 \qquad \text{Substitute } -4 \text{ for } y.$$
$$4x + 12 = -2$$
$$4x = -14 \qquad \text{Subtract 12 from both sides.}$$
$$x = -\frac{14}{4} \qquad \text{Divide both sides by 4 to isolate } y.$$
$$x = -\frac{7}{2} \qquad \text{Simplify.}$$

The solution is $\left(-\frac{7}{2}, -4 \right)$. We will leave the check to the reader.

Fractions or Decimals in a System

If any of the equations in a system of equations contains fractions or decimals, it is helpful to use the multiplication principle of equality to clear those fractions or decimals so that the equations contain only integers.

EXAMPLE 4 Solve the system of equations.

$$\begin{cases} \dfrac{1}{2}x - y = \dfrac{3}{4} & \text{(Equation 1)} \\ 0.4x - 0.3y = 1 & \text{(Equation 2)} \end{cases}$$

Solution To clear the fractions in Equation 1, we can multiply both sides by the LCD, which is 4. To clear the decimals in Equation 2, we can multiply both sides by 10.

$$\frac{1}{2}x - y = \frac{3}{4} \quad \xrightarrow{\text{Multiply by 4.}} \quad 2x - 4y = 3$$
$$0.4x - 0.3y = 1 \quad \xrightarrow[\text{Multiply by 10.}]{} \quad 4x - 3y = 10$$

Now that both equations contain only integers, it will be easier to solve the system. We will choose to eliminate x, so we multiply the first equation by -2, then combine the equations.

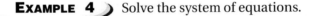

$$2x - 4y = 3 \quad \xrightarrow{\text{Multiply by } -2.} \quad -4x + 8y = -6$$
$$4x - 3y = 10 \qquad\qquad\qquad\qquad \underline{4x - 3y = 10}$$
$$0 + 5y = 4 \qquad \text{Add the rewritten equations to eliminate } x.$$
$$y = \frac{4}{5} \qquad \text{Divide both sides by 5 to isolate } y.$$

To finish, we substitute $\frac{4}{5}$ for y in one of the equations and solve for x. We will use $\frac{1}{2}x - y = \frac{3}{4}$.

Note: *We could have multiplied both sides of the equation by 20 to clear the fraction.*

$$\frac{1}{2}x - \frac{4}{5} = \frac{3}{4} \qquad \text{Substitute } \frac{4}{5} \text{ for } y.$$

$$\frac{1}{2}x = \frac{3}{4} + \frac{4}{5} \qquad \text{Add } \frac{4}{5} \text{ to both sides.}$$

$$\frac{1}{2}x = \frac{15}{20} + \frac{16}{20} \qquad \text{Write the fractions with their LCD, 20.}$$

$$\frac{1}{2}x = \frac{31}{20} \qquad \text{Add the fractions.}$$

$$x = \frac{31}{\overset{20}{\underset{10}{\cancel{20}}}} \cdot \frac{\cancel{2}^{1}}{1} \qquad \text{Multiply both sides by 2 to isolate } x.$$

$$x = \frac{31}{10}$$

The solution is $\left(\dfrac{31}{10}, \dfrac{4}{5}\right)$. We will leave the check to the reader.

YOUR TURN Solve the system of equations using elimination.

$$\begin{cases} 0.6x + y = 3 \\ \dfrac{2}{5}x + \dfrac{1}{4}y = 1 \end{cases}$$

Rewriting the Equations in the Form $Ax + By = C$

Notice in Examples 1 through 4, all of the equations were in standard form, which we learned in Chapter 4 to be $Ax + By = C$. To use the elimination method, the equations need to be written in standard form. For example, in the system that we solved in Example 1, the equations could have been given in a nonstandard form:

$$\begin{cases} y = 5 - x & \text{(Equation 1)} \\ 2x - y = 7 & \text{(Equation 2)} \end{cases}$$

Adding x to both sides of Equation 1 puts it in standard form so that we can use elimination as we did in Example 1.

$$\begin{cases} x + y = 5 & \text{(Equation 1)} \\ 2x - y = 7 & \text{(Equation 2)} \end{cases}$$

ANSWER

$\left(1, \dfrac{12}{5}\right)$ or $(1, 2.4)$

We can now summarize the elimination method with the following procedure.

PROCEDURE   Solving Systems of Equations Using Elimination

To solve a system of two linear equations using the elimination method:

1. Write the equations in standard form ($Ax + By = C$).
2. Use the multiplication principle to clear fractions or decimals.
3. Multiply one or both equations by a number (or numbers) so that they have a pair of terms that are additive inverses.
4. Add the equations. The result should be an equation in terms of one variable.
5. Solve the equation from step 4 for the value of that variable.
6. Using an equation containing both variables, substitute the value you found in step 5 for the corresponding variable and solve for the value of the other variable.
7. Check the solution in the original equations.

Inconsistent Systems and Dependent Equations

How would we recognize an inconsistent system or a system with dependent equations using the elimination method?

EXAMPLE 5   Solve the system of equations.

$$\begin{cases} 2x - y = 1 \\ 2x - y = -3 \end{cases}$$

Solution Notice that the left sides of the equations match. Multiplying one of the equations by -1, then adding the equations, will eliminate both variables.

Connection If we graph the equations in Example 5, the lines would be parallel, indicating there is no solution.

$$
\begin{array}{l}
2x - y = 1 \\
2x - y = -3
\end{array}
\quad \xrightarrow{\text{Multiply by } -1.} \quad
\begin{array}{l}
-2x + y = -1 \\
\underline{2x - y = -3} \\
0 = -4
\end{array}
\quad \begin{array}{l}\text{Add the rewritten}\\ \text{equations.}\end{array}
$$

Both variables have been eliminated and the resulting equation, $0 = -4$, is false, therefore there is no solution. This system of equations is inconsistent.

EXAMPLE 6   Solve the system of equations.

$$\begin{cases} 3x + 4y = 5 \\ 9x + 12y = 15 \end{cases}$$

Solution To eliminate x, we could multiply the first equation by -3, then combine the equations.

$$
\begin{array}{l}
3x + 4y = 5 \\
9x + 12y = 15
\end{array}
\quad \xrightarrow{\text{Multiply by } -3.} \quad
\begin{array}{l}
-9x - 12y = -15 \\
\underline{9x + 12y = 15} \\
0 = 0
\end{array}
\quad \begin{array}{l}\text{Add the rewritten}\\ \text{equations.}\end{array}
$$

Connection If graphed, both equations in Example 6 generate the same line, indicating the equations are dependent.

Both variables have been eliminated and the resulting equation, $0 = 0$, is true. This means the equations are dependent, so there are an infinite number of solutions, which are all the ordered pairs along the line $3x + 4y = 5$ (or $9x + 12y = 15$).

Solve the system of equations.

a. $\begin{cases} x - 4y = 2 \\ 5x - 20y = 10 \end{cases}$ 　　　 **b.** $\begin{cases} x + 2y = 3 \\ x + 2y = 1 \end{cases}$

OBJECTIVE 2. Solve applications using elimination. In Section 9.2, we solved problems in which the substitution method was advantageous because after translating, one of the equations had an isolated variable. In this section, we focus on problems in which elimination is advantageous. These problems translate to a system of linear equations in which every equation is in standard form ($Ax + By = C$).

EXAMPLE 7 In constructing a roof frame, a support beam is connected to a horizontal joist forming two angles. If the larger angle minus twice the smaller angle equals 6°, what are the measures of the two angles?

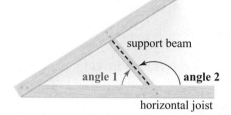

Understand We are to find two angle measurements, but it appears that we are given only one relationship. Because the joist is a straight line, it represents an angle of 180°; therefore, the two angles are supplementary, which is our second relationship.

Plan Translate the relationships to a system of equations, and then solve the system.

Execute We will let x represent the measure of angle 1 and y represent the measure of angle 2. Now we translate the two relationships to a system of two equations.

Relationship 1: The larger angle minus twice the smaller angle equals 6°.

Translation: $y - 2x = 6$

Relationship 2: From the picture we see that the angles are supplementary.

Translation: $x + y = 180$

ANSWERS

a. all ordered pairs along
$x - 4y = 2$ (dependent)
b. no solution (inconsistent)

716　CHAPTER 9　Systems of Equations

We can now solve the system using elimination. We will eliminate y.

Note: *We rearranged the terms to align the like terms vertically.*

$$\begin{cases} -2x + y = 6 \\ x + y = 180 \end{cases} \xrightarrow{\text{Multiply by } -1.} \begin{array}{l} 2x - y = -6 \\ \underline{x + y = 180} \\ 3x + 0 = 174 \end{array}$$

Add the rewritten equations to eliminate y.

$$x = 58$$

Divide both sides by 3 to isolate x.

Now we can find the value of y using $x + y = 180$.

$$58 + y = 180 \qquad \text{Substitute 58 for } x.$$
$$y = 122 \qquad \text{Subtract 58 from both sides to isolate } y.$$

Answer Angle 1 measures $58°$ and angle 2 measures $122°$.

Check Verify that the measure of the larger angle minus twice the measure of the smaller angle equals $6°$: $122 - 2(58) = 122 - 116 = 6$. Also, verify that the angles are supplementary: $58 + 122 = 180$.

YOUR TURN A museum's security system uses a laser placed in the corner just above the floor with its beam projecting across the room parallel to the floor. The beam makes two angles in the corner of the room. The difference of the measures of the two angles is $22°$. Find the angle measurements.

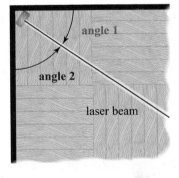

angle 1

angle 2

laser beam

ANSWER

$34°, 56°$

9.3 Exercises

For Extra Help | *MyMathLab* | MyMathLab | Videotape/DVT | InterAct Math | Tutor Center Math Tutor Center | Math XL.com

1. What advantages does the method of elimination have over graphing and substitution?

2. Suppose you are given the system of equations $\begin{cases} x - y = 2. \\ 3x + y = 5 \end{cases}$

   Which variable would you eliminate? Why?

3. Suppose you are given the system of equations $\begin{cases} 6x - 2y = 1. \\ 3x + 5y = 7 \end{cases}$

a. Which variable is easier to eliminate? Why?

b. How would you eliminate the variable you chose?

4. Suppose you are given the system of equations $\begin{cases} 5x + 4y = 8. \\ 3x - 6y = 2 \end{cases}$

a. Which variable is easier to eliminate? Why?

b. How would you eliminate the variable you chose?

5. When using elimination, how do you know if a system of equations has no solution (inconsistent system)?

6. When using elimination, how do you know if a system of equations has an infinite number of solutions (dependent equations)?

For Exercises 7–42, solve the system of equations using the elimination method.

7. $\begin{cases} x - y = 14 \\ x + y = -2 \end{cases}$

8. $\begin{cases} x + y = 9 \\ 5x - y = 3 \end{cases}$

9. $\begin{cases} 3x + y = 9 \\ 2x - y = 1 \end{cases}$

10. $\begin{cases} 3x + y = 10 \\ -2x - y = -7 \end{cases}$

11. $\begin{cases} 3x + 2y = -7 \\ 5x - 2y = -1 \end{cases}$

12. $\begin{cases} 2x - 3y = 8 \\ 4x + 3y = 16 \end{cases}$

13. $\begin{cases} 4x + 3y = 17 \\ 2x + 3y = 13 \end{cases}$

14. $\begin{cases} x - 3y = 7 \\ x - 5y = 13 \end{cases}$

15. $\begin{cases} 5x + y = 14 \\ 2x + y = 5 \end{cases}$

16. $\begin{cases} 3x - y = 7 \\ 5x - y = 15 \end{cases}$

17. $\begin{cases} 3x - 5y = -17 \\ 4x + y = -15 \end{cases}$

18. $\begin{cases} 7x - 4y = 4 \\ 5x + y = 26 \end{cases}$

19. $\begin{cases} 12x - 2y = -54 \\ 13x + 4y = -40 \end{cases}$

20. $\begin{cases} 5x + 2y = 3 \\ 7x - 6y = 13 \end{cases}$

21. $\begin{cases} 5x + 6y = 11 \\ 2x - 4y = -2 \end{cases}$

22. $\begin{cases} -6x + 7y = -2 \\ 9x - 5y = -8 \end{cases}$

23. $\begin{cases} 4x + 5y = -4 \\ 3x + 8y = -20 \end{cases}$

24. $\begin{cases} 3x - 4y = -14 \\ 5x + 7y = 45 \end{cases}$

25. $\begin{cases} 5x + 6y = 2 \\ 10x + 3y = -2 \end{cases}$

26. $\begin{cases} 8x + 6y = 5 \\ 2x + y = 1 \end{cases}$

27. $\begin{cases} 10x + 5y = 3.5 \\ 3x - 4y = -0.6 \end{cases}$

28. $\begin{cases} 2x - y = -0.8 \\ 3x - 2y = -1 \end{cases}$

29. $\begin{cases} \dfrac{3}{2}x - \dfrac{4}{3}y = \dfrac{11}{3} \\ \dfrac{1}{4}x - \dfrac{2}{3}y = -\dfrac{7}{6} \end{cases}$

30. $\begin{cases} \dfrac{3}{4}x - \dfrac{2}{7}y = -5 \\ \dfrac{5}{8}x + \dfrac{1}{4}y = -\dfrac{3}{4} \end{cases}$

31. $\begin{cases} 0.7x - \dfrac{1}{2}y = 9.4 \\ 0.9x + \dfrac{7}{10}y = 0 \end{cases}$

32. $\begin{cases} \dfrac{1}{3}x + 0.2y = -0.2 \\ \dfrac{2}{3}x - 0.75y = -5 \end{cases}$

33. $\begin{cases} 2x + y = -2 \\ 8x + 4y = -8 \end{cases}$

34. $\begin{cases} 4x - 2y = 3 \\ -8x + 4y = -6 \end{cases}$

35. $\begin{cases} 2x - 5y = 4 \\ -8x + 20y = -20 \end{cases}$

36. $\begin{cases} 5x + 5y = 50 \\ x + y = 2.5 \end{cases}$

37. $\begin{cases} y = 2x - 5 \\ x - y = 9 \end{cases}$

38. $\begin{cases} x + 3y = 8 \\ 2y = x + 6 \end{cases}$

39. $\begin{cases} y = \dfrac{3}{4}x + 1 \\ 4x = 2y + 3 \end{cases}$

40. $\begin{cases} y = -\dfrac{2}{3}x - 5 \\ 3y = 2x - 27 \end{cases}$

41. $\begin{cases} 0.2x - y = -3.2 \\ y = \dfrac{3}{4}x + 1 \end{cases}$

42. $\begin{cases} y = \dfrac{1}{5}x - 2 \\ 6y = 0.3x - 1.2 \end{cases}$

For Exercises 43–44, find the mistake(s).

43. $\begin{cases} x - y = 1 \\ 9x + 8y = 77 \end{cases}$

$\begin{aligned} -9(x - y) &= 1 \\ 9x + 8y &= 77 \end{aligned} \longrightarrow \begin{aligned} -9x + 9y &= 1 \\ \underline{9x + 8y} &= \underline{77} \\ 17y &= 78 \\ y &= \dfrac{78}{17} \end{aligned}$

$x - y = 1$

$x - \left(\dfrac{78}{17}\right) = 1$

$x = \dfrac{95}{17}$

Solution: $\left(\dfrac{95}{17}, \dfrac{78}{17}\right)$

44. $\begin{cases} x + y = 1 \\ x + 2y = 2 \end{cases}$

$\begin{aligned} x + y &= 1 \\ -1(x + 2y &= 2) \end{aligned} \longrightarrow \begin{aligned} x + y &= 1 \\ \underline{-x - 2y} &= \underline{2} \\ -y &= 3 \\ y &= -3 \end{aligned}$

$x + y = 1$

$x + (-3) = 1$

$x = 4$

Solution: $(4, -3)$

For Exercises 45–62, translate the problem to a system of equations, then solve using the elimination method.

45. One number is 62 more than another. If twice the greater number is subtracted from five times the lesser number, the result is 155. What are the two numbers?

46. The difference of two numbers is 4. Three times the greater number is two more than five times the lesser. Find the numbers.

47. One number is 7 less than another number. The lesser number subtracted from three-fourths of the greater number is 3. Find the numbers.

48. One number exceeds another by 9. If two-thirds of the lesser number is increased by 22, the result equals the greater number. What are the two numbers?

49. Janet is four years younger than her brother. If the sum of their ages is 15, what are their ages?

50. The sum of Morgan and Angela's ages is 56 and their age difference is 14. What are the ages?

51. Two angles are supplementary. One of the angles is 20° more than twice the other. Find the measure of the two angles.

52. The sum of two angles is 80° and their difference is 30°. Find the measure of the two angles.

53. Two angles are complementary. One of the angles is twice the other. Find the measure of the two angles.

54. Two angles are complementary. The larger angle is 6° less than twice the smaller angle. Find the measure of the two angles.

55. What are the length and width of a rectangle if the length exceeds the width by 14 inches and the perimeter is 60 inches?

56. What are the length and width of a rectangle if the width is 27 feet less than its length and the perimeter is 398 feet?

57. During the 2002 Winter Olympic Games in Salt Lake City, Germany won the most medals, followed closely by the United States. The total medals won by both countries was 69 and the difference between the numbers of medals won by each country was 1. Find the total number of medals won by each country.

58. During the same Olympics, Germany won twice as many gold medals as Canada. If the total gold medals awarded to both countries was 18, find the number of gold medals awarded to each.

Of Interest

When the modern Olympics were revived in 1896, silver medals were given to first-place winners. Gold was considered inferior. It was not until the 1904 games that gold medals were used to award first place.
(Source: Atlanta Journal-Constitution)

59. For Super Bowl XXXVI, when New England Patriots kicker Adam Vinatieri kicked a 48-yard field goal as time expired, the Patriots won by 3 points over the Rams. If the total points scored during the game were 37, how many points did each team score?

60. In Super Bowl XXXIX, the New England Patriots won their third Super Bowl in four years. They beat the Philadelphia Eagles by 3 points. If the total points scored during the game were 45, how many points did each team score? (*Source:* NFL Internet Network)

61. Jack Nicklaus has the most victories ever won by a golfer in the Masters Tournament. He exceeds the number of Arnold Palmer's victories by 2 wins. If they have won a total of 10 tournaments, how many times has each man won?

62. In the 2004 season, Albert Pujols of the St. Louis Cardinals had three more home runs than Manny Ramirez of the Boston Red Sox. If their combined number of home runs was 89, find how many home runs each player had. (*Source:* MLB.com)

PUZZLE PROBLEM

Use algebraic methods to prove that 0.999 … is equal to 1. (Hint: Let $x = 0.\overline{9}$, then write a second equation using the multiplication principle so that when the two equations are added, the repeated 9 is eliminated.)

REVIEW EXERCISES

[3.3] **1.** Monique has $4.05 in her purse. If she has 24 coins, all dimes and quarters, how many of each does she have?

[3.4] **2.** Jay and Lisa are traveling west in separate cars on the same highway. Jay is traveling at 60 miles per hour and Lisa at 70 miles per hour. Jay passes Exit 54 at 2:30 P.M. Lisa passes the same exit at 2:36 P.M. At what time will Lisa catch up to Jay?

[3.5] **3.** The total value of two bank accounts is $1100. The annual interest earned on the checking account is 5% while the CD pays 6%. The total interest earned is $61. How much was deposited into each account?

[3.5] **4.** A chemist has 80 milliliters of 10% HCl solution in a container. He has a large amount of 25% HCl solution in another container. How much of the 25% solution should be added to the 10% solution to make a solution that is 20% HCl?

[9.1] **5.** Use graphing to solve the system $\begin{cases} x + y = -5. \\ 3x - y = 9 \end{cases}$

[9.2] **6.** Use substitution to solve the system $\begin{cases} y = -2x. \\ 2x + 5y = 10 \end{cases}$

9.4 Solving Systems of Linear Equations in Three Variables

OBJECTIVES

1. Determine if an ordered triple is a solution for a system of equations.

2. Understand the graphs of systems of three equations.

3. Solve a system of three linear equations using the elimination method.

4. Solve applications involving three unknowns using a system of equations.

OBJECTIVE 1. Determine if an ordered triple is a solution for a system of equations. In this section, we will solve systems of three linear equations with three unknowns. Solutions of these systems will be *ordered triples* with the form (x, y, z). We check solutions for these systems just as we did in Section 9.1: Replace each variable with its corresponding value, and verify that each equation is true.

EXAMPLE 1 Determine whether each ordered triple is a solution to the system of equations.

$$\begin{cases} x + y + z = 3 & \text{(Equation 1)} \\ 2x + 3y + 2z = 7 & \text{(Equation 2)} \\ 3x - 4y + z = 4 & \text{(Equation 3)} \end{cases}$$

a. $(2, 1, 0)$

Solution In all three equations, replace x with 2, y with 1, and z with 0.

Equation 1:

$x + y + z = 3$

$2 + 1 + 0 \stackrel{?}{=} 3$

$3 = 3$ True

Equation 2:

$2x + 3y + 2z = 7$

$2(2) + 3(1) + 2(0) \stackrel{?}{=} 7$

$4 + 3 + 0 \stackrel{?}{=} 7$

$7 = 7$ True

Equation 3:

$3x - 4y + z = 4$

$3(2) - 4(1) + 0 \stackrel{?}{=} 4$

$6 - 4 + 0 \stackrel{?}{=} 4$

$2 = 4$ False

Because $(2, 1, 0)$ does not solve all three equations in the system, it is not a solution for the system.

b. $(3, 1, -1)$

Solution

Equation 1:

$x + y + z = 3$

$3 + 1 + (-1) \stackrel{?}{=} 3$

$3 = 3$ True

Equation 2:

$2x + 3y + 2z = 7$

$2(3) + 3(1) + 2(-1) \stackrel{?}{=} 7$

$6 + 3 - 2 \stackrel{?}{=} 7$

$7 = 7$ True

Equation 3:

$3x - 4y + z = 4$

$3(3) - 4(1) + (-1) \stackrel{?}{=} 4$

$9 - 4 - 1 \stackrel{?}{=} 4$

$4 = 4$ True

Because $(3, 1, -1)$ solves all three equations in the system, it is a solution for the system.

YOUR TURN Determine whether the following ordered triples are solutions to the system of equations.

a. $(4, 1, 1)$ **b.** $(-1, 5, 2)$

$$\begin{cases} x + y + z = 6 & \text{(Equation 1)} \\ 3x - 2y + 3z = -7 & \text{(Equation 2)} \\ 4x - 2y + z = -12 & \text{(Equation 3)} \end{cases}$$

OBJECTIVE 2. Understand the graphs of systems of three equations.   Recall that we plot an ordered pair using two axes: x and y. Similarly, we plot an ordered triple like $(3, 4, 5)$ using three axes: x, y, and z, each of which is perpendicular to the other two, as shown.

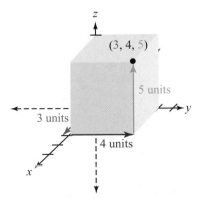

ANSWERS

a. not a solution
b. solution

In Section 9.1, we examined the graphs of the equations in systems of two equations in two unknowns to understand the number of possible solutions these systems could have. We learned that these systems could have one solution (when the lines intersect at a single point), an infinite number of solutions (when both equations produce the same line), or no solution (when the lines are parallel). Similarly, systems of three equations in three unknowns can have one solution, an infinite number of solutions, or no solution depending on how the graphs of the equations interact.

The graph of a linear equation in three variables is a *plane*, which is a flat surface much like a sheet of paper with infinite length and width. We draw a plane as a parallelogram (shown to the right), but remember that the length and width are infinite. Let's examine some of the ways three planes can be configured to produce one solution, an infinite number of solutions, or no solution.

A Single Solution

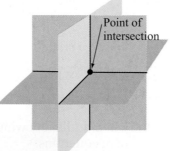

If the planes intersect at a single point, then that ordered triple is the solution to the system.

Infinite Number of Solutions

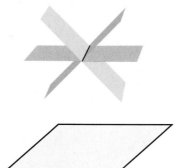

If the three planes intersect along a line, then the system has an infinite number of solutions, which are the coordinates of any point along that line.

If all three graphs are in the same plane, then the system has an infinite number of solutions, which are the coordinates of any point in the plane.

No Solution

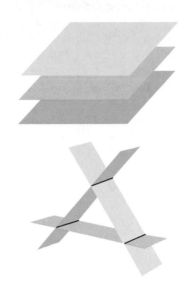

If all the planes are parallel, then the system has no solution.

Pairs of planes could also intersect, as shown. However, since all three planes do not have a common intersection, the system has no solution.

Note: *There are other ways the planes could be configured so that they do not have a common intersection.*

OBJECTIVE 3. Solve a system of three linear equations using the elimination method. Since graphing systems of three linear equations is difficult, we will first solve them using elimination. The process is much like what we learned in Section 9.3.

EXAMPLE 2 Solve the following systems of equations using the elimination method.

$$\begin{cases} x + y + z = 2 & \text{(Equation 1)} \\ 2x + y + 2z = 1 & \text{(Equation 2)} \\ 3x + 2y + z = 1 & \text{(Equation 3)} \end{cases}$$

Solution First we choose a variable and eliminate that variable from two pairs of equations. Let's eliminate y using Equations 1 and 2. We multiply Equation 1 by -1, then add the equations.

Note: *Our initial goal is to generate two equations with two variables that are the same.*

(Eq. 1) $x + y + z = 2$ $\xrightarrow{\text{Multiply by } -1}$ $\begin{array}{r} -x - y - z = -2 \\ 2x + y + 2z = 1 \\ \hline x \quad\quad + z = -1 \end{array}$

(Eq. 2) $2x + y + 2z = 1$

Add the equations.
(Equation 4)

Now we need to eliminate y from another pair of equations. We will choose Equations 1 and 3.

(Eq. 1) $x + y + z = 2$ $\xrightarrow{\text{Multiply by } -2}$ $\begin{array}{r} -2x - 2y - 2z = -4 \\ 3x + 2y + z = 1 \\ \hline x \quad\quad - z = -3 \end{array}$

(Eq. 3) $3x + 2y + z = 1$

Add the equations.
(Equation 5)

Equations 4 and 5 form a system of equations with variables x and z. We can eliminate z by adding the equations together.

$$(\text{Eq. 4}) \quad x + z = -1$$
$$(\text{Eq. 5}) \quad \underline{x - z = -3} \qquad \text{Add the equations.}$$
$$\phantom{(\text{Eq. 5}) \quad} 2x = -4$$
$$\phantom{(\text{Eq. 5}) \quad 2} x = -2 \qquad \text{Divide both sides by 2 to isolate } x.$$

To find z, we substitute -2 for x in Equation 4 or 5. We will use Equation 4.

$$x + z = -1$$
$$-2 + z = -1 \qquad \text{Substitute } -2 \text{ for } x.$$
$$z = 1 \qquad \text{Add 2 to both sides.}$$

To find y, substitute -2 for x and 1 for z in any of the original equations. We will use Equation 1.

$$x + y + z = 2$$
$$-2 + y + 1 = 2 \qquad \text{Substitute } -2 \text{ for } x \text{ and 1 for } z.$$
$$y - 1 = 2 \qquad \text{Simplify the left side.}$$
$$y = 3 \qquad \text{Add 1 to both sides to isolate } y.$$

The solution is $(-2, 3, 1)$. We can check the solution by verifying that the ordered triple satisfies each of the three original equations. We will leave the check to the reader.

In some systems, one or more variables may be missing from one or more of the equations, which can simplify solving the system.

EXAMPLE 3 Solve the following systems of equations using the elimination method.

$$\begin{cases} 2x + 3y = 3 & (\text{Equation 1}) \\ 2y - 3z = -8 & (\text{Equation 2}) \\ 4x - z = 10 & (\text{Equation 3}) \end{cases}$$

Solution Notice that a variable is missing in each of the equations. We will eliminate z using Equations 2 and 3.

Note: *We chose Equation 3 because z has a coefficient of* -1.

$$(\text{Eq. 2}) \quad 2y - 3z = -8 \qquad \qquad \qquad 2y - 3z = -8$$
$$(\text{Eq. 3}) \quad 4x - z = 10 \xrightarrow{\text{Multiply by } -3} \underline{-12x + 3z = -30} \quad \text{Add the equations.}$$
$$\phantom{(\text{Eq. 3}) \quad} -12x + 2y = -38 \quad (\text{Equation 4})$$

Because Equations 1 and 4 do not have z terms, we do not need to eliminate z from another pair of equations. We will now use Equations 1 and 4 to eliminate x.

$$\text{(Eq. 1)} \qquad 2x + 3y = 3 \xrightarrow{\text{Multiply by 6}} \quad 12x + 18y = 18$$

$$\text{(Eq. 4)} \quad -12x + 2y = -38 \qquad\qquad\qquad \underline{-12x + 2y = -38} \quad \textbf{Add the equations.}$$

$$20y = -20$$

$$y = -1 \quad \textbf{Divide both sides by 20 to isolate } y.$$

To find x, substitute -1 for y in Equation 1 or 4. We will use Equation 1.

$$2x + 3(-1) = 3 \qquad \textbf{Substitute } -1 \textbf{ for } y \textbf{ in Equation 1.}$$

$$2x - 3 = 3 \qquad \textbf{Simplify.}$$

$$2x = 6 \qquad \textbf{Add 3 to both sides.}$$

$$x = 3 \qquad \textbf{Divide both sides by 2 to isolate } x.$$

To find z, substitute in either Equation 2 or 3 (equation 1 has no z term). We will use Equation 2.

$$2(-1) - 3z = -8 \qquad \textbf{Substitute } -1 \textbf{ for } y \textbf{ in Equation 2.}$$

$$-2 - 3z = -8 \qquad \textbf{Simplify.}$$

$$-3z = -6 \qquad \textbf{Add 2 to both sides.}$$

$$z = 2 \qquad \textbf{Divide both sides by } -3 \textbf{ to isolate } z.$$

The solution is $(3, -1, 2)$. We will leave the check to the reader.

Following is a summary of the process of solving a system of three linear equations.

PROCEDURE   **Solving Systems of Three Equations Using Elimination**

To solve a system of three equations with three unknowns using elimination,

1. Write each equation in the form $Ax + By + Cz = D$.
2. Eliminate one variable from one pair of equations using the elimination method.
3. If necessary, eliminate the same variable from another pair of equations.
4. Steps 2 and 3 result in two equations with the same two variables. Solve these equations using the elimination method.
5. To find the third variable, substitute the values of the variables found in step 4 back into any of the three original equations that contain the third variable.
6. Check the ordered triple in all three of the original equations.

Let's see what happens when a system of three linear equations has no solution (inconsistent).

EXAMPLE 4) Solve the following system using elimination.

$$\begin{cases} 2x + y + 2z = -1 & \text{(Equation 1)} \\ -3x + 2y + 3z = -13 & \text{(Equation 2)} \\ 4x + 2y + 4z = 5 & \text{(Equation 3)} \end{cases}$$

Note: *We mentioned in the Connection box on page 725 that we can identify systems of dependent equations when using the elimination method. A system of three equations is dependent if during the solution process we add two equations and get $0 = 0$ (we saw this with two equations in Example 6 of Section 9.3). For example, if Equation 4 of Example 4 were $-7x - z = -18$, the sum of equations 4 and 5 would be $0 = 0$.*

Eq. 4 $\quad -7x - z = -18$
Eq. 5 $\quad \underline{7x + z = 18}$
$\quad\quad\quad\quad 0 = 0$

In such a case, we say the equations in the system are dependent and the system has an infinite number of solutions.

Solution We need to eliminate a variable from two pairs of equations. Let's eliminate y using Equations 1 and 2.

Multiply by -2

(Eq. 1) $\quad 2x + y + 2z = -1 \longrightarrow \quad -4x - 2y - 4z = 2$
(Eq. 2) $\quad -3x + 2y + 3z = -13 \qquad\qquad \underline{-3x + 2y + 3z = -13}$ **Add the equations.**
$\qquad\qquad\qquad\qquad\qquad\qquad\qquad -7x \quad\quad - z = -11$ (Equation 4)

Now we eliminate y from Equations 2 and 3.

Multiply by -1

(Eq. 2) $\quad -3x + 2y + 3z = -13 \longrightarrow \quad 3x - 2y - 3z = 13$
(Eq. 3) $\quad 4x + 2y + 4z = 5 \qquad\qquad\qquad \underline{4x + 2y + 4z = 5}$ **Add the equations.**
$\qquad\qquad\qquad\qquad\qquad\qquad\qquad\quad 7x \quad\quad + z = 18$ (Equation 5)

Equations 4 and 5 form a system in x and z. Add the equations together.

$\qquad$ (Eq. 4) $\quad -7x - z = -11$
$\qquad$ (Eq. 5) $\quad \underline{7x + z = 18}$
$\qquad\qquad\qquad\qquad 0 = 7$

All variables are eliminated and the resulting equation is false, which means this system has no solution; it is inconsistent.

YOUR TURN) Solve the following systems of equations.

a. $\begin{cases} x + 3y + 2z = 6 \\ 2x - 3y + z = -18 \\ -3x + 2y + z = 12 \end{cases}$ **b.** $\begin{cases} 2x + y + 2z = 1 \\ x + y + z = 2 \\ 4x + 2y + 4z = 6 \end{cases}$ **c.** $\begin{cases} x + 3y - 6z = 9 \\ -7y + 6z = -3 \\ -2x + y + 6z = -15 \end{cases}$

OBJECTIVE 4. Solve applications involving three unknowns using a system of equations. We follow the same procedure to solve applications involving three unknowns as we did in Sections 9.2 and 9.3 when solving applications with two unknowns.

EXAMPLE 5) At a movie theater, John buys one popcorn, one soft drink, and one candy bar, all for $7. Fred buys two popcorns, three soft drinks, and two candy bars for $16. Carla buys one popcorn, two soft drinks, and three candy bars for $12. Find the price of one popcorn, one soft drink, and one candy bar.

Understand We have three unknowns, three relationships, and we are to find the cost of each.

Plan Select a variable for each unknown, translate the relationships to a system of three equations, and then solve the system.

Execute Let x represent the cost of one popcorn, y represent the cost of one soft drink, and z represent the cost of one candy bar.

Relationship 1: One popcorn, one soft drink, and one candy bar cost $7.

$$\textbf{Translation: } x + y + z = 7$$

Relationship 2: Two popcorns, three soft drinks, and two candy bars cost $16.

$$\textbf{Translation: } 2x + 3y + 2z = 16$$

Relationship 3: One popcorn, two soft drinks, and three candy bars cost $12.

$$\textbf{Translation: } x + 2y + 3z = 12$$

$$\text{Our system: } \begin{cases} x + y + z = 7 & \text{(Equation 1)} \\ 2x + 3y + 2z = 16 & \text{(Equation 2)} \\ x + 2y + 3z = 12 & \text{(Equation 3)} \end{cases}$$

We will choose to eliminate z from two pairs of equations. We will start with Equations 1 and 2.

Multiply by -2.

(Eq. 1) $x + y + z = 7 \longrightarrow -2x - 2y - 2z = -14$

(Eq. 2) $2x + 3y + 2z = 16$ $\underline{2x + 3y + 2z = 16}$ **Add the equations.**

$y = 2$ (Equation 4)

Note: *Although we were trying to eliminate z, we ended up eliminating both x and z.*

Equation 4 gives us the value of y, indicating that a soft drink costs $2.00. Now, we choose another pair of equations and eliminate z again. We will use Equations 1 and 3.

Multiply by -3.

(Eq. 1) $x + y + z = 7 \longrightarrow -3x - 3y - 3z = -21$

(Eq. 3) $x + 2y + 3z = 12$ $\underline{x + 2y + 3z = 12}$ **Add the equations.**

$-2x - y = -9$ (Equation 5)

Since we already know $y = 2$, we can substitute for y in $-2x - y = -9$.

$-2x - \mathbf{2} = -9$ **Substitute 2 for y.**

$-2x = -7$ **Add 2 to both sides.**

$x = 3.50$ **Divide both sides by -2 to isolate x.**

Since x represents the cost of one popcorn, a popcorn costs $3.50. To find z, substitute for x and y into one of the original equations. We will use Equation 1.

$$3.5 + 2 + z = 7 \qquad \text{Substitute 3.5 for } x \text{ and 2 for } y.$$
$$5.5 + z = 7 \qquad \text{Simplify.}$$
$$z = 1.5 \qquad \text{Subtract 5.5 from both sides to isolate } z.$$

Since z represents the cost of one candy bar, a candy bar costs $1.50.

Answer   Popcorn costs $3.50, a soft drink costs $2.00, and a candy bar costs $1.50.

Check   Verify that at these prices, the amounts of money spent by John, Fred, and Carla are correct. We will leave the check to the reader.

YOUR TURN   A small aquarium is in the shape of a rectangular solid. The sum of the length, width, and height is 46 inches. The sum of twice the length, three times the width, and the height is 90 inches. The sum of the length, twice the width, and three times the height is 86 inches. Find the dimensions of the aquarium.

ANSWER

length: 20 in.,
width: 12 in.,
height: 14 in.

9.4 Exercises

For Extra Help   *MyMathLab*   MyMathLab   Videotape/DVT   InterAct Math   Math Tutor Center   Math XL.com

1. When solving a system of three equations with three variables, you eliminate a variable from one pair of equations, then eliminate the same variable from a second pair of equations. Does it matter which equations you choose?

2. When solving a system of three equations with three variables, you eliminate a variable from one pair of equations, then eliminate the same variable from a second pair of equations. Why do we eliminate the same variable from two pairs of equations?

3. Suppose you are given the system of equations $\begin{cases} x + y + z = 0 & \text{(Equation 1)} \\ 2x + 4y + 3z = 5 & \text{(Equation 2)} \\ 4x - 2y + 3z = -13 & \text{(Equation 3)} \end{cases}$

Which variable would you choose to eliminate, using which pairs of equations? Why?

4. When using elimination to solve a system of three equations with three variables, how do you know if it has no solution?

5. Two drawings were given in the text of inconsistent systems of equations with three unknowns. Make another drawing of an inconsistent system other than the ones given.

6. Where would a point be located if it solves two equations of a system of three equations in three variables but does not solve the third equation?

For Exercises 7–12, determine if the given point is a solution of the system.

7. $(3, -1, 1)$
$\begin{cases} x + y + z = 3 \\ 2x - 2y - z = 4 \\ 2x + y - 2z = 3 \end{cases}$

8. $(2, -2, 1)$
$\begin{cases} 3x + 2y + z = 3 \\ 2x - 3y - 2z = 8 \\ -2x + 4y + 3z = -9 \end{cases}$

9. $(1, 0, 2)$
$\begin{cases} 2x + 3y - 3z = -4 \\ -2x + 4y - z = -4 \\ 3x - 4y + 2z = 5 \end{cases}$

10. $(3, 4, 0)$
$\begin{cases} x + 2y + 5z = 11 \\ 3x - 2y - 4z = 1 \\ 2x + 2y + 3z = 12 \end{cases}$

11. $(2, -2, 4)$
$\begin{cases} x + 2y - z = -6 \\ 2x - 3y + 4z = 26 \\ -x + 2y - 3z = -18 \end{cases}$

12. $(0, -2, 4)$
$\begin{cases} 3x + 2y - 3z = -16 \\ 2x - 4y - z = 4 \\ -3x + 4y - 2z = -16 \end{cases}$

For Exercises 13–32, solve the systems of equations.

13. $\begin{cases} x + y + z = 5 \\ 2x + y - 2z = -5 \\ x - 2y + z = 8 \end{cases}$

14. $\begin{cases} x + y + z = 4 \\ 3x + y - z = -4 \\ 2x - 2y + 3z = 3 \end{cases}$

15. $\begin{cases} x + y + z = 2 \\ 4x - 3y + 2z = 2 \\ 2x + 3y - 2z = -8 \end{cases}$

16. $\begin{cases} x + y - z = 7 \\ -2x + 2y - z = -5 \\ 3x - 3y + 2z = 6 \end{cases}$

17. $\begin{cases} 2x + y + 2z = 5 \\ 3x - 2y + 3z = 4 \\ -2x + 3y + z = 8 \end{cases}$

18. $\begin{cases} 2x + 3y - 2z = -4 \\ 4x - 3y + z = 25 \\ x + 2y - 4z = -12 \end{cases}$

19. $\begin{cases} x + 2y - z = 1 \\ 2x + 4y - 2z = -8 \\ -0.5x - y + 0.5z = -2 \end{cases}$

20. $\begin{cases} x + y + z = 6 \\ 3x + 2y - 4z = -4 \\ 1.5x + y - 2z = 6 \end{cases}$

21. $\begin{cases} 6x + 4y - 8z = 12 \\ 3x + 2y - 4z = 6 \\ 1.5x + y - 2z = 3 \end{cases}$

22. $\begin{cases} -8x + 4y + 6z = -18 \\ 2x - y - 1.5z = 4.5 \\ 4x - 2y - 3z = 9 \end{cases}$

23. $\begin{cases} y = 2x + z - 8 \\ x = 2y - 3z + 11 \\ 2x + 3y - z = -6 \end{cases}$

24. $\begin{cases} z = -3x + y - 10 \\ 2x + 3y - 2z = 5 \\ x = 3y - 3z - 14 \end{cases}$

25. $\begin{cases} x = 4y - z + 1 \\ 3x + 2y - z = -8 \\ x + 6y + 2z = -3 \end{cases}$

26. $\begin{cases} 2x - 3y - 4z = 3 \\ y = -4x + 8z - 1 \\ x = -5y - 2z - 4 \end{cases}$

27. $\begin{cases} 4x + 2y + 3z = 9 \\ 2x - 4y - z = 7 \\ 3x - 2z = 4 \end{cases}$

28. $\begin{cases} 4x + 2y - 3z = 20 \\ 2x + 5z = -4 \\ 3x - 3y + 2z = 2 \end{cases}$

29. $\begin{cases} 3x + 2y = -2 \\ 2x - 3z = 1 \\ 0.4y - 0.5z = -2.1 \end{cases}$

30. $\begin{cases} 0.2x - 0.3z = -1.8 \\ 3x + 2y = -5 \\ 3y + 2z = 14 \end{cases}$

31. $\begin{cases} \dfrac{3}{2}x + y - z = 0 \\ 4y - 3z = -22 \\ -0.2x + 0.3y = -2 \end{cases}$

32. $\begin{cases} -0.2x - 0.3y + 0.1z = -0.3 \\ -3x + 2y = 13 \\ \dfrac{1}{4}y - \dfrac{1}{2}z = 2 \end{cases}$

For Exercises 33–64, translate the problem to a system of equations, then solve.

33. The sum of three numbers is 16. One of the numbers is 2 more than twice a second number, and 2 less than the third number. Find the numbers.

34. The sum of three numbers is 18. Twice the first number is 1 less than the third number, and twice the second number is 1 more than the third number. Find the numbers.

35. The sum of the measures of the angles in every triangle is 180°. Suppose the measure of one angle of a triangle is three times that of a second angle and the measure of the second angle is 5° less than the measure of the third. Find the measure of each angle of the triangle.

36. The measure of one angle of a triangle is 25° less than the measure of a second angle, and twice the measure of the second angle is 65° more than the measure of the third angle. Find the measure of each angle of the triangle.

37. The perimeter of a triangle is 33 inches. The sum of the length of the longest side and twice the length of the shortest side is 31 inches. Twice the length of the longest side minus both the other side lengths is 12 inches. Find the side lengths.

38. The perimeter of the figure shown is 118 centimeters. Twice x minus y is 1, and y minus z is 10 centimeters. Find the side lengths.

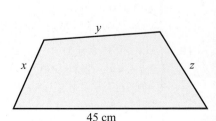

39. At a fast-food restaurant, one burger, one order of fries, and one drink cost $5; three burgers, two orders of fries, and two drinks cost $12.50; and two burgers, four order of fries, and three drinks cost $14. Find the individual cost of one burger, one order of fries, and one drink.

40. James went to the college bookstore and purchased two pens, two erasers, and one pack of paper for $6. Tamika purchased four pens, three erasers, and two packs of paper for $11.50. Jermaine purchased three pens, one eraser, and three packs of paper for $9.50. Find the cost of one pen, one eraser, and one pack of paper.

41. Tickets for a high school band concert were $3 for children, $5 for students, and $8 for adults. There was a total of 500 tickets sold and the total money received from the sale of the tickets was $2500. There were 150 fewer adult tickets sold than student tickets. Find the number of each type of ticket sold.

42. Tickets to a play were $2 for children, $4 for students, and $6 for adults. There was a total of 300 tickets sold and the total money received from the sale of the tickets was $1250. There were 35 more student tickets sold than children's tickets. Find the number of each type of ticket sold.

43. On April 2, 2005 in a game against the New Jersey Nets, Steve Francis of the Orlando Magic scored twenty-three times for a total of 33 points. The sum of the number of 3-point field goals and 2-point field goals was 5 less than the number of free throws (1 point each). Find how many 3-point field goals, 2-point field goals, and free throws he made in this game. (*Source: Orlando Sentinel*, April 3, 2005)

44. At a track meet, 10 points are awarded for each first-place finish, 5 points for each second, and 1 point for each third. Suppose a track team scored a total of 71 points and had two more firsts than seconds and one less first than third. Find the number of first-, second-, and third-place finishes for the team.

45. Tanisha has 25 coins consisting of nickels, dimes, and quarters with a total value of $3.60. If the sum of the number of nickels and dimes is 5 more than the number of quarters, how many of each type of coin does she have?

46. A vending machine accepts nickels, dimes, and quarters. When the owner checks the machine, there is a total of 56 coins worth $7.55. If the sum of the number of nickels and dimes is 2 more than twice the number of quarters, find the number of each type of coin.

47. Tommy empties his piggy bank and finds that he has 25 coins consisting of pennies, nickels, and dimes that are worth $1.36. If there are twice as many nickels as there are pennies, find the number of each type of coin.

48. At the end of each day, Josh puts all the change in his pockets into a jar. After one week he has 25 coins consisting of pennies, dimes, and quarters that are worth $2.17. If there are 3 more dimes than quarters, find the number of each type of coin.

49. Bronze is an alloy that is made of zinc, tin, and copper in a specified proportion. The amount of tin in bronze is three times the amount of zinc, and the amount of copper is 20 pounds more than 15 times the amount of tin. Suppose an order is placed for 1000 pounds of bronze. Find the number of pounds of zinc, tin, and copper. (*Source: World Almanac and Book of Facts* 2002)

50. The Fresh Roasted Almond Company sells nuts in bulk quantities. When bought in bulk, peanuts sell for $1.34 per pound, almonds for $4.36 per pound, and pecans for $5.88 per pound. Suppose a local specialty shop wants a mixture of 200 pounds that will cost $2.70 per pound. Find the number of pounds of each type of nut if the sum of the number of pounds of peanuts and almonds is nine times the number of pounds of pecans. (*Source:* Fresh Roasted Almond Company)

51. The Wunderbar delicatessen sells Black Forest ham for $11.96 per pound, breast of turkey for $8.76 per pound, and roast beef for $9.16 per pound. Suppose they make a 10-pound party tray of these three meats such that the average cost is $9.80 per pound. Find the number of pounds of each if the number of pounds of breast of turkey is equal to the sum of the number of pounds of Black Forest ham and roast beef. (*Source: The Wunderbar, Altamonte Springs*)

52. Barnie's coffee shop sells Jamaican Blue Mountain coffee for $45.99 per pound, Hawaiian Kona for $36.99 per pound, and Sulawesi Kalossi for $12.99 per pound. A 25-pound mixture of these three coffees sells for $30.27 per pound. Find the number of pounds of each if the sum of the number of pounds of Jamaica Blue Mountain and Hawaiian Kona is 5 more than the number of pounds of Sulawesi Kalossi. (*Source: Barnie's, Orlando, Florida*)

53. A total of $5000 is invested in three funds. The money market fund pays 5% annually, the income fund pays 6% annually, and the growth fund pays 8% annually. The total earnings for one year from the three funds are $340. If the amount invested in the growth fund is $500 less than twice the amount invested in the money market fund, find the amount in each of the funds.

54. A total of $8000 is invested in three stocks, which paid 4%, 6%, and 7% dividends, respectively, in one year. The amount invested in the stock that returned 6% dividends is $500 more than twice the amount that returned 4% dividends. If the total dividends in one year were $475, find the amount invested in each stock.

55. Nina inherited $100,000, which she invested in stocks, bonds, and certificates of deposit. After the first year, she had lost 6% in the stocks, earned 5% on the bonds, and earned 4% on the certificates of deposit. The sum of twice the principal in certificates of deposit and the principal in stocks is $20,000 more than the principal in bonds. Find the principal in each investment if the total income for the year was $1500.

56. An executive received a $30,000 end-of-the-year bonus. He invests the bonus in stocks, municipal bonds, and certificates of deposit. At the end of the first year, he had lost 5% in the stocks, earned 5% on the bonds, and earned 4% on the certificates of deposit, which resulted in a net return of $220. Find the principal in each of the investments if the principal in stocks is $6000 less than the sum of the principals in bonds and certificates.

57. Eggs are classified by the weight of a dozen. The three largest sizes are jumbo, extra large, and large. The total weight of one dozen of each size is 81 ounces. Three dozen jumbo eggs weigh 39 ounces more than the sum of the weight of one dozen extra large and one dozen large. The sum of the weights of two dozen extra large and one dozen large is 18 ounces more than the weight of two dozen jumbo. Find the weight of one dozen of each type of egg. (*Source: Numbers: How Many, How Long, How Far, How Much*)

58. The intensity of sound is measured in decibels. The decibel reading of a rock concert is 60 less than three times the reading of normal conversation and 10 less than that of a jet at takeoff. The decibel reading of a jet takeoff is 10 more than twice that of normal conversation. Find the decibel reading of each. (*Source: Numbers: How Many, How Long, How Far, How Much*)

59. The number of calories burned per hour bicycling is 120 more than the number burned from brisk walking. The number of calories burned per hour in climbing stairs is 180 more than from bicycling and is twice that from brisk walking. Find the number of calories burned per hour for each of the three activities. (*Source: Numbers: How Many, How Long, How Far, How Much*)

60. The human body is composed of many elements. About 98% of the human body is made up of carbon, oxygen, and hydrogen. The percentage of the human body that is hydrogen is the sum of twice the percentage of oxygen and the percentage of carbon. The percentage of oxygen is 6% more than twice the percentage of carbon. Find the percentage of the human body that is composed of each of these three elements. (*Source: Numbers: How Many, How Long, How Far, How Much*)

61. The three countries in the world with the highest death rates per 1000 are Angola, Mozambique, and Botswana. The sum of the death rates of all three countries is 75 per 1000. The sum of the death rates per 1000 of Angola and Botswana is twice the death rate of Mozambique, and the death rate per 1000 for Botswana is 2 more than the rate for Angola. Find the death rate per 1000 for each country. (*Source: Phrase Base*)

62. The three countries in the world with the highest birth rates per 1000 are Chad, Mali, and Niger. The sum of the birth rates is 145 per 1000. The sum of the birth rates per 1000 of Chad and Mali is 5 less than twice the birth rate of Niger. The birth rate per 1000 of Chad is 3 less than the birth rate of Niger. Find the birth rate per 1000 for each of these countries. (*Source: Phrase Base*)

63. An object is thrown upward with an initial velocity of v_0 from an initial height of h. The height of the object, h, is described by an equation of the form $h = at^2 + v_0 t + h_0$, where h and h_0 are in feet and t is in seconds. The following table shows the height for different values of t. Find the values of a, v_0, and h_0 and write the equation for h.

t	h
1	234
3	306
6	174

64. Use the same general equation from Exercise 61, $h = at^2 + v_0 t + h_0$, where h and h_0 are in feet and t is in seconds. The table at the right shows the distance above the ground for different values of t. Find the values of a, v_0, and h_0 and write the equation for h.

t	h
2	136
3	106
4	44

PUZZLE PROBLEM

The sum of four positive integers a, b, c, and d is 228 and a < b < c < d. The greatest number subtracted from twice the smallest number gives 18. If twice the sum of the third number and fourth number is subtracted from the product of 6 and the second number, the result is 10. If the smallest two numbers are consecutive odd integers, find all numbers.

REVIEW EXERCISES

$\begin{bmatrix} 1.7 \\ 5.3 \end{bmatrix}$ *For Exercises 1 and 2, simplify.*

1. $\frac{1}{2}(2x + 3y) - (x + 4y)$

2. $-2(x + y + 2z) + (2x + y - 2z)$

$\begin{bmatrix} 1.7 \\ 2.3 \end{bmatrix}$ **3.** Given $2x - 3y = 6$, find x if $y = 4$.

$\begin{bmatrix} 1.7 \\ 2.3 \end{bmatrix}$ **4.** Given $x + y + 2z = 7$, find x if $y = -1$ and $z = 3$.

For Exercises 5 and 6, solve.

[9.3] **5.** $\begin{cases} 3x + 4y = 6 \\ x - y = -5 \end{cases}$

[9.3] **6.** $\begin{cases} 2x + 5y = 1 \\ -3x + 2y = 11 \end{cases}$

9.5 Solving Systems of Linear Equations Using Matrices

OBJECTIVES

1. Write a system of equations as an augmented matrix.

2. Solve a system of linear equations by transforming its augmented matrix to echelon form.

OBJECTIVE 1. Write a system of equations as an augmented matrix. Although the elimination method that we learned in Section 9.2 is effective for solving systems of linear equations, we can streamline the method by manipulating just the coefficients and constants in a **matrix**.

DEFINITION *Matrix:* A rectangular array of numbers.

Following are some examples of matrices (plural of matrix):

$$\begin{bmatrix} 1 & -2 \\ -3 & 4 \end{bmatrix} \quad \begin{bmatrix} 1 & -5 & 6 \\ -2 & 4 & 0 \end{bmatrix} \quad \begin{bmatrix} -3 & 0 & 9 \\ -2 & 4 & 7 \\ 9 & -2 & 0 \end{bmatrix}$$

Matrices are made up of horizontal rows and vertical columns.

Column 1 Column 2 Column 3

$$\text{Row 1} \longrightarrow \begin{bmatrix} 1 & 4 & -2 \\ 5 & -1 & 3 \end{bmatrix} \longleftarrow \text{Row 2}$$

Column 1 Column 2

$$\begin{array}{l} \text{Row 1} \longrightarrow \\ \text{Row 2} \longrightarrow \\ \text{Row 3} \longrightarrow \end{array} \begin{bmatrix} 3 & -4 \\ -1 & 2 \\ 5 & -3 \end{bmatrix}$$

Note: *One way to remember that columns are vertical is to imagine the columns on a building, which are vertical.*

We call a matrix like $\begin{bmatrix} 2 & 5 & -1 \\ 5 & 4 & 3 \end{bmatrix}$ a 2×3 (read "2 by 3") matrix because it has two

rows and three columns; $\begin{bmatrix} 2 & 5 \\ -4 & 6 \\ 3 & -2 \end{bmatrix}$ is a 3×2 matrix because it has three rows

and two columns. Each number in a matrix is called an *element*. To solve a system of equations using matrices, we first rewrite the system as an **augmented matrix**.

DEFINITION *Augmented matrix:* A matrix made up of the coefficients and the constant terms of a system. The constant terms are separated from the coefficients by a dashed vertical line.

For example, to write the system $\begin{cases} 3x - 2y = 7 \\ 4x - 3y = 10 \end{cases}$ as an augmented matrix, we omit the variables and write $\begin{bmatrix} 3 & -2 & | & 7 \\ 4 & -3 & | & 10 \end{bmatrix}$.

EXAMPLE 1 Write $\begin{cases} 2x + 3y - 4z = 1 \\ 3x - 5y + z = -4 \\ 2x - 6y + 3z = 2 \end{cases}$ as an augmented matrix.

Solution $\begin{bmatrix} 2 & 3 & -4 & | & 1 \\ 3 & -5 & 1 & | & -4 \\ 2 & -6 & 3 & | & 2 \end{bmatrix}$

◀ **Note:** *It is helpful to think of the dashes as the equal sign of each equation.*

YOUR TURN Write the augmented matrix for $\begin{cases} x + 3y = -2 \\ -4x - 5y = 7 \end{cases}$.

OBJECTIVE 2. **Solve a system of linear equations by transforming its augmented matrix to echelon form.**
Recall that in the elimination method, we could interchange equations, add equations, or multiply equations by a number and then add the rewritten equations. Since each row of an augmented matrix contains the constants and coefficients of each equation in the system, we can interchange rows, add rows, or multiply rows by a number and then add the rewritten rows.

RULE **Row Operations**

The solution of a system is not affected by the following row operations in its augmented matrix:

1. Any two rows may be interchanged.
2. The elements of any row may be multiplied (or divided) by any nonzero real number.
3. Any row may be replaced by a row resulting from adding the elements of that row (or multiples of that row) to a multiple of the elements of any other row.

We use these row operations to solve a system of linear equations by transforming the augmented matrix of the system into an equivalent matrix that is in **echelon form**.

DEFINITION *Echelon form:* An augmented matrix whose coefficient portion has 1s on the diagonal from upper left to lower right and 0s below the 1s.

For example, $\begin{bmatrix} 1 & 3 & | & 3 \\ 0 & 1 & | & 4 \end{bmatrix}$ and $\begin{bmatrix} 1 & 2 & -4 & | & 5 \\ 0 & 1 & -5 & | & 7 \\ 0 & 0 & 1 & | & 5 \end{bmatrix}$ are in echelon form.

In the following examples we will let R_1 mean row 1, R_2 mean row 2, and so on.

ANSWER
$\begin{bmatrix} 1 & 3 & | & -2 \\ -4 & -5 & | & 7 \end{bmatrix}$

EXAMPLE 2 Solve the following linear system by transforming its augmented matrix into echelon form.

$$\begin{cases} x - 2y = -4 & \text{(Equation 1)} \\ -2x + 5y = 9 & \text{(Equation 2)} \end{cases}$$

Solution First, we write the augmented matrix: $\begin{bmatrix} 1 & -2 & | & -4 \\ -2 & 5 & | & 9 \end{bmatrix}$.

Now, we perform row operations to transform the matrix into echelon form. The element in the first row, first column is already 1, which is what we want. So we need to rewrite the matrix so that -2 in the second row, first column becomes 0. To do this, we multiply the first row by 2 and add it to the second row.

$$2R_1 + R_2 \longrightarrow \begin{bmatrix} 1 & -2 & | & -4 \\ 0 & 1 & | & 1 \end{bmatrix}$$

Note: *The result of the row operations replaces the affected row in the matrix. Other rows are rewritten unchanged.*

The resulting matrix represents the system $\begin{cases} x - 2y = -4 \\ y = 1 \end{cases}$.

Since $y = 1$, we can solve for x using substitution.

$$\begin{aligned} x - 2(1) &= -4 & &\text{Substitute 1 for } y \text{ in } x - 2y = -4. \\ x - 2 &= -4 & &\text{Simplify.} \\ x &= -2 & &\text{Add 2 to both sides.} \end{aligned}$$

The solution is $(-2, 1)$.

EXAMPLE 3 Solve the following linear system by transforming its augmented matrix into echelon form:

$$\begin{cases} 2x + 6y = 1 & \text{(Equation 1)} \\ x + 8y = 3 & \text{(Equation 2)} \end{cases}$$

Solution First, we write the augmented matrix: $\begin{bmatrix} 2 & 6 & | & 1 \\ 1 & 8 & | & 3 \end{bmatrix}$.

Now we perform row operations to get echelon form.

Connection Row operations correspond to equation manipulations in the elimination method. For example, $-1 \cdot R_1 + R_2$ corresponds to multiplying $x + 3y = \dfrac{1}{2}$ by -1 and then adding the resulting equation to $x + 8y = 3$ so that we get $5y = \dfrac{5}{2}$.

$$\frac{1}{2} \cdot R_1 \longrightarrow \begin{bmatrix} 1 & 3 & | & \frac{1}{2} \\ 1 & 8 & | & 3 \end{bmatrix}$$

We need the 2 in row 1, column 1 to be a 1, so multiply each number in row 1 by $\frac{1}{2}$.

$$-1 \cdot R_1 + R_2 \longrightarrow \begin{bmatrix} 1 & 3 & | & \frac{1}{2} \\ 0 & 5 & | & \frac{5}{2} \end{bmatrix}$$

We need the 1 in row 2, column 1 to be 0, so multiply row 1 by -1 and add it to row 2.

$$\frac{1}{5} \cdot R_2 \longrightarrow \begin{bmatrix} 1 & 3 & | & \frac{1}{2} \\ 0 & 1 & | & \frac{1}{2} \end{bmatrix}$$

We need the 5 in row 2, column 2 to be 1, so multiply row 2 by $\frac{1}{5}$.

This matrix represents the system $\begin{cases} x + 3y = \dfrac{1}{2} \\ y = \dfrac{1}{2} \end{cases}$.

Since $y = \dfrac{1}{2}$, we can solve for x using substitution.

$$x + 3\left(\dfrac{1}{2}\right) = \dfrac{1}{2} \qquad \text{Substitute } \dfrac{1}{2} \text{ for } y \text{ in } x + 3y = \dfrac{1}{2}.$$

$$x + \dfrac{3}{2} = \dfrac{1}{2} \qquad \text{Simplify.}$$

$$x = -1 \qquad \text{Subtract } \dfrac{3}{2} \text{ from both sides.}$$

The solution is $\left(-1, \dfrac{1}{2}\right)$.

YOUR TURN Solve the following system by transforming its augmented matrix to echelon form.

$$\begin{cases} 3x + 4y = 6 \\ x - 3y = -11 \end{cases}$$

Now let's use row operations to solve systems of three equations.

EXAMPLE 4 Use the echelon method to solve

$$\begin{cases} 2x + y - 2z = -3 & \text{(Equation 1)} \\ x + y + 2z = 7 & \text{(Equation 2)} \\ 4y + 3z = 5 & \text{(Equation 3)} \end{cases}$$

Solution Write the augmented matrix: $\begin{bmatrix} 2 & 1 & -2 & | & -3 \\ 1 & 1 & 2 & | & 7 \\ 0 & 4 & 3 & | & 5 \end{bmatrix}$.

Now we perform row operations to get echelon form

$$\begin{matrix} R_2 \longrightarrow \\ R_1 \longrightarrow \\ {} \end{matrix} \begin{bmatrix} 1 & 1 & 2 & | & 7 \\ 2 & 1 & -2 & | & -3 \\ 0 & 4 & 3 & | & 5 \end{bmatrix}$$

Because we need a 1 in row 1, column 1, we will interchange rows 1 and 2.

$$\begin{matrix} {} \\ -2 \cdot R_1 + R_2 \longrightarrow \\ {} \end{matrix} \begin{bmatrix} 1 & 1 & 2 & | & 7 \\ 0 & -1 & -6 & | & -17 \\ 0 & 4 & 3 & | & 5 \end{bmatrix}$$

To get a 0 in row 2, column 1, we multiply the new row 1 by -2 and add it to row 2.

$$\begin{matrix} {} \\ {} \\ 4 \cdot R_2 + R_3 \longrightarrow \end{matrix} \begin{bmatrix} 1 & 1 & 2 & | & 7 \\ 0 & -1 & -6 & | & -17 \\ 0 & 0 & -21 & | & -63 \end{bmatrix}$$

Row 3 only needs a 0 in the second column, so we multiply row 2 by 4 and add it to row 3.

$$\begin{matrix} {} \\ -1 \cdot R_2 \longrightarrow \\ -\dfrac{1}{21} \cdot R_3 \longrightarrow \end{matrix} \begin{bmatrix} 1 & 1 & 2 & | & 7 \\ 0 & 1 & 6 & | & 17 \\ 0 & 0 & 1 & | & 3 \end{bmatrix}$$

To get a 1 in row 2, column 2, we multiply row 2 by -1.

To get a 1 in row 3, column 3, we multiply row 3 by $-\dfrac{1}{21}$.

ANSWER

$(-2, 3)$

Solving Systems of Linear Equations Using Matrices **739**

The resulting matrix represents the system $\begin{cases} x + y + 2z = 7 \\ \quad\;\; y + 6z = 17. \\ \qquad\qquad z = 3 \end{cases}$

To find y, substitute 3 for z in $y + 6z = 17$

$$y + 6(\mathbf{3}) = 17 \qquad \text{Substitute 3 for } z.$$
$$y + 18 = 17 \qquad \text{Multiply 6 and 3.}$$
$$y = -1 \qquad \text{Subtract 18 from both sides.}$$

To find x, substitute -1 for y and 3 for z in $x + y + 2z = 7$.

$$x + (-1) + 2(\mathbf{3}) = 7 \qquad \text{Substitute } -1 \text{ for } y \text{ and } \mathbf{3} \text{ for } z.$$
$$x + 5 = 7 \qquad \text{Simplify the left side of the equation.}$$
$$x = 2 \qquad \text{Subtract 5 from both sides of the equation.}$$

The solution is $(2, -1, 3)$. We can check by verifying that the ordered triple satisfies all three of the original equations.

YOUR TURN Solve the following linear system by transforming its augmented matrix into echelon form.

$$\begin{cases} 3x + y + z = 2 \\ x + 2y - z = -4 \\ 2x - 2y + 3z = 9 \end{cases}$$

Calculator TIPS

To use a graphing calculator to solve the system in Example 4, press the [MATRX] *key, then use the arrow keys to highlight* **EDIT**. *You will see a list of matrices and their current sizes. Press 1 to select the first matrix, named [A]. We need our matrix to be 3 × 4, so press* [3] [ENTER], *then* [4] [ENTER]. *The cursor will now be in row 1, column 1 of a 3 × 4 matrix. You can now enter each number. Use the arrow keys to move the cursor to any position in the matrix. Note that the cursor position's row number, column number, and current value are displayed at the bottom of the screen. After entering each number for Example 4, your screen should look like this:*

MATRX [A] 3 × 4
$\begin{bmatrix} 2 & 1 & -2 & -3 \end{bmatrix}$
$\begin{bmatrix} 1 & 1 & 2 & 7 \end{bmatrix}$
$\begin{bmatrix} 0 & 4 & 3 & 5 \end{bmatrix}$

When you finish entering all the numbers, press [2nd] [MODE] *to quit editing.*

ANSWER

$(1, -2, 1)$

Now, to put the matrix into row echelon form, press MATRX *again and move the cursor to highlight **MATH** at the top of the screen. From the menu, select **A: ref (** and press* ENTER *. You will be prompted to enter the name of the desired matrix. Press* MATRX *again and select**1:[A] 3 × 4.** You will now see ref ([A]. Use the*) *key to close the parentheses and press* ENTER *. You now have the row echelon form and your screen should look like this:*

ref ([A])

$$\begin{bmatrix} 1 & .5 & -1 & -1.5 \\ 0 & 1 & .75 & 1.25 \\ 0 & 0 & 1 & 3 \end{bmatrix}$$

◀ **Note:** *The function* **ref** *stands for row echelon form.*

Notice the last row indicates that z = 3 and we can use substitution to find the other values.

*Your calculator can also show the complete solution. Instead of selecting **A: ref (**, select **B: rref (** from the math menu. Your screen will look like this:*

rref ([A])

$$\begin{bmatrix} 1 & 0 & 0 & 2 \\ 0 & 1 & 0 & -1 \\ 0 & 0 & 1 & 3 \end{bmatrix}$$

◀ **Note:** *This form is called reduced row echelon form, which is why the function is* **rref**.

Note that the first row of this form indicates that x = 2, the second row indicates that y = −1, and the third row indicates that z = 3, which is the solution for our system.

9.5 Exercises

For Extra Help — MyMathLab — Videotape/DVT — InterAct Math — Math Tutor Center — Math XL.com

1. How many rows and columns are in a 4 × 2 matrix?

2. How do you write a system of equations as an augmented matrix?

3. What does the dashed line in an augmented matrix correspond to in a system of equations?

4. Explain how solving a system of equations using row operations is similar to solving the system using elimination.

5. What is echelon form?

6. Once an augmented matrix is in echelon form, how do you find the values of the variables?

For Exercises 7–14, write the augmented matrix for the system of equations.

7. $\begin{cases} 14x + 7y = 6 \\ 7x + 6y = 8 \end{cases}$

8. $\begin{cases} 5x + 6y = 2 \\ 10x + 3y = -2 \end{cases}$

9. $\begin{cases} 7x - 6y = 1 \\ 8x - 12y = 6 \end{cases}$

10. $\begin{cases} 3x - y = 6 \\ 9x + 2y = -2 \end{cases}$

11. $\begin{cases} x - 3y + z = 4 \\ 2x - 4y + 2z = -4 \\ 6x - 2y + 5z = -4 \end{cases}$

12. $\begin{cases} 3x + 2y - 3z = 2 \\ 2x - 4y + 5z = -10 \\ 5x - 4y + z = 0 \end{cases}$

13. $\begin{cases} 4x + 6y - 2z = -1 \\ 8x + 3y = -12 \\ -y + 2z = 4 \end{cases}$

14. $\begin{cases} x - 3y = 3 \\ -3x + 4y + 9z = 3 \\ 2x + 7z = -9 \end{cases}$

For Exercises 15–18, given the matrices in echelon form, find the solution for the system.

15. $\left[\begin{array}{cc|c} 1 & 5 & -6 \\ 0 & 1 & -2 \end{array}\right]$

16. $\left[\begin{array}{cc|c} 1 & -3 & 7 \\ 0 & 1 & -5 \end{array}\right]$

17. $\left[\begin{array}{ccc|c} 1 & -4 & -8 & 6 \\ 0 & 1 & -2 & -7 \\ 0 & 0 & 1 & 1 \end{array}\right]$

18. $\left[\begin{array}{ccc|c} 1 & -2 & 6 & 16 \\ 0 & 1 & -4 & -13 \\ 0 & 0 & 1 & 3 \end{array}\right]$

For Exercises 19–24, complete the indicated row operation.

19. Replace R_2 in $\left[\begin{array}{cc|c} 1 & 3 & -1 \\ -2 & 5 & 6 \end{array}\right]$ with $2R_1 + R_2$.

20. Replace R_2 in $\left[\begin{array}{cc|c} 1 & 2 & -2 \\ 3 & 8 & -4 \end{array}\right]$ with $-3R_1 + R_2$.

21. Replace R_3 in $\left[\begin{array}{ccc|c} 1 & -2 & 4 & 6 \\ 0 & 2 & -1 & -5 \\ 0 & 8 & -6 & -3 \end{array}\right]$ with $-4R_2 + R_3$.

22. Replace R_2 in $\left[\begin{array}{ccc|c} 1 & 5 & -3 & 8 \\ -3 & 2 & -4 & -6 \\ 0 & 1 & -2 & 9 \end{array}\right]$ with $3R_1 + R_2$.

23. Replace R_1 in $\left[\begin{array}{cc|c} 4 & 8 & -10 \\ -1 & 3 & 2 \end{array}\right]$ with $\frac{1}{4}R_1$.

24. Replace R_3 in $\left[\begin{array}{ccc|c} 1 & 3 & 4 & 8 \\ 0 & 1 & -2 & -6 \\ 0 & 0 & -6 & 24 \end{array}\right]$ with $-\frac{1}{6}R_3$.

For Exercises 25–28, describe the row operation that should be performed to make the matrix closer to echelon form.

25. $\begin{bmatrix} 1 & 2 & | & -1 \\ -3 & 4 & | & 5 \end{bmatrix}$

26. $\begin{bmatrix} 1 & -3 & | & 2 \\ 5 & 6 & | & -4 \end{bmatrix}$

27. $\begin{bmatrix} 1 & -4 & 3 & | & 6 \\ 0 & 1 & -2 & | & 4 \\ 0 & 2 & -5 & | & 1 \end{bmatrix}$

28. $\begin{bmatrix} 1 & -2 & 4 & | & 7 \\ 0 & 1 & -2 & | & 3 \\ 0 & -5 & 4 & | & -9 \end{bmatrix}$

For Exercises 29–52, solve by transforming the augmented matrix into echelon form.

29. $\begin{cases} x - y = 2 \\ x + y = 2 \end{cases}$

30. $\begin{cases} x + y = 2 \\ x - y = -8 \end{cases}$

31. $\begin{cases} x + y = 3 \\ 3x - y = 1 \end{cases}$

32. $\begin{cases} x + 3y = -9 \\ -x + 2y = -11 \end{cases}$

33. $\begin{cases} x + 2y = -7 \\ 2x - 4y = 2 \end{cases}$

34. $\begin{cases} x - 3y = -13 \\ 3x + 2y = 5 \end{cases}$

35. $\begin{cases} -2x + 5y = 9 \\ x - 2y = -4 \end{cases}$

36. $\begin{cases} 2x + y = 12 \\ x - 3y = 6 \end{cases}$

37. $\begin{cases} 4x - 3y = -2 \\ 2x - 3y = -10 \end{cases}$

38. $\begin{cases} 6x - 3y = 0 \\ 2x + 4y = 0 \end{cases}$

39. $\begin{cases} 5x + 2y = 12 \\ 2x + 3y = -4 \end{cases}$

40. $\begin{cases} 4x + 3y = 14 \\ 3x - 2y = 2 \end{cases}$

41. $\begin{cases} x + y + z = 6 \\ 2x + y - 3z = -3 \\ 2x + 2y + z = 6 \end{cases}$

42. $\begin{cases} x + 2y + z = 2 \\ 3x + y - z = 3 \\ 2x + y + 2z = 7 \end{cases}$

43. $\begin{cases} x - y + z = -1 \\ 2x - 2y + z = 0 \\ x + 3y + 2z = 1 \end{cases}$

44. $\begin{cases} x + 3y - 2z = 4 \\ 2x - 3y + 2z = -7 \\ 3x + 2y - 2z = -1 \end{cases}$

45. $\begin{cases} 2x - y + z = 8 \\ x - 2y + 3z = 11 \\ 2x + 3y - z = -6 \end{cases}$

46. $\begin{cases} 2x + 3y - 2z = -21 \\ 2x - 4y + 3z = 15 \\ 3x + 2y - 3z = -24 \end{cases}$

47. $\begin{cases} 3x + 2y - 3z = 1 \\ -2x + 3y - 4z = 7 \\ 5x - 2y + z = -5 \end{cases}$

48. $\begin{cases} 2x - 2y + z = 6 \\ -2x + 4y + 3z = 4 \\ 4x - 3y - 2z = -7 \end{cases}$

49. $\begin{cases} 3x - 6y + z = -10 \\ 7y - z = 2 \\ 2x + 4z = 14 \end{cases}$

50. $\begin{cases} 6x - y + 5z = -28 \\ 4x + 2y = 10 \\ 5x + 6z = -23 \end{cases}$

51. $\begin{cases} 2x + 5y - z = 10 \\ x - y - z = 14 \\ x - 6y = 20 \end{cases}$

52. $\begin{cases} 2x + 4y - 5z = 16 \\ x - 2y - 11z = 1 \\ 4x + 5y = -6 \end{cases}$

For Exercises 53–60, solve using a matrix on a graphing calculator.

53. $\begin{cases} 2x + 5y = -14 \\ 6x + 7y = -10 \end{cases}$

54. $\begin{cases} 4x - 3y = 11 \\ 5x + 4y = 68 \end{cases}$

55. $\begin{cases} 4x - 7y = -80 \\ 9x + 5y = -14 \end{cases}$

56. $\begin{cases} 7x - 3y = 23 \\ 4x - 9y = -67 \end{cases}$

57. $\begin{cases} 3x + 2y - 5z = -19 \\ 4x - 7y + 6z = 67 \\ 5x - 6y - 4z = 18 \end{cases}$

58. $\begin{cases} 6x - 7y + 2z = -52 \\ 8x + 5y - 6z = 132 \\ 12x + 4y - 7z = 150 \end{cases}$

59. $\begin{cases} 8x - 7y + 2z = 108 \\ 6x + 11y - 10z = -74 \\ 9x - 2y + 13z = 140 \end{cases}$

60. $\begin{cases} 8x - 11y + 14z = 40 \\ 5x + 15y - 5z = 35 \\ 17x - 9y + 2z = -133 \end{cases}$

For Exercises 61–62, explain the mistake, then find the correct solution.

61. $\begin{cases} x + 3y = 13 \\ -4x - y = -26 \end{cases}$

$$\begin{bmatrix} 1 & 3 & | & 13 \\ -4 & -1 & | & -26 \end{bmatrix}$$

$4 \cdot R_1 + R_2 \longrightarrow \begin{bmatrix} 1 & 3 & | & 13 \\ 0 & 11 & | & -13 \end{bmatrix}$

$\dfrac{1}{11} \cdot R_2 \longrightarrow \begin{bmatrix} 1 & 3 & | & 13 \\ 0 & 1 & | & -\dfrac{13}{11} \end{bmatrix}$

Solution: $\left(-\dfrac{182}{11}, -\dfrac{13}{11} \right)$

62. $\begin{cases} x - y + z = 8 \\ 3x - z = -9 \\ 4y + z = -6 \end{cases}$

$$\begin{bmatrix} 1 & -1 & 1 & | & 8 \\ 3 & 0 & -1 & | & -9 \\ 4 & 0 & 1 & | & -6 \end{bmatrix}$$

$-3 \cdot R_1 + R_2 \longrightarrow \begin{bmatrix} 1 & -1 & 1 & | & 8 \\ 0 & 3 & -4 & | & -33 \\ 4 & 0 & 1 & | & -6 \end{bmatrix}$

$\begin{matrix} \dfrac{1}{3} \cdot R_2 \\ -4 \cdot R_1 + R_3 \end{matrix} \longrightarrow \begin{bmatrix} 1 & -1 & 1 & | & 8 \\ 0 & 1 & -\dfrac{4}{3} & | & -11 \\ 0 & 4 & -3 & | & -38 \end{bmatrix}$

$-4 \cdot R_2 + R_3 \longrightarrow \begin{bmatrix} 1 & -1 & 1 & | & 8 \\ 0 & 1 & -\dfrac{4}{3} & | & -11 \\ 0 & 0 & \dfrac{7}{3} & | & 6 \end{bmatrix}$

$\dfrac{3}{7} \cdot R_3 \longrightarrow \begin{bmatrix} 1 & -1 & 1 & | & 8 \\ 0 & 1 & -\dfrac{4}{3} & | & -11 \\ 0 & 0 & 1 & | & \dfrac{18}{7} \end{bmatrix}$

Solution: $\left(-\dfrac{15}{7}, -\dfrac{53}{7}, \dfrac{18}{7} \right)$

For Exercises 63–76, translate the problem to a system of equations, then solve using matrices.

63. Brad purchased three grilled chicken sandwiches and two drinks for $9.90, and Angel purchased seven grilled chicken sandwiches and four drinks for $22.30. Find the price of one grilled chicken sandwich and one drink.

64. Sharika purchased three general admission tickets and two student tickets to a college play for $55, and Yo Chen purchased two general admission tickets and four student tickets for $50. Find the cost of one general admission and one student ticket.

65. A right triangle has one angle whose measure is 90°. One of the remaining angles is 6° less than twice the measure of the other. Find the measure of each of the remaining angles.

66. The measure of the largest angle of a triangle is 16° less than the sum of the measures of the other two. Twice the measure of the middle angle is 40° more than the sum of the largest and smallest. Find the measure of each of the angles.

67. The two longest rivers in the world, the Nile and the Amazon, have a combined length of 8050 miles. The Nile is 250 miles longer than the Amazon. Find the length of both rivers.

68. The longest vehicular tunnel in the world is the Saint Gotthard Tunnel in Switzerland and the second longest is the Arlberg Tunnel in Austria. The total length of the two tunnels is 18.8 miles. The Saint Gotthard tunnel is 1.4 miles longer than the Arlberg. Find the length of each tunnel. (*Source: Webster's New World Book of Facts*)

69. Nikita invested a total of $10,000 in certificates of deposit that pay 5% annually and in a money market account that pays 6% annually. If the total interest earned in one year from the two investments is $536, find the principal that was invested in each.

70. An athlete received a $60,000 signing bonus, which he invested in three funds—a money market fund paying 5% annually, an income fund paying 5.5% annually, and a growth fund paying 7% annually. The sum of the principals invested in the money market and the growth fund equals the principal in the income fund. The total annual interest received from the three investments is $3400. Find the principal invested in each fund.

71. John bought two CDs, four tapes, and three DVDs for $164, and Tanelle bought five CDs, two tapes, and two DVDs for $160. If the sum of the cost of one CD and one tape equals the cost of one DVD, find the cost of one of each.

72. Angel bought 3 pounds of salmon, 2 pounds of tuna, and 5 pounds of cod for $63. Sara bought 6 pounds of salmon, 3 pounds of tuna, and 5 pounds of cod for $94. The cost of 1 pound of salmon and 1 pound of tuna is the same as the cost of 3 pounds of cod. Find the cost of 1 pound of each.

73. In New England's 24–21 victory over Philadelphia in Super Bowl XXXIX, New England scored touchdowns (6 points), extra points (1 point), and field goals (3 points). The number of touchdowns equaled the number of extra points. Also, the number of touchdowns was one less than four times the number of field goals. Find how many of each type of score New England had. (*Source: Super Bowl.com*)

74. In the 2005, NCAA Women's Championship Basketball game, Baylor player Emily Niemann scored eight times for a total of 19 points using a combination of 3-point field goals, 2-point field goals, and free throws (1 point). If the sum of the number of 3-point and 2-point field goals was three times the number of free throws, find the number of each type. (*Source: Orlando Sentinel,* Apr. 16, 2005)

75. An electrical circuit has three points of connection, with different voltage measurements at each of the three connections. An engineer has written the following equations to describe the voltages. Find each voltage.

$$4v_1 - v_2 = 30$$
$$-2v_1 + 5v_2 - v_3 = 10$$
$$-v_2 + 5v_3 = 4$$

76. An engineer has written the following system of equations to describe the forces in pounds acting on a steel structure. Find the forces.

$$6F_1 - F_2 = 350$$
$$9F_1 - 2F_2 - F_3 = -100$$
$$3F_2 - F_3 = 250$$

REVIEW EXERCISES

[1.5] *For Exercises 1–6, simplify.*

1. $(-3)(4) - (-4)(5)$

2. $2(2 - 8) + 3[-1 - (-6)] - 4(4 - 6)$

3. $\dfrac{(-5)(-1) - (9)(3)}{(2)(-1) - (3)(3)}$

4. $\dfrac{2\left(\dfrac{5}{4}\right) - 8\left(\dfrac{1}{4}\right)}{\left(\dfrac{1}{2}\right)\left(\dfrac{5}{4}\right) - \left(\dfrac{3}{2}\right)\left(\dfrac{1}{4}\right)}$

5. $\dfrac{(1.1)(-0.2) - (1.7)(-0.5)}{(0.2)(-0.2) - (0.5)(-0.5)}$

6. $\dfrac{1[-45 - (-9)] + 7(-9 - 2) - 2(-27 - 30)}{1(6 - 3) - 2(-9 - 2) - 2[9 - (-4)]}$

9.6 Solving Systems of Linear Equations Using Cramer's Rule

OBJECTIVES
1. Evaluate determinants of 2 × 2 matrices.
2. Evaluate determinants of 3 × 3 matrices.
3. Solve systems of equations using Cramer's Rule.

OBJECTIVE 1. Evaluate determinants of 2 × 2 matrices. In this section, we will explore another way of solving systems of linear equations using a special type of matrix called a **square matrix**.

DEFINITION *Square matrix:* A matrix that has the same number of rows and columns.

For example, $\begin{bmatrix} 2 & 3 \\ 1 & -4 \end{bmatrix}$ and $\begin{bmatrix} 1 & -3 & 5 \\ -7 & 2 & 9 \\ 0 & 4 & -6 \end{bmatrix}$ are square matrices.

Every square matrix has a *determinant*. We write the determinant of a matrix, A, as $\det(A)$ or $|A|$. The method used to find the determinant of a matrix depends upon its size.

RULE Determinant of a 2 × 2 Matrix

Warning: Be careful to note the difference between [A] and $|A|$. The notation [A] means "the matrix A" whereas $|A|$ means "the determinant of the matrix A."

If $A = \begin{bmatrix} a_1 & b_1 \\ a_2 & b_2 \end{bmatrix}$, then $\det(A) = \begin{vmatrix} a_1 & b_1 \\ a_2 & b_2 \end{vmatrix} = a_1 b_2 - a_2 b_1$.

Notice that the determinant contains diagonal products, as illustrated by the following:

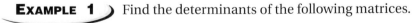

$\begin{vmatrix} a_1 & b_1 \\ a_2 & b_2 \end{vmatrix} = a_1 b_2 - a_2 b_1$

Note: *Since subtraction is not commutative, be sure to note the order of the two products.*

Learning Strategies

The products in the determinant are arranged with the downward cross product subtracting the upward cross product. An easy way to remember this order is that you have to "fall down before you can get up."

EXAMPLE 1 Find the determinants of the following matrices.

a. $A = \begin{bmatrix} 3 & -2 \\ 2 & 4 \end{bmatrix}$

Solution $\det(A) = \begin{vmatrix} 3 & -2 \\ 2 & 4 \end{vmatrix} = (3)(4) - (2)(-2) = 12 + 4 = 16$

b. $B = \begin{bmatrix} -3 & 5 \\ 4 & -2 \end{bmatrix}$

Solution $\det(B) = (-3)(-2) - (4)(5) = 6 - 20 = -14$

c. $M = \begin{bmatrix} 1 & 3 \\ 3 & 9 \end{bmatrix}$

Solution $\det(M) = (1)(9) - (3)(3) = 9 - 9 = 0$

ANSWER

-10

YOUR TURN Find the determinant of $\begin{bmatrix} 1 & -3 \\ -4 & 2 \end{bmatrix}$.

OBJECTIVE 2. Evaluate determinants of 3 × 3 matrices. There are various methods of evaluating the determinant of a 3 × 3 matrix. One of the most common methods is *expanding by minors*. Each element of a square matrix has a number called the **minor** for that element.

DEFINITION *Minor:* The determinant of the remaining matrix when the row and column in which the element is located are ignored.

EXAMPLE 2 Find the minor of 2 in $\begin{bmatrix} 2 & -3 & -6 \\ -1 & 5 & -2 \\ 3 & -4 & 1 \end{bmatrix}$.

Solution To find the minor of 2, we ignore its row and column (shown in blue) and evaluate the determinant of the remaining matrix (shown in red).

$$\begin{bmatrix} 2 & -3 & -6 \\ -1 & 5 & -2 \\ 3 & -4 & 1 \end{bmatrix}$$

$$\begin{vmatrix} 5 & -2 \\ -4 & 1 \end{vmatrix} = (5)(1) - (-4)(-2) = 5 - 8 = -3$$

ANSWER

$\begin{vmatrix} -2 & 2 \\ 4 & -1 \end{vmatrix} = -6$

YOUR TURN Find the minor of 6 in $\begin{bmatrix} 1 & -3 & 6 \\ -2 & 2 & 0 \\ 4 & -1 & 5 \end{bmatrix}$.

Note: *We can actually expand by minors along any row or column to find the determinant of a 3×3 matrix. For simplicity, we have chosen to show expanding by minors only along the first column. You may learn how to expand by minors along other rows or columns in future courses.*

▶ **RULE**   **Evaluating the Determinant of a 3×3 Matrix**

$$\begin{vmatrix} a_1 & b_1 & c_1 \\ a_2 & b_2 & c_2 \\ a_3 & b_3 & c_3 \end{vmatrix} = a_1 \begin{pmatrix} \text{minor} \\ \text{of } a_1 \end{pmatrix} - a_2 \begin{pmatrix} \text{minor} \\ \text{of } a_2 \end{pmatrix} + a_3 \begin{pmatrix} \text{minor} \\ \text{of } a_3 \end{pmatrix}$$

$$= a_1 \begin{vmatrix} b_2 & c_2 \\ b_3 & c_3 \end{vmatrix} - a_2 \begin{vmatrix} b_1 & c_1 \\ b_3 & c_3 \end{vmatrix} + a_3 \begin{vmatrix} b_1 & c_1 \\ b_2 & c_2 \end{vmatrix}$$

EXAMPLE 3   Find the determinant of $\begin{bmatrix} 2 & -3 & -4 \\ -1 & 2 & -2 \\ 3 & -4 & 1 \end{bmatrix}$.

Solution Using the rule for expanding by minors along the first column, we have

$$\begin{vmatrix} 2 & -3 & -4 \\ -1 & 2 & -2 \\ 3 & -4 & 1 \end{vmatrix} = 2 \begin{pmatrix} \text{minor} \\ \text{of } 2 \end{pmatrix} - (-1) \begin{pmatrix} \text{minor} \\ \text{of } -1 \end{pmatrix} + 3 \begin{pmatrix} \text{minor} \\ \text{of } 3 \end{pmatrix}$$

$$= 2 \begin{vmatrix} 2 & -2 \\ -4 & 1 \end{vmatrix} - (-1) \begin{vmatrix} -3 & -4 \\ -4 & 1 \end{vmatrix} + 3 \begin{vmatrix} -3 & -4 \\ 2 & -2 \end{vmatrix}$$

$$= 2(2 - 8) + 1(-3 - 16) + 3(6 + 8)$$

$$= -12 + (-19) + 42$$

$$= 11$$

YOUR TURN   Find the determinant of $\begin{bmatrix} 3 & -2 & 4 \\ -2 & 3 & 1 \\ 2 & -4 & 2 \end{bmatrix}$.

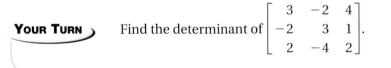

Calculator TIP

*The determinant of a matrix can be found using a graphing calculator. After entering the matrix (see p. 740) we use the **det (** function. Press the* $\boxed{\text{MATRX}}$ *key and select the* **det (** *function from the **MATH** menu. Now indicate the matrix you want to find the determinant of by pressing* $\boxed{\text{MATRX}}$ *and selecting the desired matrix from your list. After indicating the desired matrix, press* $\boxed{\text{ENTER}}$ *.*

ANSWER

26

OBJECTIVE 3. Solve systems of equations using Cramer's Rule. Now we can use **Cramer's Rule**, which uses determinants to solve systems of equations. To derive Cramer's Rule, we solve a general system of equations using the elimination method. We will show the derivation for a system of two equations in two unknowns.

$$\begin{cases} a_1x + b_1y = c_1 & \text{(Equation 1)} \\ a_2x + b_2y = c_2 & \text{(Equation 2)} \end{cases}$$

We will eliminate y by multiplying equation 1 by b_2 and equation 2 by $-b_1$.

$$
\begin{array}{ll}
a_1b_2x + b_1b_2y = b_2c_1 & \text{Multiply equation 1 by } b_2. \\
\underline{-a_2b_1x - b_1b_2y = -b_1c_2} & \text{Multiply equation 2 by } -b_1. \\
a_1b_2x - a_2b_1x \qquad = b_2c_1 - b_1c_2 & \text{Add the equations.} \\
(a_1b_2 - a_2b_1)x = b_2c_1 - b_1c_2 & \text{Factor out } x \text{ from the left side.} \\
x = \dfrac{b_2c_1 - b_1c_2}{a_1b_2 - a_2b_1} & \text{Divide by } a_1b_2 - a_2b_1.
\end{array}
$$

Notice that the numerator is $\begin{vmatrix} c_1 & b_1 \\ c_2 & b_2 \end{vmatrix}$ and the denominator is $\begin{vmatrix} a_1 & b_1 \\ a_2 & b_2 \end{vmatrix}$, so $x = \dfrac{\begin{vmatrix} c_1 & b_1 \\ c_2 & b_2 \end{vmatrix}}{\begin{vmatrix} a_1 & b_1 \\ a_2 & b_2 \end{vmatrix}}$.

If we repeat the same process and solve for y, we get $y = \dfrac{\begin{vmatrix} a_1 & c_1 \\ a_2 & c_2 \end{vmatrix}}{\begin{vmatrix} a_1 & b_1 \\ a_2 & b_2 \end{vmatrix}}$.

A similar approach is used to derive the rule for a system of three equations in three unknowns.

DEFINITION *Cramer's Rule*

The solution to the system of linear equations $\begin{cases} a_1x + b_1y = c_1 \\ a_2x + b_2y = c_2 \end{cases}$ is

$$x = \dfrac{\begin{vmatrix} c_1 & b_1 \\ c_2 & b_2 \end{vmatrix}}{\begin{vmatrix} a_1 & b_1 \\ a_2 & b_2 \end{vmatrix}} = \dfrac{D_x}{D} \quad \text{and} \quad y = \dfrac{\begin{vmatrix} a_1 & c_1 \\ a_2 & c_2 \end{vmatrix}}{\begin{vmatrix} a_1 & b_1 \\ a_2 & b_2 \end{vmatrix}} = \dfrac{D_y}{D}$$

Note: *Each denominator, D, is the determinant of a matrix containing only the coefficients in the system. To find D_x, we replace the column of x-coefficients in the coefficient matrix with the constants from the system. To find D_y, we replace the column of y-coefficients in the coefficient matrix with the constant terms, and do likewise to find D_z.*

The solution to the system of linear equations $\begin{cases} a_1x + b_1y + c_1z = d_1 \\ a_2x + b_2y + c_2z = d_2 \\ a_3x + b_3y + c_3z = d_3 \end{cases}$ is

$$x = \dfrac{\begin{vmatrix} d_1 & b_1 & c_1 \\ d_2 & b_2 & c_2 \\ d_3 & b_3 & c_3 \end{vmatrix}}{\begin{vmatrix} a_1 & b_1 & c_1 \\ a_2 & b_2 & c_2 \\ a_3 & b_3 & c_3 \end{vmatrix}} = \dfrac{D_x}{D}, \quad y = \dfrac{\begin{vmatrix} a_1 & d_1 & c_1 \\ a_2 & d_2 & c_2 \\ a_3 & d_3 & c_3 \end{vmatrix}}{\begin{vmatrix} a_1 & b_1 & c_1 \\ a_2 & b_2 & c_2 \\ a_3 & b_3 & c_3 \end{vmatrix}} = \dfrac{D_y}{D}, \quad \text{and} \quad z = \dfrac{\begin{vmatrix} a_1 & b_1 & d_1 \\ a_2 & b_2 & d_2 \\ a_3 & b_3 & d_3 \end{vmatrix}}{\begin{vmatrix} a_1 & b_1 & c_1 \\ a_2 & b_2 & c_2 \\ a_3 & b_3 & c_3 \end{vmatrix}} = \dfrac{D_z}{D}$$

EXAMPLE 4 Use Cramer's Rule to solve $\begin{cases} 2x + 3y = -5 \\ 3x - y = 9 \end{cases}$.

Solution First, we find D, D_x, and D_y.

$$D = \begin{vmatrix} 2 & 3 \\ 3 & -1 \end{vmatrix} = (2)(-1) - (3)(3) = -2 - 9 = -11$$

$$D_x = \begin{vmatrix} -5 & 3 \\ 9 & -1 \end{vmatrix} = (-5)(-1) - (9)(3) = 5 - 27 = -22$$

$$D_y = \begin{vmatrix} 2 & -5 \\ 3 & 9 \end{vmatrix} = (2)(9) - (3)(-5) = 18 + 15 = 33$$

Now we can find x and y.

$$x = \frac{D_x}{D} = \frac{-22}{-11} = 2 \qquad y = \frac{D_y}{D} = \frac{33}{-11} = -3$$

The solution is $(2, -3)$, which we can check by verifying that it satisfies both equations in the system. We will leave the check to the reader.

YOUR TURN Use Cramer's Rule to solve $\begin{cases} 3x - 2y = -16 \\ x + 3y = 2 \end{cases}$.

EXAMPLE 5 Use Cramer's Rule to solve $\begin{cases} x + 2y - 2x = -7 \\ 3x - 2y + z = 15 \\ 2x + 3y - 3z = -9 \end{cases}$.

Note: *Recall that we find the determinant of a 3×3 matrix by expanding by minors along the first column.*

Solution We need to find D, D_x, D_y, and D_z.

$$D = \begin{vmatrix} 1 & 2 & -2 \\ 3 & -2 & 1 \\ 2 & 3 & -3 \end{vmatrix} = (1)\begin{vmatrix} -2 & 1 \\ 3 & -3 \end{vmatrix} - (3)\begin{vmatrix} 2 & -2 \\ 3 & -3 \end{vmatrix} + (2)\begin{vmatrix} 2 & -2 \\ -2 & 1 \end{vmatrix}$$

$$= 1(6 - 3) - 3(-6 + 6) + 2(2 - 4)$$
$$= 3 - 0 + (-4)$$
$$= -1$$

$$D_x = \begin{vmatrix} -7 & 2 & -2 \\ 15 & -2 & 1 \\ -9 & 3 & -3 \end{vmatrix} = (-7)\begin{vmatrix} -2 & 1 \\ 3 & -3 \end{vmatrix} - (15)\begin{vmatrix} 2 & -2 \\ 3 & -3 \end{vmatrix} + (-9)\begin{vmatrix} 2 & -2 \\ -2 & 1 \end{vmatrix}$$

$$= -7(6 - 3) - 15(-6 + 6) + (-9)(2 - 4)$$
$$= -21 - 0 + 18$$
$$= -3$$

ANSWER

$(-4, 2)$

$$D_y = \begin{vmatrix} 1 & -7 & -2 \\ 3 & 15 & 1 \\ 2 & -9 & -3 \end{vmatrix} = (1)\begin{vmatrix} 15 & 1 \\ -9 & -3 \end{vmatrix} - (3)\begin{vmatrix} -7 & -2 \\ -9 & -3 \end{vmatrix} + (2)\begin{vmatrix} -7 & -2 \\ 15 & 1 \end{vmatrix}$$

$$= 1(-45 + 9) - 3(21 - 18) + 2(-7 + 30)$$
$$= -36 - 9 + 46$$
$$= 1$$

$$D_z = \begin{vmatrix} 1 & 2 & -7 \\ 3 & -2 & 15 \\ 2 & 3 & -9 \end{vmatrix} = (1)\begin{vmatrix} -2 & 15 \\ 3 & -9 \end{vmatrix} - (3)\begin{vmatrix} 2 & -7 \\ 3 & -9 \end{vmatrix} + (2)\begin{vmatrix} 2 & -7 \\ -2 & 15 \end{vmatrix}$$

$$= 1(18 - 45) - 3(-18 + 21) + 2(30 - 14)$$
$$= -27 - 9 + 32$$
$$= -4$$

Note: *After finding the values of two of the variables, you could find the value of the third variable by substituting these values back into any one of the original equations.*

$$x = \frac{D_x}{D} = \frac{-3}{-1} = 3, \qquad y = \frac{D_y}{D} = \frac{1}{-1} = -1, \qquad z = \frac{D_z}{D} = \frac{-4}{-1} = 4$$

The solution is $(3, -1, 4)$. We will leave the check to the reader.

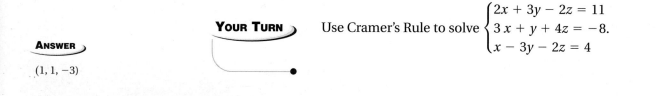

YOUR TURN   Use Cramer's Rule to solve $\begin{cases} 2x + 3y - 2z = 11 \\ 3x + y + 4z = -8. \\ x - 3y - 2z = 4 \end{cases}$

ANSWER

$(1, 1, -3)$

9.6 Exercises

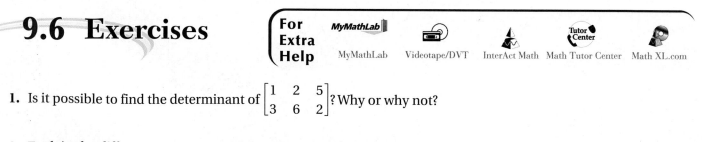

For Extra Help   MyMathLab   MyMathLab   Videotape/DVT   InterAct Math   Math Tutor Center   Math XL.com

1. Is it possible to find the determinant of $\begin{bmatrix} 1 & 2 & 5 \\ 3 & 6 & 2 \end{bmatrix}$? Why or why not?

2. Explain the difference between a matrix and a determinant.

3. How do you find the minor for an element of a 3×3 matrix?

4. How do you find D_y when solving a system of equations using Cramer's Rule?

For Exercises 5–32, find the determinant.

5. $\begin{bmatrix} 3 & 2 \\ 1 & 5 \end{bmatrix}$

6. $\begin{bmatrix} 4 & 1 \\ 3 & 7 \end{bmatrix}$

7. $\begin{bmatrix} -3 & 5 \\ 2 & 4 \end{bmatrix}$

8. $\begin{bmatrix} -5 & 4 \\ 2 & 6 \end{bmatrix}$

9. $\begin{bmatrix} 3 & 6 \\ -2 & 4 \end{bmatrix}$

10. $\begin{bmatrix} 2 & 8 \\ -3 & 2 \end{bmatrix}$

11. $\begin{bmatrix} -3 & -4 \\ 2 & 5 \end{bmatrix}$

12. $\begin{bmatrix} -3 & -5 \\ 4 & 6 \end{bmatrix}$

13. $\begin{bmatrix} -2 & -3 \\ -4 & 5 \end{bmatrix}$

14. $\begin{bmatrix} -6 & -2 \\ 3 & -4 \end{bmatrix}$

15. $\begin{bmatrix} 0 & 3 \\ -5 & 7 \end{bmatrix}$

16. $\begin{bmatrix} -5 & 0 \\ -4 & 2 \end{bmatrix}$

17. $\begin{bmatrix} 1 & 2 & 1 \\ 3 & 1 & 4 \\ 2 & 3 & 2 \end{bmatrix}$

18. $\begin{bmatrix} 3 & 1 & 4 \\ 2 & 2 & 3 \\ 1 & 4 & 3 \end{bmatrix}$

19. $\begin{bmatrix} -1 & 2 & 0 \\ -3 & 2 & 4 \\ -4 & 2 & 3 \end{bmatrix}$

20. $\begin{bmatrix} -2 & 0 & -1 \\ 3 & -2 & 4 \\ -3 & 2 & 1 \end{bmatrix}$

21. $\begin{bmatrix} 2 & 1 & -3 \\ 0 & -3 & 2 \\ 4 & 1 & -3 \end{bmatrix}$

22. $\begin{bmatrix} 3 & -2 & 4 \\ 3 & 0 & 2 \\ -4 & -2 & 2 \end{bmatrix}$

23. $\begin{bmatrix} 1 & 4 & -2 \\ 3 & 2 & 0 \\ -1 & 4 & 3 \end{bmatrix}$

24. $\begin{bmatrix} 3 & -5 & 0 \\ 2 & -4 & 1 \\ -2 & -1 & -3 \end{bmatrix}$

25. $\begin{bmatrix} 0.3 & -0.5 \\ 1.3 & -0.6 \end{bmatrix}$

26. $\begin{bmatrix} -0.4 & 1.6 \\ -4.7 & 3.1 \end{bmatrix}$

27. $\begin{bmatrix} -0.4 & 0.7 & -1.2 \\ 3.1 & 1.5 & -3.2 \\ 1.6 & -2.2 & -1.5 \end{bmatrix}$

28. $\begin{bmatrix} 1.7 & -3.2 & 4.1 \\ 5.3 & -6.2 & -1.1 \\ -1.3 & 2.3 & -4.5 \end{bmatrix}$

29. $\begin{bmatrix} \dfrac{1}{2} & -\dfrac{1}{3} \\ \dfrac{2}{5} & \dfrac{3}{5} \end{bmatrix}$

30. $\begin{bmatrix} -\dfrac{3}{4} & -\dfrac{3}{5} \\ \dfrac{3}{2} & \dfrac{2}{5} \end{bmatrix}$

31. $\begin{bmatrix} \dfrac{1}{2} & -\dfrac{3}{4} & \dfrac{2}{5} \\ \dfrac{1}{3} & \dfrac{1}{5} & -\dfrac{3}{2} \\ -\dfrac{3}{4} & \dfrac{1}{2} & \dfrac{3}{5} \end{bmatrix}$

32. $\begin{bmatrix} -\dfrac{1}{4} & -\dfrac{3}{2} & \dfrac{4}{3} \\ \dfrac{1}{5} & -\dfrac{5}{4} & \dfrac{1}{2} \\ -\dfrac{5}{3} & \dfrac{1}{4} & -\dfrac{4}{5} \end{bmatrix}$

For Exercises 33 and 34, expand by minors along the first column.

33. $\begin{bmatrix} x & y & 1 \\ 2 & -1 & 3 \\ -2 & 0 & 1 \end{bmatrix}$

34. $\begin{bmatrix} x & y & 1 \\ -3 & -2 & 4 \\ 3 & -2 & 2 \end{bmatrix}$

For Exercises 35–48, solve using Cramer's Rule.

35. $\begin{cases} x + y = -5 \\ x - 2y = -2 \end{cases}$

36. $\begin{cases} x + 3y = 1 \\ x + y = -3 \end{cases}$

37. $\begin{cases} 2x - 3y = -6 \\ x - y = -1 \end{cases}$

38. $\begin{cases} 2x + 5y = -7 \\ x - y = -7 \end{cases}$

39. $\begin{cases} -x + 2y = -12 \\ 2x - 3y = 20 \end{cases}$

40. $\begin{cases} -x + 2y = -9 \\ 4x + 5y = -3 \end{cases}$

41. $\begin{cases} 2x - y = -4 \\ -x + 3y = -3 \end{cases}$

42. $\begin{cases} x - 3y = -19 \\ 2x + y = -3 \end{cases}$

43. $\begin{cases} 8x - 3y = 10 \\ 4x + 3y = 14 \end{cases}$

44. $\begin{cases} 2x - y = -4 \\ 4x - 5y = -51 \end{cases}$

45. $\begin{cases} \dfrac{1}{2}x - \dfrac{1}{4}y = 0 \\[2mm] \dfrac{3}{4}x + \dfrac{5}{2}y = \dfrac{23}{2} \end{cases}$

46. $\begin{cases} \dfrac{2}{3}x + \dfrac{1}{4}y = \dfrac{1}{2} \\[2mm] \dfrac{3}{4}x + \dfrac{4}{3}y = -\dfrac{23}{4} \end{cases}$

47. $\begin{cases} 0.2x + 0.5y = 3.4 \\ 0.7x - 0.3y = -0.4 \end{cases}$

48. $\begin{cases} 1.2x - 0.6y = -2.4 \\ 3.1x + 1.3y = -11.9 \end{cases}$

For Exercises 49–60, solve using Cramer's Rule.

49. $\begin{cases} x + y + z = 6 \\ 2x - 4y + 2z = 6 \\ 3x + 2y + z = 11 \end{cases}$

50. $\begin{cases} 2x + y - 3z = -1 \\ x + 2y - 2z = -3 \\ -3x - 4y + z = -3 \end{cases}$

51. $\begin{cases} 3x + y - z = -4 \\ 2x - y + 2z = -7 \\ x - 3y + z = -6 \end{cases}$

52. $\begin{cases} x - y + 3z = -10 \\ 5x + 4y - z = -7 \\ 2x + y - z = -4 \end{cases}$

53. $\begin{cases} 4x + 2y + 3z = 9 \\ 2x - 4y - z = 7 \\ 3x - 2z = 4 \end{cases}$

54. $\begin{cases} 2x - y = -1 \\ 5x - 3y + 2z = 0 \\ 3x + 2y - 3z = -8 \end{cases}$

55. $\begin{cases} 3x + 2y = -12 \\ 3y + 10z = -16 \\ 6x - 2z = 3 \end{cases}$

56. $\begin{cases} 4x + 2z = 7 \\ 8x - 2y = -7 \\ 10y - 2z = -5 \end{cases}$

57. $\begin{cases} \dfrac{1}{2}x + \dfrac{1}{3}y + \dfrac{3}{4}z = \dfrac{25}{12} \\[2mm] \dfrac{2}{3}x + \dfrac{3}{2}y - \dfrac{1}{4}z = -\dfrac{37}{12} \\[2mm] \dfrac{3}{4}x + \dfrac{1}{4}y - \dfrac{2}{3}z = -\dfrac{7}{4} \end{cases}$

58. $\begin{cases} \dfrac{3}{4}x - \dfrac{1}{2}y + \dfrac{1}{2}z = \dfrac{1}{4} \\[2mm] \dfrac{2}{5}x + \dfrac{5}{2}y - \dfrac{4}{3}z = -\dfrac{51}{5} \\[2mm] \dfrac{7}{4}x + \dfrac{5}{3}y - \dfrac{3}{5}z = -\dfrac{623}{60} \end{cases}$

59. $\begin{cases} 0.3x + 0.4y - 0.6z = 2.6 \\ 0.5x - 0.2y + 0.7z = -0.8 \\ 1.4x + 1.3y - 2.2z = 9.8 \end{cases}$

60. $\begin{cases} 1.2x + 2.1y - 0.5z = 7.3 \\ 3.2x - 2.4y + 1.3z = 6.1 \\ 2.5x + 1.3y - 1.7z = 8.4 \end{cases}$

For Exercises 61–64, find x.

61. $\begin{vmatrix} 9 & x \\ -6 & 5 \end{vmatrix} = 21$

62. $\begin{vmatrix} 12 & 2 \\ x & -3 \end{vmatrix} = -22$

63. $\begin{vmatrix} 2 & -1 & 0 \\ 0 & x & 1 \\ -2 & 0 & 4 \end{vmatrix} = -38$

64. $\begin{vmatrix} 1 & -2 & 4 \\ -1 & 0 & 2 \\ 0 & x & 5 \end{vmatrix} = -28$

For Exercises 65–70, use the following. Suppose a triangle has vertices of (x_1, y_1), (x_2, y_2), and (x_3, y_3) as shown in the graph to the right. The area of the triangle is given by $A = \dfrac{1}{2}\left| \det \begin{bmatrix} x_1 & y_1 & 1 \\ x_2 & y_2 & 1 \\ x_3 & y_3 & 1 \end{bmatrix} \right|$. For example, the area of the triangle whose

Note: *This notation indicates half of the absolute value of the determinant.*

vertices are $(1, 2)$, $(3, -2)$, and $(-2, 5)$ is $A = \dfrac{1}{2}\left| \det \begin{bmatrix} 1 & 2 & 1 \\ 3 & -2 & 1 \\ -2 & 5 & 1 \end{bmatrix} \right|$. *Expanding*

about the first column we have:

$$A = \frac{1}{2}\left| 1 \det \begin{bmatrix} -2 & 1 \\ 5 & 1 \end{bmatrix} - 3 \det \begin{bmatrix} 2 & 1 \\ 5 & 1 \end{bmatrix} - 2 \det \begin{bmatrix} 2 & 1 \\ -2 & 1 \end{bmatrix} \right|$$

$$= \frac{1}{2}\left| (-2 - 5) - 3(2 - 5) - 2(2 + 2) \right|$$

$$= \frac{1}{2}\left| -7 + 9 - 8 \right| = \frac{1}{2}\left| -6 \right| = 3 \; square \; units$$

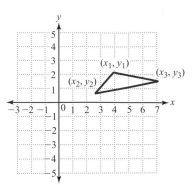

Find the area of the triangles with vertices at the given points.

65. $(2, 4)\,(4, 0)\,(6, 5)$

66. $(0, 2)\,(3, -2)\,(5, 5)$

67. $(-3, 1)\,(2, -3)\,(4, 4)$

68. $(-3, -2)\,(-1, 4)\,(3, -4)$

69. $(-4, -1)\,(1, 3)\,(3, -3)$

70. $(-3, -3)\,(2, 2)\,(4, -1)$

For Exercises 71–76, translate the problem to a system of equations, then solve using Cramer's Rule.

71. The two heaviest known meteorites to be found on Earth's surface are the Hoba West, which was found in Namibia, and the Ahnighito, which was found in Greenland. The total weight of the two meteorites is 90 tons. The Hoba West is twice as heavy as the Ahnighito. Find the weight of each. (*Source: Webster's New World Book of Facts*)

72. The two largest expenses for the average American family are federal taxes and housing, including household expenses. Together these two items total 43.3% of the average family's income. The amount spent on taxes is 12.1% more than the amount spent on housing. Find the percent spent on each. (*Source: Numbers: How Many, How Long, How Far, How Much*)

73. A restaurant makes a soup that includes garbanzo and black turtle beans. The manager purchased 10 pounds of beans at a cost of $8.80. If the garbanzo beans cost $1.00 per pound and the black turtle beans cost $0.70 per pound, how many pounds of each did he purchase?

74. The perimeter of a triangle is 21 inches and two sides are of equal length. The length of the third side is 3 inches less than the length of the two equal sides. Find the length of each side of the triangle.

75. Coinage bronze is made up of zinc, tin, and copper. The percent of tin is four times the percent of zinc. The percent of copper is nineteen times the sum of the percents of zinc and tin. Find the percent of zinc, tin, and copper in coinage bronze. (*Source: Webster's New World Book of Facts*)

76. In winning the 2005 NCAA basketball championship, North Carolina scored a total of 39 times in their 75 to 70 victory over Illinois. The sum of the number of 3-point field goals and 2-point field goals was three more than twice the number of free throws (1 point each). How many of each did they score? (*Source:* CBS Sports Line)

REVIEW EXERCISES

[2.6] *For Exercises 1 and 2, solve and graph the solution set.*

  1. $x > 2x - 3$

                                                   2. $2x + 3 \geq 10 - 5x$

[3.3]   **3.** Find two consecutive integers whose sum is 39.

[4.6] *For Exercises 4 and 5, graph.*

  4. $y > 3x + 2$

                                                   5. $x + y \leq -2$

[9.2]   **6.** Use substitution to solve $\begin{cases} x = 3 - y \\ 5x + 3y = 5 \end{cases}$.

9.7 Solving Systems of Linear Inequalities

OBJECTIVES
1. Graph the solution set of a system of linear inequalities.
2. Solve applications involving a system of linear inequalities.

OBJECTIVE 1. Graph the solution set of a system of linear inequalities. In Section 2.6, we learned to solve linear inequalities in terms of one variable. In Section 4.6, we learned how to graph linear inequalities. In this section, we will develop a graphical approach to solving *systems of linear inequalities*.

Consider the system of linear inequalities $\begin{cases} x + 2y < 6 \\ 2x - y \geq 2 \end{cases}$. First, let's graph each inequality separately.

Note: *Recall from Section 4.6 that we use a dashed line with $<$ or $>$, and points on a dashed line are not in the solution set.*

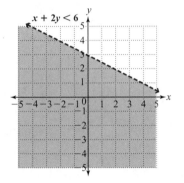

In the graph of $x + 2y < 6$, the shaded region contains all ordered pairs that make $x + 2y < 6$ true.

Note: *Recall from Section 4.6 that we determine which side of the line to shade by choosing an ordered pair and checking to see if it makes the inequality true. If it does, we shade on the side of the line containing that ordered pair. If it does not, we shade the other side of the line.*

Note: *Recall from Section 4.6 that we use a solid line with $\leq$ or $\geq$ and points on a solid line are part of the solution set.*

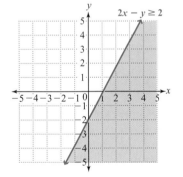

In the graph of $2x - y \geq 2$, ordered pairs in the shaded region and on the line itself make $2x - y \geq 2$ true.

Now we put the two graphs together on the same grid to determine the solution set for the system. A solution for a system of inequalities is an ordered pair that makes every inequality in the system true.

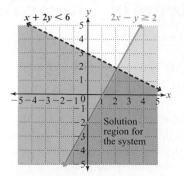

$x + 2y < 6$ $2x - y \geq 2$

Solution region for the system

On our graph containing both $x + 2y < 6$ and $2x - y \geq 2$, the region where the shading overlaps contains ordered pairs that make both inequalities true, so all ordered pairs in this region are in the solution set for the system. Also, ordered pairs on the solid line for $2x - y \geq 2$ where it touches the region of overlap are in the solution set for the system, whereas ordered pairs on the dashed line for $x + 2y < 6$ are not. This implies the following procedure.

PROCEDURE **Solving a System of Linear Inequalities**

To solve a system of linear inequalities, graph all of the inequalities on the same grid. The solution set for the system contains all ordered pairs in the region where the inequalities' solution sets overlap along with ordered pairs on the portion of any solid line that touches the region of overlap.

To check, we can select a point in the solution region such as (3, 0) and verify that it makes both inequalities true.

First inequality:	**Second inequality:**	
$x + 2y < 6$	$2x - y \geq 2$	
$3 + 2(0) \overset{?}{<} 6$	$2(3) - 0 \overset{?}{\geq} 2$	Replace x with 3 and y with 0 in both inequalities.
$3 < 6$ True.	$6 \geq 2$ True.	

Since (3, 0) makes both inequalities true, it is a solution to the system. Though we selected only one ordered pair in the solution region, remember that *every* ordered pair in that region is a solution.

EXAMPLE 1 Graph the solution set for the system of inequalities.

a. $\begin{cases} x + 3y > 6 \\ y < 2x - 1 \end{cases}$

Solution Graph the inequalities on the same grid. Because both lines are dashed, the solution set for the system contains only those ordered pairs in the region of overlap (the purple shaded region).

Note: *Ordered pairs on dashed lines are not part of the solution region.*

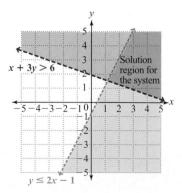

$x + 3y > 6$

Solution region for the system

$y \leq 2x - 1$

Connection Remember that for a system of two linear equations, a solution is a point of intersection of the graphs of the two equations. Similarly, for a system of linear inequalities, the solution set is a region of intersection of the graphs of the two inequalities.

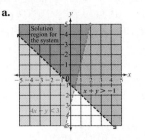
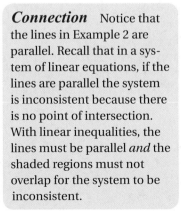
b. $\begin{cases} y < 2 \\ 5x - y \le 4 \end{cases}$

Solution Graph the inequalities on the same grid. The solution set for this system contains all ordered pairs in the region of overlap (purple shaded region) together with all ordered pairs on the portion of the solid blue line that touches the purple shaded region.

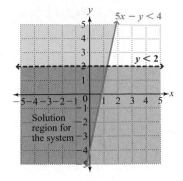

YOUR TURN Graph the solution set for the system of inequalities.

a. $\begin{cases} x + y > -1 \\ 4x - y < 3 \end{cases}$

b. $\begin{cases} x > -3 \\ 3x - 4y \le -8 \end{cases}$

Inconsistent Systems

Some systems of linear inequalities have no solution. We say these systems are inconsistent.

EXAMPLE 2 Graph the solution set of the system of inequalities.

$$\begin{cases} x - 4y \ge 8 \\ y \ge \dfrac{1}{4}x + 3 \end{cases}$$

Solution Graph the inequalities on the same grid.

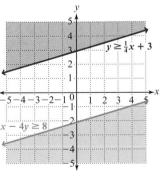

The lines appear to be parallel. We can verify by writing $x - 4y \ge 8$ in slope-intercept form and comparing the slopes.

$x - 4y \ge 8$

$-4y \ge -x + 8$ Subtract x from both sides.

$y \le \dfrac{1}{4}x - 2$ Divide both sides by -4, which changes the direction of the inequality.

The slopes are equal, so the lines are in fact parallel. Since the lines are parallel and the shaded regions do not overlap, there is no solution region for this system. The system is inconsistent.

Connection Notice that the lines in Example 2 are parallel. Recall that in a system of linear equations, if the lines are parallel the system is inconsistent because there is no point of intersection. With linear inequalities, the lines must be parallel *and* the shaded regions must not overlap for the system to be inconsistent.

EXAMPLE 3 A home-interiors store stocks two different-size prints from an artist. The manager wants to purchase at least 15 prints for the store. The artist sells the smaller print for $20 and the larger print for $30. The manager cannot spend more than $800 for the prints. Write a system of inequalities that describes the manager's order, then solve the system by graphing.

Understand We must translate to a system of inequalities, then solve the system.

Plan and Execute Let x represent the number of small prints and y represent the number of large prints ordered.

Relationship 1: The manager wants to purchase at least 15 prints.

The words *at least* indicate that the combined number of prints is to be greater than or equal to 15, so $x + y \geq 15$.

Relationship 2: The manager cannot spend more than $800.

The small prints cost $20 each, so $20x$ describes the amount spent on small prints. Similarly, $30y$ describes the amount spent on large prints. The total cannot exceed $800, so $20x + 30y \leq 800$.

$$\text{Our system: } \begin{cases} x + y \geq 15 \\ 20x + 30y \leq 800 \end{cases}$$

Answer See graph. Since the manager cannot order a negative number of prints, the solution set is confined to Quadrant I. Any ordered pair in the solution region or on a portion of either line touching the solution region is a solution for the system. However, assuming only whole prints can be purchased, only ordered pairs of whole numbers in the solution set, such as (10, 15) or (25, 10), are realistic.

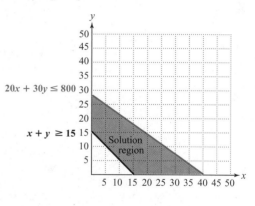

Check We will check one ordered pair in the solution region. The ordered pair (10, 15) indicates an order of 10 small prints and 15 large prints, so that has 25 prints are ordered (which is more than 15) that cost a total of $20(10) + 30(15) = \$650$ (which is less than $800).

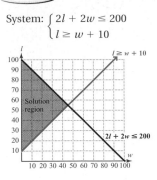
YOUR TURN In designing a building with a rectangular base, an architect decides that the perimeter cannot exceed 200 feet. Also, it is decided that the length will be at least 10 feet more than the width. Write a system of inequalities that describes the design requirements, then solve the system graphically.

Note: *Since length and width must be positive numbers, the solution region is confined to Quadrant 1.*

9.7 Exercises

For Extra Help

MyMathLab   MyMathLab   Videotape/DVT   InterAct Math   Math Tutor Center   Math XL.com

1. How would you check a possible solution to a system of inequalities?

2. How do you determine the region that is the solution set for a system of linear inequalities?

3. What circumstances would cause a system of inequalities to have no solution?

4. Write a system of linear inequalities whose solution set is the entire first quadrant.

5. Write a system of linear inequalities whose solution set is the entire third quadrant.

6. Write a system of linear inequalities whose solution set is the entire fourth quadrant.

For Exercises 7–30, graph the solution set for the system of inequalities.

7. $\begin{cases} x + y > 4 \\ x - y < 6 \end{cases}$

8. $\begin{cases} x - y > 2 \\ x + y > 4 \end{cases}$

9. $\begin{cases} x + y < -2 \\ x - y > 6 \end{cases}$

10. $\begin{cases} x - y < -5 \\ x + y < 3 \end{cases}$

11. $\begin{cases} 2x + y \geq -1 \\ x - y \leq 5 \end{cases}$

12. $\begin{cases} x - y < -5 \\ 2x - y < -7 \end{cases}$

13. $\begin{cases} x + y < 3 \\ x - 2y \geq 2 \end{cases}$

14. $\begin{cases} x + 3y \leq 6 \\ x - y > 5 \end{cases}$

15. $\begin{cases} 3x + 4y \leq -9 \\ y < 3x \end{cases}$

16. $\begin{cases} 2x + y > 9 \\ y > \dfrac{1}{4}x \end{cases}$

17. $\begin{cases} y < x \\ y > -x + 1 \end{cases}$

18. $\begin{cases} y < x \\ y < 2x - 3 \end{cases}$

19. $\begin{cases} 3x > -4y \\ 3x + 4y \leq -8 \end{cases}$

20. $\begin{cases} 2y - x \geq 6 \\ 2x - 4y \geq 5 \end{cases}$

21. $\begin{cases} y < 3x + 1 \\ 3x - y \leq 4 \end{cases}$

22. $\begin{cases} x - 2y > 3 \\ 2x - 4y \leq 20 \end{cases}$

23. $\begin{cases} x > 2 \\ 3x - 2y > 6 \end{cases}$

24. $\begin{cases} y \leq -1 \\ x + 2y > 3 \end{cases}$

25. $\begin{cases} x \geq -3y \\ y \geq 2x \end{cases}$

26. $\begin{cases} x > 2y \\ x + y > 6 \end{cases}$

27. $\begin{cases} y > 2 \\ x < -1 \end{cases}$

28. $\begin{cases} y \geq -1 \\ x < 2 \end{cases}$

29. $\begin{cases} 2x + y \geq 1 \\ x - y > -1 \\ x > 2 \end{cases}$

30. $\begin{cases} x + y \geq 1 \\ x - y \geq -5 \\ y > -2 \end{cases}$

For Exercises 31–34, explain the mistake.

31. $\begin{cases} x + y > 3 \\ x - y \leq 2 \end{cases}$

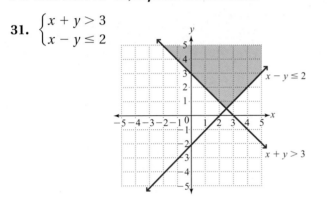

32. $\begin{cases} y < \dfrac{1}{2}x \\ y \leq -3x + 1 \end{cases}$

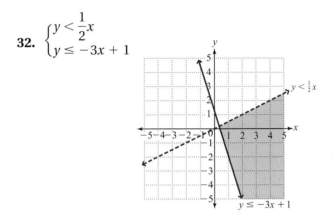

33. $\begin{cases} x < 3 \\ y > 4 \end{cases}$

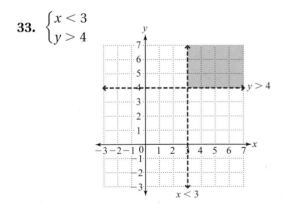

34. $\begin{cases} x > 3y \\ x + y \geq 2 \end{cases}$

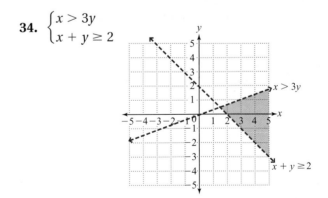

35. To be admitted to a certain college, prospective students must have a combined verbal and math score of at least 1100 on the SAT. To be admitted to the engineering program at this college, students must also have a math score of at least 500.

   a. Write a system of inequalities that describes the requirements to be admitted to the college of engineering.

   b. The maximum score on each part of the SAT is 800. Write two inequalities describing these maximum scores.

   c. Solve the system by graphing.

   d. Give two different combinations of verbal and math scores that satisfy the admission requirements to the college of engineering.

36. An architect is designing a rectangular platform for an auditorium. The client wants the length of the platform to be at least twice the width and the perimeter must not exceed 200 feet.

   a. Write a system of inequalities that describes the specifications for the platform.

   b. Solve the system by graphing.

   c. Give two different combinations of length and width that satisfy the requirements of the client.

37. A company sells two versions of software: the regular version and the deluxe version. The regular version sells for \$15.95 and the deluxe sells for \$20.95. The company gives a bonus to any salesperson who sells at least 100 units and has at least \$1700 in total sales.

   a. Write a system of inequalities that describes the requirements a salesperson must meet in order to receive the bonus.

   b. Solve the system by graphing.

   c. Give two different combinations of number of units sold that meet the requirements for receiving the bonus.

38. An investor is trying to decide how much to invest in two different stocks. She has decided that she will invest at most \$5000. The "safe" stock is very stable and is projected to return about 5% over the next year. The other stock is more risky, but could return about 9% if the company grows as expected. She wishes her total dividend at the end of the year to be at least \$300.

   a. Write a system of inequalities that describes the amount invested in the two stocks and the return on that investment.

   b. Solve the system by graphing.

c. Give two different combinations of investment amounts that return at least $300.

Collaborative Exercises   MAXIMIZING THE PROFIT

Linear programming is an area of mathematics that solves problems like the one described below. One of the fundamental elements of a linear programming model is the constraint. A constraint is simply an inequality that describes some limited resource required in the problem. For example, suppose you make two types of bicycles, style A and style B. Also, suppose that style A requires 40 minutes of welding for each bicycle and style B requires 25 minutes of welding. If you have only 600 minutes of welding time available, then the welding-time constraint would be $40A + 25B \leq 600$, where A represents the number of style A produced and B represents the number of style B produced.

An aspiring-artists' group is preparing for a fund-raising sale. They design and construct two types of antique-reproduction tables: a semicircular foyer table and a side table. The materials cost $100 for each foyer table and $200 for each side table. Both types of table require 2 hours of cutting and carving. The foyer table requires 3.5 hours to assemble, sand, and finish; and the side table requires 1.75 hours to assemble, sand, and finish. The group has $3300 to purchase all materials; 40 hours available for cutting and carving; and 63 hours available for assembling, sanding, and finishing. For each foyer table that they sell, their profit will be $350. For each side table, their profit will be $215. How many of each should they make to maximize their profit?

1. For the artists' fund-raising problem there are three basic constraints: 1. cost of materials, 2. cutting and carving, and 3. assembling-sanding-finishing. For each of these constraints, write an inequality similar to the one in the bicycle example. Let x represent the number of foyer tables and y the number of side tables.

2. Since x and y represent numbers of tables, they cannot represent negative values. Write inequalities for these two additional constraints.

3. Now graph the system of inequalities described by these five constraints.
4. The region that is the solution of the system of inequalities is called the *feasible region* and includes all points that satisfy the system. The goal is to find the optimal solution, which, according to linear programming, is one of the corner points of the feasible region. There should be five corner points: one at the origin, one on each axis, and two more in Quadrant I. Determine the coordinates of each corner point. Note that a corner point lying on an axis is the x- or y-intercept for the line passing through that point. Since two lines intersect to form a corner point that is not on an axis, make a system out of their two equations, then solve the system to find their point of intersection.

5. Next develop an algebraic expression describing the profit gained for selling x of the foyer tables and y of the side tables. This is called the *objective function*.

6. Using the objective function, test each of the points found in step 4 to see which one yields the maximum profit.

7. Using your solution, answer the following questions.
   a. How many of each type of table should the group produce?

   b. How many hours will be spent in each phase of the production?

   c. How much money will they need to purchase the materials?

   d. What amount of profit will the group receive?

REVIEW EXERCISES

[1.5] *For Exercises 1 and 2, evaluate.*

1. 3^2

2. -10^2

[1.7] 3. Simplify: $3x - 2(2x + 7)$

[5.4] 4. Multiply: $4a^2b \cdot (-2ab^4)$

[5.6] 5. Divide: $\dfrac{-25a^3b^5}{5a^9}$

[6.6] 6. Solve: $x^2 - 5x + 6 = 0$

Chapter 9 Summary

Defined Terms

Section 9.1
System of equations
(p. 688)
Solution for a system of
equations (p. 688)
Consistent system
(p. 693)

Inconsistent system
(p. 693)
Dependent equations
(p. 693)
Independent equations
(p. 693)

Section 9.5
Matrix (p. 736)
Augmented matrix
(p. 736)
Echelon form (p. 737)

Section 9.6
Square matrix (p. 747)
Minor (p. 748)
Cramer's Rule (p. 750)

Procedures, Rules, and Key Examples

Procedures/Rules	Key Example(s)

Section 9.1 Solving Systems of Linear Equations Graphically

To verify or check a solution to a system of equations:
1. Replace each variable in each equation with its corresponding value.
2. Verify that each equation is true.

Example 1: Determine whether the ordered pair is a solution to the system.

$$\begin{cases} 2x + 3y = 8 \\ y = 5x - 3 \end{cases}$$

a. $(-2, 4)$

$$2x + 3y = 8 \qquad y = 5x - 3$$
$$2(-2) + 3(4) \stackrel{?}{=} 8 \qquad 4 \stackrel{?}{=} 5(-2) - 3$$
$$8 = 8 \qquad 4 \neq -13$$

$(-2, 4)$ is not a solution to the system.

b. $(1, 2)$

$$2x + 3y = 8 \qquad y = 5x - 3$$
$$2(1) + 3(2) \stackrel{?}{=} 8 \qquad 2 \stackrel{?}{=} 5(1) - 3$$
$$8 = 8 \qquad 2 = 2$$

$(1, 2)$ is a solution to the system.

To classify a system of two linear equations in two unknowns, write the equations in slope-intercept form and compare the slopes and y-intercepts.
1. If the slopes are different, then the system is **consistent with independent equations** and has a single solution.
2. If the slopes are equal and the y-intercepts are also equal, then the system is **consistent with dependent equations** and has an infinite number of solutions.
3. If the slopes are equal with different y-intercepts, then the system is **inconsistent with independent equations** and has no solution.

Example 2: Classify $\begin{cases} x + 3y = 6 \\ 2x + 6y = -5 \end{cases}$
and discuss the number of solutions.

Solution: Write the equations in slope-intercept form.

$$x + 3y = 6 \quad \text{becomes} \quad y = -\frac{1}{3}x + 2$$

$$2x + 6y = 25 \quad \text{becomes} \quad y = -\frac{1}{3}x - \frac{25}{6}$$

continued

Procedures/Rules	Key Example(s)

Section 9.1 (continued)

Since the slopes are the same (both $-\frac{1}{3}$), but the y-intercepts are different, this system is inconsistent with independent equations. This means that there is no solution to the system.

To solve a system of linear equations graphically:
1. Graph each equation.
   a. If the lines intersect at a single point, then the coordinates of that point form the solution (consistent system with independent equations).
   b. If the lines are parallel, then there is no solution (inconsistent system with independent equations).
   c. If the lines are identical, then there are an infinite number of solutions, which are the coordinates of all the points on that line (consistent system with dependent equations).
2. Check the solution in the original equations.

Example 3: Solve $\begin{cases} 2x + 3y = 8 \\ y = 5x - 3 \end{cases}$ graphically.

Solution:

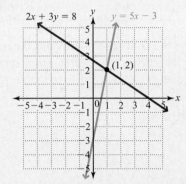

Solution: $(1, 2)$ is the solution.

Section 9.2 Solving Systems of Linear Equations by Substitution

To find the solution of a system of two linear equations using the substitution method:
1. Isolate one of the variables in one of the equations.
2. In the other equation, substitute the expression you found in step 1 for that variable.
3. Solve this new equation. (It will now have only one variable.)
4. Using one of the equations containing both variables, substitute the value you found in step 3 for that variable and solve for the value of the other variable.
5. Check the solution in the original equations.

Example 1: Solve $\begin{cases} y - 4x = -1 \\ 5x - y = 2 \end{cases}$ using substitution.

Solution: We will isolate y in the first equation.

$y = 4x - 1$ Add $4x$ to both sides.

Now we substitute $4x - 1$ in place of y in the second equation.

$$5x - y = 2$$
$$5x - (4x - 1) = 2$$
$$5x - 4x + 1 = 2 \quad \text{Distribute}$$
$$x + 1 = 2 \quad \text{Combine like terms.}$$
$$x = 1 \quad \text{Subtract 1 from both sides.}$$

Now substitute 1 for x in one of the equations. We will use $y = 4x - 1$.

$$y = 4x - 1$$
$$y = 4(1) - 1$$
$$y = 3$$

Solution: $(1, 3)$

continued

Procedures/Rules	Key Example(s)

Section 9.3 Solving Systems of Linear Equations by Elimination

To solve a system of two linear equations using the elimination method:

1. Write the equations in standard form ($Ax + By = C$).
2. Use the multiplication principle of equality to clear fractions or decimals.
3. Multiply one or both equations by a number (or numbers) so that they have a pair of terms that are additive inverses.
4. Add the equations. The result should be an equation in terms of one variable.
5. Solve the equation from step 4 for the value of that variable.
6. Using an equation containing both variables, substitute the value you found in step 4 for the corresponding variable and solve for the value of the other variable.
7. Check the solution in the original equations.

Example 1: Solve $\begin{cases} x + 2y = -4 \\ 4x + y = 5 \end{cases}$ using elimination.

Solution: We will eliminate y.

$$x + 2y = -4 \qquad\qquad x + 2y = -4$$
$$4x + y = 5 \xrightarrow{\text{Multiply by } -2} \underline{-8x - 2y = -10}$$
$$-7x + 0 = -14$$

Add equations.
Divide both sides by -7. $\quad x = 2$

Now solve for the value of y.

$$4(2) + y = 5 \qquad \text{Substitute 2 for } x.$$
$$y = -3$$

Solution: $(2, -3)$

Section 9.4 Solving Systems of Linear Equations in Three Variables

To solve a system of three equations with three unknowns:

1. If necessary, write each equation in the form $Ax + By + Cz = D$.
2. Choose a variable and choose two equations and eliminate that variable.
3. Choose two other equations and eliminate the same variable as was eliminated in step 2.
4. The two equations resulting from steps 2 and 3 form a system of two equations with two unknowns, which can be solved using the methods of Section 9.3.
5. To find the third variable, substitute the values of the variables found in step 4 back into any of the three original equations that contain that variable.
6. Check the ordered triple in all three original equations.

Example 1: Solve the following system of equations.

$$\begin{cases} x + 2y - z = -6 & \text{(Equation 1)} \\ 2x - 3y + 4z = 26 & \text{(Equation 2)} \\ -x + 2y - 3z = -18 & \text{(Equation 3)} \end{cases}$$

Eliminate x by adding Equation 1 and Equation 3.

$$x + 2y - z = -6$$
$$\underline{-x + 2y - 3z = -18}$$
$$4y - 4z = -24 \qquad \text{Divide both sides by 4.}$$
$$y - z = -6 \qquad \text{(Equation 4)}$$

Eliminate x again using Equations 2 and 3.

$$2x - 3y + 4z = 26 \qquad\qquad 2x - 3y + 4z = 26$$
$$-x + 2y - 3z = -18 \longrightarrow \underline{-2x + 4y - 6z = -36}$$
$$\text{Multiply by 2.} \qquad y - 2z = -10 \text{ (Eq.5)}$$

Use Equations 4 and 5 to solve for z.

$$y - z = -6 \qquad\qquad y - z = -6$$
$$y - 2z = -10 \xrightarrow{\text{Multiply by } -1.} \underline{-y + 2z = 10}$$
$$z = 4$$

Substitute 4 for z in Equation 4 and solve for y.

$$y - 4 = -6$$
$$y = -2$$

continued

Procedures/Rules	Key Example(s)

Section 9.4 (continued)

Substitute -2 for y and 4 for z in Equation 1 and solve for x.

$$x + 2(-2) - 4 = -6$$
$$x - 4 - 4 = -6$$
$$x = 2$$

Solution: $(2, -2, 4)$

Section 9.5 Solving Systems of Linear Equations Using Matrices

Elementary row operations:
1. Any two rows may be interchanged.
2. The elements of any row may be multiplied (or divided) by any nonzero real number.
3. Any row may be replaced by a row resulting from adding the elements of that row (or multiples of that row) to a multiple of the elements of any other row.

Elementary row transformations can be used to solve a system of linear equations by transforming the augmented matrix of the system into a matrix that is in echelon form. A matrix is in echelon form if the coefficient portion of the augmented matrix has 1s on the diagonal from upper left to lower right and 0s below the 1s.

Note: Systems of three equations with three unknowns are solved in a similar manner.

Example 1: Solve $\begin{cases} 2x + 3y = -5 \\ x + 2y = -4 \end{cases}$ using row operations.

The augmented matrix is $\begin{bmatrix} 2 & 3 & | & -5 \\ 1 & 2 & | & -4 \end{bmatrix}$

$\begin{bmatrix} 1 & 2 & | & -4 \\ 2 & 3 & | & -5 \end{bmatrix}$ We need row 1, column 1 to be a 1, so interchange the two rows.

$-2 \cdot R_1 + R_2 \longrightarrow \begin{bmatrix} 1 & 2 & | & -4 \\ 0 & -1 & | & 3 \end{bmatrix}$ We need row 2, column 1 to be 0, so multiply row 1 by -2 and add it to row 2.

$-1 \cdot R_2 \longrightarrow \begin{bmatrix} 1 & 2 & | & -4 \\ 0 & 1 & | & -3 \end{bmatrix}$ Multiply row 2 by -1 to get echelon form.

Row 2 means $y = -3$ and row 1 means $x + 2y = -4$.

$$x + 2(-3) = -4 \quad \text{Substitute } -3 \text{ for } y \text{ and solve for } x$$

$$x = 2$$

Solution: $(2, -3)$

Section 9.6 Cramer's Rule

If $A = \begin{bmatrix} a_1 & b_1 \\ a_2 & b_2 \end{bmatrix}$, then $\det(A) = \begin{vmatrix} a_1 & b_1 \\ a_2 & b_2 \end{vmatrix} = a_1 b_2 - a_2 b_1$.

Example 1: Find the determinant of $\begin{bmatrix} 3 & -4 \\ 2 & -5 \end{bmatrix}$.

Solution:
$$\begin{vmatrix} 3 & -4 \\ 2 & -5 \end{vmatrix} = 3(-5) - 2(-4) = -15 + 8 = -7$$

continued

Procedures/Rules	Key Example(s)

Section 9.6 Cramer's Rule (Continued)

The determinant of a 3×3 matrix is found by

$$\begin{vmatrix} a_1 & b_1 & c_1 \\ a_2 & b_2 & c_2 \\ a_3 & b_3 & c_3 \end{vmatrix} = a_1 \begin{pmatrix} \text{minor} \\ \text{of } a_1 \end{pmatrix} - a_2 \begin{pmatrix} \text{minor} \\ \text{of } a_2 \end{pmatrix} + a_3 \begin{pmatrix} \text{minor} \\ \text{of } a_3 \end{pmatrix}$$

$$= a_1 \begin{vmatrix} b_2 & c_2 \\ b_3 & c_3 \end{vmatrix} - a_2 \begin{vmatrix} b_1 & c_1 \\ b_3 & c_3 \end{vmatrix} + a_3 \begin{vmatrix} b_1 & c_1 \\ b_2 & c_2 \end{vmatrix}$$

Example 2: Find the determinant of

$$\begin{bmatrix} -1 & -2 & 3 \\ 1 & -3 & 4 \\ -2 & 3 & -1 \end{bmatrix}.$$

Solution:

$$\begin{vmatrix} -1 & -2 & 3 \\ 1 & -3 & 4 \\ -2 & 3 & -1 \end{vmatrix} = -1 \begin{vmatrix} -3 & 4 \\ 3 & -1 \end{vmatrix} - 1 \begin{vmatrix} -2 & 3 \\ 3 & -1 \end{vmatrix} + (-2) \begin{vmatrix} -2 & 3 \\ -3 & 4 \end{vmatrix}$$

$$= -1(3 - 12) - 1(2 - 9) + (-2)(-8 + 9)$$
$$= 9 + 7 - 2$$
$$= 14$$

Cramer's Rule

The solution to $\begin{cases} a_1x + b_1y = c_1 \\ a_2x + b_2y = c_2 \end{cases}$ is

$$x = \frac{\begin{vmatrix} c_1 & b_1 \\ c_2 & b_2 \end{vmatrix}}{\begin{vmatrix} a_1 & b_1 \\ a_2 & b_2 \end{vmatrix}} = \frac{D_x}{D} \quad \text{and} \quad y = \frac{\begin{vmatrix} a_1 & c_1 \\ a_2 & c_2 \end{vmatrix}}{\begin{vmatrix} a_1 & b_1 \\ a_2 & b_2 \end{vmatrix}} = \frac{D_y}{D}$$

Example 3: Solve $\begin{cases} 3x + 2y = -2 \\ 5x + 2y = 2 \end{cases}.$

Solution: $D = \begin{vmatrix} 3 & 2 \\ 5 & 2 \end{vmatrix} = -4$

$$x = \frac{\begin{bmatrix} -2 & 2 \\ 2 & 2 \end{bmatrix}}{-4} = \frac{-4 - 4}{-4} = \frac{-8}{-4} = 2$$

$$y = \frac{\begin{vmatrix} 3 & -2 \\ 5 & 2 \end{vmatrix}}{-4} = \frac{6 + 10}{-4} = \frac{16}{-4} = -4$$

The solution is $(2, -4)$

The solution to $\begin{cases} a_1x + b_1y + c_1z = d_1 \\ a_2x + b_2y + c_2z = d_2 \\ a_3x + b_3y + c_3z = d_3 \end{cases}$ is

$$x = \frac{\begin{vmatrix} d_1 & b_1 & c_1 \\ d_2 & b_2 & c_2 \\ d_3 & b_3 & c_3 \end{vmatrix}}{\begin{vmatrix} a_1 & b_1 & c_1 \\ a_2 & b_2 & c_2 \\ a_3 & b_3 & c_3 \end{vmatrix}} = \frac{D_x}{D}, \quad y = \frac{\begin{vmatrix} a_1 & d_1 & c_1 \\ a_2 & d_2 & c_2 \\ a_3 & d_3 & c_3 \end{vmatrix}}{\begin{vmatrix} a_1 & b_1 & c_1 \\ a_2 & b_2 & c_2 \\ a_3 & b_3 & c_3 \end{vmatrix}} = \frac{D_y}{D},$$

and $z = \frac{\begin{vmatrix} a_1 & b_1 & d_1 \\ a_2 & b_2 & d_2 \\ a_3 & b_3 & d_3 \end{vmatrix}}{\begin{vmatrix} a_1 & b_1 & c_1 \\ a_2 & b_2 & c_2 \\ a_3 & b_3 & c_3 \end{vmatrix}} = \frac{D_z}{D}$

Example 4: Solve $\begin{cases} x + 2y - z = -4 \\ 2x - 3y - 2z = 13. \\ 2x - 2y + 3z = 20 \end{cases}$

Solution: $D = \begin{vmatrix} 1 & 2 & -1 \\ 2 & -3 & -2 \\ 2 & -2 & 3 \end{vmatrix} = -35$

$$x = \frac{\begin{vmatrix} -4 & 2 & -1 \\ 13 & -3 & -2 \\ 20 & -2 & 3 \end{vmatrix}}{-35} = \frac{-140}{-35} = 4$$

$$y = \frac{\begin{vmatrix} 1 & -4 & -1 \\ 2 & 13 & -2 \\ 2 & 20 & 3 \end{vmatrix}}{-35} = \frac{105}{-35} = -3$$

$$z = \frac{\begin{vmatrix} 1 & 2 & -4 \\ 2 & -3 & 13 \\ 2 & -2 & 20 \end{vmatrix}}{-35} = \frac{-70}{-35} = 2$$

The solution is $(4, -3, 2)$

continued

Procedures/Rules	Key Example(s)

Section 9.7 Solving Systems of Linear Inequalities

To solve a system of linear inequalities, graph all of the inequalities on the same grid. The solution set for the system contains all ordered pairs in the region where the inequalities' solution sets overlap along with ordered pairs on the portion of any solid line that touches the region of overlap.

Example 1: Solve $\begin{cases} x + y > -2 \\ y \le 3x - 1 \end{cases}$.

Solution:

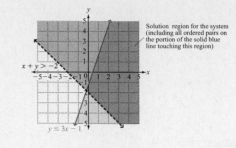

Solution region for the system (including all ordered pairs on the portion of the solid blue line touching this region)

Chapter 9   Review Exercises

For Exercises 1–5, answer true or false.

[9.1]   **1.** Systems of linear equations either have one solution or no solutions.

[9.1]   **2.** Given a system of two equations with two variables, if there is no solution then the graphs are parallel lines.

[9.3]   **3.** Any system of linear equations that can be solved, can be solved using the elimination method.

[9.5]   **4.** Solving a system of linear equations using the echelon method is like solving a system using the elimination method except the echelon method uses the coefficients and constant terms only.

[9.6]   **5.** The determinant of a matrix is another matrix.

For Exercises 6–10, complete the rule.

[9.1]   **6.** To verify or check a solution to a system of equations:

    1. _____ each variable in each equation with its corresponding value.

    2. Verify that each equation is true.

[9.1]   **7.** An inconsistent system has no solution. The graphs have the same _____ but different _____.

[9.5]   **8.** A matrix is in echelon form if _____
_____.

[9.6]   **9.** When solving a system of linear equations using Cramer's Rule, the denominator is the _____ of the matrix containing only the coefficients of the system.

[9.7]   **10.** To solve a system of two linear inequalities:

    1. Graph each inequality.

    2. The solution of the system is _____
_____.

[9.1] *For Exercises 11–12, determine whether the given ordered pair is a solution to the given system of equations.*

11. $(4, 3);$ $x - y = 1$
                 $-x + y = -1$

12. $(1, 1); 2x - y = 7$
                  $x + y = 8$

Equations and
Inequalities

Exercises 11–56

[9.1] *For Exercises 13–16:* ***a.** **Determine whether the system of equations is consistent with independent equations, inconsistent with independent equations, or consistent with dependent equations.***

b.** **How many solutions does the system have?

13.

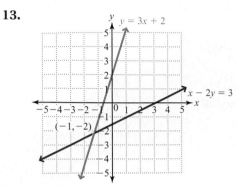

14.

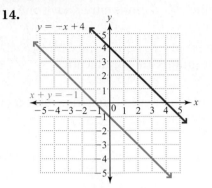

15. $\begin{cases} x - 3y = 10 \\ 2x + 3y = 5 \end{cases}$

16. $\begin{cases} x - 5y = 10 \\ 2x - 10y = 20 \end{cases}$

[9.1] *For Exercises 17–20, solve the system graphically.*

17. $\begin{cases} 4x = y \\ 3x + y = -7 \end{cases}$

18. $\begin{cases} x - y = 4 \\ 2x + 3y = 3 \end{cases}$

19. $\begin{cases} y = -2x - 3 \\ 4x + 2y = 6 \end{cases}$

20. $\begin{cases} 3x - 2y = 6 \\ y = \dfrac{3}{2}x - 3 \end{cases}$

[9.2] *For Exercises 21–24, solve the system of equations using substitution.*

21. $\begin{cases} 3x + 10y = 2 \\ x - 2y = 6 \end{cases}$

22. $\begin{cases} 2x + 5y = 8 \\ x - 10y = 9 \end{cases}$

23. $\begin{cases} 3y - 4x = 6 \\ y = \dfrac{4}{3}x + 2 \end{cases}$

24. $\begin{cases} 8x - 2y = 12 \\ y - 4x = 3 \end{cases}$

[9.3] For Exercises 25–28, solve the system of equations using elimination.

25. $\begin{cases} x + y = 4 \\ x - y = -2 \end{cases}$ **26.** $\begin{cases} 3x + 2y = 4 \\ 2x - 3y = 7 \end{cases}$ **27.** $\begin{cases} 0.25x + 0.75y = 4 \\ x - y = -4 \end{cases}$ **28.** $\begin{cases} \dfrac{1}{5}x - \dfrac{1}{3}y = 2 \\ x + y = 2 \end{cases}$

[9.4] For Exercises 29–32, solve the systems using the elimination method.

29. $\begin{cases} x + y + z = -2 \\ 2x + 3y + 4z = -10 \\ 3x - 2y - 3z = 12 \end{cases}$ **30.** $\begin{cases} x + y + z = 0 \\ 2x - 4y + 3z = -12 \\ 3x - 3y + 4z = 2 \end{cases}$

31. $\begin{cases} 3x + 4y = -3 \\ -2y + 3z = 12 \\ 4x - 3z = 6 \end{cases}$ **32.** $\begin{cases} 2x + 2y - 3z = 3 \\ x = -3y + 4z - 3 \\ z = -x - 2y + 6 \end{cases}$

[9.5] For Exercises 33–36, solve the system using the echelon method.

33. $\begin{cases} x - 4y = 8 \\ x + 2y = 2 \end{cases}$ **34.** $\begin{cases} 2x - y = -4 \\ 2x - 3y = 0 \end{cases}$

35. $\begin{cases} x + y + z = 2 \\ 2x + y + 2z = 1 \\ 3x + 2y + z = 1 \end{cases}$ **36.** $\begin{cases} x + 2y + z = -1 \\ 2x - 2y + z = -10 \\ -x + 3y - 2z = 15 \end{cases}$

[9.6] For Exercises 37–40, solve the system of equations using Cramer's Rule.

37. $\begin{cases} 9x + 4y = 18 \\ 5x - 2y = -28 \end{cases}$ **38.** $\begin{cases} 5x - y = 3 \\ 10x - 2y = 5 \end{cases}$

39. $\begin{cases} x + y + z = 5 \\ 3x - 2y + z = -3 \\ x + 3y + 4z = 10 \end{cases}$ **40.** $\begin{cases} x + 3y + 2z = 6 \\ 2x - 3y + z = -18 \\ -3x + 2y + z = 12 \end{cases}$

[9.7] For Exercises 41–44, graph the solution set for the system of inequalities.

41. $\begin{cases} x + y > -5 \\ x - y \geq -1 \end{cases}$ **42.** $\begin{cases} 2x + y \geq -4 \\ -3x + y \geq -1 \end{cases}$ **43.** $\begin{cases} 2x + y > -1 \\ 3x + y \leq 4 \end{cases}$ **44.** $\begin{cases} -3x + 4y < 12 \\ 2x - y \leq -3 \end{cases}$

45. A movie theater charges $7.00 for adults and $4.25 for children. During a recent showing of *The Matrix Reloaded*, 139 tickets were sold for a total of $720. How many of each ticket were sold?

46. At the University of Kentucky, in 2001 there were 23,221 students enrolled in both undergraduate and graduate degree programs. There are about three times as many undergraduate students as graduate students. About how many graduate students are there? (*Source:* University of Kentucky)

47. At a local automobile dealership, 409 of the cars are either black or silver. There are 35 more black cars than silver cars. How many cars of each color are there?

48. The sum of two numbers is 16. Their difference is 4. What are the two numbers?

49. If two angles are complementary, find the measures of each if the smaller angle is 10° less than one-third of the larger angle.

50. Duke is beginning to invest for his son's college fund. He put part of a $5000 account in a fund that yields 7% and the rest was placed in a fund that yields 9%. If he earned $380 in interest during the year, how much was invested in each fund?

51. Yolanda and Dee are traveling west in separate cars on the same highway. Yolanda is traveling at 60 miles per hour and Dee at 70 miles per hour. Yolanda passes the I-95 exit at 4:00 P.M. Dee passes the same exit at 4:15 P.M. At what time will Dee catch up to Yolanda?

52. Suppose a small plane is traveling with the jet stream and can travel 450 miles in 3 hours. When the plane returns, it takes 5 hours to travel the same distance against the jet stream. What is the speed of the plane in still air?

53. How much of a 10% saline solution is needed to combine with a 60% saline solution to obtain 50 milliliters of a 30% saline solution?

54. John went to the drugstore and bought some vitamin supplements that cost $8 each, some film that costs $4 per roll, and some bags of candy that cost $3 per bag. He bought a total of eight items that cost $32. The number of rolls of film was one less than the number of bags of candy. Find the number of each type of item that he bought.

55. At a swim meet 5 points are awarded for each first-place finish, 3 points for each second, and 1 point for each third. Shawnee Mission South High School scored a total of 38 points. The number of first-place finishes was one more than the number of second-place finishes. The number of third-place finishes was three times the number of second-place finishes. Find the number of first-, second-, and third-place finishes for the school.

56. A real estate course is offered in a continuing education program at a two-year college. The course is limited to 16 students. Also, to run the course, the college must receive at least $2750 in tuition. The college charges $250 per student if they live in the same county as the college and $400 if they live in any other county (or out-of-state).

   a. If x represents the number of in-county students and y represents the number of out-of-county students, write a system of inequalities that describes the number of in-county students enrolled in the course versus the number of out-of-county students and how much they paid.

   b. Solve the system.

Chapter 9 Practice Test

For Exercises 1–2, determine whether the given ordered pair is a solution to the given system of equations.

1. $(-1, 3)$; $\begin{cases} 3x + 2y = 3 \\ 4x - y = -7 \end{cases}$

2. $(2, 2, 3)$; $\begin{cases} 2x - 3y + z = 1 \\ -2x + y - 3z = 11 \\ 3x + y + 3z = 14 \end{cases}$

3. Solve $\begin{cases} x + 3y = 1 \\ -2x + y = 5 \end{cases}$ by graphing.

For Exercises 4–9, solve the system of equations using substitution or elimination. Note that some systems may be inconsistent or consistent with dependent equations.

4. $\begin{cases} 2x + y = 15 \\ y = 7 - x \end{cases}$

5. $\begin{cases} x - 2y = 1 \\ 3x - 5y = 4 \end{cases}$

6. $\begin{cases} 3x - 2y = -8 \\ 2x + 3y = -14 \end{cases}$

7. $\begin{cases} 4x + 6y = 2 \\ 6x + 9y = 3 \end{cases}$

8. $\begin{cases} x + 2y + z = 2 \\ x + 4y - z = 12 \\ 3x - 3y - 2z = -11 \end{cases}$

9. $\begin{cases} x + 2y - 3z = -9 \\ 3x - y + 2z = -8 \\ 4x - 3y + 3z = -13 \end{cases}$

For Exercises 10–11, solve the equations using the echelon method.

10. $\begin{cases} x + 2y = -6 \\ 3x + 4y = -10 \end{cases}$

11. $\begin{cases} x + y + z = 6 \\ 3x - 2y + 3z = -7 \\ 4x - 2y + z = -12 \end{cases}$

For Exercises 12–13, solve the system of equations using Cramer's Rule.

12. $\begin{cases} 3x + 4y = 14 \\ 2x - 3y = -19 \end{cases}$

13. $\begin{cases} x + y + z = -1 \\ 3x - 2y + 4z = 0 \\ 2x + 5y - z = -11 \end{cases}$

For Exercises 14, graph the solution set for the system of inequalities.

14. $\begin{cases} 2x - 3y < 1 \\ x + 2y \le -2 \end{cases}$

For Exercises 15–20, solve.

15. Excedrin surveyed workers in various professions who get headaches at least once a year on the job. There were 9 more accountants than there were waiters/waitresses in the survey who got a headache. If the combined number of accountants and waiters/waitresses that got a headache was 163, how many people in each of those professions got a headache? (*Source:* the *State* newspaper)

16. When asked what the Internet most resembled, three times as many people said a library as opposed to a highway. If 240 people were polled, how may people considered the Internet to be a library? (*Source: bLINK* magazine, winter 2001)

17. A boat traveling with the current took 3 hours to go 30 miles. The same boat went 12 miles in 3 hours against the current. What is the rate of the boat in still water?

18. Janice invested $12,000 into two funds. One of the funds returned 6% interest and the other returned 8% interest after one year. If the total interest for the year was $880, how much did she invest in each fund?

19. Tickets for the senior play at Apopka High School cost $3 for children, $5 for students, and $8 for adults. There were 800 tickets sold for a total of $4750. The number of adult tickets sold was 50 more than two times the number of children's tickets. Find the number of each type of ticket.

20. A landscaper wants to plan a bordered rectangular garden area in a yard. Since she currently has 200 feet of border materials, the perimeter needs to be at most 200 feet. She thinks it would look best if the length is at least 10 feet more than the width.

   a. If x represents the length and y represents the width, write a system of inequalities to describe the situation.

   b. Solve the system of inequalities.

Chapters 1–9 Cumulative Review Exercises

For Exercises 1–6, answer true or false.

[4.4] **1.** The graph of $y = -3x + 5$ is a line with a slope of $-3x$.

[6.4] **2.** The expression $x^2 + 16$ cannot be factored (it is prime).

[8.1] **3.** The solution set for a compound inequality involving *and* can be empty.

[9.1] **4.** The ordered pair $(1, 1)$ is a solution for the system of equations
$$\begin{cases} x = y \\ 9y - 13 = -4x \end{cases}.$$

[9.4] **5.** A solution to a system of equations that has three variables will be an ordered triple.

[9.3] **6.** To use the elimination method to solve a system of equations, all equations must be in standard form.

For Exercises 7–10, fill in the blank.

[1.7] **7.** Complete using the distributive property: $4(2x + 3)$ = _____ .

[3.1] **8.** A proportion is an equation in the form $\dfrac{a}{b}$ = _____ .

[4.2] **9.** To find a solution to an equation in two variables:

    a. Choose a value for one of the variables.

    b. Replace the corresponding _____ with your chosen value.

    c. Solve the equation for the value of the other variable.

[5.2] **10.** The degree of a monomial is the _____ of the exponents of all the variables.

Expressions

Exercises 11–26

[1.5] **For Exercises 11–13, simplify.**

11. $-(6 - 2^2)^2 + 15$ **12.** $\sqrt{36} - 6^2 \div 0$ **13.** $-|2 - 8| + 3(2)$

For Exercises 14–17, simplify.

[5.3] **14.** $(-3x - 4) - (3x + 2)$ [5.5] **15.** $(x - 3)(4x^2 - 2x + 1)$

[5.6] **16.** $(x^3 + 5x^2 + 10x) \div 5x$ [5.6] **17.** $(15m^2 - 22m + 14) \div (3m - 2)$

For Exercises 18–20, factor completely.

[6.4] **18.** $m^3 + 8$ [6.2] **19.** $25 + 10x + x^2$ [6.4] **20.** $x^4 - 81$

For Exercises 21–26, simplify.

[7.2] **21.** $\dfrac{4u^2 + 4u + 1}{u + 2u^2} \cdot \dfrac{u}{2u^2 - u - 1}$

[7.2] **22.** $\dfrac{a^2 - b^2}{x^2 - y^2} \div \dfrac{a + b}{x - y}$

[7.5] **23.** $\dfrac{\dfrac{x}{4x^2 - 1}}{\dfrac{5}{2x - 1}}$

[7.4] **24.** $\dfrac{3}{y - 2} + \dfrac{y}{y + 3}$

[7.3] **25.** $\dfrac{g}{g^2 - 9} - \dfrac{3}{g^2 - 9}$

[7.4] **26.** $\dfrac{3}{x - 2} - \dfrac{2}{x + 2}$

For Exercises 27–31, solve.

[2.3] **27.** $6x - 8 = 2(3x - 4)$

[8.1] **28.** $-2 < x + 4 < 5$

[6.6] **29.** $x^2 - 36 = 0$

Equations and Inequalities

Exercises 27–32, 34–50

[7.6] **30.** $\dfrac{5}{x - 2} - \dfrac{3}{x} = \dfrac{11}{3x}$

[8.2] **31.** $|3x + 4| + 3 = 7$

[2.4] **32.** Solve for m in the formula $E = \dfrac{1}{2} mv^2$.

[4.7] **33.** Determine if the following represents a function. List the domain and range.

The computers and related equipment that the University of Oregon recycled from May 2001 to October 2001. (*Source: The Chronicle of Higher Education*, 2/14/03)

CPUs	1163
Monitors	775
Keyboards	706
Printers	280
Peripherals	81
Macintoshes	67
Typewriters	62
Laptops	25
Televisions	21
Other	228

For Exercises 34–36, graph.

[4.2] **34.** $y = -2x$

[6.7] **35.** $f(x) = x^2 - 3$

[8.4] **36.** $f(x) = |x + 2|$

[4.5] **37.** Write the equation of a line in standard form that passes through the point $(-2, 4)$ and is perpendicular to the line $3x - 2y = 6$.

For Exercises 38–41, solve the systems.

[9.3] **38.** $\begin{cases} 4x - 3y = -2 \\ 6x - 7y = 7 \end{cases}$

[9.4] **39.** $\begin{cases} x + y + z = 5 \\ 2x + y - 2z = -5 \\ x - 2y + z = 8 \end{cases}$

[9.6] **40.** Use Cramer's Rule to solve the system:
$\begin{cases} 2x + 3y = -5 \\ 3x - y = 9 \end{cases}$

[9.7] **41.** Solve and graph the solution set: $\begin{cases} x + y < 3 \\ x - y \geq 4 \end{cases}$

For Exercises 42–50, solve.

[3.2] **42.** The average size of a single-family home in 1973 was 1660 square feet. In 2003, the average size was 2,266. What was the percent of increase in the average size of a single-family home? (*Source:* the *State* newspaper 2/1/03)

[3.3] **43.** In 2004, the median salary per week for men was $144 more than the median salary for women. If the combined median salary was $1300, what was the median salary for men and for women? (*Source: Parade Magazine* 3/2/03)

[3.5] **44.** A chemist has a bottle containing 50 ml of 15% saline solution and a bottle of 40% saline solution. She wants a 30% solution. How much of the 40% solution must be added to the 15% solution so that a 30% concentration is created?

[4.6] **45.** An event planner is trying to determine how many events she must plan in a year to make a profit of at least $50,000. She estimates that based on her current prices, she receives a profit of $250 for each small event and $500 for each large event that she plans.

   a. Select a variable for each type of event and write an inequality that describes a total profit of at least $50,000.

   b. Graph the solution region for this inequality.

   c. Give one combination of the number of each type of event that would give her a profit of exactly $50,000.

   d. Give one combination of the number of each type of event that would give her a profit of more than $50,000.

[5.6] **46.** If light travels about 3×10^8 meters per second, how long does it take the light from the Sun to travel to Pluto, which is a distance of about 5.9×10^{12} meters. Convert your answer to hours.

[7.7] **47.** If the wavelength of a wave remains constant, then the velocity, v, of a wave is inversely proportional to its period, T. In an experiment, waves are created in a pool of water so that the period is 8 seconds and the velocity is 6 centimeters per second. If the period is increased to 15 seconds, what is the velocity?

[7.7] **48.** LaTilla takes 3 hours to mow and edge the yard. Her husband can mow and edge in 2 hours. How long will it take them if they work on the yard together?

[9.3] **49.** In constructing a roof frame, a support beam is connected to a horizontal joist, forming two angles. If the greater angle is 15° less than twice the smaller angle, what are the measures of the two angles?

[9.4] **50.** Joanna purchased two pens, two erasers, and one ream of paper for $6. Willa purchased four pens, three erasers, and two reams of paper for $11.50. On the same day, Scott purchased three pens, one eraser and three reams of paper for $9.50. Find the cost of each.

10

Rational Exponents, Radicals, and Complex Numbers

"Radical simply means 'grasping things at the root.' "
—Angela Davis, U.S. political activist,
address June 25, 1987, to Spellman College

In Section 1.5, we reviewed the basic rules of square roots. In this chapter, we will explore square roots and radicals in much more detail. We will begin in Section 10.1 with evaluating and simplifying radical expressions, much like how we started with evaluating and simplifying polynomial expressions in the first part of Chapter 5. Then, in Section 10.2, we will see that exponents can be fractions and that those rational exponents mean roots. In Sections 10.3–10.5, we move to rewriting expressions containing radicals. Once we have explored how to rewrite radical expressions, we will learn to solve equations that contain radicals in Section 10.6. Finally, in Section 10.7, we will discuss a new set of numbers called *complex numbers*, which come from square roots of negative numbers.

10.1 Radical Expressions and Functions

OBJECTIVES

1. Find the nth root of a number.
2. Approximate roots using a calculator.
3. Simplify radical expressions.
4. Evaluate radical functions.
5. Find the domain of radical functions.
6. Solve applications involving radical functions.

In this section through Section 10.5, we focus on the expression portion of our algebra pyramid and explore square root and radical expressions.

The Algebra Pyramid

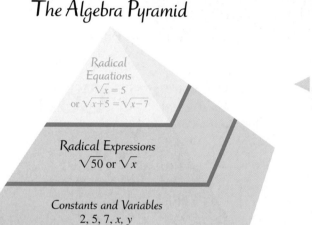

Note: *We will explore radical equations in Section 10.6.*

OBJECTIVE 1. Find the nth root of a number. In Section 1.5, we learned that a square root of a given number is a number that, when squared, equals the given number. For example, a square root of 16 is 4 since $4^2 = 16$. However, another number can be squared to equal 16: $(-4)^2 = 16$, so -4 is also a square root of 16. Consequently, 16 has two square roots: 4 and -4. We can write them more compactly as ± 4.

Similarly, a *cube root* of a given number is a number that, when cubed, equals the given number. For example, the cube root of 8 is 2 because $2^3 = 8$.

The *fourth root* of a number is a number that, when raised to the fourth power, equals the given number. For example, 3 is a fourth root of 81 because $3^4 = 81$. However, $(-3)^4 = 81$ also, so the fourth roots of 81 are ± 3. Our examples suggest the following definition of nth **root**.

DEFINITION n^{th} *root:* The number b is an nth root of a number a if $b^n = a$.

Recall from Section 1.5 that we used the symbol $\sqrt{}$, called a *radical sign*, to denote the square root of a number. To indicate roots other than square roots we use the symbol $\sqrt[n]{a}$, read "the nth root of a," where n is called the *root index* and indicates which root we are to find. If no root index is given, we assume it is 2, the square root. The number *a* is called the *radicand* and is the number or expression whose root we are to find. The entire expression is called a *radical* and any expression containing a radical is called a *radical expression*.

We have seen some numbers that have more than one root, such as 16, which has two square roots, 4 and -4. To keep things simple, the symbol $\sqrt[n]{a}$ means to find a single root. If *n* is even, then $\sqrt[n]{a}$ denotes the nonnegative root only and is called the *principal root*. For example, the principal square root of 16 is 4. Further, for $\sqrt[n]{a}$ to exist as a real number when *n* is even, *a* must be nonnegative because there is no real number that can be raised to an even power to equal a negative number. For example, $\sqrt{-4}$ does not exist as a real number because there is no real number whose square is -4.

If the root index is odd, the radicand can be positive or negative since a positive number raised to an odd power is positive and a negative number raised to an odd power is negative.

RULE   Evaluating n^{th} Roots

When evaluating a radical expression $\sqrt[n]{a}$, the sign of *a* and the index *n* will determine possible outcomes.

If *a* is nonnegative, then $\sqrt[n]{a} = b$, where $b \geq 0$ and $b^n = a$.

If *a* is negative and *n* is even, then there is no real-number root.

If *a* is negative and *n* is odd, then $\sqrt[n]{a} = b$, where *b* is negative and $b^n = a$.

EXAMPLE 1   Evaluate the following roots.

a. $\sqrt{225}$

Solution $\sqrt{225} = 15$ Because $15^2 = 225$.

Note: *It can be helpful to think of* $-\sqrt{0.81}$ *as* $-1 \cdot \sqrt{0.81} = -1 \cdot 0.9 = -0.9$.

b. $-\sqrt{0.81}$

Solution $-\sqrt{0.81} = -0.9$ Because $(-0.9)^2 = 0.81$.

c. $\sqrt{-9}$

Solution $\sqrt{-9}$ is not a real number.

Note: *In Section 10.7, we will define the square root of a negative number using a set of numbers called the* imaginary numbers.

d. $\pm\sqrt{121}$

Solution $\pm\sqrt{121} = \pm 11$ Because $(11)^2 = 121$ and $(-11)^2 = 121$.

e. $\sqrt{\dfrac{4}{9}}$

Solution $\sqrt{\dfrac{4}{9}} = \dfrac{2}{3}$ Because $\left(\dfrac{2}{3}\right)^2 = \dfrac{4}{9}$.

f. $\sqrt[3]{8}$

Solution $\sqrt[3]{8} = 2$ Because $2^3 = 8$.

g. $\sqrt[3]{-8}$

Solution $\sqrt[3]{-8} = -2$ Because $(-2)^3 = -8$.

h. $\sqrt[4]{16}$

Solution $\sqrt[4]{16} = 2$ Because $2^4 = 16$.

YOUR TURN   Evaluate the following expressions.

a. $\sqrt{0.36}$ **b.** $-\sqrt{4}$ **c.** $\sqrt{-64}$ **d.** $\pm\sqrt{169}$

e. $\sqrt{\dfrac{25}{36}}$ **f.** $\sqrt[3]{64}$ **g.** $\sqrt[3]{-27}$ **h.** $-\sqrt[4]{81}$

OBJECTIVE 2. Approximate roots using a calculator.   All the roots that we have considered so far have been rational. Some roots, like $\sqrt{3}$, are called irrational. As we learned in Section 1.1, we cannot express the exact value of an irrational number using rational numbers. In fact, writing $\sqrt{3}$ with the radical sign is the only way we can express the exact value. However, we may approximate $\sqrt{3}$ with a calculator or a table, such as the one in the endpapers of this text.

For example, a calculator shows that $\sqrt{2} \approx 1.414213562$, which we can round to various decimal places:

Approximating to two decimal places: $\sqrt{2} \approx 1.41$

Approximating to three decimal places: $\sqrt{2} \approx 1.414$

◀ **Note:** *Recall that the symbol $\approx$ means "approximately equal to."*

EXAMPLE 2   Approximate the roots using a calculator or the table in the endpapers. Round to three decimal places.

a. $\sqrt{12}$

Answer $\sqrt{12} \approx 3.464$

b. $-\sqrt{38}$

Answer $-\sqrt{38} \approx -6.164$

c. $\sqrt[3]{45}$

Answer $\sqrt[3]{45} \approx 3.557$

Calculator TIPS

To evaluate roots higher than a square root on a calculator, use the $\boxed{\sqrt[x]{\ }}$ key. First type the index, then press $\boxed{\sqrt[x]{\ }}$, then the radicand. For example, to evaluate $\sqrt[3]{45}$, type: $\boxed{3}\ \boxed{\sqrt[x]{\ }}\ \boxed{4}\ \boxed{5}\ \boxed{\text{ENTER}}$.

On some calculators, $\sqrt[x]{\ }$ is a function in a menu accessed by pressing the $\boxed{\text{MATH}}$ key.

YOUR TURN   Approximate the following irrational numbers. Round to three decimal places.

a. $\sqrt{19}$ **b.** $-\sqrt{93}$ **c.** $\sqrt[3]{63}$

OBJECTIVE 3. Simplify radical expressions.   The definition of a root can also be used to find roots with variable radicands. Recall that with an even index, the principal root is nonnegative, so at first we will assume all variables represent nonnegative values. We will use $(a^m)^n = a^{mn}$ to verify the roots.

Note: *Remember that with an even index, the principal root is nonnegative. Since we are assuming that the variables are nonnegative, our result accurately indicates the principal square root.*

Connection To raise a power to another power using $(a^m)^n = a^{mn}$, we multiply the powers. To find a root of a power, we divide the index into the exponent.

EXAMPLE 3 Find the root. Assume all variables represent nonnegative values.

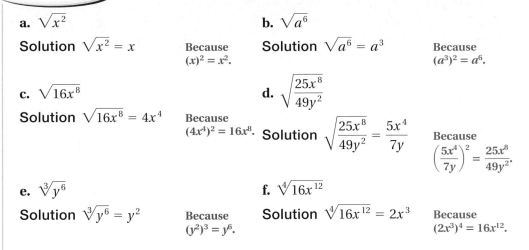

a. $\sqrt{x^2}$

Solution $\sqrt{x^2} = x$ Because $(x)^2 = x^2$.

b. $\sqrt{a^6}$

Solution $\sqrt{a^6} = a^3$ Because $(a^3)^2 = a^6$.

c. $\sqrt{16x^8}$

Solution $\sqrt{16x^8} = 4x^4$ Because $(4x^4)^2 = 16x^8$.

d. $\sqrt{\dfrac{25x^8}{49y^2}}$

Solution $\sqrt{\dfrac{25x^8}{49y^2}} = \dfrac{5x^4}{7y}$ Because $\left(\dfrac{5x^4}{7y}\right)^2 = \dfrac{25x^8}{49y^2}$.

e. $\sqrt[3]{y^6}$

Solution $\sqrt[3]{y^6} = y^2$ Because $(y^2)^3 = y^6$.

f. $\sqrt[4]{16x^{12}}$

Solution $\sqrt[4]{16x^{12}} = 2x^3$ Because $(2x^3)^4 = 16x^{12}$.

Connection In parts c and f, notice the similarity between finding the root of a product and raising a product to a power. To raise a product to a power, we raise each factor to the power. To find a root of a product, we find the root of each factor.

YOUR TURN Find the root. Assume variables represent nonnegative values.

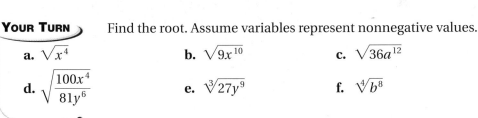

a. $\sqrt{x^4}$

b. $\sqrt{9x^{10}}$

c. $\sqrt{36a^{12}}$

d. $\sqrt{\dfrac{100x^4}{81y^6}}$

e. $\sqrt[3]{27y^9}$

f. $\sqrt[4]{b^8}$

If the variables can represent *any* real number and the index is even, then we must be careful to ensure that the principal root is nonnegative by using absolute value symbols. If the index is odd, however, we do not need absolute value signs because the root can be either positive or negative depending on the radicand's sign.

Note: *To illustrate why $\sqrt{x^2} = |x|$, suppose we were to evaluate $\sqrt{x^2}$ when $x = -3$. We would have: $\sqrt{(-3)^2} = \sqrt{9} = 3$. Notice that the root, 3, is, in fact, the absolute value of -3. If we had incorrectly stated that $\sqrt{x^2} = x$, then $\sqrt{(-3)^2}$ would have to equal -3, which is not true.*

EXAMPLE 4 Find the root. Assume variables represent any real number.

a. $\sqrt{x^2}$

Solution $\sqrt{x^2} = |x|$

b. $\sqrt{a^6}$

Solution $\sqrt{a^6} = |a^3|$

c. $\sqrt{(n+1)^2}$

Solution $\sqrt{(n+1)^2} = |n+1|$

d. $\sqrt{25y^8}$

Solution $\sqrt{25y^8} = 5y^4$

e. $\sqrt[3]{8n^3}$

Solution $\sqrt[3]{8n^3} = 2n$

f. $\sqrt[3]{(t-2)^3}$

Solution $\sqrt[3]{(t-2)^3} = t-2$

Note: In part d, we do not need absolute value because y^4, with its even exponent, will always be nonnegative. In parts e and f, we do not use absolute value because the indexes are odd.

ANSWERS

a. x^2 **b.** $3x^5$

c. $6a^6$ **d.** $\dfrac{10x^2}{9y^3}$

e. $3y^3$ **f.** b^2

YOUR TURN   Find the root. Assume variables represent any real number.

a. $\sqrt{x^4}$ **b.** $\sqrt{9x^{10}}$ **c.** $\sqrt{36a^{12}}$ **d.** $\sqrt{1.21u^6t^2}$

e. $\sqrt{\dfrac{100x^4}{81y^6}}$ **f.** $\sqrt[3]{27y^9}$ **g.** $\sqrt[4]{b^8}$

OBJECTIVE 4. Evaluate radical functions.   Now that we have learned about radical expressions, let's examine **radical functions**.

DEFINITION   *Radical function:* A function of the form $f(x) = \sqrt[n]{P}$, where P is a polynomial.

EXAMPLE 5

a. Given $f(x) = \sqrt{3x - 2}$, find $f(3)$.

Solution To find $f(3)$, substitute 3 for x and simplify.

$$f(3) = \sqrt{3(3) - 2} = \sqrt{9 - 2} = \sqrt{7}$$

b. Given $f(x) = \sqrt{2x - 6}$, find $f(0)$.

Solution To find $f(0)$, substitute 0 for x and simplify. $f(0) = \sqrt{2(0) - 6} = \sqrt{-6}$, which is not a real number.

YOUR TURN   Given $f(x) = \sqrt{2x + 5}$, find $f(-1)$.

OBJECTIVE 5. Find the domain of radical functions.   In Example 5(b), we found that $f(0)$ did not exist because $\sqrt{-6}$ is not a real number. What does this suggest about the domains of functions involving radicals? Consider the graphs of $f(x) = \sqrt{x}$ and $f(x) = \sqrt[3]{x}$, which we can generate by choosing values for x.

x	$f(x) = \sqrt{x}$
-1	Not real
0	0
1	1
4	2

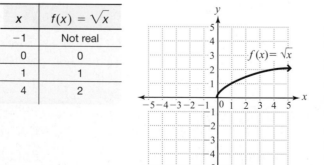

Notice that $\sqrt{x}$ is not a real number if x is negative, which means the domain for $f(x) = \sqrt{x}$ is $\{x | x \geq 0\}$. Alternatively, $\sqrt[3]{x}$, with its odd index, is a real number when x is negative, so its domain is all real numbers. Our graphs suggest the following conclusion.

x	$f(x) = \sqrt[3]{x}$
-8	-2
-1	-1
0	0
1	1
8	2

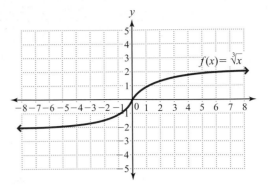

Conclusion The domain of a radical function with an even index must contain values that keep its radicand nonnegative.

EXAMPLE 6 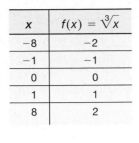 Find the domain of each of the following.

a. $f(x) = \sqrt{x - 4}$

Solution Since the index is even, the radicand must be nonnegative.

$$x - 4 \geq 0$$
$$x \geq 4 \qquad \text{Add 4 to both sides.}$$

Domain: $\{x \mid x \geq 4\}$, or $[4, \infty)$

b. $f(x) = \sqrt{-2x + 6}$

Solution The radicand must be nonnegative.

$$-2x + 6 \geq 0$$
$$-2x \geq -6 \qquad \text{Subtract 6 from both sides.}$$
$$x \leq 3 \qquad \text{Divide both sides by } -2.$$

Domain: $\{x \mid x \leq 3\}$, or $(-\infty, 3]$

Connection Recall that when we divide *both* sides of an inequality by a negative number, we reverse the direction of the inequality.

ANSWERS

a. $\{x \mid x \geq 2\}$, or $[2, \infty)$
b. $\{x \mid x \leq -3\}$, or $(-\infty, -3]$

YOUR TURN Find the domain of each of the following.

a. $f(x) = \sqrt{2x - 4}$ **b.** $f(x) = \sqrt{-3x - 9}$

OBJECTIVE 6. Solve applications involving radical functions. Often, radical functions appear in real-world situations where one variable is a function of another.

EXAMPLE 7 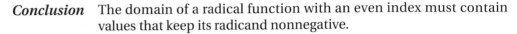 The velocity of a free-falling object is a function of the distance that it has fallen. Ignoring air resistance, the velocity of an object, v, in meters per second, can be found after falling h meters using the formula $v = -\sqrt{19.6h}$. Find the velocity of a stone that has fallen 30 meters after being dropped from a cliff.

Understand We are to find the velocity of an object after it falls 30 meters.

Plan Use the formula $v = -\sqrt{19.6h}$, replacing h with 30.

Execute $v = -\sqrt{19.6(30)}$ **Replace *h* with 30.**

$v = -\sqrt{588}$ **Multiply within the radical.**

$v \approx -24.2$ **Evaluate the square root.** ◀ **Note:** *A negative velocity indicates that the object is traveling downward.*

Answer   After falling 30 meters, the stone is traveling at a velocity of -24.2 meters per second.

Check   We can verify the calculations, which we will leave to the reader.

Your Turn   A skydiver jumps from a plane and puts her body into a dive position so that air resistance is minimized. Find her velocity after she falls 100 meters.

EXAMPLE 8   The period of a pendulum is the amount of time it takes the pendulum to swing from the point of release to the opposite extreme then back to the point of release. The period is a function of the length. The period, T, measured in seconds, can be found using the formula $T = 2\pi\sqrt{\dfrac{L}{9.8}}$, where L represents the length of the pendulum in meters. Find the period of a pendulum that is 0.5 meters long.

Understand   We are to find the period of a pendulum that is 0.5 meters long.

Plan   Use the formula $T = 2\pi\sqrt{\dfrac{L}{9.8}}$, approximating π with 3.14 and replacing L with 0.5.

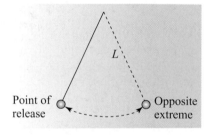

Point of release L Opposite extreme

Execute $T \approx 2(3.14)\sqrt{\dfrac{0.5}{9.8}}$ **Simplify.**

$T \approx 6.28\sqrt{0.051}$

$T \approx 6.28(0.226)$ **Approximate the square root.**

$T \approx 1.42$

Answer   The period of a pendulum that is 0.5 meters long is 1.42 seconds.

Check   We can verify the calculations, which we will leave to the reader.

Your Turn   Find the period of a pendulum that is 0.2 meters long. Round to the nearest thousandth.

Answer

-44.3 m/sec.

Answer

≈ 0.898 sec.

1. Give an example of a number with rational square roots and an example of a number with irrational square roots. What makes the second number's roots irrational?

2. How can we express the exact value of a root that is irrational?

3. Why are there two square roots for every positive real number?

4. If the index is even, what is the principal root of a number?

5. Why is an even root of any negative number not a real number?

6. If x is any real number, explain why $\sqrt{x^2} = |x|$ instead of x (with no absolute value).

For Exercises 7–14, find all square roots of the number given.

7. 36

8. 64

9. 121

10. 81

11. 196

12. 400

13. 225

14. 289

For Exercises 15–56, evaluate the roots, if possible.

15. $\sqrt{25}$

16. $\sqrt{49}$

17. $\sqrt{-64}$

18. $\sqrt{-25}$

19. $-\sqrt{25}$

20. $-\sqrt{100}$

21. $\pm\sqrt{25}$

22. $\pm\sqrt{100}$

23. $\sqrt{1.44}$

24. $\sqrt{0.36}$

25. $\sqrt{-10.64}$

26. $\sqrt{-2.25}$

27. $-\sqrt{0.0121}$

28. $-\sqrt{0.0009}$

29. $\sqrt{\dfrac{49}{81}}$

30. $\sqrt{\dfrac{64}{121}}$

31. $-\sqrt{\dfrac{144}{169}}$

32. $-\sqrt{\dfrac{25}{4}}$

33. $\sqrt[3]{27}$

34. $\sqrt[3]{64}$

35. $\sqrt[3]{-27}$

36. $\sqrt[3]{-216}$

37. $-\sqrt[3]{-216}$

38. $-\sqrt[3]{-125}$

39. $\sqrt[4]{625}$

40. $\sqrt[4]{81}$

41. $\sqrt[4]{-625}$

42. $\sqrt[4]{-81}$

43. $-\sqrt[4]{16}$

44. $-\sqrt[4]{625}$

45. $\sqrt[5]{32}$

46. $\sqrt[5]{243}$

47. $\sqrt[5]{-32}$

48. $\sqrt[5]{-243}$

49. $-\sqrt[5]{-32}$

50. $-\sqrt[5]{-3125}$

51. $\sqrt[6]{64}$

52. $\sqrt[6]{-64}$

53. $\sqrt[3]{\dfrac{8}{27}}$

54. $\sqrt[3]{-\dfrac{64}{125}}$

55. $\sqrt[4]{\dfrac{16}{81}}$

56. $\sqrt[5]{\dfrac{1}{32}}$

For Exercises 57–72, use a calculator to approximate each root to the nearest thousandth.

57. $\sqrt{5}$

58. $\sqrt{12}$

59. $-\sqrt{11}$

60. $-\sqrt{41}$

61. $\sqrt[3]{50}$

62. $\sqrt[3]{21}$

63. $\sqrt[3]{-53}$

64. $\sqrt[3]{-83}$

65. $\sqrt[4]{65}$

66. $\sqrt[4]{123}$

67. $-\sqrt[4]{85}$

68. $-\sqrt[4]{77}$

69. $\sqrt[5]{89}$

70. $\sqrt[5]{-62}$

71. $\sqrt[6]{146}$

72. $\sqrt[6]{98}$

For Exercises 73–96, simplify. Assume variables represent nonnegative values.

73. $\sqrt{b^4}$

74. $\sqrt{r^8}$

75. $\sqrt{16x^2}$

76. $\sqrt{81t^6}$

77. $\sqrt{100r^2s^6}$

78. $\sqrt{121x^8y^{10}}$

79. $\sqrt{0.25a^6b^{12}}$

80. $\sqrt{0.36r^4s^{14}}$

81. $\sqrt[3]{m^3}$

82. $\sqrt[3]{n^6}$

83. $\sqrt[3]{27a^9b^6}$

84. $\sqrt[3]{64u^{12}t^9}$

85. $\sqrt[3]{-64a^3b^{12}}$

86. $\sqrt[3]{-27r^{15}s^3}$

87. $\sqrt[3]{0.008x^{18}}$

88. $\sqrt[3]{-0.125r^6}$

89. $\sqrt[4]{a^4}$

90. $\sqrt[4]{x^{12}}$

91. $\sqrt[4]{16x^{16}}$

92. $\sqrt[4]{81t^8}$

93. $\sqrt[5]{32x^{10}}$

94. $\sqrt[5]{243x^{15}}$

95. $\sqrt[6]{x^{12}y^6}$

96. $\sqrt[7]{s^7t^{21}}$

For Exercises 97–108, simplify. Assume variables represent any real number.

97. $\sqrt{36m^2}$

98. $\sqrt{9t^6}$

99. $\sqrt{(r-1)^2}$

100. $\sqrt{(k+3)^2}$

101. $\sqrt[4]{256y^{12}}$

102. $\sqrt[4]{16x^4}$

103. $\sqrt[3]{27y^3}$

104. $\sqrt[3]{125x^6}$

105. $\sqrt{(y-3)^4}$

106. $\sqrt{(x+2)^4}$

107. $\sqrt[3]{(y-4)^3}$

108. $\sqrt[4]{(n+5)^8}$

For Exercises 109–112, find the indicated value of the function.

109. $f(x) = \sqrt{2x+4}$, find $f(0)$

110. $f(x) = \sqrt{3x+2}$, find $f(3)$

111. $f(x) = \sqrt{4x+3}$, find $f(3)$

112. $f(x) = \sqrt{-2x+3}$, find $f(-2)$

For Exercises 113–116, find the domain.

113. $f(x) = \sqrt{2x-8}$

114. $f(x) = \sqrt{3x+12}$

115. $f(x) = \sqrt{-4x+16}$

116. $f(x) = \sqrt{-2x+6}$

For Exercises 117–120: **a. Use a graphing calculator to graph the function.**
                        b. Find the domain of the function.

117. $f(x) = \sqrt{x-2}$

118. $f(x) = \sqrt{x+3}$

119. $f(x) = \sqrt[3]{x+1}$

120. $f(x) = \sqrt[4]{x-3}$

For Exercises 121 and 122, use the following. Neglecting air resistance, the velocity, v, in meters per second, of an object after falling h meters can be found using the formula $v = -\sqrt{19.6h}$.

121. Find the velocity of a rock that has been dropped from a cliff after it falls 16 meters.

122. Find the velocity of a ball that has been dropped from a roof after it falls 5 meters.

Of Interest
One of the many topics Galileo Galilei studied was pendulums. His interest in them was piqued when he noticed a swinging light fixture in the cathedral of his hometown of Pisa and timed its period using his pulse.

For Exercises 123 and 124, use the following. The period, T, of a pendulum in seconds can be found using the formula $2\pi\sqrt{\dfrac{L}{9.8}}$, where L represents the length of the pendulum in meters.

123. Find the period of a pendulum that is 3 meters long.

124. Find the period of a pendulum that is 2.5 meters long.

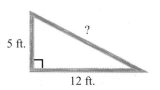

point of release / L / opposite extreme

For Exercises 125 and 126, use the following. The formula $S = \dfrac{7}{2}\sqrt{2D}$ can be used to approximate the speed, S, in miles per hour, that a car was traveling prior to braking and skidding a distance, D, in feet, on asphalt. (Source: Harris Technical Services Traffic Accident Reconstructionists.)

125. Find the speed of a car if the skid distance is 15 feet.

126. Find the speed of a car if the skid distance is 40 feet.

For Exercises 127 and 128, use the following. Given the lengths a and b of the sides of a right triangle, we can find the hypotenuse, c, using the formula $c = \sqrt{a^2 + b^2}$.

127. Three pieces of lumber are to be connected to form a right triangle that will be part of the frame for a roof. If the horizontal piece is to be 12 feet and the vertical piece is to be 5 feet, how long must the connecting piece be?

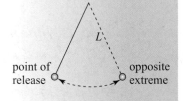

5 ft.

?

12 ft.

128. A counselor decides to create a ropes course with a zip line. She wants the line to connect from a 40-foot-tall tower to the ground at a point 100 feet from the base of the tower. Assuming the tower and ground form a right angle, find the length of the zip line.

Connection In Section 6.6, we used the Pythagorean theorem, $c^2 = a^2 + b^2$, to find missing side lengths of a right triangle. The formula $c = \sqrt{a^2 + b^2}$ comes from isolating c in that theorem.

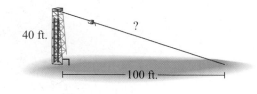

40 ft.

?

100 ft.

For Exercises 129 and 130 use the following. Two forces, F_1 and F_2, acting on an object at a 90° angle will pull the object with a resultant force, R, at an angle in between F_1 and F_2 (see the figure). The value of the resultant force can be found by the formula $R = \sqrt{F_1^2 + F_2^2}$.

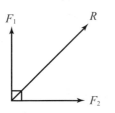

129. Find the resultant force if $F_1 = 9$ N and $F_2 = 12$ N.

Note: *The unit of force in the metric system is the newton (abbreviated N).*

130. Find the resultant force if $F_1 = 6.2$ N and $F_2 = 4.5$ N.

131. The number of earthquakes worldwide for 2000–2004 with a magnitude of 3.0–3.9 can be approximated using the function $f(x) = 1570\sqrt{x} + 4784$, where x represents the number of years after 2000. (*Hint:* The year 2000 corresponds to $x = 0$.) (*Source:* USGS Earthquake Hazards Program)

   a. Find the approximate number of earthquakes with a magnitude of 3.0–3.9 that occurred in 2004.

   b. Find the approximate number of earthquakes with a magnitude of 3.0–3.9 that occurred in 2002.

132. The number of associate degrees, in thousands, conferred by institutions of higher education for 1970 through 2004 can be approximated using the function $f(x) = 52\sqrt{x} + 252$, where x represents the number of years after 1970. (*Hint:* The year 1970 corresponds to $x = 0$.) (*Source:* National Center for Education Statistics)

   a. How many associate degrees were awarded in 1980?

   b. How many associate degrees were awarded in 2004?

PUZZLE PROBLEM

A telemarketer calls a house and speaks to the mother of three children. The telemarketer asks the mother, "How old are you?" The woman answers, "41." The marketer then asks, "How old are your children?" The woman replies, "The product of their ages is our house number, 1296, and all of their ages are perfect squares." What are the ages of the children?

REVIEW EXERCISES

[1.5] 1. Follow the order of operations to simplify.

  a. $\sqrt{16 \cdot 9}$

  b. $\sqrt{16} \cdot \sqrt{9}$

For Exercises 2–5, multiply.

[5.4] 2. $x^5 \cdot x^3$ [5.4] 3. $(-9m^3n)(5mn^2)$ [5.5] 4. $4x^2(3x^2 - 5x + 1)$ [5.5] 5. $(7y - 4)(3y + 5)$

[5.5] 6. Write an expression for the area of the figure shown.

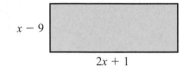

$x - 9$

$2x + 1$

10.2 Rational Exponents

OBJECTIVES

1. Evaluate rational exponents.

2. Write radicals as expressions raised to rational exponents.

3. Simplify expressions with rational number exponents using the rules of exponents.

4. Use rational exponents to simplify radical expressions.

OBJECTIVE 1. Evaluate rational exponents.

Rational Exponents with a Numerator of 1

So far we have only seen integer exponents. However, expressions can have fraction exponents also, as in $3^{1/2}$. The fraction exponent in $3^{1/2}$ is called a **rational exponent.**

DEFINITION *Rational exponents:* An exponent that is a fraction.

To discover what rational exponents mean, consider the following.

$3^{1/2} \cdot 3^{1/2} = 3^{1/2 + 1/2} = 3^1 = 3$ Use $a^m \cdot a^n = a^{m+n}$, which we learned in Section 5.4.

Notice that multiplying $\sqrt{3}$ by itself gives the same result:

$$\sqrt{3} \cdot \sqrt{3} = 3$$

These two results suggest that a base raised to the 1/2 power and the square root of that same base are equivalent expressions, or $a^{1/2} = \sqrt{a}$. Note that the denominator, 2, of the rational exponent is the index of the root (remember that square roots have an invisible index of 2). This relationship holds for other roots.

RULE **Rational Exponents with a Numerator of 1**

$a^{1/n} = \sqrt[n]{a}$, where n is a natural number other than 1.

> **Note:** *If a is negative and n is odd, then the root will be negative.*
> *If a is negative and n is even, then there is no real-number root.*

EXAMPLE 1 Rewrite using radicals, then simplify, if possible. Assume all variables represent nonnegative values.

a. $36^{1/2}$

Solution $36^{1/2} = \sqrt{36} = 6$

b. $81^{1/4}$

Solution $81^{1/4} = \sqrt[4]{81} = 3$

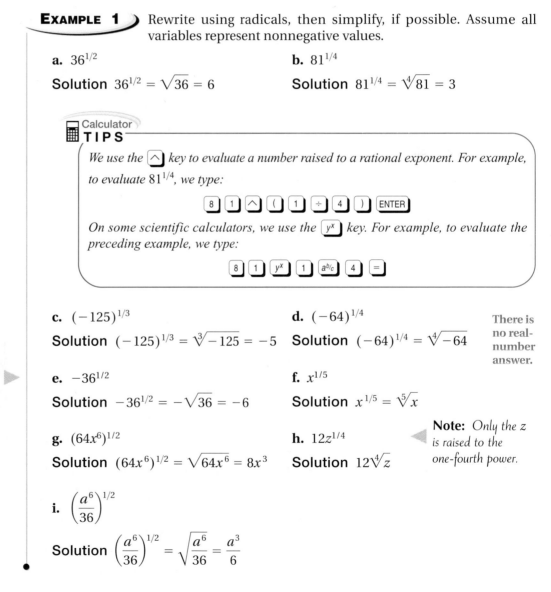

> Calculator **TIPS**
>
> *We use the* $\boxed{\wedge}$ *key to evaluate a number raised to a rational exponent. For example, to evaluate $81^{1/4}$, we type:*
>
> $\boxed{8}\ \boxed{1}\ \boxed{\wedge}\ \boxed{(}\ \boxed{1}\ \boxed{\div}\ \boxed{4}\ \boxed{)}\ \boxed{\text{ENTER}}$
>
> *On some scientific calculators, we use the* $\boxed{y^x}$ *key. For example, to evaluate the preceding example, we type:*
>
> $\boxed{8}\ \boxed{1}\ \boxed{y^x}\ \boxed{1}\ \boxed{a^{b/c}}\ \boxed{4}\ \boxed{=}$

Note: *In $(-64)^{1/4}$, the parentheses group the minus sign with the radicand, whereas in $-36^{1/2}$, the minus sign is not part of the radicand.*

c. $(-125)^{1/3}$

Solution $(-125)^{1/3} = \sqrt[3]{-125} = -5$

d. $(-64)^{1/4}$

Solution $(-64)^{1/4} = \sqrt[4]{-64}$ There is no real-number answer.

e. $-36^{1/2}$

Solution $-36^{1/2} = -\sqrt{36} = -6$

f. $x^{1/5}$

Solution $x^{1/5} = \sqrt[5]{x}$

g. $(64x^6)^{1/2}$

Solution $(64x^6)^{1/2} = \sqrt{64x^6} = 8x^3$

h. $12z^{1/4}$

Solution $12\sqrt[4]{z}$

Note: *Only the z is raised to the one-fourth power.*

i. $\left(\dfrac{a^6}{36}\right)^{1/2}$

Solution $\left(\dfrac{a^6}{36}\right)^{1/2} = \sqrt{\dfrac{a^6}{36}} = \dfrac{a^3}{6}$

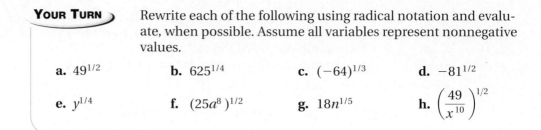

YOUR TURN

Rewrite each of the following using radical notation and evaluate, when possible. Assume all variables represent nonnegative values.

a. $49^{1/2}$ **b.** $625^{1/4}$ **c.** $(-64)^{1/3}$ **d.** $-81^{1/2}$

e. $y^{1/4}$ **f.** $(25a^8)^{1/2}$ **g.** $18n^{1/5}$ **h.** $\left(\dfrac{49}{x^{10}}\right)^{1/2}$

Rational Exponents with a Numerator Other Than 1

Let's explore how we can rewrite expressions like $8^{2/3}$ in which the numerator of the rational exponent is a number other than 1. Note that we could write the fraction $2/3$ as a product:

$$8^{2/3} = 8^{2(1/3)} \quad \text{or} \quad 8^{2/3} = 8^{(1/3)2}$$

Using the rule of exponents $(n^a)^b = n^{ab}$ in reverse, we can write:

$$8^{2/3} = 8^{2(1/3)} = (8^2)^{1/3} \quad \text{or} \quad 8^{2/3} = 8^{(1/3)2} = (8^{1/3})^2$$

Applying the rule $a^{1/n} = \sqrt[n]{a}$, we can write:

$$8^{2/3} = 8^{2(1/3)} = (8^2)^{1/3} = \sqrt[3]{8^2} \quad \text{or} \quad 8^{2/3} = 8^{(1/3)2} = (8^{1/3})^2 = (\sqrt[3]{8})^2$$

Notice that the denominator of the rational exponent becomes the index of the radical. The numerator of the rational exponent can be written either as the exponent of the radicand or as an exponent for the entire radical.

RULE **General Rule for Rational Exponents**

$a^{m/n} = \sqrt[n]{a^m} = (\sqrt[n]{a})^m$, where $a \geq 0$ and m and n are natural numbers other than 1.

EXAMPLE 2
Rewrite using radicals, then simplify, if possible. Assume all variables represent nonnegative values.

a. $8^{2/3}$

Solution $8^{2/3} = \sqrt[3]{8^2}$ Rewrite.   or $8^{2/3} = (\sqrt[3]{8})^2$ Rewrite.

            $= \sqrt[3]{64}$ Square the radicand. $= (2)^2$ Evaluate the root.

            $= 4$ Evaluate the root. $= 4$ Evaluate the exponential form.

b. $625^{3/4}$

Solution We could rewrite $625^{3/4}$ as $\sqrt[4]{625^3}$, but calculating 625^3 is tedious without a calculator, so we will use the other form of the rule.

$$625^{3/4} = (\sqrt[4]{625})^3 \quad \text{Rewrite.}$$
$$= (5)^3 \quad \text{Evaluate the root.}$$
$$= 125 \quad \text{Evaluate the exponential form.}$$

ANSWERS

a. $\sqrt{49} = 7$
b. $\sqrt[4]{625} = 5$
c. $\sqrt[3]{-64} = -4$
d. $-\sqrt{81} = -9$
e. $\sqrt[4]{y}$ **f.** $\sqrt{25a^8} = 5a^4$
g. $18\sqrt[5]{n}$ **h.** $\sqrt{\dfrac{49}{x^{10}}} = \dfrac{7}{x^5}$

Warning: In the expression $-9^{5/2}$, the negative sign tells us to find the opposite of the value of the exponential form.

$$-9^{5/2} = -(9^{5/2}) = -(\sqrt{9})^5$$

Or, some people find it helpful to think of $-9^{5/2}$ as $-1 \cdot 9^{5/2}$:

$$-9^{5/2} = -1 \cdot 9^{5/2}$$
$$= -1 \cdot (\sqrt{9})^5$$

c. $-9^{5/2}$

Solution $-9^{5/2} = -(\sqrt{9})^5 = -(3)^5 = -243$

d. $\left(\dfrac{1}{9}\right)^{5/2}$

Solution $\left(\dfrac{1}{9}\right)^{5/2} = \left(\sqrt{\dfrac{1}{9}}\right)^5 = \left(\dfrac{1}{3}\right)^5 = \dfrac{1}{243}$

e. $x^{2/3}$

Solution $x^{2/3} = \sqrt[3]{x^2}$

◀ **Note:** *Since the base is a variable, we put the exponent beneath the radical sign.*

f. $(3x - 5)^{3/5}$

Solution $(3x - 5)^{3/5} = \sqrt[5]{(3x - 5)^3}$

◀ **Note:** *The simplification is usually easier if we write $a^{m/n}$ as $(\sqrt[n]{a})^m$ if a is a constant, and as $\sqrt[n]{a^m}$ if a is a variable.*

YOUR TURN Simplify.

a. $32^{3/5}$　　**b.** $-49^{3/2}$　　**c.** $\left(\dfrac{1}{16}\right)^{3/2}$　　**d.** $y^{5/6}$　　**e.** $(3a + 4)^{4/5}$

Negative Rational Exponents

Recall from Section 5.1 that $a^{-b} = \dfrac{1}{a^b}$. For example, $2^{-3} = \dfrac{1}{2^3}$. This same rule applies to negative rational exponents.

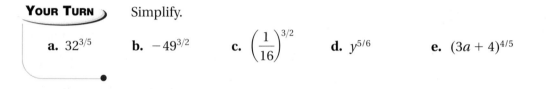

RULE Negative Rational Exponents

$a^{-m/n} = \dfrac{1}{a^{m/n}}$, where $a \neq 0$ and m and n are natural numbers with $n \neq 1$.

EXAMPLE 3 Rewrite using radicals, then simplify, if possible.

a. $81^{-1/2}$

Solution $81^{-1/2} = \dfrac{1}{81^{1/2}}$ 　　Rewrite the exponential form with a positive exponent by inverting the base and changing the sign of the exponent.

$= \dfrac{1}{\sqrt{81}}$ 　　Write the rational exponent in radical form.

$= \dfrac{1}{9}$ 　　Evaluate the square root.

ANSWERS

a. 8 　　　　**b.** -343

c. $\dfrac{1}{64}$ 　　**d.** $\sqrt[6]{y^5}$

e. $\sqrt[5]{(3a + 4)^4}$

b. $16^{-3/4}$

Solution $16^{-3/4} = \dfrac{1}{16^{3/4}}$ Rewrite the exponential form with a positive exponent by inverting the base and changing the sign of the exponent.

$= \dfrac{1}{(\sqrt[4]{16})^3}$ Write the rational exponent in radical form.

$= \dfrac{1}{(2)^3}$ Evaluate the radical.

$= \dfrac{1}{8}$ Simplify the exponential form.

Connection Though we used $a^{-m/n} = \dfrac{1}{a^{m/n}}$ in part c, we could have used $\left(\dfrac{a}{b}\right)^{-n} = \left(\dfrac{b}{a}\right)^n$, which we learned in Chapter 5.

c. $\left(\dfrac{16}{25}\right)^{-1/2}$

Solution $\left(\dfrac{16}{25}\right)^{-1/2} = \dfrac{1}{\left(\dfrac{16}{25}\right)^{1/2}}$ Rewrite the exponential form with a positive exponent by inverting the base and changing the sign of the exponent.

$= \dfrac{1}{\sqrt{\dfrac{16}{25}}}$ Write the rational exponent in radical form.

$= \dfrac{1}{\dfrac{4}{5}}$ Evaluate the square root.

$= \dfrac{5}{4}$ Simplify the complex fraction.

d. $(-64)^{-2/3}$

Solution $(-64)^{-2/3} = \dfrac{1}{(-64)^{2/3}}$ Rewrite the exponential form with a positive exponent.

$= \dfrac{1}{(\sqrt[3]{-64})^2}$ Write the rational exponent in radical form.

$= \dfrac{1}{(-4)^2}$ Evaluate the cube root.

$= \dfrac{1}{16}$ Square -4.

YOUR TURN   Simplify.

   a. $49^{-1/2}$ **b.** $-27^{-2/3}$ **c.** $\left(\dfrac{16}{81}\right)^{-3/4}$ **d.** $(-8)^{-5/3}$

Of Interest

John Wallis (1616–1703) was one of the first mathematicians to explain rational and negative exponents. Wallis wrote extensively about physics and mathematics but, unfortunately, his work was overshadowed by the work of his countryman Isaac Newton. Newton added to what Wallis began with exponents and roots and popularized the notation that we use today. (*Source:* D. E. Smith, *History of Mathematics*, Dover, 1953.)

ANSWERS

a. $\dfrac{1}{7}$ **b.** $-\dfrac{1}{9}$

c. $\dfrac{27}{8}$ **d.** $-\dfrac{1}{32}$

OBJECTIVE 2. Write radicals as expressions raised to rational exponents. In upper-level math courses, it is often necessary to change radical expressions into exponential form. To do so, we use the facts that $\sqrt[n]{a^m} = a^{m/n}$ and $(\sqrt[n]{a})^m = a^{m/n}$.

> **EXAMPLE 4** Write each of the following in exponential form.
>
> **a.** $\sqrt[4]{x^3}$ **b.** $\dfrac{1}{\sqrt[3]{x^2}}$
>
> **Solution** $\sqrt[4]{x^3} = x^{3/4}$ **Solution** $\dfrac{1}{\sqrt[3]{x^2}} = \dfrac{1}{x^{2/3}} = x^{-2/3}$
>
> **c.** $(\sqrt[5]{z})^3$ **d.** $\sqrt[6]{(3x-5)^5}$
>
> **Solution** $(\sqrt[5]{z})^3 = z^{3/5}$ **Solution** $\sqrt[6]{(3x-5)^5} = (3x-5)^{5/6}$

> **YOUR TURN** Write each of the following in exponential form.
>
> **a.** $\sqrt[5]{y^2}$ **b.** $\dfrac{1}{\sqrt[4]{x^3}}$ **c.** $(\sqrt[7]{a})^3$ **d.** $\sqrt[3]{(3y-2)^5}$

OBJECTIVE 3. Simplify expressions with rational number exponents using the rules of exponents. Rational exponents follow the same rules that we established in Chapter 5 for integer exponents, which we review here.

RULE   Rules of Exponents Summary

(Assume that no denominators are 0, that a and b are real numbers, and that m and n are integers.)

Zero as an exponent: $a^0 = 1$, where $a \neq 0$

0^0 is indeterminate.

Negative exponents: $a^{-n} = \dfrac{1}{a^n}$

$\dfrac{1}{a^{-n}} = a^n$

$\left(\dfrac{a}{b}\right)^{-n} = \left(\dfrac{b}{a}\right)^{n}$

Product rule for exponents: $a^m \cdot a^n = a^{m+n}$

Quotient rule for exponents: $a^m \div a^n = a^{m-n}$

Raising a power to a power: $(a^m)^n = a^{mn}$

Raising a product to a power: $(ab)^n = a^n b^n$

Raising a quotient to a power: $\left(\dfrac{a}{b}\right)^n = \dfrac{a^n}{b^n}$

EXAMPLE 5 Use the rules of exponents to simplify. Write answers with positive exponents.

a. $x^{1/5} \cdot x^{3/5}$

Solution $x^{1/5} \cdot x^{3/5} = x^{1/5+3/5}$ Use $a^m \cdot a^n = a^{m+n}$.

$\qquad\qquad\qquad\quad = x^{4/5}$ Add the exponents.

b. $(2a^{1/2})(4a^{1/3})$

Solution $(2a^{1/2})(4a^{1/3}) = 8a^{1/2+1/3}$ Use $a^m \cdot a^n = a^{m+n}$.

$\qquad\qquad\qquad\qquad\quad = 8a^{3/6+2/6}$ Rewrite the exponents with a common denominator of 6.

$\qquad\qquad\qquad\qquad\quad = 8a^{5/6}$ Add the exponents.

c. $\dfrac{5^{2/7}}{5^{6/7}}$

Solution $\dfrac{5^{2/7}}{5^{6/7}} = 5^{2/7-6/7}$ Use $\dfrac{a^m}{a^n} = a^{m-n}$.

$\qquad\qquad\quad = 5^{-4/7}$ Subtract the exponents.

$\qquad\qquad\quad = \dfrac{1}{5^{4/7}}$ Rewrite with a positive exponent.

d. $(-4y^{-3/5})(5y^{4/5})$

Solution $(-4y^{-3/5})(5y^{4/5}) = -20y^{-3/5+4/5}$ Use $a^m \cdot a^n = a^{m+n}$.

$\qquad\qquad\qquad\qquad\qquad = -20y^{1/5}$ Add the exponents.

e. $(7w^{3/4})^2$

Solution $(7w^{3/4})^2 = 7^2 \cdot (w^{3/4})^2$ Use $(ab)^n = a^n \cdot b^n$.

$\qquad\qquad\qquad\quad = 49 \cdot w^{(3/4)2}$ Use $(a^m)^n = a^{mn}$.

$\qquad\qquad\qquad\quad = 49w^{3/2}$ Multiply the exponents.

f. $(2a^{2/3}b^{3/5})^3$

Solution $(2a^{2/3}b^{3/5})^3 = 2^3(a^{2/3})^3(b^{3/5})^3$ Use $(ab)^n = a^n \cdot b^n$.

$\qquad\qquad\qquad\qquad\quad = 8a^{(2/3)3} b^{(3/5)3}$ Use $(a^m)^n = a^{mn}$.

$\qquad\qquad\qquad\qquad\quad = 8a^2 b^{9/5}$ Multiply the exponents.

g. $\dfrac{(3x^{4/3})^3}{x^3}$

Solution $\dfrac{(3x^{4/3})^3}{x^3} = \dfrac{3^3(x^{4/3})^3}{x^3}$ Use $(ab)^n = a^n \cdot b^n$ in the numerator.

$\qquad\qquad\qquad = \dfrac{27x^{(4/3)3}}{x^3}$ Use $(a^m)^n = a^{mn}$.

$\qquad\qquad\qquad = \dfrac{27x^4}{x^3}$ Simplify.

$\qquad\qquad\qquad = 27x^{4-3}$ Use $\dfrac{a^m}{a^n} = a^{m-n}$.

$\qquad\qquad\qquad = 27x$ Simplify.

YOUR TURN Use the rules of exponents to simplify. Write answers with positive exponents.

a. $(3x^{2/3})(-2x^{1/2})$ **b.** $\dfrac{12y^{2/7}}{3y^{5/7}}$ **c.** $(4b^{3/4})^2$ **d.** $(2x^{3/2}y^{2/3})^4$ **e.** $\dfrac{(2z^{5/3})^3}{z^2}$

OBJECTIVE 4. Use rational exponents to simplify radical expressions. Often, radical expressions can be simplified by rewriting them with rational exponents, simplifying, and then rewriting as a radical expression.

EXAMPLE 6 Rewrite as a radical with a smaller root index. Assume all variables represent nonnegative values.

a. $\sqrt[4]{49}$

Solution $\sqrt[4]{49} = 49^{1/4}$ Rewrite in exponential form.

$\qquad = (7^2)^{1/4}$ Write 49 as 7^2.

$\qquad = 7^{2 \cdot 1/4}$ Use $(a^m)^n = a^{mn}$.

$\qquad = 7^{1/2}$ Simplify.

$\qquad = \sqrt{7}$ Write in radical form.

b. $\sqrt[6]{x^4}$

Solution $\sqrt[6]{x^4} = x^{4/6}$ Rewrite in exponential form.

$\qquad = x^{2/3}$ Simplify to lowest terms.

$\qquad = \sqrt[3]{x^2}$ Write in radical form.

c. $\sqrt[6]{a^4 b^2}$

Solution $\sqrt[6]{a^4 b^2} = (a^4 b^2)^{1/6}$ Rewrite in exponential form.

$\qquad = (a^4)^{1/6}(b^2)^{1/6}$ Use $(ab)^n = a^n b^n$.

$\qquad = a^{4 \cdot 1/6} b^{2 \cdot 1/6}$ Use $(a^m)^n = a^{mn}$.

$\qquad = a^{2/3} b^{1/3}$ Simplify.

$\qquad = (a^2 b)^{1/3}$ Use $(ab)^n = a^n b^n$.

$\qquad = \sqrt[3]{a^2 b}$ Write in radical form.

YOUR TURN Rewrite as a radical with a smaller root index. Assume all variables represent nonnegative values.

a. $\sqrt[4]{36}$ **b.** $\sqrt[8]{x^6}$ **c.** $\sqrt[8]{x^6 y^2}$

Writing radicals in exponential form also allows us to multiply and divide radical expressions with different root indices.

EXAMPLE 7 Perform the indicated operations. Write the result using a radical.

a. $\sqrt{x} \cdot \sqrt[3]{x^2}$

Solution $\sqrt{x} \cdot \sqrt[3]{x^2} = x^{1/2} \cdot x^{2/3}$ **Write in exponential form.**

$= x^{1/2+2/3}$ **Use $a^m \cdot a^n = a^{m+n}$.**

$= x^{3/6+4/6}$ **Rewrite the exponents with their LCD.**

$= x^{7/6}$ **Simplify.**

$= \sqrt[6]{x^7}$ **Write in radical form.**

b. $\dfrac{\sqrt[4]{x^3}}{\sqrt[3]{x}}$

Note: *In Section 10.3, we will learn to simplify expressions like $\sqrt[6]{x^7}$ further.*

Solution $\dfrac{\sqrt[4]{x^3}}{\sqrt[3]{x}} = \dfrac{x^{3/4}}{x^{1/3}}$ **Write in exponential form.**

$= x^{3/4-1/3}$ **Use $\dfrac{a^m}{a^n} = a^{m-n}$.**

$= x^{9/12-4/12}$ **Rewrite the exponents with their LCD.**

$= x^{5/12}$ **Simplify.**

$= \sqrt[12]{x^5}$ **Write in radical form.**

c. $\sqrt{3} \cdot \sqrt[3]{2}$

Solution $\sqrt{3} \cdot \sqrt[3]{2} = 3^{1/2} \cdot 2^{1/3}$ **Write in exponential form.**

$= 3^{3/6} \cdot 2^{2/6}$ **Write exponents with their LCD.**

$= (3^3 \cdot 2^2)^{1/6}$ **Use $(ab)^n = a^n b^n$.**

$= (27 \cdot 4)^{1/6}$ **Evaluate 3^3 and 2^2.**

$= 108^{1/6}$ **Multiply.**

$= \sqrt[6]{108}$ **Rewrite as a radical.**

YOUR TURN Perform the indicated operations.

a. $\sqrt[3]{x^2} \cdot \sqrt[4]{x}$ **b.** $\dfrac{\sqrt[4]{a^3}}{\sqrt[3]{a^2}}$ **c.** $\sqrt[4]{2} \cdot \sqrt{5}$

By writing radical expressions using rational exponents, we can also find the root of a root.

EXAMPLE 8 Write $\sqrt[3]{\sqrt{x}}$ with a single radical.

Solution $\sqrt[3]{\sqrt{x}} = (x^{1/2})^{1/3}$ **Write in exponential form.**

$= x^{(1/2)(1/3)}$ **Apply $(a^m)^n = a^{mn}$.**

$= x^{1/6}$ **Simplify.**

$= \sqrt[6]{x}$ **Write as a radical.**

ANSWERS

a. $\sqrt[12]{x^{11}}$ **b.** $\sqrt[12]{a}$ **c.** $\sqrt[4]{50}$

ANSWER

$\sqrt[12]{y}$

YOUR TURN Write $\sqrt[4]{\sqrt[3]{y}}$ with a single radical.

10.2 Exercises

1. If $4^{1/3}$ were written as a radical expression, what would be the radicand?

2. If $4^{3/5}$ were written as a radical expression, what is the root index?

3. If $a^{m/n}$, with n even and m/n reduced, were written as a radical expression, what restrictions would be placed on a? Why?

4. If $a^{m/n}$, with n odd, were written as a radical expression, what restrictions would be placed on a? Why?

5. Does $\sqrt[4]{100} = \sqrt{10}$? Why or why not?

6. Does $\sqrt[3]{3} \cdot \sqrt[5]{2} = \sqrt[15]{1944}$? Why or why not?

For Exercises 7–40, rewrite each of the following using radical notation and evaluate, when possible. Assume all variables represent nonnegative values.

7. $25^{1/2}$

8. $64^{1/2}$

9. $-100^{1/2}$

10. $-64^{1/2}$

11. $27^{1/3}$

12. $125^{1/3}$

13. $(-64)^{1/3}$

14. $(-125)^{1/3}$

15. $y^{1/4}$

16. $w^{1/8}$

17. $(144x^8)^{1/2}$

18. $(9z^{10})^{1/2}$

19. $18r^{1/2}$

20. $22a^{1/2}$

21. $\left(\dfrac{x^4}{121}\right)^{1/2}$

22. $\left(\dfrac{n^8}{36}\right)^{1/2}$

23. $8^{2/3}$

24. $16^{3/4}$

25. $-81^{3/4}$

26. $-16^{5/4}$

27. $(-8)^{4/3}$

28. $(-27)^{5/3}$

29. $16^{-3/2}$

30. $8^{-4/3}$

31. $x^{4/5}$

32. $m^{5/1}$

33. $8n^{2/3}$

34. $6u^{5/6}$

35. $(-32)^{-2/5}$

36. $(-216)^{-2/3}$

37. $\left(\dfrac{1}{25}\right)^{3/2}$

38. $\left(\dfrac{1}{32}\right)^{3/5}$

39. $(2a+4)^{5/6}$

40. $(5r-2)^{5/7}$

For Exercises 41–56, write each of the following in exponential form.

41. $\sqrt[4]{25}$

42. $\sqrt[8]{33}$

43. $\sqrt[6]{z^5}$

44. $\sqrt[7]{r^5}$

45. $\dfrac{1}{\sqrt[6]{5^5}}$

46. $\dfrac{1}{\sqrt[7]{6^2}}$

47. $\dfrac{5}{\sqrt[5]{x^4}}$

48. $\dfrac{8}{\sqrt[7]{n^3}}$

49. $(\sqrt[3]{5})^7$

50. $(\sqrt[5]{8})^3$

51. $(\sqrt[7]{x})^2$

52. $(\sqrt[3]{m})^8$

53. $\sqrt[4]{(4a-5)^7}$

54. $\sqrt[7]{(5w+3)^5}$

55. $(\sqrt[5]{2r-5})^8$

56. $(\sqrt[5]{3r-6})^9$

For Exercises 57–96, use the rules of exponents to simplify. Write answers with positive exponents.

57. $x^{1/5} \cdot x^{3/5}$

58. $n^{1/4} \cdot n^{2/4}$

59. $x^{3/2} \cdot x^{-1/3}$

60. $n^{2/3} \cdot n^{-1/2}$

61. $a^{2/3} \cdot a^{3/4}$

62. $r^{5/2} \cdot r^{5/4}$

63. $(3w^{1/5})(4w^{2/5})$

64. $(5p^{1/9})(3p^{4/9})$

65. $(-3a^{2/3})(4a^{3/4})$

66. $(8c^{4/5})(-4c^{3/2})$

67. $\dfrac{7^{6/5}}{7^{3/5}}$

68. $\dfrac{3^{7/9}}{3^{2/9}}$

69. $\dfrac{x^{2/5}}{x^{4/5}}$

70. $\dfrac{y^{2/7}}{y^{5/7}}$

71. $\dfrac{x^{3/4}}{x^{1/2}}$

72. $\dfrac{x^{5/8}}{x^{1/2}}$

73. $\dfrac{r^{3/4}}{r^{2/3}}$

74. $\dfrac{m^{3/5}}{m^{1/2}}$

75. $\dfrac{x^{-3/7}}{x^{2/7}}$

76. $\dfrac{v^{-4/5}}{v^{2/5}}$

77. $\dfrac{a^{3/4}}{a^{-3/2}}$

78. $\dfrac{b^{4/3}}{b^{-2/3}}$

79. $(5s^{-2/7})(4s^{5/7})$

80. $(6u^{8/9})(-6u^{-5/9})$

81. $(-6b^{-5/4})(4b^{3/2})$

82. $(-6y^{-5/6})(-7y^{5/3})$

83. $(x^{2/3})^3$

84. $(r^{3/4})^4$

85. $(a^{5/6})^2$

86. $(n^{3/8})^4$

87. $(b^{2/3})^{3/5}$

88. $(m^{3/2})^{2/5}$

89. $(2x^{2/3}y^{1/2})^6$

90. $(3a^{1/4}b^{3/2})^4$

91. $(8q^{3/2}t^{3/4})^{1/3}$

92. $(16x^{2/3}y^{1/3})^{3/4}$

93. $\dfrac{(3a^{3/4})^4}{a^2}$

94. $\dfrac{(5v^{5/2})^2}{v^3}$

95. $\dfrac{(9z^{7/3})^{1/2}}{z^{5/6}}$

96. $\dfrac{(36x^{3/4})^{1/2}}{x^{1/8}}$

For Exercises 97–108, represent each of the following as a radical with a smaller root index. Assume all variables represent nonnegative values.

97. $\sqrt[4]{4}$

98. $\sqrt[4]{36}$

99. $\sqrt[6]{25}$

100. $\sqrt[6]{27}$

101. $\sqrt[4]{x^2}$

102. $\sqrt[6]{y^3}$

103. $\sqrt[8]{r^6}$

104. $\sqrt[10]{n^6}$

105. $\sqrt[8]{x^6y^2}$

106. $\sqrt[6]{y^2z^4}$

107. $\sqrt[10]{m^4n^6}$

108. $\sqrt[10]{a^2b^8}$

For Exercises 109–120, perform the indicated operations. Write the result using a radical.

109. $\sqrt[3]{x} \cdot \sqrt{x}$

110. $\sqrt[4]{y} \cdot \sqrt[3]{x^2}$

111. $\sqrt[4]{y^2} \cdot \sqrt[3]{y^2}$

112. $\sqrt[5]{x^4} \cdot \sqrt[3]{x^2}$

113. $\dfrac{\sqrt[3]{x^4}}{\sqrt{x}}$

114. $\dfrac{\sqrt[4]{y^3}}{\sqrt{y}}$

115. $\dfrac{\sqrt[5]{n^4}}{\sqrt[3]{n^2}}$

116. $\dfrac{\sqrt[6]{z^4}}{\sqrt{z}}$

117. $\sqrt{5} \cdot \sqrt[3]{3}$

118. $\sqrt[3]{4} \cdot \sqrt{5}$

119. $\sqrt[4]{6} \cdot \sqrt[3]{2}$

120. $\sqrt[4]{4} \cdot \sqrt[3]{2}$

For Exercises 121–124, write each as a single radical.

121. $\sqrt[3]{\sqrt[3]{x}}$

122. $\sqrt[3]{\sqrt[5]{m}}$

123. $\sqrt{\sqrt[3]{n}}$

124. $\sqrt[4]{\sqrt[5]{z}}$

REVIEW EXERCISES

[1.2]   **1.** Write the prime factorization of 240.

[1.5] *For Exercises 2 and 3, simplify using the order of operations agreement.*

2. $\sqrt{16} \cdot \sqrt{9}$

3. $\sqrt[3]{27} \cdot \sqrt[3]{125}$

[5.4] *For Exercises 4 and 5, simplify.*

4. $(2.5 \times 10^6)(3.2 \times 10^5)$

5. $\left(\dfrac{3}{4}x^3y\right)\left(-\dfrac{5}{6}xyz^2\right)$

[5.6]   **6.** Use long division to find the quotient: $\dfrac{2x^3 - 2x^2 - 19x + 18}{x + 3}$

10.3 Multiplying, Dividing, and Simplifying Radicals

OBJECTIVES

1. Multiply radical expressions.
2. Divide radical expressions.
3. Use the product rule to simplify radical expressions.

In this section, we explore some ways to simplify expressions that involve multiplication or division of radicals.

OBJECTIVE 1. Multiply radical expressions. Consider the expression $\sqrt{9} \cdot \sqrt{16}$. The usual approach would be to find the roots and then multiply those roots. However, we can also multiply the radicands and then find the root of the product.

Find the roots first: $\qquad$ Multiply the radicands first:
$$\sqrt{9} \cdot \sqrt{16} = 3 \cdot 4 = 12 \qquad \sqrt{9} \cdot \sqrt{16} = \sqrt{9 \cdot 16} = \sqrt{144} = 12$$

Both approaches give the same result, suggesting the following rule:

RULE **Product Rule for Radicals**

If $\sqrt[n]{a}$ and $\sqrt[n]{b}$ are both real numbers, then $\sqrt[n]{a} \cdot \sqrt[n]{b} = \sqrt[n]{a \cdot b}$.

EXAMPLE 1 Find the product and write the answer in simplest form. Assume all variables represent nonnegative values.

a. $\sqrt{3} \cdot \sqrt{27}$

Solution $\sqrt{3} \cdot \sqrt{27} = \sqrt{3 \cdot 27} = \sqrt{81} = 9$

b. $\sqrt{11} \cdot \sqrt{x}$

Solution $\sqrt{11} \cdot \sqrt{x} = \sqrt{11x}$

c. $\sqrt[3]{4} \cdot \sqrt[3]{2}$

Solution $\sqrt[3]{4} \cdot \sqrt[3]{2} = \sqrt[3]{4 \cdot 2} = \sqrt[3]{8} = 2$

d. $\sqrt[3]{3x} \cdot \sqrt[3]{4x}$

Solution $\sqrt[3]{3x} \cdot \sqrt[3]{4x} = \sqrt[3]{3x \cdot 4x} = \sqrt[3]{12x^2}$

e. $\sqrt[4]{5} \cdot \sqrt[4]{7x^2}$

Solution $\sqrt[4]{5} \cdot \sqrt[4]{7x^2} = \sqrt[4]{5 \cdot 7x^2} = \sqrt[4]{35x^2}$

f. $\sqrt{\dfrac{5}{x}} \cdot \sqrt{\dfrac{y}{2}}$

Solution $\sqrt{\dfrac{5}{x}} \cdot \sqrt{\dfrac{y}{2}} = \sqrt{\dfrac{5}{x} \cdot \dfrac{y}{2}} = \sqrt{\dfrac{5y}{2x}}$

g. $\sqrt{x} \cdot \sqrt{x}$

Solution $\sqrt{x} \cdot \sqrt{x} = \sqrt{x \cdot x} = \sqrt{x^2} = x$

> ◀ **Connection** Example 1(a) illustrates that the product of two irrational numbers can be a rational number.

> **Connection**
> Expressions that have a fraction in a radical, like $\sqrt{\dfrac{5y}{2x}}$, are not considered to ◀ be in simplest form. We will learn how to simplify them in Section 10.5.

Find the product and write the answer in simplest form.

a. $\sqrt{2} \cdot \sqrt{32}$ **b.** $\sqrt{5} \cdot \sqrt{a}$ **c.** $\sqrt[3]{5x} \cdot \sqrt[3]{2y}$ **d.** $\sqrt{\dfrac{2}{a}} \cdot \sqrt{\dfrac{b}{7}}$

Notice that in Example 1(g), $\sqrt{x} \cdot \sqrt{x} = x$. It is also true that $\sqrt{x} \cdot \sqrt{x} = (\sqrt{x})^2$. Therefore, $(\sqrt{x})^2 = x$. Similarly, $\sqrt[3]{x} \cdot \sqrt[3]{x} \cdot \sqrt[3]{x} = \sqrt[3]{x \cdot x \cdot x} = \sqrt[3]{x^3} = x$. It is also true that $\sqrt[3]{x} \cdot \sqrt[3]{x} \cdot \sqrt[3]{x} = (\sqrt[3]{x})^3$, so $(\sqrt[3]{x})^3 = x$. These examples suggest the following rule.

RULE Raising an nth Root to the nth Power

For any nonnegative real number a, $(\sqrt[n]{a})^n = a$.

This means that $(\sqrt[3]{4})^3 = 4$, $(\sqrt[5]{19})^5 = 19$, and $(\sqrt[4]{3x^2})^4 = 3x^2$.

OBJECTIVE 2. Divide radical expressions.   Earlier, we developed the product rule for radicals. Now, we develop a similar rule for quotients like $\dfrac{\sqrt{100}}{\sqrt{25}}$. We can follow the order of operations and divide the roots, or we can divide the radicands, then find the square root of the quotient.

Find the roots first: Divide the radicands first:

$$\frac{\sqrt{100}}{\sqrt{25}} = \frac{10}{5} = 2 \qquad \frac{\sqrt{100}}{\sqrt{25}} = \sqrt{\frac{100}{25}} = \sqrt{4} = 2$$

Both approaches give the same result, which suggests the following rule.

RULE Quotient Rule for Radicals

If $\sqrt[n]{a}$ and $\sqrt[n]{b}$ are both real numbers, then $\dfrac{\sqrt[n]{a}}{\sqrt[n]{b}} = \sqrt[n]{\dfrac{a}{b}}$, where $b \neq 0$.

As with all equations, this rule can be used going from left to right or right to left.

EXAMPLE 2 Simplify. Assume variables represent nonnegative values.

a. $\sqrt{\dfrac{7}{36}}$

Solution $\sqrt{\dfrac{7}{36}} = \dfrac{\sqrt{7}}{\sqrt{36}} = \dfrac{\sqrt{7}}{6}$

b. $\dfrac{\sqrt{108}}{\sqrt{3}}$

Solution $\dfrac{\sqrt{108}}{\sqrt{3}} = \sqrt{\dfrac{108}{3}} = \sqrt{36} = 6$

ANSWERS

a. 8   **b.** $\sqrt{5a}$

c. $\sqrt[3]{10xy}$ **d.** $\sqrt{\dfrac{2b}{7a}}$

c. $\sqrt[3]{\dfrac{9}{x^3}}$

Solution $\sqrt[3]{\dfrac{9}{x^3}} = \dfrac{\sqrt[3]{9}}{\sqrt[3]{x^3}} = \dfrac{\sqrt[3]{9}}{x}$

d. $\dfrac{\sqrt[3]{15}}{\sqrt[3]{5}}$

Solution $\dfrac{\sqrt[3]{15}}{\sqrt[3]{5}} = \sqrt[3]{\dfrac{15}{5}} = \sqrt[3]{3}$

e. $\sqrt[4]{\dfrac{y}{81}}$

Solution $\sqrt[4]{\dfrac{y}{81}} = \dfrac{\sqrt[4]{y}}{\sqrt[4]{81}} = \dfrac{\sqrt[4]{y}}{3}$

YOUR TURN   Simplify. Assume variables represent nonnegative values.

a. $\sqrt{\dfrac{x}{49}}$ **b.** $\dfrac{\sqrt{75}}{\sqrt{5}}$ **c.** $\sqrt[4]{\dfrac{6}{x^4}}$ **d.** $\dfrac{\sqrt[3]{32}}{\sqrt[3]{4}}$

OBJECTIVE 3. Use the product rule to simplify radical expressions.   Several conditions exist in which a radical is not considered to be in simplest form. One such condition is if a radicand has a factor that can be written to a power greater than or equal to the index. For example, $\sqrt[3]{81}$ is not in simplest form because the perfect cube 27 is a factor of 81. Our first step in simplifying $\sqrt[3]{81}$ would be to rewrite it as $\sqrt[3]{27 \cdot 3}$ so that we can then use the product rule for radicals.

Note: *We will explore other conditions that require simplification in future sections.*

PROCEDURE   Simplifying nth Roots

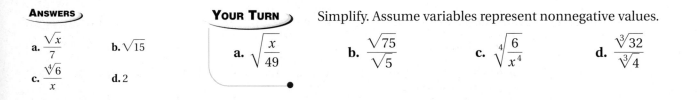

1. Write the radicand as a product of the greatest possible perfect nth power and a number or expression that has no perfect nth power factors.
2. Use the product rule $\sqrt[n]{ab} = \sqrt[n]{a} \cdot \sqrt[n]{b}$, where a is a perfect nth power.
3. Find the nth root of the perfect nth power radicand.

Connection   A list of perfect powers may be helpful.

Perfect squares:
1, 4, 9, 16, 25, 36, 49, 64, 81, 100, . . .

Perfect cubes:
1, 8, 27, 64, 125, 216, . . .

Perfect fourth powers:
1, 16, 81, 256, 625, . . .

EXAMPLE 3   Simplify.

a. $\sqrt{18}$

Solution $\sqrt{18} = \sqrt{9 \cdot 2}$ The greatest perfect square factor of 18 is 9, so we write 18 as $9 \cdot 2$.

$= \sqrt{9} \cdot \sqrt{2}$ Use the product rule of roots to separate the factors into two radicals.

$= 3\sqrt{2}$ Simplify the square root of 9.

b. $5\sqrt{72}$

Solution $5\sqrt{72} = 5 \cdot \sqrt{36 \cdot 2}$ The greatest perfect square factor of 72 is 36, so we write 72 as 36 · 2.

$$= 5 \cdot \sqrt{36} \cdot \sqrt{2}$$ Use the product rule of roots to separate the factors into two radicals.

$$= 5 \cdot 6 \cdot \sqrt{2}$$ Simplify the square root of 36.

$$= 30\sqrt{2}$$ Multiply 5 · 6.

Note: *We can use perfect n^{th} power factors other than the greatest perfect n^{th} power factor. For example, in simplifying $\sqrt{72}$, instead of $\sqrt{72} = \sqrt{36 \cdot 2} = 6\sqrt{2}$, we could write*

$$\sqrt{72} = \sqrt{4 \cdot 18} = 2\sqrt{18} = 2\sqrt{9 \cdot 2} = 2 \cdot 3\sqrt{2} = 6\sqrt{2}$$

Notice that using the greatest perfect n^{th} factor saves steps.

c. $\sqrt[3]{40}$

Solution $\sqrt[3]{40} = \sqrt[3]{8 \cdot 5}$ The greatest perfect cube factor of 40 is 8.

$$= \sqrt[3]{8} \cdot \sqrt[3]{5}$$ Use the product rule of roots.

$$= 2\sqrt[3]{5}$$ Simplify the cube root of 8.

d. $4\sqrt[4]{162}$

Solution $4\sqrt[4]{162} = 4\sqrt[4]{81 \cdot 2}$ The greatest perfect fourth power factor of 162 is 81.

$$= 4\sqrt[4]{81} \cdot \sqrt[4]{2}$$ Use the product rule of roots.

$$= 4 \cdot 3 \cdot \sqrt[4]{2}$$ Simplify the fourth root of 81.

$$= 12\sqrt[4]{2}$$

YOUR TURN   Simplify.

   a. $\sqrt{150}$ **b.** $6\sqrt{80}$ **c.** $\sqrt[3]{108}$ **d.** $3\sqrt[4]{80}$

Using Prime Factorization

If the greatest perfect n^{th} power of a particular radicand is not obvious, then try using the prime factorization of the radicand. Each prime factor that appears twice will have a square root equal to one of the two factors, each prime factor that appears three times will have a cube root that is one of the three factors, and so on. The remaining factors stay in the radical sign.

EXAMPLE 4   Simplify the following radicals using prime factorizations.

   a. $\sqrt{375}$

Solution $\sqrt{375} = \sqrt{5 \cdot 5 \cdot 5 \cdot 3}$ Write 375 as the product of its prime factors.

$$= 5\sqrt{3 \cdot 5}$$ The square root of the pair of 5s is 5.

$$= 5\sqrt{15}$$ Multiply the prime factors in the radicand.

ANSWERS

a. $5\sqrt{6}$ **b.** $24\sqrt{5}$

c. $3\sqrt[3]{4}$ **d.** $6\sqrt[4]{5}$

b. $\sqrt[3]{324}$

Solution $\sqrt[3]{324} = \sqrt[3]{3 \cdot 3 \cdot 3 \cdot 3 \cdot 2 \cdot 2}$ Write 324 as the product of its prime factors.

$= 3\sqrt[3]{3 \cdot 2 \cdot 2}$ The cube root of the three 3s is 3.

$= 3\sqrt[3]{12}$ Multiply the prime factors in the radicand.

c. $\sqrt[4]{240}$

Solution $\sqrt[4]{240} = \sqrt[4]{2 \cdot 2 \cdot 2 \cdot 2 \cdot 3 \cdot 5}$ Write 240 as the product of its prime factors.

$= 2\sqrt[4]{3 \cdot 5}$ The fourth root of the four 2s is 2.

$= 2\sqrt[4]{15}$ Multiply the prime factors in the radicand.

YOUR TURN   Simplify the following radicals using prime factorizations.

a. $\sqrt{294}$ **b.** $\sqrt[3]{324}$ **c.** $\sqrt[4]{486}$

Simplifying Radicals with Variables

We can use either procedure to simplify radicals whose radicands contain variables. Since $\sqrt[n]{a^m} = a^{m/n}$, we can find an exact root if n divides into m evenly. For example, $\sqrt{x^4} = x^{4/2} = x^2$ and $\sqrt[3]{x^{12}} = x^{12/3} = x^4$. Therefore, to find the nth root we rewrite the radicand as a product in which one factor has the greatest possible exponent divisible by the index n.

EXAMPLE 5   Simplify. Assume variables represent nonnegative values.

a. $\sqrt{x^7}$

Solution $\sqrt{x^7} = \sqrt{x^6 \cdot x}$ The greatest number smaller than 7 that is divisible by 2 is 6, so write x^7 as $x^6 \cdot x$.

$= \sqrt{x^6} \cdot \sqrt{x}$ Use the product rule of roots.

$= x^3\sqrt{x}$ Simplify: $\sqrt{x^6} = x^3$.

b. $3\sqrt{24a^5b^9}$

Solution $3\sqrt{24a^5b^9} = 3\sqrt{4 \cdot 6 \cdot a^4 \cdot a \cdot b^8 \cdot b}$ Write 24 as $4 \cdot 6$, a^5 as $a^4 \cdot a$, and b^9 as $b^8 \cdot b$.

$= 3\sqrt{4a^4b^8 \cdot 6ab}$ Regroup the factors so that perfect squares are together.

$= 3\sqrt{4a^4b^8} \cdot \sqrt{6ab}$ Use the product rule of roots.

$= 3 \cdot 2a^2b^4 \cdot \sqrt{6ab}$ Simplify: $\sqrt{4} = 2$, $\sqrt{a^4} = a^2$, and $\sqrt{b^8} = b^4$.

$= 6a^2b^4\sqrt{6ab}$ Multiply: $3 \cdot 2 = 6$.

c. $y^2\sqrt[3]{y^8}$

Solution $y^2\sqrt[3]{y^8} = y^2\sqrt[3]{y^6 \cdot y^2}$ The greatest number smaller than 8 that is divisible by 3 is 6, so write y^8 as $y^6 \cdot y^2$.

$= y^2\sqrt[3]{y^6} \cdot \sqrt[3]{y^2}$ Use the product rule of roots.

$= y^2 \cdot y^2 \cdot \sqrt[3]{y^2}$ Simplify: $\sqrt[3]{y^6} = y^2$.

$= y^4\sqrt[3]{y^2}$ Multiply: $y^2 \cdot y^2 = y^4$.

ANSWERS

a. $7\sqrt{6}$ **b.** $3\sqrt[3]{12}$

c. $3\sqrt[4]{6}$

d. $\sqrt[5]{64x^9y^{12}}$

Solution $\sqrt[5]{64x^9y^{12}} = \sqrt[5]{32 \cdot 2 \cdot x^5 \cdot x^4 \cdot y^{10} \cdot y^2}$ Write 64 as $32 \cdot 2$, x^9 as $x^5 \cdot x^4$, and y^{12} as $y^{10} \cdot y^2$.

$= \sqrt[5]{32x^5y^{10} \cdot 2x^4y^2}$ Regroup the factors.

$= \sqrt[5]{32x^5y^{10}} \cdot \sqrt[5]{2x^4y^2}$ Use the product rule of roots.

$= 2xy^2\sqrt[5]{2x^4y^2}$ Simplify: $\sqrt[5]{32} = 2$, $\sqrt[5]{x^5} = x$, and $\sqrt[5]{y^{10}} = y^2$.

YOUR TURN   Simplify. Assume variables represent nonnegative values.

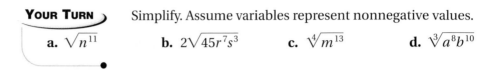

a. $\sqrt{n^{11}}$ **b.** $2\sqrt{45r^7s^3}$ **c.** $\sqrt[4]{m^{13}}$ **d.** $\sqrt[3]{a^8b^{10}}$

After using the product or quotient rules, it is often necessary to simplify the results.

EXAMPLE 6   Find the product or quotient and simplify the results. Assume variables represent nonnegative values.

a. $\sqrt{3} \cdot \sqrt{6}$

Solution $\sqrt{3} \cdot \sqrt{6} = \sqrt{18}$ Use the product rule of roots to multiply.

$= \sqrt{9 \cdot 2}$ Write 18 as $9 \cdot 2$.

$= 3\sqrt{2}$ Simplify: $\sqrt{9} = 3$.

b. $5\sqrt{3x^3} \cdot 3\sqrt{15x^2}$

Solution $5\sqrt{3x^3} \cdot 3\sqrt{15x^2} = 5 \cdot 3\sqrt{3x^3} \cdot \sqrt{15x^2}$ Regroup the factors.

$= 15\sqrt{45x^5}$ Multiply.

$= 15\sqrt{9 \cdot 5 \cdot x^4 \cdot x}$ Write 45 as $9 \cdot 5$ and x^5 as $x^4 \cdot x$.

$= 15 \cdot 3x^2\sqrt{5x}$ Simplify: $\sqrt{9} = 3$ and $\sqrt{x^4} = x^2$.

$= 45x^2\sqrt{5x}$ Multiply.

c. $\dfrac{\sqrt{288}}{\sqrt{6}}$

Solution $\dfrac{\sqrt{288}}{\sqrt{6}} = \sqrt{\dfrac{288}{6}}$ Use the quotient rule of roots.

$= \sqrt{48}$ Divide the radicand.

$= \sqrt{16 \cdot 3}$ Write 48 as $16 \cdot 3$.

$= 4\sqrt{3}$ Simplify: $\sqrt{16} = 3$.

ANSWERS

a. $n^5\sqrt{n}$ **b.** $6r^3s\sqrt{5rs}$

c. $m^3\sqrt[4]{m}$ **d.** $a^2b^3\sqrt[3]{a^2b}$

d. $\dfrac{8\sqrt{756a^8b^5}}{2\sqrt{7a^4b^2}}$

$\text{Solution } \dfrac{8\sqrt{756a^8b^5}}{2\sqrt{7a^4b^2}} = 4\sqrt{\dfrac{756a^8b^5}{7a^4b^2}}$ **Divide coefficients and use the quotient rule of radicals.**

$\qquad\qquad\qquad = 4\sqrt{108a^4b^3}$ **Divide the radicand.**

$\qquad\qquad\qquad = 4\sqrt{36a^4b^2 \cdot 3b}$ **Rewrite the radicand with a perfect square factor.**

$\qquad\qquad\qquad = 4 \cdot 6a^2b\sqrt{3b}$ **Find the square roots.**

$\qquad\qquad\qquad = 24a^2b\sqrt{3b}$ **Multiply.**

YOUR TURN   Find the product or quotient and simplify the results. Assume variables represent nonnegative values.

a. $\sqrt{6} \cdot \sqrt{15}$ **b.** $4\sqrt{14x^3} \cdot 2\sqrt{6x^4}$ **c.** $\dfrac{\sqrt{1296}}{\sqrt{12}}$ **d.** $\dfrac{14\sqrt{315x^{11}y^8}}{2\sqrt{5x^6y^5}}$

ANSWERS
a. $3\sqrt{10}$ **b.** $16x^3\sqrt{21x}$
c. $6\sqrt{3}$ **d.** $21x^2y\sqrt{7xy}$

10.3 Exercises

1. For $\sqrt{8} \cdot \sqrt{18}$, explain the difference between using the product rule for radicals and multiplying the approximate roots of 8 and 18.

2. Explain why $\sqrt{28}$ is not in simplest form.

3. Describe in your own words how to simplify a cube root containing a radicand with a perfect cube factor.

4. Explain why the expression $3x^2\sqrt[3]{x^5}$ is not in simplest form.

For Exercises 5–32, find the product and write the answer in simplest form. Assume variables represent nonnegative values.

5. $\sqrt{2} \cdot \sqrt{32}$ **6.** $\sqrt{3} \cdot \sqrt{12}$ **7.** $\sqrt{3x} \cdot \sqrt{27x^5}$ **8.** $\sqrt{8y^3} \cdot \sqrt{2y}$

9. $\sqrt{6xy^3} \cdot \sqrt{24xy}$ **10.** $\sqrt{50u^3v^2} \cdot \sqrt{2uv^4}$ **11.** $\sqrt{2} \cdot \sqrt{7}$ **12.** $\sqrt{6} \cdot \sqrt{11}$

13. $\sqrt{15} \cdot \sqrt{x}$

14. $\sqrt{17} \cdot \sqrt{y}$

15. $\sqrt[3]{3} \cdot \sqrt[3]{9}$

16. $\sqrt[3]{4} \cdot \sqrt[3]{16}$

17. $\sqrt[3]{5y} \cdot \sqrt[3]{2y}$

18. $\sqrt[3]{6m} \cdot \sqrt[3]{2m}$

19. $\sqrt[4]{3} \cdot \sqrt[4]{7}$

20. $\sqrt[4]{8} \cdot \sqrt[4]{5}$

21. $\sqrt[4]{12w^3} \cdot \sqrt[4]{6w}$

22. $\sqrt[4]{21r^3} \cdot \sqrt[4]{7r}$

23. $\sqrt[4]{3x^2y} \cdot \sqrt[4]{5xy^2}$

24. $\sqrt[4]{ab^2} \cdot \sqrt[4]{ab}$

25. $\sqrt[5]{6x^3} \cdot \sqrt[5]{5x}$

26. $\sqrt[5]{3m^2} \cdot \sqrt[5]{8m^2}$

27. $\sqrt[6]{4x^2y^3} \cdot \sqrt[6]{2x^3y}$

28. $\sqrt[6]{ab^3} \cdot \sqrt[6]{7a^3b^2}$

29. $\sqrt{\dfrac{7}{2}} \cdot \sqrt{\dfrac{3}{5}}$

30. $\sqrt{\dfrac{5}{2}} \cdot \sqrt{\dfrac{11}{3}}$

31. $\sqrt{\dfrac{6}{x}} \cdot \sqrt{\dfrac{y}{5}}$

32. $\sqrt{\dfrac{a}{3}} \cdot \sqrt{\dfrac{7}{b}}$

For Exercises 33–48, use the quotient rule to simplify. Assume variables represent nonnegative values.

33. $\sqrt{\dfrac{25}{36}}$

34. $\sqrt{\dfrac{49}{64}}$

35. $\sqrt{\dfrac{10}{9}}$

36. $\sqrt{\dfrac{15}{81}}$

37. $\dfrac{\sqrt{196}}{\sqrt{4}}$

38. $\dfrac{\sqrt{243}}{\sqrt{3}}$

39. $\dfrac{\sqrt{15}}{\sqrt{5}}$

40. $\dfrac{\sqrt{21}}{\sqrt{3}}$

41. $\sqrt[3]{\dfrac{4}{w^6}}$

42. $\sqrt[3]{\dfrac{7}{v^3}}$

43. $\sqrt[3]{\dfrac{5y^2}{27x^9}}$

44. $\sqrt[3]{\dfrac{5a}{8r^6}}$

45. $\dfrac{\sqrt[3]{320}}{\sqrt[3]{5}}$

46. $\dfrac{\sqrt[3]{162}}{\sqrt[3]{6}}$

47. $\sqrt[4]{\dfrac{3u^3}{16x^8}}$

48. $\sqrt[4]{\dfrac{3x^2}{81y^4}}$

For Exercises 49–84, simplify. Assume variables represent nonnegative values.

49. $\sqrt{98}$

50. $\sqrt{48}$

51. $\sqrt{128}$

52. $\sqrt{180}$

53. $6\sqrt{80}$

54. $4\sqrt{50}$

55. $5\sqrt{112}$

56. $3\sqrt{245}$

57. $\sqrt{a^3}$

58. $\sqrt{d^5}$

59. $\sqrt{x^2y^4}$

60. $\sqrt{a^6b^2}$

61. $\sqrt{x^6y^8z^{10}}$

62. $\sqrt{p^4q^8r^8}$

63. $rs^2\sqrt{r^9s^5}$

64. $a^2b^3\sqrt{a^{11}b^3}$

65. $3\sqrt{72x^5}$

66. $6\sqrt{75d^3}$

67. $\sqrt[3]{32}$

68. $\sqrt[3]{54}$

69. $\sqrt[3]{x^7}$

70. $\sqrt[3]{b^{11}}$

71. $\sqrt[3]{x^6 y^5}$

72. $\sqrt[3]{m^{13} n^9}$

73. $\sqrt[3]{128 z^8}$

74. $\sqrt[3]{48 h^{14}}$

75. $2\sqrt[3]{24}$

76. $4\sqrt[3]{250}$

77. $\sqrt[4]{80}$

78. $\sqrt[4]{162}$

79. $3x^2 \sqrt[4]{243 x^9}$

80. $3a^4 \sqrt[4]{48 a^7}$

81. $\sqrt[5]{486 x^{16}}$

82. $\sqrt[5]{160 n^{18}}$

83. $\sqrt[6]{x^8 y^{14} z^{11}}$

84. $\sqrt[7]{a^{16} b^9 c^{12}}$

For Exercises 85–94, find the product and write the answer in simplest form. Assume variables represent nonnegative values.

85. $\sqrt{3} \cdot \sqrt{21}$

86. $\sqrt{5} \cdot \sqrt{15}$

87. $5\sqrt{10} \cdot 3\sqrt{14}$

88. $2\sqrt{6} \cdot 5\sqrt{21}$

89. $\sqrt{y^3} \cdot \sqrt{y^2}$

90. $\sqrt{m^7} \cdot \sqrt{m^4}$

91. $x\sqrt{x^2 y^3} \cdot y^2 \sqrt{x^4 y^4}$

92. $x\sqrt{x^5 y^2} \cdot y\sqrt{xy^3}$

93. $4\sqrt{6c^3} \cdot 3\sqrt{10c^5}$

94. $6\sqrt{15c^2} \cdot 2\sqrt{10c^5}$

For Exercises 95 and 96, write an expression in simplest form for the area of the figure.

95.

96.

![Figure for exercises 95 and 96: a rectangle with height $4\sqrt{3}$ and base $5\sqrt{6}$, and a parallelogram with height $3\sqrt{5}$ and base $6\sqrt{10}$.]

For Exercises 97–108, find the quotient and write the answer in simplest form. Assume variables represent nonnegative values.

97. $\dfrac{\sqrt{48}}{\sqrt{6}}$

98. $\dfrac{\sqrt{54}}{\sqrt{3}}$

99. $\dfrac{9\sqrt{160}}{3\sqrt{8}}$

100. $\dfrac{10\sqrt{280}}{2\sqrt{10}}$

101. $\dfrac{\sqrt{c^5 d^4}}{\sqrt{cd^3}}$

102. $\dfrac{\sqrt{m^6 n^5}}{\sqrt{m^3 n^3}}$

103. $\dfrac{8\sqrt{45 a^5}}{2\sqrt{5a}}$

104. $\dfrac{6\sqrt{48 n^7}}{3\sqrt{3 n^3}}$

105. $\dfrac{12\sqrt{72 c^5}}{4\sqrt{6 c^2}}$

106. $\dfrac{15\sqrt{48 a^7}}{5\sqrt{2 a^2}}$

107. $\dfrac{36\sqrt{96 x^6 y^{11}}}{4\sqrt{3 x^2 y^4}}$

108. $\dfrac{54\sqrt{240 r^9 s^{10}}}{9\sqrt{5 r^6 s^4}}$

For Exercises 109–114, find the products and write the answers in simplest form. Assume variables represent nonnegative values.

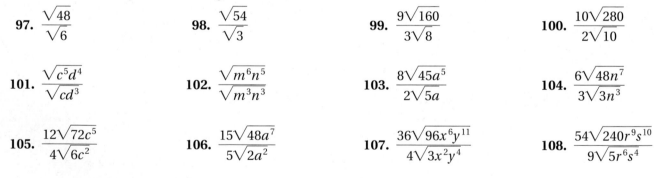

109. $\sqrt{\dfrac{3}{7}} \cdot \sqrt{\dfrac{8}{7}}$

110. $\sqrt{\dfrac{8}{5}} \cdot \sqrt{\dfrac{6}{5}}$

111. $\sqrt{\dfrac{a^3}{2}} \cdot \sqrt{\dfrac{a^5}{2}}$

112. $\sqrt{\dfrac{c^7}{6}} \cdot \sqrt{\dfrac{c^5}{6}}$

113. $\sqrt{\dfrac{3x^5}{2}} \cdot \sqrt{\dfrac{15x^5}{8}}$

114. $\sqrt{\dfrac{5y^3}{3}} \cdot \sqrt{\dfrac{10y^3}{27}}$

For Exercises 115 and 116, write an expression in simplest form for the area of the figure.

115.

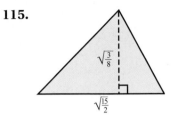

116.

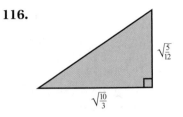

REVIEW EXERCISES

[1.7] *For Exercises 1 and 2, use a = 6 and b = 8.*

1. Does $\sqrt{a^2 + b^2} = \sqrt{a^2} + \sqrt{b^2} = a + b$?

2. Does $\sqrt{(a + b)^2} = a + b$?

3. Simplify: $6x^2 - 4x - 3x + 2x^2$

[5.5] *For Exercises 4–6, multiply.*

4. $(2a - 3b)(4a + 3b)$

5. $(3m + 5n)(3m - 5n)$

6. $(2x - 3y)^2$

10.4 Adding, Subtracting, and Multiplying Radical Expressions

OBJECTIVES

1. Add or subtract like radicals.
2. Use the distributive property in expressions containing radicals.
3. Simplify radical expressions that contain mixed operations.

OBJECTIVE 1. Add or subtract like radicals. Recall that like terms such as $-8a^2$ and $3a^2$ have identical variables and identical exponents. Similarly, **like radicals** have identical radicands and indexes.

DEFINITION *Like radicals:* Radical expressions with identical radicands and identical indexes.

The radicals $3\sqrt{2}$ and $5\sqrt{2}$ are like. The radicals $3\sqrt{2}$ and $5\sqrt{3}$ are unlike because their radicands are different. The radicals $3\sqrt{2}$ and $5\sqrt[3]{2}$ are unlike because their indexes are different.

Adding or subtracting like radicals is essentially the same as combining like terms. Remember that the distributive property is at work when we combine like terms because we factor out the common variable, which leaves us a sum or difference of the coefficients.

Like terms: Like radicals:

$$3x + 4x = (3 + 4)x \qquad 3\sqrt{5} + 4\sqrt{5} = (3 + 4)\sqrt{5}$$
$$= 7x \qquad\qquad\qquad\qquad = 7\sqrt{5}$$

Our example suggests the following procedure for adding like radical expressions.

PROCEDURE   Adding Like Radicals

To add or subtract like radicals, add or subtract the coefficients and keep the radicals the same.

EXAMPLE 1   Simplify.

a. $7\sqrt{5} + 2\sqrt{5}$

Solution $7\sqrt{5} + 2\sqrt{5} = (7 + 2)\sqrt{5}$
$$= 9\sqrt{5}$$

b. $6\sqrt[3]{4} + \sqrt[3]{4}$

Solution $6\sqrt[3]{4} + \sqrt[3]{4} = (6 + 1)\sqrt[3]{4}$
$$= 7\sqrt[3]{4}$$

c. $6x\sqrt[4]{3x} - 2x\sqrt[4]{3x}$

Solution $6x\sqrt[4]{3x} - 2x\sqrt[4]{3x} = (6x - 2x)\sqrt[4]{3x}$
$$= 4x\sqrt[4]{3x}$$

d. $8\sqrt{3} + 6\sqrt{2} - 5\sqrt{3} + 3\sqrt{2}$

Solution $8\sqrt{3} + 6\sqrt{2} - 5\sqrt{3} + 3\sqrt{2} = 8\sqrt{3} - 5\sqrt{3} + 6\sqrt{2} + 3\sqrt{2}$ **Regroup the terms.**
$$= (8 - 5)\sqrt{3} + (6 + 3)\sqrt{2}$$
$$= 3\sqrt{3} + 9\sqrt{2}$$

Warning: Never add radicands! Consider $\sqrt{9} + \sqrt{16}$. It might be tempting to add the radicands to get $\sqrt{25}$, but $\sqrt{9} + \sqrt{16} \neq \sqrt{25}$. If we calculate each side separately, we see why they are not equivalent.

$$\sqrt{9} + \sqrt{16} \neq \sqrt{25}$$
$$3 + 4 \neq 5$$
$$7 \neq 5$$

YOUR TURN   Simplify.

a. $5\sqrt{6} + 2\sqrt{6}$

b. $5\sqrt[4]{5x^2} + \sqrt[4]{5x^2}$

c. $7y\sqrt[3]{7} - 12y\sqrt[3]{7}$

d. $9x\sqrt{5x} - 2y\sqrt{3y} - 6x\sqrt{5x} + 7y\sqrt{3y}$

ANSWERS

a. $7\sqrt{6}$
b. $6\sqrt[4]{5x^2}$
c. $-5y\sqrt[3]{7}$
d. $3x\sqrt{5x} + 5y\sqrt{3y}$

In a problem involving addition or subtraction of radicals, if the radicals are not like, it may be possible to simplify one or more of the radicals so that they are like.

EXAMPLE 2 Simplify. Assume variables represent nonnegative quantities.

a. $7\sqrt{3} + \sqrt{12}$

Solution
$$\begin{aligned}
7\sqrt{3} + \sqrt{12} &= 7\sqrt{3} + \sqrt{4 \cdot 3} && \text{Factor out the perfect square factor, 4.}\\
&= 7\sqrt{3} + \sqrt{4} \cdot \sqrt{3} && \text{Use the product rule to separate the radicals.}\\
&= 7\sqrt{3} + 2\sqrt{3} && \text{Simplify.}\\
&= 9\sqrt{3} && \text{Combine like radicals.}
\end{aligned}$$

b. $3\sqrt[3]{24} - 2\sqrt[3]{3}$

Solution
$$\begin{aligned}
3\sqrt[3]{24} - 2\sqrt[3]{3} &= 3\sqrt[3]{8 \cdot 3} - 2\sqrt[3]{3} && \text{Rewrite 24 as } 8 \cdot 3.\\
&= 3\sqrt[3]{8} \cdot \sqrt[3]{3} - 2\sqrt[3]{3} && \text{Use the product rule.}\\
&= 3 \cdot 2\sqrt[3]{3} - 2\sqrt[3]{3} && \text{Simplify } \sqrt[3]{8}.\\
&= 6\sqrt[3]{3} - 2\sqrt[3]{3} && \text{Multiply.}\\
&= 4\sqrt[3]{3} && \text{Combine like radicals.}
\end{aligned}$$

c. $\sqrt{48x^3} + \sqrt{12x^3}$

Solution
$$\begin{aligned}
\sqrt{48x^3} + \sqrt{12x^3} &= \sqrt{16x^2 \cdot 3x} + \sqrt{4x^2 \cdot 3x} && \text{Rewrite } 48x^3 \text{ as } 16x^2 \cdot 3x \text{ and } 12x^3 \text{ as } 4x^2 \cdot 3x.\\
&= \sqrt{16x^2} \cdot \sqrt{3x} + \sqrt{4x^2} \cdot \sqrt{3x} && \text{Use the product rule.}\\
&= 4x\sqrt{3x} + 2x\sqrt{3x} && \text{Find } \sqrt{16x^2} \text{ and } \sqrt{4x^2}.\\
&= 6x\sqrt{3x} && \text{Combine like radicals.}
\end{aligned}$$

d. $3\sqrt[4]{32x^5} + \sqrt[4]{162x^9}$

Solution $3\sqrt[4]{32x^5} + \sqrt[4]{162x^9} = 3\sqrt[4]{16 \cdot 2 \cdot x^4 \cdot x} + \sqrt[4]{81 \cdot 2 \cdot x^8 \cdot x}$ Write 32 as $16 \cdot 2$, x^5 as $x^4 \cdot x$, 162 as $81 \cdot 2$ and x^9 as $x^8 \cdot x$.

Note: *Although the radicals are like, we cannot add the coefficients because they are not like.*

$$\begin{aligned}
&= 3 \cdot 2x\sqrt[4]{2x} + 3x^2\sqrt[4]{2x} && \text{Find the roots.}\\
&= 6x\sqrt[4]{2x} + 3x^2\sqrt[4]{2x} && \text{Multiply.}\\
&= (6x + 3x^2)\sqrt[4]{2x} && \text{Combine like radicals.}
\end{aligned}$$

ANSWERS

a. $-10\sqrt{6}$
b. $14x^2\sqrt{2x}$
c. $10a^2\sqrt[3]{2a}$

YOUR TURN Simplify. Assume variables represent nonnegative quantities.

a. $4\sqrt{24} - 6\sqrt{54}$ **b.** $4\sqrt{50x^5} - 2\sqrt{18x^5}$ **c.** $6a\sqrt[3]{54a^4} - 2\sqrt[3]{128a^7}$

OBJECTIVE 2. Use the distributive property in expressions containing radicals. Products involving sums and differences of radicals are found in much the same way as products of polynomials. We will use the distributive property, multiply binomials using FOIL, and square a binomial (all from Section 5.5).

EXAMPLE 3) Find the product. Assume variables represent nonnegative values.

a. $\sqrt{3}(\sqrt{3} + \sqrt{15})$

Solution $\sqrt{3}(\sqrt{3} + \sqrt{15}) = \sqrt{3} \cdot \sqrt{3} + \sqrt{3} \cdot \sqrt{15}$ **Use the distributive property.**

$\qquad\qquad\qquad\qquad\quad = \sqrt{3 \cdot 3} + \sqrt{3 \cdot 15}$ **Use the product rule.**

$\qquad\qquad\qquad\qquad\quad = \sqrt{9} + \sqrt{45}$ **Multiply.**

$\qquad\qquad\qquad\qquad\quad = 3 + 3\sqrt{5}$ **Find $\sqrt{9}$ and simplify $\sqrt{45}$.**

b. $2\sqrt{6}(3 + 5\sqrt{5})$

Solution $2\sqrt{6}(3 + 5\sqrt{5}) = 2\sqrt{6} \cdot 3 + 2\sqrt{6} \cdot 5\sqrt{5}$ **Use the distributive property.**

$\qquad\qquad\qquad\qquad\quad = 2 \cdot 3\sqrt{6} + 2 \cdot 5\sqrt{6 \cdot 5}$ **Use the product rule.**

$\qquad\qquad\qquad\qquad\quad = 6\sqrt{6} + 10\sqrt{30}$ **Find the products.**

c. $(2 + \sqrt{3})(\sqrt{5} - \sqrt{6})$

Solution $(2 + \sqrt{3})(\sqrt{5} - \sqrt{6}) = 2\sqrt{5} - 2\sqrt{6} + \sqrt{3} \cdot \sqrt{5} - \sqrt{3} \cdot \sqrt{6}$ **Use FOIL.**

$\qquad\qquad\qquad\qquad\qquad\qquad = 2\sqrt{5} - 2\sqrt{6} + \sqrt{15} - \sqrt{18}$ **Use the product rule.**

$\qquad\qquad\qquad\qquad\qquad\qquad = 2\sqrt{5} - 2\sqrt{6} + \sqrt{15} - 3\sqrt{2}$ **Simplify $\sqrt{18}$.**

d. $(3\sqrt{x} + \sqrt{y})(2\sqrt{x} - 5\sqrt{y})$

Solution $(3\sqrt{x} + \sqrt{y})(2\sqrt{x} - 5\sqrt{y}) =$

$\qquad 3\sqrt{x} \cdot 2\sqrt{x} - 3\sqrt{x} \cdot 5\sqrt{y} + \sqrt{y} \cdot 2\sqrt{x} - \sqrt{y} \cdot 5\sqrt{y}$ **Use FOIL.**

$\qquad\qquad = 6x - 15\sqrt{xy} + 2\sqrt{xy} - 5y$ **Use the product rule.**

$\qquad\qquad = 6x - 13\sqrt{xy} - 5y$ **Combine like radicals.**

e. $(5 + \sqrt{3})^2$

Solution $(5 + \sqrt{3})^2 = 5^2 + 2 \cdot 5\sqrt{3} + (\sqrt{3})^2$ **Use $(a + b)^2 = a^2 + 2ab + b^2$.**

$\qquad\qquad\qquad\quad = 25 + 10\sqrt{3} + 3$ **Simplify.**

$\qquad\qquad\qquad\quad = 28 + 10\sqrt{3}$ **Add 25 and 3.**

YOUR TURN) Find the product. Assume variables represent nonnegative quantities.

a. $2\sqrt{11}(3 - 3\sqrt{6})$ **b.** $(2\sqrt{a} + 3\sqrt{b})(\sqrt{a} - 3\sqrt{b})$ **c.** $(3 + \sqrt{5})^2$

Radicals in Conjugates

Radical expressions can be conjugates. Like binomial conjugates, conjugates involving radicals differ only in the sign separating the terms. For example, $7 - \sqrt{3}$ and $7 + \sqrt{3}$ are conjugates. Let's explore what happens when we multiply conjugates containing radicals.

ANSWERS)

a. $6\sqrt{11} - 6\sqrt{66}$

b. $2a - 3\sqrt{ab} - 9b$

c. $14 + 6\sqrt{5}$

EXAMPLE 4 Find the product.

a. $(2 + \sqrt{3})(2 - \sqrt{3})$

Solution $(2 + \sqrt{3})(2 - \sqrt{3}) = 2^2 - (\sqrt{3})^2$ Use $(a + b)(a - b) = a^2 - b^2$ (from Section 5.5).

$= 4 - 3$ Simplify.

$= 1$

b. $(\sqrt{5} - 3\sqrt{3})(\sqrt{5} + 3\sqrt{3})$

Solution $(\sqrt{5} - 3\sqrt{3})(\sqrt{5} + 3\sqrt{3}) = (\sqrt{5})^2 - (3\sqrt{3})^2$ Use $(a + b)(a - b) = a^2 - b^2$.

$= 5 - 9 \cdot 3$ Simplify.

$= 5 - 27$

$= -22$

Note: *The product of conjugates always results in a rational number. This will be useful in Section 10.5 when rationalizing denominators.*

YOUR TURN Find the product.

a. $(4 + \sqrt{10})(4 - \sqrt{10})$

b. $(3\sqrt{5} + 2\sqrt{6})(3\sqrt{5} - 2\sqrt{6})$

OBJECTIVE 3. Simplify radical expressions that contain mixed operations. Now let's use the order of operations to simplify radical expressions that have more than one operation.

EXAMPLE 5 Simplify.

a. $\sqrt{2} \cdot \sqrt{10} + \sqrt{3} \cdot \sqrt{15}$

Solution $\sqrt{2} \cdot \sqrt{10} + \sqrt{3} \cdot \sqrt{15} = \sqrt{2 \cdot 10} + \sqrt{3 \cdot 15}$ Use the product rule.

$= \sqrt{20} + \sqrt{45}$ Multiply.

$= \sqrt{4 \cdot 5} + \sqrt{9 \cdot 5}$ Rewrite 20 as $4 \cdot 5$ and 45 as $9 \cdot 5$.

$= \sqrt{4} \cdot \sqrt{5} + \sqrt{9} \cdot \sqrt{5}$ Use the product rule.

$= 2\sqrt{5} + 3\sqrt{5}$ Find $\sqrt{9}$ and $\sqrt{4}$.

$= 5\sqrt{5}$ Combine like radicals.

b. $\dfrac{\sqrt{54}}{\sqrt{3}} + \sqrt{32}$

Solution $\dfrac{\sqrt{54}}{\sqrt{3}} + \sqrt{32} = \sqrt{\dfrac{54}{3}} + \sqrt{16 \cdot 2}$ Use the quotient rule and rewrite 32 as $16 \cdot 2$.

$= \sqrt{18} + \sqrt{16} \cdot \sqrt{2}$ Divide and use the product rule.

$= \sqrt{9 \cdot 2} + 4\sqrt{2}$ Rewrite 18 as $9 \cdot 2$ and find $\sqrt{16}$.

$= \sqrt{9} \cdot \sqrt{2} + 4\sqrt{2}$ Use the product rule.

$= 3\sqrt{2} + 4\sqrt{2}$ Find $\sqrt{9}$.

$= 7\sqrt{2}$ Combine like radicals.

ANSWERS

a. 6   b. 21

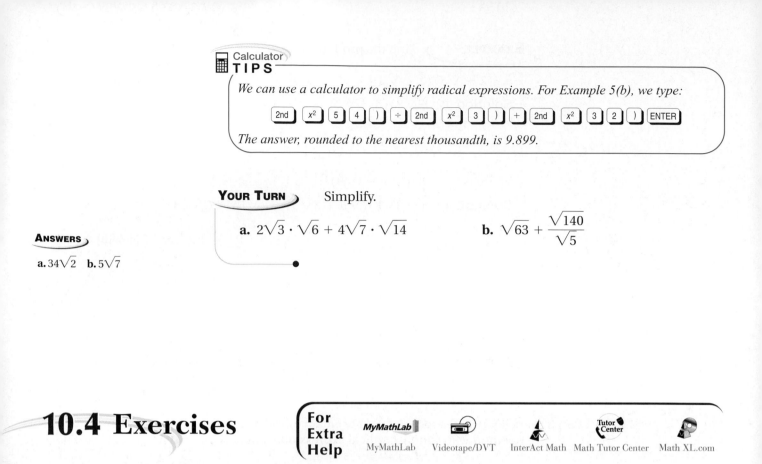

YOUR TURN   Simplify.

a. $2\sqrt{3} \cdot \sqrt{6} + 4\sqrt{7} \cdot \sqrt{14}$

b. $\sqrt{63} + \dfrac{\sqrt{140}}{\sqrt{5}}$

ANSWERS

a. $34\sqrt{2}$ **b.** $5\sqrt{7}$

10.4 Exercises

For Extra Help

MyMathLab   Videotape/DVT   InterAct Math   Math Tutor Center   Math XL.com

1. What must be identical in like radicals? What can be different?

2. Add $3x + 2x$ and then add $3\sqrt{2} + 2\sqrt{2}$. Discuss the similarities in the process.

3. Multiply $(x + 3)(x + 2)$ and then multiply $(\sqrt{5} + 3)(\sqrt{5} + 2)$. Discuss the similarities in the process.

4. Why is $\sqrt{a} + \sqrt{b} \neq \sqrt{a + b}$? Use examples if necessary.

For Exercises 5–18, simplify. Assume variables represent nonnegative values.

5. $9\sqrt{6} - 15\sqrt{6}$

6. $2\sqrt{10} - 13\sqrt{10}$

7. $7\sqrt{a} + 2\sqrt{a}$

8. $5\sqrt{y} + 7\sqrt{y}$

9. $4\sqrt{5} - 2\sqrt{6} + 8\sqrt{5} - 6\sqrt{6}$

10. $-6\sqrt{2} + 5\sqrt{7} + 3\sqrt{2} - 2\sqrt{7}$

11. $3a\sqrt{5a} - 4b\sqrt{7b} + 8a\sqrt{5a} + 2b\sqrt{7b}$

12. $12n\sqrt{2n} - 14m\sqrt{5m} - 8n\sqrt{2n} + 18m\sqrt{5m}$

13. $6x\sqrt[3]{9} - 3x\sqrt[3]{9}$

14. $4y\sqrt[3]{3} - y\sqrt[3]{3}$

15. $6x^2\sqrt[4]{5x} + 12x^2\sqrt[4]{5x}$

16. $3y^3\sqrt[4]{8y} - 9y^3\sqrt[4]{8y}$

17. $3x\sqrt{5x} + 4x\sqrt[3]{5x}$

18. $4z\sqrt[4]{2z} - 7z\sqrt{2z}$

For Exercises 19–38, simplify the radicals and then find the sum or difference. Assume variables represent nonnegative values.

19. $\sqrt{48} - \sqrt{75}$

20. $\sqrt{80} - \sqrt{20}$

21. $\sqrt{80y} - \sqrt{125y}$

22. $\sqrt{27x} + \sqrt{75x}$

23. $\sqrt{80} - 4\sqrt{45}$

24. $\sqrt{20} - 2\sqrt{180}$

25. $3\sqrt{96} - 2\sqrt{54}$

26. $5\sqrt{63} + 2\sqrt{28}$

27. $6\sqrt{48a^3} - 2\sqrt{75a^3}$

28. $4\sqrt{98y^5} - 7\sqrt{128y^5}$

29. $\sqrt{150} - \sqrt{54} + \sqrt{24}$

30. $\sqrt{20} + \sqrt{125} - \sqrt{80}$

31. $2\sqrt{8} - 3\sqrt{48} + 2\sqrt{98} - \sqrt{75}$

32. $3\sqrt{216} - \sqrt{147} - 4\sqrt{96} - \sqrt{108}$

33. $\sqrt[3]{16} + \sqrt[3]{54}$

34. $\sqrt[3]{24} + \sqrt[3]{81}$

35. $4\sqrt[3]{135x^5} - 6x\sqrt[3]{320x^2}$

36. $3a^2\sqrt[3]{500a^4} + 6a\sqrt[3]{108a^7}$

37. $-4\sqrt[4]{x^9} + 2\sqrt[4]{16x^7}$

38. $6y\sqrt[4]{81y^5} - 2y\sqrt[4]{16y^6}$

For Exercises 39–46 use the distributive property. Assume variables represent nonnegative values.

39. $\sqrt{2}(3 + \sqrt{2})$

40. $\sqrt{5}(4 - \sqrt{5})$

41. $\sqrt{3}(\sqrt{3} - \sqrt{15})$

42. $\sqrt{6}(\sqrt{6} + \sqrt{2})$

43. $\sqrt{5}(\sqrt{3} + 2\sqrt{15})$

44. $\sqrt{7}(\sqrt{5} - 3\sqrt{14})$

45. $4\sqrt{3x}(2\sqrt{3x} - 4\sqrt{6x})$

46. $6\sqrt{2y}(3\sqrt{2y} + 2\sqrt{10y})$

For Exercises 47–66, multiply (use FOIL). Assume variables represent nonnegative values.

47. $(3 + \sqrt{5})(4 - \sqrt{2})$

48. $(3 + \sqrt{7})(7 - \sqrt{3})$

49. $(3 + \sqrt{x})(2 + \sqrt{x})$

50. $(5 - \sqrt{a})(2 + \sqrt{a})$

51. $(2 + 3\sqrt{3})(3 + 5\sqrt{2})$

52. $(7 - 3\sqrt{5})(2 - 2\sqrt{10})$

53. $(\sqrt{2} + \sqrt{3})(\sqrt{3} + \sqrt{5})$

54. $(\sqrt{5} + \sqrt{2})(\sqrt{2} + \sqrt{7})$

55. $(\sqrt{x} + \sqrt{y})(\sqrt{x} - 2\sqrt{y})$

56. $(2\sqrt{a} + \sqrt{b})(\sqrt{a} + \sqrt{b})$

57. $(4\sqrt{2} + 2\sqrt{5})(3\sqrt{7} - 3\sqrt{3})$

58. $(8\sqrt{2} - 2\sqrt{3})(2\sqrt{5} + 3\sqrt{10})$

59. $(2\sqrt{a} + 3\sqrt{b})(4\sqrt{a} - \sqrt{b})$

60. $(\sqrt{m} - 4\sqrt{n})(2\sqrt{m} - 3\sqrt{n})$

61. $(\sqrt[3]{4} + 5)(\sqrt[3]{4} - 8)$

62. $(\sqrt[3]{9} + 5)(\sqrt[3]{9} - 2)$

63. $(\sqrt[3]{9} + \sqrt[3]{4})(\sqrt[3]{3} - \sqrt[3]{2})$

64. $(\sqrt[3]{5} + \sqrt[3]{9})(\sqrt[3]{25} - \sqrt[3]{3})$

65. $(\sqrt[3]{x} + 2)(\sqrt[3]{x^2} - 2\sqrt[3]{x} + 4)$

66. $(\sqrt[3]{r} - 3)(\sqrt[3]{r^2} + 3\sqrt[3]{r} + 9)$

For Exercises 67–74, find the product.

67. $(4 + \sqrt{6})^2$

68. $(6 + \sqrt{3})^2$

69. $(1 - \sqrt{2})^2$

70. $(5 - \sqrt{2})^2$

71. $(2 + 2\sqrt{3})^2$

72. $(3 + 2\sqrt{5})^2$

73. $(2\sqrt{3} + 3\sqrt{2})^2$

74. $(4\sqrt{2} - 5\sqrt{6})^2$

For Exercises 75–88, multiply the conjugates. Assume variables represent nonnegative values.

75. $(2 + \sqrt{3})(2 - \sqrt{3})$

76. $(3 + \sqrt{5})(3 - \sqrt{5})$

77. $(\sqrt{2} + 4)(\sqrt{2} - 4)$

78. $(\sqrt{7} - 6)(\sqrt{7} + 6)$

79. $(6 + \sqrt{x})(6 - \sqrt{x})$

80. $(5 + \sqrt{y})(5 - \sqrt{y})$

81. $(\sqrt{3} + \sqrt{2})(\sqrt{3} - \sqrt{2})$

82. $(\sqrt{5} - \sqrt{3})(\sqrt{5} + \sqrt{3})$

83. $(\sqrt{x} + \sqrt{y})(\sqrt{x} - \sqrt{y})$

84. $(\sqrt{a} - \sqrt{b})(\sqrt{a} + \sqrt{b})$　　　**85.** $(4 + 2\sqrt{3})(4 - 2\sqrt{3})$　　　**86.** $(5 + 3\sqrt{3})(5 - 3\sqrt{3})$

87. $(3\sqrt{7} + \sqrt{13})(3\sqrt{7} - \sqrt{13})$　　**88.** $(4\sqrt{5} - 3\sqrt{2})(4\sqrt{5} + 3\sqrt{2})$

For Exercises 89–96, simplify.

89. $\sqrt{3} \cdot \sqrt{15} + \sqrt{8} \cdot \sqrt{10}$　　　**90.** $\sqrt{6} \cdot \sqrt{8} + \sqrt{5} \cdot \sqrt{15}$　　　**91.** $3\sqrt{3} \cdot \sqrt{18} - 4\sqrt{18} \cdot \sqrt{12}$

92. $3\sqrt{2} \cdot 2\sqrt{40} - 5\sqrt{12} \cdot \sqrt{15}$　　**93.** $\dfrac{\sqrt{40}}{\sqrt{5}} + \sqrt{50}$　　　**94.** $\dfrac{\sqrt{60}}{\sqrt{5}} + \sqrt{48}$

95. $\dfrac{\sqrt{540}}{\sqrt{3}} - 4\sqrt{125}$　　　　**96.** $\dfrac{\sqrt{288}}{\sqrt{6}} - 6\sqrt{108}$

For Exercises 97 and 98, find the perimeter of the shape.

97.

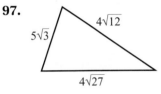

98.

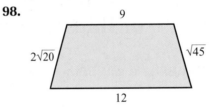

99. Crown molding, which is placed at the top of a wall, is to be installed around the perimeter of the room shown.

   a. Write an expression in simplest form for the perimeter of the room.

   b. Use a calculator to approximate the perimeter, rounded to the nearest tenth.

   c. If crown molding costs $1.89 per foot length, how much will the crown molding cost for this room?

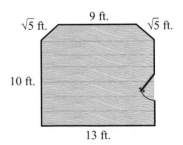

100. A tabletop is to be fitted with veneer strips along the sides.

   a. Write an expression in simplest form for the perimeter of the tabletop.

   b. Use a calculator to approximate the perimeter, rounded to the nearest tenth.

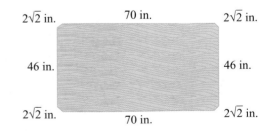

REVIEW EXERCISES

[4.4]   **1.** For the equation $5y - 2x = 10$, find the slope and the y-intercept.

[4.4]   **2.** Graph: $y = -\dfrac{1}{3}x + 4$

[5.5]   **3.** What is the conjugate of $2x - 5$?

[5.5]   **4.** Multiply: $(4x + 3)(4x - 3)$

[10.3] **5.** What factor could multiply $\sqrt{8}$ in order to equal $\sqrt{16}$?

[10.3] **6.** What factor could multiply $\sqrt[3]{2}$ in order to equal $\sqrt[3]{8}$?

10.5   Rationalizing Numerators and Denominators of Radical Expressions

OBJECTIVES

1. **Rationalize denominators.**
2. **Rationalize denominators that have a sum or difference with a square root term.**
3. **Rationalize numerators.**

We are now ready to formalize the conditions for a radical expression that is in simplest form. A radical expression is in simplest form if

1. All rational roots have been found.
2. There are no perfect nth factors of radicands of the form $\sqrt[n]{a^n}$.
3. All possible products, quotients, sums, and differences have been found.
4. There are no radicals in the denominator of a fraction.

   In this section, we explore how to simplify expressions that have a radical in the denominator of a fraction, as in $\dfrac{1}{\sqrt{2}}$.

OBJECTIVE 1. Rationalize denominators. If the denominator of a fraction contains a radical, our goal will be to *rationalize* the denominator, which means to rewrite the expression so that it has a rational number in the denominator. In general, we multiply the fraction by a well-chosen 1 so that the radical is eliminated. We determine that 1 by finding a factor that multiplies the nth root in the denominator so that its radicand is a perfect nth power.

Square Root Denominators

In the case of a square root in the denominator, we multiply it by a factor that makes the radicand a perfect square, which allows us to eliminate the square root. For example, to rationalize $\dfrac{1}{\sqrt{2}}$, we could multiply by $\dfrac{\sqrt{2}}{\sqrt{2}}$ because the product's denominator is the square root of a perfect square.

$$\frac{1}{\sqrt{2}} = \frac{1}{\sqrt{2}} \cdot \frac{\sqrt{2}}{\sqrt{2}} = \frac{\sqrt{2}}{\sqrt{4}} = \frac{\sqrt{2}}{2}$$

▲

Note: *We are not changing the value of* $\dfrac{1}{\sqrt{2}}$ *because we are multiplying it by* 1 *in the form of* $\dfrac{\sqrt{2}}{\sqrt{2}}$.

Any factor that produces a perfect square radicand will work. For example, we could have multiplied $\dfrac{1}{\sqrt{2}}$ by $\dfrac{\sqrt{8}}{\sqrt{8}}$.

$$\frac{1}{\sqrt{2}} = \frac{1}{\sqrt{2}} \cdot \frac{\sqrt{8}}{\sqrt{8}} = \frac{\sqrt{8}}{\sqrt{16}} = \frac{\sqrt{4 \cdot 2}}{4} = \frac{2\sqrt{2}}{4} = \frac{\sqrt{2}}{2}$$

Notice, however, that multiplying by $\dfrac{\sqrt{2}}{\sqrt{2}}$ required fewer steps to simplify than multiplying by $\dfrac{\sqrt{8}}{\sqrt{8}}$.

EXAMPLE 1 ⟩ Rationalize the denominator. Assume variables represent non-negative values.

a. $\dfrac{2}{\sqrt{5}}$

Solution $\dfrac{2}{\sqrt{5}} = \dfrac{2}{\sqrt{5}} \cdot \dfrac{\sqrt{5}}{\sqrt{5}}$ **Multiply by** $\dfrac{\sqrt{5}}{\sqrt{5}}$.

$\phantom{\dfrac{2}{\sqrt{5}}} = \dfrac{2\sqrt{5}}{\sqrt{25}}$ **Simplify.**

$\phantom{\dfrac{2}{\sqrt{5}}} = \dfrac{2\sqrt{5}}{5}$

b. $\sqrt{\dfrac{5}{8}}$

Solution $\sqrt{\dfrac{5}{8}} = \dfrac{\sqrt{5}}{\sqrt{8}}$ Use the quotient rule of square roots to separate the numerator and denominator into two radicals.

$$= \dfrac{\sqrt{5}}{\sqrt{8}} \cdot \dfrac{\sqrt{2}}{\sqrt{2}}$$ Multiply by $\dfrac{\sqrt{2}}{\sqrt{2}}$.

Note: *We chose to multiply by* $\dfrac{\sqrt{2}}{\sqrt{2}}$ *because it leads to a smaller perfect square than other choices, like* $\dfrac{\sqrt{8}}{\sqrt{8}}$.

$$= \dfrac{\sqrt{10}}{\sqrt{16}}$$ Simplify.

$$= \dfrac{\sqrt{10}}{4}$$

Multiplying by $\dfrac{\sqrt{8}}{\sqrt{8}}$ *produces the same final answer, but requires more steps.*

Warning: Though it may be tempting, we cannot divide out the 4 and 10 because 10 is a radicand, whereas 4 is not. We *never* divide out factors common to a radicand and a number not under a radical.

c. $\dfrac{3}{\sqrt{2x}}$

Solution $\dfrac{3}{\sqrt{2x}} = \dfrac{3}{\sqrt{2x}} \cdot \dfrac{\sqrt{2x}}{\sqrt{2x}}$ Multiply by $\dfrac{\sqrt{2x}}{\sqrt{2x}}$.

$$= \dfrac{3\sqrt{2x}}{\sqrt{4x^2}}$$ Simplify.

$$= \dfrac{3\sqrt{2x}}{2x}$$

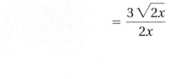 **YOUR TURN**   Rationalize the denominator. Assume variables represent nonnegative values.

a. $\dfrac{1}{\sqrt{7}}$ **b.** $\sqrt{\dfrac{7}{12}}$ **c.** $\dfrac{3}{\sqrt{10x}}$

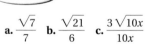 **ANSWERS**

a. $\dfrac{\sqrt{7}}{7}$ **b.** $\dfrac{\sqrt{21}}{6}$ **c.** $\dfrac{3\sqrt{10x}}{10x}$

n^{th}-Root Denominators

If the denominator contains a higher-order root, such as a cube root, then we multiply appropriately to get a perfect cube radicand in the denominator so that we can eliminate the radical. For example, $\dfrac{2}{\sqrt[3]{5}} = \dfrac{2}{\sqrt[3]{5}} \cdot \dfrac{\sqrt[3]{25}}{\sqrt[3]{25}} = \dfrac{2\sqrt[3]{25}}{\sqrt[3]{125}} = \dfrac{2\sqrt[3]{25}}{5}$. We summarize as follows.

PROCEDURE   **Rationalizing Denominators**

To rationalize a denominator containing a single n^{th} root, multiply the fraction by a 1 so that the product's denominator has a radicand that is a perfect n^{th} power.

EXAMPLE 2 Rationalize the denominator. Assume variables represent nonnegative values.

a. $\dfrac{3}{\sqrt[3]{2}}$

Solution $\dfrac{3}{\sqrt[3]{2}} = \dfrac{3}{\sqrt[3]{2}} \cdot \dfrac{\sqrt[3]{4}}{\sqrt[3]{4}}$ Since $\sqrt[3]{2} \cdot \sqrt[3]{4} = \sqrt[3]{8} = 2$, we multiply the fraction by $\dfrac{\sqrt[3]{4}}{\sqrt[3]{4}}$.

$= \dfrac{3\sqrt[3]{4}}{\sqrt[3]{8}}$ Simplify.

$= \dfrac{3\sqrt[3]{4}}{2}$

b. $\dfrac{\sqrt[3]{a}}{\sqrt[3]{b}}$

Solution $\dfrac{\sqrt[3]{a}}{\sqrt[3]{b}} = \dfrac{\sqrt[3]{a}}{\sqrt[3]{b}} \cdot \dfrac{\sqrt[3]{b^2}}{\sqrt[3]{b^2}}$ Since $\sqrt[3]{b} \cdot \sqrt[3]{b^2} = \sqrt[3]{b^3} = b$, we multiply the fraction by $\dfrac{\sqrt[3]{b^2}}{\sqrt[3]{b^2}}$.

$= \dfrac{\sqrt[3]{ab^2}}{\sqrt[3]{b^3}}$ Simplify.

$= \dfrac{\sqrt[3]{ab^2}}{b}$

c. $\sqrt[3]{\dfrac{5}{9a^2}}$

Solution $\sqrt[3]{\dfrac{5}{9a^2}} = \dfrac{\sqrt[3]{5}}{\sqrt[3]{9a^2}}$ Use the quotient rule to separate the numerator and denominator.

$= \dfrac{\sqrt[3]{5}}{\sqrt[3]{9a^2}} \cdot \dfrac{\sqrt[3]{3a}}{\sqrt[3]{3a}}$ Since $\sqrt[3]{9a^2} \cdot \sqrt[3]{3a} = \sqrt[3]{27a^3} = 3a$, multiply the fraction by $\dfrac{\sqrt[3]{3a}}{\sqrt[3]{3a}}$.

$= \dfrac{\sqrt[3]{15a}}{\sqrt[3]{27a^3}}$ Simplify.

$= \dfrac{\sqrt[3]{15a}}{3a}$

d. $\dfrac{5}{\sqrt[4]{3}}$

Solution $\dfrac{5}{\sqrt[4]{3}} = \dfrac{5}{\sqrt[4]{3}} \cdot \dfrac{\sqrt[4]{27}}{\sqrt[4]{27}}$ Since $\sqrt[4]{3} \cdot \sqrt[4]{27} = \sqrt[4]{81} = 3$, multiply the fraction by $\dfrac{\sqrt[4]{27}}{\sqrt[4]{27}}$.

$= \dfrac{5\sqrt[4]{27}}{\sqrt[4]{81}}$ Simplify.

$= \dfrac{5\sqrt[4]{27}}{3}$

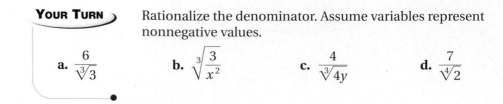

YOUR TURN Rationalize the denominator. Assume variables represent nonnegative values.

a. $\dfrac{6}{\sqrt[3]{3}}$ **b.** $\sqrt[3]{\dfrac{3}{x^2}}$ **c.** $\dfrac{4}{\sqrt[3]{4y}}$ **d.** $\dfrac{7}{\sqrt[4]{2}}$

OBJECTIVE 2. Rationalize denominators that have a sum or difference with a square root term. In Example 4 of Section 10.4, we saw that the product of two conjugates containing square roots does not contain any radicals. Consequently, if the denominator of a fraction contains a sum or difference with a square root term, we can rationalize the denominator by multiplying the fraction by a 1 made up of the conjugate of the denominator. For example, to rationalize $\dfrac{5}{7 - \sqrt{3}}$, we multiply by $\dfrac{7 + \sqrt{3}}{7 + \sqrt{3}}$. Because $7 - \sqrt{3}$ and $7 + \sqrt{3}$ are conjugates, their product will not contain any radicals, so the denominator will be rationalized.

$$\frac{5}{7 - \sqrt{3}} = \frac{5}{7 - \sqrt{3}} \cdot \frac{7 + \sqrt{3}}{7 + \sqrt{3}} = \frac{5(7 + \sqrt{3})}{(7)^2 - (\sqrt{3})^2} = \frac{35 + 5\sqrt{3}}{49 - 3} = \frac{35 + 5\sqrt{3}}{46}$$

PROCEDURE Rationalizing a Denominator Containing a Sum or Difference

To rationalize a denominator containing a sum or difference with at least one square root term, multiply the fraction by a 1 whose numerator and denominator are the conjugate of the denominator.

EXAMPLE 3 Rationalize the denominator and simplify. Assume variables represent nonnegative values.

a. $\dfrac{9}{\sqrt{2} + 7}$

Solution $\dfrac{9}{\sqrt{2} + 7} = \dfrac{9}{\sqrt{2} + 7} \cdot \dfrac{\sqrt{2} - 7}{\sqrt{2} - 7}$ The conjugate of $\sqrt{2} + 7$ is $\sqrt{2} - 7$, so we multiply by $\dfrac{\sqrt{2} - 7}{\sqrt{2} - 7}$.

$= \dfrac{9(\sqrt{2} - 7)}{(\sqrt{2})^2 - (7)^2}$ Multiply. In the denominator, use the rule $(a + b)(a - b) = a^2 - b^2$.

$= \dfrac{9\sqrt{2} - 63}{2 - 49}$ Simplify.

$= \dfrac{9\sqrt{2} - 63}{-47}$ We can simplify the negative denominator by factoring out -1 in the numerator and denominator.

$= \dfrac{-1(63 - 9\sqrt{2})}{-1(47)}$ After factoring out the -1, the signs of the terms change. Since 63 is now positive, we write it first.

$= \dfrac{63 - 9\sqrt{2}}{47}$ Divide out the common factor -1.

ANSWERS

a. $2\sqrt[3]{9}$ **b.** $\dfrac{\sqrt[3]{3x}}{x}$

c. $\dfrac{2\sqrt[3]{2y^2}}{y}$ **d.** $\dfrac{7\sqrt[4]{8}}{2}$

b. $\dfrac{2\sqrt{3}}{\sqrt{6} - \sqrt{2}}$

Solution $\dfrac{2\sqrt{3}}{\sqrt{6} - \sqrt{2}} = \dfrac{2\sqrt{3}}{\sqrt{6} - \sqrt{2}} \cdot \dfrac{\sqrt{6} + \sqrt{2}}{\sqrt{6} + \sqrt{2}}$ The conjugate of $\sqrt{6} - \sqrt{2}$ is $\sqrt{6} + \sqrt{2}$, so we multiply by $\dfrac{\sqrt{6} + \sqrt{2}}{\sqrt{6} + \sqrt{2}}$.

$$= \dfrac{2\sqrt{3}(\sqrt{6} + \sqrt{2})}{(\sqrt{6})^2 - (\sqrt{2})^2}$$

$$= \dfrac{2\sqrt{18} + 2\sqrt{6}}{6 - 2}$$ Multiply in the numerator and evaluate the exponents in the denominator.

$$= \dfrac{2\sqrt{9 \cdot 2} + 2\sqrt{6}}{4}$$ Simplify $\sqrt{18}$ by factoring out a perfect square factor in 18.

$$= \dfrac{2 \cdot 3\sqrt{2} + 2\sqrt{6}}{4}$$ Simplify $\sqrt{9 \cdot 2}$ by finding the square root of 9.

Note: *We cannot add* $3\sqrt{2}$ *and* $\sqrt{6}$ *because their radicands do not match.*

$$= \dfrac{2(3\sqrt{2} + \sqrt{6})}{4}$$ Factor out the common 2 factor in the numerator.

$$= \dfrac{3\sqrt{2} + \sqrt{6}}{2}$$ Divide out the common factor 2.

c. $\dfrac{6}{\sqrt{x} - 5}$

Solution $\dfrac{6}{\sqrt{x} - 5} = \dfrac{6}{\sqrt{x} - 5} \cdot \dfrac{\sqrt{x} + 5}{\sqrt{x} + 5}$ The conjugate of $\sqrt{x} - 5$ is $\sqrt{x} + 5$, so we multiply by $\dfrac{\sqrt{x} + 5}{\sqrt{x} + 5}$.

$$= \dfrac{6(\sqrt{x} + 5)}{(\sqrt{x})^2 - (5)^2}$$

$$= \dfrac{6\sqrt{x} + 30}{x - 25}$$ Multiply in the numerator and evaluate the exponents in the denominator.

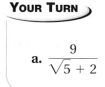

YOUR TURN   Rationalize the denominator and simplify. Assume variables represent nonnegative values.

a. $\dfrac{9}{\sqrt{5} + 2}$ **b.** $\dfrac{\sqrt{2}}{\sqrt{5} - \sqrt{3}}$ **c.** $\dfrac{3}{\sqrt{x} + 4}$

ANSWERS

a. $9\sqrt{5} - 18$

b. $\dfrac{\sqrt{10} + \sqrt{6}}{2}$

c. $\dfrac{3\sqrt{x} - 12}{x - 16}$

OBJECTIVE 3. Rationalize numerators.   In later mathematics courses, you may need to rationalize the numerator. We use the same procedure that we use in rationalizing denominators.

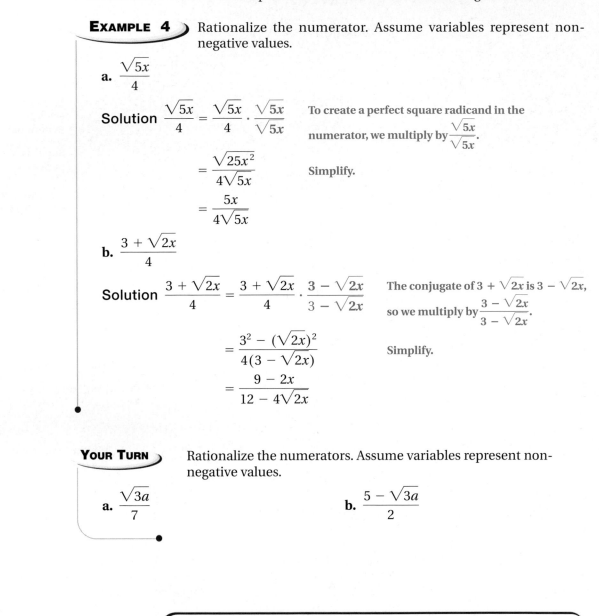

EXAMPLE 4 Rationalize the numerator. Assume variables represent non-negative values.

a. $\dfrac{\sqrt{5x}}{4}$

Solution $\dfrac{\sqrt{5x}}{4} = \dfrac{\sqrt{5x}}{4} \cdot \dfrac{\sqrt{5x}}{\sqrt{5x}}$ To create a perfect square radicand in the numerator, we multiply by $\dfrac{\sqrt{5x}}{\sqrt{5x}}$.

$= \dfrac{\sqrt{25x^2}}{4\sqrt{5x}}$ Simplify.

$= \dfrac{5x}{4\sqrt{5x}}$

b. $\dfrac{3 + \sqrt{2x}}{4}$

Solution $\dfrac{3 + \sqrt{2x}}{4} = \dfrac{3 + \sqrt{2x}}{4} \cdot \dfrac{3 - \sqrt{2x}}{3 - \sqrt{2x}}$ The conjugate of $3 + \sqrt{2x}$ is $3 - \sqrt{2x}$, so we multiply by $\dfrac{3 - \sqrt{2x}}{3 - \sqrt{2x}}$.

$= \dfrac{3^2 - (\sqrt{2x})^2}{4(3 - \sqrt{2x})}$ Simplify.

$= \dfrac{9 - 2x}{12 - 4\sqrt{2x}}$

ANSWERS

a. $\dfrac{3a}{7\sqrt{3a}}$

b. $\dfrac{25 - 3a}{10 + 2\sqrt{3a}}$

YOUR TURN Rationalize the numerators. Assume variables represent non-negative values.

a. $\dfrac{\sqrt{3a}}{7}$

b. $\dfrac{5 - \sqrt{3a}}{2}$

10.5 Exercises

For Extra Help   *MyMathLab*   MyMathLab   Videotape/DVT   InterAct Math   Tutor Center Math Tutor Center   Math XL.com

1. Explain why each of the following expressions is not in simplest form.

a. $\sqrt{\dfrac{3}{16}}$

b. $\dfrac{5}{\sqrt{3}}$

2. Although $\dfrac{1}{\sqrt{3}}$ and $\dfrac{\sqrt{3}}{3}$ are equal, explain why $\dfrac{\sqrt{3}}{3}$ is considered simplest form.

3. Explain in your own words how to rationalize a denominator that is the square root of a number $\left(\text{like } \dfrac{2}{\sqrt{3}} \text{ or } \dfrac{\sqrt{5}}{\sqrt{7}}\right)$.

4. Explain in your own words how to rationalize a denominator that is a sum or difference with a square root term $\left(\text{like } \dfrac{3}{5 + \sqrt{2}} \text{ or } \dfrac{2}{\sqrt{x} - \sqrt{y}}\right)$.

For Exercises 5–28, rationalize the denominator. Assume variables represent non-negative values.

5. $\dfrac{1}{\sqrt{3}}$

6. $\dfrac{1}{\sqrt{5}}$

7. $\dfrac{3}{\sqrt{8}}$

8. $\dfrac{5}{\sqrt{12}}$

9. $\sqrt{\dfrac{36}{7}}$

10. $\sqrt{\dfrac{81}{5}}$

11. $\sqrt{\dfrac{5}{12}}$

12. $\sqrt{\dfrac{11}{18}}$

13. $\dfrac{\sqrt{7x^2}}{\sqrt{50}}$

14. $\dfrac{\sqrt{3x^2}}{\sqrt{32}}$

15. $\dfrac{\sqrt{8}}{\sqrt{56}}$

16. $\dfrac{\sqrt{10}}{\sqrt{20}}$

17. $\dfrac{5}{\sqrt{3a}}$

18. $\dfrac{11}{\sqrt{7b}}$

19. $\sqrt{\dfrac{3m}{11n}}$

20. $\sqrt{\dfrac{5r}{6s}}$

21. $\dfrac{10}{\sqrt{5x}}$

22. $\dfrac{20}{\sqrt{10a}}$

23. $\dfrac{\sqrt{6x}}{\sqrt{32x}}$

24. $\dfrac{\sqrt{10a}}{\sqrt{18a}}$

25. $\dfrac{3}{\sqrt{x^3}}$

26. $\dfrac{5}{\sqrt{b^5}}$

27. $\dfrac{8x^2}{\sqrt{2x}}$

28. $\dfrac{14a^3}{\sqrt{7a}}$

For Exercises 29 and 30, explain the mistake, then simplify correctly.

29. $\dfrac{\sqrt{3}}{\sqrt{2}} = \dfrac{\sqrt{3}}{\sqrt{2}} \cdot \dfrac{2}{2} = \dfrac{2\sqrt{3}}{2}$

30. $\sqrt{\dfrac{7}{3}} = \dfrac{\sqrt{7}}{\sqrt{3}} \cdot \dfrac{\sqrt{3}}{\sqrt{3}} = \dfrac{\sqrt{21}}{9}$

For Exercises 31–50, rationalize the denominators. Assume variables represent nonnegative values.

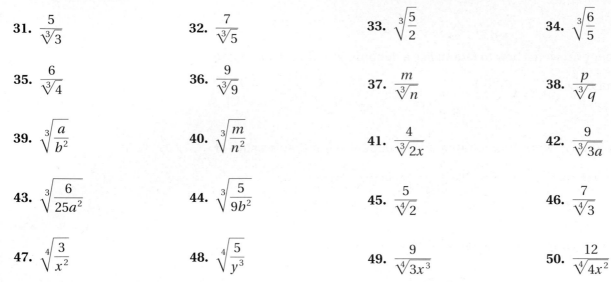

31. $\dfrac{5}{\sqrt[3]{3}}$

32. $\dfrac{7}{\sqrt[3]{5}}$

33. $\sqrt[3]{\dfrac{5}{2}}$

34. $\sqrt[3]{\dfrac{6}{5}}$

35. $\dfrac{6}{\sqrt[3]{4}}$

36. $\dfrac{9}{\sqrt[3]{9}}$

37. $\dfrac{m}{\sqrt[3]{n}}$

38. $\dfrac{p}{\sqrt[3]{q}}$

39. $\sqrt[3]{\dfrac{a}{b^2}}$

40. $\sqrt[3]{\dfrac{m}{n^2}}$

41. $\dfrac{4}{\sqrt[3]{2x}}$

42. $\dfrac{9}{\sqrt[3]{3a}}$

43. $\sqrt[3]{\dfrac{6}{25a^2}}$

44. $\sqrt[3]{\dfrac{5}{9b^2}}$

45. $\dfrac{5}{\sqrt[4]{2}}$

46. $\dfrac{7}{\sqrt[4]{3}}$

47. $\sqrt[4]{\dfrac{3}{x^2}}$

48. $\sqrt[4]{\dfrac{5}{y^3}}$

49. $\dfrac{9}{\sqrt[4]{3x^3}}$

50. $\dfrac{12}{\sqrt[4]{4x^2}}$

For Exercises 51–72, rationalize the denominator and simplify. Assume variables represent nonnegative values.

51. $\dfrac{\sqrt{3}}{\sqrt{2}+1}$

52. $\dfrac{3}{\sqrt{5}+2}$

53. $\dfrac{4}{2-\sqrt{3}}$

54. $\dfrac{2}{4-\sqrt{15}}$

55. $\dfrac{5}{\sqrt{2}+\sqrt{3}}$

56. $\dfrac{7}{\sqrt{6}+\sqrt{7}}$

57. $\dfrac{4}{1-\sqrt{5}}$

58. $\dfrac{6}{1-\sqrt{7}}$

59. $\dfrac{\sqrt{3}}{\sqrt{3}-1}$

60. $\dfrac{\sqrt{5}}{1-\sqrt{5}}$

61. $\dfrac{2\sqrt{3}}{\sqrt{3}-4}$

62. $\dfrac{2\sqrt{5}}{\sqrt{5}-4}$

63. $\dfrac{4\sqrt{3}}{\sqrt{7}+\sqrt{2}}$

64. $\dfrac{\sqrt{8}}{\sqrt{4}+\sqrt{3}}$

65. $\dfrac{8\sqrt{2}}{4\sqrt{2}-\sqrt{3}}$

66. $\dfrac{5\sqrt{3}}{\sqrt{2}+4\sqrt{6}}$

67. $\dfrac{6\sqrt{y}}{\sqrt{y}+1}$

68. $\dfrac{4\sqrt{x}}{\sqrt{x}+1}$

69. $\dfrac{3\sqrt{t}}{\sqrt{t}+2\sqrt{u}}$

70. $\dfrac{2\sqrt{m}}{\sqrt{n}-\sqrt{m}}$

71. $\dfrac{\sqrt{2y}}{\sqrt{x}-\sqrt{6y}}$

72. $\dfrac{\sqrt{14h}}{\sqrt{2h}+\sqrt{k}}$

For Exercises 73–84, rationalize the numerator. Assume variables represent non-negative values.

73. $\dfrac{\sqrt{3}}{2}$

74. $\dfrac{\sqrt{5}}{8}$

75. $\dfrac{\sqrt{2x}}{5}$

76. $\dfrac{\sqrt{7y}}{3}$

77. $\dfrac{\sqrt{8n}}{6}$

78. $\dfrac{\sqrt{20t}}{8}$

79. $\dfrac{2 + \sqrt{3}}{5}$

80. $\dfrac{5 - \sqrt{2}}{6}$

81. $\dfrac{\sqrt{5x} - 6}{9}$

82. $\dfrac{\sqrt{2x} + 7}{3}$

83. $\dfrac{5\sqrt{n} + \sqrt{6n}}{2n}$

84. $\dfrac{4\sqrt{k} - \sqrt{10k}}{5k}$

85. Given $f(x) = \dfrac{5\sqrt{2}}{x}$, find each of the following. Express your answer in simplest form.

  a. $f(\sqrt{6})$

  b. $f(\sqrt{10})$

  c. $f(\sqrt{22})$

86. Given $g(x) = \dfrac{\sqrt{2}}{x - 1}$, find each of the following. Express your answer in simplest form.

  a. $g(\sqrt{5})$

  b. $g(3\sqrt{2})$

  c. $g(2\sqrt{6})$

87. Graph $f(x) = \dfrac{1}{\sqrt{x}}$, then graph $g(x) = \dfrac{\sqrt{x}}{x}$.

  a. What do you notice about the two graphs? What does this indicate about the two functions?

  b. Simplify $f(x)$ by rationalizing the denominator. What do you notice?

88. Graph $f(x) = -\dfrac{1}{\sqrt{x}}$, then graph $g(x) = -\dfrac{\sqrt{x}}{x}$.

  a. What do you notice about the two graphs? What does this indicate about the two functions?

  b. Simplify $f(x)$ by rationalizing the denominator. What do you notice?

89. In Section 10.1, we used the formula $T = 2\pi\sqrt{\dfrac{L}{9.8}}$ to determine the period of a pendulum where T is the period in seconds and L is the length in meters.

    a. Rewrite the formula so that the denominator is rationalized.

    b. Rewrite the formula so that the numerator is rationalized.

90. The formula $t = \sqrt{\dfrac{h}{16}}$ can be used to find the time, t, in seconds for an object to fall a distance of h feet.

    a. Rewrite the formula so that the denominator is rationalized.

    b. Rewrite the formula so that the numerator is rationalized.

91. The formula $s = \sqrt{\dfrac{3V}{h}}$ can be used to find the length, s, of each side of the base of a pyramid having a square base, volume V in cubic feet, and height h in feet.

    a. Rationalize the denominator in the formula.

    b. The volume of the Great Pyramid at Giza in Egypt is approximately 83,068,742 cubic feet and its height is 449 feet. Find the length of each side of its base.

92. The formula $s = 2\sqrt{\dfrac{A}{6\sqrt{3}}}$ can be used to find the length, s, of each side of a regular hexagon having an area A.

    a. Rationalize the denominator in the formula.

s

    b. If A is 100 square meters, write an expression in simplest form for the side length.

    c. Use a calculator to approximate the side lengths, rounded to three decimal places.

93. In AC circuits, voltage is often expressed as a *root-mean-square*, or *rms*, value. The formula for calculating the rms voltage, V_{rms}, given the maximum voltage, V_m, value is $V_{rms} = \dfrac{V_m}{\sqrt{2}}$.

  a. Rationalize the denominator in the formula.

  b. Given a maximum voltage of 163 V, write an expression for the rms voltage.

  c. Use a calculator to approximate the rms voltage rounded to the nearest tenth.

94. The velocity, in meters per second, of a particle can be determined by the formula $v = \sqrt{\dfrac{2E}{m}}$, where E represents the kinetic energy, in joules, of the particle and m represents the mass, in kilograms, of the particle.

  a. Rationalize the denominator in the formula.

  b. A particle with a mass of 1×10^{-6} kilograms has 2.4×10^7 joules of kinetic energy. Write an expression for its velocity.

  c. Use a calculator to approximate the velocity rounded to the nearest tenth.

95. The resistance in a circuit is found to be $\dfrac{5\sqrt{2}}{3 + \sqrt{6}}$ Ω. Rationalize the denominator.

96. Two charged particles, q_1 and q_2, are separated by a distance of 8 centimeters. The values of the charges are $q_1 = 3 \times 10^{-6}$ coulombs and $q_2 = 1 \times 10^{-6}$ coulombs. The charged particles each exert an electrical field. At a point in between the two particles x centimeters away from q_1, the electric fields will cancel each other so that the value of the fields at the point x is 0.

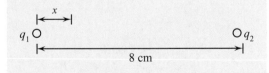

  a. Use the formula $x = \dfrac{l}{1 + \sqrt{\dfrac{q_2}{q_1}}}$ to find the distance from q_1 at which the electric field is canceled, where l is the distance separating the particles. Write the distance with a rationalized denominator.

  b. Use a calculator to approximate the distance rounded to the nearest tenth.

REVIEW EXERCISES

[10.1] **1.** Simplify: $\pm\sqrt{28}$

[6.4] **2.** Factor: $x^2 - 6x + 9$

[2.3] *For Exercises 3 and 4, solve.*

3. $2x - 3 = 5$

4. $2x - 3 = -5$

[6.6] *For Exercises 5 and 6, solve and check.*

5. $x^2 - 36 = 0$

6. $x^2 - 5x + 6 = 0$

$\mathbf{10.6}$ Radical Equations and Problem Solving

OBJECTIVE

1. Use the power rule to solve radical equations.

We now explore how to solve **radical equations**.

DEFINITION *Radical equation:* An equation containing at least one radical expression whose radicand has a variable.

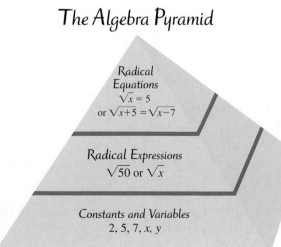

The Algebra Pyramid

Radical
Equations
$\sqrt{x} = 5$
or $\sqrt{x+5} = \sqrt{x-7}$

Radical Expressions
$\sqrt{50}$ or $\sqrt{x}$

Constants and Variables
2, 5, 7, x, y

Note: *We move upward on the Algebra Pyramid, making equations out of the expressions we've studied.*

OBJECTIVE 1. Use the power rule to solve radical equations. To solve radical equations we will use a new principle of equality called the *power rule*.

RULE Power Rule

If $a = b$, then $a^n = b^n$.

Isolated Radicals

First, we consider equations such as $\sqrt{x} = 9$ in which the radical is isolated. When we use the power rule, we raise both sides of the equation to the same power as the root index, then use the principle $(\sqrt[n]{x})^n = x$, which eliminates the radical, leaving us with its radicand.

EXAMPLE 1 Solve.

a. $\sqrt{x} = 9$

Solution $(\sqrt{x})^2 = (9)^2$ Since the root index is 2, we square both sides.

$$x = 81$$

Check $\sqrt{81} \overset{?}{=} 9$

$$9 = 9 \quad \text{True}$$

b. $\sqrt[3]{y} = -2$

Solution $(\sqrt[3]{y})^3 = (-2)^3$ Since the root index is 3, we cube both sides.

$$y = -8$$

Check $\sqrt[3]{-8} \overset{?}{=} -2$

$$-2 = -2 \quad \text{True}$$

Extraneous Solutions

In Section 7.4, we learned that some equations have *extraneous solutions*, which are apparent solutions that actually do not make the original equation true. Using the power rule can sometimes lead to extraneous solutions, so it is important to check solutions. For example, watch what happens when we use the power rule to solve the equation $\sqrt{x} = -9$.

Warning: As we shall see in the check, this result is not a solution. $\longrightarrow$

$(\sqrt{x})^2 = (-9)^2$ Square both sides.

$$x = 81$$

By checking 81 in the original equation we see that it is extraneous.

$$\sqrt{81} \overset{?}{=} -9$$

$$9 = -9 \qquad \text{This equation is false, so 81 is extraneous.}$$

In fact, $\sqrt{x} = -9$ has no real number solution because if x is a real number, then $\sqrt{x}$ must be positive (or 0).

EXAMPLE 2 Solve.

a. $\sqrt{x - 7} = 8$

Solution $(\sqrt{x - 7})^2 = (8)^2$ **Square both sides.**

$$x - 7 = 64$$

$$x = 71 \quad \text{Add 7 to both sides.}$$

Check $\sqrt{71 - 7} \overset{?}{=} 8$

$$\sqrt{64} \overset{?}{=} 8$$

$$8 = 8 \quad \text{True}$$

b. $\sqrt{2x + 1} = -3$

Solution $(\sqrt{2x + 1})^2 = (-3)^2$ **Square both sides.**

$$2x + 1 = 9$$

$$2x = 8 \qquad \text{Subtract 1 from both sides.}$$

$$x = 4 \qquad \text{Divide both sides by 2.}$$

Check $\sqrt{2(4) + 1} \overset{?}{=} -3$

$$\sqrt{9} \overset{?}{=} -3$$

$$3 = -3 \quad \text{False, so 4 is extraneous. This equation has no real-number solution.}$$

Note: *We can see that this equation has no real-number solution because a principal square root cannot be equal to a negative number. However, we will go through the steps to confirm this.*

YOUR TURN Solve.

a. $\sqrt{x - 5} = 6$ **b.** $\sqrt[3]{x + 2} = 3$ **c.** $\sqrt{x + 3} = -7$

Radicals on Both Sides of the Equation

As we will see in Example 3, the power rule can be used to solve equations with radicals on both sides of the equal sign.

EXAMPLE 3 Solve.

a. $\sqrt{6x - 1} = \sqrt{x + 2}$

Solution $(\sqrt{6x - 1})^2 = (\sqrt{x + 2})^2$ **Square both sides.**

$$6x - 1 = x + 2 \qquad \text{Subtract } x \text{ from both sides and add 1 to both sides.}$$

$$5x = 3 \qquad \text{Divide both sides by 5.}$$

$$x = \frac{3}{5}$$

Check $\quad\sqrt{6\left(\dfrac{3}{5}\right) - 1} \overset{?}{=} \sqrt{\dfrac{3}{5} + 2}$

$$\sqrt{\dfrac{18}{5} - \dfrac{5}{5}} \overset{?}{=} \sqrt{\dfrac{3}{5} + \dfrac{10}{5}}$$

$$\sqrt{\dfrac{13}{5}} = \sqrt{\dfrac{13}{5}} \qquad \text{True}$$

b. $\sqrt[3]{5x - 2} = \sqrt[3]{3x + 2}$

Solution $\quad(\sqrt[3]{5x - 2})^3 = (\sqrt[3]{3x + 2})^3 \qquad$ Cube both sides.

$$5x - 2 = 3x + 2$$

$$2x = 4 \qquad\qquad\qquad \text{Subtract } 3x \text{ from both sides and add 2 to both sides.}$$

$$x = 2 \qquad\qquad\qquad \text{Divide both sides by 2.}$$

Check $\quad\sqrt[3]{5(2) - 2} \overset{?}{=} \sqrt[3]{3(2) + 2}$

$$\sqrt[3]{10 - 2} \overset{?}{=} \sqrt[3]{6 + 2}$$

$$\sqrt[3]{8} \overset{?}{=} \sqrt[3]{8}$$

$$2 = 2 \qquad\qquad \text{True}$$

YOUR TURN Solve.

$\quad$ **a.** $\sqrt{8x + 5} = \sqrt{2x + 7}$ $\qquad\qquad$ **b.** $\sqrt[3]{6x + 9} = \sqrt[3]{10x - 3}$

Multiple Solutions

Radical equations may have multiple solutions if, after using the power rule, we are left with a quadratic form.

EXAMPLE 4 Solve. $x + 2 = \sqrt{5x + 16}$

Solution $\quad(x + 2)^2 = (\sqrt{5x + 16})^2 \qquad$ Square both sides.

$$x^2 + 4x + 4 = 5x + 16 \qquad \text{Use FOIL on the left-hand side.}$$

$$x^2 - x - 12 = 0 \qquad\qquad \text{Since the equation is quadratic, we set it equal to 0 by subtracting } 5x \text{ and 16 from both sides.}$$

$$(x + 3)(x - 4) = 0 \qquad \text{Factor.}$$

$$x + 3 = 0 \quad \text{or} \quad x - 4 = 0 \qquad \text{Use the zero-factor theorem.}$$

$$x = -3 \qquad\qquad x = 4$$

Checks $\quad -3 + 2 \overset{?}{=} \sqrt{5(-3) + 16} \qquad 4 + 2 \overset{?}{=} \sqrt{5(4) + 16}$

$$-1 \overset{?}{=} \sqrt{-15 + 16} \qquad\qquad 6 \overset{?}{=} \sqrt{20 + 16}$$

$$-1 \overset{?}{=} \sqrt{1} \qquad\qquad\qquad 6 \overset{?}{=} \sqrt{36}$$

$$-1 = 1 \quad \text{False} \qquad\qquad 6 = 6 \quad \text{True}$$

Because -3 does not check, it is an extraneous solution. The only solution is 4.

YOUR TURN Solve.

$$\sqrt{9x + 7} = x + 3$$

Radicals Not Isolated

Now we consider radical equations in which the radical term is not isolated. In such equations, we must first isolate the radical term.

EXAMPLE 5 Solve.

a. $\sqrt{x + 1} - 2x = x + 1$

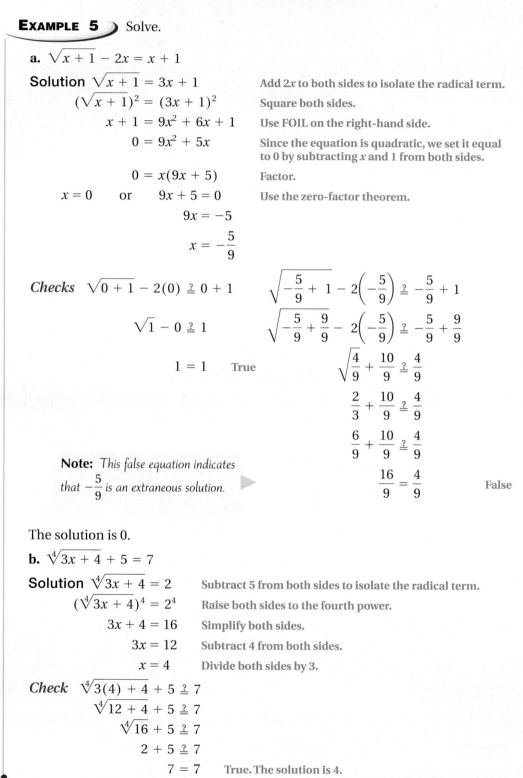

Solution $\sqrt{x + 1} = 3x + 1$ Add $2x$ to both sides to isolate the radical term.

$(\sqrt{x + 1})^2 = (3x + 1)^2$ Square both sides.

$x + 1 = 9x^2 + 6x + 1$ Use FOIL on the right-hand side.

$0 = 9x^2 + 5x$ Since the equation is quadratic, we set it equal to 0 by subtracting x and 1 from both sides.

$0 = x(9x + 5)$ Factor.

$x = 0$ or $9x + 5 = 0$ Use the zero-factor theorem.

$9x = -5$

$x = -\dfrac{5}{9}$

Checks $\sqrt{0 + 1} - 2(0) \stackrel{?}{=} 0 + 1$

$\sqrt{1} - 0 \stackrel{?}{=} 1$

$1 = 1$ **True**

$\sqrt{-\dfrac{5}{9} + 1} - 2\left(-\dfrac{5}{9}\right) \stackrel{?}{=} -\dfrac{5}{9} + 1$

$\sqrt{-\dfrac{5}{9} + \dfrac{9}{9}} - 2\left(-\dfrac{5}{9}\right) \stackrel{?}{=} -\dfrac{5}{9} + \dfrac{9}{9}$

$\sqrt{\dfrac{4}{9}} + \dfrac{10}{9} \stackrel{?}{=} \dfrac{4}{9}$

$\dfrac{2}{3} + \dfrac{10}{9} \stackrel{?}{=} \dfrac{4}{9}$

$\dfrac{6}{9} + \dfrac{10}{9} \stackrel{?}{=} \dfrac{4}{9}$

$\dfrac{16}{9} = \dfrac{4}{9}$ **False**

Note: *This false equation indicates that* $-\dfrac{5}{9}$ *is an extraneous solution.*

The solution is 0.

b. $\sqrt[4]{3x + 4} + 5 = 7$

Solution $\sqrt[4]{3x + 4} = 2$ Subtract 5 from both sides to isolate the radical term.

$(\sqrt[4]{3x + 4})^4 = 2^4$ Raise both sides to the fourth power.

$3x + 4 = 16$ Simplify both sides.

$3x = 12$ Subtract 4 from both sides.

$x = 4$ Divide both sides by 3.

Check $\sqrt[4]{3(4) + 4} + 5 \stackrel{?}{=} 7$

$\sqrt[4]{12 + 4} + 5 \stackrel{?}{=} 7$

$\sqrt[4]{16} + 5 \stackrel{?}{=} 7$

$2 + 5 \stackrel{?}{=} 7$

$7 = 7$ **True. The solution is 4.**

YOUR TURN) Solve.

a. $\sqrt{5x^2 + 6x - 7} + 3x = 5x + 1$ b. $\sqrt[4]{3x + 6} - 7 = -4$

Using the Power Rule Twice

Some equations may require that we use the power rule twice in order to eliminate all radicals.

EXAMPLE 6) Solve. $\sqrt{x + 21} = \sqrt{x} + 3$

Solution $(\sqrt{x + 21})^2 = (\sqrt{x} + 3)^2$ Since one of the radicals is isolated, we square both sides.

$$x + 21 = (\sqrt{x} + 3)(\sqrt{x} + 3)$$

$$x + 21 = x + 3\sqrt{x} + 3\sqrt{x} + 9$$ Use FOIL on the right-hand side.

$$x + 21 = x + 6\sqrt{x} + 9$$ Combine like terms.

$$12 = 6\sqrt{x}$$ Subtract x and 9 from both sides to isolate the remaining radical expression.

$$2 = \sqrt{x}$$ Divide both sides by 6.

$$(2)^2 = (\sqrt{x})^2$$ Square both sides.

$$4 = x$$

Check $\sqrt{4 + 21} \overset{?}{=} \sqrt{4} + 3$

$\sqrt{25} \overset{?}{=} 2 + 3$

$5 = 5$ True. The solution is 4.

Our examples suggest the following procedure.

PROCEDURE   **Solving Radical Equations**

To solve a radical equation:

1. Isolate the radical. (If there is more than one radical term, then isolate one of the radical terms.)
2. Raise both sides of the equation to the same power as the root index.
3. If all radicals have been eliminated, then solve. If a radical term remains, then isolate that radical term and raise both sides to the same power as its root index.
4. Check each solution. Any apparent solution that does not check is an extraneous solution.

YOUR TURN) Solve.

$$\sqrt{2x + 1} = \sqrt{x} + 1$$

10.6 Exercises

For Extra Help · MyMathLab · MyMathLab · Videotape/DVT · InterAct Math · Tutor Center · Math Tutor Center · Math XL.com

1. Explain why we must check all potential solutions to radical equations.

2. What is an extraneous solution?

3. Explain why there is no real-number solution for the radical equation $\sqrt{x} = -6$.

4. Show why $(\sqrt{a})^2 = a$, assuming $a \geq 0$.

5. Given the radical equation $\sqrt{x + 2} + 3x = 4x - 1$, what would be the first step in solving the equation? Why?

6. Give an example of a radical equation that would require you to use the power rule twice in solving the equation. Explain why the principle would have to be used twice.

For Exercises 7–28, solve.

7. $\sqrt{x} = 2$

8. $\sqrt{y} = 5$

9. $\sqrt{k} = -4$

10. $\sqrt{x} = -1$

11. $\sqrt[3]{y} = 3$

12. $\sqrt[3]{m} = 4$

13. $\sqrt[3]{z} = -2$

14. $\sqrt[3]{p} = -5$

15. $\sqrt{n - 1} = 4$

16. $\sqrt{x + 3} = 7$

17. $\sqrt{t - 7} = 2$

18. $\sqrt{m + 8} = 1$

19. $\sqrt{3x - 2} = 4$

20. $\sqrt{2x + 5} = 3$

21. $\sqrt{2x + 17} = 4$

22. $\sqrt{3y - 2} = 5$

23. $\sqrt{2n - 8} = -3$

24. $\sqrt{5x - 1} = -6$

25. $\sqrt[3]{x - 3} = 2$

26. $\sqrt[3]{k + 2} = 4$

27. $\sqrt[3]{3y - 2} = -2$

28. $\sqrt[3]{2x + 4} = -3$

For Exercises 29–38 solve. First isolate the radical term.

29. $\sqrt{u-3} - 10 = 1$

30. $\sqrt{y+1} - 4 = 2$

31. $\sqrt{y-6} + 2 = 9$

32. $\sqrt{r-5} + 6 = 10$

33. $\sqrt{6x-5} - 2 = 3$

34. $\sqrt{8x+4} - 2 = 4$

35. $\sqrt[3]{n+3} - 2 = -4$

36. $\sqrt[3]{x-4} + 2 = 3$

37. $\sqrt[4]{x-2} - 2 = -4$

38. $\sqrt[4]{m+3} + 2 = 5$

For Exercises 39–64, solve.

39. $\sqrt{3x-2} = \sqrt{8-2x}$

40. $\sqrt{m+2} = \sqrt{2m-3}$

41. $\sqrt{4x-5} = \sqrt{6x+5}$

42. $\sqrt{3x-4} = \sqrt{5x+2}$

43. $\sqrt[3]{2r+2} = \sqrt[3]{3r-1}$

44. $\sqrt[3]{3h-4} = \sqrt[3]{h+4}$

45. $\sqrt[4]{4x+4} = \sqrt[4]{5x+1}$

46. $\sqrt[4]{2x+4} = \sqrt[4]{3x-2}$

47. $\sqrt{k+4} = k-8$

48. $\sqrt{5n-1} = 4-2n$

49. $y-1 = \sqrt{2y-2}$

50. $3+x = \sqrt{7+3x}$

51. $\sqrt{3x+10} - 4 = x$

52. $\sqrt{4y+1} + 5 = y$

53. $\sqrt{10n+4} - 3n = n+1$

54. $\sqrt{6x-1} - 6x = 2-9x$

55. $\sqrt[3]{5x+2} + 2 = 5$

56. $\sqrt[3]{4x-1} - 4 = -1$

57. $\sqrt[3]{n^2-2n+5} = 2$

58. $\sqrt[3]{y^2+y+7} = 3$

59. $1 + \sqrt{x} = \sqrt{2x+1}$

60. $\sqrt{t+2} - 1 = \sqrt{t}$

61. $\sqrt{3x+1} + \sqrt{3x} = 2$

62. $\sqrt{x+5} - \sqrt{x} = 1$

63. $\sqrt{x+8} - 2 = \sqrt{x}$

64. $\sqrt{5x-1} - 1 = \sqrt{x+2}$

For Exercises 65–68, explain the mistake, then solve correctly.

65. $\sqrt{x} = -9$
$(\sqrt{x})^2 = (-9)^2$
$x = 81$

66. $\sqrt{x} = -2$
$(\sqrt{x})^2 = (-2)^2$
$x = 4$

67. $\sqrt{x+3} = x - 3$
$x + 3 = x^2 - 9$
$0 = x^2 - x - 12$
$0 = (x + 3)(x - 4)$
$x = -3, 4$

68. $\sqrt{x+3} = 4$
$(\sqrt{x+3})^2 = 4^2$
$x + 9 = 16$
$x + 9 - 9 = 16 - 9$
$x = 7$

For Exercises 69–72, given the period of a pendulum, find the length of the pendulum. Use the formula $T = 2\pi\sqrt{\dfrac{L}{9.8}}$, where T is the period in seconds and L is the length in meters.

69. 2π seconds

70. 6π seconds

71. π seconds

72. $\dfrac{\pi}{3}$ seconds

For Exercises 73–76, find the distance an object has fallen. Use the formula $t = \sqrt{\dfrac{h}{16}}$, where t is the time in seconds that it takes an object to fall a distance of h feet.

73. 0.3 seconds

74. 0.5 seconds

75. 3 seconds

76. $\dfrac{1}{4}$ seconds

For Exercises 77–80, find the skid distance after braking hard at a given speed. Use the formula $S = \dfrac{7}{2}\sqrt{2D}$, where D represents the skid distance in feet on asphalt and S represents the speed of the car in miles per hour. (Source: Harris Technical Services, Traffic Accident Reconstructionists.)

77. 30 miles per hour

78. 60 miles per hour

79. 45 miles per hour

80. 75 miles per hour

For Exercises 81–84, use the following. Two forces, F_1 and F_2, acting on an object at a 90° angle will pull the object with a resultant force, R, at an angle in between F_1 and F_2 (see the figure). The value of the resultant force can be found by the formula $R = \sqrt{F_1^2 + F_2^2}$.

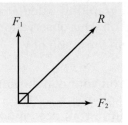

81. Find F_1 if $R = 5$ N and $F_2 = 3$ N.

82. Find F_1 if $R = 10$ N and $F_2 = 8$ N.

83. Find F_2 if $R = 3\sqrt{5}$ N and $F_1 = 3$ N.

84. Find F_2 if $R = 2\sqrt{13}$ N and $F_1 = 6$ N.

85. a. Complete the table of ordered pairs for the equation $y = \sqrt{x}$.

x	0	1	4	9	16	?
y	0	1	2	?	?	5

    b. Graph the ordered pairs in the coordinate plane, then connect the points to make a curve.

    c. Does the graph extend below or to the left of the origin? Explain.

    d. Does the curve represent a function? Explain.

86. a. Complete the table of ordered pairs for the equation $y = \sqrt{x} + 2$.

x	0	1	4	9	16	?
y	2	3	4	?	?	7

    b. Graph the ordered pairs in the coordinate plane, then connect the points to make a curve.

    c. Does the graph extend below or to the left of the origin? Explain.

    d. Does the curve represent a function? Explain.

87. Use a graphing calculator to graph $y = \sqrt{x}$, $y = 2\sqrt{x}$, and $y = 3\sqrt{x}$. Based on your observations, as you increase the size of the coefficient, what happens to the graph?

88. Use a graphing calculator to compare the graph of $y = \sqrt{x}$ to the graph of $y = -\sqrt{x}$. What effect does the negative sign have on the graph?

89. Use a graphing calculator to graph $y = \sqrt{x}$, $y = \sqrt{x} + 2$, and $y = \sqrt{x} - 2$. Based on your observations, what effect does adding or subtracting a constant have on the graph of $y = \sqrt{x}$?

90. Based on your conclusions from Exercises 87–89, without graphing, describe the graph of $y = -2\sqrt{x} + 3$. Use a graphing calculator to graph the equation to confirm your description.

Collaborative Exercises BUILDING TIME

You are constructing a grandfather clock that operates using a pendulum. The period of the pendulum is the time required to complete one full swing. The formula giving the relationship between the length,

L (in meters), and the period, T (in seconds), is $T = 2\pi\sqrt{\dfrac{L}{9.8}}$.

1. Find the length of the pendulum if the period is 1 second.

2. Suppose that you want the pendulum to complete one period in 2 seconds. Would you need to increase or decrease the length of the pendulum found in Exercise 1? Use the same formula to determine the length of the pendulum so that the period is 2 seconds.

3. Based on the results of Exercises 1 and 2, what can you conclude about the required length of the pendulum as the period increases?

REVIEW EXERCISES

[5.1] *For Exercises 1–3, evaluate.*

1. 3^4

2. $(-0.2)^3$

3. $\left(\dfrac{2}{5}\right)^{-4}$

For Exercises 4–6, use the rules of exponents to simplify.

[5.4]   **4.** $(x^3)(x^5)$

[5.4]   **5.** $(n^4)^6$

[5.6]   **6.** $\dfrac{y^7}{y^3}$

10.7 Complex Numbers

OBJECTIVES

1. Write imaginary numbers using i.
2. Perform arithmetic operations with complex numbers.
3. Raise i to powers.

We have said that the square root of a negative number is not a real number. In this section, we will learn about the *imaginary number system*, in which square roots of negative numbers are expressed using a notation involving the letter i.

OBJECTIVE 1. Write imaginary numbers using i. Using the product rule of square roots, any square root of a negative number can be rewritten as a product of a real number and an **imaginary unit**, which we express as i.

DEFINITION *Imaginary unit:* The number represented by i, where $i = \sqrt{-1}$ and $i^2 = -1$.

A number that can be expressed as a product of a real number and the imaginary unit is called an **imaginary number**.

DEFINITION *Imaginary number:* A number that can be expressed in the form bi, where b is a real number and i is the imaginary unit.

Of Interest

When the idea of finding square roots of negative numbers was first introduced, members of the established mathematics community said this type of number existed, but only in the imagination of those finding them. From then on they've been called "imaginary numbers."

Note: *In a product with the imaginary unit, it is customary to write integer factors first, then i, then square root factors.*

EXAMPLE 1 Write each imaginary number as a product of a real number and i.

a. $\sqrt{-9}$

Solution $\sqrt{-9} = \sqrt{-1 \cdot 9}$ Factor out -1 in the radicand.

$\qquad\qquad = \sqrt{-1} \cdot \sqrt{9}$ Use the product rule of square roots.

$\qquad\qquad = i \cdot 3$

$\qquad\qquad = 3i$

b. $\sqrt{-54}$

Solution $\sqrt{-54} = \sqrt{-1 \cdot 54}$ Factor out -1 in the radicand.

$\qquad\qquad = \sqrt{-1} \cdot \sqrt{54}$ Use the product rule of square roots.

$\qquad\qquad = i\sqrt{9 \cdot 6}$ Use the product rule again to simplify further.

$\qquad\qquad = 3i\sqrt{6}$

Example 1 suggests the following procedure.

PROCEDURE **Rewriting Imaginary Numbers**

To write an imaginary number $\sqrt{-n}$ in terms of the imaginary unit i:
1. Separate the radical into two factors, $\sqrt{-1} \cdot \sqrt{n}$.
2. Replace $\sqrt{-1}$ with i.
3. Simplify $\sqrt{n}$.

YOUR TURN Write each imaginary number as a product of a real number and i.

a. $\sqrt{-64}$ **b.** $\sqrt{-18}$ **c.** $\sqrt{-48}$

OBJECTIVE 2. Perform arithmetic operations with complex numbers. We now have two distinct sets of numbers, the set of real numbers and the set of imaginary numbers. There is yet another set of numbers, called the set of **complex numbers**, that contains both the real and the imaginary numbers.

DEFINITION ***Complex number:*** A number that can be expressed in the form $a + bi$, where a and b are real numbers and i is the imaginary unit.

When written in the form $a + bi$, a complex number is said to be in *standard form*. Following are some examples of complex numbers written in standard form:

$$2 + 3i \qquad 4 - 7i \qquad -2.1 - i\sqrt{5}$$

Note that if $a = 0$, then the complex number is purely an imaginary number, such as:

$$-6i \qquad 4i\sqrt{3}$$

If $b = 0$, then the complex number is a real number.

The following Venn Diagram shows how the set of complex numbers contains both the real numbers and imaginary numbers.

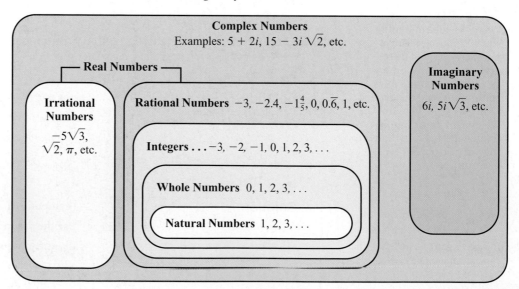

We can perform arithmetic operations with complex numbers. In general, we treat the complex numbers just like polynomials where i is like a variable.

Adding and Subtracting Complex Numbers

EXAMPLE 2 Add or subtract.

a. $(-8 + 7i) + (5 - 19i)$

Solution We add complex numbers just like we add polynomials by combining like terms.

$$(-8 + 7i) + (5 - 19i) = -3 - 12i$$

b. $(-1 + 13i) - (8 - 3i)$

Solution We subtract complex numbers just like we subtract polynomials by writing an equivalent addition and changing the signs in the second complex number.

$$(-1 + 13i) - (8 - 3i) = (-1 + 13i) + (-8 + 3i)$$
$$= -9 + 16i$$

Multiplying Complex Numbers

We multiply complex numbers in the same way that we multiply monomials and binomials. However, we must be careful when simplifying because these products may contain i^2, which is equal to -1.

EXAMPLE 3 Multiply.

a. $(9i)(-8i)$

Solution Multiply in the same way that we multiply monomials.

$$(9i)(-8i) = -72i^2$$
$$= -72(-1) \qquad \text{Replace } i^2 \text{ with } -1.$$
$$= 72$$

b. $(5i)(4 - 7i)$

Solution Multiply in the same way that we multiply a binomial by a monomial.

$$(5i)(4 - 7i) = 20i - 35i^2 \qquad \text{Distribute.}$$
$$= 20i - 35(-1) \qquad \text{Replace } i^2 \text{ with } -1.$$
$$= 20i + 35$$
$$= 35 + 20i \qquad \text{Write in standard form.}$$

c. $(8 - 3i)(2 + i)$

Solution Multiply in the same way that we multiply binomials.

$$(8 - 3i)(2 + i) = 16 + 8i - 6i - 3i^2 \qquad \text{Use FOIL.}$$
$$= 16 + 2i - 3(-1) \qquad \text{Combine like terms and replace } i^2 \text{ with } -1.$$
$$= 16 + 2i + 3 \qquad \text{Simplify.}$$
$$= 19 + 2i \qquad \text{Write in standard form.}$$

d. $(6 - 5i)(6 + 5i)$

Solution Note that these complex numbers are conjugates.

$$(6 - 5i)(6 + 5i) = 36 + 30i - 30i - 25i^2 \qquad \text{Use FOIL.}$$
$$= 36 - (-25) \qquad \text{Combine like terms and replace } i^2 \text{ with } -1.$$
$$= 36 + 25 \qquad \text{\textbf{Note:} } \textit{The product of these two complex}$$
$$= 61 \qquad \textit{numbers is a real number.}$$

The complex numbers that we multiplied in Example 3(d) are called **complex conjugates**.

DEFINITION

> ***Complex conjugates:*** The complex conjugate of a complex number $a + bi$ is $a - bi$.

Some other examples of complex conjugates follow:

$$4 + i \quad \text{and} \quad 4 - i$$
$$9 - 7i \quad \text{and} \quad 9 + 7i$$

Example 3(d) also illustrates the fact that the product of complex conjugates will always be a real number.

YOUR TURN Multiply.

a. $(-4i)(-7i)$ **b.** $(-3i)(6 - i)$ **c.** $(8 - 3i)(5 + 7i)$ **d.** $(7 + 3i)(7 - 3i)$

Dividing Complex Numbers

We have established that the product of complex conjugates is a real number. We can use that fact when the divisor is a complex number. The process is similar to rationalizing denominators.

EXAMPLE 4 Write in standard form.

a. $\dfrac{6 + 5i}{7 - 2i}$

Connection Recall that we rationalize denominators in order to clear undesired square root expressions from a denominator. The imaginary unit i represents a square root expression, $\sqrt{-1}$, which is why we rationalize denominators that contain i.

Solution $\dfrac{6 + 5i}{7 - 2i} = \dfrac{6 + 5i}{7 - 2i} \cdot \dfrac{7 + 2i}{7 + 2i}$

Multiply numerator and denominator by the complex conjugate of the denominator, which is $7 + 2i$.

$$= \frac{42 + 12i + 35i + 10i^2}{49 - 4i^2}$$

$$= \frac{42 + 47i + 10(-1)}{49 - 4(-1)} \qquad \text{Simplify.}$$

$$= \frac{42 + 47i - 10}{49 + 4}$$

$$= \frac{32 + 47i}{53}$$

$$= \frac{32}{53} + \frac{47}{53}i \qquad \text{Write in standard form.}$$

b. $\dfrac{9}{2i}$

Solution Think of $2i$ as $0 + 2i$ so the conjugate is $0 - 2i$, which is $-2i$.

$$\dfrac{9}{2i} = \dfrac{9}{2i} \cdot \dfrac{-2i}{-2i} \qquad \text{Multiply numerator and denominator by } -2i.$$

$$= \dfrac{-18i}{-4i^2}$$

$$= \dfrac{-18i}{-4(-1)} \qquad \text{Replace } i^2 \text{ with } -1.$$

$$= \dfrac{-18i}{4}$$

$$= -\dfrac{9}{2}i$$

Note: *Remember the goal is to eliminate* i *from the denominator. Multiplying by* $\dfrac{i}{i}$ *would have worked, too.*

$$\dfrac{9}{2i} = \dfrac{9}{2i} \cdot \dfrac{i}{i} = \dfrac{9i}{2i^2} = \dfrac{9i}{2(-1)} = -\dfrac{9}{2}i$$

YOUR TURN Write in standard form.

a. $\dfrac{6 + i}{4 - 5i}$

b. $\dfrac{8}{3i}$

OBJECTIVE 3. Raise i to powers. We have defined i as $\sqrt{-1}$ and seen that $i^2 = -1$. Raising i to other powers leads to an interesting pattern.

$$i^1 = i \qquad\qquad\qquad i^5 = i^4 \cdot i = 1 \cdot i = i$$
$$i^2 = -1 \qquad\qquad\qquad i^6 = i^4 \cdot i^2 = 1 \cdot (-1) = -1$$
$$i^3 = i^2 \cdot i = -1 \cdot i = -i \qquad\qquad i^7 = i^4 \cdot i^3 = 1 \cdot (-i) = -i$$
$$i^4 = (i^2)^2 = (-1)^2 = 1 \qquad\qquad i^8 = (i^4)^2 = 1^2 = 1$$

Note: *If i's exponent is an even number not divisible by 4, then the result is -1. If i's exponent is divisible by 4, then the result is 1. If the exponent is an odd number that precedes a number divisible by 4, then the result is $-i$.*

If we continue the pattern, we get:

$$i^1 = i \qquad\qquad i^5 = i \qquad\qquad i^9 = i$$
$$i^2 = -1 \qquad\qquad i^6 = -1 \qquad\qquad i^{10} = -1$$
$$i^3 = -i \qquad\qquad i^7 = -i \qquad\qquad i^{11} = -i$$
$$i^4 = 1 \qquad\qquad i^8 = 1 \qquad\qquad i^{12} = 1$$

Notice that i to any integer power is i, -1, $-i$, or 1. This pattern allows us to find i to any integer power. We will use the fact that since $i^4 = 1$, then $(i^4)^n = 1$ for any integer value of n.

EXAMPLE 5 Simplify.

a. i^{25}

Solution $i^{25} = i^{24} \cdot i$ Write i^{25} as $i^{24} \cdot i$ since 24 is the largest multiple of 4 that is smaller than 25.

$$= (i^4)^6 \cdot i \qquad \text{Write } i^{24} \text{ as } (i^4)^6 \text{ since } i^4 = 1.$$

$$= 1^6 \cdot i \qquad \text{Replace } i^4 \text{ with 1.}$$

$$= 1 \cdot i$$

$$= i$$

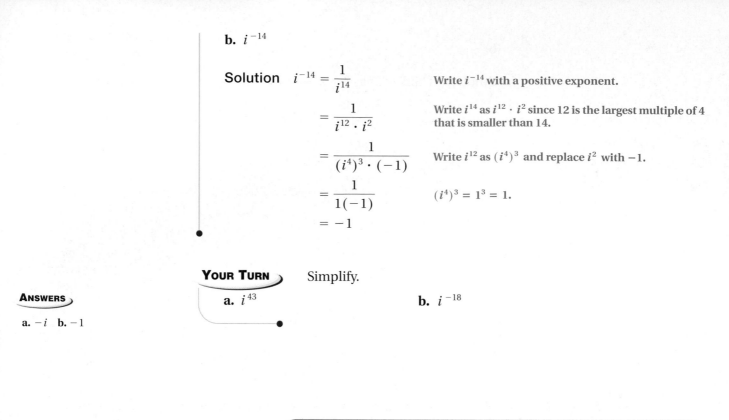

b. i^{-14}

Solution $i^{-14} = \dfrac{1}{i^{14}}$ Write i^{-14} with a positive exponent.

$= \dfrac{1}{i^{12} \cdot i^2}$ Write i^{14} as $i^{12} \cdot i^2$ since 12 is the largest multiple of 4 that is smaller than 14.

$= \dfrac{1}{(i^4)^3 \cdot (-1)}$ Write i^{12} as $(i^4)^3$ and replace i^2 with -1.

$= \dfrac{1}{1(-1)}$ $(i^4)^3 = 1^3 = 1.$

$= -1$

YOUR TURN Simplify.

a. i^{43} **b.** i^{-18}

ANSWERS

a. $-i$ **b.** -1

10.7 Exercises

For Extra Help MyMathLab MyMathLab Videotape/DVT InterAct Math Math Tutor Center Math XL.com

1. As an imaginary number, what does i represent?

2. Is every real number a complex number? Explain.

3. Is every complex number an imaginary number? Explain.

4. Explain how to add complex numbers.

5. Explain how to subtract complex numbers.

6. Is the expression $\dfrac{5 - 4i}{3}$ in standard form for a complex number? Explain.

For Exercises 7–22, write the imaginary number using i.

7. $\sqrt{-36}$ **8.** $\sqrt{-81}$ **9.** $\sqrt{-5}$ **10.** $\sqrt{-7}$

11. $\sqrt{-8}$ **12.** $\sqrt{-12}$ **13.** $\sqrt{-28}$ **14.** $\sqrt{-32}$

15. $\sqrt{-27}$ **16.** $\sqrt{-72}$ **17.** $\sqrt{-125}$ **18.** $\sqrt{-80}$

19. $\sqrt{-63}$ **20.** $\sqrt{-45}$ **21.** $\sqrt{-245}$ **22.** $\sqrt{-810}$

For Exercises 23–38, add or subtract.

23. $(9 + 3i) + (-3 + 4i)$

24. $(5 - 3i) + (4 - 5i)$

25. $(6 + 2i) + (5 - 8i)$

26. $(4 + i) + (7 - 6i)$

27. $(-4 + 6i) - (3 + 5i)$

28. $(8 - 3i) - (-1 - 2i)$

29. $(15 - 3i) - (-7i)$

30. $(19 + 6i) - (-2i)$

31. $(12 + 3i) + (-15 - 13i)$

32. $(14 - 7i) + (-8 - 9i)$

33. $(-5 - 9i) - (-5 - 9i)$

34. $(-4 + 2i) - (-4 + 2i)$

35. $(10 + i) - (2 - 13i) + (6 - 5i)$

36. $(-14 + 2i) + (6 + i) + (19 + 10i)$

37. $(5 - 2i) - (9 - 14i) + (16i)$

38. $(-12i) - (4 - 7i) - (18 + 4i)$

For Exercises 39–54, multiply.

39. $(8i)(3i)$ **40.** $(4i)(-7i)$ **41.** $(-8i)(5i)$ **42.** $(9i)(i)$

43. $2i(6 - 7i)$ **44.** $6i(9 + i)$ **45.** $-8i(4 + 9i)$ **46.** $-7i(5 - 8i)$

47. $(6 + i)(3 - i)$ **48.** $(5 - 2i)(4 + i)$ **49.** $(8 + 5i)(5 - 2i)$ **50.** $(6 - 3i)(7 + 8i)$

51. $(8 + i)(8 - i)$ **52.** $(4 + 9i)(4 - 9i)$ **53.** $(10 + i)^2$ **54.** $(5 - 3i)^2$

For Exercises 55–74, write in standard form.

55. $\dfrac{2}{i}$ **56.** $\dfrac{6}{-i}$ **57.** $\dfrac{4}{5i}$ **58.** $\dfrac{6}{7i}$

59. $\dfrac{6}{2i}$ **60.** $\dfrac{8}{4i}$ **61.** $\dfrac{2 - i}{2i}$ **62.** $\dfrac{3 + i}{3i}$

63. $\dfrac{4 + 2i}{4i}$

64. $\dfrac{5 - 3i}{4i}$

65. $\dfrac{7}{2 + i}$

66. $\dfrac{5}{6 + i}$

67. $\dfrac{2i}{3 - 7i}$

68. $\dfrac{-4i}{5 + i}$

69. $\dfrac{5 - 9i}{1 - i}$

70. $\dfrac{3 + i}{2 - i}$

71. $\dfrac{3 + i}{2 + 3i}$

72. $\dfrac{1 + 3i}{5 + 2i}$

73. $\dfrac{1 + 6i}{4 + 5i}$

74. $\dfrac{5 - 6i}{2 - 9i}$

For Exercises 75–90, find the powers of i.

75. i^{19}

76. i^{25}

77. i^{46}

78. i^{51}

79. i^{38}

80. i^{41}

81. i^{60}

82. i^{52}

83. i^{-20}

84. i^{-32}

85. i^{-26}

86. i^{-38}

87. i^{-21}

88. i^{-45}

89. i^{-35}

90. i^{-44}

REVIEW EXERCISES

[6.4]   **1.** Factor: $4x^2 - 12x + 9$

[10.1]   **2.** Simplify: $\pm\sqrt{48}$

For Exercises 3 and 4, solve.

[2.3]   **3.** $3x - 2 = 4$

[6.6]   **4.** $2x^2 - x - 6 = 0$

[4.3]   **5.** Find the *x*- and *y*-intercepts for $2x + 3y = 6$.

[6.7]   **6.** Graph: $y = x^2 - 3$

Chapter 10 Summary

Defined Terms

Section 10.1
n^{th} root (p. 784)
Radical function (p. 788)

Section 10.2
Rational exponent (p. 796)

Section 10.4
Like radicals (p. 817)

Section 10.6
Radical equation (p. 838)

Section 10.7
Imaginary unit (p. 849)
Imaginary number (p. 849)
Complex number (p. 850)
Complex conjugate (p. 852)

Procedures, Rules, and Key Examples

Procedures/Rules	Key Example(s)

Section 10.1 Radical Expressions and Functions

If a is positive, then $\sqrt[n]{a} = b$, where $b \geq 0$ and $b^n = a$.

If a is negative and n is even, then there is no real-number root.

If a is negative and n is odd, then $\sqrt[n]{a} = b$, where b is negative and $b^n = a$.

Example 1: Simplify.

a. $\sqrt{36} = 6$

b. $\sqrt{-25}$ is not a real number

c. $-\sqrt{121} = -11$

d. $\sqrt{\dfrac{49}{100}} = \dfrac{7}{10}$

e. $\sqrt{16x^6} = 4|x^3|$

f. $\sqrt[3]{-27} = -3$

Note: *If we assume that the variables represent nonnegative numbers, then the solution for part e would be* $\sqrt{16x^6} = 4x^3$.

To evaluate a radical function, replace the variable with the indicated value and simplify.

Example 2: If $f(x) = \sqrt{4x + 3}$, find $f(3)$.
Solution:
$$f(3) = \sqrt{4(3) + 3} = \sqrt{12 + 3} = \sqrt{15}$$

For radical functions with an even index, the radicand must be nonnegative.

Example 3: Find the domain of
$$f(x) = \sqrt{2x - 4}.$$

Solution: Since $2x - 4$ must be nonnegative, we solve $2x - 4 \geq 0$.
$$2x - 4 \geq 0$$
$$2x \geq 4$$
$$x \geq 2$$

Section 10.2 Rational Exponents

$a^{1/n} = \sqrt[n]{a}$, where n is a natural number other than 1.

Note: If a is negative and n is odd, then the root will be negative. If a is negative and n is even, then there is no real-number solution.

Example 1: Evaluate.

a. $25^{1/2} = \sqrt{25} = 5$

b. $(-27)^{1/3} = \sqrt[3]{-27} = -3$

c. $(-16)^{1/4} = \sqrt[4]{-16}$, which is not a real number.

continued

Procedures/Rules	Key Example(s)

Section 10.2 Rational Exponents (continued)

$a^{m/n} = \sqrt[n]{a^m} = (\sqrt[n]{a})^m$, where $a \geq 0$ and m and n are natural numbers other than 1.

$a^{-m/n} = \dfrac{1}{a^{m/n}}$, where $a \neq 0$ and m and n are natural numbers with $n \neq 1$.

Example 2: Write the following in exponential form.

a. $\sqrt[3]{x^2} = x^{2/3}$

b. $(\sqrt[6]{x})^5 = x^{5/6}$

Example 3: Evaluate.

$$16^{-3/4} = \frac{1}{16^{3/4}} = \frac{1}{(\sqrt[4]{16})^3} = \frac{1}{2^3} = \frac{1}{8}$$

Example 4: Write as a radical expression with a smaller root index.

a. $\sqrt[4]{25} = 25^{1/4} = (5^2)^{1/4}$
$$= 5^{2 \cdot 1/4} = 5^{1/2} = \sqrt{5}$$

b. $\sqrt[8]{x^6} = x^{6/8} = x^{3/4} = \sqrt[4]{x^3}$

Example 5: Simplify.

a. $\sqrt[4]{x} \cdot \sqrt{x} = x^{1/4} \cdot x^{1/2} = x^{1/4+1/2}$
$$= x^{1/4+2/4}$$
$$= x^{3/4} = \sqrt[4]{x^3}$$

b. $\dfrac{\sqrt[3]{x^2}}{\sqrt[4]{x}} = \dfrac{x^{2/3}}{x^{1/4}} = x^{(2/3)-(1/4)}$
$$= x^{(8/12)-(3/12)} = x^{5/12} = \sqrt[12]{x^5}$$

c. $\sqrt{5} \cdot \sqrt[3]{2} = 5^{1/2} \cdot 2^{1/3} = 5^{3/6} \cdot 2^{2/6}$
$$= (5^3 \cdot 2^2)^{1/6} = (125 \cdot 4)^{1/6}$$
$$= 500^{1/6} = \sqrt[6]{500}$$

Example 6: Simplify $\sqrt[3]{\sqrt[5]{x}}$ to a single radical.

Solution: $\sqrt[3]{\sqrt[5]{x}} = (x^{1/5})^{1/3} = x^{(1/5) \cdot (1/3)}$
$$= x^{1/15} = \sqrt[15]{x}$$

Section 10.3 Multiplying, Dividing, and Simplifying Radicals

Product rule for radicals:
If $\sqrt[n]{a}$ and $\sqrt[n]{b}$ are both real numbers, then $\sqrt[n]{a} \cdot \sqrt[n]{b} = \sqrt[n]{a \cdot b}$.

Example 1: Simplify. Assume variables represent nonnegative values.

a. $\sqrt{2} \cdot \sqrt{50} = \sqrt{2 \cdot 50} = \sqrt{100} = 10$

b. $\sqrt{3x} \cdot \sqrt{12x^3} = \sqrt{3x \cdot 12x^3}$
$$= \sqrt{36x^4} = 6x^2$$

c. $\sqrt[3]{5x} \cdot \sqrt[3]{3x} = \sqrt[3]{15x^2}$

d. $\sqrt[4]{27x^3} \cdot \sqrt[4]{3x} = \sqrt[4]{27x^3 \cdot 3x}$
$$= \sqrt[4]{81x^4} = 3x$$

continued

Procedures/Rules	Key Example(s)

Section 10.3 Multiplying, Dividing, and Simplifying Radicals (continued)

Raising an nth root to the nth power:

For any nonnegative real number a, $(\sqrt[n]{a})^n = a$.

Quotient rule for radicals:

If $\sqrt[n]{a}$ and $\sqrt[n]{b}$ are both real numbers, then $\dfrac{\sqrt[n]{a}}{\sqrt[n]{b}} = \sqrt[n]{\dfrac{a}{b}}$, where $b \neq 0$.

Simplifying an nth root.
1. Write the radicand as the product of the greatest possible n^{th} power and a number or expression that has no perfect n^{th} power factors.
2. Use the product rule $\sqrt[n]{ab} = \sqrt[n]{a} \cdot \sqrt[n]{b}$, where a is the perfect n^{th} power.
3. Find the nth root of the perfect n^{th} power radicand.

Example 2: Simplify $(\sqrt[4]{x^3})^4$.
Solution: $(\sqrt[4]{x^3})^4 = x^3$

Example 3: Simplify.

a. $\dfrac{\sqrt{45}}{\sqrt{5}} = \sqrt{\dfrac{45}{5}} = \sqrt{9} = 3$

b. $\dfrac{\sqrt[3]{128x^7}}{\sqrt[3]{2x}} = \sqrt[3]{\dfrac{128x^7}{2x}} = \sqrt[3]{64x^6} = 4x^2$

c. $\sqrt{\dfrac{18x^5}{8x^3}} = \sqrt{\dfrac{9x^2}{4}} = \dfrac{\sqrt{9x^2}}{\sqrt{4}} = \dfrac{3x}{2}$

d. $\sqrt[4]{\dfrac{8}{x^4}} = \dfrac{\sqrt[4]{8}}{\sqrt[4]{x^4}} = \dfrac{\sqrt[4]{8}}{x}$

Example 4: Simplify.

a. $\sqrt{50} = \sqrt{25 \cdot 2} = \sqrt{25} \cdot \sqrt{2} = 5\sqrt{2}$

b. $\sqrt{48x^3} = \sqrt{16x^2 \cdot 3x}$
$= \sqrt{16x^2} \cdot \sqrt{3x} = 4x\sqrt{3x}$

c. $\sqrt[4]{48a^9} = \sqrt[4]{16a^8 \cdot 3a}$
$= \sqrt[4]{16a^8} \cdot \sqrt[4]{3a} = 2a^2\sqrt[4]{3a}$

Example 5: Perform the operation and simplify.
$$\sqrt{6} \cdot \sqrt{10} = \sqrt{60} = \sqrt{4 \cdot 15}$$
$$= \sqrt{4} \cdot \sqrt{15} = 2\sqrt{15}$$

Section 10.4 Adding, Subtracting, and Multiplying Radical Expressions

To add or subtract like radicals, add or subtract the coefficients and keep the radicals the same.

Example 1: Simplify.
a. $2\sqrt{x} - 7\sqrt{x} = (2 - 7)\sqrt{x} = -5\sqrt{x}$
b. $4\sqrt[3]{5} + 2\sqrt[3]{5} = (4 + 2)\sqrt[3]{5} = 6\sqrt[3]{5}$

Example 2: Simplify the radicals, then combine like radicals.
$9\sqrt{3} - \sqrt{12} + 6\sqrt{75}$
$= 9\sqrt{3} - \sqrt{4 \cdot 3} + 6\sqrt{25 \cdot 3}$
$= 9\sqrt{3} - 2\sqrt{3} + 6 \cdot 5\sqrt{3}$
$= 9\sqrt{3} - 2\sqrt{3} + 30\sqrt{3}$
$= 37\sqrt{3}$

Example 3: Multiply then simplify.
$(\sqrt{3} - 4)(\sqrt{3} + 6)$
$= \sqrt{3} \cdot \sqrt{3} + \sqrt{3} \cdot 6 - 4 \cdot \sqrt{3} - 4 \cdot 6$ **Use FOIL.**

$= \sqrt{9} + 6\sqrt{3} - 4\sqrt{3} - 24$ **Multiply.**
$= 3 + 6\sqrt{3} - 4\sqrt{3} - 24$ **Simplify $\sqrt{9}$.**
$= -21 + 2\sqrt{3}$ **Combine like radicals.**
continued

Procedures/Rules	Key Example(s)

Section 10.4 Adding, Subtracting and Multiplying Radical Expressions (continued)

Example 4: Perform the following operations.

a. $\sqrt{6} \cdot \sqrt{8} + \sqrt{5} \cdot \sqrt{15}$
$$= \sqrt{48} + \sqrt{75}$$
$$= \sqrt{16 \cdot 3} + \sqrt{25 \cdot 3}$$
$$= 4\sqrt{3} + 5\sqrt{3} = 9\sqrt{3}$$

b. $\dfrac{\sqrt{60}}{\sqrt{5}} + \sqrt{48} = \sqrt{\dfrac{60}{5}} + \sqrt{48}$
$$= \sqrt{12} + \sqrt{48}$$
$$= \sqrt{4 \cdot 3} + \sqrt{16 \cdot 3}$$
$$= 2\sqrt{3} + 4\sqrt{3} = 6\sqrt{3}$$

Section 10.5 Rationalizing Numerators and Denominators of Radical Expressions

Summary of Simplest Form

A radical expression is in simplest form if:
1. All rational roots have been found.
2. There are no perfect nth factors of radicands of the form $\sqrt[n]{a^n}$.
3. All possible products, quotients, sums, and differences have been found.
4. There are no radicals in the denominator of a fraction.

To rationalize a denominator containing a single nth root, multiply the fraction by a 1 so that the product's denominator has a radicand that is a perfect nth power.

To rationalize a denominator containing a sum or difference with at least one square root term, multiply the fraction by a 1 whose numerator and denominator are the conjugate of the denominator.

The numerator of a rational expression is rationalized exactly the same way a denominator is rationalized.

Example 1: Rationalize the denominator.

a. $\dfrac{7}{\sqrt{3}} = \dfrac{7}{\sqrt{3}} \cdot \dfrac{\sqrt{3}}{\sqrt{3}} = \dfrac{7\sqrt{3}}{3}$

b. $\dfrac{4}{\sqrt[3]{3}} = \dfrac{4}{\sqrt[3]{3}} \cdot \dfrac{\sqrt[3]{9}}{\sqrt[3]{9}} = \dfrac{4\sqrt[3]{9}}{\sqrt[3]{27}} = \dfrac{4\sqrt[3]{9}}{3}$

c. $\dfrac{6}{4 - \sqrt{5}} = \dfrac{6}{4 - \sqrt{5}} \cdot \dfrac{4 + \sqrt{5}}{4 + \sqrt{5}}$
$$= \dfrac{6 \cdot 4 + 6 \cdot \sqrt{5}}{16 - 5}$$
$$= \dfrac{24 + 6\sqrt{5}}{11}$$

Example 2: Rationalize the numerator.

a. $\dfrac{\sqrt{6}}{3} = \dfrac{\sqrt{6}}{3} \cdot \dfrac{\sqrt{6}}{\sqrt{6}} = \dfrac{\sqrt{36}}{3\sqrt{6}}$
$$= \dfrac{6}{3\sqrt{6}} = \dfrac{2}{\sqrt{6}}$$

b. $\dfrac{2 - \sqrt{3}}{5} = \dfrac{2 - \sqrt{3}}{5} \cdot \dfrac{2 + \sqrt{3}}{2 + \sqrt{3}}$
$$= \dfrac{4 - 3}{5(2 + \sqrt{3})} = \dfrac{1}{10 + 5\sqrt{3}}$$

continued

Procedures/Rules	Key Example(s)

Section 10.6 Radical Equations and Problem Solving

Power Rule: If $a = b$, then $a^n = b^n$.

To solve a radical equation:
1. Isolate the radical. (If there is more than one radical term, then isolate one of the radical terms.)
2. Raise both sides of the equation to the same power as the root index.
3. If all radicals have been eliminated, then solve. If a radical term remains, then isolate that radical term and raise both sides to the same power as its root index.
4. Check each solution. Any apparent solution that does not check is an extraneous solution.

Example 1: Solve $\sqrt{x - 5} = 7$.

$$(\sqrt{x - 5})^2 = 7^2 \quad \text{Square both sides.}$$
$$x - 5 = 49$$
$$x = 54 \quad \text{Add 5 on both sides.}$$

Check: $\sqrt{54 - 5} = 7$
$$\sqrt{49} = 7 \quad \text{True}$$

Example 2: Solve $\sqrt{n + 14} = n + 2$.

$$(\sqrt{n + 14})^2 = (n + 2)^2 \quad \text{Square both sides.}$$

$$n + 14 = n^2 + 4n + 4$$
$$0 = n^2 + 3n - 10 \quad \text{Subtract } n \text{ and } 14 \text{ on both sides.}$$
$$0 = (n + 5)(n - 2) \quad \text{Factor.}$$
$$n + 5 = 0 \text{ or } n - 2 = 0 \quad \text{Use the zero-factor theorem.}$$
$$n = -5 \qquad n = 2$$

Check:
$$\sqrt{-5 + 14} = -5 + 2 \qquad \sqrt{2 + 14} = 2 + 2$$
$$\sqrt{9} = -3 \quad \text{False} \qquad \sqrt{16} = 4 \quad \text{True}$$

The only solution is 2 (-5 is an extraneous solution).

Example 3: Solve $3 + \sqrt{t} = \sqrt{t + 21}$.

$$(3 + \sqrt{t})^2 = (\sqrt{t + 21})^2 \quad \text{Square both sides.}$$

$$9 + 6\sqrt{t} + t = t + 21$$
$$6\sqrt{t} = 12 \quad \text{Subtract 9 and } t \text{ on both sides.}$$
$$\sqrt{t} = 2 \quad \text{Divide both sides by 6.}$$
$$(\sqrt{t})^2 = 2^2 \quad \text{Square both sides.}$$
$$t = 4$$

We will leave the check to the reader.

continued

Procedures/Rules	Key Example(s)

Section 10.7 Complex Numbers

To write an imaginary number $\sqrt{-n}$ in terms of the imaginary unit i:
1. Separate the radical into two factors $\sqrt{-1} \cdot \sqrt{n}$.
2. Replace $\sqrt{-1}$ with i.
3. Simplify $\sqrt{n}$.

Example 1: Write using the imaginary unit.

a. $\sqrt{-36} = \sqrt{-1} \cdot \sqrt{36} = 6i$

b. $\sqrt{-32} = \sqrt{-1} \cdot \sqrt{32}$
$$= i \cdot 4\sqrt{2} = 4i\sqrt{2}$$

To add complex numbers, combine like terms.

Example 2: Add $(5 - 6i) + (9 + 2i)$.
$$(5 - 6i) + (9 + 2i) = 14 - 4i$$

To subtract complex numbers, change the signs of the second complex numbers, then combine like terms.

Example 3: Subtract $(7 - i) - (3 + 5i)$.
$$(7 - i) - (3 + 5i) = (7 - i) + (-3 - 5i)$$
$$= 4 - 6i$$

To multiply complex numbers, follow the same procedures as for multiplying monomials or binomials (FOIL). Remember that $i^2 = -1$.

Example 4: Multiply $(6 + 5i)(2 - 3i)$.
$$(6 + 5i)(2 - 3i) = 12 - 18i + 10i - 15i^2$$
$$= 12 - 8i - 15(-1)$$
$$= 12 - 8i + 15$$
$$= 27 - 8i$$

To divide complex numbers, rationalize the denominator using the complex conjugate.

Example 5: Write $\dfrac{2 - 3i}{4 + 5i}$ in standard form.

$$\frac{2 - 3i}{4 + 5i} = \frac{2 - 3i}{4 + 5i} \cdot \frac{4 - 5i}{4 - 5i}$$
$$= \frac{8 - 10i - 12i + 15i^2}{16 - 25i^2}$$
$$= \frac{8 - 22i + 15(-1)}{16 - 25(-1)}$$
$$= \frac{-7 - 22i}{41}$$
$$= -\frac{7}{41} - \frac{22}{41}i$$

Powers of i.
$$i^1 = i$$
$$i^2 = -1$$
$$i^3 = -i$$
$$i^4 = 1$$

Example 6: Simplify.
a. $i^{19} = i^{16} \cdot i^3 = (i^4)^4 \cdot (-i)$
$$= 1^4(-i) = 1(-i) = -i$$

b. $i^{-21} = \dfrac{1}{i^{21}} = \dfrac{1}{i^{20} \cdot i} = \dfrac{1}{(i^4)^5 \cdot i} = \dfrac{1}{1^5 \cdot i}$

$$= \frac{1}{i} = \frac{1}{i} \cdot \frac{-i}{-i} = \frac{-i}{-i^2}$$

$$= \frac{-i}{-(-1)} = \frac{-i}{1} = -i$$

Chapter 10 Review Exercises

For Exercises 1–5, answer true or false.

[10.1] **1.** Every positive number has two square roots: a positive root and a negative root.

[10.1] **2.** $\sqrt{x^2} = x$, where x is any real number.

[10.5] **3.** The expression $\dfrac{3}{\sqrt{2}}$ is in simplest form.

[10.6] **4.** Every radical equation has a single solution.

[10.6] **5.** It is necessary to check all potential solutions to radical equations.

For Exercises 6–10, complete the rule.

[10.4] **6.** Like radicals are two radical expression with identical _____ and indexes.

[10.4] **7.** To add or subtract like radicals, add or subtract the _____ and keep the radicals the same.

[10.5] **8.** To rationalize a denominator that is a single square root,

    1. _____ both the numerator and denominator of the fraction by the same square root as appears in the denominator.

    2. Simplify.

[10.5] **9.** To rationalize a denominator containing a sum or difference with a square root term, multiply the numerator and denominator by the _____ of the denominator.

[10.6] **10.** An apparent solution to a radical equation that does not solve the equation is considered _____ .

[10.1] **For Exercises 11 and 12, find all square roots of the given number.**

11. 121

12. 49

[10.1] **For Exercises 13–16, evaluate the square root.**

13. $\sqrt{169}$

14. $-\sqrt{49}$

15. $\sqrt{-36}$

16. $\sqrt{\dfrac{1}{25}}$

Expressions

Exercises 11–96

[10.1] *For Exercises 17 and 18, use a calculator to approximate each root to the nearest thousandth.*

17. $\sqrt{7}$ **18.** $\sqrt{90}$

[10.1] *For Exercises 19–26, simplify. Assume variables represent nonnegative numbers.*

19. $\sqrt{49x^8}$ **20.** $\sqrt{144a^6b^{12}}$ **21.** $\sqrt{0.16m^2n^{10}}$ **22.** $\sqrt[3]{x^{15}}$

23. $\sqrt[3]{-64r^9s^3}$ **24.** $\sqrt[4]{81x^{12}}$ **25.** $\sqrt[5]{32x^{15}y^{20}}$ **26.** $\sqrt[7]{x^{14}y^7}$

[10.1] *For Exercises 27–28, simplify. Assume variables can represent any real number.*

27. $\sqrt{81x^2}$ **28.** $\sqrt[4]{(x-1)^8}$

[10.2] *For Exercises 29–32, rewrite using radical notation and evaluate. Assume all variables represent nonnegative quantities.*

29. $(-64)^{1/3}$ **30.** $(24a^4)^{1/2}$ **31.** $\left(\dfrac{1}{32}\right)^{3/5}$ **32.** $(5r-2)^{5/7}$

[10.2] *For Exercises 33–38, write in exponential form.*

33. $\sqrt[8]{33}$ **34.** $\dfrac{8}{\sqrt[7]{n^3}}$ **35.** $(\sqrt[5]{8})^3$

36. $(\sqrt[3]{m})^8$ **37.** $(\sqrt[4]{3xw})^3$ **38.** $\sqrt[3]{(a+b)^4}$

[10.2] *For Exercises 39–44, use the rules of exponents to simplify.*

39. $x^{2/3} \cdot x^{4/3}$ **40.** $(4m^{1/4})(8m^{5/4})$ **41.** $\dfrac{y^{3/5}}{y^{4/5}}$

42. $\dfrac{b^{2/5}}{b^{-3/5}}$ **43.** $(k^{2/3})^{3/4}$ **44.** $(2xy^{1/5})^{3/4}$

[10.3] *For Exercises 45–56, simplify. Assume variables represent nonnegative values.*

45. $\sqrt{3} \cdot \sqrt{27}$ **46.** $\sqrt{5x^5} \cdot \sqrt{20x^3}$ **47.** $\sqrt[3]{2} \cdot \sqrt[3]{4}$ **48.** $\sqrt[4]{7} \cdot \sqrt[4]{6}$

49. $\sqrt[5]{3x^2y^3} \cdot \sqrt[5]{5x^2y}$

50. $\sqrt{\dfrac{49}{121}}$

51. $\sqrt[3]{-\dfrac{27}{8}}$

52. $\sqrt{72}$

53. $4b\sqrt{27b^7}$

54. $5\sqrt[3]{108}$

55. $2\sqrt[3]{40x^{10}}$

56. $2x^4\sqrt[4]{162x^7}$

[10.3] *For Exercises 57–60, find the products and write the answer in simplified form. Assume variables represent nonnegative values.*

57. $4\sqrt{6} \cdot 7\sqrt{15}$

58. $\sqrt{x^9} \cdot \sqrt{x^6}$

59. $4\sqrt{10c} \cdot 2\sqrt{6c^4}$

60. $a\sqrt{a^3b^2} \cdot b^2\sqrt{a^5b^3}$

[10.3] *For Exercises 61–64, find the quotients and write the answers in simplified form. Assume all variables represent nonnegative values.*

61. $\dfrac{\sqrt{48}}{\sqrt{6}}$

62. $\dfrac{9\sqrt{160}}{3\sqrt{8}}$

63. $\dfrac{8\sqrt{45a^5}}{2\sqrt{5a}}$

64. $\dfrac{36\sqrt{96x^6y^{11}}}{4\sqrt{3x^2y^4}}$

For Exercises 65–72, find the sum or difference.

65. $-5\sqrt{n} + 2\sqrt{n}$

66. $3y^3\sqrt[4]{8y} - 9y^3\sqrt[4]{8y}$

67. $\sqrt{45} + \sqrt{20}$

68. $4\sqrt{24} - 6\sqrt{54}$

69. $\sqrt{150} - \sqrt{54} + \sqrt{24}$

70. $4\sqrt{72x^2y} - 2x\sqrt{128y} + 5\sqrt{32x^2y}$

71. $\sqrt[3]{250x^4y^5} + \sqrt[3]{128x^4y^5}$

72. $\sqrt[4]{48} + \sqrt[4]{243}$

[10.4] *For Exercises 73–82, find the products.*

73. $\sqrt{5}(\sqrt{3} + \sqrt{2})$

74. $\sqrt[3]{7}(\sqrt[3]{3} + 2\sqrt[3]{7})$

75. $3\sqrt{6}(2 - 3\sqrt{6})$

76. $(\sqrt{2} - \sqrt{3})(\sqrt{5} + \sqrt{7})$

77. $(\sqrt[3]{2} - 4)(\sqrt[3]{4} + 2)$

78. $(\sqrt[4]{6x} + 2)(\sqrt[4]{2x} - 1)$

79. $(\sqrt{5a} + \sqrt{3b})(\sqrt{5a} - \sqrt{3b})$

80. $(2\sqrt{3} - \sqrt{5})(2\sqrt{3} + \sqrt{5})$

81. $(\sqrt{2} + 1)^2$

82. $(\sqrt{2} - \sqrt{5})^2$

[10.5] *For Exercises 83–92, rationalize the denominator and simplify.*

83. $\dfrac{1}{\sqrt{2}}$

84. $\dfrac{3}{\sqrt[3]{3}}$

85. $\sqrt{\dfrac{4}{7}}$

86. $\dfrac{\sqrt[4]{5x^2}}{\sqrt[4]{2}}$

87. $\sqrt[3]{\dfrac{17}{3y^2}}$

88. $\dfrac{\sqrt[4]{9}}{\sqrt[4]{27}}$

89. $\dfrac{4}{\sqrt{2} - \sqrt{3}}$

90. $\dfrac{1}{4 + \sqrt{3}}$

91. $\dfrac{1}{2 - \sqrt{n}}$

92. $\dfrac{2\sqrt{3}}{3\sqrt{2} - 2\sqrt{3}}$

[10.5] *For Exercises 93–96, rationalize the numerator.*

93. $\dfrac{\sqrt{10}}{6}$

94. $\dfrac{\sqrt{3x}}{5}$

95. $\dfrac{2 - \sqrt{3}}{8}$

96. $\dfrac{2\sqrt{t} + \sqrt{3t}}{5t}$

Equations and Inequalities

Exercises 97–1

[10.6] *For Exercises 97–108, solve.*

97. $\sqrt{x} = 9$

98. $\sqrt{y} = -3$

99. $\sqrt{w - 1} = 3$

100. $\sqrt[3]{3x - 2} = -2$

101. $\sqrt[4]{x - 2} - 3 = -1$

102. $\sqrt{y + 1} = \sqrt{2y - 4}$

103. $\sqrt{x - 6} = x + 2$

104. $\sqrt[3]{3x + 10} - 4 = 5$

105. $\sqrt[4]{x + 8} = \sqrt[4]{2x + 1}$

106. $\sqrt{5n - 1} = 4 - 2n$

107. $1 + \sqrt{x} = \sqrt{2x + 1}$

108. $\sqrt{3x + 1} = 2 - \sqrt{3x}$

[10.7] *For Exercises 109 and 110, write the imaginary number using i.*

109. $\sqrt{-9}$

110. $\sqrt{-20}$

Expressions

Exercises 109–125

[10.7] *For Exercises 111 and 112, add or subtract.*

111. $(3 + 2i) + (5 - 8i)$

112. $(7 - 3i) - (-2 + 4i)$

[10.7] *For Exercises 113–116, multiply and simplify.*

113. $(3i)(4i)$

114. $2i(4 - i)$

115. $(6 + 2i)(4 - i)$

116. $(5 - i)^2$

[10.7] *For Exercises 117–122, write in standard form.*

117. $\dfrac{5}{i}$

118. $\dfrac{3}{-i}$

119. $\dfrac{4}{3i}$

120. $\dfrac{7+i}{5i}$

121. $\dfrac{3}{2+i}$

122. $\dfrac{5+i}{2-3i}$

[10.7] *For Exercises 123 and 124, simplify.*

123. i^{20}

124. i^{15}

[10.3] 125. Write an expression in simplest form for the area of the figure.

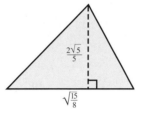

For Exercises 126 and 127, use the following. The speed of a car can be determined by the length of the skid marks using the formula $S = 2\sqrt{2L}$, where L represents the length of the skid mark in feet and S represents the speed of the car in miles per hour. (Source: Harris Technical Services, Traffic Accident Reconstructionists)

Equations and Inequalities

Exercises 126–132

[10.1] 126. a. Write an expression for the exact speed of a car if the length of the skid marks measures 40 feet.

   b. Approximate the speed to the nearest tenth.

[10.6] 127. Find the length of skid marks if a driver brakes hard at a speed of 50 miles per hour.

For Exercises 128 and 129, use the formula $T = 2\pi\sqrt{\dfrac{L}{9.8}}$, where T is the period of a pendulum in seconds and L is the length of the pendulum in meters.

[10.1] 128. Find the period of a pendulum with a length of 2.45 meters.

[10.6] 129. Suppose the period of the pendulum is $\dfrac{\pi}{3}$ seconds. Find the length.

For Exercises 130–131, use the formula $t = \sqrt{\dfrac{h}{16}}$*, where t is the time in seconds that it takes on object to fall a distance of h feet.*

[10.1] **130. a.** Write an expression in simplest form of the time that an object takes to fall 40 feet.

     b. Approximate the time to the nearest hundredth.

[10.6] **131.** Find the distance an object falls in 0.3 seconds.

[10.1] **132.** Three pieces of lumber are to be connected to form a right triangle that will be part of the roof frame for a small storage building. If the horizontal piece is to be 4 feet and the vertical piece is to be 3 feet, how long must the connecting piece be?

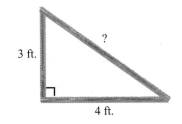

3 ft.

?

4 ft.

Chapter 10 Practice Test

For Exercises 1 and 2, evaluate the square root.

1. $\sqrt{36}$

2. $\sqrt{-49}$

For Exercises 3–12, simplify. Assume variables represent nonnegative values.

3. $\sqrt{81x^2y^5}$

4. $\sqrt[3]{54}$

5. $\sqrt[4]{4x} \cdot \sqrt[4]{4x^5}$

6. $-\sqrt[3]{-27r^{15}}$

7. $\dfrac{\sqrt{5}}{\sqrt{45}}$

8. $\dfrac{\sqrt[4]{1}}{\sqrt[4]{81}}$

9. $6\sqrt{7} - \sqrt{7}$

10. $(\sqrt{3} - 1)^2$

11. $x^{2/3} \cdot x^{-4/3}$

12. $(\sqrt[3]{2} - 4)(\sqrt[3]{4} + 2)$

For Exercises 13 and 14, write in exponential form.

13. $\sqrt[5]{8x^3}$

14. $\sqrt[3]{(2x + 5)^2}$

For Exercises 15 and 16, rationalize the denominator and simplify. Assume variables represent nonnegative values.

15. $\dfrac{1}{\sqrt[3]{4}}$

16. $\dfrac{\sqrt{x}}{\sqrt{x} + \sqrt{y}}$

For Exercises 17 and 18, solve the equation.

17. $\sqrt{3x - 2} = 8$

18. $\sqrt[4]{x + 8} = \sqrt[4]{2x + 1}$

For Exercises 19–22, simplify and write the answer in standard form $(a + bi)$.

19. $(2 - i) - (4 + 3i)$

20. $(2i)(-i)$

21. $(4 - i)(4 + i)$

22. $\dfrac{2}{4 - 3i}$

23. Write an expression in simplest form for the area of the figure.

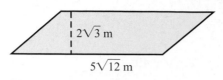

$$2\sqrt{3}\text{ m}$$
$$5\sqrt{12}\text{ m}$$

24. The formula $t = \sqrt{\dfrac{h}{16}}$ describes the amount of time t, in seconds, that an object falls a distance of h feet.

   a. Write an expression in simplest form for the exact amount of time an object falls a distance of 12 feet.

   b. Find the distance an object falls in 2 seconds.

25. The formula $S = \dfrac{7}{4}\sqrt{D}$ can be used to approximate the speed, S, in miles per hour, that a car was traveling prior to braking and skidding a distance D, in feet, on ice. (*Source:* Harris Technical Services, Traffic Accident Reconstructionists)

   a. Find the speed of a car if the skid length measures 40 feet long.

   b. Find the length of the skid marks that a car traveling 30 miles per hour would make if it brakes hard and skids to a halt.

CHAPTER 11

Quadratic Equations

"All our progress is unfolding like the vegetable bud."

—Ralph Waldo Emerson,
(1803–1882), U.S. essayist, poet, philosopher

"For beautiful variety no crop can be compared with this. Here is not merely the plain yellow of the grains, but nearly all the colors that we know . . ."

—Henry David Thoreau

In Section 6.6, we introduced quadratic equations and solved them by factoring. However, not all quadratic equations can be solved by factoring, so we need more powerful methods. In this chapter, we explore three other methods for solving quadratic equations, each of which will use square roots. In the first method, we will solve quadratic equations using a new principle of equality, called the *square root principle*. Like factoring, this first method has some limitations, so we will develop a more powerful method called *completing the square*. Although we can complete the square to solve any quadratic equation, it is rather tedious. So we will develop a third method that uses a general formula, called the *quadratic formula,* to solve any quadratic equation. We will then use these methods to solve equations that are quadratic in form and that are quadratic inequalities. Finally, we will add to what we learned in Sections 6.7 and 8.4 and learn more about the graphs of quadratic functions.

11.1 Completing the Square

OBJECTIVES
1. Use the square root principle to solve quadratic equations.
2. Solve quadratic equations by completing the square.

OBJECTIVE 1. Use the square root principle to solve quadratic equations. In Section 6.4, we solved quadratic equations such as $x^2 = 25$ by subtracting 25 from both sides, factoring, and then using the zero-factor theorem. Let's recall that process:

$$x^2 - 25 = 0$$
$$(x - 5)(x + 5) = 0 \qquad \text{Factor}$$
$$x - 5 = 0 \quad \text{or} \quad x + 5 = 0 \qquad \text{Use the zero-factor theorem.}$$
$$x = 5 \qquad\qquad x = -5 \qquad \text{Isolate } x \text{ in each equation.}$$

Another approach to solving $x^2 = 25$ involves square roots. Notice that the solutions to this equation must be numbers that can be squared to equal 25. These numbers are the square roots of 25, which are 5 and -5. This suggests a new rule called the *square root principle*.

Note: *The expression $\pm\sqrt{a}$ is read "plus or minus the square root of a."*

RULE The Square Root Principle

If $x^2 = a$, where a is a real number, then $x = \sqrt{a}$ or $x = -\sqrt{a}$.
It is common to indicate the positive and negative solutions by writing $\pm\sqrt{a}$.

For example, if $x^2 = 64$, then $x = \sqrt{64} = 8$ or $x = -\sqrt{64} = -8$. Or, we could simply write $x = \pm\sqrt{64} = \pm 8$.

Solve Equations in the Form $x^2 = a$

The square root principle is especially useful for solving equations in the form $x^2 = a$ when a is not a perfect square.

EXAMPLE 1 Solve. $x^2 = 40$

Solution $x^2 = 40$

$$x = \pm\sqrt{40} \qquad \text{Use the square root principle.}$$
$$x = \pm\sqrt{4 \cdot 10} \qquad \text{Simplify by factoring out a perfect square.}$$
$$x = \pm 2\sqrt{10}$$

Connection We learned how to simplify square roots of numbers that have perfect square factors in Section 10.3.

Note: *Remember the $\pm$ symbol means that the two solutions are $2\sqrt{10}$ and $-2\sqrt{10}$.*

We can check the two solutions using the original equation:

Check $2\sqrt{10}$: $(2\sqrt{10})^2 \stackrel{?}{=} 40$ *Check* $-2\sqrt{10}$: $(-2\sqrt{10})^2 \stackrel{?}{=} 40$

              $4 \cdot 10 = 40$ True $4 \cdot 10 = 40$ True

For the remaining examples, the checks will be left to the reader.

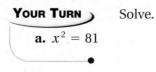

YOUR TURN Solve.

a. $x^2 = 81$

b. $x^2 = 50$

Solve Equations in the Form $ax^2 + b = c$

If an equation is in the form $ax^2 + b = c$, we use the addition and multiplication principles of equality to isolate x^2 and then use the square root principle.

EXAMPLE 2 Solve.

a. $x^2 + 2 = 100$

Solution $x^2 + 2 = 100$

$$x^2 = 98 \qquad \text{Subtract 2 from both sides to isolate } x^2.$$
$$x = \pm\sqrt{98} \qquad \text{Use the square root principle.}$$
$$x = \pm\sqrt{49 \cdot 2} \qquad \text{Simplify by factoring out a perfect square.}$$
$$x = \pm 7\sqrt{2}$$

b. $9x^2 = 27$

Solution $9x^2 = 27$

$$x^2 = 3 \qquad \text{Divide both sides by 9 to isolate } x^2.$$
$$x = \pm\sqrt{3} \qquad \text{Use the square root principle.}$$

c. $5x^2 + 2 = 62$

Solution $5x^2 + 2 = 62$

$$5x^2 = 60 \qquad \text{Subtract 2 from both sides.}$$
$$x^2 = 12 \qquad \text{Divide both sides by 5.}$$
$$x = \pm\sqrt{12} \qquad \text{Use the square root principle.}$$
$$x = \pm\sqrt{4 \cdot 3} \qquad \text{Simplify by factoring out a perfect square.}$$
$$x = \pm 2\sqrt{3}$$

YOUR TURN Solve.

a. $x^2 - 6 = 42$

b. $3x^2 = 60$

c. $2x^2 + 11 = 65$

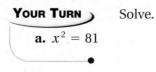

ANSWERS

a. ± 9 b. $\pm 5\sqrt{2}$

ANSWERS

a. $\pm 4\sqrt{3}$ b. $\pm 2\sqrt{5}$ c. $\pm 3\sqrt{3}$

Solve Equations in the Form $(ax + b)^2 = c$

In an equation in the form $(ax + b)^2 = c$, notice the expression $ax + b$ is squared. We can use the square root principle to eliminate the square.

EXAMPLE 3 Solve.

a. $(x + 6)^2 = 49$

Solution $(x + 6)^2 = 49$

$\qquad x + 6 = \pm\sqrt{49}$ Use the square root principle.

$\qquad x + 6 = \pm 7$

$\qquad\qquad x = -6 \pm 7$ Subtract 6 from both sides.

$x = -6 + 7 \quad$ or $\quad x = -6 - 7$ Simplify by separating the two solutions.

$x = 1 \qquad\qquad\quad x = -13$

b. $(5x - 1)^2 = 10$

Solution $(5x - 1)^2 = 10$

$\qquad 5x - 1 = \pm\sqrt{10}$ Use the square root principle.

$\qquad\quad 2x = 1 \pm \sqrt{10}$ Add 1 to both sides to isolate $5x$.

$\qquad\quad\; x = \dfrac{1 \pm \sqrt{10}}{5}$ Divide both sides by 5.

c. $(8x + 6)^2 = -32$

Solution $(8x + 6)^2 = -32$

Note: *The square root of a negative number is an imaginary number, which we can rewrite using* i.

$\qquad 8x + 6 = \pm\sqrt{-32}$ Use the square root principle.

$\qquad\quad 8x = -6 \pm \sqrt{-16 \cdot 2}$ Subtract 6 from both sides and simplify the square root.

$\qquad\quad 8x = -6 \pm 4i\sqrt{2}$ Rewrite the imaginary number using i notation.

$\qquad\quad\; x = \dfrac{-6 \pm 4i\sqrt{2}}{8}$ Divide both sides by 8.

$\qquad\quad\; x = -\dfrac{6}{8} \pm \dfrac{4\sqrt{2}}{8}i$ Write the complex number in standard form.

$\qquad\quad\; x = -\dfrac{3}{4} \pm \dfrac{\sqrt{2}}{2}i$ Simplify.

ANSWERS

a. 8 or −2

b. $\dfrac{-1 \pm \sqrt{14}}{2}$ **c.** $\dfrac{5}{4} \pm \dfrac{\sqrt{3}}{2}i$

YOUR TURN Solve.

a. $(x - 3)^2 = 25$ **b.** $(2x + 1)^2 = 14$ **c.** $(4x - 5)^2 = -12$

OBJECTIVE 2. Solve quadratic equations by completing the square. To make use of the square root principle, we need one side of the equation to be a perfect square and the other side to be a constant, as in $(x + 2)^2 = 5$. But, suppose we are given an equation like $x^2 + 6x = 2$ whose left-hand side is an "incomplete" square. We can use the addition principle of equality to add an appropriate number (9 in this case) to both sides so that the left-hand side becomes a perfect square. We call this process *completing the square*.

$\qquad x^2 + 6x + 9 = 2 + 9$ Adding 9 to both sides completes the square on the left side.

We can now factor the left-hand side of the equation, then use the square root principle to finish solving the equation.

$$(x + 3)^2 = 11$$ **Write the left-hand side in factored form.**

$$x + 3 = \pm\sqrt{11}$$ **Use the square root principle to eliminate the square.**

$$x = -3 \pm \sqrt{11}$$ **Subtract 3 from both sides to isolate x.**

How do we know what number to add to complete the square?

Notice that squaring half of the coefficient of x gives the constant term in the completed square.

$$x^2 + 6x + 9$$

Half of 6 is 3, which when squared is 9.

Also notice that half of the coefficient of x is the constant in the factored form.

$$x^2 + 6x + 9 = (x + 3)^2$$

Half of 6 is 3, which is the constant in the factored form.

Connection   The product of every perfect square in the form $(x + b)^2$ is a trinomial in the form $x^2 + 2bx + b^2$. Notice that half of x's coefficient, $2b$, is b, which is then squared to equal the last term.

Equations in the Form $x^2 + bx = c$

To solve a quadratic equation by completing the square, we need the equation in the form $x^2 + bx = c$, so that the coefficient of x^2 is 1. Once the equation is in the form $x^2 + bx = c$, we complete the square and then use the square root principle.

EXAMPLE 4   Solve by completing the square.

a. $x^2 + 12x + 15 = 0$

Solution   We first write the equation in the form $x^2 + bx = c$.

Note: *We found* 36
by squaring half of 12.

$$\left(\frac{12}{2}\right)^2 = 6^2 = 36$$

$$x^2 + 12x = -15$$ **Subtract 15 from both sides to get the form $x^2 + bx = c$.**

$$x^2 + 12x + 36 = -15 + 36$$ **Complete the square by adding 36 to both sides.**

$$(x + 6)^2 = 21$$ **Factor.**

$$x + 6 = \pm\sqrt{21}$$ **Use the square root principle.**

$$x = -6 \pm \sqrt{21}$$ **Subtract 6 from both sides to isolate x.**

b. $x^2 - 7x + 8 = 5$

Solution $x^2 - 7x = -3$ **Subtract 8 from both sides to get the form $x^2 + bx = c$.**

Note: *We found* $\dfrac{49}{4}$

by squaring half of -7.

$$\left(\dfrac{-7}{2}\right)^2 = \dfrac{49}{4}$$

$$x^2 - 7x + \frac{49}{4} = -3 + \frac{49}{4} \qquad \text{Complete the square by adding } \frac{49}{4} \text{ to both sides.}$$

$$\left(x - \frac{7}{2}\right)^2 = \frac{37}{4} \qquad \text{Factor the left side and simplify the right.}$$

$$x - \frac{7}{2} = \pm\sqrt{\frac{37}{4}} \qquad \text{Use the square root principle.}$$

$$x = \frac{7}{2} \pm \sqrt{\frac{37}{4}} \qquad \text{Add } \frac{7}{2} \text{ to both sides to isolate } x.$$

$$x = \frac{7}{2} \pm \frac{\sqrt{37}}{2} \qquad \text{Simplify the square root.}$$

$$x = \frac{7 \pm \sqrt{37}}{2} \qquad \text{Combine the fractions.}$$

YOUR TURN   Solve by completing the square.

    a. $x^2 + 8x - 29 = 0$ **b.** $x^2 - 9x - 6 = 5$

Equations in the Form $ax^2 + bx = c$, where $a \neq 1$

So far our equations have been in the form $x^2 + bx = c$, where the coefficient of the x^2 term is 1. To solve an equation such as $2x^2 + 12x = 3$, we will need to divide both sides of the equation by 2 (or multiply both sides by $\dfrac{1}{2}$) so that the coefficient of x^2 is 1.

$$\frac{2x^2 + 12x}{2} = \frac{3}{2} \qquad \text{Divide both sides by 2.}$$

$$x^2 + 6x = \frac{3}{2}$$

We can now solve by completing the square.

$$x^2 + 6x + 9 = \frac{3}{2} + 9 \qquad \text{Add 9 to both sides to complete the square.}$$

$$(x + 3)^2 = \frac{21}{2} \qquad \text{Factor the left side and simplify the right.}$$

$$x + 3 = \pm\sqrt{\frac{21}{2}} \qquad \text{Use the square root principle.}$$

$$x = -3 \pm \sqrt{\frac{21}{2}} \qquad \text{Subtract 3 from both sides to isolate } x.$$

$$x = -3 \pm \frac{\sqrt{21}}{\sqrt{2}} \cdot \frac{\sqrt{2}}{\sqrt{2}} \qquad \text{Rationalize the denominator.}$$

$$x = -3 \pm \frac{\sqrt{42}}{2}$$

ANSWERS

a. $-4 \pm 3\sqrt{5}$ **b.** $\dfrac{9 \pm 5\sqrt{5}}{2}$

We can now write a procedure for solving any quadratic equation by completing the square.

PROCEDURE **Solving Quadratic Equations by Completing the Square**

To solve a quadratic equation by completing the square:

1. Write the equation in the form $x^2 + bx = c$.

2. Complete the square by adding $\left(\dfrac{b}{2}\right)^2$ to both sides.

3. Write the completed square in factored form.

4. Use the square root principle to eliminate the square.

5. Isolate the variable.

6. Simplify as needed.

Connection In Section 11.4, we will use the process of completing the square to rewrite quadratic functions in a form that allows us to easily determine features of the graph.

EXAMPLE 5 Solve by completing the square.

a. $16x^2 - 24x = 1$

Solution

$$\dfrac{16x^2 - 24x}{16} = \dfrac{1}{16}$$ Divide both sides by 4.

$$x^2 - \dfrac{3}{2}x = \dfrac{1}{16}$$ Simplify.

$$x^2 - \dfrac{3}{2}x + \dfrac{9}{16} = \dfrac{1}{16} + \dfrac{9}{16}$$ Add $\dfrac{9}{16}$ to both sides to complete the square.

$$\left(x - \dfrac{3}{4}\right)^2 = \dfrac{10}{16}$$ Factor the left side and simplify the right.

$$x - \dfrac{3}{4} = \pm\sqrt{\dfrac{10}{16}}$$ Use the square root principle.

$$x = \dfrac{3}{4} \pm \dfrac{\sqrt{10}}{4}$$ Add $\dfrac{3}{4}$ to both sides and simplify the square root.

$$x = \dfrac{3 \pm \sqrt{10}}{4}$$ Combine the fractions.

Note: *To complete the square, we square half of* $\dfrac{3}{2}$:

$$\left(\dfrac{1}{2} \cdot \dfrac{3}{2}\right)^2 = \left(\dfrac{3}{4}\right)^2 = \dfrac{9}{16}$$

b. $2x^2 + 9x + 7 = -5$

Solution

$$2x^2 + 9x = -12$$ Subtract 7 from both sides.

$$\dfrac{2x^2 + 9x}{2} = \dfrac{-12}{2}$$ Divide both sides by 2.

$$x^2 + \dfrac{9}{2}x = -6$$ Simplify.

$$x^2 + \dfrac{9}{2}x + \dfrac{81}{16} = -6 + \dfrac{81}{16}$$ Add $\dfrac{81}{16}$ to both sides to complete the square.

$$\left(x + \dfrac{9}{4}\right)^2 = -\dfrac{15}{16}$$ Factor the left side and simplify the right.

$$x + \dfrac{9}{4} = \pm\sqrt{-\dfrac{15}{16}}$$ Use the square root principle.

$$x = -\dfrac{9}{4} \pm \dfrac{\sqrt{15}}{4}i$$ Subtract $\dfrac{9}{4}$ from both sides and write the complex number in standard form.

ANSWERS

a. $\dfrac{-3 \pm 2\sqrt{2}}{3}$ or

$-1 \pm \dfrac{2\sqrt{2}}{3}$

b. $4 \pm 3i$

YOUR TURN Solve by completing the square.

a. $9x^2 + 18x = -1$ **b.** $x^2 - 8x + 11 = -14$

11.1 Exercises

For Extra Help

MyMathLab   Videotape/DVT   InterAct Math   Math Tutor Center   Math XL.com

1. Explain why there are two solutions to an equation in the form $x^2 = a$.

2. Write a formula for the solutions of $ax^2 - b = c$ by solving for x.

3. Write a formula for the solutions of $(ax - b)^2 = c$ by solving for x.

4. Given an equation in the form $x^2 + bx = c$, explain how to complete the square.

5. Consider the equation $x^2 - 7x + 12 = 0$. Which is a better method for solving the equation, factoring and then using the zero-factor theorem or completing the square? Explain.

6. Consider the equation $x^2 + 4x + 5 = 0$. Which is a better method for solving the equation, factoring and then using the zero-factor theorem or completing the square? Explain.

For Exercises 7–16, solve and check.

7. $x^2 = 49$

8. $x^2 = 144$

9. $y^2 = \dfrac{4}{25}$

10. $t^2 = \dfrac{1}{64}$

11. $n^2 = 0.81$

12. $p^2 = 1.21$

13. $z^2 = 45$

14. $m^2 = 72$

15. $w^2 = -25$

16. $c^2 = -49$

For Exercises 17–36, solve and check. Begin by using the addition or multiplication principles of equality to isolate the squared term.

17. $n^2 - 7 = 42$

18. $y^2 - 5 = 59$

19. $y^2 - 7 = 29$

20. $k^2 + 5 = 30$

21. $4n^2 = 36$

22. $5y^2 = 125$

23. $25t^2 = 9$

24. $16d^2 = 49$

25. $4h^2 = -16$

26. $-3k^2 = 27$

27. $\dfrac{5}{6}x^2 = \dfrac{3}{8}$

28. $-\dfrac{2}{3}m^2 = -\dfrac{8}{5}$

29. $2x^2 + 5 = 21$

30. $3x^2 + 5 = 80$

31. $5y^2 - 7 = -97$

32. $4n^2 + 20 = -76$

33. $\dfrac{3}{4}y^2 - 5 = 3$

34. $\dfrac{25}{9}m^2 - 2 = 6$

35. $0.2t^2 - 0.5 = 0.012$

36. $0.5p^2 + 1.28 = 1.6$

For Exercises 37–52, solve and check. Use the square root principle to eliminate the square.

37. $(x + 8)^2 = 49$

38. $(y + 7)^2 = 144$

39. $(5n - 3)^2 = 16$

40. $(6h - 5)^2 = 81$

41. $(m - 7)^2 = -12$

42. $(t - 5)^2 = -28$

43. $(4k - 1)^2 = 40$

44. $(3x - 7)^2 = 50$

45. $(m - 8)^2 = -1$

46. $(t - 2)^2 = -4$

47. $\left(y - \dfrac{3}{4}\right)^2 = \dfrac{9}{16}$

48. $\left(x + \dfrac{4}{9}\right)^2 = \dfrac{25}{81}$

49. $\left(\dfrac{5}{9}d - \dfrac{1}{2}\right)^2 = \dfrac{1}{36}$

50. $\left(\dfrac{3}{4}h + \dfrac{4}{5}\right)^2 = \dfrac{1}{100}$

51. $(0.4x + 3.8)^2 = 2.56$

52. $(0.8n - 6.8)^2 = 1.96$

For Exercises 53–56, explain the mistake, then find the correct solutions.

53. $x^2 - 15 = 34$
$$x^2 = 49$$
$$x = \sqrt{49}$$
$$x = 7$$

54. $x^2 = 20$
$$x = \sqrt{20}$$
$$x = 2\sqrt{5}$$

55. $(x - 5)^2 = -6$
$$x - 5 = \pm\sqrt{6}$$
$$x = 5 \pm \sqrt{6}$$

56. $(x - 1)^2 = -12$
$$x - 1 = \pm\sqrt{-12}$$
$$x = 1 \pm 2\sqrt{3}$$

For Exercises 57–62, solve, then use a calculator to approximate the irrational solutions rounded to three places.

57. $x^2 = 96$

58. $t^2 = 56$

59. $y^2 - 15 = 5$

60. $x^2 - 22 = 6$

61. $(n - 6)^2 = 15$

62. $(m + 3)^2 = 10$

For Exercises 63–74: a. Add a term to the expression to make it a perfect square.
b. Factor the perfect square.

63. $x^2 + 14x$

64. $c^2 + 8c$

65. $n^2 - 10n$

66. $a^2 - 4a$

67. $y^2 - 7y$

68. $m^2 - 11m$

69. $s^2 - \dfrac{2}{3}s$

70. $y^2 - \dfrac{4}{5}y$

71. $m^2 + \dfrac{1}{7}m$

72. $v^2 + \dfrac{1}{5}v$

73. $p^2 + 9p$

74. $z^2 - 15z$

For Exercises 75–90, solve the equation by completing the square.

75. $w^2 + 2w = 15$

76. $p^2 + 8p = 9$

77. $y^2 + 10y = -16$

78. $x^2 + 10x = -24$

79. $r^2 - 2r + 50 = 0$

80. $c^2 - 6c + 45 = 0$

81. $k^2 = 9k - 18$

82. $a^2 = 7a - 10$

83. $n^2 + 9n - 20 = 16$

84. $u^2 - 5u - 9 = -13$

85. $b^2 - 2b - 11 = 5$

86. $z^2 + 2z = 6$

87. $h^2 - 6h + 3 = -26$

88. $j^2 - 4j + 25 = -3$

89. $u^2 + \dfrac{1}{2}u = \dfrac{3}{2}$

90. $y^2 + \dfrac{1}{3}y = \dfrac{2}{3}$

For Exercises 91–104, solve the equation by completing the square. Begin by writing the equation in the form $x^2 + bx = c$.

91. $4x^2 + 12x = 7$

92. $8m^2 - 10m = 3$

93. $2n^2 - n - 3 = 0$

94. $f^2 - 12f - 45 = 0$

95. $6w^2 - 6 = -5w$

96. $2t^2 - 5 = -3t$

97. $2g^2 + g - 11 = -5$

98. $4l^2 + l - 30 = 30$

99. $2x^2 = 4x + 3$

100. $2x^2 = 6x - 3$

101. $5k^2 + k - 2 = 0$

102. $3s^2 - 4s = 2$

103. $3a^2 - 1 = -6a$

104. $3x^2 - 4 = -8x$

For Exercises 105–106, explain the mistake.

105.
$$3x^2 + 4x = 2$$
$$3x^2 + 4x + 4 = 2 + 4$$
$$(3x + 2)^2 = 6$$
$$3x + 2 = \pm\sqrt{6}$$
$$3x = -2 \pm \sqrt{6}$$
$$x = \frac{-2 \pm \sqrt{6}}{3}$$

106.
$$x^2 + 6x = 7$$
$$x^2 + 6x + 9 = 7 + 9$$
$$(x + 3)^2 = 16$$
$$x + 3 = \sqrt{16}$$
$$x + 3 = 4$$
$$x = -3 + 4$$
$$x = 1$$

For Exercises 107–122, solve.

107. A square sheet of metal has an area of 196 square inches. What is the length of each side?

108. A severe thunderstorm warning is issued by the National Weather Service for a square area covering 14,400 square miles. What is the length of each side of the square area?

109. A tank is to be made for an aquarium so that the length is 8 feet, the width is twice the height, and the volume is 144 cubic feet. Find the height and width of the tank.

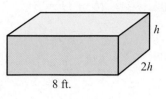

110. The length of a swimming pool is to be three times the width. If the depth is to be a constant 4 feet and the volume is to be 4800 cubic feet, find the length and width.

111. The Arecibo radio telescope is like a giant satellite dish that covers a circular area of approximately $23,256.25\pi$ square meters. Find the diameter of the dish.

Of Interest

The Arecibo radio telescope, the world's largest radio telescope, analyzes radiation and signals emitted by objects in space. It was built in a natural depression near Arecibo, Puerto Rico.

112. A field is planted in a circular pattern. A watering device is to be constructed with pipe in a line extending from the center of the field to the edge of the field. The pipe is set upon wheels so that it can rotate around the field and cover the entire field with water. If the area of the field is 7225π square feet, how long will the watering device be?

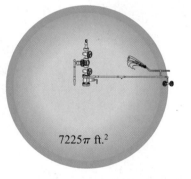

7225π ft.^{2}

113. The corner of a sheet of plywood has been cut out as shown. If the area of the piece that was removed was 256 square inches, find x.

Removed piece

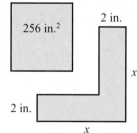

256 in.^{2}

2 in.

2 in.

x

x

114. The area of the hole in the washer shown is 25π square millimeters. Find r, the radius of the washer.

4 mm

r

115. A rectangular basement is 6 feet longer than it is wide. If the area of the room is 315 square feet, what are the dimensions of the room?

116. The length of a rectangular garden is 25 feet more than the width. If the area of the garden is 3150 square feet, what are its dimensions?

117. An LCD computer monitor measures 20 inches across its diagonal. If the width is 4 inches less than the length, find the dimensions of the monitor.

118. Two identical right triangles are placed together to form the frame of a roof. If the base of each triangle is 7 feet longer than its height and its hypotenuse is 13 feet, what are the dimensions of the base and height?

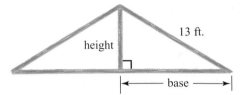

119. The metal panel shown is to have a total area of 1050 square inches. Find the length x.

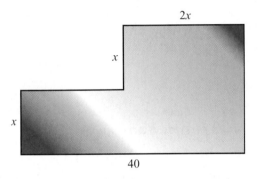

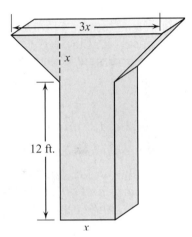

120. A side view of a concrete bridge support footing is shown. The side of the footing is to have a total area of 110 square feet. Find the length of x.

121. A plastic panel is to have a rectangular hole cut as shown.

    a. Find l so that the area remaining after the hole is cut is 1230 square centimeters.

    b. Find the length and width of the plastic panel.

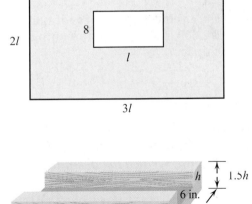

122. A 6-inch-wide groove with a height of h is to be cut into a wood block as shown.

    a. Find h so that the volume remaining in the block after the groove is cut is 360 cubic inches.

    b. Find the height and width of the block.

For Exercises 123–126, if an object is dropped, the formula $d = 16t^2$ describes the distance d in feet that the object falls in t seconds.

123. Suppose a cover for a ceiling light falls from a 9-foot ceiling. How long does it take for the cover to hit the floor?

124. A construction worker tosses a scrap piece of lumber from the roof of a house. How long does it take the piece of lumber to reach the ground 25 feet below?

125. A toy rocket is launched straight up and reaches a height of 180 feet in 1.5 seconds then plummets back to the ground. Determine the total time of the rocket's round-trip.

126. A lead weight is dropped from a tower at a height of 150 feet. How much time passes for the weight to be halfway to the ground? Write an exact answer, then approximate to the nearest tenth of a second.

For Exercises 127 and 128, use the following information. In physics, if an object is in motion it has kinetic energy. The formula $E = \dfrac{1}{2}mv^2$ is used to calculate the kinetic energy E of an object with a mass m and velocity v. If the mass is measured in kilograms and the velocity is in meters per second, then the kinetic energy will be in units called joules (J).

127. Suppose an object with a mass of 50 kilograms has 400 joules of kinetic energy. Find its velocity.

128. In a crash test, a vehicle with a mass of 1200 kilograms is found to have kinetic energy of 117,600 joules just before impact. Find the velocity of the vehicle just before impact.

PUZZLE PROBLEM

Without using a calculator, which of the following four numbers is a perfect square? (Hint: Write a list of smaller perfect squares and look for a pattern)

9,456,804,219,745,618

2,512,339,789,576,516

7,602,985,471,286,543

4,682,715,204,643,182

REVIEW EXERCISES

[6.4] *For Exercises 1 and 2, factor.*

1. $x^2 - 10x + 25$

2. $x^2 + 6x + 9$

[10.1] *For Exercises 3 and 4, simplify. Assume variables represent nonnegative values.*

3. $\sqrt{36x^2}$

4. $\sqrt{(2x + 3)^2}$

[1.5, 10.1, 10.7] *For Exercises 5 and 6, simplify.*

5. $\dfrac{-4 + \sqrt{8^2 - 4(2)(5)}}{2(2)}$

6. $\dfrac{-6 - \sqrt{6^2 - 4(5)(2)}}{2(5)}$

11.2 Solving Quadratic Equations Using the Quadratic Formula

OBJECTIVES

1. Solve quadratic equations using the quadratic formula.
2. Use the discriminant to determine the number of real solutions that a quadratic equation will have.
3. Find the x- and y-intercepts of a quadratic function.
4. Solve applications using the quadratic formula.

In this section, we will solve quadratic equations using a formula called the *quadratic formula*. This formula is much easier to use than completing the square.

OBJECTIVE 1. Solve quadratic equations using the quadratic formula. To derive the quadratic formula, we begin with the general form of the quadratic equation, $ax^2 + bx + c = 0$, and assume that $a \neq 0$. We will follow the procedure for solving a quadratic equation by completing the square.

$$ax^2 + bx + c = 0$$

$$ax^2 + bx = -c \qquad \text{Subtract } c \text{ from both sides.}$$

$$\frac{ax^2 + bx}{a} = \frac{-c}{a} \qquad \text{Divide both sides by } a \text{ so that the coefficient of } x^2 \text{ will be 1.}$$

$$x^2 + \frac{b}{a}x = -\frac{c}{a} \qquad \text{Simplify.}$$

$$x^2 + \frac{b}{a}x + \frac{b^2}{4a^2} = -\frac{c}{a} + \frac{b^2}{4a^2} \qquad \text{Complete the square.}$$

Note: *To complete the square, we square half of $\dfrac{b}{a}$.*

$$\left(\frac{1}{2} \cdot \frac{b}{a}\right)^2 = \left(\frac{b}{2a}\right)^2$$

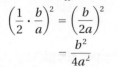

$$\frac{b^2}{4a^2}$$

$$\left(x + \frac{b}{2a}\right)^2 = -\frac{4ac}{4a^2} + \frac{b^2}{4a^2}$$

On the left side, write factored form. On the right side, rewrite $-\dfrac{c}{a}$ with the common denominator $4a^2$ so that we can combine the rational expressions.

$$\left(x + \frac{b}{2a}\right)^2 = \frac{b^2 - 4ac}{4a^2}$$

Combine the rational expressions. For simplicity, we rearrange the order of the b^2 and $-4ac$ terms in the numerator.

$$x + \frac{b}{2a} = \pm\sqrt{\frac{b^2 - 4ac}{4a^2}}$$

Use the square root principle to eliminate the square.

$$x + \frac{b}{2a} = \pm\frac{\sqrt{b^2 - 4ac}}{2a}$$

Simplify the square root in the denominator.

$$x = -\frac{b}{2a} \pm \frac{\sqrt{b^2 - 4ac}}{2a}$$

Subtract $\dfrac{b}{2a}$ from both sides to isolate x.

$$x = \frac{-b \pm \sqrt{b^2 - 4ac}}{2a}$$

Combine rational expressions.

This final equation is the *quadratic formula* and can be used to solve any quadratic equation simply by replacing a, b, and c with the corresponding numbers from the given equation.

PROCEDURE   Using the Quadratic Formula

Note: *A quadratic equation must be in the form $ax^2 + bx + c = 0$ to identify a, b, and c for use in the quadratic formula.*

▶

To solve a quadratic equation in the form $ax^2 + bx + c = 0$, where $a \neq 0$, use the quadratic formula:

$$x = \frac{-b \pm \sqrt{b^2 - 4ac}}{2a}$$

EXAMPLE 1 ▶ Solve.

a. $3x^2 + 10x - 8 = 0$

Solution This equation is in the form $ax^2 + bx + c = 0$, where $a = 3$, $b = 10$, and $c = -8$, so we can use the quadratic formula, $x = \dfrac{-b \pm \sqrt{b^2 - 4ac}}{2a}$.

$$x = \frac{-10 \pm \sqrt{10^2 - 4(3)(-8)}}{2(3)}$$

Replace a with 3, b with 10, and c with -8.

$$x = \frac{-10 \pm \sqrt{196}}{6}$$

Simplify in the radical and the denominator.

$$x = \frac{-10 \pm 14}{6}$$

Evaluate $\sqrt{196}$.

$$x = \frac{-10 + 14}{6} \quad \text{or} \quad x = \frac{-10 - 14}{6}$$

Now we split up the $\pm$ to calculate the two solutions.

$$x = \frac{4}{6} \qquad\qquad x = \frac{-24}{6}$$

$$x = \frac{2}{3} \qquad\qquad x - -4$$

Connection Notice that the radicand 196 is a perfect square, which causes the solutions to be two rational numbers. As we consider more examples in this section, note how the radicand determines the type of solutions.

b. $x^2 + 8x - 15 = 14$

Solution $x^2 + 8x - 29 = 0$ | Subtract 14 from both sides to get the form $ax^2 + bx + c = 0$.

$$x = \frac{-8 \pm \sqrt{8^2 - 4(1)(-29)}}{2(1)}$$ | Use the quadratic formula, replacing a with 1, b with 8, and c with -29.

$$x = \frac{-8 \pm \sqrt{180}}{2}$$ | Simplify in the radical and the denominator.

$$x = \frac{-8 \pm \sqrt{36 \cdot 5}}{2}$$ | Use the product rule of radicals.

$$x = \frac{-8 \pm 6\sqrt{5}}{2}$$ | Simplify the radical.

$$x = \frac{-8}{2} \pm \frac{6\sqrt{5}}{2}$$ | Separate into two rational expressions in order to simplify.

$$x = -4 \pm 3\sqrt{5}$$ | Simplify by dividing out the 2.

> **Connection** Notice that the radicand 180 is not a perfect square, which causes the solutions to be irrational numbers.

c. $9x^2 + 4 = -12x$

Solution $9x^2 + 12x + 4 = 0$ | Add 12x to both sides to get the form $ax^2 + bx + c = 0$.

$$x = \frac{-12 \pm \sqrt{12^2 - 4(9)(4)}}{2(9)}$$ | Use the quadratic formula, replacing a with 9, b with 12, and c with 4.

$$x = \frac{-12 \pm \sqrt{0}}{18}$$ | Simplify in the radical and the denominator.

$$x = -\frac{2}{3}$$ | Simplify the radical and write the fraction in lowest terms.

> **Connection** Notice that the radicand is 0, which causes this equation to have only one solution.

d. $3x^2 + 9 = -8x + 2$

Solution $3x^2 + 8x + 7 = 0$ | Add 8x to and subtract 2 from both sides to get the form $ax^2 + bx + c = 0$.

$$x = \frac{-8 \pm \sqrt{8^2 - 4(3)(7)}}{2(3)}$$ | Use the quadratic formula, replacing a with 3, b with 8, and c with 7.

$$x = \frac{-8 \pm \sqrt{-20}}{6}$$ | Simplify in the radical and the denominator.

$$x = \frac{-8 \pm 2i\sqrt{5}}{6}$$ | Simplify the radical.

$$x = \frac{-4 \pm i\sqrt{5}}{3}$$ | Simplify to lowest terms.

> **Connection** Notice that the negative radicand causes the solutions to be complex numbers.

Note: *In standard form, the two complex solutions are* $-\frac{4}{3} + \frac{\sqrt{5}}{3}i$ *and* $-\frac{4}{3} - \frac{\sqrt{5}}{3}i.$

ANSWERS

a. 2 or $-\frac{3}{4}$

b. $\dfrac{-3 \pm 2\sqrt{6}}{2}$

c. 4

d. $-2 \pm i\sqrt{2}$

YOUR TURN   Solve using the quadratic formula.

a. $4x^2 - 5x - 6 = 0$ **b.** $4x^2 = 15 - 2x$

c. $x^2 - 8x + 16 = 0$ **d.** $x^2 + 8 = 2 - 4x$

Choosing a Method for Solving Quadratic Equations

We have learned several methods for solving quadratic equations. The following table summarizes the methods and the conditions that would make each method the best choice.

Methods for Solving Quadratic Equations

Method	When the Method Is Beneficial
1. Factoring (Section 6.6)	Use when the quadratic equation can be easily factored.
2. Square root principle (Section 11.1)	Use when the quadratic equation can be easily written in the form $ax^2 = c$ or $(ax + b)^2 = c$.
3. Completing the square (Section 11.1)	Rarely the best method, but important for future topics.
4. Quadratic formula (Section 11.2)	Use when factoring is not easy.

OBJECTIVE 2. Use the discriminant to determine the number of real solutions that a quadratic equation will have. In the Connection boxes for Example 1, we pointed out how the radicand in the quadratic formula affects the solutions to a given quadratic equation. The expression $b^2 - 4ac$, which is the radicand, is called the **discriminant**.

DEFINITION **Discriminant:** The radicand $b^2 - 4ac$ in the quadratic formula.

We use the discriminant to determine the number and type of solutions to a quadratic equation.

PROCEDURE **Using the Discriminant**

Note: *When the discriminant is 0, the solution is* $-\dfrac{b \pm \sqrt{0}}{2a} = -\dfrac{b}{2a}$. *This real-number solution is a rational number if* $-\dfrac{b}{2a}$ *is rational.*

Given a quadratic equation in the form $ax^2 + bx + c = 0$, where $a \neq 0$, to determine the number and type of solutions it has, evaluate the discriminant $b^2 - 4ac$.

If the **discriminant is positive**, then the equation has two real-number solutions. They will be rational if the discriminant is a perfect square and irrational otherwise.

If the **discriminant is 0**, then the equation has one real solution.

If the **discriminant is negative**, then the equation has two nonreal complex solutions.

EXAMPLE 2 Use the discriminant to determine the number and type of solutions.

a. $3x^2 - 7x = 8$

Solution $3x^2 - 7x - 8 = 0$ Write the equation in the form $ax^2 + bx + c = 0$.

Evaluate the discriminant, $b^2 - 4ac$.

$(-7)^2 - 4(3)(-8)$ Replace a with 3, b with -7, and c with -8.

$= 49 + 96$ **Warning:** 145 is the value of the discriminant,

$= 145$ not a solution for the equation $3x^2 - 7x - 8 = 0$.

Because the discriminant is positive, this equation has two real-number solutions. Since 145 is not a perfect square, the solutions are irrational.

b. $x^2 = \dfrac{4}{5}x - \dfrac{4}{25}$

Solution $x^2 - \dfrac{4}{5}x + \dfrac{4}{25} = 0$ Write the equation in the form $ax^2 + bx + c = 0$.

It is easier to evaluate the disciminant if the numbers are integers, so we multiply the equation through by the LCD, which is 25.

$$25 \cdot x^2 - \dfrac{25}{1} \cdot \dfrac{4}{5}x + \dfrac{25}{1} \cdot \dfrac{4}{25} = 25 \cdot 0$$

$$25x^2 - 20x + 4 = 0$$

Note: *Since the discriminant is 0, the solution is*
$$-\dfrac{b}{2a} = -\dfrac{(-20)}{2(25)} = \dfrac{20}{50} = \dfrac{2}{5}$$
which is a rational number.

Now evaluate the discriminant, $b^2 - 4ac$.

$$(-20)^2 - 4(25)(4) \quad \text{Replace } a \text{ with 25, } b \text{ with } -20, \text{ and } c \text{ with 4.}$$
$$= 400 - 400$$
$$= 0$$

Because the discriminant is zero, this equation has only one real solution.

Note: *We could avoid calculations with decimal numbers by multiplying the original equation through by 10 so that it becomes:*
$$5x^2 - 8x + 25 = 0$$

c. $0.5x^2 - 0.8x + 2.5 = 0$

Solution $(-0.8)^2 - 4(0.5)(2.5)$ In $b^2 - 4ac$, replace a with 0.5, b with -0.8, and c with 2.5.
$$= 0.64 - 5$$
$$= -4.36$$

Because the discriminant is negative, there are two nonreal complex solutions.

YOUR TURN   Use the discriminant to determine the number and type of solutions for the equation. If the solutions are real, state whether they are rational or irrational.

a. $5x^2 + 8x - 9 = 0$ **b.** $\dfrac{5}{2}x^2 + \dfrac{5}{6} = \dfrac{4}{3}x$ **c.** $x^2 - 0.12x = -0.0036$

OBJECTIVE 3. Find the x- and y-intercepts of a quadratic function. In Chapter 4, we learned that x-intercepts are points where a graph intersects the x-axis. Since an x-intercept is on the x-axis, the y-coordinate of the point will be 0. In Section 6.7, we graphed quadratic functions, which have the form $f(x) = ax^2 + bx + c$ (or $y = ax^2 + bx + c$) and learned their graphs are parabolas. Notice when we replace y with 0 to find the x-intercepts in $y = ax^2 + bx + c$, we have the quadratic equation $ax^2 + bx + c = 0$, which has two, one, or no real-number solutions. As a result, quadratic functions in the form $y = ax^2 + bx + c$ have two, one, or no x-intercepts. The following graphs illustrate the possibilities.

ANSWERS

a. Discriminant = 244; two irrational solutions

b. Discriminant = $-\dfrac{59}{9}$; two nonreal complex solutions

c. Discriminant = 0; one rational solution

Two x-intercepts: If $0 = ax^2 + bx + c$ has two real-number solutions, which occurs if $b^2 - 4ac > 0$, then the graph of $y = ax^2 + bx + c$ has two x-intercepts and looks like one of the following graphs.

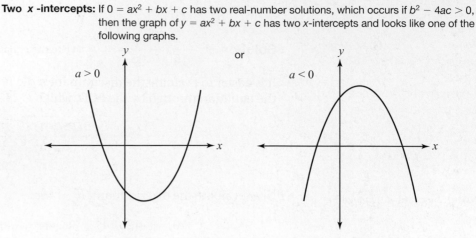

One x-intercept: If $0 = ax^2 + bx + c$ has one real-number solution, which occurs if $b^2 - 4ac = 0$, then the graph of $y = ax^2 + bx + c$ has one x-intercept and looks like one of the following graphs.

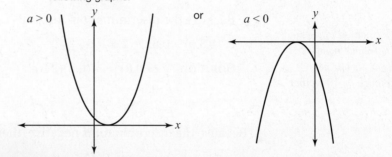

No x-intercepts: If $0 = ax^2 + bx + c$ has no real-number solution, which occurs if $b^2 - 4ac < 0$, then the graph of $y = ax^2 + bx + c$ has no x-intercepts and looks like one of the following graphs.

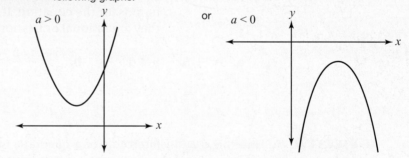

We have also learned that a y-intercept is where a graph intersects the y-axis. To find y-intercepts, we let $x = 0$ and solve for y. Notice when we replace x with 0 in $y = ax^2 + bx + c$, we have $y = a(0)^2 + b(0) + c = c$, so the y-intercept is $(0, c)$.

EXAMPLE 3 Find the x- and y-intercepts of $y = x^2 - 2x - 8$, then graph.

Solution For the x-intercepts, letting $y = 0$ gives the equation $0 = x^2 - 2x - 8$, which we will solve using the quadratic formula.

$$x = \frac{(-2) \pm \sqrt{(-2)^2 - 4(1)(-8)}}{2(1)} = \frac{2 \pm \sqrt{36}}{2} = 1 \pm 3 = 4 \quad \text{or} \quad -2$$

x-intercepts: $(4, 0)$ and $(-2, 0)$

Connection We could have solved $0 = x^2 - 2x - 8$ by factoring:

$$0 = x^2 - 2x - 8$$
$$0 = (x - 4)(x + 2)$$
$$x - 4 = 0 \quad \text{or} \quad x + 2 = 0$$
$$x = 4 \qquad\qquad x = -2$$

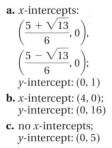

Calculator TIPS

The ZERO function on a graphing calculator can be used to find the coordinates of x-intercepts. After graphing the parabola on the calculator, select ZERO from the CALC menu. You will then be prompted to enter the left and right bound of one of the intercepts. Move the cursor along the parabola until it is to the left of one of the intercepts and press [ENTER]. *Then move the cursor along the parabola to the right of the same intercept and press* [ENTER] *again. When prompted for the guess, press* [ENTER] *once more. The coordinates of the intercept will appear at the bottom of the screen.*

Since $c = -8$, the y-intercept is $(0, -8)$.

Note: *We could also have calculated the y-intercept:*
$$y = (0)^2 + 2(0) - 8 = -8$$

Now we can graph. We know the graph opens upwards because $a = 1$, which is positive.

Note: *Knowing the intercepts and whether the parabola opens up or down gives us a pretty good sense of its position. However, if we knew the exact location of the vertex, we could get a more accurate graph. We will learn how to find the coordinates of the vertex in Section 11.4.*

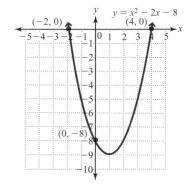

YOUR TURN Find the x- and y-intercepts. Verify on a graphing calculator.

a. $y = 3x^2 - 5x + 1$ **b.** $y = x^2 - 8x + 16$ **c.** $y = 2x^2 - 3x + 5$

OBJECTIVE 4. Solve applications using the quadratic formula. In physics, the general formula for describing the height of an object after it has been thrown upwards is $h = \frac{1}{2}gt^2 + v_0t + h_0$, where g represents the acceleration due to gravity, t is the time in flight, v_0 is the initial velocity, and h_0 is the initial height. For Earth, the acceleration due to gravity is -32.2 ft./sec.^{2} or -9.8 m/sec.^{2}

EXAMPLE 4 In an extreme-games competition, a motorcyclist jumps with an initial velocity of 70 feet per second from a ramp height of 25 feet, landing on a ramp with a height of 15 feet. Find the time the motorcyclist is in the air.

Understand We are given the initial velocity, initial height, and final height and we are to find the time the bike is in the air.

Plan Use the formula $h = \frac{1}{2}gt^2 + v_0t + h_0$ with $h = 15$, $v_0 = 70$, and $h_0 = 25$. Since the units are in feet, we will use -32.2 ft./sec.^{2} for g.

Execute

$15 = \frac{1}{2}(-32.2)t^2 + 70t + 25$ Substitute the values.

$15 = -16.1t^2 + 70t + 25$

$0 = -16.1t^2 + 70t + 10$ Subtract 15 from both sides to get the form $ax^2 + bx + c = 0$.

$x = \dfrac{-70 \pm \sqrt{70^2 - 4(-16.1)(10)}}{2(-16.1)}$ Use the quadratic formula, replacing a with -16.1, b with 70, and c with 10.

$x = \dfrac{-70 \pm \sqrt{5544}}{-32.2}$ Simplify in the radical and the denominator.

$x \approx -0.138$ or 4.486 Approximate the two irrational solutions.

ANSWERS

a. x-intercepts:
$\left(\dfrac{5 + \sqrt{13}}{6}, 0\right)$,
$\left(\dfrac{5 - \sqrt{13}}{6}, 0\right)$;
y-intercept: $(0, 1)$

b. x-intercept: $(4, 0)$;
y-intercept: $(0, 16)$

c. no x-intercepts;
y-intercept: $(0, 5)$

Answer Since the time cannot be negative, the motorcycle is in the air approximately 4.486 seconds.

Check We can check by evaluating the original formula using $t = 4.486$ seconds to see if the motorcycle indeed lands at a height of 15 feet. We will leave this check to the reader.

YOUR TURN A ball is thrown from an initial height of 1.5 meters with an initial velocity of 8 meters per second. Find the time for the ball to land on the ground ($h = 0$). Approximate the time to the nearest thousandth.

ANSWER

1.802 sec.

11.2 Exercises

1. Explain the general plan for deriving the quadratic formula.

2. Discuss the advantages and disadvantages of using the quadratic formula to solve quadratic equations.

3. Are there quadratic equations that cannot be solved using factoring? Explain.

4. What part of the quadratic formula is the discriminant?

5. Under what conditions will a quadratic equation have only one solution? Modify the quadratic formula to describe this single solution.

6. If x-intercepts for a quadratic function are imaginary numbers, what does that indicate about its graph?

For Exercises 7–14, rewrite each quadratic equation in the form $ax^2 + bx + c = 0$, then identify a, b, and c.

7. $x^2 - 3x + 7 = 0$

8. $x^2 + 6x - 15 = 0$

9. $3x^2 - 9x = 4$

10. $6x^2 + 10x = -15$

11. $1.5x^2 = x - 0.2$

12. $x = 0.8x^2 + 4.5$

13. $\frac{3}{4}x = -\frac{1}{2}x^2 + 6$

14. $\frac{5}{6}x^2 = \frac{1}{4}x - 6$

For Exercises 15–38, solve using the quadratic formula.

15. $x^2 + 9x + 20 = 0$ **16.** $x^2 - 4x - 21 = 0$ **17.** $x^2 - x - 1 = 0$ **18.** $x^2 + 3x - 5 = 0$

19. $4x^2 + 5x = 6$ **20.** $2x^2 - x = 3$ **21.** $x^2 - 9x = 0$ **22.** $x^2 + 2x = 0$

23. $x^2 - 8x = -16$ **24.** $x^2 + 14x = -49$ **25.** $x^2 + 2 = 2x$ **26.** $x^2 + 5 = 4x$

27. $4x^2 - 3x = 1$ **28.** $5x^2 - 4x = 1$ **29.** $3x^2 + 10x + 5 = 0$ **30.** $3x^2 - x - 3 = 0$

31. $-4x^2 + 6x = 5$ **32.** $-5x^2 = 3 - 4x$ **33.** $18x^2 + 2 = -15x$ **34.** $10x^2 - 12 = -7x$

35. $3x^2 - 4x = -3$ **36.** $-3x^2 + 6x = 8$ **37.** $6x^2 - 3x = 4$ **38.** $6x^2 - 6 - 13x = 0$

For Exercises 39–50, solve using the quadratic formula. If the solutions are irrational, give an exact answer and an approximate answer rounded to the nearest thousandth. (Hint: You might first clear the fractions or decimals by multiplying both sides by an appropriately chosen number.)

39. $2x^2 + 0.1x = 0.03$ **40.** $4x^2 + 8.6x - 2.4 = 0$ **41.** $x^2 + \dfrac{1}{2}x - 3 = 0$

42. $x^2 + \dfrac{1}{3}x - \dfrac{2}{3} = 0$ **43.** $x^2 - \dfrac{49}{36} = 0$ **44.** $x^2 - 0.0144 = 0$

45. $\dfrac{1}{2}x^2 + \dfrac{3}{2} = x$ **46.** $\dfrac{1}{5}x^2 + 2 = \dfrac{1}{2}x$ **47.** $x^2 - 0.5 = -0.06x$

48. $\dfrac{1}{3}x^2 - \dfrac{1}{4}x - \dfrac{1}{24} = 0$ **49.** $1.2x^2 - 0.6x = -0.5$ **50.** $2.4x^2 + 4.5 = 6.3x$

For Exercises 51–54, explain the mistake, then solve correctly.

51. Solve $3x^2 - 7x + 1 = 0$ using the quadratic formula.
$$\frac{-7 \pm \sqrt{(-7)^2 - (4)(3)(1)}}{2(3)} = \frac{-7 \pm \sqrt{49 - 12}}{6}$$
$$= \frac{-7 \pm \sqrt{37}}{6}$$

52. Solve $2x^2 - 6x - 5 = 0$ using the quadratic formula.
$$\frac{6 \pm \sqrt{(-6)^2 - (4)(2)(5)}}{2(2)} = \frac{6 \pm \sqrt{36 - 40}}{4}$$
$$= \frac{3}{2} \pm \frac{1}{2}i$$

53. Solve $x^2 - 2x + 3 = 0$ using the quadratic formula.

$$\frac{-(-2) \pm \sqrt{(-2)^2 - (4)(1)(3)}}{2(1)} = \frac{2 \pm \sqrt{4 - 12}}{2}$$

$$= \frac{2 \pm \sqrt{-8}}{2}$$

54. Solve $x^2 - 8 = 0$ using the quadratic formula.

$$\frac{-(-8) \pm \sqrt{(-8)^2 - (4)(1)(0)}}{2(1)} = \frac{8 \pm \sqrt{64 - 0}}{2}$$

$$= \frac{8 \pm \sqrt{64}}{2}$$

$$= \frac{8 \pm 8}{2}$$

$$= 8, 0$$

For Exercises 55–64, use the discriminant to determine the number and type of solutions for the equation. If the solution(s) are real, state whether they are rational or irrational.

55. $x^2 + 10x = -25$

56. $2x^2 - 8x + 8 = 0$

57. $\frac{1}{4}x^2 - 4x = -4$

58. $\frac{1}{2}x^2 + 18 = 6x$

59. $x^2 + 4x + 9 = 0$

60. $2x^2 - 3x - 2 = 0$

61. $x^2 - x + 3 = 0$

62. $x^2 + 4x = -5$

63. $x^2 - 6x + 6 = 0$

64. $3x^2 = 13x - 8$

For Exercises 65–74, indicate which of the following methods is the best choice for solving the given equation: factoring, using the square root principle, or using the quadratic formula. Then solve the equation.

65. $x^2 - 81 = 0$

66. $x^2 = 44$

67. $4x^2 + 48 = 0$

68. $2x^2 = 5x - 2$

69. $x^2 - 6x + 13 = 0$

70. $(x + 7)^2 = 40$

71. $x^2 + 6x = 0$

72. $x^2 + 14x + 45 = 0$

73. $x^2 = 8x - 19$

74. $x^2 - 2x = 149$

For Exercises 75–82, find the x- and y-intercepts.

75. $y = x^2 - x - 2$

76. $y = x^2 - 3x + 2$

77. $y = -x^2 - 2x + 8$

78. $y = x^2 - x - 30$

79. $y = 2x^2 + 15x - 8$

80. $y = -15x^2 + x + 6$

81. $y = -2x^2 + 3x - 6$

82. $y = -3x^2 - 2x - 5$

For Exercises 83–90, translate to a quadratic equation then solve using the quadratic formula.

83. A positive integer squared plus five times its consecutive integer is equal to 71. Find the integers.

84. The square of a positive integer minus half of its consecutive integer is equal to 162. Find the integers.

85. A right triangle has side lengths that are three consecutive integers. Use the Pythagorean theorem to find the lengths of those sides. (Remember that the hypotenuse in a right triangle is always the longest side.)

86. A right triangle has side lengths that are consecutive even integers. Use the Pythagorean theorem to find the lengths of those sides.

87. The length of a rectangular fence gate is 3.5 feet less than three times the width. Find the length and width of the gate if the area is 34 square feet.

88. A small access door for an attic storage area is designed so that the width is 3.25 feet more than the length. Find the length and width if the area is 10.5 square feet.

89. An architect is experimenting with two different shapes of a room, as shown.

   a. Find x so that the rooms have the same area.

   b. Complete the dimensions for the L-shaped room.

90. A cylinder is to be made so that its volume is equal to that of a sphere with a radius of 9 inches. If a cylinder is to have a height of 4 inches, find its radius.

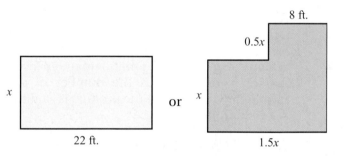

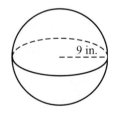

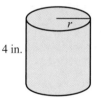

For Exercises 91–94, use the formula $h = \frac{1}{2}gt^2 + v_0t + h_0$, where g represents the acceleration due to gravity, t is the time in flight, v_0 is the initial velocity, and h_0 is the initial height. For Earth, the acceleration due to gravity is -32.2 ft./sec.^{2} or -9.8 m/sec.^{2} Approximate irrational answers to the nearest hundredth.

91. In 2001, Robby Knievel jumped the Grand Canyon on a motorcycle. Suppose that on take off, his motorcycle had a vertical velocity of 48 feet per second and at the end of the launch ramp his altitude was 485 feet above sea level. If he landed on a ramp that was 460 feet above sea level, how long was he in flight?

92. At the 2003 extreme-games competition, Brian Deegan successfully performed the first 360-degree flip ever attempted in a competition. Suppose his motorcycle had a vertical velocity of 12.8 meters per second from the end of a ramp that was 8 meters high and he landed on a ramp at a point 4 meters above the ground. How long was he in the air?

93. A platform diver dives from a platform that is 10 meters above the water.

    a. Write the equation that describes his height during the dive. (Assume his initial velocity is 0.)

    b. Find the time it takes for the diver to be at 5 meters above the water.

    c. Find the time for the diver to enter the water.

94. In a cliff-diving championship in Acapulco, a diver dives from a cliff at a height of 70 feet.

    a. Write the equation that describes his height during the dive. (Assume his initial velocity is 0.)

    b. Find the time it takes for the diver to be at 50 feet above the water.

    c. Find the time for the diver to enter the water.

95. The expression $0.5n^2 + 2.5n$ describes the gross income from the sale of a particular software product, where n is the number of units sold in thousands. The expression $4.5n + 16$ describes the cost of producing the n units. Find the number of units that must be produced and sold for the company to break even. (To *break even* means that the gross income and cost are the same.)

96. An economist and marketing manager discover that the expression $2n^2 + 5n$ models the price of a CD based on the demand for it, where n is the number of units (in millions) that the market demands. The expression $-0.5n^2 + 17.1$ describes the price of the CD based on the number of units (also in millions) supplied to the market.

    a. Find the number of units that need to be demanded and supplied so that the price based on demand is equal to the price based on supply. When the number of units demanded by the market is the same as the number of units supplied to the market, the product is said to be at equilibrium.

    b. What is the price of the CD at equilibrium?

97. For the equation $2x^2 - 5x + c = 0$,

    a. Find c so that the equation has only one rational number solution.

    b. Find the range of values of c for which the equation has two real-number solutions.

    c. Find the range of values of c for which the equation has no real-number solution.

98. For the equation $4x^2 + 6x + c = 0$,

    a. Find c so that the equation has only one rational number solution.

    b. Find the range of values of c for which the equation has two real-number solutions.

    c. Find the range of values of c for which the equation has no real-number solution.

99. For the equation $ax^2 + 12x + 8 = 0$,

    a. Find a so that the equation has only one rational number solution.

    b. Find the range of values of a for which the equation has two real-number solutions.

    c. Find the range of values of a for which the equation has no real-number solution.

100. For the equation $3x^2 + bx + 8 = 0$,

    a. Find b so that the equation has only one rational number solution.

    b. Find all positive values of b for which the equation has two real-number solutions.

    c. Find all positive values of b for which the equation has no real-number solution.

PUZZLE PROBLEM

In the equation shown, A, B, C, D, and E are five consecutive positive integers where $A < B < C < D < E$. What are they?

$$A^2 + B^2 + C^2 = D^2 + E^2$$

REVIEW EXERCISES

For Exercises 1 and 2, factor.

[6.2]   **1.** $u^2 - 9u + 14$

[6.3]   **2.** $3u^2 - 2u - 16$

For Exercises 3 and 4, solve.

[6.6]   **3.** $5u^2 + 13u = 6$

[7.6]   **4.** $\dfrac{7}{3u} = \dfrac{5}{u} - \dfrac{1}{u - 5}$

For Exercises 5 and 6, simplify.

[5.4]   **5.** $(x^2)^2$

[10.2]   **6.** $(x^{1/3})^2$

11.3 Solving Equations That Are Quadratic in Form

OBJECTIVES

1. Solve equations by rewriting them in quadratic form.

2. Solve equations that are quadratic in form by using substitution.

3. Solve applications problems using equations that are quadratic in form.

Many equations that are not quadratic equations are **quadratic in form** and can be solved using the methods for solving quadratic equations.

DEFINITION   An equation is **quadratic in form** if it can be rewritten as a quadratic equation $au^2 + bu + c = 0$, where $a \neq 0$ and u is a variable or an expression.

Equations with Rational Expressions

In Section 7.6, we solved equations containing rational expressions by multiplying both sides of the equation by the LCD. Those rewritten equations are often quadratic. Remember that equations containing rational expressions sometimes have extraneous solutions.

Connection Recall that a solution for an equation with rational expressions is extraneous if it causes one or more of the denominators to equal 0.

EXAMPLE 1 Solve. $\dfrac{3}{x+1} = 1 - \dfrac{3}{x(x+1)}$

Solution $x(x+1) \cdot \dfrac{3}{x+1} = x(x+1)\left(1 - \dfrac{3}{x(x+1)}\right)$ Multiply both sides by the LCD $x(x+1)$.

$x(x+1) \cdot \dfrac{3}{x+1} = x(x+1) \cdot 1 - x(x+1) \cdot \dfrac{3}{x(x+1)}$ Distribute $x(x+1)$.

$3x = x^2 + x - 3$ Simplify both sides.

$0 = x^2 - 2x - 3$ Subtract $3x$ from both sides to get the quadratic form $ax^2 + bx + c = 0$.

$0 = (x-3)(x+1)$ Factor.

$x - 3 = 0$ or $x + 1 = 0$ Use the zero-factor theorem.

$x = 3$ $x = -1$ Solve each equation.

Checks $x = 3$ $x = -1$

$\dfrac{3}{3+1} = 1 - \dfrac{3}{3(3+1)}$ $\dfrac{3}{-1+1} = 1 - \dfrac{3}{-1(-1+1)}$

$\dfrac{3}{4} = 1 - \dfrac{3}{12}$ $\dfrac{3}{0} = 1 - \dfrac{3}{0}$ ◀ **Note:** *The expression* $\dfrac{3}{0}$ *is undefined, so* -1 *is extraneous.*

$\dfrac{3}{4} = \dfrac{3}{4}$ **True**

The solution is 3 (-1 is extraneous).

YOUR TURN Solve. $\dfrac{x}{x-2} = \dfrac{6}{x} + \dfrac{4}{x(x-2)}$

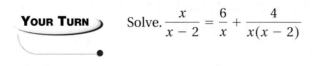

Equations Containing Radicals

In Section 10.6, we found that after using the power rule on equations containing radicals the result was often a quadratic equation. Remember that these radical equations sometimes have extraneous solutions.

ANSWER

4 (2 is extraneous)

EXAMPLE 2 Solve.

a. $\sqrt{x} + x = 6$

Solution $\sqrt{x} = 6 - x$ **Subtract x from both sides to isolate the radical.**

$(\sqrt{x})^2 = (6 - x)^2$ **Square both sides.**

$x = 36 - 12x + x^2$ **Simplify both sides.**

$0 = x^2 - 13x + 36$ **Subtract x from both sides to write in quadratic form.**

$0 = (x - 4)(x - 9)$ **Factor.**

$x - 4 = 0$ or $x - 9 = 0$ **Use the zero-factor theorem.**

$x = 4$ $x = 9$ **Solve each equation.**

Checks $x = 4$ $x = 9$

$\sqrt{4} + 4 \stackrel{?}{=} 6$ $\sqrt{9} + 9 \stackrel{?}{=} 6$

$2 + 4 = 6$ **True** $3 + 9 = 6$ **False, so 9 is extraneous.**

The solution is 4 (9 is extraneous).

b. $\sqrt{x - 1} = 2x - 1$

Solution $(\sqrt{x - 1})^2 = (2x - 1)^2$ **Square both sides of the equation.**

$x - 1 = 4x^2 - 4x + 1$ **Simplify both sides.**

$0 = 4x^2 - 5x + 2$ **Subtract x and add 1 on both sides to write in quadratic form.**

We cannot factor $4x^2 - 5x + 2$, so we will use the quadratic formula.

$$x = \frac{-(-5) \pm \sqrt{(-5)^2 - 4(4)(2)}}{2(4)} = \frac{5 \pm \sqrt{-7}}{8} = \frac{5 \pm i\sqrt{7}}{8}$$

Check   Because the solutions are complex, we will not check them.

Connection   Recall that a solution for an equation containing radicals is extraneous if it makes the original equation false.

Note: *Written in standard form, these solutions are* $\frac{5}{8} + \frac{\sqrt{7}}{8}i$ *and* $\frac{5}{8} - \frac{\sqrt{7}}{8}i$.

ANSWERS

a. 4 (1 is extraneous)

b. $\dfrac{2 \pm 2i\sqrt{2}}{3}$

YOUR TURN   Solve.

a. $x - \sqrt{x - 2} = 0$ **b.** $\sqrt{6x - 11} = 3x - 1$

OBJECTIVE 2. Solve equations that are quadratic in form using substitution. Recall that an equation that is quadratic in form can be written as $au^2 + bu + c = 0$, where $a \neq 0$ and u can be an *expression*. We now explore a method for solving these equations where we substitute u for an expression. To use substitution, it is important to note a pattern with the exponents of the terms in a quadratic equation. Notice that the degree of the first term, au^2, is 2, the degree of the middle term, bu, is 1, and the third term is a constant. If a trinomial has one term with an expression raised to the second power, a second term with that same expression raised to the first power, and a third term that is a constant, then the equation is quadratic in form and we can use substitution to solve it.

Consider the equation $x^4 - 13x^2 + 36 = 0$. Notice we can rewrite the equation as $(x^2)^2 - 13(x^2) + 36 = 0$, so it is quadratic in form. By substituting u for each x^2,

Note: *Given an equation containing a trinomial, if the degree of the first term is twice the degree of the middle term, and the third term is a constant, then try using substitution.*

we have a "friendlier" form of quadratic equation, which we can then solve by factoring, completing the square, or the quadratic formula.

$$(x^2)^2 - 13(x^2) + 36 = 0$$
$$u^2 - 13u + 36 = 0 \qquad \text{Substitute } u \text{ for } x^2.$$
$$(u - 9)(u - 4) = 0 \qquad \text{Factor.}$$
$$u - 9 = 0 \quad \text{or} \quad u - 4 = 0 \qquad \text{Use the zero-factor theorem.}$$
$$u = 9 \qquad\qquad u = 4$$

Note that these solutions are for u, not x. We must substitute x^2 back in place of u to finish solving for x.

Note: *The equation* $x^4 - 13x^2 + 36 = 0$ *has four solutions:* 3, −3, 2, *and* −2.

$$x^2 = 9 \qquad x^2 = 4 \qquad \text{Substitute } x^2 \text{ for } u.$$
$$x = \pm\sqrt{9} \qquad x = \pm\sqrt{4}$$
$$x = \pm 3 \qquad x = \pm 2$$

Our example suggests the following procedure.

PROCEDURE **Using Substitution to Solve Equations That Are Quadratic in Form**

To solve equations that are quadratic in form using substitution,

1. Rewrite the equation so that it is in the form $au^2 + bu + c = 0$.
2. Solve the quadratic equation for u.
3. Substitute for u and solve.
4. Check the solutions.

EXAMPLE 3 Solve.

a. $(x + 2)^2 - 2(x + 2) - 8 = 0$

Solution If we substitute u for $x + 2$, we have an equation in the form $au^2 + bu + c = 0$.

$$(x + 2)^2 - 2(x + 2) - 8 = 0$$
$$u^2 - 2u - 8 = 0 \qquad \text{Substitute } u \text{ for } x + 2.$$
$$(u - 4)(u + 2) = 0 \qquad \text{Factor.}$$
$$u - 4 = 0 \quad \text{or} \quad u + 2 = 0 \qquad \text{Use the zero-factor theorem.}$$
$$u = 4 \qquad\qquad u = -2 \qquad \text{Solve each equation for } u.$$
$$x + 2 = 4 \qquad x + 2 = -2 \qquad \text{Substitute } x + 2 \text{ for } u.$$
$$x = 2 \qquad\qquad x = -4 \qquad \text{Solve each equation for } x.$$

Check Verify that 2 and −4 make $(x + 2)^2 - 2(x + 2) - 8 = 0$ true. We will leave this check to the reader.

b. $x^{2/3} - x^{1/3} - 6 = 0$

Solution Since $x^{2/3} - x^{1/3} - 6 = 0$ can be written as $(x^{1/3})^2 - x^{1/3} - 6 = 0$, it is quadratic in form. We will substitute u for $x^{1/3}$.

$(x^{1/3})^2 - x^{1/3} - 6 = 0$	**Rewrite in quadratic form.**
$u^2 - u - 6 = 0$	**Substitute u for $x^{1/3}$.**
$(u - 3)(u + 2) = 0$	**Factor.**
$u - 3 = 0$ or $u + 2 = 0$	**Use the zero-factor theorem.**
$u = 3$ $\qquad u = -2$	**Solve each equation for u.**
$x^{1/3} = 3$ $\qquad x^{1/3} = -2$	**Substitute $x^{1/3}$ for u.**
$(x^{1/3})^3 = 3^3$ $\qquad (x^{1/3})^3 = (-2)^3$	**Cube both sides of the equations.**
$x = 27$ $\qquad x = -8$	**Simplify.**

Checks $\quad x = 27$ $\qquad\qquad\qquad\qquad\qquad x = -8$

$27^{2/3} - 27^{1/3} - 6 \overset{?}{=} 0$ $\qquad\qquad (-8)^{2/3} - (-8)^{1/3} - 6 \overset{?}{=} 0$

$(\sqrt[3]{27})^2 - \sqrt[3]{27} - 6 \overset{?}{=} 0$ $\qquad\qquad (\sqrt[3]{-8})^2 - \sqrt[3]{-8} - 6 \overset{?}{=} 0$

$3^2 - 3 - 6 \overset{?}{=} 0$ $\qquad\qquad\qquad\quad (-2)^2 - (-2) - 6 \overset{?}{=} 0$

$9 - 3 - 6 \overset{?}{=} 0$ $\qquad\qquad\qquad\qquad\quad 4 + 2 - 6 \overset{?}{=} 0$

$0 = 0 \quad$ True $\qquad\qquad\qquad\qquad\qquad 0 = 0 \quad$ True

YOUR TURN Solve.

a. $x^4 - 10x^2 + 9 = 0$ $\qquad\qquad\qquad$ **b.** $(n - 2)^2 + 4(n - 2) - 12 = 0$

c. $x^{2/3} - x^{1/3} - 2 = 0$

OBJECTIVE 3. Solve applications problems using equations that are quadratic in form.

EXAMPLE 4 The average speed of a car is 10 miles per hour more than the average speed of a bus. The bus takes 1 hour longer than the car to travel 200 miles. Find how long it takes the car to travel 200 miles.

Understand Both the car and the bus travel 200 miles, but it takes the bus 1 hour longer than the car. The rate of the car is 10 miles per hour more than the rate of the bus.

Plan We will use a table to organize the information, then write an equation, which we can solve.

Execute We will let x represent the time for the car to travel 200 miles. Since the bus travels 1 more hour, $x + 1$ describes the time for the bus to travel 200 miles.

Vehicle	d	t	r
bus	200	$x + 1$	$\dfrac{200}{x + 1}$
car	200	x	$\dfrac{200}{x}$

We use $r = \dfrac{d}{t}$ to describe each rate.

Since the rate of the car is 10 miles per hour faster than the rate of the bus, we can say (the rate of the car) = (the rate of the bus) + 10.

$$\frac{200}{x} = \frac{200}{x+1} + 10$$

$$x(x+1)\left(\frac{200}{x}\right) = x(x+1)\left(\frac{200}{x+1} + 10\right) \qquad \text{Multiply both sides by the LCD } x(x+1).$$

$$(x+1)(200) = x(x+1)\left(\frac{200}{x+1}\right) + x(x+1)(10) \qquad \text{Simplify both sides.}$$

$$200x + 200 = 200x + 10x^2 + 10x \qquad \text{Continue simplifying.}$$

$$0 = 10x^2 + 10x - 200 \qquad \text{Subtract } 200x \text{ and } 200 \text{ from both sides to get quadratic form.}$$

$$0 = x^2 + x - 20 \qquad \text{Divide both sides by 10.}$$

$$0 = (x+5)(x-4) \qquad \text{Factor.}$$

$$x + 5 = 0 \quad \text{or} \quad x - 4 = 0 \qquad \text{Use the zero-factor theorem.}$$

$$x = -5 \qquad\qquad x = 4 \qquad \text{Solve each equation.}$$

Answer   Since time cannot be negative, it takes the car 4 hours to travel 200 miles.

Check   The rate of the car is $\dfrac{200}{x} = \dfrac{200}{4} = 50$ miles per hour, so in 4 hours the car travels $50(4) = 200$ miles. The bus travels 200 miles in $4 + 1 = 5$ hours at a rate of $\dfrac{200}{4+1} = \dfrac{200}{5} = 40$ miles per hour, so in 5 hours the bus travels $5(40) = 200$ miles.

Notice also that the car's rate is 10 miles per hour more than the bus's.

EXAMPLE 5   Bobby and Pam manufacture and install blinds. Working together, they can install blinds in every window of an average-sized house in $1\frac{1}{3}$ hours. Working alone, Bobby takes 2 hours longer than Pam to do the same installation. How long does each take to install blinds in an average-sized house working alone?

Understand   We are given the time it will take Bobby and Pam working together and are asked to find how long it would take each individually. We also know that it will take Bobby 2 hours longer than it would take Pam.

Plan   We will use a table to organize the information, write an equation, and then solve.

Execute   We will let x represent the number of hours for Pam to install working alone. Since Bobby takes 2 more hours, his time working alone is represented by $x + 2$.

Worker	Time to Complete the Job Alone	Rate of Work	Time at Work	Portion of Job Completed
Pam	x	$\dfrac{1}{x}$	$\dfrac{4}{3}$	$\dfrac{4}{3x}$
Bobby	$x + 2$	$\dfrac{1}{x + 2}$	$\dfrac{4}{3}$	$\dfrac{4}{3(x + 2)}$

Multiplying the rate of work and the time at work gives an expression of the amount of the job completed.

The total job in this case is 1 average-sized house, so we can write an equation that combines their individual expressions for work completed and set this sum equal to 1.

(Portion Pam does) + (Portion Bobby does) = 1 (the entire job)

$$\frac{4}{3x} + \frac{4}{3(x + 2)} = 1$$

$$3x(x + 2)\left(\frac{4}{3x} + \frac{4}{3(x + 2)}\right) = 3x(x + 2)(1) \qquad \text{Multiply both sides by the LCD } 3x(x + 2).$$

$$3x(x + 2)\left(\frac{4}{3x}\right) + 3x(x + 2)\left(\frac{4}{3(x + 2)}\right) = 3x^2 + 6x \qquad \text{Simplify both sides.}$$

$$4(x + 2) + 4x = 3x^2 + 6x \qquad \text{Continue simplifying.}$$

$$4x + 8 + 4x = 3x^2 + 6x \qquad \text{Distribute.}$$

$$8x + 8 = 3x^2 + 6x \qquad \text{Combine like terms.}$$

$$0 = 3x^2 - 2x - 8 \qquad \text{Subtract } 8x \text{ and } 8 \text{ from both sides.}$$

$$0 = (3x + 4)(x - 2) \qquad \text{Factor.}$$

$$3x + 4 = 0 \quad \text{or} \quad x - 2 = 0 \qquad \text{Use the zero-factor theorem.}$$

$$3x = -4 \qquad\qquad x = 2 \qquad \text{Solve each equation.}$$

$$x = -\frac{4}{3}$$

Answer   Since negative time makes no sense in the context of this problem, it takes Pam 2 hours to install working alone and Bobby $x + 2 = 2 + 2 = 4$ hours working alone.

Check   Since Pam can install in 2 hours, she can install $\dfrac{1}{2}$ of an average-sized house in 1 hour.

Since Bobby takes 4 hours to do the same work, he can install $\dfrac{1}{4}$ of an average-sized house in 1 hour. Together, they can install $\dfrac{1}{2} + \dfrac{1}{4} = \dfrac{2}{4} + \dfrac{1}{4} = \dfrac{3}{4}$ of an average-sized house in 1 hour, so in $1\dfrac{1}{3}$ hours they can install blinds in $1\dfrac{1}{3} \cdot \dfrac{3}{4} = \dfrac{4}{3} \cdot \dfrac{3}{4} = 1$ average-sized house.

YOUR TURN Solve.

a. The average speed of a passenger train is 25 miles per hour more than the average speed of a car. The time required for the car to travel 300 miles is 2 hours more than the time required for the train. Find the average speed of the car.

b. Terri and Tommy run a flea market. It takes Terri 2 hours longer to put the merchandise out than it does Tommy. If they can put the merchandise out together in $1\frac{7}{8}$ hours, how long would it take each working alone?

ANSWERS

a. 50 mph

b. It takes Tommy 3 hr. and Terri 5 hr.

11.3 Exercises

For Extra Help

MyMathLab Videotape/DVT InterAct Math Math Tutor Center Math XL.com

1. What does it mean to say that an equation is quadratic in form?

2. Is $x^{3/4} - x^{1/4} - 6 = 0$ quadratic in form? Why or why not?

3. If you solved $3\left(\dfrac{x+2}{3}\right)^2 + 13\left(\dfrac{x+2}{3}\right) - 10 = 0$ using substitution, what would your substitution be?

4. If you solved $x^4 - 7x^2 + 12 = 0$ using substitution, what would your substitution be?

5. If Alisha can clean her house in x hours, what part of her house can she clean in 1 hour?

6. If a car travels 400 miles in x hours, how many miles does it travel in 1 hour?

For Exercises 7–18, solve the equations with rational expressions.

7. $\dfrac{1}{x} + \dfrac{1}{x+2} = \dfrac{3}{4}$

8. $\dfrac{1}{x} + \dfrac{2}{x-3} = \dfrac{5}{6}$

9. $\dfrac{60}{x+2} = \dfrac{60}{x} - 5$

10. $\dfrac{120}{x+2} = \dfrac{120}{x} - 5$

11. $\dfrac{1}{p-4} + \dfrac{1}{4} = \dfrac{8}{p^2-16}$

12. $\dfrac{1}{y-5} - \dfrac{10}{y^2-25} = -\dfrac{1}{5}$

13. $\dfrac{6}{2y+5} = \dfrac{2}{y+5} + \dfrac{1}{5}$

14. $\dfrac{6}{2x+3} = \dfrac{2}{x-6} + \dfrac{4}{3}$

15. $1 + 2x^{-1} - 8x^{-2} = 0$ **16.** $1 + 5x^{-1} + 6x^{-2} = 0$ **17.** $3 + 13x^{-1} - 10x^{-2} = 0$ **18.** $2 - x^{-1} - 15x^{-2} = 0$

For Exercises 19–30, solve the equations with radical expressions.

19. $x - 8\sqrt{x} + 15 = 0$

20. $x - 3\sqrt{x} + 2 = 0$

21. $2x - 5\sqrt{x} - 7 = 0$

22. $3x + 4\sqrt{x} - 4 = 0$

23. $\sqrt{2a + 5} = 3a - 3$

24. $\sqrt{3b + 1} = 5b - 3$

25. $\sqrt{2m - 8} - m - 1 = 0$

26. $\sqrt{8 - 12x} - 2x + 3 = 0$

27. $\sqrt{4x + 1} = \sqrt{x + 2} + 1$

28. $\sqrt{3x + 7} = \sqrt{2x + 3} + 1$

29. $\sqrt{2x + 1} - \sqrt{3x + 4} = -1$

30. $\sqrt{2x - 1} - \sqrt{4x + 5} = -2$

For Exercises 31–52, solve using substitution.

31. $x^4 - 10x^2 + 9 = 0$

32. $x^4 - 13x^2 + 36 = 0$

33. $4x^4 - 13x^2 + 9 = 0$

34. $9x^4 - 13x^2 + 4 = 0$

35. $x^4 + 5x^2 - 36 = 0$

36. $x^4 - 3x^2 - 4 = 0$

37. $(x + 2)^2 + 6(x + 2) + 8 = 0$

38. $(x - 3)^2 + 2(x - 3) - 15 = 0$

39. $2(x + 3)^2 - 9(x + 3) - 5 = 0$

40. $3(x - 1)^2 + 4(x - 1) - 4 = 0$

41. $\left(\dfrac{x - 1}{2}\right)^2 + 8\left(\dfrac{x - 1}{2}\right) + 15 = 0$

42. $\left(\dfrac{x + 2}{3}\right)^2 - 5\left(\dfrac{x + 2}{3}\right) - 6 = 0$

43. $2\left(\dfrac{x + 2}{2}\right)^2 + \left(\dfrac{x + 2}{2}\right) - 3 = 0$

44. $3\left(\dfrac{x - 1}{3}\right)^2 + 10\left(\dfrac{x - 1}{3}\right) - 8 = 0$

45. $x^{2/3} - 5x^{1/3} + 6 = 0$

46. $x^{2/3} + 3x^{1/3} - 10 = 0$

47. $2x^{2/3} - 3x^{1/3} - 2 = 0$

48. $3x^{2/3} - 4x^{1/3} - 4 = 0$

49. $x^{1/2} - 7x^{1/4} + 12 = 0$

50. $x^{1/2} - 4x^{1/4} + 3 = 0$

51. $5x^{1/2} + 8x^{1/4} - 4 = 0$

52. $2x^{1/2} - x^{1/4} - 3 = 0$

For Exercises 53–64, solve.

53. The average rate of a bus is 15 miles per hour more than the average rate of a truck. The truck takes 1 hour longer than the bus to travel 180 miles. How long does it take the bus to travel 180 miles?

54. A charter business has two types of small planes, a jet and a twin propeller plane. The prop-plane's average air speed is 120 miles per hour less than the jet's. The prop-plane takes 1 hour longer to travel 720 miles than the jet. How long does it take the prop-plane to travel the 720 miles?

55. The average speed of the winner of the Boston Marathon in 2003 was 3.8 miles per hour faster than the person who finished 1 hour behind. Find the time it took the winner to run the 26-mile course.

56. In the 2003 Tour de France, Lance Armstrong's average speed was 0.0081 kilometers per hour faster than the second-place finisher, who finished 1 minute behind Armstrong. Find Armstrong's time to complete the 3415-kilometer race.

57. Suppose the time required for a bus to travel 360 miles is 3 hours more than the time required for a motorcycle to travel 300 miles. If the average rate of the motorcycle is 15 miles per hour more than the average rate of the bus, find the average rate of the bus.

58. Suppose the time required for a truck to travel 400 miles is 3 hours more than the time required for a car to travel 300 miles. If the average rate of the car is 10 miles per hour more than the average rate of the truck, find the average rate of the truck.

59. After training hard for a year, a novice cyclist discovers that she has increased her average rate by 6 miles per hour and can travel 36 miles in 1 less hour than a year ago.
    a. What was her old average rate? What is her new average rate?
    b. What was her old time to travel 36 miles? What is her new time for 36 miles?

60. A high school track coach determines that his fastest long-distance runner runs 2 miles per hour faster than his slowest runner. The slower runner takes a half-hour longer to run 12 miles than the faster runner.
    a. Find the average rates of both runners.
    b. Find the time for both runners to run 12 miles.

61. Billy and Jody are commercial fishermen. Working alone it takes Billy 2 hours longer to run the hoop nets than it takes Jody working alone. Together they can run the hoop nets in $2\frac{2}{5}$ hours. How long does it take each working alone?

62. Using a riding mower, Fran can mow the grass at the campground in 4 hours less time than it takes Donnie using a push mower. Together they can mow the grass in $2\frac{2}{3}$ hours. How long does it take each working alone?

63. A newspaper has two presses, one of which is older than the other. Working alone, the newer press can print all the copies for a typical day in a half-hour less time than the older press. When running at the same time, they print all of a typical day's copies in 2 hours. How long would it take to print a day's worth if they worked alone?

64. A school copy center prints the school newsletter in a half-hour using two copiers running at the same time. If the center uses only one copier, the faster of the two copiers takes 6 minutes less time than the other copier to print all of the newsletters. How long does each copier take to print the newsletters working alone?

On a guitar, the frequency of the vibrating string is related to the tension on the string. Suppose two strings of the same diameter and length are placed on an instrument and wound to different tensions. The formula $\dfrac{F_1^2}{F_2^2} = \dfrac{T_1}{T_2}$ describes the relationship between their frequency and tension.

65. One string is wound on a guitar to a tension of 50 pounds. A second string of the same diameter is wound to 60 pounds. If the string with the greater tension has a frequency that is 40 vibrations per second greater than the other string, what are the frequencies of the two strings? Round to the nearest ten.

66. Two strings of the same diameter are wound on a banjo to different tensions: one at 80 pounds and the other at 90 pounds. The frequency of the string under less tension is 20 vibrations per second less than the other string. What are the frequencies of the two strings? Round to the nearest ten.

> **Of Interest**
> In addition to tension, the diameter and length of a string also affect its frequency. Increasing the diameter or length of a string under the same tension decreases its frequency.

REVIEW EXERCISES

[1.7] **1.** Evaluate $-\dfrac{b}{2a}$ when a is 2 and b is 8.

[4.3] **2.** Find the x- and y-intercepts for $2x + 3y = 6$.

[4.7] **3.** If $f(x) = 2x^2 - 3x + 1$, find $f(-1)$.

[8.4] *For Exercises 4–6, graph.*

4. $f(x) = |x - 3| + 1$

5. $f(x) = x^2 - 3$

6. $f(x) = 3x^2$

11.4 Graphing Quadratic Functions

OBJECTIVE

1. Graph quadratic functions of the form $f(x) = ax^2$.
2. Graph quadratic functions of the form $f(x) = ax^2 + k$.
3. Graph quadratic functions of the form $f(x) = a(x - h)^2$.
4. Graph quadratic functions of the form $f(x) = a(x - h)^2 + k$.
5. Graph quadratic functions of the form $f(x) = ax^2 + bx + c$.
6. Solve applications involving parabolas.

In Sections 6.7 and 8.4, we learned that quadratic functions have the form $f(x) = ax^2 + bx + c$, where a, b, and c are real numbers and $a \neq 0$. By plotting lots of ordered pairs, we found that the graphs of these functions are parabolas that open up if $a > 0$ and down if $a < 0$. By replacing x with 0, we found that the y-intercept is $(0, c)$. By replacing y (or $f(x)$) with 0, we found the x-intercepts. We also learned that every parabola will have symmetry about a line called its *axis of symmetry* and the axis of symmetry always passes through a point called the *vertex*, which is the lowest point on a parabola that opens upwards or the highest point on a parabola that opens downwards. Look at the following graphs.

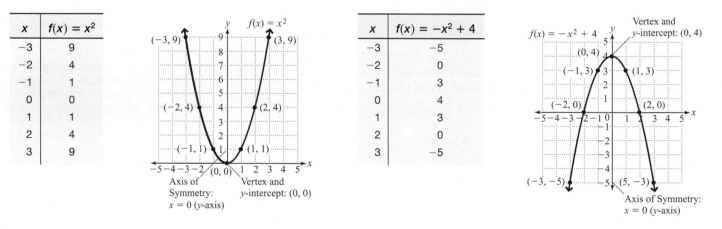

x	$f(x) = x^2$
−3	9
−2	4
−1	1
0	0
1	1
2	4
3	9

x	$f(x) = -x^2 + 4$
−3	−5
−2	0
−1	3
0	4
1	3
2	0
3	−5

Notice that the graph of $f(x) = x^2$ opens upwards because $a = 1$, which is positive, whereas the graph of $f(x) = -x^2 + 4$ opens downwards because $a = -1$, which is negative. In this section, we will learn more about the graphs of quadratic functions. Later in the section, we will revisit the form $f(x) = ax^2 + bx + c$ and learn how to determine a parabola's vertex and axis of symmetry from the values of a and b.

OBJECTIVE 1. Graph quadratic functions of the form $f(x) = ax^2$. First we consider $f(x) = ax^2$ to discover more about how a affects the parabola. We will see that a not only affects whether the parabola opens up or down but also affects the width of the parabola.

EXAMPLE 1 Compare the graphs of each function.

a. $f(x) = \frac{1}{4}x^2$, $g(x) = \frac{1}{3}x^2$, $h(x) = \frac{1}{2}x^2$, $k(x) = x^2$, and $m(x) = 2x^2$

Solution We will graph all five functions on the same grid.

Note: *All these graphs open up because **a** is positive. Also notice that as **a** increases, the parabolas appear narrower.*

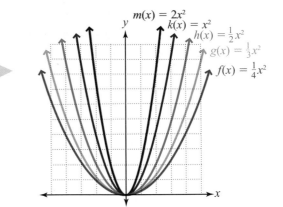

b. $f(x) = -\frac{1}{4}x^2$, $g(x) = -\frac{1}{3}x^2$, $h(x) = -\frac{1}{2}x^2$, $k(x) = -x^2$, and $m(x) = -2x^2$

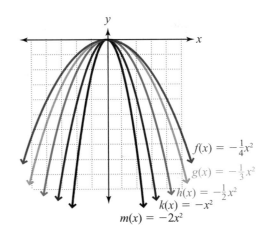

Note: *All these graphs open down because **a** is negative. Also notice that as $|a|$ increases, the parabolas appear narrower.*

Example 1 suggests the following conclusions about functions of the form $f(x) = ax^2$.

Conclusion Given a function in the form $f(x) = ax^2$, the axis of symmetry is $x = 0$ (the y-axis) and the vertex is $(0, 0)$. Also, the greater the absolute value of a, the narrower the parabola appears (or, the smaller the absolute value of a, the wider the parabola appears).

YOUR TURN Graph the following functions.

a. $f(x) = -4x^2$

b. $f(x) = \frac{3}{4}x^2$

ANSWERS

a.

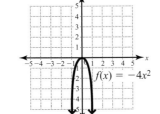

b.

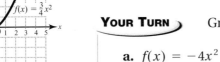

OBJECTIVE 2. Graph quadratic functions of the form $f(x) = ax^2 + k$. We have learned that the vertex of a parabola of the form $f(x) = ax^2$ is at $(0, 0)$ and the axis of symmetry is $x = 0$ (the y-axis). Now let's consider quadratic functions of the form $f(x) = ax^2 + k$, where k is a constant. We will see that for the same value of a, the graphs of $f(x) = ax^2 + k$ and $f(x) = ax^2$ have the same width and shape but the graph of $f(x) = ax^2 + k$ is shifted up or down k units on the y-axis from the origin so that the vertex is at $(0, k)$.

EXAMPLE 2 Graph the following functions.

a. $g(x) = 2x^2 + 3$ and $h(x) = 2x^2 - 4$

Solution We will graph both functions on the same grid.

Note: *The vertex of the "basic" function $f(x) = 2x^2$ is the origin $(0, 0)$. The vertex of $g(x) = 2x^2 + 3$ is shifted up 3 and the vertex of $h(x) = 2x^2 - 4$ is shifted down 4 from the origin.*

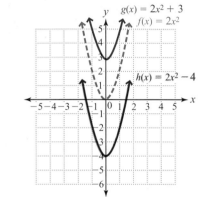

b. $g(x) = -x^2 + 4$ and $h(x) = -x^2 - 2$

Solution

Note: *The vertex of the "basic" function $f(x) = -x^2$ is the origin $(0, 0)$. The vertex of $g(x) = -x^2 + 4$ is shifted up 4 and the vertex of $h(x) = -x^2 - 2$ is shifted down 2 from the origin.*

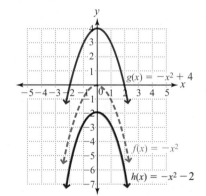

Example 2 suggests the following conclusion about the constant k in functions of the form $f(x) = ax^2 + k$.

Conclusion Given a function in the form $f(x) = ax^2 + k$, if $k > 0$, then the graph of $f(x) = ax^2$ is shifted k units *up* from the origin. If $k < 0$, then the graph of $f(x) = ax^2$ is shifted k units *down* from the origin. The new position of the vertex is $(0, k)$. The axis of symmetry is $x = 0$ (the y-axis).

YOUR TURN Graph the following functions.

a. $f(x) = 3x^2 - 2$ **b.** $f(x) = -\dfrac{1}{2}x^2 + 4$

ANSWERS

a.

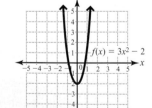

$f(x) = 3x^2 - 2$

b.

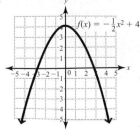

$f(x) = -\frac{1}{2}x^2 + 4$

OBJECTIVE 3. Graph quadratic functions of the form $f(x) = a(x - h)^2$. We now consider the form $f(x) = a(x - h)^2$ and we will see that the constant h causes the parabola to shift right or left.

> **EXAMPLE 3** Graph. $m(x) = 2(x - 3)^2$ and $n(x) = 2(x + 4)^2$

Solution

Note: *The vertex of the "basic" function $f(x) = 2x^2$ is $(0, 0)$ and the axis of symmetry is $x = 0$. The graph of $m(x) = 2(x - 3)^2$ has the same shape as $f(x) = 2x^2$, only the vertex and axis of symmetry are shifted right 3 units from the origin along the x-axis so that they are $(3, 0)$ and $x = 3$. Similarly, the vertex and axis of symmetry of $n(x) = 2(x + 4)^2$ are shifted left 4 units from the origin along the x-axis so that they are $(-4, 0)$ and $x = -4$.*

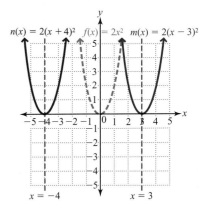

Example 3 suggests the following conclusion about h in $f(x) = a(x - h)^2$.

Conclusion Given a function in the form $f(x) = a(x - h)^2$, if $h > 0$, then the graph of $f(x) = ax^2$ is shifted h units *right* from the origin. If $h < 0$, then the graph of $f(x) = ax^2$ is shifted h units *left* from the origin. The new position of the vertex is $(h, 0)$ and the axis of symmetry is $x = h$.

> **YOUR TURN** Graph the following functions.
>
> **a.** $f(x) = -2(x + 1)^2$ **b.** $g(x) = \dfrac{1}{3}(x - 4)^2$

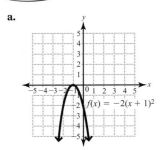

OBJECTIVE 4. Graph quadratic functions of the form $f(x) = a(x - h)^2 + k$. Examples 2 and 3 suggest that the graph of a function of the form $f(x) = a(x - h)^2 + k$ has the same shape as $f(x) = ax^2$, but the vertex is shifted from the origin to (h, k) and the axis of symmetry is shifted from $x = 0$ to $x = h$. We can summarize all that we have learned with the following rule.

Parabola with Vertex (h, k).

The graph of a function in the form $f(x) = a(x - h)^2 + k$ is a parabola with vertex at (h, k). The equation of the axis of symmetry is $x = h$. The parabola opens upwards if $a > 0$ and downwards if $a < 0$. The larger the $|a|$, the narrower the graph.

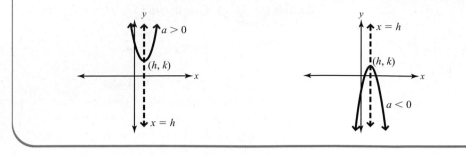

Now let's graph quadratic equations of the form $f(x) = a(x - h)^2 + k$.

EXAMPLE 4 Given $f(x) = 2(x - 3)^2 + 1$, determine whether the graph opens up or down, find the vertex and axis of symmetry, and draw the graph.

Solution We see that $f(x) = 2(x - 3)^2 + 1$ is in the form $f(x) = a(x - h)^2 + k$, where $a = 2$, $h = 3$, and $k = 1$. Since a is positive 2, the parabola opens upwards. The vertex is at (3, 1) and the axis of symmetry is $x = 3$. To complete the graph, we find a few points on either side of the axis of symmetry.

Note: *We find these additional points by choosing x-values on either side of the axis of symmetry and by using the equation to find the corresponding y-values.*

x	y
2	3
1	9
4	3
5	9

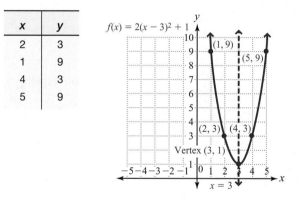

opens downwards, vertex: (1, 3), axis: $x = 1$

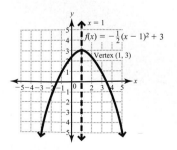

YOUR TURN Given the equation $f(x) = -\dfrac{1}{2}(x - 1)^2 + 3$, determine whether the graph opens upwards or downwards, find the vertex and axis of symmetry, and draw the graph.

OBJECTIVE 5. Graph quadratic functions of the form $f(x) = ax^2 + bx + c$. The advantage of the form $f(x) = a(x-h)^2 + k$ is that we can "see" the vertex, axis of symmetry, and whether the parabola opens upwards or downwards by looking at the equation. If an equation is in the form $f(x) = ax^2 + bx + c$, it can be transformed into $f(x) = a(x-h)^2 + k$ by completing the square.

EXAMPLE 5 Write $f(x) = x^2 - 2x - 8$ in the form $f(x) = a(x-h)^2 + k$. Then determine whether the graph opens upwards or downwards, find the vertex and axis of symmetry, and draw the graph.

Solution

$y = x^2 - 2x - 8$	Replace $f(x)$ with y to make manipulations more workable.
$y + 8 = x^2 - 2x$	Add 8 to both sides of the equation to isolate the x terms.
$y + 8 + 1 = x^2 - 2x + 1$	Add 1 to both sides of the equation to complete the square.
$y + 9 = (x-1)^2$	Write the right side as a square.
$y = (x-1)^2 - 9$	Subtract 9 from both sides to solve for y.

Since $a = 1$ and 1 is positive, the graph opens upwards. The vertex is at $(1, -9)$ and the axis of symmetry is $x = 1$. Plot a few points on either side of the axis of symmetry and graph. We will include the x- and y-intercepts, which we found in Example 3 of Section 11.2

Note: *Remember, to find x-intercepts, we replace $f(x)$ (or y) with 0 and solve for x. To find y-intercepts, we replace x with 0 and solve for y.*

ANSWERS

x-intercepts: $(-1, 0), (3, 0)$
y-intercept: $(0, -3)$
$f(x) = (x-1)^2 - 4$,
opens upwards, vertex: $(1, -4)$,
axis: $x = 1$

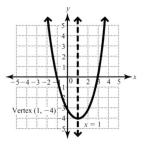

	x	$f(x)$
x-intercept ⟶	-2	0
	-1	-5
y-intercept ⟶	0	-8
vertex ⟶	1	-9
	2	-8
	3	-5
x-intercept ⟶	4	0

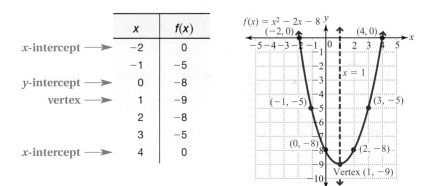

YOUR TURN For $f(x) = x^2 - 2x - 3$, find the x- and y-intercepts, write the equation in the form $f(x) = a(x-h)^2 + k$, determine whether the graph opens upwards or downwards, find the vertex and axis of symmetry, and draw the graph.

We can also find the vertex using a formula, which we can derive by completing the square on the general form $f(x) = ax^2 + bx + c$.

Note: *Adding $\left[\frac{1}{2}\left(\frac{b}{a}\right)\right]^2$ or $\frac{b^2}{4a^2}$ completes the square inside the parentheses. Since those parentheses are multiplied by a, we add $a \cdot \frac{b^2}{4a^2}$ to the left side to keep the equation balanced.*

$y = ax^2 + bx + c$	Replace $f(x)$ with y.
$y - c = ax^2 + bx$	Isolate the x terms.
$y - c = a\left(x^2 + \dfrac{b}{a}x\right)$	Factor out a.
$y - c + \dfrac{ab^2}{4a^2} = a\left(x^2 + \dfrac{b}{a}x + \dfrac{b^2}{4a^2}\right)$	Complete the square.
$y - \dfrac{4a^2c - ab^2}{4a^2} = a\left(x + \dfrac{b}{2a}\right)^2$	Find the LCD of the two terms on the left. Factor on the right.

$$y - \frac{a(4ac - b^2)}{4a^2} = a\left(x + \frac{b}{2a}\right)^2 \qquad \text{Remove a common factor of } a \text{ on the left side.}$$

$$y - \frac{4ac - b^2}{4a} = a\left(x + \frac{b}{2a}\right)^2 \qquad \text{Reduce to lowest terms.}$$

$$y = a\left(x + \frac{b}{2a}\right)^2 + \frac{4ac - b^2}{4a} \qquad \text{Add } \frac{4ac - b^2}{4a} \text{ to both sides.}$$

Notice that if we rewrite the last line above as $y = a\left(x - \left(-\dfrac{b}{2a}\right)\right)^2 + \dfrac{4ac - b^2}{4a}$, we see that $h = -\dfrac{b}{2a}$ and $k = \dfrac{4ac - b^2}{4a}$ so that the vertex is $\left(-\dfrac{b}{2a}, \dfrac{4ac - b^2}{4a}\right)$. Notice this also means that the axis of symmetry is $x = -\dfrac{b}{2a}$. Also, though the y-coordinate of the vertex is $\dfrac{4ac - b^2}{4a}$, it is usually easier to find it by substituting the x-coordinate into the original function.

PROCEDURE Finding the Vertex of a Quadratic Function in the Form $f(x) = ax^2 + bx + c$

Given an equation in the form $f(x) = ax^2 + bx + c$, to determine the vertex,

1. Find the x-coordinate using the formula $x = -\dfrac{b}{2a}$.

2. Find the y-coordinate by evaluating $f\left(-\dfrac{b}{2a}\right)$.

EXAMPLE 6 For the function $f(x) = 2x^2 - 8x + 5$, find the coordinates of the vertex.

Solution First, find the x-coordinate of the vertex using $-\dfrac{b}{2a}$.

Note: Using $\dfrac{4ac - b^2}{4a}$ also gives the y-coordinate:

$$\frac{4(2)(5) - (-8)^2}{4(2)} = \frac{-24}{8} = -3$$

However, substituting the x-value, 2, into the original function is easier.

$$x\text{-coordinate of the vertex:} \quad -\frac{b}{2a} = -\frac{(-8)}{2(2)} = \frac{8}{4} = 2$$

Now find the y-coordinate by evaluating $f(2)$.

$$y\text{-coordinate of the vertex:} \quad f(2) = 2(2)^2 - 8(2) + 5 = 8 - 16 + 5 = -3$$

The vertex is $(2, -3)$.

ANSWER

$(-3, -46)$

YOUR TURN For the equation $f(x) = 3x^2 + 18x - 19$, find the vertex.

OBJECTIVE 6. Solve applications involving parabolas. Because the vertex of a parabola is the highest point in parabolas that open down or the lowest point in parabolas that open up, we can find a maximum or minimum value in applications involving quadratic functions.

EXAMPLE 7 A toy rocket is launched straight up with an initial velocity of 40 feet per second. The equations $h = -16t^2 + 40t$ describes the height, h, of the rocket t seconds after being launched.

a. Find the maximum height that the rocket reaches.

b. Find the amount of time that the rocket is in the air.

Solution

a. Since the graph of $h = -16t^2 + 40t$ is a parabola that opens down ($a = -16$), the maximum height occurs at its vertex.

$$t\text{-coordinate of the vertex: } -\frac{b}{2a} = -\frac{40}{2(-16)} = -\frac{40}{-32} = \frac{5}{4} = 1.25 \text{ seconds}$$

To find the h-coordinate of the vertex we replace t with 1.25 in $h = -16t^2 + 40t$.

$$h = -16(1.25)^2 + 40(1.25) = 25 \text{ feet}$$

Note: *We could have found the vertex by writing the equation in $y = a(x - h)^2 + k$ form, which would be $h = -16(t - 1.25)^2 + 25$.*

The vertex is (1.25, 25), so the maximum height is 25 feet, which occurs 1.25 seconds after the rocket is launched.

b. The time the rocket is in the air is from launch until it returns to the ground. At launch and upon returning to the ground, the rocket's height is 0, so we need to find t when $h = 0$.

$$
\begin{aligned}
0 &= -16t^2 + 40t && \text{Replace } h \text{ with 0.} \\
0 &= -8t(2t - 5) && \text{Factor out a common factor of } -8t. \\
-8t = 0 \quad &\text{or} \quad 2t - 5 = 0 && \text{Use the zero-factor theorem.} \\
t = 0 \quad & \qquad \quad 2t = 5 \\
& \qquad \quad \ \ t = 2.5
\end{aligned}
$$

This means the height is 0 when $t = 0$ and when $t = 2.5$ seconds, so the rocket is in the air for 2.5 seconds.

Note: *The graph of $h = -16t^2 + 40t$ is not the flight path of the rocket. Also, because the situation involves a rocket being launched from the ground (0 feet) and returning to the ground, only the portion of the graph in the first quadrant is realistic.*

The graph of $h = -16t^2 + 40t$ shows that at 0 seconds the rocket is at 0 feet, which is when it is launched. Its height increases to a maximum of 25 feet 1.25 seconds after launch. Then the height decreases back to 0 feet 2.5 seconds after launch.

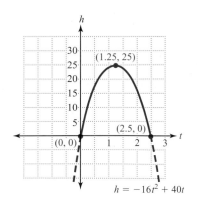

YOUR TURN A soccer player kicks the ball straight up with an initial velocity of 56 feet per second. The equation $h = -16t^2 + 56t$ describes the height, h, of the ball t seconds after being kicked.

a. After how many seconds is the ball at its maximum height?

b. What is the maximum height that the ball reaches?

c. How long is the ball in the air?

ANSWERS

a. 1.75 sec.

b. 49 ft.

c. 3.5 sec.

If an object is thrown or shot upwards and outwards, gravity causes its path, or trajectory, to be in the shape of a parabola.

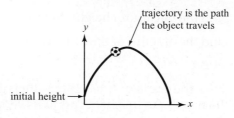

trajectory is the path the object travels

initial height

Note: *x represents the horizontal distance the object travels, and y represents the vertical distance the object travels.*

EXAMPLE 8 The equation $y = -0.8x^2 + 2.4x + 6$ when $x \geq 0$ and $y \geq 0$ describes the trajectory of a ball thrown upwards and outwards from an initial height of 6 feet.

a. What is the maximum height that the ball reaches?

b. How far does the ball travel horizontally?

c. Graph the trajectory of the ball.

Solution

a. The maximum height will be the y-coordinate of the vertex.

$$x\text{-coordinate of the vertex: } -\frac{b}{2a} = -\frac{(2.4)}{2(-0.8)} = 1.5$$

$$y\text{-coordinate of the vertex: } y = -0.8(1.5)^2 + 2.4(1.5) + 6 = 7.8$$

The vertex is $(1.5, 7.8)$, so the maximum height is 7.8 feet.

b. To find how far the ball travels, we need to find the x-value when it hits the ground, which is where $y = 0$.

$0 = -0.8x^2 + 2.4x + 6$ **Substitute 0 for y in the equation.**

$x = \dfrac{-(2.4) \pm \sqrt{(2.4)^2 - 4(-0.8)(6)}}{2(-0.8)}$ **Substitute $a = -0.8$, $b = 2.4$, and $c = 6$ into the quadratic formula.**

$= \dfrac{-2.4 \pm \sqrt{24.96}}{-1.6} \approx -1.62 \text{ or } 4.62$

The negative value does not make sense in the context of this problem, so the distance the ball travels must be approximately 4.62 feet.

Note: *Because the equation describes the trajectory when $x \geq 0$ and $y \geq 0$, we only consider the part of the graph in the first quadrant.*

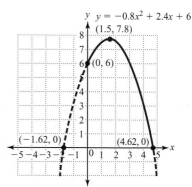

11.4 Exercises

For Extra Help | MyMathLab | Videotape/DVT | InterAct Math | Math Tutor Center | Math XL.com

1. In an equation in the form $f(x) = a(x - h)^2 + k$, what determines whether the parabola opens upwards or downwards?

2. In an equation in the form $f(x) = a(x - h)^2 + k$, what are the coordinates of the vertex?

3. Describe the axis of symmetry in a parabola that opens up or down. What is its equation?

4. Suppose the y-intercept for a given quadratic equation is $(0, -5)$. If the vertex is at $(3, -2)$, what can you conclude about the x-intercepts?

5. If the solutions for a quadratic equation are imaginary numbers, then what does that indicate about the corresponding quadratic function?

6. Given that the x-coordinate of the vertex is $-\dfrac{b}{2a}$, what must be true about a and b if the x-coordinate of the vertex is -1?

For Exercises 7–16, state whether the parabola opens upwards or downwards.

7. $f(x) = 2x^2$

8. $g(x) = -3x^2$

9. $h(x) = -x^2 + 3$

10. $k(x) = 4x^2 - 5$

11. $f(x) = 2(x - 1)^2 - 4$

12. $h(x) = -0.8(x + 3)^2 - 1$

13. $g(x) = -\dfrac{2}{3}x^2 + 4x - 1$

14. $f(x) = -x^2 - 8x - 9$

15. $k(x) = 3x - 0.5x^2$

16. $g(x) = -2x + \dfrac{3}{4}x^2$

For Exercises 17–34, find the coordinates of the vertex and write the equation of the axis of symmetry.

17. $f(x) = 5(x - 2)^2 - 3$

18. $h(x) = 2(x - 1)^2 + 4$

19. $g(x) = -2(x + 1)^2 - 5$

20. $k(x) = -(x + 4)^2 - 3$

21. $k(x) = x^2 + 2$

22. $g(x) = x^2 - 6$

23. $h(x) = 3(x - 7)^2$

24. $f(x) = 5(x + 1)^2$

25. $f(x) = -0.5x^2$

26. $k(x) = 2.5x^2$

27. $f(x) = x^2 - 4x + 8$

28. $g(x) = x^2 + 6x + 8$

29. $k(x) = 2x^2 + 16x + 27$

30. $f(x) = 3x^2 - 6x + 1$

31. $g(x) = -3x^2 + 2x + 1$

32. $h(x) = 5x^2 - 3x + 2$

33. $k(x) = -0.5x^2 - 0.4x + 1$

34. $f(x) = 0.2x^2 - 0.3x - 5$

For Exercises 35–54: a. State whether the parabola opens upwards or downwards.
b. Find the coordinates of the vertex.
c. Write the equation of the axis of symmetry.
d. Graph.

35. $h(x) = -3x^2$

36. $f(x) = 2x^2$

37. $k(x) = \dfrac{1}{4}x^2$

38. $g(x) = -\dfrac{1}{3}x^2$

39. $f(x) = 4x^2 - 3$

40. $h(x) = -x^2 + 4$

41. $g(x) = -0.5x^2 + 2$

42. $k(x) = 0.5x^2 - 3$

43. $f(x) = (x - 3)^2 + 2$

44. $h(x) = (x - 2)^2 - 1$

45. $k(x) = -2(x + 1)^2 - 3$

46. $g(x) = -(x + 3)^2 + 1$

47. $h(x) = \dfrac{1}{3}(x - 2)^2 - 1$

48. $f(x) = -\dfrac{1}{4}(x - 3)^2 - 2$

For Exercises 49–54: a. Find the x- and y-intercepts.
b. Write the equation in the form $f(x) = a(x - h)^2 + k$.
c. State whether the parabola opens upwards or downwards.
d. Find the coordinates of the vertex.
e. Write the equation of the axis of symmetry.
f. Graph.

49. $h(x) = x^2 + 6x + 9$

50. $k(x) = 3x^2 - 6x + 1$

51. $f(x) = 2x^2 + 6x + 3$

52. $k(x) = -x^2 - 2x + 3$

53. $g(x) = -3x^2 + 6x - 5$

54. $h(x) = -3x^2 - 6x + 4$

For Exercises 55–66, solve.

55. A toy rocket is launched with an initial velocity of 45 meters per second. The equation $h = -4.9t^2 + 45t$ describes the height, h, of the rocket in meters t seconds after being launched.

a. After how many seconds does the rocket reach its maximum height?

b. What is the maximum height the rocket reaches?

c. How long is the rocket in the air?

56. A ball is drop-kicked straight up with an initial velocity of 36 feet per second. The equation $h = -16t^2 + 36t$ describes the height, h, of the ball in feet t seconds after being kicked.

a. After how many seconds does the ball reach its maximum height?

b. What is the maximum height the ball reaches?

c. How long is the ball in the air?

57. The equation $y = -0.8x^2 + 2.4x + 6$ models the trajectory of a ball thrown upwards and outwards from a height of 6 feet (assume that $x \geq 0$ and $y \geq 0$).

a. What is the maximum height that the ball reaches?

b. How far does the ball travel horizontally?

c. Graph the trajectory of the ball.

58. The javelin toss is an event in track and field in which participants try to throw a javelin the farthest. Suppose the equation $y = -0.02x^2 + 1.3x + 8$ models the trajectory of one particular throw (assume that x and y represent distances in meters and that $x \geq 0$ and $y \geq 0$).

  a. What is the maximum height that the javelin reaches?

  b. How far does the javelin travel horizontally?

  c. Graph the trajectory of the javelin.

59. A record company discovers that the number of CDs sold each week after release follows a parabolic pattern. The function $n(t) = -200t^2 + 4000t$ describes the number, n, of CDs an artist sold each of t weeks after the release of the album.

  a. Which week had the greatest number of CDs sold?

  b. How many CDs sold that week?

60. The function $n(t) = -3t^2 + 42t$ describes the number, n, of tickets sold for a play each of t days after tickets went on sale.

  a. What day had the greatest number of tickets sold?

  b. How many tickets sold that day?

61. A farmer has enough materials to build a fence with a total length of 400 feet. He wants the enclosed space to be rectangular and also wants to maximize the area enclosed. Find the length and width so that the area is maximized.

62. An architect wants the length and width of a rectangular building to be a total of 150 feet long. She also wants to maximize the rectangular area that the building occupies. Find the length and width so that the area is maximized.

63. The function $C(n) = n^2 - 110n + 5000$ describes a company's cost, C, of producing n units of its product. Find the number of units the company should produce to minimize its cost.

64. The function $P(t) = 0.001t^2 - 0.24t + 59.90$ roughly models a particular stock's closing price, P, each of t days of trading during one year.

  a. After how many days is the price at its lowest?

  b. What was the lowest price during that year?

65. One integer is 12 more than another. If their product is minimized, find the integers and their product.

66. The greater of two integers minus the smaller gives a result of 20. Find the two integers so that their product is minimized. Also find their product.

PUZZLE PROBLEM

Given the vertex and the x- and y-intercepts of a parabola, reconstruct the equation.

  a. *Vertex* $(-3, -1)$
     Intercepts $(-2, 0), (-4, 0), (0, 8)$

  b. *Vertex* $(-1, 0)$
     Intercepts $(-1, 0), (0, 1)$

  c. *Vertex* $(-2, 1)$
     Intercept $(0, 5)$

  d. *Vertex* $(1, 0)$
     Intercepts $(1, 0), (0, 1)$

Collaborative Exercises (ARCH SPAN)

The Gateway Arch, located in St. Louis, Missouri, was built from 1963 to 1965 and is the nation's tallest memorial. The equation $h(x) = -0.0063492063x^2 + 630$ can be used to approximate the height of the arch, where x represents the distance from its axis of symmetry and h(x) represents its height above the ground.

1. Using the equation, find the maximum height of the structure.

2. The span of the arch at a given height is the horizontal distance between the two opposing points on the parabola at that height. Find the span of the arch at ground level. (*Hint:* Think of ground level as the x-axis. The height is 0 along the x-axis.)

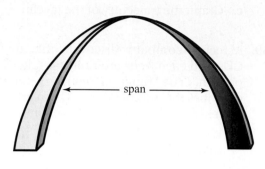

3. How does the span at ground level relate to the maximum height of the arch?

4. The arch has foundations 60 feet below ground level. What is the span of the arch between its foundations?

REVIEW EXERCISES

[1.5]   1. Evaluate: $-|-2 + 3 \cdot 4| - 4^0$

[6.6]   2. Solve $x^2 + 2x = 15$ by factoring.

[11.1]   3. Solve $x^2 = -20$ using the square root principle.

For Exercises 4–6, solve, then graph the solution set.

[2.6]   4. $4x - 8 \leq 2x + 1$

[8.3]   5. $|x - 3| \geq 8$

[9.7]   6. $\begin{cases} x + y > 2 \\ 2x - 3y \leq 6 \end{cases}$

11.5 Solving Nonlinear Inequalities

OBJECTIVES
1. Solve quadratic and other inequalities.
2. Solve rational inequalities.

OBJECTIVE 1. Solve quadratic and other inequalities. Now that we have solved quadratic equations, we can learn how to solve **quadratic inequalities**.

DEFINITION *Quadratic inequality:* An inequality that can be written in the form $ax^2 + bx + c > 0$ or $ax^2 + bx + c < 0$, where $a \neq 0$.

Note: *In this definition, the symbols $<$ and $>$ can be replaced with $\leq$ or $\geq$.*

For example, $2x^2 - x - 3 < 0$ and $2x^2 - x - 3 \geq 0$ are quadratic inequalities. Let's see what we can learn about these inequalities by looking at the graph of the corresponding quadratic function: $y = 2x^2 - x - 3$. Note that by letting $y = 0$, we have the corresponding equation $2x^2 - x - 3 = 0$ and solving this equation gives the x-intercepts $x = -1$ and $x = \frac{3}{2}$. Also notice that those x-intercepts divide the x-axis into three intervals: $(-\infty, -1)$, $\left(-1, \frac{3}{2}\right)$, and $\left(\frac{3}{2}, \infty\right)$.

Notice the parabola is above the x-axis in intervals $(-\infty, -1)$ and $\left(\frac{3}{2}, \infty\right)$, shown in blue. Evaluating $y = 2x^2 - x - 3$ using any x-value in these intervals produces a y-value that is greater than 0, so these intervals are the solution sets for $2x^2 - x - 3 > 0$.

The parabola is below the x-axis in the interval $\left(-1, \frac{3}{2}\right)$, shown in red. Evaluating $y = 2x^2 - x - 3$ using any x-value in this interval produces a y-value that is less than 0, so this interval is the solution set for $2x^2 - x - 3 < 0$.

The following number lines indicate various solution sets based on the inequality.

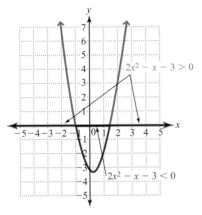

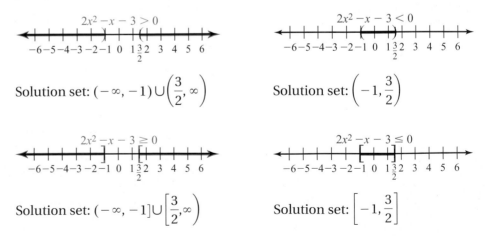

Solution set: $(-\infty, -1) \cup \left(\frac{3}{2}, \infty\right)$

Solution set: $\left(-1, \frac{3}{2}\right)$

Solution set: $(-\infty, -1] \cup \left[\frac{3}{2}, \infty\right)$

Solution set: $\left[-1, \frac{3}{2}\right]$

Based on these observations, we use the following procedure to solve quadratic inequalities.

PROCEDURE   **Solving Quadratic Inequalities**

1. Solve the related equation $ax^2 + bx + c = 0$.
2. Plot the solutions of $ax^2 + bx + c = 0$ on a number line. These solutions will divide the number line into intervals.
3. Choose a test number from each interval and substitute the number into the inequality. If the test number makes the inequality *true*, then all numbers in that interval will solve the inequality. If the test number makes the inequality *false*, then no numbers in that interval will solve the inequality.
4. State the solution set of the inequality: It is the union of all the intervals that solve the inequality. If the inequality symbols are $\leq$ or $\geq$, then the values from step 2 are included. If the symbols are $<$ or $>$, they are not solutions.

EXAMPLE 1 Solve $x^2 - x < 6$. Write the solution set using interval notation, then graph the solution set on a number line.

Solution $x^2 - x = 6$ Write the related equation.

$\qquad x^2 - x - 6 = 0$ Subtract 6 from both sides to get quadratic form.

$\qquad (x - 3)(x + 2) = 0$ Factor.

$\qquad x - 3 = 0 \quad \text{or} \quad x + 2 = 0$ Use the zero-factor theorem.

$\qquad x = 3 \qquad\qquad x = -2$ These are the x-intercepts of the graph of $y = x^2 - x - 6$.

Plot -2 and 3 on a number line (x-axis), which divides it into three intervals.

$\xleftarrow{\qquad}$ $-6\,-5\,-4\,-3\,-2\,-1\ 0\ 1\ 2\ 3\ 4\ 5\ 6$ $\xrightarrow{\qquad}$
$(-\infty, -2)\quad (-2, 3)\qquad (3, \infty)$

▷ **Note:** -2 and 3 are not part of the solution set because the inequality symbol is $<$.

Choose a test number from each interval and substitute that value into $x^2 - x < 6$.

For $(-\infty, -2)$, we choose $x = -3$.	For $(-2, 3)$, we choose $x = 0$.	For $(3, \infty)$, we choose $x = 4$.
$(-3)^2 - (-3) < 6$	$0^2 - 0 < 6$	$4^2 - 4 < 6$
$9 + 3 < 6$	$0 < 6$	$16 - 4 < 6$
$12 < 6$		$12 < 6$

This is false, so $(-\infty, -2)$ is not in the solution set. | This is true, so $(-2, 3)$ is in the solution set. | This is false, so $(3, \infty)$ is not in the solution set.

Since $(-2, 3)$ is the only interval that has solutions to the inequality, it is the solution set. Following is the graph of the solution set.

$\xleftarrow{\qquad}$ $-6\,-5\,-4\,-3\,-2\,-1\ 0\ 1\ 2\ 3\ 4\ 5\ 6$ $\xrightarrow{\qquad}$

Connection Notice the solution set for $x^2 - x - 6 < 0$ corresponds to the interval where the graph of $f(x) = x^2 - x - 6$ is below the x-axis and the x-intercepts are the endpoints of the interval.

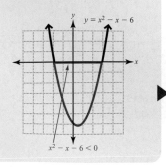

EXAMPLE 2 Solve $x^2 + 3x \geq 0$. Write the solution set using interval notation, then graph the solution set.

Solution $x^2 + 3x = 0$ **Write the related equation.**

$\qquad\qquad x(x + 3) = 0$ **Factor.**

$\qquad\quad x = 0 \quad \text{or} \quad x + 3 = 0$ **Use the zero-factor theorem.**

$\qquad\qquad\qquad\qquad\qquad x = -3$

Plot -3 and 0 on a number line and note the intervals.

Note: *Since we have already determined that -3 and 0 are in the solution set, we will not include them in any of our test intervals, which is why we use parentheses for each interval.*

$(-\infty, -3)\ (-3, 0) \qquad\qquad (0, \infty)$

Note: *-3 and 0 are included in the solution set because the inequality symbol is $\geq$.*

Choose a test number from each interval and test in $x^2 + 3x \geq 0$.

For $(-\infty, -3)$, we choose $x = -4$.

For $(-3, 0)$, we choose $x = -1$.

For $(0, \infty)$, we choose $x = 1$.

$(-4)^2 + 3(-4) \geq 0$

$\quad 16 - 12 \geq 0$

$\qquad\quad 4 \geq 0$

$(-1)^2 + 3(-1) \geq 0$

$\quad 1 - 3 \geq 0$

$\qquad -2 \geq 0$

$1^2 + 3(1) \geq 0$

$\quad 1 + 3 \geq 0$

$\qquad 4 \geq 0$

This is true, so $(-\infty, -3)$ is in the solution set.

This is false, so $(-3, 0)$ is not in the solution set.

This is true, so $(0, \infty)$ is in the solution set.

Connection Notice that the solution set for $x^2 + 3x \geq 0$ corresponds to the intervals where the graph of $f(x) = x^2 + 3x$ is above the x-axis and the x-intercepts are endpoints in the intervals.

The solution set is $(-\infty, -3] \cup [0, \infty)$, which is graphed next.

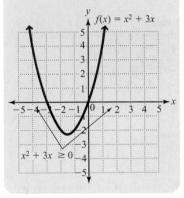

EXAMPLE 3 Solve. Write the solution set using interval notation, then graph the solution set.

a. $(x - 1)^2 > -2$

Solution Since $(x - 1)^2$ is always 0 or positive; it is always greater than -2, so every real number is a solution.

Solution set: $\mathbb{R}$ or $(-\infty, \infty)$

b. $(x - 1)^2 < -2$

Solution Since $(x - 1)^2$ is never negative, its value can never be less than -2, so there are no real solutions for $(x - 1)^2 < -2$.

Solution set: $\varnothing$

Connection Look at the graph of $f(x) = (x - 1)^2$.

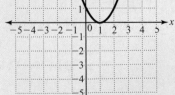

Notice that $f(x)$ is always 0 or positive, so $(x - 1)^2 > -2$ is true for all real x-values and $(x - 1)^2 < -2$ is false for all real x-values.

YOUR TURN

Solve. Write the solution set using interval notation, then graph the solution set on a number line.

a. $x^2 - 2x - 8 \le 0$ **b.** $x^2 - 4x > 0$ **c.** $(3x - 8)^2 < -5$

Other Polynomial Inequalities

We use a similar procedure for expressions with more than two factors. In the following example, we condense the procedure to a table.

EXAMPLE 4 Solve $(x + 4)(x + 1)(x - 3) \le 0$. Write the solution set in interval notation, then graph the solution set on a number line.

Solution $(x + 4)(x + 1)(x - 3) = 0$ **Write the related equation.**

$x + 4 = 0$ or $x + 1 = 0$ or $x - 3 = 0$ **Set each factor equal to 0.**

$\qquad x = -4 \qquad\qquad x = -1 \qquad\qquad x = 3$ **Solve the equations.**

Plot -4, -1, and 3 on a number line and note the intervals.

Note: -4, -1, and 3 are included in the solution set because the inequality symbol is $\le$. Remember that we test intervals around these values, so they are not included in those intervals.

Interval	$(-\infty, -4)$	$(-4, -1)$	$(-1, 3)$	$(3, \infty)$
Test Number	-5	-2	0	4
Test Results	$-32 \le 0$	$10 \le 0$	$-12 \le 0$	$40 \le 0$
True or False	True	False	True	False

Note: *Because the inequality is $\le$, -4, -1, and 3 are included in the solution set.*

Therefore, the solution set is $(-\infty, -4] \cup [-1, 3]$, which is graphed next.

YOUR TURN

Solve. Write the solution set using interval notation, then graph the solution set on a number line.

$$(x + 3)(x - 2)(x + 5) \le 0$$

OBJECTIVE 2. Solve rational inequalities. Now we consider solving **rational inequalities**.

DEFINITION   *Rational inequality:* An inequality containing a rational expression.

For example, $\dfrac{x + 2}{x - 3} > 4$ is a rational inequality. Recall that in solving a polynomial inequality, we divided the number line into intervals using x-values we found from solving its related equation. With rational inequalities, we not only find values that solve the related equation but we must also consider values that make the rational expression undefined (denominator $= 0$).

1. Find all values that make any denominator equal to 0. These values must be excluded from the solution set.
2. Solve the related equation.
3. Plot the numbers found in steps 1 and 2 on a number line.
4. Choose a test number from each interval and determine whether it solves the inequality.
5. The solution set is the union of all the regions whose test number solves the inequality. If the inequality symbol is ≤ or ≥, include the values found in step 1. The solution set never includes the values found in step 2 because they make a denominator equal to 0.

EXAMPLE 5 Solve. Write the solution set using interval notation, then graph the solution set on a number line.

a. $\dfrac{x - 4}{x + 2} \geq 0$

Solution First, we find the values that make the denominator equal to 0.

$$x + 2 = 0$$
$$x = -2$$

Now, solve the related equation.

$$\frac{x - 4}{x + 2} = 0 \qquad \textbf{Write the related equation.}$$

$$(x + 2)\frac{x - 4}{x + 2} = (x + 2)(0) \qquad \textbf{Multiply both sides by } x + 2.$$

$$x - 4 = 0 \qquad \textbf{Simplify.}$$
$$x = 4$$

Plot −2 and 4 on a number line and note the intervals.

$$
\begin{array}{c}
\text{number line with open circles at } -2 \text{ and } 4 \\
-6\;-5\;-4\;-3\;-2\;-1\;\;0\;\;1\;\;2\;\;3\;\;4\;\;5\;\;6 \\
(-\infty, -2) \qquad (-2, 4) \qquad (4, \infty)
\end{array}
$$

Note: *Remember, since* −2 *makes* $\dfrac{x - 4}{x + 2}$ *undefined, it is not included in the solution set.*

Again, we use a table.

Interval	$(-\infty, -2)$	$(-2, 4)$	$(4, \infty)$
Test Number	−3	0	5
Test Results	$7 \geq 0$	$-2 \geq 0$	$\frac{1}{7} \geq 0$
True or False	True	False	True

The solution set is $(-\infty, -2) \cup [4, \infty)$ and is graphed next.

$$
\begin{array}{c}
\text{number line} \\
-6\;-5\;-4\;-3\;-2\;-1\;\;0\;\;1\;\;2\;\;3\;\;4\;\;5\;\;6
\end{array}
$$

b. $\dfrac{x+5}{x-1} < 4$

Solution Find the values that make the denominator equal to 0.

$$x - 1 = 0$$
$$x = 1$$

Solve the related equation.

$$\frac{x+5}{x-1} = 4 \qquad \qquad \textbf{Write the related equation.}$$

$$(x-1)\frac{x+5}{x-1} = (x-1)(4) \qquad \textbf{Multiply both sides by the LCD.}$$

$$x + 5 = 4x - 4 \qquad \qquad \textbf{Simplify.}$$

$$-3x = -9 \qquad \qquad \textbf{Subtract } 4x \textbf{ and 5 from both sides.}$$

$$x = 3$$

Note: *Since 1 makes* $\dfrac{x+5}{x-1}$ *undefined, it is not included in the solution set. Also, 3 is not in the solution set because* $\dfrac{x+5}{x-1} = 4$ *when* $x = 3$.

Plot 1 and 3 on a number line.

```
 ←+—+—+—+—+—+—+—+—Ø—+—Ø—+—+—+→
  −6−5−4−3−2−1  0  1  2  3  4  5  6
     (−∞, 1)        (1, 3)   (3, ∞)
```

Interval	$(-\infty, 1)$	$(1, 3)$	$(3, \infty)$
Test Number	0	2	4
Test Results	$-5 < 4$	$7 < 4$	$3 < 4$
True or False	True	False	True

The solution set is $(-\infty, 1) \cup (3, \infty)$ and is graphed next.

```
 ←+—+—+—+—+—+—+—)—+—(—+—+—+→
  −6−5−4−3−2−1  0  1  2  3  4  5  6
```

YOUR TURN   Solve. Write the solution set using interval notation, then graph the solution set on a number line.

a. $\dfrac{x+3}{x-1} < 0$ **b.** $\dfrac{x-2}{x+4} \leq 3$

ANSWERS

a. $(-3, 1)$

```
 ←+—(—+—+—+—)—+→
  −4−3−2−1  0  1  2
```

b. $(-\infty, -7] \cup (-4, \infty)$

```
 ←+—+—]—+—+—(—+→
  −9−8−7−6−5−4−3−2
```

11.5 Exercises

For Extra Help

MyMathLab

MyMathLab

Videotape/DVT

InterAct Math

Tutor Center

Math Tutor Center

Math XL.com

1. Explain how you would find the solution set of $x^2 + 2x - 15 \geq 0$.

2. If the graph of $y = ax^2 + bx + c$ intersects the x-axis at -2 and 2, what regions would you check to solve $ax^2 + bx + c > 0$?

3. Is it possible to have a quadratic inequality whose solution set is the empty set? Explain.

4. Is it possible to have a quadratic inequality whose solution set is one number? Explain.

5. The quadratic inequality $(x - 2)(x + 5) < 0$ and the rational inequality $\dfrac{x - 2}{x + 5} < 0$ have the same solution sets. Explain how this is possible.

6. If you were to solve $\dfrac{x + 2}{(x + 5)(x - 3)} \geq 0$, what are the regions that you would need to test?

For Exercises 7–14, the graph of a quadratic function is given. Use the graph to solve each equation and inequality. For solution sets that involve intervals, use interval notation.

7. **a.** $x^2 + 6x + 5 = 0$

    b. $x^2 + 6x + 5 < 0$

    c. $x^2 + 6x + 5 > 0$

8. **a.** $x^2 - 4 = 0$

    b. $x^2 - 4 > 0$

    c. $x^2 - 4 < 0$

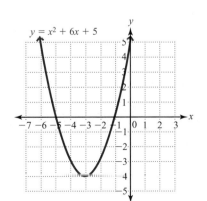

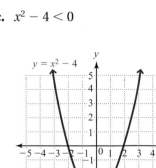

9. a. $-x^2 + 2x + 3 = 0$

  b. $-x^2 + 2x + 3 \leq 0$

  c. $-x^2 + 2x + 3 \geq 0$

10. a. $-x^2 + 5x = 0$

  b. $-x^2 + 5x \leq 0$

  c. $-x^2 + 5x \geq 0$

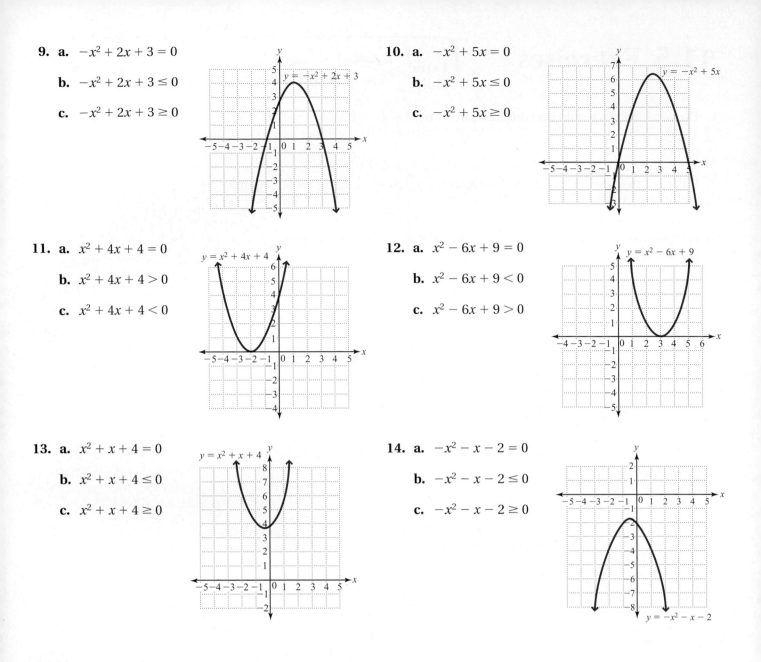

11. a. $x^2 + 4x + 4 = 0$

  b. $x^2 + 4x + 4 > 0$

  c. $x^2 + 4x + 4 < 0$

12. a. $x^2 - 6x + 9 = 0$

  b. $x^2 - 6x + 9 < 0$

  c. $x^2 - 6x + 9 > 0$

13. a. $x^2 + x + 4 = 0$

  b. $x^2 + x + 4 \leq 0$

  c. $x^2 + x + 4 \geq 0$

14. a. $-x^2 - x - 2 = 0$

  b. $-x^2 - x - 2 \leq 0$

  c. $-x^2 - x - 2 \geq 0$

For Exercises 15–44, solve. Write the solution set using interval notation, then graph the solution set on a number line.

15. $(x + 4)(x + 2) < 0$

16. $(x - 3)(x + 1) < 0$

17. $(x - 2)(x - 5) > 0$

18. $(x + 4)(x - 3) > 0$

19. $x^2 + 5x + 4 < 0$

20. $x^2 + 6x + 5 < 0$

21. $x^2 - 4x + 3 > 0$

22. $x^2 - 8x + 7 > 0$

23. $b^2 - 6b + 8 \leq 0$

24. $c^2 - 9c + 14 \leq 0$

25. $y^2 - 5 \geq 4y$

26. $z^2 - 3 \leq 2z$

27. $a^2 - 3a < 10$

28. $b^2 + 4b \geq 21$

29. $y^2 + 6y + 9 \geq 0$

30. $x^2 + 10x + 25 \geq 0$

31. $c^2 - 4c + 5 \leq 0$

32. $2a^2 - 3a + 5 \leq 0$

33. $4r^2 + 21r + 5 > 0$

34. $4s^2 + 5s - 6 < 0$

35. $x^2 - 5x > 0$

36. $x^2 + 3x > 0$

37. $x^2 \leq 6x$

38. $x^2 \geq 3x$

39. $(x - 3)^2 \geq -1$

40. $(2x - 1)^2 > -4$

41. $(x - 4)(x + 2)(x + 4) \geq 0$

42. $(x + 5)(x - 3)(x + 1) \geq 0$

43. $(x + 2)(x + 6)(x - 1) < 0$

44. $(x - 3)(x + 3)(x + 1) < 0$

For Exercises 45–50, solve using the quadratic formula.

45. $3c^2 + 4c - 1 < 0$

46. $5a^2 - 10a + 2 < 0$

47. $4r^2 + 8r - 3 \geq 0$

48. $-2y^2 + 6y + 5 < 0$

49. $-0.2a^2 - 1.6a - 2 \leq 0$

50. $-0.1x^2 - 1.2x + 4 > 0$

For Exercises 51–68, solve the rational inequalities. Write the solution set using interval notation, then graph the solution set on a number line.

51. $\dfrac{a + 4}{a - 1} > 0$

52. $\dfrac{m - 6}{m + 2} \geq 0$

53. $\dfrac{n + 1}{n + 5} \leq 0$

54. $\dfrac{b - 2}{b + 3} < 0$

55. $\dfrac{6}{x + 4} > 0$

56. $\dfrac{3}{x - 3} < 0$

57. $\dfrac{c}{c + 3} < 3$

58. $\dfrac{m}{m - 2} < 2$

59. $\dfrac{a + 5}{a - 4} > 4$

60. $\dfrac{j + 4}{j - 4} > 5$

61. $\dfrac{p + 2}{p - 3} \geq 4$

62. $\dfrac{c + 1}{c - 4} \leq 6$

63. $\dfrac{(k + 3)(k - 2)}{k - 5} \leq 0$

64. $\dfrac{(m + 1)(m - 3)}{m + 4} \geq 0$

65. $\dfrac{(2x - 1)^2}{x} \geq 0$

66. $\dfrac{(4x + 3)^2}{x} < 0$

67. $\dfrac{x^2 - 7x + 10}{x + 1} < 0$

68. $\dfrac{x^2 - 2x - 8}{x - 1} \geq 0$

For Exercises 69–72, solve.

69. If a ball is thrown upward with an initial velocity of 80 feet per second from the top of a building 96 feet high, then the height, h, above the ground after t seconds is given by $h = -16t^2 + 80t + 96$, where h is in feet.

    a. After how many seconds will the ball hit the ground? (*Hint:* Think about the value of h when the ball is on the ground.)

    b. After how many seconds is the ball 192 feet above the ground?

c. Find the interval of time when the ball is more than 192 feet above the ground.

d. Use the answers to parts a, b, and c to find the intervals of time when the ball is less than 192 feet above the ground.

70. If an object is dropped from the top of a cliff that is 144 feet high, the equation giving the height, h, above the ground is $h = 144 - 16t^2$, where h is in feet and t is in seconds.

a. After how many seconds will the object hit the ground?

b. After how many seconds is the object 80 feet above the ground?

c. Find the interval of time when the object is more than 80 feet above the ground.

d. Find the interval of time when the ball is less than 80 feet above the ground.

71. In the parallelogram shown, the height is to be 2 inches less than x. The base is to be 6 inches more than x.

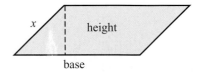

a. Find the range of values for x so that the area of the parallelogram is at least 20 square inches.

b. Find the range of values for the base and the height.

72. An auditorium is to be designed roughly in the shape of a box. The height is set to be 40 feet. It is desired that the length of the space be 20 feet more than the width.

a. Find the range of values for the width so that the volume of the space is at least 140,000 cubic feet.

b. Find the range of values for the length.

For Exercises 73 and 74, use the formula for the slope of a line: $m = \dfrac{y_2 - y_1}{x_2 - x_1}$.

73. Suppose a line is to be drawn in the coordinate plane so that it passes through the point at $(2, 5)$. Find the range of values for the second point (x_2, y_2) so that $x_2 = y_2$ and the slope of the line is at most $\dfrac{1}{2}$.

74. Suppose a line is to be drawn in the coordinate plane so that it passes through the point at $(-3, 2)$. Find the range of values for the second point (x_2, y_2) so that $x_2 = y_2$ and the slope of the line is at least $\dfrac{1}{4}$.

REVIEW EXERCISES

1. Solve: $\sqrt{x-2} = 4$

For Exercises 2 and 3, add a term to the expression to make it a perfect square then factor the perfect square.

2. $x^2 - 6x$

3. $x^2 + 5x$

4. Find the x-intercepts for $y = x^2 + 4x - 12$.

5. What are the coordinates of the vertex of the graph of $f(x) = x^2 - 6x + 5$?

6. Does the graph of $f(x) = -x^2 + 2x - 3$ open upwards or downwards? Why?

Chapter 11 Summary

Defined Terms

Section 11.2
Discriminant (p. 888)

Section 11.3
Equations quadratic in form (p. 897)

Section 11.5
Quadratic Inequality (p. 923)
Rational Inequality (p. 926)

Procedures, Rules, and Key Examples

Procedures/Rules	Key Examples

Section 11.1 Completing the Square

The square root principle:

If $x^2 = a$, where a is a real number, then $x = \sqrt{a}$ or $x = -\sqrt{a}$.

It is common to indicate the positive and negative solutions by writing $\pm \sqrt{a}$.

To solve a quadratic equation by completing the square,
1. Write the equation in the form $x^2 + bx = c$.
2. Complete the square by adding $\left(\dfrac{b}{2}\right)^2$ to both sides.
3. Write the completed square in factored form.
4. Use the square root principle to eliminate the square.
5. Isolate the variable.
6. Simplify as needed.

Example 1: Solve.

a. $x^2 = 36$

Solution: $x = \pm\sqrt{36}$
$x = \pm 6$

b. $(x - 3)^2 = 12$

Solution: $x - 3 = \pm\sqrt{12}$
$x = 3 \pm 2\sqrt{3}$

Example 2: Solve by completing the square.

a. $x^2 + 6x - 7 = 8$

$\quad x^2 + 6x = 15$ Add 7 to both sides.

$x^2 + 6x + 9 = 15 + 9$ Complete the square.

$\quad\quad (x + 3)^2 = 24$ Factor.

$\quad\quad\quad x + 3 = \pm\sqrt{24}$ Use the square root principle.

$\quad\quad\quad\quad x = -3 \pm 2\sqrt{6}$ Subtract 3 from both sides and simplify the square root.

b. $3x^2 - 9x - 2 = 10$

$\quad 3x^2 - 9x = 12$ Add 2 to both sides.

$\quad\quad x^2 - 3x = 4$ Divide both sides by 3.

$x^2 - 3x + \dfrac{9}{4} = 4 + \dfrac{9}{4}$ Complete the square.

$\left(x - \dfrac{3}{2}\right)^2 = \dfrac{25}{4}$ Factor.

$x - \dfrac{3}{2} = \pm\sqrt{\dfrac{25}{4}}$ Use the square root principle.

$x = \dfrac{3}{2} \pm \dfrac{5}{2}$ Add $\dfrac{3}{2}$ to both sides and simplify the square root.

$x = \dfrac{3}{2} + \dfrac{5}{2} = 4$ or $x = \dfrac{3}{2} - \dfrac{5}{2} = -1$

continued

Procedures/Rules	Key Examples

Section 11.2 Solving Quadratic Equations Using the Quadratic Formula

To solve a quadratic equation in the form $ax^2 + bx + c = 0$, where $a \neq 0$, use the quadratic formula:

$$x = \frac{-b \pm \sqrt{b^2 - 4ac}}{2a}$$

Given a quadratic equation in the form $ax^2 + bx + c = 0$, where $a \neq 0$, to determine the number and type of solutions it has, evaluate the discriminant $b^2 - 4ac$.

If the **discriminant is positive**, then the equation has two real-number solutions. They will be rational if the discriminant is a perfect square and irrational otherwise.

If the **discriminant is 0**, then the equation has one real solution. The solution is rational if $-\dfrac{b}{2a}$ is a rational number.

If the **discriminant is negative**, then the equation has two nonreal complex solutions.

Example 1: Solve: $3x^2 - 4x + 2 = 0$

Solution: In the quadratic formula, replace a with 3, b with -4, and c with 2.

$$x = \frac{-(-4) \pm \sqrt{(-4)^2 - 4(3)(2)}}{2(3)}$$

$$= \frac{4 \pm \sqrt{16 - 24}}{6} = \frac{4 \pm \sqrt{-8}}{6}$$

$$= \frac{4 \pm 2\sqrt{-2}}{6} = \frac{2}{3} \pm \frac{\sqrt{2}}{3}i$$

Example 2: Use the discriminant to determine the number and type of solutions for $5x^2 - 7x + 8 = 0$.

Solution: In the discriminant, replace a with 5, b with -7, and c with 8.

$$(-7)^2 - 4(5)(8) = 49 - 160$$
$$= -111$$

Since the discriminant is negative, the equation has two nonreal complex solutions.

Section 11.3 Solving Equations That Are Quadratic in Form

If the equation involves rational expressions, multiply both sides by the LCD. Be sure to check for extraneous solutions.

Example 1: Solve $\dfrac{6}{x - 2} + \dfrac{6}{x - 1} = 5$.

Solution: First, multiply both sides by the LCD, $(x - 2)(x - 1)$.

$6(x - 1) + 6(x - 2) = 5(x - 2)(x - 1)$

$12x - 18 = 5x^2 - 15x + 10$ Simplify.

$0 = 5x^2 - 27x + 28$ Write in $ax^2 + bx + c = 0$ form.

$0 = (5x - 7)(x - 4)$ Factor.

$5x - 7 = 0$ or $x - 4 = 0$ Set each factor equal to 0, then solve each equation.

$x = \dfrac{7}{5}$ $x = 4$

Check: Verify that $\dfrac{7}{5}$ and 4 solve the original equation.

continued

Procedures/Rules	Key Examples

Section 11.3 (continued)

If an equation has a radical, isolate the radical and then square both sides. Continue the process as needed until all radicals have been eliminated. Solve the resulting equation and check for extraneous solutions.

Example 2: Solve $\sqrt{4x + 4} = 2x - 2$.

$$4x + 4 = 4x^2 - 8x + 4 \quad \text{Square both sides.}$$
$$0 = 4x^2 - 12x \quad \text{Write in } ax^2 + bx + c = 0 \text{ form.}$$
$$0 = 4x(x - 3) \quad \text{Factor.}$$
$$4x = 0 \quad \text{or} \quad x - 3 = 0 \quad \text{Set each factor equal to 0, then}$$
$$x = 0 \quad\quad\quad x = 3 \quad \text{solve each equation.}$$

Check:

$$x = 0 \quad\quad\quad\quad\quad x = 3$$
$$\sqrt{4(0) + 4} = 2(0) - 2 \quad \sqrt{4(3) + 4} = 2(3) - 2$$
$$\sqrt{0 + 4} = 0 - 2 \quad\quad \sqrt{12 + 4} = 6 - 2$$
$$2 = -2 \quad \text{False} \quad\quad 4 = 4 \quad \text{True}$$

$x = 3$ is the only solution.

To solve equations that are quadratic in form using substitution,
1. Rewrite the equation so that it is in the form $au^2 + bu + c = 0$.
2. Solve the quadratic equation for u.
3. Substitute for u and solve.
4. Check the solutions.

Example 3: Solve
$$(a + 2)^2 + 7(a + 2) + 12 = 0.$$

$$u^2 + 7u + 12 = 0 \quad \text{Substitute } u \text{ for } a + 2.$$
$$(u + 3)(u + 4) = 0 \quad \text{Factor.}$$
$$u + 3 = 0 \quad \text{or} \quad u + 4 = 0 \quad \text{Set each factor equal to 0, then solve each equation.}$$
$$u = -3 \quad\quad\quad u = -4$$

$$a + 2 = -3 \quad \text{or} \quad a + 2 = -4 \quad \text{Substitute } a + 2 \text{ for } u,$$
$$a = -5 \quad\quad\quad a = -6 \quad \text{solve for } a.$$

Check: Verify that -5 and -6 satisfy the original equation.

Section 11.4 Graphing Quadratic Functions

The graph of a function in the form $f(x) = a(x - h)^2 + k$ is a parabola with vertex at (h, k). The equation of the axis of symmetry is $x = h$. The parabola opens up if $a > 0$ and down if $a < 0$.

Example 1: For $f(x) = -3(x + 1)^2 - 2$,
a. Determine whether the graph opens upwards or downwards.
b. Find the vertex.
c. Write the equation of the axis of symmetry.
d. Graph.

Answers:
a. downwards   b. $(-1, -2)$ c. $x = -1$
d.

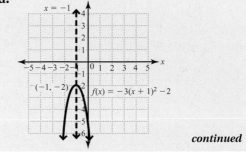

continued

Procedures/Rules	Key Examples

Section 11.4 (continued)

Given an equation in the form $f(x) = ax^2 + bx + c$, to determine the vertex of the corresponding parabola,

1. Find the x-coordinate using the formula $x = -\dfrac{b}{2a}$.

2. Find the y-coordinate by evaluating $f\left(-\dfrac{b}{2a}\right)$.

Note: *As an alternative approach to Example 2, we could have transformed* $f(x) = 2x^2 - 12x + 19$ *to* $f(x) = 2(x - 3)^2 + 1$ *by completing the square.*

Example 2: For $f(x) = 2x^2 - 12x + 19$,

a. Determine whether the graph opens upwards or downwards.

b. Find the vertex.

c. Write the equation of the axis of symmetry.

d. Graph.

Answers:

a. upwards

b. For the x-coordinate, use $-\dfrac{b}{2a}$, replacing a with 2 and b with -12.

$$x = -\frac{(-12)}{2(2)} = 3$$

For the y-coordinate, evaluate $f(3)$.
$$f(3) = 2(3)^2 - 12(3) + 19$$
$$= 18 - 36 + 19 = 1$$

Vertex: $(3, 1)$

c. $x = 3$

d.

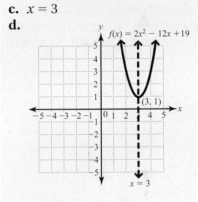

Section 11.5 Solving Nonlinear Inequalities

Solving quadratic inequalities.

1. Solve the related equation $ax^2 + bx + c = 0$.

2. Plot the solutions of $ax^2 + bx + c = 0$ on a number line. These solutions will divide the number line into intervals.

3. Choose a test number from each interval and substitute the number into the inequality. If the test number makes the inequality true, then all numbers in that interval will solve the inequality. If the test number makes the inequality false, then no numbers in that interval will solve the inequality.

4. State the solution set of the inequality: It is the union of all the intervals that solve the inequality. If the inequality symbols are $\leq$ or $\geq$, then the values from step 2 are included. If the symbols are $<$ or $>$, they are not solutions.

Example 1: Solve $x^2 + 2x - 15 \geq 0$.

$x^2 + 2x - 15 = 0$ Write the related equation.

$(x + 5)(x - 3) = 0$ Factor.

$x + 5 = 0$ or $x - 3 = 0$ Set each

$x = -5$ $x = 3$ factor equal to 0, then solve each equation.

Plot -5 and 3 on a number line and label the intervals.

$-8\ -7\ -6\ -5\ -4\ -3\ -2\ -1\ \ 0\ \ 1\ \ 2\ \ 3\ \ 4\ \ 5\ \ 6$

$(-\infty, -5)$ $(-5, 3)$ $(3, \infty)$

continued

Procedures/Rules	Key Examples

Section 11.5 (continued)

Choose a number from each interval and test it in the original inequality.

Interval	$(-\infty, -5)$	$(-5, 3)$	$(3, \infty)$
Test value	-6	0	4
Result	$9 \geq 0$	$-15 \geq 0$	$9 \geq 0$
True/False	True	False	True

The solution set is $(-\infty, -5] \cup [3, \infty)$.

Solving rational inequalities.
1. Find all values that make any denominator equal to 0. These values must be excluded from the solution set.
2. Solve the related equation.
3. Plot the numbers found in steps 1 and 2 on a number line and label the regions.
4. Choose a test number from each interval and determine whether it solves the inequality.
5. The solution set is the union of all the regions whose test number solves the inequality. If the inequality symbol is $\leq$ or $\geq$, include the values found in step 1. The solution set never includes the values found in step 2 because they make a denominator equal to 0.

Example 2: Solve $\dfrac{x + 5}{x - 1} > 4$.

$\dfrac{x + 5}{x - 1} = 4$ Write the related equation.

$(x - 1)\dfrac{x + 5}{x - 1} = (x - 1)(4)$

      Multiply both sides by the LCD.

$x + 5 = 4x - 4$ Simplify.

$9 = 3x$ Subtract x and add 5 on both sides.

$3 = x$ Divide by 3.

Find the value(s) that make any denominator equal to 0.

$x - 1 = 0$

$x = 1$

Plot 1 and 3 on the number line and label the regions.

Choose a number from each interval and test it in the original inequality.

Interval	$(\infty, 1)$	$(1, 3)$	$(3, \infty)$
Test value	0	2	4
Result	$-5 > 4$	$7 > 4$	$3 > 4$
True/False	False	True	False

The solution set is $(1, 3)$.

Formulas

The quadratic formula: Given an equation in the form $ax^2 + bx + c = 0$, $x = \dfrac{-b \pm \sqrt{b^2 - 4ac}}{2a}$

Given an equation in the form $f(x) = ax^2 + bx + c$, the x-coordinate of the vertex of a parabola is $-\dfrac{b}{2a}$

and the equation of the axis of symmetry is $x = -\dfrac{b}{2a}$.

Chapter 11 Review Exercises

For Exercises 1–5, answer true or false.

[11.1] **1.** The notation $\sqrt{a}$ has the same meaning as the notation $\pm\sqrt{a}$.

[11.3] **2.** Equations that are quadratic in form never have extraneous solutions.

$\begin{bmatrix}11.1\\11.2\end{bmatrix}$ **3.** Every quadratic equation has a real-number solution.

[11.2] **4.** If the solutions for a quadratic equation in the form $ax^2 + bx + c = 0$ are complex, the discriminant can be used to find those solutions.

[11.4] **5.** Given an equation in the form $y = ax^2 + bx + c$, if $a > 0$, then the parabola opens up.

For Exercises 6–10, complete the rule.

[11.1] **6.** If $x^2 = a$ and $a \geq 0$, then $x = $ _____ or _____ .

[11.1] **7.** Given an equation in the form $x^2 + bx = c$, to complete the square we add the constant term _____ to both sides of the equation.

[11.2] **8.** Given an equation in the form $ax^2 + bx + c = 0$, the quadratic formula is $x = $ _____ .

[11.2] **9.** Given an equation in the form $ax^2 + bx + c = 0$, the discriminant is _____ .

[11.4] **10.** Given a function in the form $f(x) = a(x - h)^2 + k$, the coordinates of the vertex are _____ and the axis of symmetry is _____ .

Equations and Inequalities

Exercises 11–66

[11.1] *For Exercises 11–18, solve and check.*

11. $x^2 = 16$

12. $y^2 = \dfrac{1}{36}$

13. $k^2 + 2 = 30$

14. $3x^2 = 42$

15. $5h^2 + 24 = 9$

16. $(x + 7)^2 = 25$

17. $(x - 9)^2 = -16$

18. $\left(m + \dfrac{3}{5}\right)^2 = \dfrac{16}{25}$

[11.1] *For Exercises 19–22, solve by completing the square.*

19. $m^2 + 8m = -7$ **20.** $u^2 - 6u - 12 = 100$ **21.** $2b^2 - 6b + 7 = 0$ **22.** $u^2 + \dfrac{1}{4}u = \dfrac{3}{4}$

[11.2] *For Exercises 23–26, solve using the quadratic formula.*

23. $p^2 - 5 = -2p$ **24.** $3x^2 - 2x + 1 = 0$

25. $2t^2 + t - 5 = 0$ **26.** $2x^2 + 0.1x - 0.03 = 0$

[11.2] *For Exercises 27–30, use the discriminant to determine the number and type of solutions for the equation.*

27. $b^2 - 4b - 12 = 0$ **28.** $6z^2 - 7z + 5 = 0$

29. $k^2 + 6k + 9 = 0$ **30.** $0.8x^2 + 1.2x + 0.3 = 0$

[11.3] *For Exercises 31–34, solve.*

31. $\dfrac{1}{y} + \dfrac{1}{y + 3} = \dfrac{2}{3}$ **32.** $\dfrac{1}{u} + \dfrac{1}{u - 5} = \dfrac{10}{u^2 - 25}$

33. $6 - 5x^{-1} + x^{-2} = 0$ **34.** $2 - 3x^{-1} - x^{-2} = 0$

[11.3] *For Exercises 35–38, solve.*

35. $14\sqrt{x} + 45 = 0$ **36.** $\sqrt{4m} = 3m - 1$

37. $\sqrt{6r + 13} = 2r + 1$ **38.** $\sqrt{21t + 2} + t = 2 + 4t$

[11.3] *For Exercises 39–44, solve the equations using substitution.*

39. $x^4 - 5x^2 + 6 = 0$ **40.** $2m^4 - 3m^2 + 1 = 0$

41. $6(x + 5)^2 - 5(x + 5) + 1 = 0$ **42.** $\left(\dfrac{x - 1}{3}\right)^2 + 10\left(\dfrac{x - 1}{3}\right) + 9 = 0$

43. $p^{2/3} - 11p^{1/3} + 24 = 0$ **44.** $5a^{1/2} + 13a^{1/4} - 6 = 0$

For Exercises 45–50:   *a. Find the x- and y-intercepts.*
                                           b. State whether the parabola opens upwards or downwards.
                                           c. Find the coordinates of the vertex.
                                           d. Write the equation of the axis of symmetry.
                                           e. Graph.

45. $f(x) = -2x^2$ **46.** $g(x) = \dfrac{1}{2}x^2 + 1$

47. $h(x) = -\dfrac{1}{3}(x - 2)^2$ **48.** $k(x) = 4(x + 3)^2 - 2$

For exercises, 49 and 50:   *a. Write the equation in the form f(x) = a(x − h)² + k.*
                                           b. Find the x- and y-intercepts.
                                           c. State whether the parabola opens upwards or downwards.
                                           d. Find the coordinates of the vertex.
                                           e. Write the equation of the axis of symmetry.
                                           f. Graph.

49. $m(x) = x^2 + 2x - 1$ **50.** $p(x) = -0.5x^2 + 4x - 6$

For Exercises 51–54, solve the following inequalities.

51. $(x + 5)(x - 3) > 0$ **52.** $n^2 - 6n \le -8$

53. $x^2 + 9x + 14 < 0$ **54.** $(x + 3)(x - 1)(x - 2) \ge 0$

For Exercises 55–58, solve the rational inequalities.

55. $\dfrac{a + 3}{a - 1} \ge 0$ **56.** $\dfrac{r}{r + 2} < 2$

57. $\dfrac{n - 3}{n - 4} \le 5$ **58.** $\dfrac{(k + 2)(k - 3)}{k - 5} < 0$

59. A crop circle appeared July 7, 2003 in Windham Hill, England, and covered $22,500\pi$ square feet. Find the radius of the circle.

60. Using the formula $E = \dfrac{1}{2}mv^2$, where E represents the kinetic energy in joules of an object with a mass of m kilograms and a velocity of v meters per second, find the velocity of an object with a mass of 50 kilograms and 400 joules of kinetic energy.

61. The length of a small rectangular shed is 4 feet more than its width. If the area is 285 square feet, what are the dimensions of the shed?

62. A right circular cylinder is to be constructed so that its volume is equal to that of a sphere with a radius of 9 inches. If the cylinder is to have a height of 4 inches, find the radius of the cylinder.

63. A ramp is constructed so that it is a right triangle with a base that is 9 feet longer than its height. If the hypotenuse is 17 feet, find the dimensions of the base and height.

64. An acrobat is launched upwards from one end of a lever with an initial velocity of 24 feet per second. The function $h = -16t^2 + 24t$ describes the height, h, of the acrobat t seconds after being launched.

   a. After how many seconds does the acrobat reach maximum height?

   b. What is the maximum height that the acrobat reaches?

   c. How long is the acrobat in the air?

   d. Graph the function.

65. The longest punt on record in the NFL was by Steve O'Neal in a game between the New York Jets and Denver Broncos on September 21, 1969. The function $y = -0.03x^2 + 2.16x$ models the trajectory of the punt. (Note that x and y are distances in yards.)

   a. Find the maximum height the punt reached.

   b. Find the distance of the punt.

Of Interest

The yardage you found is how far the punt carried in the air. It was actually recorded as a 98-yard punt, which includes the amount of roll after the punt landed and was downed.

66. In a triangle, the height is to be 2 inches less than x. The base is to be 4 inches more than x.

   a. Find the range of values for x so that the area of the triangle is at least 56 square inches.

   b. Find the range of values for the base and the height.

Chapter 11 Practice Test

For 1 and 2, use the square root principle to solve and check.

1. $x^2 = 81$

2. $(x - 3)^2 = 20$

For 3 and 4, solve by completing the square.

3. $x^2 - 8x = -4$

4. $3m^2 - 6m = 5$

For 5 and 6, solve using the quadratic formula.

5. $2x^2 + x - 6 = 0$

6. $x^2 - 8x + 15 = 0$

For 7–12, solve using any method.

7. $u^2 - 16 = -6u$

8. $x^2 = 81$

9. $4w^2 + 6w + 3 = 0$

10. $2x^2 + 4x = 0$

11. $x^2 + 16 = 0$

12. $3k^2 = -5k$

For 13–16, solve.

13. $\dfrac{1}{x + 2} + \dfrac{1}{x} = \dfrac{5}{12}$

14. $3 - x^{-1} - 2x^{-2} = 0$

15. $9\sqrt{x} + 8 = 0$

16. $\sqrt{x + 8} - x = 2$

For 17 and 18, solve the equations using substitution.

17. $9a^4 + 26a^2 - 3 = 0$

18. $(x + 1)^2 + 3(x + 1) - 4 = 0$

19. For $f(x) = -x^2 + 6x - 4$,

    a. Find the x- and y-intercepts.

    b. Write the equation in the form $f(x) = a(x - h)^2 + k$.

    c. State whether the parabola opens upwards or downwards.

    d. Find the coordinates of the vertex.

    e. Write the equation of the axis of symmetry.

    f. Graph.

For 20 and 21: a. Solve the inequality.
                 b. Graph the solution set on a number line.

20. $(x + 1)(x - 4) \leq 0$

21. $\dfrac{x + 2}{x - 1} > 0$

For 22–25, solve.

22. A ball is thrown downward from a window in a tall building. The distance, d, traveled by the ball is given by the equation $d = 16t^2 + 32t$, where t is the time traveled in seconds. How long will it take the ball to fall 180 feet?

23. A rectangular parking lot needs to have an area of 400 square feet. The length is to be 20 feet more than the width. Find the dimensions of the parking lot.

24. An archer shoots an arrow in a field. Suppose the equation $y = -0.02x^2 + 1.3x + 8$ models the trajectory of the arrow (assume that x and y represent distances in meters and that $x \geq 0$ and $y \geq 0$).

    a. What is the maximum height that the arrow reaches?

    b. How far does the arrow travel horizontally?

    c. Graph the trajectory of the arrow.

25. A zoo is planning to install a new aquarium tank. The tank is to be in the shape of a box with a height of 12 feet. It is desired that the length be 15 feet more than the width.

    a. Find the range of values for the width so that the volume of the space is at most 12,000 cubic feet.

    b. Find the range of values for the length.

12

Exponential and Logarithmic Functions

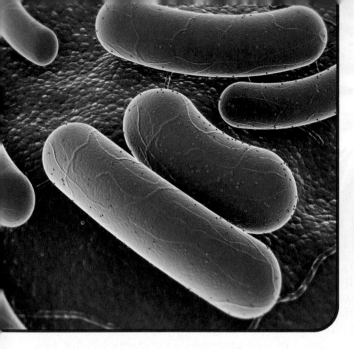

"The mathematics of uncontrolled growth are frightening. A single cell of the bacterium *E. coli* would, under ideal circumstances, divide every twenty minutes. . . . one cell becomes two, two become four, four become eight and so on. In this way it can be shown that in a single day, one cell of *E. coli* could produce a super-colony equal in size and weight to the entire planet Earth."

—Michael Crichton, *The Andromeda Strain* (1971)

We begin this chapter with composite and inverse functions. Exponential and logarithmic functions, the subject of the rest of the chapter, are inverses of each other. These functions are used in modeling population growth, electrical circuits, sound decibels, pH, and the Richter scale.

12.1 Composite and Inverse Functions

OBJECTIVES

1. Find the composition of two functions.
2. Show that two functions are inverses.
3. Show that a function is one-to-one.
4. Find the inverse of a function.
5. Graph a given function's inverse function.

OBJECTIVE 1. Find the composition of two functions. In Section 8.5, we learned to perform the basic operations with functions. We now explore a new operation. Recall that a function pairs elements from an "input" set called the *domain* with elements in an "output" set called the *range*. If the output elements of one function are used as input elements in a second function, the resulting function is the **composition** of the two functions. Composition occurs when one quantity depends on a second quantity, which, in turn, depends on a third quantity.

We can visualize the composition of two functions as a machine with two stages, which are the two functions. For example, some coffee machines have two stages. We put in roasted coffee beans, which are ground in the first stage (first function). The grounds are then fed into the second stage (second function), which runs hot water over the grounds to produce coffee.

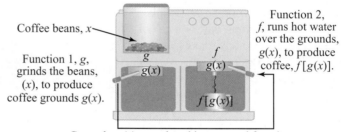

Coffee beans, x

Function 1, g, grinds the beans, (x), to produce coffee grounds $g(x)$.

Function 2, f, runs hot water over the grounds, $g(x)$, to produce coffee, $f[g(x)]$.

Grounds, $g(x)$, are placed into second function.

DEFINITION

Note: *The notation $(g \circ f)(x)$ is read "g composed with f of x," or simply "g of f of x."*

Warning: Do not confuse $(f \circ g)(x)$ with $(f \cdot g)(x)$.

Composition of functions: If f and g are functions, then the composition of f and g is defined as $(f \circ g)(x) = f[g(x)]$ for all x in the domain g for which $g(x)$ is in the domain of f. The composition of g and f is defined as $(g \circ f)(x) = g[f(x)]$ for all x in the domain of f for which $f(x)$ is in the domain of g.

For example, suppose a pebble is dropped into a calm lake causing ripples to form in concentric circles. If the radius of the outer ripple increases at a rate of 2 feet per second, then we can describe the length of the radius as a function of t by using $r(t) = 2t$, where t is in seconds. The area of the circle formed by the outer ripple is also a function described by $A(r) = \pi r^2$. If we substitute the output of $r(t)$, which is $2t$ for r, into $A(r)$, then we form the composite function $(A \circ r)(t)$.

$$
\begin{aligned}
(A \circ r)(t) &= A[r(t)] \\
&= \pi (2t)^2 && \text{Substitute } 2t \text{ for } r \text{ in } A(r) = \pi r^2. \\
&= 4\pi t^2 && \text{Simplify.}
\end{aligned}
$$

The composite function $(A \circ r)(t)$ gives the area of the circle as a function of time.

EXAMPLE 1 If $f(x) = 2x - 3$ and $g(x) = x^2 + 1$, find the following.

a. $(f \circ g)(2)$

Solution $(f \circ g)(2) = f[g(2)]$ **Definition of composition.**

$\qquad = f(5)$ $g(2) = 2^2 + 1 = 4 + 1 = 5.$

$\qquad = 2(5) - 3$ **In $f(x)$, replace x with 5.**

$\qquad = 7$ **Simplify.**

Note: We first find $g(2)$, which is 5, then substitute that result into f.

b. $(g \circ f)(2)$

Solution $(g \circ f)(2) = g[f(2)]$ **Definition of composition.**

$\qquad = g(1)$ $f(2) = 2(2) - 3 = 4 - 3 = 1.$

$\qquad = 1^2 + 1$ **In $g(x)$, replace x with 1.**

$\qquad = 2$ **Simplify.**

c. $(f \circ g)(x)$

Solution $(f \circ g)(x) = f[g(x)]$ **Definition of composition.**

$\qquad = f(x^2 + 1)$ **Replace $g(x)$ with $x^2 + 1$.**

$\qquad = 2(x^2 + 1) - 3$ **In $f(x)$, replace x with $x^2 + 1$.**

$\qquad = 2x^2 + 2 - 3$ **Simplify.**

$\qquad = 2x^2 - 1$

Note: We could have found $(f \circ g)(2)$ from part a by first finding $(f \circ g)(x) = 2x^2 - 1$, then substituting 2 for x:
$(f \circ g)(2) = 2(2)^2 - 1$
$\qquad = 8 - 1$
$\qquad = 7$

d. $(g \circ f)(x)$

Solution $(g \circ f)(x) = g[f(x)]$ **Definition of composition.**

$\qquad = g(2x - 3)$ **Replace $f(x)$ with $2x - 3$.**

$\qquad = (2x - 3)^2 + 1$ **In $g(x)$, replace x with $2x - 3$.**

$\qquad = 4x^2 - 12x + 9 + 1$ **Simplify.**

$\qquad = 4x^2 - 12x + 10$

Note: We could have found $(g \circ f)(2)$ from part b by first finding $(g \circ f)(x) = 4x^2 - 12x + 10$ then substituting 2 for x:
$(g \circ f)(2) = 4(2)^2 - 12(2) + 10$
$\qquad = 16 - 24 + 10$
$\qquad = 2$

Note: Generally, $(f \circ g)(x) \neq (g \circ f)(x)$ as in Examples 1(c) and (d), but there are special functions for which $(f \circ g)(x) = (g \circ f)(x)$.

YOUR TURN If $f(x) = x^2 + 2$ and $g(x) = 3x + 5$, find the following:

a. $f[g(-3)]$ **b.** $g[f(2)]$ **c.** $f[g(x)]$

OBJECTIVE 2. Show that two functions are inverses. Operations, such as addition and subtraction, that undo each other are called *inverse operations*. For example, if we begin with a number x and add a second number y, we have $x + y$. Now subtract y (the number we just added) from that result and we have $x + y - y = x$, which is the beginning number.

Likewise, two functions that undo each other under composition are called **inverse functions**.

ANSWERS
a. $f[g(-3)] = 18$
b. $g[f(2)] = 23$
c. $f[g(x)] = 9x^2 + 30x + 27$

DEFINITION *Inverse functions:* Two functions f and g are inverses if and only if $(f \circ g)(x) = x$ for all x in the domain of g and $(g \circ f)(x) = x$ for all x in the domain of f.

Loosely speaking, f and g are inverse functions if you evaluate f for a value x in its domain, substitute that result into g and evaluate, and you get x again and vice versa.

PROCEDURE Inverse Functions

To determine whether two functions f and g are inverses of each other,

1. Show that $f[g(x)] = x$ for all x in the domain of g.
2. Show that $g[f(x)] = x$ for all x in the domain of f.

EXAMPLE 2 Verify that f and g are inverses.

a. $f(x) = 3x + 2, g(x) = \dfrac{x - 2}{3}$

Solution We need to show that $f[g(x)] = x$ and $g[f(x)] = x$.

$$f[g(x)] = f\left(\dfrac{x - 2}{3}\right) \qquad \text{Substitute } \dfrac{x - 2}{3} \text{ for } g(x).$$

$$= 3\left(\dfrac{x - 2}{3}\right) + 2 \qquad \text{Substitute } \dfrac{x - 2}{3} \text{ for } x \text{ in } f(x).$$

$$= x - 2 + 2 \qquad \text{Simplify.}$$

$$= x \qquad \text{So } f[g(x)] = x.$$

$$g[f(x)] = g(3x + 2) \qquad \text{Substitute } 3x + 2 \text{ for } f(x).$$

$$= \dfrac{3x + 2 - 2}{3} \qquad \text{Replace } x \text{ with } 3x + 2 \text{ in } g(x).$$

$$= \dfrac{3x}{3} \qquad \text{Simplify.}$$

$$= x \qquad \text{So } g[f(x)] = x.$$

Since $f[g(x)] = x$ and $g[f(x)] = x$, f and g are inverses.

b. $f(x) = x^3 + 5$ and $g(x) = \sqrt[3]{x - 5}$

Solution We need to show that $f[g(x)] = x$ and $g[f(x)] = x$.

$$f[g(x)] = f(\sqrt[3]{x - 5}) \qquad \text{Replace } g(x) \text{ with } \sqrt[3]{x - 5}.$$

$$= (\sqrt[3]{x - 5})^3 + 5 \qquad \text{Replace } x \text{ with } \sqrt[3]{x - 5} \text{ in } f(x).$$

$$= x - 5 + 5 \qquad \text{Simplify.}$$

$$= x \qquad \text{So } f[g(x)] = x.$$

$$g[f(x)] = g(x^3 + 5) \qquad \text{Replace } f(x) \text{ with } x^3 + 5.$$

$$= \sqrt[3]{x^3 + 5 - 5} \qquad \text{Replace } x \text{ with } x^3 + 5 \text{ in } g(x).$$

$$= \sqrt[3]{x^3} \qquad \text{Simplify.}$$

$$= x \qquad \text{So } g[f(x)] = x.$$

Since $f[g(x)] = x$ and $g[f(x)] = x$, f and g are inverses.

Verify that $f(x) = 5x + 6$ and $g(x) = \dfrac{x - 6}{5}$ are inverse functions.

Calculator TIPS

We can verify that $f[g(x)] = g[f(x)] = x$ without actually finding the composition functions by using a graphing utility. We illustrate using $f(x) = \dfrac{x + 3}{x}$ and $g(x) = \dfrac{3}{x - 1}$, which are inverse functions for $x \neq 0$ and $x \neq 1$. Enter $f(x) = Y_1 = \dfrac{x + 3}{x}$, $g(x) = Y_2 = \dfrac{3}{x - 1}$. Since $f[g(x)] = Y_1(Y_2)$, enter $Y_3 = Y_1(Y_2)$. To enter Y_1 and Y_2 in the $\boxed{Y=}$ menu on the TI-83 plus, put the cursor at Y_3, press $\boxed{VARS}$, select Y-VARS, then select Function and press $\boxed{ENTER}$. Select your choice and press $\boxed{ENTER}$. Graph Y_3 without graphing Y_1 or Y_2. To graph Y_3 only, press $\boxed{Y=}$, put the cursor on the = sign of Y_1, and press $\boxed{ENTER}$. Repeat for Y_2. Press $\boxed{GRAPH}$. Trace to verify that the graph is $y = x$.

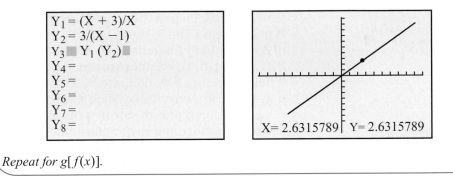

$Y_1 = (X + 3)/X$
$Y_2 = 3/(X - 1)$
$Y_3 \blacksquare Y_1 (Y_2) \blacksquare$
$Y_4 =$
$Y_5 =$
$Y_6 =$
$Y_7 =$
$Y_8 =$

X= 2.6315789   Y= 2.6315789

Repeat for $g[f(x)]$.

Before finding inverse functions, we need to determine what types of functions have inverses. Consider the following:

$$f = \{(1, 2), (3, 4), (5, 6), (x, y)\}; \text{ Domain} = \{1, 3, 5, x\} \text{ and range} = \{2, 4, 6, y\}$$
$$g = \{(2, 1), (4, 3), (6, 5), (y, x)\}; \text{ Domain} = \{2, 4, 6, y\} \text{ and range} = \{1, 3, 5, x\}$$

Note the ordered pairs in g are the ordered pairs of f with x and y interchanged. Since $(1, 2)$ is in f, by definition $f(1) = 2$. Since $(2, 1)$ is in g, $g(2) = 1$. Therefore, $f[g(2)] = f(1) = 2$ and $g[f(1)] = g(2) = 1$. Similarly, $f[g(4)] = f(3) = 4$ and $g[f(3)] = g(4) = 3$. In particular, $f[g(y)] = f(x) = y$ and $g[f(x)] = g(y) = x$ for all real numbers x and y. Consequently, f and g are inverses.

ANSWER

$f[g(x)] = f\left[\dfrac{x - 6}{5}\right]$

$\quad = 5\left(\dfrac{x - 6}{5}\right) + 6$

$\quad = x - 6 + 6$

$\quad = x$

$g[f(x)] = g[5x + 6]$

$\quad = \dfrac{5x + 6 - 6}{5}$

$\quad = \dfrac{5x}{5}$

$\quad = x$

Since $f[g(x)] = x$ and $g[f(x)] = x$, f and g are inverses.

Warning: Do not confuse f^{-1} with raising an expression to the negative 1 power. It is usually clear from the context that f^{-1} means an inverse function.

In general, to find the inverse of a function, we reverse the ordered pairs by interchanging x and y. To indicate the inverse of the function f, we use a special notation f^{-1} instead of g. The following figure illustrates inverse functions.

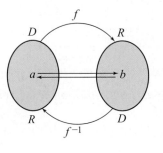

Note: *Notice that f sends a to b and f^{-1} sends b back to a. Hence, $f^{-1}[f(a)] = a$. Also, f^{-1} sends b to a and f sends a back to b. Hence, $f[f^{-1}(b)] = b$. We also see that the domain of f is the range of f^{-1} and the range of f is the domain of f^{-1}.*

Before we can find inverse functions, we need to explore one more concept.

OBJECTIVE 3. Show that a function is one-to-one.   If we merely reverse the ordered pairs of a function, we do not always get an inverse function. Consider the following sets in which the ordered pairs of B are the ordered pairs of A with x and y interchanged:

$$A = \{(1, 2), (2, 4), (3, 2), (-2, 5)\}$$
$$B = \{(2, 1), (4, 2), (2, 3), (5, -2)\}$$

Note that A represents a function, but B does not because 2 in the domain of B is paired with 1 and 3. How did this happen? Since the ordered pairs of B are those of A with x and y interchanged, two ordered pairs of A having the same y-value of 2 [(1, 2) and (3, 2)] became two ordered pairs with the same x-value in B. Therefore, if the inverse of a function is to be a function, no two ordered pairs of the function may have the same y-value. Such a function is called a **one-to-one function**.

A function is one-to-one if each value in the domain corresponds to only one value in the range and each value in the range corresponds to only one value in the domain. In terms of x and y, two different x-values must result in two different y-values, which suggests the following formal definition.

DEFINITION   ***One-to-one function:*** *A function f is one-to-one if for any two numbers a and b in its domain, when $f(a) = f(b)$, $a = b$ and when $a \neq b$, $f(a) \neq f(b)$.*

It is possible to determine if a function is one-to-one by looking at its graph. If two ordered pairs have the same y-value, then the corresponding points lie on the same horizontal line. Consequently, if a function is one-to-one, then the graph cannot be intersected by any horizontal line in more than one point.

RULE   **Horizontal Line Test for One-to-One Functions**

Given a function's graph, the function is one-to-one if every horizontal line that can intersect the graph does so at one and only one point.

EXAMPLE 3 Determine whether the following are graphs of one-to-one functions.

a. $f(x) = x^3$ **b.** $f(x) = x^2 + 2$ **c.** $f(x) = \sqrt{36 - x^2}$

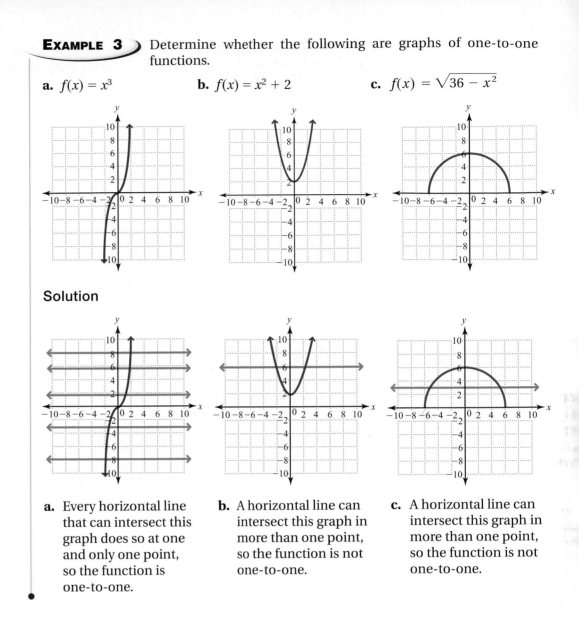

Solution

a. Every horizontal line that can intersect this graph does so at one and only one point, so the function is one-to-one.

b. A horizontal line can intersect this graph in more than one point, so the function is not one-to-one.

c. A horizontal line can intersect this graph in more than one point, so the function is not one-to-one.

YOUR TURN Determine whether the following are graphs of one-to-one functions.

a.

b.

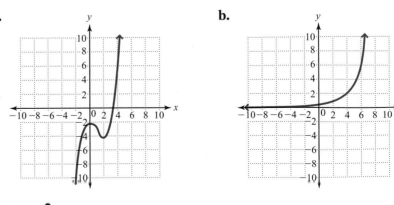

ANSWERS

a. no **b.** yes

12.1 Composite and Inverse Functions **953**

OBJECTIVE 4. Find the inverse of a function. Interchanging the ordered pairs of a function that is not one-to-one results in a relation that is not a function. Consequently, we have the following.

RULE **Existence of Inverse Functions**

A function has an inverse function if and only if the function is one-to-one.

We have already seen that in order to find the inverse of a function, we need to interchange the x- and y-values of all the ordered pairs of f. That is, we replace x with y and y with x in the equation defining the function. However, we are not finished. We now have the inverse in the form $f(y) = x$, so we must solve this equation for y in order to have y as a function of x. The steps are summarized as follows.

PROCEDURE **Finding the Inverse Function of a One-to-One Function**

1. If necessary, replace $f(x)$ with y.
2. Replace all x's with y's and y's with x's.
3. Solve the equation from step 2 for y.
4. Replace y with $f^{-1}(x)$.

EXAMPLE 4 Find $f^{-1}(x)$ for each of the following one-to-one functions.

a. $f(x) = 2x + 4$

The domain and range of f is the set of all real numbers, so the domain and range of f^{-1} will also be the set of all real numbers.

Solution
$$y = 2x + 4 \qquad \text{Replace } f(x) \text{ with } y.$$
$$x = 2y + 4 \qquad \text{Replace } x \text{ with } y \text{ and } y \text{ with } x.$$
$$\frac{x - 4}{2} = y \qquad \text{Solve for } y.$$
$$f^{-1}(x) = \frac{x - 4}{2} \qquad \text{Replace } y \text{ with } f^{-1}(x).$$

In order to verify that we have found the inverse, we need to show that $f[f^{-1}(x)] = x$ and $f^{-1}[f(x)] = x$.

$$f[f^{-1}(x)] = f\left(\frac{x-4}{2}\right) = 2\left(\frac{x-4}{2}\right) + 4 = x - 4 + 4 = x$$

$$f^{-1}[f(x)] = f^{-1}(2x + 4) = \frac{(2x+4) - 4}{2} = \frac{2x + 4 - 4}{2} = \frac{2x}{2} = x$$

Since $f[f^{-1}(x)] = x$ and $f^{-1}[f(x)] = x$, they are inverses.

b. $f(x) = x^3 + 2$

The domain and range of f is the set of all real numbers, so the domain and range of f^{-1} will also be the set of all real numbers.

Solution $y = x^3 + 2$ Replace $f(x)$ with y.

$\qquad\qquad x = y^3 + 2$ Replace x with y and y with x.

$\qquad\qquad x - 2 = y^3$ Begin solving for y by subtracting 2 from both sides.

$\qquad\qquad \sqrt[3]{x - 2} = y$ Take the cube root of each side to solve for y.

$\qquad\qquad f^{-1}(x) = \sqrt[3]{x - 2}$ Replace y with $f^{-1}(x)$.

We can verify that f and f^{-1} are inverses as in part a.

c. $f(x) = \sqrt{x - 3}$

The domain of f is $[3, \infty)$ and the range is $[0, \infty)$. Therefore, the domain of f^{-1} is $[0, \infty)$ and the range of f^{-1} is $[3, \infty)$.

Solution $y = \sqrt{x - 3}$ Replace $f(x)$ with y.

$\qquad\qquad x = \sqrt{y - 3}$ Replace x with y and y with x.

$\qquad\qquad x^2 = y - 3$ Begin solving for y by squaring both sides.

$\qquad\qquad x^2 + 3 = y$ Add 3 to both sides to solve for y.

The domain of $y = x^2 + 3$ is all real numbers, but the domain of f^{-1} is $[0, \infty)$. Therefore, we write $f^{-1}(x) = x^2 + 3, x \geq 0$.

In order to verify that these are inverses, we need to show that $f[f^{-1}(x)] = x$ and $f^{-1}[f(x)] = x$.

$$f[f^{-1}(x)] = f(x^2 + 3) = \sqrt{(x^2 + 3) - 3} = \sqrt{x^2 + 3 - 3} = \sqrt{x^2}$$

Since the domain of f^{-1} is $[0, \infty)$, $\sqrt{x^2} = x$. So, $f[f^{-1}(x)] = x$.

$$f^{-1}[f(x)] = f^{-1}(\sqrt{x - 3}) = (\sqrt{x - 3})^2 + 3 = x - 3 + 3 = x$$

Since $f[f^{-1}(x)] = x$ and $f^{-1}[f(x)] = x$, they are inverses.

YOUR TURN   Find $f^{-1}(x)$ for each of the following one-to-one functions.

a. $f(x) = 5x + 2$ **b.** $f(x) = \dfrac{x + 3}{x}$

OBJECTIVE 5. Graph a given function's inverse function. In the following figure, we have plotted pairs of points whose coordinates are interchanged and graphed the line $y = x$.

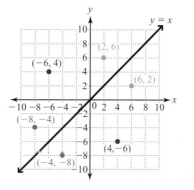

If the graph were folded along the line $y = x$, the points whose coordinates are interchanged would fall on top of each other. These points, therefore, are symmetric with respect to the line $y = x$. It can be shown that the graphs of any two points of the form (a, b) and (b, a) are symmetric with respect to the graph of $y = x$. Similarly, since the ordered pairs of f and f^{-1} have interchanged coordinates, the graphs of f and f^{-1} are symmetric with respect to the line $y = x$.

ANSWERS

a. $f^{-1}(x) = \dfrac{x - 2}{5}$

b. $f^{-1}(x) = \dfrac{3}{x - 1}$

RULE **Graphs of Inverse Functions**

The graphs of f and f^{-1} are symmetric with respect to the graph of $y = x$.

Learning Strategy

If you are a tactile learner, imagine placing a mirror on the line $y = x$. The graphs of f and f^{-1} are reflections in the line $y = x$.

Following are the graphs of f and f^{-1} for Example 4 along with the graph of $y = x$.

a. $f(x) = 2x + 4$

$f^{-1}(x) = \dfrac{x - 4}{2}$

b. $f(x) = x^3 + 2$

$f^{-1}(x) = \sqrt[3]{x - 2}$

c. $f(x) = \sqrt{x - 3}$

$f^{-1}(x) = x^2 + 3, \quad x \geq 0$

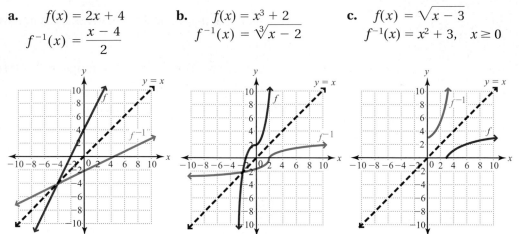

EXAMPLE 5 Sketch the inverse of the functions whose graphs are shown in parts a and b.

a.

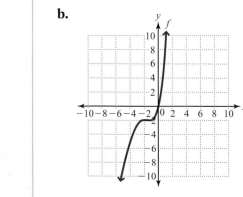

Solution Draw the line $y = x$ and reflect the graph in the line.

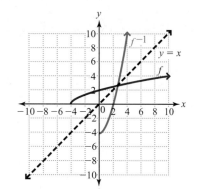

b.

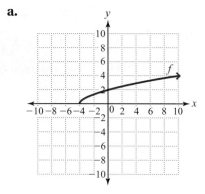

Solution Draw the line $y = x$ and reflect the graph in the line.

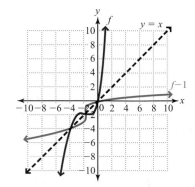

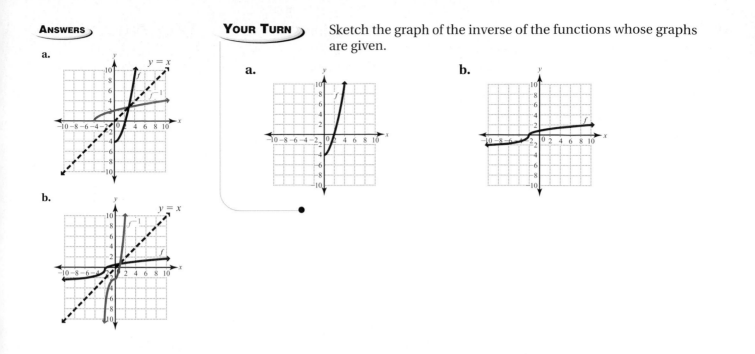
YOUR TURN

Sketch the graph of the inverse of the functions whose graphs are given.

a.

b.

12.1 Exercises

For Extra Help

MyMathLab   MyMathLab   Videotape/DVT   InterAct Math   Math Tutor Center   Math XL.com

1. Does $f \circ g = g \circ f$ for any two functions f and g? Why or why not?

2. What must be shown to prove that two functions, f and g, are inverses?

3. How are the domains and ranges of inverse functions related? Why?

4. If the coordinates of the ordered pairs of a function are interchanged, is the result always the inverse function? Why or why not?

5. How are horizontal lines used to determine whether or not a function is one-to-one?

6. How is the graph of $(-2, 3)$ related to the graph of $(3, -2)$?

For Exercises 7–18, if $f(x) = 3x + 5$, $g(x) = x^2 + 3$, and $h(x) = \sqrt{x + 1}$, find each composition.

7. $(f \circ g)(0)$

8. $(g \circ f)(0)$

9. $(h \circ f)(1)$

10. $(h \circ g)(3)$

11. $(f \circ g)(-2)$ **12.** $(g \circ f)(-1)$ **13.** $(f \circ g)(x)$ **14.** $(g \circ f)(x)$

15. $(f \circ h)(x)$ **16.** $(h \circ g)(x)$ **17.** $(h \circ f)(0)$ **18.** $(g \circ h)(0)$

For Exercises 19–28, find $(f \circ g)(x)$ and $(g \circ f)(x)$.

19. $f(x) = 2x - 2, g(x) = 3x + 4$ **20.** $f(x) = 4x + 7, g(x) = 3x - 2$

21. $f(x) = x + 2, g(x) = x^2 + 1$ **22.** $f(x) = x^2 - 3, g(x) = x + 5$

23. $f(x) = x^2 + 3x - 4, g(x) = 3x$ **24.** $f(x) = -2x, g(x) = x^2 - 2x + 4$

25. $f(x) = \sqrt{x + 2}, g(x) = 2x - 5$ **26.** $f(x) = 5x + 2, g(x) = \sqrt{x - 5}$

27. $f(x) = \dfrac{x + 1}{x}, g(x) = \dfrac{x - 3}{x}$ **28.** $f(x) = \dfrac{x + 4}{x}, g(x) = \dfrac{2 - x}{x}$

For Exercises 29–30, answer each question.

29. If the domain of f is $[3, \infty)$ and the range is $[0, \infty)$, what are the domain and range of f^{-1}?

30. If f and g are inverse functions and $f(2) = 5$, then $g(5) = $ _____.

For Exercises 31–34, determine whether the following functions f and g are inverses.

31. $f = \{(1, 2), (-1, -3), (3, 4), (2, -5)\}, g = \{(2, 1), (-3, -1), (4, 3), (-5, 2)\}$

32. $f = \{(4, -3), (1, 4), (5, 2), (-3, 1), (-1, 3)\}, g = \{(-3, 4), (4, 1), (2, 5), (1, -3), (3, -1)\}$

33. $f = \{(-2, -2), (3, -3), (-4, 4), (-6, -6)\}, g = \{(-2, -2), (-3, 3), (-4, 4), (-6, -6)\}$

34. $f = \{(5, 5), (-4, 4), (2, -2), (-7, -7)\}, g = \{(-5, -5), (4, -4), (-2, 2), (7, 7)\}$

For Exercises 35–48, determine if f and g are inverses by determining whether $(f \circ g)(x) = x$ and $(g \circ f)(x) = x$.

35. $f(x) = x + 5, g(x) = x - 5$ **36.** $f(x) = x - 1, g(x) = x + 1$

37. $f(x) = 6x, g(x) = \dfrac{x}{6}$ **38.** $f(x) = -\dfrac{x}{3}, g(x) = -3x$

39. $f(x) = 2x - 3$, $g(x) = \dfrac{x - 3}{2}$

40. $f(x) = \dfrac{x - 4}{5}$, $g(x) = 5x + 4$

41. $f(x) = x^3 - 4$, $g(x) = \sqrt[3]{x + 4}$

42. $f(x) = x^5 + 3$, $g(x) = \sqrt[5]{x - 3}$

43. $f(x) = x^2$, $g(x) = \sqrt{x}$

44. $f(x) = x^4$, $g(x) = \sqrt[4]{x}$

45. $f(x) = x^2$, $x \geq 0$; $g(x) = \sqrt{x}$

46. $f(x) = x^2 + 2$, $x \geq 0$; $g(x) = \sqrt{x - 2}$

47. $f(x) = \dfrac{3}{x + 5}$, $g(x) = \dfrac{3 - 5x}{x}$

48. $f(x) = \dfrac{x}{x - 3}$, $g(x) = \dfrac{3x}{x - 1}$

For Exercises 49–50, answer each question.

49. Is $f(x) = \dfrac{1}{x}$ its own inverse? Explain.

50. Is $f(x) = x$ its own inverse? Explain.

For Exercises 51–62, sketch the graph of the inverse of each of the following functions.

51. **52.** **53.** **54.**

55. **56.** **57.** **58.**

59. **60.** **61.** **62.**

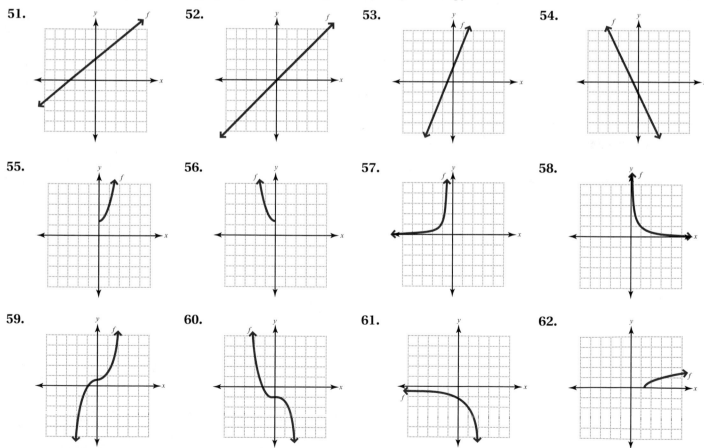

For Exercises 63–66, determine whether the function is one-to-one.

63.
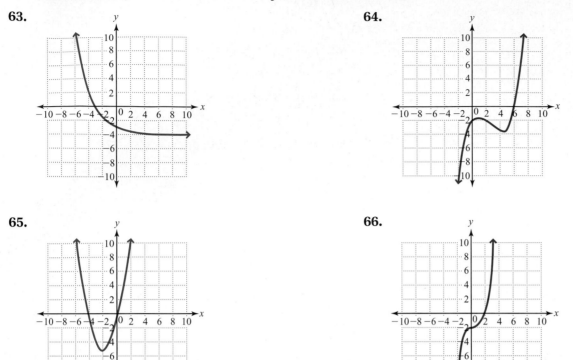

64.

65.

66.

For Exercises 67–88, find $f^{-1}(x)$ for each of the following one-to-one functions f.

67. $f = \{(-3, 2), (-1, -3), (0, 4), (4, 6)\}$

68. $f = \{(-4, 2), (-1, -3), (2, 3), (5, 7)\}$

69. $f = \{(7, -2), (9, 2), (-4, 1), (3, 3)\}$

70. $f = \{(1, 2), (4, 5), (-3, 2), (-9, 0)\}$

71. $f(x) = x + 6$

72. $f(x) = x - 4$

73. $f(x) = 2x + 3$

74. $f(x) = -3x + 2$

75. $f(x) = x^3 + 2$

76. $f(x) = x^3 - 3$

77. $f(x) = \dfrac{2}{x + 2}$

78. $f(x) = \dfrac{-3}{x - 3}$

79. $f(x) = \dfrac{x + 2}{x - 3}$

80. $f(x) = \dfrac{x - 4}{x + 2}$

81. $f(x) = \sqrt{x - 2}$

82. $f(x) = \sqrt{2x - 4}$

83. $f(x) = 2x^3 + 4$

84. $f(x) = 4x^3 - 5$

85. $f(x) = \sqrt[3]{x + 2}$

86. $f(x) = \sqrt[3]{x - 5}$

87. $f(x) = 2\sqrt[3]{2x + 4}$

88. $f(x) = 3\sqrt[3]{4x - 3}$

For Exercises 89–92, solve each problem.

89. If a salesperson worked for $100 per week plus 5% commission on sales, the weekly salary is $y = 0.05x + 100$, where y represents the salary and x represents the sales.

   a. Find the inverse function.

   b. What does each variable of the inverse function represent?

   c. Use the inverse function to find the sales for a week in which the salary was $350.

90. A painting contractor purchased 24 gallons of paint for the interior and exterior of a house. If he paid $12 per gallon for the interior paint and $18 per gallon for the exterior paint, then the amount that he paid is $y = 12x + 18(24 - x)$, where y represents the amount paid and x represents the number of gallons of interior paint.

   a. Find the inverse function.

   b. What does each variable of the inverse function represent?

   c. Use the inverse function to find the number of gallons of interior paint if he paid $372 total for the 24 gallons of paint.

91. An office building installed 105 fans, some of which measure 36 inches and the remainder measure 54 inches. If the 36-inch fans cost $45 each and the 54-inch fans cost $65 each, then the cost of the fans is $y = 45x + 65(105 - x)$, where y represents the cost and x represents the number of 36-inch fans.

   a. Find the inverse function.

   b. What does each variable of the inverse function represent?

   c. Use the inverse function to find the number of 36-inch fans if the total cost was $5225.

92. If a Toyota Prius averages 50 miles per hour on a trip, then $y = 50x$, where y represents the number of miles traveled and x represents the number of hours.

   a. Find the inverse function.

   b. What does each variable of the inverse function represent?

   c. Use the inverse function to find the number of hours to travel 210 miles.

For Exercises 93–98, answer the question.

93. If $(-4, 5)$ is an ordered pair on the graph of g, what are the coordinates of an ordered pair on the graph of g^{-1}?

94. If $f(2) = 4$ and $f^{-1}(a) = 2$, find a.

95. A linear function is of the form $f(x) = ax + b$, $a \neq 0$. Find $f^{-1}(x)$.

96. The square root function is defined by $f(x) = \sqrt{x}$. Find $f^{-1}(x)$. Be careful!

97. The graph of an even function is always symmetric with respect to the y-axis. Is the inverse of an even function also a function? Explain.

98. The graph of an odd function is always symmetric with respect to the origin. Is the inverse of an odd function also a function? Explain.

REVIEW EXERCISES

[1.2] **1.** Write 32 as 2 to a power.

For Exercises 2–5, evaluate each expression.

[1.5] **2.** 2^3 [1.5] **3.** $\left(-\dfrac{1}{3}\right)^3$ [5.1] **4.** 4^{-2} [10.2] **5.** $4^{3/2}$

[11.4] **6.** What are the coordinates of the vertex of $g(x) = (x - 3)^2 + 1$?

12.2 Exponential Functions

OBJECTIVES

1. Define and graph exponential functions.
2. Solve equations of the form $b^x = b^a$ for x.
3. Use exponential functions to solve application problems.

OBJECTIVE 1. Define and graph exponential functions. Previously we defined rational number exponents. For example, we know that $b^3 = b \cdot b \cdot b$, $b^{2/3} = \sqrt[3]{b^2}$, $b^0 = 1$, and $b^{-3} = \dfrac{1}{b^3}$. It can be shown that irrational number exponents have meaning as well, and we can approximate expressions like $2^{\sqrt{3}}$ and 5^π by using rational approximations for the exponents. Therefore, the exponential expression b^x has meaning if x is any real number (rational or irrational), so we can define the **exponential function** as follows.

DEFINITION *Exponential function:* If $b > 0$, $b \neq 1$, and x is any real number, then the exponential function is $f(x) = b^x$.

Note: *The definition of the exponential function has two restrictions on b. If $b = 1$, then $f(x) = b^x = 1^x = 1$, which is a linear function. If $b < 0$, then we could get values for which the function is not defined as a real number. For example, if $f(x) = (-4)^x$, then*

$$f\left(\frac{1}{2}\right) = (-4)^{1/2} = \sqrt{-4},$$

which is not a real number.

We graph exponential functions by plotting enough points to determine the graph's shape. We will find that the graph has one typical shape if $b > 1$ and a different typical shape if $0 < b < 1$.

EXAMPLE 1 Graph.

a. $f(x) = 2^x$ and $g(x) = 4^x$

Solution Choose some values of x and find the corresponding values of $f(x)$ and $y(x)$ (which are the y-values).

x	−3	−2	−1	0	1	2	3
f(x)	$\frac{1}{8}$	$\frac{1}{4}$	$\frac{1}{2}$	1	2	4	8
g(x)	$\frac{1}{64}$	$\frac{1}{16}$	$\frac{1}{4}$	1	4	16	64

Plotting the ordered pairs for each function gives smooth curves typical of the graphs of $f(x) = b^x$ with $b > 1$. Comparing the graphs of $f(x) = 2^x$ and $g(x) = 4^x$, we can see that the greater the value of b, the steeper the graph.

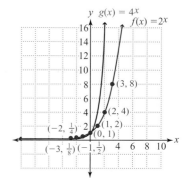

b. $h(x) = \left(\frac{1}{2}\right)^x$ and $k(x) = \left(\frac{1}{4}\right)^x$

Solution Choose some values of x and find the corresponding values of $h(x)$ and $k(x)$.

x	−3	−2	−1	0	1	2	3
h(x)	8	4	2	1	$\frac{1}{2}$	$\frac{1}{4}$	$\frac{1}{8}$
k(x)	64	16	4	1	$\frac{1}{4}$	$\frac{1}{16}$	$\frac{1}{64}$

Note: *We left the points and coordinate labels off the graphs of $g(x)$ and $k(x)$ to avoid additional clutter. From all the graphs in Example 1, we see that b^x is never negative.*

After plotting the ordered pairs for each function, we see graphs that are typical of exponential functions in the form $f(x) = b^x$ with $0 < b < 1$. Comparing the graphs of $h(x) = \left(\frac{1}{2}\right)^x$ and $k(x) = \left(\frac{1}{4}\right)^x$, we see that the smaller the value of b, the steeper the graph.

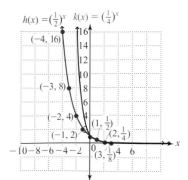

We summarize the graphs of $f(x) = b^x$ as follows.

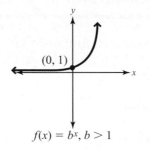

$f(x) = b^x, b > 1$

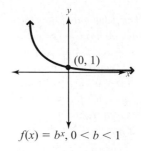

$f(x) = b^x, 0 < b < 1$

Note: *Any function that increases or decreases along its entire domain is one-to-one because no horizontal line will intersect its graph at more than one point.*

The function is increasing from left to right and one-to-one. The graph always passes through (0, 1). For negative values of x, the graph approaches the x-axis but never touches it. The larger the values of b, the steeper the graph. The domain is $(-\infty, \infty)$ and the range is $(0, \infty)$.

The function is decreasing from left to right and one-to-one. The graph always passes through (0, 1). For positive values of x, the graph approaches the x-axis but never touches it. The smaller the values of x, the steeper the graph. The domain is $(-\infty, \infty)$ and the range is $(0, \infty)$.

More complicated exponential functions are graphed in the same manner.

EXAMPLE 2   Graph. $f(x) = 3^{2x-1}$

Solution Find some ordered pairs, plot them, and draw the graph.

x	$y = f(x)$
-1	$3^{-3} = \dfrac{1}{27}$
0	$3^{-1} = \dfrac{1}{3}$
1	$3^1 = 3$
2	$3^3 = 27$

YOUR TURN   Graph. $f(x) = 2^{x+1}$

ANSWER

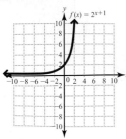

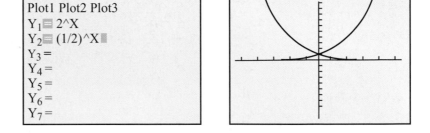

Calculator
TIPS

To graph $y = 2^x$ on a graphing calculator, we enter $Y_1 = 2^x$. Set the window to $[-5, 5]$ for x and $[-10, 10]$ for y and press GRAPH *. We graph $y = \left(\dfrac{1}{2}\right)^x$ in the same window by entering $Y_2 = \left(\dfrac{1}{2}\right)^x$. Notice that if we folded the graph along the y-axis, the graphs would lie on top of each other, indicating they are symmetric with respect to the y-axis.*

OBJECTIVE 2. Solve equations of the form $b^x = b^a$ for x. Earlier we solved equations containing expressions having a variable base and a constant exponent, like $x^2 = 4$. We will now solve equations that contain expressions having a constant base and a variable exponent, like $2^x = 16$. We need a new method to solve these equations. Previously we noted that the exponential function is one-to-one, so for each x-value there is a unique y-value and vice versa. Consequently, we have the following rule.

RULE **The One-to-One Property of Exponentials**
Given $b > 0$ and $b \neq 1$, if $b^x = b^y$, then $x = y$.

To solve some types of exponential equations, we use the following procedure.

PROCEDURE **Solving Exponential Equations**
1. If necessary, write both sides of the equation as a power of the same base.
2. If necessary, simplify the exponents.
3. Set the exponents equal to each other.
4. Solve the resulting equation.

EXAMPLE 3 Solve.

a. $3^x = 81$

Solution $3^x = 3^4$ Write 81 as 3^4 so both sides have the same base.
$\qquad\quad x = 4$ Set the exponents equal to each other.

Check $3^x = 81$
$\qquad\quad 3^4 = 81$ Replace x with 4.
$\qquad\quad 81 = 81$ True, so $x = 4$ is the solution.

b. $4^x = 32$

Solution $(2^2)^x = 2^5$ Write 4 as 2^2 and 32 as 2^5 so both sides have the same base.

$2^{2x} = 2^5$ Simplify $(2^2)^x$ by applying $(a^m)^n = a^{mn}$.

$2x = 5$ Set the exponents equal to each other.

$x = \dfrac{5}{2}$ Solve for x.

Note: *We leave the checks for parts b–d to the reader.*

c. $8^{x+4} = 4^{2x+3}$

Solution $(2^3)^{x+4} = (2^2)^{2x+3}$ Write 8 as 2^3 and 4 as 2^2 so both sides have the same base.

$2^{3x+12} = 2^{4x+6}$ Simplify the exponents by applying $(a^m)^n = a^{mn}$.

$3x + 12 = 4x + 6$ Set the exponents equal to each other.

$6 = x$ Solve for x.

d. $\left(\dfrac{1}{5}\right)^x = 25$

Solution $(5^{-1})^x = 5^2$ Write $\dfrac{1}{5}$ as 5^{-1} and 25 as 5^2 so both sides have the same base.

$5^{-x} = 5^2$ Simplify $(5^{-1})^x$ by applying $(a^m)^n = a^{mn}$.

$-x = 2$ Set the exponents equal to each other.

$x = -2$ Solve for x.

YOUR TURN   Solve.

a. $2^{x+1} = 8$ **b.** $27^{x-2} = 9^x$ **c.** $\left(\dfrac{1}{2}\right)^x = 16$

OBJECTIVE 3. Use exponential functions to solve application problems.

Compound Interest Formula

Exponential functions occur in many areas, especially the sciences and business. If P dollars are invested at an annual interest rate of r (written as a decimal) compounded n times per year for t years, the accumulated amount A in the account is given by the formula $A = P\left(1 + \dfrac{r}{n}\right)^{nt}$.

EXAMPLE 4   Find the accumulated amount in an account if $5000 is deposited at 6% compounded quarterly for 10 years. Round the answer to the nearest cent.

Understand   We are asked to find A given that $P = \$5000$, $r = 0.06$, $n = 4$, and $t = 10$.

Plan   Use the formula $A = P\left(1 + \dfrac{r}{n}\right)^{nt}$.

ANSWERS

a. 2   **b.** 6   **c.** -4

Execute $A = P\left(1 + \dfrac{r}{n}\right)^{nt}$

$A = 5000\left(1 + \dfrac{0.06}{4}\right)^{4(10)}$ **Substitute 5000 for *P*, 0.06 for *r*, 4 for *n*, and 10 for *t*.**

$A = 5000(1 + 0.015)^{40}$ **Simplify.**

$A = 5000(1.015)^{40}$

$A = 9070.09$ **Evaluate using a calculator and round to the nearest cent (hundredths place).**

Answer After 10 years, the accumulated amount in the account is $9070.09

Check Verify that the principal is $5000 if the accumulated amount is $9070.09 after the principal is compounded quarterly for 10 years.

$$9070.09 = P\left(1 + \dfrac{0.06}{4}\right)^{4(10)}$$

$$9070.09 = P(1.015)^{40}$$

$$\dfrac{9070.09}{(1.015)^{40}} = P$$

$$4999.998874 = P$$

Since the accumulated amount, 9070.09, was rounded, it is reasonable to expect our calculated value for the principal to be slightly different from $5000.

Half-Life

The *half-life* of a radioactive substance is the amount of time it takes until only half the original amount of the substance remains. Suppose we begin with 100 grams of a substance that has a half-life of 10 days. After 10 days (one half-life) 50 grams will remain, after 20 days (two half-lives) 25 grams will remain, after 30 days (three half-lives) 12.5 grams will remain, and so forth. The formula $A = A_0\left(\dfrac{1}{2}\right)^{t/h}$ gives the amount remaining, where A_0 is the initial amount, t is the time, and h is the half-life.

EXAMPLE 5 The isotope ^{45}Ca has a half-life of 165 days. How many grams of a 50-gram sample of ^{45}Ca will remain after 825 days? (*Source: CRC Handbook of Chemistry and Physics*, 62nd edition, CRC Press, 1981)

Understand Given a 50-gram sample of ^{45}Ca, we are to find the amount remaining after 825 days.

Plan Use the formula $A = A_0\left(\dfrac{1}{2}\right)^{t/h}$.

Execute $A = A_0\left(\dfrac{1}{2}\right)^{t/h}$

$A = 50\left(\dfrac{1}{2}\right)^{825/165}$ Substitute 50 for A_0, and **825** for t and 165 for h.

$A = 50\left(\dfrac{1}{2}\right)^{5}$ Simplify.

$A = 1.5625$ Evaluate using a calculator.

Answer 1.5625 grams of ^{45}Ca will remain after 825 days.

Check Verify that the sample was 50 grams if 1.5625 grams remain after 825 days. We will leave this check to the reader.

Aging

EXAMPLE 6 The number of people in the United States age 65 or over (in millions) is given in the following table.

Year	Number 65 or over (in millions)
1900	3.1
1910	4.0
1920	4.9
1930	6.7
1940	9.0
1950	12.4
1960	16.7
1970	20.1
1980	25.5
1990	31.4
2000	35.6

Source: U.S. Bureau of the Census

The data can be approximated by the function $y = 3.17(1.026)^x$, where x is the number of years after 1900 and y is the number of people in millions. Use the model to estimate the number of people in the United States age 65 or over in the year 2020. Round the answer to the nearest tenth of a million.

Understand We are given the function $y = 3.17(1.026)^x$, which approximates the data in the table. We are to find the number of people age 65 or over in the year 2020.

Plan Since x represents the number of years after 1900, we first subtract 1900 from 2020 to find the value of x that corresponds to the year 2020. We can then use $y = 3.17(1.026)^x$.

Execute $x = 2020 - 1900 = 120$ Subtract 1900 from 2020 to find the value of x that corresponds to the year 2020.

$y = 3.17(1.026)^{120}$ Substitute 120 for x in $y = 3.17(1.026)^x$.

$y = 69.0$ million Evaluate using a calculator and round to the nearest tenth.

Answer According to the function, about 69 million people will be age 65 or over in the year 2020.

Check Verify that if $y = 69$ in the function $y = 3.17(1.026)^x$, then $x = 120$.

$69 = 3.17(1.026)^x$ Substitute 69 for y in $y = 3.17(1.026)^x$.

To solve the above equation, we need to use logarithms, which we have not learned yet. We will learn about logarithms and how to solve equations like the one above in the rest of this chapter.

Connection Following is a graph of the function $y = 3.17(1.026)^x$ with the point corresponding to the solution of Example 6 indicated.

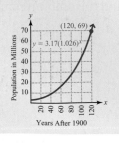

Your Turn

a. If $3000 is invested at 4% compounded semiannually (twice per year), how much money is in the account at the end of 8 years? Round to the nearest cent.

b. The radioactive isotope ^{61}Cr has a half-life of 26 days. How much of a 10-gram sample would remain after 208 days? Give the answer to the nearest thousandth of a gram. (*Source: CRC Handbook of Chemistry and Physics*, 62nd edition, CRC Press, 1981)

c. The following table shows the number of computers (in millions) in use in the United States for selected years since 1984:

Year	Number of computers (in millions)
1985	21.5
1988	40.8
1989	47.6
1991	62.0
1992	68.2
1993	76.5
1994	85.8
1995	96.2
2000	160.5

The data can be approximated by $y = 23.14(1.138)^x$, where x is the number of years after 1984 and y is the number of computers (in millions). Use the model to approximate the number of computers in use in the United States in 2010.

Answers

a. $4118.36   b. 0.039 g

c. 666.9 million

12.2 Exercises

1. Are the graphs of $f(x) = 2^x$ and $g(x) = \left(\dfrac{1}{2}\right)^x$ symmetric with respect to the y-axis? Explain.

2. As x gets larger, which graph is steeper: $f(x) = 3^x$ or $g(x) = 1.5^x$? Why?

3. Find the domain and range of $f(x) = 2^{x-4}$.

4. Is $f(x) = (-2)^x$ an exponential function? Why or why not?

5. For $b > 0$ and $b \neq 1$, if $b^x = b^y$, why does $x = y$?

6. If we have 10 grams of a substance that has a half-life of 50 days, how many grams will be present after 200 days?

For Exercises 7–24 graph.

7. $f(x) = 3^x$

8. $f(x) = 4^x$

9. $f(x) = 4^x - 3$

10. $f(x) = 3^x + 1$

11. $f(x) = \left(\dfrac{1}{3}\right)^x$

12. $f(x) = \left(\dfrac{1}{4}\right)^x$

13. $f(x) = \left(\dfrac{2}{3}\right)^x + 2$

14. $f(x) = \left(\dfrac{3}{2}\right)^x - 1$

15. $f(x) = -2^x$

16. $f(x) = -3^x$

17. $f(x) = 2^{x-2}$

18. $f(x) = 3^{x+1}$

19. $f(x) = 3^{-x}$

20. $f(x) = 2^{-x}$

21. $f(x) = 2^{2x-3}$

22. $f(x) = 3^{2x+1}$

23. $f(x) = 3^{-x+2}$

24. $f(x) = 2^{-x-1}$

For Exercises 25–42, solve each equation.

25. $2^x = 8$

26. $3^x = 81$

27. $8^x = 32$

28. $27^x = 81$

29. $16^x = 4$

30. $36^x = 216$

31. $5^x = \dfrac{1}{25}$

32. $3^x = \dfrac{1}{27}$

33. $\left(\dfrac{1}{3}\right)^x = 9$

34. $\left(\dfrac{1}{5}\right)^x = 125$

35. $\left(\dfrac{2}{3}\right)^x = \dfrac{8}{27}$

36. $\left(\dfrac{3}{2}\right)^x = \dfrac{9}{4}$

37. $\left(\dfrac{1}{2}\right)^x = 16$

38. $\left(\dfrac{1}{3}\right)^x = 27$

39. $25^{x+1} = 125$

40. $9^{x+2} = 81$

41. $8^{2x-1} = 32^{x-3}$

42. $5^{2x+1} = 125^{2x}$

For Exercises 43–46, use a graphing utility.

43. a. Graph $f(x) = 2^x$ and $g(x) = 2^{x+2}$ in the window $[-5, 5]$ for x and $[-1, 10]$ for y.
   b. How does the graph of g compare with the graph of f?

44. a. Graph $f(x) = 3^x$ and $g(x) = 3^x - 3$ in the window $[-3, 3]$ for x and $[-3, 9]$ for y.
   b. How does the graph of g compare with the graph of f?

45. a. Graph $f(x) = 2^x$ and $g(x) = 2^{-x}$ in the window $[-5, 5]$ for x and $[-1, 10]$ for y.
   b. How does the graph of g compare with the graph of f?

46. a. Graph $f(x) = 2^x$ and $g(x) = -2^x$ in the window $[-5, 5]$ for x and $[-5, 5]$ for y.
   b. How does the graph of g compare with the graph of f?

For Exercises 47–48, use the following.

Under ideal conditions a culture of *E. coli* bacteria doubles in size every 20 minutes. If A_0 is the initial amount and t is the number of minutes passed, the amount present is A, where $A = A_0(2)^{t/20}$.

47. If a culture of *E. coli* began with 100 cells, how many cells would be present after 120 minutes?

48. If a culture of *E. coli* currently has 500 cells, how many cells were present 90 minutes earlier? (*Hint:* Let $A_0 = 500$ and the time will be negative.)

49. Under ideal conditions, human beings could double their population every 50 years. If A_0 is the initial population and t is the number of years passed, the current population is A, where $A = A_0(2)^{t/50}$. If there were 6 billion humans on Earth in the year 2000, how many would there be in the year 2500 if their growth is uncontrolled?

50. Using the information from Exercise 49, how many human beings were on the Earth in 1900?

For Exercises 51–52, use the formula $A = P\left(1 + \dfrac{r}{n}\right)^{nt}$.

51. If $10,000 is deposited into an account paying 8% interest compounded quarterly, how much would be in the account after 12 years?

52. If $15,000 is deposited into an account paying 6% interest compounded semi-annually, how much would be in the account after 9 years?

For Exercises 53–54, use the formula $A = A_0\left(\dfrac{1}{2}\right)^{t/h}$ from Example 6.

53. Einsteinium (^{254}Es) has a half-life of 270 days. How much of a 5-gram sample would remain after 2160 days? Give the answer to the nearest thousandth of a gram. (*Source: CRC Handbook of Chemistry and Physics*, 62nd Edition, CRC Press, 1981)

54. Nobelium (^{257}No) has a half-life of 23 seconds. How much of a 100-gram sample would remain after 275 seconds? Give the answer to the nearest thousandth of a gram. (*Source: CRC Handbook of Chemistry and Physics*, 62nd Edition, CRC Press, 1981)

55. Since the first Super Bowl in 1967, ticket prices for the game have risen exponentially. Ticket prices can be approximated by the function $P(t) = 9.046(1.114)^t$, where $t = 1$ is the year 1967 and ticket prices are in dollars. Estimate the price of a Super Bowl ticket in the year 2010. (*Source: Orlando Sentinel*, January 21, 2003, p. A8)

56. The average cost of a 30-second commercial during the Super Bowl has increased exponentially over the years and can be approximated by $C(t) = 55.66(1.114)^t$, where t is the number of years after 1967 and C is the cost in thousands of dollars. Estimate the cost of a 30-second commercial in 1975; in 1985; 2005. (*Source: Orlando Sentinel*, January 21, 2003, p. A8)

57. Chlorine is frequently used to disinfect swimming pools. The concentration should remain between 1.5 and 2.5 parts per million. On a warm, sunny day, 30% of the chlorine can dissipate into the air or combine with other chemicals. If the initial amount of chlorine is 2.5 million parts per million, the function $f(x) = 2.5(0.7)^x$ models the amount of chlorine after x days. How much chlorine is in the pool after 2 days? (*Source:* D. Thomas, *Swimming Pool Operations Handbook*, 1972)

58. It is estimated that the value of a car depreciates 20% per year for the first five years. If the original price of a car is P, the value, A, of a car after t years is given by $A = P(.8)^t$. If a car originally cost $25,960, find the value of the car after 3 years to the nearest dollar.

59. Between 1971 and 2004 the number of transistors that can be placed on a single chip has grown significantly, as indicated in the table shown.

The data can be approximated by the function $T(x) = 0.001757(1.39)^x$, where x is the number of years after 1970 and $T(x)$ is the number of transistors in millions. If the current trend continues, estimate the number of transistors that could be put on a single chip in 2010. (*Source:* Intel)

Year	Chip	Transistors (millions)
1971	4004	0.0023
1986	386DX	0.275
1989	486DX	1.2
1993	Pentium	3.3
1995	P6	5.5
1997	Pentium II	7.5
1999	Pentium III	9.5
2000	Pentium IV	42
2004	Pentium IV Prescott	125

60. Suppose that Dave fixes a cup of coffee with cream and places it on the counter to cool. The temperature of the coffee at various times is given in the table shown.

The data can be approximated by the function $T(t) = 146.9(.989)^t$, where t is time in seconds and T is the temperature in °F. Estimate the temperature of the coffee after 1 minute.

t (seconds)	T (°F)
0.2	155.8
8.4	133.2
16.6	117.9
24.8	107.9
33	100.7
41.1	94.9
49.3	90.5

PUZZLE PROBLEM

What is the greatest number that can be written using three numerals?

REVIEW EXERCISES

[5.1] *For Exercises 1–3, evaluate.*

1. 5^3

2. $\left(\dfrac{1}{3}\right)^{-2}$

3. 4^{-3}

[10.2] **4.** Write $5^{2/3}$ in radical notation.

[12.1] **5.** If $f(x) = 3x + 4$, find $f^{-1}(x)$.

[12.1] **6.** Graph the inverse of the function whose graph is shown.

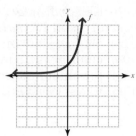

12.3 Logarithmic Functions

OBJECTIVES

1. Convert between exponential and logarithmic forms.
2. Solve logarithmic equations by changing to exponential form.
3. Graph logarithmic functions.
4. Solve applications involving logarithms.

OBJECTIVE 1. Convert between exponential and logarithmic forms. In Section 12.2, we defined the exponential function as $f(x) = b^x$ with $b > 0$ and $b \neq 1$. Since the exponential function is a one-to-one function, it has an inverse. Following is the graph of $f(x) = 2^x$ and its inverse, which we find by reflecting the graph about the line $y = x$.

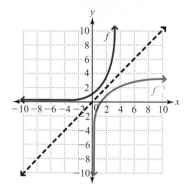

To find the inverse of $f(x) = 2^x$, we use the procedure from Section 12.1.

$$f(x) = 2^x$$

$\qquad y = 2^x$ **Replace $f(x)$ with y.**

$\qquad x = 2^y$ **Replace y with x and x with y.**

The next step is to solve for y, but we haven't learned how to isolate a variable that is an exponent. To rewrite $x = 2^y$, we define a **logarithm**.

Logarithm: If $b > 0$ and $b \neq 1$, then $y = \log_b x$ is equivalent to $x = b^y$.

If we apply this definition to $x = 2^y$ in the preceding equation, we get $y = \log_2 x$. Replacing y with $f^{-1}(x)$, we get $f^{-1}(x) = \log_2 x$, so the inverse function for $f(x) = 2^x$ is $f^{-1}(x) = \log_2 x$. To generalize, exponential functions and *logarithmic functions* are inverses. Consequently, if $y = \log_b x$, the domain is $(0, \infty)$ and the range is $(-\infty, \infty)$ with $b > 0$ and $b \neq 1$.

The expression $\log_b x$ is read "the logarithm base b of x" and *is the exponent to which b must be raised to get x*. Compare the two forms.

The exponent is the logarithm.

$$x = b^y \qquad y = \log_b x$$

Note: $y = \log_b x$ means y is the power to which we raise b to get x.

The base is the base of the logarithm.

The definition of a logarithm allows us to convert from one form to another. The following table contains pairs of equivalent forms.

Logarithmic Form	Exponential Form
$\log_2 8 = 3$	$2^3 = 8$
$\log_{10} \dfrac{1}{10} = -1$	$10^{-1} = \dfrac{1}{10}$
$\log_5 1 = 0$	$5^0 = 1$
$\log_{16} 4 = \dfrac{1}{2}$	$16^{1/2} = 4$
$\log_{1/2} 8 = -3$	$\left(\dfrac{1}{2}\right)^{-3} = 8$

Note: *The logarithmic equations are in the form $\log_b x = y$. The values of x are positive numbers only and the values of y are both positive and negative.*

EXAMPLE 1 Write in logarithmic form.

a. $3^4 = 81$

Solution $\log_3 81 = 4$ The base of the exponent is the base of the logarithm, and the exponent is the logarithm.

b. $\left(\dfrac{1}{2}\right)^{-3} = 8$

Solution $\log_{1/2} 8 = -3$ The base of the exponent is the base of the logarithm, and the exponent is the logarithm.

c. $9^{1/2} = 3$

Solution $\log_9 3 = \dfrac{1}{2}$ The base of the exponent is the base of the logarithm, and the exponent is the logarithm.

ANSWERS

a. $\log_4 64 = 3$ **b.** $\log_2 \dfrac{1}{8} = -3$

c. $\log_{27} 3 = \dfrac{1}{3}$

YOUR TURN Write in logarithmic form.

a. $4^3 = 64$ **b.** $2^{-3} = \dfrac{1}{8}$ **c.** $27^{1/3} = 3$

EXAMPLE 2 Write in exponential form.

Learning Strategy

If you are a visual learner, try visualizing the following "loop" for rewriting logarithms in exponential form.

equals

$\log_6 36 = 2$

6 raised to 2

a. $\log_6 36 = 2$

Solution $6^2 = 36$ The base of the logarithm is the base of the exponent, and the logarithm is the exponent.

b. $\log_{16} 2 = \dfrac{1}{4}$

Solution $16^{1/4} = 2$ The base of the logarithm is the base of the exponent, and the logarithm is the exponent.

c. $\log_{1/2} 16 = -4$

Solution $\left(\dfrac{1}{2}\right)^{-4} = 16$ The base of the logarithm is the base of the exponent, and the logarithm is the exponent.

YOUR TURN Write in exponential form.

a. $\log_4 64 = 3$ **b.** $\log_{16} 2 = \dfrac{1}{4}$ **c.** $\log_5 \dfrac{1}{125} = -3$

OBJECTIVE 2. Solve logarithmic equations by changing to exponential form. A logarithmic equation in the form $\log_b x = y$ could have b, x, or y as an unknown.

PROCEDURE Solving Logarithmic Equations

To solve an equation of the form $\log_b x = y$, where b, x, or y is a variable, write the equation in exponential form, $b^y = x$, and then solve for the variable.

EXAMPLE 3 Solve.

a. $\log_b 16 = 2$

Solution $b^2 = 16$ Write in exponential form.

$\quad\quad\quad b = \pm 4$ Find the positive and negative square roots of 16.

$\quad\quad\quad b = 4$ The base must be positive and not 1, so $b = 4$.

b. $\log_3 \dfrac{1}{27} = y$

Solution $3^y = \dfrac{1}{27}$ Write in exponential form.

ANSWERS

a. $4^3 = 64$ **b.** $16^{1/4} = 2$

c. $5^{-3} = \dfrac{1}{125}$

$\quad\quad\quad 3^y = \dfrac{1}{3^3}$ Write 27 as 3^3.

$\quad\quad\quad 3^y = 3^{-3}$ Write $\dfrac{1}{3^3}$ as 3^{-3}.

$\quad\quad\quad y = -3$ Set the exponents equal to each other.

c. $\log_{25} x = \dfrac{1}{2}$

Solution $25^{1/2} = x$ Write in exponential form.

$\qquad\qquad\quad 5 = x$ $25^{1/2} = \sqrt{25} = 5.$

d. $\log_{36} \sqrt[4]{6} = y$

Solution $36^y = \sqrt[4]{6}$

$\qquad\quad (6^2)^y = 6^{1/4}$ Write 36 as 6^2 and $\sqrt[4]{6}$ as $6^{1/4}$.

$\qquad\quad\ 6^{2y} = 6^{1/4}$ $(6^2)^y = 6^{2y}$.

$\qquad\qquad 2y = \dfrac{1}{4}$ Set the exponents equal to each other.

$\qquad\qquad\ y = \dfrac{1}{8}$ Solve for y.

YOUR TURN Solve.

a. $\log_b \dfrac{1}{9} = -2$ **b.** $\log_5 \dfrac{1}{25} = y$ **c.** $\log_{27} x = \dfrac{1}{3}$

If an equation containing a logarithm is not in the form $\log_b a = c$, then try using the addition or multiplication principles of equality to rewrite the equation in that form.

EXAMPLE 4 Solve. $5 - 3\log_2 x = -7$

Solution Use the addition principle of equality and multiplication principle of equality to write the equation in the form $\log_b a = c$. We can then change the equation to exponential form to solve for x.

$\qquad -3\log_2 x = -12$ Subtract 5 from both sides.

$\qquad\qquad \log_2 x = 4$ Divide both sides by -3 to isolate the logarithm.

$\qquad\qquad\qquad x = 2^4$ Write in exponential form.

$\qquad\qquad\qquad x = 16$ Simplify.

YOUR TURN Solve.

a. $9 + \log_3 x = 13$ **b.** $5\log_n 36 = 10$ **c.** $12 - 7\log_4 t = -9$

The definition of logarithms leads to the following two properties.

RULE For any real number b, where $b > 0$ and $b \neq 1$,

1. $\log_b b = 1$ **2.** $\log_b 1 = 0$

Based on the definition of a logarithm, $\log_b b = 1$ because $b^1 = b$, and $\log_b 1 = 0$ because $b^0 = 1$.

ANSWERS

a. 3   **b.** 2   **c.** 3

ANSWERS

a. 81   **b.** 6   **c.** 64

EXAMPLE 5 Find the value.

a. $\log_5 5$

Solution $\log_5 5 = 1$

b. $\log_e e$

Solution $\log_e e = 1$

c. $\log_{10} 1$

Solution $\log_{10} 1 = 0$

YOUR TURN Find the value.

a. $\log_{10} 10$ **b.** $\log_{\sqrt{3}} 1$

OBJECTIVE 3. Graph logarithmic functions. To graph logarithmic functions, which have the form $f(x) = \log_b x$, where $b > 0$ and $b \neq 1$, we will first change to exponential form so it will be easier to find ordered pairs.

PROCEDURE Graphing Logarithmic Functions

To graph a function in the form $f(x) = \log_b x$,

1. Replace $f(x)$ with y and then write the logarithm in exponential form $x = b^y$.
2. Find ordered pairs that satisfy the equation by assigning values to y and finding x.
3. Plot the ordered pairs and draw a smooth curve through the points.

EXAMPLE 6 Graph.

a. $f(x) = \log_2 x$

Solution $y = \log_2 x$ **Replace $f(x)$ with y.**
$2^y = x$ **Write in exponential form.**

Choose values for y and find x.

y	0	1	2	3	-1	-2
x	$2^0 = 1$	$2^1 = 2$	$2^2 = 4$	$2^3 = 8$	$2^{-1} = \dfrac{1}{2}$	$2^{-2} = \dfrac{1}{4}$

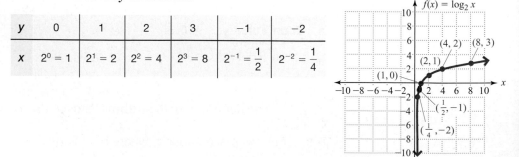

b. $f(x) = \log_{1/2} x$

Solution $y = \log_{1/2} x$ Replace $f(x)$ with y.

$$x = \left(\frac{1}{2}\right)^y$$ Write in exponential form.

Choose values for y and find x.

y	0	1	2	3	-1	-2	-3
x	$\left(\frac{1}{2}\right)^0 = 1$	$\left(\frac{1}{2}\right)^1 = \frac{1}{2}$	$\left(\frac{1}{2}\right)^2 = \frac{1}{4}$	$\left(\frac{1}{2}\right)^3 = \frac{1}{8}$	$\left(\frac{1}{2}\right)^{-1} = 2$	$\left(\frac{1}{2}\right)^{-2} = 4$	$\left(\frac{1}{2}\right)^{-3} = 8$

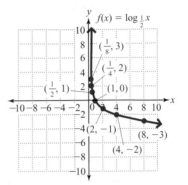

Following is a summary of the key features of the graphs of logarithmic functions.

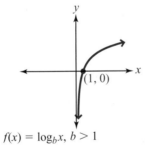

$f(x) = \log_b x,\; b > 1$

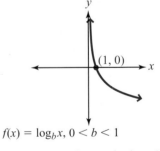

$f(x) = \log_b x,\; 0 < b < 1$

The graph passes through $(1, 0)$, approaches the y-axis, and increases. The domain is $(0, \infty)$. The range is $(-\infty, \infty)$.

The graph passes through $(1, 0)$, approaches the y-axis, and decreases. The domain is $(0, \infty)$. The range is $(-\infty, \infty)$.

ANSWER

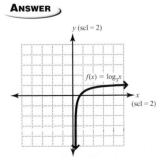

YOUR TURN   Graph.

$$f(x) = \log_3 x$$

EXAMPLE 7 The function $P = 95 - 30 \log_2 x$ models the percent, P, of students who recall the important features of a classroom lecture over time, where x is the number of days that have elapsed since the lecture was given. What percent of the students recall the important features of a lecture 8 days after it was given? (*Source: Psychology for the New Millennium*, 8th Edition, Spencer A. Rathos, Thomson Publishing Company)

Understand We are given a function that models the percent, P, of students who recall the important features of a lecture x days after it is given. We are to find the percent of students who recall the important features of a lecture 8 days after it was given.

Plan Evaluate $P = 95 - 30 \log_2 x$ where $x = 8$.

Execute $P = 95 - 30 \log_2 8$ Substitute 8 for x in $P = 95 - 30 \log_2 x$.

$\qquad\qquad P = 95 - 30(3)$ $\log_2 8 = 3$. Since $2^3 = 8$.

$\qquad\qquad P = 95 - 90$

$\qquad\qquad P = 5$ Simplify.

Answer 5% of the students remember the important features of a lecture 8 days after it is given.

Check Verify that 8 days after a lecture is given when 5% of the students recall the important features of the lecture.

$\qquad\qquad 5 = 95 - 30 \log_2 x$ Replace P with 5 in $P = 95 - 30 \log_2 x$.

$\qquad\quad -90 = -30 \log_2 x$ Subtract 95 from both sides.

$\qquad\qquad 3 = \log_2 x$ Divide both sides by -30.

$\qquad\qquad 2^3 = x$ Write in exponential form.

$\qquad\qquad 8 = x$ Simplify. It checks.

YOUR TURN Refer to Example 7.

a. Find the percent of students who remember the important features of a lecture 2 days after it was given.

b. When the number of days was decreased by a third (from 8 to 2), was the amount retained also decreased by one-third?

ANSWERS

a. 65% b. no

12.3 Exercises

For Extra Help

MyMathLab

MyMathLab   Videotape/DVT   InterAct Math   Math Tutor Center   Math XL.com

1. If $f(x) = 3^x$, what is $f^{-1}(x)$? Why?

2. If $f(x) = \log_b x$, what are the restrictions on b?

3. If $f(x) = \log_b x$, what are the restrictions on x?

4. What is the relationship between the graphs of $f(x) = 5^x$ and $g(x) = \log_5 x$? Why?

5. Why is a logarithm an exponent?

6. The $\log_b x$ is the _____ to which _____ must be raised to get _____ .

For Exercises 7–22, write in logarithmic form.

7. $2^5 = 32$

8. $4^3 = 64$

9. $10^3 = 1000$

10. $10^4 = 10{,}000$

11. $e^4 = x$

12. $e^{-2} = z$

13. $5^{-3} = \dfrac{1}{125}$

14. $6^{-3} = \dfrac{1}{216}$

15. $10^{-2} = \dfrac{1}{100}$

16. $10^{-4} = \dfrac{1}{10000}$

17. $625^{1/4} = 5$

18. $343^{1/3} = 7$

19. $\left(\dfrac{1}{4}\right)^2 = \dfrac{1}{16}$

20. $\left(\dfrac{3}{4}\right)^3 = \dfrac{27}{64}$

21. $7^{1/2} = \sqrt{7}$

22. $9^{1/3} = \sqrt[3]{9}$

For Exercises 23–38, write in exponential form.

23. $\log_3 81 = 4$

24. $\log_4 64 = 3$

25. $\log_4 \dfrac{1}{16} = -2$

26. $\log_3 \dfrac{1}{243} = -5$

27. $\log_{10} 100 = 2$

28. $\log_{10} \dfrac{1}{100} = -2$

29. $\log_e a = 5$

30. $\log_e y = -4$

31. $\log_e \frac{1}{e^4} = -4$

32. $\log_e \frac{1}{e} = -1$

33. $\log_{1/8} \frac{1}{64} = 2$

34. $\log_{1/4} \frac{1}{256} = 4$

35. $\log_{1/5} 25 = -2$

36. $\log_{1/3} 81 = -4$

37. $\log_7 \sqrt{7} = \frac{1}{2}$

38. $\log_6 \sqrt[4]{6} = \frac{1}{4}$

For Exercises 39–58, solve.

39. $\log_2 x = 3$

40. $\log_3 x = 4$

41. $\log_5 x = -2$

42. $\log_6 x = -1$

43. $\log_3 81 = y$

44. $\log_2 32 = y$

45. $\log_5 \frac{1}{5} = y$

46. $\log_4 \frac{1}{16} = y$

47. $\log_b 1000 = 3$

48. $\log_b 81 = 4$

49. $\log_m \frac{1}{16} = -4$

50. $\log_n \frac{1}{36} = -2$

51. $\log_{1/2} x = 2$

52. $\log_{1/5} x = 3$

53. $\log_{1/3} h = -5$

54. $\log_{1/6} k = -2$

55. $\log_{1/3} \frac{1}{9} = y$

56. $\log_{1/4} \frac{1}{64} = y$

57. $\log_{1/2} 64 = t$

58. $\log_{1/5} 125 = u$

For Exercises 59–70, solve.

59. $\log_2 x + 4 = 8$

60. $\log_3 x + 2 = -1$

61. $\log_{1/4} h - 2 = 1$

62. $\log_{1/2} k - 3 = -1$

63. $3 \log_b 16 = 12$

64. $-2 \log_b 81 = -8$

65. $\frac{1}{3} \log_5 c = 1$

66. $\frac{1}{4} \log_2 d = 2$

67. $3 \log_t 9 + 6 = 12$

68. $2 \log_u 125 + 8 = 14$

69. $\frac{1}{2} \log_4 m - 2 = -3$

70. $\frac{1}{3} \log_5 n - 4 = -5$

For Exercises 71–74, graph.

71. $f(x) = \log_4 x$

72. $f(x) = \log_5 x$

73. $f(x) = \log_{1/3} x$

74. $f(x) = \log_{1/4} x$

75. Why does $\log_b b = 1$ for any value of $b > 0$ and $b \neq 1$?

76. In the definition of a logarithm, $y = \log_b x$, why must $b \neq 1$?

77. Why does $\log_b 1 = 0$?

78. If $f(x) = b^x$, the domain is $(-\infty, \infty)$ and the range is $(0, \infty)$. What are the domain and range of $f(x) = \log_b x$? Why?

79. The percent of adult height attained by a 5- to 15-year-old girl can be approximated by $f(x) = 62 + 35 \log_{10}(x - 4)$, where x is the age in years and $f(x)$ is the percent. At age 14, what percent of her adult height has a girl reached?

80. The percent of adult height attained by a 5- to 15-year-old boy can be approximated by $f(x) = 29 + 48.8 \log_{10}(x + 1)$, where x is the age in years and $f(x)$ is the percent. At age 9, what percent of his adult height has a boy reached?

81. In Example 7 we learned that the percent of students who recall the important features of a lecture is given by $P = 95 - 30 \log_2 x$, where P is the percent and x is the number of days that have elapsed since the lecture was given. After how many days will 35% of the students recall the important features?

82. Using the formula from Exercise 81, find the percent of the students who recall the important features of a lecture 4 days after it was given.

Collaborative Exercises EXPLORING GRAPHS OF LOGARITHMS

Using a graphing calculator, draw the graph of $f(x) = \log x$ and $g(x) = \log(10x)$ in the window $[0, 100]$ for x and $[-2, 4]$ for y.

1. Press the [TRACE] key and using up or down arrow keys move from one graph to the other and observe the y-values of several points with the same x-values. What do you observe?

2. How is the graph of $g(x) = \log(10x)$ related to the graph of $f(x) = \log x$? Why?

3. In general, how is the graph of $g(x) = \log(kx)$ for $k > 0$ related to the graph of $f(x) = \log x$?

4. Repeat step 1 using $f(x) = \log x$ and $g(x) = \log \dfrac{x}{10}$.

5. How is the graph of $g(x) = \log \dfrac{x}{10}$ related to the graph of $f(x) = \log x$? Why?

6. In general, how is the graph of $g(x) = \log \dfrac{x}{k}$ for $k > 0$ related to the graph of $f(x) = \log x$?

REVIEW EXERCISES

[5.1]   **1.** Write $\dfrac{1}{x^6}$ as x to a negative power.

For Exercises 2–5, simplify using the rules of exponents.

[5.4]   **2.** $x^4 \cdot x^2$ **[5.4]**   **3.** $(x^3)^5$ **[5.6]**   **4.** $\dfrac{x^6}{x^3}$ **[5.6]**   **5.** $\dfrac{(x^3)^2 \cdot x^4}{x^5}$

[10.2]   **6.** Write $\sqrt[4]{x^3}$ in exponential form.

12.4   Properties of Logarithms

OBJECTIVES

1. Apply the inverse properties of logarithms.

2. Apply the product, quotient, and power properties of logarithms.

OBJECTIVE 1. Apply the inverse properties of logarithms.   Earlier, we developed two properties of logarithms, $\log_b b = 1$ and $\log_b 1 = 0$, which were based on the definition of a logarithm. The fact that $f[f^{-1}(x)] = x$ and $f^{-1}[f(x)] = x$ gives us the following properties of logarithms.

> **RULE**   **Inverse Properties of Logarithms**
>
> For any real numbers b and x, where $b > 0$, $b \neq 1$ and $x > 0$,
>
> **1.** $b^{\log_b x} = x$ **2.** $\log_b b^x = x$

To prove $b^{\log_b x} = x$, let $f(x) = b^x$ so $f^{-1}(x) = \log_b x$.

$f[f^{-1}(x)] = x$ **Composition of a function with its inverse is x.**

$f[\log_b x] = x$ **Replace $f^{-1}(x)$ with $\log_b x$.**

$b^{\log_b x} = x$ **Replace x with $\log_b x$ in $f(x)$.**

To prove $\log_b b^x = x$, use the definition of a logarithm. $\text{Log}_a b = c$ means $a^c = b$, so $\log_b b^x = x$, because $b^x = b^x$.

For example, using $b^{\log_b x} = x$ we have $3^{\log_3 8} = 8$, $6^{\log_6 x} = x$, and $e^{\log_e 4} = 4$. Using $\log_b b^x = x$, we have $\log_3 3^6 = 6$, $\log_5 5^a = a$, and $\log_e e^{x^2} = x^2$.

YOUR TURN   Find the value.

a. $8^{\log_8 4}$ **b.** $\log_3 3^{-2}$

ANSWERS

a. 4   **b.** -2

OBJECTIVE 2. **Apply the product, quotient, and power properties of logarithms.** Logarithms were invented to perform operations on very large or very small numbers. With the invention of handheld calculators, they are no longer used for this purpose. However, we still use the properties of logarithms, which are based on the fact that logarithms are exponents.

RULE **Further Properties of Logarithms**

Warning: There is no rule for the logarithms of sums or differences. The $\log_b(x + y) \neq \log_b x + \log_b y$.

Note: *When using the product and quotient rules, all the bases of all the logarithms must be the same.*

For real numbers x, y, and b, where $x > 0$, $y > 0$, $b > 0$, and $b \neq 1$,

Product Rule of Logarithms: $\log_b xy = \log_b x + \log_b y$
(The logarithm of the product of two numbers is equal to the sum of the logarithms of the numbers.)

Quotient Rule of Logarithms: $\log_b \dfrac{x}{y} = \log_b x - \log_b y$
(The logarithm of the quotient of two numbers is equal to the difference of the logarithms of the numbers.)

Power Rule of Logarithms: $\log_b x^r = r \log_b x$
(The logarithm of a number raised to a power is equal to the exponent times the logarithm of the number.)

To prove $\log_b xy = \log_b x + \log_b y$, let $M = \log_b x$ and $N = \log_b y$.

$x = b^M$ and $y = b^N$	Write each logarithmic equation in exponential form.
$xy = b^M \cdot b^N$	Multiply the left and right sides of the exponential forms.
$xy = b^{M+N}$	Add the exponents.
$\log_b xy = M + N$	Write in logarithmic form.
$\log_b xy = \log_b x + \log_b y$	Substitute $\log_b x$ for M and $\log_b y$ for N and the proof is complete.

The proof of $\log_b \dfrac{x}{y} = \log_b x - \log_b y$ is left as an exercise. The proof of $\log_b x^r = r \log_b x$ will be a collaborative exercise.

EXAMPLE 1 Use the product rule of logarithms to write the expression as a sum of logarithms.

a. $\log_{10} xyz$

Solution $\log_{10} xyz = \log_{10} x + \log_{10} y + \log_{10} z$

b. $\log_b x(x + 3)$

Solution $\log_b x(x + 3) = \log_b x + \log_b(x + 3)$ ◀ **Note:** $\log_b(x + 3) \neq \log_b x + \log_b 3.$

EXAMPLE 2 Use the product rule of logarithms in the form $\log_b x + \log_b y = \log_b xy$ to write the expression as a single logarithm.

a. $\log_3 8 + \log_3 2$

Solution $\log_3 8 + \log_3 2 = \log_3 8 \cdot 2 = \log_3 16$

b. $\log_8 5 + \log_8 x + \log_8(2x - 3)$

Solution $\log_8 5 + \log_8 x + \log_8(2x - 3) = \log_8 5x(2x - 3)$
$$= \log_8(10x^2 - 15x)$$

YOUR TURN Use the product rule of logarithms to write the expression as a sum of logarithms.

a. $\log_a 6x$ **b.** $\log_2 x(3x + 2)$

Use the product rule of logarithms to write each of the following as a single logarithm.

c. $\log_6 7 + \log_6 9$ **d.** $\log_7 2 + \log_7 x + \log_7(2x + 4)$

EXAMPLE 3 Use the quotient rule of logarithms to write the expression as a difference of logarithms.

a. $\log_5 \dfrac{5}{11}$

Solution $\log_5 \dfrac{5}{11} = \log_5 5 - \log_5 11$

$\qquad\qquad\qquad = 1 - \log_5 11$ Remember, $\log_5 5 = 1$.

b. $\log_4 \dfrac{x}{x - 5}$

Note: $\log_4(x - 5) \neq \log_4 x - \log_4 5.$

Solution $\log_4 \dfrac{x}{x - 5} = \log_4 x - \log_4 (x - 5)$

EXAMPLE 4 Use the quotient rule of logarithms in the form $\log_b x - \log_b y = \log_b \dfrac{x}{y}$ to write the expression as a single logarithm.

a. $\log_9 3 - \log_9 x$

Solution $\log_9 3 - \log_9 x = \log_9 \dfrac{3}{x}$

b. $\log_{10} x - \log_{10}(x^2 + 4)$

Solution $\log_{10} x - \log_{10}(x^2 + 4) = \log_{10} \dfrac{x}{x^2 + 4}$

ANSWERS

a. $\log_a 6 + \log_a x$
b. $\log_2 x + \log_2(3x + 2)$
c. $\log_6 63$
d. $\log_7(4x^2 + 8x)$

ANSWERS

a. $1 - \log_9 10$
b. $\log_5 x - \log_5(x + 2)$
c. 1
d. $\log_4 \dfrac{x^2 + 2}{x + 1}$

YOUR TURN Use the quotient rule of logarithms to write the expression as a difference of logarithms.

a. $\log_9 \dfrac{9}{10}$ **b.** $\log_5 \dfrac{x}{x + 2}$

Use the quotient rule of logarithms to write each of the following as a single logarithm.

c. $\log_3 15 - \log_3 5$ **d.** $\log_4(x^2 + 2) - \log_4(x + 1)$

EXAMPLE 5 Use the power rule of logarithms to write the expression as a multiple of a logarithm.

a. $\log_4 a^6$

Solution $\log_4 a^6 = 6 \log_4 a$

b. $\log_b \sqrt[4]{x^3}$

Solution $\log_b \sqrt[4]{x^3} = \log_b x^{3/4}$ Write $\sqrt[4]{x^3}$ as $x^{3/4}$.

$$= \frac{3}{4} \log_b x$$

c. $\log_b \dfrac{1}{x^3}$

Solution $\log_b \dfrac{1}{x^3} = \log_b x^{-3}$ Write $\dfrac{1}{x^3}$ as x^{-3}.

$$= -3 \log_b x$$

EXAMPLE 6 Use the power rule of logarithms in the form $r \log_b x = \log_b x^r$ to write the expression as a logarithm of a quantity to a power. Leave answers in simplest form without negative or fractional exponents.

a. $5 \log_4 x$

Solution $5 \log_4 x = \log_4 x^5$

b. $-2 \log_b 5$

Solution $-2 \log_b 5 = \log_b 5^{-2}$

$$= \log_b \frac{1}{5^2}$$ Write 5^{-2} as $\dfrac{1}{5^2}$.

$$= \log_b \frac{1}{25}$$ $5^2 = 25$.

c. $\dfrac{2}{3} \log_7 y$

Solution $\dfrac{2}{3} \log_7 y = \log_7 y^{2/3}$

$$= \log_7 \sqrt[3]{y^2}$$ Write $y^{2/3}$ as $\sqrt[3]{y^2}$.

YOUR TURN Use the power rule of logarithms to write the expression as a multiple of a logarithm.

a. $\log_a z^4$ **b.** $\log_a \sqrt[3]{x^2}$

Use the power rule of logarithms to write the expression as a logarithm of a quantity to a power.

c. $5 \log_7 x$ **d.** $-3 \log_b z$

ANSWERS

a. $4 \log_a z$ **b.** $\dfrac{2}{3} \log_a x$

c. $\log_7 x^5$ **d.** $\log_b \dfrac{1}{z^3}$

Often it is necessary to use more than one rule to simplify a logarithmic expression.

EXAMPLE 7) Write the expression as a sum or difference of multiples of logarithms.

a. $\log_b \dfrac{z^3}{yz}$

Solution $\log_b \dfrac{z^3}{yz} = \log_b z^3 - \log_b yz$ Use the quotient rule.

$= 3\log_b z - (\log_b y + \log_b z)$ **Use the power rule and product rule. Note the use of parentheses.**

$= 3\log_b z - \log_b y - \log_b z$ **Remove the parentheses.**

b. $\log_3 \sqrt{\dfrac{a^3}{b}}$

Solution $\log_3 \sqrt{\dfrac{a^3}{b}} = \log_3 \left(\dfrac{a^3}{b}\right)^{1/2}$ Write $\sqrt{\dfrac{a^3}{b}}$ as $\left(\dfrac{a^3}{b}\right)^{1/2}$.

$= \dfrac{1}{2}\log_3 \left(\dfrac{a^3}{b}\right)$ **Use the power rule.**

$= \dfrac{1}{2}(\log_3 a^3 - \log_3 b)$ **Use the quotient rule.**

$= \dfrac{1}{2}(3\log_3 a - \log_3 b)$ **Use the power rule.**

$= \dfrac{3}{2}\log_3 a - \dfrac{1}{2}\log_3 b$ **Distribute $\dfrac{1}{2}$.**

c. $\log_5 5^2 b^3$

Solution $\log_5 5^2 b^3 = \log_5 5^2 + \log_5 b^3$ Use the product rule.

$= 2\log_5 5 + 3\log_5 b$ Use the power rule.

$= 2 + 3\log_5 b$ $\log_5 5 = 1$.

EXAMPLE 8) Write the expression as a single logarithm. Leave answers in simplest form without negative or fractional exponents.

a. $4\log_b 2 - 2\log_b 3$

Solution $4\log_b 2 - 2\log_b 3 = \log_b 2^4 - \log_b 3^2$ Use the power rule.

$= \log_b \dfrac{2^4}{3^2}$ Use the quotient rule.

$= \log_b \dfrac{16}{9}$ Simplify.

b. $\frac{1}{2}(\log_2 5 - \log_2 b)$

Solution $\frac{1}{2}(\log_2 5 - \log_2 b) = \frac{1}{2}\log_2 \frac{5}{b}$ Use the quotient rule.

$$= \log_2\left(\frac{5}{b}\right)^{1/2}$$ Use the power rule.

$$= \log_2\sqrt{\frac{5}{b}}$$ Write $\left(\frac{5}{b}\right)^{1/2}$ as $\sqrt{\frac{5}{b}}$.

c. $\log_a(x + 2) + \log_a(x - 3)$

Solution $\log_a(x + 2) + \log_a(x - 3) = \log_a(x + 2)(x - 3)$ Apply $\log_b xy = \log_b x + \log_b y$.

$$= \log_a(x^2 - x - 6)$$ Multiply.

ANSWERS

a. $\log_a \frac{16}{x^3}$ **b.** $\log_a \sqrt[3]{xy^2}$

c. $\log_6(x^2 - 2x)$

YOUR TURN   Write the expression as a single logarithm.

a. $2\log_a 4 - 3\log_a x$ **b.** $\frac{1}{3}(\log_a x + 2\log_a y)$ **c.** $\log_6 x + \log_6(x - 2)$

12.4 Exercises

1. Why does $b^{\log_b x} = x$?

2. How can you use the facts that $\log_b x^r = r\log_b x$ and $\log_b b = 1$ to show that $\log_3 81 = 4$?

3. $\log_b(x + y) = $ _____.

4. Why does $\log_a \sqrt[3]{x^4} = \frac{4}{3}\log_a x$?

For Exercises 5–16, find the value.

5. $8^{\log_8 2}$ **6.** $5^{\log_5 6}$ **7.** $a^{\log_a r}$ **8.** $b^{\log_b a}$

9. $a^{\log_a 4x}$ **10.** $b^{\log_b 5a}$ **11.** $\log_3 3^5$ **12.** $\log_7 7^3$

13. $\log_b b^a$ **14.** $\log_c c^x$ **15.** $\log_a a^{7x}$ **16.** $\log_b b^{6y}$

For Exercises 17–24, use the product rule to write the expression as a sum of logarithms.

17. $\log_2 5y$

18. $\log_3 4z$

19. $\log_a pq$

20. $\log_b rs$

21. $\log_4 mnp$

22. $\log_4 pqr$

23. $\log_a x(x-5)$

24. $\log_b x(2x+6)$

For Exercises 25–36, use the product rule to write the expression as a single logarithm.

25. $\log_3 5 + \log_3 8$

26. $\log_6 4 + \log_6 7$

27. $\log_4 2 + \log_4 8$

28. $\log_9 3 + \log_9 27$

29. $\log_a 7 + \log_a m$

30. $\log_b 2 + \log_b n$

31. $\log_4 a + \log_4 b$

32. $\log_6 r + \log_6 m$

33. $\log_a 2 + \log_a x + \log_a(x+5)$

34. $\log_a 4 + \log_a y + \log_a(y-5)$

35. $\log_4(x+1) + \log_4(x+3)$

36. $\log_6(x-3) + \log_6(x+3)$

For Exercises 37–46, use the quotient rule to write the expression as a difference of logarithms.

37. $\log_2 \dfrac{7}{9}$

38. $\log_4 \dfrac{5}{12}$

39. $\log_a \dfrac{x}{5}$

40. $\log_b \dfrac{y}{3}$

41. $\log_a \dfrac{a}{b}$

42. $\log_b \dfrac{a}{b}$

43. $\log_a \dfrac{x}{x-3}$

44. $\log_b \dfrac{y}{2y+5}$

45. $\log_4 \dfrac{2x-3}{4x+5}$

46. $\log_6 \dfrac{4x-1}{x^2+3}$

For Exercises 47–58, use the quotient rule to write the expression as a single logarithm.

47. $\log_6 24 - \log_6 3$

48. $\log_8 12 - \log_8 3$

49. $\log_2 24 - \log_2 12$

50. $\log_3 48 - \log_3 16$

51. $\log_a x - \log_a 3$

52. $\log_a r - \log_a 5$

53. $\log_4 p - \log_4 q$

54. $\log_a m - \log_a n$

55. $\log_b x - \log_b(x-4)$

56. $\log_b y - \log_b(y-5)$

57. $\log_x(x^2-x) - \log_x(x-1)$

58. $\log_a(a^2+2a) - \log_a(a+2)$

For Exercises 59–70, use the power rule to write the expression as a multiple of a logarithm.

59. $\log_7 4^3$

60. $\log_3 5^4$

61. $\log_a x^7$

62. $\log_b y^8$

63. $\log_a \sqrt{3}$

64. $\log_a \sqrt{6}$

65. $\log_3 \sqrt[3]{x^2}$

66. $\log_4 \sqrt[5]{a^3}$

67. $\log_a \dfrac{1}{6^2}$

68. $\log_a \dfrac{1}{5^4}$

69. $\log_a \dfrac{1}{y^2}$

70. $\log_a \dfrac{1}{x^5}$

For Exercises 71–82, use the power rule to write the expression as a logarithm of a quantity to a power. Leave answers in simplest form without negative or fractional exponents.

71. $4 \log_3 5$

72. $5 \log_3 4$

73. $-3 \log_2 x$

74. $-4 \log_3 y$

75. $\dfrac{1}{2} \log_5 4$

76. $\dfrac{1}{3} \log_3 8$

77. $\dfrac{3}{4} \log_a x$

78. $\dfrac{5}{6} \log_b y$

79. $\dfrac{2}{3} \log_a 8$

80. $\dfrac{3}{4} \log_a 16$

81. $-\dfrac{1}{2} \log_3 x$

82. $-\dfrac{1}{3} \log_2 y$

For Exercises 83–94, write the expression as the sum or difference of multiples of logarithms.

83. $\log_a \dfrac{x^3}{y^4}$

84. $\log_a \dfrac{x^6}{y^5}$

85. $\log_3 a^4 b^2$

86. $\log_7 m^3 n^5$

87. $\log_a \dfrac{xy}{z}$

88. $\log_a \dfrac{pq}{r}$

89. $\log_x \dfrac{a^2}{bc^3}$

90. $\log_x \dfrac{c^2}{m^2 n}$

91. $\log_4 \sqrt{\dfrac{x^3}{y}}$

92. $\log_5 \sqrt[3]{\dfrac{a^5}{b^2}}$

93. $\log_a \sqrt{\dfrac{x^2 y}{z^3}}$

94. $\log_3 \sqrt[3]{\dfrac{c^2 d^3}{m^4}}$

For Exercises 95–108, write the expression as a single logarithm.

95. $3 \log_3 2 - 2 \log_3 4$

96. $2 \log_4 5 - 3 \log_4 2$

97. $4 \log_b x + 3 \log_b y$

98. $5 \log_b a + 4 \log_b c$

99. $\dfrac{1}{2}(\log_a 5 - log_a 7)$

100. $\dfrac{1}{2}(\log_4 6 - \log_4 5)$

101. $\frac{2}{3}(\log_a x^2 + \log_a y)$

102. $\frac{3}{4}(\log_a m^3 + \log_a n^2)$

103. $\log_b x + \log_b(3x - 2)$

104. $\log_a y + \log_a(2y - 4)$

105. $3\log_a(x - 2) - 4\log_a(x + 1)$

106. $2\log_b(x + 4) - 3\log_b(2x - 1)$

107. $2\log_a x + 4\log_a z - 3\log_a w - 6\log_a u$

108. $5\log_a b + 2\log_a c - 4\log_a d - \log_a e$

Collaborative Exercises   PROVING THE QUOTIENT AND POWER RULES

1. Using the proof of $\log_b xy = \log_b x + \log_b y$ as a guide, prove $\log_b \dfrac{x}{y} = \log_b x - \log_b y$.

2. The proof that $\log_b x^r = r\log_b x$ was not given. For r a positive integer, the following use of the product rule suggests (but does not prove) this rule:

$$\log_b x^3 = \log_b x \cdot x \cdot x = \log_b x + \log_b x + \log_b x = 3\log_b x$$

Using the preceding as a guide, prove that $\log_b x^5 = 5\log_b x$.

3. Complete the following steps to prove $\log_b x^r = r\log_b x$ for all values of r:
Let $N = \log_b x$.
Rewrite in exponential form.
Raise both sides to the r power.
Simplify the exponents, if necessary.
Rewrite in logarithmic form.
Substitute for N.

REVIEW EXERCISES

For Exercises 1 and 2, simplify.

[5.4] **1.** $(10^{9.5})(10^{-12})$

[5.6] **2.** $\dfrac{10^{-3}}{10^{-12}}$

[12.2] **3.** If $10,000 is deposited at 6% compounded quarterly for five years, how much will be in the account?

[12.3] **4.** Write $10^{1.6990} = 50$ in logarithmic form.

[12.3] *For Exercises 5 and 6, write in exponential form.*

5. $\log_{10} 45 = 1.6532$

6. $\log_e 0.25 = -1.3863$

12.5 Common and Natural Logarithms

OBJECTIVES

1. Define common logarithms and evaluate them using a calculator.
2. Solve applications using common logarithms.
3. Define natural logarithms and evaluate them using a calculator.
4. Solve applications using natural logarithms.

OBJECTIVE 1. Define common logarithms and evaluate them using a calculator. Of all the possible bases of logarithms, two are the most useful. As previously mentioned, logarithms were invented to do computations on very large and very small numbers. Since ours is a base-10 number system, base-10 logarithms were commonly used for this purpose. Consequently, base-10 logarithms are called **common logarithms** and $\log_{10} x$ is written as $\log x$, where the base 10 is understood. Base-10 logarithms are found in engineering, economics, social sciences, and in the natural sciences.

DEFINITION *Common logarithms:* Logarithms with a base of 10. $\text{Log}_{10}\, x$ is written as $\log x$.

Connection Since logarithmic and exponential functions are inverse functions, if $f(x) = 10^x$, then $f^{-1}(x) = \log x$ and if $g(x) = \log x$, then $g^{-1}(x) = 10^x$.

To evaluate common logarithms, we use a calculator. We will round all results to four places.

Calculator TIPS

On most calculators, to evaluate common logarithms, press the $\boxed{\text{LOG}}$ *key, enter the number, and then press* $\boxed{\text{ENTER}}$.

EXAMPLE 1 Use a calculator to approximate each common logarithm to four decimal places.

a. $\log 23$

Solution $\log 23 \approx 1.3617$

b. $\log 0.00236$

Solution $\log 0.00236 \approx -2.6271$

Connection Remember that $\log 23 \approx 1.3617$ means $10^{1.3617} \approx 23$, which provides a way to check. Because we rounded the decimal value, evaluating $10^{1.3617}$ may not give exactly 23.

YOUR TURN Use a calculator to approximate each common logarithm to four decimal places.

a. $\log 436$ **b.** $\log 0.0724$

ANSWERS

a. 2.6395 **b.** -1.1403

OBJECTIVE 2. Solve applications using common logarithms. Common logarithms can be used to calculate sound intensity and runway length.

EXAMPLE 2

a. Sound intensity can be measured in watts per unit of area or, more commonly, in decibels. The function $d = 10 \log \frac{I}{I_0}$ is used to calculate sound intensity, where d represents the intensity in decibels, I represents the intensity in watts per unit of area, and I_0 represents the faintest audible sound to the average human ear, which is 10^{12} watts per square meter. A motorcycle has a sound intensity of about $10^{-2.5}$ watts per square meter. Find the decibel reading for the motorcycle.

Understand We are given the function $d = 10 \log \frac{I}{I_0}$, values for I and I_0, and we are to find the decibel reading, which is d.

Plan Using $d = 10 \log \frac{I}{I_0}$, substitute $10^{-2.5}$ for I and 10^{-12} for I_0 and then solve for d.

Execute

$d = 10 \log \dfrac{10^{-2.5}}{10^{-12}}$ Substitute for I and I_0 in $d = 10 \log \frac{I}{I_0}$.

$d = 10 \log 10^{\,9.5}$ Subtract exponents $[-2.5 - (-12) = 9.5]$.

$d = 10(9.5)$ Use $\log_b b^x = x$.

$d = 95$ Simplify.

Answer The motorcycle has a decibel reading of 95 decibels.

Check If the sound intensity is 95 decibels, verify that I is $10^{-2.5}$.

Note: *The abbreviation for decibels is dB.*

$95 = 10 \log \dfrac{I}{10^{-12}}$ Substitute 95 for d.

$9.5 = \log \dfrac{I}{10^{-12}}$ Divide both sides by 10 to isolate the logarithm.

$10^{9.5} = \dfrac{I}{10^{-12}}$ Write in exponential form.

$10^{9.5} \cdot 10^{-12} = I$ Multiply both sides by 10^{-12}.

$10^{-2.5} = I$ Simplify by adding the exponents. It checks.

b. The minimum length of airport runway needed for a plane to take off is related to the weight of the plane. For some planes, the minimum runway length may be modeled by the function $y = 3 \log x$, where x is the plane's weight in thousands of pounds and y is the length of the runway in thousands of feet. Find the minimum length of a runway needed by a Boeing 737 whose maximum takeoff weight is 174,200 pounds.

Understand We are given the function $y = 3 \log x$, where y represents the runway length in thousands of feet required for a plane with a takeoff weight of x, in thousands of pounds. We are given that a plane's takeoff weight is 174,200 pounds and must find the runway length required for that plane.

Plan Since x is the plane's weight in thousands of pounds, we first need to divide 174,200 by 1000 to find the value of x corresponding to 174,200 pounds. We can then use $y = 3 \log x$.

Execute $x = \dfrac{174{,}200}{1{,}000} = 174.2$

$y = 3 \log 174.2$ **Substitute 174.2 for x in $y = 3 \log x$.**

$y = 6.723$ **Evaluate using a calculator.**

Answer Since y is in thousands of feet, the minimum runway length is $(6.723)(1000) = 6723$ ft.

Check Verify that a runway length of 6723 feet corresponds to a plane whose takeoff weight is 174,200. Since y is in thousands of feet, we first divide 6723 by 1000 to find the value of y: $y = 6723/1000 = 6.723$.

$6.723 = 3 \log x$ **Substitute 6.723 for y in $y = 3 \log x$.**

$2.241 = \log x$ **Divide both sides by 3 to isolate the logarithm.**

$10^{2.241} = x$ **Write in exponential form.**

$174.2 \approx x$ **Simplify. We rounded to the nearest tenth.**

Since x is in thousands of feet, $174.2(1000) = 174{,}200$ ft., which checks.

YOUR TURN

   a. The sound intensity of a rock band often exceeds $10^{-0.5}$ watts per square meter. Find the decibel rating of this band.

   b. Find the minimum runway length needed for a B-52 Stratofortress whose maximum takeoff weight is 488,000 pounds.

OBJECTIVE 3. Define natural logarithms and evaluate them using a calculator. The number e is an irrational number whose approximate value is 2.7182818285. It is a universal constant like π. In the natural sciences, base-e logarithms are much more prevalent than base-10. Since base-e logarithms occur in so many "natural" situations, they are called **natural logarithms**. The notation for $\log_e x$ is $\ln x$, which is read "el en x."

DEFINITION *Natural logarithms:* Base-e logarithms are called natural logarithms and $\log_e x$ is written as $\ln x$. Note that $\ln e = 1$.

Calculator TIPS

On most calculators, to evaluate natural logarithms, press the [ln] *key, enter the number, and then press* (ENTER).

ANSWERS

a. $d = 115$ **b.** 8065 ft.

To find natural logarithms, we use a calculator.

EXAMPLE 3 Use a calculator to approximate each natural logarithm to four decimal places.

   a. $\ln 83$

   Solution $\ln 83 \approx 4.4188$

   b. $\ln 0.0055$

   Solution $\ln 0.0055 \approx -5.2030$

> **Connection** In Example 3, $\ln 83 \approx 4.4188$ means that $e^{4.4188} \approx 83$. Similarly, $\ln 0.0055 = -5.2030$ means that $e^{-5.2030} \approx 0.0055$.

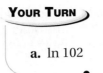

YOUR TURN Use a calculator to approximate each natural logarithm to four decimal places.

    a. ln 102   **b.** ln 0.0573

OBJECTIVE 4. Solve applications using natural logarithms. If money is deposited into an account and the interest is compounded continuously, then the time t (in years) that it would take an investment of P dollars to grow into A dollars at an interest rate r (written as a decimal) is given by $t = \dfrac{1}{r} \ln \dfrac{A}{P}$.

EXAMPLE 4 An amount of $5000 is deposited into an account earning 5% annual interest compounded continuously. How many years would it take until the account has reached $10,000?

Understand We are to find the time it takes for $5000 to grow to $10,000 if it is compounded continuously at 5%.

Plan In $t = \dfrac{1}{r} \ln \dfrac{A}{P}$, replace P with 5000, r with 0.05, A with 10,000, and then simplify.

Execute $t = \dfrac{1}{0.05} \ln \dfrac{10{,}000}{5000}$ Substitute for P, r, and A.

             $t = 20 \ln 2$ Divide

             $t \approx 13.86$ Multiply using a calculator.

Answer It would take about 13.86 years for $5000 to grow to $10,000 if it is compounded continuously.

Check We can use the formula to verify that if $5000 is compounded continuously for 13.86 years, the amount will be $10,000.

$$13.86 = \frac{1}{0.05} \ln \frac{A}{5000}$$ Substitute for t, r, and P.

$$0.693 = \ln \frac{A}{5000}$$ Multiply both sides by 0.05.

$$0.693 = \ln A - \ln 5000$$ Use $\log_b \dfrac{x}{y} = \log_b x - \log_b y$.

$$0.693 + \ln 5000 = \ln A$$ Add ln 5000 to both sides.

$$9.21 \approx \ln A$$ Add 0.693 and the approximate value of ln 5000.

$$e^{9.21} \approx A$$ Write in exponential form.

$$9996.6 \approx A$$ Calculate using a calculator.

Because 13.86 is not the exact time and we rounded the sum of 0.693 and ln 5000, it is reasonable to expect that the amount would be very close to but not exactly $10,000, so $9996.60 is reasonable.

ANSWERS

a. 4.6250   **b.** −2.8595

ANSWER

11.5 yr.

YOUR TURN How long would it take $2000 to grow into $5000 at 8% interest if the interest is compounded continuously?

12.5 Exercises

1. If $y = \log_b x$, what values are allowed for b? Why?

2. Is $\log x$ positive or negative for $0 < x < 1$? Why?

3. Is $\ln x$ positive or negative for $x > 1$? Why?

4. Give the symbol and meaning for **a.** common logarithms, **b.** natural logarithms.

5. If $e^{0.91629} = 2.5$, find $\ln 2.5$.

6. Without using a calculator, find the exact value of $\log_{10} 10^{\sqrt{5}}$.

For Exercises 7–26, use a calculator to approximate each logarithm to four decimal places.

7. $\log 64$

8. $\log 82$

9. $\log 0.0067$

10. $\log 0.00087$

11. $\log 247.8$

12. $\log 785.4$

13. $\log(1.5 \times 10^4)$

14. $\log(5.7 \times 10^7)$

15. $\log(1.6 \times 10^{-6})$

16. $\log(7.5 \times 10^{-5})$

17. $\ln 9.34$

18. $\ln 5.33$

19. $\ln 79.2$

20. $\ln 765.4$

21. $\ln 0.034$

22. $\ln 0.923$

23. $\ln(5.4 \times e^4)$

24. $\ln(7.3 \times e^6)$

25. $\log e$

26. $\ln 10$

27. Use your calculator to find $\log 0$. What happened? Why?

28. Use your calculator to find $\ln -1$. What happened? Why?

For Exercises 29–40, find the exact value of each logarithm using $\log_b b^x = x$.

29. $\log 100$

30. $\log 1000$

31. $\log \dfrac{1}{10}$

32. $\log \dfrac{1}{1000}$

33. $\log \sqrt[3]{10}$

34. $\log \sqrt{10}$

35. $\log 0.001$

36. $\log 0.00001$

37. $\ln e^3$

38. $\ln e^5$

39. $\ln \sqrt{e}$

40. $\ln \sqrt[5]{e}$

For Exercises 41–44, use the formula $d = 10 \log \dfrac{I}{I_0}$, where $I_0 = 10^{-12}$ watts/m².

41. The sound intensity of a firecracker is 10^{-3} watts per square meter. What is the decibel reading for the firecracker?

42. The sound intensity of a race car is 10^{-1} watts per square meter. What is the decibel reading for the race car?

43. If a noisy office has a decibel reading of 60, what is the sound intensity?

44. If loud thunder has a decibel reading of 90, what is the sound intensity?

For Exercises 45–48, use the following:

In chemistry, the pH of a substance determines whether it is a base (pH > 7) or an acid (pH < 7). To find the pH of a solution, we use the formula pH = $-\log [H_3O^+]$, where $[H_3O^+]$ is the hydronium ion concentration in moles per liter. Note that the pH is unitless.

45. Find the pH of vinegar if $[H_3O^+] = 1.6 \times 10^{-3}$ moles per liter. (*Source: CRC Handbook of Chemistry and Physics*, 62nd Edition, CRC Press, 1981)

46. Find the pH of maple syrup if $[H_3O^+] = 2.3 \times 10^{-7}$ moles per liter. (*Source: CRC Handbook of Chemistry and Physics*, 62nd Edition, CRC Press, 1981)

47. Find the hydronium ion concentration of sauerkraut, which has a pH of 3.5.

48. Find the hydronium ion concentration of blood, which has a pH of 7.4.

For Exercises 49 and 50, use the formula $t = \dfrac{1}{r} \ln \dfrac{A}{P}$.

49. Find how long it will take $2000 to grow to $5000 at 4% interest if the interest is compounded continuously.

50. Find how long it will take $5000 to grow to $8000 at 5% interest if the interest is compounded continuously.

51. The magnitude of an earthquake is given by the Richter scale, whose formula is $R = \log \dfrac{I}{I_0}$, where I is the intensity of the earthquake and I_0 is the intensity of a minimal earthquake and is used for comparison purposes. The 1906 San Francisco earthquake had a magnitude of 7.8, and the 1964 Alaska earthquake had a magnitude of 8.4. Compare the intensity of the two earthquakes. (*Hint:* Express the intensity of each in terms of I_0.)

52. Using the formula from Exercise 51, compare the intensity of the 1949 Queen Charlotte Islands earthquake, whose magnitude was 8.1, with the 1700 earthquake at Cascadia, whose magnitude was estimated at 9.0.

53. During an earthquake, energy is released in various forms. The amount of energy radiated from the earthquake as seismic waves is given by $\log E_s = 11.8 + 1.5\,M$, where E_s is measured in ergs and M is the magnitude of the earthquake as given by the Richter scale. Vancouver Island had an earthquake whose magnitude was 7.3. How much energy was released in the form of seismic waves?

54. Using the formula from Exercise 53, find the energy released in the form of seismic waves from the Double Springs Flat earthquake of 1994, whose magnitude was 6.1 on the Richter scale.

55. Since 1995, the number of adult deaths from AIDS in the United States has been declining and can be approximated by $y = 49{,}971.5 - 21{,}298.87\ln x$, where y is the number of deaths and $x = 1$ corresponds to 1995. (*Source:* CDC Division of HIV/AIDS Prevention, *Surveillance Report*, Vol. 13, No. 2)

  a. Estimate the number of adult AIDS deaths in 2000.

  b. Find the year when the number of adult deaths by AIDS was 26,572.

56. The number of adult deaths by AIDS in the United States as a percent of the number of diagnosed cases peaked in 1995 at 73.4%. Since then, the percent has declined and can be approximated by $y = (0.7216 - 0.1911 \ln x)100$, where y is the percent of deaths and $x = 1$ corresponds to 1995. (*Data Source:* CDC Division of HIV/AIDS Prevention, *Surveillance Report*, Vol. 13, No. 2)

  a. Find the percent of deaths among diagnosed cases in 1999.

  b. In what year did 35% of the diagnosed cases of AIDS result in death?

57. Using the formula from Example 2b, $y = 3 \log x$, find the minimum runway length needed for a Boeing 717 whose maximum weight is 110,000 pounds at takeoff.

58. Walking speeds in various cities is a function of the population, since as populations increase, so does the pace of life. Average walking speeds can be modeled by the function $W = 0.35 \ln P + 2.74$, where P is the population, in thousands, and W is the walking speed in feet per second. Find the average walking speed in Chicago, whose population was approximately 8,500,000 in 2002.

REVIEW EXERCISES

For Exercises 1–4, solve.

[2.3] **1.** $3x - (7x + 2) = 12 - 2(x - 4)$

[6.6] **2.** $x^2 + 2x = 15$

[7.6] **3.** $\dfrac{5}{x} + \dfrac{3}{x + 1} = \dfrac{23}{3x}$

[10.6] **4.** $\sqrt{5x - 1} = 7$

5. Write $\log_3(2x + 1) - \log_3(x - 1)$ as a single logarithm.

6. Write $\log_3(2x + 5) = 2$ in exponential form.

12.6 Exponential and Logarithmic Equations with Applications

OBJECTIVES

1. Solve equations that have variables as exponents.

2. Solve equations containing logarithms.

3. Solve applications involving exponential and logarithmic functions.

4. Use the change-of-base formula.

In Section 12.2, we solved equations that could be put in the form $b^x = b^y$. For example, if $3^x = 81$, then $3^x = 3^4$, so $x = 4$. In order to use this form, we had to write both sides of the equation as the same base raised to a power. In this section, we will solve equations like $3^x = 16$, which is the same as $3^x = 2^4$, where the bases are not the same. To solve this and other exponential and logarithmic equations, we need the following properties.

RULE **Properties for Solving Exponential and Logarithmic Equations**

For any real numbers b, x, and y, where $b > 0$ and $b \neq 1$,

1. If $b^x = b^y$, then $x = y$.
2. If $x = y$, then $b^x = b^y$.
3. For $x > 0$ and $y > 0$, if $\log_b x = \log_b y$, then $x = y$.
4. For $x > 0$ and $y > 0$, if $x = y$, then $\log_b x = \log_b y$.
5. For $x > 0$, if $\log_b x = y$, then $b^y = x$.

These properties are true because the exponential and logarithmic functions are one-to-one.

OBJECTIVE 1. Solve equations that have variables as exponents. To solve equations that have variables as exponents, we use property 4, which says that if two positive numbers are equal, then so are their logarithms.

EXAMPLE 1 Solve. $5^x = 16$

Solution $\log 5^x = \log 16$ Use if $x = y$, then $\log_b x = \log_b y$ (property 4).

$x \log 5 = \log 16$ Use $\log_b x^r = r \log_b x$.

$x = \dfrac{\log 16}{\log 5}$ Divide both sides of the equation by log 5.

The exact solution is $x = \dfrac{\log 16}{\log 5}$. Using a calculator, we find $x \approx 1.7227$ correct to four decimal places.

Check $5^x = 16$

$5^{1.7227} = 16$ Substitute 1.7727 for x.

$15.9998 \approx 16$ The answer is correct.

In Example 1, we took the common logarithm of both sides, but we could have used natural logarithms (or any other base) instead. If one side of the equation contains a power of e, natural logs are preferred so that we can use the fact that $\log_e e^x = x$ or, more simply, $\ln e^x = x$.

EXAMPLE 2 Solve. $e^{4x} = 23$

Solution $\ln e^{4x} = \ln 23$ Use if $x = y$, then $\log_b x = \log_b y$.

$4x = \ln 23$ Use $\log_b b^x = x$.

$x = \dfrac{\ln 23}{4}$ Divide both sides by 4.

$x \approx 0.7839$

We will leave the check to the reader.

YOUR TURN Solve the following for x. Round answers to four decimal places.

   a. $6^{2x} = 42$ **b.** $e^{3x} = 5$

OBJECTIVE 2. Solve equations containing logarithms. Now that we have explored additional properties of logarithms, we can modify the procedure for solving equations with logarithms that we learned in Section 12.3.

ANSWERS

a. 1.0430   **b.** 0.5365

To solve equations containing logarithms, use the properties of logarithms to simplify each side of the equation and then use one of the following.

If the simplification results in an equation in the form $\log_b x = \log_b y$, use the fact that $x = y$, and then solve for the variable.

If the simplification results in an equation in the form $\log_b x = y$, write the equation in exponential form, $b^y = x$, and then solve for the variable (as we did in section 12.3).

EXAMPLE 3 Solve.

a. $\log_5 x + \log_5(x + 3) = \log_5 4$

Solution $\log_5 x(x + 3) = \log_5 4$ Use $\log_b xy = \log_b x + \log_b y$ to simplify the left side.

$$x(x + 3) = 4$$ The equation is in the form $\log_b x = \log_b y$, so $x = y$.

$$x^2 + 3x = 4$$ Simplify.

$$x^2 + 3x - 4 = 0$$ Write in $ax^2 + bx + c = 0$ form.

$$(x + 4)(x - 1) = 0$$ Factor.

$$x + 4 = 0 \quad \text{or} \quad x - 1 = 0$$ Set each factor equal to 0.

$$x = -4 \quad \text{or} \quad x = 1$$ Possible solutions.

If -4 is substituted into the original equation, we have $\log_5(-4) + \log_5(-1) = \log_5 4$, but logarithms are defined for positive numbers only. So $x = -4$ is not a solution. A check will show that $x = 1$ is a solution.

b. $\log_3(2x + 5) = 2$

Solution $3^2 = 2x + 5$ The equation is in the form $\log_b x = y$, so write it in exponential form, $b^y = x$.

$$9 = 2x + 5$$

$$4 = 2x$$

$$2 = x$$ Solve for x.

We will let the reader check that $x = 2$ is a solution.

c. $\log_2(5x + 1) - \log_2(x - 1) = 3$

Solution $\log_2 \dfrac{5x + 1}{x - 1} = 3$ Use $\log_b \dfrac{x}{y} = \log_b x - \log_b y$ to simplify the left side.

$$\dfrac{5x + 1}{x - 1} = 2^3$$ The equation is in the form $\log_b x = y$, so $b^y = x$.

$$5x + 1 = 8x - 8$$ Multiply both sides by $x - 1$.

$$9 = 3x$$ Isolate the x term.

$$x = 3$$ Solve for x.

We will let the reader check that $x = 3$ is a solution.

YOUR TURN Solve.

a. $\ln x + \ln(x + 2) = \ln 8$

b. $\log_3(4x + 2) - \log_3(x - 2) = 2$

OBJECTIVE 3. Solve applications involving exponential and logarithmic functions. A wide variety of problems from business and the sciences can be solved with exponential or logarithmic functions. Earlier, we used the formula for compound interest, $A = P\left(1 + \dfrac{r}{n}\right)^{nt}$, to find A when given P, r, n, and t. Using logarithms, it is also possible to find t when given A, P, r, and n.

EXAMPLE 4 How long will it take $6000 invested at 6% interest compounded quarterly to grow to $10,000?

Understand   We are given $A = 10,000$, $P = 6000$, $r = 0.06$, and $n = 4$.

Plan   Use $A = P\left(1 + \dfrac{r}{n}\right)^{nt}$.

Execute

$$10,000 = 6000\left(1 + \frac{.06}{4}\right)^{4t} \qquad \text{Substitute for } A, P, r, \text{ and } n.$$

$$\frac{5}{3} = (1.015)^{4t} \qquad \begin{array}{l}\text{Divide both sides by 6000 and simplify inside}\\\text{the parentheses.}\end{array}$$

$$\log \frac{5}{3} = \log 1.015^{4t} \qquad \text{Use if } x = y, \text{ then } \log_b x = \log_b y.$$

$$\log \frac{5}{3} = 4t \log 1.015 \qquad \text{Use } \log_b x^r = r \log_b x.$$

$$\frac{\log \dfrac{5}{3}}{4 \log 1.015} = t \qquad \text{Divide both sides by } 4 \log 1.015$$

$$8.58 \approx t \qquad \text{Evaluate using a calculator.}$$

Answer   It would take about 8.58 years.

Check   We can use the formula to verify that if $6000 is compounded quarterly for 8.58 years, the amount will be $10,000.

$$A = 6000\left(1 + \frac{0.06}{4}\right)^{4(8.58)} \qquad \text{Substitute for } P, r, n, \text{ and } t.$$

$$A = 10,001.52 \qquad \text{Calculate.}$$

Because 8.58 is not the exact time, it is reasonable to expect that the amount would be very close to, but not exactly, $10,000, so $10,001.52 is reasonable.

Many banks compound interest continuously. The formula for interest compounded continuously is $A = Pe^{rt}$, where A is the amount in the account, P is the amount deposited, r is the interest rate as a decimal, and t is the time in years.

EXAMPLE 5 If $5000 is deposited into an account at 5% interest compounded continuously, how much will be in the account after 9 years?

Understand   We are given $P = \$5000$, $r = 0.05$, and $t = 9$ and are asked to find A.

Plan   Use $A = Pe^{rt}$.

Execute $A = 5000e^{0.05(9)}$ Substitute for P, r, and t.

$\qquad\qquad A = 5000e^{0.45}$

$\qquad\qquad A = \$7841.56$ Evaluate using a calculator.

Answer   There will be $7841.56 in the account.

Check Use the formula $t = \dfrac{1}{r} \ln \dfrac{A}{P}$ that we learned in Example 4 of Section 12.5 to verify that it takes 9 years for $5000 to grow to $7841.56 if it is compounded continuously at 5%.

$$t = \frac{1}{0.05} \ln \frac{7841.56}{5000} \qquad \text{Substitute for } r, A, \text{ and } P.$$

$$t \approx 9 \qquad \text{Calculate. It checks.}$$

Connection Solving $A = Pe^{rt}$ for t gives the formula $t = \dfrac{1}{r} \ln \dfrac{A}{P}$.

$$\frac{A}{P} = e^{rt} \qquad \text{Divide both sides by } P \text{ to isolate } e^{rt}.$$

$$\log_e \frac{A}{P} = rt \qquad \text{Write in log form.}$$

$$\frac{1}{r} \ln \frac{A}{P} = t \qquad \text{Divide both sides by } r \text{ and by definition, } \log_e \frac{A}{P} \text{ is } \ln \frac{A}{P}.$$

EXAMPLE 6 Since the 1970s, Orlando, Florida, has been one of the fastest-growing cities in the United States. The following table shows the population of Orlando for selected years.

Year	Population in millions
1975	0.6896
1980	0.805
1985	0.996
1990	1.225
1995	1.428
2000	1.645
2003	1.803

The data can be approximated by $y = 0.6965e^{0.0344t}$, where y is the population in millions and t is the number of years after 1975. (*Source:* U.S. Bureau of the Census)

a. Assuming the population continues to grow in the same manner, use the model to estimate the population of Orlando in 2010.

Understand We are given the function $y = 0.6965e^{0.0344t}$, where y represents the population of Orlando t years after 1975.

Plan Since t represents the number of years after 1975, we first subtract 1975 from 2010 to determine the value of t that corresponds to 2010. We can then substitute that value into the function.

Execute $t = 2010 - 1975 = 35.$

$$y = 0.6965e^{0.0344t} \qquad \text{Formula.}$$

$$y = 0.6965e^{0.0344(35)} \qquad \text{Substitute 35 for } t.$$

$$y = 2.322 \qquad \text{Evaluate using a calculator.}$$

Answer Since y is in millions, the estimated population is 2,322,000 in 2010.

Check Use $y = 0.6965e^{0.0344t}$ to verify that if the population is 2,322,000 ($y = 2.322$), then the year is 2010 ($t = 35$).

$$2.322 = 0.6965e^{0.0344t} \qquad \text{Substitute 2.322 for } y.$$

$$\frac{2.322}{0.6965} = e^{0.0344t} \qquad \text{Divide both sides by 0.6965.}$$

$$\ln\frac{2.322}{0.6965} = \ln e^{0.0344t} \qquad \text{Use if } x = y, \text{ then } \log_b x = \log_b y.$$

$$\ln\frac{2.322}{0.6965} = 0.0344t \qquad \text{Use } \log_b b^x = x.$$

$$\frac{\ln\frac{2.322}{0.6965}}{0.0344} = t \qquad \text{Divide both sides by 0.0344.}$$

$$35 \approx t \qquad \text{Calculate.}$$

Note: *Our manipulations suggest the following formula for calculating t given y:*

$$t = \frac{1}{0.0344}\ln\frac{y}{0.6965}$$

b. Find the year in which the population will be 2,500,000.

Understand We are to find t given the population.

Plan Since y represents the population in millions, we must first divide 2,500,000 by 1,000,000 to find the value of y that corresponds to 2,500,000. We can then use $y = 0.6965e^{0.0344t}$ and follow the same steps as in our check for part a. Or, as we will do here, use the formula $t = \dfrac{1}{0.0344}\ln\dfrac{y}{0.6965}$ that we discovered as a result of checking part a.

Execute
$$y = \frac{2,500,000}{1,000,000} = 2.5 \qquad \begin{array}{l}\text{Divide 2,500,000 by 1,000,000 to find the} \\ \text{value of } y \text{ that corresponds to 2,500,000.}\end{array}$$

$$t = \frac{1}{0.0344}\ln\frac{2.5}{0.6965} \approx 37.2 \qquad \begin{array}{l}\text{Substitute 2.5 for } y \text{ in } t = \dfrac{1}{0.0344}\ln\dfrac{y}{0.6965} \\ \text{and then calculate.}\end{array}$$

Answer Since t represents the number of years after 1975, the population will reach 2,500,000 in $1975 + 37.2 = 2012.2$, which means during the year 2012.

Check Use $y = 0.6965e^{0.0344t}$ to verify that in the year 2012 ($t = 37.2$), the population will be 2,500,000 ($y = 2.5$). We will leave this check to the reader.

YOUR TURN

a. How long will it take $8000 invested at 4% annual interest compounded semiannually to grow to $12,000?

b. Between April 2, 2003, and April 8, 2003, the reported number of SARS (severe acute respiratory syndrome) cases was increasing at a rate of 3.1% per day. As of April 2, there were 2671 reported cases of SARS. If the reported number of SARS cases continues to increase at this rate, the function $A = 2671e^{0.031t}$ models the spread of SARS, where A represents the number of cases and t is the number of days after April 2. How many cases would there be on April 10?

c. How many days after April 2 would it take for there to be 8000 cases?

ANSWERS

a. 10.24 yr.

b. 3423 cases

c. Approximately 35.39 days.

Exponential growth and decay can be represented by the equation $A = A_0 e^{kt}$, where A is the amount present, A_0 is the initial amount, t is the time, and k is a constant that is determined by the substance. There is exponential growth indicated if $k > 0$ and exponential decay if $k < 0$. Recall that the half-life of a substance is the amount of time until only one-half the original amount is present.

Plutonium-239 is frequently used as fuel in nuclear reactors to generate electricity. One of the problems with using plutonium is disposing of the radioactive waste, which is extremely dangerous for a very long period of time.

EXAMPLE 7 A nuclear reactor contains 10 kilograms of radioactive plutonium ^{239}P. Plutonium disintegrates according to the formula $A = A_0 e^{-0.0000284t}$.

a. How much will remain after 10,000 years?

Understand We are given $A_0 = 10$ kg and $t = 10,000$ years and asked to find A.

Plan Use $A = A_0 e^{-0.0000284t}$

Execute $A = 10e^{-0.0000284(10,000)}$ **Substitute 10 for A_0 and 10,000 for t.**

$A = 10e^{-0.284}$ **Simplify.**

$A \approx 7.53$ **Evaluate using a calculator.**

Answer About 7.53 kilograms (or about $\frac{3}{4}$ of the original amount) will remain after 10,000 years.

Check Use $A = A_0 e^{-0.0000284t}$ to verify that it takes 10,000 years for 10 kilograms of ^{239}P to disintegrate to 7.53 kg.

$$7.53 = 10e^{-0.0000284t} \qquad \textbf{Substitute 7.53 for } A \textbf{ and 10 for } A_0.$$

$$\frac{7.53}{10} = e^{-0.0000284t} \qquad \textbf{Divide both sides by 10.}$$

$$\ln \frac{7.53}{10} = \ln e^{-0.0000284t} \qquad \textbf{Use if } x = y, \textbf{ then } \log_b x = \log_b y.$$

$$\ln \frac{7.53}{10} = -0.0000284t \qquad \textbf{Use } \log_b b^x = x.$$

$$\frac{\ln \frac{7.53}{10}}{-0.0000284} = t \qquad \textbf{Divide both sides by } -0.0000284.$$

$$9989.1 \approx t \qquad \textbf{Calculate.}$$

Note: *Our manipulations suggest the following formula for calculating t given A and A_0:*

$$t = \frac{1}{-0.0000284} \ln \frac{A}{A_0}$$

Because 7.53 is not the exact amount remaining, we should expect that the corresponding time calculation would not be exactly 10,000 years, so 9989.1 years is reasonable.

b. Find the half-life of ^{239}P.

Understand After one half-life, 5 kg will remain, so we must find t when $A = 5$.

Plan Use $A = A_0 e^{-0.0000284t}$ when $A = 5$ and $A_0 = 10$ and follow the same steps as in our check of part a. Or, as we will do, use the formula $t = \frac{1}{-0.0000284} \ln \frac{A}{A_0}$, which we discovered as a result of checking part a.

Execute $t = \dfrac{1}{-0.0000284} \ln \dfrac{5}{10} \approx 24{,}406.59$ Substitute 5 for A and 10 for A_0 and then calculate.

Answer The half-life of ^{239}P is about 24,400 years.

Check Use $A = A_0 e^{-0.0000284t}$ to verify that in 24,406.59 years, 5 kg out of an initial 10 kg of ^{239}P will remain. We will leave this check to the reader.

YOUR TURN Carbon-14 is a radioactive form of carbon that is present in all living things. Archaeologists and paleontologists frequently use carbon-14 dating in estimating the age of organic fossils. Carbon-14 disintegrates according to the formula $A = A_0 e^{-0.000121t}$.

a. If a sample contains 5 grams of carbon-14, how much will be present after 1500 years?

b. What is the half-life of carbon-14?

OBJECTIVE 4. Use the change-of-base formula. Sometimes applications involve logarithms other than common or natural logarithms. For example, earlier we were given the formula $P = 95 - 30 \log_2 x$, where P is the percent of students who recall the important features of a lecture after x days. To find the percent after 5 days, we need to calculate $\log_2 5$. Most calculators have only base-10 and base-e logarithms, so to calculate $\log_2 5$ using a calculator, we need to write $\log_2 5$ in terms of common or natural logarithms using the change-of-base formula. To derive the change-of-base formula, we let $y = \log_a x$.

$a^y = x$	Write $y = \log_a x$ in exponential form.
$\log_b a^y = \log_b x$	Take $\log_b$ of both sides.
$y \log_b a = \log_b x$	Use $\log_b x^r = r \log_b x$.
$y = \dfrac{\log_b x}{\log_b a}$	Divide both sides by $\log_b a$.
$\log_a x = \dfrac{\log_b x}{\log_b a}$	Substitute $\log_a x$ for y.

RULE Change-of-Base Formula

In general, if $a > 0$, $a \neq 1$, $b > 0$, $b \neq 1$, and $x > 0$, then $\log_a x = \dfrac{\log_b x}{\log_b a}$.

In terms of common and natural logarithms, $\log_a x = \dfrac{\log x}{\log a} = \dfrac{\ln x}{\ln a}$.

EXAMPLE 8 Use the change-of-base formula to calculate $\log_5 19$. Round the answer to four decimal places.

Note: *We could have used* ln *rather than* log.

$\log_5 19 = \dfrac{\ln 19}{\ln 5} \approx 1.8295$

Solution $\log_5 19 = \dfrac{\log 19}{\log 5} \approx 1.8295$ Use $\log_a x = \dfrac{\log_b x}{\log_b a}$, then evaluate using a calculator.

Check $5^{1.8295} = 19.0005 \approx 19$, so the answer is correct.

EXAMPLE 9 Use $P = 95 - 30 \log_2 x$ and the change-of-base formula to find the percent of students who retained the main points of a lecture after 5 days.

Solution $P = 95 - 30 \log_2 5$ Substitute 5 for x in $P = 95 - 30 \log_2 x$.

$$P = 95 - 30 \frac{\ln 5}{\ln 2}$$ Use $\log_a x = \dfrac{\log_b x}{\log_b a}$.

$$P \approx 95 - 30(2.3219)$$ Evaluate $\dfrac{\ln 5}{\ln 2}$ using a calculator.

$$P \approx 25.34$$ Simplify.

About 25% of the students remember the main points of a lecture 5 days later.

YOUR TURN

a. Find $\log_6 25$.

b. Use the formula from Example 9 to find the percent of students who retain the main points of a lecture 8 days later.

ANSWERS

a. 1.7965 **b.** 5%

12.6 Exercises

For Extra Help

MyMathLab   Videotape/DVT   InterAct Math   Math Tutor Center   Math XL.com

1. If $\log_a m = \log_a n$, then _____.

2. When solving $10^{x+2} = 45$, would natural or common logarithms be the better choice? Why?

3. What principle is used to solve $\log(x - 3) = \log(3x - 13)$?

4. What principle is used to solve $100 = (5)^{2n}$?

5. In solving $\log x + \log(x + 2) = \log 15$, we get possible solutions of $x = 3$ and $x = -5$. Why must $x = -5$ be rejected?

6. Suppose you were solving an equation that contained $\log(5 - x)$. You did all the algebraic work correctly and you obtained $x = -3$ as an apparent solution. Would you have to reject $x = -3$ as a solution? Why or why not?

For Exercises 7–18, solve. Round answers to four decimal places.

7. $2^x = 9$ **8.** $3^x = 20$ **9.** $5^{2x} = 32$ **10.** $4^{3x} = 13$

11. $5^{x+3} = 10$ **12.** $7^{x+1} = 41$ **13.** $8^{x-2} = 6$ **14.** $5^{x-3} = 12$

15. $4^{x+2} = 5^x$ **16.** $6^{x+4} = 10^x$ **17.** $2^{x+1} = 3^{x-2}$ **18.** $5^{x-3} = 3^{x+1}$

For Exercises 19–26, solve. Round answers to four decimal places.

19. $e^{3x} = 5$ **20.** $e^{2x} = 7$ **21.** $e^{0.03x} = 25$ **22.** $e^{0.002x} = 12$

23. $e^{-0.022x} = 5$ **24.** $e^{-0.0032x} = 8$ **25.** $\ln e^{4x} = 24$ **26.** $\ln e^{5x} = 35$

For Exercises 27–60, solve. Give exact answers.

27. $\log_2(x - 3) = 3$ **28.** $\log_3(x + 4) = 2$ **29.** $\log_4(4x - 8) = 2$ **30.** $\log_3(3x + 9) = 3$

31. $\log_4 x^2 = 2$ **32.** $\log_2 x^2 = 6$ **33.** $\log_6(x^2 + 5x) = 2$ **34.** $\log_4(x^2 + 6x) = 2$

35. $\log(3x - 2) = \log(2x + 5)$ **36.** $\log(4x + 1) = \log(2x + 7)$ **37.** $\ln(4x + 6) = \ln(2x - 8)$

38. $\ln(5x + 6) = \ln(3x - 8)$ **39.** $\log_9(x^2 + 4x) = \log_9 12$ **40.** $\log_8(x^2 + x) = \log_8 30$

41. $\log_4 x + \log_4 8 = 2$ **42.** $\log_6 x + \log_6 4 = 2$ **43.** $\log_5 x - \log_5 2 = 1$

44. $\log_5 x - \log_5 3 = 2$ **45.** $\log_3 x + \log_3(x + 6) = 3$ **46.** $\log_2 x + \log_2(x - 3) = 2$

47. $\log_3(2x + 15) + \log_3 x = 3$ **48.** $\log_2(3x - 2) + \log_2 x = 4$ **49.** $\log_2(7x + 3) - \log_2(2x - 3) = 3$

50. $\log_3(3x + 3) - \log_3(x - 3) = 2$ **51.** $\log(x + 2) - \log(2x - 2) = 0$ **52.** $\log_3(2x + 1) - \log_3(x - 1) = 1$

53. $\log_8 2x + \log_8 6 = \log_8 10$ **54.** $\log_5 3x + \log_5 2 = \log_5 17$ **55.** $\ln x + \ln(2x - 1) = \ln 10$

56. $\ln x + \ln(3x - 5) = \ln 12$ **57.** $\log x - \log(x - 5) = \log 6$ **58.** $\log x - \log(x - 2) = \log 3$

59. $\log_6(3x + 4) - \log_6(x - 2) = \log_6 8$ **60.** $\log_7(5x + 2) - \log_7(x - 2) = \log_7 9$

61. How long will it take $5000 invested at 5% compounded quarterly to grow to $8000? Round your answer to the nearest tenth of a year.

62. How long will it take $7000 invested at 6% compounded monthly to grow to $12,000? Round your answer to the nearest tenth of a year.

63. If $8000 is deposited into an account at 6% annual interest compounded continuously:

    a. How much money will be in the account after 15 years?

    b. How long will it take the $8000 to grow into $14,000?

64. If $4000 is deposited into an account at 5% annual interest compounded continuously:

    a. How much money will be in the account after 10 years?

    b. How long will it take the $4000 to grow into $10,000?

65. The probability that a person will have an accident while driving at a given blood alcohol level is approximated by $P(b) = e^{21.5b}$, where b is the blood alcohol level ($0 \leq b \leq 0.4$) and P is the percent probability of having an accident.

    a. What is the probability of an accident if the blood alcohol level is 0.08, which is legally drunk in many states?

    b. Estimate the blood alcohol level when the probability of an accident is 50%.

66. Atmospheric pressure (in pounds per square inch, psi) is a function of the altitude above sea level and can be modeled by $P(a) = 14.7e^{-0.21a}$, where P is the pressure and a is the altitude above sea level in miles.

    a. Find the atmospheric pressure at the peak of Mt. McKinley, Alaska, which is 3.85 miles above sea level.

    b. If the atmospheric pressure at the peak of Mt. Everest in Nepal, is 4.68 pounds per square inch, find the height of Mt. Everest.

67. The population of a mosquito colony increases at a rate of 4% per day. If the initial number of mosquitoes is 500, the number present, A, at the end of t days is given by $A = 500e^{0.04t}$.

    a. How many mosquitoes are present after 2 weeks?

    b. Find the number of days until 10,000 mosquitoes are present.

68. A cake is removed from the oven at a temperature of 210°F and is left to cool on a counter where the room temperature is 70°F. The cake cools according to the function $T = 70 + 140e^{-0.0231t}$, where T is the temperature of the cake and t is in minutes.

    a. What is the temperature of the cake after 40 minutes?

    b. After how many minutes will the cake's temperature be 75°F?

69. The world population in 2005 was about 6.4 billion and is increasing at a rate of 1% per year. The world population after 2005 can be approximated by the equation $A = A_0 e^{0.01t}$, where A_0 is the world population in 2005 and t is the number of years after 2005.

    a. Assuming the growth rate follows the same trend, find the world population in 2020.

    b. In what year will the world population reach 7 billion?

70. In 2004, Africa had a population of 885 million and a natural growth rate of 2.4% (2.4 times the world's growth rate). The population growth can be approximated by $A = A_0 e^{0.024t}$, where A_0 is the population in millions in 2004 and t is the number of years after 2004.

    a. Excluding immigration, what would the population of Africa be in 2020?

    b. Excluding immigration, in what year would Africa's population reach 1 billion?

71. The barometric pressure x miles from the eye of a hurricane is approximated by the function $P = 0.48 \ln(x + 1)$, where P is inches of mercury. (*Source:* A. Miller and R. Anthes, *Meteorology*, 4th edition, Merrill Publishing)

    a. Find the barometric pressure 50 miles from the center of the hurricane.

    b. Find the distance from the center where the pressure is 1.5 inches of mercury.

72. The first two-year college, Joliet Junior College in Chicago, was founded in 1901. The number of two-year colleges in the United States grew rapidly, especially during the 1960s, but growth has tapered off. The total number of two-year colleges in the United States since 1960 can be approximated by the function $y = 175.6 \ln x + 513$, where x is the number of years after 1960 and y is the number of two-year colleges. Use the model to estimate the number of two-year colleges in the United States in 2010. (*Source:* American Association of Two-Year Colleges)

73. The percent, $f(x)$, of adult height attained by a boy who is x years old is modeled by $f(x) = 29 + 48.8 \log(x + 1)$, where $5 \le x \le 15$.

    a. At age 10, about what percent of his adult height has a boy reached?

    b. At what age will a boy attain 75% of his adult height?

74. The annual depreciation rate r of a car purchased for P dollars and worth A dollars after t years can be found by the formula $\log (1 - r) = \frac{1}{t} \log \frac{A}{P}$. Find the depreciation rate of a car that is purchased for $22,500 and is sold 3 years later for $10,000.

75. The following table shows the number of registered cocaine users in the United Kingdom from 1987 to 1995.

Year	Number of Users
1987	431
1988	462
1989	527
1990	633
1991	882
1992	1131
1993	1375
1994	1636
1995	1809

The data can be approximated by the exponential function $y = 387.7(1.222)^t$, where y is the number of registered users and t is the number of years after 1987. If the current trend continues, use the model to estimate the number of users in 2010.

76. The number of females working in automotive repair has been increasing, as shown in the following table.

Year	Number of Female Technicians
1988	556
1989	614
1990	654
1991	737
1992	849
1993	1086
1994	1329
1995	1592

The data can be approximated by $y = 507(1.116)^t$, where y is the number of technicians and t is the number of years after 1988. If the trend continues in the same manner, use the equation to estimate the number of female technicians in 2012. (*Source:* National Institute for Automotive Service Excellence)

77. Housing prices in Volusia County, Florida, have risen exponentially. The average sales price of existing houses sold in Volusia County can be approximated by $S(t) = 40.2(1.1)^t$, where t is the number of years after 1990 and S is the price in thousands of dollars.

a. Estimate the average price of existing homes sold in 2005.

b. Find the number of years after 1990 when the average price of an existing home will be $250,000.

78. Since the first Super Bowl in 1967, ticket prices have risen exponentially according to the formula $P(t) = 9.046(1.114)^t$, where $t = 1$ is the year 1967 and ticket prices are in dollars. (*Source: Orlando Sentinel,* January 21, 2003, p. A8)

a. Estimate the price of a Super Bowl ticket in 2012.

b. Find the year when Super Bowl prices were $230.

79. The thickness of a runway's pavement is determined by the weight of the planes that will be using it. The relationship between runway thickness, x, in inches and gross airplane weight, y, in thousands of pounds can be approximated by $y = 18.29(1.279)^x$. If a partially loaded B-52 Stratofortress weighs 350,000 pounds, how thick should a runway be for it to safely use the runway? (*Source:* Federal Aviation Administration)

80. The gap between available organs for liver transplants and people who need them has widened in recent years. The equation $y = 2329(1.2406)^x$, where y is the number of persons waiting and x is the number of years after 1988, approximated the gap. Assuming the same trend continues, use the equation to estimate the year in which the number of people waiting for a transplant is 200,000.

For Exercises 81–88, use the change-of-base formula to find the logarithms. Round answers to four decimal places.

81. $\log_4 12$

82. $\log_5 23$

83. $\log_8 3$

84. $\log_9 5$

85. $\log_{1/2} 5$

86. $\log_{1/3} 4$

87. $\log_{1/4} \dfrac{3}{5}$

88. $\log_{1/3} \dfrac{4}{7}$

REVIEW EXERCISES

For Exercises 1 and 2, simplify.

[1.5] **1.** $(3x + 2) - (2x - 1)$

[1.7] **2.** $(3\sqrt{5} + 2)(2\sqrt{5} - 1)$

For Exercises 3–6, use $f(x) = x^2 + 4$ and $g(x) = 2x - 1$.

[1.7] **3.** Find $f(-5)$.

[1.7] **4.** Find $f + g$

[1.7] **5.** Find $(f \circ g)$

[3.2] **6.** Find $(f \circ g)(2)$

Chapter 12 Summary

Defined Terms

Section 12.1
Composition of functions (p. 948)
Inverse functions (p. 950)
One-to-one function (p. 952)

Section 12.2
Exponential function (p. 963)

Section 12.3
Logarithm (p. 975)

Section 12.5
Common logarithms (p. 993)
Natural logarithms (p. 995)

Procedures, Rules, and Key Examples

Procedures/Rules	Key Examples

Section 12.1 Composite and Inverse Functions

Composition of Functions

$(f \circ g)(x) = f[g(x)]$ for all x in the domain of g such that $g(x)$ is in the domain of f.

$(g \circ f)(x) = g[f(x)]$ for all x in the domain of f such that $f(x)$ is in the domain of g.

Example 1: If $f(x) = x^2 - 3$ and $g(x) = 3x + 4$, find (a) $(f \circ g)(x)$ and (b) $(g \circ f)(x)$.

a. $(f \circ g)(x) = f[g(x)]$

$= f(3x + 4)$ Substitute $3x + 4$ for $g(x)$.

$= (3x + 4)^2 - 3$ Replace x with $3x + 4$ in $f(x)$.

$= 9x^2 + 24x + 13$ Square $3x + 4$ and add like terms.

b. $(g \circ f)(x) = g[f(x)]$

$= g(x^2 - 3)$ Substitute $x^2 - 3$ for $f(x)$.

$= 3(x^2 - 3) + 4$ Replace x with $x^2 - 3$ in $g(x)$.

$= 3x^2 - 5$ Distribute 3 and add.

Inverse Functions

To determine whether two functions f and g are inverses of each other,

1. Show that $f[g(x)] = x$ for all x in the domain of g.
2. Show that $g[f(x)] = x$ for all x in the domain of f.

Example 2: Show that $f(x) = 5x - 6$ and $g(x) = \dfrac{x + 6}{5}$ are inverse functions.

$$f[g(x)] = f\left(\frac{x + 6}{5}\right)$$

$$= 5\left(\frac{x + 6}{5}\right) - 6$$

$$= x + 6 - 6 = x$$

$$g[f(x)] = g(5x - 6)$$

$$= \frac{5x - 6 + 6}{5}$$

$$= \frac{5x}{5} = x$$

Since $f[g(x)] = g[f(x)] = x$, f and g are inverse functions.

continued

Procedures/Rules	Key Examples

Section 12.1 Composite and Inverse Functions (continued)

Horizontal Line Test for One-to-One Functions

Given a function's graph, the function is one-to-one if every horizontal line that can intersect the graph does so at one and only one point.

Note: Only one-to-one functions have inverses that are functions.

Example 3: Determine whether the functions whose graphs follow are one-to-one.

a.

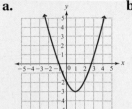

b.

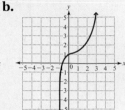

Solution: A horizontal line can intersect this graph in more than one point, so the function is not one-to-one.

Solution: Every horizontal line that can intersect this graph does so at one and only one point, so the function is one-to-one.

Finding the Inverse of a One-to-One Function
1. If necessary, replace $f(x)$ with y.
2. Replace all x's with y's and all y's with x's.
3. Solve the equation from step 2 for y.
4. Replace y with $f^{-1}(x)$.

Example 4: If $f(x) = 3x - 7$, find $f^{-1}(x)$.

Solution:

$$f(x) = 3x - 7$$
$$y = 3x - 7 \quad \text{Replace } f(x) \text{ with } y.$$
$$x = 3y - 7 \quad \text{Interchange } x \text{ and } y.$$
$$\frac{x + 7}{3} = y \quad \text{Solve for } y.$$
$$f^{-1}(x) = \frac{x + 7}{3} \quad \text{Replace } y \text{ with } f^{-1}(x).$$

Verify by showing that $f[f^{-1}(x)] = x$ and $f^{-1}[f(x)] = x$.

Graphs of Inverse Functions

The graphs of f and f^{-1} are symmetric with respect to the graph of $y = x$.

Example 5: Sketch the inverse of the function whose graph follows.

Solution: Draw the line $y = x$ and reflect the graph in the line.

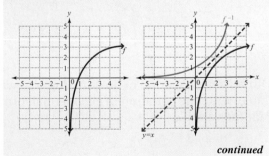

continued

Procedures/Rules	Key Examples

Section 12.2 Exponential Functions
Graphs of Exponential Functions

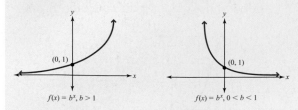

$f(x) = b^x, b > 1$ $\qquad$ $f(x) = b^x, 0 < b < 1$

The One-to-One Property of Exponentials
Given $b > 0$ and $b \neq 1$, if $b^x = b^y$, then $x = y$.

Solving Exponential Equations
1. If necessary, write both sides as a power of the same base.
2. If necessary, simplify the exponents.
3. Set the exponents equal to each other.
4. Solve the resulting equation.

Example 1: Graph $f(x) = 2^x$.

Choose some values for x and find the corresponding values of $f(x)$, which are the y-values.

x	-2	-1	0	1	2
$f(x)$	$\dfrac{1}{4}$	$\dfrac{1}{2}$	1	2	4

Plot the points and draw the graph.

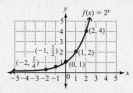

Example 2: Solve $8^{x-3} = 16^{2x}$.

Solution: $8^{x-3} = 16^{2x}$

$(2^3)^{x-3} = (2^4)^{2x}$ $\quad$ Rewrite 8 and 16 as powers of 2.

$2^{3x-9} = 2^{8x}$ $\quad$ Multiply exponents.

$3x - 9 = 8x$ $\quad$ Set exponents equal.

$-\dfrac{9}{5} = x$ $\quad$ Solve for x.

Section 12.3 Logarithmic Functions
Definition of Logarithm
If $b > 0$ and $b \neq 1$, then $y = \log_b x$ is equivalent to $x = b^y$.

To solve an equation of the form $\log_b x = y$, where b, x, or y is a variable, write the equation in exponential form, $b^y = x$ and then solve for the variable.

Example 1: Write in logarithmic form.

a. $3^4 = 81$ $\qquad$ **b.** $\left(\dfrac{1}{4}\right)^{-3} = 64$

Solution: $\qquad\qquad$ Solution:
$\log_3 81 = 4$ $\qquad\quad$ $\log_{1/4} 64 = -3$

Example 2: Write in exponential form.

a. $\log_5 125 = 3$ $\qquad$ **b.** $\log_3 \dfrac{1}{27} = -3$

Solution: $5^3 = 125$ $\qquad$ Solution: $3^{-3} = \dfrac{1}{27}$

Example 3: Solve.

a. $\log_2 \dfrac{1}{16} = x$

continued

Procedures/Rules	Key Examples

Procedures/Rules

Section 12.3 Logarithmic Functions (continued)

Two Properties of Logarithms
For any real number b, where $b > 0$ and $b \neq 1$,
1. $\log_b b = 1$
2. $\log_b 1 = 0$

To graph a function in the form $f(x) = \log_b x$,
1. Replace $f(x)$ with y and then write the logarithm in exponential form, $x = b^y$.
2. Find ordered pairs that satisfy the equation by assigning values to y and finding x.
3. Plot the ordered pairs and draw a smooth curve through the points.

Section 12.4 Properties of Logarithms

Inverse Properties of Logarithms
For any real numbers b and x, where $b > 0$, $b \,?\, 1$, and $x > 0$.
1. $b^{\log_b x} = x$
2. $\log_b b^x = x$

Further Properties of Logarithms
For real numbers x, y, and b, where $x > 0$, $y > 0$, $b > 0$, and $b \neq 1$,
Product Rule: $\log_b xy = \log_b x + \log_b y$

Quotient Rule: $\log_b \dfrac{x}{y} = \log_b x - \log_b y$

Power Rule: $\log_b x^r = r \log_b x$

Key Examples

Solution: $2^x = \dfrac{1}{16}$ Change into exponential form.

$2^x = \dfrac{1}{2^4}$ $16 = 2^4$

$2^x = 2^{-4}$ $\dfrac{1}{2^4} = 2^{-4}$

$x = -4$

b. $\log_x 49 = 2$

Solution: $x^2 = 49$ Change into exponential form.

$x = \pm 7$ The square roots of 49 are ± 7.

$x = 7$ The base must be positive.

Example 4: Find the following logarithms:
a. $\log_8 8$ **b.** $\log_5 1$
Solution: $\log_8 8 = 1$ Solution: $\log_5 1 = 0$

Example 5: Graph $y = \log_3 x$.
Solution: Write as $3^y = x$ and let y have values.

x	y
$\dfrac{1}{9}$	-2
$\dfrac{1}{3}$	-1
1	0
3	1
9	2

Example 1: Find the value of each.
a. $6^{\log_6 5}$ **b.** $3^{\log_3 8}$
Solution: $6^{\log_6 5} = 5$ Solution: $3^{\log_3 8} = 8$

Example 2: Use the product rule for the following:
a. Write $\log_4 4x$ as the sum of logarithms.
Solution:

$\log_4 4x = \log_4 4 + \log_4 x$ Use $\log_b xy = \log_b x + \log_b y$.

$= 1 + \log_4 x$ $\log_4 4 = 1$

b. Write $\log_7 4 + \log_7 2$ as a single logarithm.
Solution:

$\log_7 4 + \log_7 2 = \log_7 4 \cdot 2$ Use $\log_b xy = \log_b x + \log_b y$.

$= \log_7 8$

continued

Procedures/Rules	Key Examples

Section 12.4 Properties of Logarithms (continued)

Example 3: Use the quotient rule for the following:

a. Write $\log_a \dfrac{w}{7}$ as the difference of logarithms.

Solution:
$$\log_a \frac{w}{7} = \log_a w - \log_a 7 \qquad \text{Use } \log_b \frac{x}{y} = \log_b x - \log_b y.$$

b. Write $\log_7(x + 5) - \log_7(x - 3)$ as a single logarithm.

Solution:
$$\log_7(x + 5) - \log_7(x - 3) = \log_7 \frac{x + 5}{x - 3}$$
$$\text{Use } \log_b \frac{x}{y} = \log_b x - \log_b y.$$

Example 4: Use the power rule for the following:

c. Write $\log_b \sqrt[5]{x^3}$ as a multiple of a logarithm.

Solution:
$$\log_b \sqrt[5]{x^3} = \log_b x^{3/5} = \frac{3}{5} \log_b x \qquad \text{Use } \log_b x^r = r \log_b x.$$

d. Write $-3 \log_2 y$ as a logarithm of a quantity to a power.

Solution:
$$-3 \log_2 y = \log_2 y^{-3} \qquad \text{Use } \log_b x^r = r \log_b x.$$
$$= \log_2 \frac{1}{y^3}$$

Example 5: Write $\log_a \dfrac{x^2 y}{z^4}$ as the sum or difference of multiples of logarithms.

Solution:
$$\log_a \frac{x^2 y}{z^4} = \log_a x^2 y - \log_a z^4 \qquad \text{Quotient rule.}$$
$$= \log_a x^2 + \log_a y - \log_a z^4 \qquad \text{Product rule.}$$
$$= 2 \log_a x + \log_a y - 4 \log_a z \qquad \text{Power rule.}$$

Example 6: Write $\dfrac{1}{4}(2 \log_5 x - 3 \log_5 y)$ as a single logarithm.

Solution: $\dfrac{1}{4}(2 \log_5 x - 3 \log_5 y)$
$$= \frac{1}{4}(\log_5 x^2 - \log_5 y^3) \qquad \text{Power rule.}$$
$$= \frac{1}{4} \log_5 \frac{x^2}{y^3} \qquad \text{Quotient rule.}$$
$$= \log_5 \left(\frac{x^2}{y^3}\right)^{1/4} \qquad \text{Power rule.}$$
$$= \log_5 \sqrt[4]{\frac{x^2}{y^3}}$$

continued

Procedures/rules	Key Examples

Section 12.5 Common and Natural Logarithms

Base-10 logarithms are common logarithms and $\log_{10} x$ is written as $\log x$.

Common logarithms can be evaluated using the $\boxed{\text{LOG}}$ key on a calculator.

Base-e logarithms are called natural logarithms and $\log_e x$ is written as $\ln x$.

Natural logarithms can be evaluated using the $\boxed{\text{ln}}$ key on a calculator.

Example 1: Evaluate using a calculator and round to 4 decimal places.
a. log 356
Solution: $\log 356 \approx 2.5514$

b. log 0.0059
Solution: $\log 0.0059 \approx -2.2291$

Example 2: Evaluate using a calculator and round to four decimal places.
a. ln 72
Solution: $\ln 72 \approx 4.2767$
b. ln 0.097
Solution: $\ln 0.097 \approx -2.3330$

Section 12.6 Exponential and Logarithmic Equations with Applications

Properties for Solving Exponential and Logarithmic Equations
For any real numbers b, x, and y, where $b > 0$ and $b \neq 1$,
1. If $b^x = b^y$, then $x = y$.
2. If $x = y$, then $b^x = b^y$.
3. For $x > 0$ and $y > 0$, if $\log_b x = \log_b y$, then $x = y$.
4. For $x > 0$ and $y > 0$, if $x = y$, then $\log_b x = \log_b y$.
5. For $x > 0$, if $\log_b x = y$, then $b^y = x$.

Example 1: Solve $3^x = 4$.
Solution:

$$\log 3^x = \log 4 \qquad \text{Use if } x = y, \text{ then } \log_b x = \log_b y.$$
$$x \log 3 = \log 4 \qquad \text{Power rule.}$$
$$x = \frac{\log 4}{\log 3} \qquad \text{Divide by log 3.}$$
$$x \approx 1.2619 \qquad \text{Evaluate using a calculator.}$$

Example 2: Solve $e^{3x} = 12$.
Solution: Since the base of the exponent is e, take the natural logarithm of both sides.
$$\ln e^{3x} = \ln 12 \qquad \text{Use if } x = y, \text{ then } \log_b x = \log_b y.$$
$$3x = \ln 12 \qquad \text{Use } \log_b b^x = x.$$
$$x = \frac{\ln 12}{3} \approx 0.8283 \qquad \text{Divide by 3 and approximate.}$$

Solving Equations Containing Logarithms
To solve equations containing logarithms, use the properties of logarithms to simplify each side of the equation and then use one of the following.

If the simplification results in an equation in the form $\log_b x = \log_b y$, use the fact that $x = y$, and then solve for the variable.

If the simplification results in an equation in the form $\log_b x = y$, write the equation in exponential form, $b^y = x$, and then solve for the variable (as we did in Section 12.3).

Example 3: Solve
$\log_3 x + \log_3(x - 3) = \log_3 10$.
Solution:
$$\log_3 x(x - 3) = \log_3 10 \qquad \text{Use the product rule.}$$
$$x(x - 3) = 10 \qquad \text{Use if } \log_b x = \log_b y, \text{ then } x = y.$$
$$x^2 - 3x - 10 = 0 \qquad \text{Multiply then subtract 10.}$$
$$(x - 5)(x + 2) = 0 \qquad \text{Factor.}$$
$$x - 5 = 0 \quad \text{or} \quad x + 2 = 0 \qquad \text{Set each factor equal to 0.}$$
$$x = 5 \quad \text{or} \quad x = -2 \qquad \text{Solve each equation.}$$

We must reject -2 because it results in $\log_3(-2)$ and $\log_3(-5)$ in the original equation, which do not exist.

continued

Procedures/Rules	Key Examples

Section 12.6 Exponential and Logarithmic Equations with Applications (continued)

Example 4: Solve $\log_2(2x + 2) - \log_2(x - 4) = 2$.

Solution:

$\log_2 \dfrac{2x + 2}{x - 4} = 2$ Use the quotient rule.

$\dfrac{2x + 2}{x - 4} = 2^2$ Use if $\log_b x = y$, then $b^y = x$.

$\dfrac{2x + 2}{x - 4} = 4$ $2^2 = 4$.

$2x + 2 = 4x - 16$ Multiply by $x - 4$.

$18 = 2x$

$9 = x$ Solve for x.

Example 5: How long will it take $12,000 invested at 4% compounded quarterly to grow to $15,000?

Solution: Use $A = P\left(1 + \dfrac{r}{n}\right)^{nt}$ with $A = 15{,}000$, $P = 12{,}000$, $r = 4\% = 0.04$, and $n = 4$. Find t.

$A = P\left(1 + \dfrac{r}{n}\right)^{nt}$

$15{,}000 = 12{,}000\left(1 + \dfrac{0.04}{4}\right)^{4t}$ Substitute.

$\dfrac{5}{4} = (1.01)^{4t}$ Divide by 12,000.

$\log \dfrac{5}{4} = \log 1.01^{4t}$ Apply if $x = y$, then $\log_b x = \log_b y$.

$\log \dfrac{5}{4} = 4t \log 1.01$ Power rule.

$\dfrac{\log \dfrac{5}{4}}{4 \log 1.01} = t$ Divide by 4 log 1.01.

$5.61 \approx t$ Evaluate.

It would take about 5.6 years.

continued

Procedures/Rules	Key Examples

Section 12.6 Exponential and Logarithmic Equations with Applications (continued)

Example 6: If $3000 is invested in an account paying 4.5% compounded continuously:

a. How much will be in the account after 10 years?

Solution: Use $A = Pe^{rt}$ with $P = 3000$, $r = 0.045$, and $t = 10$.

$A = Pe^{rt}$

$A = 3000e^{(0.045)(10)}$ Substitute.

$A = 3000e^{0.45}$ $10(0.045) = 0.45$.

$A = 4704.94$ Evaluate using a calculator.

b. How long would it take for the $3000 to grow to $5000?

Solution: Use $A = Pe^{rt}$ with $A = 5000$, $P = 3000$, and $r = 0.045$. Find t.

$A = Pe^{rt}$

$5000 = 3000e^{0.045t}$ Substitute

$\dfrac{5}{3} = e^{0.045t}$ Divide by 3000.

$\ln \dfrac{5}{3} = \ln e^{0.045t}$ Use if $x = y$, then $\log_b x = \log_b y$.

$\ln \dfrac{5}{3} = 0.045t$ Use if $x = y$, then $\log_b x = \log_b y$.

$\dfrac{\ln \dfrac{5}{3}}{0.045} = t$ Divide by 0.045.

$11.35 \approx t$ Evaluate using a calculator.

It would take about $11\dfrac{1}{3}$ years.

Exponential Growth and Decay
Exponential growth and decay can be represented by the equation $A = A_0 e^{kt}$, where A_0 is the initial amount, t is the time, and k is a constant that is determined by the substance. If $k > 0$, there is exponential growth, and if $k < 0$, there is exponential decay. The half-life of a substance is the amount of time that must pass until one-half of the original substance remains.

Example 7: The element bismuth has an isotope, ^{200}Bi, that disintegrates according to the formula $A = A_0 e^{-0.0198t}$, where t is in minutes.

a. How much of a 100-gram sample remains after 120 minutes?

Solution: We are given $A_0 = 100$ and $t = 120$.

$A = A_0 e^{-0.0198t}$

$A = 100e^{-0.0198(120)}$ Substitute for A_0 and t.

$A = 9.29$ Evaluate using a calculator.

There would be about 9.29 grams left after 120 minutes.

continued

Procedures/Rules	Key Examples

Section 12.6 Exponential and Logarithmic Equations with Applications (continued)

b. Find the half-life of ^{200}Bi.

Solution: After one half-life, 50 of the original 100 grams remain, so $A = 50$.

$$A = A_0 e^{-0.0198t}$$
$$50 = 100e^{-0.0198t} \quad \text{Substitute for } A \text{ and } A_0.$$

$$\frac{1}{2} = e^{-0.0198t} \quad \text{Divide by 100.}$$

$$\ln \frac{1}{2} = \ln e^{-0.0198t} \quad \text{Use if } x = y, \text{ then } \log_b x = \log_b y.$$

$$\ln \frac{1}{2} = -0.0198t \quad \text{Use } \log_b b^x = x.$$

$$\frac{\ln \frac{1}{2}}{-0.0198} = t \quad \text{Divide by } -0.0198$$

$$35 = t \quad \text{Evaluate using a calculator.}$$

The half-life is 35 minutes.

Change-of-Base Formula
In general, if $a > 0$, $a \neq 1$, $b > 0$, $b \neq 1$, and $x > 0$, then
$$\log_a x = \frac{\log_b x}{\log_b a}.$$

In terms of common and natural logarithms, $\log_a x = \dfrac{\log x}{\log a} = \dfrac{\ln x}{\ln a}$.

Example 8: Find $\log_5 16$.

Solution: Use $\log_a x = \dfrac{\log x}{\log a} = \dfrac{\ln x}{\ln a}$, then evaluate using a calculator.

$$\log_5 16 = \frac{\log 16}{\log 5} \approx 1.7227$$

or

$$\log_5 16 = \frac{\ln 16}{\ln 5} \approx 1.7227$$

Chapter 12 Review Exercises

For Exercises 1–6, answer true or false.

[12.1] 1. If f and g are inverse functions and $f(2) = 5$, then $g(5) = 2$.

[12.2] 2. If b, x, and y are real numbers and $b^x = b^y$, then $x = y$.

[12.3] 3. If $y = \log_a(x - 4)$, it is possible for x to equal 3.

[12.3] 4. If $\log_a b = c$, then $c^a = b$.

[12.3] 5. Logarithms are exponents.

[12.5] 6. Base-e logarithms are called common logarithms.

Equations and Inequalities

Exercises 7–62

[12.1] **For Exercises 7–10, find each composition if $f(x) = 3x + 4$ and $g(x) = x^2 - 2$.**

7. $(f \circ g)(3)$ 8. $(g \circ f)(3)$ 9. $f[g(0)]$ 10. $g[f(0)]$

[12.1] **For Exercises 11–14, find $(f \circ g)(x)$ and $(g \circ f)(x)$.**

11. $f(x) = 3x - 6$, $g(x) = 2x + 3$ 12. $f(x) = x^2 + 4$, $g(x) = 3x - 7$

13. $f(x) = \sqrt{x - 3}$, $g(x) = 2x - 1$ 14. $f(x) = \dfrac{x + 3}{x}$, $g(x) = \dfrac{x - 4}{x}$

[12.1] **For Exercises 15 and 16, determine if each is the graph of a one-to-one function. If the function is one-to-one, sketch the graph of the inverse function.**

15.
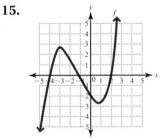

16.

[12.1] *For Exercises 17–20, determine whether f and g are inverse functions.*

17. $f(x) = 2x - 9$, $g(x) = \dfrac{x + 9}{2}$

18. $f(x) = x^3 + 6$, $g(x) = \sqrt[3]{x - 6}$

19. $f(x) = x^2 - 3$, $x \geq 0$; $g(x) = \sqrt{x + 2}$

20. $f(x) = \dfrac{x}{x + 4}$, $g(x) = \dfrac{-4x}{x + 1}$

[12.1] *For Exercises 21–24, find $f^{-1}(x)$ for each of the following one-to-one functions.*

21. $f(x) = 5x + 4$

22. $f(x) = x^3 + 6$

23. $f(x) = \dfrac{4}{x + 5}$

24. $f(x) = \sqrt[3]{3x + 2}$

[12.1] 25. If $(4, -6)$ is an ordered pair on the graph of f, what ordered pair is on the graph of f^{-1}?

[12.1] 26. Fill in the blanks. If $f(a) = b$, then $f^{-1}(__) = $ ____.

[12.2] *For Exercises 27–30, graph.*

27. $f(x) = 3^x$

28. $f(x) = \left(\dfrac{1}{2}\right)^x$

29. $f(x) = 2^{x-3}$

30. $f(x) = 3^{-x+2}$

[12.2] *For Exercises 31–38, solve each equation.*

31. $5^x = 625$

32. $9^x = 27$

33. $6^x = \dfrac{1}{36}$

34. $\left(\dfrac{3}{4}\right)^x = \dfrac{16}{9}$

35. $3^{x-2} = 9$

36. $5^{x+2} = 25^x$

37. $\left(\dfrac{1}{3}\right)^{-x} = 27$

38. $8^{3x-2} = 16^{4x}$

[12.2] 39. The median doubling time for a malignant tumor is about 100 days. If there are 500 cells initially, then the number of cells, A, after t days is given by $A = A_0 2^{t/100}$. How many cells are present after one year?

[12.2] 40. If $25,000 is deposited into an account paying 6% interest compounded monthly, how much will be in the account after eight years?

[12.2] 41. The radioactive isotope ^{82}R has a half-life of 107 days. How much of a 50-gram sample remains after 300 days? (Use $A = A_0\left(\dfrac{1}{2}\right)^{t/h}$.)

[12.2] 42. Cocaine use in the United Kingdom skyrocketed between 1987 and 1995 and can be approximated by the exponential function $N(t) = 387.7(1.222)^t$, where $N(t)$ is the number of addicts and t is the number of years after 1987. Estimate the number of addicts in 1991.

[12.3] *For Exercises 43–46, write in logarithmic form.*

43. $7^3 = 343$

44. $4^{-3} = \dfrac{1}{64}$

45. $\left(\dfrac{3}{2}\right)^4 = \dfrac{81}{16}$

46. $11^{1/3} = \sqrt[3]{11}$

[12.3] *For Exercises 47–50, write in exponential form.*

47. $\log_9 81 = 2$

48. $\log_{1/5} 125 = -3$

49. $\log_a 16 = 4$

50. $\log_e c = b$

[12.3] *For Exercises 51–58, solve.*

51. $\log_3 x = -4$

52. $\log_{1/2} x = -2$

53. $\log_2 32 = x$

54. $\log_{1/4} 16 = x$

55. $\log_x 81 = 4$

56. $\log_x \dfrac{1}{1000} = 3$

57. $\log_{3/4} \dfrac{9}{16} = x$

58. $\log_{121} x = \dfrac{1}{2}$

[12.3] *For Exercises 59–60, graph.*

59. $f(x) = \log_4 x$

60. $f(x) = \log_{1/3} x$

[12.3] 61. If $f(x) = \log_b x$, the domain is $(0, \infty)$, and the range is $(-\infty, \infty)$, what are the domain and range of $g(x) = b^x$? Why?

[12.3] 62. The formula for the number of decibels in a sound is $d = 10 \log \dfrac{I}{I_0}$. Find the decibel reading of a sound whose intensity is $I = 1000\, I_0$.

[12.4] *For Exercises 63 and 64, find the value.*

63. $3^{\log_3 8}$

64. $\log_9 9^6$

[12.4] *For Exercises 65 and 66, use the product rule to write the expression as a sum of logarithms.*

65. $\log_6 6x$

66. $\log_4 x(2x - 5)$

[12.4] *For Exercises 67 and 68, use the product rule to write the expression as a single logarithm.*

67. $\log_3 4 + \log_3 8$

68. $\log_5 3 + \log_5 x + \log_5(x - 2)$

[12.4] *For Exercises 69 and 70, use the quotient rule to write the expression as a difference of logarithms.*

69. $\log_b \dfrac{x}{5}$

70. $\log_a \dfrac{3x - 2}{4x + 3}$

[12.4] *For Exercises 71 and 72, use the quotient rule to write the expression as a single logarithm.*

71. $\log_5 24 - \log_5 4$

72. $\log_2(x + 5) - \log_2(2x - 3)$

[12.4] *For Exercises 73–76, use the power rule to write the expression as a multiple of a logarithm.*

73. $\log_3 7^4$

74. $\log_a \sqrt[3]{x}$

75. $\log_4 \dfrac{1}{a^4}$

76. $\log_a \sqrt[5]{a^4}$

[12.4] *For Exercises 77 and 78, use the power rule to write the expression as the logarithm of a quantity to a power. Simplify the answer, if possible.*

77. $4 \log_a x$

78. $\dfrac{3}{5} \log_a y$

[12.4] *For Exercises 79–82, write the expression as the sum or differences of multiples of logarithms.*

79. $\log_a x^2 y^3$

80. $\log_a \dfrac{c^4}{d^3}$

81. $\log_a \dfrac{x^2 y^3}{z^4}$

82. $\log_a \sqrt{\dfrac{a^3}{b^4}}$

[12.4] *For Exercises 83–86, write the expression as a single logarithm.*

83. $2 \log_a b + 4 \log_a c$

84. $3 \log_a 4 - 2 \log_a 3$

85. $\dfrac{1}{4}(2 \log_a x + 3 \log_a y)$

86. $4 \log_a(x + 5) + 2 \log_a(x - 3)$

[12.5] *For Exercises 87–90, use a calculator to approximate each logarithm to four decimal places.*

87. $\log 326$ **88.** $\log 0.0035$ **89.** $\ln 0.043$ **90.** $\ln 92$

[12.5] *For Exercises 91 and 92, find the exact value without using a calculator.*

91. $\log 0.00001$ **92.** $\ln \sqrt[4]{e}$

[12.5] 93. The sound intensity of a clap of thunder was $10^{-3.5}$ watts per square meter. What is the decibel reading? (Use $d = 10 \log \dfrac{I}{I_0}$, where $I_0 = 10^{-12}$ watts/m^2.)

Equations and Inequalities

Exercises 93–114

[12.5] 94. Using pH $= -\log [H_3O^+]$ (the hydronium ion concentration), find the pH of an apple whose $[H_3O^+]$ is 0.001259.

[12.5] 95. How long will it take $6000 to grow to $10,000 if it is invested at 3% annual interest compounded continuously? (Use $A = Pe^{rt}$.)

[12.5] 96. Using $R = \log \dfrac{I}{I_0}$, for Richter scale readings, compare the intensity of an earthquake whose Richter scale reading was 7.8 with one whose reading was 6.8. What do you notice? (*Hint:* Solve for I in terms of I_0.)

[12.6] *For Exercises 97–102, solve. Round answers to four decimal places.*

97. $9^x = 32$ **98.** $3^{5x} = 19$ **99.** $6^{2x-1} = 22$ **100.** $4^{2x-3} = 5^{x+1}$

101. $e^{4x} = 11$ **102.** $e^{-0.003x} = 5$

[12.6] *For Exercises 103–110, solve. Give exact answers.*

103. $\log_3(x - 4) = 2$

104. $\log_2(x^2 + 2x) = 3$

105. $\log(3x - 8) = \log(x - 2)$

106. $\log 5 + \log x = 2$

107. $\log_3 x - \log_3 4 = 2$

108. $\log_2 x + \log_2(x - 6) = 4$

109. $\log_4 x + \log_4(6x - 9) = \log_4 15$

110. $\log_3(5x + 2) - \log_3(x - 2) = 2$

[12.6] 111. How long will it take $15,000 invested at 3% compounded monthly to grow to $18,000? Round the answer to the nearest tenth of a year. (Use $A = P\left(1 + \dfrac{r}{n}\right)^{nt}$.)

[12.6] 112. If $7000 is deposited at 7% annual interest compounded continuously:

    a. How much will be in the account after eight years? (Use $A = Pe^{rt}$.)

    b. How long will it take until there is $12,000 in the account?

[12.6] 113. The population of an ant colony is 400 and is increasing at the rate of 3% per month. Use $A = A_0 e^{0.03t}$ to answer the following:

    a. How many ants will be in the colony after one year?

    b. After how many months will there be 800 ants in the colony?

[12.6] 114. The number of diagnosed AIDS cases among adults in the United States since 1992 has declined and can be approximated by the function $y = 87{,}419 - 24{,}647.9 \ln x$, where y is the number of diagnosed cases and x is the number of years after 1992. (*Data Source:* CDC Division of HIV/AIDS Prevention, *Surveillance Report*, Vol. 13, No. 2)

    a. Estimate the number of cases of diagnosed AIDS in 2000.

    b. In what year were 43,256 new cases of AIDS diagnosed?

[12.6] *For Exercises 115 and 116, use the change-of-base formula to approximate each logarithm to four decimal places.*

115. $\log_4 15$ **116.** $\log_{1/2} 6$

Expressions

Exercises 115–116

Chapter 12 Practice Test

1. If $f(x) = x^2 - 6$ and $g(x) = 3x - 5$, find $f[g(x)]$.

2. Use $f(x) = 4x - 3$, to answer a–c.
   a. Find $f^{-1}(x)$.

   b. Verify that $f[f^{-1}(x)] = x$ and $f^{-1}[f(x)] = x$.

   c. What is the relationship between the graphs of $f(x)$ and $f^{-1}(x)$?

3. The graph of a function f is shown to the right. Graph f^{-1}.

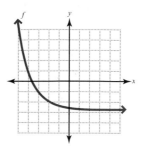

4. Graph $f(x) = 2^{x-1}$.

5. Solve $32^{x-2} = 8^{2x}$.

6. A sum of $20,000 is invested at 5% compounded quarterly.
   a. Find the amount in the account after 15 years.

   b. Find the number of years until there is $30,000 in the account.

7. The isotope ^{98}Nb has a half-life of 30 minutes.
   a. How much of a 100-gram sample remains after 3.6 hours?

   b. How long will it take until there are only 20 grams remaining?

8. The number of people in the United States aged 65 or over has increased rapidly since 1900 and can be approximated by $y = 3.17(1.026)^x$, where y is the number in millions aged 65 or older and x is the number of years after 1900. (*Source:* U.S. Bureau of the Census)
   a. Find the number aged 65 or older in 1960. In 2005.

   b. In what year were there 31.4 million people aged 65 or older?

9. Write $\log_{1/3} 81 = -4$ in exponential form.

For Exercises 10 and 11, solve. Give exact answers.

10. $\log_6 \dfrac{1}{216} = x$

11. $\log_x 625 = 4$

12. Graph $f(x) = \log_2 x$.

For Exercises 13 and 14, write as the sum or difference of multiples of logarithms.

13. $\log_b \dfrac{x^4 y^2}{z}$

14. $\log_b \sqrt[4]{\dfrac{x^5}{y^7}}$

15. Write $\dfrac{3}{4}(2 \log_b x + 3 \log_b y)$ as a single logarithm.

16. The isotope ^{119}Sn disintegrates according to the function $A = A_0 e^{-0.0028t}$, where t is the time in days.

   a. How much of a 300-gram sample remains after 500 days?

   b. What is the half-life of ^{119}Sn?

For Exercises 17–19, solve. If necessary, use a calculator to approximate to four decimal places.

17. $6^{x-3} = 19$

18. $\log_3 x + \log_3(x + 6) = 3$

19. $\log(4x + 2) - \log(3x - 2) = \log 2$

20. The number of children younger than 13 diagnosed with AIDS has declined since 1992 and can be modeled by $y = 1141.7 - 428.64 \ln x$, where y is the number of diagnosed cases and $x = 1$ is 1992. (*Data Source:* CDC Divisions of HIV/AIDS Prevention, *Surveillance Report*, Vol. 13, no. 2)

   a. Find the number of diagnosed cases in 1998.

   b. In what year were 155 cases diagnosed?

Conic Sections

"The universe . . . is written in the language of mathematics, and its characters are triangles, circles, and other geometrical figures."

—Galileo Galilei (1564–1642), Italian mathematician, astronomer, physicist, and philosopher. *Il Saggiatore* (1623).

"Mankind is not a circle with a single center but an ellipse with two focal points of which facts are one and ideas the other."

—Victor Hugo (1802–1883),
French poet, dramatist, novelist. *Les Miserables* (1862)

In this chapter, we expand our understanding of equations, inequalities, and graphs and learn about conic sections. We will see that the conic sections are parabolas, circles, ellipses, and hyperbolas. These shapes are called conic sections because they are formed by the intersection of a cone with a plane. Later, we will study the equations of conic sections, nonlinear systems, and nonlinear inequalities.

13.1 The Parabola and Circle

OBJECTIVES

1. Graph parabolas of the form $x = a(y - k)^2 + h$.
2. Find the distance between two points.
3. Graph circles of the form $(x - h)^2 + (y - k)^2 = r^2$.
4. Find the equation of a circle with a given center and radius.
5. Graph circles of the form $x^2 + y^2 + dx + ey + f = 0$.

The intersection of a plane with a cone will be a circle, ellipse, parabola, or hyperbola. For that reason, these curves are called **conic sections** or **conics**.

DEFINITION   *Conic Section:* A curve in a plane that is the result of intersecting the plane with a cone. More specifically, a circle, ellipse, parabola, or hyperbola.

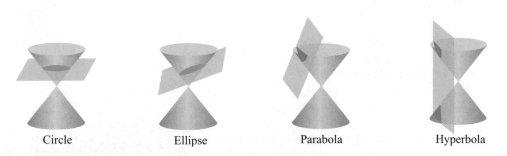

| Circle | Ellipse | Parabola | Hyperbola |

Recall from Section 11.4 that we graphed parabolas in the form $y = a(x - h)^2 + k$. The graph opened upwards if $a > 0$, downwards if $a < 0$, had a vertex at (h, k), and $x = h$ as the axis of symmetry.

EXAMPLE 1   For $y = 2(x - 3)^2 + 1$, determine whether the graph opens upwards or downwards, find the vertex and axis of symmetry, and draw the graph.

Solution The graph opens upwards since $a = 2$ and 2 is positive. We compare the equation with the form $y = a(x - h)^2 + k$ and observe that the vertex is at the point with coordinates (3, 1) and the axis of symmetry is $x = 3$. Plot a few points on either side of the axis of symmetry by letting x have values on either side of 3 and finding y.

x	y
2	3
1	9
4	3
5	9

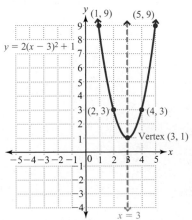

YOUR TURN   For $y = -2(x - 1)^2 + 3$, determine whether the graph opens upwards or downwards, find the vertex and axis of symmetry, and draw the graph.

OBJECTIVE 1. Graph parabolas of the form $x = a(y - k)^2 + h$.   If we interchange x and y in the equations of parabolas that open upwards and downwards, we get the equations of parabolas that open to the left or right. To keep the vertex at (h, k), we also interchange h and k.

RULE   **Equations of Parabolas Opening Left or Right**

The graph of an equation in the form $x = a(y - k)^2 + h$ is a parabola with vertex at (h, k). The parabola opens to the right if $a > 0$ and to the left if $a < 0$. The equation of the axis of symmetry is $y = k$.

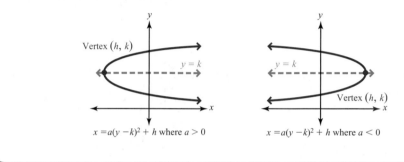

$x = a(y - k)^2 + h$ where $a > 0$ $x = a(y - k)^2 + h$ where $a < 0$

EXAMPLE 2   For each equation, determine whether the graph opens left or right, find the vertex and axis of symmetry, and draw the graph.

a. $x = -2(y + 3)^2 - 2$

Solution This parabola opens to the left because $a = -2$, which is negative. Rewrite the equation as $x = -2(y - (-3))^2 - 2$. Comparing this equation with $x = a(y - k)^2 + h$, we see that $h = -2$ and $k = -3$. The vertex is at the point with coordinates $(-2, -3)$ and the axis of symmetry is $y = -3$. To graph, plot a few points on either side of the axis of symmetry by letting y equal values on either side of -3 and finding x.

ANSWER

opens downwards; vertex: $(1, 3)$; axis of symmetry: $x = 1$

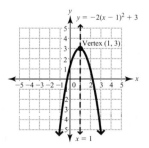

x	y
-4	-2
-10	-1
-4	-4
-10	-5

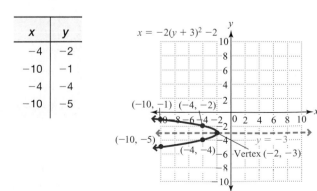

b. $x = 3y^2 + 12y + 8$

Solution This parabola opens to the right because $a = 3$, which is positive. To find the vertex and axis of symmetry we need to write the equation in the form $x = a(y - k)^2 + h$.

$x = 3y^2 + 12y + 8$	**Original equation.**
$x - 8 = 3y^2 + 12y$	**Subtract 8 from both sides.**
$x - 8 = 3(y^2 + 4y)$	**Factor out the common factor, 3.**
$x - 8 + 12 = 3(y^2 + 4y + 4)$	**Complete the square. Note that we added $3 \cdot 4 = 12$ to both sides of the equation.**
$x + 4 = 3(y + 2)^2$	**Simplify the left side and factor the right.**
$x = 3(y + 2)^2 - 4$	**Subtract 4 from both sides of the equation.**

The vertex is at the point with coordinates $(-4, -2)$ and the axis of symmetry is $y = -2$.

To complete the graph, let y equal values on either side of -2 and find x.

x	y
-1	-1
8	0
-1	-3
8	-4

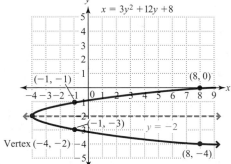

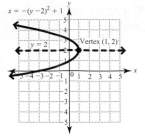
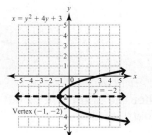
YOUR TURN For each equation, determine whether the graph opens left or right, find the vertex and axis of symmetry, and draw the graph.

a. $x = -(y - 2)^2 + 1$

b. $x = y^2 + 4y + 3$

OBJECTIVE 2. Find the distance between two points. To derive the other conic's general equations, we need to be able to find the distance between any two points in the coordinate plane. Consider the two points (x_1, y_1) and (x_2, y_2) shown in the following graph.

Note: *If y_2 were 6 and y_1 were 2, the distance between would be $6 - 2 = 4$. Therefore, to calculate the length of the vertical leg we calculate $y_2 - y_1$.*

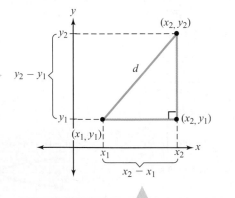

Since the distance between the two points is measured along a line segment that is the hypotenuse of a right triangle, we can use the Pythagorean theorem (see below) to find the distance.

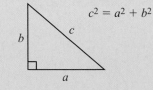

$$c^2 = a^2 + b^2$$

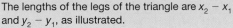

The lengths of the legs of the triangle are $x_2 - x_1$ and $y_2 - y_1$, as illustrated.

Note: *If x_2 were 4 and x_1 were 1, the distance between would be $4 - 1 = 3$. Therefore, to calculate the length of the horizontal leg we calculate $x_2 - x_1$.*

Now we can use the Pythagorean theorem, replacing a with $x_2 - x_1$, b with $y_2 - y_1$, and c with d.

$$d^2 = (x_2 - x_1)^2 + (y_2 - y_1)^2$$

$$d = \pm \sqrt{(x_2 - x_1)^2 + (y_2 - y_1)^2}$$ **Use the square root principle to isolate d.**

Because d is a distance, it must be positive, so we use only the positive value.

RULE The Distance Formula

The distance, d, between two points with coordinates (x_1, y_1) and (x_2, y_2) can be found using the formula

$$d = \sqrt{(x_2 - x_1)^2 + (y_2 - y_1)^2}$$

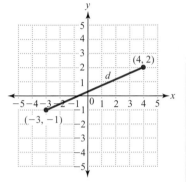

Note: *The distance formula holds no matter what quadrants the points are in.*

EXAMPLE 3 Find the distance between $(4, 2)$ and $(-3, -1)$. If the distance is an irrational number, also give a decimal approximation rounded to three places.

Solution $d = \sqrt{(x_2 - x_1)^2 + (y_2 - y_1)^2}$ **Use the distance formula.**

$d = \sqrt{(-3 - 4)^2 + (-1 - 2)^2}$ **Let $(4, 2) = (x_1, y_1)$ and $(-3, -1) = (x_2, y_2)$.**

$d = \sqrt{(-7)^2 + (-3)^2}$

$d = \sqrt{49 + 9}$ **Note:** *It does not matter which ordered pair is (x_1, y_1) and*

$d = \sqrt{58}$ *which is (x_2, y_2). Consider Example 3 again with $(-3, -1)$ as*

$d \approx 7.616$ *(x_1, y_1) and $(4, 2)$ as (x_2, y_2):*

$$d = \sqrt{(4 - (-3))^2 + (2 - (-1))^2}$$
$$= \sqrt{(7)^2 + (3)^2} = \sqrt{49 + 9}$$
$$= \sqrt{58} \approx 7.616$$

YOUR TURN Determine the distance between the given points. If the distance is an irrational number, also give a decimal approximation rounded to three places.

a. $(8, 2)$ and $(3, -4)$ **b.** $(6, -5)$ and $(0, -1)$

ANSWERS

a. $\sqrt{61} \approx 7.810$ **b.** $2\sqrt{13} \approx 7.211$

OBJECTIVE 3. Graph circles of the form $(x - h)^2 + (y - k)^2 = r^2$. The second conic section that we will consider is the **circle** with **radius** r.

DEFINITIONS *Circle:* A set of points that are equally distant from a central point. The central point is the center.
Radius: The distance from the center of a circle to any point on the circle.

If the center of a circle is (h, k) and the radius is r, we can use the distance formula to derive the equation of the circle. If (x, y) is any point on the circle, the distance between (x, y) and (h, k) must be the radius, r.

$$\sqrt{(x_2 - x_1)^2 + (y_2 - y_1)^2} = r$$

$$\sqrt{(x - h)^2 + (y - k)^2} = r \quad \text{Substitute } (x, y) \text{ for } (x_2, y_2) \text{ and } (h, k) \text{ for } (x_1, y_1).$$

$$(x - h)^2 + (y - k)^2 = r^2 \quad \text{Square both sides.}$$

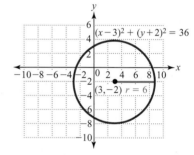

RULE   **Standard Form of the Equation of a Circle**

The equation of a circle with center at (h, k) and radius r is
$(x - h)^2 + (y - k)^2 = r^2$.

Note: *If the center of a circle is at the origin, then $(h, k) = (0, 0)$ and the equation of the circle becomes $(x - 0)^2 + (y - 0)^2 = r^2$, which simplifies to $x^2 + y^2 = r^2$.*

EXAMPLE 4   Find the center and radius of each circle and draw the graph.

a. $(x - 3)^2 + (y + 2)^2 = 36$

Solution $(x - 3)^2 + (y - (-2))^2 = 6^2$ **Write in the form $(x - h)^2 + (y - k)^2 = r^2$.**

Since $h = 3$ and $k = -2$, the center is $(3, -2)$.
Since $36 = 6^2$, the radius is 6.

Note: *To find the radius, we evaluate the square root of 36. Because the radius is a distance, we give only the principal square root.*

$$r = \sqrt{36} = 6$$

ANSWERS

a. center: $(3, -5)$; radius: 2

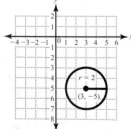

b. center: $(-1, 1)$; radius: $3\sqrt{2}$

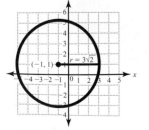

b. $(x + 4)^2 + (y + 1)^2 = 28$

Solution $(x - (-4))^2 + (y - (-1))^2 = (\sqrt{28})^2$ **Write in the form $(x - h)^2 + (y - k)^2 = r^2$.**

Since $h = -4$ and $k = -1$, the center is $(-4, -1)$.
For this radius, $r = \sqrt{28} = 2\sqrt{7} \approx 5.292$.

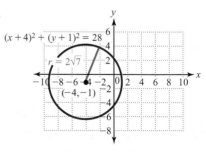

YOUR TURN   Find the center and radius of each circle and draw the graph.

a. $(x - 3)^2 + (y + 5)^2 = 4$ **b.** $(x + 1)^2 + (y - 1)^2 = 18$

OBJECTIVE 4. Find the equation of a circle with a given center and radius. We can also use the standard form of a circle to write equations of circles.

EXAMPLE 5 Write the equation of each circle in standard form.

a. center: $(-4, 2)$; radius: 8

Solution Since the center is at $(-4, 2)$, $h = -4$ and $k = 2$.

$$(x - h)^2 + (y - k)^2 = r^2$$
$$(x - (-4))^2 + (y - 2)^2 = 8^2 \quad \text{Substitute for } h, k, \text{ and } r.$$
$$(x + 4)^2 + (y - 2)^2 = 64 \quad \text{Simplify}$$

b. center: $(0, 0)$; radius: 2

Solution Since the center is at $(0, 0)$, the standard form of the equation is $x^2 + y^2 = r^2$.

$$x^2 + y^2 = 2^2 \quad \text{Substitute for } r.$$
$$x^2 + y^2 = 4$$

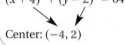
ANSWERS

a. $(x - 4)^2 + (y + 2)^2 = 25$
b. $x^2 + y^2 = 81$

YOUR TURN Write the equation of each circle in standard form.

a. center: $(4, -2)$; radius: 5   **b.** center: $(0, 0)$; radius: 9

OBJECTIVE 5. Graph circles of the form $x^2 + y^2 + dx + ey + f = 0$. If the equation of a circle is not given in standard form, we complete the square to write the equation in the form $(x - h)^2 + (y - k)^2 = r^2$.

EXAMPLE 6 Find the center and radius of the circle whose equation is $x^2 + y^2 - 6x + 8y + 9 = 0$ and draw the graph.

Solution $x^2 + y^2 - 6x + 8y = -9$ Subtract 9 from both sides to isolate the variable terms.

$$(x^2 - 6x) + (y^2 + 8y) = -9 \quad \text{Group the terms in } x \text{ and } y.$$

$$(x^2 - 6x + 9) + (y^2 + 8y + 16) = -9 + 9 + 16 \quad \begin{array}{l}\text{Complete the square for } x \text{ and } y \text{ by}\\ \text{adding 9 and 16 to both sides of the}\\ \text{equation.}\end{array}$$

$$(x - 3)^2 + (y + 4)^2 = 16 \quad \text{Factor and simplify.}$$

The center is at $(3, -4)$ and the radius is 4.

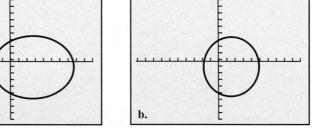

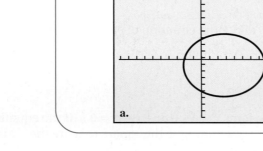

Calculator TIPS

The graph of a circle fails the vertical line test, so the equation of a circle is not a function. To graph equations that are not functions, we solve for y and graph the two resulting functions on the same screen.

EXAMPLE Graph $(x - 2)^2 + (y + 1)^2 = 16$.

Solution We solve $(x - 2)^2 + (y + 1)^2 = 16$ for y.

$$(x - 2)^2 + (y + 1)^2 = 16$$
$$(y + 1)^2 = 16 - (x - 2)^2 \qquad \text{Subtract } (x - 2)^2 \text{ from both sides.}$$
$$y + 1 = \pm\sqrt{16 - (x - 2)^2} \qquad \text{Apply the square root principle.}$$
$$y = -1 \pm \sqrt{16 - (x - 2)^2} \qquad \text{Subtract 1 from both sides.}$$

This equation defines two functions. On your graphing calculator define $Y_1 = -1 + \sqrt{16 - (x - 2)^2}$ and $Y_2 = -1 - \sqrt{16 - (x - 2)^2}$ and graph both in the window $[-8, 8]$ for x and $[-8, 8]$ for y. The resulting graph is labeled "**a**" below; note that the graph does not look like the graph of a circle. To make the graph look like a circle, select ZSquare from the [ZOOM] menu. This results in the graph labeled "**b**" below.

a.　　　　　　　　　　b.

ANSWER

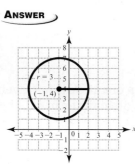

center: $(-1, 4)$; radius: 3

YOUR TURN Find the center and radius of the circle whose equation is $x^2 + y^2 + 2x - 8y + 8 = 0$ and draw the graph.

13.1 Exercises

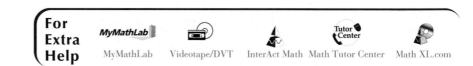

For Extra Help MyMathLab Videotape/DVT InterAct Math Math Tutor Center Math XL.com

1. Is $x^2 + 2x - 3 + y = 0$ the equation of a circle or a parabola? Explain.

2. Find the coordinates of the vertex of the parabola $x = 3y^2 - 12y + 2$.

3. In which direction does the graph of the parabola in Exercise 2 open? Why?

4. Explain how the Pythagorean theorem is used to derive the distance formula.

5. Complete the following: The set of all points that are 4 units from the point whose coordinates are $(-3, 2)$ is a _____ with center at _____ and radius of _____ .

6. How is the distance formula related to the equation of a circle?

For Exercises 7–28, find the direction the parabola opens, the coordinates of the vertex, the equation of the axis of symmetry, and draw the graph.

7. $y = (x - 1)^2 + 2$

8. $y = (x + 2)^2 - 3$

9. $y = -x^2 - 2x + 3$

10. $y = -x^2 + 4x - 1$

11. $x = (y + 2)^2 - 2$

12. $x = (y + 3)^2 + 2$

13. $x = -(y - 1)^2 + 3$

14. $x = -(y - 3)^2 + 3$

15. $x = 2(y + 2)^2 - 4$

16. $x = 3(y + 3)^2 - 4$

17. $x = -3(y + 2)^2 - 5$

18. $x = -2(y - 4)^2 + 1$

19. $x = y^2 + 4y + 3$

20. $x = y^2 - 2y - 8$

21. $x = -y^2 + 6y - 5$

22. $x = -y^2 - 4y - 3$

23. $x = 2y^2 + 8y + 3$

24. $x = 2y^2 - 4y + 1$

25. $x = 3y^2 - 6y + 3$

26. $x = 3y^2 - 12y + 9$

27. $x = -2y^2 + 4y + 5$

28. $x = -3y^2 - 6y - 2$

For Exercises 29–32, match the equation with the correct graph.

29. $x = (y + 3)^2 - 2$

a.

b.

30. $y = (x + 3)^2 - 2$

31. $y = 2x^2 - 8x + 5$

c.

d.

32. $x = 2y^2 - 8y + 5$

For Exercises 33–44, find the distance between the two points.

33. $(-4, 2)$ and $(-1, 6)$

34. $(5, -1)$ and $(1, 2)$

35. $(-8, -4)$ and $(-3, 8)$

36. $(-3, 2)$ and $(3, -6)$

37. $(-8, -10)$ and $(4, -5)$

38. $(4, -6)$ and $(10, 2)$

39. $(2, 4)$ and $(4, 8)$

40. $(-3, 2)$ and $(1, -4)$

41. $(-5, 2)$ and $(3, -2)$

42. $(3, -4)$ and $(7, 2)$

43. $(6, -2)$ and $(1, -5)$

44. $(3, -6)$ and $(-2, 1)$

For Exercises 45–48, the coordinates of the center of a circle and a point on the circle are given. Find the radius of the circle.

45. center: $(4, 2)$, point on the circle: $(8, -1)$

46. center: $(-4, 6)$, point on the circle: $(2, -2)$

47. center: $(2, -6)$, point on the circle: $(10, -1)$

48. center: $(3, -4)$, point on the circle: $(6, 8)$

For Exercises 49–64, find the center and radius and draw the graph.

49. $(x - 2)^2 + (y - 1)^2 = 4$

50. $(x - 1)^2 + (y - 3)^2 = 25$

51. $(x + 3)^2 + (y + 2)^2 = 81$

52. $(x + 5)^2 + (y + 4)^2 = 36$

53. $(x - 5)^2 + (y + 3)^2 = 49$

54. $(x + 6)^2 + (y - 2)^2 = 1$

55. $(x + 1)^2 + (y - 3)^2 = 18$

56. $(x - 6)^2 + (y + 1)^2 = 12$

57. $(x + 4)^2 + (y + 2)^2 = 32$

58. $(x - 6)^2 + (y + 5)^2 = 8$

59. $x^2 + y^2 - 2x - 6y - 39 = 0$

60. $x^2 + y^2 + 8x - 6y + 16 = 0$

61. $x^2 + y^2 + 10x - 4y - 35 = 0$

62. $x^2 + y^2 + 12x + 10y + 60 = 0$

63. $x^2 + y^2 + 14x - 4y + 49 = 0$

64. $x^2 + y^2 + 8x - 10y + 16 = 0$

For Exercises 65–68, match the equation with the correct graph.

65. $(x - 2)^2 + (y + 3)^2 = 25$

66. $(x + 2)^2 + (y - 3)^2 = 25$

67. $x^2 + y^2 + 8x + 2y - 8 = 0$

68. $x^2 + y^2 + 2x - 8y - 8 = 0$

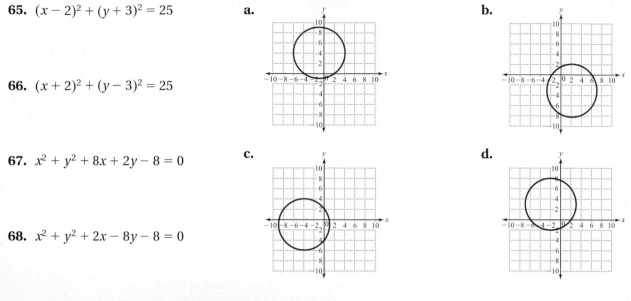

For Exercises 69–72, graph using a graphing calculator.

69. $x^2 + y^2 = 49$

70. $x^2 + y^2 = 36$

71. $(x - 2)^2 + (y + 3)^2 = 25$

72. $(x + 4)^2 + (y - 1)^2 = 9$

For Exercises 73–80, the center and radius of a circle are given. Write the equation of the circle in standard form.

73. center: $(8, -1)$; radius: 6

74. center: $(-3, 2)$; radius: 6

75. center: $(-4, -3)$; radius: 5

76. center: $(-6, -5)$; radius: 2

77. center: $(6, -2)$; radius: $\sqrt{14}$

78. center: $(-3, -3)$; radius: $\sqrt{26}$

79. center: $(-5, 2)$; radius: $3\sqrt{5}$

80. center: $(6, 2)$; radius: $2\sqrt{6}$

For Exercises 81–84, the center of a circle and a point on the circle are given. Write the equation of the circle in standard form.

81. center: $(2, 4)$, point on the circle: $(5, 8)$

82. center: $(-4, 3)$, point on the circle: $(4, 9)$

83. center: $(2, 4)$, point on the circle: $(7, 16)$

84. center: $(-6, 8)$, point on the circle: $(-12, 0)$

85. Write the equation of the set of all points that are a distance of 8 units from $(2, -5)$.

86. Write the equation of the set of all points that are a distance of 10 units from $(-3, -7)$.

87. If a rock is thrown vertically upward from the top of a building 112 feet high with an initial velocity of 96 feet per second, the height, h, above ground level after t seconds is given by $h = -16t^2 + 96t + 112$, where h is in feet and t is in seconds.

   a. What is the maximum height the rock will reach?

   b. How many seconds will it take the rock to reach its maximum height?

   c. How many seconds will it take the rock to hit the ground?

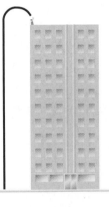

88. The path of a shell fired from ground level is in the shape of the parabola $y = 4x - x^2$, where x and y are given in kilometers.

   a. How high does the shell go?

   b. How many seconds does it take the shell to reach its maximum height?

   c. How far from its firing point does the shell land?

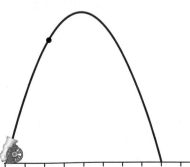

89. Arches in the shapes of parabolas are often used in construction. Find the equation of a parabolic arc that is 18 feet high at its highest point and 30 feet wide at the base, as illustrated in the following figure. Place the origin at the midpoint of the base.

18 ft.

30 ft.

90. The cross sections of satellite dishes are in the shape of parabolas. Find the equation of a dish that is 6 feet across and 1 foot deep if the vertex is at the origin and the parabola is opening upward.

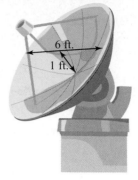

6 ft.

1 ft.

91. The percent of deaths by age per million miles driven can be approximated by the equation $y = 0.0038x^2 - 0.3475x + 8.316$, where x is the age and y is the percent. Find the percent of deaths per million miles for drivers 17 years old. (*Source: National Highway Traffic Safety Administration*)

92. The number of drivers involved in fatal accidents for a given blood alcohol content (BAC) can be approximated by $y = -8862.5x^2 + 26622.6x + 332$, where x is the BAC and y is the number of drivers involved in fatal accidents. Find the number of drivers involved in fatal accidents who had a BAC of 0.20. (*Source: National Highway Traffic Safety Administration*)

93. Bill and Don are fishing in the Gulf of Mexico in separate boats that are equipped with radios with a range of 20 miles. If we put Bill's radio at the origin of a coordinate system, what is the equation of all possible locations of Don's boat where the radios would be at their maximum range.

94. A toy plane is attached to a string pinned to the ceiling so that the plane flies in a circle. If the string is 4 feet long, write an equation that describes the path of the plane if the pin is at the origin.

95. A Ferris wheel has a diameter of 200 feet and the bottom of the Ferris wheel is 10 feet above the ground. Find the equation of the wheel if the origin is placed on the ground directly below the center of the wheel, as illustrated.

y

200 ft.

10 ft.

x

96. The Fermilab Tunnel houses the world's largest superconducting synchrotron. A cross section of the tunnel is a circle with radius of 1000 meters. Find the equation of a cross section of the tunnel if the center of the circle is at the center of the tunnel.

97. If a satellite is placed in a circular orbit of 230 kilometers above the Earth, what is the equation of the path of the satellite if the origin is placed at the center of the Earth (the radius of the Earth is approximately 6370 kilometers)?

98. The minute hand of Big Ben is 14 feet long. What is the equation of the circle swept out by the tip of the hand as it makes one complete revolution?

PUZZLE PROBLEM

Using only a pencil, a rough circle and its center can both be drawn without the point of the pencil losing contact with the paper resulting in a picture like the one shown. Explain how.

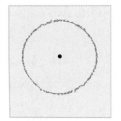

REVIEW EXERCISES

[4.1] **1.** Following is a coordinate system with the point (2, 3) plotted. Plot the points that are 4 units to the left and right of (2, 3) and give their coordinates.

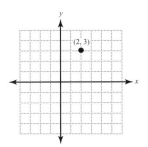

[4.4] **2.** Graph $y = \frac{2}{3}x$ and $y = -\frac{2}{3}x$ on the same set of axes.

[6.4] **3.** Factor: $25x^2 - 9y^2$

[11.1] **4.** Solve: $\frac{x^2}{16} = 1$

[4.3, 11.1] **5.** Find the x- and y-intercepts of $9x^2 + 16y^2 = 144$.

[4.3, 11.1] **6.** Find the x- and y-intercepts of $25x^2 + 9y^2 = 225$

13.2 Ellipses and Hyperbolas

1. Graph ellipses.
2. Graph hyperbolas.

OBJECTIVE 1. Graph ellipses. Suppose you drive two nails in a board and tie a string to the two nails. Now take a pencil, pull the string taut, and draw a figure around the two nails.

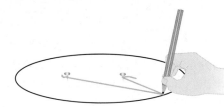

The figure you've drawn is called an **ellipse**. The locations of the two nails are the *focal points*.

DEFINITION   *Ellipse:* The set of all points the sum of whose distance from two fixed points is constant.

Ellipses occur in many situations. The orbits of the planets about the Sun are elliptical with the Sun at one focal point. The orbits of satellites about the Earth are also elliptical. The cams of compound bows are elliptical, which allows a decrease in the amount of effort required to hold the bow at full draw.

In the definition of an ellipse, the two fixed points are the *foci* (plural of *focus*) and the point halfway between the foci is the *center*. The following figure shows the graph of an ellipse with foci at $(c, 0)$ and $(-c, 0)$, x-intercepts at $(a, 0)$ and $(-a, 0)$, and y-intercepts at $(0, b)$ and $(0, -b)$. Consequently, the center is at the origin, $(0, 0)$. It can be shown that $c^2 = a^2 - b^2$ for ellipses in which $a > b$.

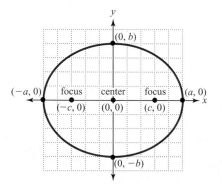

Using the distance formula, it can be shown that an ellipse with these characteristics has the following equation.

RULE **The Equation of an Ellipse Centered at (0, 0)**

The equation of an ellipse with center $(0, 0)$, x-intercepts $(a, 0)$ and $(-a, 0)$, and y-intercepts $(0, b)$ and $(0, -b)$ is

$$\frac{x^2}{a^2} + \frac{y^2}{b^2} = 1$$

EXAMPLE 1 Graph each ellipse and label the x- and y-intercepts.

a. $\dfrac{x^2}{16} + \dfrac{y^2}{9} = 1$

Solution The equation can be rewritten as $\dfrac{x^2}{4^2} + \dfrac{y^2}{3^2} = 1$, so $a = 4$ and $b = 3$. Consequently, the x-intercepts are $(4, 0)$ and $(-4, 0)$ and the y-intercepts are $(0, 3)$ and $(0, -3)$.

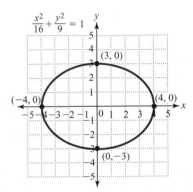

b. $25x^2 + 9y^2 = 225$

Solution We first need to write the equation in standard form: $\dfrac{x^2}{a^2} + \dfrac{y^2}{b^2} = 1$.

$$\frac{25x^2}{225} + \frac{9y^2}{225} = \frac{225}{225} \qquad \text{Divide both sides by 225.}$$

$$\frac{x^2}{9} + \frac{y^2}{25} = 1 \qquad \text{Simplify both sides.}$$

We now see that this is an equation of an ellipse with $a = 3$ and $b = 5$, so the x-intercepts are $(3, 0)$ and $(-3, 0)$ and the y-intercepts are $(0, 5)$ and $(0, -5)$.

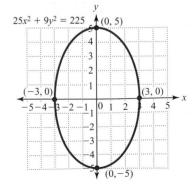

If the center of an ellipse is not at the origin, it can be shown that the equation has the following form.

The equation of an ellipse with center (h, k) is $\dfrac{(x-h)^2}{a^2} + \dfrac{(y-k)^2}{b^2} = 1$. The ellipse passes through two points that are a units to the left and right of the center, and two points that are b units above and below the center.

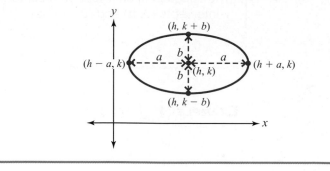

ANSWERS

a. $\dfrac{x^2}{36} + \dfrac{y^2}{9} = 1$

b. $\dfrac{(x+1)^2}{25} + \dfrac{(y-2)^2}{36} = 1$

EXAMPLE 2 Graph the ellipse. Label the center and the points above, below, to the left, and to the right of the center. $\dfrac{(x-2)^2}{36} + \dfrac{(y+3)^2}{16} = 1$

Solution Since $h = 2$, and $k = -3$, the center of the ellipse is $(2, -3)$. Also, we see that $a = 6$, which means the ellipse passes through two points 6 units to the right and left of $(2, -3)$. These points are $(8, -3)$ and $(-4, -3)$. Since $b = 4$, the ellipse passes through two points 4 units above and below the center. These points are $(2, 1)$ and $(2, -7)$.

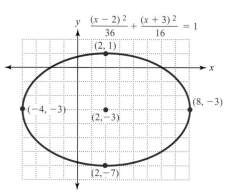

YOUR TURN Graph each ellipse. Label the center and the points above, below, to the left, and to the right of the center.

a. $\dfrac{x^2}{36} + \dfrac{y^2}{9} = 1$

b. $\dfrac{(x+1)^2}{25} + \dfrac{(y-2)^2}{36} = 1$

OBJECTIVE 2. Graph hyperbolas. The last conic is the **hyperbola**. Applications of hyperbolas include the LORAN tracking system and the orbits of some comets.

DEFINITION *Hyperbola:* The set of all points the difference of whose distances from two fixed points remains constant.

Just as with the ellipse, the two fixed points are the *foci* and the point halfway between the foci is the *center.*

RULE **Equations of Hyperbolas in Standard Form**

The equation of a hyperbola with center (0, 0), x-intercepts (a, 0) and (−a, 0), and no y-intercepts is $\dfrac{x^2}{a^2} - \dfrac{y^2}{b^2} = 1$.

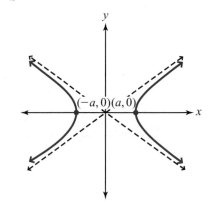

The equation of a hyperbola with center (0, 0), y-intercepts (0, b) and (0, −b), and no x-intercepts is $\dfrac{y^2}{b^2} - \dfrac{x^2}{a^2} = 1$.

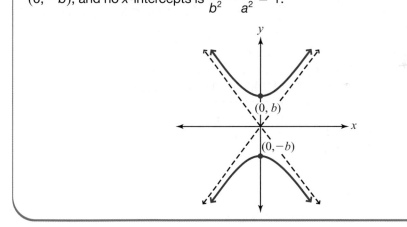

The dashed lines that intersect at the center of a hyperbola are *asymptotes* and are not a part of the graph, but are used as an aid in graphing. An asymptote is a line that the graph approaches, but does not cross, as the graph goes away from the origin. The rectangle whose vertices are (a, b), (−a, b), (a, −b), and (−a, −b) is called the *fundamental rectangle*, and the asymptotes are the extended diagonals of the fundamental rectangle, as in the following illustration.

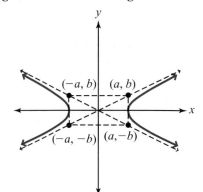

PROCEDURE Graphing a Hyperbola in Standard Form

1. Find the intercepts. If the x^2 term is positive, the x-intercepts are $(a, 0)$ and $(-a, 0)$ and there are no y-intercepts. If the y^2 term is positive, the y-intercepts are $(0, b)$ and $(0, -b)$ and there are no x-intercepts.
2. Draw the fundamental rectangle. The vertices are (a, b), $(-a, b)$, $(a, -b)$, and $(-a, -b)$.
3. Draw the asymptotes, which are the extended diagonals of the fundamental rectangle.
4. Draw the graph so that each branch passes through an intercept and approaches the asymptotes the farther they are from the origin.

EXAMPLE 3 Graph each hyperbola. Also show the fundamental rectangle with its corner points labeled, the asymptotes, and the intercepts.

a. $\dfrac{x^2}{16} - \dfrac{y^2}{9} = 1$

Solution

1. This equation can be written as $\dfrac{x^2}{4^2} - \dfrac{y^2}{3^2} = 1$, so $a = 4$ and $b = 3$. Since the x^2 term is positive, the graph has x-intercepts at $(4, 0)$ and $(-4, 0)$.
2. The fundamental rectangle has vertices at $(4, 3)$, $(-4, 3)$, $(4, -3)$, and $(-4, -3)$.
3. Draw the asymptotes.
4. Sketch the graph so it passes through the x-intercepts and then approaches the asymptotes.

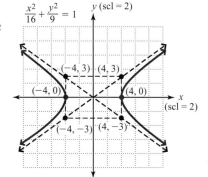

b. $\dfrac{y^2}{4} - \dfrac{x^2}{9} = 1$

Solution

1. This equation can be written as $\dfrac{y^2}{2^2} - \dfrac{x^2}{3^2} = 1$, so $b = 2$ and $a = 3$. Since the y^2 term is positive, the graph has y-intercepts at $(0, 2)$ and $(0, -2)$.
2. The fundamental rectangle has vertices at $(3, 2)$, $(-3, 2)$, $(3, -2)$, and $(-3, -2)$.
3. Draw the asymptotes.
4. Sketch the graph so it passes through the y-intercepts and then approaches the asymptotes.

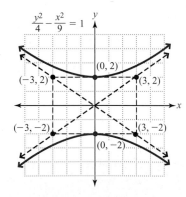

ANSWER

$\dfrac{x^2}{25} - \dfrac{y^2}{16} = 1$

YOUR TURN Graph the hyperbola. Also show the fundamental rectangle with its corner points labeled, the asymptotes, and the intercepts.

$$\dfrac{x^2}{25} - \dfrac{y^2}{16} = 1$$

Graph the ellipse $\dfrac{x^2}{9} + \dfrac{y^2}{16} = 1$. *We solve for y just as when graphing a circle.*

$$\frac{x^2}{9} + \frac{y^2}{16} = 1$$

$$\frac{y^2}{16} = 1 - \frac{x^2}{9} \qquad \textbf{Subtract } \frac{x^2}{9} \textbf{ from both sides.}$$

$$y^2 = 16\left(1 - \frac{x^2}{9}\right) \qquad \textbf{Multiply both sides by 16.}$$

$$y = \pm 4\sqrt{1 - \frac{x^2}{9}} \qquad \textbf{Apply the square root principle.}$$

Define $Y_1 = 4\sqrt{1 - \dfrac{x^2}{9}}$ *and* $Y_2 = -4\sqrt{1 - \dfrac{x^2}{9}}$ *and graph in a window* $[-5, 5]$ *for x and* $[-5, 5]$ *for y and square the window.*

Following is a summary of the conic sections.

Standard Forms

Parabola

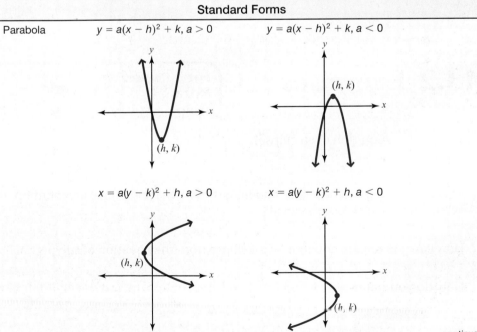

$y = a(x - h)^2 + k, a > 0$	$y = a(x - h)^2 + k, a < 0$
$x = a(y - k)^2 + h, a > 0$	$x = a(y - k)^2 + h, a < 0$

continued

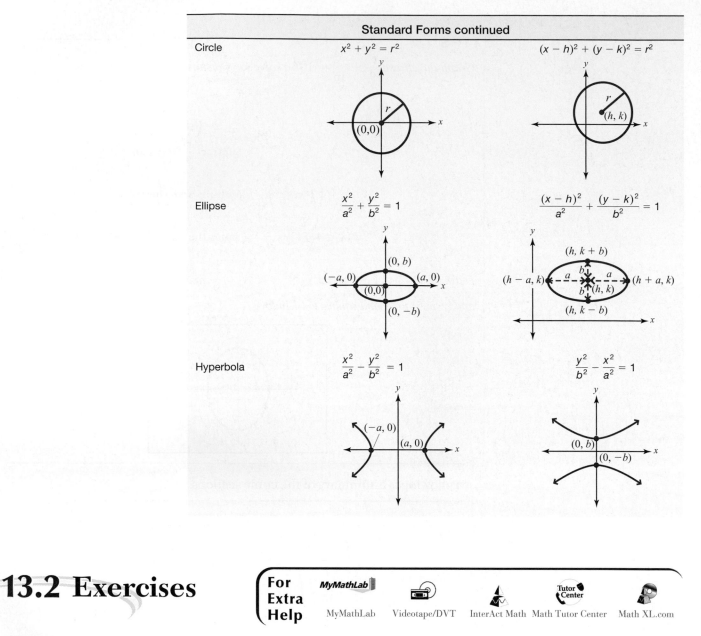

Standard Forms continued

Circle $\qquad x^2 + y^2 = r^2 \qquad\qquad (x - h)^2 + (y - k)^2 = r^2$

Ellipse $\qquad \dfrac{x^2}{a^2} + \dfrac{y^2}{b^2} = 1 \qquad\qquad \dfrac{(x - h)^2}{a^2} + \dfrac{(y - k)^2}{b^2} = 1$

Hyperbola $\qquad \dfrac{x^2}{a^2} - \dfrac{y^2}{b^2} = 1 \qquad\qquad \dfrac{y^2}{b^2} - \dfrac{x^2}{a^2} = 1$

13.2 Exercises

1. What is the definition of an ellipse?

2. In the definitions of the ellipse and hyperbola, two fixed points are mentioned. What are these two points called?

3. How can you tell the equation of an ellipse from the equation of a hyperbola?

4. How do you determine whether a hyperbola opens up and down or left and right?

5. Which one conic has an equation that has either x^2 or y^2, but not both?

6. What do the graphs of $\dfrac{x^2}{9} - \dfrac{y^2}{16} = 1$ and $\dfrac{x^2}{9} + \dfrac{y^2}{16} = 1$ have in common?

For Exercises 7–18, graph each ellipse. Label the center and the points above, below, to the left, and to the right of the center.

7. $\dfrac{x^2}{25} + \dfrac{y^2}{4} = 1$

8. $\dfrac{x^2}{81} + \dfrac{y^2}{64} = 1$

9. $\dfrac{x^2}{36} + y^2 = 1$

10. $\dfrac{x^2}{9} + \dfrac{y^2}{16} = 1$

11. $4x^2 + 9y^2 = 36$

12. $25x^2 + 4y^2 = 100$

13. $36x^2 + 4y^2 = 144$

14. $x^2 + 9y^2 = 36$

15. $\dfrac{(x-1)^2}{49} + \dfrac{(y+3)^2}{25} = 1$

16. $\dfrac{(x+3)^2}{25} + \dfrac{(y-2)^2}{64} = 1$

17. $\dfrac{(x-4)^2}{4} + \dfrac{(y+3)^2}{36} = 1$

18. $\dfrac{(x+5)^2}{9} + \dfrac{(y+3)^2}{25} = 1$

19. $\dfrac{x^2}{25} + \dfrac{y^2}{4} = 1$

20. $\dfrac{y^2}{12} + \dfrac{x^2}{8} = 1$

21. $\dfrac{(y-2)^2}{36} + \dfrac{(x+4)^2}{25} = 1$

22. $\dfrac{(x+1)^2}{16} + \dfrac{(y+2)^2}{4} = 1$

23. If the x-intercepts of an ellipse are $(-3, 0)$ and $(3, 0)$ and the y-intercepts are $(0, 4)$ and $(0, -4)$, what is the equation of the ellipse?

24. If the x-intercepts of an ellipse are $(-6, 0)$ and $(6, 0)$ and the y-intercepts are $(0, 3)$ and $(0, -3)$, what is the equation of the ellipse?

For Exercises 25–32, graph each hyperbola. Also show the fundamental rectangle with its corner points labeled, the asymptotes, and the intercepts.

25. $\dfrac{x^2}{9} - \dfrac{y^2}{4} = 1$

26. $\dfrac{x^2}{16} - \dfrac{y^2}{25} = 1$

27. $\dfrac{y^2}{36} - \dfrac{x^2}{9} = 1$

28. $\dfrac{y^2}{4} - \dfrac{x^2}{25} = 1$

29. $9x^2 - y^2 = 36$

30. $x^2 - 4y^2 = 16$

31. $16y^2 - 4x^2 = 64$

32. $9y^2 - 25x^2 = 225$

For Exercises 33–34, graph using a graphing calculator.

33. $\dfrac{x^2}{36} - \dfrac{y^2}{4} = 1$

34. $\dfrac{y^2}{9} - \dfrac{x^2}{25} = 1$

35. If a hyperbola opens left and right and the vertices of the fundamental rectangle are $(3, 2)$, $(3, -2)$, $(-3, 2)$ and $(-3, -2)$, what is the equation of the hyperbola?

36. If a hyperbola opens up and down and the vertices of the fundamental rectangle are $(5, 3)$, $(5, -3)$, $(-5, 3)$ and $(-5, -3)$, what is the equation of the hyperbola?

For Exercises 37–40, match the equation with the graph.

37. $\dfrac{x^2}{9} + \dfrac{y^2}{25} = 1$

38. $\dfrac{x^2}{9} - \dfrac{y^2}{25} = 1$

39. $\dfrac{x^2}{25} + \dfrac{y^2}{9} = 1$

40. $\dfrac{x^2}{25} - \dfrac{y^2}{9} = 1$

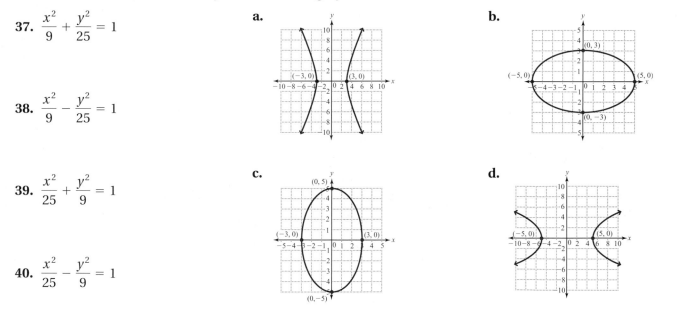

For Exercises 41–44, determine whether the graph of the equation is a circle, parabola, ellipse, or hyperbola. Do not draw the graph.

41. $9x^2 + 16y^2 = 144$

42. $16y^2 - 9x^2 = 144$

43. $x^2 + y^2 - 6x + 8y - 75 = 0$

44. $2x^2 - 12x + 23 - y = 0$

For Exercises 45–58, indicate whether the graph of the given equation is a circle, parabola, ellipse, or hyperbola, then draw the graph.

45. $(x - 2)^2 + (y + 2)^2 = 49$

46. $x^2 + y^2 = 81$

47. $y = 2(x + 1)^2 + 3$

48. $x = (y + 3)^2 - 4$

49. $\dfrac{x^2}{36} + \dfrac{y^2}{16} = 1$

50. $\dfrac{x^2}{36} + \dfrac{y^2}{81} = 1$

51. $\dfrac{x^2}{4} - \dfrac{y^2}{25} = 1$

52. $\dfrac{y^2}{49} - \dfrac{x^2}{4} = 1$

53. $y = 2x^2 + 8x + 6$

54. $x = y^2 + 6$

55. $\dfrac{x^2}{16} - \dfrac{y^2}{16} = 1$

56. $x = -2y^2 + 4y - 3$

57. $\dfrac{(x + 3)^2}{9} + \dfrac{(y + 2)^2}{36} = 1$

58. $\dfrac{(x - 1)^2}{4} + (y + 3)^2 = 1$

For Exercises 59–66, solve.

59. A bridge over a waterway has an arch in the form of half an ellipse. The equation of the ellipse is $400x^2 + 256y^2 = 102{,}400$.

  a. A sailboat, the top of whose mast is 18 feet above the water, is approaching the arch. Will the mast clear the bridge? Why or why not?

  b. How wide is the base of the arch?

60. A highway passes beneath an overpass that is in the shape of half an ellipse. The overpass is 15 feet high at the center and 40 feet wide at the base.

  a. What is the equation of the ellipse?

  b. A truck that is 10 feet wide and carrying a load that is 14 feet high is approaching the bridge. If the truck goes down the middle of the road, will the load clear the bridge? Why or why not?

61. The comet Epoch has an orbit that is in the shape of an ellipse with the Sun at one of the foci. The equation is approximately $\dfrac{x^2}{3.6^2} + \dfrac{y^2}{2.88^2} = 1$, where x and y are in astronomical units (an astronomical unit is 93,000,000 miles). Sketch the graph of the comet Epoch. (*Source: Orbital Motion*, A. E. Roy, Institute of Physics Publishing, London)

62. The planet Pluto has an orbit that is in the shape of an ellipse with the Sun at one of the foci. The equation is approximately $\dfrac{x^2}{39.4^2} + \dfrac{y^2}{38.2^2} = 1$, where x and y are measured in astronomical units. Sketch the graph of the orbit of Pluto. (*Hint:* Make each unit on the x- and y-axes equal to 10.) (*Source: Orbital Motion*, A. E. Roy, Institute of Physics Publishing, London)

63. Compound bows have elliptical cams that decrease the amount of effort required to hold the bow at full draw. If a cam on a bow is 4 inches from top to bottom and 3 inches across, what is the equation of the ellipse if the origin is at the center of the cam?

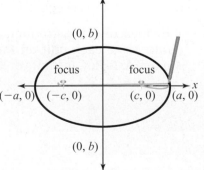

64. One of the most popular exercise machines is an elliptical trainer in which your foot moves in an elliptical path. On a typical machine, the length of the stride is about 19 inches and the height varies from 3 to 5 inches depending on the settings. Write the equation for the path that your foot takes if the total length of the stride is 19 inches and the total height is 4 inches. (*Source:* Precor National Headquarters)

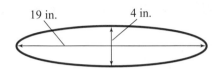

65. If a source of light or sound is placed at one focal point of an elliptic reflector, the light or sound is reflected through the other focal point. This principle is used in a lithotripter that uses sound waves to crush kidney stones by placing a source of sound at one focal point and the kidney stone at the other. If the elliptic reflector is based on the ellipse $\dfrac{x^2}{25} + \dfrac{y^2}{9} = 1$, how many units from the center should the kidney stone be placed? (*Hint:* $c^2 = a^2 - b^2$.)

66. The same principle used in Exercise 65 is also used in whispering rooms. The room is in the shape of an elliptic reflector where one person speaks at one focal point and the other places his or her ear at the other and can hear sounds as faint as a whisper. If a whispering room is based on the ellipse $\dfrac{x^2}{169} + \dfrac{y^2}{144} = 1$, how many units from the center of the ellipse would the two people stand? (*Hint:* $c^2 = a^2 - b^2$.)

Collaborative Exercises THE ELLIPTICAL TABLECLOTH

Recall that an ellipse can be drawn by fixing the ends of a string to the foci and then tracing out the ellipse. By considering the following figures we can determine an expression for the length of the string and also a relationship between a, b, and c.

1. Use the figure to write a formula for the length of the string.

2. In the following figure, notice that the string forms an isosceles triangle and the y-axis splits that triangle into two identical right triangles.

   a. Find the length of the hypotenuse of each of those right triangles.

   b. What expression describes the length of the string?

   c. Use the Pythagorean theorem to write a formula relating *a*, *b*, and *c*.

   d. Solve the formula for *c*.

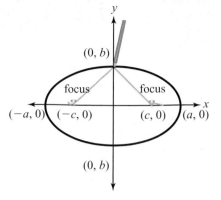

Suppose we are to make an elliptical tablecloth for an elliptical table that is 76 inches long and 58 inches wide. The tablecloth is to drape 6 inches over the edge of the table all the way around plus we need an additional inch for the hem. We have a large rectangular piece of cloth to make the tablecloth. To trace the ellipse on the cloth, we need to know the string length and the location of the foci.

  3. Find the dimensions of the elliptical tablecloth taking into account the amount it needs to drape and also the hem.

  4. How long must the string be in order to trace the ellipse?

  5. How far from the center are the foci located?

PUZZLE PROBLEM

Suppose an ellipse is drawn using the method described on p. 000. Assuming the ellipse is centered at the origin, if the nails are 8 inches apart and the string is 10 inches in length, what are the x- and y-intercepts?

REVIEW EXERCISES

[2.4]   **1.** Solve $3x^2 + y = 6$ for *y*.

[9.1]   **2.** Solve the following system using the graphical method. $\begin{cases} x + y = 3 \\ 2x + y = 4 \end{cases}$

[9.2]   **3.** Solve the following system using the substitution method. $\begin{cases} 2x + y = 1 \\ 3x + 4y = -6 \end{cases}$

[9.3]   **4.** Solve the following system using the elimination method. $\begin{cases} 2x + 3y = 6 \\ 3x - 4y = -25 \end{cases}$

For Exercises 5 and 6, solve.

[6.6]   **5.** $3x^2 + 10x - 8 = 0$

[11.1]   **6.** $4x^2 = 36$

13.3 Nonlinear Systems of Equations

OBJECTIVES

1. Solve nonlinear systems of equations using substitution.
2. Solve nonlinear systems of equations using elimination.

We solved systems of linear equations in Chapter 9. Now we will solve **nonlinear systems** of equations.

DEFINITION | *Nonlinear system of equations:* A system of equations that contains at least one nonlinear equation.

The types of equations in a nonlinear system determine the number of solutions that are possible for the system. For example, a system containing a quadratic equation and a linear equation can have 0, 1, or 2 points of intersection and therefore 0, 1, or 2 solutions, as shown by the following figures.

Connection In Chapter 4, we learned that the number of solutions for a system of linear equations depended on the relative positions of the graphs of each of the equations. We see a similar relationship between the graphs and the number of solutions here.

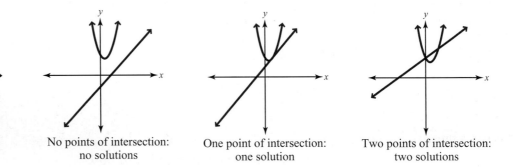

No points of intersection: no solutions

One point of intersection: one solution

Two points of intersection: two solutions

OBJECTIVE 1. Solve nonlinear systems of equations using substitution. When solving nonlinear systems, if one equation is linear, or one of the variables in an equation is isolated, the substitution method is usually preferred.

EXAMPLE 1 Solve using the substitution method.

a. $\begin{cases} y = 2(x + 2)^2 - 2 \\ 2x + y = -2 \end{cases}$ ◀ **Note:** *The graphs are a parabola and a line, so there will be 0, 1, or 2 solutions.*

Solution It is easier to solve the linear equation for one of its variables and substitute into the nonlinear equation. Since the coefficient of y is 1, we will solve $2x + y = -2$ for y.

$$y = -2x - 2 \qquad \text{Subtract } -2x \text{ from both sides.}$$
$$-2x - 2 = 2(x + 2)^2 - 2 \qquad \text{Substitute } -2x - 2 \text{ for } y \text{ in } y = 2(x + 2)^2 - 2.$$
$$-2x - 2 = 2(x^2 + 4x + 4) - 2 \qquad \text{To solve this quadratic equation, we need to write it in the form } ax^2 + bx + c = 0. \text{ First, we square } x + 2.$$
$$-2x - 2 = 2x^2 + 8x + 6 \qquad \text{Distribute 2, then combine like terms.}$$
$$0 = 2x^2 + 10x + 8 \qquad \text{Add } 2x \text{ and 2 to both sides.}$$
$$0 = x^2 + 5x + 4 \qquad \text{Divide both sides by 2.}$$
$$0 = (x + 1)(x + 4) \qquad \text{Factor.}$$
$$0 = x + 1 \quad \text{or} \quad 0 = x + 4 \qquad \text{Use the zero-factor theorem.}$$
$$-1 = x \quad \text{or} \quad -4 = x \qquad \text{Solve each equation.}$$

To find y, substitute -1 and -4 for x in $y = -2x - 2$.

$$y = -2(-1) - 2 \qquad\qquad y = -2(-4) - 2$$
$$y = 2 - 2 \qquad\qquad\qquad y = 8 - 2$$
$$y = 0 \qquad\qquad\qquad\quad y = 6$$

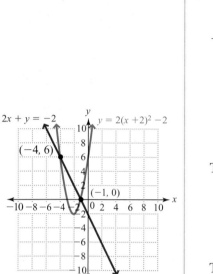

The solutions are $(-1, 0)$ and $(-4, 6)$, as verified by the graph to the left.

b. $\begin{cases} y = \sqrt{x + 2} \\ x^2 + y^2 = 8 \end{cases}$

Solution Since $y = \sqrt{x + 2}$ is already solved for y, substitute $\sqrt{x + 2}$ for y in $x^2 + y^2 = 8$.

$$x^2 + (\sqrt{x + 2})^2 = 8 \qquad \text{Substitute } \sqrt{x + 2} \text{ for } y \text{ in } x^2 + y^2 = 8.$$
$$x^2 + x + 2 = 8 \qquad \text{Square } \sqrt{x + 2}.$$
$$x^2 + x - 6 = 0 \qquad \text{Since the equation is now quadratic, we subtract 8 from both sides to get the form } ax^2 + bx + c = 0.$$
$$(x + 3)(x - 2) = 0 \qquad \text{Factor}$$
$$x + 3 = 0 \quad \text{or} \quad x - 2 = 0 \qquad \text{Use the zero-factor theorem.}$$
$$x = -3 \quad \text{or} \quad x = 2 \qquad \text{Solve each equation.}$$

To find y, substitute -3 and 2 for x in $y = \sqrt{x + 2}$.

$$y = \sqrt{-3 + 2} \qquad\qquad y = \sqrt{2 + 2}$$
$$y = \sqrt{-1} \qquad\qquad\qquad y = \sqrt{4}$$
$$\qquad\qquad\qquad\qquad\qquad y = 2$$

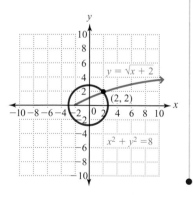

Since $\sqrt{-1}$ is imaginary, the only real solution is $(2, 2)$, as verified by the graph to the left.

Note: *A system containing a root function and a circle, as in Example 1b, can have 0, 1, or 2 solutions, as shown.*

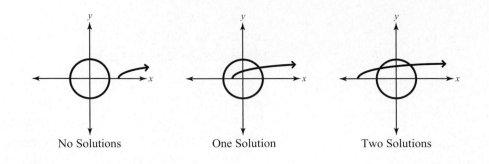

No Solutions   One Solution   Two Solutions

YOUR TURN   Solve using the substitution method.

a. $\begin{cases} 2x - y = 1 \\ x^2 + y^2 = 1 \end{cases}$ b. $\begin{cases} xy = 3 \\ 4x - y = 1 \end{cases}$

OBJECTIVE 2. Solve nonlinear systems of equations using elimination. If neither equation contains a radical expression or both equations contain the same powers of the variables, the elimination method can be used.

EXAMPLE 2   Solve using the elimination method. $\begin{cases} 9x^2 - 4y^2 = 20 \\ x^2 + y^2 = 8 \end{cases}$

Solution $\begin{cases} 9x^2 - 4y^2 = 20 & \text{(Equation 1)} \\ x^2 + y^2 = 8 & \text{(Equation 2)} \end{cases}$

To eliminate y^2, multiply equation 2 by 4 and add the equations.

$$9x^2 - 4y^2 = 20 \qquad\qquad\qquad 9x^2 - 4y^2 = 20$$
$$x^2 + y^2 = 8 \xrightarrow{\text{Multiply by 4}} \underline{4x^2 + 4y^2 = 32}$$
$$13x^2 \qquad\quad = 52 \qquad \textbf{Add the equations.}$$
$$x^2 = 4 \qquad \textbf{Divide both sides by 13.}$$
$$x = \pm 2 \qquad \textbf{Find the square roots of 4.}$$

To find y, substitute 2 and -2 for x in one of the original equations. We will use $x^2 + y^2 = 8$.

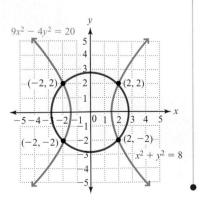

$$2^2 + y^2 = 8 \qquad\qquad (-2)^2 + y^2 = 8$$
$$4 + y^2 = 8 \qquad\qquad 4 + y^2 = 8$$
$$y^2 = 4 \qquad\qquad\qquad y^2 = 4$$
$$y = \pm 2 \qquad\qquad\qquad y = \pm 2$$

Therefore, $(2, 2)$ and $(2, -2)$ are solutions.   Therefore, $(-2, 2)$ and $(-2, -2)$ are solutions.

This system has four solutions: $(2, 2)$, $(2, -2)$, $(-2, 2)$, and $(-2, -2)$, as verified by the graph to the left.

Note: *If the graphs are a hyperbola and a circle both centered at the origin as in Example 2, there will be 0, 2, or 4 solutions, as shown.*

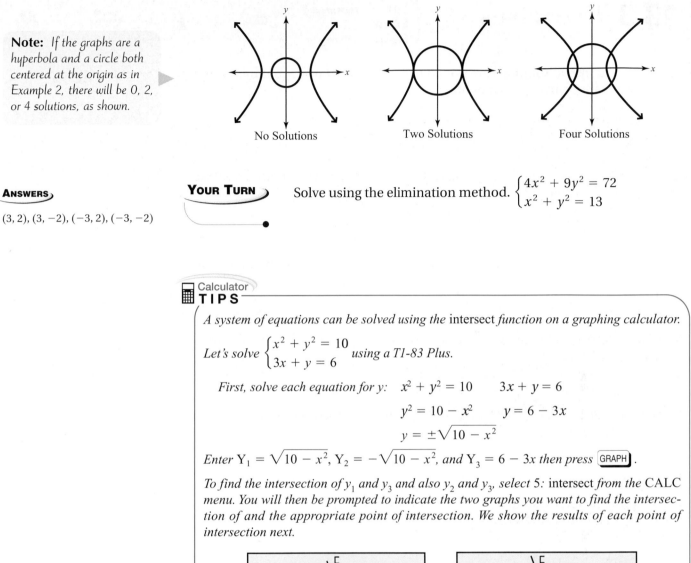

No Solutions   Two Solutions   Four Solutions

YOUR TURN   Solve using the elimination method. $\begin{cases} 4x^2 + 9y^2 = 72 \\ x^2 + y^2 = 13 \end{cases}$

Calculator TIPS

A system of equations can be solved using the intersect *function on a graphing calculator.*

Let's solve $\begin{cases} x^2 + y^2 = 10 \\ 3x + y = 6 \end{cases}$ *using a T1-83 Plus.*

First, solve each equation for y:

$$x^2 + y^2 = 10 \qquad 3x + y = 6$$
$$y^2 = 10 - x^2 \qquad y = 6 - 3x$$
$$y = \pm\sqrt{10 - x^2}$$

Enter $Y_1 = \sqrt{10 - x^2}$, $Y_2 = -\sqrt{10 - x^2}$, *and* $Y_3 = 6 - 3x$ *then press* [GRAPH].

To find the intersection of y_1 *and* y_3 *and also* y_2 *and* y_3, *select* 5: intersect *from the* CALC *menu. You will then be prompted to indicate the two graphs you want to find the intersection of and the appropriate point of intersection. We show the results of each point of intersection next.*

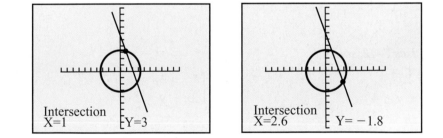

Intersection
X=1   Y=3

Intersection
X=2.6   Y=−1.8

1. **a.** How many real solutions are possible for a system of two equations whose graphs are a circle and a parabola?

   b. Draw a figure illustrating such a system with two solutions.

2. **a.** How many real solutions are possible for a system of two equations whose graphs are a line and a hyperbola?

   b. Draw a figure illustrating such a system with one solution.

3. Which method (substitution or elimination) would you use to solve the system $\begin{cases} 2x + y = 8 \\ 4x^2 + 3y^2 = 24 \end{cases}$? Why? (Do not attempt to solve the system.)

4. Which method (substitution or elimination) would you use to solve the system $\begin{cases} 3x^2 - 4y^2 = 12 \\ 2x^2 + 3y^2 = 24 \end{cases}$? Why? (Do not attempt to solve the system.)

5. Without solving the system, what is the number of possible solutions of the system $\begin{cases} 3x - y = 6 \\ 4x^2 + 9y^2 = 36 \end{cases}$?

6. Is it possible for a system of two equations whose graphs are an ellipse and a hyperbola (both centered at the origin) to have three solutions? Why or why not?

For Exercises 7–42, solve.

7. $\begin{cases} y = 2x^2 \\ 2x + y = 4 \end{cases}$

8. $\begin{cases} x^2 + 2y = 1 \\ 2x + y = 2 \end{cases}$

9. $\begin{cases} y = x^2 + 4x + 4 \\ 3x - y = -6 \end{cases}$

10. $\begin{cases} y = 6x - x^2 \\ 2x - y = -3 \end{cases}$

11. $\begin{cases} x^2 + y^2 = 25 \\ x - y = -1 \end{cases}$

12. $\begin{cases} x^2 + y^2 = 25 \\ x - 7y = -25 \end{cases}$

13. $\begin{cases} x^2 + y^2 = 13 \\ 2x + y = 7 \end{cases}$

14. $\begin{cases} x^2 + y^2 = 10 \\ 3x + y = 6 \end{cases}$

15. $\begin{cases} x^2 + 2y^2 = 4 \\ x + y = 5 \end{cases}$

16. $\begin{cases} 3x^2 + y^2 = 9 \\ 2x + y = 11 \end{cases}$

17. $\begin{cases} y = 2x^2 + 3 \\ 2x - y = -3 \end{cases}$

18. $\begin{cases} y = -3x^2 - 2 \\ 3x + y = -2 \end{cases}$

19. $\begin{cases} y = (x - 3)^2 + 2 \\ y = -(x - 2)^2 + 3 \end{cases}$

20. $\begin{cases} y = 2(x + 4)^2 + 2 \\ y = -2(x + 3)^2 + 4 \end{cases}$

21. $\begin{cases} x^2 + y^2 = 20 \\ x^2 - y^2 = 12 \end{cases}$

22. $\begin{cases} x^2 + y^2 = 48 \\ x^2 - y^2 = 24 \end{cases}$

23. $\begin{cases} 9x^2 + 4y^2 = 145 \\ x^2 + y^2 = 25 \end{cases}$

24. $\begin{cases} 4x^2 + 9y^2 = 72 \\ x^2 + y^2 = 13 \end{cases}$

25. $\begin{cases} 4x^2 - y^2 = 15 \\ x^2 + y^2 = 5 \end{cases}$

26. $\begin{cases} 9x^2 - 4y^2 = 32 \\ x^2 + y^2 = 5 \end{cases}$

27. $\begin{cases} x^2 + 3y^2 = 36 \\ x = y^2 - 6 \end{cases}$

28. $\begin{cases} 4x^2 + y^2 = 16 \\ y = x^2 - 4 \end{cases}$

29. $\begin{cases} 9x^2 - 16y^2 = 144 \\ 4x^2 + 9y^2 = 36 \end{cases}$

30. $\begin{cases} 9x^2 + 4y^2 = 36 \\ 4x^2 - 9y^2 = 36 \end{cases}$

31. $\begin{cases} y = x^2 \\ x^2 + y^2 = 20 \end{cases}$

32. $\begin{cases} 16x^2 + y^2 = 128 \\ y = 2x^2 \end{cases}$

33. $\begin{cases} 25x^2 - 16y^2 = 400 \\ x^2 + 4y^2 = 16 \end{cases}$

34. $\begin{cases} 9y^2 - 25x^2 = 225 \\ 4y^2 + 25x^2 = 100 \end{cases}$

35. $\begin{cases} xy = 2 \\ 4x^2 + y^2 = 8 \end{cases}$

36. $\begin{cases} xy = 4 \\ 2x^2 - y^2 = 4 \end{cases}$

37. $\begin{cases} y = x^2 - 2x - 3 \\ y = -x^2 + 6x + 7 \end{cases}$

38. $\begin{cases} y = \dfrac{1}{3}x^2 - 2 \\ 3x^2 + 9y^2 = 36 \end{cases}$

39. $\begin{cases} 4x^2 + 5y^2 = 36 \\ 4x^2 - 3y^2 = 4 \end{cases}$

40. $\begin{cases} 4x^2 + 7y^2 = 64 \\ 4x^2 - 3y^2 = 24 \end{cases}$

41. $\begin{cases} x = -y^2 + 2 \\ x^2 - 5y^2 = 4 \end{cases}$

42. $\begin{cases} x = -y^2 + 2 \\ 9x^2 - 45y^2 = 36 \end{cases}$

43. Create a system of two equations whose graphs are a circle and a line for which there is no solution. Include the graphs.

44. Create a system of two equations whose graphs are a circle and a hyperbola for which there are exactly two solutions. Include the graphs.

45. The sum of the squares of two integers is 34 and the difference of their squares is 16. Find the integers.

46. The difference of the squares of two integers is 32 and their product is 12. Find the integers.

47. A computer keyboard has an area of 144 square inches and a perimeter of 52 inches. Find the length and width.

48. A rectangular living room has an area of 48 square meters and a perimeter of 28 meters. Find the length and width.

49. If p is in dollars and x is in hundreds of units, the demand function for a certain style of chair is given by $p = -3x^2 + 120$ and the supply function is given by $p = 11x + 28$. The *market equilibrium* occurs when the number produced is equal to the number demanded. Find the number of chairs and the price per chair when market equilibrium is reached.

50. If y is in dollars and x is the number of cell phones manufactured (in thousands), a cell phone manufacturer has determined that the cost y to manufacture x cell phones is given by $y = 5x^2 + 30x + 50$ and the revenue from the sales is given by $y = 13x^2$. The *break-even point* is the point (x, y) for which the cost equals the revenue. Find the number of units necessary to break even by solving the system.

For Exercises 51–54, use a graphing calculator to verify the results of the exercise given.

51. Exercise 17   **52.** Exercise 18   **53.** Exercise 19   **54.** Exercise 20

REVIEW EXERCISES

[4.6] *For Exercises 1 and 2, determine if the ordered pair is a solution for $x + 3y \le 6$.*

1. $(0, 0)$ **2.** $(3, 4)$

[4.6] *For Exercises 3–6, graph.*

3. $y \ge 3$ **4.** $y < 2x + 3$ **5.** $2x + 3y < -6$ **6.** $x \ge -2$

13.4 Nonlinear Inequalities and Systems of Inequalities

OBJECTIVES

1. Graph nonlinear inequalities.
2. Graph the solution set of a system of nonlinear inequalities.

OBJECTIVE 1. Graph nonlinear inequalities. In Section 3.4, we graphed linear inequalities like $x + 2y > 6$ by first graphing the line corresponding to $x + 2y = 6$ and then shading the appropriate region on one side of that boundary line. Recall that we used a dashed line for $<$ and $>$ and a solid line for $\leq$ and $\geq$. We determined which side to shade by using a test point on one side of the boundary.

We use a similar procedure to graph nonlinear inequalities such as $x^2 + y^2 \leq 25$. The boundary is the graph of $x^2 + y^2 = 25$, which is a circle with center at $(0, 0)$ and radius of 5. The solution set of $x^2 + y^2 \leq 25$ contains all ordered pairs on the circle (boundary) along with all ordered pairs inside it or all ordered pairs outside it. To determine which of those two regions is correct, we choose an ordered pair from one of the regions to test in the inequality. Let's test $(0, 0)$.

$0^2 + 0^2 \leq 25$ **Substitute $(0, 0)$ into $x^2 + y^2 \leq 25$.**

$0 \leq 25$ **Simplify. The inequality is true.**

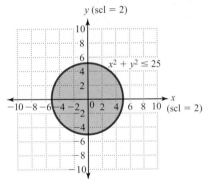

Since $0 \leq 25$ is true, all ordered pairs in the region containing $(0, 0)$ are in the solution set, so we shade that region.

Ordered pairs in the region outside the circle are not in the solution set because they do not solve the inequality. To illustrate, let's test $(6, 8)$.

$6^2 + 8^2 \leq 25$ **Substitute $(6, 8)$ into $x^2 + y^2 \leq 25$.**

$100 \leq 25$ **Simplify. The inequality is false.**

Note: *If the inequality had been $x^2 + y^2 < 25$, we would have drawn a dashed circle.*

Since $100 \leq 25$ is false, the region containing $(6, 8)$ is not in the solution set, so we do not shade that region.

Our example suggests the following procedure for graphing nonlinear inequalities.

PROCEDURE   Graphing Nonlinear Inequalities

1. Graph the related equation. If the inequality symbol is $\leq$ or $\geq$, draw the graph as a solid curve. If the inequality symbol is $>$ or $<$, draw a dashed curve.
2. The graph divides the coordinate plane into at least two regions. Test an ordered pair from each region by substituting it into the inequality. If the ordered pair satisfies the inequality, then shade the region containing that ordered pair.

EXAMPLE 1 Graph the inequality.

a. $y > 2(x - 1)^2 + 3$

Solution We first graph the related equation, $y = 2(x - 1)^2 + 3$.

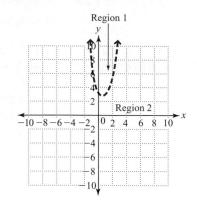

Note: *The parabola is dashed because the inequality is $>$. Also, notice that the graph divides the coordinate plane into two regions: Region 1 above the boundary and Region 2 below.*

We now test an ordered pair from each region.

Note: *To make computations easier, choose ordered pairs with 0 for at least one of the coordinates.*

Region 1: We choose (0, 6).

$6 > 2(0 - 1)^2 + 3$	**Substitute.**
$6 > 2(-1)^2 + 3$	**Simplify.**
$6 > 5$	

True, so Region 1 is in the solution set.

Region 2: We choose (0, 0).

$0 > 2(0 - 1)^2 + 3$	**Substitute.**
$0 > 2(-1)^2 + 3$	**Simplify.**
$0 > 5$	

False, so Region 2 is not in the solution set.

Since ordered pairs in only Region 1 solve the inequality, we shade only that region. The solution set contains all ordered pairs in Region 1. Ordered pairs on the parabola are not in the solution set.

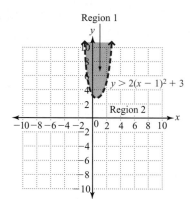

b. $\dfrac{x^2}{9} - \dfrac{y^2}{25} \leq 1$

Solution Graph the related equation, $\dfrac{x^2}{9} - \dfrac{y^2}{25} = 1$.

Note: *The hyperbola is solid because the inequality is $\leq$. Also, notice that the hyperbola divides the coordinate plane into three regions.*

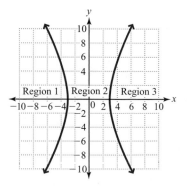

We test an ordered pair from each region.

Region 1: We choose $(-5, 0)$.

$\dfrac{(-5)^2}{9} - \dfrac{0^2}{25} \le 1$ **Substitute.**

$\dfrac{25}{9} \le 1$ **Simplify.**

False, so Region 1 is not in the solution set.

Region 2: We choose $(0, 0)$.

$\dfrac{0^2}{9} - \dfrac{0^2}{25} \le 1$ **Substitute.**

$0 \le 1$ **Simplify.**

True, so Region 2 is in the solution set.

Region 3: We choose $(5, 0)$.

$\dfrac{(5)^2}{9} - \dfrac{0^2}{25} \le 1$ **Substitute.**

$\dfrac{25}{9} \le 1$ **Simplify.**

False, so Region 3 is not in the solution set.

Since Region 2 is the only region containing ordered pairs that solve the inequality, we shade only that region. The solution set contains all ordered pairs on the hyperbola along with all ordered pairs in Region 2.

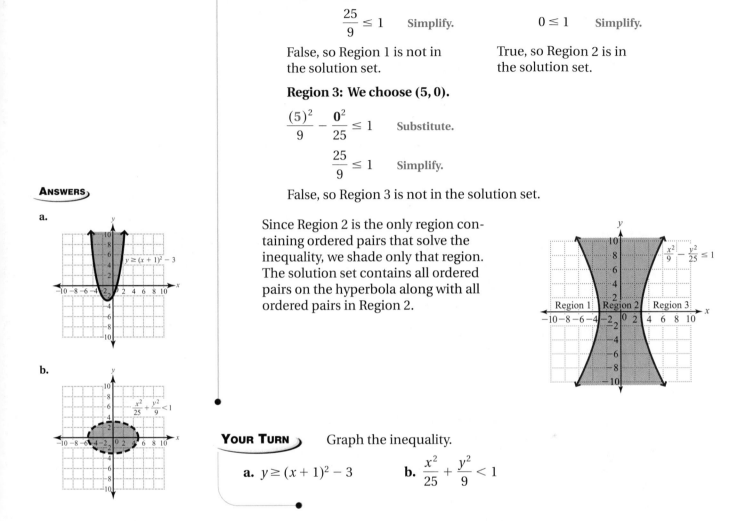

YOUR TURN   Graph the inequality.

a. $y \ge (x+1)^2 - 3$ **b.** $\dfrac{x^2}{25} + \dfrac{y^2}{9} < 1$

OBJECTIVE 2. Graph the solution set of a system of nonlinear inequalities.   In Section 9.7, we solved systems of linear inequalities by graphing each inequality on the same grid. The solution set contained all ordered pairs in the region where the inequalities' solution sets overlapped together with ordered pairs on the portion of any solid line touching the region of overlap. We use a similar procedure for systems of nonlinear inequalities.

EXAMPLE 2   Graph the solution set of the system of inequalities.

a. $\begin{cases} y \ge x^2 - 4 \\ 2x - y < 2 \end{cases}$

Solution We begin by graphing $y \ge x^2 - 4$ and $2x - y < 2$.

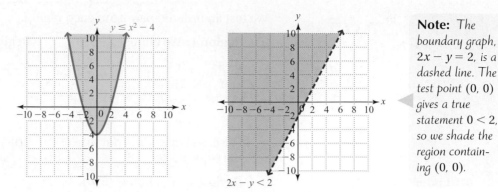

$y \le x^2 - 4$

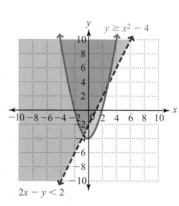

Note: *The boundary graph, $y = x^2 - 4$, is a parabola opening up with vertex at $(0, -4)$. The test point $(0, 0)$ gives the true statement $0 \ge -4$, so we shade the region containing $(0, 0)$.*

Note: *The boundary graph, $2x - y = 2$, is a dashed line. The test point $(0, 0)$ gives a true statement $0 < 2$, so we shade the region containing $(0, 0)$.*

Note: *Remember, ordered pairs on dashed lines or curves are not in the solution set for a system of inequalities.*

If we place both graphs on the same grid, their intersection (purple shading) is the solution region for the system. In addition to ordered pairs in the purple shaded region, the solution set also contains all ordered pairs on the portions of the parabola that touch the purple shaded region.

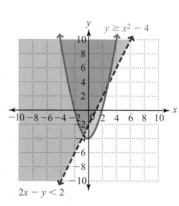

$y \ge x^2 - 4$

$2x - y < 2$

b. $\begin{cases} x^2 + y^2 \le 49 \\ \dfrac{x^2}{16} - \dfrac{y^2}{9} < 1 \\ y \ge 2x + 2 \end{cases}$

Solution Graph each inequality on the same coordinate system.

The graph of $x^2 + y^2 \le 49$ is a circle and its interior. The graph of $\dfrac{x^2}{16} - \dfrac{y^2}{9} < 1$ is the region between the branches of the hyperbola with the curve dashed. The graph of $y \ge 2x + 2$ is a solid line and the region above the line. The solution set for the system contains all ordered pairs in the region where the three graphs overlap (purple shaded region) together with all ordered pairs on the portion of the circle and the line that touches the purple shaded region.

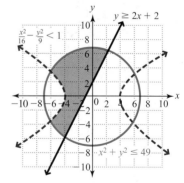

$y \ge 2x + 2$

$\dfrac{x^2}{16} - \dfrac{y^2}{9} < 1$

$x^2 + y^2 \le 49$

ANSWERS

a.

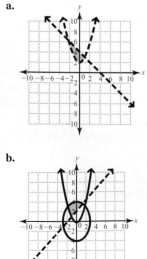

b.

YOUR TURN Graph the solution set of the system of inequalities.

a. $\begin{cases} y > x^2 + 2 \\ x + y < 4 \end{cases}$

b. $\begin{cases} \dfrac{x^2}{9} + \dfrac{y^2}{16} \le 1 \\ y \ge x^2 \\ -x + y > 2 \end{cases}$

13.4 Exercises

For Extra Help

MyMathLab   Videotape/DVT   InterAct Math   Math Tutor Center   Math XL.com

1. Do the ordered pairs that lie on the boundary curve solve the inequality $9x^2 + 4y^2 > 36$? Why or why not? Would the graph be drawn solid or dashed? Why?

2. Discuss how graphing linear inequalities like $x + 2y \geq 4$ is similar to graphing a nonlinear inequality like $y \geq x^2 + 3$.

3. Draw the graph of a system of nonlinear inequalities for which there is no solution and explain why there is no solution.

4. Describe the procedure you would use to graph $\dfrac{x^2}{9} - \dfrac{y^2}{16} > 1$.

For Exercises 5–24, graph the inequality.

5. $x^2 + y^2 \geq 4$

6. $x^2 + y^2 \leq 9$

7. $y > x^2$

8. $y < -x^2$

9. $y < (x - 1)^2 + 2$

10. $y > 2(x + 2)^2 - 3$

11. $\dfrac{x^2}{16} + \dfrac{y^2}{9} \leq 1$

12. $\dfrac{x^2}{4} + \dfrac{y^2}{9} \geq 1$

13. $\dfrac{x^2}{25} - \dfrac{y^2}{4} > 1$

14. $\dfrac{x^2}{36} - \dfrac{y^2}{16} < 1$

15. $x^2 + y^2 < 16$

16. $x^2 + y^2 > 25$

17. $y \geq -2(x-3)^2 + 3$

18. $y < -(x+3)^2 - 2$

19. $\dfrac{x^2}{49} + \dfrac{y^2}{25} > 1$

20. $\dfrac{x^2}{16} + \dfrac{y^2}{9} \leq 1$

21. $\dfrac{y^2}{25} - \dfrac{x^2}{4} \leq 1$

22. $\dfrac{y^2}{36} - \dfrac{x^2}{25} > 1$

23. $y < x^2 + 4x - 5$

24. $y > x^2 - 6x - 6$

For Exercises 25–50, graph the solution set of each system of inequalities.

25. $\begin{cases} y \geq x^2 \\ x + y \leq 3 \end{cases}$

26. $\begin{cases} y < -x^2 \\ 2x - y < 4 \end{cases}$

27. $\begin{cases} x - 2y < -4 \\ x^2 + y^2 < 16 \end{cases}$

28. $\begin{cases} 3x + 2y \ge -6 \\ x^2 + y^2 \le 25 \end{cases}$

29. $\begin{cases} x^2 + y^2 \ge 9 \\ x^2 + y^2 \le 25 \end{cases}$

30. $\begin{cases} x^2 + y^2 > 4 \\ x^2 + y^2 > 9 \end{cases}$

31. $\begin{cases} y > x^2 + 1 \\ 2x + y < 3 \end{cases}$

32. $\begin{cases} 3x - y \le 2 \\ y \le -x^2 + 2 \end{cases}$

33. $\begin{cases} y < -x^2 + 3 \\ y > x^2 - 2 \end{cases}$

34. $\begin{cases} y > -x^2 + 2 \\ y < x^2 + 5 \end{cases}$

35. $\begin{cases} \dfrac{x^2}{25} + \dfrac{y^2}{9} \le 1 \\ x^2 + y^2 \ge 4 \end{cases}$

36. $\begin{cases} \dfrac{x^2}{9} + \dfrac{y^2}{4} \le 1 \\ x^2 + y^2 \le 4 \end{cases}$

37. $\begin{cases} \dfrac{x^2}{25} + \dfrac{y^2}{9} < 1 \\ y > x^2 + 1 \end{cases}$

38. $\begin{cases} \dfrac{x^2}{9} + \dfrac{y^2}{4} < 1 \\ y < -x^2 + 2 \end{cases}$

39. $\begin{cases} \dfrac{x^2}{9} - \dfrac{y^2}{4} \le 1 \\ \dfrac{x^2}{25} + \dfrac{y^2}{9} \le 1 \end{cases}$

40. $\begin{cases} \dfrac{x^2}{16} - \dfrac{y^2}{9} \ge 1 \\ \dfrac{x^2}{36} + \dfrac{y^2}{16} \le 1 \end{cases}$

41. $\begin{cases} \dfrac{x^2}{4} - \dfrac{y^2}{4} > 1 \\ y > 2 \end{cases}$

42. $\begin{cases} \dfrac{x^2}{9} - \dfrac{y^2}{9} > 1 \\ x > 3 \end{cases}$

43. $\begin{cases} 3x + 2y \le 6 \\ x - y > -3 \\ x + 6y \ge 2 \end{cases}$

44. $\begin{cases} 2x + 3y < 6 \\ x - 2y > -4 \\ x + 5y \ge -4 \end{cases}$

45. $\begin{cases} \dfrac{x^2}{16} + \dfrac{y^2}{4} \le 1 \\ x^2 + y^2 \le 9 \\ y \le x \end{cases}$

46. $\begin{cases} \dfrac{x^2}{9} + \dfrac{y^2}{16} < 1 \\ x^2 + y^2 < 9 \\ y > x + 1 \end{cases}$

47. $\begin{cases} \dfrac{x^2}{49} - \dfrac{y^2}{16} \leq 1 \\ \dfrac{x^2}{64} + \dfrac{y^2}{36} \leq 1 \\ 2x - y \leq -3 \end{cases}$

48. $\begin{cases} \dfrac{x^2}{9} + \dfrac{y^2}{49} \leq 1 \\ y \geq x^2 + 3 \\ x + y \leq 6 \end{cases}$

49. $\begin{cases} y < 2x^2 + 8 \\ 2x + y > 3 \\ x - y < 4 \end{cases}$

50. $\begin{cases} y < x^2 + 2 \\ 2x + y < 4 \\ 2x - y > -5 \end{cases}$

REVIEW EXERCISES

[1.5]   **1.** Simplify: $(3 \cdot 1 + 2) + (3 \cdot 2 + 2) + (3 \cdot 3 + 2) + (3 \cdot 4 + 2)$

[1.7]   **2.** Evaluate: $\dfrac{n}{2}(a_1 + a_n)$ if $n = 12$, $a_1 = -8$, and $a_n = 60$

[1.7]   **3.** Evaluate: $a_1 + (n - 1)d$ if $a_1 = -12$, $n = 25$, and $d = -3$

[1.7] *For Exercises 4 and 5, evaluate the expression for $n = 1, 2, 3,$ and 4.*

4. $2n^2 - 3$

5. $\dfrac{(-1)^n}{3n - 2}$

[3.2]   **6.** 5% of 550 is what number?

Chapter 13 Summary

Defined Terms

Section 13.1
Conic Section (p. 1032)
Circle (p. 1035)
Radius (p. 1035)

Section 13.2
Ellipse (p. 1046)
Hyperbola (p. 1048)

Section 13.3
Nonlinear system of equations (p. 1060)

Procedures, Rules, and Key Examples

Procedures/Rules	Key Example(s)

Section 13.1 The Parabola and Circle

The graph of an equation in the form $x = a(y - k)^2 + h$ is a parabola with vertex at (h, k). The parabola opens to the right if $a > 0$ and to the left if $a < 0$. The equation of the axis of symmetry is $y = k$.

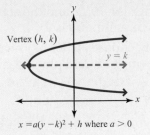

$x = a(y - k)^2 + h$ where $a > 0$

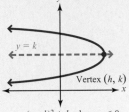

$x = a(y - k)^2 + h$ where $a < 0$

Example 1: For $x = -2(y - 3)^2 - 2$, determine whether the graph opens left or right, find the vertex and the axis of symmetry, and draw the graph.

Solution: Since $a = -2$ and $-2 < 0$, the graph opens to the left. The vertex is $(-2, 3)$ and the equation of the axis of symmetry is $y = 3$.

Choose values for y near the axis of symmetry and plot the graph.

x	y
−10	1
−4	2
−4	4
−10	5

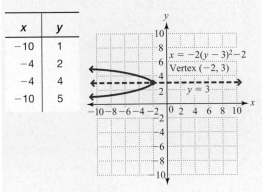

$x = -2(y - 3)^2 - 2$
Vertex $(-2, 3)$
$y = 3$

continued

| Procedures/Rules | Key Example(s) |

Section 13.1 (continued)

Example 2: For $x = y^2 - 4y - 5$, determine whether the graph opens left or right, find the vertex and axis of symmetry, and draw the graph.

Solution: Since $a = 1$ and $1 > 0$ the graph opens to the right. To find the vertex and axis of symmetry, we will complete the square.

$$x = y^2 - 4y - 5$$
$$x + 5 = y^2 - 4y \qquad \text{Add 5 to both sides.}$$
$$x + 5 + 4 = y^2 - 4y + 4 \qquad \text{Add 4 to both sides.}$$
$$x + 9 = (y - 2)^2 \qquad \text{Simplify both sides.}$$
$$x = (y - 2)^2 - 9 \qquad \text{Subtract 9 from both sides.}$$

The vertex is $(-9, 2)$ and the equation of the axis of symmetry is $y = 2$.

To graph, choose values of y near the axis of symmetry and plot the graph.

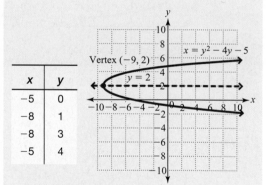

x	y
−5	0
−8	1
−8	3
−5	4

The distance, d, between two points with coordinates (x_1, y_1) and (x_2, y_2), can be found using the formula

$$d = \sqrt{(x_2 - x_1)^2 + (y_2 - y_1)^2}$$

Example 3: Find the distance between the points whose coordinates are $(-4, 3)$ and $(2, -1)$.

Solution: Let $(x_1, y_1) = (-4, 3)$ and $(x_2, y_2) = (2, -1)$.

$$d = \sqrt{(x_2 - x_1)^2 + (y_2 - y_1)^2}$$
$$d = \sqrt{(2 - (-4))^2 + (-1 - 3)^2}$$
$$\qquad\qquad\qquad\qquad \text{Substitute.}$$
$$d = \sqrt{52} \qquad\qquad \text{Simplify.}$$
$$d = \sqrt{4 \cdot 13} \qquad\quad \text{Rewrite 52 as } 4 \cdot 13.$$
$$d = 2\sqrt{13}$$

continued

Procedures/Rules	Key Example(s)
Section 13.1 (continued) The equation of a circle with center at (h, k) and radius r is $(x - h)^2 + (y - k)^2 = r^2$.	**Example 4:** Find the center and radius of the circle whose equation is $(x - 2)^2 + (y + 4)^2 = 25$ and draw the graph. Solution: Rewrite the equation as $$(x - 2)^2 + (y - (-4))^2 = 5^2.$$ Since $h = 2$ and $k = -4$, the center is $(2, -4)$ and $r = 5$. 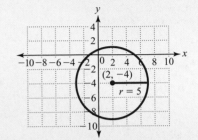

Example 5: Write the equation of the circle whose center is $(5, -6)$ and whose radius is 8.
Solution: Substitute for h, k, and r in
$$(x - h)^2 + (y - k)^2 = r^2.$$
$$(x - 5)^2 + (y - (-6))^2 = 8^2$$
$$(x - 5)^2 + (y + 6)^2 = 64 \qquad \textbf{Simplify.}$$

To graph circles in the form $x^2 + y^2 + dx + ey + f = 0$, complete the square in x and y to put the equation in the form $(x - h)^2 + (y - k)^2 = r^2$.

Example 6: Find the center and radius of the circle whose equation is $x^2 + y^2 + 8x - 2y + 8 = 0$ and draw the graph.
Solution: Complete the square in x and y.
$$x^2 + 8x + y^2 - 2y = -8 \qquad \begin{array}{l}\textbf{Group the}\\\textbf{terms in } x\\\textbf{and } y \textbf{ and}\\\textbf{subtract 8.}\end{array}$$
$$x^2 + 8x + \mathbf{16} + y^2 - 2y + \mathbf{1} = -8 + \mathbf{16} + \mathbf{1}$$
$$\textbf{Complete the square.}$$
$$(x + 4)^2 + (y - 1)^2 = 9$$
$$\begin{array}{l}\textbf{Factor on the left and}\\\textbf{simplify on the right.}\end{array}$$
The center is $(-4, 1)$ and the radius is 3.

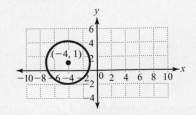

continued

Procedures/Rules	Key Example(s)

Section 13.2 Ellipses and Hyperbolas

The equation of an ellipse with center $(0, 0)$, x-intercepts $(a, 0)$ and $(-a, 0)$, and y-intercepts $(0, b)$ and $(0, -b)$, is $\dfrac{x^2}{a^2} + \dfrac{y^2}{b^2} = 1$.

Example 1: Graph $\dfrac{x^2}{49} + \dfrac{y^2}{36} = 1$.

Solution: The equation can be rewritten as $\dfrac{x^2}{7^2} + \dfrac{y^2}{6^2} = 1$, so $a = 7$ and $b = 6$. Consequently, the x-intercepts are $(7, 0)$ and $(-7, 0)$ and the y-intercepts are $(0, 6)$ and $(0, -6)$.

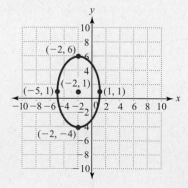

The equation of an ellipse with center (h, k) is $\dfrac{(x - h)^2}{a^2} + \dfrac{(y - k)^2}{b^2} = 1$.

The ellipse passes through two points that are a units to the left and right of the center, and two points that are b units above and below the center.

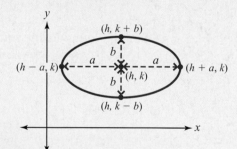

Example 2: Graph $\dfrac{(x + 2)^2}{9} + \dfrac{(y - 1)^2}{25} = 1$.

Solution: The equation can be rewritten as $\dfrac{(x - (-2))^2}{3^2} + \dfrac{(y - 1)^2}{5^2} = 1$, so $h = -2$, $k = 1$, $a = 3$, and $b = 5$. The center is $(-2, 1)$. To find other points on the ellipse, go 3 units left and right of the center and 5 units up and down from the center.

continued

Procedures/Rules	Key Example(s)

Section 13.2 (continued)

The equation of a hyperbola with center $(0, 0)$, x-intercepts $(a, 0)$ and $(-a, 0)$, and no y-intercepts, is $\dfrac{x^2}{a^2} - \dfrac{y^2}{b^2} = 1$.

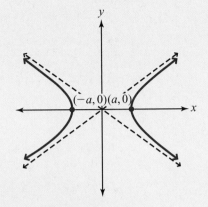

The equation of a hyperbola with center $(0, 0)$, y-intercepts $(0, b)$ and $(0, -b)$, and no x-intercepts, is $\dfrac{y^2}{b^2} - \dfrac{x^2}{a^2} = 1$.

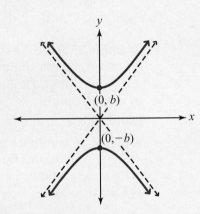

To graph a hyperbola:
1. Find the intercepts. If the x^2 term is positive, the x-intercepts are $(a, 0)$ and $(-a, 0)$ and there are no y-intercepts. If the y^2 term is positive, the y-intercepts are $(0, b)$ and $(0, -b)$ and there are no x-intercepts.
2. Draw the fundamental rectangle. The vertices are (a, b), $(-a, b)$, $(a, -b)$, and $(-a, -b)$.
3. Draw the asymptotes, which are the extended diagonals of the fundamental rectangle.
4. Draw the graph so that each branch passes through an intercept and approaches the asymptotes the farther they are from the origin.

Example 3: Graph $\dfrac{x^2}{36} - \dfrac{y^2}{16} = 1$.

Solution: The equation can be rewritten as $\dfrac{x^2}{6^2} - \dfrac{y^2}{4^2} = 1$, so $a = 6$ and $b = 4$. Since the x^2 term is positive, the x-intercepts are $(6, 0)$ and $(-6, 0)$ and there are no y-intercepts. The fundamental rectangle has vertices $(6, 4)$, $(-6, 4)$, $(6, -4)$, and $(-6, -4)$.

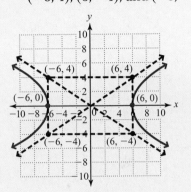

continued

Procedures/Rules	Key Example(s)

Section 13.3 Nonlinear Systems of Equations

If a nonlinear system has a linear equation, the substitution method is usually preferred. Solve the linear equation for one of its variables and substitute for that variable in the other equation.

Example 1: Solve the system
$$\begin{cases} y = (x-3)^2 + 2 \\ 2x - y = -4 \end{cases}$$ by substitution.

(**Note:** This is a parabola and a line.)

Solution:

$$y = 2x + 4$$ Solve $2x - y = -4$ for y.

$$2x + 4 = (x-3)^2 + 2$$ Substitute $2x + 4$ for y in $y = (x-3)^2 + 2$.

$$2x + 4 = x^2 - 6x + 11$$ Simplify the right side.

$$0 = x^2 - 8x + 7$$ Write in the form $ax^2 + bx + c = 0$.

$$0 = (x-7)(x-1)$$ Factor.

$$x - 7 = 0 \quad \text{or} \quad x - 1 = 0$$ Set each factor equal to 0.

$$x = 7 \quad \text{or} \quad x = 1$$ Solve each equation.

To find y, substitute 7 and 1 into $y = 2x + 4$.

$$y = 2(7) + 4 \qquad y = 2(1) + 4$$
$$y = 18 \qquad\qquad y = 6$$

Solutions: $(7, 18)$, $(1, 6)$

If neither equation is linear, the elimination method is often preferred.

Example 2: Solve the system
$$\begin{cases} 4x^2 + 5y^2 = 36 \ \text{(Equation 1)} \\ x^2 + y^2 = 8 \quad\ \text{(Equation 2)} \end{cases}$$
by elimination.

(**Note:** This is an ellipse and a circle.)

Solution: To eliminate x^2, multiply Equation 2 by -4 and add the equations.

$$4x^2 + 5y^2 = 36 \qquad\qquad 4x^2 + 5y^2 = 36$$
$$x^2 + y^2 = 8 \longrightarrow \underline{-4x^2 - 4y^2 = -32}$$
$$\text{Multiply by } -4 \qquad\qquad y^2 = 4$$
$$y = \pm 2$$

Substitute 2 and -2 for y into either equation to find x. Let's use $x^2 + y^2 = 8$.

$$x^2 + 2^2 = 8 \qquad x^2 + (-2)^2 = 8$$
$$x^2 + 4 = 8 \qquad x^2 + 4 = 8$$
$$x^2 = 4 \qquad\quad x^2 = 4$$
$$x = \pm 2 \qquad\quad x = \pm 2$$

Solutions: $(2, 2)$, $(2, -2)$, $(-2, 2)$, and $(-2, -2)$

Procedures/Rules	Key Example(s)

Section 13.4 Nonlinear Inequalities and Systems of Inequalities

Graphing nonlinear inequalities.

1. Graph the related equation. If the inequality symbol is $\leq$ or $\geq$, draw the graph as a solid curve. If the inequality symbol is $<$ or $>$, draw a dashed curve.
2. The graph divides the coordinate plane into at least two regions. Test an ordered pair from each region by substituting it into the inequality. If the ordered pair satisfies the inequality, then shade the region that contains that ordered pair.

Example 1: Graph $y \geq (x - 2)^2 - 1$.

Solution: The related equation is a parabola with vertex $(2, -1)$ and opening upward. Since the inequality is $\geq$, all ordered pairs on the parabola are in the solution set, so we draw it with a solid curve.

We will choose $(0, 0)$ as a test point.

$0 \geq (0 - 2)^2 - 1$

$0 \geq 3$ which is false, so $(0, 0)$ is not in the solution set.

Choose $(2, 0)$ as a test point.

$0 \geq (2 - 2)^2 - 1$

$0 \geq -1$ which is true, so $(2, 0)$ is in the solution set. Shade the region that contains $(2, 0)$.

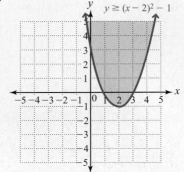

To graph a system of inequalities, graph the solution sets of the individual inequalities on the same grid. The solution set of the system is the intersection of the solution sets of the individual inequalities.

Example 2: Graph the solution set of

$$\begin{cases} \dfrac{x^2}{4} + \dfrac{y^2}{16} \leq 1 \\ x^2 + y^2 \geq 9 \end{cases}.$$

Solution: The graph of $\dfrac{x^2}{4} + \dfrac{y^2}{16} \leq 1$ is an ellipse and all points inside the ellipse. The graph of $x^2 + y^2 \geq 9$ is a circle and all points outside the circle. So, the solution set of the system is the set of all points inside the ellipse and outside the circle.

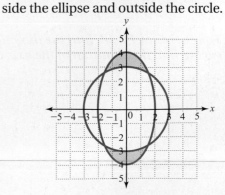

Chapter 13 Review Exercises

For Exercises 1–6, answer true or false.

[13.1] **1.** The graph of $x = -3(y - 2)^2 + 5$ opens down.

[13.1] **2.** The graph of $(x + 2)^2 + (y - 3)^2 = 36$ is a circle with center at $(-2, 3)$ and radius 6.

[13.2] **3.** A parabola has asymptotes.

[13.2] **4.** Choose any two points on an ellipse. The sum of their distances from the foci is the same.

[13.3] **5.** It is possible for a circle and a parabola to intersect in three points.

[13.4] **6.** The graph of the system $\begin{cases} x^2 + y^2 > 36 \\ 2x + y > 3 \end{cases}$ lies inside the circle and above the line.

For Exercises 7–10, fill in the blanks.

[13.1] **7.** The distance between the points whose coordinates are $(-3, 4)$ and $(3, -4)$ is _____.

[13.1] **8.** The graph of $x^2 + y^2 + 8x - 2y + 13 = 0$ is a circle with center at _____ and radius of _____.

[13.2] **9.** The graph of $\dfrac{x^2}{16} + \dfrac{y^2}{36} = 1$ is a(n) _____ with x-intercepts at _____.

[13.2] **10.** The fundamental rectangle for the graph of $\dfrac{x^2}{25} - \dfrac{y^2}{16} = 1$ has vertices at _____.

Equations and
Inequalities

Exercises 11–64

[13.1] **For Exercises 11–14, find the direction the parabola opens, the coordinates of the vertex, and the equation of the axis of symmetry. Draw the graph.**

11. $y = 2(x - 3)^2 - 5$

12. $y = -2(x + 2)^2 + 3$

13. $x = -(y - 2)^2 + 4$

14. $x = 2(y + 3)^2 - 2$

[13.1] For Exercises 15–18, find the direction the parabola opens, the coordinates of the vertex, the equation of the axis of symmetry, and draw the graph.

15. $x = y^2 + 6y + 8$

16. $x = 2y^2 - 8y - 6$

17. $x = -y^2 - 2y + 3$

18. $x = -3y^2 - 12y - 9$

[13.1] For Exercises 19 and 20, find the distance between the two points.

19. $(-1, -2)$ and $(-5, 1)$

20. $(2, -5)$ and $(-2, 3)$

[13.1] 21. The center of a circle is at $(6, -4)$ and the circle passes through $(-2, 2)$. What is the radius of the circle?

[13.1] 22. The center of a circle is at $(-6, 8)$ and the circle passes through $(3, -4)$. What is the equation of the circle?

[13.1] For Exercises 23–26, find the center and radius and draw the graph.

23. $(x - 3)^2 + (y + 2)^2 = 25$

24. $(x + 5)^2 + (y - 1)^2 = 4$

25. $x^2 + y^2 - 4x + 8y + 11 = 0$

26. $x^2 + y^2 + 10x + 2y + 22 = 0$

[13.1] *For Exercises 27–28, the center and radius of a circle are given. Write the equation of each circle in standard form.*

27. Center: $(6, -8)$, $r = 9$

28. Center: $(-3, -5)$, $r = 10$

[13.1] 29. If a heavy object is thrown vertically upward with an initial velocity of 32 feet per second from the top of a building 128 feet high, its height above the ground after t seconds is given by $h = -16t^2 + 32t + 128$, where h is the height in feet and t is the time in seconds.

  a. What is the maximum height the object will reach?

  b. How many seconds will it take the object to reach its maximum height?

  c. How many seconds will it take for the object to strike the ground?

[13.1] 30. To lay out the border of a circular flower bed, sticks are tied to each end of a rope that is 20 feet long. One person holds one of the sticks station-ary while another person uses the other stick to trace out a circular path while keeping the rope taut. If the center of the circle is at the stationary stick, what is the equation of the circle?

[13.2] *For Exercises 31–36, graph each ellipse. Label the center and the points above, below, to the left, and to the right of the center.*

31. $\dfrac{x^2}{49} + \dfrac{y^2}{25} = 1$

32. $\dfrac{x^2}{9} + \dfrac{y^2}{25} = 1$

33. $4x^2 + 9y^2 = 36$

34. $25x^2 + 9y^2 = 225$

35. $\dfrac{(x + 2)^2}{16} + \dfrac{(y - 3)^2}{4} = 1$

36. $\dfrac{(x - 1)^2}{9} + \dfrac{(y + 4)^2}{25} = 1$

[13.2] *For Exercises 37–40, graph each hyperbola. Also show the fundamental rectangle with its corner points labeled, the asymptotes, and the intercepts.*

37. $\dfrac{x^2}{25} - \dfrac{y^2}{16} = 1$

38. $\dfrac{x^2}{36} - \dfrac{y^2}{9} = 1$

39. $y^2 - 9x^2 = 36$

40. $25y^2 - 9x^2 = 225$

[13.2] 41. A bridge over a canal is in the shape of half an ellipse. If the highest point of the bridge is 15 feet above the water and the base of the bridge is 50 feet across, what is the equation of the ellipse, half of which forms the bridge?

[13.2] 42. The cam of a compound bow is elliptical and is 4 inches long and 3.25 inches wide, as shown in the figure. What is the equation of the ellipse forming the shape of the cam?

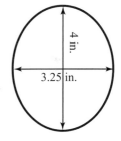

[13.3] *For Exercises 43–50, solve.*

43. $\begin{cases} y = 2x^2 - 3 \\ 2x - y = -1 \end{cases}$

44. $\begin{cases} x^2 + y^2 = 17 \\ x - y = 3 \end{cases}$

45. $\begin{cases} 25x^2 + 3y^2 = 100 \\ 2x - y = -3 \end{cases}$

46. $\begin{cases} x^2 + y^2 = 64 \\ x^2 - y^2 = 64 \end{cases}$

47. $\begin{cases} x^2 + y^2 = 25 \\ 25y^2 - 16x^2 = 256 \end{cases}$

48. $\begin{cases} x^2 + 5y^2 = 36 \\ 4x^2 - 7y^2 = 36 \end{cases}$

49. $\begin{cases} y = x^2 - 2 \\ 4x^2 + 5y^2 = 36 \end{cases}$

50. $\begin{cases} y = x^2 - 1 \\ 4y^2 - 5x^2 = 16 \end{cases}$

[13.3] 51. The sum of the squares of two integers is 89 and the difference of their squares is 39. Find the integers.

[13.3] 52. A rectangular rug has a perimeter of 36 feet and an area of 80 square feet. Find the length and width of the rug.

53. Exercise 43.

54. Exercise 49.

[13.4] *For Exercises 55–58, graph the inequality.*

55. $x^2 + y^2 \leq 64$

56. $y < 2(x - 3)^2 + 4$

57. $\dfrac{y^2}{25} + \dfrac{x^2}{49} > 1$

58. $\dfrac{x^2}{25} - \dfrac{y^2}{36} < 1$

[13.4] *For Exercises 59–64, graph the solution set of the system of inequalities.*

59. $\begin{cases} y \geq x^2 - 3 \\ 2x + y < 2 \end{cases}$

60. $\begin{cases} x + 2y < 4 \\ x^2 + y^2 \leq 25 \end{cases}$

61. $\begin{cases} y > x^2 - 2 \\ y < -x^2 + 1 \end{cases}$

62. $\begin{cases} \dfrac{x^2}{9} + \dfrac{y^2}{25} \leq 1 \\ x^2 + y^2 \geq 9 \end{cases}$

63. $\begin{cases} \dfrac{y^2}{4} - \dfrac{x^2}{9} \leq 1 \\ \dfrac{x^2}{9} + \dfrac{y^2}{25} \leq 1 \end{cases}$

64. $\begin{cases} \dfrac{y^2}{4} + \dfrac{x^2}{16} \leq 1 \\ x^2 + y^2 \leq 9 \\ y \leq x + 1 \end{cases}$

Chapter 13 Practice Test

For Exercises 1–3, find the direction the parabola opens, the coordinates of the vertex, the equation of the axis of symmetry, and draw the graph.

1. $y = 2(x + 1)^2 - 4$

2. $x = -2(y - 3)^2 + 1$

3. $x = y^2 + 4y - 3$

4. Find the distance between the points whose coordinates are $(2, -3)$ and $(6, -5)$.

For Exercises 5 and 6, find the center and radius, and draw the graph.

5. $(x + 4)^2 + (y - 3)^2 = 36$

6. $x^2 + y^2 + 4x - 10y + 20 = 0$

7. Write the equation of the circle with center $(2, -4)$ and that passes through the point $(-4, 4)$.

For Exercises 8–11, graph the equation and label relevant points. If the graph is a hyperbola, show the fundamental rectangle with its corner points labeled, the asymptotes, and the intercepts.

8. $16x^2 + 36y^2 = 576$

9. $\dfrac{(x + 2)^2}{4} + \dfrac{(y - 1)^2}{25} = 1$

10. $\dfrac{x^2}{49} - \dfrac{y^2}{25} = 1$

11. $\dfrac{y^2}{9} - \dfrac{x^2}{16} = 1$

For Exercises 12–15, solve.

12. $\begin{cases} y = (x+1)^2 + 2 \\ 2x + y = 8 \end{cases}$

13. $\begin{cases} 3x - y = 4 \\ x^2 + y^2 = 34 \end{cases}$

14. $\begin{cases} x^2 + y^2 = 13 \\ 3x^2 + 4y^2 = 48 \end{cases}$

15. $\begin{cases} x^2 - 2y^2 = 1 \\ 4x^2 + 7y^2 = 64 \end{cases}$

For Exercises 16 and 17, graph.

16. $y \le -2(x+3)^2 + 2$

17. $\dfrac{x^2}{4} + \dfrac{y^2}{9} > 1$

18. Graph the solution set. $\begin{cases} y \ge x^2 - 4 \\ \dfrac{x^2}{9} + \dfrac{y^2}{16} \le 1 \end{cases}$

19. An arch is in the shape of a parabola as shown in the figure. If we place the origin, O, as indicated, find the height of the arch and the distance across the base if the equation is $y = -\dfrac{1}{2}x^2 + 18$.

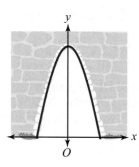

20. The path of a person's foot using an elliptical exercise machine on a particular setting is an ellipse. If the length of the stride is 19 inches and the height of the stride is 2 inches, what is the equation of the path of the foot?

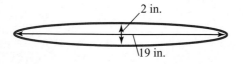

2 in.

19 in.

Chapters 1–13 Cumulative Review Exercises

For Exercises 1–6, answer true or false.

[3.1]   **1.** If $\dfrac{a}{b} = \dfrac{c}{d}$, then $ad = bc$ where $b \neq 0$ and $d \neq 0$.

[12.1]   **2.** A function must be one-to-one to have an inverse that is a function.

[12.3]   **3.** Logarithms are exponents.

[4.7]   **4.** The vertical line test can help determine if a graph is a function.

[5.1]   **5.** A positive base raised to a negative exponent simplifies to a negative number.

[13.2]   **6.** A circle has asymptotes.

For Exercises 7–10, fill in the blank.

[13.1]   **7.** The equation of a circle with center at (h, k) and radius of r is _____.

[10.7]   **8.** $\sqrt{-1} =$ _____

[13.1]   **9.** The formula for finding the distance between two points (x_1, y_1) and (x_2, y_2) is _____ .

[9.6]   **10.** If $A = \begin{bmatrix} a_1 & b_1 \\ a_2 & b_2 \end{bmatrix}$, then $\det(A) = \begin{vmatrix} a_1 & b_1 \\ a_2 & b_2 \end{vmatrix} =$ _____.

For Exercises 11–16, simplify.

[1.5]   **11.** $6 - (-3) + 2^{-4}$

[10.7] **12.** $\sqrt{-20}$

[7.2]   **13.** $7x \cdot \dfrac{8}{21x^2}$

Expressions

Exercises 11–18

[7.5]   **14.** $\dfrac{\dfrac{1}{x}}{3 + \dfrac{1}{x^2}}$

[7.1]   **15.** $\dfrac{x - y}{4y - 4x}$

[5.4]   **16.** $(5m^3 n^{-2})^{-3}$

For Exercises 17 and 18, factor completely.

[6.4]   **17.** $k^4 - 81$

[6.1]   **18.** $ax + bx + ay + by$

For Exercises 19–28, solve.

[9.3] **19.** $\begin{cases} 2x + y = 3 \\ 2x - y = 5 \end{cases}$

[11.2] 20. $4x^2 - 2x + 1 = 0$

[8.2] **21.** $|2x + 1| = |x - 3|$

[10.6] 22. $\sqrt{x - 3} = 7$

[11.3] 23. $u^4 - 3u^2 - 4 = 0$

[11.3] 24. $9 + 24x^{-1} + 16x^{-2} = 0$

[12.2] 25. $9^x = 27$

[13.3] 26. $\begin{cases} x + y = 5 \\ x^2 + y^2 = 4 \end{cases}$

[12.3] 27. $\log_3 x = -2$

[12.3] 28. $\log_3(x + 1) - \log_3 x = 2$

For Exercises 29 and 30, solve the inequality, then: a. Graph the solution set.
b. Write the solution set in
set-builder notation.
c. Write the solution set in
interval notation.

[8.3] **29.** $|3x + 1| > 4$

[11.5] 30. $5m^2 - 3m < 0$

[8.5] **31.** Given $f(x) = 2x + 1$ and $g(x) = x^2 - 1$, find

    a. $(f + g)(x)$ **b.** $(f - g)(x)$ **c.** $(f \cdot g)(x)$

    d. $(f/g)(x)$ **e.** $(f \circ g)(x)$ **f.** $(g \circ f)(x)$

[12.1] 32. Determine whether $f(x) = \dfrac{x - 4}{5}$ and $g(x) = 5x + 4$ are inverses.

[4.5] **33.** Write the equation of a line perpendicular to $3x + 4y = 8$ and passing through the point $(3, -8)$. Write the equation in slope-intercept form and in standard form.

[12.3] 34. Convert $5^3 = 125$ into logarithmic form.

[12.4] 35. Write as a sum or difference of multiples of logarithms: $\log_5 x^5 y$

[13.1] 36. Write the equation of a circle in standard form with a center at $(2, 4)$ passing through the point $(7, 16)$.

For Exercises 37–41, graph.

[9.7] 37. $\begin{cases} y < -x + 2 \\ y \geq x - 4 \end{cases}$

[12.3] 38. $f(x) = \log_3 x$

[13.2] 39. $\dfrac{x^2}{4} + \dfrac{y^2}{9} = 1$

[13.2] 40. $\dfrac{x^2}{16} - \dfrac{y^2}{36} = 1$

[13.4] 41. $\begin{cases} y < x + 2 \\ x^2 + y^2 < 9 \end{cases}$

For Exercises 42–50, solve.

[3.2] 42. The percentage of infants regularly sleeping in a bed with parents or caregivers increased from 5.5% in 1993 to 12.8% in 2000. By how much did the percent increase? (*Source:* Time, January 27, 2003)

[2.4] 43. The volume of a cylinder can be found using the formula $V = \pi r^2 h$.

   a. Solve the formula for h.

   b. Suppose the volume of a cylinder is 15π cubic inches. Find its height.

[3.3 or 9.2] **44.** During 2001, shrimp overtook tuna as the top-selling seafood in the United States. It is estimated that each person consumed a combined total of 6.3 pounds of these two seafoods and that each person ate $\frac{1}{2}$ pound more of shrimp than tuna. How many pounds of each were consumed? (*Source: Good Housekeeping*, March 2003)

[7.7] **45.** Karen and Eric operate a landscaping company. Karen can mow, edge, and trim hedges on a quarter acre lot in 2 hours. Eric takes 1.5 hours to do the same work. How much time would it take them working together?

[3.5] **46.** A total of $5000 is invested in three funds. The money market fund pays 5%, the income fund pays 6%, and the growth fund pays 3%. The total annual interest from the three accounts is $255 and the amount in the growth fund is $500 less than the amount invested in the money market account. Find the amount invested in each of the funds.

[13.2] **47.** A bridge over a canal is in the shape of half an ellipse. If the highest point of the bridge is 20 feet above the water and the base of the bridge is 150 feet across, what is the equation of the ellipse, half of which forms the bridge?

[12.5] **48.** What is the pH of a solution whose concentration of hydrogen ions is 4.23×10^{-8} moles per liter?

[12.2] **49.** If $15,000 is deposited into an account paying 4% interest compounded monthly, how much will be in the account after 3 years?

[12.5] **50.** The magnitude of an earthquake was defined in 1935 by Charles Richter as $M = \log\frac{I}{I_0}$. We let I be the intensity of the earthquake measured by the amplitude of a seismograph reading taken 100 km from the epicenter of the earthquake, and we let I_0 be the intensity of a "standard earthquake" whose amplitude is 10^{-4} cm. On January 26, 2001, an earthquake in Rann of Kutch, Gujarat was measured with an intensity of $10^{2.9}$. What was the magnitude of this earthquake?

R.1 Review of Expressions

OBJECTIVES
1. Evaluate an expression.
2. Apply the distributive property.
3. Combine like terms.

OBJECTIVE 1. Evaluate an expression

PROCEDURES   Evaluating an Algebraic Expression
1. Replace the variables with their corresponding given numbers.
2. Calculate the numerical expression using the order of operations.

Order of Operations
1. Grouping symbols: parentheses ( ), brackets [ ], braces { }, absolute value | |, and radicals $\sqrt{}$ beginning with the innermost and working outward.
2. Exponents/Roots from left to right, in order as they occur.
3. Multiplication/Division from left to right, in order as they occur.
4. Addition/Subtraction from left to right, in order as they occur.

EXAMPLE 1 Evaluate. $|3a + 2(b^3 - 2c)| - \sqrt{a^2 + c^2}$; $a = 3, b = -2, c = 4$

Solution $|3a + 2(b^3 - 2c)| - \sqrt{a^2 + c^2}$

$= |3(3) + 2[(-2)^3 - 2[(4)]| - \sqrt{3^2 + 4^2}$ Replace a with 3, b with -2, and c with 4.

$= |3(3) + 2[-8 - 2(4)]| - \sqrt{9 + 16}$ Calculate the exponential forms.

$= |3(3) + 2[-8 - 8]| - \sqrt{25}$ Multiply inside the brackets and add under the radical sign.

$= |3(3) + 2[-16]| - 5$ Add inside the brackets and $\sqrt{25} = 5$.

$= |9 - 32| - 5$ Multiply.

$= |-23| - 5$ Add inside absolute value symbols.

$= 23 - 5$ $|-23| = 23$.

$= 18$ Subtract.

YOUR TURN   Evaluate the expressions using $x = -2$, $y = 4$, and $z = -3$.

ANSWERS

a. 30   b. -51 c. $\dfrac{32}{21}$

d. 244   e. 2   f. -955

a. $3y - 2(3x + z)$

b. $-y^2 - (3z^2 - 4x)$

c. $\dfrac{2x^3 - 3y^2}{3x^3 - 2z^2}$

d. $|4x^2 - 3y(z^3 + 2y)|$

e. $\sqrt{y} - |-3x + 2z|$

f. $\sqrt{y^2 + z^2} - 2|5xy^2z|$

For additional review, see Section 1.7, page 78. For additional exercises, see Exercises 7–32 on pages 85–86.

OBJECTIVE 2. Apply the distributive property.

RULE The Distributive Property of Multiplication over Addition

$$a(b + c) = ab + ac$$

EXAMPLE 2 Use the distributive property to write an equivalent expression.

a. $3(2a + 4) = 3 \cdot 2a + 3 \cdot 4$
$$= 6a + 12$$

b. $-4(r - 5) = -4 \cdot r - (-4) \cdot 5$
$$= -4r - (-20)$$
$$= -4r + 20$$

c. $\dfrac{3}{4}\left(2b + \dfrac{5}{6}\right) = \dfrac{3}{4} \cdot 2b + \dfrac{3}{4} \cdot \dfrac{5}{6}$
$$= \dfrac{3}{2}b + \dfrac{5}{8}$$

YOUR TURN Use the distributive property to write an equivalent expression.

a. $4(3x + 6)$ b. $-7(-3a + 5b)$ c. $2.3x(4.2x - 1.7)$ d. $\dfrac{5}{8}\left(6x - \dfrac{4}{3}\right)$

For additional review, see Section 1.7, page 80. For additional exercises, see Exercises 41–48 on page 87.

OBJECTIVE 3. Combine like terms.

DEFINITION *Like terms:* Variable terms that have the same variable(s) raised to the same exponents, or constant terms.

PROCEDURE Combining Like Terms

ANSWERS

a. $12x + 24$ b. $21a - 35b$
c. $9.66x^2 - 3.91x$ d. $\dfrac{15}{4}x - \dfrac{5}{6}$

To combine like terms, add or subtract the coefficients and keep the variables and exponents the same. If an algebraic expression contains more than one set of like terms, we can collect like terms before combining.

EXAMPLE 3 Combine like terms.

a. $8x + 3x = 11x$ **We think: Eight x's plus three x's equals eleven x's.**

b. $9xy + 5x - 3xy + x$

$= 9xy - 3xy + 5x + x$ **Collect like terms.**

$= 6xy + 6x$ **Combine like terms.**

c. $\dfrac{5}{8}x^2 + 4.6y - \dfrac{3}{4}x^2 - 3.8y$

$= \dfrac{5}{8}x^2 - \dfrac{3}{4}x^2 + 4.6y - 3.8y$ **Collect like terms.**

$= \dfrac{5}{8}x^2 - \dfrac{3(2)}{4(2)}x^2 + 0.8y$ **Write the coefficients of x^2 as equivalent fractions with their LCD, 12, and subtract $3.8y$ from $4.6y$.**

$= \dfrac{5}{8}x^2 - \dfrac{6}{8}x^2 + 0.8y$

$= -\dfrac{1}{8}x^2 + 0.8y$ **Subtract $\dfrac{6}{8}x^2$ from $\dfrac{5}{8}x^2$.**

YOUR TURN   Combine like terms.

a. $-13a + 4a$ **b.** $8y - 4k + 2 - 4y + 3 - 9k$

c. $6.3t - 2.8u - 4.7t - 5.7u$ **d.** $\dfrac{2}{3}x^2y - \dfrac{3}{5}xy^2 + \dfrac{1}{4}x^2y - \dfrac{7}{10}xy^2$

e. $\dfrac{1}{3}p + 5.1q - \dfrac{3}{5}p + 6.9q$

ANSWERS

a. $-9a$ **b.** $4y - 13k + 5$

c. $1.6t - 8.5u$ **d.** $\dfrac{11}{12}x^2y - \dfrac{13}{10}xy^2$

e. $-\dfrac{4}{15}p + 12q$

For additional review, see Section 1.7, pages 81–84. For additional exercises, see Exercises 59–78 on pages 87–88.

R.2   Review of Solving Linear Equations

OBJECTIVES

1. Verify solutions to equations.

2. Solve linear equations using the addition principle

3. Solve linear equations using the multiplication principle.

4. Solve equations using both the addition and multiplication principles.

DEFINITION **Solution:** A number that makes an equation true when it replaces the variable in the equation.

PROCEDURE Checking a Possible Solution

To determine whether a value is a solution to a given equation, replace the variable(s) in the equation with the value. If the resulting equation is true, then the value is a solution.

EXAMPLE 1 Check to see if the given number is a solution for the given equation.

a. $5(u - 2) + 7(3 - u) = 9u;\ u = 1$

Solution $5(u - 2) + 7(3 - u) = 9u$

$5(1 - 2) + 7(3 - 1) \stackrel{?}{=} 9(1)$ Replace u with 1.

$5(-1) + 7(2) \stackrel{?}{=} 9$ Simplify inside the parentheses.

$-5 + 14 \stackrel{?}{=} 9$ Multiply.

$9 = 9$ The equation is true, so 1 is a solution.

b. $6x^2 - 10 = 11x;\ x = 2$

Solution $6x^2 - 10 = 11x$

$6(2)^2 - 10 \stackrel{?}{=} 11(2)$ Replace x with 2.

$6(4) - 10 \stackrel{?}{=} 22$ Square 2 and multiply $11(2)$.

$24 - 10 \stackrel{?}{=} 22$ Multiply.

$14 = 22$ The equation is false, so 2 is not a solution.

YOUR TURN Check to see if the given number is a solution for the given equation.

a. $9y - 5 - 8y = 5y + 10 - 3y;\ y = -15$ **b.** $6(p - 4) - 3p = 18 - 3p;\ p = 6$

c. $x(x + 3) = 40;\ x = 5$ **d.** $8x^2 - 15 = 14x;\ x = -\dfrac{3}{4}$

e. $6x^2 - 7x = 3;\ x = \dfrac{1}{3}$

For additional review, see Section 2.1, pages 102–104. For additional exercises, see Exercises 5–20 on page 112.

ANSWERS

a. yes **b.** no **c.** yes **d.** yes **e.** no

OBJECTIVE 2. Solve linear equations using the addition principle.

DEFINITIONS **Linear Equation:** An equation in which each variable term contains a single variable raised to an exponent of 1.
Linear Equation in One Variable: An equation that can be written in the form $ax + b = c$ where a, b, and c are real numbers and $a \neq 0$.

To solve equations of the form $x + a = b$, or those that can be simplified to that form, we use the Addition Principle of Equality.

RULE **The Addition Principle of Equality**
If $a = b$ then $a + c = b + c$ is true for all real numbers a, b, and c.

PROCEDURE **Solving Linear Equations Requiring the Addition Principle Only**
1. Simplify both sides of the equation as needed.
   a. Distribute to clear parentheses.
   b. Combine like terms.
2. Use the addition principle so that all variable terms are on one side of the equation and all constants are on the other side. Then combine like terms.

Tip: Clear the variable term that has the lesser coefficient to avoid negative coefficients.

EXAMPLE 2 Solve and check.

a. $-5.7a + 4.6 + 6.7a = 16.5 - 9.2$

Solution $-5.7a + 4.6 + 6.7a = 16.5 - 9.2$

$$a + 4.6 = 7.3 \qquad \text{Combine like terms.}$$
$$\underline{-4.6 \quad -4.6} \qquad \text{Subtract 4.6 from both sides.}$$
$$a = 2.7$$

Check $-5.7(2.7) + 4.6 + 6.7(2.7) = 16.5 - 9.2$ Replace a with 2.7.
$$-15.39 + 4.6 + 18.09 = 7.3 \qquad \text{Multiply.}$$
$$7.3 = 7.3 \qquad \text{True, so 2.7 is a solution.}$$

b. $3 = 2(4x + 5) - 7(x + 2)$

Solution $3 = 2(4x + 5) - 7(x + 2)$

$$3 = 8x + 10 - 7x - 14 \qquad \text{Distribute.}$$
$$3 = x - 4 \qquad \text{Combine like terms.}$$
$$\underline{+4 \qquad +4} \qquad \text{Add 4 to both sides.}$$
$$7 = x$$

Check The check is left to the reader.

YOUR TURN   Solve and check.

a. $7 + 5 = 6r + 8 - 5r$ **b.** $16x - 5(3x - 2) = 12$

c. $-2(4.7y - 5.4) + 10.4y = 3.1$ **d.** $-4(5x - 4) + 7(3x - 2) - 6 = 8 - 2$

For additional review, see Section 2.2, pages 118–126. For additional exercises, see Exercises 7–66 on pages 128–129.

OBJECTIVE 3. Solve linear equations using the multiplication principle.

To solve equations of the form $ax = b$, we use the Multiplication Principle of Equality.

RULE   **The Multiplication Principle of Equality**

If $a = b$, then $ac = bc$ for all real numbers a, b, and c, where $c \neq 0$.

PROCEDURE   **Solving Equations Requiring the Multiplication Principle Only**

To use the multiplication principle to clear a coefficient:

1. Determine the coefficient you want to clear.
2. Multiply both sides by the multiplicative inverse of that coefficient, or divide both sides by the coefficient.

EXAMPLE 3   Solve and check.

a. $-1.2z = 4.8$

Solution $-1.2z = 4.8$

$$\frac{-1.2z}{-1.2} = \frac{4.8}{-1.2}$$ Divide both sides by the coefficient, -1.2.

$$z = -4$$ Simplify both sides.

Check $-1.2(-4) = 4.8$ Substitute -4 for z.

$ 4.8 = 4.8$ True, so -4 is a solution.

b. $\dfrac{8}{5}x = -32$

Solution $\dfrac{8}{5}x = -32$

$$\frac{\overset{1}{\cancel{5}}}{\cancel{8}} \cdot \frac{\overset{1}{\cancel{8}}}{\cancel{5}} x = \frac{5}{\cancel{8}}(-\overset{4}{\cancel{32}})$$ Multiply both sides by the multiplicative inverse of the coefficient, $\dfrac{5}{8}$.

$$x = -20$$ Simplify both sides.

Check   The check is left to the reader.

ANSWERS

a. 4   **b.** 2
c. -7.7 **d.** 10

YOUR TURN Solve and check.

a. $5x = -20$

b. $3.4a = -10.2$

c. $\dfrac{3}{4}x = 9$

d. $\dfrac{4}{5}r = -\dfrac{12}{25}$

e. $\dfrac{w}{-4} = -6$

For additional review, see Section 2.3, pages 132–133. For additional exercises, see Exercises 7–18 on page 141.

OBJECTIVE 4. **Solve equations using both the addition and multiplication principles.** To solve equations that can be written in the form $ax + b = c$, we use both the Addition and Multiplication Principles of Equality along with the following procedure.

PROCEDURE Solving Linear Equations

To solve linear equations in one variable:

1. Simplify both sides of the equation as needed.
    a. Distribute to clear parentheses.
    b. Clear fractions or decimals by multiplying through by the LCD. In the case of decimals, the LCD is the power of 10 with the same number of zero digits as decimal places in the number with the most decimal places. (Clearing fractions and decimals is optional.)
    c. Combine like terms.
2. Use the addition principle so that all variable terms are on one side of the equation and all constant terms are on the other side. (Clear the variable term with the lesser coefficient.) Then combine like terms.
3. Use the multiplication principle to clear the remaining coefficient.

EXAMPLE 4 Solve and check.

a. $3x + 4 = -5x + 28$

Solution

$$3x + 4 = -5x + 28$$

Add 5x to both sides (-5 is the lesser coefficient).

$$\underline{+5x \qquad\quad +5x}$$
$$8x + 4 = \quad 0 + 28$$

$$8x + 4 = 28$$

Subtract 4 from both sides to isolate the 8x term.

$$\underline{-4 \quad -4}$$
$$\dfrac{8x}{8} = \dfrac{24}{8}$$

Divide both sides by 8 to clear the 8 coefficient.

$$x = 3$$

ANSWERS

a. -4 **b.** -3 **c.** 12

d. $-\dfrac{3}{5}$ **e.** 24

Check The check is left to the reader.

b. $4(b - 4) + 2(b + 2) = 2(b - 2)$

Solution $4(b - 4) + 2(b + 2) = 2(b - 2)$

$$4b - 16 + 2b + 4 = 2b - 4 \qquad \text{Distribute to clear parentheses.}$$

$$6b - 12 = 2b - 4 \qquad \text{Combine like terms.}$$

$$6b - 12 = 2b - 4 \qquad \text{Subtract } 2b \text{ from both sides (2 is the lesser coefficient.)}$$

$$\underline{-2b \qquad\quad -2b}$$

$$4b - 12 = 0 - 4$$

$$4b - 12 = -4 \qquad \text{Add 12 to both sides to isolate the } 4b \text{ term.}$$

$$\underline{+12 \quad +12}$$

$$4b + 0 = 8$$

$$\frac{4b}{4} = \frac{8}{4} \qquad \text{Divide both sides by 4 to clear the 4 coefficient.}$$

$$b = 2$$

Check The check is left to the reader.

YOUR TURN   Solve and check.

a. $-7x - 6 = 8$

b. $2.3y - 1.6 - 0.8y = 4.4$

c. $3a - 4a + 9 = -12a + 18 - 6a$

d. $14(w - 2) + 13 = 4w + 5$

e. $0.3n + 0.2(4n - 1) = 0.4 + 0.5n$

f. $5(a - 1) - 9(a - 2) = -3(2a + 1) - 2$

ANSWERS

a. -2 **b.** 4 **c.** $\dfrac{9}{17}$
d. 2 **e.** 1 **f.** -9

For additional review, see Section 2.3, pages 132–138. For additional exercises, see Exercises 19–72 on pages 141–142.

R.3 Review of Graphing Linear Equations

OBJECTIVES

1. Plot points in the coordinate plane.
2. Find solutions for equations in two unknowns.
3. Graph linear equations by plotting solutions.
4. Graph linear equations using intercepts.
5. Graph vertical and horizontal lines.

PROCEDURE   Plotting a Point

To graph or plot a point given its coordinates:

1. Beginning at the origin, (0, 0) move to the right or left along the *x*-axis the amount indicated by the first coordinate.
2. From that position on the *x*-axis, move up or down the amount indicated by the second coordinate.
3. Draw a dot to represent the point described by the coordinates.

EXAMPLE 1   Plot the point described by the coordinates.

a. $(4, 5)$ **b.** $(0, 3)$ **c.** $(-2, -3)$

Solution

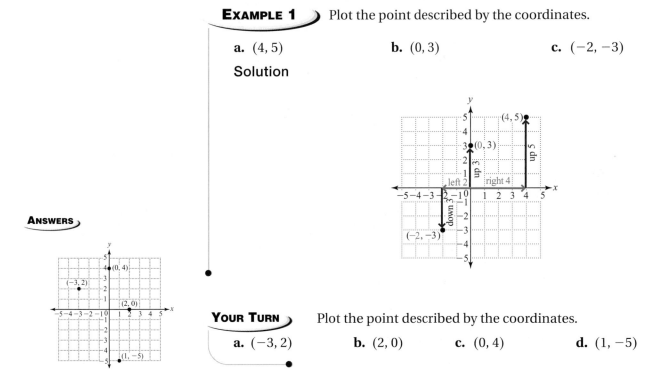

ANSWERS

YOUR TURN   Plot the point described by the coordinates.

a. $(-3, 2)$ **b.** $(2, 0)$ **c.** $(0, 4)$ **d.** $(1, -5)$

For additional review, see Section 4.1, page 270. For additional exercises, see Exercises 9–12 on page 274.

PROCEDURE   Finding Solutions to Equations with Two Variables

To find a solution to an equation in two variables:

1. Choose a value for one of the variables (any value).
2. Replace the corresponding variable with your chosen value.
3. Solve the equation for the value of the other variable.

EXAMPLE 2 Find three solutions for the equation $2x + y = 2$.

Solution

For the first solution, we will choose x to be 0.

$$2x + y = 2$$
$$2(0) + y = 2$$
$$y = 2$$

Solution $(0, 2)$

For the second solution, we will choose x to be -1.

$$2x + y = 2$$
$$2(-1) + y = 2$$
$$-2 + y = 2$$
$$y = 4$$

Solution $(-1, 4)$

For the third solution, we will choose y to be -2.

$$2x + y = 2$$
$$2x + (-2) = 2$$
$$2x = 4$$
$$x = 2$$

Solution $(2, -2)$

YOUR TURN Find three solutions for each equation. (Answers may vary.)

a. $x + 4y = 8$

b. $3x + 2y = 6$

c. $y = -2x - 2$

d. $y = -\dfrac{2}{3}x + 4$

For additional review, see Section 4.2, pages 278–280. For additional exercises, see Exercises 23–58 on pages 284–286.

OBJECTIVE 3. Graph linear equations by plotting solutions.

PROCEDURE Graphing Linear Equations

To graph a linear equation:

1. Find at least two solutions to the equation.
2. Plot the solutions as points in the rectangular coordinate system.
3. Connect the points to form a straight line.

Even though two points uniquely determine a line, it is a good idea to plot a third point as a check.

ANSWERS

(Remember, answers may vary.)

a. $(0, 2)\ (8, 0)\ (4, 1)$

b. $(0, 3)\ (2, 0)\ (4, -3)$

c. $(0, -5)\ (-2, -1)\ (2, -9)$

d. $(0, 4)\ (3, 2)\ (-3, 6)$

i.

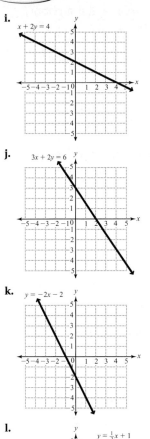

j.

k.

l.

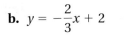

EXAMPLE 3 Find three solutions for the given equations and graph.

a. $2x + y = 2$

b. $y = -\dfrac{2}{3}x + 2$

Solution

In Example 2, we found three solutions for this equation: $(0, 2)$, $(-1, 4)$, and $(2, -2)$. We will plot these points and draw a straight line through them.

Solution

Using the methods of Example 2, three solutions of the equation are $(0, 2)$, $(-3, 4)$ and $(3, 0)$. We will plot these points and draw a straight line through them.

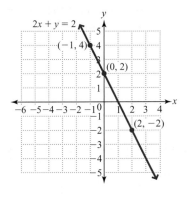

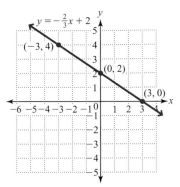

YOUR TURN Find three solutions for the given equations and graph.

a. $x + 2y = 4$

b. $3x + 2y = 6$

c. $y = -2x - 2$

d. $y = \dfrac{1}{3}x + 1$

For additional review, see Section 4.2, pages 278–281. For additional exercises, see Exercises 23–52 on pages 284–285.

OBJECTIVE 4. Graph linear equations using intercepts. Another method of graphing linear equations involves finding the intercepts.

DEFINTIONS *x-intercept:* A point where a graph intersects the *x*-axis.
y-intercept: A point where a graph intersects the *y*-axis.

PROCEDURE Finding the *x*- and *y*-intercepts

To find an *x*-intercept:

1. Replace *y* with 0 in the given equation.
2. Solve for *x*.

To find a *y*-intercept:

1. Replace *x* with 0 in the given equation.
2. Solve for *y*.

RULES Intercepts for $y = mx$

If an equation can be written in the form $y = mx$, where m is a real number other than 0, then the x- and y-intercepts are at the origin, $(0, 0)$.

y-intercept for $y = mx + b$

If an equation is in the form $y = mx + b$, where m and b are real numbers, then the y intercept is $(0, b)$.

Intercepts for $y = c$

The graph of an equation in the form $y = c$, where c is a real number constant, has no x-intercept and the y intercept is $(0, c)$.

Intercepts for $x = c$

The graph of an equation in the form $x = c$, where c is a real number constant, has no y-intercept and the x intercept is $(c, 0)$.

EXAMPLE 4 Graph using the x- and y-intercepts.

a. $3x + 4y = 12$

Solution Find the intercepts, then graph.

For the x-intercept, replace y with 0 and solve for x.

$$3x + 4(0) = 12$$
$$3x + 0 = 12$$
$$x = 4$$

x-intercept: $(4, 0)$

For the y-intercept, replace x with 0 and solve for y.

$$3(0) + 4y = 12$$
$$0 + 4y = 12$$
$$4y = 12$$
$$y = 3$$

y-intercept: $(0, 3)$

It is a good idea to find a third solution as a check. Letting $x = -4$, we solve the equation for y to find that $y = 6$, so a third solution is $(-4, 6)$. Plotting the intercepts and our third solution gives the graph to the right.

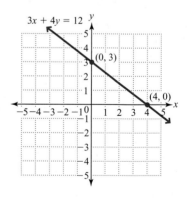

b. $y = -\dfrac{1}{4}x$

Solution The equation is in the form $y = mx$, so the x- and y- intercepts are $(0, 0)$. We need at least one more solution, but will find a third solution as a check.

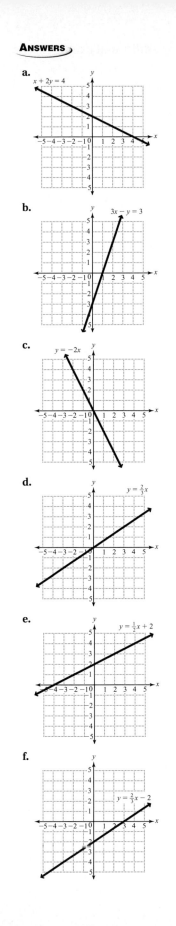

a. $x + 2y = 4$

b. $3x - y = 3$

c. $y = -2x$

d. $y = \frac{2}{3}x$

e. $y = \frac{1}{2}x + 2$

f. $y = \frac{2}{3}x - 2$

Second solution:

Choose x to be 4.

$$y = -\frac{1}{4}(4)$$

$$y = -1$$

Second solution: $(4, -1)$

Third solution:

Choose x to be -4.

$$y = -\frac{1}{4}(-4)$$

$$y = 1$$

Third solution: $(-4, 1)$

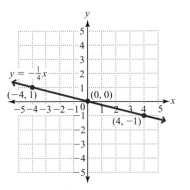

c. $y = \frac{1}{3}x - 1$

Solution The equation is in the form $y = mx + b$, so the y-intercept is $(0, -1)$.

For the x-intercept, let $y = 0$ and solve for x.

$$0 = \frac{1}{3}x - 1$$

$$1 = \frac{1}{3}x$$

$$3(1) = 3\left(\frac{1}{3}x\right)$$

$$3 = x$$

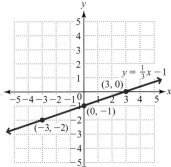

x-intercept: $(3, 0)$

As a check we will choose $x = -3$ to find a third solution, which is $(-3, -2)$.

YOUR TURN Graph using the x- and y-intercepts.

a. $x + 2y = 4$

b. $3x - y = 3$

c. $y = -2x$

d. $y = \frac{2}{3}x$

e. $y = \frac{1}{2}x + 2$

f. $y = \frac{2}{3}x - 2$

For additional review, see Section 4.3, pages 289–294. For additional exercises, see Exercises 23–42 on pages 295–296.

OBJECTIVE 5. Graph vertical and horizontal lines.

RULES   **Horizontal Lines**

The graph of $y = c$, where c is a real number constant, is a horizontal line parallel to the x-axis with a y-intercept of $(0, c)$.

Vertical Lines

The graph of $x = c$, where c is a real number constant, is a vertical line parallel to the y-axis with an x-intercept of $(c, 0)$.

EXAMPLE 5   Graph.

a. $y = 2$

Since the equation is in the form of $y = c$, this is a horizontal line with a y-intercept of $(0, 2)$

b. $x = -3$

Since the equation is in the form of $x = c$, this is a vertical line with an x-intercept of $(-3, 0)$.

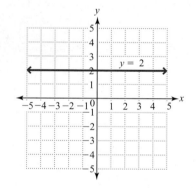

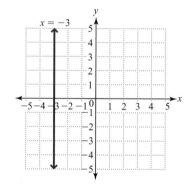

ANSWERS

a.

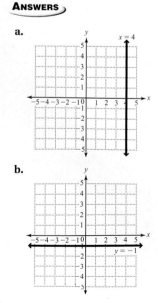

b.

YOUR TURN   Graph.

a. $x = 4$

b. $y = -1$

For additional review, see Section 4.2, pages 282–283 and Section 4.3, pages 291–292. For additional exercises, see Exercises 53–56 on page 286 and Exercises 43–50 on pages 296–297.

R.4 Review of Polynomials

OBJECTIVES

1. Add and subtract polynomials.
2. Multiply polynomials.
3. Divide polynomials.

OBJECTIVE 1. Add and subtract polynomials.

PROCEDURE **Adding and Subtracting Polynomials**

To add polynomials, combine like terms.

To subtract polynomials:

1. Change the operation symbol from a minus sign to a plus sign.
2. Change the subtrahend (second polynomial) to its additive inverse. To get the additive inverse we change the sign of each term in the polynomial.
3. Combine like terms.

EXAMPLE 1 Add or subtract.

a. $(5x^3 - 3x^2 + 6x - 7) + (-2x^3 + 2x^2 + 3x - 5)$

Solution $(5x^3 - 3x^2 + 6x - 7) + (-2x^3 + 2x^2 + 3x - 5)$

$= 3x^3 - x^2 + 9x - 12$ Combine like terms.

b. $(3n^3 - 5n^2 - 7n + 8) - (6n^3 + 3n^2 - 8n - 2)$

Solution $(3n^3 - 5n^2 - 7n + 8) - (6n^3 + 3n^2 - 8n - 2)$

$= (3n^3 - 5n^2 - 7n + 8) + (-6n^3 - 3n^2 + 8n + 2)$ Change the operation symbol from − to + and change the subtrahend to its additive inverse.

$= -3n^3 - 8n^2 + n + 10$ Combine like terms.

YOUR TURN Add or subtract.

a. $(4a^2 - 7a - 4) + (2a^2 + 9a - 6)$

b. $(-3a^3 + 8a^2 - 7a - 3) + (4a^3 + 6a^2 - 9a + 7)$

c. $(6r^2 - 2r + 7) - (4r^2 - 6r + 6)$

d. $(7m^3 - m^2 + 4m - 7) - (-5m^3 + 4m^2 - 9m + 3)$

ANSWERS

a. $6a^2 + 2a - 10$
b. $a^3 + 14a^2 - 16a + 4$
c. $2r^2 + 4r + 1$
d. $12m^3 - 5m^2 + 13m - 10$

For additional review, see Section 5.3, pages 400–405. For additional exercises, see Exercises 5–58 on pages 406–408.

OBJECTIVE 2. Multiply polynomials.

PROCEDURE   Multiplying Monomials

To multiply monomials:
1. Multiply coefficients.
2. Add the exponents of the like bases.
3. Write any unlike variable bases unchanged in the product.

EXAMPLE 2   Multiply. $(-5x^3yz^4)(2x^5y^2)$

Solution $(-5x^3yz^4)(2x^5y^2) = -5 \cdot 2x^{3+5}\, y^{1+2}\, z^4$
$$= -10x^8y^3z^4$$

YOUR TURN   Multiply.

a. $(3a^3b^2)(-6a^5b^6c^2)$ **b.** $(-2p^2q^4)(-6p^7q^3r^5)(3q^4r^5)$

c. $(-2.3m^8n^4)(0.4m^2n^3)$ **d.** $\left(\dfrac{2}{3}x^3y^5\right)\left(\dfrac{9}{5}x^4y^2\right)$

For additional review, see Section 5.4, pages 412–413. For additional exercises, see Exercises 7–32 on page 418.

PROCEDURE   Multiplying a Polynomial by a Monomial

To multiply a polynomial by a monomial, use the distributive property to multiply each term in the polynomial by the monomial.

EXAMPLE 3   Multiply. $3x^2(2x^3 - 3x^2 + 5)$

ANSWERS

a. $-18a^8b^8c^2$ **b.** $36p^9q^{11}r^{10}$

c. $-9.2m^{10}n^7$ **d.** $\dfrac{6}{5}x^7y^7$

$3x^2(2x^3 - 3x^2 + 5)$
$= 3x^2 \cdot 2x^3 + 3x^2 \cdot (-3x^2) + 3x^2 \cdot 5$ Multiply each term in the polynomial by $3x^2$.
$= 6x^5 - 9x^4 + 15x^2$ Multiply.

ANSWERS

a. $24a^5 - 16a^4 + 12a^3$
b. $-12x^5y^3 - 24x^3y^4 + 16x^7y - 20x^2y$
c. $-12x^7 - \dfrac{1}{2}x^5 + \dfrac{1}{3}x^4 - \dfrac{1}{5}x^3$
d. $9.45a^5b^4c - 15.75a^6b^7c^3 + 18a^3b^2c$

YOUR TURN   Multiply.

a. $4a^3(6a^2 - 4a + 3)$ **b.** $-4x^2y(3x^3y^2 + 6xy^3 - 4x^5 + 5)$

c. $4x^2\left(-3x^5 - \dfrac{1}{8}x^3 + \dfrac{1}{12}x^2 - \dfrac{1}{20}x\right)$ **d.** $4.5a^3b^2c(2.1a^2b^2 - 3.5a^3b^5c^2 + 4)$

For additional review, see Section 5.5, pages 423–424. For additional exercises, see Exercises 7–34 on page 431.

PROCEDURE Multiplying Polynomials

To multiply two polynomials:
1. Multiply every term in the second polynomial by every term in the first polynomial.
2. Combine like terms.

EXAMPLE 4 Multiply.

a. $(3x - 4)(2x + 3)$

Solution $(3x - 4)(2x + 3)$

$$= 3x \cdot 2x + 3x \cdot 3 + (-4) \cdot 2x + (-4) \cdot 3$$ Multiply every term in the second polynomial by every term in the first.

$$= 6x^2 + 9x - 8x - 12$$ Multiply.

$$= 6x^2 + x - 12$$ Combine like terms.

b. $(a + 3)(2a^2 - 3a + 2)$

Solution $(a + 3)(2a^2 - 3a + 2)$

$$= a \cdot 2a^2 + a \cdot (-3a) + a \cdot 2 + 3 \cdot 2a^2$$ Multiply every term in the second
$$+ 3 \cdot (-3a) + 3 \cdot 2$$ polynomial by every term in the first.

$$= 2a^3 - 3a^2 + 2a + 6a^2 - 9a + 6$$ Multiply.

$$= 2a^3 + 3a^2 - 7a + 6$$ Combine like terms.

YOUR TURN Multiply.

a. $(3y + 5)(y - 4)$

b. $(3p - 4q)(2p - 3q)$

c. $(a + 2)(4a^2 + 3a - 5)$

d. $(3y - 2)(2y^2 - 4y - 5)$

For additional review, see Section 5.5, pages 425–428. For additional exercises, see Exercises 37–70 on pages 432–433.

DEFINITION *Conjugates:* Binomials that differ only in the sign separating the terms.

PROCEDURE Multiplying Conjugates

If a and b are real numbers, variables, or expressions, then $(a + b)(a - b) = a^2 - b^2$

ANSWERS

a. $3y^2 - 7y - 20$

b. $6p^2 - 17pq + 12q^2$

c. $4a^3 + 11a^2 + a - 10$

d. $6y^3 - 16y^2 - 7y + 10$

EXAMPLE 5 Multiply. $(4a - 3)(4a + 3)$

Solution $(4a - 3)(4a + 3) = (4a)^2 - 3^2$ Use $(a + b)(a - b) = a^2 - b^2$.
$$= 16a^2 - 9$$

YOUR TURN Multiply.

a. $(2y + 5)(2y - 5)$ **b.** $(3m - 7n)(3m + 7n)$

For additional review, see Section 5.5, pages 428–429. For additional exercises, see Exercises 79–88 on page 433.

PROCEDURES   Squaring a Binomial

$$(a + b)^2 = a^2 + 2ab + b^2$$
$$(a - b)^2 = a^2 - 2ab + b^2$$

EXAMPLE 6 Multiply. $(3a - 2b)^2$

Solution $(3a - 2b)^2 = (3a)^2 - 2(3a)(2b) + (2b)^2$ Use $(a - b)^2 = a^2 - 2ab + b^2$.
$$= 9a^2 - 12ab + 4b^2$$

YOUR TURN Square the following binomials.

a. $(7x + 2y)^2$ **b.** $(5a - 3b)^2$

For additional review, see Section 5.5, pages 429–430. For additional exercises, see Exercises 89–100 on page 433.

OBJECTIVE 3. Divide polynomials.

PROCEDURES   Dividing a Polynomial by a Monomial

To divide a polynomial by a monomial, divide each term of the polynomial by the monomial.

ANSWERS

a. $4y^2 - 25$ **b.** $9m^2 - 49n^2$

ANSWERS

a. $49x^2 + 28xy + 4y^2$
b. $25a^2 - 30ab + 9b^2$

EXAMPLE 7 Divide. $\dfrac{16x^2y^4 - 12x^4y + 40x^3}{4x^2y}$

Solution $\dfrac{16x^2y^4 - 12x^4y + 40x^3}{4x^2y} = \dfrac{16x^2y^4}{4x^2y} - \dfrac{12x^4y}{4x^2y} + \dfrac{40x^3}{4x^2y}$
     Divide each term of the polynomial by the monomial.

$$= 4y^3 - 3x^2 + \dfrac{10x}{y}$$

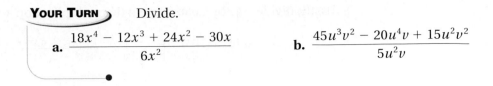

YOUR TURN   Divide.

a. $\dfrac{18x^4 - 12x^3 + 24x^2 - 30x}{6x^2}$ b. $\dfrac{45u^3v^2 - 20u^4v + 15u^2v^2}{5u^2v}$

For additional review, see Section 5.6, pages 435–439. For additional exercises, see Exercises 59–76 on page 448.

PROCEDURE   **Dividing a Polynomial by a Polynomial**

To divide a polynomial by a polynomial use long division. If there is a remainder, write the result in the following form:

$$\text{quotient} + \frac{\text{remainder}}{\text{divisor}}$$

Be sure both the divisor and dividend are written in descending order before using long division.

EXAMPLE 8   Divide. $\dfrac{8x^3 - 22x + 8}{2x - 3}$

Solution   Write as a long division with a place holder for the x^2 term. Divide the first term of the dividend by the first term of the divisor: $8x^3 \div 2x = 4x^2$. Then multiply the quotient, $4x^2$, by the divisor $(2x - 3)$ and subtract.

$$
\begin{array}{r}
4x^2 \\
2x - 3 \overline{)8x^3 + 0x^2 - 22x + 8} \\
-(8x^3 - 12x^2)
\end{array}
$$
Change signs.

$$
\begin{array}{r}
4x^2 \\
2x - 3 \overline{)8x^3 + 0x^2 - 22x + 8} \\
-8x^3 + 12x^2 \\
\hline
12x^2 - 22x
\end{array}
$$
Combine like terms and bring down the next term.

Now divide $12x^2$ by $2x$ and repeat the multiplication and subtraction steps.

$$
\begin{array}{r}
4x^2 + 6x \\
2x - 3 \overline{)8x^3 + 0x^2 - 22x + 8} \\
-8x^3 + 12x^2 \\
\hline
12x^2 - 22x \\
-(12x^2 - 18x)
\end{array}
$$
Change signs.

$$
\begin{array}{r}
4x^2 + 6x \\
2x - 3 \overline{)8x^3 + 0x^2 - 22x + 8} \\
-8x^3 + 12x^2 \\
\hline
12x^2 - 22x \\
-12x^2 + 18x \\
\hline
-4x + 8
\end{array}
$$
Combine like terms and bring down the next term.

Lastly, divide $-4x$ by $2x$ and repeat the multiplication and subtraction steps.

$$
\begin{array}{r}
4x^2 + 6x - 2 \\
2x - 3\overline{)8x^3 + 0x^2 - 22x + 8} \\
\underline{-8x^3 + 12x^2} \\
12x^2 - 22x \\
\underline{-12x^2 + 18x} \\
-4x + 8 \\
\underline{-(-4x + 6)}
\end{array}
\qquad
\begin{array}{r}
4x^2 + 6x - 2 \\
2x - 3\overline{)8x^3 + 0x^2 - 22x + 8} \\
\underline{-8x^3 + 12x^2} \\
12x^2 - 22x \\
\underline{-12x^2 + 18x} \\
-4x + 8 \\
\underline{+4x - 6} \\
2
\end{array}
$$

Change signs. Combine like terms.

Answer $4x^2 + 6x - 2 + \dfrac{2}{2x - 3}$

YOUR TURN   Divide.

a. $\dfrac{16x - 12x^2 + 3x^3 - 12}{x - 2}$

b. $\dfrac{2x^3 - 14x + 5}{2x - 4}$

ANSWERS

a. $3x^2 - 6x + 4 + \dfrac{-4}{x - 2}$

b. $x^2 + 2x - 3 + \dfrac{-7}{2x - 4}$

For additional review, see Section 5.6, pages 440–443. For additional exercises, see Exercises 81–104 on pages 448–449.

R.5   Review of Factoring

OBJECTIVES

1. Write a polynomial as a product of a monomial GCF and a polynomial.
2. Factor by grouping.
3. Factor trinomials of the form $x^2 + bx + c$.
4. Factor trinomials of the form $ax^2 + bx + c$, where $a \neq 1$.
5. Factor special products.

OBJECTIVE 1. Write a polynomial as a product of a monomial GCF and a polynomial.

DEFINITION   *Factored form:* A number or expression written as a product.

The first step in factoring any polynomial is to factor out any monomial GCF.

PROCEDURES   **Prime Factorization Method for Finding the GCF**

To find the greatest common factor of a given set of numbers:

1. Write the prime factorization of each given number in exponential form.
2. Create a factorization for the GCF that includes only those prime factors common to all the factorizations, each raised to its smallest exponent in the factorization.
3. Multiply the factors in the factorization created in step 2.

   Note: If there are no common prime factors, then the GCF is 1.

Factoring a Monomial GCF Out of a Polynomial

To factor a monomial GCF out of a given polynomial:

1. Find the GCF of the terms that make up the polynomial.
2. Rewrite the given polynomial as a product of the GCF and parentheses that contain the result of dividing the given polynomial by the GCF.

$$\text{Given polynomial} = \text{GCF}\left(\frac{\text{Given polynomial}}{\text{GCF}}\right)$$

EXAMPLE 1   Factor.

  a. $18x^3y^4 + 9x^2y^4 - 12xy^3$

Solution 1. Find the GCF of $18x^3y^4$, $9x^2y^4$, and $12xy^3$.
  The GCF of $18x^3y^4$, $9x^2y^4$, and $12xy^3$, is $3xy^3$

  2. Write $18x^3y^4$, $9x^2y^4$ and $-12xy^3$ as the product of the GCF and parentheses containing the quotient of $18x^3y^4 + 9x^2y^4 - 12xy^3$ and the GCF $3xy^3$.

$$18x^3y^4 + 9x^2y^4 - 12xy^3 = 3xy^3\left(\frac{18x^3y^4 + 9x^2y^4 - 12xy^3}{3xy^3}\right)$$

$$= 3xy^3\left(\frac{18x^3y^4}{3xy^3} + \frac{9x^2y^4}{3xy^3} - \frac{12xy^3}{3xy}\right) \qquad \textbf{Separate the terms.}$$

$$= 3xy^3(6x^2y + 3xy - 4) \qquad \textbf{Divide.}$$

  b. $-40x^4y^5 - 32x^3y^4 + 16x^2y^3z$

Solution The GCF of $40x^4y^5$, $32x^3y^4$, and $16x^2y^3z$ is $8x^2y^3$. Because the first term is negative, we will factor out the negative of the GCF, $-8x^2y^3$.

$$-40x^4y^5 - 32x^3y^4 + 16x^2y^3z = -8x^2y^3\left(\frac{-40x^4y^5 - 32x^3y^4 + 16x^2y^3z}{-8x^2y^3}\right)$$

$$= -8x^2y^3\left(\frac{-40x^4y^5}{-8x^2y^3} - \frac{32x^3y^4}{-8x^2y^3} + \frac{16x^2y^3z}{-8x^2y^3}\right)$$

$$= -8x^2y^3(5x^2y^2 + 4xy - 2z)$$

YOUR TURN Factor by removing the GCF.

a. $18c^3d^4 - 12c^2d^6$ **b.** $16c^3d^2 - 24c^2d^4 + 36cd^2$

Factor by removing a negative GCF.

c. $-10x^3 - 15x^2 + 25x$ **d.** $-24m^2n - 28mn + 16mn^2$

For additional review, see Section 6.1, pages 460–466. For additional exercises, see Exercises 23–66 on pages 468–469.

OBJECTIVE 2. Factoring by grouping.

PROCEDURE   Factoring by Grouping

To factor a four-term polynomial by grouping:

1. Factor out any monomial GCF (other than 1) that is common to all four terms.
2. Group together pairs of terms and factor the GCF out of each pair.
3. If there is a common binomial factor, then factor it out.
4. If there is no common binomial factor, then interchange the middle two terms and repeat the process. If there is still no common binomial factor, then the polynomial cannot be factored by grouping.

EXAMPLE 2 Factor by grouping.

a. $6a^2 - 9ay + 10a - 15y$

Solution $6a^2 - 9ay + 10a - 15y = (6a^2 - 9ay) + (10a - 15y)$

$$= 3a(2a - 3y) + 5(2a - 3y) \quad \text{Factor } 3a \text{ from } 6a^2 - 9ay \text{ and } 5 \text{ from } 10a - 15y.$$

$$= (2a - 3y)(3a + 5) \quad \text{Factor out } 2a - 3y.$$

b. $3x - xr - 21 + 7r$

Solution $3x - xr - 21 + 7r = (3x - xr) + (-21 + 7r)$

$$= x(3 - r) - 7(3 - r) \quad \text{Factor } x \text{ from } 3x - xr \text{ and } -7 \text{ from } -21 + 7r.$$

$$= (3 - r)(x - 7) \quad \text{Factor out } 3 - r.$$

ANSWERS

a. $6c^2d^4(3c - 2d^2)$
b. $4cd^2(4c^2 - 6cd^2 + 9)$
c. $-5x(2x^2 + 3x - 5)$
d. $-4mn(6m + 7 - 4n)$

YOUR TURN Factor by grouping.

a. $a^3 + 4a^2 + 3a + 12$ **b.** $8x^2 - 2xy + 12x - 3y$

c. $6ac + 4ad - 9bc - 6bd$ **d.** $15ac - 5ad - 3bc + bd$

ANSWERS

a. $(a + 4)(a^2 + 3)$
b. $(4x - y)(2x + 3)$
c. $(3c + 2d)(2a - 3b)$
d. $(3c - d)(5a - b)$

For additional review, see Section 6.1, pages 466–467. For additional exercises, see Exercises 67–90 on page 469.

OBJECTIVE 3. Factor trinomials of the form $x^2 + bx + c$.

PROCEDURE Factoring $x^2 + bx + c$

To factor a trinomial of the form $x^2 + bx + c$:

1. Find two numbers with a product equal to c and a sum equal to b.
2. The factored trinomial will have the form:
$(x + \text{first number})(x + \text{second number})$

Note: The signs in the binomial factors can be minus signs, depending on the signs of b and c.

EXAMPLE 3 ▶ Factor.

a. $r^2 - 15r + 36$

Solution We must find a pair of numbers whose product is 36 and whose sum is -15. If two numbers have a positive product and a negative sum, they must both be negative. Following is a table listing the products and sums:

Product	Sum
$(-1)(-36) = 36$	$-1 + (-36) = -37$
$(-2)(-18) = 36$	$-2 + (-18) = -20$
$(-3)(-12) = 36$	$-3 + (-12) = -15$

◀ *This is the correct combination, so -3 and -12 are the second terms in each binomial of the factored form.*

Answer $r^2 - 15r + 36 = (r - 3)(r - 12)$

Check We can check by multiplying the binomial factors to see if their product is the original polynomial.

$(r - 3)(r - 12) = r^2 - 12r - 3r + 36$ **Multiply the factors using FOIL.**
$= r^2 - 15r + 36$ **The product is the original polynomial.**

b. $3m^2 + 33mn - 126n^2$

Solution First factor out the monomial GCF, 3.

$$3m^2 + 33mn - 126n^2 = 3(m^2 + 11mn - 42n^2)$$

Now factor $m^2 + 11mn - 42n^2$. Since n is in the last term, it is helpful to think of the middle term as $11nm$ and view the "coefficient" to be $11n$. We must find a pair of terms whose product is $-42n^2$ and whose sum is $11n$. Since the product is negative, the terms must have different signs. Since the sum is positive, the term whose coefficient has the greatest absolute value will be positive. These terms are $-3n$ and $14n$, so $m^2 + 11mn - 42n^2 = (m - 3n)(m + 14n)$.

Answer $3m^2 + 33mn - 126n^2 = 3(m - 3n)(m + 14n)$

Check The check is left to the reader.

ANSWERS
a. $(x + 2)(z + 9)$
b. $(x - 5y)(x - 7y)$
c. $(y + 7)(y - 6)$
d. $2(a + 12b)(a - 2b)$

YOUR TURN ▶ Factor.

a. $z^2 + 11z + 18$.   **b.** $x^2 - 12xy + 35y^2$ **c.** $y^2 + y - 42$ **d.** $2a^2 + 20ab - 48b^2$

For additional review, see Section 6.2, pages 471–474. For additional exercises, see Exercises 5–70 on pages 475–476.

OBJECTIVE 4. Factoring trinomials of the form $ax^2 + bx + c$ where $a \neq 1$. If $a \neq 1$, then we factor by using trial and error, checking our trials using FOIL.

PROCEDURE Factoring by Trial and Error

To factor a trinomial of the form $ax^2 + bx + c$, where $a \neq 1$, by trial and error:

1. Look for a monomial GCF. If there is one, factor it out.
2. Write a pair of *first* terms whose product is ax^2.
3. Write a pair of *last* terms whose product is c.
4. Verify that the sum of the *inner* and *outer* products is bx (the middle term of the trinomial).

   If the sum of the inner and outer products is not bx, then try the following:

   a. Exchange the first terms of the binomials from step 3, then repeat step 4.
   b. Exchange the last terms of the binomials from step 3, then repeat step 4.
   c. Repeat steps 2–4 with a different combination of first and last terms.

EXAMPLE 4 Factor. $6x^2 + x - 12$

Solution

The product of the *first* terms must be $6x^2$.
These could be x and $6x^2$ or $2x$ and $3x$.

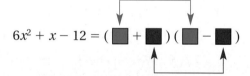

$$6x^2 + x - 12 = (\blacksquare + \blacksquare)(\blacksquare - \blacksquare)$$

The product of the *last* terms must be -12. Because -12 is negative, the last terms in the binomials must have different signs. We have already written the appropriate signs, so these factor pairs could be 1 and 12, 2 and 6, or 3 and 4.

Now we multiply binomials with various combinations of these first and last terms until we find a combination whose inner and outer products combine to equal x.

$$(3x + 1)(2x - 12) = 6x^2 - 34x - 12$$
$$(3x + 12)(2x + 1) = 6x^2 + 27x + 12$$
$$(3x + 2)(2x - 6) = 6x^2 - 14x - 12$$
$$(3x + 6)(2x - 2) = 6x^2 + 6x - 12$$
$$(3x + 3)(2x - 4) = 6x^2 - 6x - 12$$
$$(3x + 4)(2x - 3) = 6x^2 - x - 12$$

Incorrect combinations.

$$(3x - 4)(2x + 3) = 6x^2 + x - 12 \longleftarrow \text{Correct combination.}$$

ANSWERS

a. $(3x - 2)(3x - 4)$
b. $(3c - 2d)(2c - 3d)$
c. $3(3c - 5d)(2c - 3d)$
d. $(3a + 4b)(4a - 3b)$

YOUR TURN Factor.

a. $9x^2 - 18x + 8$

b. $6c^2 - 13cd + 6d^2$

c. $18c^2 - 57cd + 45d^2$

d. $12a^2 + 7ab - 12b^2$

For additional review, see Section 6.3, pages 478–481. For additional exercises, see Exercises 11–42 on page 484.

OBJECTIVE 5. Factor special products.

RULE   **Factoring Perfect Square Trinomials**

$$a^2 + 2ab + b^2 = (a + b)^2$$
$$a^2 - 2ab + b^2 = (a - b)^2$$

EXAMPLE 5   Factor.

a. $x^2 - 10x + 25$

Solution This trinomial is a perfect square because it has the form $a^2 - 2ab + b^2$ where

$$a = x \text{ and } b = 5.$$
$$x^2 - 10x + 25 = (x - 5)^2 \qquad \text{Use } a^2 - 2ab + b^2 = (a - b)^2 \text{ with } a = x \text{ and } b = 5.$$

b. $16u^2 + 24ut + 9t^2$

Solution This is a perfect square trinomial with $a = 4u$ and $b = 3t$.

$$16u^2 + 24ut + 9t^2 = (4u + 3t)^2 \qquad \text{Use } a^2 - 2ab + b^2 = (a - b)^2 \text{ with } a = 4u \text{ and } b = 3t.$$

YOUR TURN   Factor.

a. $a^2 - 16a + 64$ **b.** $25x^2y - 10xy + y$

c. $4x^2 + 20xy + 25y^2$ **d.** $36m^2 - 84mn + 49n^2$

For additional review, see Section 6.4, pages 487–489. For additional exercises, see Exercises 7–24 on page 493.

RULE   **Factoring a Difference of Squares**

$$a^2 - b^2 = (a + b)(a - b)$$ **Warning:** A sum of squares, $a^2 + b^2$, is prime and cannot be factored.

ANSWERS
a. $(a - 8)^2$ **b.** $y(5x - 1)^2$
c. $(2x + 5y)^2$ **d.** $(6m - 7n)^2$

EXAMPLE 6   Factor. $16y^2 - 9$

Solution $16y^2 - 9 = (4y)^2 - 3^2 = (4y + 3)(4y - 3)$ Use $a^2 - b^2 = (a + b)(a - b)$ with $a = 4y$ and $b = 3$.

ANSWERS
a. $(2x + 5y)(2x - 5y)$
b. $(y^2 + 9)(y + 3)(y - 3)$
c. prime

YOUR TURN   Factor.

a. $4x^2 - 25y^2$ **b.** $y^4 - 81$ **c.** $36x^2 + 49y^2$

For additional review, see Section 6.4, pages 489–490. For additional exercises, see Exercises 25–44 on page 493.

RULE   Factoring the Sum or Difference of Cubes

$$a^3 + b^3 = (a + b)(a^2 - ab + b^2)$$
$$a^3 - b^3 = (a - b)(a^2 + ab + b^2)$$

EXAMPLE 7   Factor. $27x^3 - 8$

Solution $27x^3 - 8 = (3x)^3 - 2^3 = (3x - 2)((3x)^2 + (3x)(2) + 2^2)$

Use $a^3 - b^3 = (a - b)(a^2 + ab + b^2)$ with $a = 3x$ and $b = 2$.

$$= (3x - 2)(9x^2 + 6x + 4)$$

YOUR TURN   Factor.

a. $r^3 - 125$

b. $27x^3 + 64$

ANSWERS

a. $(r - 5)(r^2 + 5r + 25)$
b. $(3x + 4)(9x^2 - 12x + 16)$

For additional review, see Section 6.4, pages 490–492. For additional exercises, see Exercises 45–64 on pages 493–494.

Appendix A Arithmetic Sequences and Series

OBJECTIVES

1. Find the terms of a sequence when given the general term.
2. Define and write arithmetic sequences, find their common difference, and find a particular term.
3. Define and write series, find partial sums, and use summation notation.
4. Write arithmetic series and find their sums.

The word **sequence** is used in mathematics much as it is in everyday life. For example, the classes that you attend on a given day occur in a particular order or sequence. As such, a sequence is an ordered list. Suppose we have a bacteria colony that has an initial population of 10,000 and increases at a rate of 10% each day.

On the second day, we would have
$10,000 + 0.10(10,000) = 10,000 + 1000 = 11,000$.
On the third day we have $11,000 + 0.10(11,000) = 11,000 + 1100 = 12,100$.
On the fourth day we have $12,100 + 0.10(12,100) = 12,100 + 1210 = 13,310$.
On the fifth day we have $13,310 + 0.10(13,310) = 13,310 + 1331 = 14,641$.

The number of bacteria present after each day forms a sequence that we can summarize as follows.

Days	1	2	3	4	5
Number of Bacteria	10,000	11,000	12,100	13,310	14,641

Based on this, we make the following definition.

DEFINITION *Sequence:* A function list whose domain is 1, 2, 3, . . . , *n*.

The *domain* of our bacteria example is the numbers of the days {1, 2, 3, 4, 5}. Each number in the *range* of a sequence is called a *term*, so the terms of our bacteria sequence are the numbers of bacteria: 10,000, 11,000, 12,100, 13,310, 14,641. Sequences can be finite or infinite depending on the number of terms. Our bacteria example is a **finite sequence** because it has a finite number of terms. The sequence 2, 4, 6, 8, 10, . . . is an **infinite sequence** because it has an infinite number of terms.

DEFINITION *Finite sequence:* A function with a domain that is the set of natural numbers from 1 to *n*.
Infinite sequence: A function with a domain that is the set of natural numbers.

OBJECTIVE 1. Find the terms of a sequence when given the general term. Since a sequence is a function, we could describe sequences with functional notation. Instead, we use a different notation that emphasizes the fact that the domain is a subset of the natural numbers. We think of the terms of a sequence as $a_1, a_2, a_3, \ldots, a_n$, where the subscript gives the number of the term. Thus, a_n is the nth or *general term* of the sequence and we represent a sequence by giving a formula for a_n. Consider the following sequence:

Term	a_1	a_2	a_3	a_4	a_5	a_n
Term of Sequence	1	4	9	16	25	n^2

We represent this sequence by writing $a_n = n^2$, which means $a_1 = 1^2$, $a_2 = 2^2$, etc.

EXAMPLE 1) Find the first three terms of the following sequences and the 25th term.

a. $a_n = 3n - 1$

Solution We let $n = 1, 2, 3,$ and 25 and evaluate.

$$a_1 = 3(1) - 1 = 3 - 1 = 2 \qquad \text{Let } n = 1 \text{ and evaluate.}$$
$$a_2 = 3(2) - 1 = 6 - 1 = 5 \qquad \text{Let } n = 2 \text{ and evaluate.}$$
$$a_3 = 3(3) - 1 = 9 - 1 = 8 \qquad \text{Let } n = 3 \text{ and evaluate.}$$
$$a_{25} = 3(25) - 1 = 75 - 1 = 74 \qquad \text{Let } n = 25 \text{ and evaluate.}$$

The first three terms of the sequence are 2, 5, and 8. The 25th term is 74.

b. $a_n = \dfrac{(-1)^n}{n^2 + 1}$

Solution We let $n = 1, 2, 3,$ and 25 and evaluate.

$$a_1 = \frac{(-1)^1}{1^2 + 1} = \frac{-1}{1 + 1} = -\frac{1}{2} \qquad \text{Let } n = 1 \text{ and evaluate.}$$

$$a_2 = \frac{(-1)^2}{2^2 + 1} = \frac{1}{4 + 1} = \frac{1}{5} \qquad \text{Let } n = 2 \text{ and evaluate.}$$

$$a_3 = \frac{(-1)^3}{3^2 + 1} = \frac{-1}{9 + 1} = -\frac{1}{10} \qquad \text{Let } n = 3 \text{ and evaluate.}$$

$$a_{25} = \frac{(-1)^{25}}{25^2 + 1} = \frac{-1}{625 + 1} = -\frac{1}{626} \qquad \text{Let } n = 25 \text{ and evaluate.}$$

The first three terms of the sequence are $-\dfrac{1}{2}, \dfrac{1}{5},$ and $-\dfrac{1}{10}.$ The 25th term is $-\dfrac{1}{626}.$

OBJECTIVE 2. Define and write arithmetic sequences, find their common difference, and find a particular term. In the sequence $-3, 1, 5, 9, 13, \ldots$, notice that each term after the first is found by adding 4 to the previous term. This is an example of an **arithmetic sequence** or *arithmetic progression*. Any two successive terms of an arithmetic sequence differ by the same amount, which is called the **common difference** and is denoted by d. To find d, choose any term (except the first) and subtract the previous term.

DEFINITION *Arithmetic sequence:* A sequence in which each term after the first is found by adding the same number to the previous term.
Common difference of an arithmetic sequence: The value d found by $d = a_n - a_{n-1}$ where a_n is any value in the sequence and a_{n-1} is the previous value.

EXAMPLE 2 Write the first four terms of the following arithmetic sequences.

a. The first term is 2 and the common difference is 5.

Solution Begin with 2 and find each successive term by adding 5 to the previous term.

$$a_1 = 2, a_2 = 2 + 5 = 7, a_3 = 7 + 5 = 12, a_4 = 12 + 5 = 17$$

The first four terms of the sequence are 2, 7, 12, 17.

b. $a_1 = -1, d = -2$

Solution Begin with -1 and find each successive term by adding -2 to the previous term.

$$a_1 = -1, a_2 = -1 - 2 = -3, a_3 = -3 - 2 = -5, a_4 = -5 - 2 = -7$$

The first four terms of the sequence are $-1, -3, -5, -7$.

EXAMPLE 3 Find the common difference, d, for the following arithmetic sequence.

a. $-4, -1, 2, 5, 8, \ldots$

Solution Pick any term (except the first) and subtract the term before it.

$$d = -1 - (-4) = -1 + 4 = 3$$

or we could use $d = 5 - 2 = 3$, etc.

b. $8, 6, 4, 2, 0, \ldots$

Solution Pick any term (except the first) and subtract the term before it.

$$d = 6 - 8 = -2$$

or we could use $d = 2 - 4 = -2$, etc.

If the first term of an arithmetic sequence is a_1 and the common difference is d, then the arithmetic sequence can be written as a_1, $a_1 + d$, $a_1 + 2d$, $a_1 + 3d$, $a_1 + 4d$, ... Note that the coefficient of d is one less than the number of the term, so we have the following rule.

RULE   **The nth Term of an Arithmetic Sequence**

The formula for finding the nth term of an arithmetic sequence is $a_n = a_1 + (n - 1)d$, where a_1 is the first term and d is the common difference.

EXAMPLE 4 Find the 23rd term and an expression for the nth term of an arithmetic sequence in which $a_1 = -8$ and $d = 3$.

Solution

$$a_n = a_1 + (n - 1)d$$
$$a_{23} = -8 + (23 - 1)(3) \qquad \text{Substitute 23 for } n, -8 \text{ for } a_1, 3 \text{ for } d.$$
$$a_{23} = 58 \qquad \text{Evaluate.}$$
$$n^{\text{th}} \text{ term: } a_n = -8 + (n - 1)3 \qquad \text{Substitute for } a_1 \text{ and } d.$$
$$a_n = -11 + 3n \qquad \text{Simplify.}$$

Calculator TIPS

Some graphing calculators have a sequence mode that allows you to display the terms of a sequence in a table. To generate the sequence $a_n = n^2$ on a TI-83, press $\boxed{\text{MODE}}$ *and select seq then follow the steps below.*

Press $\boxed{\text{Y =}}$ *and enter as below.*

Press $\boxed{\text{2nd}}$ $\boxed{\text{WINDOW}}$ *to set the table as below.*

Press $\boxed{\text{2nd}}$ $\boxed{\text{GRAPH}}$ *to display the table.*

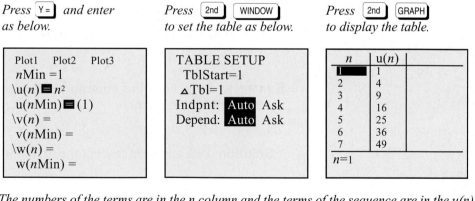

The numbers of the terms are in the n column and the terms of the sequence are in the u(n) column.

If we know the first term and one other term, we can use the formula for the nth term of an arithmetic sequence to find the common difference. Consequently, we can find the sequence.

EXAMPLE 5 The first term of an arithmetic sequence is 1 and the 20th term (a_{20}) is 58. Find the common difference and the first four terms of the sequence.

Solution We will use the fact that we know a_{20} and the formula for the nth term to find d.

$$a_n = a_1 + (n - 1)d$$
$$a_{20} = 1 + (\mathbf{20} - 1)d \qquad \text{Substitute 1 for } a_1 \text{ and 20 for } n.$$
$$58 = 1 + 19d \qquad \text{Substitute 58 for } a_{20} \text{ and solve.}$$
$$3 = d$$

The first four terms are $1, 1 + 3 = 4, 4 + 3 = 7, 7 + 3 = 10$.

OBJECTIVE 3. Define and write series, find partial sums, and use summation notation. When the terms of a sequence are added, the sum is called a **series**.

DEFINITION *Series:* The sum of the terms of a sequence.

Given a sequence $a_1, a_2, a_3, \ldots, a_n$, then $a_1 + a_2 + a_3 + \cdots + a_n$ is the corresponding series. Finite series correspond to finite sequences and infinite series correspond to infinite sequences. If an expression for the nth term is known, we can write the series in *summation notation* using the capital Greek letter *sigma* (Σ) as follows:

$$a_1 + a_2 + a_3 + \cdots + a_n = \sum_{i=1}^{n} a_i$$

The i is called the *index of summation*; 1 is the *lower limit* of i and n is the *upper limit* of i. If the upper limit is a natural number, the series is finite, and if the upper limit is ∞, the series is infinite. To find a finite series from summation notation, first replace the index of summation (usually i) with its lower limit and evaluate that term, replace i with the lower limit plus 1 and evaluate that term, and so on until you reach the upper limit. This gives the series. To find the sum, add the terms of the series.

EXAMPLE 6

Warning: Do not confuse the use of i in sigma notation, as in $\sum_{i=1}^{5} (3i + 2)$, with its use as the imaginary unit, as in $7 \pm 5i$. In sigma notation, i is a variable representing natural numbers, whereas as the imaginary unit it represents $\sqrt{-1}$.

a. Write the terms of $\sum_{i=1}^{5} (3i + 2)$ and find the sum of the series.

Solution Replace i with $1, 2, 3, 4,$ and 5. Then evaluate.

$$\sum_{i=1}^{5} (3i + 2) = (3 \cdot 1 + 2) + (3 \cdot 2 + 2) + (3 \cdot 3 + 2) + (3 \cdot 4 + 2) + (3 \cdot 5 + 2)$$
$$= 5 + 8 + 11 + 14 + 17$$
$$= 55$$

b. Write the first five terms of the infinite series $\sum_{i=1}^{\infty} 2i^2$.

Solution Replace i with $1, 2, 3, 4,$ and 5. Then evaluate.

$$\sum_{i=1}^{\infty} 2i^2 = 2 \cdot 1^2 + 2 \cdot 2^2 + 2 \cdot 3^2 + 2 \cdot 4^2 + 2 \cdot 5^2 + \cdots$$
$$= 2 + 8 + 18 + 32 + 50 + \cdots$$

OBJECTIVE 4. Write arithmetic series and find their sums. If the terms of an arithmetic sequence are added, it is called an **arithmetic series**.

DEFINITION *Arithmetic series:* The sum of the terms in an arithmetic sequence.

Consequently, an arithmetic series has the form $a_1 + (a_1 + d) + (a_1 + 2d) + (a_1 + 3d) + \ldots + [a_1 + (n - 1)d]$.

Adding a finite number of terms of an infinite series gives a *partial sum*. The symbol S_n is used to indicate the sum of the first n terms. For example, S_5 means add the first five terms. Let's derive a formula for S_n for an arithmetic series as follows:

$$S_n = a_1 + (a_1 + d) + (a_1 + 2d) + (a_1 + 3d) + \ldots + a_n$$

We also need to include the terms between a_1 and $3d$ and a_n, so we write another version of S_n beginning with the last term, a_n, and subtracting the common difference, d, from the previous term.

$$S_n = a_n + (a_n - d) + (a_n - 2d) + (a_n - 3d) + \ldots + a_1$$

To describe the entire sum, we add our two versions of S_n together.

$$
\begin{array}{ll}
S_n = a_1 & + (a_1 + d) + (a_1 + 2d) + (a_1 + 3d) + \cdots + a_n \\
+\ S_n = a_n & + (a_n - d) + (a_n - 2d) + (a_n - 3d) + \cdots + a_1 \\
\hline
2S_n = (a_1 + a_n) + (a_1 + a_n) + (a_1 + a_n) + (a_1 + a_n) + \cdots + (a_1 + a_n)
\end{array}
$$

Since S_n has n terms, there are n terms of $(a_1 + a_n)$, so

$$2S_n = n(a_1 + a_n), \text{ or}$$

$$S_n = \frac{n}{2}(a_1 + a_n)$$

RULE **Partial Sum S_n of an Arithmetic Series**

The sum of the first n terms of an arithmetic series, S_n, called the nth partial sum, is given by

$$S_n = \frac{n}{2}(a_1 + a_n)$$

where n is the number of terms, a_1 is the first term, and a_n is the nth term.

EXAMPLE 7 Find the sum of the first 20 terms (S_{20}) of the arithmetic series $-6 - 2 + 2 + 6 + \cdots$.

Solution To find the S_{20}, we first need to find the 20th term. Since $d = 4$,

$$a_{20} = -6 + (20 - 1)(4) = 70$$

$$S_n = \frac{n}{2}(a_1 + a_n) \qquad \text{Rule for } n^{\text{th}} \text{ partial sum.}$$

$$S_{20} = \frac{20}{2}(-6 + 70) \qquad \text{Substitute 20 for } \textbf{\textit{n}}, -6 \text{ for } a_1, \text{ and 70 for } a_n.$$

$$S_{20} = 10(64) = 640 \qquad \text{Evaluate.}$$

Appendix A
Exercises

For Extra Help

MyMathLab

MyMathLab   Videotape/DVT   InterAct Math   Math Tutor Center   Math XL.com

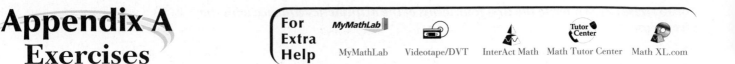

1. What set of numbers makes up the domain of a sequence?

2. A sequence with an unlimited number of terms is called a(n) _____ sequence.

3. What is a series?

4. What is an arithmetic sequence? How do you find the common difference?

5. The series $-10 - 5 + 0 + 5 + 10 + \cdots$ is an example of a(n) _____ series with a common difference of _____.

6. What does S_n represent?

For Exercises 7–14, write the first four terms of the sequence and the indicated term.

7. $a_n = 2n + 1$, 20th term

8. $a_n = 3n - 4$, 18th term

9. $a_n = n^2 + 2$, 15th term

10. $a_n = n^2 - 3$, 12th term

11. $a_n = \dfrac{n}{n + 2}$, 22nd term

12. $a_n = \dfrac{2n}{n + 3}$, 10th term

13. $a_n = \dfrac{(-1)^n}{n^2 + 1}$, 15th term

14. $a_n = \dfrac{(-1)^n}{n^2 - 3}$, 26th term

For Exercises 15–20, find the common difference, d, for each arithmetic sequence.

15. $2, 7, 12, 17, \ldots$

16. $3, 11, 19, 27, \ldots$

17. $25, 22, 19, 16, \ldots$

18. $42, 36, 30, 24, \ldots$

19. $-12, -5, 2, 9, \ldots$

20. $-24, -27, -30, -33, \ldots$

For Exercises 21–26, find the indicated term and an expression for the nth term of the given arithmetic sequence.

21. a_{14} if $a_1 = 14$ and $d = 4$

22. a_{24} if $a_1 = 16$ and $d = 6$

23. a_{28} if $a_1 = -8$ and $d = -3$

24. a_{30} if $a_1 = -7$ and $d = -6$

25. a_{34} of $-5, -1, 3, 7, \ldots$

26. a_{21} of $8, 15, 22, 29, \ldots$

For Exercises 27–32, write the first four terms of the arithmetic sequence with the given characteristics.

27. $a_1 = -6, d = 7$

28. $a_1 = -2, d = 6$

29. $a_1 = -5, d = -3$

30. $a_1 = -13, d = -5$

31. $a_1 = 7, a_{18} = 75$

32. $a_1 = 6, a_{22} = 153$

33. Find the first term of an arithmetic sequence if $a_{45} = 143$ and $d = 3$.

34. Find the first term of an arithmetic sequence if $a_{39} = 181$ and $d = 6$.

35. Find the common difference of an arithmetic sequence if the first term is -110 and the 29th term is 2.

36. Find the common difference of an arithmetic sequence if the first term is 78 and the 15th term is -76.

For Exercises 37–40, write the first four terms of the arithmetic sequence with the given d and a_n.

37. $d = 7, a_8 = 41$

38. $d = 4, a_{11} = 27$

39. $d = -3, a_7 = 9$

40. $d = -6, a_{10} = -36$

For Exercises 41–48, write the series and find the sum.

41. $\displaystyle\sum_{i=1}^{6} i^2$

42. $\displaystyle\sum_{i=1}^{3} 3i^2$

43. $\displaystyle\sum_{i=1}^{4} (2i - 5)$

44. $\displaystyle\sum_{i=1}^{5} (4i + 1)$

45. $\displaystyle\sum_{i=1}^{3} (3i^2 - 4)$

46. $\displaystyle\sum_{i=1}^{4} (-2i^2 + 5)$

47. $\displaystyle\sum_{i=3}^{6} (4i - 3)$

48. $\displaystyle\sum_{i=2}^{5} (4i - 2)$

For Exercises 49–56, find the given S_n for the arithmetic series.

49. If $a_1 = 10$ and $d = 4$, find S_{25}.

50. If $a_1 = -8$ and $d = 2$, find S_{30}.

51. If $a_1 = 12$ and $d = -3$, find S_{22}.

52. If $a_1 = 18$ and $d = -2$, find S_{30}.

53. $3 + 9 + 15 + 21 + \cdots$. Find S_{15}.

54. $5 + 9 + 13 + 17 + \cdots$. Find S_{25}.

55. $54 + 46 + 38 + 30 + \cdots$. Find S_{12}.

56. $44 + 39 + 34 + 29 + \cdots$. Find S_{18}.

For Exercises 57–64, answer each question.

57. Find the sum of the first 100 natural numbers.

58. Find the sum of the even integers 2 through 200.

59. A concert hall has 60 seats in the first row, 64 in the second, 68 in the third, and so on.

   a. How many seats are in the 22nd row?

   b. How many seats are in the concert hall if there are 35 rows?

60. During the year 2003, many car dealers were offering loans at 0% interest. Johanna bought a car for $24,000 and made monthly payments of $400. Using 24,000 as the first term, write the first five terms of an arithmetic sequence that gives the amount that she still owes at the end of each month. How much will she owe at the end of the twenty-fifth month?

61. Fence posts are arranged in a triangular stack with 25 on the bottom row, 24 on the next, 23 on the next, and so forth until there is a single post on the top.

   a. How many posts are on the 10th row from the bottom?

   b. How many posts are in the stack?

62. Carlos is doing sit-ups to flatten his stomach. He did 25 the first night and plans to add 1 each night.

   a. Write the first five terms of an arithmetic sequence that gives the number of sit-ups that he does each night.

   b. Find the number he does on the 30th night.

   c. Find the total number of sit-ups that he has done after 30 nights.

63. Tanisha takes a job that pays $28,000 the first year and a raise of $1500 per year.

   a. Write the first five terms of an arithmetic sequence that gives her salary at the end of each year.

   b. What will her salary be for the 10th year?

   c. What are her total earnings for her first 10 years?

64. You have a choice of two jobs. Job A has a starting salary of $25,000 with raises of $900 per year, and job B has a starting salary of $28,000 with raises of $600 per year.

   a. Which job will pay the most during the 10th year?

   b. Which job will pay the most total amount during the first 15 years?

Appendix B Geometric Sequences and Series

OBJECTIVES

1. Write a geometric sequence and find its common ratio and a specified term.
2. Find partial sums of geometric series.
3. Find the sums of infinite geometric series.
4. Solve applications using geometric series.

OBJECTIVE 1. Write a geometric sequence and find its common ratio and a specified term. In Appendix A, we generated an arithmetic sequence by adding the same number to each term to get the next term. In this section, we will multiply each term by the same number to generate a **geometric sequence**. The number we multiply by is called the **common ratio** and is denoted as r. We can find r by dividing any term (except the first) by the term before it.

DEFINITION *Geometric sequence:* A sequence in which every term after the first is found by multiplying the previous term by the same number, called the common ratio,

r, where $r = \dfrac{a_n}{a_{n-1}}$.

Note: *In geometric sequences, the first term is usually denoted as a rather than a_1.*

If the first term is a and the common ratio is r, a geometric sequence has the form $a, ar, ar^2, ar^3, ar^4, \ldots, ar^{n-1}$. Thus, the general term of a geometric series is $a_n = ar^{n-1}$.

We notice that the exponent of r is one less than the number of the term, so we have the following rule.

RULE The nth Term of a Geometric Sequence

The formula for finding the nth term of a geometric sequence is $a_n = ar^{n-1}$, where a is the first term and r is the common ratio.

EXAMPLE 1 A geometric sequence has a first term of 3 and a common ratio of 4.

a. Write the first five terms.

Solution To find each term, multiply the term before it by 4. So,

$$a_1 = 3, a_2 = 3(4) = 12, a_3 = 12(4) = 48, a_4 = 48(4) = 192, a_5 = 192(4) = 768$$

The first five terms of the sequence are 3, 12, 48, 192, 768.

b. Find the 10th term.

Solution $a_n = ar^{n-1}$ Formula for a_n.

$a_{10} = 3(4)^{10-1}$ Substitute **10** for n, **3** for a, and **4** for r.

$a_{10} = 786,432$ Evaluate.

c. Find the general term.

Solution $a_n = ar^{n-1}$ Formula for a_n.

$a_n = 3(4)^{n-1}$ Substitute for a and r.

EXAMPLE 2 Given the geometric sequence 32, -16, 8, -4, . . . , find the common ratio r.

Solution To find r, divide any term except the first by the term before it.

$$r = \frac{-16}{32} = -\frac{1}{2} \quad \text{or} \quad r = \frac{8}{-16} = -\frac{1}{2}, \text{etc.}$$

Geometric sequences often occur in populations and other applications.

EXAMPLE 3 The population of rabbits in a large pen increases at a rate of 12% per month. If there are currently 50 rabbits, find the population after 15 months.

Solution Let P_0 be the initial number of rabbits, P_1 the number after 1 month, P_2 the number after 2 months, etc. The number of rabbits at the end of each month is the number at the beginning of the month plus an increase of 12% of that number. So the number at the end of each month can be found as follows:

$$P_1 = P_0 + 0.12P_0 = 1.12P_0$$
$$P_2 = P_1 + 0.12P_1 = 1.12P_1 = 1.12(1.12P_0) = (1.12)^2 P_0$$
$$P_3 = P_2 + 0.12P_2 = 1.12P_2 = 1.12[(1.12)^2 P_0] = (1.12)^3 P_0$$

The number of rabbits at the end of each month can be represented by the sequence $P_0, P_1, P_2, P_3, \ldots P_n = P_0, 1.12P_0, (1.12)^2 P_0, (1.12)^3 P_0, \ldots, (1.12)^{n-1} P_0$, which is a geometric sequence whose first term is P_0 and $r = 1.12$. Consequently, the nth term is $P_0(1.12)^{n-1}$. The number of rabbits at the end of 15 months is the 16th term of the sequence, so

$$P_{16} = P_0(1.12)^{16-1}$$
$$P_{16} = 50(1.12)^{15}$$
$$P_{16} = 273.68$$

Answer There are about 274 rabbits at the end of 15 months.

Note: It can be shown using this procedure that if a population is growing at p% per unit time and the initial population is P_0, then the population at the end of each unit of time forms a geometric sequence whose first term is P_0 and whose common ratio is $(1 + p)$, where p is p% written as a decimal. So the geometric sequence is $P_0, P_0(1 + p), P_0(1 + p)^2, P_0(1 + p)^3, \ldots, P_0(1 + p)^{n-1}$. The population after n time periods is the $(n + 1)$st term of the sequence, which is $P_0(1 + p)^n$.

OBJECTIVE 2. Find partial sums of geometric series. If we add the terms of an arithmetic sequence, we get an arithmetic series. Likewise, if we add the terms of a geometric sequence, we get a **geometric series**.

DEFINITION *Geometric series:* The sum of the terms of a geometric sequence.

A geometric series is of the form $a + ar + ar^2 + \cdots + ar^{n-1}$ if the series is finite, and $a + ar + ar^2 + \cdots$ if the series is infinite.

In the geometric series $3 + 6 + 12 + 24 + \cdots$, we have $a = 3$ and $r = 2$. In $243 - 81 + 27 - 9 + \cdots$, we have $a = 243$ and $r = -\dfrac{1}{3}$. Just as with arithmetic series, we can find a formula for partial sum, S_n. Begin with

$$S_n = a + ar + ar^2 + \cdots + ar^{n-1}$$
$$\underline{-rS_n = \quad - ar - ar^2 - \cdots - ar^{n-1} - ar^n} \quad \text{Multiply both sides of the equation by } -r.$$
$$S_n - rS_n = a - ar^n \quad\quad\quad\quad\quad\quad \text{Add the equations.}$$
$$S_n(1 - r) = a(1 - r^n) \quad\quad\quad\quad\quad \text{Factor both sides.}$$
$$S_n = \frac{a(1 - r^n)}{1 - r} \quad\quad\quad\quad\quad\quad \text{Divide both sides by } 1 - r.$$

RULE The Partial Sum, S_n, of a Geometric Series

The sum of the first n terms of a geometric series, S_n, called the nth partial sum, is given by

$$S_n = \frac{a(1 - r^n)}{1 - r}$$

where n is the number of terms, a is the first term, and r is the common ratio ($r \neq 1$).

EXAMPLE 4 Find the sum of the first 10 terms of the geometric series. $1 - 3 + 9 - 27 + \cdots$

Solution We first find r: $r = \dfrac{-3}{1} = -3$.

$$S_n = \frac{a(1 - r^n)}{1 - r} \quad\quad\quad \text{Formula for } S_n.$$

$$S_{10} = \frac{1(1 - (-3)^{10})}{1 - (-3)} \quad \text{Substitute 10 for } n, 1 \text{ for } a, \text{ and } -3 \text{ for } r.$$

$$S_{10} = -14,762 \quad\quad\quad\quad \text{Evaluate.}$$

OBJECTIVE 3. Find the sums of infinite geometric series. If the common ratio satisfies $|r| > 1$, the partial sums become infinitely large as n becomes infinitely large. However, if $|r| < 1$, the partial sums approach a value as n becomes infinitely large. This value is called the *limit* of the partial sums and is the sum of the infinite series.

For example, for the series $2 + 1 + \dfrac{1}{2} + \dfrac{1}{4} + \cdots$, we have

$$S_5 = \frac{2\left(1 - \left(\frac{1}{2}\right)^5\right)}{1 - \frac{1}{2}} = 3.875$$

$$S_{10} = \frac{2\left(1 - \left(\frac{1}{2}\right)^{10}\right)}{1 - \frac{1}{2}} \approx 3.996$$

$$S_{15} = \frac{2\left(1 - \left(\frac{1}{2}\right)^{15}\right)}{1 - \frac{1}{2}} \approx 3.9998779$$

Notice that the greater the value of n, the closer the sum is to 4. The partial sums approach 4 because as n gets larger, $\left(\dfrac{1}{2}\right)^n$ approaches 0. In the preceding example, $\left(\dfrac{1}{2}\right)^5 = 0.03125$, $\left(\dfrac{1}{2}\right)^{10} \approx 0.000977$, and $\left(\dfrac{1}{2}\right)^{15} \approx 0.0000305$. In general, if $|r| < 1$, then r^n approaches 0 as n becomes large. Consequently, if $|r| < 1$ the formula $S_n = \dfrac{a(1 - r^n)}{1 - r}$ becomes $S_\infty = \dfrac{a(1 - 0)}{1 - r} = \dfrac{a}{1 - r}$ as n becomes infinitely large. We denote the sum as S_∞ rather than S_n.

RULE **The Sum of an Infinite Geometric Series**

If $|r| < 1$, the sum of an infinite geometric series, S_∞, is given by the formula

$$S_\infty = \frac{a}{1 - r}$$

where a is the first term and r is the common ratio. If $|r| \geq 1$, S_∞ does not exist.

EXAMPLE 5 Find the sum of the infinite geometric series.

$$2 + 1 + \frac{1}{2} + \frac{1}{4} + \cdots$$

Solution We know that $a = 2$ and $r = \dfrac{1}{2}$. Since $|r| < 1$, the sum exists.

$$S_\infty = \frac{a}{1 - r}$$

$$S_\infty = \frac{2}{1 - \frac{1}{2}} \qquad \text{Substitute 2 for } a \text{ and } \frac{1}{2} \text{ for } r.$$

$$S_\infty = 4 \qquad \text{Evaluate.}$$

Infinite geometric series also provides us with a method of changing a repeating decimal into a fraction.

EXAMPLE 6 Write as a fraction. $0.\overline{37}$

Solution First write $0.\overline{37}$ as an infinite geometric series as follows:

$$0.\overline{37} = 0.37 + 0.0037 + 0.000037 + \cdots$$

$$0.\overline{37} = \frac{37}{100} + \frac{37}{10000} + \frac{37}{1000000} + \cdots$$

The last line is an infinite geometric series with $a = \dfrac{37}{100}$ and $r = \dfrac{1}{100}$. Since $|r| < 1$, the sum of this series exists.

$$S_\infty = \frac{a}{1 - r}$$

$$S_\infty = \frac{\dfrac{37}{100}}{1 - \dfrac{1}{100}} \qquad \text{Substitute } \dfrac{37}{100} \text{ for } a \text{ and } \dfrac{1}{100} \text{ for } r.$$

$$S_\infty = \frac{\dfrac{37}{100}}{\dfrac{99}{100}} \qquad \text{Simplify.}$$

$$S_\infty = \frac{37}{99}$$

Answer $0.\overline{37} = \dfrac{37}{99}$

OBJECTIVE 4. Solve applications using geometric series.

EXAMPLE 7 To save for their child's college education, the McBride family put $1000 into a savings account the first year, and each year thereafter deposited 10% more than the previous year.

a. Write the first five terms of the geometric sequence that gives the amount of money deposited into the account each year.

Solution From the note following Example 3, this is a geometric series in which $a = 1000$ and $r = (1 + 0.10) = 1.1$

$$a_1 = 1000, \ a_2 = 1000(1.1) = 1100, \ a_3 = 1000(1.1)^2 = 1210,$$
$$a_4 = 1000(1.10)^3 = 1331, \ a_5 = 1000(1.10)^4 = 1464.10$$

The first five terms of the sequence are 1000, 1100, 1210, 1331, 1464.10.

b. How much money will have been deposited into the account at the end of 15 years?

Solution The series $1000 + 1100 + 1210 + 1331 + 1464.10 + \cdots$ is geometric with $a = 1000$ and $r = 1.1$.

$$S_n = \frac{a(1 - r^n)}{1 - r}$$

$$S_{15} = \frac{1000(1 - 1.1^{15})}{1 - 1.1} \qquad \text{Substitute 15 for } n, 1000 \text{ for } a, \text{ and } 1.1 \text{ for } r.$$

$$S_{15} = 31{,}772.48 \qquad \text{Evaluate.}$$

Answer They will have deposited $31,772.48 in the account after 15 years.

Appendix B
Exercises

1. If you divide a term of a geometric sequence by the term before it, what do you get?

2. What is the form of a geometric sequence?

3. Can you find the sum of the first n terms of the geometric series $2 + 6 + 18 + \ldots$ using the formula $S_n = \dfrac{a(1 - r^n)}{1 - r}$? Why or why not?

4. Can you find the sum of the infinite geometric series $2 + 6 + 18 + \cdots$ using the formula $S_\infty = \dfrac{a}{1 - r}$? Why or why not?

5. The fourth partial sum, S_4, of $2 + 6 + 18 + \ldots$ is ___ + ___ + ___ + ___ = ___.

6. Write $0.\overline{23}$ as an infinite geometric series. What are a and r?

For Exercises 7–14: a. Find the common ratio, r, for the given geometric sequence.
                 b. Find the indicated term.
                 c. Find an expression for the general term, a_n.

7. $1, 3, 9, 27, \ldots$; 9th term

8. $7, 14, 28, 56, \ldots$; 11th term

9. $-2, 4, -8, 16, \ldots$; 15th term

10. $-8, 24, -72, 216, \ldots$; 9th term

11. $243, 81, 27, 9, \ldots$; 10th term

12. $8, 4, 2, 1, \ldots$; 8th term

13. $128, -32, 8, -2, \ldots$; 7th term

14. $125, -25, 5, -1, \ldots$; 9th term.

For Exercises 15–18: a. Write the first five terms of the geometric sequence satisfying the given conditions.
b. Find the indicated term.

15. $a = -4, r = -3$; 8th term

16. $a = -6, r = -2$; 10th term

17. $a = 243, r = \dfrac{1}{3}$; 7th term

18. $a = 256, r = -\dfrac{1}{2}$; 10th term

For Exercises 19–24: a. Use the formula for the nth term to find r.
b. Write the first four terms of the geometric sequence.

19. $a = 1, r > 0, a_5 = 16$

20. $a = 3, a_4 = -24$

21. $a = -6, r > 0, a_5 = -96$

22. $a = -4, a_4 = -32$

23. $a = 128, a_4 = 2$

24. $a = -243, a_6 = 1$

25. Find the first term of a geometric sequence in which $r = 3$ and the fifth term is -486.

26. Find the first term of a geometric sequence in which $r = -\dfrac{1}{2}$ and the fifth term is $-\dfrac{1}{8}$.

For Exercises 27–32, find the sum of the first n terms of each geometric series for the given value of n.

27. $3 + 9 + 27 + 81 + \cdots, n = 11$

28. $-1 - 2 - 4 - 8 - \cdots, n = 9$

29. $32 + 16 + 8 + 4 + \cdots, n = 9$

30. $81 + 27 + 9 + 3 + \cdots, n = 7$

31. $128 - 32 + 8 - 2 + \cdots, n = 9$

32. $625 - 125 + 25 - 5 + \cdots, n = 8$

For Exercises 33–38, find the sum of the infinite geometric series, if possible. If it is not possible, explain why.

33. $27 + 9 + 3 + \cdots$

34. $8 + 4 + 2 + \cdots$

35. $15 - 9 + \dfrac{27}{5} - \cdots$

36. $15 - 10 + \dfrac{20}{3} - \cdots$

37. $9 + 12 + 16 + \cdots$

38. $16 + 20 + 25 + \cdots$

For Exercises 39–42, write each repeating decimal as a fraction.

39. $0.\overline{4}$

40. $0.\overline{7}$

41. $0.\overline{17}$

42. $0.\overline{25}$

For Exercises 43 and 44, answer each question.

43. If $a_n = 500(1.04)^n$:
   a. Find the first five terms of the sequence.

   b. Find the 10th term of the sequence.

44. If $a_n = 350(1.06)^n$:
   a. Find the first five terms of the sequence.

   b. Find the 8th term of the sequence.

For Exercises 45–50, solve.

45. A population of mink is increasing at a rate of 8% per month. The current mink population is 100.

   a. Using 100 as the first term, find the first four terms of the geometric sequence that gives the number of mink at the beginning of each month.

   b. Find the number of mink present at the beginning of the eighth month.

   c. Find the expression for the general term, a_n.

46. The generation time (the time required for the number present to double) for a particular bacteria is 1 hour. Suppose initially one bacteria was present.

    a. Using 1 as the first term, write the first five terms of the geometric sequence giving the number of bacteria present after each hour.

    b. Find an expression for the number present after the nth hour.

    c. How many bacteria are present after 1 day?

47. Suppose you took a job for a month (20 working days) that paid $0.01 the first day and your salary doubled each day.

    a. Write the first five terms of the geometric sequence that gives your salary each day.

    b. Find an expression for the amount earned on the nth day.

    c. How much would you earn on the 20th day?

    d. What are your total earnings for the month?

48. Damarys deposits $200 in the bank and each month thereafter deposits 5% more than the month before. She does this for 1 year.

    a. Find an expression for the amount she deposits on the nth month.

    b. Write the first four terms of the geometric sequence that gives the amount of her deposit each month.

    c. How much did she deposit on the 10th month?

    d. How much does she deposit for the year?

49. A new boat costs $20,000 and depreciates by 7% each year. What will the boat be worth in 8 years?

50. The isotope $_{15}P^{33}$ has a half-life of 25 days. A sample has 400 grams.

    a. Find the first five terms of the geometric sequence that gives the amount present at the end of each half-life.

    b. Find an expression for the amount present after the nth half-life.

    c. Find the amount present after the 10th half-life.

Appendix C The Binomial Theorem

OBJECTIVES

1. Expand a binomial using Pascal's triangle.
2. Evaluate factorial notation and binomial coefficients.
3. Expand a binomial using the binomial theorem.
4. Find a particular term of a binomial expansion.

In Section 5.5 we learned to square binomials. In this section, we will learn to raise binomials to natural-number powers.

OBJECTIVE 1. Expand a binomial using Pascal's triangle. We begin by writing out $(a + b)^n$, where n is a natural number, and look for patterns. These products are called *binomial expansions.*

$$(a + b)^0 = 1$$
$$(a + b)^1 = a + b$$
$$(a + b)^2 = a^2 + 2ab + b^2$$
$$(a + b)^3 = a^3 + 3a^2b + 3ab^2 + b^3$$
$$(a + b)^4 = a^4 + 4a^3b + 6a^2b^2 + 4ab^3 + b^4$$
$$(a + b)^5 = a^5 + 5a^4b + 10a^3b^2 + 10a^2b^3 + 5ab^4 + b^5$$

Conclusions: Several patterns can be observed from the preceding expansions.

1. The first term in the expansion, a, is raised to the same power as the binomial and the power of a decreases by 1 in each successive term. Note that the last term does not contain an a.
2. The exponent of b is 0 in the first term and increases by 1 on each successive term.
3. The sum of the exponents of the variables of each term equals the exponent of the binomial.
4. The number of terms in the expansion is one more than the exponent of the binomial.

Now consider the coefficients of the terms in these expansions.

Coefficients of Expansions

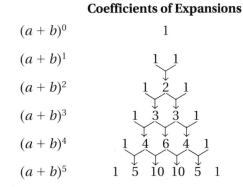

$(a + b)^0$	1
$(a + b)^1$	1 1
$(a + b)^2$	1 2 1
$(a + b)^3$	1 3 3 1
$(a + b)^4$	1 4 6 4 1
$(a + b)^5$	1 5 10 10 5 1

If we arrange the coefficients of each expansion in a triangular array, we see an interesting pattern. Each row begins and ends with 1. Each number inside a row is the sum of the two numbers in the row above it. For example, each 10 in the bottom row comes from adding the 4 and 6 directly above.

This triangular array of numbers is called *Pascal's triangle* in honor of the French mathematician Blaise Pascal. Using these observations, the expansion of $(a + b)^6$ would have seven terms and the variable portions of the terms would be a^6, a^5b, a^4b^2, a^3b^3, a^2b^4, ab^5, b^6. By continuing the pattern in Pascal's triangle, we can find the coefficients for each of those terms.

$(a + b)^5$ 1   5   10   10   5   1

$(a + b)^6$ 1   6   15   20   15   6   1

Using the coefficients from the last line of Pascal's triangle and the variables previously listed gives us

$$(a + b)^6 = a^6 + 6a^5b + 15a^4b^2 + 20a^3b^3 + 15a^2b^4 + 6ab^5 + b^6$$

OBJECTIVE 2. Evaluate factorial notation and binomial coefficients. Although Pascal's triangle is easy to use, it isn't practical, especially for binomials raised to large powers. Consequently, another method called the *binomial theorem* is often used. Before introducing the binomial theorem, we need **factorial notation**.

DEFINITION   *Factorial notation:* For any natural number n, the symbol $n!$ (read "n factorial") means $n(n - 1)(n - 2) \ldots 3 \cdot 2 \cdot 1$. $0!$ is defined to be 1, so $0! = 1$.

EXAMPLE 1   Evaluate the following factorials.

a. $5!$

Solution $5! = 5 \cdot 4 \cdot 3 \cdot 2 \cdot 1 = 120$

b. $7!$

Solution $7! = 7 \cdot 6 \cdot 5 \cdot 4 \cdot 3 \cdot 2 \cdot 1 = 5040$

Sometimes we may not write all the factors of a factorial. In such cases, the last desired factor is written as a factorial. Below are some alternate ways to write $7!$

$$7! = 7 \cdot 6 \cdot 5 \cdot 4 \cdot 3 \cdot 2 \cdot 1 = 7 \cdot 6! \quad \text{or} \quad 7! = 7 \cdot 6 \cdot 5! \quad \text{or} \quad 7! = 7 \cdot 6 \cdot 5 \cdot 4!$$

The coefficients of a binomial expansion can be expressed in terms of factorials using a special notation called the **binomial coefficient**. We will see exactly how the binomial coefficient is used later.

DEFINITION   *Binomial coefficient:* A number written as $\binom{n}{r}$ and defined as $\dfrac{n!}{r!(n - r)!}$.

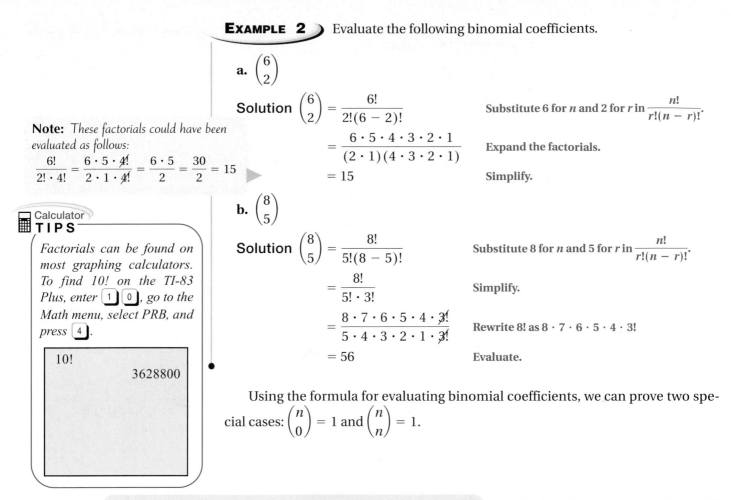

EXAMPLE 2 Evaluate the following binomial coefficients.

a. $\dbinom{6}{2}$

Solution $\dbinom{6}{2} = \dfrac{6!}{2!(6-2)!}$ Substitute 6 for n and 2 for r in $\dfrac{n!}{r!(n-r)!}$.

$= \dfrac{6 \cdot 5 \cdot 4 \cdot 3 \cdot 2 \cdot 1}{(2 \cdot 1)(4 \cdot 3 \cdot 2 \cdot 1)}$ Expand the factorials.

$= 15$ Simplify.

Note: *These factorials could have been evaluated as follows:*

$\dfrac{6!}{2! \cdot 4!} = \dfrac{6 \cdot 5 \cdot \cancel{4!}}{2 \cdot 1 \cdot \cancel{4!}} = \dfrac{6 \cdot 5}{2} = \dfrac{30}{2} = 15$

b. $\dbinom{8}{5}$

Solution $\dbinom{8}{5} = \dfrac{8!}{5!(8-5)!}$ Substitute 8 for n and 5 for r in $\dfrac{n!}{r!(n-r)!}$.

$= \dfrac{8!}{5! \cdot 3!}$ Simplify.

$= \dfrac{8 \cdot 7 \cdot 6 \cdot 5 \cdot 4 \cdot \cancel{3!}}{5 \cdot 4 \cdot 3 \cdot 2 \cdot 1 \cdot \cancel{3!}}$ Rewrite 8! as $8 \cdot 7 \cdot 6 \cdot 5 \cdot 4 \cdot 3!$

$= 56$ Evaluate.

Calculator TIPS

Factorials can be found on most graphing calculators. To find 10! on the TI-83 Plus, enter `1` `0`, *go to the Math menu, select PRB, and press* `4`.

```
10!
          3628800
```

Using the formula for evaluating binomial coefficients, we can prove two special cases: $\dbinom{n}{0} = 1$ and $\dbinom{n}{n} = 1$.

OBJECTIVE 3. Expand a binomial using the binomial theorem. Let's look at the expansion of $(a + b)^6$ and make another observation.

$$(a + b)^6 = a^6 + 6a^5b + 15a^4b^2 + 20a^3b^3 + 15a^2b^4 + 6ab^5 + b^6$$

Look at the third term of the expansion, $15a^4b^2$. The coefficient is 15 and from Example 2a, we see that $\dbinom{6}{2} = 15$. The coefficient of the fourth term is 20 and $\dbinom{6}{3} = 20$. In both cases, the n value of the binomial coefficient, $\dbinom{n}{r}$, is the exponent of the binomial and the r value is the exponent of b. This observation leads to the **binomial theorem**.

DEFINITION ***The Binomial theorem:*** For any positive integer n,

$$(a + b)^n = \dbinom{n}{0}a^n + \dbinom{n}{1}a^{n-1}b + \dbinom{n}{2}a^{n-2}b^2 + \dbinom{n}{3}a^{n-3}b^3 + \cdots + \dbinom{n}{n}b^n$$

EXAMPLE 3 Expand each of the following binomials using the binomial theorem.

a. $(a + b)^6$

Solution

$$(a + b)^6 = \binom{6}{0}a^6 + \binom{6}{1}a^5b + \binom{6}{2}a^4b^2 + \binom{6}{3}a^3b^3 + \binom{6}{4}a^2b^4 + \binom{6}{5}ab^5 + \binom{6}{6}b^6$$

$$= \frac{6!}{0!6!}a^6 + \frac{6!}{1!5!}a^5b + \frac{6!}{2!4!}a^4b^2 + \frac{6!}{3!3!}a^3b^3 + \frac{6!}{4!2!}a^2b^4 + \frac{6!}{5!1!}ab^5 + \frac{6!}{6!0!}b^6$$

$$= a^6 + 6a^5b + 15a^4b^2 + 20a^3b^3 + 15a^2b^4 + 6ab^5 + b^6$$

Note: *This is the same result we got using Pascal's triangle earlier. Also note that*
$$\binom{6}{3} = \frac{6!}{3!3!} = \frac{6 \cdot 5 \cdot 4 \cdot 3!}{3!3!} = \frac{6 \cdot 5 \cdot 4 \cdot \cancel{3}!}{3 \cdot 2 \cdot 1 \cdot \cancel{3}!} = \frac{\cancel{6} \cdot 5 \cdot 4}{\cancel{3} \cdot \cancel{2} \cdot 1} = 20, \text{ etc.}$$

b. $(a - 3b)^5$

Solution Write $(a - 3b)^5$ as $[a + (-3b)]^5$

$$[a + (-3b)]^5 = \binom{5}{0}a^5 + \binom{5}{1}a^4(-3b) + \binom{5}{2}a^3(-3b)^2 +$$

$$\binom{5}{3}a^2(-3b)^3 + \binom{5}{4}a(-3b)^4 + \binom{5}{5}(-3b)^5$$

$$= a^5 + 5a^4(-3b) + 10a^3(9b^2) + 10a^2(-27b^3) + 5a(81b^4) + (-243b^5)$$

$$= a^5 - 15a^4b + 90a^3b^2 - 270a^2b^3 + 405ab^4 - 243b^5$$

Note: $\binom{5}{2} = \frac{5!}{2!3!} = \frac{5 \cdot 4 \cdot 3!}{2 \cdot 1 \cdot 3!} = \frac{5 \cdot 4 \cdot \cancel{3}!}{2 \cdot 1 \cdot \cancel{3}!} = \frac{5 \cdot 4}{2} = \frac{20}{2} = 10, \text{ etc.}$

OBJECTIVE 4. Find a particular term of a binomial expansion. Sometimes, it is necessary to find only a specific term of a binomial expansion without writing out the entire expansion. Look again at the binomial expansion. Note that the third term (which we will call the $(2 + 1)^{\text{st}}$ term) is $\binom{n}{2}a^{n-2}b^2$ and the fourth term (which we will call the $(3 + 1)^{\text{st}}$ term) is $\binom{n}{3}a^{n-3}b^3$. Similarly, the $(m + 1)^{\text{st}}$ term is $\binom{n}{m}a^{n-m}b^m$. These observations lead to the following.

RULE **Finding the $(m + 1)^{\text{st}}$ Term of a Binomial Expansion**

The $(m + 1)^{\text{st}}$ term of the expansion $(a + b)^n$ is $\binom{n}{m}a^{n-m}b^m$.

EXAMPLE 4 Find the indicated term of each of the following binomial expansions.

a. $(a + b)^{11}$, seventh term

Solution Use the formula for the $(m + 1)^{\text{st}}$ term with $n = 11$ and $m = 6$ (to find the seventh term).

$$\binom{n}{m}a^{n-m}b^m = \binom{11}{6}a^{11-6}b^6 = 462a^5b^6$$

b. $(2x - 5y)^8$, fourth term

Solution Write $(2x - 5y)^8$ as $[2x + (-5y)]^8$. Use the formula for the $(m + 1)^{\text{st}}$ term with $n = 8$, $m = 3$ (to find the fourth term), $a = 2x$, and $b = -5y$.

$$\binom{n}{m}a^{n-m}b^m = \binom{8}{3}(2x)^{8-3}(-5y)^3 = 56(32x^5)(-125y^3) = -224{,}000x^5y^3$$

Appendix C
Exercises

For Extra Help   MyMathLab   Videotape/DVT   InterAct Math   Math Tutor Center   Math XL.com

1. How many terms are in the expansion of $(5a + b)^{12}$?

2. What is the sum of the exponents on x and y for any term in the expansion of $(x + y)^9$?

3. How is the symbol 8! read?

4. What is the meaning of 8!? What is its value?

5. What is the exponent of b in the sixth term of the expansion of $(a + b)^{13}$? What is the exponent of a in that term?

6. On a term in the expansion of $(x + y)^n$ the exponent of x is 4. Find the exponent of y.

For Exercises 7–18, evaluate each expression.

7. $4!$

8. $7!$

9. $(4!)(3!)$

10. $(3!)(2!)$

11. $(6!)(5!)$

12. $(4!)(7!)$

13. $\dfrac{8!}{10!}$

14. $\dfrac{7!}{9!}$

15. $\dfrac{10!}{9!}$

16. $\dfrac{12!}{11!}$

17. $\dfrac{8!}{6!(8-6)!}$

18. $\dfrac{10!}{6!(10-6)!}$

For Exercises 19–26, evaluate each binomial coefficient.

19. $\dbinom{7}{3}$

20. $\dbinom{5}{2}$

21. $\dbinom{10}{4}$

22. $\dbinom{6}{5}$

23. $\dbinom{7}{7}$

24. $\dbinom{4}{4}$

25. $\dbinom{8}{0}$

26. $\dbinom{9}{0}$

For Exercises 27–38, use the binomial theorem to expand each of the following.

27. $(a + b)^5$

28. $(a + b)^7$

29. $(x - y)^4$

30. $(x - y)^3$

31. $(2a + b)^3$

32. $(a + 2b)^3$

33. $(x - 2y)^5$

34. $(x - 3y)^4$

35. $(2m + 3n)^6$

36. $(3c + 2d)^5$

37. $(3x - 4y)^4$

38. $(4a - b)^7$

For Exercises 39–46, find the indicated term of each binomial expansion.

39. $(x + y)^8$, fifth term

40. $(a + b)^9$, fourth term

41. $(a - b)^{10}$, third term

42. $(m - n)^7$, second term

43. $(4x + y)^9$, sixth term

44. $(3a + b)^{11}$, seventh term

45. $(3m - 2n)^7$, fourth term

46. $(5x - 3y)^{12}$, fifth term

Appendix D Permutations and Combinations

OBJECTIVES
1. Use the multiplication principle.
2. Calculate permutations.
3. Calculate combinations.

In this section, we will be counting the number of different ways in which tasks can be performed. For example, in a race with 8 people, we might count the possible number of first, second, and third place finishers. Or in a club with 25 members, how many different ways can we choose a committee of 5 to plan a party? Before we answer these questions, we need to establish the *multiplication principle*.

OBJECTIVE 1. Use the multiplication principle. Murlene goes into an ice cream shop to buy some fat-free frozen yogurt. For a container, she has a choice of a cup, sugar cone, or waffle cone. Her choices in flavors are vanilla, chocolate, strawberry, or peach. In how many ways can she choose a container and flavor? We use a tree diagram to answer this question.

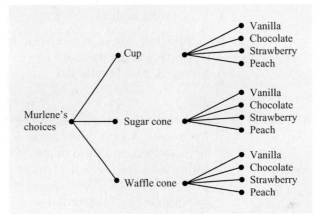

From the diagram, we see that for each of the 3 choices of containers there are 4 choices of flavors, so there are $3 \cdot 4 = 12$ different choices of containers and flavors. Our example suggests the multiplication principle.

RULE The Multiplication Principle

> To find the number of ways in which successive tasks can be performed, multiply the number of ways in which each task can be performed. Symbolically, if task A can be performed in m ways and task B can be performed in n ways, then the two tasks can be performed in mn ways.

EXAMPLE 1

a. How many different three-digit numbers can be formed using the digits 2, 3, 5, and 7 if no digit may be repeated?

Solution We have four choices of the first digit (2, 3, 5 or 7), three for the second (can't use the one chosen as the first digit), and two for the third, so the total number of possible different three digit numbers is $4 \cdot 3 \cdot 2 = 24$.

b. In how many different ways (orders) can 6 people sit on a bench?

Solution Six choices are possible for the first person. After that person is seated, 5 choices are possible for the second person. Similarly, there are 4 choices for the third, 3 for the fourth, 2 for the fifth and 1 for the sixth. Therefore, 6 people can sit on a bench in $6 \cdot 5 \cdot 4 \cdot 3 \cdot 2 \cdot 1 = 720$ different ways.

c. A race has 8 participants. How many different first-, second-, and third-place finishers are possible?

Solution Any one of the 8 could finish first. Any one of the remaining 7 could finish second. Similarly, any one of the remaining 6 could finish third, so there are $8 \cdot 7 \cdot 6 = 336$ different first-, second-, and third-place finishers possible.

d. How many different phone numbers are possible for a given area code and prefix such as (920) 423- ___ ___ ___ ___ ?

Solution The telephone numbers have four more digits. Each digit has 10 possible choices, so there are $10 \cdot 10 \cdot 10 \cdot 10 = 10,000$ possible phone numbers in area code 920 and prefix 423.

OBJECTIVE 2. Calculate permutations. Examples 1a–c illustrate **permutations**. In a permutation, the *order* in which things occur is important. For example, if points are awarded for first, second, and third place, the order in which the participants finish in a race is important. However, if a troupe of dancers has 15 members and 4 are selected to perform a routine, the order in which dancers are selected is not important. Therefore, this is not a permutation.

DEFINITION **Permutation:** An ordered arrangement of objects. The permutation of n things taken r at a time is often denoted as $_nP_r$.

In Example 1b, we found $_6P_6 = 720$. In Example 1c, we found $_8P_3 = 336$. Notice from Example 1b that $_6P_6 = 6 \cdot 5 \cdot 4 \cdot 3 \cdot 2 \cdot 1 = 6!$ Instead of finding $_8P_3 = 336$ as $8 \cdot 7 \cdot 6$, we could have used $\dfrac{8 \cdot 7 \cdot 6 \cdot 5 \cdot 4 \cdot 3 \cdot 2 \cdot 1}{5 \cdot 4 \cdot 3 \cdot 2 \cdot 1} = \dfrac{8!}{(8-3)!} = \dfrac{8!}{5!}$. This leads to the following formula.

RULE Evaluating $_nP_r$.

The formula for the number of permutations of n things taken r at a time is

$$_nP_r = \frac{n!}{(n-r)!}$$

EXAMPLE 2

a. A horse race has 10 horses entered. How many different first-, second-, and third-place finishers are possible?

Solution We need to find the number of permutations of 10 horses taken 3 at a time, $_{10}P_3$ so we use the formula $_nP_r = \dfrac{n!}{(n-r)!}$.

$$_{10}P_3 = \frac{10!}{(10-3)!} \qquad \text{Substitute 10 for } n \text{ and 3 for } r \text{ in } _nP_r = \frac{n!}{(n-r)!}.$$

$$= \frac{10 \cdot 9 \cdot 8 \cdot 7!}{7!} \qquad \text{Replace 10! with } 10 \cdot 9 \cdot 8 \cdot 7! \text{ and } (10-3)! = 7!.$$

$$= 10 \cdot 9 \cdot 8 \qquad \text{Divide the 7!s.}$$

$$= 720 \qquad \text{Evaluate}$$

There are 720 different possibilities for first-, second-, and third-place finishers.

b. Johann is going on a short trip and selects 4 CDs to play while driving. In how many different orders can he play the 4 CDs?

Solution We need to find the number of permutations of 4 things taken 4 at a time, $_4P_4$. Use $_nP_r = \dfrac{n!}{(n-r)!}$ with $n = 4$ and $r = 4$.

$$_4P_4 = \frac{4!}{(4-4)!} = \frac{4!}{0!} = \frac{4 \cdot 3 \cdot 2 \cdot 1}{1} = 24$$

From Example 2, we can derive the following.

RULE **Three Special Permutation Formulas**

$$_nP_n = n! \qquad _nP_0 = 1 \qquad _nP_1 = n$$

Also note that Example 2 could have been done using the multiplication principle.

Calculator TIPS

Most graphing calculators can evaluate $_nP_r$. To evaluate $_{15}P_5$ on the TI-83 Plus, press 1 5 *, MATH, select PRB, select 2, press* 5 *, and press* ENTER *. The screen appears as follows.*

```
15 nPr 5
                360360
```

OBJECTIVE 3. Calculate combinations. Suppose a student governance association has 12 members and selects a committee of 4 to organize the upcoming election. The order in which the members of the committee are selected is not important, since an individual is either on the committee or not. So this situation is not a permutation. It is an example of a **combination**.

DEFINITION **Combination:** An unordered arrangement of objects. The combination of n things taken r at a time is often denoted as ${}_nC_r$.

To determine the number of different ways the committee of 4 can be formed from the 12 members, let's first calculate the number if order did matter (that is, a permutation).

$$_{12}P_4 = \frac{12!}{(12-4)!} = \frac{12!}{8!} = \frac{12 \cdot 11 \cdot 10 \cdot 9 \cdot \cancel{8!}}{\cancel{8!}} = 12 \cdot 11 \cdot 10 \cdot 9 = 11{,}880$$

Since order is not important, many of these committees have the same members. For example, a committee made up of Aricellis, Ebony, Hector, and Carol is the same as the committee made up of Ebony, Aricellis, Carol, and Hector. Since each committee has 4 members, there are 4! = 24 different orders in which the same committee can be written. To eliminate these duplicate committees, divide 11,880 by 24, which gives 495 different committees.

Summarizing, to find the number of combinations of 12 things taken 4 at a time, we found $\dfrac{_{12}P_4}{4!} = \dfrac{\dfrac{12!}{(12-4)!}}{4!} = \dfrac{12!}{4!(12-4)!}$, which suggests the following rule.

PROCEDURE Evaluating ${}_nC_r$

Connection ${}_nC_r = \dbinom{n}{r}$, which is the binomial coefficient.

▶

The formula for the number of combinations of n things taken r at a time, is

$$_nC_r = \frac{n!}{r!(n-r)!}$$

Note that ${}_nC_n = 1$, ${}_nC_0 = 1$, and ${}_nC_1 = n$.

EXAMPLE 3 In a BMX dirt bike race, 4 of the 8 bikes in a preliminary heat can advance to the next heat. How many different groups of 4 can advance?

Solution Since the order in which the top 4 finishers is not important (all 4 go on to the next heat), this is a combination of 8 things taken 4 at a time, or ${}_8C_4$, so we use the formula ${}_nC_r = \dfrac{n!}{r!(n-r)!}$ with $n = 8$ and $r = 4$.

$$_8C_4 = \frac{8!}{4!(8-4)!} = \frac{8!}{4!(4!)} = \frac{8 \cdot 7 \cdot 6 \cdot 5 \cdot \cancel{4!}}{4 \cdot 3 \cdot 2 \cdot 1(\cancel{4!})} = \frac{8 \cdot 7 \cdot 6 \cdot 5}{4 \cdot 3 \cdot 2 \cdot 1} = 70$$

There are 70 possible combinations of four riders who can advance to the next heat.

Appendix D
Exercises

1. If one task can be performed in x ways and a second task can be performed in y ways, what does $x \cdot y$ represent?

2. How do permutations differ from combinations?

3. A family is going on vacation and can visit any 5 of 12 amusement parks. To determine the number of different groups of 5 parks they could visit, would you use the formula for $_nP_r$ or $_nC_r$? Why?

4. Fredrica has a choice of 5 pies and a choice of 3 different flavors of ice cream. What formula or principle would you use to determine the total number of different choices of pie and ice cream?

5. The combination to a combination lock consists of 3 different numbers chosen from a possible 50. To determine the number of possible combinations would you use the formula for $_nP_r$ or $_nC_r$? Why?

6. In deriving the formula $_nC_r$, why did we calculate $\dfrac{_nP_r}{r!}$?

For Exercises 7–18, evaluate each.

7. $_7P_2$

8. $_9P_5$

9. $_6P_6$

10. $_5P_5$

11. $_3P_0$

12. $_8P_0$

13. $_6C_4$

14. $_{10}C_4$

15. $_5C_5$

16. $_9C_9$

17. $_7C_0$

18. $_8C_0$

19. John works as a salesperson and has to wear a sports coat and tie to work each day. Assuming all his clothing matches, find the number of possible outfits that he can wear if he has 4 coats, 6 pairs of pants, 5 shirts, and 3 ties.

20. The executive board of a corporation has 8 members. If one member is to be in charge of public relations, a different member in charge of marketing, and a third member in charge of research, how many different ways can these 3 positions be filled?

21. Jorge plans to take one course each in the subject areas of literature, computers, history, and humanities. If he has a choice of 4 literature courses, 3 computer courses, 5 history courses, and 2 humanities courses that fit his schedule, in how many different ways can he build his schedule?

22. At a sandwich shop, you have a choice of 3 breads, one of 3 dressings, with or without lettuce, with or without tomato, with or without pickles, and 5 meats. How many different sandwiches are possible?

23. How many different three-digit numbers can be formed from the digits 4, 6, and 7 (a) if no digit may be repeated? (b) If the digits may be repeated?

24. A club consists of 8 women and 5 men. In how many different ways can the club elect a president and a secretary if the president must be a woman and the secretary must be a man?

25. Some states' license plates have 3 numbers followed by 3 letters. How many different license plates are possible in one of these states?

26. Postal Zip codes have 5 digits. How many Zip codes are possible?

27. A telephone calling card requires a four-digit PIN (personal identification number). How many different PINs are possible? How many different PINs have no repeated digits?

28. A singer has 12 songs to record on a CD. In how many different orders can the songs be recorded?

29. In how many different orders can the 11 members of the starting offensive lineup of a football team be introduced?

30. A television executive has decided on 4 shows to air on Thursday evenings. In how many different orders can the shows be shown?

31. To win a state lottery, you must correctly select 5 numbers, from 1 to 49 inclusive. Since the numbers are drawn using ping-pong balls, no number can be repeated and the order in which the numbers occur is not important. What is the total number of possible winning tickets?

32. Basketball teams in the National Basketball Association have 12 members. How many different starting lineups of 5 players are possible?

33. A singer has been asked to perform 3 songs at a benefit. If she has 18 songs to choose from, how many different groups of 3 songs are possible?

34. A poet is putting together a collection for publication. He has 30 poems to choose from, but can include only 20 in the collection. How many different collections are possible?

Appendix E Probability

OBJECTIVES
1. Find sample spaces and events.
2. Calculate probabilities.

We often hear the word *probability* in everyday language. For example, a weather forecaster might say that the probability of rain is 40% (0.40). In this usage, probability measures the likelihood that the event will occur. In this section, we will develop a more precise way of defining probability.

OBJECTIVE 1. Find sample spaces and events. To talk about probability, we need to define an **experiment** and **outcomes**.

DEFINITIONS *Experiment:* Any act or process whose result is not known in advance.
Outcomes: The possible results of an experiment.

If an experiment is tossing a coin, then an outcome would be getting a head. If an experiment is rolling two dice (plural of *die*), then an outcome would be getting a total of 7. If an experiment is drawing a card from a deck of cards, then getting the King of Hearts is an outcome. Associated with every experiment is the **sample space**.

DEFINITION *Sample space:* The set of all possible outcomes of an experiment. The sample space is denoted by S.

EXAMPLE 1 Find the sample spaces for each of the following experiments.

a. Toss a single coin.

Solution $S = \{H, T\}$, where H means heads; T tails

b. Toss two coins (or one coin twice).

Solution $S = \{HH, HT, TH, TT\}$, where HT means heads on the first toss and tails on the second

c. Roll a single die.

Solution $S = \{1, 2, 3, 4, 5, 6\}$

d. Roll two dice.

Solution It is convenient to think of the dice as having different colors. Suppose one is blue and the other red. The sample space contains ordered pairs of the form (B, R), where B represents the number on the blue (first) die and R represents the number on the red (second) die.

$$
\begin{aligned}
S = \{&(1, 1)\,(1, 2)\,(1, 3)\,(1, 4)\,(1, 5)\,(1, 6) \\
&(2, 1)\,(2, 2)\,(2, 3)\,(2, 4)\,(2, 5)\,(2, 6) \\
&(3, 1)\,(3, 2)\,(3, 3)\,(3, 4)\,(3, 5)\,(3, 6) \\
&(4, 1)\,(4, 2)\,(4, 3)\,(4, 4)\,(4, 5)\,(4, 6) \\
&(5, 1)\,(5, 2)\,(5, 3)\,(5, 4)\,(5, 5)\,(5, 6) \\
&(6, 1)\,(6, 2)\,(6, 3)\,(6, 4)\,(6, 5)\,(6, 6)\}
\end{aligned}
$$

Note that there are 36 outcomes in the sample space.

To calculate probabilities, we need to discuss **events**.

DEFINITION   *Event:* Any subset of a sample space.

EXAMPLE 2 Find the events that correspond with the following outcomes.

a. Having exactly one head on the toss of two coins

Solution $E = \{\text{HT, TH}\}$

b. Rolling a number greater than 2 with a single roll of a die

Solution $E = \{3, 4, 5, 6\}$

c. Rolling two dice for a total of 7

Solution $E = \{(1, 6)\,(2, 5)\,(3, 4)\,(4, 3)\,(5, 2)\,(6, 1)\}$

OBJECTIVE 2. Calculate probabilities.   We are now ready to give a more formal definition of probability of an event.

DEFINITION   *Probability of an Event* **E:** Given an experiment, the probability of an event E, written as $P(E)$, is the number of outcomes in event E divided by the number of outcomes in the sample space S of the experiment. That is,

$$
P(E) = \frac{\text{The number of outcomes in } E}{\text{The number of outcomes in } S}
$$

From the definition, we can conclude that the probability of an event must be between 0 and 1 inclusive, that is, $0 \le P(E) \le 1$. If an event cannot occur, then the number of outcomes in the event is 0, so the probability of that event is 0. For example, the probability of rolling a die and getting a 7 is 0 since this cannot occur. The event is {7}, which can occur in 0 ways, and the sample space is $S = \{1, 2, 3, 4, 5, 6\}$, so $P(E) = \dfrac{0}{6} = 0$. If an event is certain to occur, then the number of outcomes in the event is the same as the number of outcomes in the sample space, so the probability is 1. For example, the probability of rolling a die and getting a number less than 7 is 1 since this event is certain to occur. The event is $E = \{1, 2, 3, 4, 5, 6\}$ and the sample space is $S = \{1, 2, 3, 4, 5, 6\}$, so $P(E) = \dfrac{6}{6} = 1$.

Many probability problems involve a standard deck of cards, which consists of 52 cards with 4 suits of 13 cards each. The suits are diamonds and hearts, which are red, and spades and clubs, which are black. Each suit has the cards 2, 3, 4, 5, 6, 7, 8, 9, 10, Jack, Queen, King, and Ace. The cards Jack, Queen, and King are called face cards.

EXAMPLE 3 Find the probabilities of each of the following events.

a. Drawing a King from a standard deck on a single draw

Solution Since there are 52 cards in the deck, there are 52 possible outcomes in the sample space. Since there are 4 Kings in the deck, the event has 4 outcomes. Therefore, $P(E) = \dfrac{4}{52} = \dfrac{1}{13}$.

b. Rolling a total of 7 on a single roll of two dice

Solution We found in Example 1d that the sample space for rolling two dice has 36 outcomes, and in Example 2c we found 6 different ways of getting a total of 7. Therefore, $P(E) = \dfrac{6}{36} = \dfrac{1}{6}$.

c. A club has 20 members, of which 8 are men and 12 are women. If a committee of 5 is chosen to revise the bylaws, what is the probability that all 5 are women?

Solution $P(E) = \dfrac{\text{Number of ways of selecting 5 women from 12}}{\text{Number of ways of selecting 5 members from 20}}$

The number of ways of selecting 5 women from 12:

$$_{12}C_5 = \frac{12!}{5!(12-5)!} = \frac{12!}{5!7!} = 792$$

The number of ways of selecting 5 members from 20:

$$_{20}C_5 = \frac{20!}{5!(20-5)!} = \frac{20!}{5!15!} = 15{,}504$$

Therefore, $P(E) = \dfrac{792}{15{,}504} = \dfrac{33}{646}$.

Appendix E
Exercises

1. What is the sample space for an experiment?

2. What is an event?

3. The probability of any event lies between what two numbers inclusively?

4. What is the probability of an event that cannot happen?

5. Define the probability of an event.

6. How many face cards are in a standard deck of cards?

For Exercises 7–10, find the sample space for each experiment.

7. Selecting a vowel from the English alphabet

8. A jar contains 1 red, 1 blue, and 1 white ball; draw a ball, replace it, and draw again

9. Tossing 3 coins (or 1 coin 3 times)

10. Jar 1 contains a blue marble and a white marble; jar 2 contains a red marble and a black marble; select a marble from jar 1 and then select a marble from jar 2

For Exercises 11–14, find the number of elements in the sample space of the experiment.

11. Selecting a two-digit number

12. Rolling a die and tossing a coin

13. Choosing a committee of 4 from a club with 16 members

14. Five-card hands using a standard deck of cards

For Exercises 15–18, find the event that corresponds with the following outcomes.

15. Having exactly 1 head if 3 coins are tossed

16. Rolling a total of 4 on a single roll of 2 dice

17. Having a number greater than 6 on a single roll of a die

18. Rolling a total of 13 on a single roll of 2 dice

For Exercises 19 and 20, a single die is rolled. Find the probability of each of the following events.

19. Having a 5

20. Having a number greater than 3

For Exercises 21–24, two coins are tossed. Find the probability of each of the following events.

21. Having exactly 1 head

22. Having exactly 2 tails

23. Having at least 1 tail

24. Having more than 2 heads

For Exercises 25–28, two dice are rolled. Find the probability of each of the following events.

25. Having a total of 4

26. Having a total of 8

27. Having a total less than 4

28. Having a total greater than 7

For Exercises 29–32, a single card is drawn from a standard deck. Find the probability of each of the following events.

29. Drawing a 10

30. Drawing the Jack of spades

31. Drawing a face card

32. Drawing a number less than 5

For Exercises 33–34, use the spinner shown to find the probability of each of the following events.

33. The spinner stopping on red

34. The spinner stopping on blue

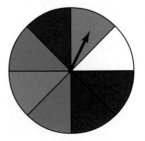

For Exercises 35–40, we have a jar that contains 1 blue, 1 red, and 1 pink ball. Find the probability of each of the following events.

35. Drawing a pink ball on a single draw

36. Drawing a blue ball on a single draw

37. Draw a ball, do not replace it, and then draw a second ball. What is the probability that both balls are blue?

38. Draw a ball, do not replace it, and then draw a second ball. What is the probability that both balls are red?

39. Draw a ball, replace it, and draw again. What is the probability that both balls are red?

40. Draw a ball, replace it, and draw again. What is the probability that the first ball is blue and the second ball is pink?

41. A farm pond is stocked with 80 warmouth perch, 140 bluegills, and 200 shellcrackers. If we assume that each fish is equally likely to be caught, what is the probability that the first fish caught is a shellcracker?

42. The quality-control department of a company that manufactures car batteries has determined that 3% of the batteries produced are defective. If one battery that is manufactured by this company is selected at random, what is the probability that it is defective?

43. A committee has 12 members of which 8 are Democrats and 4 are Republicans. A subcommittee of 3 is appointed. What is the probability that all 3 are Republicans [if each number is equally likely to be chosen]?

44. A club has 20 members of which 13 are women and 7 are men. If a committee of 6 is appointed, what is the probability that all 6 are women [if each number is equally likely to be chosen]?

45. A jar contains 14 balls of which 8 are pink and 6 are purple. If 4 balls are drawn at random, what is the probability that all 4 are pink?

46. A box contains 18 pieces of paper of which 9 are blue and 9 are white. If 4 pieces are selected at random, what is the probability that all 4 are blue?

Answers

Chapter 1

1.1 Exercises
1. a collection of objects   **3.** Rational numbers can be expressed as a ratio of integers and irrational numbers cannot.   **5.** true   **7.** true   **9.** False, all real numbers are either rational or irrational.   **11.** Sunday, Monday, Tuesday, Wednesday, Thursday, Friday, Saturday   **13.** $\{a, e, i, o, u\}$ **15.** $\{5, 10, 15, 20, \ldots\}$ **17.** $\{9, 11, 13, 15, \ldots\}$ **19.** $\{-6, -5, -4, -3\}$ **21.** rational   **23.** irrational   **25.** irrational   **27.** rational   **29.** rational

31.

33.

35.

37.

39. 23   **41.** 2   **43.** 5.7   **45.** $3\frac{1}{8}$ **47.** 0   **49.** $>$ **51.** $>$ **53.** $>$ **55.** $<$ **57.** $>$ **59.** $>$ **61.** $<$ **63.** $=$ **65.** $<$ **67.** $=$ **69.** $>$ **71.** $>$ **73.** $<$ **75.** $<$

77. $-4.7, -2.56, 5.4, \left|-7\frac{1}{2}\right|, |8.3|$

79. $-0.6, -0.44, 0, |-0.02|, 0.4, \left|1\frac{2}{3}\right|, 3\frac{1}{4}$

1.2 Exercises
1. Multiply or divide the numerator and denominator by the same nonzero number.   **3.** No, 4 is not a prime number. The correct prime factorization would be $2 \cdot 2 \cdot 2 \cdot 2 \cdot 7$.   **5.** No, for example 3 and 7 are whole numbers that are not composite (they are prime numbers).   **7.** True, prime numbers are natural numbers and every natural number is a whole number.   **9.** True, every composite number can be written as a product of prime numbers.

11. $\frac{3}{10}$ **13.** $\frac{1}{3}$ **15.** $\frac{3}{4}$ **17.** $\frac{15}{16}$ **19.** 12   **21.** -35 **23.** 16   **25.** 15   **27.** $\frac{12}{15}$ and $\frac{10}{15}$ **29.** $\frac{8}{18}$ and $\frac{21}{18}$ **31.** $-\frac{44}{72}$ and $-\frac{51}{72}$

33. $-\frac{9}{144}$ and $-\frac{60}{144}$ **35.** $2 \cdot 2 \cdot 11$ **37.** $2 \cdot 2 \cdot 3 \cdot 3$

39. $2 \cdot 2 \cdot 2 \cdot 2 \cdot 2 \cdot 2$ **41.** $2 \cdot 5 \cdot 5 \cdot 5$ **43.** $\frac{4}{5}$ **45.** $\frac{7}{11}$

47. $-\frac{3}{4}$ **49.** $-\frac{8}{15}$ **51.** Incorrect, you may only divide out factors, not addends.   **53.** $\frac{7}{8}$ **55.** $\frac{12}{23}$ **57.** $\frac{5}{21}$ **59.** $\frac{2}{3}$

61. $\frac{13}{25}$ **63.** $\frac{31}{50}$ **65. a.** 1975   **b.** $\frac{277}{500}$

67. $\frac{14}{69}$ **69.** $\frac{6}{73}$ **71.** $\frac{2}{3}$ **73.** $\frac{8}{15}$ **75.** $\frac{4}{15}$

Review Exercises
1. It is an expression because it has no = sign.   **2.** {Mercury, Venus, Earth, Mars}   **3.** It is a rational number because it can be written as a ratio of the integers 8 and 10.   **4.**   **5.** 27   **6.** $=$

1.3 Exercises
1. The commutative property of addition changes the order while the associative property of addition changes the grouping.   **3.** Adding 0 to a number or expression does not change the identity of the number or expression.   **5.** When adding two numbers that have the same sign, add their absolute values and keep the same sign.   **7.** To write a subtraction statement as an equivalent addition statement, change the operation symbol from a minus sign to a plus sign and change the subtrahend to its additive inverse.   **9.** commutative property of addition   **11.** additive identity   **13.** additive inverse   **15.** associative property of addition   **17.** commutative property of addition   **19.** commutative property of addition   **21.** additive inverse   **23.** commutative property of addition   **25.** 21   **27.** -6

29. 14   **31.** -4 **33.** $\frac{3}{4}$ **35.** $-\frac{2}{3}$ **37.** $-\frac{2}{3}$ **39.** $\frac{11}{12}$

41. $-\frac{3}{4}$ **43.** $-\frac{9}{14}$ **45.** 0.26   **47.** 6.52   **49.** -4.38 **51.** -5

53. 12   **55.** 0   **57.** $-\frac{5}{6}$ **59.** 0.29   **61.** x **63.** $-\frac{m}{n}$

65. -9 **67.** -13 **69.** 7   **71.** -6 **73.** $\frac{13}{10}$ **75.** $-\frac{18}{35}$

77. 0   **79.** 0.36   **81.** -5.75 **83.** -1.6 **85.** -6 **87.** -56

89. 1.62   **91.** $-\frac{11}{24}$ **93.** 2   **95.** -3 **97.** \$583 million profit

99. -268.2 N; The negative indicates the beam is moving downward.   **101.** 373.16   **103.** \$8.87   **105.** $-196 - (-208)$; 12   **107. a.** $18.6 - 18.8$ **b.** -0.2 **c.** The negative difference indicates that the mean composite score in 1989 was less than the score in 1986.   **109.** 2000; $\approx \$16{,}000$

Review Exercises
1. {Washington, Adams, Jefferson, Madison}   **2.** 7   **3.** 5   **4.** $<$ **5.** $2 \cdot 2 \cdot 5 \cdot 5$ **6.** $\frac{6}{7}$

1.4 Exercises
1. positive   **3.** negative   **5.** There is no quotient that we can multiply by 0 to get 5.   **7.** Their product is 1.   **9.** An odd number of negatives in division will result in a negative answer.   **11.** distributive property   **13.** multiplicative identity   **15.** associative property of multiplication   **17.** multiplicative property of 0   **19.** commutative property of multiplication   **21.** associative property of multiplication   **23.** commutative property of multiplication   **25.** distributive property   **27.** -18

29. 20　**31.** 54　**33.** 720　**35.** -48　**37.** $-\dfrac{1}{4}$　**39.** $\dfrac{1}{2}$

41. $-\dfrac{4}{9}$　**43.** -38　**45.** 6.72　**47.** -6.536　**49.** $\dfrac{3}{2}$　**51.** $-\dfrac{2}{5}$

53. $-\dfrac{1}{5}$　**55.** There is no multiplicative inverse of 0.　**57.** -3

59. 14　**61.** 2　**63.** undefined　**65.** indeterminate　**67.** 9

69. $-\dfrac{2}{3}$　**71.** $\dfrac{3}{4}$　**73.** $-\dfrac{14}{15}$　**75.** 2.1　**77.** 91.8　**79.** $-46.\overline{6}$

81. $\dfrac{1}{2}$　**83.** $-\$8680$　**85.** 612.8　**87.** -402.5 lb; the force is downward　**89.** 169,202.2245 kg　**91.** -51.2 V　**93.** 3600 W

Review Exercises
1. irrational

2.

3. 6.8

4. -22　**5.** $-\dfrac{2}{3}$　**6.** $-\dfrac{11}{24}$

1.5 Exercises
1. two cubed　**3.** Squaring a number means to multiply the number by itself, finding its square root means to find a number whose square is the given number.　**5.** We can multiply and then find the square root of the product, or we can find the square root of each factor and then multiply those square roots.　**7.** Base: 7; Exponent: 2; "seven squared"　**9.** Base: -5; Exponent: 3; "negative five cubed"　**11.** Base: 2; Exponent: 7; "additive inverse of two to the seventh power"　**13.** 81　**15.** -64　**17.** -8　**19.** 1

21. $\dfrac{1}{25}$　**23.** $-\dfrac{27}{64}$　**25.** 0.008　**27.** -68.921　**29.** ± 11

31. No real-number square roots exist.　**33.** ± 14　**35.** ± 16

37. 4　**39.** 12　**41.** 0.7　**43.** not a real number　**45.** $\dfrac{8}{9}$

47. 5　**49.** 19　**51.** -11　**53.** -40　**55.** 7　**57.** 5　**59.** -7

61. -28　**63.** 26.8　**65.** -100.8　**67.** -53　**69.** $-10\dfrac{4}{5}$

71. 78　**73.** $-\dfrac{253}{300}$　**75.** 22　**77.** 12　**79.** 0　**81.** 4　**83.** 7

85. undefined　**87.** associative property of multiplication; The multiplication was not performed from left to right.　**89.** distributive property; The parentheses were not simplified first.　**91.** Mistake: Multiplied before dividing. Correct: 1　**93.** Mistake: Found the square roots of the subtrahend and minuend in square root of a difference. Correct: 24　**95.** 77.2　**97.** 96　**99.** $\$1.99$　**101.** $\$53.80$

Review Exercises
1. It is an expression because it has no = sign.　**2.** $\{0, 1, 2, 3, 4, 5, 6, 7, 8, 9, 10\}$

3.

4. 24　**5.** 3　**6.** $-\dfrac{1}{9}$

1.6 Exercises
1. sum, plus, added　**3.** product, times, twice　**5.** less than, subtracted from　**7.** $4x$　**9.** $4x + 16$

11. $5 + y$　**13.** $-4 \div y^3$ or $\dfrac{-4}{y^3}$　**15.** $8p - 4$　**17.** $\dfrac{m}{14}$

19. $x^4 + 5$　**21.** $7w - \dfrac{1}{5}$　**23.** $-3(n - 2)$　**25.** $(2 - l)^3$

27. $mn - 5$　**29.** $4 \div n - 2$ or $\dfrac{4}{n} - 2$　**31.** $-27 - (a + b)$

33. $0.6 - 4(y - 2)$　**35.** $(p - q) - (m + n)$

37. $\sqrt{y} - mn$　**39.** $n - 3(n - 6)$　**41.** Mistake: Multiplied x by 9 instead of the sum of x and y. Correct: $9(x + y)$

43. Mistake: Order is incorrect. Correct: $m^2 - 4$　**45.** $5 + w$

47. $3w$　**49.** $\dfrac{1}{2}d$　**51.** $42 - n$　**53.** $t + \dfrac{1}{4}$　**55.** $2w + 2l$

57. $\dfrac{1}{3}\pi r^2 h$　**59.** $\sqrt{(x_2 - x_1)^2 + (y_2 - y_1)^2}$　**61.** mc^2

63. one-half the product of the base and height　**65.** the product of the length, width, and height　**67.** product of the length and width added to the product of the length and height added to the product of the width and height, all doubled　**69.** the ratio of the difference of y_2 and y_1 to the difference of x_2 and x_1

Review Exercises
1. Any real number that can be expressed in the form $\dfrac{a}{b}$, where a and b are integers and $b \neq 0$.

2. additive identity　**3.** associative property of multiplication　**4.** -28　**5.** 12　**6.** -9

1.7 Exercises
1. To evaluate an expression: 1. Replace the variables with their corresponding given numbers. 2. Calculate the numerical expression using the order of operations.　**3.** An expression is undefined when the denominator is equal to 0.　**5.** A coefficient is the numerical factor in a term.　**7.** 9　**9.** 3.2　**11.** 23　**13.** $\dfrac{11}{2}$　**15.** 4.6　**17.** -14

19. 33　**21.** -12　**23.** -21　**25.** -48　**27.** $\dfrac{38}{13}$

29. a. 25　**b.** -12　**31. a.** 2　**b.** $\dfrac{2}{3}$　**33.** -5　**35.** $-1, 3$

37. 0　**39.** $\dfrac{2}{3}$　**41.** $6a + 12$　**43.** $-32 + 24y$　**45.** $\dfrac{7}{16}c - 14$

47. $0.6n - 1.6$　**49.** coefficient: -6　**51.** coefficient: 1

53. coefficient: -1　**55.** coefficient: $-\dfrac{2}{3}$　**57.** coefficient: $\dfrac{1}{5}$

59. $8y$　**61.** $-\dfrac{1}{3}b^2$　**63.** $-1.9n$　**65.** $-26c$　**67.** $5x - 1$

69. $-10x + 4y - 1$　**71.** $-9m + 5n + y - 11$

73. $\dfrac{5}{6}c + \dfrac{5}{2}d + \dfrac{1}{7}$　**75.** $\dfrac{25}{4}a + \dfrac{18}{35}b^2 + 4$　**77.** $-1.3x - 0.4$

79. a. $14 + (6n - 8n)$　**b.** $14 - 2n$　**c.** 20

Review Exercises
1. $-3, -2.5, 4.2, 4\dfrac{5}{8}, |-6|$　**2.** $\dfrac{6}{35}$

3. -22　**4.** 6.1　**5.** $\dfrac{3}{2}$　**6.** 0

Chapter 1 Review Exercises 1. false 2. false

3. false 4. true 5. true 6. true 7. positive 8. prime
9. keep 10. even 11. {March, May} 12. {a, l, g, e, b, r}
13. {2, 4, 6, ...} 14. {..., −6, −4, −2, 0, 2, 4, ...}

15. [number line: −4, −3⅖, −3] 16. [number line: 2, 2¼, 3]

17. [number line: 8, 8.2, 9] 18. [number line: −5, −4.6, −4]

19. 6.3 20. 8.46 21. $2\frac{1}{6}$ 22. $4\frac{3}{8}$ 23. = 24. <

25. < 26. = 27. $\frac{11}{15}$ 28. $\frac{3}{8}$ 29. 12 30. 25 31. 5

32. −10 33. $3 \cdot 17$ 34. $2 \cdot 2 \cdot 5 \cdot 5$ 35. $2 \cdot 2 \cdot 3 \cdot 3 \cdot 3$

36. $2 \cdot 2 \cdot 3 \cdot 7$ 37. $\frac{19}{25}$ 38. $\frac{1}{2}$ 39. $\frac{25}{36}$ 40. $\frac{2}{3}$

41. associative property of addition 42. distributive
property 43. commutative property of multiplication
44. commutative property of addition 45. commutative
property of multiplication 46. associative property of

multiplication 47. −3.8 48. 15.1 49. $\frac{11}{15}$ 50. $\frac{5}{12}$

51. 91 52. −48 53. −5 54. 10 55. −36 56. 9

57. 52 58. 9 59. 160 60. 1 61. $-\frac{28}{3}$ 62. −8.6

63. −1 64. 27 65. 7 66. −103 67. 3 68. 24

69. ±7 70. No real-number square roots exist.

71. $14 - 2n$ 72. $\frac{y}{7}$ 73. $y - 2(y + 4)$ 74. $7(m - n)$

75. 61 76. 52 77. −8 78. 7 79. 5 80. −4 81. −6

82. −3, 4 83. $5x + 30$ 84. $-15n + 24$ 85. $2y + \frac{3}{4}$

86. $2.7m - 1.26$ 87. $-3x - 3y - 15$ 88. $4y^2 + 7y$
89. $2xy$ 90. $-2x^3 - 6x^2$ 91. $-8m$ 92. $-14x - 6y + 6$
93. −$503.59 94. a. −11.15 b. loss of $11.61
95. a. $8,984 b. $2322 loss or − $2,322 96. −152 V

97. ≈ 6.2 98. $54,158$\frac{1}{3}$ or $54,158.33

Chapter 1 Practice Test 1. [number line: −4, −3$\frac{2}{7}$, −3]

2. 3.67 3. $2 \cdot 2 \cdot 5 \cdot 5$ 4. $\frac{1}{3}$ 5. 12 6. distributive

property 7. commutative property of addition 8. 4

9. $1\frac{17}{24}$ or $\frac{41}{24}$ 10. 0.6 11. $-\frac{5}{4}$ 12. −64 13. ±9

14. −8 15. 2 16. 5 17. 10 18. $2(m + n)$ 19. $3w - 5$

20. $250 21. 5.58$\overline{3}$ 22. −11 23. 10 24. $-20y - 45$

25. $5.6x + 1.3$

Chapter 2

2.1 Exercises 1. An equation has an equal symbol.

3. 1. Replace the variable(s) in the equation with the value.
2. If the resulting equation is true, then the value is a solution.
5. no 7. yes 9. no 11. yes 13. no 15. no 17. yes
19. yes 21. yes 23. yes 25. no 27. no 29. no
31. a. 66 ft. b. 6 c. $53.94 33. ≈12.57 km 35. 7000 ft.²
37. $975 39. a. 78 ft.² b. 8 bales c. $36 41. a. 260 in.²
b. $3900 43. $6926.40 45. 82.25 ft.² 47. ≈6.9 ft.²
49. 47.5 m³ 51. ≈1005.3 cm³ 53. 83,068,741.$\overline{6}$ ft.³
55. 1747.5 mi. 57. 64.5 mph 59. ≈54.3 mph 61. −52 V
63. 93.2°F 65. 2795°F 67. −60°C

Review Exercises 1. 59 2. −34 3. $\frac{1}{3}x + 2$

4. $-4w + 6$ 5. $8x - 5$ 6. $5.4y + 5.5$

2.2 Exercises 1. The addition principle of equality says

that we can add (or subtract) the same amount on both sides
of an equation without affecting its solution(s). 3. Add 9 to
both sides of the equation. 5. Use the addition principle of
equality to get the variable terms together on the same side of
the equal sign. 7. yes 9. no 11. no 13. yes 15. yes

17. no 19. yes 21. yes 23. 9 25. −25 27. $-\frac{3}{40}$

29. $-\frac{11}{24}$ 31. −8.2 33. 6.2 35. 21 37. 7 39. 13

41. 6 43. −1 45. −2 47. 13.2 49. all real numbers
51. no solution 53. −42 55. −10 57. 16 59. 4
61. −8.8 63. no solution 65. all real numbers
67. $1947 + x = 2373$; $426 69. $16 + x = 42$; 26 mi.
71. $23 + x = 28$; 5 73. $x + 12.4 + 16.3 + 27.2 = 67.2$;
11.3 cm 75. $x + 19 + 10 = 54$; 25 ft.

77. $x + \frac{1}{3} + \frac{2}{5} = 1$; $\frac{4}{15}$

Review Exercises 1. −33.2 2. −9 3. 37 4. −19

5. $8x^2 - 5x - 9$ 6. $12x - 20$

2.3 Exercises 1. We can multiply (or divide) both sides

of an equation by the same amount without affecting its solu-
tion(s). 3. Multiply both sides of the equation by the multiplicative inverse of that coefficient. 5. Multiply both sides
of the equation by a common multiple of the denominators.
Using the LCD results in an equation with the smallest

integers possible. 7. 4 9. 3 11. 16 13. −20 15. $-\frac{7}{6}$

17. $-\frac{4}{3}$ 19. 7 21. 5 23. $-\frac{3}{2}$ 25. 16 27. 0 29. −5

31. −7 33. 16 35. 2 37. $\frac{20}{3}$ 39. −2 41. no solution

43. 2 45. $-\frac{15}{2}$ 47. all real numbers 49. −9 51. all real

numbers 53. $\frac{15}{8}$ 55. $\frac{48}{13}$ 57. 1 59. $-\frac{2}{3}$ 61. −8 63. 9

65. no solution **67.** -9 **69.** 12 **71.** 6.5 **73.** Mistake: The minus sign was dropped from in front of the 11. Correct: 2 **75.** Mistake: In the second line, the minus sign was not distributed to the 2. Correct: $\dfrac{13}{3}$ **77.** 14 ft. **79.** 19 cm **81.** 52 ft. **83.** 110 ft. **85.** 2 in. **87.** ≈ 6374.8 km **89.** ≈ 4.8 m **91.** -9.5 A **93.** 4 m/sec.^{2} **95.** 945.5 kg **97.** 413.6 m/sec. **99.** \$286.67

Review Exercises
1. 70 **2.** $3.8x^2 - 7.63x + 1.5$ **3.** $-\dfrac{9}{2}x - 6$ **4.** 20 **5.** -25 **6.** -10

2.4 Exercises
1. Add 2 to both sides. Since 2 is subtracted from w, adding 2 eliminates the 2, leaving w isolated. **3.** Subtract $2x$ from both sides. Subtracting $2x$ isolates the $3y$ term, which contains the variable we wish to isolate.

5. $t = 4u + v$ **7.** $y = -\dfrac{x}{5}$ **9.** $x = \dfrac{b+3}{2}$ **11.** $m = \dfrac{y-b}{x}$

13. $y = \dfrac{8 - 3x}{4}$ **15.** $Y = \dfrac{mn}{4} - f$ **17.** $c = \dfrac{m - np - 12d}{6}$

19. $y = \dfrac{15 - 5x}{3}$ **21.** $n = c\left(\dfrac{3}{4} + 5a\right)$ **23.** $p = \dfrac{A - P}{tr}$

25. $a = P - b - c$ **27.** $l = \dfrac{A}{w}$ **29.** $d = \dfrac{C}{\pi}$ **31.** $r^2 = \dfrac{3V}{\pi h}$

33. $b = \dfrac{2A}{h}$ **35.** $w = \dfrac{P - 2l}{2}$ **37.** $l = \dfrac{2S - na}{n}$

39. $w = \dfrac{\pi r^2 h - V}{lh}$ **41.** $C = R - P$ **43.** $r = \dfrac{I}{Pt}$

45. $C = nP$ **47.** $r = \dfrac{A - P}{Pt}$ **49.** $t = \dfrac{d}{r}$ **51.** $d = \dfrac{W}{F}$

53. $t = \dfrac{W}{P}$ **55.** $t = \dfrac{v - v_0}{-32}$ or $t = \dfrac{v_0 - v}{32}$

57. $C = \dfrac{5}{9}(F - 32)$ **59.** $m = \dfrac{FR^2}{GM}$ **61.** Mistake: Subtracted the coefficient 7 instead of dividing. Correct: $t = \dfrac{54 - 3n}{7}$

63. Mistake: Multiplied 5 by -2. Correct: $m = \dfrac{5nk + 2}{3}$

Review Exercises
1. $>$ **2.** 0.36 **3.** -16 **4.** -6 **5.** $4 + 7n$ **6.** $3(x + 2) - 9$

2.5 Exercises
1. sum, plus, added **3.** of, times, twice **5.** The order of the subtraction is different. **7.** $6 - x = -3; 9$ **9.** $11m = -99; -9$ **11.** $y - 5 = -1; 4$ **13.** $m \div 4 = 1.6; 6.4$

15. $\dfrac{4}{5}x = \dfrac{5}{8}; \dfrac{25}{32}$ **17.** $7 + 3w = 34; 9$ **19.** $5a - 9 = 76; 17$

21. $8(8 + t) = 160; 12$ **23.** $-3(x - 2) = 12; -2$

25. $\dfrac{1}{3}(g + 2) = 1; 1$ **27.** $3m - 11 = m + 5; 8$

29. $10 = \dfrac{r}{4} - 1; 44$ **31.** $3(a + 4) + 2a = -3; -3$

33. $(2d - 8) - (d - 12) = 11; 7$ **35.** $5\left(x + \dfrac{1}{3}\right) = 3x + \dfrac{2}{3};$

$-\dfrac{1}{2}$ **37.** $2x - 4 = 16 + 3x; -20$ **39.** $\dfrac{2}{5}x = \dfrac{1}{2}x - 2; 20$

41. $\dfrac{n - 3}{3} = n \div 6; 6$ **43.** $-4(2 - x) = 2(4 - 3x) - 6; 1$

45. Four times a number added to three is seven. **47.** Six times the sum of a number and four is equal to the product of negative 10 and the number. **49.** One-half the difference of a number and three will result in two-thirds the difference of the number and eight. **51.** Five-hundredths of a number added to six-hundredths of the difference of the number and eleven is twenty-two. **53.** The sum of two-thirds, three-fourths, and one-half of the same number will equal ten. **55.** Mistake: Subtraction translated in reverse order. Correct: $n - 10 = 40$ **57.** Mistake: Multiplied 5 times the unknown number instead of the difference, which requires parentheses. Correct: $5(x - 6) = -2$ **59.** Mistake: Subtracted the unknown number instead of the sum, which requires parentheses. Correct: $2t - (t + 3) = -6$ **61.** Translation: $P = 2(l + w)$ **a.** 84 ft. **b.** 61 cm **c.** $37\dfrac{1}{4}$ in. **63.** Translation: $P = b + 2s$ **a.** 27 in. **b.** 2.7 m **c.** $34\dfrac{1}{4}$ in.

65. Translation: $I = Prt$ **a.** \$400 **b.** \$7.50 **c.** \$30 **67.** Translation: $t = \dfrac{e - d}{153.8}$ **a.** ≈ 49 sec. **b.** ≈ 52 sec. **c.** ≈ 62 sec.

Review Exercises
1. True, 19 is equal to 19. **2.** $=$ **3.** $>$ **4.** -12 **5.** -25 **6.** -30

2.6 Exercises
1. Any number that can replace the variable(s) in the inequality and make it true. **3.** Any number greater than -5 and -5 itself. **5. a.** $\{x \mid x \geq -3\}$ **b.** $[-3, \infty)$

c. **7. a.** $\{h \mid h < 6\}$ **b.** $(-\infty, 6)$

c. **9. a.** $\left\{n \mid n < -\dfrac{2}{3}\right\}$

b. $\left(-\infty, -\dfrac{2}{3}\right)$ **c.**

11. a. $\{t \mid t \geq 2.4\}$ **b.** $[2.4, \infty)$ **c.**

13. a. $\{x \mid -3 < x < 6\}$ **b.** $(-3, 6)$

c. **15. a.** $\{n \mid 0 \leq n \leq 5\}$ **b.** $[0, 5]$

c. **17. a.** $m < 25$ **b.** $\{m \mid m < 25\}$

c. $(-\infty, 25)$ **d.** **19. a.** $y \geq 2$

b. $\{y \mid y \geq 2\}$ **c.** $[2, \infty)$ **d.**

21. a. $x \leq 4$ **b.** $\{x \mid x \leq 4\}$ **c.** $(-\infty, 4]$

d. [number line: −1 0 1 2 3 4 5]   **23. a.** $x \ge 6$ **b.** $\{x \mid x \ge 6\}$

c. $[6, \infty)$ **d.** [number line: 3 4 5 6 7 8 9]   **25. a.** $m > 8$

b. $\{m \mid m > 8\}$ **c.** $(8, \infty)$ **d.** [number line: 4 5 6 7 8 9 10 11]

27. a. $y > 3$ **b.** $\{y \mid y > 3\}$ **c.** $(3, \infty)$

d. [number line: 0 1 2 3 4 5 6]   **29. a.** $x > -4$

b. $\{x \mid x > -4\}$ **c.** $(-4, \infty)$ **d.** [number line: −5 −4 −3 −2 −1 0 1]

31. a. $a < 1$ **b.** $\{a \mid a < 1\}$ **c.** $(-\infty, 1)$

d. [number line: 0 1 2 3 4]   **33. a.** $f \le -2$ **b.** $\{f \mid f \le -2\}$

c. $(-\infty, -2]$ **d.** [number line: −4 −3 −2 −1 0]   **35. a.** $u \le 2$

b. $\{u \mid u \le 2\}$ **c.** $(-\infty, 2]$ **d.** [number line: −2 −1 0 1 2 3]

37. a. $c > -12$ **b.** $\{c \mid c > -12\}$ **c.** $(-12, \infty)$

d. [number line: −14 −13 −12 −11 −10 −9]   **39. a.** $w \ge 3$ **b.** $\{w \mid w \ge 3\}$

c. $[3, \infty)$ **d.** [number line: 0 1 2 3 4 5]   **41. a.** $x \le -13$

b. $\{x \mid x \le -13\}$ **c.** $(-\infty, -13]$ **d.** [number line: −15 −14 −13 −12 −11 −10]

43. a. $n < 7$ **b.** $\{n \mid n < 7\}$ **c.** $(-\infty, 7)$

d. [number line: 4 5 6 7 8 9]   **45. a.** $l < 10$ **b.** $\{l \mid l < 10\}$

c. $(-\infty, 10)$ **d.** [number line: 5 6 7 8 9 10 11]   **47. a.** $t \le 3.\overline{3}$

b. $\{t \mid t \le 3.\overline{3}\}$ **c.** $(-\infty, 3.\overline{3}]$

d. [number line: 2 3 3.3̄ 3.6̄ 4 5]   **49.** $4 + n > 24$

$n > 20$ **51.** $\dfrac{4}{9}x \le -8$; $x \le -18$ **53.** $8y - 36 \ge 60$;

$y \ge 12$ **55.** $5 + \dfrac{1}{2}a \le 2$; $a \le -6$ **57.** $4x - 8 < 2x$;

$x < 4$ **59.** $25 \ge 6x + 7$; $x \le 3$ **61.** $x \le 10$ ft.

63. $h \ge 15$ in.   **65.** $t \ge 6\dfrac{4}{13}$ hr.   **67.** $i \le 1.5$ A

69. $R \ge \$1{,}475{,}000$ **71.** $x \ge 98$

Review Exercises   **1.** $\{1, 2, 3, 4, 5, 6, 7, 8, 9, 10\}$
2. commutative   **3.** undefined   **4.** $6 + 2(x - 5)$ **5.** 16
6. -14

Chapter 2 Review Exercises   **1.** true   **2.** true
3. false   **4.** false   **5.** true   **5.** true   **7.** identify   **8.** linear
9. 1. Simplify   **a.** Distribute   **b.** multiplying; LCD.

c. Combine   **2.** addition   **3.** multiplication   **10.** Replace the variable(s) in the equation with the value. If the resulting equation is true, then the value is a solution.   **11.** yes   **12.** no
13. no   **14.** yes   **15.** no   **16.** yes   **17.** -5 **18.** 18   **19.** 6
20. 5   **21.** -2 **22.** $\dfrac{9}{5}$ **23.** no solution   **24.** all real numbers   **25.** 2   **26.** 18   **27.** $\dfrac{9}{4}$ **28.** $-\dfrac{1}{2}$ **29.** $\dfrac{3}{4}$ **30.** 5.25
31. $x = 1 - y$ **32.** $a^2 = c^2 - b^2$ **33.** $d = \dfrac{C}{\pi}$ **34.** $m = \dfrac{E}{c^2}$
35. $h = \dfrac{2A}{b}$ **36.** $m = \dfrac{Fr}{v^2}$ **37.** $h = \dfrac{3V}{\pi r^2}$ **38.** $m = \dfrac{y - b}{x}$
39. $w = \dfrac{P - 2l}{2}$ **40.** $h = \dfrac{2A}{a + b}$ **41.** $6x = -18$; $x = -3$
42. $\dfrac{1}{2}m - 3 = \dfrac{1}{4}m - 9$; $m = -24$ **43.** $2(v + 4) - 1 = 1$; $v = -3$ **44.** $2(y - 1) + 3y = 6y - 20$; $y = 18$
45. a. $x < -3$ **b.** $\{x \mid x < -3\}$ **c.** $(-\infty, -3)$
d. [number line: −4 −3 −2 −1 0 1 2]   **46. a.** $x \ge -1$ **b.** $\{x \mid x \ge -1\}$
c. $[-1, \infty)$ **d.** [number line: −3 −2 −1 0 1 2 3]   **47. a.** $z \le -1$
b. $\{z \mid z \le -1\}$ **c.** $(-\infty, -1]$ **d.** [number line: −3 −2 −1 0 1 2 3]
48. a. $v > -4$ **b.** $\{v \mid v > -4\}$ **c.** $(-4, \infty)$
d. [number line: −5 −4 −3 −2 −1 0 1]   **49. a.** $m > 4$ **b.** $\{m \mid m > 4\}$
c. $(4, \infty)$ **d.** [number line: 0 1 2 3 4 5 6]   **50. a.** $c \ge -9$
b. $\{c \mid c \ge -9\}$ **c.** $[-9, \infty)$ **d.** [number line: −10 −9 −8 −7 −6 −5 −4]
51. $13 > 3 - 10p$; $p > -1$ **52.** $3 + 2z < 3(z - 5)$;
$z > 18$ **53.** $-\dfrac{1}{2}x \ge 4$; $x \le -8$ **54.** $-2 \le 1 - \dfrac{1}{4}k$;
$k \le 12$ **55.** 19.75 ft.^{2} **56.** 102.6 in.^{2} **57.** \$23.75   **58.** 80 in.
59. 7.1 in.   **60.** 1.2 mi.   **61.** 15 in.   **62.** $d \le 3.18$ ft.
63. $t \ge 5\dfrac{5}{6}$ hr.   **64.** $R \ge \$825{,}000$

Chapter 2 Practice Test   **1.** Equation because there is an equals symbol.   **2.** Nonlinear because there is a variable raised to an exponent other than 1.   **3.** yes   **4.** no
5. -3 **6.** 13.2   **7.** 3   **8.** no solution   **9.** $-\dfrac{2}{3}$ **10.** 4
11. $x = \dfrac{7 - y}{2}$ **12.** $r = \dfrac{I}{Pt}$ **13.** $h = \dfrac{2A}{b}$ **14.** $F = \dfrac{9C + 160}{5}$
or $F = \dfrac{9}{5}C + 32$ **15.** Answer: Set-builder notation: $\{x \mid x \ge 3\}$;
Interval notation: $[3, \infty)$; Graph: [number line: 0 1 2 3 4 5 6 7]

16. Answer: Set-builder notation: $\{x \mid -1 \le x < 4\}$; Interval notation: $[-1, 4)$; Graph:

17. a. $m < 3$ **b.** $\{m \mid m < 3\}$ **c.** $(-\infty, 3)$

d.

18. a. $x \ge -8$

b. $\{x \mid x \ge -8\}$ **c.** $[-8, \infty)$ **d.**

19. a. $p > -\dfrac{2}{5}$ **b.** $\left\{p \mid p > -\dfrac{2}{5}\right\}$ **c.** $\left(-\dfrac{2}{5}, \infty\right)$

d.

20. a. $l \le -\dfrac{5}{3}$

b. $\left\{l \mid l \le -\dfrac{5}{3}\right\}$ **c.** $\left(-\infty, -\dfrac{5}{3}\right]$ **d.**

21. $\dfrac{2}{3}n - \dfrac{1}{6} = 2n$; $n = -\dfrac{1}{8}$ **22.** $-7x + 12 = 5$; $x = 1$

23. $5(n - 2) - 3 = 10 - 4(n - 1)$; $n = 3$ **24.** $1 - n > 2n$;

$n < \dfrac{1}{3}$ **25.** 20 ft. **26.** 68.25 in.^{2} **27.** 7 cm **28.** \$530

29. $l \le 3$ ft. **30.** $x \ge 101$

Chapter 3

3.1 Exercises
1. The quantity preceding *to* (longest side) is written in the numerator, and the quantity following the word *to* (shortest side) is written in the denominator.
3. $ad = bc$ **5.** To solve proportion problems: **a.** Set up the given ratio any way you wish. **b.** Set up the ratio with the unknown so that it logically corresponds to the way you set up the given ratio. **c.** Solve using cross products. **7. a.** $\dfrac{4}{15}$ **b.** $\dfrac{6}{11}$

c. $\dfrac{2}{1}$ **9.** $\dfrac{1}{2}$ **11.** $\dfrac{2}{3}$ **13.** $\approx \dfrac{\$16.47}{\$1}$; The price of the stock is \$16.47 for every \$1 earned in 2004. **15.** The 24-oz. container is better because it costs less per ounce (the unit ratio of price to quantity is less). **17. a.** $\approx \dfrac{0.83}{1}$ **b.** $\approx \dfrac{0.89}{1}$ **c.** The ratio of convictions to total cases was greater in 2002 than in 1994.

19. a. $\approx \dfrac{0.70}{1}$; Women earn \$0.70 for every \$1 men earn.

b. $\approx \dfrac{0.75}{1}$ **c.** The ratio of women's median income to men's median income was greater in 2001 than in 1990. **21.** yes
23. no **25.** no **27.** yes **29.** no **31.** 10.5 **33.** -14

35. 10.5 **37.** -8 **39.** 115 **41.** $3\dfrac{3}{5}$ **43.** 4 **45.** 7

47. $183\dfrac{1}{3}$ mi. **49.** \$1350.65 **51.** \$9450

53. ≈ 422.25 pounds **55.** 4.875 in. **57.** $2\dfrac{1}{2}$ tsp.
59. 380 highway miles **61.** ≈ 15.8 min. **63.** 225 lb.
65. \$85.91 **67.** ≈ 66 deer **69.** 617,647 **71.** 11.25 cm
73. $a = 4\dfrac{8}{19}$ ft.; $b = 5\dfrac{1}{4}$ ft.; $c = 5\dfrac{17}{19}$ ft. **75.** 300 m
77. 26.4 m

Review Exercises **1.** Yes, 0.58 is a rational number because it can be written as a ratio of integers, $\dfrac{58}{100}$. **2.** $\dfrac{3}{25}$
3. 3.06 **4.** 0.452 **5.** 320 **6.** 0.4

3.2 Exercises
1. To write a percent as a fraction or decimal: **a.** Write the percent as a ratio with 100 in the denominator. **b.** Simplify to the desired form. **3.** multiplication
5. In a word-for-word translation, the division yields a decimal number that must be written as a percent. When using the proportion method, the decimal number is multiplied by 100 which gives the percent. **7.** $0.2, \dfrac{1}{5}$ **9.** $0.15, \dfrac{3}{20}$

11. $0.148, \dfrac{37}{250}$ **13.** $0.0375, \dfrac{3}{80}$ **15.** $0.455, \dfrac{91}{200}$ **17.** $0.\overline{3}, \dfrac{1}{3}$
19. 60% **21.** 37.5% **23.** 83.3% **25.** 66.7% **27.** 96%
29. 80% **31.** 9% **33.** 120% **35.** 2.8% **37.** 405.1%
39. 28 **41.** 3.1 **43.** 5.92 **45.** 103.2 **47.** 50 **49.** 80
51. 62.4 **53.** 20% **55.** 105% **57.** $66.\overline{6}$% **59.** 43
61. \$4800, \$400 **63.** \$898.80 **65.** 5783.5 tg **67.** \$42.50
69. \$300 **71.** ≈ 45.3% **73.** ≈ 70% **75.** ≈ 74.8%
77. MA: ≈ 66.6%, NH: ≈ 20.2%, VT: ≈ 22.3%; Massachusetts
79. 23.7% **81.** 20% **83.** \$6.22; 130.67 **85.** \$94.90; \$854.10
87. \$1950, 75% **89.** ≈ 34.5% **91.** $34.1\overline{6}$% **93.** ≈ 40.3%
95. 1960% **97.** \$29.97

Review Exercises **1.** -1 **2.** $-3x - 21$
3. $-xy + 4x - 3y$ **4.** $3(n + 9) = n - 11$; -19 **5.** 2.1
6. 10.5 cm

3.3 Exercises
1. "The second number:" $5n$. **3.** There is always a difference of 2 between consecutive even or odd numbers. **5.** 11, 17 **7.** 6, 8 **9.** 13, 39 **11.** Marie is 16, Susan is 4. **13.** Arianna is 50, Emma is 25. **15.** grapefruit: 22 g, orange: 27 g **17.** Big Mac: 580 cal., large fry: 520 cal.
19. 25 watt: 215 lumens, 60 watt: 880 lumens **21.** length is 70 ft., width is 50 ft. **23.** length is 11 cm, width is 7 cm
25. length is 7.5 ft., width is 5 ft. **27.** 26 m, 26 m, 17 m

29. $53\dfrac{1}{3}°$, $126\dfrac{2}{3}°$ **31.** 35°, 55° **33.** 73, 74 **35.** 76, 78, 80
37. 9, 11, 13, 15 **39.** 69, 70, 71 **41.** 15, 17, 19
43. 220 student tickets, 530 general public tickets
45. ten \$5 bills, six \$10 bills **47.** \$5, \$50 **49.** \$3, \$8

Review Exercises **1.** yes **2.** yes **3.** -16 **4.** -8
5. 3 **6.** 6.75 mi.

3.4 Exercises
1. They will travel the same amount of time. **3.** $45t + 24t = 10$ **5.**

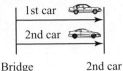

1st car
2nd car
Bridge   2nd car catches up

7. 4 hr. **9.** 3.5 hr. **11.** 2 hr. **13.** 2 hr. **15.** 9:36 A. M. **17.** Car 1: 22 mph, Car 2: 33 mph **19.** 4.5 hr.; 247.5 mi. **21.** 5:30 P.M. **23.** 5:00 P.M. **25. a.** 75 mph **b.** 300 mi. **27.** 6 hr. **29.** 4 hr. **31.** 79 mph **33.** 3 mph, 67 mph

Review Exercises
1. commutative property of multiplication **2.** 38 **3.** -43 **4.** $8x + 7y + 6$ **5.** $90.\overline{90}$ **6.** 700

3.5 Exercises
1. a. $1500, $3600, $6800 **b.** 10,000 − Known amount **c.** 10,000 − P **3.** The sum of the expressions for interest is set equal to the total interest. **5. a.** 120, 200, 300 **b.** Multiply the amount of solution 1 by 2. **c.** $2n$ **d.** $n + 2n$ or $3n$ **7.** $4600 at 2%, $13,800 at 4% **9.** $1600 at 2%, $2400 at 3% **11.** $1250 at 8%, $1900 at 12% **13.** $1800 in plan 1, $2700 in plan 2 **15.** $940 at 5%, $660 at 4% **17.** $1000 at 5.5%, $4000 at 7% **19.** $4200 at 6%, $8400 at −2% **21.** $4000 in plan A, $6000 in plan B **23.** 30 ml **25.** 1.8 gal. **27.** 1.2 qt. **29.** $166\frac{2}{3}$ lb. of 26%, $133\frac{1}{3}$ lb. of 35% **31.** 52.5 ml **33.** 50 g of 6%, 25 g of 18% **35.** 56 lb. at $1.20, 24 lb. at $0.90 **37.** 3 lb. at $1.65, 2 lb. at $1.25 **39.** 6 lb. of peanuts and 4 lb. of cashews

Review Exercises
1. < **2.** $\frac{41}{108}$ **3.** -64 **4.** $12x$ **5.** -2 **6.** 72

Chapter 3 Review Exercises
1. true **2.** true **3.** false **4.** false **5.** false **6.** false **7. 1.** 100 **8.** congruent **9.** complementary **10.** proportion **11.** $\frac{3}{8}$ **12.** $\approx \frac{\$0.085}{1\,oz.}$ **13.** no **14.** yes **15.** yes **16.** no **17.** 25 **18.** 22 **19.** $78\frac{3}{4}$ **20.** 9.5625 **21.** -120 **22.** $21.1\overline{6}$ **23.** 300 mi. **24.** 5 ft. **25.** $9.10 **26.** $2587.20 **27.** ≈ 211.4 mi. **28.** $270.03 **29.** 12 in. **30.** 43.2 m **31.** $x = 12$ ft., $y = 6$ ft. **32.** $a = 4\frac{5}{7}$ in., $b = 4\frac{2}{3}$ in., $c = 13\frac{1}{2}$ in. **33.** 0.15, $\frac{3}{20}$ **34.** 0.825, $\frac{33}{40}$ **35.** 0.125, $\frac{1}{8}$ **36.** 0.0245, $\frac{49}{2000}$ **37.** 0.1, $\frac{1}{10}$ **38.** $0.\overline{3}, \frac{1}{3}$ **39.** 40% **40.** $33\frac{1}{3}$% or $33.\overline{3}$% **41.** $36\frac{4}{11}$% or $36.\overline{36}$% **42.** 35% **43.** 120% **44.** 201.6% **45.** 14.56 **46.** 289.7 **47.** 32.5%

48. 56.25% **49.** 478 **50.** $805.59 **51.** ≈ 60.5% **52.** ≈ 60.5% **53.** 2.5, 12.5 **54.** $-2, -1, 0$ **55.** length is 45 ft., width is 15 ft. **56.** 17 cm, 19 cm, 19 cm **57.** 106°, 74° **58.** 31°, 59° **59.** 3 regular, 2 control top **60.** Pepsi is $46, Coca-Cola is $58 **61.** 3.5 hr. **62.** 2.6 hr. **63.** 4:00 **64.** 75 mph **65.** $20,000 at 8%, $40,000 at 10% **66.** $10,200 at 7%, $30,600 at −2% **67.** 40 L of 90%, 60 L of 40% **68.** 60 oz.

Chapter 3 Practice Test
1. $\frac{2}{5}$ **2.** $\approx \frac{\$0.247}{1\,oz.}$ **3.** $3\frac{1}{3}$ **4.** 2.25 **5.** ≈ 49 gal. **6.** 12.96 cm **7.** 0.22, $\frac{11}{50}$ **8.** $0.0\overline{3}, \frac{1}{30}$ **9.** 320% **10.** 40% **11.** 20% **12.** 14.35 **13.** 175 **14.** 185.5 or about 186 **15.** ≈ 73.5% **16.** 12.5% **17.** 8 **18.** 54, 56 **19.** 5 km, 11 km **20.** 102°, 78° **21.** Target: $53, Walmart: $47 **22.** 0.5 hr. **23.** 0.625 hr. **24.** $3000 in plan A, $1500 in plan B **25.** 30 oz.

Chapters 1–3 Cumulative Review Exercises
1. false **2.** true **3.** false **4.** false **5.** true **6.** true **7.** itself **8.** LCD **9.** 100 **10.** cross **11.** $\{1, 2, 3, \ldots\}$ (answers may vary) **12.**

$-3\frac{1}{4}$ number line marked from −4 to 0

13. $2^3 \cdot 3 \cdot 5$ **14.** 3 **15.** distributive property **16.** Set the cross products equal to each other and solve for the variable. **17.** $\frac{4}{9}$ **18.** $\frac{2}{9}$ **19.** not a real number **20.** -6.5 **21.** 7 **22.** $\frac{1}{10}$ **23.** -14 **24.** $\frac{11}{10}$ **25.** -26.48 **26.** undefined **27.** $-\frac{1}{2}x + 27$ **28.** additive inverse **29.** commutative property of addition **30.** associative property of addition **31.** additive inverse **32.** -5 **33.** 9 **34.** $\frac{2}{5}$ **35.** $\frac{32}{19}$ **36.** all real numbers **37.** 5 **38.** $b = \frac{2A}{h}$ **39.** $r = \frac{A - P}{Pt}$ **40.** $c = P - a - b$ **41.** $t = \frac{d}{r}$ **42.** 12 **43.** 6.345 **44.** $-3, -1, 1$ **45.** $t \geq 88$ **46.** 20 L **47.** 15 dimes, 20 quarters **48.** 4:00 **49.** 3.75 lb. of cashews, 6.25 lb. of peanuts **50.** 300 mi.

Chapter 4
4.1 Exercises
1. horizontal-axis coordinate
3.

coordinate plane with quadrants labeled II (−, +), I (+, +), III (−, −), IV (+, −)

5. $A(4, 3), B(-3, 1), C(0, -2), D(2, -5)$ **7.** $A(-4, 0), B(2, 4), C(-2, -3), D(4, -4)$

9.

11.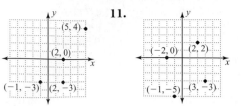

13. II **15.** I

17. IV **19.** III **21.** *y*-axis **23.** *x*-axis **25.** linear
27. nonlinear **29.** linear **31.** nonlinear **33.** linear
35. Answers may vary. Some possible answers are $(-1, -5)$, $(0, -3)$, $(2, 1)$. **37.** *A*: $(-6, -4)$, *B*: $(-3, 2)$, *C*: $(5, 2)$, *D*: $(8, -4)$ **39. a.** $(-3, 1)$, $(2, 1)$, $(1, -3)$, $(-4, -3)$
b. $(0, 2)$, $(5, 2)$, $(4, -2)$, $(-1, -2)$ **c.** $(x + 3, y + 1)$
41. a.   **b.** 20 units **c.** 21 square units

Review Exercises

1.
$$\begin{array}{c}\leftarrow\!\!+\!\!+\!\!\diamond\!\!+\!\!+\!\!+\!\!+\!\!+\!\!+\!\!\rightarrow\\ -4\;-3\;-2\;-1\;\;0\;\;\;1\;\;\;2\end{array}$$
2. 2

3. 12 **4.** $\dfrac{15}{2}$ **5.** $-\dfrac{9}{10}$ **6.** 2

4.2 Exercises

1. a. Replace the variables in the equation with the corresponding coordinates. **b.** Verify that the equation is true. **3. a.** Choose a value for one of the variables. **b.** Replace the corresponding variable with your chosen value. **c.** Solve the equation for the value of the other variable. **5.** A minimum of two ordered pairs are needed because two points determine a line. **7.** yes **9.** yes **11.** no
13. yes **15.** no **17.** yes **19.** no **21.** no
23. $(3, -5)$, $(5, -3)$, $(6, -2)$

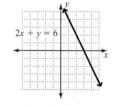

25. $(1, 4)$, $(3, 0)$, $(5, -4)$

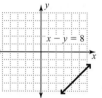

27. $(0, 4)$, $(6, 0)$, $(3, 2)$

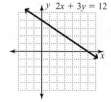

29. $(0, 4)$, $(-3, 0)$, $(3, 8)$

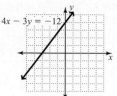

31. $(-1, -1)$, $(0, 0)$, $(1, 1)$

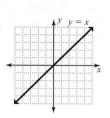

33. $(-1, -2)$, $(0, 0)$, $(1, 2)$

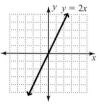

35. $(-1, 5)$, $(0, 0)$, $(1, -5)$

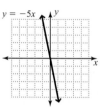

37. $(0, -3)$, $(2, -1)$, $(4, 1)$

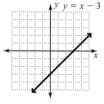

39. $(0, -2)$, $(4, -6)$, $(-4, 2)$

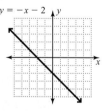

41. $(0, -5)$, $(1, -3)$, $(2, -1)$

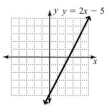

43. $(-1, 6), (0, 4), (1, 2)$

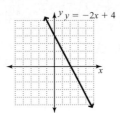

45. $(-2, -1), (0, 0), (2, 1)$

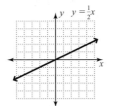

47. $(0, 0), (-3, 2), (3, -2)$

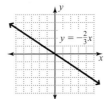

49. $(-3, 6), (0, 4), (3, 2)$

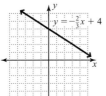

51. $(1, -4), (2, 0), (3, 4)$

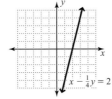

53. $(-1, -5), (0, -5), (1, -5)$

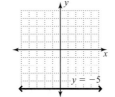

55. $(7, -1), (7, 0), (7, 1)$

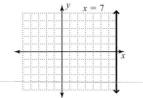

57. $(-1, -2.9), (0, -2.5), (1, -2.1)$

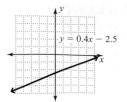

59. As a gets larger, the graph gets steeper.
61. The graph will shift up b units on the y-axis.
63. a. $160 **b.** 4 hr. **c.**

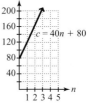

d. The c-intercept represents the initial charge, which is $80.
65. a. $18 **b.** 500 copies **c.**

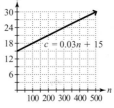

67. a. 412.5 gal. **b.** ≈ 71.4 min. **c.** ≈ 143 min.
d.

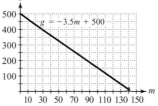

e. The g-intercept represents the original amount of water in the hot tub, which is 500 gallons. **f.** The m-intercept represents the amount of time it takes for the hot tub to empty (0 gallons of water).
69. a. $1200 **b.** 4 yr. **c.** 8 yr. **d.**

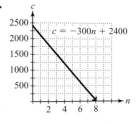

Review Exercises **1.** -64.4 **2.** 31.5 in. **3.** 5.67 sq. in.
4. 3 **5.** $x = \dfrac{v - k}{m}$

4.3 Exercises **1.** The point where a graph intersects the *x*-axis. **3.** 1. Replace *y* with 0 in the given equation. 2. Solve for *x*. **5.** The graph of any equation in the form $y = mx$ is always a line that passes through the origin, which means the *x*- and *y*-intercepts are both at $(0, 0)$. **7.** $(4, 0)$, $(0, -4)$

9. $(3, 0)$, $(0, 2)$ **11.** $\left(-\dfrac{10}{3}, 0\right)$, $\left(0, \dfrac{5}{2}\right)$ **13.** $(2, 0)$, $(0, -6)$

15. $\left(-\dfrac{5}{2}, 0\right)$, $(0, 5)$ **17.** $(0, 0)$ for both. **19.** $(0, 0)$ for both.

21. $(5, 0)$, no *y*-intercept.

23.
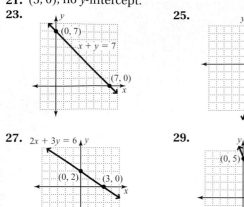

25.

27.
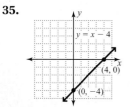

29.

31.

33.

35.
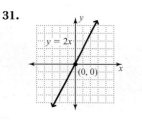

37.

39.
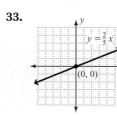

41.

43.
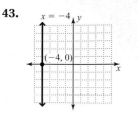

45.
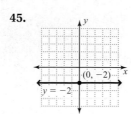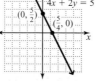

47.

49.
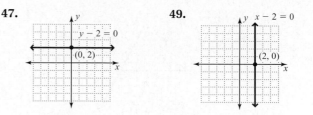

51. d, because the *x*-coordinate of the *x*-intercept is positive and the *y*-coordinate of the *y*-intercept is negative. **53.** $(-2.7, 0)$, $(2.7, 0)$, $(0, -2.7)$, $(0, 2.7)$ **55. a.** $15x + 20y$

b. $15x + 20y = 2000$ **c.** $\left(133\dfrac{1}{3}, 0\right)$, $(0, 100)$ **d.** The number

of units required to meet the goal if only one size is sold.
e. Answers may vary. Some posibilities are $(20, 85)$, $(40, 70)$, or $(60, 55)$. **f.**

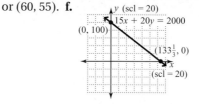

Review Exercises **1.** $w = \dfrac{P - 2l}{2}$

2. $4 + 3x = 5(x + 6)$; -13 **3.** $x + x + 2 + x + 4 = 141$; $45, 47, 49$ **4.** $37.5°, 52.5°$ **5.** I **6.** no

4.4 Exercises **1.** How steep a line is. **3.** Downhill, because the slope is negative. **5.** Horizontal, because the rise is 0 for any amount of run.

7.

9.

11.

13.

15. $m = 2$ $(0, 3)$

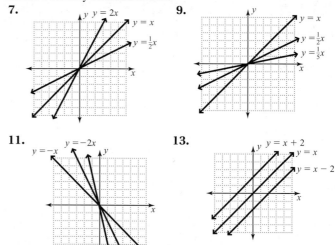

17. $m = -2$ $(0, -1)$

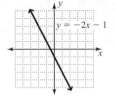
$y = -2x - 1$

19. $m = \dfrac{1}{3}$ $(0, 2)$

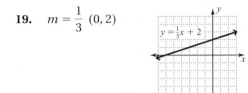
$y = \frac{1}{3}x + 2$

21. $m = \dfrac{3}{4}$ $(0, -2)$

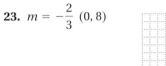
$y = \frac{3}{4}x - 2$

23. $m = -\dfrac{2}{3}$ $(0, 8)$

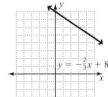
$y = -\frac{2}{3}x + 8$

25. $m = -2$ $(0, 4)$

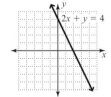
$2x + y = 4$

27. $m = 3$ $(0, 1)$

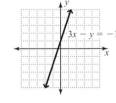
$3x - y = -1$

29. $m = -\dfrac{2}{3}$ $(0, 2)$

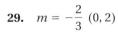
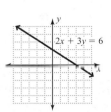
$2x + 3y = 6$

31. $m = -\dfrac{2}{3}$ $\left(0, \dfrac{7}{3}\right)$

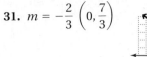

$2x + 3y = 7$

33. $m = \dfrac{2}{3}$ $\left(0, \dfrac{8}{3}\right)$

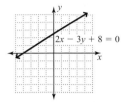
$2x - 3y + 8 = 0$

35. $m = -\dfrac{3}{2}$ $\left(0, \dfrac{5}{2}\right)$

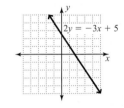
$2y = -3x + 5$

37. $m = 3$ $(0, -5)$

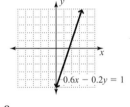
$0.6x - 0.2y = 1$

39. $\dfrac{5}{2}$ **41.** $-\dfrac{2}{3}$ **43.** $\dfrac{2}{7}$ **45.** -1 **47.** -3 **49.** $-\dfrac{5}{4}$ **51.** 0

53. undefined   **55.** d   **57.** a   **59.** f   **61.** $m = 2$; $(0, 3)$;
$y = 2x + 3$ **63.** m is undefined; no y-intercept; $x = 2$.

65. $y = -2$ **67.** $y = 0$

69. a.

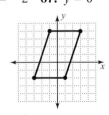

b. 3 and 0

c. They are the same.   **71.** $0.58\overline{3}, \dfrac{7}{12}$ **73.** $\dfrac{293}{230} \approx 1.27$

75. a.

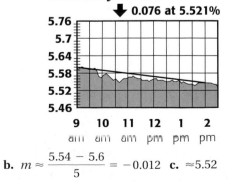

b. $m \approx \dfrac{5.54 - 5.6}{5} = -0.012$ **c.** ≈ 5.52

Review Exercises **1.** $11x + 7$ **2.** $-2x - 14$

3. $\dfrac{21}{2}$ **4.** $x = \dfrac{C - By}{A}$ **5.**

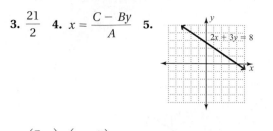

$2x + 3y = 8$

6. $\left(\dfrac{7}{3}, 0\right), \left(0, -\dfrac{7}{5}\right)$

4.5 Exercises **1.** The slope-intercept form would be easiest because you are given the values m and b. **3.** Find the slope. **5.** Solve for y. **7.** $y = 2x + 3$ **9.** $y = -3x - 9$

11. $y = \dfrac{3}{4}x + 5$ **13.** $y = -\dfrac{2}{5}x + \dfrac{7}{8}$ **15.** $y = 0.8x - 5.1$

17. $y = -3x$ **19.** $y = 2x - 2$ **21.** $y = -\dfrac{6}{5}x + 3$

23. $y = 3x - 13$ **25.** $y = -x - 2$ **27.** $y = \dfrac{2}{3}x - 1$

29. $y = -\dfrac{3}{4}x - \dfrac{23}{4}$ **31.** $y = -\dfrac{4}{3}x + \dfrac{4}{3}$ **33.** $y = 2x$

35. $y = -2x + 5$ **37.** $y = \dfrac{5}{2}x - \dfrac{13}{2}$ **39.** $y = 7x$ **41.** $x = 8$

43. $y = -1$ **45.** $y = 3x + 0.8$ **47.** $5x - 2y = 4$

49. $3x - 2y = 6$ **51.** $4x + 11y = -27$ **53.** $3x - 8y = 37$
55. $x = -1$ **57.** $y = -1$ **59.** parallel **61.** perpendicular
63. parallel **65.** perpendicular **67.** perpendicular
69. a. $y = 4x - 14$ **b.** $4x - y = 14$ **71. a.** $y = -2x + 3$

b. $2x + y = 3$ **73. a.** $y = \dfrac{1}{3}x + \dfrac{1}{3}$ **b.** $x - 3y = -1$

75. a. $y = -2x + 1$ **b.** $2x + y = 1$ **77. a.** $y = \dfrac{3}{4}x + \dfrac{11}{2}$

b. $3x - 4y = -22$ **79. a.** $y = -\dfrac{1}{2}x - 3$ **b.** $x + 2y = -6$

81. a. $y = -4x + 8$ **b.** $4x + y = 8$ **83. a.** $y = \dfrac{1}{2}x - 5$

b. $x - 2y = 10$ **85. a.** $y = \dfrac{3}{2}x - \dfrac{17}{2}$ **b.** $3x - 2y = 17$

87. a. $y = -\dfrac{2}{3}x + \dfrac{5}{3}$ **b.** $2x + 3y = 5$

89. a. $c = 0.3n + 7.75$ **b.** $19.75
c. **91. a.**

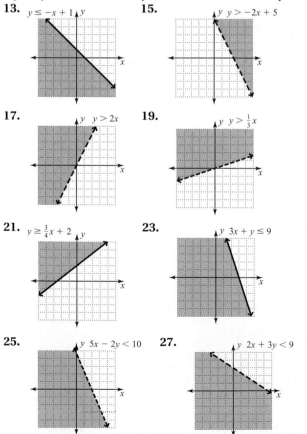

$c = 0.3n + 7.75$

$(10, 2.78)$

$(0, 1.89)$

b. 0.089 **c.** $p = 0.089n + 1.89$ **d.** 2.513 million

e. 3.848 million **93. a.**

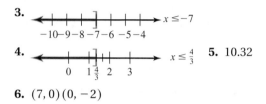

$(0, 36)$ $(5, 28.5)$

b. -1.5

c. $p = -1.5n + 36$ **d.** $(0, 36)$; The p-intercept indicates the initial price. **e.** $(24, 0)$; The n-intercept indicates that on the 24th day, the price will be $0. **f.** $24

Review Exercises **1.** $<$ **2.** $\dfrac{36}{5}$

3.

$x \le -7$

$-10\ -9\ -8\ -7\ -6\ -5\ -4$

4.

$x \le \dfrac{4}{3}$ **5.** 10.32

$0 \quad 1\frac{4}{3}\ 2 \quad 3$

6. $(7, 0)(0, -2)$

4.6 Exercises **1.** To determine whether an ordered pair is a solution for an inequality, replace the variables with the corresponding coordinates and see if the resulting inequality is true. If so, the ordered pair is a solution. **3.** A solid line indicates the equality is included. A dashed line indicates the equality is excluded. **5.** yes **7.** no **9.** no **11.** yes

13. $y \le -x + 1$ **15.** $y \quad y > -2x + 5$

17. $y \quad y > 2x$ **19.** $y \quad y > \dfrac{1}{3}x$

21. $y \ge \dfrac{3}{4}x + 2$ **23.** $y \quad 3x + y \le 9$

25. $y \quad 5x - 2y < 10$ **27.** $y \quad 2x + 3y < 9$

29. $2y + 4x \le 3$ **31.** $y + 3x \ge 0$

33. $x > -4$ **35.** $y \le 2$

37. $x + 2 \ge 0$ **39.** $y - 3 < 0$

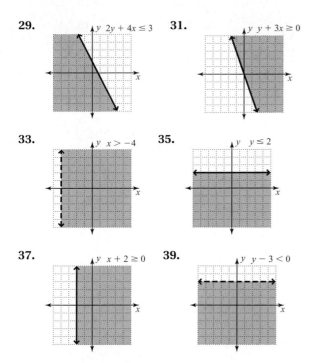

41. a. x represents the number of large bottles sold, y represents the number of small bottles sold.
b.

$2x + 1.5\,y \ge 280{,}000$
y (scl = 20,000)
x (scl = 20,000)

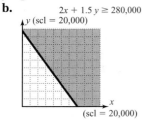

c. The boundary line represents combinations of bottle sizes sold to produce a revenue of exactly \$280,000 (break even). The shaded region represents combinations that produce a revenue greater than \$280,000 (profit). **d.** Answers will vary; some combinations are (80,000, 80,000), (20,000, 160,000).
e. Answers will vary; some combinations are (150,000, 0), (90,000, 100,000). **f.** No, only combinations of natural numbers are possible because we cannot sell negative numbers of bottles nor fractions of a bottle. **43. a.** $5.5x + 12.5y \le 100$
b.

$5.5x + 12.5y \le 100$
y (scl = 2)
x (scl = 2)

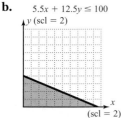

c. The combinations of plant purchases that would cost exactly \$100. **d.** The combinations of plants sold that would cost less than \$100. **e.** Answers will vary; some combinations are (2, 4), (6, 2) or (8, 4). **f.** Yes, (0, 8). Since she cannot purchase negative or fractional numbers of plants, the

combination must be natural numbers and (0, 8) is the only combination of natural numbers that costs exactly \$100.
45. a. $2l + 2w \le 180$ **b.**

$2l + 2w \le 180$
w (scl = 10)
l
(scl = 10)

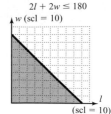

c. The combination of lengths and widths that would yield a perimeter of exactly 180 feet. **d.** The combinations of length and width that yield a perimeter that is less than 180 feet.
e. Answers will vary; some combinations are (10, 80), (20, 70), or (40, 50). **f.** Answers will vary; some combinations are (10, 10), (20, 20), or (30, 30).

Review Exercises
1. $\{a, e, i, o, u\}$ **2.** 1 **3.** -13
4. 20 ft. by 24 ft. **5.** $\left(\dfrac{8}{3}, 0\right), (0, -4)$ **6.**

$-x + y = 7$

4.7 Exercises
1. The domain is a set containing all x-values. **3.** Trail along the x-axis and determine what values of x have a corresponding y-value. **5.** Draw or imagine a vertical line through every point. If each vertical line intersects the graph in at most one point, the relation is a function.
7. Domain: $\{2, 3, 1, -1\}$ Range: $\{1, -2, 4, -1\}$ **9.** Domain: $\{4, 2, 3, -4\}$ Range: $\{1, -5, 0\}$ **11.** Domain: $\{0, -3, 2, 8\}$ Range: $\{0, -3, 9, -16\}$ **13.** Yes. Every element in the domain is assigned to one element in the range.
15. No. An element in the domain is assigned to more than one element in the range (8 is assigned to both 2 and -1).
17. Yes. Every element in the domain is asigned to one element in the range. **19.** No. An element in the domain is assigned to more than one element in the range. **21.** No. It fails the vertical line test. **23.** Yes. It passes the vertical line test. **25.** yes **27.** no **29.** yes **31.** no **33.** Domain: $\{x \,|\, 1900 \le x \le 2001\}$ Range: $\{y \,|\, 13\text{ million} \le y \le 63\text{ million}\}$ Function **35.** Domain $\{1900, 1910, 1920, 1930, 1940, 1950, 1960, 1970, 1980, 1990, 2000\}$ Range: $\{4.7, 5.4, 6.2, 6.9, 8.0, 8.8, 10.4, 11.6, 11.6, 13.2, 13.6, 14.7\}$ Function **37.** Domain: $\{x \,|\, -4 \le x \le 4\}$ Range: $\{y \,|\, -3 \le y \le 2\}$ Function
39. Domain: $\{x \,|\, x \le 0\}$ Range: all real numbers Not a function **41.** Domain: all real numbers Range: $\{y \,|\, y \ge -1\}$ Function **43. a.** -5 **b.** -3 **c.** -9 **d.** -4 **45. a.** 7 **b.** 5 **c.** 11 **d.** 17 **47. a.** $\sqrt{3}$ **b.** not a real number **c.** $\sqrt{2}$ **d.** $\sqrt{a + 3}$ **49. a.** 0 **b.** not a real number **c.** 0 **b.** $\sqrt{5}$ **51. a.** -1 **b.** undefined **c.** $\dfrac{1}{2}$ **d.** $\dfrac{1}{3}$
53. a. -3 **b.** -3.19 **c.** $a^2 - 2a - 3$ **d.** $a^2 - 8a + 12$

55. a. -2 **b.** $-\dfrac{5}{3}$ **c.** $-\dfrac{7}{3}$ **d.** -3 **57. a.** $-\dfrac{3}{4}$ **b.** $-\dfrac{4}{3}$

c. $-\dfrac{1}{6}$ **d.** -6 **59. a.** undefined **b.** $\dfrac{1}{\sqrt{3}}$ **c.** not a real

number **d.** $\dfrac{1}{\sqrt{a+2}}$ **61. a.** 2 **b.** 4 **c.** 6 **d.** 10

63. a. 3 **b.** 4 **c.** 2 **65. a.** 0 **b.** -2 **c.** undefined
67.

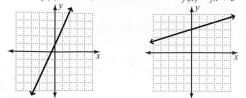

$f(x) = 2x + 1$

69.

$f(x) = \frac{1}{3}x + 3$

71.

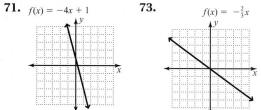

$f(x) = -4x + 1$

73.

$f(x) = -\frac{2}{3}x$

75.

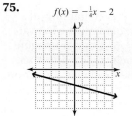

$f(x) = -\frac{1}{4}x - 2$

77. a. $350 **b.** $450

c. The cost of producing 30 items is $750.   **79. a.** $13,500
b. $9000   **c.** The value of the car after ten years is $3000.

Review Exercises   **1.** Commutative property of addi-
tion.   **2.** $-\dfrac{13}{30}$ **3.** $-4x - 2$ **4.** 8   **5.** $\dfrac{36}{5}$ **6.** 35 people

Chapter 4 Review Exercises   **1.** true   **2.** true
3. false   **4.** false   **5.** true   **6.** true   **7. 1.** y **2.** x.   **8. 1.** x
2. y **9.** y-intercept   **10.** uphill; downhill   **11.** $A(4,3)$,
$B(-2,4)$, $C(-4,-2)$, $D(2,-1)$ **12.** $A(3,1)$, $B(0,4)$,
$C(-3,0)$, $D(0,-2)$ **13.**

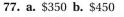

• (3, 4)
◆(0, 1)
(−2, 0)
(−1, −5)

14.

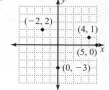

$(-2,2)$ $(4,1)$ $(5,0)$ $(0,-3)$

15. II   **16.** III   **17.** I   **18.** IV

19. yes   **20.** yes   **21.** no   **22.** no   **23.** no   **24.** no
25. $(0,0)$, $(1,-2)$, $(-1,2)$

$y = -2x$

26. $(0,0)$, $(1,1)$, $(-1,-1)$

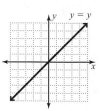

$y = y$

27. $(0,7)$, $(1,6)$, $(3,4)$

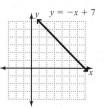

$y = -x + 7$

28. $(5,0)$, $(1,1)$, $(-3,2)$

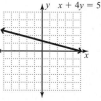

$x + 4y = 5$

29. $(0,2)$, $(3,0)$, $(-3,4)$

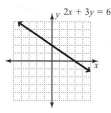

$2x + 3y = 6$

30. $(0,3)$, $(3,2)$, $(-3,4)$

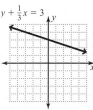

$y + \frac{1}{3}x = 3$

31. $(12,0)$, $(0,6)$ **32.** $(0,0)$ for both   **33.** $(3.1,0)$, $(0,-2)$

34. $\left(-\dfrac{15}{2}, 0\right)$, $(0, 5)$ **35.** $(-2, 0)$, no y-intercept

36. no x-intercept, $(0, 7)$ **37.**

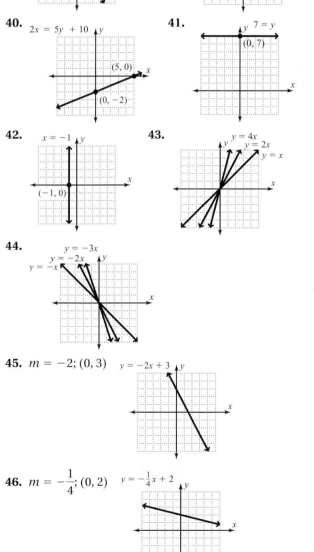

38.   **39.** $y = -\dfrac{1}{5}x + 3$

40.   **41.**

42.   **43.**

44.

45. $m = -2$; $(0, 3)$

46. $m = -\dfrac{1}{4}$; $(0, 2)$

47. $m = 3$; $(0, 0)$

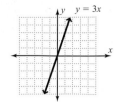

48. $m = -1$; $(0, 5)$

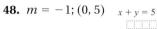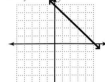

49. $m = \dfrac{1}{3}$; $\left(0, -\dfrac{7}{3}\right)$

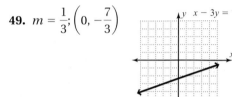

50. $m = \dfrac{2}{3}$; $\left(0, -\dfrac{8}{3}\right)$

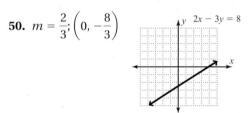

51. 3   **52.** 1   **53.** $-\dfrac{1}{3}$ **54.** $-\dfrac{3}{5}$ **55.** 0   **56.** undefined

57. $y = -x + 7$ **58.** $y = -\dfrac{1}{5}x - 8$ **59.** $y = 0.2x + 6$

60. $y = x$ **61.** $y = -2x + 9$ **62.** $y = x - 9$

63. $y = \dfrac{1}{3}x + \dfrac{5}{3}$ **64.** $y = -\dfrac{2}{5}x + \dfrac{21}{5}$ **65.** $y = 6.2x + 9.4$

66. $y = -0.4x + 1.6$ **67.** $y = -x + 4$, $x + y = 4$

68. $y = \dfrac{5}{4}x - \dfrac{37}{4}$, $5x - 4y = 37$ **69.** $y = -\dfrac{1}{2}x + \dfrac{7}{2}$,

$x + 2y = 7$ **70.** $y = -\dfrac{1}{5}x - \dfrac{13}{5}$, $x + 5y = -13$

71. $y = 2x - 2$ **72.** $y = -\dfrac{2}{3}x + \dfrac{17}{3}$ **73.** $y = -\dfrac{5}{3}x - \dfrac{11}{3}$

74. $y = \dfrac{5}{2}x + 9$ **75. a.** \$2200   **b.** $(0, 1000)$; The p-intercept
indicates the person's gross pay if he has \$0 in sales during a
month.

c.

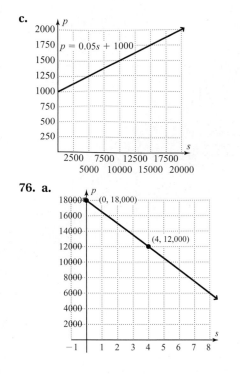

$p = 0.05s + 1000$

76. a.

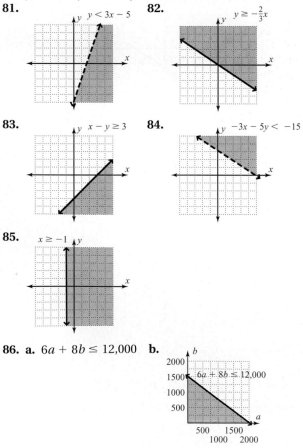

(0, 18,000)

(4, 12,000)

b. -1500　**c.** $v = -1500n + 18,000$　**d.** 2006　**e.** 2012
77. yes　**78.** yes　**79.** yes　**80.** no
81.　　　　　　　　　**82.**

$y < 3x - 5$　　　$y \geq -\frac{2}{3}x$

83.　　　　　　　　　**84.**

$x - y \geq 3$　　　$-3x - 5y < -15$

85.

$x \geq -1$

86. a. $6a + 8b \leq 12,000$　**b.**

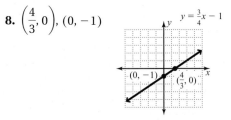

$6a + 8b \leq 12,000$

c. Answers may vary; some combinations are (400, 1200), (800, 900), (1200, 600).　**d.** Answers may vary; some combinations are (500, 100), (1000, 200), (200, 1100).　**87.** Domain: $\{2, -1, 3, -3\}$; Range: $\{3, 5, 6, -4\}$　**88.** Domain: {Volkswagen, Honda, Toyota, Subaru, Nissan}; Range: $\{52.2, 49.7, 49.0, 47.8, 45.8\}$　**89.** no　**90.** yes　**91.** Domain: all real numbers; Range: all real numbers; It is a function.
92. Domain: $\{x | x \leq 0\}$; Range: all real numbers; It is not a function.　**93.** Domain: $\{x | -4 \leq x \leq 5\}$; Range: $\{1, 2, 3\}$; It is a function.　**94. a.** 1　**b.** 3　**c.** 3　**d.** 2　**95. a.** 0　**b.** 2
c. -10　**96. a.** 2　**b.** undefined　**c.** $\frac{5}{8}$

Chapter 4 Practice Test

1. $A(2, 4)$ $B(-3, 2)$ $C(0, -5)$ $D(4, -2)$　**2.**

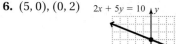

3. I　**4.** no

$(-3, 2)$
$(2, 0)$
$(4, -1)$
$(-1, -5)$ $(0, -3)$

5. $(4, 0), (0, -2)$

$x - 2y = 4$

$(4, 0)$
$(0, -2)$

6. $(5, 0), (0, 2)$　$2x + 5y = 10$

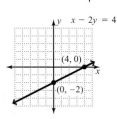

7. $(0, 0)$　$y = -2x$

$(0, 0)$

8. $\left(\frac{4}{3}, 0\right), (0, -1)$　$y = \frac{3}{4}x - 1$

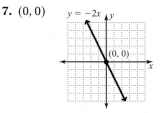

$(0, -1)$ $\left(\frac{4}{3}, 0\right)$

9. $m = \frac{3}{4}$; $(0, 11)$　**10.** $m = \frac{5}{3}$; $\left(0, -\frac{8}{3}\right)$　**11.** $-\frac{10}{3}$　**12.** 0

13. $y = \frac{3}{5}x + 4$ **14.** $y = -\frac{2}{5}x + \frac{11}{5}$ **15.** $2x - 3y = 1$
16. $x - 3y = 10$ **17. a.**   **b.** 20.225

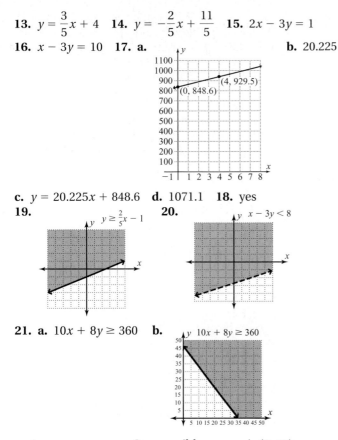

c. $y = 20.225x + 848.6$ **d.** 1071.1   **18.** yes
19.   **20.**

21. a. $10x + 8y \geq 360$ **b.**

c. Answers may vary. One possible answer is (0, 45).
d. Answers may vary. One possible answer is (10, 35).

22. yes   **23.** no   **24. a.** -2 **b.** undefined   **c.** $-\frac{9}{7}$

25. a. Domain: $\{x \mid -3 \leq x \leq 3\}$; Range: $\{-2, 1, 2\}$ **b.** 2

Chapter 5

5.1 Exercises   **1.** If a is a real number, where $a \neq 0$ and n is a natural number, then $a^{-n} = \frac{1}{a^n}$.   **3.** The result is positive. We simplify an expression with a negative exponent by inverting the expression and changing the sign of the exponent. $\left(2^{-3} = \frac{1}{2^3} = \frac{1}{8}\right)$ **5.** To write a number greater than 1 in scientific notation: **a.** Move the decimal point so that the number is greater than or equal to 1 but less than 10.   **b.** Write the decimal number multiplied by 10^n, where n is the number of places between the new decimal position and the original decimal position.   **c.** Delete 0s to the right of the last nonzero digit.   **7.** 1   **9.** -8 **11.** -125 **13.** -16

15. $\frac{81}{256}$ **17.** -0.008 **19.** 5.29   **21.** $\frac{1}{a^5}$ **23.** $\frac{1}{8}$ **25.** $-\frac{1}{64}$

27. $-\frac{1}{10}$ **29.** -125 **31.** $\frac{n^5}{m^5}$ **33.** $\frac{49}{25}$ **35.** b^6 **37.** 64

39. $\frac{-1}{7,962,624}$ **41.** $\approx -77,160.49$ **43.** $-\frac{512}{125}$

45. Mistake: The expression was interpreted to mean $(-3)^4$, but it means the additive inverse of 3^4. Correct: -81
47. Mistake: The negative sign in the exponent was passed to the product instead of inverting the original expression as it should have. Correct: $\frac{1}{25}$ **49.** 16,500,000,000
51. 299,800,000   **53.** 6,378,000   **55.** 92,920,000
57. 0.0000025   **59.** 0.0000000000000000000000000000167
61. 0.00000000295
63. 0.0000000000000000000000000006645   **65.** 2.5×10^{13}
67. 5.3×10^9 **69.** 2.342×10^7 **71.** 1.3×10^9
73. 5.86×10^{-7} m   **75.** 1.67×10^{-27} **77.** 6.25×10^{-5}
79. 5×10^{-8} **81.** 8.95×10^5, 7.2×10^6, 7.5×10^6, 9×10^7, 1.3×10^8 **83.** The wavelength of ultraviolet light is about 38.8 times the wavelength of x-rays.

Review Exercises   **1.** $-4y + 12$ **2.** $3\frac{1}{3}$ **3.** $266\frac{2}{3}$
4. slope $\frac{2}{3}$, y-intercept: $(0, -2)$ **5.** $\frac{1}{5}$ **6.** 43

5.2 Exercises   **1.** The sum of the exponents on all variables in a monomial.   **3.** The greatest degree of any of the terms in the polynomial.   **5.** Add the coefficients and leave the variables and exponents the same.   **7.** monomial
9. monomial   **11.** not   **13.** not   **15.** monomial
17. monomial   **19.** c: -5, d: 3   **21.** c: -1, d: 5
23. c: -9, d: 0   **25.** c: 4.2, d: 4   **27.** c: 16, d: 3   **29.** c: 1, d: 1
31. trinomial   **33.** monomial   **35.** none   **37.** binomial
39. monomial   **41.** none   **43.** 4   **45.** 5   **47.** 7   **49.** 60
51. -8 **53.** -4.8 **55. a.** 46 ft.   **b.** 26.96 ft.   **57. a.** 30 units   **b.** 43 units   **59. a.** ≈ 1273.4 ft.^{3} **b.** ≈ 537.2 ft.^{3}
61. a. 26 million   **b.** 45.95 million   **c.** 40.45 million
d. 12.8 million; Answers will vary, but the prediction is most likely not reasonable because of the increase in world population and unforeseen social factors that might affect travel.
63. $5x^7 - 3x^5 + 7x^4 - 8x + 14$ **65.** $-8r^6 + 18r^5 + 7r^3 - 3r^2 + 4r$ **67.** $-12w^4 - w^3 + 11w^2 + 5w + 20$
69. $5x - 3$ **71.** $-4a + 5$ **73.** $-2x^2 + 6x$ **75.** $-5x^2 + 14$
77. $5y^2 - 9y + 22$ **79.** $-2k^4 + 6k^3 - 3k^2 + 6k$
81. $-10a^4 + 7a^2 + a + 16$ **83.** $-21v^5 + 8v^3 + 35v^2 - 20$
85. $4a - 2b$ **87.** $-4y^2 + 3y$ **89.** $-9x^2 + 9y$
91. $y^6 + 5yz^4 - 6yz - 3z^5 - 7z^3 - 10$
93. $-4w^2z - wz^2 + 12w^2 - 1$

Review Exercises   **1.** -179 **2.** 12   **3.** $\frac{71}{500}$ **4.** 600 L
5. -1 **6.** 2×10^{11}

5.3 Exercises   **1.** To add polynomials, combine like terms.   **3.** Change the signs of each term in the polynomial.
5. $0x + 1$ **7.** $10y + 12$ **9.** $7x$ **11.** $3x^2 - 2x + 12$
13. $6z^2 - 11z - 2$ **15.** $5r^2 - 7r - 1$ **17.** $9y^2 + 3$
19. $5a^3 + 3a^2 + 3a - 3$ **21.** $3p^3 - p^2 + 7p + 9$

23. $-8w^4 + 3w^3 - 7w^2 + 13w - 9$
25. $-3a^3b^2 - 2ab^2 - 8a^2b + 4ab + 3b^2 - 8$
27. $-7u^4 + u^2v^3 + 7u^3v + 7u - 13v^2 - 3$
29. $-12mnp - 16m^2n^2 - 12mn^2p + 14m^2n - 31n + 9$
31. $5x - 2$ **33.** $14a - 12$ **35.** $4x + 2$ **37.** $13a^2 - 1$
39. $2x^2 - 9x + 2$ **41.** $2z^2 - 7z + 9$ **43.** $12a^2 + 10a - 15$
45. $-u^3 - 2u^2 - 6u + 8$ **47.** $-11w^5 + 3w^4 + 5w^2 + 2$
49. $-11p^4 + 3p^3 - 13p^2 + 5p - 6$
51. $11v^4 + 3v^3 + 23v^2 - 5v - 9$ **53.** $2xy$
55. $-6xy + 6xz + 2yz$ **57.** $19p^3q^2 + 2pq^2 - pq + 16q^2 + 2$
59. $32.1x + 35.45y + 6.01z$ **61.** $4.22a + 1.63b + 12.13c$
63. Mistake: Only the sign of the first term in the subtrahend was changed instead of all three terms. Correct: $-4x^2 - 2x + 5$

Review Exercises   **1.** -682 **2.** -32 **3.** 0.0001
4. 128 **5.** $58,900,000$ **6.** 300 m^2

5.4 Exercises   **1.** product rule for exponents
3. To multiply monomials: **a.** Multiply coefficients.
b. Add the exponents of the like bases.   **c.** Write any unlike variable bases unchanged in the product.   **5.** Multiply the exponents and keep the base the same.   **7.** a^7 **9.** 2^{13}

11. a^5b **13.** $6x^3$ **15.** $-8m^3n^2$ **17.** $-\dfrac{5}{28}s^3t^8$

19. $2.76a^3b^5c^6$ **21.** $-6r^5s^{10}t^7$ **23.** $-12x^4t^2z^9$

25. $90x^6y^5z^3$ **27.** $-\dfrac{1}{4}a^6b^3c^2$ **29.** $w^3x^3y^5$ **31.** $5q^5r^{10}s^{21}$

33. $35x^7y$ Mistake: The exponents of x were multiplied instead of added. Correct: $35x^7y$ **35.** $54x^4y^6z$ Mistake: The exponents of x were not added. Correct: $54x^4y^6z$

37. $5w^2$ **39.** $4.5w^3$ **41. a.** height $= \dfrac{1}{2}b$, top side $= \dfrac{1}{4}b$

b. $A = \dfrac{5}{16}b^2$ **43.** 1.2×10^9 **45.** 2.4×10^{14}

47. 2.4985×10^{19} **49.** 4.2×10^4 square miles
51. 5.8404×10^8 mi.   **53.** 2.9817×10^{-19} joules   **55.** x^6
57. 3^8 **59.** h^8 **61.** $-y^6$ **63.** x^3y^3 **65.** $27x^6$

67. $-8x^9y^6$ **69.** $\dfrac{1}{64}m^6n^3$ **71.** $-0.008x^6y^{15}z^3$

73. $64p^{24}q^{24}r^{12}$ **75.** $18x^7$ **77.** r^5s^4 **79.** $100a^6b^4$

81. $\dfrac{1}{1024}a^{10}b^{15}c^5$ **83.** $3.24u^6v^3$ **85.** $324u^6v^3$

87. Mistake: The exponents were added instead of multiplied. Correct: $25x^6$ **89.** Mistake: The coefficient was multiplied by 4 instead of being raised to the 4th power. Also, the exponents were not multiplied properly. Correct: $16x^{16}y^4$
91. $\approx 1.412 \times 10^{27} \text{ m}^3$ **93.** 2.16×10^7 joules

Review Exercises   **1.** $-4x - 4z$ **2.** -16 **3.** -61
4. $4, 5, 6$ **5.** 6 **6.** $\$615,000$

5.5 Exercises   **1.** distributive property   **3.** They differ only in the sign separating the terms.   **5.** The middle term is missing.   **7.** $5x + 15$ **9.** $-12x + 24$ **11.** $28n^2 - 8n$

13. $9a^2 - 9ab$ **15.** $\dfrac{1}{4}m^2 - \dfrac{5}{8}mn$ **17.** $-0.3p^5 + 0.6p^2q^2$

19. $20x^3 + 10x^2 - 25x$ **21.** $-21x^5 + 15x^3 - 3x^2$
23. $-r^4s^4 - 3r^3s^3 + r^2s^2$ **25.** $-4a^5b^2 + 12a^2b^3 - 6a^3b^3$
$+ 2a^4b^4$ **27.** $12a^3b^3c^3 - 6a^2b^2c^2 + 15abc$

29. $4b^6 - b^4 + \dfrac{1}{5}b^3 - \dfrac{1}{2}b^2 + 16b$

31. $-r^4t^3 + 3.2r^3t^4 - 1.64r^2t^5 + 0.8rt^2$
33. $25x^6y - 15x^5y^2 + 50x^3y^9$ **35. a.** $x + 5$ **b.** $2x$
c. $2x(x + 5)$ **d.** $2x^2 + 10x$ **e.** They describe the same area.   **37.** $x^2 + 7x + 12$ **39.** $x^2 - 5x - 14$
41. $y^2 - 9y + 18$ **43.** $6y^2 + 19y + 10$
45. $15m^2 + 11m - 12$ **47.** $12t^2 - 26t + 10$
49. $15q^2 + 11qt - 12t^2$ **51.** $14y^2 + 34xy + 12x^2$
53. $a^4 - a^2b - 2b^2$ **55. a.** $x + 3$ **b.** $x + 2$
c. $(x + 2)(x + 3)$ **d.** $x^2 + 5x + 6$
e. They describe the same area.   **57.** $x^3 - 3x + 2$
59. $3a^3 - 4a^2 - a + 2$ **61.** $6c^3 + c^2 - 11c - 6$
63. $15p^3 + 26p^2q + 58pq^2 + 20q^3$
65. $12y^3 - 24y^2z + 33yz^2 - 12z^3$
67. $x^4 + 3x^3y + 2x^2y^2 + xy^3 - y^4$
69. $x^4 + 14x^3 + 69x^2 + 140x + 100$ **71.** $x + 3$
73. $4x + 2y$ **75.** $4d + 3c$ **77.** $-3j + k$ **79.** $x^2 - 25$
81. $4m^2 - 25$ **83.** $x^2 - y^2$ **85.** $64r^2 - 100s^2$ **87.** $4x^2 - 9$
89. $x^2 + 6x + 9$ **91.** $16t^2 - 8t + 1$ **93.** $m^2 + 2mn + n^2$
95. $4u^2 + 12uv + 9v^2$ **97.** $81w^2 - 72wz + 16z^2$
99. $81 - 90y + 25y^2$ **101.** $10xyz + 2y^2z$ **103.** $h^2 + 4h$
105. $w^2 + 3w$ **107.** $4x^3 + 24x^2 + 20x$ **109.** $4w^3 + 6w^2$
111. $V = 3.14r^3 + 6.28r^2$, or $\pi r^3 + 2\pi r^2$

Review Exercises   **1.** -5.25 **2.** -1 **3.** $r = \dfrac{5 - d}{m}$

4. $9, 36$ **5.** $6,304,000$ **6.** x^7

5.6 Exercises   **1.** When dividing exponential forms that have the same base, we can subtract the divisor's exponent from the dividend's exponent and keep the same base.
3. The decimal factors and powers of 10 can be separated into a product of two fractions. This allows us to calculate the decimal division and divide the powers of 10 separately.
5. To divide a polynomial by a monomial, divide each term in the polynomial by the monomial.   **7.** $\dfrac{1}{a^3}$ **9.** $\dfrac{1}{2^5}$ **11.** 3

13. $\dfrac{1}{4^6}$ **15.** y^5 **17.** $\dfrac{1}{x^2}$ **19.** a^2 **21.** $\dfrac{1}{w^3}$ **23.** $\dfrac{1}{a^7}$ **25.** r^{10}

27. y^8 **29.** $\dfrac{1}{p^3}$ **31.** 1 **33.** 3.75×10^{-2} **35.** 6.67×10^{-8}

37. 7.1×10^{10} **39.** 2.734×10^4 **41.** 500 sec. or $8\dfrac{1}{3}$ min.

43. $\$27,692.86$ **45.** $\approx 9.3 \times 10^{-4}$ kg   **47.** $5x^3$ **49.** $\dfrac{-4}{x^3}$

51. $-3m^2n^4$ **53.** $\dfrac{-8q^3}{5}$ **55.** $\dfrac{4y^4}{3x^5}$ **57.** $\dfrac{4c}{3a^2}$ **59.** $a + 2b$

61. $4x^2 - 2x$ **63.** $3x + \dfrac{2}{x}$ **65.** $xy - 3y^2$ **67.** $-2c + 8ab$

69. $3x^2 + 2x - 1$ **71.** $2x - 4 + \dfrac{3}{x}$ **73.** $12uv^3 + \dfrac{4v^4}{u} - 5v$

75. $y^2 + y^4 - 3y^5 + \dfrac{1}{y^2}$ **77.** $\dfrac{5mn}{2}$ **79. a.** $3x - 5 + \dfrac{4}{x}$

b. area = 18, length = 6, width = 3 **81.** $x + 4$ **83.** $m + 5$

85. $x - 12 + \dfrac{4}{x - 5}$ **87.** $3x^2 + 15x + 2$

89. $2x^2 + 3x + 6 + \dfrac{17}{x - 2}$ **91.** $p^2 - 5p + 25$ **93.** $3x + 2$

95. $2y - 2 + \dfrac{23}{7y + 3}$ **97.** $4k^2 - 1 + \dfrac{2}{k + 2}$

99. $7u^2 - 11u + 12 - \dfrac{30}{3u + 2}$ **101.** $b^2 - 3b + 5 + \dfrac{3}{b - 3}$

103. $y^2 - 2y + 3$ **105. a.** $x + 4$ **b.** length = 7 ft.

height = 8 ft. volume = 336 ft.² **107.** $\dfrac{36m^4}{n^4}$ **109.** $\dfrac{x^9}{64y^6}$

111. $256r^9s^7$ **113.** m^{16} **115.** $\dfrac{1}{x^{14}}$ **117.** $\dfrac{y^{27}}{x^9}$ **119.** x^{14}

121. $\dfrac{27y^6}{x^3z^6}$ **123.** $\dfrac{1}{x^{10}}$ **125.** $\dfrac{4x^4z^3}{y^5}$ **127.** $\dfrac{2}{3ab^2}$ **129.** $\dfrac{1}{2x^{11}}$

Review Exercises
1. $x + \dfrac{1}{2}y$ **2.** 340 ft.² **3.** -24

4. No. it does not pass the vertical line test. **5.** 1.67×10^{-24}
6. 29,980,000,000

Chapter 5 Review Exercises
1. false **2.** false
3. true **4.** false **5.** false **6.** false **7.** combine
8. a. coefficients **b.** exponents **9.** $(a^m)^n = a^{mn}$

10. $2ab$ **11.** $\dfrac{8}{125}$ **12.** -16 **13.** $\dfrac{1}{25}$ **14.** $\dfrac{1}{13}$

15. 4,500,000,000 **16.** 13,800,000
17. 0.00000000000000000000001663
18. 0.00000000000000000000000002006
19. 1.661×10^{-24} **20.** 9.63×10^{-23} **21.** 3×10^8
22. 6.37×10^6 **23.** yes **24.** yes **25.** no **26.** no
27. c: 6 d: 4 **28.** c: 27 d: 0 **29.** c: -2.6 d: 4 **30.** c: -1 d: 1
31. binomial **32.** monomial **33.** trinomial
34. none of these **35.** $9; -2x^9 + 21x^5 - 19x^3 - 3x + 15$
36. $4; -j^4 + 22j^2 + 5j - 19$ **37.** $5; 4u^5 + 13u^4 - 18u^3$
$- u - 21$ **38.** $8; 6v^8 + 21v^4 + 16v^3 - v^2 - 2v - 19$
39. $4y^5 + 4y^4 + 8$ **40.** $18l^5 + 5l^4 - 2l^3 + 6l^2 + 7l + 20$
41. $-2m^3 + 7m^2 + m + 3$ **42.** $-3y^2 + 7y - 15$
43. $a^2bc - ab^2c - 6$ **44.** $-11x^2yz - 2xyz^2$
45. $-4c^3 - cd^2 + 11cd + 2d + 8$
46. $-8a^5 + 3abc^3 - 8abc + 8$
47. $-3j^4 + 3jk^3 + k^2 + 10jk + 12$
48. $3mn^2 - 8n^2 - 9mn + 7$ **49.** $-4ab - 18a - 19c + 8$
50. $11y - 10$ **51.** $10x^2 + 3x - 4$ **52.** $3m - 8$
53. $-2n^3 + 3n^2 + 4n + 3$ **54.** 9 **55.** $-x^3 + 4x^2 + 17$
56. $-p^3 + 10p^2 - 10p + 9$ **57.** $-16x^3 - 4x - 2$
58. $8x^2 - xy + 6y^2$ **59.** $-3m^2 + 8mn - 2$ **60.** m^5
61. $10a^9$ **62.** $-6x^6y^6$ **63.** $25x^4y^2$ **64.** $72u^{10}$ **65.** $32x^{10}$
66. x^8 **67.** u^7 **68.** $\dfrac{1}{s^6}$ **69.** x^6 **70.** $\dfrac{b^6}{8a^9}$ **71.** $\dfrac{4j^8}{9hk^2}$

72. $\dfrac{-5}{x^2}$ **73.** 7 **74.** 1 **75.** $2x - 42$ **76.** $-16x + 8$
77. $4a^2 - 12a$ **78.** $-12b^3 + 24b^2 + 6b$
79. $-8m^7 - 12m^5 + 4m^4 + 20m^3$
80. $-12a^2bc^2 + 16a^2b^4c^2 - 8a^2b^2c^4 - 32abc^2$
81. $x^2 + 4x - 5$ **82.** $y^2 - 11y + 24$ **83.** $12m^2 + 28m - 5$
84. $15a^2 - 4ab - 4b^2$ **85.** $4x^2 - 1$ **86.** $x^3 - x^2 - x - 15$
87. $3y^3 - 7y^2 + 14y - 4$ **88.** $a^4 - a^3b - 2a^2b^2 - 3ab^3$
$- b^4$ **89.** $16 - x^2$ **90.** $x^2 - 12x + 36$ **91.** $9r^2 - 30r + 25$
92. $36s^2 - 4r^2$ **93. a.** $20x^2 - 28x$ **b.** 96 ft.²
94. $2.25 \times 10^8\ \text{m}^2$ **95.** $2.94 \times 10^{-4}\ \text{V}$ **96.** $x - 5$

97. $-y^2 + 2y - 3$ **98.** $1 + 10st^3 - 2t^4$ **99.** $a^2 - \dfrac{1}{c} + 2ac$

100. $2 + b$ **101.** $\dfrac{2z}{5y^2} - \dfrac{xz}{y} + \dfrac{2}{y}$ **102. a.** $9x - 10$

b. 24 in., 26 in., and 624 in.² **103.** < 4.2 yr. **104.** $z + 5$

105. $3m - 2 + \dfrac{1}{m - 5}$ **106.** $y - 4 - \dfrac{4}{2y + 1}$

107. $2m - 2 + \dfrac{1}{3m - 2}$ **108.** $x^2 - x + 1 - \dfrac{2}{x + 1}$

109. $4x^2 + 6x + 9$ **110.** $2s^2 + 2s - \dfrac{3}{s - 1}$

111. $5y^3 - 3y + 2$ **112.** $4x - 5$

Chapter 5 Practice Test
1. $\dfrac{1}{8}$ **2.** $\dfrac{9}{4}$ **3.** 0.006201

4. 2.75×10^8 **5.** 3 **6.** 4 **7.** 56 **8.** $-5x^4 + 7x^2 + 6x^2$
$- 5x^4 + 12 - 6x^3 + x^2 - 10x^4 - 6x^3 + 14x^2 + 12$
9. $8x^2 + x - 4$ **10.** $5x^4 - 3x^2 - x + 8$ **11.** $12x^7$
12. $-2a^6b^4c^7$ **13.** $16x^2y^6$ **14.** $3x^3 - 12x^2 + 15x$
15. $-24t^5u + 48t^3u^3$ **16.** $n^2 + 3n - 4$ **17.** $4x^2 - 12x + 9$
18. $x^3 - 2x^2 - 5x + 6$ **19.** $6n^2 - 7n - 20$ **20.** x^5

21. $\dfrac{1}{x^5}$ **22.** $\dfrac{x^9y^4}{9}$ **23.** $4x^3 + 3x$ **24.** $x - 4$

25. $5x - 4 + \dfrac{6}{3x - 2}$

Chapter 6
6.1 Exercises
1. To list all of the possible factors of a number, we can divide by 1, 2, 3, and so on, writing each divisor and quotient pair as a product until we have all possible combinations. **3.** The factorization for the GCF includes only those prime factors common to all the factorizations, each raised to its smallest exponent in the factorizations of the given terms. **5.** four-term polynomials **7.** 1, 3, 9 **9.** 1, 3, 11, 33
11. 1, 2, 3, 6, 9, 18 **13.** 1, 2, 4, 8, 16 **15.** 1, 2, 4, 11, 22, 44
17. 1, 2, 3, 4, 5, 6, 10, 12, 15, 20, 30, 60 **19.** 1, 2, 4, 7, 8, 14, 28, 56 **21.** 1, 2, 3, 5, 6, 9, 10, 15, 18, 30, 45, 90 **23.** 3
25. 8 **27.** 6 **29.** $2xy$ **31.** $5h^2$ **33.** $3a^4b$ **35.** $5(c - 4)$
37. $4(2x - 3)$ **39.** $x(x - 1)$ **41.** $6z^4(3z^2 - 2)$
43. $3p^3(6 - 5p^2)$ **45.** $3ab(2a - b)$ **47.** $7uv(2v - 1)$
49. $25x(y - 2z + 4)$ **51.** $xy(x + y + x^2y^2)$
53. $4ab^2c(7b - 9a)$ **55.** $-4pq(5p + 6 - 4q)$

57. $3mnp(n^4p + 6n^2 - 2)$ **59.** $21a^2b^2(5a - 3b + 4a^4b^2)$
61. $3x(6x^3 - 3x^2 + 10x - 4)$ **63.** $(a - 3)(2 + y)$
65. $(4x - 3)(ay - 2)$ **67.** $(x + 2)(b + c)$
69. $(m - n)(a - b)$ **71.** $(x + 2)(x^2 - 3)$
73. $(1 - m)(1 + m^2)$ **75.** $(3x + 5)(y + 2)$
77. $(4b - 1)(b + 1)$ **79.** $(3 + b)(2 - a)$
81. $(x + 2y)(3a + 4b)$ **83.** $(y - s)(x^2 - r)$
85. $(w + 3z)(w + 5)$ **87.** $(s + y)(3t - 2)$
89. $(x^2 + y^2)(a - 5)$ **91.** $2(a + b)(a + 3)$
93. $3m(m + 1)(n - 4m)$ **95.** $3a(y - 4)(a + 3)$
97. $2x(x + 3y)(x + 5)$ **99.** $6x(x + 6)$ **101.** $3x(7x + 9)$
103. $12\pi r^2(3r + 2)$

Review Exercises   **1.** 3,740,000,000   **2.** 4.56×10^7
3. $x^2 + 7x + 10$ **4.** $x^2 - x - 12$ **5.** $x^2 - 8x + 15$
6. $x^2 - 7x$

6.2 Exercises   **1.** Both are positive.   **3.** The number
with the greater absolute value is positive and the other number is negative.   **5.** 5   **7.** 2   **9.** 6   **11.** 2
13. $(r + 3)(r + 1)$ **15.** $(x - 7)(x - 1)$
17. $(z - 3)(z + 1)$ **19.** $(y + 2)(y + 3)$
21. $(u - 2)(u - 4)$ **23.** $(u + 3)(u - 2)$
25. $(a + 3)(a + 4)$ **27.** $(y - 6)^2$ **29.** $(w - 4)(w + 3)$
31. $(x - 6)(x + 5)$ **33.** $(n - 6)(n - 5)$
35. $(r - 3)(r - 6)$ **37.** $(x - 8)(x + 3)$
39. $(p - 9)(p + 4)$ **41.** prime   **43.** prime   **45.** $(p + 3q)^2$
47. $(a - 9b)(a + 3b)$ **49.** $(x - 12y)(x - 2y)$
51. $(r - 6s)(r + 5s)$ **53.** $4(x - 3)(x - 7)$
55. $2m(m - 1)(m - 6)$ **57.** $3b(a - 8)(a + 3)$
59. $4(x - 3)^2$ **61.** $n^2(n + 2)(n + 3)$
63. $7u^2(u + 5)(u + 1)$ **65.** $6b^2c(a - 2)(a - 4)$
67. $3(x - 2y)(x - 5y)$ **69.** $2b(a - 6b)(a + 3b)$
71. $(x - 6)(x + 1)$ **73.** $(x - 4)(x + 1)$ **75.** 4, 20
77. 7, 8, 13   **79.** 6, 10, 12   **81.** 9, 16, 21, 24, 25
83. length $h + 4$; width: $h + 2$ **85.** $w - 6$ and $w - 10$

Review Exercises   **1.** commutative property of
multiplication   **2.** $10x^2 + 17x + 3$ **3.** $15y^2 - 2y - 24$
4. $72n^3 - 60n^2 + 8n$ **5.** 1, 2, 3, 4, 6, 8, 12, 16, 24, 48   **6.** 1

6.3 Exercises   **1.** monomial GCF in all the terms
3. Write a pair of last terms whose product is c.   **5.** 2, 4
7. $t, 4t$ **9.** $7, 3x$ **11.** $(2j + 1)(j + 2)$ **13.** $(2y - 5)(y + 1)$
15. $(3m - 4)(m - 2)$ **17.** $(6a + 7)(a + 1)$
19. prime   **21.** $(4a - 3)(a - 4)$ **23.** $(2x + 3)(3x + 5)$
25. $(2d - 3)(8d + 5)$ **27.** $(3p - q)(p - 4q)$
29. $(2a - 5b)(6a - 5b)$ **31.** $(k + h)(5k - 12h)$
33. $(8m - 3n)(m - 3n)$ **35.** $y(8x^2 - 4x - 1)$
37. $3(2y + 1)(2y + 3)$ **39.** $2a(3b - 7)(b - 1)$
41. $2v^2(3w + 2)(w + 1)$ **43.** $(a + 1)(3a + 1)$
45. $(2t - 1)(t - 1)$ **47.** $(3x - 7)(x + 1)$
49. $(y - 2)(2y + 3)$ **51.** $(2a - 1)(5a - 7)$
53. $(4r + 3)(2r - 3)$ **55.** $2(k - 1)(2k - 3)$
57. $(5x - 2)(4x - 3)$ **59.** $(2k + j)(3k + 2j)$
61. $(x - 5y)(5x - y)$ **63.** $(5s - 2t)(2s + t)$
65. $(3u - 2v)(2u + 3v)$ **67.** $2m(2m + 3)(3m - 2)$

69. $6y(3x - 2)(2x - 3)$ **71.** $4x(2x - 3y)(3x + 2y)$
73. Mistake: Using FOIL to check, we see that the first and
last terms check but the inner and outer terms combine to
give $31x$ instead of $13x$. Correct: $(2x + 1)(3x + 5)$
75. Mistake: Did not factor completely. Correct:
$4(n + 2)(n + 1)$ **77.** $(5x - 2)(3x - 1)$ **79.** $2w + 1$ and
$3w - 2$ **81.** 7, 8, or 13   **83.** 4, 12, or 44   **85.** 1, 11, 19, or 41

Review Exercises   **1.** $16x^2$ **2.** $8y^3$ **3.** $9x^2 - 25$
4. $x^3 - x^2 + 2x - 8$ **5.** $4x^2 - 20x + 25$ **6.** $36y^2 + 12y + 1$

6.4 Exercises   **1.** The first and last terms are perfect
squares and twice the product of their roots equals the
middle term.   **3.** conjugates   **5.** The trinomial factor.
7. $(x + 7)^2$ **9.** $(b - 4)^2$ **11.** not a perfect square
13. $(5u - 3)^2$ **15.** $(10w + 1)^2$ **17.** $(y + z)^2$
19. $(2p - 7q)^2$ **21.** $(4g + 3h)^2$ **23.** $4(2t + 5)^2$
25. $(x + 2)(x - 2)$ **27.** $(4 + y)(4 - y)$
29. $(p + q)(p - q)$ **31.** $(5u + 4)(5u - 4)$
33. $(3x + b)(3x - b)$ **35.** $(8m - 5n)(8m + 5n)$
37. $2(5x + 4y)(5x - 4y)$ **39.** $4(x + 5y)(x - 5y)$
41. $(x^2 + y^2)(x + y)(x - y)$ **43.** $(x^2 + 4)(x + 2)(x - 2)$
45. $(n - 3)(n^2 + 3n + 9)$ **47.** $(x - 1)(x^2 + x + 1)$
49. $(3k - 2)(9k^2 + 6k + 4)$
51. $(c - 4d)(c^2 + 4cd + 16d^2)$
53. $(3x - 4y)(9x^2 + 12xy + 16y^2)$
55. $(x + 3)(x^2 - 3x + 9)$ **57.** $(m + n)(m^2 - mn + n^2)$
59. $(3k + 2)(9k^2 - 6k + 4)$
61. $(5x + 4y)(25x^2 - 20xy + 16y^2)$
63. $(2p + qz)(4p^2 - 2pqz + q^2z^2)$
65. $2(x + 5)(x - 5)$ **67.** $\left(4x + \dfrac{5}{7}\right)\left(4x - \dfrac{5}{7}\right)$

69. $2u(u + 1)(u - 1)$ **71.** $y^3(y + 4b)(y - 4b)$
73. $2x(5x + 1)(5x - 1)$ **75.** $3(y - 2z)(y^2 + 2yz + 4z^2)$
77. $\left(c - \dfrac{2}{3}\right)\left(c^2 + \dfrac{2}{3}c + \dfrac{4}{9}\right)$
79. $2c(2c + d)(4c^2 - 2cd + d^2)$
81. $(2a - b + c)(2a - b - c)$
83. $(1 - 3x + 3y)(1 + 3x - 3y)$
85. $(x + 3y + 3z)(x^2 - 3xy - 3xz + 9y^2 + 18yz + 9z^2)$
87. $(4d - x - y)(16d^2 + 4dx + 4dy + x^2 + 2xy + y^2)$
89. 24   **91.** 36   **93.** 25   **95.** 16
97. $10(x + 2)(x - 2)$ **99.** $(2x - 3)(4x^2 + 6x + 9)$

Review Exercises   **1.** 1, 2, 3, 4, 6, 9, 12, 18, 36
2. $2 \cdot 2 \cdot 5 \cdot 5$ **3.** $4x^3 + 2x^2 - 3x - 26$
4. $10n^4 - 18n^3 + 12n^2 - 16n - 6$ **5.** $16x^2 - 40x + 25$
6. $5x^2 - 6x + 7 + \dfrac{3}{x + 2}$

6.5 Exercises   **1.** a monomial GCF   **3.** sum of squares
5. Try to factor by grouping if there are four terms. Grouping
may also be used as a method to factor trinomials.
7. $3xy(y + 2x)$ **9.** $(x + y)(7a - 1)$ **11.** $2(x + 4)(x \quad 4)$
13. $(a + b)(x + y)$ **15.** $abc(12a^2b + 3abc + 5c^2)$

17. $(x + 3)(x + 5)$ **19.** $(x^2 + 4)(x + 2)(x - 2)$
21. prime   **23.** $(5x - 1)(3x + 2)$ **25.** $a(x + 2)^2$
27. $6ab(1 - 6b)$ **29.** prime   **31.** $(x + 7)(x - 7)$
33. $(p + 6)(p - 5)$ **35.** $u(u + 1)(u - 1)$
37. $2(b + 3)(b + 4)$ **39.** $3r^2(2 - 5r)$
41. $7(u - 5)(u + 3)$ **43.** $h^2(h^2 + h + 1)$
45. $5(p + 4)(p - 4)$ **47.** $(q^2 - 8)(q^2 + 2)$
49. $(3w + 2)(w - 1)$ **51.** $(3v + 2)(4v + 5)$
53. $2(1 + 5x)(1 - 5x)$ **55.** $20k^2(2 + l)(2 - l)$
57. $(2j + 3)(j - 1)$ **59.** $2(5 - t)^2$ **61.** $(x - 2y)(3a + 4b)$
63. $(x + 1)(x - 1)^2$ **65.** $(2x - 7)^2$
67. $2(x^2 + 9)(x + 3)(x - 3)$ **69.** $(a + 2b)(a + b)$
71. $(3 + 2m)(3 - 2m)$ **73.** $x(x^2 + 2y)(x^2 - 2y)$
75. prime   **77.** $(b + 5)(b^2 - 5b + 25)$
79. $3(y - 2)(y^2 + 2y + 4)$ **81.** $2x(3 - y)(9 + 3y + y^2)$
83. $l = 2x + 1$ $w = 3x - 7$ **85.** $l = 2x$ $w = 3x + 5$
$h = 3x + 5$ **87.** $4(5 + 2t)(5 - 2t)$ **89.** current: $3i + 4$,
resistance: $2r + 5$

Review Exercises   **1.** 224 ft.^2 **2.** $4(x + 3) = 20; 2$
3. 17, 18, 19   **4.** $x - 4 - \dfrac{4}{2x + 1}$ **5.** $6x^2$

6.6 Exercises   **1.** If a and b are real numbers and
$ab = 0$, then $a = 0$ or $b = 0$.   **3.** Write the equation in
standard form $(ax^2 + bx + c = 0)$.   **5.** $-5, -2$ **7.** $-3, 4$
9. $-\dfrac{3}{2}, \dfrac{4}{3}$ **11.** $0, 7$ **13.** $-2, 1, 0$ **15.** $0, 2$ **17.** $0, 4$
19. $0, -\dfrac{5}{3}$ **21.** $-2, 2$ **23.** $-3, 2$ **25.** $-9, 3$ **27.** $-3, 7$
29. $1, 2$ **31.** $-3, 3$ **33.** $-7, -2$ **35.** $0, 4, -3$
37. $0, -\dfrac{5}{2}, \dfrac{2}{3}$ **39.** $\dfrac{2}{3}, 4$ **41.** $-4, \dfrac{15}{4}$ **43.** $3, 4$ **45.** $-2, 7$
47. $-\dfrac{7}{2}$ **49.** 5, 11   **51.** 13, 52   **53.** 13, 14   **55.** 12 m by
21 m   **57.** 172 ft. by 54 ft.   **59.** 14 ft.   **61.** 9 in.
63. 1.25 sec.   **65.** 10%   **67.** 15   **69.** 25   **71.** 37 cm
73. 30 blocks

Review Exercises   **1.** No; It is not linear because the
variable x has an exponent other than 1.   **2.** Yes
3. $(-10, 0), (0, 4)$ **4.** $3x - y = -7$ **5. a.** 2 **b.** 6
6.

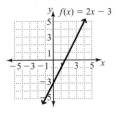

6.7 Exercises   **1.** If a vertical line is drawn through the
vertex, the sides of the parabola are mirror images.
3. The y-coordinate is c.   **5.** $(2, -1)$ **7.** $(3, 0)$ **9.** $(0, -4)$

11.

13.

15.

17.

19.

21.

23.

25.

27.

29.

31.

33.

35.

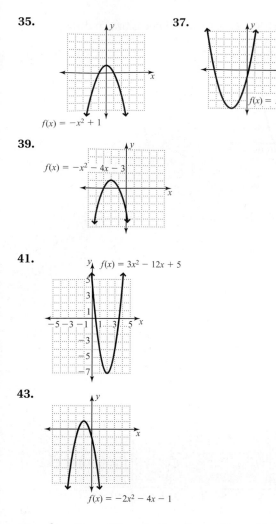

$f(x) = -x^2 + 1$

37.

$f(x) = x^2 + 4x - 1$

39.

$f(x) = -x^2 - 4x - 3$

41.

$f(x) = 3x^2 - 12x + 5$

43.

$f(x) = -2x^2 - 4x - 1$

45. a function **47.** not a function **49. a.** $(0, 0)$ **b.** $(0, 2)$ **c.** c is the y-intercept **d.** The original graph moved downward 3 units. **51. a.** $y = x^2$ opens up and $y = -x^2$ opens down. The sign of a indicates whether the parabola will open up or down. **b.** yes **c.** These graphs are "wider." **d.** The smaller the absolute value of a, the wider the parabola.

Review Exercises
1. $3m^3n - 12m^2n^2 + 15mn$
2. $2x^2 - 5x - 3$ **3.** $(3n - 5)(n - 4)$ **4.** $(x + 5)(x - 5)$
5. $0, 3$ **6.** $-6, 3$

Chapter 6 Review Exercises
1. false **2.** false
3. false **4.** true **5.** true **6. 1.** GCF **2.** $\dfrac{\text{given polynomial}}{\text{GCF}}$
7. 1. c; b **8.** $(a + b)(a - b)$ **9.** $(a - b)(a^2 + ab + b^2)$
10. $(a - b)^2$ **11.** $1, 2, 3, 4, 5, 6, 10, 12, 15, 20, 30, 60$
12. $1, 3, 9$ **13.** $1, 3, 9, 27, 81$ **14.** $1, 2, 4, 11, 22, 44$
15. $1, 3, 5, 9, 15, 45$ **16.** $1, 7, 49$ **17.** $1, 2, 3, 5, 6, 10, 15, 30$
18. $1, 3, 11, 33$ **19.** 3 **20.** 17 **21.** 6 **22.** 5 **23.** $5y^2$
24. $4x$ **25.** $3mn$ **26.** $2k$ **27.** $2(2x - 1)$
28. $5m(1 - 7m^2)$ **29.** $y(y^2 - 1)$

30. $abc(1 + abc - a^2b^2c^2)$ **31.** $xy(x + y + x^2y^2)$
32. $21a^2b^2(5a - 3b + 4a^4b^2)$
33. $10k(10k^3 + 12k^4 - 1 + 4k^2)$ **34.** $18ab^2c(b - 2a)$
35. $(x + y)(a + b)$ **36.** $(x + 2)(a + b)$
37. $(y + 2)(y^2 + 3)$ **38.** $(y + 4)(y^3 - b)$
39. $(x + 1)(y + 1)$ **40.** $(x^2 + y^2)(a - 5)$
41. $(2b^2 + 1)(b - 1)$ **42.** $(u - 3)(u + 4v)$
43. $(x - 4)(x + 3)$ **44.** $(x + 5)(x + 9)$
45. $(n - 2)(n - 4)$ **46.** prime **47.** $(h + 3)(h + 48)$
48. $(y - 12)(y + 2)$ **49.** $4(x - 3)^2$ **50.** $3(m - 9)(m - 2)$
51. prime **52.** $(2u + 1)(u + 2)$ **53.** $(3m - 1)(m - 3)$
54. $(5k - 12h)(k + h)$ **55.** $2(3a - 4)(a - 2)$
56. $y(8x^2 - 4x - 1)$ **57.** $(3p - q)(p - 4q)$
58. $8(3x + y)(x - 3y)$ **59.** $(v - 4)^2$ **60.** $(u + 3)^2$
61. $(2x + 5)^2$ **62.** $(3y - 2)^2$ **63.** $(x + 2)(x - 2)$
64. $(5 + y)(5 - y)$ **65.** $(x - 1)(x^2 + x + 1)$
66. $(x + 3)(x^2 - 3x + 9)$ **67.** $(3b + 2)(2b - 1)$
68. $4ab(1 - 6b)$ **69.** prime **70.** $3(x + y)(x - y)$
71. $(x^2 + 9)(x + 3)(x - 3)$ **72.** $7(u - 5)(u + 3)$
73. $(2x - 3y)(4x^2 + 6xy + 9y^2)$ **74.** $2(1 + 5y)(1 - 5y)$
75. $(m - 2n)(3a + 4b)$ **76.** $3(m^2 + 3m + 9)$ **77.** $-3, 4$
78. $-2, 2$ **79.** $2, 3$ **80.** $-3, 7$ **81.** $-3, 3$ **82.** $-6, 3$

83. -3 **84.** $-\dfrac{1}{3}, 4$ **85.** $5x^2 = 2x; 0, \dfrac{2}{5}$

86. $x^2 + (x + 1)^2 = 61; 5, 6$ **87.** $x \cdot 4x = 100; -5, 5$
88. $(w + 6)w = 91; 13$ in. by 7 in. **89.** $x(x + 1) = 110;$
$10, 11$ **90.** $3^2 + b^2 = 5^2; 4$
91. **92.**

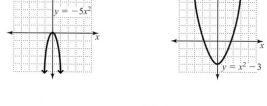

93. **94.**

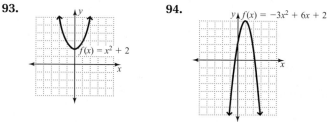

Chapter 6 Practice Test
1. $1, 2, 4, 5, 8, 10, 16, 20,$
$40, 80$ **2.** $5m$ **3.** $5(y - 6)$ **4.** $2y(3x^2 - y)$
5. $(x - y)(a - b)$ **6.** $(m + 3)(m + 4)$ **7.** $(r - 4)^2$
8. $(2y + 5)^2$ **9.** $(q - 2)(3q - 4)$ **10.** $(3x - 4)(2x - 5)$
11. $a(x + 3)(x - 8)$ **12.** $2n(n + 4)(5n - 1)$
13. $(c + 5)(c - 5)$ **14.** prime **15.** $2(1 + 5u)(1 - 5u)$
16. $-4(x + 2)(x - 2)$ **17.** $(m + 5)(m^2 - 5m + 25)$
18. $(x - 2)(x^2 + 2x + 4)$ **19.** $-3, 0$ **20.** $6, -2$

21. $\dfrac{3}{2}, -5$ **22.** $8, 9$ **23.** 9 ft., 4 ft.

24.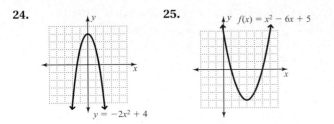

25. $f(x) = x^2 - 6x + 5$

Chapters 1–6 Cumulative Review Exercises
1. false **2.** true **3.** false **4.** false **5.** false **6.** false **7. a.** GCF **e.** grouping **8.** highest **9.** $(a + b)(a - b)$ **10. a.** 0 **b.** factored **11.** An x-intercept is a point where a graph intersects the x-axis. A y-intercept is a point where a graph intersects the y-axis. **12.** a polynomial containing two terms **13.** 42 **14.** -89 **15.** $12m^2 - 6m - 7$ **16.** $5x^2 + 4x + 2$ **17.** $18x^4y$ **18.** $48x^7$ **19.** $2y^2 - 15y - 8$
20. $4x^2 + 20x + 25$ **21.** u **22.** $4h + 1 - \dfrac{2}{h}$ **23.** $x + 4$
24. $x^2 - 2x + 4 - \dfrac{56}{x + 2}$ **25.** $(x + 11)(x - 11)$
26. $10(x^2 + 4)$ **27.** $(x - 4)^2$ **28.** $(2x - 3)(x + 4)$
29. $(x - 5)(x^2 + 5)$ **30.** $(n - 2)(7m + 6)$
31. $(2x - 3)(4x^2 + 6x + 9)$ **32.** $(3x + 2)(x + 2)$
33. 10 **34.** $\dfrac{17}{12}$ **35.** 13 **36.** 13.2 **37.** $-10, 10$ **38.** $-4, 7$
39. a. $\{x | x \le 4\}$ **b.** $(-\infty, 4]$
c.

$$\xleftarrow{\hspace{0.5em}} \overset{\hspace{0.5em}-5\;-4\;-3\;-2\;-1\;\;0\;\;1\;\;2\;\;3\;\;4\;\;5}{\hspace{0.5em}} \rightarrow$$

40. a. $\{x | x > -2\}$ **b.** $(-2, \infty)$
c.

$$\xleftarrow{\hspace{0.5em}} \overset{\hspace{0.5em}-5\;-4\;-3\;-2\;-1\;\;0\;\;1\;\;2\;\;3\;\;4\;\;5}{\hspace{0.5em}} \rightarrow$$

41. $x = \dfrac{7 + 3y}{2}$ **42.** no **43.** $(3, 0), (0, -6)$ **44.** 0
45.

$y = 3x + 2$

46.

$2x - y = 6$

47. ≈ 686 **48.** 6, 8, 10 **49.** \$2500 in 5%, \$5500 in 8%
50. $33\dfrac{1}{3}$ ml

Chapter 7

7.1 Exercises
1. We replace the variables with given values, and then simplify the numerical expression.
3. Write an equation that has the denominator set equal to zero, and solve the equation. **5.** y is not a factor in the numerator, $x + y$. **7. a.** $\dfrac{16}{45}$ **b.** -1 **c.** 0 **9. a.** 1

b. undefined **c.** -3 **11. a.** $\dfrac{11}{8}$ **b.** $-\dfrac{1}{7}$
c. $-\dfrac{24}{49}$ or ≈ -0.49 **13. a.** $-\dfrac{4}{3}$ **b.** $-\dfrac{3}{2}$ **c.** $-\dfrac{341}{30}$ or $-11.3\overline{6}$
15. 3 **17.** $-3, 3$ **19.** $5, 9$ **21.** $0, 3$ **23.** $-1, \dfrac{5}{2}$ **25.** $\dfrac{3x}{4y}$
27. $-\dfrac{4m^2}{n}$ **29.** $\dfrac{3}{2x^2z^2}$ **31.** $-\dfrac{3m^2n}{m + 2n}$ **33.** $\dfrac{x + 5}{2}$ **35.** $\dfrac{3}{5}$
37. $\dfrac{2}{3}$ **39.** $\dfrac{5}{3}$ **41.** $\dfrac{x}{4}$ **43.** x **45.** $\dfrac{5}{x}$ **47.** $\dfrac{x - y}{x + y}$ **49.** $\dfrac{x}{x - 2}$
51. $\dfrac{t - 3}{t + 2}$ **53.** $\dfrac{2a - 3}{3a - 2}$ **55.** $\dfrac{x - 4}{x + 2}$ **57.** $\dfrac{p + q}{m - n}$
59. $\dfrac{2u^2 - 1}{u + 1}$ **61.** -4 **63.** $-y - 2$ **65.** $-\dfrac{1}{w + 1}$
67. $\dfrac{3}{x^2 - 2x + 4}$ **69.** $\dfrac{x + y}{x}$ **71.** $-\dfrac{1}{m^2 + 3m + 9}$
73. Mistake: Placed remaining x in numerator. Correct: $-\dfrac{1}{2x}$
75. Mistake: Divided out a part of a multiterm polynomial. Correct: Can't reduce **77.** Mistake: $y - x$ and $x - y$ are not identical factors until -1 is factored out of one of them. Correct: $-x$ **79.** The first two are normal weight and the third is overweight. **81.** $533.\overline{3}\pi, 666.\overline{6}\pi, 666.\overline{6}\pi, 833.\overline{3}\pi$
83. a. $h = \dfrac{3V}{\pi r^2}$ **b.** ≈ 9 cm **85.** $\dfrac{s - 3}{s + 4}$

Review Exercises
1. 1 **2.** $-\dfrac{3}{4}$ **3.** $\dfrac{9}{20}$ **4.** $\dfrac{28}{15}$
5. $(x - y)(a + b)$ **6.** $(3x - 2)(x + 5)$

7.2 Exercises
1. To eliminate any common factors.
3. Their product is 1. **5.** A fraction with a ratio equivalent to 1. Examples will vary. **7.** $\dfrac{2m^2}{5n^2}$ **9.** $\dfrac{s^2}{r^3}$ **11.** y^2 **13.** $-\dfrac{1}{mn^2s}$
15. $\dfrac{9r}{s}$ **17.** $\dfrac{mn^2}{2}$ **19.** $\dfrac{9}{10}$ **21.** $\dfrac{4}{15}$ **23.** $(2x - 5)(3x + 4)$
25. $\dfrac{r^2}{(r + 6)(r - 3)}$ **27.** $\dfrac{q + 1}{2q(q - 1)}$ **29.** $\dfrac{w - 1}{(w + 1)(w - 5)}$
31. $\dfrac{n + 1}{n}$ **33.** 1 **35.** $\dfrac{(d - 5)(a + 2)}{(b - 4)(d + 1)}$ **37.** $\dfrac{1}{y}$ **39.** $-\dfrac{4y^2}{3x}$
41. $1 - 3x$ **43.** $\dfrac{1}{x^2y^2}$ **45.** $\dfrac{5}{4m^2n}$ **47.** $\dfrac{y^2}{5z^2}$ **49.** $\dfrac{-32x^6}{45}$
51. $\dfrac{4a}{9}$ **53.** $\dfrac{8}{5}$ **55.** $\dfrac{a - b}{x + y}$ **57.** $\dfrac{x + 8}{x - 8}$ **59.** $x - 4$
61. $\dfrac{a + 2b}{a - 2b}$ **63.** $\dfrac{2k + 1}{k(2k - 1)}$ **65.** $\dfrac{x + 2}{(x + 1)(3x - 1)}$
67. $\dfrac{b - 4}{c - 3}$ **69.** $-b - 5$ **71.** $\dfrac{7 + y}{y(y + 4)}$ **73.** $-\dfrac{x^3}{x - 3}$
75. $-\dfrac{x}{y}$ **77.** Mistake: Did not invert the divisor. Correct: $\dfrac{50}{27}$
79. Mistake: Divided out y, which is illegal because it is not a factor in $y - 2$. Correct: $\dfrac{(x - 3)(y - 2)}{15y}$ **81.** 222 in.

83. 10,560 yd. **85.** 288 in. **87.** 88 oz. **89.** 12.7 T
91. 96,000 oz. **93.** 260.48 ft./sec. **95.** $\approx$ 19.03 mi./hr.
97. 939,785,328 km **99.** $\approx 5.87 \times 10^{12}$ mi./yr.
101. 259.55 £ **103.** \$604.03 **105.** 932.85 €

Review Exercises
1. 1 **2.** $-\dfrac{1}{2}$ **3.** $2x$ **4.** $\dfrac{1}{4}y - 4$
5. $8x^2 - \dfrac{2}{3}x - 4$ **6.** $-\dfrac{1}{5}n^2 + 4$

7.3 Exercises
1. The numerator may not contain like terms. **3.** Write an equivalent addition with the additive inverse of the subtrahend. **5.** $\dfrac{x}{2}$ **7.** $\dfrac{a^2}{3}$ **9.** $\dfrac{3}{a}$ **11.** $-\dfrac{x}{7y^2}$
13. $\dfrac{9}{n+5}$ **15.** $\dfrac{3r}{r+2}$ **17.** $\dfrac{q+1}{q}$ **19.** $\dfrac{t+11}{5u}$ **21.** 3
23. $\dfrac{11}{w+1}$ **25.** $\dfrac{3m-1}{m-1}$ **27.** $\dfrac{-11t+3}{2t-3}$ **29.** $\dfrac{6w-5}{w+5}$
31. $\dfrac{1}{g+3}$ **33.** $z-3$ **35.** $\dfrac{1}{s+2}$ **37.** $\dfrac{y-3}{y}$ **39.** $\dfrac{x+2}{x-4}$
41. $\dfrac{t-1}{t+7}$ **43.** $\dfrac{3x+9}{x^2-25}$ **45.** $\dfrac{3x-2}{x-5}$ **47.** $\dfrac{x-3}{x+5}$
49. Mistake: Combined terms that are not like. Correct: $\dfrac{4y+1}{5}$ **51.** Mistake: Did not change sign of -5. Correct: 2
53. $\dfrac{5x+9}{2}$ **55.** $\dfrac{3x+2}{2}$

Review Exercises
1. $\dfrac{11}{10}$ **2.** $\dfrac{11}{24}$ **3.** $1\dfrac{6}{35}$
4. $\dfrac{11}{15}x + \dfrac{7}{12}$ **5.** $2x^2 - 12x - \dfrac{11}{24}$ **6.** $2y^2 - \dfrac{1}{20}y - 10$

7.4 Exercises
1. 1. Factor each denominator. 2. For each unique factor, compare the number of times it appears in each factorization. Write a factored form that includes each factor the greatest number of times it appears in the denominator factorizations. 3. Factor the denominators. **5.** Rewrite the numerator as an equivalent addition sentence.
7. $15a^2$; $\dfrac{10a}{15a^2}$, $\dfrac{9}{15a^2}$ **9.** $90m^3n^3$; $\dfrac{48}{90m^3n^3}$, $\dfrac{35mn^2}{90m^3n^3}$
11. $(y+2)(y-2)$; $\dfrac{3y-6}{(y+2)(y-2)}$, $\dfrac{4y^2+8y}{(y+2)(y-2)}$
13. $12(y-5)$; $\dfrac{9y}{12(y-5)}$, $\dfrac{14y}{12(y-5)}$
15. $6x(x+3)$; $\dfrac{5x}{6x(x+3)}$, $\dfrac{12}{6x(x+3)}$
17. p^2-4; $\dfrac{2}{p^2-4}$, $\dfrac{p^2+p-6}{p^2-4}$
19. $3(u+2)^2$; $\dfrac{12u}{3(u-2)^2}$, $\dfrac{2u+4}{3(u+2)^2}$
21. $\dfrac{8t^3-8t^2-t+1}{(t+6)(t-6)(t-1)}$, $\dfrac{t^2+6t}{(t+6)(t-6)(t-1)}$

23. $\dfrac{x^2+4x+4}{(x+4)(x-1)(x+2)}$, $\dfrac{x^2-4x+3}{(x+4)(x-1)(x+2)}$
25. $\dfrac{2x-4}{(x-1)(x-2)}$, $\dfrac{3}{(x-1)(x-2)}$, $\dfrac{5x^2-5x}{(x-1)(x-2)}$
27. $\dfrac{-31}{18}$ **29.** $\dfrac{8}{3c}$ **31.** $\dfrac{5x^2-2}{x^3}$ **33.** $\dfrac{8y-2}{(y+2)(y-4)}$
35. $\dfrac{x+10}{(x-2)(x+2)}$ **37.** $\dfrac{2x+14}{(x+5)(x-5)}$ **39.** $\dfrac{10p-8}{p^2-2p+1}$
41. $\dfrac{-z^2-5z-3}{(z+4)(z-3)}$ **43.** $\dfrac{2n^2-2n+18}{(n-4)(n+3)}$
45. $\dfrac{5t+16}{(t+4)(t+6)}$ **47.** $\dfrac{3}{2(c-5)}$ **49.** $\dfrac{y+2}{y}$
51. $\dfrac{r^2-6r-1}{(r+1)(r-1)^2}$ **53.** $\dfrac{q+1}{q(q-3)}$ **55.** $\dfrac{1}{(w-1)(w-2)}$
57. $\dfrac{v^2+16v+51}{(v+4)(v-4)(v+3)}$ **59.** $\dfrac{-3}{t-3}$ **61.** $\dfrac{4-x}{2x-1}$
63. $\dfrac{y+6}{3(y-2)}$ **65.** Mistake: Did not find a common denominator. Correct: $\dfrac{6-x^2}{2x}$ **67.** Mistake: Did not change the sign of 7. Correct: $\dfrac{-x-7}{x+2}$ **69.** Mistake: Added denominators.
Correct: $\dfrac{7w}{x}$ **71.** $\dfrac{5t}{6}$ **73.** $\dfrac{4x+6}{x^2+3x}$ **75.** $\dfrac{4ah+5bh}{8}$
77. $\dfrac{6x^2+8x+2}{(2x-1)(2x+1)}$ **79.** $\dfrac{2a^2+2b^2}{(a+b)^2}$ **81.** xy **83.** $\dfrac{m^2+1}{m^4}$
85. $\dfrac{x-5}{3(x+2)}$

Review Exercises
1. $\dfrac{3}{8}$ **2.** $\dfrac{m^2}{108}$ **3.** $4(x+2)(x-2)$
4. $3y(2y+1)(y-5)$ **5.** $\dfrac{x-1}{x+2}$ **6.** $\dfrac{1}{x^2-3x+9}$

7.5 Exercises
1. $\dfrac{4}{x+3}$ is the numerator. $\dfrac{3}{x}$ is the denominator. **3.** $\dfrac{\dfrac{2x}{x-3}}{\dfrac{4-x}{2x-6}}$ **5.** In this case, multiplying the numerator and denominator by their LCD, 12, (method 2) is preferable because doing so requires fewer steps than simplifying $\dfrac{x}{6} - \dfrac{3}{4}$ and $\dfrac{2}{3} + x$ and then dividing the simplified expressions (method 1). (Answers may vary.) **7.** $\dfrac{1}{2}$ **9.** $\dfrac{mp}{nq}$
11. 5 **13.** $\dfrac{5a+2}{3a+4}$ **15.** $\dfrac{1-y}{1+y}$ **17.** $\dfrac{4r-3}{2-4r}$ **19.** $\dfrac{x}{3+x}$
21. xy **23.** $\dfrac{x}{x-2}$ **25.** $\dfrac{12x+6}{3x^2-2x}$ **27.** $\dfrac{a}{b(a-3)}$

29. $\dfrac{2y(y-5)}{5}$ **31.** u **33.** $\dfrac{1}{2a-1}$ **35.** $\dfrac{1}{3}$ **37.** 2

39. $y-1$ **41.** $\dfrac{1}{n-1}$ **43.** $3-y$ **45.** $\dfrac{x-1}{x}$

47. $\dfrac{-2}{x(x+h)}$ **49.** Mistake: Divided out $\dfrac{1}{2}$, which is not a

common factor. Correct: $\dfrac{2x-1}{2y-1}$ **51.** Mistake: Did not write

as an equivalent multiplication. Correct: $\dfrac{m^2}{18}$ **53.** Mistake:

Did not distribute c in the numerator. Correct: c^2+1
55. Mistake: Added the x's instead of multiplying them.

Correct: $\dfrac{18}{x^2}$ **57.** $34\dfrac{2}{7}$ mph **59.** $\dfrac{3(5n-2)}{4(n-1)}$

61. a. $\dfrac{R_1 R_2}{R_2+R_1}$ **b.** ≈ 9.1 **c.** 8

Review Exercises **1.** -7 **2.** 3 **3.** 2.5 **4.** 4
5. 4.8 mph **6.** 0.09 hr. or 5.4 min.

7.6 Exercises **1.** Multiply both sides of the equation by
the LCD of the rational expressions. **3.** An extraneous solution causes an expression in the equation to be undefined.
5. 4 and -2; Substituting either of those numbers for x
causes an expression in the equation to be undefined. **7.** no
9. yes **11.** yes **13.** 2 **15.** 10 **17.** 15 **19.** 3 **21.** 1

23. 10 **25.** 8 **27.** $\dfrac{1}{2}$ **29.** -1 **31.** $-\dfrac{3}{4}$ **33.** 9 **35.** $-2, \dfrac{1}{3}$

37. -5 **39.** $-\dfrac{11}{3}$, (-1 is extraneous) **41.** $4, -2$

43. $-\dfrac{1}{4}, 5$ **45.** $-2, 9$ **47.** 1, (2 is extraneous) **49.** $0, \dfrac{9}{2}$

51. $-3, -\dfrac{2}{3}$ **53.** -2 **55.** 1 **57.** $6, -1$ **59.** no solution

61. 5, (1 is extraneous) **63.** -3 is the only answer, 3 is extraneous. **65.** In an expression, the LCD is used to combine the
numerator over a single denominator. In an equation, the
LCD is used to eliminate the denominator. **67.** 400 **69.** 20
and 30 **71.** ≈ 2.1 ft. **73.** $f = 15$ mm, $i = 20$ mm

Review Exercises **1.** ≈ 68.6 mph **2.** $\approx \$0.084$/oz.
3. 6 hr. **4.** 72, 74 **5.** The length is 11 mm and the width is
7 mm. **6.** 0.15 hr. (9 min.)

7.7 Exercises **1.** $\dfrac{1}{x}$ **3.** $\dfrac{100}{r}$ **5.** increases **7.** $4\dfrac{4}{5}$ days

9. 12 min. **11.** $2\dfrac{2}{5}$ hr. **13.** $2\dfrac{8}{11}$ hr. **15.** 6 mph, 4 mph

17. 340 mph **19.** 60 mph, 20 mph **21.** 16 mph, 20 mph
23. 30 **25.** 4.5 **27.** \$25.76 **29.** 15 gal. **31.** 144.9 lb.
33. 660 vps **35.** 420 in.³ **37.** 5 **39.** 14 **41.** 30 psi
43. 4 A **45.** ≈ 4.39 million pounds **47.** $2.\overline{3}$ cm/sec.

49. a.

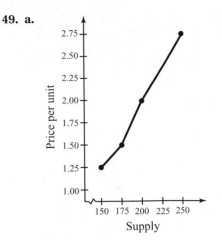

b. Supply and price are directly proportional. As supply
increased, so did the price per unit. **c.** Yes; each value in the
domain (supply) is paired with one value in the range (price
per unit).

Review Exercises **1.** 5, 6 **2.** no

3.

4. $\left(-\dfrac{3}{2}, 0\right)$, $(0, 6)$ **5.** $\dfrac{17}{4}$ **6.** $3a-4$

Chapter 7 Review Exercises **1.** false **2.** true
3. true **4.** true **5.** true **6. 1.** multiplication **2.** Divide
7. 1. numerator; denominator **8.** complex
9. 1. numerator; denominator **10. 1.** LCD **11. a.** $\dfrac{17}{6}$

b. ≈ 0.077 **c.** $\dfrac{3}{4}$ **12. a.** 3 **b.** 15 **c.** 1.875 **13.** 5

14. $-2, 2$ **15.** $-6, 1$ **16.** $0, -6$ **17.** $\dfrac{x+5}{3}$ **18.** $\dfrac{2}{3}$ **19.** $\dfrac{x}{4}$

20. $\dfrac{y}{3}$ **21.** $\dfrac{a+b}{a-b}$ **22.** $\dfrac{3}{(x^2+9)(x-3)}$ **23.** $-(x+y)$

24. $-\dfrac{1}{w+3}$ **25.** $\dfrac{7x^2}{4y^2}$ **26.** $-\dfrac{3}{5mnp^4}$ **27.** $\dfrac{3}{5}$

28. $\dfrac{x^2}{(x+4)(x-3)}$ **29.** $\dfrac{n+1}{n}$ **30.** $\dfrac{2}{r-1}$

31. $\dfrac{m+1}{4m(m-1)}$ **32.** $\dfrac{w+4}{w-4}$ **33.** $\dfrac{y^3}{abx}$ **34.** $\dfrac{5}{6}$ **35.** 2

36. $\dfrac{5a}{2}$ **37.** $\dfrac{x+y}{c-d}$ **38.** $\dfrac{5(b+4)(b-4)}{6}$ **39.** $\dfrac{2j+1}{j(2j-1)}$

40. $\dfrac{1}{(u+1)^2}$ **41.** $\dfrac{x}{3}$ **42.** $\dfrac{3r}{r+3}$ **43.** $-\dfrac{1}{3m}$ **44.** $\dfrac{3x+1}{x+1}$

45. 1 **46.** $\dfrac{s}{r^2-s^2}$ **47.** $\dfrac{11x^2-21x+7}{2x+1}$ **48.** $q-3$

49. $\dfrac{6c^2}{12c^3}, \dfrac{12}{12c^3}$ **50.** $\dfrac{5x}{x^2y^3}, \dfrac{3y}{x^2y^3}$ **51.** $\dfrac{4m+4}{(m-1)(m+1)},$ $\dfrac{4my-4y}{(m-1)(m+1)}$ **52.** $\dfrac{3h}{4h(h-2)}, \dfrac{20}{4h(h-2)}$

53. $\dfrac{x}{(x+1)(x-1)}, \dfrac{x^2+x-2}{(x+1)(x-1)}$

54. $\dfrac{2p+4}{(p-2)(p+2)^2}, \dfrac{p^2-2p}{(p-2)(p+2)^2}$ **55.** $-\dfrac{23}{36}$

56. $\dfrac{-x-27}{(x+3)(x-3)}$ **57.** $\dfrac{8y-22}{(y-2)(y-4)}$

58. $\dfrac{4x-3}{(x+2)(x-2)}$ **59.** $\dfrac{-3t^2+21t}{2(t-3)^2}$ **60.** $\dfrac{18-a^2}{(a+3)(a+4)}$

61. $\dfrac{15}{2y-1}$ **62.** $\dfrac{3x+57}{3(x+4)(x-4)}$ **63.** $\dfrac{ay}{bx}$ **64.** -1

65. $\dfrac{5(x-2)}{x(3x+5)}$ **66.** $\dfrac{x}{x-3}$ **67.** $\dfrac{x}{5(2x-1)}$ **68.** 1

69. $\dfrac{y^2+x^2}{7xy}$ **70.** $-h$ **71.** $\dfrac{38}{13}$ **72.** 6 **73.** 1 **74.** 4

75. $3, 4$ **76.** -1 **77.** -8, (4 is extraneous) **78.** 23
79. 31.875 min. **80.** 8 yd./sec. and 10 yd./sec. **81.** 30
82. 2 **83.** 10 gal. **84.** 4.8 A

Chapter 7 Practice Test **1.** 7 **2.** $-4, 4$ **3.** $-\dfrac{3}{x-4}$

4. $\dfrac{1}{x+y}$ **5.** $\dfrac{6f}{f^2g^3}, \dfrac{2g^3}{f^2g^3}$ **6.** $\dfrac{5x(x+3)}{(x+3)^2(x-3)},$

$\dfrac{2(x-3)}{(x+3)^2(x-3)}$ **7.** $\dfrac{5ab^2x^2}{12}$ **8.** $-\dfrac{7a^6b^7y^2}{40x^2}$ **9.** $-4x$

10. $\dfrac{12x^2}{(x+1)(2x+1)}$ **11.** $\dfrac{7x}{2x+3}$ **12.** $\dfrac{1}{x+3}$

13. $\dfrac{x^2+5x-6}{x(x+2)(x-2)}$ **14.** $\dfrac{9x+3}{(x-1)(x+2)}$

15. $\dfrac{3x+13}{(x-4)(x+1)}$ **16.** $\dfrac{x-2}{2x}$ **17.** $\dfrac{2y+1}{3y-1}$ **18.** $\dfrac{2}{x+y}$

19. $\dfrac{25}{9}$ **20.** -1 **21.** -1, (2 is extraneous) **22.** 2, 4

23. 1.2 hr. **24.** 6 mph, 4 mph **25.** 529.2 N

Chapter 8

8.1 Exercises **1.** Two inequalities joined by either "and" or "or." **3.** For two sets A and B, the intersection of A and B, symbolized by $A \cap B$, is a set containing only elements that are in both A and B. **5.** We graph the region of overlap of the two inequalities. **7. a.** $\{1, 3, 5\}$ **b.** $\{1, 3, 5, 7, 9\}$ **9. a.** $\{7\}$ **b.** $\{5, 6, 7, 8, 9\}$ **11. a.** $\varnothing$ **b.** $\{a, c, d, g, o, t\}$ **13. a.** $\{x, y, z\}$ **b.** $\{w, x, y, z\}$ **15.** $-4 < x < 5$ **17.** $-2 < y \le 0$

19. $-7 < w < 3$ **21.** $0 \le u \le 2$

23. (number line graph)

25. (number line graph)

27. (number line graph)

29. (number line graph)

31. a. (number line graph) **b.** $\{x \mid -3 < x < -1\}$ **c.** $(-3, -1)$

33. a. (number line graph) **b.** $\{x \mid 3 < x < 6\}$ **c.** $(3, 6]$

35. a. (number line graph) **b.** $\{\ \}$ or $\varnothing$ **c.** no interval notation

37. a. (number line graph) **b.** $\{x \mid -5 \le x < 3\}$ **c.** $[-5, 3)$

39. a. (number line graph) **b.** $\{\ \}$ or $\varnothing$ **c.** no interval notation

41. a. (number line graph) **b.** $\{x \mid -7 < x < -3\}$ **c.** $(-7, -3)$

43. a. (number line graph) **b.** $\{x \mid -1 \le x \le 2\}$ **c.** $[-1, 2]$

45. a. (number line graph) **b.** $\left\{x \,\middle|\, -\dfrac{2}{3} \le x < 2\right\}$ **c.** $\left[-\dfrac{2}{3}, 2\right)$

47. a. (number line graph) **b.** $\{x \mid 0 \le x < 3\}$ **c.** $[0, 3)$

49. a. (number line graph) **b.** $\{x \mid 0 \le x \le 3\}$ **c.** $[0, 3]$

51. (number line graph)

53. (number line graph)

55. (number line graph)

57. (number line graph)

59. a. (number line graph) **b.** $\{y \mid y < -9 \text{ or } y > 5\}$ **c.** $(-\infty, -9) \cup (5, \infty)$

61. a.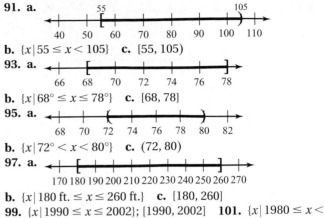
number line with open parenthesis at -2 and open parenthesis at 1
$-6\ -5\ -4\ -3\ -2\ -1\ 0\ 1\ 2\ 3\ 4\ 5\ 6$
b. $\{r \mid r < -2 \text{ or } r > 1\}$ **c.** $(-\infty, -2) \cup (1, \infty)$

63. a.
$-6\ -5\ -4\ -3\ -2\ -1\ 0\ 1\ 2\ 3\ 4\ 5\ 6\ 7\ 8$
b. $\{w \mid w \le -1 \text{ or } w \ge 7\}$ **c.** $(-\infty, -1] \cup [7, \infty)$

65. a.
$-6\ -5\ -4\ -3\ -2\ -1\ 0\ 1\ 2\ 3\ 4\ 5\ 6$
b. $\{k \mid k \le 2 \text{ or } k \ge 3\}$ **c.** $(-\infty, 2] \cup [3, \infty)$

67. a.
$-\frac{6}{5}\ -1\ -\frac{4}{5}\ -\frac{3}{5}\ -\frac{2}{5}\ -\frac{1}{5}\ 0\ \frac{1}{5}\ \frac{2}{5}\ \frac{3}{5}\ \frac{4}{5}\ 1\ \frac{6}{5}$
b. $\left\{ x \le x \le -\frac{3}{5} \text{ or } x \ge \frac{1}{5} \right\}$ **c.** $\left(-\infty, -\frac{3}{5} \right] \cup \left[\frac{1}{5}, \infty \right)$

69. a.
$-6\ -5\ -4\ -3\ -2\ -1\ 0\ 1\ 2\ 3\ 4\ 5\ 6$
b. $\left\{ c \mid c > 3 \right\}$ **c.** $(3, \infty)$

71. a.
$\frac{8}{3}$
$-6\ -5\ -4\ -3\ -2\ -1\ 0\ 1\ 2\ 3\ 4\ 5\ 6$
b. $\left\{ m \mid m \le 1 \text{ or } m \ge \frac{8}{3} \right\}$ **c.** $(-\infty, 1] \cup \left[\frac{8}{3}, \infty \right)$

73. a.
$-6\ -5\ -4\ -3\ -2\ -1\ 0\ 1\ 2\ 3\ 4\ 5\ 6$
b. $\{x \mid x \le -4\}$ **c.** $(-\infty, -4)$

75. a.
$-6\ -5\ -4\ -3\ -2\ -1\ 0\ 1\ 2\ 3\ 4\ 5\ 6$
b. $\{x \mid x \text{ is a real number}\}$ or $\mathbb{R}$ **c.** $(-\infty, \infty)$

77. a.
$-7\ -6\ -5\ -4\ -3\ -2\ -1\ 0\ 1\ 2\ 3\ 4\ 5$
b. $\{x \mid -7 < x < 3\}$ **c.** $(-7, 3)$

79. a.
$-6\ -5\ -4\ -3\ -2\ -1\ 0\ 1\ 2\ 3\ 4\ 5\ 6$
b. $\{x \mid x < -2 \text{ or } x > 2\}$ **c.** $(-\infty, -2) \cup (2, \infty)$

81. a.
$-6\ -5\ -4\ -3\ -2\ -1\ 0\ 1\ 2\ 3\ 4\ 5\ 6$
b. $\{x \mid -3 < x < -2\}$ **c.** $(-3, -2)$

83. a.
$-6\ -5\ -4\ -3\ -2\ -1\ 0\ 1\ 2\ 3\ 4\ 5\ 6$
b. $\{x \mid 3 \le x \le 6\}$ **c.** $[3, 6]$

85. a.
$-6\ -5\ -4\ -3\ -2\ -1\ 0\ 1\ 2\ 3\ 4\ 5\ 6$
b. $\{x \mid 1 < x < 3\}$ **c.** $(1, 3)$

87. a.
$-6\ -5\ -4\ -3\ -2\ -1\ 0\ 1\ 2\ 3\ 4\ 5\ 6$
b. $\{x \mid x > 1\}$ **c.** $(1, \infty)$

89. a.
$333.\overline{3}$
$250\ 300\ 350\ 400\ 450\ 500\ 550$
b. $\{x \mid 333.\overline{3} \le x \le 500\}$ **c.** $[333.\overline{3}, 500]$

91. a.
$55 \qquad 105$
$40\ 50\ 60\ 70\ 80\ 90\ 100\ 110$
b. $\{x \mid 55 \le x < 105\}$ **c.** $[55, 105)$

93. a.
$66\ 68\ 70\ 72\ 74\ 76\ 78$
b. $\{x \mid 68° \le x \le 78°\}$ **c.** $[68, 78]$

95. a.
$68\ 70\ 72\ 74\ 76\ 78\ 80\ 82$
b. $\{x \mid 72° < x < 80°\}$ **c.** $(72, 80)$

97. a.
$170\ 180\ 190\ 200\ 210\ 220\ 230\ 240\ 250\ 260\ 270$
b. $\{x \mid 180 \text{ ft.} \le x \le 260 \text{ ft.}\}$ **c.** $[180, 260]$
99. $\{x \mid 1990 \le x \le 2002\}$; $[1990, 2002]$ **101.** $\{x \mid 1980 \le x < 1988\}$; $[1980, 1988)$ **103.** $\{x \mid 1989 \le x < 2000 \text{ or } 2001 \le x \le 2002\}$; $[1989, 2000) \cup [2001, 2002]$

Review Exercises
1. No, the absolute value of zero is zero and zero is neither negative or positive. **2.** -11
3. -16 **4.** 6 **5.** $\frac{1}{5}$ **6.** -2

8.2 Exercises
1. The units a number is from zero.
3. If $|n| = a$, where n is a variable or an expression and $a \ge 0$, then $n = a$ or $n = -a$. **5.** Separate the absolute value equation into two equations: $ax + b = cx + d$ and $ax + b = -(cx + d)$. **7.** $-2, 2$ **9.** no solution **11.** $-11, 5$
13. $2, 3$ **15.** $1, \frac{7}{5}$ **17.** $-\frac{2}{3}, \frac{10}{3}$ **19.** no solution **21.** $\frac{3}{4}$
23. $-4, 4$ **25.** $-1, 3$ **27.** $-12, 4$ **29.** $-\frac{3}{5}, 1$ **31.** $-4, 1$
33. $-2, 6$ **35.** $-\frac{9}{2}, \frac{15}{2}$ **37.** $0, 20$ **39.** $-2, 4$ **41.** $\frac{1}{3}, 7$
43. $-5, -\frac{3}{5}$ **45.** all real numbers **47.** 1 **49.** $-3, 13$
51. $-6, 10$ **53.** $\frac{5}{6}, \frac{11}{6}$ **55.** $-\frac{7}{2}, 2$

Review Exercises
1. $x > -7$ **2.** $x \ge -3$
3. $n - 2 < 5; x < 7$ **4.** $(-2, 0]$
5.
$-6\ -5\ -4\ -3\ -2\ -1\ 0\ 1\ 2\ 3\ 4\ 5\ 6$
6. $-2 < x < \frac{14}{3}$

8.3 Exercises
1. $x \ge -a$ and $x \le a$ **3.** Shade between the values of $-a$ and a. **5.** when $a \le 0$
7. a.
$-6\ -5\ -4\ -3\ -2\ -1\ 0\ 1\ 2\ 3\ 4\ 5\ 6$
b. $\{x \mid -5 < x < 5\}$ **c.** $(-5, 5)$
9. a.
$-12\ -10\ -8\ -6\ -4\ -2\ 0\ 2\ 4$
b. $\{x \mid -10 \le x \le 4\}$ **c.** $[-10, 4]$

11. a.
b. $\{s \mid -6 < s < 0\}$ **c.** $(-6, 0)$
13. a.
b. $\{m \mid -2 < m < 7\}$ **c.** $(-2, 7)$
15. a.
b. $\left\{k \mid \dfrac{4}{3} \le k \le 2\right\}$ **c.** $\left[\dfrac{4}{3}, 2\right]$
17. a.
b. $\{x \mid -5 \le x \le 5\}$ **c.** $[-5, 5]$
19. a.
b. $\{w \mid 0 < w < 6\}$ **c.** $(0, 6)$
21. a.
b. $\{c \mid c < -12 \text{ or } c > 12\}$ **c.** $(-\infty, -12) \cup (12, \infty)$
23. a.
b. $\{y \mid y \le -9 \text{ or } y \ge 5\}$ **c.** $(-\infty, -9] \cup [5, \infty)$
25. a.
b. $\{p \mid p < -2 \text{ or } p > 14\}$ **c.** $(-\infty, -2) \cup (14, \infty)$
27. a.
b. $\{x \mid x \le -6 \text{ or } x \ge 2\}$ **c.** $(-\infty, -6] \cup [2, \infty)$
29. a.
b. $\left\{n \mid n < -\dfrac{5}{2} \text{ or } n > 0\right\}$ **c.** $\left(-\infty, -\dfrac{5}{2}\right) \cup (0, \infty)$
31. a.
b. $\{v \mid v \le -1 \text{ or } v \ge 1\}$ **c.** $(-\infty, -1] \cup [1, \infty)$
33. a.
b. $\{y \mid y < -3 \text{ or } y > -1\}$ **c.** $(-\infty, -3) \cup (-1, \infty)$
35. a.
b. $\{m \mid m < -5 \text{ or } m > 1\}$ **c.** $(-\infty, -5) \cup (1, \infty)$
37. a.
b. $\{x \mid 0 < x < 4\}$ **c.** $(0, 4)$
39. a.
b. $\{r \mid r \text{ is a real number}\}$ **c.** $(-\infty, \infty)$

41. a.
b. $\{x \mid -4 < x < -2\}$ **c.** $(-4, -2)$
43. a.
b. $\{x \mid 0 < x < 1\}$ **c.** $(0, 1)$
45. a.
b. $\varnothing$ **c.** no interval notation
47. a.
b. $\left\{k \mid -2 \le k \le \dfrac{14}{3}\right\}$ **c.** $\left[-2, \dfrac{14}{3}\right]$
49. a.
b. $\{x \mid x < 4 \text{ or } x > 20\}$ **c.** $(-\infty, 4) \cup (20, \infty)$
51. a.
b. $\{y \mid -6.4 \le y \le 12.8\}$ **c.** $[-6.4, 12.8]$
53. a.
b. $\varnothing$ **c.** no solution
55. $|x + 1| > 1$ **57.** $|x - 3| \le 2$ **59.** $|x| < 3$
61. $|x| >$ any negative number

Review Exercises
1. true **2.** true **3.** -2 **4.** $\dfrac{15}{2}$

5. $y = \dfrac{C - Ax}{B}$ **6.** 46, 47, 48

8.4 Exercises
1. A function is a relation in which every value in the domain is paired with exactly one value in the range.
3. *Domain*: The set of all input values for a relation. *Range*: The set of all output values for a relation. **5.** The graph of every quadratic function is a parabola. Find many ordered pairs, plot them, and then draw a smooth U-shaped curve through those points.
7. Domain: {1, 2, 3, 4, 5}; Range: {Australia, Italy, U.K., France, Ireland}; It is a function.
9. Domain: {2%, 3%, 5%, 14%, 15%, 26%, 33%};

Range:
$$\left\{\begin{array}{l}\text{Work at home} \\ \text{Midnight to 4:59 A.M.} \\ \text{10 A.M. to 11:59 A.M.} \\ \text{9 A.M. to 9:59 A.M.} \\ \text{Noon to 11:59 P.M.} \\ \text{8 A.M. to 8:59 A.M.} \\ \text{5 A.M. to 6:59 A.M.} \\ \text{7 A.M. to 7:59 A.M.}\end{array}\right\}$$

It is not a function.
11. Domain: {−3, 0, 1, 2}; Range: {1, 2, 5, 8}; It is a function.
13. Domain: {−3, 2, 4} ; Range: {1, 2, 5, 8}; It is not a function.
15. Domain: {1970, 1975, 1980, 1985, 1990, 1995, 2000};

Range: {21.4, 21.8, 22.7, 23.7, 24.2, 24.5, 24.9}; It is a function.
17. Domain: all real numbers; Range: all real numbers; It is a function. **19.** Domain: all real numbers; Range: $\{y \mid y \geq -4\}$; It is a function. **21.** Domain: $\{x \mid -4 \leq x < 4\}$; Range: $\{1, 2, 3\}$; It is a function. **23.** Domain: $\{x \mid x \geq -4\}$; Range: all real numbers; It is not a function. **25.** 15 **27.** $\frac{2}{9}$ **29.** 5

31. $\frac{28}{5}$ **33.** 3 **35.** 2.928 **37.** undefined **39.** 6 **41. a.** 2 **b.** 1 **c.** 0 **43. a.** 1 **b.** 9 **c.** 4 **45. a.** -3 **b.** 1 **c.** -1

47. **49.**

51. **53.**

55. **57.**

59. **61.**

63. 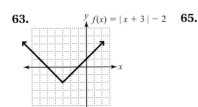 **65.**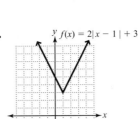

67. **69.**

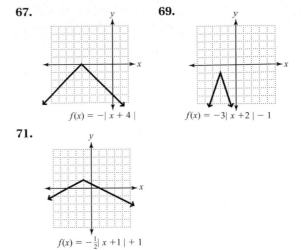

$f(x) = -\mid x + 4 \mid$ $f(x) = -3 \mid x + 2 \mid - 1$

71.

$f(x) = -\frac{1}{2} \mid x + 1 \mid + 1$

Review Exercises
1. $5x^2 + 6$ **2.** $7x^2 - 3x - 7$ **3.** $6x^2 - 5x - 4$ **4.** $4x^2 - 16x + 16$ **5.** $2x^2 + 3x - 5$ **6.** $x^3 + 3x^2 + 9x + 27$

8.5 Exercises
1. $(p + q)(x) = p(x) + q(x)$ **3.** Use FOIL to multiply the binomials. **5.** $5x - 1$; $-x + 1$ **7.** $3x - 8$; $-x - 2$ **9.** $x^2 - 3x + 8$; $x^2 - 5x + 6$ **11.** $3x^2 - 5x + 2$; $x^2 - 5x - 8$ **13.** $-2x^2 + 2x + 7$; $-8x^2 + 6x + 9$ **15.** $x^2 - 5x$ **17.** $x^2 - x - 2$ **19.** $3x^2 + 8x + 4$ **21.** $x^4 - x^3 - 11x^2 + 25x - 14$ **23.** $6x^4 - 11x^3 + 4x^2 + 2x - 1$ **25.** $x - 3, x \neq 0$ **27.** $2x - 1 + \frac{2}{x}, x \neq 0$ **29.** $x - 5, x \neq 9$

31. $x + 1, x \neq \frac{5}{2}$ **33.** $2x - 15 + \frac{68}{x + 5}, x \neq -5$

35. a. $7x + 2$ **b.** $3x - 4$ **c.** $10x^2 + 13x - 3$ **d.** $\frac{5x - 1}{2x + 3}, x \neq \frac{-3}{2}$ **37. a.** $2x^2 - 2x + 4$ **b.** $2x^2 - 4x + 6$ **c.** $2x^3 - 5x^2 + 8x - 5$ **d.** $2x - 1 + \frac{4}{x - 1}, x \neq 1$ **39. a.** -12 **b.** 11 **c.** -6 **d.** 2 **41. a.** 16 **b.** 54 **c.** 7 **d.** -8 **43. a.** $4x + 6$ **b.**

f(x)	g(x)	$(f + g)(x)$
-11	1	-10
-5	3	-2
1	5	6
7	7	14
13	9	22

c.

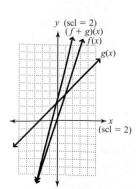

d. The output values for $f(x)$ added to the output values of $g(x)$ are equal to the output values of $(f+g)(x)$.

45. a. $-x - 14$

b.

$h(x)$	$k(x)$	$(h-k)(x)$
-17	-7	-10
-13	-1	-12
-9	5	-14
-5	11	-16
-1	17	-18

c.

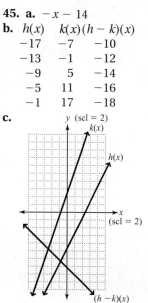

d. The output values for $k(x)$ subtracted from the output values of $h(x)$ are equal to the output values of $(h-k)(x)$.

47. a. $t(x) = 4x + 4$ **b.** \$804   **c.**

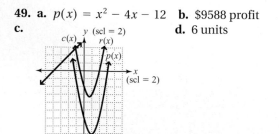

49. a. $p(x) = x^2 - 4x - 12$ **b.** \$9588 profit

c.

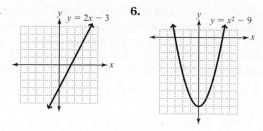

d. 6 units

51. a. $A(x) = 6x^2 + 4x$ **b.** 66 ft.^{2}
53. a. $h(x) = 4x + 1$ **b.** 17 cm

Review Exercises   **1.** $(3-y)(x-y)$
2. $-3(x+3)(x-3)$ **3.** $-4, 4$ **4.** no solution

Chapter 8 Review Exercises   **1.** true
2. true   **3.** false   **4.** true   **5.** true   **6.** true   **7.** $\cap$
8. $2.$ $ax + b = -c, ax + b = c$ **9.** the range   **10.** $f(x) + g(x)$
11. Intersection: $\{1, 5, 9\}$; Union: $\{1, 5, 7, 9\}$
12. Intersection: $\{\ \}$ or $\varnothing$; Union: $\{1, 2, 3, 4, 5, 6, 7\}$
13. a.

b. $\{x | -3 < x < -2\}$ **c.** $(-3, -2)$
14. a.

b. $\{x | -5 \le x \le -3\}$ **c.** $[-5, -3]$
15. a.

b. $\{x | -7 < x < 3\}$ **c.** $(-7, 3)$
16. a.

b. $\{x | 1 < x \le 4\}$ **c.** $(1, 4]$
17. a.

b. $\{x | -2 \le x < 0\}$ **c.** $[-2, 0)$
18. a.

b. $\{\ \}$ or $\varnothing$ **c.** no interval notation
19. a.

b. $\{w | w \le -6 \text{ or } w \ge -2\}$ **c.** $(-\infty, -6] \cup [-2, \infty)$
20. a.

b. $\{x | x \text{ is a real number}\}$ **c.** $(-\infty, \infty)$
21. a.

b. $\left\{ m | m < \dfrac{5}{2} \text{ or } m > 5 \right\}$ **c.** $\left(-\infty, \dfrac{5}{2} \right) \cup (5, \infty)$
22. a.

b. $\left\{ x | x \le -\dfrac{4}{3} \text{ or } x \ge 2 \right\}$ **c.** $\left(-\infty, -\dfrac{4}{3} \right] \cup [2, \infty)$
23. a.

b. $\{x \mid x \le -9 \text{ or } x \ge -4\}$ **c.** $(-\infty, -9] \cup [-4, \infty)$

24. a. 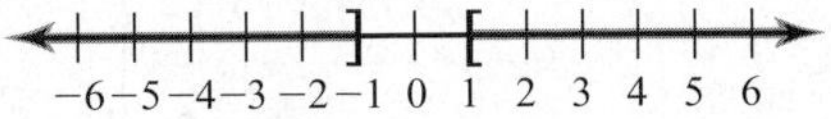
−6 −5 −4 −3 −2 −1 0 1 2 3 4 5 6

b. $\{w \mid w \le -1 \text{ or } w \ge 1\}$ **c.** $(-\infty, -1] \cup [1, \infty)$

25. a. 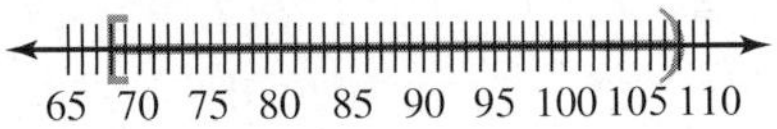
65 70 75 80 85 90 95 100 105 110

b. $\{x \mid 68 \le x < 108\}$ **c.** $[68, 108)$

26. a. 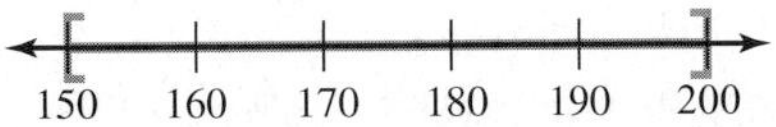
150 160 170 180 190 200

b. $\{x \mid 150 \le x \le 200\}$ **c.** $[150, 200]$ **27.** $-4, 4$ **28.** $-3, 11$ **29.** $-1, 2$ **30.** no solution **31.** $-7, 15$ **32.** $-3, 3$ **33.** $0, \frac{8}{3}$ **34.** $-1, 11$ **35. a.** 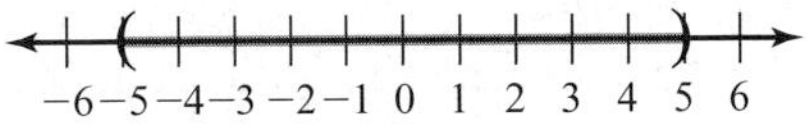
−6 −5 −4 −3 −2 −1 0 1 2 3 4 5 6

b. $\{x \mid -5 < x < 5\}$ **c.** $(-5, 5)$

36. a. 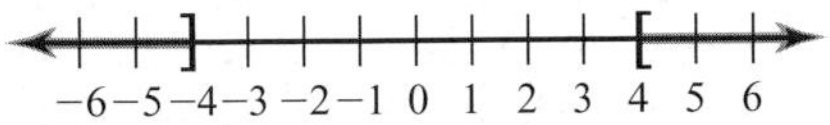
−6 −5 −4 −3 −2 −1 0 1 2 3 4 5 6

b. $\{p \mid p \le -4 \text{ or } p \ge 4\}$ **c.** $(-\infty, -4] \cup [4, \infty]$

37. a. 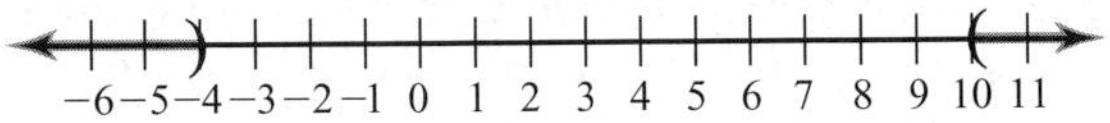
−6 −5 −4 −3 −2 −1 0 1 2 3 4 5 6 7 8 9 10 11

b. $\{x \mid x < -4 \text{ or } x > 10\}$ **c.** $(-\infty, -4) \cup (10, \infty)$

38. a. 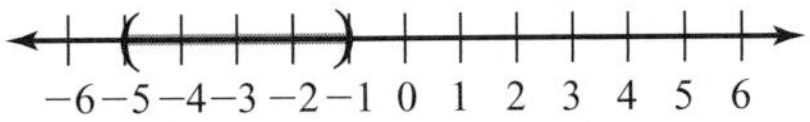
−6 −5 −4 −3 −2 −1 0 1 2 3 4 5 6

b. $\{m \mid -5 < m < -1\}$ **c.** $(-5, -1)$

39. a. 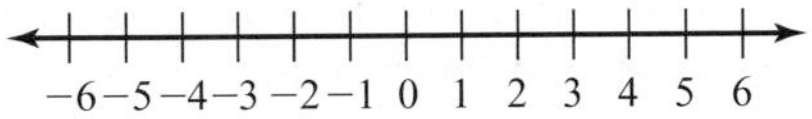
−6 −5 −4 −3 −2 −1 0 1 2 3 4 5 6

b. $\{\ \}$ or $\varnothing$ **c.** no interval notation

40. a. 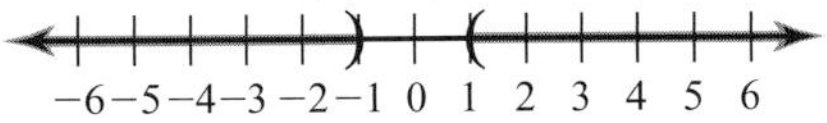
−6 −5 −4 −3 −2 −1 0 1 2 3 4 5 6

b. $\{b \mid b < -1 \text{ or } b > 1\}$ **c.** $(-\infty, -1) \cup (1, \infty)$

41. a. 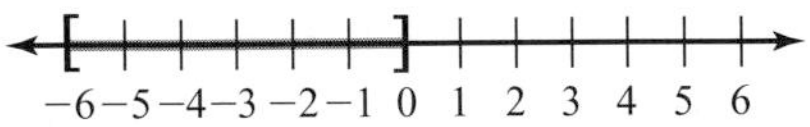
−6 −5 −4 −3 −2 −1 0 1 2 3 4 5 6

b. $\{m \mid -6 \le m \le 0\}$ **c.** $[-6, 0]$

42. a. 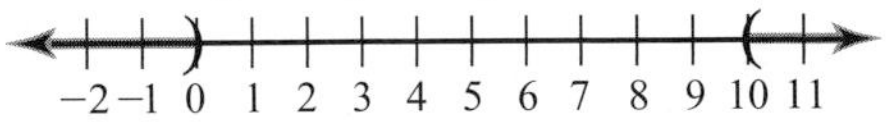
−2 −1 0 1 2 3 4 5 6 7 8 9 10 11

b. $\{t \mid t < 0 \text{ or } t > 10\}$ **c.** $(-\infty, 0) \cup (10, \infty)$

43. a. 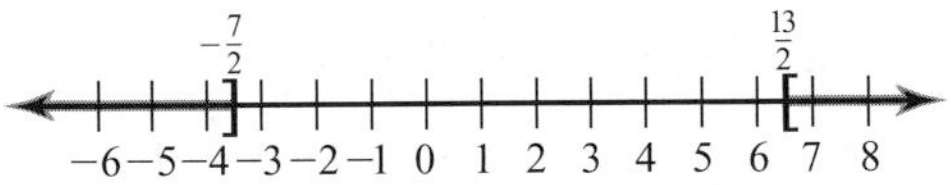
−7/2
13/2
−6 −5 −4 −3 −2 −1 0 1 2 3 4 5 6 7 8

b. $\left\{k \mid k \le -\frac{7}{2} \text{ or } k \ge \frac{13}{2}\right\}$ **c.** $\left(-\infty, -\frac{7}{2}\right] \cup \left[\frac{13}{2}, \infty\right)$

44. a. 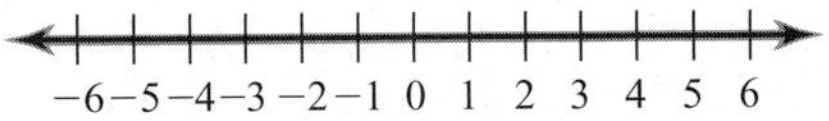
−6 −5 −4 −3 −2 −1 0 1 2 3 4 5 6

b. $\{x \mid x \text{ is a real number}\}$ **c.** $(-\infty, \infty)$

45. Domain: {1, 2, 3, 4, 5, 6}; Range: {Superman #204, New Avengers #1, Superman Batman #8, Identity Crisis #1, Astonishing X-Men #1, Superman #205}; It is a function.

46. Domain: {1, 2, 3, 4}
Range: {Harvard University (MA), Stanford University (CA), University of Pennsylvania (Wharton), Massachusetts Institute of Technology (Sloan), Northwestern University (Kellogg) (IL), Columbia University (NY)}; It is not a function.

47. Domain: {2, 3, 4}; Range: $\{-3, 1, 4\}$; It is not a function.
48. Domain: $\{-2, 0, 1\}$; Range: {3, 4}; It is a function.
49. Domain: all real numbers; Range: $\{y \mid y \ge -4\}$; It is a function. **50.** Domain: all real numbers; Range: $\{y \mid y \le 4\}$; It is a function. **51.** 17 **52.** -9

53. 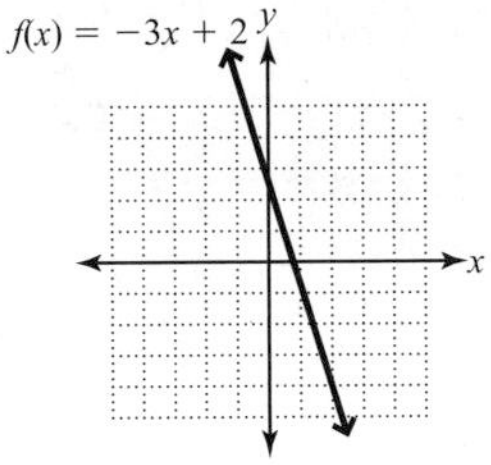
f(x) = −3x + 2
y
x

54. 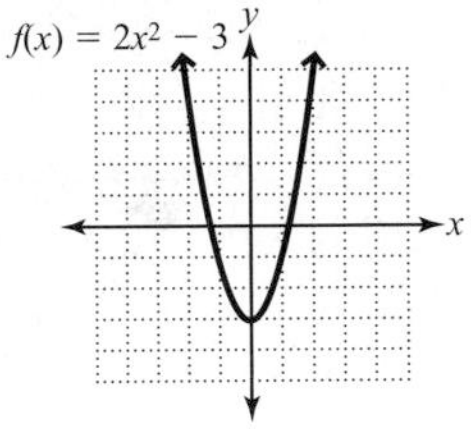
f(x) = 2x² − 3
y
x

55. 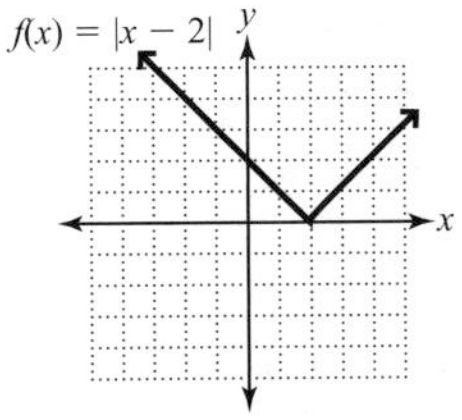
f(x) = |x − 2|
y
x

56. 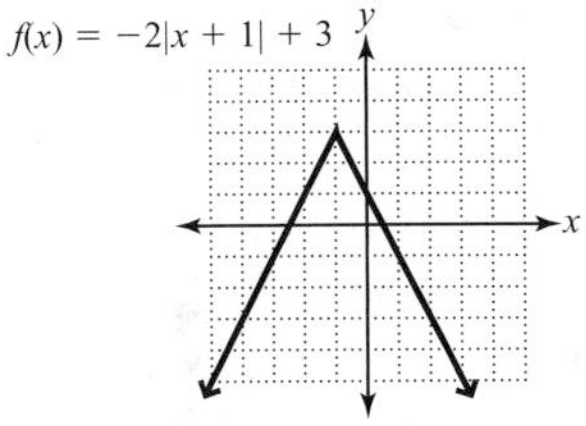
f(x) = −2|x + 1| + 3
y
x

57. a. $x + 10$ **b.** $-7x + 8$ **c.** $-12x^2 + 33x + 9$
58. a. $7x$ **b.** $-9x - 4$ **c.** $-8x^2 - 18x - 4$
59. a. $3x^2 + 7$ **b.** $3x^2 - 2x + 3$ **c.** $3x^3 + 5x^2 + 3x + 10$
60. a. $x^2 + x$ **b.** $x^2 - 5x - 2$ **c.** $3x^3 - 5x^2 - 5x - 1$
61. $2x^2 - 5x + 4, x \ne 0$ **62.** $x^2 - 6x + 3 + \frac{-2}{2x - 5}, x \ne \frac{5}{2}$
63. a. $c(x) = 4x + 4$ **b.** $904 **c.** 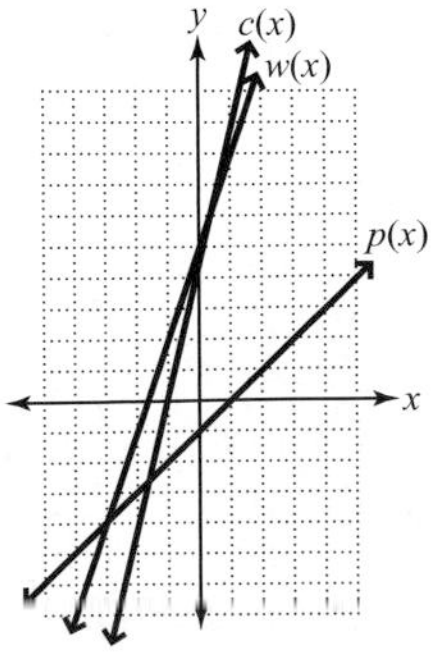
y
c(x)
w(x)
p(x)
x

64. a. $h(x) = 2x - 1$ **b.** 17 in.

Chapter 8 Practice Test   **1.** $\{e, h, o\}$; $\{e, h, m, o, s, u\}$

2. $-8, 2$ **3.** $-3, 6$ **4.** $-8, \dfrac{2}{3}$ **5.** no solution

6. a.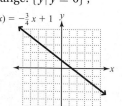
number line with open paren at -7 and bracket at 3, from -9 to 5
b. $\{x \mid -7 < x \leq 3\}$ **c.** $(-7, 3]$

7. a. number line with bracket at -3 and open paren at -2, from -6 to 6
b. $\{x \mid -3 \leq x < -2\}$ **c.** $[-3, -2)$

8. a. number line with bracket at -1 and bracket at 2, from -3 to 4
b. $\{x \mid x \leq -1 \text{ or } x \geq 2\}$ **c.** $(-\infty, -1] \cup [2, \infty)$

9. a. number line with open paren at 1, from -1 to 5
b. $\{n \mid n > 1\}$ **c.** $(1, \infty)$

10. a. number line with open paren at -13 and open paren at 5, from -14 to 8
b. $\{x \mid -13 < x < 5\}$ **c.** $(-13, 5)$

11. a. number line with open paren at -1 and open paren at 3, from -6 to 6
b. $\{x \mid x < -1 \text{ or } x > 3\}$ **c.** $(-\infty, -1) \cup (3, \infty)$

12. a. number line with open paren at -7 and open paren at -1, from -8 to 6
b. $\{x \mid -7 < x < -1\}$ **c.** $(-7, -1)$

13. a. number line from -6 to 6
b. $\{ \ \}$ or $\varnothing$ **c.** no interval notation

14. a. number line from -6 to 6
b. $\{x \mid x \text{ is a real number}\}$ **c.** $(-\infty, \infty)$

15. a. number line with bracket at $-\frac{1}{3}$ and bracket at 3, from -6 to 6
b. $\left\{ x \mid -\dfrac{1}{3} \leq x \leq 3 \right\}$ **c.** $\left[-\dfrac{1}{3}, 3 \right]$

16. a. number line with bracket at 8 and bracket at 10, from 0 to 12
b. $\{x \mid 8 \leq x \leq 10\}$ **c.** $[8, 10]$

17. Domain: all real numbers; Range: $\{y \mid y \leq 0\}$; It is a function.   **18.** 18   **19.** $f(x) = -\frac{3}{4}x + 1$
graph of line

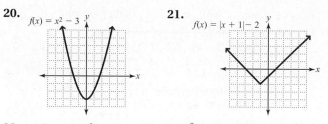

20. $f(x) = x^2 - 3$
graph of parabola

21. $f(x) = |x + 1| - 2$
graph of absolute value

22. a. $7x - 7$ **b.** $5x - 11$ **c.** $6x^2 + 3x - 18$
23. a. $4x^2 - 2$ **b.** $-2x^2 + 2$ **c.** $3x^4 - 2x^2$
24. $5x^2 + 9x - 7$ **25. a.** $h(x) = 9x + 2$ **b.** 137 cm

Chapter 9

9.1 Exercises   **1.** 1. Replace each variable in each equation with its corresponding value.   2. Verify that each equation is true.   **3.** The graphs are parallel.   **5.** Yes, the graphs are identical.   **7.** yes   **9.** no   **11.** yes   **13.** no   **15.** yes   **17.** $(4, 1)$ **19.** $(3, 2)$ **21.** $(-3, 4)$ **23.** $(-1, 3)$ **25.** $(3, -4)$ **27.** $(4, -3)$ **29.** $(2, -3)$ **31.** $(-2, 3)$ **33.** $(5, -2)$ **35.** no solution (inconsistent with independent equations)   **37.** all ordered pairs along $2x + 3y = 2$ (consistent with dependent equations)   **39. a.** inconsistent with independent equations   **b.** no solution   **41. a.** consistent with independent equations   **b.** one solution   **43. a.** consistent with dependent equations   **b.** infinite number of solutions   **45. a.** consistent with independent equations   **b.** one solution   **47. a.** inconsistent with independent equations   **b.** no solution   **49. a.** consistent with dependent equations   **b.** infinite number of solutions   **51. a.** consistent with independent equations   **b.** one solution   **53. a.** 5000 units   **b.** \$4000   **c.** $n > 5000$ **55. a.** the One-Rate plan   **b.** 375 min.   **c.** more than 375 min.

Review Exercises   **1.** $-3y - 35$ **2.** $y = \dfrac{10 - x}{3}$

3. $x = 30 + 6y$ **4.** $\dfrac{40\pi}{3}$ in.^{3} or ≈ 41.89 in.^{3} **5.** 11, 73

6. $2x + 17$

9.2 Exercises   **1.** When using the graphing method, if either coordinate in the solution is a fraction, we may have to guess the value. Substitution requires no guessing.   **3.** y in the first equation because it can be isolated without using division.   **5.** The resulting equation will no longer contain variables and becomes a false equation.   **7.** $(4, 2)$ **9.** $(12, -4)$ **11.** $(-10, -24)$ **13.** $(-3, 2)$ **15.** $(2, 9)$ **17.** $(-1, -3)$ **19.** $(3, 4)$ **21.** $(1, 1)$ **23.** $\left(3, -\dfrac{1}{2} \right)$ **25.** $(1, -2)$ **27.** $(4, -2)$ **29.** $\left(-\dfrac{2}{5}, \dfrac{2}{3} \right)$ **31.** all ordered pairs along $y = -2x$ (consistent with dependent equations)   **33.** no solution (inconsistent system)   **35.** $(2, 5)$ **37.** $(1, 2)$ **39.** Mistake: 3 was not distributed to $3y$.

Correct: $\left(\dfrac{27}{11}, -\dfrac{13}{11}\right)$ **41.** 23, 17 **43.** 2, −1 **45.** −6, −2
47. Jon is 20, Tony is 7 **49.** $58,600 **51.** length: 11 m, width: 5 m **53.** length: 14 ft., width: 10 ft.
55. length: 18 m, width: 9 m

Review Exercises

1. $5x - 4y$ **2.** $-3x + 21y$ **3.** $9x$
4. $13y$ **5.** 80 **6.** $-\dfrac{3}{2}$

9.3 Exercises

1. The elimination method is advantageous over graphing when the solution involves fractions. The elimination method is advantageous over substitution when no coefficients are 1. **3. a.** x, because multiplying only the second equation by -2 will allow the x to be eliminated, whereas to eliminate y, we would need to multiply each equation by a number. **b.** Multiply $3x + 5y = 7$ by -2 then add the resulting equation to $6x - 2y = 1$. **5.** Both variables have been eliminated and the resulting equation is false.
7. $(6, -8)$ **9.** $(2, 3)$ **11.** $(-1, -2)$ **13.** $(2, 3)$
15. $(3, -1)$ **17.** $(-4, 1)$ **19.** $(-4, 3)$ **21.** $(1, 1)$
23. $(4, -4)$ **25.** $\left(-\dfrac{2}{5}, \dfrac{2}{3}\right)$ **27.** $(0.2, 0.3)$ **29.** $(6, 4)$
31. $(7, -9)$ **33.** all ordered pairs along $2x + y = -2$ (dependent) **35.** no solution (inconsistent)
37. $(-4, -13)$ **39.** $\left(2, \dfrac{5}{2}\right)$ **41.** $(4, 4)$ **43.** Mistake: Did not multiply the right side of $x - y = 1$ by -9. Correct: $(5, 4)$
45. 155, 93 **47.** 16, 9 **49.** 5.5 yr., 9.5 yr. **51.** $126\dfrac{2}{3}°, 53\dfrac{1}{3}°$
53. $60°, 30°$ **55.** 22 in., 8 in. **57.** Germany: 35, U.S.: 34
59. Patriots: 20, Rams: 17 **61.** Nicklaus: 6, Palmer: 4

Review Exercises

1. 13 dimes, 11 quarters
2. 3:12 P.M. **3.** $500 in checking, $600 in CD **4.** 160 ml
5. $(1, -6)$ **6.** $\left(-\dfrac{5}{4}, \dfrac{5}{2}\right)$

9.4 Exercises

1. It does not matter which pair of equations you choose as long as the second pair is different from the first pair. **3.** Answers may vary. A good choice is to eliminate z using equations 1 and 2 and 1 and 3. Multiplying Equation 1 by -3 and then adding this result to both Equations 2 and 3 takes very few steps. **5.** Two parallel planes intersecting a third plane. **7.** no **9.** no **11.** yes
13. $(2, -1, 4)$ **15.** $(-1, 0, 3)$ **17.** $(-1, 1, 3)$ **19.** no solution **21.** infinite solutions (dependent equations)
23. $(2, -3, 1)$ **25.** $\left(-2, -\dfrac{1}{2}, 1\right)$ **27.** $(2, -1, 1)$
29. $(2, -4, 1)$ **31.** $(4, -4, 2)$ **33.** 6, 2, 8 **35.** $105°, 35°, 40°$
37. 8 in., 10 in., and 15 in. **39.** Burger: $2.50, Fry: $1.50, Drink: $1.00 **41.** 150 children, 250 students, 100 adults
43. 14 free throws, 8 field goals, 1 three pointer
45. 8 nickels, 7 dimes, 10 quarters **47.** 6 pennies, 12 nickels,

7 dimes **49.** zinc: 20 lb., tin: 60 lb., copper: 920 lb.
51. ham: 3 lb., turkey: 5 lb., beef: 2 lb. **53.** MM: $1666.67, IF: $500, GF: $2833.33 **55.** $30,000 in stocks, $50,000 in bonds, $20,000 in certificates **57.** jumbo: 30 oz., extra large: 27 oz., large: 24 oz **59.** bicycling: 420, walking: 300, stair: 600
61. Angola: 24, Mozambique: 25, Botswana: 26 **63.** $a = -16$, $v_0 = 100$, $h_0 = 150$; $h = -16t^2 + 100t + 150$

Review Exercises

1. $-\dfrac{5}{2}y$ **2.** $-y - 6z$ **3.** 9 **4.** 2
5. $(-2, 3)$ **6.** $\left(-\dfrac{53}{19}, \dfrac{25}{19}\right)$

9.5 Exercises

1. 4 rows, 2 columns **3.** The dashed line corresponds to the equal signs in the equations.
5. A matrix is in echelon form when the coefficient portion of the augmented matrix has 1s on the diagonal from upper left to lower right and 0s below the 1s. **7.** $\left[\begin{array}{cc:c} 14 & 7 & 6 \\ 7 & 6 & 8 \end{array}\right]$

9. $\left[\begin{array}{cc:c} 7 & -6 & 1 \\ 8 & -12 & 6 \end{array}\right]$ **11.** $\left[\begin{array}{ccc:c} 1 & -3 & 1 & 4 \\ 2 & -4 & 2 & -4 \\ 6 & -2 & 5 & -4 \end{array}\right]$

13. $\left[\begin{array}{ccc:c} 4 & 6 & -2 & -1 \\ 8 & 3 & 0 & -12 \\ 0 & -1 & 2 & 4 \end{array}\right]$ **15.** $(4, -2)$ **17.** $(-6, -5, 1)$

19. $\left[\begin{array}{cc:c} 1 & 3 & -1 \\ 0 & 11 & 4 \end{array}\right]$ **21.** $\left[\begin{array}{ccc:c} 1 & -2 & 4 & 6 \\ 0 & 2 & -1 & -5 \\ 0 & 0 & -2 & 17 \end{array}\right]$

23. $\left[\begin{array}{c:cc} 1 & 2 & -2.5 \\ -1 & 3 & 2 \end{array}\right]$ **25.** Replace R_2 with $3R_1 + R_2$.
27. Replace R_3 with $-2R_2 + R_3$. **29.** $(2, 0)$ **31.** $(1, 2)$
33. $(-3, -2)$ **35.** $(-2, 1)$ **37.** $(4, 6)$ **39.** $(4, -4)$
41. $(15, -15, 6)$ **43.** $(2, 1, -2)$ **45.** $(2, -3, 1)$
47. $(-1, -1, -2)$ **49.** $(-3, 1, 5)$ **51.** $(8, -2, -4)$
53. $(3, -4)$ **55.** $(-6, 8)$ **57.** $(4, -3, 5)$ **59.** $(7, -6, 5)$
61. Mistake: In the second step, $4R_1 + R_2$ is calculated incorrectly. Correct: The correct calculation is $\left[\begin{array}{cc:c} 1 & 3 & 13 \\ 0 & 11 & 26 \end{array}\right]$. The solution is $(65/11, 26/11)$. **63.** chicken sandwich: $2.50, drink: $1.20 **65.** $58°, 32°$ **67.** Nile: 4150 mi., Amazon: 3900 mi. **69.** $6400 in CD, $3600 in money market **71.** CD: $16; tape: $12, DVD: $28 **73.** 3 touchdowns, 3 extra points, 1 field goal **75.** $v_1 = 9$, $v_2 = 6$, $v_3 = 2$

Review Exercises

1. 8 **2.** 11 **3.** 2 **4.** 2 **5.** 3
6. -1

9.6 Exercises

1. No, because it is not a square matrix.
3. Eliminate the elements in the same row and column as the element. **5.** 13 **7.** -22 **9.** 24 **11.** -7 **13.** -22
15. 15 **17.** 1 **19.** -12 **21.** -14 **23.** -58 **25.** 0.47

27. 14.451 **29.** $\dfrac{13}{30}$ **31.** $-\dfrac{317}{2400}$ **33.** $-x - 8y - 2$

35. $(-4, -1)$ **37.** $(3, 4)$ **39.** $(4, -4)$ **41.** $(-3, -2)$
43. $(2, 2)$ **45.** $(2, 4)$ **47.** $(2, 6)$ **49.** $(2, 1, 3)$
51. $(-2, 1, -1)$ **53.** $(2, -1, 1)$ **55.** $\left(\dfrac{2}{3}, -7, \dfrac{1}{2}\right)$

57. $(1, -2, 3)$ **59.** $(2, 2, -2)$ **61.** $x = -4$ **63.** $x = -5$
65. 9 **67.** 21.5 **69.** 19 **71.** Hoba is 60 tons, Ahnighito is
30 tons. **73.** 6 lb. of garbanzo, 4 lb. of black turtle
75. 1% zinc, 4% tin, 95% copper

Review Exercises **1.** $x < 3$;

2. $x \geq 1$;

3. 19, 20 **4.**   **5.**

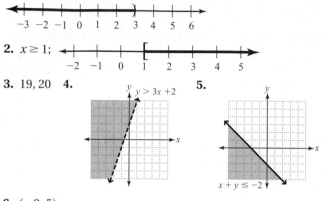

6. $(-2, 5)$

9.7 Exercises **1.** Select a point in the solution region
and verify that it makes both inequalities true. **3.** Parallel
lines with shaded regions that do not overlap. **5.** $\begin{cases} x < 0 \\ y < 0 \end{cases}$

7.   **9.**

11.   **13.**

15.   **17.**

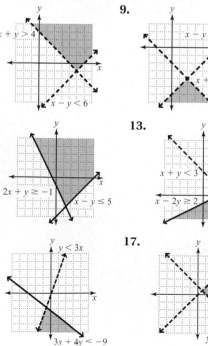

A-34 Answers

19.   **21.**

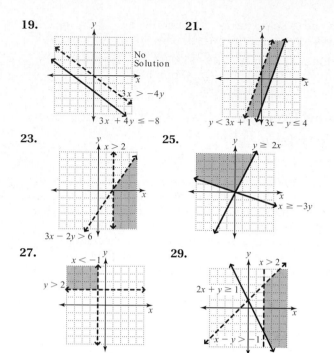

23.   **25.**

27.   **29.**

31. $x + y > 3$ should be a dashed line. **33.** The wrong area
is shaded. It should be the region containing the point $(1, 5)$.

35. a. $\begin{cases} V + M \geq 1100 \\ M \geq 500 \end{cases}$ **b.** $V \leq 800$; $M \leq 800$

c. 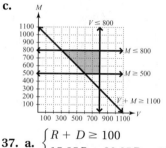 **d.** Answers will vary. Some
possible (V, M) pairs are $(500,$
$700)$, $(600, 500)$, or $(700, 600)$.

37. a. $\begin{cases} R + D \geq 100 \\ 15.95R + 20.95D \geq 1700 \end{cases}$

b. 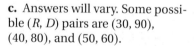 **c.** Answers will vary. Some possible (R, D) pairs are $(30, 90)$,
$(40, 80)$, and $(50, 60)$.

Review Exercises **1.** 9 **2.** -100 **3.** $-x - 14$
4. $-8a^3b^5$ **5.** $\dfrac{-5b^5}{a^6}$ **6.** 2, 3

Chapter 9 Review Exercises **1.** false **2.** true
3. true **4.** true **5.** false **6.** 1. Replace **7.** slope,
y-intercepts **8.** the coefficient portion of the augmented
matrix has 1s on the diagonal from the upper left to lower

right and 0s below the 1s   **9.** determinant   **10.** 2. the inter-section (overlap) of the two solutions   **11.** yes   **12.** no
13. a. consistent with independent equations   **b.** one
14. a. inconsistent with independent equations   **b.** zero
15. a. consistent with independent equations   **b.** one
16. a. consistent with dependent equations   **b.** infinite solutions along $x - 5y = 10$ **17.** $(-1, -4)$ **18.** $(3, -1)$
19. no solution   **20.** all ordered pairs along $y = \dfrac{3}{2}x - 3$
21. $(4, -1)$ **22.** $\left(5, -\dfrac{2}{5}\right)$ **23.** all ordered pairs along
$y = \dfrac{4}{3}x + 2$ **24.** no solution   **25.** $(1, 3)$ **26.** $(2, -1)$
27. $(1, 5)$ **28.** $(5, -3)$ **29.** $(1, 0, -3)$ **30.** no solution
31. $(3, -3, 2)$ **32.** $(4.2, 0, 1.8)$ **33.** $(4, -1)$ **34.** $(-3, -2)$
35. $(-2, 3, 1)$ **36.** $(-1, 2, -4)$ **37.** $(-2, 9)$ **38.** no solution
39. $(2, 4, -1)$ **40.** $(-2, 4, -2)$ **41.**

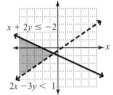

42. 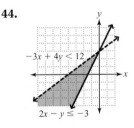   **43.**

44. 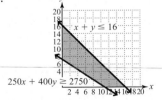   **45.** 47 adult, 92 children.

46. 5805   **47.** 222 black cars, 187 silver cars.   **48.** 10, 6
49. 75°, 15°   **50.** $3500 at 7% and $1500 at 9%.   **51.** 5:45 P.M.
52. 120 mph   **53.** 30 ml   **54.** 1 vitamin supplement, 3 rolls of film, 4 bags of candy.   **55.** 4 first place, 3 second place, 9 third place.

56. a. $x + y \le 16$, $250x + 400y \ge 2750$, $x \ge 0$, $y \ge 0$

Chapter 9 Practice Test   **1.** yes   **2.** no
3. $(-2, 1)$ **4.** $(8, -1)$ **5.** $(3, 1)$

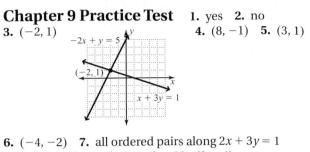

6. $(-4, -2)$ **7.** all ordered pairs along $2x + 3y = 1$
8. $(-2, 3, -2)$ **9.** $(-4, 2, 3)$ **10.** $(2, -4)$
11. $(-1, 5, 2)$ **12.** $(-2, 5)$ **13.** $(-4, 0, 3)$
14. 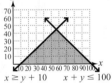   **15.** accountants: 86, waiters/waitresses: 77

16. 180   **17.** 7 mph   **18.** $4000 at 6%, $8000 at 8%
19. children: 150, students: 300, adult: 350   **20. a.** $x \ge 0$, $y \ge 0$, $x + y \le 100$, $x \ge y + 10$ **b.**

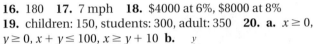

Chapters 1–9 Cumulative Review Exercises
1. false   **2.** true   **3.** true   **4.** true   **5.** true   **6.** false
7. $8x + 12$ **8.** $\dfrac{c}{d}$ **9. b.** variable   **10.** sum   **11.** 11
12. undefined   **13.** 0   **14.** $-6x - 6$ **15.** $4x^3 - 14x^2 + 7x - 3$
16. $\dfrac{1}{5}x^2 + x + 2$ **17.** $5m - 4 + \dfrac{6}{3m - 2}$
18. $(m + 2)(m^2 - 2m + 4)$ **19.** $(x + 5)^2$
20. $(x^2 + 9)(x + 3)(x - 3)$ **21.** $\dfrac{1}{u - 1}$ **22.** $\dfrac{a - b}{x + y}$
23. $\dfrac{x}{10x + 5}$ **24.** $\dfrac{y^2 + y + 9}{(y - 2)(y + 3)}$ **25.** $\dfrac{1}{g + 3}$
26. $\dfrac{x + 10}{(x - 2)(x + 2)}$ **27.** all real numbers   **28.** $-6 < x < 1$
29. ± 6 **30.** 8   **31.** $-\dfrac{8}{3}, 0$ **32.** $m = \dfrac{2E}{v^2}$ **33.** The data does represent a function. Domain = {CPUs, Monitors, Keyboards, Printers, Peripherals, Macintoshes, Typewriters, Laptops, Televisions, Other}; Range = {1163, 775, 706, 280, 81, 67, 62, 25, 21, 228}

34.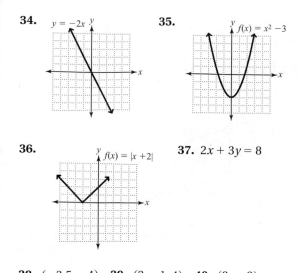

$y = -2x$

35. $f(x) = x^2 - 3$

36. $f(x) = |x + 2|$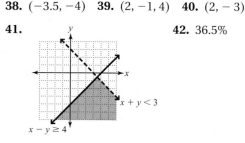

37. $2x + 3y = 8$

38. $(-3.5, -4)$ **39.** $(2, -1, 4)$ **40.** $(2, -3)$

41.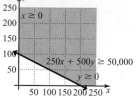

$x + y < 3$

$x - y \geq 4$

42. 36.5%

43. women: \$578, men: \$722 **44.** 75 ml **45. a.** x: number of small events, y: number of large events; $250x + 500y \geq 50,000$, where $x \geq 0$, $y \geq 0$ **b.**

$250x + 500y \geq 50,000$

$x \geq 0$

$y \geq 0$

c. 100 small events, 50 large events **d.** 50 small events, 150 large events **46.** ≈ 5.5 hr. **47.** 3.2 cm/sec. **48.** 1.2 hr., or 1 hr. and 12 min. **49.** $115°, 65°$ **50.** pens at \$2.00, erasers at \$0.50, reams of paper at \$1.00

Chapter 10

10.1 Exercises
1. Answers will vary. For example, 9 has rational square roots whereas 17 has irrational square roots. The square roots of 17 are irrational because they cannot be expressed in the form $\frac{a}{b}$ where a and b are integers and $b \neq 0$.
3. Squaring a number or its additive inverse results in the same positive number. **5.** You cannot raise a number to an even power and get a negative value. **7.** ± 6 **9.** ± 11
11. ± 14 **13.** ± 15 **15.** 5 **17.** not a real number **19.** -5
21. ± 5 **23.** 1.2 **25.** not a real number **27.** -0.11
29. $\frac{7}{9}$ **31.** $-\frac{12}{13}$ **33.** 3 **35.** -3 **37.** 6 **39.** 5 **41.** not a

real number **43.** -2 **45.** 2 **47.** -2 **49.** 2 **51.** 2
53. $\frac{2}{3}$ **55.** $\frac{2}{3}$ **57.** 2.236 **59.** -3.317 **61.** 3.684
63. -3.756 **65.** 2.839 **67.** -3.036 **69.** 2.454 **71.** 2.295
73. b^2 **75.** $4x$ **77.** $10rs^3$ **79.** $0.5a^3b^6$ **81.** m **83.** $3a^3b^2$
85. $-4ab^4$ **87.** $0.2x^6$ **89.** a **91.** $2x^4$ **93.** $2x^2$ **95.** x^2y
97. $6|m|$ **99.** $|r - 1|$ **101.** $4|y^3|$ **103.** $3y$ **105.** $(y - 3)^2$
107. $y - 4$ **109.** 2 **111.** $\sqrt{15}$ **113.** $\{x | x \geq 4\}$ or $[4, \infty)$
115. $\{x | x \leq 4\}$ or $(-\infty, 4]$

117. a.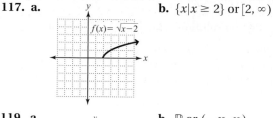

$f(x) = \sqrt{x - 2}$

b. $\{x | x \geq 2\}$ or $[2, \infty)$

119. a.

$f(x) = \sqrt[3]{x + 1}$

b. $\mathbb{R}$ or $(-\infty, \infty)$

121. ≈ -17.709 m/sec. **123.** ≈ 3.476 sec. **125.** ≈ 19.170 mph **127.** 13 ft. **129.** 15 N **131. a.** 7924 **b.** ≈ 7004

Review Exercises **1. a.** 12 **b.** 12 **2.** x^8 **3.** $-45m^4n^3$
4. $12x^4 - 20x^3 + 4x^2$ **5.** $21y^2 + 23y - 20$ **6.** $2x^2 - 17x - 9$

10.2 Exercises
1. 4 **3.** a must be positive, because with an even index the radicand must be positive. **5.** Yes, because $100^{1/4} = (10^2)^{1/4} = 10^{1/2}$. **7.** $\sqrt{25} = 5$
9. $-\sqrt{100} = -10$ **11.** $\sqrt[3]{27} = 3$ **13.** $\sqrt[3]{-64} = -4$
15. $\sqrt[4]{y}$ **17.** $\sqrt{144x^8} = 12x^4$ **19.** $18\sqrt{r}$ **21.** $\sqrt{\frac{x^4}{121}} = \frac{x^2}{11}$
23. $\sqrt[3]{8^2} = 4$ **25.** $(-\sqrt[4]{81})^3 = -27$ **27.** $(\sqrt[3]{-8})^4 = 16$
29. $\frac{1}{(\sqrt{16})^3} = \frac{1}{64}$ **31.** $\sqrt[5]{x^4}$ **33.** $8\sqrt[3]{n^2}$ **35.** $\frac{1}{\sqrt[5]{-32^2}} = \frac{1}{4}$
37. $\left(\sqrt{\frac{1}{25}}\right)^2 = \frac{1}{125}$ **39.** $\sqrt[6]{(2a + 4)^5}$ **41.** $25^{1/4}$ **43.** $z^{5/6}$
45. $5^{-5/6}$ **47.** $5x^{-4/5}$ **49.** $5^{7/3}$ **51.** $x^{2/7}$ **53.** $(4a - 7)^{7/4}$
55. $(2r - 5)^{8/5}$ **57.** $x^{4/5}$ **59.** $x^{7/6}$ **61.** $a^{17/12}$ **63.** $12w^{3/5}$
65. $-12a^{17/12}$ **67.** $7^{3/5}$ **69.** $\frac{1}{x^{2/5}}$ **71.** $x^{1/4}$ **73.** $r^{1/12}$
75. $\frac{1}{x^{5/7}}$ **77.** $a^{9/4}$ **79.** $20s^{3/7}$ **81.** $-24b^{1/4}$ **83.** x^2 **85.** $a^{5/3}$
87. $b^{2/5}$ **89.** $64x^4y^3$ **91.** $2q^{1/2}t^{1/4}$ **93.** $81a$ **95.** $3z^{1/3}$
97. $\sqrt{2}$ **99.** $\sqrt[3]{5}$ **101.** $\sqrt{x}$ **103.** $\sqrt[4]{r^3}$ **105.** $\sqrt[4]{x^3y}$
107. $\sqrt[5]{m^2n^3}$ **109.** $\sqrt[6]{x^5}$ **111.** $\sqrt[6]{y^7}$ **113.** $\sqrt[6]{x^5}$ **115.** $\sqrt[15]{n^2}$
117. $\sqrt[6]{1125}$ **119.** $\sqrt[12]{3456}$ **121.** $\sqrt[9]{x}$ **123.** $\sqrt[6]{n}$

Review Exercises 1. $2^4 \cdot 3 \cdot 5$ 2. 12 3. 15
4. 8×10^{11} 5. $-\frac{5}{8}x^4y^2z^2$ 6. $2x^2 - 8x + 5 + \frac{3}{x+3}$

10.3 Exercises 1. Multiplying the approximate roots gives $\sqrt{8} \cdot \sqrt{18} \approx 2.828 \cdot 4.243 = 11.999204$, which is tedious and inexact. Using the product rule for radicals gives $\sqrt{8} \cdot \sqrt{18} = \sqrt{8 \cdot 18} = \sqrt{144} = 12$, which is fast and exact. 3. Rewrite the expression as a product of two radicals, the first containing the perfect cube and the second containing no perfect cubes. Then simplify the first radical. For example, $\sqrt[3]{54x^8}$ can be rewritten as $\sqrt[3]{27x^6} \cdot \sqrt[3]{2x^2}$, then simplified to $3x^2 \cdot \sqrt[3]{2x^2}$. 5. 8 7. $9x^3$ 9. $12xy^2$ 11. $\sqrt{14}$ 13. $\sqrt{15x}$
15. 3 17. $\sqrt[3]{10y^2}$ 19. $\sqrt[4]{21}$ 21. $w\sqrt[4]{72}$ 23. $\sqrt[4]{15x^3y^3}$
25. $\sqrt[5]{30x^4}$ 27. $\sqrt[6]{8x^5y^4}$ 29. $\sqrt{\frac{21}{10}}$ 31. $\sqrt{\frac{6y}{5x}}$ 33. $\frac{5}{6}$
35. $\frac{\sqrt{10}}{3}$ 37. 7 39. $\sqrt{3}$ 41. $\frac{\sqrt[3]{4}}{w^2}$ 43. $\frac{\sqrt[3]{5y^2}}{3x^3}$ 45. 4
47. $\frac{\sqrt[4]{3u^3}}{2x^2}$ 49. $7\sqrt{2}$ 51. $8\sqrt{2}$ 53. $24\sqrt{5}$ 55. $20\sqrt{7}$
57. $a\sqrt{a}$ 59. xy^2 61. $x^3y^4z^5$ 63. $r^5s^4\sqrt{rs}$ 65. $18x^2\sqrt{2x}$
67. $2\sqrt[3]{4}$ 69. $x^2\sqrt[3]{x}$ 71. $x^2y^3\sqrt[3]{y^2}$ 73. $4z^2\sqrt[3]{2z^2}$ 75. $4\sqrt[3]{3}$
77. $2\sqrt[4]{5}$ 79. $9x^4\sqrt[4]{3x}$ 81. $3x^3\sqrt[5]{2x}$ 83. $xy^2z\sqrt[6]{x^2y^2z^5}$
85. $3\sqrt{7}$ 87. $30\sqrt{35}$ 89. $y^2\sqrt{y}$ 91. $x^4y^5\sqrt{y}$
93. $24c^4\sqrt{15}$ 95. $60\sqrt{2}$ 97. $2\sqrt{2}$ 99. $6\sqrt{5}$ 101. $c^2\sqrt{d}$
103. $12a^2$ 105. $6c\sqrt{3c}$ 107. $36x^2y^3\sqrt{2y}$ 109. $\frac{2\sqrt{6}}{7}$
111. $\frac{a^4}{2}$ 113. $\frac{3x^5\sqrt{5}}{4}$ 115. $\frac{3}{8}\sqrt{5}$

Review Exercises 1. no 2. yes 3. $8x^2 - 7x$
4. $8a^2 - 6ab - 9b^2$ 5. $9m^2 - 25n^2$ 6. $4x^2 - 12xy + 9y^2$

10.4 Exercises 1. Like radicals have the same index and the same radicand, but their coefficients may be different. 3. $(x+3)(x+2) = x^2 + 2x + 3x + 6 = x^2 + 5x + 6$, $(\sqrt{5}+3)(\sqrt{5}+2) = \sqrt{5}^2 + 2\sqrt{5} + 3\sqrt{5} + 6 = 5 + 5\sqrt{5} + 6 = 11 + 5\sqrt{5}$. In both cases we use the FOIL method and then simplify the expansion. 5. $-6\sqrt{6}$
7. $9\sqrt{a}$ 9. $12\sqrt{5} - 8\sqrt{6}$ 11. $11a\sqrt{5a} - 2b\sqrt{7b}$
13. $3x\sqrt[3]{9}$ 15. $18x^2\sqrt[4]{5x}$ 17. Cannot combine because the radicals are not like. 19. $4\sqrt{3} - 5\sqrt{3} = -\sqrt{3}$
21. $4\sqrt{5y} - 5\sqrt{5y} = -\sqrt{5y}$ 23. $4\sqrt{5} - 12\sqrt{5} = -8\sqrt{5}$
25. $12\sqrt{6} - 6\sqrt{6} = 6\sqrt{6}$ 27. $24a\sqrt{3a} - 10a\sqrt{3a} = 14a\sqrt{3a}$ 29. $5\sqrt{6} - 3\sqrt{6} + 2\sqrt{6} = 4\sqrt{6}$
31. $4\sqrt{2} - 12\sqrt{3} + 14\sqrt{2} - 5\sqrt{3} = 18\sqrt{2} - 17\sqrt{3}$
33. $2\sqrt[3]{2} + 3\sqrt[3]{2} = 5\sqrt[3]{2}$ 35. $12x\sqrt[3]{5x^2} - 24x\sqrt[3]{5x^2} = -12x\sqrt[3]{5x^2}$ 37. $-4x^2\sqrt[4]{x} + 4x\sqrt[4]{x^3}$ 39. $3\sqrt{2} + 2$
41. $3 - 3\sqrt{5}$ 43. $\sqrt{15} + 10\sqrt{3}$ 45. $24x - 48x\sqrt{2}$
47. $12 - 3\sqrt{2} + 4\sqrt{5} - \sqrt{10}$ 49. $6 + 5\sqrt{x} + x$
51. $6 + 10\sqrt{2} + 9\sqrt{3} + 15\sqrt{6}$ 53. $\sqrt{6} + \sqrt{10} + 3 + \sqrt{15}$
55. $x - \sqrt{xy} - 2y$ 57. $12\sqrt{14} - 12\sqrt{6} + 6\sqrt{35} - 6\sqrt{15}$
59. $8a + 10\sqrt{ab} - 3b$ 61. $2\sqrt[3]{2} - 3\sqrt[3]{4} - 40$

63. $1 - \sqrt[3]{18} + \sqrt[3]{12}$ 65. $x + 8$ 67. $22 + 8\sqrt{6}$
69. $3 - 2\sqrt{2}$ 71. $16 + 8\sqrt{3}$ 73. $30 + 12\sqrt{6}$ 75. 1
77. -14 79. $36 - x$ 81. 1 83. $x - y$ 85. 4 87. 50
89. $7\sqrt{5}$ 91. $-15\sqrt{6}$ 93. $7\sqrt{2}$ 95. $-14\sqrt{5}$ 97. $25\sqrt{3}$
99. a. $42 + 2\sqrt{5}$ ft. b. 46.5 ft. c. \$87.89

Review Exercises 1. Slope is $\frac{2}{5}$, y-intercept is (0, 2).
2. 3. $2x + 5$ 4. $16x^2 - 9$ 5. $\sqrt{2}$

$y = -\frac{1}{3}x + 4$

6. $\sqrt[3]{4}$

10.5 Exercises 1. a. $\sqrt{16}$ is rational. b. There is a radical, $\sqrt{3}$, in the denominator. 3. Multiply the fraction by a 1 so that the product's denominator has a radicand that is a perfect square. 5. $\frac{\sqrt{3}}{3}$ 7. $\frac{3\sqrt{2}}{4}$ 9. $\frac{6\sqrt{7}}{7}$ 11. $\frac{\sqrt{15}}{6}$
13. $\frac{x\sqrt{14}}{10}$ 15. $\frac{\sqrt{7}}{7}$ 17. $\frac{5\sqrt{3a}}{3a}$ 19. $\frac{\sqrt{33mn}}{11n}$ 21. $\frac{2\sqrt{5x}}{x}$
23. $\frac{\sqrt{3}}{4}$ 25. $\frac{3\sqrt{x}}{x^2}$ 27. $4x\sqrt{2x}$ 29. Mistake: The product of $\sqrt{2}$ and 2 is not 2. Correct: $\frac{\sqrt{6}}{2}$. 31. $\frac{5\sqrt[3]{9}}{3}$ 33. $\frac{\sqrt[3]{20}}{2}$
35. $3\sqrt[3]{2}$ 37. $\frac{m\sqrt[3]{n^2}}{n}$ 39. $\frac{\sqrt[3]{ab}}{b}$ 41. $\frac{2\sqrt[3]{4x^2}}{x}$ 43. $\frac{\sqrt[3]{30a}}{5a}$
45. $\frac{5\sqrt[4]{8}}{2}$ 47. $\frac{\sqrt[4]{3x^2}}{x}$ 49. $\frac{3\sqrt[4]{27x}}{x}$ 51. $\sqrt{6} - \sqrt{3}$
53. $8 + 4\sqrt{3}$ 55. $5\sqrt{3} - 5\sqrt{2}$ 57. $-1 - \sqrt{5}$ 59. $\frac{3 + \sqrt{3}}{2}$
61. $\frac{-6 - 8\sqrt{3}}{13}$ 63. $\frac{4\sqrt{21} - 4\sqrt{6}}{5}$ 65. $\frac{64 + 8\sqrt{6}}{29}$
67. $\frac{6y - 6\sqrt{y}}{y - 1}$ 69. $\frac{3t - 6\sqrt{tu}}{t - 4u}$ 71. $\frac{\sqrt{2xy} + 2y\sqrt{3}}{x - 6y}$
73. $\frac{3}{2\sqrt{3}}$ 75. $\frac{2x}{5\sqrt{2x}}$ 77. $\frac{2n}{3\sqrt{2n}}$ 79. $\frac{1}{10 - 5\sqrt{3}}$
81. $\frac{5x - 36}{9\sqrt{5x} + 54}$ 83. $\frac{19n}{10n\sqrt{n} - 2n\sqrt{6n}}$ 85. a. $\frac{5\sqrt{3}}{3}$ b. $\sqrt{5}$
c. $\frac{5\sqrt{11}}{11}$ 87. a. The graphs are identical. The functions are identical. b. $f(x) = g(x)$ 89. a. $T = 2\pi\sqrt{9.8L}/9.8$
b. $T = 2\pi L/\sqrt{9.8L}$ 91. a. $s = \frac{\sqrt{3Vh}}{h}$ b. 745 ft.
93. a. $V_{rms} = \frac{\sqrt{2}}{2}V_m$ or $\frac{\sqrt{2}V_m}{2}$ b. $V_{rms} = \frac{163\sqrt{2}}{2}$ c. ≈ 115.3
95. $\frac{15\sqrt{2} - 10\sqrt{3}}{3}\ \Omega$

10.6 Exercises
1. Some of the answers may be extraneous. 3. The principal square root of a number cannot equal a negative. 5. Subtract $3x$ from both sides to isolate the radical. This allows us to use the squaring principle of equality to eliminate the radical. 7. 4 9. no real-number solution 11. 27 13. -8 15. 17 17. 11 19. 6 21. $-\dfrac{1}{2}$ 23. no real-number solution 25. 11 27. -2 29. 124 31. 55 33. 5 35. -11 37. no real-number solution 39. 2 41. no real-number solution 43. 3 45. 3 47. 12 (5 is an extraneous solution) 49. 1, 3 51. $-3, -2$ 53. $\dfrac{1}{2}\left(-\dfrac{3}{8}\text{ is an extraneous solution}\right)$ 55. 5 57. $-1, 3$ 59. 0, 4 61. $\dfrac{3}{16}$ 63. 1 65. Mistake: You cannot take the principal square root of a number and get a negative. Correct: No real-number solution. 67. Mistake: The binomial $x-3$ was not squared correctly. Correct: $x=6$ with $x=1$ an extraneous solution. 69. 9.8 m 71. 2.45 m 73. 1.44 ft. 75. 144 ft. 77. 36.73 ft. 79. 82.65 ft. 81. 4 N 83. 6 N 85. a. 3, 4, 25 b.

c. No. The x-values must be 0 or positive because real square roots exist only when $x \geq 0$. The y-values must be 0 or positive because by definition the principal square root is either 0 or positive. d. Yes, because it passes the vertical line test. 87. The graph becomes steeper from left to right. 89. The graph rises or lowers according to the value of the constant.

10.7 Exercises
1. $\sqrt{-1}$ 3. No. The set of complex numbers contains both the real and the imaginary numbers. 5. We subtract complex numbers just like we subtract polynomials by writing an equivalent addition and changing the signs in the second complex number. 7. $6i$ 9. $i\sqrt{5}$ 11. $2i\sqrt{2}$ 13. $2i\sqrt{7}$ 15. $3i\sqrt{3}$ 17. $5i\sqrt{5}$ 19. $3i\sqrt{7}$ 21. $7i\sqrt{5}$ 23. $6+7i$ 25. $11-6i$ 27. $-7+i$ 29. $15+4i$ 31. $-3-10i$ 33. 0 35. $14+9i$ 37. $-4+28i$ 39. -24 41. 40 43. $14+12i$ 45. $72-32i$ 47. $19-3i$ 49. $50+9i$ 51. 65 53. $99+20i$ 55. $-2i$ 57. $-\dfrac{4i}{5}$ 59. $-3i$ 61. $-\dfrac{1}{2}-i$ 63. $\dfrac{1}{2}-i$ 65. $\dfrac{14}{5}-\dfrac{7}{5}i$ 67. $-\dfrac{7}{29}+\dfrac{3}{29}i$ 69. $7-2i$ 71. $\dfrac{9}{13}-\dfrac{7}{13}i$ 73. $\dfrac{34}{41}+\dfrac{19}{41}i$ 75. $-i$ 77. -1 79. -1 81. 1 83. 1 85. -1 87. $-i$ 89. i

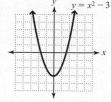

Chapter 10 Review Exercises
1. true 2. false 3. false 4. false 5. true 6. radicands 7. coefficients 8. 1. Multiply 9. conjugate. 10. extraneous 11. ± 11 12. ± 7 13. 13 14. -7 15. $6i$ 16. $\dfrac{1}{5}$ 17. 2.646 18. 9.487 19. $7x^4$ 20. $12a^3b^6$ 21. $0.4mn^5$ 22. x^5 23. $-4r^3s$ 24. $3x^3$ 25. $2x^3y^4$ 26. x^2y 27. $9|x|$ 28. $(x-1)^2$ 29. -4 30. $2a^2\sqrt{6}$ 31. $\dfrac{1}{8}$ 32. $\sqrt{(5r-2)^5}$ 33. $33^{1/8}$ 34. $8n^{-3/7}$ 35. $8^{3/5}$ 36. $m^{8/3}$ 37. $(3xw)^{3/4}$ 38. $(a+b)^{4/3}$ 39. x^2 40. $32m^{3/2}$ 41. $y^{-1/5}$ 42. b 43. $k^{1/2}$ 44. $2^{3/4}x^{3/4}y^{3/20}$ 45. 9 46. $10x^4$ 47. 2 48. $\sqrt[4]{42}$ 49. $\sqrt[5]{15x^4y^4}$ 50. $\dfrac{7}{11}$ 51. $-\dfrac{3}{2}$ 52. $6\sqrt{2}$ 53. $12b^4\sqrt{3b}$ 54. $15\sqrt[3]{4}$ 55. $4x^3\sqrt[3]{5x}$ 56. $6x^5\sqrt[4]{2x^3}$ 57. $84\sqrt{10}$ 58. $x^7\sqrt{x}$ 59. $16c^2\sqrt{15c}$ 60. $a^5b^4\sqrt{b}$ 61. $2\sqrt{2}$ 62. $6\sqrt{5}$ 63. $12a^2$ 64. $36x^2y^3\sqrt{2y}$ 65. $-3\sqrt{2}$ 66. $-6y^3\sqrt[4]{8y}$ 67. $5\sqrt{5}$ 68. $-10\sqrt{6}$ 69. $4\sqrt{6}$ 70. $28x\sqrt{2y}$ 71. $9xy\sqrt[3]{2xy^2}$ 72. $5\sqrt[3]{3}$ 73. $\sqrt{15}+\sqrt{10}$ 74. $\sqrt[3]{21}+2\sqrt[3]{49}$ 75. $6\sqrt{6}-54$ 76. $\sqrt{10}+\sqrt{14}-\sqrt{15}-\sqrt{21}$ 77. $-6+2\sqrt[3]{2}-4\sqrt[3]{4}$ 78. $\sqrt[4]{12x^2}-\sqrt[4]{6x}+2\sqrt[4]{2x}-2$ 79. $5a-3b$ 80. 7 81. $3+2\sqrt{2}$ 82. $7-2\sqrt{10}$ 83. $\dfrac{\sqrt{2}}{2}$ 84. $\sqrt[3]{9}$ 85. $\dfrac{2\sqrt{7}}{7}$ 86. $\dfrac{\sqrt[4]{40x^2}}{2}$ 87. $\dfrac{\sqrt[3]{153y}}{3y}$ 88. $\dfrac{\sqrt[4]{27}}{3}$ 89. $-4\sqrt{2}-4\sqrt{3}$ 90. $\dfrac{4-\sqrt{3}}{13}$ 91. $\dfrac{2+\sqrt{n}}{4-n}$ 92. $\sqrt{6}+2$ 93. $\dfrac{5}{3\sqrt{10}}$ 94. $\dfrac{3x}{5\sqrt{3x}}$ 95. $\dfrac{1}{16+8\sqrt{3}}$ 96. $\dfrac{1}{10\sqrt{t}-5\sqrt{3t}}$ 97. $x=81$ 98. no real-number solution 99. $w=10$ 100. $x=-2$ 101. $x=18$ 102. $y=5$ 103. no real-number solution 104. $\dfrac{719}{3}$ 105. 7 106. $1\left(\dfrac{17}{4}\text{ is extraneous}\right)$ 107. 0, 4 108. $\dfrac{3}{16}$ 109. $3i$ 110. $2i\sqrt{5}$ 111. $8-6i$ 112. $9-7i$ 113. -12 114. $2+8i$ 115. $26+2i$ 116. $24-10i$ 117. $-5i$ 118. $3i$ 119. $-\dfrac{4i}{3}$ 120. $\dfrac{1}{5}-\dfrac{7}{5}i$ 121. $\dfrac{6}{5}-\dfrac{3}{5}i$

122. $\frac{7}{13} + \frac{17}{13}i$ **123.** 1   **124.** $-i$ **125.** $\frac{\sqrt{6}}{4}$ ft.
126. a. $8\sqrt{5}$ mph   **b.** 17.9 mph   **127.** 312.5 ft.
128. π sec.   **129.** 0.272 m   **130. a.** $\frac{\sqrt{10}}{2}$ sec.   **b.** 1.58 sec.
131. 1.44 ft.   **132.** 5 ft.

Chapter 10 Practice Test   **1.** 6   **2.** $7i$ **3.** $9xy^2\sqrt{y}$
4. $3\sqrt[3]{2}$ **5.** $2x\sqrt[4]{x^2}$, or $2x\sqrt{x}$ **6.** $3r^5$ **7.** $\frac{1}{3}$ **8.** $\frac{1}{3}$ **9.** $5\sqrt{7}$
10. $4 - 2\sqrt{3}$ **11.** $x^{-2/3}$ **12.** $-6 + 2\sqrt[3]{2} - 4\sqrt[3]{4}$
13. $8^{1/5}x^{3/5}$ **14.** $(2x + 5)^{2/3}$ **15.** $\frac{\sqrt[3]{2}}{2}$ **16.** $\frac{x - \sqrt{xy}}{x - y}$
17. 22   **18.** 7   **19.** $-2 - 4i$ **20.** $2 + 0i$, or 2
21. $17 + 0i$, or 17   **22.** $\frac{8}{25} + \frac{6}{25}i$ **23.** 60 m²   **24. a.** $\frac{\sqrt{3}}{2}$
b. 64 ft.   **25. a.** $3.5\sqrt{10}$ **b.** 293.88 ft., or 14,400/49 ft.

Chapter 11
11.1 Exercises   **1.** $x^2 = a$ has two solutions because squaring $\sqrt{a}$ and $-\sqrt{a}$ gives a for every real number a.
3. $x = \frac{b \pm \sqrt{c}}{a}$ **5.** Since $x^2 - 7x + 12$ is easy to factor, factoring is better as it will require fewer steps than completing the square.   **7.** ± 7 **9.** $\pm\frac{2}{5}$ **11.** ± 0.9
13. $\pm 3\sqrt{5}$ **15.** $\pm 5i$ **17.** ± 7 **19.** ± 6 **21.** ± 3 **23.** $\pm\frac{3}{5}$
25. $\pm 2i$ **27.** $\pm\frac{3\sqrt{5}}{10}$ **29.** ± 4 **31.** $\pm 3i\sqrt{2}$ **33.** $\pm\frac{4\sqrt{6}}{3}$
35. ± 1.6 **37.** $-15, -1$ **39.** $-\frac{1}{5}, \frac{7}{5}$ **41.** $7 \pm 2i\sqrt{3}$
43. $\frac{1 \pm 2\sqrt{10}}{4}$ **45.** $8 \pm i$ **47.** $0, \frac{3}{2}$ **49.** $\frac{3}{5}, \frac{6}{5}$
51. $-13.5, -5.5$ **53.** Mistake: Gave only the positive solution. Correct: ± 7 **55.** Mistake: Changed -6 to 6. Correct: $5 \pm i\sqrt{6}$ **57.** $\pm 4\sqrt{6} \approx \pm 9.798$ **59.** $\pm 2\sqrt{5} \approx \pm 4.472$
61. $6 \pm \sqrt{15} \approx 9.873, 2.127$ **63. a.** $x^2 + 14x + 49$
b. $(x + 7)^2$ **65. a.** $n^2 - 10n + 25$ **b.** $(n - 5)^2$
67. a. $y^2 - 7y + \frac{49}{4}$ **b.** $\left(y - \frac{7}{2}\right)^2$ **69. a.** $s^2 - \frac{2}{3}s + \frac{1}{9}$
b. $\left(s - \frac{1}{3}\right)^2$ **71. a.** $m^2 + \frac{1}{7}m + \frac{1}{196}$ **b.** $\left(m + \frac{1}{14}\right)^2$
73. a. $p^2 + 9p + \frac{81}{4}$ **b.** $\left(p + \frac{9}{2}\right)^2$ **75.** $-5, 3$ **77.** $-8, -2$
79. $1 \pm 7i$ **81.** $3, 6$ **83.** $-12, 3$ **85.** $1 \pm \sqrt{17}$
87. $3 \pm 2i\sqrt{5}$ **89.** $-\frac{3}{2}, 1$ **91.** $-\frac{7}{2}, \frac{1}{2}$ **93.** $-1, \frac{3}{2}$
95. $-\frac{3}{2}, \frac{2}{3}$ **97.** $-2, \frac{3}{2}$ **99.** $\frac{2 \pm \sqrt{10}}{2}$ **101.** $\frac{-1 \pm \sqrt{41}}{10}$

103. $\frac{-3 \pm 2\sqrt{3}}{3}$ **105.** Did not divide by 3 so that x^2 has a coefficient of 1. Then wrote an incorrect factored form.
107. 14 in.   **109.** 3 ft., 6 ft.   **111.** 305 m   **113.** 18 in.
115. 15 ft. by 21 ft.   **117.** 16 ft. by 12 ft.   **119.** 15 in.
121. a. 15 cm   **b.** 4.5 cm by 30 cm   **123.** $\frac{3}{4}$ sec.
125. 4.9 sec.   **127.** 4 m/sec.

Review Exercises   **1.** $(x - 5)^2$ **2.** $(x + 3)^2$ **3.** $6x$
4. $2x + 3$ **5.** $\frac{-2 + \sqrt{6}}{2}$ or $-1 + \frac{\sqrt{6}}{2}$ **6.** $\frac{-3 - i}{5}$ or $-\frac{3}{5} - \frac{1}{5}i$

11.2 Exercises   **1.** Follow the procedure for solving a quadratic equation by completing the square on $ax^2 + bx + c = 0$.   **3.** Yes, if they have irrational or nonreal complex solutions.   **5.** When the discriminant is zero. The quadratic formula becomes $x = -\frac{b}{2a}$.   **7.** $a = 1, b = -3, c = 7$
9. $a = 3, b = -9, c = -4$ **11.** $a = 1.5, b = -1, c = 0.2$ or $a = -1.5, b = 1, c = -0.2$ **13.** $a = -\frac{1}{2}, b = -\frac{3}{4}, c = 6$ or $a = \frac{1}{2}, b = \frac{3}{4}, c = -6$ **15.** $-5, -4$ **17.** $\frac{1 \pm \sqrt{5}}{2}$
19. $-2, \frac{3}{4}$ **21.** $0, 9$ **23.** 4   **25.** $1 \pm i$ **27.** $-\frac{1}{4}, 1$
29. $\frac{-5 \pm \sqrt{10}}{3}$ **31.** $\frac{3 \pm i\sqrt{11}}{4}$ **33.** $-\frac{1}{6}, -\frac{2}{3}$ **35.** $\frac{2 \pm i\sqrt{5}}{3}$
37. $\frac{3 \pm \sqrt{105}}{12}$ **39.** $-0.15, 0.1$ **41.** $-2, \frac{3}{2}$ **43.** $\pm\frac{7}{6}$
45. $1 \pm i\sqrt{2}$ **47.** $\frac{-3 \pm \sqrt{5009}}{100} \approx -0.738, 0.678$
49. $\frac{3 \pm i\sqrt{51}}{12}$ **51.** Mistake: Did not evaluate $-b$.
Correct: $\frac{7 \pm \sqrt{37}}{6}$ **53.** Mistake: The result was not completely simplified. Correct: $1 \pm i\sqrt{2}$ **55.** one rational
57. one rational   **59.** two nonreal complex   **61.** two nonreal complex   **63.** two irrational   **65.** square root principle or factoring; ± 9 **67.** square root principle; $\pm 2i\sqrt{3}$
69. quadratic formula; $3 \pm 2i$ **71.** factoring; $0, -6$
73. quadratic formula; $4 \pm i\sqrt{3}$ **75.** $(2, 0), (-1, 0), (0, -2)$
77. $(-4, 0), (2, 0), (0, 8)$ **79.** $\left(\frac{1}{2}, 0\right), (-8, 0), (0, -8)$
81. no x-intercepts, $(0, -6)$ **83.** $x^2 + 5(x + 1) = 71; 6, 7$
85. $x^2 + (x + 1)^2 = (x + 2)^2; 3, 4, 5$ **87.** $w(3w - 3.5) = 34$; width: 4 ft., length: 8.5 ft.   **89. a.** $22x = 1.5x^2 + 4x$; 12 ft.
b. 18 ft., 12 ft., 10 ft., 6 ft., 8 ft., 18 ft. (from bottom clockwise)
91. ≈ 3.43 sec.   **93. a.** $h = -4.9t^2 + 10$ **b.** ≈ 1.01 sec.
c. ≈ 1.43 sec.   **95.** 8000 units   **97. a.** $\frac{25}{8}$ **b.** $c < \frac{25}{8}$
c. $c > \frac{25}{8}$ **99. a.** $\frac{9}{2}$ **b.** $a < \frac{9}{2}$ **c.** $a > \frac{9}{2}$

Review Exercises **1.** $(u-7)(u-2)$
2. $(3u-8)(u+2)$ **3.** $\frac{2}{5}, -3$ **4.** 8 **5.** x^4 **6.** $x^{2/3}$

11.3 Exercises **1.** It can be rewritten as a quadratic

equation. **3.** $u = \frac{x+2}{3}$ **5.** $\frac{1}{x}$ **7.** $-\frac{4}{3}, 2$ **9.** $-6, 4$

11. -8 (4 is extraneous) **13.** $-7.5, 5$ **15.** $-4, 2$ **17.** $-5, \frac{2}{3}$

19. $9, 25$ **21.** $\frac{49}{4}$ (1 is extraneous) **23.** $2\left(\frac{2}{9} \text{ is extraneous}\right)$

25. $\pm 3i$ **27.** $2\left(-\frac{2}{9} \text{ is extraneous}\right)$ **29.** $0, 4$ **31.** $\pm 3, \pm 1$

33. $\pm 1.5, \pm 1$ **35.** $\pm 2, \pm 3i$ **37.** $-6, -4$ **39.** $-3.5, 2$

41. $-9, -5$ **43.** $-5, 0$ **45.** $8, 27$ **47.** $-\frac{1}{8}, 8$ **49.** $81, 256$

51. $\frac{16}{625}$ (16 is extraneous) **53.** 3 hr. **55.** ≈ 2.16 hr.

57. 45 mph **59. a.** Old rate = 12 mph; new rate = 18 mph
b. Old time = 3 hr.; new time = 2 hr. **61.** Jody: 4 hr.;
Billy: 6 hr. **63.** ≈ 4.27 hr., 3.77 hr. **65.** 420 vps; 460 vps

Review Exercises **1.** -2 **2.** $(3, 0), (0, 2)$ **3.** 6
4.

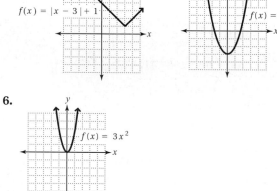

$f(x) = |x-3|+1$ $f(x) = x^2 - 3$

6.

$f(x) = 3x^2$

11.4 Exercises **1.** The sign of a. **3.** The axis of
symmetry is the vertical line through the vertex. If $y = ax^2 + bx + c$, then the equation of symmetry is $x = -\frac{b}{2a}$.
5. There are no x-intercepts. **7.** upwards **9.** downwards
11. upwards **13.** downwards **15.** downwards
17. V: $(2, -3)$; Axis: $x = 2$ **19.** V: $(-1, -5)$; Axis: $x = -1$
21. V: $(0, 2)$; Axis: $x = 0$ **23.** V: $(7, 0)$; Axis: $x = 7$
25. V: $(0, 0)$; Axis: $x = 0$ **27.** V: $(2, 4)$; Axis: $x = 2$

29. V: $(-4, -5)$; Axis: $x = -4$ **31.** V: $\left(\frac{1}{3}, \frac{4}{3}\right)$; Axis: $x = \frac{1}{3}$

33. V: $(-0.4, 1.08)$; Axis: $x = -0.4$ **35. a.** downwards
b. $(0, 0)$ **c.** $x = 0$
d.

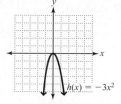

$h(x) = -3x^2$

37. a. upwards **b.** $(0, 0)$ **c.** $x = 0$
d. $k(x) = \frac{1}{4}x^2$

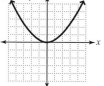

39. a. upwards **b.** $(0, -3)$ **c.** $x = 0$
d. $f(x) = 4x^2 - 3$

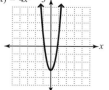

41. a. downwards **b.** $(0, 2)$ **c.** $x = 0$
d. $g(x) = -0.5x^2 + 2$

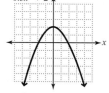

43. a. upwards **b.** $(3, 2)$ **c.** $x = 3$
d. $f(x) = (x-3)^2 + 2$

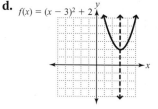

45. a. downwards **b.** $(-1, -3)$ **c.** $x = -1$
d.

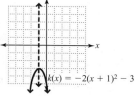

$k(x) = -2(x+1)^2 - 3$

47. a. upwards **b.** $(2, -1)$ **c.** $x = 2$

d. $h(x) = \frac{1}{3}(x - 2)^2 - 1$

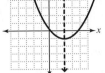

49. a. $(-3, 0), (0, 9)$ **b.** $h(x) = (x + 3)^2$ **c.** upwards

d. $(-3, 0)$ **e.** $x = -3$

f. $h(x) = x^2 + 6x + 9$

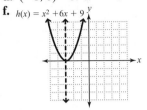

51. a. $\left(\dfrac{-3 + \sqrt{3}}{2}, 0\right), \left(\dfrac{-3 - \sqrt{3}}{2}, 0\right), (0, 3)$

b. $f(x) = 2\left(x + \dfrac{3}{2}\right)^2 - \dfrac{3}{2}$ **c.** upwards **d.** $\left(-\dfrac{3}{2}, -\dfrac{3}{2}\right)$

e. $x = -\dfrac{3}{2}$ **f.** $f(x) = 2x^2 + 6x + 3$

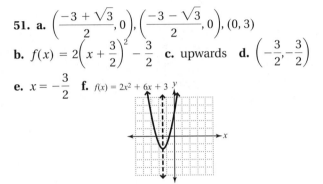

53. a. no x-intercepts, $(0, -5)$ **b.** $g(x) = -3(x - 1)^2 - 2$

c. downwards **d.** $(1, -2)$ **e.** $x = 1$

f.

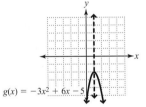

$g(x) = -3x^2 + 6x - 5$

55. a. 4.59 sec. **b.** 103.32 m **c.** 9.18 sec.

57. a. 7.8 ft. **b.** ≈ 4.62 ft. **c.**

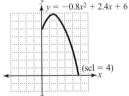

$y = -0.8x^2 + 2.4x + 6$

(scl = 4)

59. a. The greatest number of CDs were sold in the tenth week. **b.** 20,000 **61.** The area is maximized if the length and width are 100 ft. **63.** 55 units **65.** $-6, 6$; The product is -36.

Review Exercises **1.** -11 **2.** $x = -5, 3$

3. $x = \pm 2i\sqrt{5}$

4. $x \le 4.5$

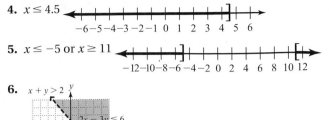

5. $x \le -5$ or $x \ge 11$

6. $x + y > 2$

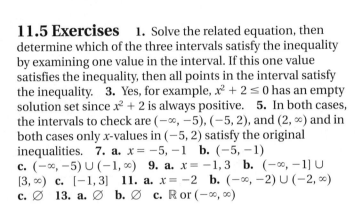

$2x - 3y \le 6$

11.5 Exercises

1. Solve the related equation, then determine which of the three intervals satisfy the inequality by examining one value in the interval. If this one value satisfies the inequality, then all points in the interval satisfy the inequality. **3.** Yes, for example, $x^2 + 2 \le 0$ has an empty solution set since $x^2 + 2$ is always positive. **5.** In both cases, the intervals to check are $(-\infty, -5)$, $(-5, 2)$, and $(2, \infty)$ and in both cases only x-values in $(-5, 2)$ satisfy the original inequalities. **7. a.** $x = -5, -1$ **b.** $(-5, -1)$

c. $(-\infty, -5) \cup (-1, \infty)$ **9. a.** $x = -1, 3$ **b.** $(-\infty, -1] \cup [3, \infty)$ **c.** $[-1, 3]$ **11. a.** $x = -2$ **b.** $(-\infty, -2) \cup (-2, \infty)$

c. $\varnothing$ **13. a.** $\varnothing$ **b.** $\varnothing$ **c.** $\mathbb{R}$ or $(-\infty, \infty)$

15. $(-4, -2)$

17. $(-\infty, 2) \cup (5, \infty)$

19. $(-4, -1)$

21. $(-\infty, 1) \cup (3, \infty)$

23. $[2, 4]$

25. $(-\infty, -1] \cup [5, \infty)$

27. $(-2, 5)$

29. $(-\infty, \infty)$, or $\mathbb{R}$

31. no solution, $\varnothing$

33. $(-\infty, -5) \cup (-0.25, \infty)$

$-6\,-5\,-4\,-3\,-2\,-1\ 0\ 1\ 2\ 3\ 4\ 5\ 6$

35. $(-\infty, 0) \cup (5, \infty)$

$-6\,-5\,-4\,-3\,-2\,-1\ 0\ 1\ 2\ 3\ 4\ 5\ 6$

37. $[0, 6]$

$-6\,-5\,-4\,-3\,-2\,-1\ 0\ 1\ 2\ 3\ 4\ 5\ 6$

39. $(-\infty, \infty)$, or $\mathbb{R}$

$-6\,-5\,-4\,-3\,-2\,-1\ 0\ 1\ 2\ 3\ 4\ 5\ 6$

41. $[-4, -2] \cup [4, \infty)$

$-6\,-5\,-4\,-3\,-2\,-1\ 0\ 1\ 2\ 3\ 4\ 5\ 6$

43. $(-\infty, -6) \cup (-2, 1)$

$-9\,-8\,-7\,-6\,-5\,-4\,-3\,-2\,-1\ 0\ 1\ 2\ 3$

45. $\left(\dfrac{-2 - \sqrt{7}}{3}, \dfrac{-2 + \sqrt{7}}{3} \right)$

47. $\left(-\infty, \dfrac{-2 - \sqrt{7}}{2} \right] \cup \left[\dfrac{-2 + \sqrt{7}}{2}, \infty \right)$

49. $(-\infty, -4 - \sqrt{6}] \cup [-4 + \sqrt{6}, \infty)$

51. $(-\infty, -4) \cup (1, \infty)$

$-6\,-5\,-4\,-3\,-2\,-1\ 0\ 1\ 2\ 3\ 4\ 5\ 6$

53. $(-5, -1]$

$-6\,-5\,-4\,-3\,-2\,-1\ 0\ 1\ 2\ 3\ 4\ 5\ 6$

55. $(-4, \infty)$

$-6\,-5\,-4\,-3\,-2\,-1\ 0\ 1\ 2\ 3\ 4\ 5\ 6$

57. $\left(-\infty, -\dfrac{9}{2} \right) \cup (-3, \infty)$

$-6\,-5\,-4\,-3\,-2\,-1\ 0\ 1\ 2\ 3\ 4\ 5\ 6$

59. $(4, 7)$

$-1\ 0\ 1\ 2\ 3\ 4\ 5\ 6\ 7\ 8\ 9\ 10\ 11$

61. $\left(3, \dfrac{14}{3} \right]$

$-6\,-5\,-4\,-3\,-2\,-1\ 0\ 1\ 2\ 3\ 4\ 5\ 6$

63. $(-\infty, -3] \cup [2, 5)$

$-6\,-5\,-4\,-3\,-2\,-1\ 0\ 1\ 2\ 3\ 4\ 5\ 6$

65. $(0, \infty)$

$-6\,-5\,-4\,-3\,-2\,-1\ 0\ 1\ 2\ 3\ 4\ 5\ 6$

67. $(-\infty, -1) \cup (2, 5)$

$-6\,-5\,-4\,-3\,-2\,-1\ 0\ 1\ 2\ 3\ 4\ 5\ 6$

69. a. 6 sec.   **b.** At 2 sec. and again at 3 sec.   **c.** $(2, 3)$ sec.   **d.** $[0, 2) \cup (3, 6]$ **71. a.** $x \geq 4$ **b.** base ≥ 10, height ≥ 2 **73.** $2 < x \leq 8$

Review Exercises   **1.** 18   **2.** $9, (x - 3)^2$

3. $\dfrac{25}{4}, \left(x + \dfrac{5}{2} \right)^2$ **4.** $(-6, 0), (2, 0)$ **5.** $(3, -4)$

6. Downwards, because the coefficient of x^2 is negative.

Chapter 11 Review Exercises   **1.** false   **2.** false

3. false   **4.** false   **5.** true   **6.** $+\sqrt{a}, -\sqrt{a}$ **7.** $\left(\dfrac{b}{2} \right)^2$

8. $\dfrac{-b \pm \sqrt{b^2 - 4ac}}{2a}$ **9.** $b^2 - 4ac$ **10.** V: (h, k); $x = h$

11. ± 4 **12.** $\pm \dfrac{1}{6}$ **13.** $\pm 2\sqrt{7}$ **14.** $\pm \sqrt{14}$ **15.** $\pm i\sqrt{3}$

16. $-12, -2$ **17.** $9 \pm 4i$ **18.** $-\dfrac{7}{5}, \dfrac{1}{5}$ **19.** $-7, -1$

20. $-8, 14$ **21.** $\dfrac{3 \pm i\sqrt{5}}{2}$ **22.** $-1, \dfrac{3}{4}$ **23.** $-1 \pm \sqrt{6}$

24. $\dfrac{1 \pm i\sqrt{2}}{3}$ **25.** $\dfrac{-1 \pm \sqrt{41}}{4}$ **26.** $-\dfrac{3}{20}, \dfrac{1}{10}$ **27.** $D = 64$;

two rational   **28.** $D = -71$; two nonreal complex

29. $D = 0$; one rational   **30.** $D = 0.48$; two irrational

31. $\pm \dfrac{3\sqrt{2}}{2}$ **32.** $-\dfrac{5}{2}$ (5 is extraneous)   **33.** $\dfrac{1}{3}, \dfrac{1}{2}$

34. $\dfrac{3 \pm \sqrt{17}}{4}$ **35.** no solution   **36.** $1 \left(\dfrac{1}{9} \text{ is extraneous} \right)$

37. $2 \left(-\dfrac{3}{2} \text{ is extraneous} \right)$ **38.** $\dfrac{1}{3}, \dfrac{2}{3}$ **39.** $\pm \sqrt{2}, \pm \sqrt{3}$

40. $\pm 1, \pm \dfrac{\sqrt{2}}{2}$ **41.** $-\dfrac{9}{2}, -\dfrac{14}{3}$ **42.** $-26, -2$ **43.** $27, 512$

44. $\dfrac{16}{625}$ (1 is extraneous)   **45. a.** $(0, 0)$ **b.** downwards

c. $(0, 0)$ **d.** $x = 0$ **e.**

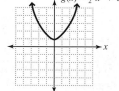

$f(x) = -2x^2$

46. a. no x-intercepts, $(0, 1)$ **b.** upwards   **c.** $(0, 1)$

d. $x = 0$ **e.**

y $g(x) = \frac{1}{2} x^2 + 1$

47. a. $(2, 0), \left(0, -\dfrac{4}{3} \right)$ **b.** downwards   **c.** $(2, 0)$ **d.** $x = 2$

e.

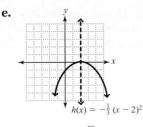

$h(x) = -\frac{1}{3}(x-2)^2$

48. a. $\left(-3 + \frac{\sqrt{2}}{2}, 0\right), \left(-3 - \frac{\sqrt{2}}{2}, 0\right), (0, 34)$ **b.** upwards

c. $(-3, -2)$ **d.** $x = -3$ **e.**

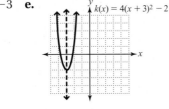

$k(x) = 4(x+3)^2 - 2$

49. a. $(-1 + \sqrt{2}, 0), (-1 - \sqrt{2}, 0), (0, -1)$
b. $m(x) = (x+1)^2 - 2$ **c.** upwards   **d.** $(-1, -2)$
e. $x = -1$ **f.**

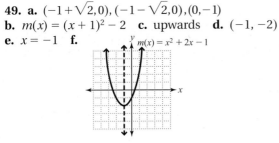

$m(x) = x^2 + 2x - 1$

50. a. $(2, 0), (6, 0), (0, -6)$ **b.** $p(x) = -0.5(x+4)^2 + 2$
c. downwards   **d.** $(4, 2)$ **e.** $x = 4$ **f.**

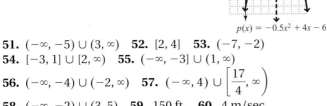

$p(x) = -0.5x^2 + 4x - 6$

51. $(-\infty, -5) \cup (3, \infty)$ **52.** $[2, 4]$ **53.** $(-7, -2)$
54. $[-3, 1] \cup [2, \infty)$ **55.** $(-\infty, -3] \cup (1, \infty)$
56. $(-\infty, -4) \cup (-2, \infty)$ **57.** $(-\infty, 4) \cup \left[\frac{17}{4}, \infty\right)$
58. $(-\infty, -2) \cup (3, 5)$ **59.** 150 ft.   **60.** 4 m/sec.
61. 15 ft. by 19 ft.   **62.** $9\sqrt{3}$ **63.** 6.65 ft. and 15.65 ft.
64. a. 0.75 sec.   **b.** 9 ft.   **c.** 1.5 sec.
d.

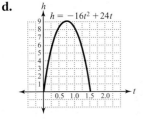

$h = -16t^2 + 24t$

65. a. 38.88 yd.   **b.** 72 yd.   **66. a.** $x \geq 10$ in.   **b.** height $\geq$ 8 in., base $\geq$ 14 in.

Chapter 11 Practice Test   **1.** ± 9 **2.** $3 \pm 2\sqrt{5}$

3. $4 \pm 2\sqrt{3}$ **4.** $\frac{3 \pm 2\sqrt{6}}{3}$ **5.** $-2, \frac{3}{2}$ **6.** 3, 5   **7.** $-8, 2$

8. ± 9 **9.** $\frac{-3 \pm i\sqrt{3}}{4}$ **10.** $0, -2$ **11.** $\pm 4i$ **12.** $0, -\frac{5}{3}$

13. $-\frac{6}{5}, 4$ **14.** $-\frac{2}{3}, 1$ **15.** no solution   **16.** 1 (-4 is extraneous)   **17.** $\pm\frac{1}{3}, \pm i\sqrt{3}$ **18.** $-5, 0$

19. a. $(3 + \sqrt{5}, 0)(3 - \sqrt{5}, 0), (0, -4)$
b. $f(x) = -(x-3)^2 + 5$ **c.** downwards   **d.** $(3, 5)$
e. $x = 3$ **f.**

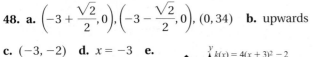

$(3, -\sqrt{5}, 0)$ $(3, +\sqrt{5}, 0)$ $(0, -4)$ $f(x) = -x^2 + 6x - 4$

20. $[-1, 4]$

-6 -5 -4 -3 -2 -1 0 1 2 3 4 5 6

21. $(-\infty, -2) \cup (1, \infty)$

-6 -5 -4 -3 -2 -1 0 1 2 3 4 5 6

22. 2.5 sec.   **23.** 12.36 ft. by 32.36 ft.   **24. a.** 29.125 m
b. 70.66 m   **c.**

$y = -0.02x^2 + 1.3x + 8$

25. a. $0 < w \leq 25$ **b.** $15 < l \leq 40$

Chapter 12

12.1 Exercises   **1.** No, $f \circ g = g \circ f$ if f and g are inverses of each other and for a few other functions.   **3.** The domain of a function is the range of its inverse; the range of a function is the domain of its inverse. This occurs because the ordered pairs of f and f^{-1} have x and y interchanged.   **5.** If any horizontal line intersects the graph in more than one point, it is not one-to-one.   **7.** 14   **9.** 3   **11.** 26   **13.** $3x^2 + 14$
15. $3\sqrt{x+1} + 5$ **17.** $\sqrt{6}$ **19.** $(f \circ g)(x) = 6x + 6$;
$(g \circ f)(x) = 6x - 2$ **21.** $(f \circ g)(x) = x^2 + 3; (g \circ f)(x) = x^2 + 4x + 5$ **23.** $(f \circ g)(x) = 9x^2 + 9x - 4; (g \circ f)(x) = 3x^2 + 9x - 12$ **25.** $(f \circ g)(x) = \sqrt{2x - 3}; (g \circ f)(x) = 2\sqrt{x+2} - 5$ **27.** $(f \circ g)(x) = \frac{2x - 3}{x - 3}; (g \circ f)(x) = \frac{1 - 2x}{x + 1}$

29. Domain is $[0, \infty)$ and range is $[3, \infty)$.   **31.** yes   **33.** no
35. yes   **37.** yes   **39.** no   **41.** yes   **43.** no   **45.** yes
47. yes   **49.** Yes, since $(f \circ g)(x) = (g \circ f)(x) = x$.

51.

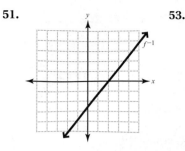

53.

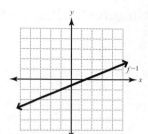

55.

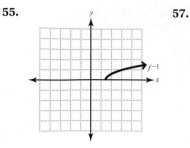

57.

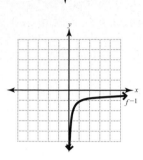

59.

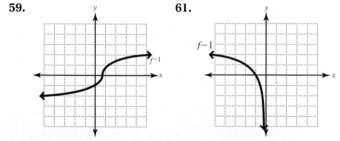

61.

63. yes **65.** no **67.** $f^{-1} = \{(2, -3), (-3, -1), (4, 0), (6, 4)\}$ **69.** $f^{-1} = \{(-2, 7), (2, 9), (1, -4), (3, 3)\}$

71. $f^{-1}(x) = x - 6$ **73.** $f^{-1}(x) = \dfrac{x - 3}{2}$

75. $f^{-1}(x) = \sqrt[3]{x - 2}$ **77.** $f^{-1}(x) = \dfrac{2 - 2x}{x}$

79. $f^{-1}(x) = \dfrac{3x + 2}{x - 1}$ **81.** $f^{-1}(x) = x^2 + 2, x \geq 0$

83. $f^{-1}(x) = \sqrt[3]{\dfrac{x - 4}{2}}$ **85.** $f^{-1}(x) = x^3 - 2$

87. $f^{-1}(x) = \dfrac{x^3}{16} - 2$ **89. a.** $y = \dfrac{x - 100}{0.05}$ **b.** x represents the salary, y represents the sales **c.** $5000

91. a. $y = \dfrac{6825 - x}{20}$ **b.** x represents the cost, y represents the number of 36-in. fans **c.** 80 36-in. fans **93.** $(5, -4)$

95. $f^{-1}(x) = \dfrac{x - b}{a}$ **97.** No, the inverse of an even function is not a function. If the graph of a function is symmetric with respect to the y-axis, the function is not one-to-one.

Review Exercises **1.** $32 = 2^5$ **2.** 8 **3.** $-\dfrac{1}{27}$ **4.** $\dfrac{1}{16}$

5. 8 **6.** $(3, 1)$

12.2 Exercises **1.** Yes, the graphs are symmetric. The point $(1, 2)$ on $f(x)$ corresponds to the point $(-1, 2)$ on $g(x)$, etc. **3.** The domain is all real numbers, the range is the interval $(0, \infty)$. **5.** Because the exponential function is one-to-one.

7.

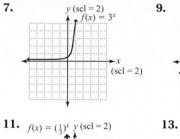

9.

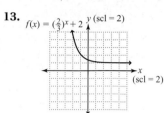

11.

13.

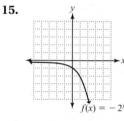

15.

17.

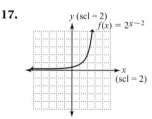

19.

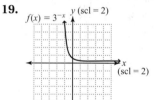

21.

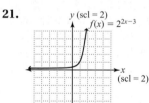

23.

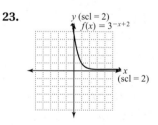

25. 3 **27.** $\dfrac{5}{3}$ **29.** $\dfrac{1}{2}$ **31.** -2 **33.** -2 **35.** 3 **37.** -4

39. $\dfrac{1}{2}$ **41.** -12 **43. .a.**

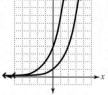

b. The graph of g is the graph of f shifted 2 units to the left.
45. a.

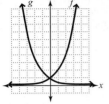

b. The graph of g is the graph of f reflected in the y-axis.
47. 6400 cells   **49.** 6144 billion people   **51.** $25,870.70
53. 0.020 g   **55.** $1046   **57.** 1.225 parts per million
59. ≈ 923 million transistors

Review Exercises   **1.** 125   **2.** 9   **3.** $\dfrac{1}{64}$ **4.** $\sqrt[3]{5^2}$ or

$(\sqrt[3]{5})^2$ **5.** $f^{-1}(x) = \dfrac{x-4}{3}$ **6.**

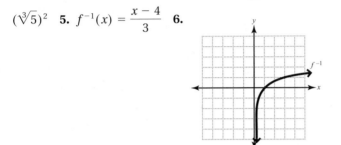

12.3 Exercises   **1.** $f^{-1}(x) = \log_3 x$, because logarithmic and exponential functions are inverses.   **3.** $x > 0$
5. Logarithms are exponents because logarithms are inverses of exponential functions.   **7.** $\log_2 32 = 5$ **9.** $\log_{10} 1000 = 3$

11. $\log_e x = 4$ **13.** $\log_5 \dfrac{1}{125} = -3$ **15.** $\log_{10} \dfrac{1}{100} = -2$

17. $\log_{625} 5 = \dfrac{1}{4}$ **19.** $\log_{1/4} \dfrac{1}{16} = 2$ **21.** $\log_7 \sqrt{7} = \dfrac{1}{2}$

23. $3^4 = 81$ **25.** $4^{-2} = \dfrac{1}{16}$ **27.** $10^2 = 100$ **29.** $e^5 = a$

31. $e^{-4} = \dfrac{1}{e^4}$ **33.** $\left(\dfrac{1}{8}\right)^2 = \dfrac{1}{64}$ **35.** $\left(\dfrac{1}{5}\right)^{-2} = 25$

37. $7^{1/2} = \sqrt{7}$ **39.** 8   **41.** $\dfrac{1}{25}$ **43.** 4   **45.** -1 **47.** 10

49. 2   **51.** $\dfrac{1}{4}$ **53.** 243   **55.** 2   **57.** -6 **59.** 16   **61.** $\dfrac{1}{64}$

63. 2   **65.** 125   **67.** 3   **69.** $\dfrac{1}{16}$

71.   **73.**

75. $\log_b b = 1$ because $b^1 = b$ **77.** Because $b^0 = 1$.
79. 97%   **81.** After 4 days.

Review Exercises   **1.** x^{-6} **2.** x^6 **3.** x^{15} **4.** x^3
5. x^5 **6.** $x^{3/4}$

12.4 Exercises   **1.** Because the exponential and logarithmic functions are inverses.   **3.** $\log_b(x + y)$; there is no rule for the logarithm of a sum.   **5.** 2   **7.** r **9.** $4x$
11. 5   **13.** a **15.** $7x$ **17.** $\log_2 5 + \log_2 y$
19. $\log_a p + \log_a q$ **21.** $\log_4 m + \log_4 n + \log_4 p$
23. $\log_a x + \log_a(x - 5)$ **25.** $\log_3(40)$ **27.** $\log_4(16)$
29. $\log_a 7m$ **31.** $\log_4 ab$ **33.** $\log_a(2x^2 + 10x)$
35. $\log_4(x^2 + 4x + 3)$ **37.** $\log_2 7 - \log_2 9$
39. $\log_a x - \log_a 5$ **41.** $1 - \log_a b$ **43.** $\log_a x - \log_a(x - 3)$
45. $\log_4(2x - 3) - \log_4(4x + 5)$ **47.** $\log_6 8$ **49.** $\log_2 2 = 1$

51. $\log_a \dfrac{x}{3}$ **53.** $\log_4 \dfrac{p}{q}$ **55.** $\log_b \dfrac{x}{x - 4}$ **57.** $\log_x x = 1$

59. $3 \log_7 4$ **61.** $7 \log_a x$ **63.** $\dfrac{1}{2} \log_a 3$ **65.** $\dfrac{2}{3} \log_3 x$

67. $-2 \log_a 6$ **69.** $-2 \log_a y$ **71.** $\log_3 5^4$ **73.** $\log_2 \dfrac{1}{x^3}$

75. $\log_5 2$ **77.** $\log_a \sqrt[4]{x^3}$ **79.** $\log_a 4$ **81.** $\log_3 \dfrac{1}{\sqrt{x}}$

83. $3 \log_a x - 4 \log_a y$ **85.** $4 \log_3 a + 2 \log_3 b$
87. $\log_a x + \log_a y - \log_a z$ **89.** $2 \log_x a - \log_x b - 3 \log_x c$
91. $\dfrac{3}{2} \log_4 x - \dfrac{1}{2} \log_4 y$ **93.** $\log_a x + \dfrac{1}{2} \log_a y - \dfrac{3}{2} \log_a z$

95. $\log_3 \dfrac{1}{2}$ **97.** $\log_b x^4 y^3$ **99.** $\log_a \sqrt{\dfrac{5}{7}}$ **101.** $\log_a \sqrt[3]{(x^2 y)^2}$

103. $\log_b(3x^2 - 2x)$ **105.** $\log_a \dfrac{(x - 2)^3}{(x + 1)^4}$ **107.** $\log_a \dfrac{x^2 z^4}{w^3 u^6}$

Review Exercises   **1.** $10^{-2.5}$ **2.** 10^9 **3.** $13,468.55
4. $\log_{10} 50 = 1.6990$ **5.** $10^{1.6532} = 45$ **6.** $e^{-1.3863} = 0.25$

12.5 Exercises   **1.** $b > 0$, $b \neq 1$. If $y = \log_b x$, then $b^y = x$.
If $b < 0$ and $y = \dfrac{1}{2}$ or any other fraction whose denominator is
even, then x is imaginary. If $b = 1$, then $x = 1$ for all values of y
and the graph is a vertical line.   **3.** Positive. If $y = \ln x$, then
$e^y = x$. If $x > 1$, then $e^y > 1$, which is true if $y > 0$.   **5.** 0.91629
7. 1.8062   **9.** -2.1739 **11.** 2.3941   **13.** 4.1761
15. -5.7959 **17.** 2.2343   **19.** 4.3720   **21.** -3.3814
23. 5.6864   **25.** 0.4343   **27.** Error results, because the
domain of $\log_a x$ is $(0, \infty)$, so log 0 is undefined.   **29.** 2

31. -1 **33.** $\dfrac{1}{3}$ **35.** -3 **37.** 3   **39.** $\dfrac{1}{2}$ **41.** 90 dB

43. 10^{-6} watts/m^2 **45.** 2.796   **47.** $10^{-3.5}$ moles/L
49. 22.9, or about 23 yr.   **51.** The 1964 Alaska earthquake
was about four times as severe as the 1906 San Francisco
earthquake.   **53.** $10^{22.75}$ ergs   **55. a.** About 11,809 deaths.
b. 1997   **57.** 6124 ft.

Review Exercises   **1.** -11 **2.** $-5, 3$ **3.** 8   **4.** 10
5. $\log_3 \dfrac{2x + 1}{x - 1}$ **6.** $3^2 = 2x + 5$

12.6 Exercises

1. $m = n$ **3.** If $\log_b x = \log_b y$, then $x = y$. **5.** The solution $x = -5$ is rejected because substituting into the equation gives $\log(-5)$ and $\log(-3)$, which are both undefined. **7.** 3.1699 **9.** 1.0767 **11.** -1.5693
13. 2.8617 **15.** 12.4251 **17.** 7.1285 **19.** 0.5365
21. 107.2959 **23.** -73.1563 **25.** 6 **27.** 11 **29.** 6 **31.** 4
33. $4, -9$ **35.** 7 **37.** no solution **39.** $-6, 2$ **41.** 2 **43.** 10
45. 3 **47.** $\dfrac{3}{2}$ **49.** 3 **51.** 4 **53.** $\dfrac{5}{6}$ **55.** $\dfrac{5}{2}$ **57.** 6 **59.** 4
61. 9.5 yr. **63. a.** $19,676.82 **b.** 9.3 yr. **65. a.** 5.58%
b. 0.18 **67. a.** 875 mosquitoes **b.** 75 days
69. a. ≈ 7.44 billion **b.** ≈ 9 yr. after 2005, in 2014.
71. a. 1.89 in. of mercury **b.** 21.8 mi. **73. a.** 79.8%
b. 7.76, so about age 8. **75.** $\approx 39{,}006$ **77. a.** $167,925
b. ≈ 19.1 after 1990, in 2009. **79.** ≈ 12 in. **81.** 1.7925
83. 0.5283 **85.** -2.3219 **87.** 0.3685

Review Exercises

1. $x + 3$ **2.** $28 + \sqrt{5}$ **3.** 29
4. $x^2 + 2x + 3$ **5.** $4x^2 - 4x + 5$ **6.** 13

Chapter 12 Review Exercises

1. true **2.** false
3. false **4.** false **5.** true **6.** false **7.** 25 **8.** 167 **9.** -2
10. 14 **11.** $(f \circ g)(x) = 6x + 3$, $(g \circ f)(x) = 6x - 9$
12. $(f \circ g)(x) = 9x^2 - 42x + 53$, $(g \circ f)(x) = 3x^2 + 5$
13. $(f \circ g)(x) = \sqrt{2x - 4}$, $(g \circ f)(x) = 2\sqrt{x - 3} - 1$
14. $(f \circ g)(x) = \dfrac{4x - 4}{x - 4}$, $(g \circ f)(x) = \dfrac{3 - 3x}{x + 3}$ **15.** no

16. yes

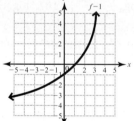

17. yes **18.** yes **19.** no **20.** no **21.** $f^{-1}(x) = \dfrac{x - 4}{5}$
22. $f^{-1}(x) = \sqrt[3]{x - 6}$ **23.** $f^{-1}(x) = \dfrac{4 - 5x}{x}$
24. $f^{-1}(x) = \dfrac{x^3 - 2}{3}$ **25.** $(-6, 4)$ **26.** $f^{-1}(b) = a$

27. **28.**

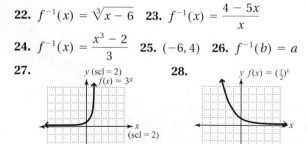

29. **30.**

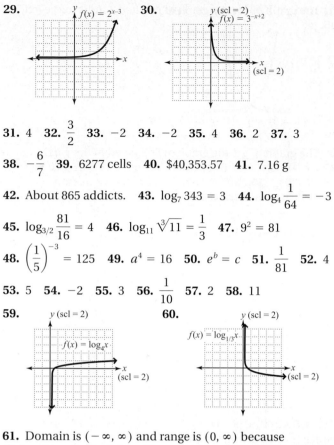

31. 4 **32.** $\dfrac{3}{2}$ **33.** -2 **34.** -2 **35.** 4 **36.** 2 **37.** 3
38. $-\dfrac{6}{7}$ **39.** 6277 cells **40.** $40,353.57 **41.** 7.16 g
42. About 865 addicts. **43.** $\log_7 343 = 3$ **44.** $\log_4 \dfrac{1}{64} = -3$
45. $\log_{3/2} \dfrac{81}{16} = 4$ **46.** $\log_{11} \sqrt[3]{11} = \dfrac{1}{3}$ **47.** $9^2 = 81$
48. $\left(\dfrac{1}{5}\right)^{-3} = 125$ **49.** $a^4 = 16$ **50.** $e^b = c$ **51.** $\dfrac{1}{81}$ **52.** 4
53. 5 **54.** -2 **55.** 3 **56.** $\dfrac{1}{10}$ **57.** 2 **58.** 11
59. **60.**

61. Domain is $(-\infty, \infty)$ and range is $(0, \infty)$ because $f(x) = \log_b x$ and $g(x) = b^x$ are inverses. **62.** 30 dB **63.** 8
64. 6 **65.** $1 + \log_6 x$ **66.** $\log_4 x + \log_4(2x - 5)$
67. $\log_3(32)$ **68.** $\log_5(3x^2 - 6x)$ **69.** $\log_b x - \log_b 5$
70. $\log_a(3x - 2) - \log_a(4x + 3)$ **71.** $\log_5 6$
72. $\log_2 \dfrac{x + 5}{2x - 3}$ **73.** $4 \log_3 7$ **74.** $\dfrac{1}{3} \log_a x$ **75.** $-4 \log_4 a$
76. $\dfrac{4}{5}$ **77.** $\log_a x^4$ **78.** $\log_a \sqrt[5]{y^3}$ **79.** $2 \log_a x + 3 \log_a y$
80. $4 \log_a c - 3 \log_a d$ **81.** $2 \log_a x + 3 \log_a y - 4 \log_a z$
82. $\dfrac{3}{2} - 2 \log_a b$ **83.** $\log_a b^2 c^4$ **84.** $\log_a \dfrac{4^3}{3^2}$ **85.** $\log_a \sqrt[4]{x^2 y^3}$
86. $\log_a(x + 5)^4(x - 3)^2$ **87.** 2.5132 **88.** -2.4559
89. -3.1466 **90.** 4.5218 **91.** -5 **92.** $\dfrac{1}{4}$ **93.** 85 dB
94. pH $= 2.9$ **95.** ≈ 17 yr. **96.** The 7.8 earthquake is 10 times as severe. **97.** 1.5773 **98.** 0.5360 **99.** 1.3626
100. 4.9592 **101.** 0.5995 **102.** -536.4793 **103.** 13
104. $-4, 2$ **105.** 3 **106.** 20 **107.** 36 **108.** 8 **109.** $\dfrac{5}{2}$
110. 5 **111.** 6.1 yr. **112. a.** $12,254.71 **b.** 7.7 yr.
113. a. 573 ants **b.** ≈ 23 months **114. a.** 36,165 cases
b. 1998 **115.** 1.9534 **116.** -2.5850

Chapter 12 Practice Test

1. $f[g(x)] = 9x^2 - 30x + 19$ **2. a.** $f^{-1}(x) = \dfrac{x+3}{4}$

b. $f[f^{-1}(x)] = f\left[\dfrac{x+3}{4}\right] = 4\left(\dfrac{x+3}{4}\right) - 3 = x + 3 - 3 = x$

$f^{1}[f(x)] = f^{-1}(4x - 3) = \dfrac{4x - 3 + 3}{4} = \dfrac{4x}{4} = x$

c. The graphs are symmetric about the graph of $y = x$.

3. 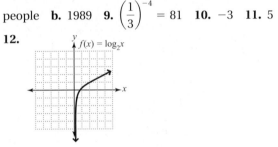   **4.**

5. -10 **6. a.** \$42,143.63   **b.** About 8.2 yr.   **7. a.** 0.68 g
b. About 1.16 hr.   **8. a.** 14.8 million people, 46.9 million

people   **b.** 1989   **9.** $\left(\dfrac{1}{3}\right)^{-4} = 81$ **10.** -3 **11.** 5

12.

13. $4 \log_b x + 2 \log_b y - \log_b z$ **14.** $\dfrac{5}{4} \log_b x - \dfrac{7}{4} \log_b y$
15. $\log_b \sqrt[4]{x^6 y^9}$ **16. a.** About 74 g   **b.** About 247.6 days.
17. 4.6433   **18.** 3   **19.** 3   **20. a.** 307 cases   **b.** 2001

Chapter 13

13.1 Exercises
1. It is a parabola, because only one
variable is squared.   **3.** Parabola opens to the right, because
the variable y is squared and $a = 3$, which is positive.
5. circle; $(-3, 2)$; 4   **7.** opens upwards; vertex: $(1, 2)$; axis of
symmetry: $x = 1$

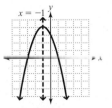

9. opens downwards; vertex: $(-1, 4)$;
axis of symmetry: $x = -1$

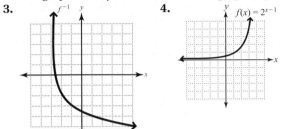

11. opens right; vertex: $(-2, -2)$;
axis of symmetry: $y = -2$

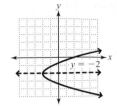

13. opens left; vertex: $(3, 1)$;
axis of symmetry: $y = 1$

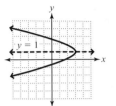

15. opens right; vertex: $(-4, -2)$;
axis of symmetry: $y = -2$

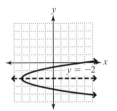

17. opens left; vertex: $(-5, -2)$;
axis of symmetry: $y = -2$

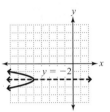

19. opens right; vertex: $(-1, -2)$;
axis of symmetry: $y = -2$

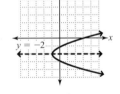

21. opens left; vertex: $(4, 3)$;
axis of symmetry: $y = 3$

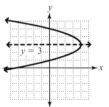

23. opens right; vertex: $(-5, -2)$;
axis of symmetry: $y = -2$

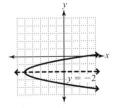

25. opens right; vertex: (0, 1);
axis of symmetry: $y = 1$

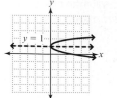

27. opens left; vertex: (7, 1);
axis of symmetry: $y = 1$

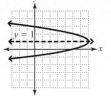

29. c **31.** a **33.** 5 **35.** 13 **37.** 13 **39.** $2\sqrt{5}$ **41.** $4\sqrt{5}$
43. $\sqrt{34}$ **45.** 5 **47.** $\sqrt{89}$
49. center: (2,1); radius: 2

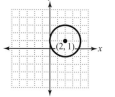

51. center: $(-3, -2)$; radius: 9

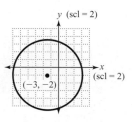

53. center: $(5, -3)$; radius: 7

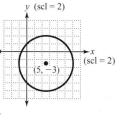

55. center: $(-1, 3)$; radius: $3\sqrt{2}$

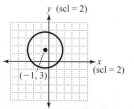

57. center: $(-4, -2)$; radius: $4\sqrt{2}$

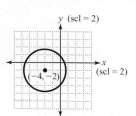

59. center: (1, 3); radius: 7

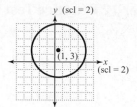

61. center: $(-5, 2)$; radius: 8

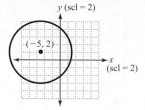

63. center: $(-7, 2)$; radius: 2

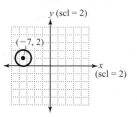

65. b **67.** c **69.**

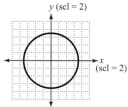

71.

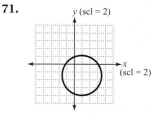

73. $(x - 8)^2 + (y + 1)^2 = 36$ **75.** $(x + 4)^2 + (y + 3)^2 = 25$
77. $(x - 6)^2 + (y + 2)^2 = 14$ **79.** $(x + 5)^2 + (y - 2)^2 = 45$
81. $(x - 2)^2 + (y - 4)^2 = 25$ **83.** $(x - 2)^2 + (y - 4)^2 = 169$
85. $(x - 2)^2 + (y + 5)^2 = 64$ **87. a.** 256 ft. **b.** 3 sec.

c. 7 sec. **89.** $y = -\dfrac{2}{25}x^2 + 18$ or $y = -0.08x^2 + 18$

91. 3.5% **93.** $x^2 + y^2 = 400$ **95.** $x^2 + (y - 110)^2 = 10,000$
97. $x^2 + y^2 = 43,560,000$

Review Exercises **1.**

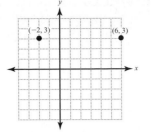

23. $\dfrac{x^2}{9} + \dfrac{y^2}{16} = 1$ **25.**

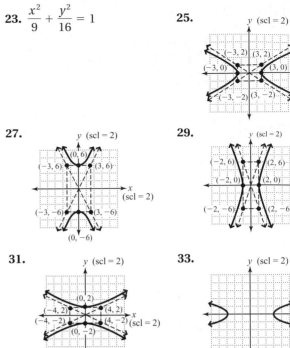

2.

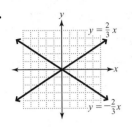

3. $(5x + 3y)(5x - 3y)$ **4.** ± 4

27.   **29.**

31.   **33.**

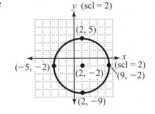

5. $(4, 0), (-4, 0), (0, 3), (0, -3)$ **6.** $(3, 0), (-3, 0), (0, 5), (0, -5)$

13.2 Exercises   **1.** An ellipse is the set of all points the sum of whose distances from two fixed points is constant. **3.** The equation of an ellipse is a sum; the equation of a hyperbola is a difference.   **5.** The parabola has only one squared term.

35. $\dfrac{x^2}{9} - \dfrac{y^2}{4} = 1$ **37.** c   **39.** b   **41.** ellipse   **43.** circle

45. circle

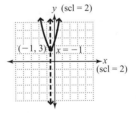

7.   **9.**

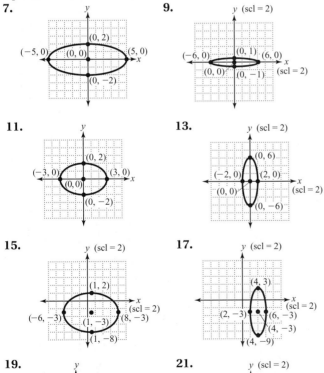

47. parabola

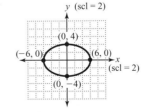

11.   **13.**

15.   **17.**

49. ellipse

19.   **21.**

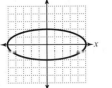
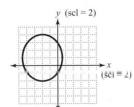

51. hyperbola

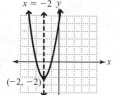

53. parabola

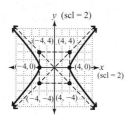

55. hyperbola

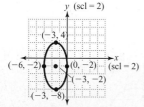

57. ellipse

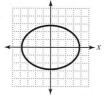

59. a. Yes, the sailboat will clear the bridge. The height of the bridge at the center is 20 feet and the boat's mast is only 18 feet above the water.   **b.** The bridge is 32 feet wide at the base of the arch.   **61.**

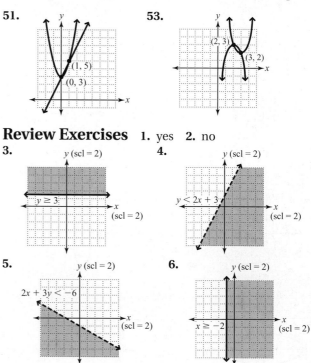

63. $\dfrac{x^2}{2.25} + \dfrac{y^2}{4} = 1$ **65.** 4 units

Review Exercises   **1.** $y = -3x^2 + 6$ **2.** $(1, 2)$

3. $(2, -3)$ **4.** $(-3, 4)$ **5.** $-4, \dfrac{2}{3}$ **6.** ± 3

13.3 Exercises   **1. a.** 0, 1, 2, 3, or 4 solutions are possible.   **b.** Answers may vary.

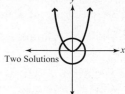

Two Solutions

3. Substitution, because the first equation has a linear term and is easy to solve for y.   **5.** The graph of the first equation is a line and the second is an ellipse, so the system could have 0, 1, or 2 solutions.   **7.** $(-2, 8), (1, 2)$ **9.** $(-2, 0), (1, 9)$ **11.** $(-4, -3), (3, 4)$ **13.** $(3.6, -0.2), (2, 3)$ **15.** no solution   **17.** $(0, 3), (1, 5)$ **19.** $(3, 2), (2, 3)$ **21.** $(4, 2), (4, -2),$ $(-4, -2), (-4, 2)$ **23.** $(3, 4), (-3, 4), (-3, -4), (3, -4)$ **25.** $(2, 1), (2, -1), (-2, -1), (-2, 1)$ **27.** $(-6, 0), (3, 3),$ $(3, -3)$ **29.** no solution   **31.** $(2, 4), (-2, 4)$ **33.** $(4, 0),$ $(-4, 0)$ **35.** $(1, 2), (-1, -2)$ **37.** $(5, 12), (-1, 0)$ **39.** $(2, 2),$ $(2, -2), (-2, -2), (-2, 2)$ **41.** $(-7, 3), (-7, -3), (2, 0)$
43. Answers will vary, but one possible system is $\begin{cases} x + y = 4 \\ x^2 + y^2 = 1 \end{cases}$.

45. $(5, 3), (5, -3), (-5, -3), (-5, 3)$ **47.** 8 in. by 18 in.

49. At equilibrium there should be 400 chairs at \$72 per chair.

51.   **53.**

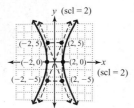

Review Exercises   **1.** yes   **2.** no

3.   **4.**

5.   **6.**

13.4 Exercises

1. No, the coordinates of the points on the curve make $9x^2 + 4y^2 = 36$. The curve would be a dashed curve because the points are not included in the solution.

3. One possible example is the system $\begin{cases} x^2 + y^2 < 1 \\ x^2 + y^2 > 4 \end{cases}$. The graph of $x^2 + y^2 < 1$ is the region inside the circle with radius 1. The graph of $x^2 + y^2 > 4$ is the region outside the circle with radius 2. Thus, there is no common region of solution.

5.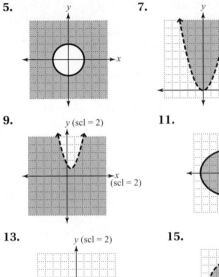

7.

9.

11.

13.

15.

17.

19.

21.

23.

25.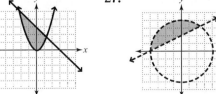

27.

29.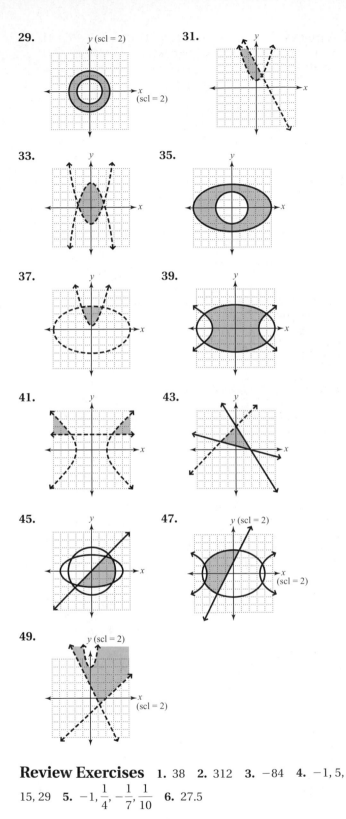

31.

33.

35.

37.

39.

41.

43.

45.

47.

49.

Review Exercises

1. 38   **2.** 312   **3.** -84 **4.** $-1, 5,$ $15, 29$ **5.** $-1, \dfrac{1}{4}, -\dfrac{1}{7}, \dfrac{1}{10}$ **6.** 27.5

Chapter 13 Review Exercises

1. false   **2.** true   **3.** false   **4.** true   **5.** true   **6.** false   **7.** 10   **8.** $(-4, 1); r = 2$

9. ellipse; $(4, 0)$ and $(-4, 0)$ **10.** $(5, 4), (5, -4), (-5, -4),$ $(-5, 4)$ **11.** opens upwards; vertex: $(3, -5)$; axis of symmetry: $x = 3$

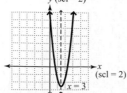

12. opens downwards; vertex: $(-2, 3)$; axis of symmetry: $x = -2$

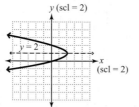

13. opens left; vertex: $(4, 2)$; axis of symmetry: $y = 2$

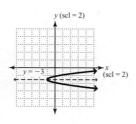

14. opens right; vertex: $(-2, -3)$; axis of symmetry: $y = -3$

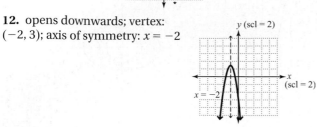

15. opens right; vertex: $(-1, -3)$; axis of symmetry: $y = -3$

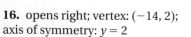

16. opens right; vertex: $(-14, 2)$; axis of symmetry: $y = 2$

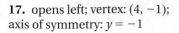

17. opens left; vertex: $(4, -1)$; axis of symmetry: $y = -1$

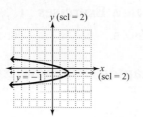

18. opens left; vertex: $(3, -2)$; axis of symmetry: $y = -2$

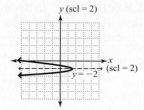

19. 5   **20.** $4\sqrt{5}$ **21.** 10   **22.** $(x + 6)^2 + (y - 8)^2 = 225$
23. center: $(3, -2)$; radius: 5

24. center: $(-5, 1)$; radius: 2

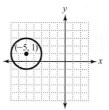

25. center: $(2, -4)$; radius: 3

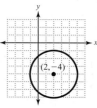

26. center: $(-5, -1)$; radius: 2

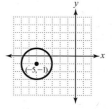

27. $(x - 6)^2 + (y + 8)^2 = 81$ **28.** $(x + 3)^2 + (y + 5)^2 = 100$ **29. a.** 144 ft.   **b.** 1 sec.   **c.** 4 sec.   **30.** $x^2 + y^2 = 400$
31.

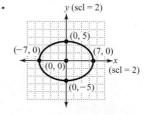

32.

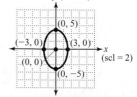

33.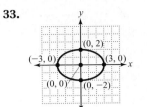

34.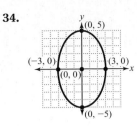

35.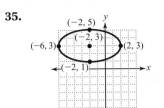

36.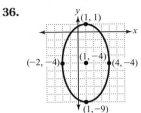

37.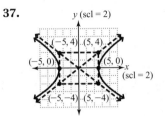

38.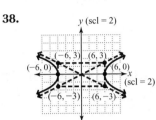

39.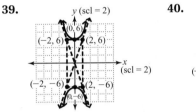

40.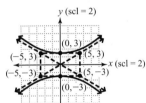

41. $\dfrac{x^2}{625} + \dfrac{y^2}{225} = 1$ **42.** $\dfrac{x^2}{1.625^2} + \dfrac{y^2}{4} = 1$ **43.** $(2, 5),$
$(-1, -1)$ **44.** $(-1, -4), (4, 1)$ **45.** $(-73/37, -35/37), (1, 5)$
46. $(8, 0), (-8, 0)$ **47.** $(3, 4), (3, -4), (-3, -4), (-3, 4)$
48. $(4, 2), (4, -2), (-4, -2), (-4, 2)$ **49.** $(2, 2), (-2, 2)$
50. $(2, 3), (-2, 3)$ **51.** $(8, 5), (8, -5), (-8, -5), (-8, 5)$
52. 8 ft. by 10 ft. **53.**

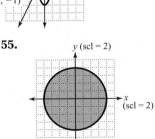

54.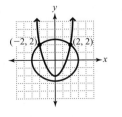

55.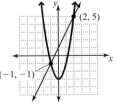

56.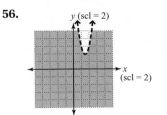

57.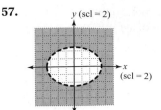

58.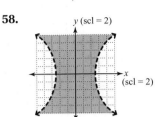

59.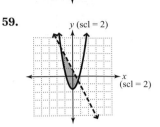

60.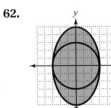

61.

62.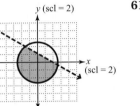

63.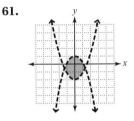

64.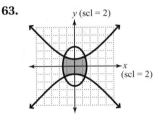

Chapter 13 Practice Test

1. opens upwards; vertex: $(-1, -4)$;
axis of symmetry: $x = -1$

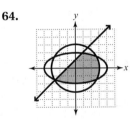

2. opens left; vertex: $(1, 3)$;
axis of symmetry: $y = 3$

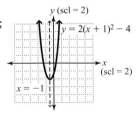

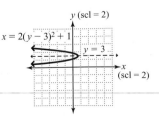

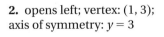

3. opens right; vertex: $(-7, -2)$; axis of symmetry: $y = -2$

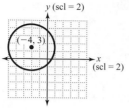

4. $2\sqrt{5}$ **5.** center: $(-4, 3)$; radius: 6

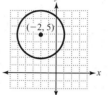

6. center: $(-2, 5)$, radius: 3

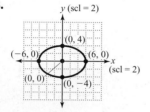

7. $(x - 2)^2 + (y + 4)^2 = 100$ **8.**

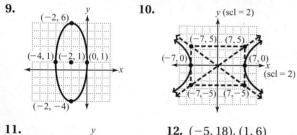

9.

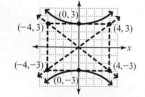

10.

11.

12. $(-5, 18), (1, 6)$

13. $(3, 5), (-3/5, -29/5)$ **14.** $(2, 3), (2, -3), (-2, -3), (-2, 3)$ **15.** $(3, 2), (3, -2), (-3, -2), (-3, 2)$

16.

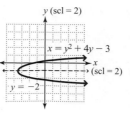

17.

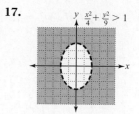

18.

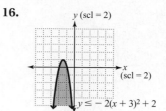

19. Height is 18, distance across the base is 12.

20. $\dfrac{x^2}{9.5^2} + y^2 = 1$

Chapters 1–13 Cumulative Review Exercises

1. true **2.** true **3.** true **4.** true **5.** false **6.** false
7. $(x - h)^2 + (y - k)^2 = r^2$ **8.** i
9. $d = \sqrt{(x_2 - x_1)^2 + (y_2 - y_1)^2}$ **10.** $a_1 b_2 - a_2 b_1$
11. $9\dfrac{1}{16}$ **12.** $2i\sqrt{5}$ **13.** $\dfrac{8}{3x}$ **14.** $\dfrac{x}{3x^2 + 1}$ **15.** $-\dfrac{1}{4}$
16. $\dfrac{n^6}{125m^9}$ **17.** $(k^2 + 9)(k + 3)(k - 3)$

18. $(x + y)(a + b)$ **19.** $(2, -1)$ **20.** $\dfrac{1 \pm i\sqrt{3}}{4}$ **21.** $-4, \dfrac{2}{3}$

22. 52 **23.** $\pm 2, \pm i$ **24.** $-\dfrac{4}{3}$ **25.** 1.5 **26.** no solution

27. $\dfrac{1}{9}$ **28.** $\dfrac{1}{8}$ **29. a.**

b. $\left\{x \mid x < -\dfrac{5}{3} \text{ or } x > 1\right\}$ **c.** $\left(-\infty, -\dfrac{5}{3}\right) \cup (1, \infty)$

30. a. **b.** $\left\{m \mid 0 < m < \dfrac{3}{5}\right\}$

c. $\left(0, \dfrac{3}{5}\right)$ **31. a.** $x^2 + 2x$ **b.** $-x^2 + 2x + 2$

c. $2x^3 + x^2 - 2x - 1$ **d.** $\dfrac{2x + 1}{x^2 - 1}; x \neq \pm 1$ **e.** $2x^2 - 1$

f. $4x^2 + 4x$ **32.** yes **33.** $y = \dfrac{4}{3}x - 12; 4x - 3y = 36$

34. $\log_5 125 = 3$ **35.** $5 \log_5 x + \log_5 y$

36. $(x - 2)^2 + (y - 4)^2 = 13^2$ **37.**

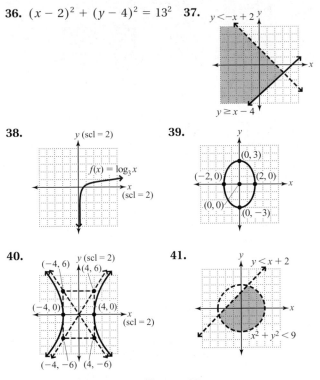

38. **39.**

40. **41.**

42. 7.3% **43. a.** $h = \dfrac{V}{\pi r^2}$ **b.** $\dfrac{15}{r^2}$ **44.** 2.9 lb. of tuna and 3.4 lb. of shrimp **45.** 0.857 hr. or approximately 51 min. **46.** \$1500 at 5%, \$2500 at 6%, \$1000 at 3%

47. $\dfrac{x^2}{5625} + \dfrac{y^2}{400} = 1$ **48.** 7.37 **49.** \$16,909.08 **50.** 6.9

Appendix A

Appendix A Exercises

1. The domain of a sequence is the set of natural numbers 1 to n. **3.** A series is the sum of the terms of a sequence. **5.** arithmetic; 5 **7.** 3, 5, 7, 9, 41

9. 3, 6, 11, 18, 227 **11.** $\dfrac{1}{3}, \dfrac{1}{2}, \dfrac{3}{5}, \dfrac{2}{3}, \dfrac{11}{12}$

13. $-\dfrac{1}{2}, \dfrac{1}{5}, -\dfrac{1}{10}, \dfrac{1}{17}, -\dfrac{1}{226}$ **15.** 5 **17.** -3 **19.** 7

21. $a_n = 14 + (n-1)(4)$, $a_{14} = 66$ **23.** $a_n = -8 + (n-1)(-3)$, $a_{28} = -89$ **25.** $a_n = -5 + (n-1)(4)$, $a_{34} = 127$ **27.** $-6, 1, 8, 15$ **29.** $-5, -8, -11, -14$
31. 7, 11, 15, 19 **33.** $a_1 = 11$ **35.** $d = 4$ **37.** $-8, -1, 6, 13$
39. 27, 24, 21, 18 **41.** $1 + 4 + 9 + 16 + 25 + 36 = 91$
43. $-3 - 1 + 1 + 3 = 0$ **45.** $-1 + 8 + 23 = 30$
47. $9 + 13 + 17 + 21 = 60$ **49.** 1450 **51.** -429 **53.** 675
55. 120 **57.** 5050 **59. a.** 144 seats **b.** 4480 seats
61. a. 16 posts **b.** 325 posts **63. a.** \$28,000 \$29,500 \$31,000 \$32,500 \$34,000 **b.** \$41,500 **c.** \$347,500

Appendix B

Appendix B Exercises

1. The common ratio, r.
3. Yes. The first n terms form a finite geometric series.
5. $2 + 6 + 18 + 54 = 80$ **7. a.** 3 **b.** $a_9 = 6561$

c. $a_n = 1(3)^{n-1}$ **9. a.** -2 **b.** $a_{15} = -32,768$
c. $a_n = -2(-2)^{n-1}$ **11. a.** $\dfrac{1}{3}$ **b.** $a_{10} = \dfrac{1}{81}$
c. $a_n = 243\left(\dfrac{1}{3}\right)^{n-1}$ **13. a.** $-\dfrac{1}{4}$ **b.** $a_7 = \dfrac{1}{32}$
c. $a_n = 128\left(-\dfrac{1}{4}\right)^{n-1}$ **15. a.** $-4, 12, -36, 108, -324$

b. 8748 **17. a.** 243, 81, 27, 9, 3 **b.** $\dfrac{1}{3}$ **19. a.** 2

b. 1, 2, 4, 8 **21. a.** 2 **b.** $-6, -12, -24, -48$ **23. a.** $\dfrac{1}{4}$

b. 128, 32, 8, 2 **25.** -6 **27.** 265,719 **29.** 63.875
31. 102.4 **33.** 40.5 **35.** 9.375 **37.** The sum does not exist, because $|r| \geq 1$. **39.** $\dfrac{4}{9}$ **41.** $\dfrac{17}{99}$ **43. a.** 520, 540.8, 562.432, 584.92928, 608.3264512 **b.** 740.122142459 **45. a.** 100, 108, 117, 126 **b.** 171 mink **c.** $a_n = 100(1.08)^{n-1}$ **47. a.** \$0.01, \$0.02, \$0.04, \$0.08, \$0.16 **b.** $a_n = 0.01(2)^{n-1}$ **c.** \$5242.88 **d.** \$10,485.75 **49.** \$11,191.64

Appendix C

Appendix C Exercises

1. 13 **3.** 8 factorial **5.** 5, 8
7. 24 **9.** 144 **11.** 86,400 **13.** $\dfrac{1}{90}$, or $0.0\overline{1}$ **15.** 10 **17.** 28

19. 35 **21.** 210 **23.** 1 **25.** 1 **27.** $a^5 + 5a^4b + 10a^3b^2 + 10a^2b^3 + 5ab^4 + b^5$ **29.** $x^4 - 4x^3y + 6x^2y^2 - 4xy^3 + y^4$
31. $8a^3 + 12a^2b + 6ab^2 + b^3$ **33.** $x^5 - 10x^4y + 40x^3y^2 - 80x^2y^3 + 80xy^4 - 32y^5$ **35.** $64m^6 + 576m^5n + 2160m^4n^2 + 4320m^3n^3 + 4860m^2n^4 + 2916mn^5 + 729n^6$ **37.** $81x^4 - 432x^3y + 864x^2y^2 - 768xy^3 + 256y^4$ **39.** $70x^4y^4$ **41.** $45a^8b^2$
43. $32,256x^4y^5$ **45.** $22,680m^4n^3$

Appendix D

Appendix D Exercises

1. The number of ways the two tasks can be performed one after the other.
3. $_nC_r$ because the order of selection does not matter.
5. $_nP_r$ because the order matters. **7.** 42 **9.** 720 **11.** 1
13. 15 **15.** 1 **17.** 1 **19.** 360 outfits **21.** 120 ways
23. a. 6 **b.** 27 **25.** 17,576,000 **27.** 10,000; 5040
29. 39,916,800 **31.** 1,906,884 **33.** 816

Appendix E

Appendix E Exercises

1. The sample space is the set of all possible outcomes. **3.** 0 and 1 **5.** The probability of an event is the ratio of the number of ways the event can occur to the number of outcomes in the sample space.
7. {a, e, i, o, u} **9.** {HHH, HHT, HTH, THH, THT, HTT, TTH, TTT} **11.** 90 **13.** 1820 **15.** {HTT, THT, TTH} **17.** {} or $\varnothing$
19. $\dfrac{1}{6}$ **21.** $\dfrac{1}{2}$ **23.** $\dfrac{3}{4}$ **25.** $\dfrac{1}{12}$ **27.** $\dfrac{1}{12}$ **29.** $\dfrac{1}{13}$ **31.** $\dfrac{3}{13}$
33. $\dfrac{3}{8}$ **35.** $\dfrac{1}{3}$ **37.** 0 **39.** $\dfrac{1}{9}$ **41.** $\dfrac{10}{21}$ **43.** 0.018 **45.** 0.070

GLOSSARY

Absolute value: A given number's distance from 0 on a number line.

Additive inverses: Two numbers whose sum is 0.

Area: The total number of square units that fill a figure.

Arithmetic sequence: A sequence in which each term after the first is found by adding the same number to the previous term.

Arithmetic series: The sum of the terms in an arithmetic sequence.

Augmented matrix: A matrix made up of the coefficients and the constant terms of a system. The constant terms are separated from the coefficients by a dashed vertical line.

Axis: A number line used to locate a point in a plane.

Axis of symmetry: A line that divides a graph into two symmetrical halves.

Base: The number that is repeatedly multiplied.

Binomial: A polynomial containing two terms.

Binomial coefficient: A number written as $\binom{n}{r}$ and defined as $\dfrac{n!}{r!(n-r)!}$.

Binomial theorem: For any positive integer n,

$$(a+b)^n = \binom{n}{0}a^n + \binom{n}{1}a^{n-1}b + \binom{n}{2}a^{n-2}b^2 + \binom{n}{3}a^{n-3}b^3 + \cdots + \binom{n}{n}b^n.$$

Circle: A set of points that are equally distant from a central point. The central point is the center.

Circumference: The distance around a circle.

Coefficient: The numerical factor in a term.

Combination: An unordered arrangement of objects. The combination of n things taken r at a time is often denoted as $_nC_r$.

Common difference of an arithmetic sequence: The value $\mathbf{d}$ found by $d = a_n - a_{n-1}$ where a_n is any value in the sequence and a_{n-1} is the previous value.

Common logarithms: Logarithms with a base of 10. $\text{Log}_{10}\, x$ is written as $\log x$.

Complementary angles: Two angles are complementary if the sum of their measures is 90°.

Complex conjugates: The complex conjugate of a complex number $a + bi$ is $a - bi$.

Complex number: A number that can be expressed in the form $a + bi$, where a and b are real numbers and i is the imaginary unit.

Complex rational expression: A rational expression that contains rational expressions in the numerator or denominator.

Compound inequality: Two inequalities joined by either "and" or "or."

Composition of functions: If f and g are functions, then the composition of f and g is defined as $(f \circ g)(x) = f[g(x)]$ for all x in the domain g for which $g(x)$ is in the domain of f. The composition of g and f is defined as $(g \circ f)(x) = g[f(x)]$ for all x in the domain of f for which $f(x)$ is in the domain of g.

Congruent angles: Angles that have the same measure. The symbol for congruent is $\cong$.

Conjugates: Binomials that differ only in the sign separating the terms.

Conic section: A curve in a plane that is the result of intersecting the plane with a cone. More specifically, a circle, ellipse, parabola, or hyperbola.

Consecutive integers: $\ldots -3, -2, -1, 0, 1, 2, 3, \ldots$

Consistent system of equations: A system of equations that has at least one solution.

Constant: A symbol that does not vary in value.

Cramer's Rule: The solution to the system of linear equations $\begin{cases} a_1x + b_1y = c_1 \\ a_2x + b_2y = c_2 \end{cases}$ is

$$x = \frac{\begin{vmatrix} c_1 & b_1 \\ c_2 & b_2 \end{vmatrix}}{\begin{vmatrix} a_1 & b_1 \\ a_2 & b_2 \end{vmatrix}} = \frac{D_x}{D} \quad \text{and} \quad x = \frac{\begin{vmatrix} a_1 & c_1 \\ a_2 & c_2 \end{vmatrix}}{\begin{vmatrix} a_1 & b_1 \\ a_2 & b_2 \end{vmatrix}} = \frac{D_y}{D}$$

The solution to the system of linear equations

$$\begin{cases} a_1x + b_1y = c_1z = d_1 \\ a_2x + b_2 = c_2z = d_2 \\ a_3x + b_3y = c_3z = d_3 \end{cases} \text{is}$$

$$x = \frac{\begin{vmatrix} d_1 & b_1 & c_1 \\ d_2 & b_2 & c_2 \\ d_3 & b_3 & c_3 \end{vmatrix}}{\begin{vmatrix} a_1 & b_1 & c_1 \\ a_2 & b_2 & c_2 \\ a_3 & b_3 & c_3 \end{vmatrix}} = \frac{D_x}{D}, \; y = \frac{\begin{vmatrix} a_1 & d_1 & c_1 \\ a_2 & d_2 & c_2 \\ a_3 & d_3 & c_3 \end{vmatrix}}{\begin{vmatrix} a_1 & b_1 & c_1 \\ a_2 & b_2 & c_2 \\ a_3 & b_3 & c_3 \end{vmatrix}} = \frac{D_y}{D}, \text{and } z = \frac{\begin{vmatrix} a_1 & b_1 & d_1 \\ a_2 & b_2 & d_2 \\ a_3 & b_3 & d_3 \end{vmatrix}}{\begin{vmatrix} a_1 & b_1 & c_1 \\ a_2 & b_2 & c_2 \\ a_3 & b_3 & c_3 \end{vmatrix}} = \frac{D_z}{D}.$$

Note: Each denominator, D, is the determinant of a matrix containing only the coefficients in the system. To find D_x, we replace the column of x-coefficients in the coefficient matrix with the constants from the system. To find D_y, we replace the column of y-coefficients in the coefficient matrix with the constant terms, and do likewise to find D_z.

Degree: The sum of the exponents of all variables in a monomial.

Degree of a polynomial: The greatest degree of any of the terms in the polynomial.

Dependent linear equations in two unknowns: Equations with identical graphs.

Diameter: The distance across a circle through its center.

Direct variation: Two variables, y and x, are in direct variation if $y = kx$, where k is a constant.

Discriminant: The radicand $b^2 - 4ac$ in the quadratic formula.

Domain: The set of all input values for a relation.

Echelon form: An augmented matrix whose coefficient portion has 1's on the diagonal from upper left to lower right and 0's below the 1's.

Ellipse: The set of all points the sum of whose distance from two fixed points is constant.

Equation: A mathematical relationship that contains an equal sign.

Equation quadratic in form: An equation is quadratic in form if it can be rewritten as a quadratic equation $au^2 + bu + c = 0$, where $a \neq 0$ and u is a variable or expression.

Event: Any subset of a sample space.

Experiment: Any act or process whose result is not known in advance.

Exponent: A symbol written to the upper right of a base number that indicates how many times to use the base as a factor.

Exponential function: If $b > 0$, $b \neq 1$, and x is any real number, then the exponential function is $f(x) = b^x$.

Expression: A constant, variable, or any combination of constants, variables, and arithmetic operations that describes a calculation.

Extraneous solution: An apparent solution that does not solve its equation.

Factored form: A number or expression written as a product of factors.

Factorial notation: For any natural number n, the symbol $n!$ (read "n factorial") means $n(n-1)(n-2)\ldots 3 \cdot 2 \cdot 1$. 0! is defined to be 1, so $0! = 1$.

Factors: If $a \cdot b = c$, then a and b are factors of c.

Finite sequence: A function with a domain that is the set of natural numbers from 1 to n.

Formula: An equation that describes a mathematical relationship.

Fraction: A quotient of two numbers or expressions a and b having the form $\frac{a}{b}$, where $b \neq 0$.

Function: A relation in which every value in the domain is paired with exactly one value in the range.

Geometric sequence: A sequence in which every term after the first is found by multiplying the previous term by the same number, called the common ratio, r, where $r = \frac{a_n}{a_{n-1}}$.

Geometric series: The sum of the terms of a geometric sequence.

Greatest common factor (GCF): The largest natural number that divides all given numbers with no remainder.

Hyperbola: The set of all points the difference of whose distances from two fixed points remains constant.

Identity: An equation that has every real number as a solution (excluding any numbers that cause an expression in the equation to be undefined).

Imaginary number: A number that can be expressed in the form bi, where b is a real number and i is the imaginary unit.

Imaginary unit: The number represented by i, where $i = \sqrt{-1}$, and $i^2 = -1$.

Inconsistent system of equations: A system of equations that has no solution.

Independent linear equations in two unknowns: Equations that have different graphs.

Inequality: A mathematical relationship that contains an inequality symbol ($\neq$, $<$, $>$, $\leq$, or $\geq$).

Infinite sequence: A function with a domain that is the set of natural numbers.

Intersection: For two sets A and B, the intersection of A and B, symbolized by $A \cap B$, is a set containing only elements that are in both A and B.

Inverse functions: Two functions f and g are inverses if and only if $(f \circ g)(x) = x$ for all x in the domain of g and $(g \circ f)(x) = x$ for all x in the domain of f.

Inverse variation: Two variables, y and x, are in inverse variation if $y = \frac{k}{x}$, where k is a constant.

Irrational number: Any real number that is not rational.

Least common denominator (LCD): The least common multiple of the denominators of a given set of fractions.

Least common multiple (LCM): The smallest number that is a multiple of each number in a given set of numbers.

Like radicals: Radical expressions with identical radicands and identical indexes.

Like terms: Variable terms that have the same variable(s) raised to the same exponents, or constant terms.

Linear equation: An equation in which each variable term contains single variable raised to an exponent of 1.

Linear equations in one variable: An equation that can be written on the form $ax + b = c$, where a, b, and c are real numbers and $a \neq 0$.

Linear inequality: An inequality containing expressions in which each variable term contains a single variable with an exponent of 1.

Logarithm: If $b > 0$ and $b \neq 1$, then $y = \log_b x$ is equivalent to $x = b^y$.

Lowest terms: Given a fraction $\dfrac{a}{b}$ and $b \neq 0$, if the only factor common to both a and b is 1, then the fraction is in lowest terms.

Matrix: A rectangular array of numbers.

Minor: The determinant of the remaining matrix when the row and column in which the element is located are ignored.

Monomial: An expression that is a constant or a product of a constant and variables that are raised to whole-number powers.

Multiple: A multiple of a given integer n is the product of n and an integer.

Multiplicative inverses: Two numbers whose product is 1.

Natural logarithm: Base-e logarithms are called natural logarithms and $\log_e x$ is written as $\ln x$. Note that $\ln e = 1$.

Nonlinear system of equations: A system of equations that contains at least one nonlinear equation.

nth root: The number b is the nth root of a number a if $b^n = a$.

One-to-one function: A function f is one-to-one if for any two numbers a and b in its domain, when $f(a) = f(b)$, $a = b$ and when $a \neq b$, $f(a) \neq f(b)$.

Outcomes: The possible results of an experiment.

Percent: A ratio representing some part out of 100.

Perimeter: The distance around a figure.

Permutation: An ordered arrangement of objects. The permutation of n things taken r at a time is often denoted as $_nP_r$.

Polynomial: A monomial or an expression that can be written as a sum of monomials.

Polynomial in one variable: A polynomial in which every variable term has the same variable.

Prime factorization: A factorization that contains only prime factors.

Prime number: A natural number that has exactly two different factors, 1 and the number itself.

Probability of an Event E: Given an experiment, the probability of an event E, written as $P(E)$, is the number of outcomes in event E divided by the number of outcomes in the sample space S of the experiment. That is,

$$P(E) = \frac{\text{The number of outcomes in } E}{\text{The number of outcomes in } S}$$

Proportion: An equation in the form $\dfrac{a}{b} = \dfrac{c}{d}$, where $b \neq 0$ and $d \neq 0$.

Quadrant: One of the four regions created by the intersection of the axes in the coordinate plane.

Quadratic equation in one variable: An equation that can be written in the form $ax^2 + bx + c = 0$, where a, b, and c are all real numbers and $a \neq 0$.

Quadratic equation in two variables: An equation that can be written in the form $y = ax^2 + bx + c$, where a, b, and c are real numbers and $a \neq 0$.

Quadratic inequality: An inequality that can be written in the form $ax^2 + bx + c > 0$ or $ax^2 + bx + c < 0$, where $a \neq 0$.

Radical equation: An equation containing at least one radical expression whose radicand has a variable.

Radical function: A function of the form $f(x) = \sqrt[n]{P}$, where P is a polynomial.

Radius: The distance from the center of a circle to any point on the circle.

Range: The set of all output values for a relation.

Ratio: A comparison of two quantities using a quotient.

Rational exponent: An exponent that is a fraction.

Rational expression: An expression that can be written in the form $\dfrac{P}{Q}$, where P and Q are polynomials and $Q \neq 0$.

Rational inequality: An inequality containing a rational expression.

Rational number: Any real number that can be expressed in the form $\frac{a}{b}$, where a and b are integers and $b \neq 0$.

Relation: A set of ordered pairs.

Sample space: The set of all possible outcomes of an experiment. The sample space is denoted by S.

Scientific notation: A number expressed in the form $a \times 10^n$, where a is a decimal number with $1 \leq |a| < 10$ and n is an integer.

Sequence: A function list whose domain is 1, 2, 3, ..., n.

Series: The sum of the terms of a sequence.

Set: A collection of objects.

Similar figures: Figures with congruent angles and proportional side lengths.

Slope: The ratio of the vertical change between any two points on a line to the horizontal change between these points.

Solution: A number that makes an equation true when it replaces the variable in the equation.

Solution for a system of equations: An ordered set of numbers that makes all equations in the system true.

Square matrix: A matrix that has the same number of rows and columns.

Supplementary angles: Two angles are supplementary if the sum of their measures is 180°.

System of equations: A group of two or more equations.

Terms: Expressions that are the addends in an expression that is a sum.

Trinomial: A polynomial containing three terms.

Unit ratio: A ratio with a denominator of 1.

Union: For two sets A and B, the union of A and B, symbolized by $A \cup B$, is a set containing every element in A or B.

Variable: A symbol that can vary in value.

Vertex: The lowest point on a parabola that opens upwards or highest point on a parabola that opens downwards.

Volume: The total number of cubic units that fill a space.

***x*-intercept:** A point where a graph intersects at the x-axis.

***y*-intercept:** A point where a graph intersects at the y-axis.

Index

Work problems, 602–603
World Series, 260

X

x-axis, 268
x-intercept, 289, 290
  finding, 289–292, 917
    of quadratic function, 889–891

Y

y-axis, 268
y-intercept, 289, 290
  equation of line given its, 315
  finding, 289–292
    of quadratic function, 889–891
  for $y = mx + b$, 291

Z

Zero
  division involving, 48–49
  as exponent, 375–376, 444, 801
  multiplicative property of, 41
  solving equations with two or more
    factors equal to, 503–504
Zero-factor theorem, 502–504,
  593–595
Zero slope, 304
Zero subscript, 149

Applications Index

Astronomy/Aerospace

Automotive

Biology

Business

Chemistry

Cross section of a tunnel, 1044
Current and resistance, 501
Current, 145, 176
Designing a lens, 600
Designing a steel plate, 448
Electrical circuit, 55, 590, 746
Estimating the distance across a river, 204
Forces on a steel structure, 746
Grade of a road, 198
Hoover Dam supplying electricity, 203
Nuclear reactor, 1006
Resistance of an electrical circuit, 55, 145, 590, 599, 613, 625
Resultant force, 37
Transistors on a chip, 973
Value of resistors, 600
Voltage of a current, 117
Wavelength, 608, 613, 782
Width of a rectangular steel plate, 580

Environment
Acceptable air quality standards, 219
Area of forest remaining, 288
Atmospheric pressure, 1010
Barometric pressure in the eye of a hurricane, 1011
Erosion rate for Niagara Falls, 203
Longest rivers, 745
Magnitude of an earthquake, 795, 999, 1027, 1092
Rainfall, 64, 350
Recycling, 673
Severe thunderstorm warning, 882
Snowfall measurements, 100
Temperatures, 117, 350
U.S. greenhouse gas emissions, 216

Finance
Balance of an account, 30
Checking account balance, 130, 217
Coins in a purse, 75
Compound interest, 966–967, 969, 972, 992, 1003, 1005, 1010, 1020, 1024, 1027, 1029, 1092
Continuous compounding, 996, 998, 1003–1004, 1010, 1021, 1027, 1028
Credit card balance, 30, 100, 126, 130
Credit card debt, 54
Debt, 54, 216
Deposits, 721, APP-19
Down payment for a car, 130
Exchanging money, 201
Growth of an account, 1003–1004, 1005, 1010, 1020, 1021, 1024, 1027, 1028

Household budget, 40
Interest earned, 162, 244
Interest rate, 513
Investment, 218, 244–245, 246, 248, 249, 250, 251, 255, 261, 263, 513, 536, 734, 745, 764–765, 776, 778, 1003, 1092
Loans, APP-9
Money needed to buy a stereo, 126
Mortgage payments, 216
Qualifying for a loan, 216
Retirement plan, 261
Savings account, APP-15–APP-16

Geometry
Angle measurement, 225, 226, 234, 260, 263, 299, 717, 720, 732, 745, 776
Area of a circular region, 115, 433
Area of a parallelogram, 934
Area of a rectangular region, 113, 114, 115, 450, 501
Area of a room, 109, 113, 433, 470, 895
Area of a shaded region, 178, 470, 495
Area of a state, 420
Area of an entertainment center, 185
Area of the front face of a box, 675
Area of the side of a house, 115, 470
Base and height of a triangle, 512, 883, 944
Circumference of a circular region or object, 113
Diagonal of a rectangular shaped object, 514
Diameter of a circular region or object, 145, 185, 882
Dimensions of a box-shaped object, 139–140, 144, 145, 449, 730, 882, 944
Dimensions of a cylindrical object, 145
Dimensions of a parallelogram, 457
Dimensions of a rectangular region, 138–139, 144, 175, 185, 187, 223–224, 233, 234, 260, 263, 336, 457, 477, 486, 512, 528, 533, 590, 601, 681, 705–706, 709, 720, 882, 883, 895, 921, 946, 1065, 1066, 1085
Dimensions of a room, 144, 149, 508, 882, 895, 1070
Dimensions of a trapezoid, 130, 144, 185, 187
Dimensions of a triangle, 944
Height of a box, 139, 145, 175, 675, 684, 686
Height of a cylinder, 145
Height of a trapezoid, 512

Height of a triangle, 590
Length of a side of a triangle, 130, 144, 234, 260, 262, 528–529, 732, 756, 895
Length of the side of a regular hexagon, 836
Length of the side of a square, 882
Length of the sides of the base of a pyramid, 836
Maximum area, 921
Maximum length of a garden, 171–172
Maximum perimeter of a building, 335
Minimum surface area, 172
Perimeter of a rectangular region, 108, 113, 161
Perimeter of a semicircle, 161
Perimeter of an isosceles triangle, 161
Radius of a circular region, 175, 477, 508, 512, 882, 944
Radius of a cylinder, 895, 944
Radius of a sphere, 145
Side lengths of a figure, 732
Surface area of a box, 161, 396
Surface area of a swimming pool, 410
Volume of a box-shaped object, 116, 178, 434, 449, 501, 934, 946
Volume of a cone, 116, 434, 550, 698
Volume of a cylindrical object, 116, 434, 1091
Volume of a pyramid, 116, 550
Volume of a shaded region, 495
Volume of a sphere, 116
Volume of a storage tank, 397
Volume of an object, 470
Volume of metal remaining in a block, 397

Government
Club electing a president and secretary, APP-32
Defendants in cases concluded in U.S., 199, 643
Democrats and Republicans, APP-38
National debt, 447
Public elementary and secondary school revenue, 217–218
Student governance association, APP-30
U.S. House of Representatives, 23
U.S. Senate, 23

Health
Americans saying they are in excellent health, 262
Body mass index (BMI), 162, 548

Velocity of an object, 273, 317
Voltage in a circuit, 396, 448, 457, 501, 607–608, 837
Wavelength of light, 384, 385, 386

Sports/Entertainment

Acrobat, 944
Archer, 946
Basketball lineups, APP-32
Basketball scores, 733, 746, 756
Batting average, 217
Bicycling, 242, 643
Boston Marathon, 906
Cliff diving championship, 896
Concert hall seating, APP-9
Concert ticket sales, 733
Cyclist training, 906
Daytona 500, 341, 562
Dirt bike race, APP-30
Event planner, 781
Extreme games competition, 891–892, 895
Ferris wheel, 1044
Field goal attempts, 217
Finishers in a race, APP-28
Fishing, 1044
Football, 625, APP-32
Games won by the Boston Red Sox, 260
High school athletics, 324
Hiking, 111
Home runs, 721
Horse race, APP-29
Javelin toss, 921
Jogging, 75, 242, 601
Jumping the Grand Canyon on a motorcycle, 895
Longest punt, 944
Masters tournament, 349, 721
Monopoly game, 235
Movie ticket sales, 673, 697, 776
Olympic medals, 720
Platform diver, 896
Poem collection, APP-32
Ropes course, 794
Running, 236, 238, 263, 562, 591, 627, 907

Singer performing at a benefit, APP-32
Singer recording songs, APP-32
Skydiving, 162, 790
Snow skiing, 185
Soccer player kicking ball, 915
Super Bowl scores, 721, 745
Super Bowl ticket prices, 972, 1012
Swim meet, 776
Ticket sales to a play, 235, 733, 744, 778, 921
Top-five money earners in the concert circuit, 349
Top-five cable TV networks, 349
Tour de France, 116, 906
Track, 610, 733
Triathlon training, 243
TV shows, APP-32
Visiting amusement parks, APP-31
Volleyball court, 709
Western conference in the WNBA, 263
Yahtzee, 130

Statistics/Demographics

Cocaine use in the United Kingdom, 1012, 1025
Countries with the highest birth rates, 735
Countries with the highest death rates, 735
Foreign-born population, 350
Number of Catholics in the United States, 350
Number of computers in the United States, 969
Number of people aged 65 or older in the United States, 968, 1029
Number of single fathers, 219
People per square mile in the United States, 447
Percent of deaths by age per million miles driven, 1044
Percentage of infants regularly sleeping in a bed, 1191
Poll or survey results, 54, 131, 259, 262, 536

Population of Africa, 1011
Population of Orlando, Florida, 1004–1005
Population, 219, 312, 385, 972
Probability of an accident while driving at a given blood alcohol level, 1010
U.S. households headed by married couples, 217
U.S. population 21 years of age or older, 217
Women waiting to have first child, 664
World population, 1011

Transportation

Airline fuel prices, 213
Average rate, 111, 116, 175, 589, 590, 601, 611, 904, 906
Distance apart, 241, 242, 261, 591
Distance from original location, 514
Distance traveled, 116, 130, 201, 202, 242, 258, 266
Gear on a bicycle, 198
Map scale, 201
Minimum runway length, 994–995, 999
Motorcycle deaths, 664
Plane's speed in still air, 610, 776
Rate of a boat in still water, 778
Rate, 607, 610, 611
Speed, 242, 243, 261, 610, 611, 901–902, 906
Speeding ticket, 642
Time for a person or vehicle to overtake another, 241, 242, 261, 266
Time for a vehicle to catch up to another vehicle, 238–239, 240, 242, 255, 263, 721, 776
Time for vehicles or people to meet, 236–237, 254
Travel time, 75, 144, 152, 175, 185, 609, 906, 961
Traveling for leisure, 372
Walking speeds, 1000

Photograph Credits

1 Getty RF; 21 © 2003 Gorton's; 23 Lawrence Lawry/Getty Images; 25 Comstock Images; 34, 72, 399 StockTrek/GettyImages; 38 © Image Source/elektraVision/PictureQuest; 40 (tl) Library of Congress, Prints and Photographs Division [reproduction number LC-USZ62-117116 DLC]; (tr) Library of Congress, Prints and Photographs Division [reproduction number LC-USZ62-13002 DLC]; (bl) Library of Congress, Prints and Photographs Division [reproduction number LC-USZ62-117117 DLC]; (br) Library of Congress, Prints and Photographs Division [reproduction number LC-USZ62-13004 DLC]; 55 © Flip Nicklin/Minden Pictures; 75, 609 Karl Weatherly/Getty Images; 76, 269, 509 Hulton/Archive by Getty Images; 101, 267, 720, 783 PhotoDisc; 109 Photo Courtesy of The Carpet and Rug Institute; 110, 580 David Young-Wolf/PhotoEdit; 111, 514 Doug Menuez/Getty Images; 113 Fermilab Photo; 116 (t), 973 Corbis RF; 116 (b) Mark Mainz/Getty Images; 117, 906 (b) © Royalty-Free/Corbis; 130, 605 Nick Rowe/Getty Images; 146 Jack Hollingsworth/Getty Images; 162 PhotoLink/Getty Images; 185 Tim Barnett/Getty Images; 189 PhotoDisc Blue; 198 Jeff Maloney/Getty Images; 200 Steve Mason/Getty Images; 201, 447 Hishman F. Ibrahim/Getty Images; 203 Jeremy Woodhouse/Getty Images; 212 © Ron Chapple/ThinkStock/PictureQuest; 213 © John Neubauer/PhotoEdit; 217 Icon SMI/Corbis; 235 Ryan McVay/Getty Images; 236 Ian Wyatt/Getty Images; 238 Sami Sarkis/Getty Images; 246, 392 David Buffington/Getty Images; 286 Dave Thompson/Life File/Getty Images; 287 SW Productions/Getty Images; 311 Jamie Squire/Allsport; 335 John Dakers/Life File/Getty Images; 373, 642, 891, 999 Corbis; 378 PAL_AAEUFML0 Photo Researchers, Inc; 384 (t) © DigitalVision/PictureQuest; (b) Nick Koudis/Getty Images; 385 EyeWire/Getty Images; 386 AFP/Getty Images; 459 Blend Images (GettyRF); 508 Photomondo/Getty Images; 513 © Bill Aron/PhotoEdit; 537, 871 PhotoDisc Red; 548, 562 © Tony Freeman/PhotoEdit; 549 Rob Melnychuk/ Getty Images; 562 Buddy Baker; 625 © Tom Carter/PhotoEdit 629, 687, 896, 1011 Digital Vision; 643 Jon Devitch; 721 Mark J. Terrill/AP Wideworld Photos; 728 PhotoDisc RF; 745 PhotoDisc BS14085; 746 AP Wideworld Photos; 755 Jonathan Blair/Corbis; 800 Bettmann/Corbis; 882 NOAA; 895 AFP/Getty Images Editorial; 906 (t) Reuters/Corbis; 947 Getty Images; 961 (t) Taxi/Getty (b) Toyota Pressroom; 994 George Hall/Corbis; 1010 Lifeloc Technologies; 1012 Benelux Press/Getty Images; 1031 Brand X Pictures; 1045 Andrew Ward/Life File/Getty Images; 1057 Kevin Fleming/Corbis

POWERS AND ROOTS

n	n^2	n^3	$\sqrt{n}$	$\sqrt[3]{n}$	$\sqrt{10n}$	n	n^2	n^3	$\sqrt{n}$	$\sqrt[3]{n}$	$\sqrt{10n}$
1	1	1	1.000	1.000	3.162	51	2,601	132,651	7.141	3.708	22.583
2	4	8	1.414	1.260	4.472	52	2,704	140,608	7.211	3.733	22.804
3	9	27	1.732	1.442	5.477	53	2,809	148,877	7.280	3.756	23.022
4	16	64	2.000	1.587	6.325	54	2,916	157,464	7.348	3.780	23.238
5	25	125	2.236	1.710	7.071	55	3,025	166,375	7.416	3.803	23.452
6	36	216	2.449	1.817	7.746	56	3,136	175,616	7.483	3.826	23.664
7	49	343	2.646	1.913	8.367	57	3,249	185,193	7.550	3.849	23.875
8	64	512	2.828	2.000	8.944	58	3,364	195,112	7.616	3.871	24.083
9	81	729	3.000	2.080	9.487	59	3,481	205,379	7.681	3.893	24.290
10	100	1,000	3.162	2.154	10.000	60	3,600	216,000	7.746	3.915	24.495
11	121	1,331	3.317	2.224	10.488	61	3,721	226,981	7.810	3.936	24.698
12	144	1,728	3.464	2.289	10.954	62	3,844	238,328	7.874	3.958	24.900
13	169	2,197	3.606	2.351	11.402	63	3,969	250,047	7.937	3.979	25.100
14	196	2,744	3.742	2.410	11.832	64	4,096	262,144	8.000	4.000	25.298
15	225	3,375	3.873	2.466	12.247	65	4,225	274,625	8.062	4.021	25.495
16	256	4,096	4.000	2.520	12.649	66	4,356	287,496	8.124	4.041	25.690
17	289	4,913	4.123	2.571	13.038	67	4,489	300,763	8.185	4.062	25.884
18	324	5,832	4.243	2.621	13.416	68	4,624	314,432	8.246	4.082	26.077
19	361	6,859	4.359	2.688	13.784	69	4,761	328,509	8.307	4.102	26.268
20	400	8,000	4.472	2.714	14.142	70	4,900	343,000	8.367	4.121	26.458
21	441	9,261	4.583	2.759	14.491	71	5,041	357,911	8.426	4.141	26.646
22	484	10,648	4.690	2.802	14.832	72	5,184	373,248	8.485	4.160	26.833
23	529	12,167	4.796	2.844	15.166	73	5,329	389,017	8.544	4.179	27.019
24	576	13,824	4.899	2.884	15.492	74	5,476	405,224	8.602	4.198	27.203
25	625	15,625	5.000	2.924	15.811	75	5,625	421,875	8.660	4.217	27.386
26	676	17,576	5.099	2.962	16.125	76	5,776	438,976	8.718	4.236	27.568
27	729	19,683	5.196	3.000	16.432	77	5,929	456,533	8.775	4.254	27.749
28	784	21,952	5.292	3.037	16.733	78	6,084	474,552	8.832	4.273	27.928
29	841	24,389	5.385	3.072	17.029	79	6,241	493,039	8.888	4.291	28.107
30	900	27,000	5.477	3.107	17.321	80	6,400	512,000	8.944	4.309	28.284
31	961	29,791	5.568	3.141	17.607	81	6,561	531,441	9.000	4.327	28.460
32	1,024	32,768	5.657	3.175	17.889	82	6,724	551,368	9.055	4.344	28.636
33	1,089	35,937	5.745	3.208	18.166	83	6,889	571,787	9.110	4.362	28.810
34	1,156	39,304	5.831	3.240	18.439	84	7,056	592,704	9.165	4.380	28.983
35	1,225	42,875	5.916	3.271	18.708	85	7,225	614,125	9.220	4.397	29.155
36	1,296	46,656	6.000	3.302	18.974	86	7,396	636,056	9.274	4.414	29.326
37	1,369	50,653	6.083	3.332	19.235	87	7,569	658,503	9.327	4.431	29.496
38	1,444	54,872	6.164	3.362	19.494	88	7,744	981,472	9.381	4.448	29.665
39	1,521	59,319	6.245	3.391	19.748	89	7,921	704,969	9.434	4.465	29.833
40	1,600	64,000	6.325	3.420	20.000	90	8,100	729,000	9.487	4.481	30.000
41	1,681	68,921	6.403	3.448	20.248	91	8,281	753,571	9.539	4.498	30.166
42	1,764	74,088	6.481	3.476	20.494	92	8,464	778,688	9.592	4.514	30.332
43	2,849	79,507	6.557	3.503	20.736	93	8,649	804,357	9.644	4.531	30.496
44	2,936	85,184	6.633	3.530	20.976	94	8,836	830,584	9.695	4.547	30.659
45	2,025	91,125	6.708	3.557	21.213	95	9,025	857,375	9.747	4.563	30.882
46	2,116	97,336	6.782	3.583	21.148	96	9,216	884,736	9.798	4.579	30.984
47	2,209	103,823	6.856	3.609	21.679	97	9,409	912,673	9.849	4.595	31.145
48	2,304	110,592	6.928	3.534	21.909	98	9,604	941,192	9.899	4.610	31.305
49	2,401	117,649	7.000	3.659	22.136	99	9,801	970,299	9.950	4.626	31.464
50	2,500	125,000	7.071	3.684	22.361	100	10,000	1,000,000	10.000	4.642	31.623